BEEKEEPING

A Compressive guide to bees and beekeeping

Dr. D P ABROL

Professor & Head
Division of Entomology, Faculty of Agriculture,
Sher-e-Kashmir University of Agricultural Sciences and Technology,
Chatha, Jammu 180 009 (J.K.) India

Published by:

Scientific Publishers (India)
5 A, New Pali Road, P.O. Box 91
Jodhpur 342 001 (India)

Branch Office
Scientific Publishers (India)
4806/24, Ansari Road, Daryaganj
New Delhi - 110 002 (India)

E-mail: info@scientificpub.com
Website: www.scientificpubonline.com

Print : 2026

ISBN: 978-81-7233-669-1 (HB)
eISBN: 978-93-86237-62-0

Printed in India

Foreword

Honeybees are wonderful social insects which have fascinated the humanity since the prehistoric times. The biological evolution of honeybees is much older than human civilization, dating back to 50,000,000 years ago or so. Commercial beekeeping (apiculture) is perhaps the only industry exploiting the domesticated species of honeybees for enhance crop productivity, honey from nectar of (floral and extra floral nectaries) from plants, beeswax and several other products of medicinal value, and help in conservation of global biodiversity through propagation of plant species as well as enablight speciation of new flora in nature. A wide human interest to investigate their biology of this marvellous insect in the quest of their interesting ways of life and methods of their colony management led to deploy such knowledge in social evolution of human society and commercial exploitation through their domenstication.

`Beekeeping - a comprehensive guide on bees and beekeeping' by Dr. D.P. Abrol, a distinguished entomologist, is yet another knowledge resource on the subject, towards enthusing entrepreneurs and practitioners of apiculture. Ever since the discovery of movable frame hives, many advances have been made in the field of apiculture or beekeeping particularly in the west with the European honeybee *Apis mellifera* L. However, in India exploitation of the rich floral diversity and congenial climates for exploitation of beekeeping of this bee species is yet to be an agricultural activity. Recognising the yeomen role of pollinators in the enhancement of crop productivity and quality of commodities, Indian Council of Agricultural Research recognized the science of pollination during the XIth five year plan period. It is in this context, that this book offers practical knowledge on beekeeping and apicultural technology to provide package of information as guide for efficient commercial apiculture. The voluminous book is a testimony to the enormous research work that has been done on this subject. The book, suitable for practical and class room reference, is an excellent presentation of worldwide comprehensive picture of beekeeping. It contains a wealth of knowledge that would prove extremely useful to students, teachers, researchers and beekeepers all over the world. I am sure that the book will enlighten us with the latest technological developments in beekeeping and shall be useful to agricultural and applied scientists, extension workers, policy makers and would motivate young minds to the wonderful world of beekeeping to improve rural economy and conservation of biodiversity.

Dr. Abrol has made serious efforts to write the book in a very popular and lucid language and in a form of running story which will provide enjoyable reading. Several recommendations/suggestions for boosting apiculture in India in a big way have found a place in this book.

It is very apt for its launch at a time when the National Bee Board in liaison with National Horticulture Mission and National Agricultural Development Project of the Ministry of Agriculture, Government of India.

I congratulate the author for this excellent book.

T.P. Rajendran

Assistant Director General (Plant Protection)
Indian Council of Agricultural Research
215, Krishi Bhawan, Dr. Rajendra Prasad Road,
New Delhi 110 114

Acknowledgments

A vast spectrum of people has helped in one way or the other in the writing of this book which would have remained a distant dream without their active help and support. The list is so long that it would not be possible to thank each of them individually. This book is outcome of my personal experiences and the contributions of several workers which have been incorporated. I express my humble and profound thank to all of them whose hard work has enabled me to compile the suitable information in such a manner that it would be useful to those interested in bees and beekeeping. The illustrations and figures are either original or redrawn from other sources which have been cited individually in the figure legends. All the authors whose work has been used/refereed deserve special appreciation and heartiest acknowledgments.

I owe deep sense of gratitude and indebtedness to my teachers who had a particular impact on my thinking about Science and Bees. I am greatly indebted to My Guide Late Professor Dr. R.P. Kapil, the then Dean Post Graduate Studies, CCS Haryana Agricultural University, Hisar who introduced me into the fascinating world of bees and beekeeping Professors Dr. R.C. Sihag, Dr. B.N. Putatunda, Dr. K.L. Jain, Dr. S.K. Garg, Dr. S.C. Paschan, Dr. N.K. Yadava, Dr.V.P. Sablok, Dr. R.B. Mathur, and late Professor Dr. Mahavir Gupta; Late Professor Dr. (Mrs.) Sudha Mathur had special impact on my thinking about the science of beekeeping.

Professor Dr. Raghavendra Gadagkar Chairman Centre for Ecological Sciences, Indian Institute of Sciences Bangalore deserves special thanks as he has always been a source of inspiration, needed help, guidance and encouragement throughout my career. Without his active support and inspiration I could have never ventured to compile such a book.

I am fortunate enough to have had the opportunity to spend past 20 years in different but complimentary intellectual environments where I had the opportunity to work with Professor Dr. M. Amin Masoodi former Director Research SKUAST Kashmir who was always available for help and get me out of the hole as when needed. Dr. F.A. Zaki Professor & Head Entomology SKUAST Kashmir deserves special thanks for his encouraging attitude and working atmosphere provided during my services in SKUAST Kashmir. I am also thankful to my students in beekeeping, bee biology and social insect classes through the years and hope this book will stimulate them to ask more questions.

I express a deep sense of gratitudes to my university authorities for the excellent working atmosphere provided in the University for Smooth sailing of my work and needed encouragement for compiling such a voluminous book.

I am also extremely thankful to M/S Scientific Publishers (India), Jodhpur especially Mr. Pawan Kumar Sharma who took great pains and keen interest in publication of this book in a very impressive way.

Last but not the least my sincere thanks are due to my wife Professor Dr. Asha Abrol, son Rajat and daughter Vitasta for their endurance and help while writing this book.

Dr. D.P. Abrol

Preface

India is a country of diverse climates, cultures, mysteries and amazing phenomenon. The land resources are being limited but increased agricultural production is to be obtained through intensive farming i.e. higher cropping intensity, better seed and greater use of fertilizers. New cropping patterns are likely to create new problems, new pests may appear or pests now considered minor may become major. In some crops, any amount of fertilizer, irrigation or pesticide use may not give even a fraction of yield unless pollinated by bees. Honeybees play an important role in the pollination of large portion of the angiosperms of the world and maintain natural vegetation needed for survival of the ecosystems and the world as whole. Honeybees have always fascinated the mankind since the times immemorial. One comes across a number of folklores praising the honeybees diligence, usefulness and sacrifice. These winged creatures find mention in all the religious epics of the world. The carving of honeybees, their combs and hives can be found in tombs, coffins, crowns and maces of both ancient and modern empires. Honey and bees wax are used in many rituals, ceremonies and festivals and on many occasions in the life of individuals. In earlier times, honey was considered as a commodity of trade exchange between different nations of the world.

Honeybee civilization is much older than the human civilization. The first bee evolved in the tertiary period (Eocene) about 50,000,000 years ago. Beekeeping in India has been practiced from times immemorial and their mention can be found in epics such as Rigveda (4500-1500 BC), Upanishad, Mahabharta and Ramayana (400 BC).

People of all ages carried on a tide of interest into a relationship with this marvelous insect have always wanted to know whether beekeeping could be carried in their circumstances? The beekeeping is possible in all those areas which have sufficient floral resources. However, it is essential for a prospective beekeeper to know the handling of bees before handling the honeybee colonies.

In this book I have tried to give the beginner the basic information, he will require to start up his colonies. Here I have tried to put over the latest scientific advancements as well as my own experience and philosophy of beekeeping and practical methods that have proved very productive and rewarding. The last few years have seen many strides in beekeeping and it has been my endeavor to capture the richness and flavour of many approaches which will be most rewarding to the prospective and enthusiastic, new recruits to the science of beekeeping. In this book I have pointed the ways towards the running of small scale units to large enterprises. If what is written in this book is able to motivate and smoothen the path of some future beekeepers and hobbyists, I shall feel highly rewarded.

The success of beekeeping depends upon understanding of the biology and behaviour of honeybees, their management techniques including knowledge of their diseases and enemies and latest equipment for handling them. In this book an attempt has been made to put the authentic and up to date information in a concise and popular style. It is intended to serve as a reference and guide book for students of agriculture, teachers, scientists, horticulturists, plant breeders, geneticists, extension workers and all those who are interested in bees and beekeeping either as hobby or profession. Efforts have been made to present the book in the form of a continuous story. In addition to basic concepts of beekeeping some topics of general interest have also been included.

This book is one of the complete readings in apiculture dealing with the theoretical and practical approaches. The book contains 34 chapters covering all aspects of beekeeping which include important events and pioneers in beekeeping, history of beekeeping, type of bees, their evolution and biodiversity, honeybee species, their biology, form and function, beginning beekeeping, establishing apiary, bee appliances, bee behaviour, bee pasturage, management of colonies and their manipulations, mass rearing of queens, honeybee nutrition, two queen management system, migratory beekeeping, breeding of honeybees, use of honeybees and wild bees for pollination, bee products and their value addition, diseases and enemies of bees, breeding bees for disease resistance, inbreeding depression in honeybees, quarantine control of bee diseases, artificial insemination, pesticide toxicity to bees, biotechnological potential of bees, honeybees as biosensors, ancillary industries, marketing of bee products, future strategies for development of beekeeping, bibliography and glossary of beekeeping terms etc. Besides this, information is also provided on books, CDs, Vedio and important websites.

"The simple language and lucid treatment of the subject shall make this book easily readable and highly useful. This book shall serve as a reference book for students, teachers, and researchers and for all those interested in bees and beekeeping. This book will be useful to all those who wish to make beekeeping their hobby or as profession, entrepreneurs and even layman. Besides, the information provided in this book will be useful to pollination biologists, students, teachers, scientists of agriculture, animal behaviour, botany, conservation, biology, ecology, entomology, environmental biology, forestry, genetics, plant breeding, horticulture, toxicology, zoology, seed growers and seed agencies." It will be highly useful to motivate the young generation to fascinating world of honeybees and adopt beekeeping as a profession. The extension workers and policy planners will find this book as a guide for their problems and evolving strategies.

There is certainly a room for more honey production as the vast areas of forage remain unexplored and unexploited or underutilized with little benefit to man or nature. Taking honey from an area removes very little other than carbon dioxide and water and is the sort of exploitation of the environment that most will be prepared to forgive.

In case the tone and temper of the book is appealing to the lovers of beekeeping I shall feel highly elevated and properly rewarded.

Although there are large number of standard and good books available on the subject but they are not in accordance with the local needs of the people and are very costly. Furthermore, much technological advancements have been made, for instance, the number of honeybees species has increased from 4 to 9 and some new diseases/pests have been discovered over the years which warrant their documentation for the benefit of beekeepers. Hence, a book fulfilling the basic needs and latest technological advancements was considered to be of immense importance.

This book is not a literary work of mine but is the mirror of facts and fundamentals of beekeeping. If linguistic errors might have crept they may be ignored, as language is the only means of relaying ideas and they do not defeat my purpose if I am able to let others know how the beekeeping could be carried out successfully.

Finally the criticism and suggestions for further improvement of the book shall be most welcome and highly appreciated.

Jammu

D.P. Abrol

Contents

Chapter 1

Introduction

A honey bee is no *ordinary* insect, but rather an extraordinary insect! Honey bees are amazing creatures! A honeybee is one of the most fascinating and marvelous insect whose usefulness is known to the mankind since the prehistoric times. Honey harvesting dates back to 7000 B.C. and is perhaps the only industry which besides the production of sweet honey, bees wax and bee venom also help in increased crop production through pollination of crops, continuation and propagation of many plants growing in nature thereby maintaining the stability of ecosystem, environmental quality and biodiversity. These tiny winged creatures find mention in almost all the epics of the world. A honeybee colony has fascination of its own, the poets, the naturalists and thinkers have always admired them for their industriousness, self sacrifice, unity, calmness of spirit, toleration, equitable division of labour in their colonies and a spirit of social service. If one wishes to acquire happiness, needed diversion, develop a philosophy of self-sufficiency, self preservation and worship the supreme art of love then one must learn beekeeping. The face of the humanity would have been quite different if the humanity could learn socialization from bees.

The question what is this book about? The answer to this question is very simple. It is about bees and beekeeping, about the instincts governing the activity of a bee colony at different stages of its development, about the ways to master these instincts and how to use them in practical ways.

Honeybees play an important role as natural pollinators for a wide variety of crops as well as plants growing in the wild. While visiting flowers to collect nectar, the bees transfer pollen from one flower to another. Three quarters of the world's cultivated crops are pollinated by different species of bees, and honeybees are the most effective and reliable pollinators. They also play an often unrecognized role in maintaining the vegetation cover: more pollination means more seed, more young plants and eventually more biomass, providing food and habitats for birds, insects and other animals.

Beekeeping is crucially important for agricultural well-being; it represents and symbolizes the natural biological interdependence that comes from insects, pollination and production of seed. Useful small-scale efforts to encourage beekeeping interventions can be found throughout the world, helping people to strengthen livelihoods and ensuring maintenance of habitat and biodiversity. Beekeeping fits in well as a subsidiary activity alongside many other livelihood endeavors because bees are capable of harvesting nectar and pollen in plants in the same natural resources which otherwise go waste. There is no competition with other insects or animals for these resources that otherwise would be inaccessible to people. Beekeeping ensures the continuation of natural assets through pollination of wild and cultivated plants. Flowering plants and bees are interdependent: one cannot exist without the other. As bees visit flowers, they collect food and their pollination activities ensure future generations of food plants, available for future generations of bees and for people too. It is a perfect self-sustaining activity. Pollination is difficult to quantify, but if it could be measured it would be the most economically significant value of beekeeping.

Honey bees produce or collect a variety of products which include honey, beeswax, pollen, royal jelly, and propolis. Beekeeping is an agricultural activity that does not demand separate land or compete with other farm animals for fodder and other materials. Beekeeping is possible in areas where abundant flowering plants providing nectar and pollen are available. Most of the forest sites, agricultural farms and fruit orchards can be selected as areas for beekeeping. It can provide employment to all members of the family -- old, physically handicapped, grown up children, young men and women. Beekeeping can be done in one's own backyard in his rural house. Beekeeping helps rural populations to become self- reliant It favours diversification of the local economy -- from manufacture of materials necessary for beekeeping by rural craftsmen, like carpenters, blacksmiths, and tailors. Apiculture augments national productivity through increasing oilseed, pulse, spice, fruit and other crops by bee pollination. The economics of beekeeping vary with the level of the industry, type of vegetation and beekeeping potential, and the infra-structural facilities available. Beekeeping is recognized as a low input and high output activity, suitable for rural, tribal and other weaker sections of population. The peculiarity of this agrobased industry is that it does not require any raw material from the artisans like other industries. The raw material is in the form of nectar and pollen from flowers which is freely available in nature. Beekeeping is a decentralized, forest and rural agrobased industry. Beekeeping can be started by any one who takes interest may be skilled, unskilled, man woman, old and young, working or retired persons. A technology that is simple, easy accessible and at the same time demanding the least capital investment is suitable to this type of industry. Beekeeping may even be started with a single colony which can be increased to thousands of them. It can provide unemployed and underemployed persons with full employment.

Honey is also a sweet base for a number of medicines and bee venom is used in many pharmaceutical applications, especially to cure rheumatic diseases. It is a natural dehydrant and excellent for those on slimming diets. As a proven anti-ageing agent and natural rejuvenator honey has no equal.

In India, production of honey is very low compared to China - the highest producer - which exports 80,000 tonnes annually compared to India's 7,000 tonnes. Its consumption is also very low in India. Honey production in the country is only about 27,000 tonnes a year. Only about 20 to 25% of the beekeeping potential is being exploited at present in the country. Punjab, Haryana, Uttar Pradesh, Bihar and West Bengal are the major honey producting states. Germany is the world's largest consumer, importing 90,000 tonnes of honey products annually. The major problem in expansion of beekeeping is that there is complete lack of beekeeping benefits among the farmers and honey consumers. In India, honey is still not being utilized as food and only utilized in rituals. The per capita consumption of honey in Switzerland is 2.0 kg compared to a dismal 3-5 g in India (Table 1). In the world market the demand for honey is around one million tonnes. There is an immense possibility for India to increase its export share from 7,000 tonnes to three lakh tonnes if more people invest in bee colonies.

Table 1. Honey consumption pattern in different countries of the world.

S.No.	Country	Honey consumption per person per year
1	India	3-5 g
2	Greece	1.8 kg
3	Australia	1.6 kg
4	Germany	1.5-2.0 kg
5	Italy, Spain, France and Hungary	0.6-0.9 kg
6	UK	0.4 kg
7	China	100 g
8	Russia	500 g
9	Serbia, Ukraine	600 g
10	USA	800-1000 g
11	Japan	300 g
12	Poland	400 g
13	Switzerland	2.0 kg
14	Nepal	3 g
15	Canada	700 g
16	Sweden	700 g
17	Finland	580 g
18	Russia	350-400 g
19	Ukraine	1.2 kg

The present trend in the beekeeping industry gives for raising the number of colonies to around four million and, thereby, increasing honey production to one lakh tonnes in another decade. A major portion of the honey produced in the country is used in medicines and only a small quantity finds its place on the table as food. Bee stings have been used as a medicine for many decades in Europe and Russia, especially in the treatment of muscular diseases. Bee wax is a high value product and its consumers are cosmetics, candles and paint industries.

Although honey bees can be managed to produce large quantities of these products, they are especially valued for the major role they play in pollination. While other insects and animals also are pollinators but their actions or numbers are not under immediate control. Honey bee colonies, however can be placed wherever and whenever they are needed. Furthermore, honey bees have additional advantages over other pollinators, as they could be available in large numbers and have instinctive pollen hoarding behavior. In the absence of pollinating service of honey bees, the cost of many fruits, vegetables, legumes, and seeds would be many times more what it is today.

The honeybee industry produces a diverse range of valuable commodities including honey, beeswax, propolis and royal jelly. This contribution is small, however, compared to the importance to pollination services provided by the industry. Honeybee pollination is essential for some crops, while for others it raises yield and quality. In addition to the crops, a wide range of pastures, including Lucerne and clover, are pollinated by honeybees hence this estimate understates the potential value of the pollination services.

Honey has a long and distinguished history in the human diet. For thousands of years honey hunters have plundered the hives of wild bees for their precious honey and beeswax – a practice still common today. The most widely used honey bees are the European *Apis mellifera,* which have now been introduced worldwide. Tropical Africa has a native *Apis mellifera,* which is slightly smaller than the European *Apis mellifera,* and is more likely to fly off the comb and to sting. They are also more likely to abandon their hives if disturbed, and in some areas the colonies migrate seasonally. In Asia, there are three main native tropical species, *Apis cerana, Apis dorsata, and Apis florea. Apis cerana* is the only species that can be managed in hives, but the single combs of the other two are collected by honey hunters.

The modern bee keeping became possible after the discovery of movable frame hive in 1851 by Revd. L.L. Langshoth. In India beekeeping was introduced in 1882 in Bengal. Revd. Newton introduced beekeeping to South India in 1911. But still India is much behind USA, Canada, Europe, Australia and New Zealand in beekeeping.

In addition to playing a crucial role in pollination and thereby improving crop yields, honeybees contribute in a balanced way to rural development efforts leading to secure and sustainable livelihoods.

It is generally known that bees are needed to pollinate our crops but it is not well known that the economic value of bee pollination is several times more the value of the world-wide production of honey. About 80% of our food crops are pollinated by animal pollinators. These are mainly bees. It is estimated that one third of what we eat and drink is produced through service supplied by pollinators. They appear not only to be extremely important for traditionally grown and well-known crops, but they are also essential for economically promising tropical and less common crops.

Besides honeybees, the specific pollinator role played by Non-*Apis* bees, such as bumble bees, solitary bees and tropical stingless bees is of immense significance. A great majority of angiosperm (flowering) plants depend upon animals for their pollination. Of the animals that visit plants and are responsible for the spread of the pollen, a great majority belongs to the insects, for example flies (Diptera), beetles (Coleoptera), butterflies (Lepidoptera), but most important, bees (Hymenoptera: Apoidea). Bees because of their morphological adaptations for the collection of pollen are considered to be the most efficient pollinators.

Certain groups of bees are able to perform specialized pollen collecting behavior, e.g. "buzz-pollination". In a wide range of angiosperm families, pollen can only be released when the stamens are shaken by vibrating bees. This buzz-pollination is performed by bumblebees, carpenter bees and by stingless bees of the genus *Melipona,* but not by honeybees. The production of crops, that need to be pollinated in enclosed environments like greenhouses, and therefore in the absence of natural pollinators, implies a new dimension for the application of bees as pollinators. In view of the available management technology and the actual pollination value, the honeybees are considered to be the most significant tool for seed production. However, other species, like the bumblebee and several solitary bee species are also being used for the pollination of greenhouse crops and ornamentals (Estes *et al.,* 1983).

Despite an increasing recognition of their important role in pollination, the population and diversity of honey bees is declining due to the habitat loss through land use changes, increasing monoculture and negative impacts of pesticides and herbicides.

During the last few years, there have been an impending "pollination crisis" in certain parts of the world due to decline of honeybee colonies due to attack by pests and diseases combined with a general increase in the area of bee-pollinated crops. In some countries the demand for pollination is increasing, at the very time that the supply of managed pollinators is decreasing. This pollination crisis is further raising the interest in management, culture and conservation of pollinating bees. Modern intensive agriculture and certain ways for managing our environment may have important consequences for the ecological position and the conservation of bees in this environment. Certain developments are considered to be detrimental for beekeeping. The use of agro-chemicals and of genetically modified crops are much discussed in this respect. Often it is difficult to reach a proper opinion about risks and benefits of these technologies in relation to beekeeping. This is a very important field for most beekeepers in the world. This includes the obtaining of knowledge concerning special characteristics of honey types from certain regions. It is important to document information on the relative importance of bee food plants and honey producing plants in the different countries of the world. This includes the analysis of the botanical origin of honeys, pollen and even propolis. Melissopalynology, the microscopic identification of pollen grains, is a classic powerful tool for the study of bee-food plant relations. The beekeeping has a tremendous scope for the development of ancillary industries. The untapped potential of beekeeping yet remains to be explored for increasing opportunities for gainful employment and income in rural areas. In summary, honeybee is perhaps the most studied insect.

Many biologists and naturalists have unfolded strange and amazing facts about honeybees which include; discovery of dance language (von Frisch, 1953); queen substance (Butler, 1954); social behavior (Ribbands, 1953; Michener, 1974); caste differentiation (Haydak, 1943, Beetsma, 1983); artificial insemination (Watson, 1927, 1928) and utility of bees as pollinators (Free, 1993; Kapil and Jain, 1980; Abrol, 1997).

There are several reasons why honey bees are perhaps one of the most studied insects (probably next to *Drosophila* in terms of amount of money spent and number of papers published). Honey bees play a critical role in agriculture. The most important role honey bees play is actually not honey production, but pollination. The value of crops that require pollination by honey bees, in the United States alone, is estimated to be around $24 billion each year and commercial bee pollination was valued around $10 billion annually. There is also a trend to consume more bee-pollinated crops (such as fruits and vegetables), making honey bees more and more important in agriculture. Honey bees also produce honey and beeswax, which are valued at $285 million in US annually. Besides that bees also produce pollen, propolis, royal jelly, and bee venom that are playing increasing roles in health food and alternative medicine. Bee stings are routinely used for treatment of arthritis, multiple sclerosis and other auto-immune diseases. Honey bees are studied extensively, also because they are fascinating organisms. Bees have captured mankind's attention since as early as Aristotle. Not only because they produce honey and honey is the earliest sweetener human beings have found, but because of their industriousness (working to their death), selflessness (producing honey for humans and dying to defend their home), and most importantly, their social organization. Honey bees, like other social insects, show "division of labor" whereby different workers specialize on different tasks. In some sense, the complexity of their society rivals that of our own. Who governs their day-to-day chores? How do workers know what to do in a city bustling with tens of thousands of individuals? Clearly these have been the questions of humankind since long time ago, as evidenced in the Bible: "Ants are creatures of little strength, yet they store up their food in the summer. Locusts have no king, yet they advance together in ranks" [Proverbs 30: 25-27]. Honey bees are increasingly being used as a model system to study other aspects of biology. Besides their intricate social organization, honey bees are easily maintained, are cost effective in terms of obtaining large numbers of insects, and their genetics can be precisely controlled. Honey bee workers take 21 days to develop from eggs to adults, this compares favorably with other insects commonly used in classrooms (such as cockroaches, grasshoppers). Since a queen can lay as many as 1,500 eggs a day, large numbers of bees can be obtained easily. Honey bees are probably the only insect that has "artificial insemination" technique successfully invented. This is used extensively both commercially and in research to speed up the selection process and to control the exact genetic makeup of a colony.

The honey bees are not domesticated in true sense but one has to understand and adjust his methods to gain maximum from the hard toil of honeybees. To be a successful beekeeper one must learn about honeybees and beekeeping, about the instincts governing the activity of a bee colony at different stages of its development, about the ways to master these instincts and how to use them in practical beekeeping.

Chapter 2

History and importance of beekeeping

History of beekeeping

Ancient civilizations considered bees as mystical emissaries from the spirit world, and both the Greeks and Egyptians carved their likeness onto tombs. Poets-from Homer to Yeats and beyond-have sung their praises, and scientists have studied them with such fascination that more is known about bees than any creature besides man. Most of all, farmers know their magic because without bees, it would be next to impossible to grow the food crops on which all of us depend upon.

Honey bees evolved about 40 million years ago, and as soon as human beings came along, they started robbing their hives. Rock paintings dating from as far back as 7,000 B.C. show honey-gatherers at work. No one knows when people first started keeping honey bees, but it's clear that the ancient Egyptians, Greeks and Romans were all avid beekeepers, using long clay cylinders as hives and, like their modern counterparts, utilizing smoke to manage bees. Beekeeping was a fairly passive endeavor then because little was known about bees or how to husband them, but by the 1600s that began to change as scientific knowledge of bees advanced. Among the discoveries that surprised early entomologists was that the large bee that seemed to rule the hive was in fact a queen, not a king. The 1800s brought a number of breakthroughs, including the discovery of "bee space" by L.L. Langstroth, called "the father of modern beekeeping." Langstroth discovered "bee space"- the small distance (about 3/8 inch).

Evidence of bees goes back as far as 140-60 million B.C. where archaeologists found evidence of pollen and insects. Flowers evolved in relation to insects including bees. A cave-drawing of a man climbing a rope to collect honey was found which was painted in 7000 B.C. in Spain. Egyptians did their beekeeping in clay pots and were the first people to operate migratory beekeeping by putting the clay tubes containing the bees onto barges on the Nile. Egyptians and Greeks preserved human bodies in honey, and believed that bees were the tears of the sun god Rah and there are many examples of carvings of bees on ancient Egyptian buildings. The Romans had beekeepers and knew a lot about bees.

Figure 1. Honey seeker depicted on 6000 year old cave painting near Valencia, Spain

A recent find in the German Eifel region (Anonymous 1993a) indicates that the earliest honey bee existed about 50 million years ago. Man, in comparison evolved only recently and man's history on earth may not have been more than a couple of million years. There are indications that honey was one of the natural foods the early man tasted. His association with honey bees had been as a hunter for thousands of years. Some of the earliest records of this hunting have been bequeathed to us by the Paleolithic and Mesolithic man in the form of cave paintings. Mathpal (1978, 1984) recorded several paintings of hunting rockbee hives by the pre-historic man that according to him might be as old as 15,000 to 11,000 B.C. Similar cave paintings have also been found elsewhere in the world. In fact, the first of such records was from Spain dating back to 6000 B.C. Paintings in rock shelters also occur in Zimbabwe and other African countries. Almost all the paintings outside India depict the hive bees (*Apis mellifera* L.), which build several combs in dark enclosures. The area now comprising Israel and the Palestine autonomous region is often referred to as "the land of milk and honey." (Exodus 3:8). In India, the paintings often depict the rockbees which build one large comb in the open, on arboreal or terrestrial supports.

Painting on shelter walls might have been an expression of artistic talent and creative impulse as a part of some religious or cult practice. Before the men went on a hunting trip, they probably performed a ritual in which the object of their hunting was drawn. (Wakankar and Brooks 1965, 1976). There is evidence that the prehistoric man used honey in their cooking. Beekeeping grew as Christianity came through Europe, mainly because of the use of candles and also the interest in mead. Mention of bees are found in almost all the epics of the world e.g. Vedas, Ramayan and Quran. For example, in Koran prophet

Mohammed said "honey is a remedy for every illness". In Hindu philosophy bee is shown in lotus flower resembling god Vishnu.

In ancient times bee were used as a weapon against the enemy and Bee hives were thrown at attacking armies. Beekeeping in real sense started when in the 19th century Langstroth worked on creating a hive with frames in it. Many other beekeeping pieces of equipment were invented at this time including the honey spinner, the queen excluder and smoker. In 1891 Porter invented the Porter Bee escape. The modern bee keeping became possible after the discovery of movable frame hive in 1851 by Rerd. L.L. Langshoth. In India beekeeping was introduced in 1882 in Bengal. Rerd. Newton introduced beekeeping to south India in 1911. But still India is much behind USA, Canada, Europe, Australia and Newzealand in beekeeping.

Honey has a long and distinguished history in the human diet. For thousands of years honey hunters have plundered the hives of wild bees for their precious honey and beeswax – a practice still common today. The most widely used honey bees are the European *Apis mellifera,* which have now been introduced worldwide. In Asia, there are three main native tropical species, *Apis cerana, Apis dorsata, and Apis florea; cerana* is the only species that can be managed in hives, but the single combs of the other two are collected by honey hunters.

Development of modern equipment

For thousands of years, colonies of honey bees were kept in wooden boxes, straw skeps, pottery vessels, and other containers. Honeycomb built in such hives could not be removed and manipulated like the movable combs of today.

Advances in beekeeping

Beekeeping changed most between 1500 and 1851. Until then, hives were removed and squashed by their keepers without an understanding of the life cycle of the bee or the structure of the hive. In the fall the bees were killed and the contents of the hive were strained to remove the honey. Only a few hives were kept over winter to repopulate the apiary in the spring. The first revolutionary discovery involved understanding the life cycle of the bee. In 1586, Luis Méndez de Torres first described the queen bee as a female that laid eggs. In 1609, Charles Butler in Feminine Monarchie identified the drones as male bees and in 1637, Richard Remnant in Discourse or Historie of Bees recognized that the worker bees were females. The understanding of the life cycle helped beekeepers care for their colonies.

The next advancement was needed to allow beekeepers to remove the honey, while doing the least amount of damage to the hive and bees. Sometimes, basket hives could be combined by driving the bees from one hive into the other. Honey could be harvested from the empty basket hive and the workers driven into the new hive would be accepted although their queen would be killed. Other approaches involved adding jars or caps as extensions to the top of the hive for honey storage. The extensions could be removed without disturbing the brood.

The understanding of the life cycle and the ability to harvest honey without disturbing the hive made beekeeping easier. However, for 200 years hives could not be managed because it was impossible to observe what was going on inside. Various kinds of frames and hive box were developed, but one problem remained. The bees would always build comb between the frame and the hive so that the comb could only be removed by cutting it out.

The first step toward a movable frame hive was the use of hives which resemble Top-Bar-Hives. These consisted of baskets or boxes with slats or bars across the top and a sloped container to serve as the hive. The bees did not attach comb to the sides of the hive and the beekeeper could remove each frame and examine the brood, comb and honey stores without bothering the bees.

In 1851, the American Lorenzo Langstroth designed the movable frame hive most commonly used today which carries his name. His design has four-sided frames which hang inside a hive box allowing a 3/8 inch space between the frames and the box. This distance, known as bee-space is the width of a path wide enough for a bee to pass through, but too wide to be filled with wax or propolis. The hive is made of units or "supers" that may vary in height, but with identical outside dimensions so they may be stacked. This hive design greatly facilitated hive management and beekeeping entered a new era.

In 1852, L.L. Langstroth, patented a hive with movable frames that is still used today. The principle upon which Langstroth based his hive is the space kept open in the hive to allow bees passage between and around combs. This space is about three eighths of an inch wide; space that is less than this is sealed with propolis and wax, while space wider is filled with comb. Langstroth is called "the father of modern beekeeping." Modern methods of beekeeping came very rapidly following Langstroth's patent. Other inventions soon followed that made large-scale, commercial beekeeping possible. Wax-comb foundation, invented in 1857, made possible the consistent production of straight, high-quality combs of pre dominantly worker cells. Pellett (1938) gives a detailed account of the development of wax-comb foundation. The invention of the centrifugal honey extractor in 1865, and its subsequent improvements, made possible large-scale production of extracted honey. The bee smoker, as now used by beekeepers, evolved from a pan used to contain some burning, freely smoking material, the smoke of which could be blown across the open hive to control the bees. The all-important bee veil gradually evolved from pieces of coarse cloth that were wrapped about the head of the beekeeper.

Bees in written history

The mention of bees or honey in Greek and Roman mythology is common. Aristotle in his History of Animals knew about bees locating flowers: ".As for insects, both winged and wingless, they can detect the presence of scented objects a far off, as for instance bees."

In Greek culture, honey is considered food for the Gods and refereed to as Ambrosia, the golden nectar. Civilizations across the world have considered honey sacred and magical, because of its purity. It was made into an intoxicating drink...**Mead** (which was one of the things the bride's father was supposed to supply to the groom for a full month after marriage- thus starting the tradition of **honey moon** in ancient Babylon). Mead was otherwise consumed during festivals and also offered to Gods. The fact that honey does not go bad even for centuries, if properly preserved, has also served to make it into everybody's idea of a divine food. Archeologists found a sealed jar of honey in a tomb in Egypt in 1800, and on opening, it was fund to be tasting perfect, even though it was thousands of years old. The importance of honey in various fields of daily life was appreciated for centuries, across civilizations. In discovering the new world, when the Spaniards reached the Americas in the sixteenth century, they found that the `natives' of the continent had already develop bee-culture and were already consuming honey. In fact they had distinct species of bees that were cultivated there. In 1638, the White invaders went ahead and introduced European honey bees to the New England colonies, and these were subsequently called `white Man's bees'. Scientifically speaking due to its higher monosaccharide composition, honey has greater power of sweetening than even sugar. But it contains fewer carbohydrates on weight basis (304 calories per 100 grams while refined sugar contains 400 calories per 100 grams). In addition it contains 38.5% fructose, 31% glucose, 17% water, and a number of other nutrients like carbohydrates, minerals, proteins, amino-acids besides.

As a medicinal base, almost all civilizations, old and new, have used honey for centuries. Ayurveda treats it as food for health while even ancient medicine systems used it both an external applicator for some conditions, as well as an oral medicine. In India it is believed that cooking honey will break down amino acids and cause acidity in the system. Besides, the age of the honey is also important in the Indian food system. Young honey (less than six months old), is healing for people with pitta characteristics. Fresh honey is actually best for everyone, but then it can be consumed with foods that counter its negative properties like cold and dry basic nature. Accompanied, for example, with sesame seeds, it can help warm the system. With yoghurt, it serves to enhance the astringent properties. Adding honey to boiled milk, serves to render warmth and drying qualities to the milk. In all forms, at all times, one may be sure that there are very few people who can resists the temptation of this warm, golden liquid...that has not been called the nectar fit for Gods for nothing.

During the 20th century, the dimensions of bee hives and frames became more standardized, thus eliminating the various sizes that were so confusing 100 or more years ago. The 10-frame movable comb hive is now used throughout the world wherever beekeeping is seriously practiced. Most beekeepers use full-depth standard hive bodies for brood chambers; some also use them for honey supers, while others use shallow or half-depth bodies. Development of strong colonies for major nectar flows rests upon such fundamentals as hive room, adequate stores, and high-quality queen bees. Commercial and part-time beekeepers control swarming in their colonies, but beginners still have difficulties. Drugs (antibiotics) are now available for the control of foulbrood and nosema disease. Artificial insemination of queen bees, that is, controlled mating, is being used commercially to a limited extent. The practice of renting colonies for the pollination of certain crops has increased markedly in this century, although management of colonies for such purposes needs to be improved.

History of beekeeping in India

Bees and Beekeeping in India is not new, the winged creatures find mention in almost all the religious epics such as Vedas, the Ramayana, the Quran and many other ancient books. The ancient Vedic hymans often refer to bees, their flower relationship with honey (Dave, 1954, 1955) "let our plants be enriched with profuse nectars" Says one of these proverbs seeking abundance of milk and honey. The Indian system of medicines like Ayurveda and Sidha heavily rely on honey as a carrier which enhances the properties of drugs.

The knowledge base on bees, honey and honey collection had been expanding through the ages. During the Vedic period, a large information base was available on different species of honey bees, their habits, different flowers that offered nectar and pollen to bees, types of honeys and several other aspects. Dave (1954, 1955) presented a convincing argument of this advanced knowledge base that was available during the Vedic period and summarized the main scientific and technological aspects as follows.

Honey had been variously praised as a nutritious food and valuable medicine in Rigveda and several later works. It is in the Ayurveda, a part of Atharva Veda, however, that honey found a place of honour, as a valuable nature's gift to man. References to honey bees, their nesting, food gathering and stinging behaviours are repeatedly found in ancient post-Vedic literature like the Brahmanas, Upanishads, Puranas and Ramayana.

Susruta emphasized the aspect of application of honey in surgery, particularly its curative properties on wounds, etc. He recognized eight varieties of honey, depending upon the honey bee producing it. Specific medicinal properties of individual honeys were given.

Honey and honey bees have a long history in India. Honey was the first sweet food tasted by the ancient Indian inhabiting rock shelters and forests. He hunted bee hives for this gift of god. India has some of the oldest records of honey industry in the form of paintings by prehistoric man in the rock shelters. With the development of civilization, honey acquired a unique status in the lives of the ancient Indians. They regarded honey as a magical substance that controlled the fertility of women, cattle, as also their lands and crops. The recent past has witnessed a revival of the industry in the rich forest regions along the sub-Himalayan mountain ranges and the Western Ghats, where it has been practised in its simplest form.

Although honey and honeybees are known to human beings since time immemorial, beekeeping unlike several other industries is not a traditional enterprise though it has a very fascinating history of its own. Collection of honey from wild bee colonies by smoking away the bees and squeezing away their combs for honey was traditional since time immemorial. Efforts were made to introduce *Apis mellifera* L. the European honey since 1880 but for various reasons these experiments did not meet success. Historical account of beekeeping has earlier been given by several authors (Singh1962, 1964; Mishra, 1995., Abrol, 1997). In India, first attempt to keep bees in moveable frame hives were made in 1882 in erstwhile Bengal and in 1883 and 1884 in erstwhile Punjab but with little success. In 1883, Government of India published the information on beekeeping collected from the provinces under the title 'A collection of papers on beekeeping in India'. A year later, John Douglas (an employee of telegraph department) published a 'Handbook of beekeeping in India' on the basis of his experiences in keeping Indian honeybees in modern beehives. Twenty five years later, in 1907, the work was taken up by imperial entomologist at Pusa institute and it continued up to 1919. Besides a few papers a bulletin entitled 'Beekeeping" was published by C.C. Ghosh. Sir Louis Dane took keen interest in beekeeping and formed Punjab beekeepers association with head quarters at Shimla and issued a publication entitled 'A Guide for Beekeeping" in 1916. The history of beekeeping in India dates back to 1910 when in South Father Rev. Newton devised a hive suitable for *Apis cerana*. The hive, which was named after him as **Newton Hive** is still popular for keeping indigenous honeybee *Apis cerana* in south India. During 1911-1917, rev. father trained number of people in southern India and helped them to establish beekeeping as an economically viable proposition. Travancore state took up beekeeping work in 1917 and Mysore in1925. During 1928, the royal commission on agriculture recommended beekeeping as cottage industry which gave a fillip for the development of beekeeping in right earnest. Efforts to establish beekeeping were made in many states like Madras (1931), Punjab (1933), Coorg (1934) and Uttar Pradesh (1938). Other states and provinces made efforts to establish beekeeping industry but no land mark achievement was made. In 1938, beekeepers of India organized themselves and formed All India beekeepers association during 1938-39. During winter of 1938-39, this association started publishing **Indian Bee Journal** which still has the distinction of being the only journal in the country exclusively devoted to the cause of beekeeping. The association held awareness programmes and conferences for the promotion of beekeeping all over the country. Pt. R.N. Mutto was its founder chairman. Following the demise of Pt. R.N. Mutto, Sh. S.G. Shende took charge and helped the rehabilitation of the association (AIBA). For his untiring contributions and devotion to the development of beekeeping Sh. Shendeji was given the honour of being the life president of the association by its board of management.

The noted pioneers in beekeeping who contributed in the development of beekeeping include Sh. S.G. Shende and Sh. S.K. Kallapur (Western Peninsula), Swami Shambhavanda (Karnataka), Sh. R.N. Mutto (Central Himalaya), Smt.

Rama Devi and Manmohan Chaudhary (Orissa), Sh. A.M. Shah and Sh. Razdan (Jammu Kashmir), Sh. O.P. Sharma and Sh. P.L. Sharma (Himachal Pradesh), Dr. Sardar Singh, Dr. A.S. Atwal and Dr. N.P. Goyal (Punjab), Dr. R.P. Kapil (Haryana).

Indian council of agricultural research established central beekeeping research station in Punjab in 1945 and six years later at Coimbatore in Tamilnadu, Baptala (Andhra Pradesh) and Sundernagar (Himachal Pradesh). After independence the Government of India took a policy decision to rejuvenate the rural industries and constituted All India Khadi and Village Industries Board to take up the task of beekeeping development in the country. This board was later reconstituted as Khadi and Village Industries commission (KVIC) - A statuary body under the Ministry of industry. It was only after the establishment of KVIC at the central level and Khadi and Village industries board at state level, that beekeeping industry receives serious attention for its development. Some states like Jammu and Kashmir, Karnataka, Uttar Pradesh and Himachal Pradesh established Department of beekeeping under the state ministry of industries/agriculture. Further, considering the importance of applied and basic research in apiculture, KVIC established central bee research institute CBRI at Pune, during November 1962. Subsequently, CBRI established some regional centre in various parts or the country. The apiculture research in the right earnest started when Indian council of Agricultural research (ICAR) previously called as imperial council of Agricultural Research started funding the beekeeping projects in the states, central institutions and other organizations. From 1950 ICAR has been funding various research projects in beekeeping. In 1980, the ICAR started All India coordinated research projects (AICP) on honeybee research and training having centers throughout the country.

Introduction and establishment of *Apis mellifera* in India

Although efforts to introduce *Apis mellifera* in India dates back to as early as 1880s but sizeable importations of the Italian and carniolan bees from Europe and Australia were made in Pune (Maharashtra) in 1920s by Ghosh (1920). He imported three colonies from Italy and maintained a small apiary of *A. mellifera*, but failure of queen bee mating resulted in ultimate finish of the apiary till 1931. Then Thompson in 1940 made introductions in Kerala but these hives could not be established. He also reported other unsuccessful attempts in the country (Thompson, 1940, 1944). Baldry in Mahabaleshwar (Maharashtra) maintained *A.mellifera* apiary for 6 years and later these colonies were shifted to Coimbatore where these dwindled and perished. Beadnell (1935) brought many consignments of *A. mellifera* and tried to maintain an apiary at Niligiri hills but did not succeed. Rahman and Singh (1940, 1945) also made unsuccessful attempts. Two consignments each of two and three colonies were imported and in the exchanges made *Apis cerana* and *A. mellifera* even did not accept each others brood, workers and queens. *Apis mellifera* colonies were imported in Kashmir in 1951(Vats, 1953). Queens were introduced into *A.c.indica* colonies which did well till 1959. There was, however, no report afterwards and so these presumed to have been perished.

After a gap, Professor A.S Atwal renowned entomologist of Punjab Agricultural University Ludhiana, with his associates made renewed efforts during sixties to introduce *A. mellifera* in the country first at Nagrota-Bagwan (now in Himachal Pradesh) by following and standardizing a new approach of Interspecific Queen Introduction Technique and then at Ludhiana (Atwal, 1964; Atwal and Sharma, 1968; Atwal and Goyal, 1973). The technique involved importing disease free gravid queens of exotic bee and introducing them singly into the dequeened colonies of *A.cerana* so that after the acceptance of exotic queens, the workers of *A.cerana* reared brood of exotic honeybee resulting in the gradual replacement of workers of *A. cerana* with the *A. mellifera* as the former died with age. The colonies of exotic bee so obtained were not gaining strength at bee research station at Nagrota-Bagwan. Some colonies were then introduced into Punjab plains at Punjab Agricultural University campus Ludhiana, colonies progressed well and reared drone brood in plenty because of long sunny days and assured cultivated bee flora in the plains. By 1972-73, the stock was multiplied and there were 45 colonies of *A. mellifera* at Nagrota-Bagwan and 120 at Ludhiana (Atwal and Goyal, 1973; Goyal, 1974). Need based researches were undertaken to acclimatize and establish the exotic bee at the Punjab Agricultural University (PAU) campus and its out apiaries and to develop sound seasonal bee management technology for *A. mellifera*.

Apis mellifera proved superior from the colony development, honey yield and behavioural point of view as well. In the meantime, some of the scientists in the country opposed the introduction of *A. melifera* with a plea that it would introduce deadly bee disease and parasites prevalent in the countries from where the exotic queen bees were imported, and would thus endanger the indigenous *Apis* fauna. Because of a hue and cry in the Indian Council of Agricultural Research (ICAR), it was almost decided to destroy the whole stock of imported bees with PAU. However, PAU pleaded for the disease freeness of *A. mellifera* stock with it, with fully authenticate record. Ultimately, ICAR appointed a high level committee comprising of Dr. Sardar Singh (the then plant protection advisor of Govt. of India) and Dr. C.V. Thakur (the then Director of CBRI, Pune) to examine *A. mellifera* stock with the PAU, Ludhiana and bee research station, Nagrota-Bagwan. The research data presented by Dr. N.P. Goyal to the expert Sardar Singh panel in 1974 were instrumental in deciding the fate of further continuation and funding the research project on the establishment and spread of *A. mellifera* in India by ICAR (Gatoria *et al.*, 2003). The committee found the stock disease free at both the stations and gave the verdict that research on *A. mellifera*, confined to PAU, Ludhiana should be continued and in future no further stock of any race of *A. mellifera* be imported.

Following the successful introduction and establishment of *A.mellifera* at Ludhiana and Nagrota-Bagwan, ICAR sanctioned Operational research project on the establishment of Italian honey bee in the plains of Punjab in 1976. During this year, PAU also started regular training courses for beekeepers and distributed some colonies of this exotic species to the farmers free of cost. By 1980, *A. mellifera* became very popular among the trained farmers and got successfully established in Punjab and Himachal Pradesh (Chahal *et al.*, 1981). It spread to other states was, however, restricted upto 1986 when the ICAR allowed its extension to other states following devastation of indigenous *Apis cerana* bees due to Thai sac brood viral disease (TSBV). Beekeeping with *Apis mellifera* resulted in complete change in the scenario of beekeeping development in the country and a further ban on import of any race of *A. mellifera* was imposed by the ICAR.

During 1993, ministry of agriculture, department of agriculture and cooperation laid special emphasis on beekeeping as an important component for increasing crop productivity and started a National scheme on beekeeping. Under this Scheme, beekeeping research and development projects were sanctioned to various SAUs/ Agriculture Departments' Govt and Non-Govt organizations.

Until 1960, beekeeping remained confined to *A cerana* and that too in the north hill region, southern states and north eastern region. Keeping in view the area, topography, and the population in India, the progress in beekeeping seems to be meagre. The National Commission on Agriculture (NCA) in its Report to the Government of India, laid down a target of establishing 6 million bee colonies by the year 2000 A.D. The Commission arrived at this figure by assuming that new bee colonies would be established through rearing of queen bees and increasing the existing number of bee colonies in the country. During 1971-72, beekeeping was present in about 30,000 villages. Looking to the Central Bee Research and Training Institute, Pune that had produced about 1,500 new queen bees under its several field centers, the NCA presumed that one centre could produce 2000 queen bees, and that one such centre should cater to the needs of 5000 villages. As there were 0.6 million villages in India, there should be 120 queen bee production centres. Each centre would produce 2000 queen bees in a year and in 25 years (NCA made its recommendations for development of beekeeping during the period 1975-2000), the 120 centres would produce 6 million queen bees. Obviously this estimate did not consider the actual potential of bee forage availability that sustains the bee population. However, this Herculean task of achieving six million colonies and a production of 60,000 tonnes of honey annually by 2000 A.D. got an unforeseen jolt during 1980, when a deadly viral disease called Thai sac brood diseases appeared in *A. cerana* colonies.

Though *A. mellifera* was successfully established in 1962 but was extended to farmers of Punjab in 1976. This long gap perhaps lead to the great delay in the spread and development of beekeeping in Punjab. Further, restrictions on its spread in the later years to other states proved still more damaging to the growth and development of beekeeping with *A. mellifera* in India. India is unique in having a variety of bee fauna, bee flora, climatic conditions and agricultural and horticultural practices. These present unlimited opportunities for production of honey and other nutritive, industrial and medicinal materials from bee colonies, as also of crops through pollination by bees. Recent trends in deforestation, reduction in the area of the natural habitats of honey bees and bee plants, pollution of atmosphere, and of the sources of bee food by insecticides, high costs of inputs, non-availability of timber for bee box manufacture, levying of taxes on honey at different stages of the market, etc. are disconcerting, and can have serious long term repercussions on the industry, and ultimately on even the agricultural economy of the country. Despite, serious efforts made by the Govt. of India, the growth of beekeeping in India is not yet encouraging and the number of colonies as also the honey production remains miserably low.

The country has about 75 million hectares notified as forests, constituting 19.47% of the total geographical area. Of this, about 41 million hectares are classified as reserved; 21 million hectares are protected forests, and the remaining about 13 million hectares are unclassified forests (Anonymous 1993b). Table 1a gives the state-wise area under different forest types. This area under natural vegetation constitutes the main resource of raw material for the honey bees.

Table 1a. State-wise area (million hectares) under different forest types in India

States	Forest types														
	1	2	3	4	5	6	7	8	9	10	11	12	13	14,15, 15	Total
Andhra Pradesh			3.00	0.17	1.68	1.76	0.03								6.64
Arunachal Pradesh	1.22	-	-	-	-	-	-	1.89	-	1.51	-	-	0.54		5.16
Assam, Meghalaya	1.19	0.30	2.91	-	-	-	-	0.10	0.07	-	-	-	-	-	4.57
Bihar	-	-	0.29	-	2.79	-	-	-	-	-	-	-	-	-	3.08
Gujarat	-	-	0.36	0.05	0.41	0.13	-	-	-	-	-	-	-	-	0.95
Haryana	-	-	-	-	0.08	0.03	-	-	0.03	-	-	-	-	-	0.14
Himachal Pradesh	-	-	0.01	-	0.21	-	-	-	0.73	0.09	-	0.75	0.07	0.30	2.16
Jammu & Kashmir	-	-	-	-	0.08	-	-	-	0.28	0.07	-	0.92	0.14	0.59	2.08
Karnataka	0.60	0.59	0.83	-	0.41	1.10	-	-	-	-	-	-	-	-	3.53
Kerala	0.55	0.26	0.22	-	-	-	-	0.01	-	-	*	-	-	-	1.04
Madhya Pradesh	-	-	7.26	-	9.76	0.27	-	0.02	-	-	-	-	-	-	6.68
Manipur	0.19	-	0.27	-	-	-	-	-	0.14	-	0.01	-	-	-	0.61
Nagaland	0.08	0.05	0.02	-	-	-	-	-	0.09	0.07	-	-	-	-	0.31
Orissa	-	0.28	4.49	0.11	1.82	-	-	0.11	-	-	-	-	-	-	6.81

Punjab	-	-	-	-	0.14	0.03	-	-	0.02	0.01	-	-	-	-	0.20
Rajasthan	-	-	-	-	3.28	0.47	-	0.02	-	-	-	-	-	-	3.77
Tamilnadu	0.02	-	0.26	0.06	1.58	-	0.04	0.01	-	-	0.02	-	-	-	1.99
Tripura	-	-	0.63	-	-	0.23	-	-	-	-	-	-	-	-	0.86
Uttar Pradesh	-	-	1.15	-	1.28	-	-	-	0.51	-	-	1.06	0.02	0.36	4.38
West Bengal	-	-	0.46	0.28	0.43	0.20	-	0.01	-	-	0.01	-	-	0.01	1.40
Total	3.85	1.84	23.25	0.67	28.17	5.24	0.07	0.29	3.76	0.17	1.62	2.73	0.23	1.80	73.6

Source: Department of Forests, Government of Maharashtra, Pune.

#: Forest types. : 1. Tropical wet evergreen forests; 2. Tropical semi-evergreen forests; 3. Tropical moist deciduous forests 4. Littoral and swamp forests; 5. Tropical dry deciduous forests; 6. Tropical thorn forests; 7. Tropical dry evergreen forests; 8. Sub-tropical broad leaved hill forests; 9. Sub-tropical pine forests; 10. Sub-tropical dry evergreen forests; 11. Montane wet temperate forests; 12. Himalayan moist temperate forests; 13. Himalayan dry temperate forests 14. Sub-alpine forests; 15. Moist alpine scrub; 16. Dry alpine scrub. * Less than 10,000 ha.

Table 2. State-wise beekeeping potential (million bee colonies) in forest areas in India

State	Forest types											Total
	1	2	3	4	5	6	7	8	9	10	11	
Andhra Pradesh	-	-	12.00	0.08	3.36	0.12	-	-	-	-	-	16.16
Arunachal Pradesh	9.76	-	-	-	-	-	-	1.89	-	12.08	-	23.73
Assam, Meghalaya	9.52	2.40	11.64	-	-	-	0.80	0.07	-	-	-	24.43
Bihar	-	-	1.16	-	5.58	-	-	-	-	-	-	6.74
Gujarat	-	-	1.44	0.20	0.82	-	-	-	-	-	-	2.46
Haryana	-	-	-	-	0.16	-	-	0.03	-	-	-	0.19
Himachal Pradesh	-	-	0.04	-	0.42	-	-	0.73	0.72	-	6.00	7.91
Jammu & Kashmir	-	-	-	-	0.16	-	-	0.28	0.56	-	7.36	8.36
Karnataka	4.80	4.72	3.32	-	0.82	-	-	-	-	-	-	13.66
Kerala	4.40	2.08	0.88	-	-	-	0.08	-	-	-	-	7.44
Madhya Pradesh	-	-	29.04	-	19.52	-	0.16	-	-	-	-	48.72
Manipur	-	2.88	4.28	-	8.44	-	0.08	-	-	-	-	15.68
Nagaland	1.52	-	1.08	-	-	-	-	0.14	-	0.08	-	2.82
Orissa	0.64	0.40	0.08	-	-	-	-	0.09	-	0.56	-	1.77
Punjab	-	2.24	17.96	0.44	3.64	-	0.88	-	-	-	-	25.16

Rajasthan	-	-	-	-	0.28	-	-	0.02	0.08	-	-	0.38
Tamilnadu	-	-	-	-	6.56	-	0.16	-	-	-	-	6.72
Tripura	0.16	-	1.04	0.24	3.16	0.16	0.08	-	-	0.16	-	5.00
Uttar Pradesh	-	-	2.52	-	-	-	-	-	-	-	-	2.52
West Bengal	-	-	4.60	-	2.56	-	-	0.51	-	-	8.48	16.15
Total	-	-	1.84	1.12	0.86	-	0.08	-	-	0.08	-	3.98
	30.80	14.72	92.92	2.68	56.34	0.28	2.32	3.76	1.36	12.96	21.84	239.98

#: Forest types. : 1. Tropical wet evergreen forests; 2. Tropical semi-evergreen forests; 3. Tropical moist deciduous forests; 4. Littoral and swamp forests; 5. Tropical dry deciduous forests; 6. Tropical dry evergreen forests; 7. Sub-tropical broad leaved hill forests; 8. Sub-tropical pine forests; 9. Sub-tropical dry evergreen forests; 10. Montane wet temperate forests; 11. Himalayan moist temperate forests.

Data presented in table 2 gives the estimated state-wise beekeeping potential of the natural vegetation in the country, based on the above assumptions. Considering this potential, the natural vegetation under forests can sustain about 240 million bee colonies. At present India has just over a million bee colonies (Taori, 1993). It has to be kept in mind here that there are at least three types of other honey bees in the wild. Population of rockbees is quite large in several forests having good bee forage. Rockbee colonies gather large quantities of pollen and nectar, and in production of honey, they may be equal to, if not better than, the hive bees. The dwarf bee is the dominant species in some localities, and works thoroughly on small patches of vegetation to gather pollen and nectar. The total honey potential estimated above is shared by the wild bees as well. This can significantly affect the number of hive bee colonies estimated above.

Most serious problem with Indian beekeeping has been the decreasing honey plants during the past few decades but during the recent years national and state governments, environmental protection agencies, social forestry, local and social organizations have helped to expand the planted areas of bee forages, prevented deforestation and thus stopped the reduction of bee flora. It has improved the prospects of expanding apiculture. It is estimated that even 1/4th of the bee flora in country is not yet fully utilized. The spread of diseases is yet another most crucial factor that at times have played havoc with beekeeping in India resulting in great annihilation of entire bee stocks. Therefore, besides proper management of bee diseases, plantation of bee flora, strategies for development of beekeeping need to be reoriented and channelized to achieve fast progress. Beekeeping with *A. mellifera* has now almost spread to al the states of the country and indigenous *A. cerana* is again reviving. Therefore, there is a concerted need to develop multiply and spread both the species for honey production pollination ecosystem stability, sustainability and providing employment opportunities to unemployed youth and a source of subsidiary income to the farmers. The other wild nesting honeybees and pollinating insects also need protection for their proper utilization in honey production and crop pollination.

There is an immense scope to develop beekeeping in areas with natural and cultivated vegetations, providing gainful employment to rural and tribal populations. About 120 million bee colonies can be established providing gainful employment to nearly 5 million rural or tribal farmers. Realization of this potential can result in an annual production of 1.2 million tons of honey. Honey can be exported and valuable foreign exchange can be earned. There is scope for further improvement by breeding better strains of bees.

Chapter 3

Type of bees

The bees belonging to superfamily Apoidea, were a group of wasps that had abandoned their habits of provisioning their nests with insects and spiders, and instead fed their larvae with pollen and nectar collected from flowers or with glandular secretions derived from flowers. Bees, like ants, are essentially a highly specialized form of wasp. While the first definite fossil bees date from only forty million years ago, there is genetic and partial fossil evidence that they evolved alongside flowers, at least 140 million years ago. They developed a highly specialized symbiosis, the bees obtaining an almost perfectly reliable food source in return for helping propogate the genes of the first flowering plants more efficiently than any land plants ever before.

Table 3. Families, subfamilies, principal tribes, and the distribution of bees (superdfamily Apoidea) (Based on Michener, 1974)

S.N.	Family	Subfamily	Distribution
1	Colletidae		Worldwide
2	Oxaeidae		New world
3	Halictidae	Dufoureinae	Holarctic; African & Oriental regions; Chile
		Nominae	Old world tropics; South temperate regions; Holarctic
		t. Augochlorini	South and central America, some in Canada
		t. Halictini	Worldwide but less abundant in subtropics
4	Andrenidae	Andreninae	Chiefly Holarctic some in Africa & S.America
		Panurginae	Africa, Eurasia, New World
5	Mellitidae	Ctenoplectrinae	Paleotropics
		Macropidnae	Holarctic
		Melittinae	Holarctic, Africa
		Dasypodinae	Holarctic, Africa
6	Fidelidae		S. Africa & Chile
7	Megachilidae	Lithurginae	Worldwide (tropical and warm regions)
		Megachilinae	

		t. megachilini	All continents
		t. anthidini	All continents
8	Anthophoridae	Nomadinae	Worldwide
		Anthophorinae	
		t. Exomalopsini	Neotropics
		t. Ancylini	Mediterranean & eastward into Asia
		t. Tetrapedini	Neotropics
		t. Melitomini	Western hemisphere
		T. Canepnorulini	S. America
		t. Eucerinodini	S. America
		t. Eucerini	All continents(except Australia)
		t. Anthophorini	Worldwide
		t. Centridini	Americas (tropical and warm parts)
9	Apidae	Bombinae	
		t. Euglossini	Neotropics
		t. Bombini	Holarctic
		Apinae	
		t. Meliponini	Tropics worldwide
		t. Apini	Eurasia & Africa (introduced to all parts of the world)

Bees are a large and diversified group, with some more than 21,000 species found on every continent except Antarctica. Bees are adapted for feeding on nectar and pollen, the former primarily as an energy source, and the latter for protein and other nutrients. Most pollen is used for food for the brood. Bees have a long proboscis that enables them to obtain the nectar from flowers. Bees have antennae almost universally made up of thirteen segments in males and twelve in females. They all have two pairs of wings, the back pair being the smaller of the two and are usually categorized into nine families (Table 3, Figure 2). Michener (1974) provides a detailed systematics with brief description as given below:

Colletidae

This is the most primitive family of bees. They nest in the soil, and in the holes of logs and pithy stems. They are found worldwide but better represented in the southern hemisphere. None are parasocial or eusocial.

Oxaeidae

This is a small group of large bees from American tropical and subtropical regions. They nest in deep burrows in the soil. None are parasocial or eusocial.

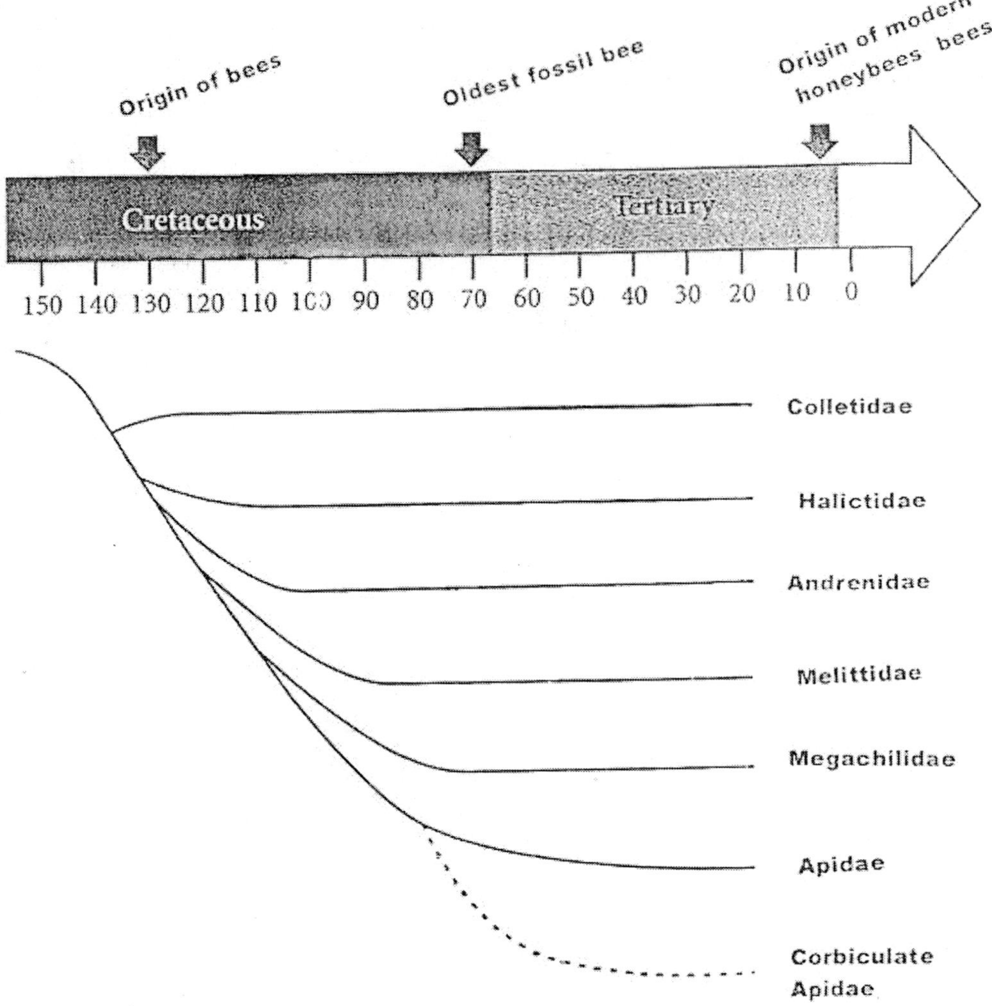

Figure 2. The phylogeny of the bee families and their approximate divergene times (After Engle 2001a; Grimaldi and Engel, 2005).

Halictidae

This is an enormous group of small to middle sized bees found worldwide. They are commonly called sweet bees which nest in burrows of soil or rotting wood. Their social behaviour ranges from communal, quasisocial, and semisocial to primitively eusocial.

Andrenidae

This is a large group of bees found in all continents except Australia. Their nests are burrows in the soil, and a few of the species nest in colonies. The bees are mainly solitary or communal but a few are parasocial.

Melittidae

This small but diverse family occurs in all continents except Australia and South America. The bees nest in burrows in soil or wood, no parasocial or eusocial species are known

Fideliidae

This family is found only in South Africa and Chile. The few genera studied nest in burrows in soil. No communal or eusocial species occur.

Megachilidae

This is a very large family found worldwide. Most nest in wood. No eusocial species occur. Many are solitary, and some are communal. The familiar leaf cutting bees belong to this family.

Anthophoridae

This is also a very large and widely distributed family. They nest in soil or wood. So far as is known most of these bees are solitary, but a few communal.

Apidae

This is only a moderate family of worldwide distribution. It is unique in having all the high eusocial bees. Solitary as well as parasocial species are also present. The Genus *Apis* belongs to this family.

Bee societies

The bee societies have a range of organization from unorganized two or three bees in a burrow to large highly organized colonies of the honeybees.

They are briefly described as below:

Solitary bees

Each female makes a nest or several of them independent of other nests. The female leaves after provisioning the cells with nectar or pollen, and ovipositing in them. Hence, there is usually no contact between generations.

Nest aggregations

In many solitary forms, especially the soil burrowing bees, the nests are grouped, from a few nests, small cluster, or enormous numbers.

Parasocial colonies

This is a collective term for communal, quasisocial and semisocial groups. The adults of these groups consist of single generation only.

(a) Communal colonies

A communal colony is made up of group of females from the same generation using a single nest, each making provisioning and ovipositing in her own cells.

(b) Quasisocial colonies

A small communal colony made up of a group of females of about the same age and generation that co-operatively construct and provision cells.

(c) Semisocial colonies

These are like the quasisocial colonies, but are distinguished by showing division of labour among the females; both egg layers and worker like females exist, although they are derived from the same generation.

(d) Subsocial colonies

These colonies are family groups with one adult female and a number of immature offspring that are cared for by the mother. She leaves or dies when the young reach maturity.

(e) Eusocial colonies

In all previous colonies, the interactions among the adults, if occur are within the same generation. However, the eusocial bees live in colonies which are family groups consisting of adult individuals of two generations, mothers and daughters. Usually, only one queen exists in these colonies, and the bulk of workers are daughters. Division of labour is often well characterized in recognizable castes.

(f) Primitively eusocial colonies

In these colonies, castes of different females are generally indistinguishable externally. Swarming is not exhibited, and the gynes (the equivalent of queens) can live alone and establish colonies as lone individuals. Often, the only food stored is in brood cells.

(g) Highly eusocial colonies

These are the well known colonies where some of the most complex social interactions are exhibited. The female castes differ strongly between each other behaviourally, physiologically, and also in size. The gynes (queens) are unable to survive alone, and new colonies are established through swarming, with the aid of workers from the parent colony. Colonies can sustain themselves for long periods with food stored in non brood cells. Communication concerning food sources and nest sites is moistly well developed; this feature perhaps has reached its acme in the true honeybee, *Apis*.

Stages in the sociality of bees

Bees may be solitary, or may live in various sorts of communities. The most advanced of these are eusocial colonies, found among the honeybees, bumblebees, and stingless bees. Sociality is believed to have evolved separately many times within the bees. In some species, groups of cohabiting females may be sisters, and if there is a division of labor within the group, then they are considered semisocial. If, in addition to a division of labor, the group consists of a mother and her daughters, then the group is called eusocial. The mother is considered the "queen" and the daughters are "workers". These castes may be purely behavioral alternatives, in which case the system is considered "primitively eusocial" (similar to many paper wasps), and if the castes are morphologically discrete, then the system is "highly eusocial".

There are many more species of primitively eusocial bees than highly eusocial bees, but they have been rarely studied, and the biology of most such species is almost completely unknown. The vast majority of such species are in the family Halictidae, or "sweat bees". Colonies are typically small, with a dozen or fewer workers, and the only physical difference between queens and workers is average size, if they differ at all. Most species have a single-season colony cycle, even in the tropics, and only mated females (future queens, or "gynes") hibernate (called diapause). A few species have long active seasons, and attain colony sizes in the hundreds. The orchid bees include a number of primitively eusocial species, as well, with similar biology. Certain species of allodapine bees (relatives of carpenter bees) also have primitively eusocial colonies, with unusual levels of interaction between the adult bees and the developing brood; this is called "progressive provisioning", when a larva's food is supplied gradually as it develops, and this system is also seen in honeybees and some bumblebees.

Highly eusocial bees live in colonies, each of which has a single queen, together with workers and drones. When home is provided for a colony, the structure is called a hive. A honeybee hive can typically contain up to about 40,000 individual bees at their annual peak, which occurs in the spring, but usually have fewer.

Bumblebees

Bumblebees are eusocial in a manner quite similar to the eusocial vespidae such as hornets, as the queen initiates a nest on her own. Bumblebee colonies typically have from 50 to 200 individual bees at peak population, which occurs in mid to late summer. Nest architecture is simple, limited by the size of the nest cavity (pre-existing), and colonies are rarely perennial.

Stingless bees

Stingless bees are very diverse in behavior, but all are highly eusocial. They practice mass provisioning, nest architecture is complex, and colonies are typically perennial.

Honeybees

The true honeybees have the most complex social behavior among the bees. The European honeybee, *Apis mellifera* is the best-known bee species, and one of the best-known of all insects; it is treated in extensive detail elsewhere.

Solitary and communal bees

Most other bees, including familiar species of bee such as the Eastern carpenter bee (*Xylocopa virginica*), alfalfa leafcutter bee (*Megachile rotundata*), orchard mason bee (*Osmia lignaria*) and the hornfaced bee (*Osmia cornifrons*) are solitary in the sense that every female is fertile. There are no worker bees for these species. Solitary bees typically produce neither honey nor beeswax. They are immune from acarine and *Varroa* mites, but have their own unique parasites, pests and diseases. Solitary bees are important pollinators, as pollen is gathered for provisioning the nests with food for their brood. Often it is mixed with nectar to form a paste-like consistency. Many solitary bees have very advanced types of pollen carry structures on their bodies. Most solitary bees are wild, with a few species being increasingly cultured for pollination. Solitary bees are often oligoleges, in that they only visit one or more species of plant. In a few cases only one species of bee can pollinate a plant species, and some plants are endangered because their pollinator is dying off.

Solitary bees create nests in hollow reeds, holes in wood, or in tunnels in the ground. The female typically creates a compartment with an egg and some provisions for the resulting larva, and then seals it off. A nest may consist of numerous compartments, usually the last (the closest to the entrance) being eggs that will become males. The adult does not care for the brood, and usually dies after making one or more nests. The males emerge first and are ready for mating when the females emerge. Solitary bees are usually stingless or very unlikely to sting. While solitary females each make individual nests, some species are gregarious, preferring to make nests near others of the same species, giving the appearance to the casual observer that they are social. In some species, however, multiple females share a common nest, but each makes and provisions her own cells independently; this type of group is called "communal", and is not uncommon. The primary advantage appears to be that a nest entrance is easier to defend when there are multiple females using that same entrance on a regular basis.

Cleptoparasitic bees

Cleptoparasitic bees, commonly referred to as "cuckoo bees" because their behavior is similar to that of cuckoo birds, occur in several bee families (e.g., subfamily Nomadinae). Females of these bees lack pollen-collecting structures (the scopa) and do not construct their own nests. Rather, they typically enter the nests of pollen-collecting species, and lay their eggs in cells provisioned by the host bee. When the cuckoo bee larva hatches, it consumes the host larva's pollen ball, and kills and eats the host larva itself. In a few cases where the hosts are social species, the cleptoparasite remains in the host nest and lays many eggs, sometimes even killing the host queen and replacing her. Many cleptoparasitic bees are closely related to, and resemble, their hosts (i.e., the subgenus *Psithyrus*, which are parasitic bumble bees that infiltrate nests of species in the subgenus *Bombus*).

Chapter 4

Evolution and Biodiversity of honeybees

Bees of all kinds belong to the order of insects known as Hymenoptera, literally "membrane wings". This order, comprising some 100,000 species, also includes wasps, ants, ichneumons and sawflies. Of the 25,000 or more described species of bees, the majority are solitary bees most of which lay their eggs in tunnels, which they excavate themselves. In some species small numbers of females may share a single tunnel system, and in other cases there may be a semi/social organization involving a hierarchical order among the females, These bees provide a supply of food (honey and pollen) for the larvae, but there is no progressive feeding of the larvae by the adult bees.

Honeybees belong to the family of social bees which includes bumble bees and the tropical stingless bees of the genus Meliponinae. The social bees nest in colonies headed by a single fertile female, the queen, which is generally the only egg layer in the colony. Foraging for nectar and other tasks such as feeding the queen and the larvae, cleaning brood cells and removing debris, are carried out by a caste of females, the Workers. Honey and pollen is stored, and larvae are reared in cells made from wax secreted by the worker bees.

Typical colonies may amount to no more than a few dozen insects, and may be annual as in the case of bumble bee colonies, or they may number several tens of thousands and persist for a number of years, as in the case of honeybees and species of Meliponinae.

The sub-family Apinae or honeybees, comprises a single genus, *Apis,* which is characterised by the building of vertical combs of hexagonal cells constructed bilaterally from a midrib, using only the wax secreted by the worker bees. The cells are multifunctional, being used repeatedly for rearing the larvae and for the storage of honey and pollen. Progressive feeding of the larvae is carried out by young bees with food produced by glands in the head of the bee from honey and pollen.

Two attributes of honeybees which have been essential to their evolution and biology are their clustering behaviour and, particularly in the case of the cavity-nesting species, their ability to cool the nest by evaporation of water collected outside. These attributes enable the colonies to achieve a marked degree of temperature regulation within the nest irrespective of the external temperature. The genus *Apis* was thus enabled to colonise a wide variety of environments, ranging from tropical to cool temperate. The Meliponinae which lack this capability are confined to tropical regions.

Another behavioural character of honeybees is the communication of information about food sources and the recruitment of foragers by "dance language". The accurate dissemination of information concerning direction and distance of forage areas leads to efficient exploitation of food sources.

Whereas representatives of most types of bee were indigenous to all the continents, bees belonging to the genus *Apis* were originally to be found only in the Old World, namely Asia, Africa and Europe. This suggests that the genus appeared much later than the other types. The genus comprises four species: *Apis florea*, the Little Honeybee; *Apis dorsata*, the Giant Honeybee; *Apis cerana*, the Eastern Honeybee; and *Apis mellifera*, the Western Honeybee. (Some authors include *Apis laboriosa* and *Apis andreniformis* as separate species, but it is likely that these are geographical subspecies of *Apis dorsata* and *Apis florea* respectively which show greater physical variations than the other subspecies and are possibly in a more advanced stage of speciation.

Apis florea and *Apis dorsata* build single comb nests in the open, *florea* in low bushes and *dorsata* in trees. Like other tropical honeybees they are prone to migrations, at times over considerable distances. These migrations may be seasonal or in some cases may be a defence against predators and parasites. Although unsuitable for apicultural use, both these species make a major contribution to the supply of honey and wax in the countries in their territorial range. Human predation usually involves destruction of the nest including the brood, but in some areas collection of honey is practised without destruction of the nest, and some honey gatherers even provide nest sites to which they transfer the whole colony.

The lifestyle of *Apis cerana* is similar to that of the Western Honeybees, and like *Apis mellifera* it is used in apiculture with modern moveable comb hives. The numerical strength of *cerana* colonies is usually much less, and honey yields are smaller. It is therefore being rapidly replaced by imported *mellifera* races, chiefly *A.m. ligustica*.

Bees of the genus *Apis* are not the only bees which contribute to the World's supply of honey and wax. Some species of Meliponinae form very large colonies and store sufficient honey to make their exploitation worthwhile.

The origin of honeybees

It is thought that bees originally evolved from hunting wasps which acquired a taste for nectar and decided to become vegetarians. Fossil evidence is sparse but bees probably appeared on the planet about the same time as flowering plants in the Cretaceous period, 146 to 74 million years ago. The oldest known fossil bee, a stingless bee named *Trigona prisca*, was found in the Upper Cretaceous of New Jersey, U.S.A., and dates from 96 to 74 million years ago. It is indistinguishable from modern *Trigona*. The precursor of the honeybees may have been living about this time, but fossils of the true *Apis* type were first discovered in the Lower Miocene (22 to 25 million years ago) of Western Germany. A bee resembling *Apis dorsata* but much smaller (about the size of a present day *mellifera)* was present in the Upper Miocene (about 12 million years ago). It is thought that *Apis florea* and *Apis dorsata* may have existed as separate species as early as the Oligocene period. It has not been possible to estimate when bees of the *mellifera/cerana* type first appeared on Earth. *Mellifera* and *cerana* must have acquired separate identities during the latter part of the Tertiary era. The two species were apparently physically separated at the time of the last glaciation, and there was no subsequent contact between them until that brought about by human intervention in recent times. In the post glacial period *mellifera* and *cerana* (and to a less extent *dorsata* and *florea*) have shown similar evolution into geographical subspecies, or races.

The development of subspecies

Although it has long been known that there are many kinds of honeybee, and these have been the subject of scientific study for more than two centuries, only in recent years has a comprehensive classification been attempted which takes into account not only differences in physical characters between subspecies and their present geographical distribution, but also the geological evidence pointing to their origins, and to the course of their subsequent evolution and distribution.

Like the stingless bees, honeybees first evolved in tropical conditions. The fossil record shows that at the time the area of land that is now Europe had a tropical climate. As the climate became cooler, the open nesting types would not have been able to survive except by migrating to the tropical region of Southern Asia. For the greater part of the Tertiary era Africa was isolated from Europe by sea, and no Tertiary types of honeybee reached Africa even after a land bridge was established. It is likely that the development of advanced thermal homeostasis in honeybees which permitted the occupation of cool temperate zones therefore occurred in Southern Asia, possibly in the Himalayan region. Once established, the cavity nesting *cerana-mellifera* type would spread East and West, eventually occupying both tropic and cool temperate zones.

A physical separation into two groups probably took place as a result of the glaciations which occurred during the Pleistocene period (1 million to 25,000 years ago) and desert and semi-desert then kept the two groups separate during intervening warm periods. Thus *mellifera* and *cerana*, although originating from a common stock, evolved into distinct species. The ultimate Western boundary of the *cerana* territory was in Afghanistan some 600 km to the East of the nearest *mellifera* colonies in Iran. The *cerana* territory comprised the Indian

Subcontinent South of the great mountain ranges, Ceylon, Malaysia and Indo-china, and the East Indies including the Celebes, Timor and the Philippines. In Eastern Asia it reached latitude 46, and occupied Japan except for the island of Hokkaido.

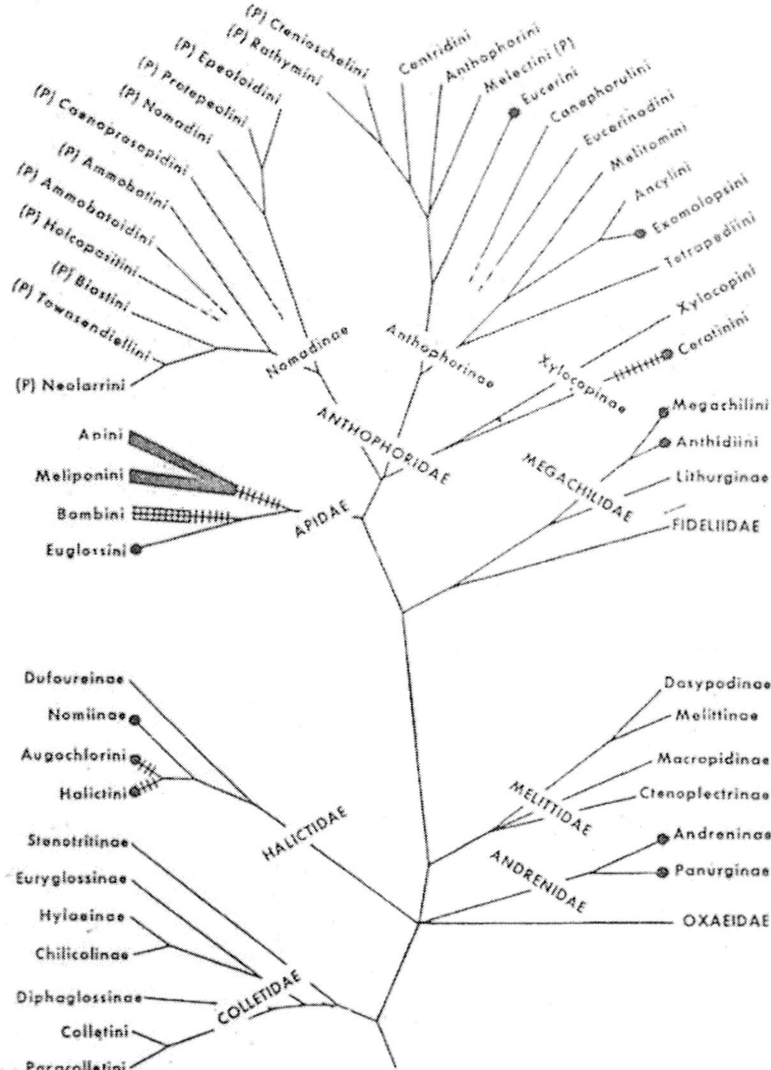

Figure 3a. Dendrogram showing relations among the major groups of bees. The lines are subjectively determined lines of descent. Lines ending in *black dots* represent taxa in which some species live in parasocial colonies, at least part of the time. Lines with *crossbars* represent taxa containing primitively eusocial species. When the crossbars are connected at the sides (Bombini), all of the nonparasitic species are primitively eusocial. *Heavy bars* represent taxa all species of which are highly eusocial. Tribes which are entirely parasitic and therefore, could not contain colonial species are marked (P) (Original drawing by Barry Siler and C.D. Michener).

Mellifera spread westwards through Asia Minor to colonise the Balkans and the Mediterranean region, and southwards through the Arabian peninsula to occupy Central and Southern Africa. Similarities between neighbouring subspecies suggest that the Iberian peninsula and Southern France were colonised from North Africa

How far *mellifera* bees may have penetrated into Northern and Western Europe during the warm intervals between the glaciations of the Pleistocene period can only be a matter of conjecture; what is certain is that no honeybees could have existed North of the Mediterranean region, the Iberian peninsula and South Western France at the time of the most recent Ice Age. Although at its maximum extent in Western Europe some 18,000 years ago, the ice sheet only reached as far as Northern Britain, the area for hundreds of miles to the South was inhospitable tundra.

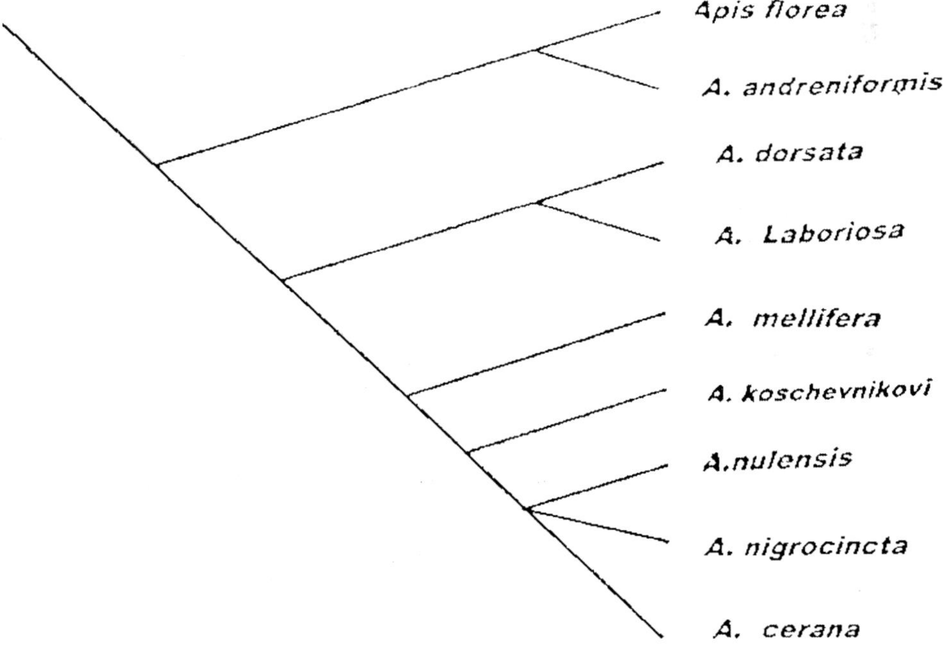

Figure 3b. A phylogeny of the genus *Apis*. A complete phylogey of all species of honey bee is not yet available. The figure shows the most likely topology in the author's opinion. The general topology is based on Arias and Sheppard (2005) and Engle and Schultz (1997). Although *A. cerana, A. nigrocincta,* and *A. nuluensis* are good species according to our criteria, their relative positions in the *Apis* phylogeny are not well resolved (Arias and Sheppard 2005; Smith *et al.,* 2003; Tanaka, Roubik *et al.,* 2001).

In the warm period which followed the Ice Age (starting about 14,000 years ago) the ice sheet gradually retreated and the tundra was replaced by forests of birch, pine, hazel, elm and broad-leaved oak. The Western honeybee was once more able to extend its domain in Europe. In the East advance beyond the Caucasian region proved impossible, owing to the lack of suitable nesting sites in the steppes of Southern Russia The bees of the Balkan area spread northwards to occupy the Eastern Alpine valleys, Central Europe as far as the 50th parallel of latitude, and the Western shores of the Black Sea. In the West the bees which had found refuge in Southern France during the Ice Age spread across Europe North of the Alps eventually occupying an area from the Atlantic seaboard to the Ural Mountains. The northern most limit of the territory may have been in Southern Norway; honeybee remains dating from *ca* 1,200 have been found in an archaeological dig in Oslo although honeybees had not been reported in Norway prior to the 19th Century. The mountain ranges of the Alps and the Pyrenees obstructed the northward movement of the bees in the Italian and Iberian peninsulas. However, in colonising this vast territory, stretching from the Urals to the Cape of Good Hope, *Apis mellifera* had to adapt itself to a large variety of habitats and climates ranging from the Continental climate of Eastern Europe with its harsh Winters, late Springs and hot, dry Summers, through Alpine, cool temperate, maritime, Mediter-ranean, semi-desert and tropical environments. This adaptation was achieved by natural selection, producing some two dozen subspecies or races. All the subspecies of the *mellifera* group can interbreed given the right conditions, but the crosses show hybridity characters.

Although *cerana* bees must have shared a common ancestor with *mellifera*, they have evolved into separate species. It is not possible to cross *cerana* with *mellifera* even using instrumental insemination, because the two species are now genetically incompatible, and viable eggs do not result from the cross fertilisation Other differences include their differing reactions to diseases, infestations and predators. *Cerana* can tolerate *Varroa* and has developed an effective defence strategy against the Giant Hornet, against which *mellifera* bees have no defence. *Cerana* is however, highly susceptible to the acarine mite, which arrived with the introduction of *mellifera* bees into *cerana* territory. It is also highly susceptible to sac brood and foul brood, but not markedly so to *nosema*.

The different races of *A. mellifera* can generally be differentiated in physiological terms. Bees from warmer climates tend to be smaller in size and lighter in colour than those adapted to the colder regions, although this rule is not invariable. The effect of altitude seems to be similar to that of increasing latitude. Accurate differentiation between races of similar appearance requires precise morphometric examination of representative samples of bees. There are also differences between races in natural history and biology. Some subspecies are more prone to swarming than others, some produce large numbers of young queens when swarming, others only a few. Tropical honeybees frequently "abscond" or migrate, sometimes due to lack forage through drought or other causes, perhaps as a defence against predators. Heavy predation is also a likely cause of the vigorous defence reaction of some races, for example, the bees of tropical Africa.

The bees of the warmer regions do not need to cluster as tightly as those confined to the nest through long, cold winters. Brood rearing is adapted to take maximum advantage of the local flora. Where bees of the same race have occupied different kinds of habitat, they have formed local strains which have accommodated themselves to the different conditions. Similarly, honeybees of different races which have occupied similar habitats have evolved similar behavioural characters. Even the "dance language" by which honeybees communicate information about the location of food sources may differ in detail between races (referred to as "honeybee dialects") as different races may be conditioned to foraging over different distances from the nest. The behavioural characters of the different races and strains, brood rearing pattern, foraging behaviour, clustering, etc., are fixed genetically, so that a colony cannot readily adapt itself when transferred to a different kind of environment.

The Dark European Honeybee, *Apis mellifera mellifera*, is fairly uniform over its whole range, having had but a comparatively short time in which regional varieties could evolve, but even in this race differences can be observed between strains. In France, where the bee has been domiciled longest, there are distinct differences in brood rearing pattern between the *mellifera* bees of the Landes district in the Southwest, the bees of the Paris area, and those of Corsica. The Landes bees are typical "heather bees", conditioned to a principal nectar flow in late Summer and early Autumn. In the Paris area there is no Summer nectar flow and the bees show early Spring brood activity. Exchange of colonies between the Landes and Paris resulted in poor performance in both cases. In Corsica the *mellifera* bees follow a Mediterranean pattern with little or no brood production in Summer and a second peak in Autumn.

The effect of transferring bees to environments to which they are not adapted is graphically illustrated by experience in the tropic zone of South America. European honeybees have been kept in Brazil for centuries, yet failed to establish a feral population in the country. When a few queens of a tropical race from Africa were introduced into the country, in a matter of a few years feral colonies of hybrids, "africanised bees" had crossed the Amazon rain forest and moved North and South completely eliminating the European bees.

The behavioural patterns which have evolved in the different races have ensured the survival of the various subspecies in their native habitats, and some of these patterns may be repeated in different races. There is one race which, although of small economic importance, possesses an apparently unique biological character which renders it of great importance in the study of the genetics of honeybees. In all other races, when a colony is rendered queen less, laying workers may appear which are capable of laying drone eggs only. In *A.m.capensis*, the Cape Bee, when a colony is deprived of its queen, a laying worker appears within a few days which, for a period, is able to lay predominantly diploid worker eggs. From these eggs true queens capable of being mated can be raised, re-establishing queen rightness in the colony.

The present situation

Apiculture has been practised in Europe and Asia throughout recorded history. For most of the time the honeybees kept in any country would be indigenous to the locality. In the New World countries, where the true honeybees, Apinae, were originally absent, the early settlers imported the bees with which they were familiar. Thus, Iberian bees were taken to Brazil and North European bees to North America, Australia and New Zealand. Whereas the Iberian bees were unsuited to the tropical climate of South America and failed to establish a feral population, the North European bees adapted well to the harsher conditions and feral colonies quickly established themselves over a wide area. Indeed, colonisation by honeybees far outstripped that by the settlers. In New Zealand and Tasmania feral and managed colonies of *A.m.mellifera* have existed in a pure state in spite of massive importations of Italian bees.

In most parts of the World, especially where beekeeping is practised on a commercial scale, the Italian bee has proved the most popular, owing to its docility, its rapid build-up, and its ability to rear brood continuously until late in the season as long as food is available. It is therefore, pre-eminently suitable for those countries where long, continuous nectar flows occur from late spring onwards. Where the nectar flows are intermittent or are interrupted by bad weather, feeding may be necessary during the barren periods, and also in spring and autumn.

The other race which has been exported world-wide is the Carniolan, *A.m. carnica*. In Germany, the native dark bee had been completely marginalized by the large scale introduction of foreign bees, chiefly *ligustica* and *carnica*, and the honeybee population was generally unproductive and aggressive.

In other North European countries there has been a tendency to move over to Carniolan bees, although in recent years an increasing interest has been shown in re-establishing the North European Dark Bee, *A.m. mellifera*, in most countries in which it is autochthonous (the original sub-species).

Which bee?

Twenty five thousand different kinds of bee have been described, divided into eleven Families, numerous subfamilies, tribes and genera, and still more numerous species and subspecies. Honeybees belong to the family Apidae, which includes other social bees such as bumble bees (Bombinae), and stingless bees (Meliponinae). The subfamily Apinae, consists of one tribe Apini, comprising one genus, *Apis*. There are four species within the genus: *florea, dorsata, cerana* and *mellifera,* but only the last two are suitable for apiculture in modern, moveable comb hives. Two dozen geographic races of the Western Honeybee, *Apis mellifera*, have been recognised, adapted to a range of environments from the cold Continental climate of Eastern Europe, through the moist temperate climate of the Atlantic seaboard, the warmth of the Mediterranean, and the heat of the tropics and semi-deserts. Only Four of these races need be considered for apiculture in a cool temperate climate. namely *A.m. ligustica, A.m. carnica, A.m. caucasica* and the native bee of the British Isles, *A m. mellifera.*

It was formerly believed not only by ordinary beekeepers but by some notable scientists, that improvements in the desirable attributes of honeybees, productivity, docility, resistance to disease, for example, could be achieved by crossbreeding different races. It is well known in other fields of bioculture that a first or second cross of two different breeds or strains will often produce progeny which are superior to either progenitor in some desirable character. It is also known that such hybrids are generally unsuitable for further breeding as the results are frequently unpredictable and generally inferior particularly if continued through several generations. So it is with honeybees; first or second crosses sometimes produce colonies which give exceptional performance, "hybrid vigour", but succeeding generations seldom repeat this performance. Moreover, crossing of any of the four races mentioned is likely to result in hybrids with very undesirable characters, namely excessive stinginess and a predilection to "following". It is now widely accepted that the best way to get improvement in bee stocks is by selective breeding within a single subspecies.

Apis mellifera ligustica

The Italian honeybee is the most widely distributed of all honeybees, and has proved adaptable to most climates from subtropical to cool temperate, but it is less satisfactory in humid tropical regions. It is very prolific but brood rearing starts late and lasts long into late summer or autumn, irrespective of nectar flow. It is therefore at its greatest advantage in those regions where favourable weather prevails throughout the summer, and there is a long, uninterrupted supply of nectar. It is less satisfactory where the main nectar flow occurs in spring, or where the weather is uncertain, as in the cool maritime regions. In poorer districts a honey crop may only be obtainable at the expense of heavy autumn feeding.

A.m. ligustica has been described as having a low swarming tendency with few queen cells. Italian bees, having been conditioned to the warmer climate of the central Mediterranean, are less able to cope with the "hard" winters and cool, wet springs of more northern latitudes. Their bodies are smaller and their overhairs shorter than those of the darker races, and they do not form such tight winter clusters. More food has to be consumed to compensate for the greater heat loss from the cluster. The tendency to raise brood late in autumn also increases food consumption. They are unable to retain faeces in the gut for long periods and require more frequent cleaning flights than the dark bees; they are more likely to be lured out of the hive by bright winter sunshine.

There is no clear evidence that *ligustica* is any more resistant to acarine than *mellifera*; no epidemic corresponding to Isle of Wight disease was ever reported from Northern Europe. Moreover, acarine is undoubtedly a problem among the Italian bees of the United States of America *ligustica* also appears to be less tolerant of *Nosema* than *mellifera*. Ligustica tends to forage over shorter distances than either Carnica or Mellifera, and may therefore be less effective in poorer nectar flows. It apparently lacks the ability to ripen heather honey before sealing. Italian bees are much more prone to drifting and robbing than the other principal races of Europe. It has a reputation for gentleness, but hybrids with the darker races can be especially vicious.

Apis mellifera carnica

The Carniolan bee of Slovenia and Austria is the nearest relative of the Italian, but it is larger and darker, the characteristic yellow rings of *ligustica* being replaced by dark bands. The Carnica territory covers a large area of south-eastern Europe, and there are numerous regional variations. The characteristic brood rhythm is a rapid build-up in spring, followed by a slow decline and an early cessation of brood rearing in the autumn. It is particularly suited to an early Spring honey flow. Like *A.m. mellifera* it can survive hard winters with a small winter cluster. Carniolan bees are said to be more prone to swarming than Italian bees, but this tendency can be reduced by selective breeding. In recent years, selective breeding has also been used with great effect in both Austria and Germany to improve the productivity of the bees. *A.m.carnica* are reputed to have better homing ability than any of the other major races, and are much less prone to drifting (and presumably to robbing). They are sparing in the use of propolis. Carniolan bees have a well deserved reputation for gentleness and quietness on the comb, but their hybrids with both *mellifera* and Ligustica are said to be particularly vicious.

Apis mellifera caucasica

The Caucasian bee closely resembles A.*m.carnica* in general appearance, and may not be easily distinguished from the latter except by morphometric examination (longer proboscis, cubital index about 2 on average). Indeed, it has been alleged that many bees sold as "Caucasians" were in fact Caucasica-Carnica hybrids.

A.m.caucasica is autochthonous (the original sub-species) to the mountain range and southern valleys of the Caucasus, and to the eastern end of the Black Sea coast in Anatolia. The climate varies from humid subtropical on the coast to cool temperate in the mountains, and local strains reflect the different climates, the bees from the mountains being larger and darker, with longer over hair, than those from the lowland region.

The Caucasian bee is noteworthy for the length of its proboscis, being the longest of all the *mellifera* races. One might expect that this would give it an advantage over shorter-tongued races from a foraging point of view, but this does not seem to be borne out in practice.

Brood rearing generally starts late and the spring build-up is slow, leading to a medium population size in summer and autumn. Swarming tendency is said to be low, and the number of swarm cells moderate. Caucasian bees are said to be at their best in protracted slight nectar flows; they seem to be unable to cope with short heavy flows, most of which is stored in the brood chamber rather than the supers. Honey cells are "wet" capped, i.e. there is no air space between the honey and the capping, and this may lead to "weeping" of the comb.

Caucasian bees are notorious for their heavy use of propolis, especially at the hive entrance. In winter, the entrance may be almost completely closed by a curtain of resin, leaving only a few small holes for ventilation and flight activity. Caucasian bees have poor resistance to *Nosema* disease and this may lead to heavy winter losses. *A.m. caucasica* is described as having a "high level of gentleness", however, they have been reported to show poor wintering qualities.

The native bee: *Apis mellifera mellifera*

The "*A. mellifera*" (1758) or "*A. mellifica*" (1761) of Linnaeus is but one small section of the Dark European Honeybee whose natural territory included the island of Corsica and ranged from the Pyrenees over Europe north of the Alps to the Ural Mountains in the East, and included Great Britain and Ireland and southern Sweden. Although there is no historical record of honeybees in Norway before 1775, it is known from archaeological evidence that *A.m.mellifera* was present in southern Norway round about 1200 A.D.

It is well adapted to survive in a harsh climate. It is thrifty in its use of stores; brood rearing is reduced when the nectar flow is interrupted. It forages over longer distances than the Italian bee and can make better use of meagre food resources. It will be observed foraging both earlier and later than *A.m.ligustica*, and will fly in dull and drizzly weather which would keep Italian bees indoors. It may also be that mating can take place at lower temperatures than in the case of the southern races. Although less prolific than Italians, the workers live longer and there is a higher ratio of foraging bees to hive bees. The wintering capabilities of the Dark bee are excellent; although colony size is at all times moderate, and the winter cluster is small, heat is conserved by the tightness of the cluster and the large bodies and long over hair of the bees. The "winter" bees of the northern race have the ability to retain faeces in the gut for long periods, due apparently to a greater production of catalase by the rectal gland in autumn. They are thus less dependent on cleaning flights. They are also less likely to be lured out of the hive by bright winter sunshine than Italian bees.

A.m.mellifera forms a compact brood nest with pollen stored as close to the brood as possible, sometimes below as well as above the brood. Honey is stored outside the pollen circle.

The swarming behaviour of *A.m. mellifera* is variable, depending on the region. *A.m.mellifera* makes abundant use of propolis to seal up small fissures and small gaps, and may even construct curtains at the hive entrance in the manner of Caucasian bees, although in general it is not as free in its use of the resin as the latter. Pure strains of *A.m. mellifera* have been found to be docile and easily handled. Hybrids with other races are often highly productive, but they frequently show a fierce temperament and proneness to "following", highly objectionable characters in densely populated countries. They have been reported to show nervous behaviour when the hive is disturbed. It usually manifests itself by the bees running to the bottom of the comb where they hang in a cluster when a frame is removed from the brood chamber.

The gentle behaviour of the major races of honeybee may be due, to selection for this quality over many generations; as beekeepers tend to destroy the worst tempered bees and retain the gentler colonies.

What of the future?

The most urgent problem in apiculture, throughout the world, is that of protecting the Western honeybee against extermination by the varroa mite. The only proven method at the present time is by using acaricides such as Bayvarol and Apistan, but these become less effective as immune strains of the mite evolve, and there must be constant research to develop new products. Research on the biology of the mite is proceeding, and alternative methods of treatment are being sought. The ultimate hope is that varroa-resistant strains of bees may evolve, but at best this is likely to be a very long term solution to the problem. If, as is supposed, the separation of the *Cerana* and *Mellifera* species occurred in (relatively) recent times, the gene which enabled *Cerana* to develop a defence against *Varroa* may still be lurking somewhere among the genes of the *Mellifera* races. There is a danger that the development of resistance among apiary stocks might be concealed by the normal anti-*Varroa* treatments and that a resistant strain might be lost through the death of the queens.

There have been many changes in the flora on which bees have depended during the past 10,000 years due to intensification in agriculture and industrialization.

In the last few years there have been extremes of weather which in the short term can be regarded as exceptional. It has been suggested that this is part of the "greenhouse effect" of manmade pollution causing global warming. It is too early to separate the climatic changes due to pollution from the short term fluctuations and long term trends in the Earth's climate. However, careful measurements over many years have shown changes from which inferences may be drawn. For example, the carbon dioxide content of the atmosphere increased from 290 parts per million in 1850 to 315 parts per million 1958, and since then has increased further to 350 parts in 1990. The increase up to 1950 was attributed partly to deforestation and partly to combustion of fuels, mainly coal. The increase since 1950 is thought to be due almost entirely to the combustion of fuels. Methane, another "greenhouse gas" has also shown a significant increase, although the concentration of this gas is much lower. Another and clearer indicator of global warming is the retreat of glaciers throughout the world, and of the arctic and Antarctic icefields which has been taking place for many years. There is little doubt that global warming is already taking place and if this trend continues the impact cannot be predicted.

One further matter to which beekeepers should be giving urgent attention is the need to improve the social acceptability of honeybees. Too many of our colonies become bad tempered and aggressive when disturbed, some even attack people without this provocation. Such bees are a menace to the public and a nuisance to the beekeeper. A swarm of bees can be a terrifying sight to anyone unaccustomed to them, and sensational films and newspaper articles about "killer" bees can only affect adversely the public's perception of beekeeping. It has been pointed out that the principal European races of honeybee in the pure form have the reputation for gentleness, and this character can be preserved and enhanced by selective breeding. The adoption of such a policy would have the added advantage that selective breeding could then be practised to pursue other desirable aims such as greater productivity, lower swarming tendency and better disease resistance.

Chapter 5

Diversity of honeybee species

Introduction

The insects constitute the largest class not only of the animal kingdom but also of the whole living world. The number of insects is much higher than all other species of the animal kingdom put together. Estimates of both the absolute number of insect species and their relative proportion vary from 7, 00, 000 to 1, 5000,000 species and from 70 to 90% of all known species of animal kingdom. The discovery of insects extends back to Upper Carboniferous period which is about 350 million years ago. Some changes in insect fauna were noticed in the Permian, Mesozoic, Triassic and Jurassic periods. When flowering plants became established in the Cretaceous period many insects were found associated with them. Of the several insect orders, the order Hymenoptera is the largest group of insects which includes at least a quarter million species including all social insects such as bees, wasps and ants (except termites which belong to order isoptera) and other insects. The name refers to the wings of the insects, which are membranous (Gr. hymen, membrane + ptera, wing) with the hindwings "married" (Hymen, Greek god of marriage + ptera, wing) to the forewings by a series of hooks called hamuli. The order is characterized by the presence of membranous wings, propodeum between thorax and abdomen and holoptic eyes. The propodeum is the first abdominal segment in Apocrita Hymenoptera (wasps, bees and ants). It is fused with the thorax to form the mesosoma. It is a single large sclerite, not subdivided, and bears a pair of spiracles. It is strongly constricted posteriorly to form the articulation of the petiole, and gives apocritans their distinctive shape). Females typically have a special ovipositor for inserting eggs into their hosts or otherwise inaccessible places, often modified into a stinger. The young develop through complete metamorphosis - that is, they have a worm-like larval stage and an inactive pupal stage before they mature. The order is divided into 2 suborders.

Suborder Symphyta

It includes sawflies, horntails, and parasitic wood wasps. The group appears to be paraphyletic, as it is often believed that the family Orussidae may be the group from which the Apocrita arose. They have an unconstricted junction between the thorax and abdomen, and the larvae of free-living forms are herbivorous, have legs, prolegs (on every segment, unlike Lepidoptera), and ocelli.

Suborder Apocrita

This sub order includes wasps, bees, and ants. This suborder is characterized by a constriction between the first and second abdominal segments called a wasp-waist (petiole), also involving the fusion of the first abdominal segment to the thorax and is called the propodeum. Also, the larvae of all Apocrita do not have legs, prolegs, or ocelli. The ovipositor of the female either extends freely or is retracted, and may be developed into a stinger for both defence and paralyzing prey. Larvae are legless and blind, and either feed inside a host (plant or animal) or in a nest cell provisioned by their mother.

The Apocrita has historically been split into two groups, the "Parasitica" and the Aculeata. The Parasitica comprising the majority of Hymenopteran insects living as parasitoids on "every other species of insect", and many non-insects. Most species are small, with the ovipositor adapted for piercing. In some hosts the parasitoids induce metamorphosis prematurely, and in others it is prolonged. There are even species ("hyperparasites") that are parasitoids on other parasitoids. The Parasitica lay their eggs inside or on another insect (egg, larva *or* pupa) and their larvae grow and develop within or on that host. The host is nearly always killed. Many parasitic Hymenoptera are used as biological control agents to control pests, such as caterpillars, true bugs and hoppers, flies, and weevils.

The Aculeata is a monophyletic group that includes those species in which the female's ovipositor is modified into a "stinger" to inject venom rather than eggs. Groups include the familiar ants, bees and various types of parasitic and predatory wasps; it also includes all of the social Hymenoptera.

Honeybees and their relatives

The present bee fauna dates back to Cretaceous period which is more than 70 million years ago (Evans, 1969). The first appearance of bees is closely tied with a change in food from insect prey to pollen and nectar obtained from flowers of angiosperms. The abundance of nectar and pollen as a readily available source of larval food was a contributing factor in the change of some wasps from a predatory existence (Sphecoidea) to that of collecting nectar and pollen. The geologically oldest and most completely preserved honey bees were found in Baltic amber in East Prussia and date to the upper Eocene period or roughly 50 million years ago (Zander and Weiss, 1964). The bees belong to superfamily Apoidea, were a group of wasps that had abandoned their habits of provisioning their nests with insects and spiders, and instead fed their larvae with pollen and nectar collected from flowers or with glandular secretions ultimately derived from flowers. These bees have morphological characters which partially point to the present day Meliponini and partially to the Apini. Eusociality in bees originated from solitary living and later subsocial living. In solitary bees, a female lays eggs in her nest after gathering sufficient food for them. The mother dies before seeing her offspring. Mass provisioning of food to young ones is the rule. In the sub social insects, a female starts a colony by laying eggs and taking care of the young ones. Her daughters later join her mother in hive maintenance. The mother then specializes in egg laying and daughters in the care of the brood. Eusocial insects are characterized by the following attributes.

Species diversity of *Apis* in Southeast Asia

Honeybees have settled almost all over the planet. They live both in regions with cold climates and long severe winters and in the tropics where winters never occur and the summer temperatures are usually higher. Bees adaptability to different climates and environments has proved to be genuinely amazing. As a result of specific climatic conditions and peculiarities of nectariferous flora, there developed various breeds of honeybees during the course of their evolutionary history.

Based on morphological and behavioural analyses and the aid of different genetic techniques, the classification of the true honeybees has obtained great achievements in the last two decades of the twentieth century. The number of recognized honeybee species has been reduced since the descriptions of Maa (Maa, 1953), because many types are now seen as subspecies (Otis, 1997). The tribe, Apini, consists of only one small monophyletic genus, *Apis* that comprises nine honeybee species:

Apis mellifera, A. cerana, A. koschevnikovi, A. nigrocincta, A. nuluensis, A. dorsata, A. laboriosa, A. florea and *A. andreniformis* (Otis, 1997; Tingek *et al*., 1996). Out of these nine species the five initial species nest in cavities have a number of combs. The last four that nest in the open have a single comb. *Apis* species are divided into three lineages: the cavity-nesting bees, *Apis mellifera, A. cerana, A. koschevnikovi, A. nigrocincta* and *A. nuluensis*; and open nesting the dwarf bees, *A. florea* and *A. andreniformis*; the giant bees, *A. dorsata* and *A. laboriosa*. Of the nine species, only *A. mellifera* and *A. cerana* have been "domesticated" for a long time (Koeniger, 1976). *A. mellifera* is the most studied and economically exploited species. All *Apis* species, except for *A. mellifera*, are native to Southeast Asia. This region is a centre of *Apis* diversity and makes scientists pay great attention to the recently recognized species such as *A. nigrocincta* and *A. nuluensis*. In Vietnam, five native honey bee species are found, namely *A. cerana* Fabricius 1793, *A. dorsata* Fabricius 1793, *A. florea* Fabricius 1787, *A. andreniformis* F. Smith 1958 (Ha and Lap, 1992) and *A. laboriosa* F Smith 1871 (Trung *et al*., 1996). They distribute in different areas of the country from north to south, especially in mountain forest areas. This rich species-diversity promotes prospective sources for basic research on biology, ecology and behaviour in honeybees. Furthermore, this diversity is of interest to applied researchers because of the great importance of pollination in Vietnamese agricultural and in the forestry ecosystem.

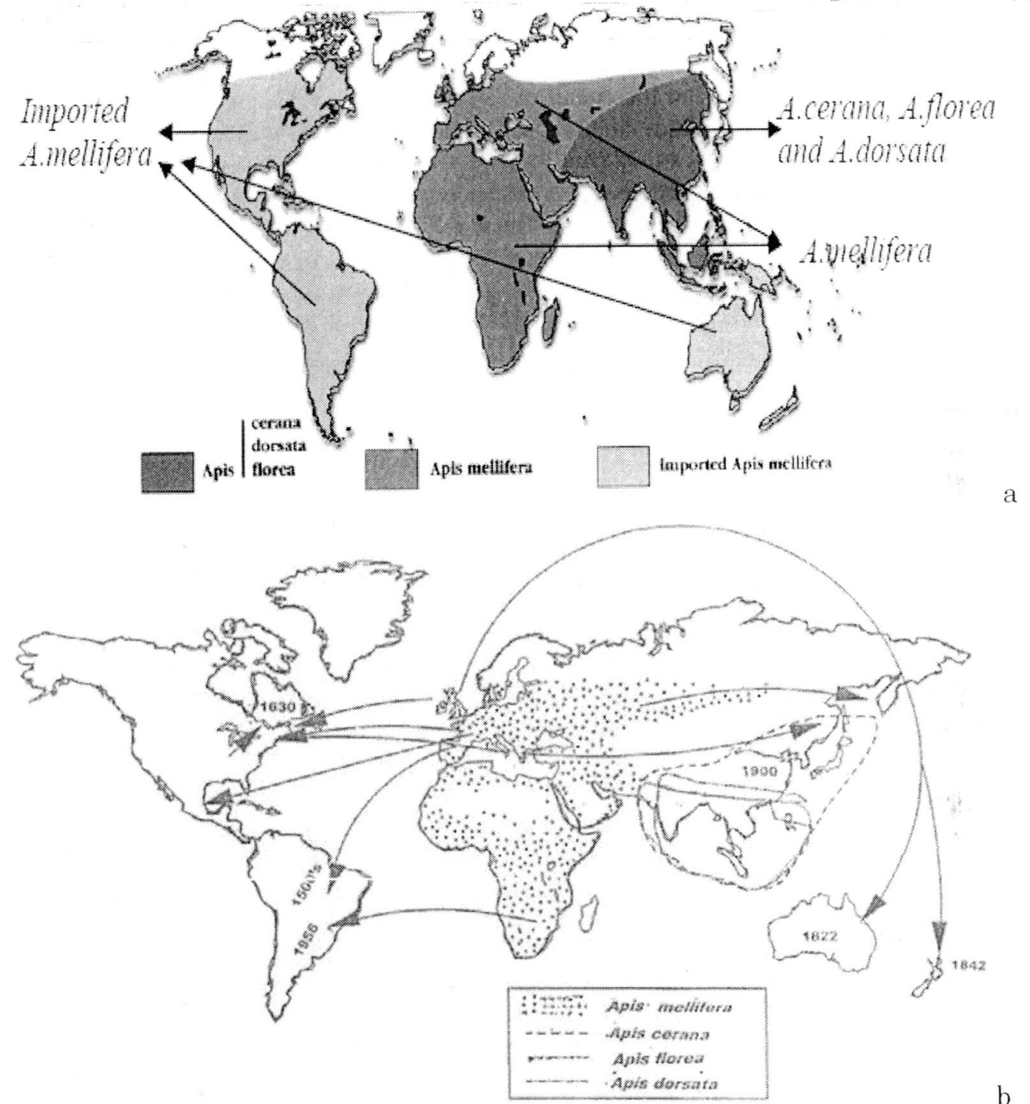

Figure 4(a) Origin and distribution of various species of honeybees **(b)** species of honeybees and major movements of *Apis mellifera*.

There are about 9 different known species of bees that make honey. The most commonly recognized honey bee species, *Apis mellifera* Linnaeus, is native to Africa and Europe, and subdivided into about 24 subspecies.

Scientific classification

Kingdom:	Animalia
Phylum:	Arthropoda

Class:	Insecta
Order:	Hymenoptera
Suborder:	Apocrita
Family:	Apidae
Subfamily:	Apinae
Tribe:	Apini
Genus	*Apis*

Species

Apis andreniformis

Apis cerana, or eastern honeybee

Apis dorsata, or giant honeybee

Cliff bee, *A. laboriosa*

Apis florea

Apis koschevnikovi

Apis mellifera, or western honeybee

Apis nigrocincta

Apis cerana nuluensis

Origin and distribution of the genus Apis

Honeybees as a group appear to have their center of origin in Southeast Asia (including the Philippines), as all but one of the extant species are native to that region, including the most primitive living species (*Apis florea* and *A. andreniformis*). The first *Apis* bees appear in the fossil record in deposits dating about 40 million years ago during the Eocene period; that these fossils are from Europe does not necessarily indicate that Europe is where the genus originated, as the likelihood of fossils being found in Southeast Asia is very small, even if that is the true origin. At about 30 million years before present they appear to have developed social behavior and structurally are virtually identical with modern honeybees. Among the extant members of the genus, the more ancient species make single, exposed combs, while the more recently-evolved species nest in cavities and have multiple combs, which greatly facilitated their domestication.

Honeybee species

Species whose nests are single combs

1. The dwarf honeybee *Apis florea*
2. The giant honeybee *Apis dorsata*

Figure 5. Showing different honeybees A, *Apis mellifera*. B, *Apis dorsata* C, *Apis florea*, D, *Apis cerana* Nests of *Apis dorsata* on branch of tree. A view of Apis dorsata on a branch of tree. A view of *Apis dorsata* on a building. A view of nest of *Apis florea* on a branch of plant. A view of nest of *Apis cerana* in earthen pot.

3. Cliff bee *A. laboriosa*

4. *A. andreniformes*

Species whose nests have parallel combs

1. The oriental honeybee *Apis cerana*

2. The common, or European, honeybee *Apis mellifera*

3. *Apis koschevnikovi*

4. *Apis nigrocincta*

5. *Apis cerana nuluensis*

A. Dwarf Honeybee *Apis florea* Fabricius, 1787

The distribution area of *A. florea* is generally confined to warm climates. In the west, the species is present in the warmer parts of Oman, Iran and Pakistan, through the Indian sub-continent and Sri Lanka. It is found as far east as Indonesia, but its primary distribution centre is Southeast Asia. Rarely found at altitudes above 1500 m, the bee is absent north of the Himalayas. It is frequently found in tropical forests, in woods and even in farming areas. In Southeast Asia it is not rare to find a nest of *A. florea* in a village. As its name implies, the dwarf honeybee is the smallest species of honeybee, both in the body size of its workers and in the size of its nest. A nest of *A. florea* consists of a single comb, whose upper part expands to form a crest that surrounds the branch or other object from which the comb is suspended. Dwarf honeybees nest in the open, but not without camouflage: most nests are hung from slender branches of trees or shrubs covered with relatively dense foliage, usually from 1 to 8 metres above the ground. In Oman, where *A. florea* nests are frequently found in caves, such combs are without crests.

Combs of the dwarf honeybee are well covered with layers of workers clinging to each other' often three or four deep. About three quarters of the colony's worker population are employed in forming this living protective curtain of bees. When disturbed, this curtain shows a "shimmering" movement, the individual bees shaking their abdomens from side to side in a synchronous manner; at the same time, a hissing sound is released. If the colony is further disturbed, the worker bees raise their abdomens and take off from the curtain to attack the intruder.

The section of comb surrounding the support consists of adjoining honey-storage cells that form a crest, from whose inroad curved surface the bees take off and on which they land. The communication dance by scouts, announcing the discovery of a food source, also takes place on this platform. Adjacent to the rows of honey-storage cells is the section of comb which the workers use for storing pollen. Beneath this band of pollen-storage cells is the area where the worker brood is reared. Prior to the swarming season, drone-brood cells are added, adjoining the lower rows of the worker-brood cells. When a colony loses its queen, emergency queen-cells are built from normal cells containing young worker larvae.

To ward off ant attacks, the workers coat both ends of the nest support with sticky strips of propolis, or "plant gum", from 2.5 to 4 cm wide. *A. florea* is the only honeybee that uses this defensive technique.During the season when there is an ample supply of nectar and honey, populous colonies of the dwarf honeybee send out multiple reproductive swarms. In addition, colonies of this bee have a high degree of mobility. Disturbance by natural enemies, exposure to inclement weather and scarcity of forage are among the major causes of colonies absconding.

In comparison with other honeybee species, the amount of honey that *A. florea* workers will store in their nests is small, usually not exceeding several hundred grams per colony. In some parts of Asia, the rural people have devised a scheme for harvesting this honey. First, nests or the bees are transferred from their natural sites to the village, and then, using twine and two short twigs, the nest is clamped and attached to a small branch of a tree. The upper part of the comb, containing the honey, is cut out, and the honey is squeezed out from it. A period of about six to eight weeks is allowed for the bees to repair the comb and replenish it with honey, and then it is harvested again. This method is not always reliable, however, because most colonies will abscond either shortly after their transfer to the new site or after the first or second harvest has taken place.

Where nests of *A. florea* are abundant, several rural families can subsist on the income generated from bee hunting alone. Although the practice appears ecologically destructive, particularly insofar as it reduces a valuable population of natural pollinators, it does not always destroy the colony being hunted. Workers and laying queens of the dwarf honeybee are able to respond to nest predation quickly. The entire colony, accompanied by a laying queen, can fly several meters away to regroup, and later abscond. Some absconding colonies are able to survive to build their new combs in a nearby area.

The Dwarf Honeybee (*Apis florea*) is one of two species of small, wild honeybees of southern and southeastern Asia. It has a much wider distribution than its sister species, *Apis andreniformis*. These two species together are the most primitive of the living species of *Apis*, and this is reflected in their small colony size, and simple nest construction. The exposed combs are built on branches of shrubs and small trees, and the forager bees do not perform a waggle dance to recruit nestmates as in the domesticated *Apis mellifera*, but instead they "dance" on the horizontal upper surface where the comb wraps around the supporting branch - the dance is a straight run pointing directly to the source of pollen or nectar that the forager has been visiting. In all other *Apis* species, the comb on which foragers dance is vertical, and the dance is not actually directed towards the food source. Aside from their small size, simple exposed nests, and simplified dance language, the life cycle and behavior of this species is fairly similar to other species of *Apis*

Apis florea is the smallest of the true honey bees and is called appropriately the dwarf or the little honey bee. In India it is known commonly as *katua, bhunga, chhoti madhumakhi, chhoti mahal* (Uttar Pradesh), *phulori masa* (Maharashtra), *visanakarra pattu* (Andhra Pradesh). The bee is generally found in plains or low lands in tropics and sub-tropics. It is rarely found in altitudes above 1500 m (Muttoo 1956, Verma 1990). The nests are built in bushes, densely leaved small trees in gardens and orchards, eaves of buildings or sheltered boxes or wall niches in urban areas and on closely placed stalks of crop plants like Sorghum (*jowar*).

There are reports of occurrence of another species of dwarf bee, closely resembling *A. florea*, but darker and a little large than it, in tropical semi-evergreen forests in the Western and Eastern Ghats. The abdomen of this type is blackish brown, while *A. florea* has white stripes alternating with orange. It has not been observed in cultivated and inhabited areas. In the nest building and other activities, it is similar to *A. florea*. Discussing on the variability in *A. florea*, Ruttner (1988) says that there are three geographic types: one found in Sri Lanka and South India; one distributed in Iran, Oman and Pakistan and a third in Thailand. It is possible that all these three types are found in India and the following account of *A. florea* applies equally well to these other types. The dwarf bee is able to survive in very hot and dry climates with ambient temperatures reaching 50°C or more. The vast salt swamps of the Rann of Kutch in Gujarat as also the dry and semi-arid plains of Rajasthan sustain large populations of this bee. The bee hive consists of a single comb, about 40 cm broad and 45 cm long. Muttoo (1956) reports that the size of the comb varies from as little as the palm of a man's hand to as much as 50 cm long. In favourable conditions the nests do grow for a long period and the comb breadth may reach 65 cm.

The comb architecture is similar to that in other *Apis* species, except for the honey storage portion, that is distinct in *A. florea* comb. The comb shows a distinct honey portion, that is situated at the top, and where the support is free from above, the honey cells are constructed around the support. This type of comb construction is very common. The cells surrounding the twig are upto 3 cm deep and extend more or less horizontally from the support on its both sides. A cylindrical comb is thus formed with the twig passing through its middle. This cylindrical section of the comb is used to store honey and is called the honey cap or crest. The dwarf bee shows a great adaptability in its nest building behaviour. When the nests are constructed in wall niches or under the eaves of houses, the crest is not in the form of a cylinder. The support is used to start a hexagonal cell pattern and the shape of the crest depends upon the nature of the support. If it is vertical as in the case of a wall niche, the crest is unilateral and is attached to the wall laterally. If on a horizontal ceiling, the crest is positioned adjacent to the vertical portion of the comb.

Beneath the crest, the comb abruptly reduces in thickness to about 16 mm, and extends vertically. The vertical part of the nest has very regular cells, 2.7 to 3.1 mm in width. During good flowering seasons, when the colony reproduces by issuing swarms, a conspicuously large celled drone brood portion is seen below the worker brood. The queen cells are constructed along the lower edge of the comb. The worker bees are appreciably smaller than the drone and queen bees. The latter two castes are almost as big as those of *A. cerana*. The number of worker cells per dm^2 was 1190 in north India, 1240 in central and 1560 in south India (Muttoo, 1956). This indicates that the bees in the south are the smallest, and the size increases as one goes northwards. The number of bees present in a good colony is about 10,000. In small colonies, it is usually half that number. The queen normally lays 350 to 700 eggs per day (Kshirsagar *et al.,* 1980, Ruttner 1988). Worker bees live longer than those of other *Apis*, usually about 60 days. In areas where *A. florea* usually occurs, the forage is available only in limited periods. Because of this the bee migrates often from one area with bee forage potential to the other. Its residence at a place varies usually from 2 to 5 months. If the forage potential of the area is good, the residence may be as long as one year. The distance to the new nesting site of a swarm may be only a few upto several 100 m. Muttoo (1956) reports that the bees undertake migrations in summer from the hot plains to the hills involving long distances of several kilometers.

The bees have a short flight range, often hardly reaching 100 meters from the nest. The maximum distance it can fly from the nest for foraging is often less than 750 m. In view of this, the honey stored by it is generally unifloral, when floral sources are plenty, as in large farms of Sorghum or sunflower, orchards of *Citrus*, litchi and guava, or plantations of the mesquite, *Prosopis chilensis* (Mol.) St. (Ramanujam and Kalpana, 1991). However, soon after the flowering season is over, the bees leave their comb and migrate to other areas with food sources, a short distance away. *A. florea* is also notorious for its absconding behaviour. The bees abandon their nest at the slightest disturbance or provocation. They build another nest a short distance away, often using for this purpose, the beeswax and honey stores of the abandoned nest. Because of its fast reproduction rate and long life, the population of the dwarf bee in any locality is maintained, in spite of the constant disturbance from its enemies and frequent absconding.

The storage capacity of the honey cap in the nest is 500 g to 1000 g or more (Muttoo 1956, Ruttner 1988). Large colonies in areas rich in bee forage are known to produce upto 4 kg of honey each. Phadke (1968) reports that honey of the dwarf bee is similar to that of the dammar bees, and has a high dextrin content and low tendency to granulate. Dextrins are 4 times that in other *Apis* honeys. Levulose content is also high. The reason for the differences in the composition of the dwarf bee honey is supposed to be the sources of forage. These usually include weeds and small flowered plants like those in the families Apiaceae, Asteraceae and Euphorbiaceae. The honey is valued in Ayurveda for its high medicinal properties. Handling *A. florea* colonies is easy. It is a common practice to cut loose the twig with the nest. The bees cling securely to the brood nest and not even single bee flies out during transportation from one site to another. The bees are mild and rather than drive away the enemy by stinging, the bees resort to absconding and constructing a new nest in a sheltered and protected place. Considering its open nesting and small size that makes it difficult to defend their nest effectively from its enemies, particularly man, this behaviour perhaps ensures the continuation of the species. Because of this behaviour, rural and tribal people, including children often plunder *A. florea* nests for the sweet honey and nutritious brood that is eaten raw.

There have been several attempts to hive this bee. Ghatge (1949) reported that these bees were amenable to gentle handling. After cutting the honey cap, the comb could be tied securely in a brood frame of *A. cerana*. The bees would immediately join the comb to the top of the frame. A super frame of *A. cerana* would be held above the brood frame. According to Ghatge, the bees would construct a honey comb and fill it with honey. There was no further report on the field application and utility of this technology. Pundir (1971) succeeded in keeping an *A. florea* colony for about two months in Ranipokhari, near Dehradun, Uttar Pradesh in a 4-frame nucleus box of *A. cerana*. The original comb of the colony was tied to one *cerana* brood frame measuring 280 x 175 mm inside. This was kept in the box between two other frames. One of these had an empty comb built by *A. cerana*. The other side frame was empty. Above the box, a top was kept leaving some space for bee flights. The colony was very strong and occupied all the three frames. During the next two weeks, the bees joined the comb to the frame and extended it to fill the frame. There were honey and pollen stores, and brood in all stages. The empty side comb was filled entirely with honey, but the third frame was left as such. The colony stayed for two months, issued three swarms and at the end deserted. This experiment indicates that a simple form of beekeeping is possible with *A. florea*.

In Latur, Maharashtra farmers who grow sunflower value *A. florea* colonies for their pollination service. To ensure pollination and good yields from sunflower, they collect *A. florea* colonies from wild shrubs and trees, take the nests to their farm lands and locate them, generally along hedges having small trees like bael (*Aegle marmelos*) that can provide adequate shelter to the nests. While collecting the colony, the honey portion is cut and consumed. The brood comb is secured between two twigs or a split bamboo piece. The twig with the nest is tied to a thin branch at the new site.

Florea beekeeping is practiced in Oman. Dutton and Free (1979) report that *florea* colonies are collected by experienced Arabs from the wild and, after removing the honey cap for consumption, the brood comb is sandwiched between two parts of a split stalk of a date palm leaf and located in wall niches or close to buildings. Honey is harvested without destroying the brood the sometimes colonies are transported during winter to especially sunny places. Ruttner (1988) says that such efforts are necessary to develop florea beekeeping in excessively hot climates, where *A. mellifera* cannot normally be kept.

A. florea has a commercial importance in the hot climate regions of Gujarat. Several hundred thousands of these colonies are found in this area. At least 300 tons of honey are collected annually from them. The dwarf bee assumes especial importance in the hot salt marshes of Gujarat, or the vast arid or semi-arid regions in Rajasthan that sustain a xerophytic vegetation, where no other honey bee can be kept profitably. It is an industrious field bee, easy to handle, and when it migrates the flight range of the swarm is generally not very large (Ruttner 1988). The technology of *florea* beekeeping is very simple and does not need even the box that is used in *A. cerana* or *A. mellifera* beekeeping.

Besides, honey production, *A. florea* can be utilized also for pollination of several agricultural and horticultural crops. The dwarf bee is an important pollinator of crops in hot and dry agricultural plains of India. Atwal (1970) and Sihag (1982) and Abrol (2006) consider *A. florea* as one of the most efficient pollinators of sarson, toria and raya varieties of mustard, lucerne, onion and cultivated Apiaceae that include ajwan, anise, carrot, cumin, dill seed and fennel.

Native from Oman is spread throughout south-east Asia as far as some of the islands of Indonesia also. In recent years it has spread to Sudan and Iraq. Apis florea builds a single-comb nest, usually fairly low down in bushes, or in the open, suspended from a branch or rock surface. Apis florea are very small bees, and their nest is small too, often no larger than a man's hand. A colony might contain 20,000 bees.

This species is a bee of the plains up to 500 m. However, seasonal migrations occur up to 1500 m and even higher (Muttoo, 1956). In spite of its small size it competes well with other the other *Apis* species (Koeniger, 1976). *A. florea* is distributed in the coasts of Persian Gulf, Pakistan, India, Srilanka, Thailand, Malaysia, Indonesia and Philippines (Palawan). Larger types in north and smaller ones in the south are seen as in other bees. Phenotypic variation among *A. florea* has not been understood well as the data on these aspects are limiting. The communication dances of *A. florea* are performed on the platform on top of the comb and point directly towards food. Another closely related species of *A. florea* is *A.andreniformis* which appears darker in colour and is found in sub-mountain regions of Srilanka, Malayan Peninsula, Thailand, Sumatra, Java and Borneo but not recorded in India.

Apis andreniformis Frederick Smith, 1858

Apis andreniformis is a species of honeybee whose native habitat is the tropical and subtropical regions of Asia. *A. andreniformis* was the second honeybee species to be recognized, and its biology, geographic distribution, and its specific status was recognized by many authors. However, it must be mentioned that the species was only recently separated from *Apis florea* since there are sites where both *A. andreniformis* and *A. florea* live conspecifically. Both species are distributed throughout tropical and subtropical Asia, including Southeast China, India, Burma, Laos, Vietnam, Malaysia, Indonesia (Java and Borneo), and the Philippines (Palawan).

The most significant morphological characteristic of the species is the black stripes legs, specifically on the tibia and on the dorsolateral (back and side) surface of the basitarsus. Additionally, the pigmentation of *A. andreniformis* is blackish, while that of *A. florea* is yellowish. Other distinguishing characteristics include a difference respective cubital indexes: *A. andreniformis* has an index of 6.37, and that *A. florea* one of 2.86. Also, the proboscis of *A. andreniformis* has a length of 2.80 mm, while that of *A. florea* is 3.27 mm. This physical difference contributes to a division in the distribution of naturally occuring nectar between the two species. Finally, there are differences in the barbs of the stinger, and in the basitarsus of the drones.

Apis florea and *Apis andreniformis* are small species of honey bees whose nests are small, single combs like *A andreniformis* nests in quiet forests, generally in darker areas where there is 25 to 30% of normal sunlight. The hive is usually made in branches of bamboo and banana plants, in shrubs, and in bushes such as coffee and tea. They can be built between 1 to 15 meters from the ground, although the average altitude is 2.5 m. The honeycomb ranges from 7 to 90 mm, however, needs further confirmation. *A. andreniformis* is generally more defensive than *A. florea*: it is known to attack when there are disturbances 3 to 4 meters from the hive.

The main parasites of both *A. andreniformis* and *A. florea* belong to Genus Euvarroa. However, *A. andreniformis* is attacked by the species *Euvarroa wongsirii*, while *Euvarroa sinhai* preys on *A. florea* and colonies of *Apis mellifera* that are imported. The two species of Euvarroa have morphological and biological differences: while *E. wongsirii* has a triangular body shape and a length of 47 to 54 micrometres, *E. sinhai* has a more circular shape and a length of 39 to 40 micrometres.

This species is quite similar to *A. fiorea*: small sized bees which nest on single combs. It has been identified in Thailand, Malaysia and the southern Chinese peninsula.

Giant or Rockbee, *Apis dorsata* F.

The distribution area of the giant honeybee is similar to that of the dwarf honeybee: it occurs from Pakistan (and, perhaps, parts of southern Afghanistan) in the west, through the Indian subcontinent and Sri Lanka to Indonesia and parts of the Philippines in the east. Its north-south distribution ranges from the southern part of China to Indonesia; it is found neither in New Guinea nor in Australia. *A.dorsata* is distributed in South China, Celebes and Timor but not Iran or the Arabian Peninsula. It is found in altitudes up to 2000 m; the mountain type named *A. laboriosa*, possibly a separate species is found even higher in Nepal (Sakagami *et al.,* 1980).

The giant honeybees of Nepal and the Himalayas have recently been reclassified as belonging to another species of *Apis*, *A. laboriosa*. It is not yet clear whether the giant honeybees of Sikkim and Assam in northern India, western Yunan Province in China, and northern Burma should he classified as *A. dorsata* or as *A. laboriosa*, but in the present state of our knowledge, it is safe to consider that all the giant bees constitute a single taxonomic identity. Although minor variations in anatomical, physiological and behavioral characteristics exist among the different geographical races of the giant honeybees, they are essentially similar in all their major biological attributes.

The giant honeybees are found predominantly in or near forests, although at times nests may be observed in towns near forest areas. The bee shares the open air, single-comb nesting habits of *Apis florea*, suspending its nest from the under surface of its support, such as a tree limb or cliff. In general, *A. dorsata* tends to nest high in the air, usually from 3 to 25 meters above the gound. In tropical forests in Thailand, many nests are suspended in Dipterocarpus trees from 12 to 25 meters high: this tree is probably preferred as a relatively safe nesting site because its smooth bark and its trunk rising for 4 to 5 meters before branching out make it very difficult of access to terrestrial predators.

Nonetheless, about three-quarters of the worker population of a colony of giant honeybees is engaged in colony defence, forming a protective curtain three to four bees thick in the same way as *Apis florea.* While birds are common predators of *A. dorsata*, the workers' large body size protects them reasonably well against ant invasion, so that the sticky bands of propolis characterizing the nests of the dwarf honeybee are not found surrounding the nests of *A. dorsata*, nor are the nests hidden by dense foliage. Nests of *A. dorsata* may occur singly or in groups; it is not uncommon to find 10-20 nests in a single tall tree, known locally as a "bee tree". In India and Thailand, tree harbouring more than 100 nests are occasionally seen in or near the tropical forest.

The single-comb nest, which does not have the crest of honey-storage cells typical of *A. florea* nests, may at times be as much as one meter in width. The organization of the comb is similar to that in the other honeybee species: honey storage at the top, followed by pollen storage, worker brood and drone brood. At the lower part of the nest is the colony's active area, known as the "mouth", where workers take off and land, and where communication dances by scouts, announcing the discovery of food sources, take place. This dance takes place on the vertical surface of the comb and during its progress, the bees must have a clear view of the sky to observe the exact location of the sun. Workers of *A. dorsata* are however able to fly at night, when the light of the moon is adequate.

In many places, the arrival of *A. dorsata* colonies is an annual event, occurring at the end of the rainy season or at the beginning of the dry season, when several species of nectaryielding plants are in bloom. This phenomenon leads to speculation that *A. dorsata* has a fixed pattern in its annual migratory route. Most professional bee-hunters know when and where the trees are to arrive, but they wait patiently until the end of the honey-flow period before taking down the nests. Observations in northern Thailand indicate that if the nests are left undisturbed, the colonies will eventually abscond or migrate when their food reserves have been depleted, usually at the end of the summer months. By the beginning of the rainy season, *A. dorsata* colonies are found deep in the lush jungles.

A. dorsata is well known for its viciousness when its nest is disturbed: the mass of defending workers can pursue attackers over long distances, sometimes more than 100 meters. Notwithstanding its ferocity, however, this bee's honey is highly prized locally, in some places commanding the best prices in local markets,

Nests of the giant honeybee have been hunted by man since antiquity, and today, organized bee hunting exists in many parts of Asia. In Thailand, bee-hunters must pay fees for permits to hunt the bee in state forests, and landowners possessing bee trees sell annual or biennial rights to hunt nests from such trees. Some professional bee-hunters prefer to work at night. Smoke is used to pacify the bees, which are then scraped from the comb. The nest is cut and placed in a cloth bag, which is lowered to an assistant on the ground. This method does not result in all colonies being killed: about a fourth of the colonies in a bee tree that has been worked over are able to reconstruct their nests.

The recent intensification of bee hunting has caused an alarm in several Asian countries. There is general concern that the total number of *A. dorsata* nests all over Asia may be on the verge of declining, partly due to shrinking forest areas, the use of toxic pesticides in foraging farm lands, and bee hunting. The giant rock bees are found in large numbers in the Himalayas. In higher reaches, Apis laboriosa is found and in the lower areas of the Terrai (foothills), Apis dorsata is commonly found. Huge quantities of honey and bees wax are sold to wholesalers from towns in this region.

In the central parts of the country honey yields are substantial from *Apis dorsata*, primarily due to good forest patches in and around sanctuaries and protected areas. *Apis dorsata* collectors are mainly tribals. Honey for health and Ayurvedic medicines has been a traditional industry in this region. The mangrove forests of the Sunderbans are an excellent habitat for *Apis dorsata*. The entire southern region is rich in *A. dorsata* populations - contributing to a large share of the total Indian honey market. In Andhra Pradesh, farmers and honey hunters in the hills of the Eastern Ghats collect honey. Significant quantity of honey is passed on to traders. Intricate technologies and practices have been going on since a long time. Honey hunting is done on rocks and trees. Any accurate estimates of the number of honey collectors is not available.

On the western edge of its distribution *Apis dorsata* is found only as far as Afghanistan but its south-east occurrence extends a long way east of Bali. Its northern distribution is limited by the Himalayas. *Apis dorsata* bees are large. Their nests consist of single combs suspended from a branch, cliff face or building. These combs can be very big, up to two metres wide and one metre from top to bottom.

There is morphometric evidence for different subspecies of *Apis dorsata* which may eventually be proven to be separate species. *A. dorsata* is the largest of the honey bees. Two subspecies of *A.dorsata* namely, *A. dorsata breviligula* with short tongue, medium forewing length found in Phillippines, beyond Meryll line in east and *A. dorsata binghami* with long tongue and long forewing in

Celebes beyond Wallace line in east have been recorded. The large combs (up to 1 m^2) are fixed on the underside of thick horizontal branches of large trees. Sixty or more nests may be found on these bee trees. Two behavioural characteristics of *A.dorsata* are remarkable. First they have a well organized mass defense reaction. An intruder once marked by the odour of a specific pheromone (2-decen-1-yl-acetate) by being stung is followed for kilometers. Second these *A. dorsata* seasonally migrates to locations 100-200 km distant every year. The timing of migration is correlated with the change in the season (rainy to dry period). Rockbees are common in plains and hills upto 3000 m altitude in several tropical, sub-tropical and temperate parts of India. Known also as the giant bee, due to the fact that it is the largest among the honey bees, in India the bee is variously called as dumna, bhandaur, bhanwar (Punjab), saranga, bhramara, pahari mahal (north India), agya masha (Maharashtra), and Konda, Thera or Pedda Pattu (Andhra Pradesh). The Vedic names of the bee (Dave 1954, 1955) are: arangara (noisy, angry, buzzing bee), chatra (bees with umbrella like comb), tugra (very ferocious bee) and twashtar (robust, bull like big nested bee).

Rockbees may belong to two distinct species : *Apis dorsata* and *A. laboriosa* Smith. The latter species is the largest among the honey bees in size. It is darker in body colour, builds bigger combs, and has larger population than the former. Truly the giant bee, it is common in the higher altitudes - between 1200 and 4100m. It is not seen in the tropical plains. *A. dorsata* is common in lower altitudes and in plains, and has a lighter orange brown or tawny body colour. There is, however, a variety distinctly darker brown in colour that occurs in some agricultural plains, e.g., Dindi reservoir in Andhra Pradesh, along with the usual lighter brown variety. However, no study has been made so far to distinguish the two. Both the species have the same local names. In view of this, the following account, although given for *A. dorsata*, also applies to and includes *A. laboriosa*. The nests of the rockbees are built in the open and are fixed underneath a broad support such as a rock overhang, branch of a tree, or the eaves of a building. An usual feature of the rockbees is aggregation of their nests on terrestrial and arboreal supports. Singh (1962) reports of a record of 156 rockbee colonies on a tree in south India. In the Nallamala forests, Andhra Pradesh, a 20 m tall *Terminalia tomentosa* tree was observed to have about 60 colonies during October 1993. The local tribals say that the tree has over 200 colonies in good seasons. Deodikar *et al.* (1977) recorded aggregations of 30 to 50 colonies in different urban localities in Karnataka, Maharashtra and Madhya Pradesh. Aggregations of about 100 colonies can be seen on a single tree or a cliff shelter on hills. These are common in areas with rich bee forage potential. In dry forests, vast agricultural fields or in urban areas, aggregations are rare. In such areas single nests or only a few nests are seen in a nesting site.

It is a common sight in thickly wooded mountainous areas to find several rock bee nests hanging down rock cliffs, rock shelters or from roofs of caves. In thick forests rockbee colonies are built on branches of lofty tress at heights upto 50 metres from the ground. In the vast farm lands near forests, nests are built on available large trees, including coconut palms, mango, peepal and banyan trees, or any terrestrial structure that offers protection like the beams underneath tall bridges on highways or railways. In the Sunderbans, the combs are built at lower levels, sometimes touching the ground (Chakrabarti and Chaudhari, 1972). This apparent difference in the nest building habit is due to the fact that the bees build their hives in places normally inaccessible to their predators and enemies. In the forests where a variety of animals can harm the nests, they are built at heights that cannot be reached by the animals. In Sunderbans the mangrove jungles are quite inaccessible. The nests are not exposed to any danger in such locations. Possibly because of this and according to the availability of suitable supports, the nests are built at lower heights. Rockbees are found even in inhabited areas including towns and cities. In such areas, multistoried buildings, large overhead water tanks and towers provide ideal nesting sites. There are also instances of several colonies nesting on ceilings and arches of temples, busy engineering workshops of factories (Deodikar *et al.*, 1977) and other buildings with regular human activity. Nests just a metre or so above the ground level, and within easy approach have also been found. As the prevailing human customs do not allow any harm to the bee hives, which seem to prefer such locations as they also get here additional protection from other wild animals.

A majority of colonies build their combs in a North-South direction. This may be due to the fact that this direction provides better protection than others against monsoon winds and allows sun light on the comb (Deodikar *et al.*, 1977, Reddy, 1983). Rockbees build a single comb often 100 cm broad and 80 cm high, but sometimes quite huge ones occur in favourable conditions, measuring upto 200 x 150 cm. The combs are semi-circular, the width corresponding to the line of attachment being greater than its height or length. The combs are vertical, attached to the support above, from where it hangs down. The comb has a midrib and two layers of cells connected to it on either side. The cell openings are slightly tilted upwards. The part where it is joined to the support is generally upto 30 cm thick and bulging from the other and lower parts of the comb. Honey is stored in this portion, usually on one side of the comb, when having an arboreal support that is not horizontal, or on both sides and nearly through the entire width, when it is attached to a terrestrial support. The bulged portion gradually decreases in thickness and merges below with the brood comb. This section is used for storage of pollen. This is followed by the normal 3.5 to 4.0 cm thick brood comb with brood in various stages of growth The brood cells are 5.28 to 5.64 mm in width (Muttoo, 1956, Thakar and Tonapi, 1961). The lower free side of the comb contains drone cells, that are not much different from worker cells, but have slightly raised walls. Sealed drone brood is thus elevated from the worker brood level. Queen cells are built along the lower edge (Deodikar *et al.*, 1977).

The nest population is 60,000 to 100,000 bee strong in well-developed colonies. In fresh swarms, the number of bees may be much less, 5,000 to 10,000. The worker bees in the nest perform two distinct functions: nest service and nest protection (Morse and Laigo, 1969, Ruttner, 1988). A thin layer of bees sitting right on the comb is engaged in the nest service. This includes the usual in-house duties like nursing the young, secretion of wax, comb construction, cleaning, and honey preparation. A large number, 80 to 90% of the worker bees, forms a curtain or 'pelt' covering the comb, but at a distance of 1 to 2 cm from it. The curtain consists of 3 to 6 layers of bees oriented uniformly, with the head of a bee below bent down under the abdomen of the above. The bees in the curtain remain steady unless disturbed. When disturbed however they exhibit a characteristic defence behaviour, called as the shimmering behaviour. This consists of (a) abdomen shaking: a collective violent lateral shaking of the slightly raised abdomen, and (b) hissing : sharp hissing sound that lasts about half a second. The sound is produced by a collective quick movement of the wings. The shimmering behaviour coupled with the massive numbers of alert bees in the defence force of the protective curtain of the colony is usually enough to frighten the mightiest of its enemies. The alerting effect of rockbee sting lasts for about 2 days. A single disturbed bee can make upto 5000 defending bees take up attack flights (Ruttner 1988).

The nest has an active "mouth" portion usually on one side of the lower free part of the comb. Here the bees are not uniformly oriented, and have their heads directed outwards. There is a regular traffic of bees landing or taking off from the comb on their foraging flights. Communication dances are visible in the "mouth" area. The size of the "mouth" area varies greatly with the amount of activity of the colony. It is believed that the "mouth" changes its location depending upon the food source and other factors, and can also be utilized to find out the "mood" of the colony. If the "mood" is right, the colony can be handled bare-handed and received no sting, even after a thorough disturbance of the colony and when "out of mood", the colony cannot even be approached, let alone handled, and many a honey collector ignorant of this behaviour, learnt about it the hard way. How to detect this "mood" however remains a mystery. Like the dwarf bee, rockbee is migratory in nature, but the stay of nests in any one place is usually long and often their movement is restricted to two times in a year. The colonies in the sub-Himalayan ranges arrive in the temperate mountain region in March - April and leave for the foot hills and plains in June. In south India, colonies migrate to farming areas in the plains in June, at the beginning of the monsoon season. They move back to the forest areas in hills in October, when rains subside. The distances covered in these migrations are quite large. Swarms are known to make short halts on the way (Ruttner 1988).

Rockbees can forage even during moonlit nights (Diwan and Salvi, 1965). Its flight range is more than 5 km (Koeniger and Vorwohl, 1979). In the normal forage conditions they have been observed to visit sources 2 to 3 km away from the nest. The bees have an average tongue length of 6.683 mm (Ruttner, 1988). These factors provide a large foraging range, both in area and variety of plant species, to the rockbees.

Honey hunting in forests is related mainly with rockbees. With their conspicuous aggregations on trees or rocks, rockbees attracted the attention of the ancient man who got in them the nature's sweet. It must have needed an irresistible urge and enormous courage to reach the nests and get this food, risking one's life and limb. Methods have slowly developed to reduce the risks. The colonies are reached by using fibre ropes and ladders to come down to the nest site from above the rock cliffs. Long poles of bamboo or other similar plants are also used when the colonies are on tall trees. The poles are tied to the tree trunk to serve as a ladder. Smoke from the ground below the colonies was given by burning dried twigs and green leaves. Sticks with green twigs are also made into brooms for making smoke to be taken up by the man harvesting the honey. The bees are driven away from the nest by the smoke. The brood comb is first removed for obtaining beeswax. Later the honey section is cut either from its place of attachment to the support, or into small chunks. The dripping honey is collected below in baskets or other containers held beneath the nests. The cut honey combs are brought down in buckets. There they are pressed usually by hand to squeeze out the honey. The pressed combs as also the brood comb is later boiled in water to extract beeswax. Although the process of honey collection appears to be destructive, the bees often survive the onslaught. The scattered bees form clusters soon after the hunting operation is completed. The cluster takes up construction of fresh comb if the flowering season is still on. Rockbees can construct combs quite rapidly. Within a short time the rehabilitated nest grows and continues its activities. If the season is over, the bees move over to safer places where they start building a new nest, as they do during normal migration.

Rockbees seem to adapt themselves to living near human societies. Large numbers of nests of rockbees are found in incredible locations like the huge workshops of machine manufacturing factories, high-rise buildings, big temples, or near sugar factories. Swarms build their nests in aggregations in these localities year after year, in spite of repeated removal of their combs for harvest of honey. Honey yield from a colony may vary from 5 to 50 kg. Muttoo (1956) reports yields of 25 to 100 kg in a single season. For a tribal honey hunter beeswax is the major produce for which he can have significant income. A normal sized comb can yield about 250 g of beeswax. Rockbee honey is similar in its physical characteristics and chemical composition to that of *A. cerana* (Phadke 1968). Rockbee honeys have the highest values for diastase, invertase and catalase enzymes, amongst the Indian honeys (Wakhle and Desai 1983). Because of the pressed method of collection of honey, the commercial samples contain large quantities of pollen and extraneous matter like dust, bee parts and brood juices.

An annual production of about 2500 tons of beeswax from the wild rock bees was reported in 1969 (Phadke *et al.,* 1969). This figure may not have changed and may be the same today. Rockbee wax has a lower melting point (59.6°C) than *A. cerana* or *A. mellifera* beeswax. Its acid value is 6.64, similar to the Indian hive bee, but lower than that of the European bee (Phadke *et al.,* 1969) Commercially it is known as the ghedda wax. Wax esters in rockbee wax constitutes 85 - 90%. Combs of deserted nests are usually quickly destroyed by wax moth. Systematic collection of the combs, soon after migration of colonies can improve beeswax yields. Swarms build combs very fast. This property can be made use of in developing a technology for increasing beeswax production. Nests of rockbees contain large quantities of pollen. Pollen is usually wasted in the process of pressing the combs during honey harvest. A few tribals eat it directly or cooked along with vegetables, etc. Work at the Central Bee Research and Training Institute, Pune showed that pollen from rockbee hives can become a commercial product, if suitable methods of its harvest are developed. This pollen can be used for preparation of pollen supplements needed as bee feed in apiculture, in dearth management. It can also be made into valuable products for human therapeutics and nutrition.

Rockbee is an important pollinator of several crops. Its long proboscis, large flight range, large number of field workers, and its habit of collecting large quantities of pollen and nectar make it the best among the honey bees for crop pollination. Its aggressive nature and migratory habit are serious drawbacks in its manipulation for planned pollination programmes. Some attempts were made to utilize rockbees for pollination of crops like safflower, which are not usually preferred by the hive bees. It was realized that the bee can be handled without fear of stings, when it is kept in artificial bee hives at ground level (Thakar 1973). It was also possible to move the rockbee colony to desired locations. Further work on rockbee pheromones, plant chemicals that can subdue the bees and designing a suitable hive to keep them at ground level, can surely lead to using this bee for crop pollination, besides commercial honey and beeswax production.

Rockbees have been associated with human societies in India since prehistoric times. The palaeolithic paintings in rock shelters depict the rockbee hives and honey hunting. A bulk of the Rigvedic literature on honey and honey bees concerns with the rockbees. The bee has been a major source of honey through the ages. In spite of this long history of man's association with rockbees, little progress is made in deliberate conservation of the species. Rockbee populations are exploited wherever possible for honey and beeswax. Improved methods of collection of these materials will help in survival and spread of the bee. Ruttner (1988) feels that the extensive honey collection by man has less impact on the species survival, than the destruction of primary forests that provide food and shelter to it

Eastern honeybee, *Apis cerana* Fabricius, 1793

Apis cerana, or the **Asiatic honeybee** (or the **Eastern honeybee**), are small honeybees of southern and southeastern Asia, such as China, India, Japan, Malaysia, Nepal, Bangladesh and Papua New Guinea. This species is the sister species of *Apis koschevnikovi*, and both are in the same subgenus as the European honeybee, *Apis mellifera*.

For ages, colonies of the oriental honeybee *Apis cerana* have provided mankind with honey and beeswax, as well as furnishing invaluable service in the pollination of agricultural crops. This bee's range of distribution is far greater than those of *A. florea* and *A. dorsata*: it is found throughout the tropical, sub-tropical and temperate zones of Asia, occurring in the Indian sub-continent and Sri Lanka in the west, through Southeast Asia, to Indonesia and the Philippines in the east. Further north, it is found in the southern USSR and China, through the Korean peninsula, to Japan. This wide range has led to important variations among the bee's geographical races: particularly between the tropical and temperate races, there are wide differences in workers' body size, nest size, colony population and swarming and absconding behaviour, The temperate and sub-tropical races appear to store greater quantities of food than the tropical races, which in turn are more mobile than the former, tending to swarm, abscond and migrate quite frequently.

In the wild, the oriental honeybees construct their multiple-comb nests in dark enclosures such as caves, rock cavities and hollow tree trunks. The normal nesting site is, in general, close to the ground, not more than 4-5 meters high. The bees' habit of nesting in the dark enables man to keep them in specially constructed vessels and for thousands of years *Apis cerana* has been kept in various kinds of hives, i.e. clay pots, logs, boxes, wall openings, etc. Despite the relatively recent introduction of movable-frame hives, colonies of *Apis cerana* kept in traditional hives are still a common sight in the villages of most Asian countries. As a result, the feral nests of the oriental honeybee in tropical Asia sustain fewer casualities in being hunted by man than those of the dwarf and giant honeybees.

The several combs of an *A. cerana* colony are built parallel to each other, and a uniform distance known as the "bee space" is respected between them. The body size of the workers of this tree is much smaller than that of the *A. dorsata* workers, and its brood comb consists of cells of two sizes: smaller for the worker brood and larger for the drone brood. The queen cells are built on the lower edge of the comb. As in the other *Apis* species, honey is stored in the upper part of the combs, but also in the outer combs, adjacent to the hive walls.

Following the invention of the movable-frame hive for the European honeybee about a century ago, traditional beekeeping with *A. cerana* has been partially replaced by this modern method in several Asian countries, and at the same time attempts have been made - with varying degrees of success - to improve hiving techniques and colony management. These beekeepers are found across the breadth and length of the country. There are rural beekeepers in the high mountains of the Himalayas who keep log hives in walls of their houses. Each family typically owns half a dozen bee-logs and honey combs are only removed for local consumption. Beekeeping is a traditional industry in West Bengal and some North Eastern states like Arunachal and Sikkim. In Karnataka and Tamil Nadu - there is a strong tradition of beekeeping with *Apis cerana*. Areas such as Coorg in Karnataka and Marthandam in Tamil Nadu are famous for their beekeeping culture. In Kerala, especially in the rubber growing areas, beekeeping is a regular activity and large quantity of honey (from extra floral nectar) is being produced.

In the wild, they prefer to nest in small spaces, such as hollowed out tree trunks. Like the European honeybee, they are sometimes domesticated and used in apiculture, mostly in wooden boxes with fixed frames. Their size is similar or somewhat smaller than *Apis mellifera*, and they also have a more prominent abdominal stripes. Their honey yield is smaller, because they form smaller colonies. Their beeswax is used to treat and heal wounds. *Apis cerana* is the natural host to the mite *Varroa destructor*, a serious pest of the European honeybee. Having coevolved with this mite, *A. cerana* exhibits more careful grooming than *A. mellifera*, and thus has an effective defense mechanism against *Varroa* that keeps the mite from devastating colonies. Other than defensive behaviors such as these, much of their behavior and biology (at least in the wild) is very similar to that of *A. mellifera*. When their hive is invaded by the Japanese giant hornet (*Vespa mandarinia*), about 500 Japanese honeybees (*A. cerana japonica*) surround the hornet and vibrate their flight muscles until the temperature is raised to 47°C (117°F), heating the hornet to death, but still under their own lethal limit (48-50°C).

Races of *Apis cerana*

There are many different races of *Apis cerana*, as could be expected from the wide range of habitats it occupies. Bees of some of the races are of the same size as some *Apis mellifera*. However, *Apis cerana* varies in size throughout its range, and tropical races are much smaller, with smaller colonies. Subspecies following Engel (1999) are as given below:

- *Apis cerana cerana* (= "sinensis") - Afghanistan, Pakistan, north India, China and north Vietnam
- *Apis cerana heimifeng*
- *Apis cerana indica* - South India, Sri Lanka, Bangladesh, Burma, Malaysia, Indonesia and the Philippines
- *Apis cerana japonica* - Japan
- *Apis cerana javana*
- *Apis cerana johni*
- *Apis cerana nuluensis*
- *Apis cerana skorikovi* (= "himalaya") - Central and east Himalayan mountains (Ruttner, 1987) *running*. Native to Asia between Afghanistan and Japan, and from Russia and China in the north to southern Indonesia. Recently introduced to Papua New Guinea.
- *Apis cerana* builds a nest consisting of a series of parallel combs, similar to *Apis mellifera*, and builds its nest within a cavity.

The distribution of *A. cerana*

A. cerana is the Asiatic honeybee or the oriental honeybee because they are only found in Asia, from Iran in the east to Pakistan in the west, and from Japan in the north to the Philippines in the south. Thus, *A. cerana* does not live only in tropical and subtropical areas of Asia, but also in colder areas as Siberia, Northern China and the high mountain area of the Himalayan region (Koeniger, 1976). *A. cerana* occurs at altitudes from 0 to 3333 m above the sea level (Rahman, 1945). For many social insects a tendency has been found that the higher the altitude and latitude the bees inhabit, the bigger the body size and the colony population (Paspari and Vargo, 1995). This relationship is the rule for *A. mellifera* (Ruttner, 1988; Ruttner, 2000; Hepburn *et al.*, 2000), for *A. cerana* (Verma *et al.*, 1994), for *A. florea* (Ruttner *et al.*, 1995), and for some stingless bee species (Pereboom and Biesmeijer, 2003). A high degree of variation in size and coloration probably reflect the ecological diversity of *A. cerana*. The influence of latitude and altitude on the size of worker bees was also found for *A. cerana* in Vietnam (Niem *et al.*, 1992). *A. cerana* colonies occur in all provinces of Vietnam (except Uminh forest) but their natural types are commonly found in mountain-forested areas such as Viet Bac, Hoang Lien Son, Truong Son; in coconut-grown provinces as such Ben Tre, Tien Giang, etc., and island-districts as Cat Ba, Phu Quoc, Con Dao (Chinh, 1996).

Studies on the taxonomy of *A. cerana* in the World

A. mellifera is the best-studied species in honeybees in particular and in social bees in general. Twenty-six subspecies and ecotypes are discriminated and have been studied in detail. Compared to *A. mellifera* there is very little research on the morphology of *A. cerana* (Verma, 1990). Based on the analysis of 34 morphological criteria of 68 samples collected from different areas of Asia, Ruttner (1988) divided *A. cerana* into four subspecies:

Apis cerana indica

This is the subspecies with the smallest body size. It lives in the south of India, in the south of Thailand, Cambodia and Vietnam, in Malaysia, in Indonesia and in The Philippines. The length of proboscis and forewing is 4.58-4.78 mm and 7.42-7.78 mm respectively (Ruttner, 1988).

Apis cerana cerana

This subspecies with the biggest body size of A. *cerana* occurs in northern parts of China, in the northwest of India, in the north of Pakistan and Afghanistan, and in the north of Vietnam. On average, the proboscis and forewing length measure 5.25 mm and 8.63 mm respectively.

Apis cerana himalayana

The body size of this subspecies is intermediate between *A. c. cerana* and *A. c. indica*. It occurs in the east of the Himalayas from Nepal to northern Thailand. On average, the proboscis and forewing length measure 5.14 and 8.03 mm respectively.

Apis cerana japonica

This subspecies is endemic in Japanese temperate climates except the island of Hokkaido. This subspecies is divided into two separate ecotypes: Honshi and Tsushima. The body size of *Apis cerana japonica* is relatively big, with an average proboscis length of 5.18 mm and an average forewing length of 8.69 mm. *A.c. japonica* gradually has been replaced by introduced *A. mellifera* (Okada, 1986).

The intraspecific classification of the Asiatic honeybee species, *A. cerana* is in a state of flux and uncertainty (Hepburn *et al.*, 2001). Next to the four subspecies distinguished by Ruttner (Ruttner, 1988), four other subspecies have been proposed: *A.c. abaensis, A.c. philippina, A.c. skorikovi* and *A.c. hainanensis* (reviewed by Hepburn *et al.*, 2001). Five of these subspecies occur in China: *A.c. indica, A.c. cerana, A.c. skorikovi, A.c. hainanensis and A.c. abaensis. A.c. cerana* is divided further into five ecotypes known as Quangdong-Quangxi, Hainan, Yunnan north, and Changbei-Shan (Zhen-Ming *et al.*, 1992). Based on multivariate morphometric analyses of 557 colonies of *A. cerana* from all of the southern mainland of Asia, Hepburn *et al.* (2001) have recently established that *A. cerana* is placed in three separable groups that are not entirely distinct morphoclusters of bees: 1. bees from the Hindu Kush, Kashmir, northern Myanmar, northern Vietnam, and southern China; 2. bees from northern India, Nepal, central Myanmar and Thailand, Cambodia, southern Vietnam and southern China; 3. bees from central and southern India, southern Myanmar, southern Thailand and peninsular Malaysia. However, the nomenclature of these intraspecific taxa of *A. cerana* still remains unadjusted in this paper. Deviations in the intraspecific classification of *A. cerana* probably reflect differences in sampling and methodology. *Apis cerana* is the East Asiatic Counterpart of *A. mellifera.* Its morphology and behaviour are so similar to *A. mellifera* that for a long time it was considered as an *A. mellifera* sub species (Buttel-Reepen, 1906). However it has several species specific characters and are genetically separated from *A. mellifera* (Ruttner and Maul, 1983). It is not true that *A. cerana* is smaller than *A. mellifera.* These species overlap in size. The northern types are generally larger than southern types. Ecological requirements of *A. cerana* are about the same as those of *A. mellifera.* This species also succeeded in colonizing forested areas in the cool temperate zone (northern China to Ussuria in East Siberia). There are four subspecies reported for *A. cerana* namely, *A. cerana cerana* in Afghanistan, Pakistan, north India, China and north Vietnam, *A. cerana indica,* in South India, Sri Lanka, Bangladesh, Burma, Malaysia, Indonesia and the Philippines, *A. cerana japonica* in Japan and *A. cerana himalaya* in Central and east Himalayan mountains (Ruttner, 1987). Thus its area of distribution is very large; it extends from West Afganistan to Japan.

Genetic variance in morphological characters of *Apis cerana* subspecies in the Himalayan region have been identified. These subspecies are named *Apis cerana, Apis cerana himalaya,* and *Apis cerana indica.* Each subspecies has further locally adapted populations called ecotypes which differ from each other in several biological and economic characters. For example, three ecotypes of subspecies *Apis cerana* himalaya that correspond to geographic distribution in (1) the Naga and Milo Hills, (2) Brahmaputra Valley and Khasi Hills, and (3) the foothills of the north-east Himalayas have been identified. In some parts of the Hindu Kush Himalaya, *Apis cerana* cerana matches the European hive bee *Apis mellifera* in commercial value and has spectacular potential for further genetic improvement.

When kept sympatrically, *A.cerana* and *A. mellifera* colonies frequently rob each other (Koeniger, 1982). In Japan, *A. cerana* (originally the only honey bee) is now replaced by imported *A. mellifera* colonies. Another cause of failing coexistence of the two species is attempted intermating which produces lethal offspring (Ruttner and Maul, 1983). Another problem is shifting of parasites from one species to the other as the geographical isolation is broken by humans. *Varroa* mite which is coadapted to *A. cerana* and is parasitic on the drone brood of this species causing no serious problem has shifted to the unadapted *A. mellifera* and is a serious pest on it. *A. cerana* colonies are smaller than that of *A. mellifera* and so are the honey yields.

STOCK IMPROVEMENT

Many of the above mentioned subspecies and ecotypes of *Apis cerana* are at present not economically viable. Therefore, to achieve stock improvement, different *Apis cerana* subspecies and ecotypes should be accumulated at a central location and superior genotypes be identified. Another important pre-requisite for stock improvement is to evolve efficient queen rearing for *Apis cerana* and also establish isolated mating stations for pure line breeding. The latter is essential because instrumental insemination in *Apis cerana* has unexpectedly turned out to be a difficult task due to very low volumes of semen ejaculated by drones. During the course of evolution, *Apis cerana* has developed certain behavioural characteristics such as frequent absconding and swarming which are essential for the survival of colonies but undesirable from a beekeeping point of view.

Asian Hive Bee, *Apis cerana* F.

The Indian hive bees are sub-species of the Asian or Oriental or Eastern hive bee, *Apis cerana* F. Along with the rockbees, this bee has been known in India since the Vedic period. In the Vedic literature this bee was described by names such as saragha, sabardugha, and puramdhi (full, prolific bee). It is called darohla, mahun (Punjab), khaira, sat lahari, satputi, mauna, pilad (Uttar Pradesh and north India), sateri masha, moholachi masha (Maharashtra) burra thene or putta theneteega (Andhra Pradesh). Natural nests of the Indian hive bee occur in tree trunks, rock crevices, ant hills, underground deserted nests of white ants, or any dark enclosure, sometimes even in the open, but quite dark spaces in forests or unused rooms in buildings. The bee occurs in a wide range of geographical areas from tropical coastal areas to the temperate Himalayan ranges at about 3000 m altitude. Colonies are found in forests or agricultural areas in the plains, and even in urban areas with good vegetation. Bees are larger than the dwarf bee, but are much small than the rockbee. This bee is similar in size to the European hive bee in similar latitudes. Two varieties of the bees had been usually recognized - the hill variety, larger and darker, usually dark brown in colour, with a cell size of about 4.8 mm, occurring in the hills and mountains; and the plains variety, smaller and lighter, generally reddish yellow in colour, with a cell size of about 4.15 mm, occurring in the plains. The hill variety is believed to be more aggressive, is more populous, builds larger nests and produces more honey than the plains variety.

Variation in the Indian Hive Bee

Like the other hive bee, *A. mellifera* L., *A. cerana* too shows an enormous variation in body size, colour, tongue length, foraging range, swarming tendency and production capacity. In spite of this variation and its rearing for several decades, the species remains even today incompletely studied and its potential unexplored. Our present knowledge of the variation in the bee and varieties that occur in India is summarized here.

Comprehensive studies on the biometry and taxonomy of *A. cerana* in India were undertaken by Kshirsagar (1976, 1983). Using over 60 biometric features of samples of 22 populations of this species from locations all over India, he differentiated seven ecotypes (Table 4). Kshirsagar and Ranade (1981) showed that the bees in the southern-most region along the coast, are the smallest, and the size increases northwards, the largest bee being found in the Kashmir Valley. A similar increase in size of the European bee towards northern latitudes has been reported (Ruttner, 1988). Besides latitude, bee size and other morphological features show variation according to the altitude. Bees in Tamilnadu plains are thus smaller than the bees in the Nilgiri hills, Tamilnadu with an altitude of about 2000 m. Considering the correlation between the body size and biology of the bee with the natural vegetation as also the distribution of natural vegetation types in the country, the seven ecotypes identified by Kshirsagar (1983) are redefined and are listed in Table 3.6.

Ruttner (1986, 1988) classified the different Asian hive bee populations into four groups, which he called as races of *A. cerana*, namely, *A. c. cerana*, *A. c. himalaya*, *A. c. indica* and *A. c. japonica*. The former three races occur in India. Their distribution is as follows.

Apis cerana cerana	North-western region including Himachal Pradesh and Jammu and Kashmir
Apis cerana himalaya	North-eastern region including the north-eastern Himalayan states
Apis cerana indica	South India, including Kerala, Tamilnadu, Karnataka and southern Andhra Pradesh

Verma and co-workers made detailed investigations on the biometry and taxonomy of Indian hive bee from 50 localities, particularly in north India.

According to these studies (Verma, 1992) the three Indian races can be further differentiated into seven sub-groups or ecotypes, two in *A. c. cerana*, three in *A. c. himalaya* and two in *A. c. indica*. Intra-specific classification of the Indian *A. cerana* into seven ecotypes indicated by Kshirsagar (1983), and redefined here (Table 4) seems to be well-founded. It is possible that by further detailed investigations, additional ecotypes and races can be found.

Deodikar *et al.* (1958) proposed that *A. cerana* originated by hybridization of a species of Trigona that had vertical multiple combs, with diploid species of *Apis*. They felt that the Indian hive bee possibly arose during Pleistocene glaciation, spread northwards through the Himalayas and gradually differentiated into the European hive bee. The centre of origin of the *Apis* is considered to be the Indo-Malayan region. Because of this there is a possibility of occurrence of several races, and varieties of *A. cerana* in this region.

Table 4. Ecotypes of *Apis cerana* F. in India

Geographic region	Latitude	Altitude	Location of sample collection	Remarks
Kashmir Valley	34°05'	1586	Srinagar, Jammu and Kashmir	Largest ecotype in the country
Western Himalayas	31°43'	761	Mandi, Himachal Pradesh	Possibly includes the next two variants
Western Sub-Himalayas	30°05'	700	Kangra, Himachal Pradesh	Possibly variant of Western Himalayas
Western Sub-Himalayan Foot Hills	30°10'	630	Ranipokhari, Uttar Pradesh	Possibly variant of Western Himalayas, and not ecotype
Eastern Himalayas	26°53'	1500	Kurseong, West Bengal	Verma (1992) proposes 3 races in this region
Indo-Gangetic Plains and Aravali Hills	29°13'	440	Haldwani, Uttar Pradesh	Mahabaleshwar included due to its high altitude
	26°06'	53	Muzaffarpur, Bihar	
	26°05'	54	Guahati, Assam	
	24°36'	1195	Mount Abu, Rajasthan	
	17°56'	1382	Mahabaleshwar, Maharashtra	
Central Peninsula	20°48'	27	Cuttack, Orissa	
	17°50'	767	Lammasingi, Andhra Pradesh	
	17°00'	670	Petlond, Maharashtra	
Western and Eastern Ghats	15°20'	700	Castle Rock, Karnataka	Kodaikanal included due to its high altitude
	14°57'	700	Yellapur, Karnataka	
	12°57'	650	Sakleshpur, Karnataka	
	10°14'	2343	Kodaikanal, Tamilnadu	
Western and Eastern Peninsular Coastal strips	14°25'	0	Kumtha, Karnataka	Smallest ecotype in the country
	11°55'	0	Pondichery, Pondichery	
	10°46'	97	Palghat, Kerala	
	08°44'	51	Tirunnelveli, Tamilnadu	
	08°05'	37	Kanya Kumari, Tamilnadu	

Source: Kshirsagar (1983)

Characteristics of the Indian hive bee

The nest consists of several parallel combs with an uniform distance between them. The nests have usually 6 to 8 combs. A wide variation occurs in the number depending upon the period of stay of the nest in the location, space available in the nest site and its shape. Sometimes only 3 to 4 combs which are narrow but about a metre long are found. In natural nests that lived for over 2 years upto 15 normal sized combs were found, that yielded over 10 kg of honey. Individual combs are uniform in thickness, unlike the combs of the dwarf or rockbee. The combs usually have honey stores in the upper part, brood in the middle and lower parts and pollen stores on the sides of the comb, adjacent to the brood. Drone cells are constructed along the lower part of the comb. These are conspicuously larger, and have raised covers with a pore in their middle. Queen cells under normal conditions are built along the lower edge. All the combs have similar functional differentiation, but usually the central combs in the nest contain brood while the outer combs have little or no brood, the entire area being utilized for honey or pollen stores. Combs are about 25 mm thick. The brood cells are about 11.5 mm deep, and vary in width usually from 4.17 to 4.83 mm, the size increasing towards north (Muttoo 1956, Deodikar *et al.* 1958, Singh 1983). Bees in the Kashmir valley may have over 4.9 mm wide cells, being at the same latitude as Peshawar, where the size is 4.87 mm (Ruttner 1988).

The egg stage in all castes lasts 3 days, the larval stage of workers 4-5 days, drones 7 days and queens 5 days; the pupal stage is 11-12 days for workers, 13-14 days for drones and 7-8 days for the queen. The total development is completed in 18-19 days by the worker, in 24 days by the drone and in 15-16 days by the queen (Muttoo 1956). There is some variation in these periods depending on the type of the bee in different parts of the country. Shah and Shah (1982) made a comparison of the Kashmir strain with other hill type in India along with the European bee. A brief account of this is given in Table 5 as it also indicates the general behaviour and performance limits of the Indian hive bee.

The Indian bees are considered as mild and are easy to handle. Its sting releases half the amount of alarm pheromone as does the sting of the European bee (Morse *et al.* 1967). Beekeepers often handle the colonies bare handed without smoking them. Perhaps this is one of the reasons for repeated handling of the colonies even by the inexperienced. This is a cause of disturbance and colonies desert. The bees have, like other Indian bee species, quite effective defence mechanisms against the usual enemies and predators. They exhibit shimmering behaviour, but since the nest is in the dark, only the sharp hissing sound is indicative of this behaviour. A knock on the hive elicits this behaviour and a 0.5 second long, clearly audible hissing results. Group defence is another strategy in which about 30 bees form a group at the entrance with the tip of their abdomen raised, on perceiving danger from an attacking hornet. Perhaps they release Nasanov pheromone to elicit the group behaviour. This along with sharp repeated hissing sounds from the nest makes the hornet abandon its attempt to attack (Ruttner 1988). The Indian bee is often blamed for its desertion or absconding tendency. Bees leave the nest when disturbed or when faced with inimical atmospheric conditions, or during periods of food scarcity. In the tropics bees are exposed to attacks of a wide range of predators and enemies. Common among them are wasps, hornets and even ants that can really be constant irritants. Absconding behaviour is not peculiar to the Indian bee; it is found equally characteristically in the tropical African and Middle Eastern races of the European hive bee. Absconding is largely a problem of management than of bee behaviour. The problem can easily be solved by understanding the reasons for absconding and removing them before the colony attempts to desert.

Table 5. Behaviour and performance of Kashmiri race and other hill type of Indian hive bee and the European hive bee

Character	Kashmiri race	Hill type	European bee
Egg laying capacity of queen (range, number of eggs laid per day)	1438 - 2033	500 - 800	871 - 1368
Population at peak of the season (number of bees)	60,000 - 70,000	18,000 - 22,000	65,000 - 70,000
Honey yield (kg/colony/year): range	15 - 35	4 - 16	3.0 - 19.5
Pollen carrying capacity, wt., g/bee	0.0187	0.0140	0.0197
Homing instinct (time in seconds taken by bees to reach their hive when a frame of bees was shaken 50 m away from their hive)	180.0	191.8	295.4
Flight range (km)	3.75	1.0	3.0 - 4.0
Weight gain in colony per day during nectar flow (g)	6.500	1.375	Not available
Mating age of queen (days: range)	4 - 6	4 - 8	6 - 13
Temperature range (°C) for bee activity	Active at 8 - 32	Active at < 8 or > 32	Very active at 21 - 35
Swarming tendency	Moderate; usually after 9 - 12 frame strength	High; even when colony is on 4 frames	Only after colony attains 12-15 frame strength
Swarming tendency	Moderate; usually after 9-12 frame strength	High; even when colony is on 4 frames	Only after colony attains 12 - 15 frame strength
Spring build-up	Quick	Moderate	Quick
Thriftiness	Thrifty	Slightly thrifty	Not thrifty
Queen introduction	Easy; even *mellifera* queens accepted	Easy; even *mellifera* queens accepted	Difficult; own queens also rejected, if reintroduced
Laying worker development	May develop after 30days	Develop within a week	Not even after a mo-nth of queenlessness
Cleanliness	Good	Poor	Good

Absconding tendency	Low	High	Low
Propolis	Not collected	Not collected	Collected
Bee response during inspection	Remain calm	Agitated; run helter-skelter; not steady	Remain calm and Steady
Nature of sting	Quite painful	Painful	Quite painful
Fanning position at hive entrance	Face away from hive entrance	Face away from hive entrance	Face towards hive entrance
Response to wasp attack	Defends effectively	Defends effectively	Cannot defend
Response to wax moth	Not serious	Easily affected	Rarely affected

The Indian hive bee does not use propolis as the European bees do. This may be an adaptation to tropical climate, where hive ventilation assumes importance. The cracks in the floor board or gaps in the hive walls or frame joints are not sealed. This may attract pests like wax moth. One of the characteristic features of the hive bee is fanning used for ventilation of the hive. During nectar flows large quantities of water have to be removed from the dilute honey in combs and ripen it. The moisture-ridden air has to be removed from the hive. For this purpose bees undertake fanning vigorously, and it is most visible at the hive entrance. The Indian hive bee fans with its head facing away from the entrance. Contrastingly, the European bee fans with its head towards the entrance. In their experiments on introducing queens of the European bee into the colonies of the Indian bee, Dhaliwal and Atwal (1970) observed the workers of both the species fanning side by side, but heads oriented in opposite directions.

Honeys from the Indian hive bee have (Phadke 1967a) about 20% water. The levulose content is about 36. 5%, dextrose about 33.4%, while the non-reducing sugars are about 3.4%. The European honeys have usually upto 17% water, about 41% levulose, and about 36% as dextrose. The high water content in Indian honey can be attributed to the tropical humid climate prevailing in the major honey producing areas. This seems to be confirmed from similar studies on honeys from Mahabaleshwar, Maharashtra with a sub-tropical climate (Phadke 1967b). Water content in samples of 8 types of unifloral honeys from here was between 17.2 to 19.0%. Cerana honeys have a high invertase content and low diastase, catalase and glucose oxidase values, compared to those of mellifera honeys. The average values found by Wakhle and Desai (1983) in 17 samples from all over India were 6.55 for diastase, 25.60 for invertase and 1.44 for catalase. Heat and long storage under tropical conditions reduce the enzyme content.

Microscopical analysis of apiary honeys (Seethalakshmi, 1983) shows generally a high pollen content. The absolute pollen count of 12 samples studied varied from 26,000 to 225,000 grains per 10 g. The pollen types in the honeys showed a wide range. The wax of the Indian hive bee has a melting point of 65°C and an acid value of 6.54 (Phadke *et al.*, 1969). It resembles the rockbee wax in having a low acid value compared to that of the European beeswax. Analyzing the factors causing the low acid value, Phadke *et al.* (1971) state that the Indian beeswaxes have only about 8% as hydrocarbons. This is nearly half that present in the European waxes. Consequently the total alcohol content is more in the

Indian waxes. In all other respects the Indian beeswax is similar to the European wax. It can therefore replace imported beeswax in pharmaceutical and cosmetic industries.

Potential of the Indian Hive Bee

Yields from the Indian hive bee are comparable to those from the European bee. In Kashmir average production of 50 kg of honey per colony is reported (Shah, 1981). There are occasional reports of yields of 50 to 60 kg of honey per colony in other regions like Bihar, West Bengal, Karnataka, Kerala and Tamilnadu, particularly when migratory beekeeping is practiced. This indicates that this bee has the potential for use for commercial beekeeping. Improvement in the hive design, adoption of suitable strain of bee for different agro-climatic conditions, improvement in management technologies can help to realize this potential. Verma (1989) lists the following advantages of beekeeping with *A. cerana*, compared to that with the exotic bee.

1. *Apis cerana* is gentle to handle, industrious and well adapted to the ecological conditions of south and Southeast Asia.
2. It is less susceptible than *A. mellilfera* to nosema disease not seriously affected by *Varroa* and is less prone to the attack of predatory wasps.
3. To control diseases, parasites and predators, beekeeping with *A. mellifera* requires chemical treatment of colonies. Chemicals are not required in beekeeping with *A. cerana*
4. The variety of geographical races/populations of *A. cerana* that exists in south and Southeast Asia provides excellent opportunities for the genetic improvement of this native species through selective breeding.
5. Through genetic engineering techniques it may be possible to introduce desirable genes from *A. cerana* into *A. mellifera.*
6. *A. cerana* is sympatric in distribution and can co-exist with the two other species of Asiatic honey bees, *A. dorsata* and *A. florea*, without any adverse ecological consequences
7. For pollination purposes, *A. cerana* is superior to *A. mellifera* in certain aspects, e.g., it is more suitable for cross-pollinating entomophilous crops grown in the small holdings of this region because of its shorter flight range and longer foraging hours than the European honey bee. Use of bee hives for pollination of agricultural and horticultural crops is another field that is gaining importance in recent years. There is an increasing demand for bee colonies by orchardists growing apples, litchi, lemon, other citrus fruits, producers of seed of cucumber and other cucurbits, cole crops, spices, onion and vegetable crops and flower seeds, as also by farmers growing sunflower. Payment by the farmer to the beekeeper for pollination service is also becoming a common practice. Bee colonies migrated to farm and orchard areas, to tide over adverse periods, can be

utilized for crop pollination. Such migrations can be doubly beneficial to the beekeeper. The overall income from apiculture can thus be quite substantial. Besides honey, the hive bees can produce beeswax, pollen, royal jelly and bee venom. Technologies exist for beeswax, pollen and bee venom production. For royal jelly too, suitable technology can easily be developed. Beekeeping for package bee production, queen rearing and supply, breeding and supply of improved strains of bees, and similar non-traditional products can be introduced. Efficient management methods for production of these items can augment the productivity of the bees. losses will be low.

Apis mellifera, Apis cerana and *Apis koschevnikovi* all build nests containing a series of parallel combs. These species usually nest in cavities.

D. The European, honeybee *Apis mellifera* Linnaeus, 1758

The **Western honeybee** or **European honeybee** (*Apis mellifera*) is a species of honeybee comprised of several subspecies or races (Table 6 Figure 23). "Mellifera" is from the Latin, and means honey-carrying - hence "*Apis mellifera*" is the honey-carrying bee. The name was coined in 1758 by Carolus Linnaeus, though in a subsequent 1761 publication, he referred to it as *mellifica*; the older name has precedence, but some Europeans still utilize the incorrect subsequent spelling.

There are many geographical races of the common honeybee *Apis mellifera*, distributed widely throughout Europe, Africa, and parts of western Asia, as well as in the Americas. All these races display similarities in their basic biological attributes, e.g. the construction of multiple-comb nests in dark cavities, colony social organization and division of labour, etc,

Table 6. Races of *Apis mellifera* (modified from Drescher and Crane, 1982; Winston, 1987).

Race	Geographical distribution	Physical characteristics	Tongue length (mm)	Temperament			Remarks
				Aggres-siveness	Swarming	absco-nding	
European races							
Apis mellifera mellifera L.(German dark bees)	Belgium, Britain, Czechoslovakia, France, Germany, Poland, Scandinavia, Switzerland, Russia	Body large,brood dark with yellow spots	5.7-6.4	Medium	Low	-	Medium honey yields, over winters well, less preferred for beekeeping due to aggressive nature, poor spring and early summer performance
Apis mellifera ligustica Spin (Italian bees)	Italy but now spread to various other countries	Smaller than *Apis mellifera mellifera* abdomen with bright yellow bands	6.3-6.6	Low	Low	Low	Over winters well,low swarming, high brood rearing, prone to robbing, good honey yielder

Apis mellifera carnica Pollman (Carniolan bees)	Austria, Northern Yugoslavia, Danube valley	Asin ligustica, grey or brown in colour	6.4-6.8	Low	Strong	-	Over winters well, rapid development in spring, slow to construct combs.
Apis mellifera caucasica Gorb. (Caucasian bees)	Caucasus USSR	Similar to carnica	Upto 7.2	Low	Low	-	High population very susceptible to **Nosema** low honey yielder, poor pover winterer.
Apis mellifera cecropia Kiesw (The Macedonian bee)		-	-	-	-	-	Little known about biology
Apis mellifera acervorum	Russia	-	-	-	-	-	Little known about biology
Apis mellifera mellifera remipes	Armenia, east Anatolia, Iran	-	-	-	-	-	Not much understood
African races							
Apis mellifera intermissa Buttel-Reepen *(Telian bees)*	North Africa from Libya to Morocco	Body long, dark pigment sparse hairs	6.4	Strong	Strong	Strong	Strong reproductive swarming
Apis mellifera major ruttner	Small pocket in RIF mountain Morocco	Body long, broad and dark with yellow markings	7.0	Medium	-	-	Biology little known
Apis mellifera lamarckii Cockerell (Egyptian bees)	Northeast Africa, Egypt and sudan along the nile valley.	Medium sized to long slender, body intense yellow colour, broad toments.	5.7	Medium	-	-	Many swarm cells
Apis mellifera nubica Ruttner	Sudan	Small bees with slender body intense yellow colour	5.4	Strong	-	-	Little known about biology
A.m. sahariensis	In north oasis near northern edge of Sahara	Bees are medium sized, slender body having yellow markings	6.3	Strong	-	-	Well adapted to environmental conditions

A.m. jementica Ruttner	Yemen	Small bees with broad having intense yellow colour of hairs	5.4	-	-	-	Little known about biology
A.m. littorea Smith	Coastal areas of Tanzania	Small bees with relatively slender body having yellow tergites	5.7	Strong	Strong	-	Intense brood production
A.m. scutellata Lepeletier (East African bees)	Ethopia, Kenya, Tanzania, Burundi, Zimbabwe, South Africa	Bees are small slender body with intense yellow colour	5.9	Strong	Strong	Strong	Intensive reproductive swarming
A.m. monticola Smith (mountain bees) Lepeletier (East African bees)	Mount regions of Tanzania,, Kenya, Ethiopia	Bees with long body, having dark colour, long hairs	6.2	Low	Strong	-	Gentle in comparison to other African races
A.m. capensis	Southern tip of africa	Small slender body, relatively dark in colour	5.9	Medium	-	-	Fast ovariole development and ability to lay parthoeno genetic females
A.m. unicolor	Madagascar	Small slender, body relatively dark in colour	5.6	Low	-	-	Fly off combs readily
A.m. adansonii Latreille (west African bees)	West Africa south of Sahara	Medium sized bees, broad body having yellow markings	6.2	-	Strong	-	Migratory swarming
Oriental races							
A.m. syrica	Turkey, Iran	Similar to ligustica	-	-	-	-	Little known about biology
A.m. anatolia	Turkey, Iran	Similar to ligustica	-	-	-	-	Little known about biology
A.m. adansonii Latreille (west African bees)	Turkey, Iran	Similar to ligustica	-	-	-	-	Little known about biology

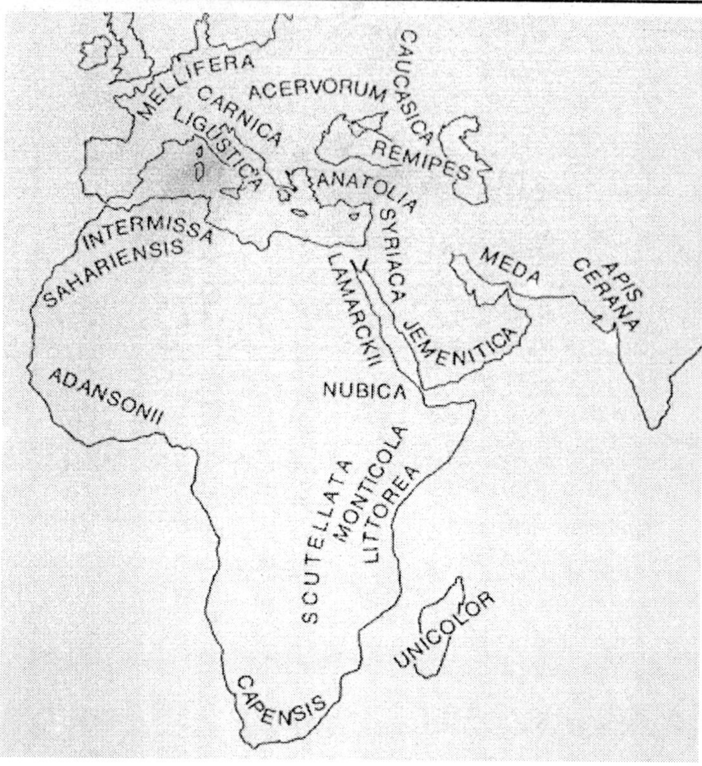

Figure 6. (a) Distribution of major honeybee species and races in europe, northern Africa and western asia

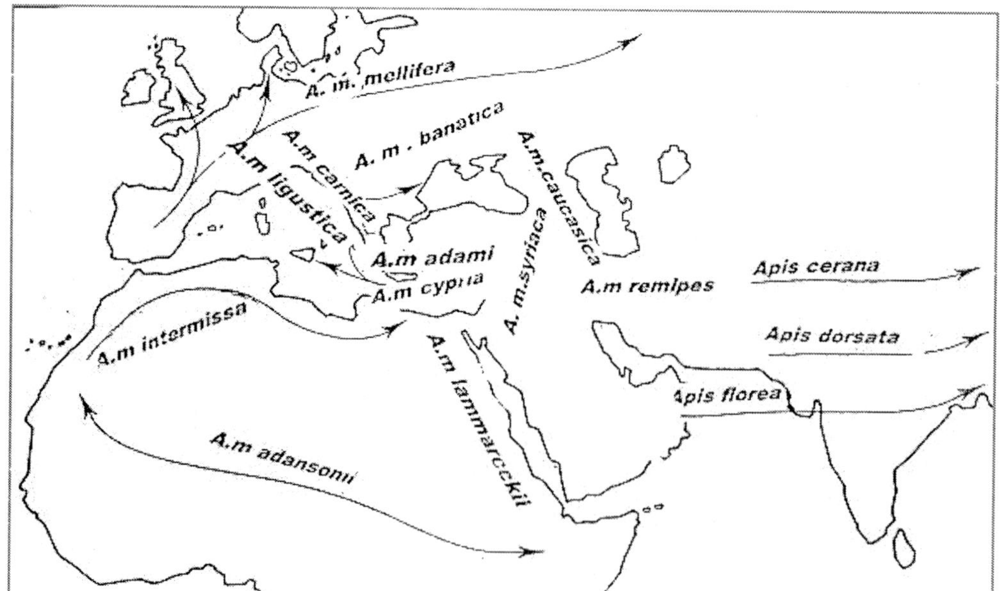

Figure 6 (b) approximate distribution of *Apis dorsata* and *A. laboriosa*. The *dorsata* populations on Luzon in the phillipines can be regardedas subspecies *A. dorsata* breviligula. The population on Sulawesi and butang is often reffered to as *A. dorsata binghami*

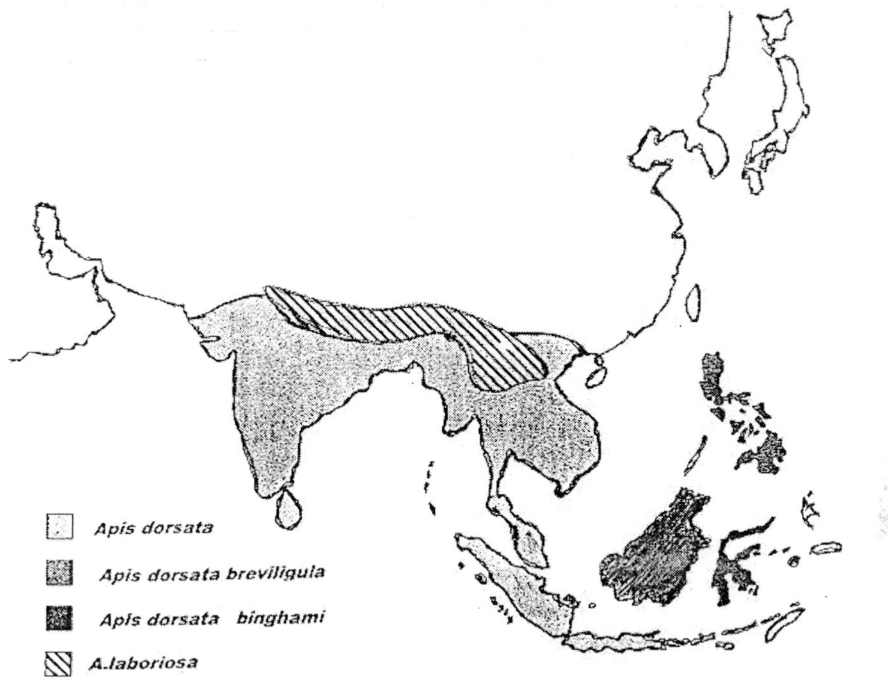

Figure 6 (c). The distribution of cavity nesting bees. *Apis cerana* is a diverse species and may have cryptic speciation especially in india, *A. nuluensis* is confined to the highlands of Borneo and its existence is known only from the Crocker range in Subah

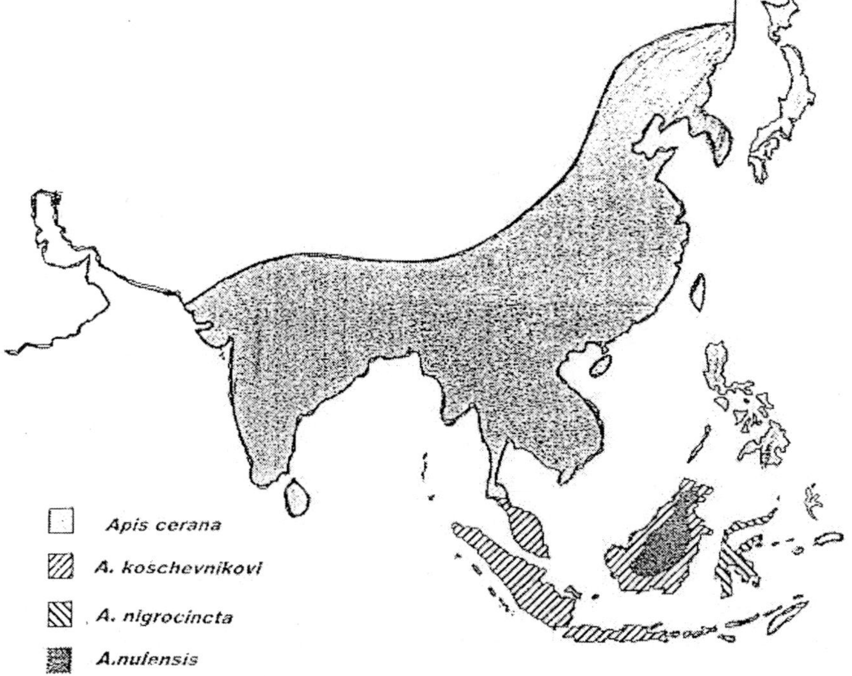

Figure 6 (d) Geographical distribution of *Apis mellifera* races.

In the wild, the natural nesting sites of *A. mellifera* are similar to those of *A. cerana*: caves, rock cavities and hollow trees. The nests are composed of multiple combs, parallel to each other, with a relatively uniform bee space. The nest usually has a single entrance. The temperate races prefer nest cavities of about 45 Litres in volume and avoid those smaller than 10, or larger than 100, litres. Colonies of the European races are composed of relatively large populations, usually between 15 000 and 60 000.

Many feral nests of *A. mellifera* in the northeastern forests of the United States have been reported to store 25 to 30 kg of honey per colony, and even more, during the nectar-flow spring season, and properly managed, commercially operated, colonies yield much more.

Anthropomorphically speaking, this behaviour of the temperate races is obviously an evolutionary advantage: without it, the colony faces starvation during the cold winter months, when food is not naturally available and the temperature is too low to permit flight activity. The shortage of natural forage and the cold temperatures prevailing from late autumn until early spring appear to play an important role in exercising rigid natural-selection pressures on the colonies. As a result, both feral and hived colonies of temperate-zone *A. mellifera* are less likely to abscond than the tropical races.

The past three centuries have seen the introduction of the common honeybee to all the habitable continents. Outside Asia, beekeeping with *A. mellifera* constitutes an integral part of modern agricultural systems, furnishing crop pollination services as well as honey and beeswax. Although this bee is one of the most studied animals, many aspects of its biology being fully known, efforts over the past few decades to introduce *A. mellifera* into Asia have encountered a number of problems, such as the inter-species transmission of bee pests and diseases. But successes have been reported from several Asian countries as regards the commercial viability and the likelihood of a profitable economic return of beekeeping with *A. mellifera.* It appears that the adaptability of the bees, appropriate beekeeping technology, better understanding of forage ecology and socio-economic suitability are among the most important factors underlying the further development of beekeeping with the common honeybee in Asia.

Apis mellifera is most widespread and the most widely studied species of honey bee. Interestingly many beekeeping texts relate only to *Apis mellifera.*

This species of honey bee is native to Africa, most of Europe and the Middle East. It has been introduced by man to the Americas, Australasia and much of the rest of the world. *Apis mellifera* usually builds its nest inside an enclosed space. The nest consists of a series of parallel combs, and this nesting pattern is followed in the design of frame hives.

There are many different races of *Apis mellifera*, some tropical, others temperate The Africanized honey bees in South and Central America are descended from tropical African *Apis mellifera.* Different races of *Apis mellifera* have different sizes of individual bees and colonies. Generally *Apis mellifera* are regarded as the medium-sized honey bees, against which other species are judged as "large" or "small".

Subspecies originating in Europe

1. *Apis mellifera ligustica*, classified by Spinola, 1806 - the Italian bee. The most commonly kept race in North America, South America and southern Europe. They are kept commercially all over the world. They are very gentle, not terribly inclined to swarm, and produce a large surplus of honey. They have few negative characteristics. Colonies tend to maintain larger populations through winter, so they require more winter stores (or feeding) than other temperate zone races. Italians are light colored, most leather colored, but some strains are golden.
2. *Apis mellifera carnica*, classified by Pollmann, 1879 - Slovenia - better known as the Carniolan honeybee - popular with beekeepers due to its extreme gentleness. The Carniolan tends to be quite dark in color, and the colonies are known to shrink to small populations over winter, and build very quickly in spring. It is a mountain bee in its native range, and is a good bee for cold climates. It does not do well in areas with long, hot summers.
3. *Apis mellifera caucasica*, classified by Pollmann, 1889 - Caucasus Mountains - This sub-species is regarded as being very gentle and fairly industrious. Some strains are excessive propolizers. It is a large honeybee of medium, sometimes grayish color.
4. *Apis mellifera remipes*, classified by [[Carl Eduard Adolph Gerst-cker [Gerst-cker]], 1862 - found in Caucasus, Iran, Caspian lake, etc.
5. *Apis mellifera mellifera*, classified by Linnaeus, 1758 - the dark bee of northern Europe also called the German Honeybee - domesticated in modern times, and taken to North America in colonial times. These small, dark-colored bees, sometimes called the German black bee, have the reputation of stinging people (and other creatures) for no good reason at all; this, however, applies to the hybrid *A. m. mellifera* x *A. m. ligustica* populations found in North America and Western Europe, not to the near-extinct "pure" *A. m. mellifera*.
6. *Apis mellifera iberiensis*, classified by Engel, 1999 - the bee from the Iberian peninsula (Spain and Portugal)
7. *Apis mellifera cecropia*, classified by Kiesenwetter, 1860 - Southern Greece
8. *Apis mellifera cypria*, classified by Pollmann, 1879 - The island of Cyprus - This sub-species has the reputation of being very fierce compared to the neighboring Italian sub-species, from which it is isolated by the Mediterranean Sea
9. *Apis mellifera sicula*, classified by Montagno, 1911 - from the Trapani province and the island of Ustica of western Sicily

Subspecies originating in Africa

Several researchers and beekeepers describe a general trait of the African subspecies which is absconding, where the Africanized honeybee colonies abscond the hive in times when food-stores are low, unlike the European colonies which tend to die in the hive.

1. *Apis mellifera scutellata*, classified by Lepeletier, 1836 - (African honeybee) Central and West Africa.
2. *Apis mellifera capensis*, classified by Eschscholtz, 1822 - the Cape bee from South Africa
3. *Apis mellifera monticola*, classified by Smith, 1961 - High altitude mountains at elevation between 1,500 and 3,100 metres of East Africa Mt. Elgon, Mt. Kilimanjaro, Mt. Kenya, Mt. Meru
4. *Apis mellifera sahariensis*, classified by Baldensperger, 1932 - from the Moroccan desert oases of Northwest Africa. This sub-species faces few predators other than humans and is therefore very gentle. Moreover, because of the low density of nectar-producing vegetation around the oases it colonizes, it forages up to five miles, much farther than sub-species from less arid regions. Other authorities say that while colonies of this species are not much inclined to sting when their hives are opened for inspection, they are, nevertheless, highly nervous.
5. *Apis mellifera intermissa*, classified by von Buttel-Reepen, 1906; Maa, 1953 - Northern part of Africa in the general area of Morocco, Libya and Tunisia. These bees are totally black. They are extremely fierce but do not attack without provocation. They are industrious and hardy, but have many negative qualities that argue against their being favored in the honey or pollination industry.
6. *Apis mellifera major*, classified by Ruttner, 1978 - from the Rif mountains of Northwest Morocco - This bee may be a brown variety of the *Apis mellifera intermissa* but there are also anatomic differences.
7. *Apis mellifera adansonii*, classified by Latreille, 1804 - originates Nigeria, Burkina Faso now hybrids also in South America, Central America and the southern USA. In an effort to address concerns by Brazilian beekeepers and to increase honey production in Brazil, Warwick Kerr, a Brazilian geneticist, was asked by Brazilian Federal and State authorities in 1956 to import about 100 pure African queens (*Apis mellifera adansonii*) to Piracicaba-Sao Paulo State in the south of Brazil. In a mishap some queens escaped. The African queens eventually mated with local drones and produced what are now known as Africanized honey bees on the American continent. The intense struggle for survival of honeybees in sub-Saharan Africa is given as the reason that this sub-species is proactive in defending the hive, and also more likely to abandon an existing hive and swarm to a more secure location. They direct more of

their energies to defensive behaviors and less of their energies to honey storage. African honeybees are leather colored, difficult to distinguish by eye from darker strains of Italian bees.

8. *Apis mellifera unicolor*, classified by Latreille, 1804 - Madagascar
9. *Apis mellifera lamarckii*, classified by Cockerell, 1906 - (Lamarck's honeybee) of the Nile valley of Egypt and Sudan. This mitotype can also be identified in honeybees from California. [2]
10. *Apis mellifera litorea*, classifed by Smith, 1961 - Low elevations of east Africa
11. *Apis mellifera nubica*, (Nubian honeybee) of Sudan
12. *Apis mellifera jemenetica*, classified by Ruttner, 1976 - Somalia, Uganda, Sudan

Subspecies originating in the Middle East and Asia

1. *Apis mellifera macedonia*, classified by Ruttner, 1988 - Northern Greece
2. *Apis mellifera ruttneri*, classified by Sheppard, Arias, Grech & Meixner, 1997
3. *Apis mellifera meda*, classified by Skorikov, 1829 - Iraq
4. *Apis mellifera adamii*, classified by Ruttner, 1977 - Crete
5. *Apis mellifera armeniaca*, Mid-East, Caucasus, Armenia
6. *Apis mellifera anatolica*, classified by Maa, 1953 - This race is typified by colonies in the central region of Anatolia in Turkey and Iraq (Range extends as far West as Armenia). It has many good characteristics but is rather unpleasant to deal with in and around the hive.
7. *Apis mellifera syriaca*, classified by Skorikov, 1829 - (Syrian honeybee) Near East and Palestine
8. *Apis mellifera yementica*, - Yemen and Oman
9. *Apis mellifera pomonella*, classified by Sheppard & Meixner, 2003- Endemic honey bees of the Tien Shan Mountains in Central Asia. This sub-species of *Apis mellifera* has a range that is the farthest East Sources

In India

European bees were successfully introduced into India in 1965. There had been several attempts to bring the exotic bees into different parts of the country through the past over 8 decades. One can assume that all these stocks perished because no reports on their performance exist. Successful establishment of the European bee was possible due to a combination of several factors. Important among these are selection of appropriate strain of the bee that could adapt to the hot agricultural climate, change in the cropping pattern that helped in ensuring a continuity in bee forage, and scientific methods of introduction of the exotic strain in an alien climate. The bee has spread to several parts of the country and at least in one or two states has replaced the indigenous bee. It is, therefore, appropriate that the status of the introduced bee is considered here.

The Mediterranean region is supposed to be the centre of origin of *Apis mellifera*, from where three important European races of the species originated - *A.m. mellifera* L., *A.m. ligustica* Spinola, and *A.m. carnica* Pollmann (Ruttner 1988). Of these, *A.m. ligustica*, also called the Italian honey bee is the most generally distributed race in the world. Giving reasons for its popularity, Ruttner (1988) writes: "It is adaptable to a wide range of climatic conditions and has a combination of behavioural traits which are needed in modern apiculture: docile and quiet on the comb, if disturbed; prolific, with a tendency to build big populations and stores of honey without swarming, and little use of propolis." It is of importance to world apiculture, because most colonies in the world belong to this race. Adam, an authority on honey bee races, writes (1983): "Indeed, I believe apiculture would never have made the progress it did without the Italian bee".

Explaining the reason for importing this race into India Atwal and Sharma (1968) stated that this race was most commonly propagated in the Western U.S.A. and over a number of years, highly prolific and productive strains suited to the comparatively warm dry climate of this region had been evolved. Climates of California in the Western U.S.A. and Punjab are similar. The bees which were ultimately established in Punjab belonged to the following: the Californian yellow, Italian yellow, English black, Starline (Mid-West Hybrid) and Caucasian Hybrid. The hybrids include genes of the other races like *A. m. mellifera*, *A. m. caucasica* and *A. m. carnica*. Except for the casual mention of the "yellow Italian" bees, there is no precise determination of the bee strain that is now used for beekeeping. The original stock at Punjab has been multiplied and the progeny distributed in the state. Later the exotic bee was taken to Himachal Pradesh, Haryana, Jammu and Kashmir, Uttar Pradesh, Bihar and West Bengal. Today the bee is kept in several areas, including Assam, Meghalaya, Karnataka, Tamilnadu and Kerala. Chahal (1993) reports that there are over 0.1 million colonies of the European bee in Punjab alone. Considering its spread for commercial beekeeping in Bihar and West Bengal, the total number of bee colonies of this bee in the country may be about 0.15 million. It is reported that no imports were made subsequent to the original experiments on introduction in Punjab. This would indicate that all these 0.15 million colonies are the progeny of the few colonies of hybrid bees.

Mishra (1993) states that the present day *A. mellifera* stock descended from varied and heterogenous blood. The progeny of this restricted population can show the effects of inbreeding depression. Performance of *A. mellifera* is specific to agro-climatic conditions. There are reports of some beekeeping firms or institutions supplying "high" bred queens and bees of the Italian race. Unfortunately the exact parentage of these is not available. If the bees are not bred for specific agro-climatic conditions, even the best "high" bred bees may not deliver the goods. The beekeeper is the sufferer in the deal.

The European bee is similar to the Indian hive bee in its biology, nesting, foraging, colony defence and other behaviour features, with minor differences. According to Chahal (1993), the European bee possesses definite superiority over the indigenous bee in Punjab. It can colonize areas where the indigenous bee is not present or cannot do well. It can yield 4 - 5 times as much honey as the Indian bee. Chahal listed some features of the two species that have economic implications in beekeeping (Table 7). However, the performance of *A. mellifera* may not be the same in other agro-climatic conditions. For example, *A. mellifera* doing well in Punjab not may not survive the new climatic and vegetational conditions in other parts of the country.

Table 7. Comparative morphometric, behavioural and economic characteristics of *Apis mellifera* and *Apis cerana*

Characteristics	*A. mellifera*	*A. cerana*
Body weight (mg)	90 -120	50 -70
Tongue length (mm)	5.7 -7.2	4.39 -5.53
Nectar load (mg)	40 -80	30 40
Pollen load (mg)	12 29	7 -14
Flight range (km)	2 -5	0.8 -2
Egg laying capacity of queen per day	800 -1800	300 -800
Colony build up at honey flow	40,000 -60,000	25000 -30,000
Swarming	Little	High tendency
Absconding	Very little	Very high tendency
Aggressiveness	Usually calm	Mostly furious
Yield under Indian conditions (kg/colony)	25 -30	4 -5

Source: Chahal (1993)

An important feature in which *A. mellifera* differs with *A. cerana* is its use of propolis. Propolis is used to seal cracks and crevices in the nest and make the hive weather proof. The frames or the inner cover is often joined to the hive body with propolis. This makes inspection or management of colonies difficult. Some races are heavy propolizers. Special techniques have to be adopted to handle the bees without unduly disturbing the nest. Propolis has antibacterial properties. It has several medicinal applications. With the introduction of beekeeping with *A. mellifera* in India it is now possible to develop suitable technology for production of propolis.

Potential of the European bee

Summarizing the prospects of beekeeping with *A. mellifera* Atwal (1987) stated that this bee revolutionized honey production and improved farmers income in Punjab and Himachal Pradesh, and contributed substantially to the stability of agro-ecosystems in the region. The European bee has replaced the indigenous bee since 1985 when Thai sacbrood disease destroyed the latter. The European bee can play a complementary role in Indian beekeeping (Goyal, 1974). It is particularly useful in vast irrigated farming and orchard areas, that are now becoming available as new developmental projects are taken up including construction of irrigation canals, dams for power generation, etc. Indigenous bees are not present in such areas. Selection of appropriate strains for different agro-climatic regions, their import, trials under quarantine conditions and their adoption in the industry are the steps necessary to take up beekeeping with the European bee in new areas.

A. mellifera does not occur in the wild. Colonies can be available only from beekeepers who keep them. This means that multiplication of the colonies has to be done from a limited stock. This has the risk of introducing inbreeding depression. Verma (1990) suggests instrumental insemination of virgin queen bees with a homogenous mixture of semen collected from a large number of drones. This increases the population size in the breeding programme and consequently brood

Table 8. Some species specific characters of the Genus *Apis* (Ruttner, 1987)

Character	*mellifera*	*cerana*	*dorsata*	*florea*
Workers				
Forewing length (mm)	8.0-9.7	7.4-9.0	12.5-14.5	6.0-6.9
Cubital index	1.65-2.95	3.1-5.1	6.1-9.8	2.8-3.7
Tomenta	Tergites 3-5	Tergites 3-6	Tergites 3-6	Tergites 3-6
Drones				
Endophallus	One pair of cornua. Bulb with chitin plates	One pair of cornua; three pairs of upper cornua	Four pairs of very long thin cornua	One pair of long cornua, distal part of endophallus elongated
Behaviour				
Nest	Several combs in cavity	Several combs in cavity	Single large comb at bottom of a branch or projecting rock	Single comb on twig encircling the branch
Distribution	Allopatric	Sympatric	Sympatric	Sympatric

Table 9. Species specific characters of *Apis mellifera* (Ruttner, 1987)

Character	Florea	dorsata	laboriosa	cerana	mellifera
Forewing length (mm)	6.0-6.9	12.5-13.5	14.2-14.8	7.27-9.02	7.64-9.70
Tomenta	Tergite 3-6	3-6	3-6	3-6	3-5
Hind wing: extension of radial vein	Variable	Present	Present	Present	Missing

Melittin:sequence of amino acids (deviation of mellifera type)	5 amino acid changed	3 amino acid changed	?	0	0
Drone					
Endophallus	1 pair of cornua bulb a thin tube	4 pairs of very long thin;cornua short bulb	?	1 pair of cornua rudiments 3 others no chitin plates. Thin pad of plumose hair	1 pair of cornua with chitin, plates
Basitarsus 3	Deep incision with plumose hair +spines	Thick pad of sturdy branched hair	?	Thin pad of plumose hair	As cerana
Behaviour					
Capping of drone cells	Solid	Solid	?	Perforated	Solid
Nest	Single comb encircling twig to form a 'dance floor' fixed with cell bases	Single big comb fixed at bottom side of branch or rock, fixed with midrib	As dorsata	Several combs in cavity fixed with midrib	As cerana
Communication	Sun-oriented dance on platform open to the sky	Sun-oriented dance on vertical comb open to the sky	?	Sun-oriented dance on vertical comb in cavity	As cerana
Distribution	Sympatric	Sympatric	?	Sympatric	allopatric

Table 10. Comparison of honeybee species

Property	*A florea*	*A dorsata*	*A laboriosa*	*A cerana*	*A mellifera*
Nest location	Twig or small branch	Large branch or rock	Cliffs or rock	Cavity	Cavity
Nest protection	Hidden in walls	High above ground	High above ground	Cavity walls	Cavity walls
Number of combs per nest	1	1	1	3-8	5-8
Nest dispersion	Solitary	Aggregated	Solitary	Solitary	Solitary
Colony population	Small<5000	<20,000	>30,000	<10,000	>10,000
Worker size(length)					
Body weight (mg)					
Cells/worker [worker/brood]	0.80	0.74		3.56	3.99
Time from egg to adult (days)	21			19	20-22
Longevity (maximum days)	>50	ND	ND	<38	30-40
Age at the start of foraging (days)	=18	ND	ND	=14	=20
Worker foraging (%)	10-23	ND	ND	7-25	ND
Trips/day/worker	6	-	-	21	
Colony aggressiveness	Timid	Fierce	Extremely fierce	-	-

Table 11. Comparison of morphological characters in different honeybee species.

Character	*A. mellifera*	*A. mellifera*	*A. dorsata*	*A. florea*
Forewing Length (mm)	8.0-9.7	7.4-9.0	12.5-14.5	6.0-6.9
Cubital index	1.65-2.95	3.1-5.1	6.1-9.8	2.8-3.7
Tomenta	Tergites 3-5	Tergites 3-6	Tergites 3-6	Tergites 3-6
Nest	Several combs in cavity	Several combs in cavity	Single large comb at bottom of branch or projecting rock	Single comb on twig encircling the branch

Apis nigrocincta

Genus: *Apis*

Species : ***Apis nigrocincta*** Frederick Smith, 1861

Apis nigrocincta is a species of honeybee that inhabits the Philippine islands of Mindanao and Sangihe as well as the Indonesian island of Celebes or Sulawesi. The species builds nests in cavities like the closely related *Apis cerana*. In fact, there are few substantial differences between the two species: the genitals of the respective drones, for instance, are identical. However, there are small morphological differences, genetic polymorphism in the mitochondrial DNA, as well as behavioral differences. In areas where *A. cerana* and *A. nigrocincta* live together, they can most immediately be distinguished by their coloration and size: *A. cerana* tends to be darker and smaller, while *A. nigrocincta* tends to be larger and have a yellowish clypeus (the lower area of the face). The architecture of the colonies is also a point of difference: the opening of the drone cell of *A. cerana* is covered in wax, under which there is a conical cocoon with a central hole or pore. In *A. nigrocincta*, however, the cell of the drone has a narrow opening, without a hard wax cap and hole. In addition, the queens of *A. nigrocincta* generally create colonies with greater numbers of drones than those of *A. cerana*. Another noticeable behavioral difference between the species is the time of day at which they prefer to gather pollen. *A. nigrocincta* contracts the parasite-caused honeybee disease Varroatosis by playing host to the species of varroa mite known as *Varroa underwoodi*. In this way, they are similar to *Apis cerana nuluensis*, which is also susceptible to the same species of parasite.

Apis cerana nuluensis

Genus: *Apis*

Species: *Apis cerana*

Subspecies: ***Apis cerana nuluensis*** Tingek, Koeniger & Koeniger, 1996

Apis cerana nuluensis is a subspecies of honeybee described in 1996 by Tingek, Koeniger & Koeniger. The geographic distribution of the subspecies is the south-east Asian island of Borneo, politically divided between Indonesia, Malaysia, and Brunei. *A.c. nuluensis* is one of a number of Asiatic honeybees, including the more obscure *Apis koschevnikovi* and *Apis nigrocincta* (the latter of which has nearby habitat on nearby Sulawesi and Mindanao islands). While this was originally described as a species, it has since been determined to represent a geographic race (subspecies) of the widespread *A. cerana* (Engel, 1999). Like many honeybees, *A.c. nuluensis* is liable to infestation by the parasitic *Varroa* mite, although in this case the particular species is *Varroa underwoodi* (In this aspect, *A.c. nuluensis* is similar to *A. nigrocincta*).

Apis koschevnikovi

Genus *Apis*

Species *Apis koschevnikovi* Buttel-Reepen, 1906

Koschevnikov's Bee, or ***Apis koschevnikovi***, is a species of honeybee which inhabits Sabah, Malaysian and Indonesian Borneo, where it lives conspecifically with other honeybee species such as *Apis cerana* (specifically *A. c. nuluensis*). The individual bees are slightly larger than *Apis cerana* found in the same locality, but otherwise the colonies are similar in size and construction. Known as red bees due to a reddish hue when clustering. The species was first described by Buttel-Reepen, who dedicated it to Koschevnikov, a 19th century pioneer of honeybee Morphology. The species was described again by Maa in 1953, this time with the name *Apis vechti*. It was finally rediscovered by Tingek *et al.* in 1988. *A. koschevnikovi* hosts a unique species of the honeybee parasite *Varroa,* named *Varroa rindereri* (Guzmán *et al.*, 1996). Although, this parasite species is quite similar to *Varroa jacobsoni* it is perfectly differentiable. It has only been reported in colonies of *A. koschevnikovi* in Borneo and seems to be specific to that species, as it has yet to be observed crossing over to colonies of *A. cerana*, even when they live in the same apiary.

The Himalayan Cliff Bee: Apis laboriosa Smith

In 1980, Sakagami *et al.*, brought attention to the worlds' largest honeybee *Apis laboriosa* Smith 1871. Earlier, Maa (1953) had given a description with key characteristics which distinguish it from other *Megapis* including less prominent ocelli, longer maler areas, much paler thoracic pubescences and much longer wax plates. A detailed morphometric analysis using more than 100 characters confirmed the distinctiveness of this bee (Sakagami *et al.*, 1980). *A. laboriosa* nests on cliffs with southerly exposures (Roubik *et al.*, 1985, Underwood, 1986) of the valleys of Himalaya mountains from Nepal east through India, Bhutan, Myanmar (Burma) and China. Active and abandoned nests have been observed at elevations ranging from 1220 to 3510 m in Nepal (Roubik *et al.*, 1985) and 930-3450 m in Tibet (Wu, 1982). In contrast, *A. dorsata* rarely nests above 1250 m (Underwood, 1990).

Not all scientists agree that *Apis laboriosa* is a species different from *Apis dorsata*. Certainly *A. laboriosa* are the largest of the honey bees. They are found in the Himalayas at higher altitudes than *A. dorsata*. *A. laboriosa* nests are similar to those of *Apis dorsata*, but *A. laboriosa* colonies are usually found together in clusters, with sometimes over 100 combs suspended from a cliff face very near to one another. The species status of *Apis laboriosa* has been contested recently. Sakagami *et al.* (1980) and McEvoy and Underwood (1988) argued that the number of quantitative differences between *Apis dorsata* and *A. laboriosa*, their sympatric distribution at the upper and lower extremes of their ranges respectively, and the lack of intermediate forms were all evidence that *A. laboriosa* should be considered a distinct species. However, the countering argument is that this evidence is not sufficient to give this morphotype species status; some of the subspecies of *A. mellifera* are as distinct from each other as are *laboriosa* and *dorsata* (Koeniger, 1976; Ruttner, 1988). Moreover, Delfinado-Baker *et al.* (1985) showed that *laboriosa* and *dorsata* share the same parasites; this is not observed in other species. On the other hand two new *Braula* species of the genus *Megabraula* were recently detected in *laboriosa* nest which may or may not occur in *dorsata* nests (Grimaldi and Underwood, 1986). Although very little is known about the biology, *laboriosa* certainly exists under extreme ecological conditions. The taxonomic rank is not determined by the morphometric distance but by the isolation and the historical age of the type. All available data on physiology and ecology indicate a step in the evolution of the limits of adaptability of an open nesting species; existing in a zone with temperate climate, surviving temperature below freezing points at least for a limited time (Roubik *et al.*, 1985). The direction of evolution points to a Bombuds type rather than to a multi-comb *Apis*: large body size, long hair, dark colour. Ruttner (1988) suggested that the taxonomic status of *laboriosa* would only be settled when the male endophallus was described as, evidence of a physical barrier to intermating. More recently, Underwood has provided data on flight timing of drones of *labioriosa*. According to him *dorsata* drones fly just before dusk, wheras *laboriosa* drones fly in early afternoon, from 1220 h to 1420h. The ocelli of drones, which are noticeably raised in most members of the *dorsata* complex and are undoubtedly an adaptation to flight at low light intensities, are flat in *laboriosa* drones which also indicates that mating flights occur during the day and not at the dusk.

Although additional data would be helpful, this difference in drone flights apperars to provide the reproductive isolation needed to consider that *laboriosa* is a valid species. *Apis laboriosa* Smith, the Himalayan cliff bee is the world's largest honeybee measuring upto 3 cm long. It has been identified in the mountainous areas of Nepal, Bhutan, India, western Chinese province of Yunnan, (Summers, 1990; Batra, 1995, Ahmad and Roy, 2000) at altitudes of between 1200 and 3500 m above sea level (Underwood, 1992). It is believed to inhabit other parts of Himalaya. It has been reported to nest mostly between 2500 and 3500 m and forages upto 4100m (Roubik *et al.*, 1985; Valli and

Summers, 1988), and that, in general, it builds brood nests under overhangs on vertical clifs (Underwood, 1986, 1990, Roubik *et al.*, 1985; Sakagami *et al.*, 1980).

The bee's large size is a sign of its adaptation to the cold and low oxygen climates in which it lives, where there is a need for increased muscle mass among other characteristics (Summers, 1990). The bees foraging area is larger than most of the other honeybee species. It can fly up to a radius of three kilometers in thin air. It produces a large quantity of honey and beeswax making honeycombs upto 3m long and 1.5 m high. As a result of its massive foraging, it plays a vital role in pollinating mountain crops and wild flora. *Apis laboriosa* live on cliffs except for the coldest months when they move to places that are more protected. From around early February to early December, the colonies are found 10 metres above the ground on the cliffs faces, with typically up to a 100 metres or more colonies, each of which builds a single nest, at one cliff site between late November and early December, at the start of cold period, the colonies abandon the cliff sides and either move to protected locations beneath rocks, logs or similar, or move to lower altitudes. The colonies tend to move to same rock faces each year and these sites have become established for honey hunting. Ahmad and Roy (2000) have also reported the existence of *Apis laboriosa* bee nests two to three metres above ground in Bhutan, while some honey hunters have even found the over wintering bee colonies in clusters without a comb on low rocks surrounded by grass.

Apis laboriosa F. Smith lives in the Himalayas on hardly accessible rock cliff sites. Its biology is little known. In autumn, the colonies migrate from the subalpine (2800 m–3600 m) and cool temperate (2000 m–2800 m) zones to the warm temperate (1200 m–2000 m) zone (Underwood, 1990b). Woyke *et al.* (2001) showed that those colonies which already resided in the lower margin of the warm zone (1250 m) did not migrate, but reared brood throughout the winter. Underwood (1990a) who monitored *A. laboriosa* during 1.5 years, mentioned only one periodic mass flight (PMF) in that species (designated by him as an "orientation flight"). The purpose of his investigation was to find out whether PMF in *A. laboriosa* represent exceptional phenomena, or whether they occur regularly. He also tried to examine the character of those flights.

Biology

Little is known of the biology of *A. laboriosa* F. Smith. This bee lives in harsh conditions in the Himalayas. The ambient temperature (ta) drops at night to 5–10°C. In winter it is low during the whole day (4.5–13.5°C), and at night ta drops below freezing. According to Underwood (1990a, b), *A. laboriosa* survives the cold winter months of December and January by huddling in energy-efficient comb-less clusters. Little heat is produced and individual bees can chill to temperatures below 4°C. However, Woyke *et al.* (2001) showed that in lower warmer Himalayan zones, the bees survive the winter on combs as normal colonies rearing the brood, in comb-right colonies. It would be interesting to compare the behaviour of *A. laboriosa* to the closely related species *A. dorsata*

Fabricius. The latter species lives in tropical conditions and often suffers from an excess of heat. However, in northern areas, or high in the mountains, the temperature drops at night to 10°C or lower. They observed dorso-ventral abdomen flipping (AF) performed intensively by *A. laboriosa* workers at temperatures below 12°C. Characteristic body movements performed by honey bees include movements performed by foragers during the waggle dance (von Frisch, 1923), vibration dance (Fletcher, 1978; Schneider *et al.*, 1986), the defence shimmering performed by *A. cerana* Fabricius (Butler, 1954; Koeniger and Fuchs, 1975) or the defence movements performed by *A. dorsata* (Seeley *et al.*, 1982). Woyke *et al.* (2001) also investigated the behaviour of *A. laboriosa* workers at different ambient temperature (ta) to detect any correlation between them.

Abdomen flipping of *Apis laboriosa* workers during different activities

A. laboriosa workers start foraging at 10°C and the number of workers leaving the nest double at 12°C (Woyke *et al.*, 2003b). They observed that during this activity the curtain mouth is created. The workers are not hanging vertically but in disorder. The foragers leave and return to this area. The recruiting dances are also performed here, but no AF within the mouth area was seen. Even at the lowest ambient temperature, at which the foraging activity just started, and the percentage of AF was the highest, no AF were performed within the mouth. However, just outside it, as well as at the rest of the curtain, AF was performed. *A. laboriosa* workers perform periodic mass flight during the day (Woyke *et al.*, 2003a). They start those flights at 12°C, when the workers perform AF. Woyke *et al.* 2003(b) observed that shortly before those flights started, as well as during that activity, the total curtain was in disorder. They did not observe any AF performed during that flight activity at any part of the curtain. In December in Chale, there were comb right colonies as well as comb-less clusters. Worker bees at the curtains of both types of bee aggregations performed AF at appropriate ambient tempera-tures. When an intruder approached the nest, or a hornet was hunting nearby, characteristic defensive circular waves occurred at the curtain. As soon as the waves passed the curtain, AFs were performed at the appropriate temperature.

Hygienic behaviour

The cavity-nesting *Apis mellifera* and *Apis cerana* bees detect, uncap, and remove diseased brood. The hygienic behaviour of open-air-nesting bees *Apis dorsata* and *Apis laboriosa* was investigated in India and Nepal. Sealed *A. dorsata* pupae were pin-killed or deep-frozen. The workers removed 37% of damaged pin-killed pupae depending on the diameter of the pins, and only 7% of the frozen undamaged pupae. Migrating *A. dorsata* and *A. laboriosa* left unopened the sealed brood in deserted combs. Thus, *A. dorsata* and *A. laboriosa* do not open undamaged cells with dead brood. This behaviour is a more efficient mechanism in preventing the spread of diseases and parasitic mites than uncapping and removing dead pupae by *A. mellifera* and *A. cerana*. It may be beneficial for migrating *A. dorsata* and *A. laboriosa* to temporarily disuse part of the comb cells in exchange for arresting the mites there and thus reducing the increase of their population.

Polyandry

Honeybee queens mate with more males and have the lowest coefficient of genetic relatedness among workers among all eusocial Hymenoptera (Tarpy and Nielsen, 2002). The Himalayan giant honeybee *A. laboriosa* F. Smith is found in the mountainous regions of Vietnam, Bhutan, China India and Nepal. The first specimen of *A. laboriosa* was reported in Yunnan, China. More recently, Sakagami *et al.* (1980) reported that *A. laboriosa* shows many morphological differences from *A. dorsata* Fabricius. Furthermore, Underwood (1986) was the first to report on certain behaviors of *A. laboriosa*. Specifically, he reports that this species constructs a large single comb in an open space in high cliffs at altitudes ranging from 1200 to 3600 m, makes a seasonal migration depending on the availability of nectar and pollen resources, and tends to colonize one site in a reproductive season. However, the comparative levels of polyandry in giant honeybee colonies are unknown within the genus *Apis*. Takahashi and Nakamura (2003) in their study used microsatellite DNA analysis to estimate queen mating frequency and the genetic relatedness among workers of two colonies of the Himalayan giant honeybee *A. laboriosa*. The queen of Colony 1 was found to have mated with 23 drones and the queen of Colony 2 with 21. The number of effective matings revealed 22.12 in the Colony 1 and 17.01 in Colony 2. The genetic relatedness of nestmate workers were 0.272 in Colony 1 and 0.278 in Colony 2. Their study demonstrated that queens of two *A. laboriosa* colonies mate 20 times, a similar level of polyandry to that found in *A. dorsata* (Moritz *et al.*, 1995; Oldroyd *et al.*, 1996). However, the level of polyandry of *A. mellifera carnica* queens ranged from 6 to 24 patrilines within species (Neumann *et al.*, 1999; Tarpy and Nielsen, 2002). It is possible that the mating frequency of *A. laboriosa* queens has an equally large range, but more colonies need to be analyzed to estimate more precisely the levels of polyandry in *A. laboriosa*.

Each honey bee colony comprises a group of worker and drone bees with one queen, who live together to supply each other's needs and cooperate to raise the off springs. Each colony builds one nest consisting of a single large wax comb with a thick honey storage area at the top and thinner brood portion below. Pollen is stored in a band separating the honey storage from the brood comb. The honey portion is usually around 15 cm thick, depending upon the overall size of the comb and the nectar flow potential of the area. Essentially there are three different types of *Apis laboriosa* honey. Spring honey collected at high altitudes – the so called red honey, spring honey collected at mid and lower altitudes, and autumn honey from any site. The red honey is valued for its special properties. It results from the collection of nectar from the white *Rhododendron* (Ericaceae family), from *Acontium* spp., and from *Entada scanders*, which only grow at higher altitudes of around 2000 m and more. Thus honey has intoxicating and relaxing qualities which, however, decrease with storage time. It is not consumed locally because of its high price, honey hunters prefer to sell it for highly need cash crop. Hive bees are not kept at this altitude, and *Apis laboriosa* is the only honeybee species that collects this intoxicating nectar. The whole sale price of freshly harvested honey is generally around US$ 13 to 19 per kilogram, five times the price of hive honey. Large amounts of *Apis laboriosa* spring honey are exported from Nepal to Japan, Korea and Hongkong. The red honey is prized in Korea for its medicinal value and fetches a premium price, nowadays Korean companies buy much of this honey in advance. Traditional honey hunting still occurs in South Africa, Southern Arabia, India (the Himalayan regions), and Nepal. The honey hunters in Nepal are mainly from a tribe of Gurungs in central Nepal. The honey hunters in Nepal harvest honey from nests made on cliffs by *Apis laboriosa*, the world's largest honeybee. Harvesters climb down rope ladde-rs, which have been lowered from the top of the cliff, and using bamboo poles ma-neuver baskets into position beneath the combs. The comb is then cut to fall into the basket. Fires are set at the base of the cliff to pacify the bees to some extent.

Two other morphotypes of giant honeybees have been recognized by some authors as full species: *Apis binghami* Cockerell 1906 of Sulawes, nearby Sula island, and Butang, and *Apis brevligula* Maa 1953 of the Phippines (Luzon, Mindoro, and probably throughout excluding Palawan). The two are quite similar in colour, both being uniformly black with distinct white bands on their abdomen. Both have raised ocelli, as is also the case with *A. dorsata* and are known to forage nocturnally. Some of the differences noted by Maa (1953) and summarized by Ruttner (1988) indicate that *A. breviligula* is slightly shorter but with broader abdomen and substantially shorter mouth parts than *A. binghami* Nesting aggregations which are common for *dorsata* have not been observed for either *brevligula* (Morse and Laigo, 1969; Starr *et al.*, 1987) or *binghami*.

It could be argued that these two morphotypes should be considered separate species because they have isolated populations with distinct morpho-logical features. However, as their distributions are allopatric from that of *A. dorsata*, their status as species is likely to remain an arbitrary decision.

Indigenous Bees

India has four species of *Apis*, i.e., the true honey bees - *Apis florea, A. cerana, A. dorsata* and *A. laboriosa* Smith - and several species of stingless bees of the genus *Trigona*. There are indications that at least an additional species close to *A. florea* occurs. Of the *A. cerana,* three sub-species have so far been identified. Kshirsagar (1976) and Verma (1992) indicate that a total of at least seven races under three recognized sub-species occur. Similar variation is sure to be present in other species as well. In spite of this variety, all the *Apis* species have a remarkably similar life cycle. The biology of the stingless bees is also almost the same as that of the *Apis*. All these bees have a well developed social organization and have similar caste differentiation, division of labour, foraging, defence and reproductive behaviours.

Stingless Bees, *Trigona* spp.

Commonly known as dammar bees, these are quite small in size and look like small mosquitoes or flies. In view of this they are called musuru pattu in Telugu. They are distributed in tropics and sub-tropics, and even in temperate regions. They build their nests in dark enclosures like cavities in branches or trunks of trees, ant hills, termite tunnels in the ground, wall crevices or any abandoned receptacle like logs, pots and tins.

The nests of *Trigona*, unlike those of *Apis*, are clusters of small uniform globular cells of wax. These pots are the cells in which the young are reared. The pots are closely stacked touching each other or separated, each cell or cluster of cells being connected with others by girders or pillars of wax. Pollen and honey are stored in conspicuously large oval cells that are constructed close to the brood cell clusters or at their periphery quite apart from them. No clear separation of honey pots is found. Because of this the honey collected from dammar bees is rich in pollen that gets into it from pollen pots interspersed among the honey pots.

The brood cells of *Trigona* (*Tetragona*) *iridipennis* Smith from Castle Rock, Karnataka (Kshirsagar and Chauhan, 1977) were on average 3.12 mm broad and 4.09 mm high. The honey pots were about 6.64 mm broad and about the same in height. The pollen cells were similar to honey cells in breadth, but were 6.78 mm in height.

After selecting a suitable nesting site, a new swarm closes all cracks, openings and crevices in the nest with a glue like very sticky propolis. The nest is connected to the outside atmosphere only through a small opening that is less than a centimetre in diameter. The opening is often made into a tunnel using wax and propolis. The tunnel opening is shut by propolis every evening after all the outdoor work is completed for the day. It is opened every morning enabling the foragers to resume their field duties. Handling the dammar bees is relatively very easy. Unlike the true honey bees, these bees do not have stinging as the defence mechanism, but developed an equally effective biting behaviour, in defending their nest. When disturbed the bees attack the enemy in large numbers, usually selecting sensitive organs like eyes, nose and ears as their target. Biting with the mandibles is quite irritating for several of the enemies of dammar bees, but can be easily protected against by man. The pots with honey and pollen stores are scooped out and pressed in a cloth to draw out honey. Honey yields from dammar bees are very low, often a few grams to some 500 g. however, from forest areas in different parts of the country up to 2 kilograms of honey are collected from each nest of well grown colonies.

Honeys of the dammar bee are dark amber in colour, show a highly positive polarization. In the 5 samples of dammar bee honeys from Castle Rock, Karnataka and Indore, Madhya Pradesh, Phadke (1968) found levulose to be about 32 per cent and dextrose only 20 per cent. However, dextrins were nearly 6 per cent. The proportion of ash was also quite high, indicating a high mineral content. Due to the low dextrose content, honeys did not granulate even after 3 years after collection. The peculiar chemical composition and physical properties of the honey are attributed to the characteristic floral range the dammar bees have. The stingless bees have been almost completely neglected. There is no comprehensive study on the Indian dammar bees. *Trigona* (*Tetragona*) *iridipennis* is the most common dammar bee in the country. Other species were reported during 1940s and 1950s in general or from some parts of the country. Besides *Trigona iridipennis* that is common all over India, three other species occur in the Khasi hills, Meghalaya (Pugh 1947). Neto (1949) recorded a total of three species from India: *T.* (*Tetragona*) *iridipennis*, *T.* (*T.*) *ruficornis* and *T.* (*Lepidotrigona*) *arcifera*. *T.* (*T.*) *ruficornis* occurs in Haldwani, Uttar Pradesh (Neto, 1951). The tribals in Andhra Pradesh assert that there are at least five different kinds of the dammar bees : *pedda* (big) *musuru*, *putta* (ant hill) *musuru*, *kaki* (crow) *musuru*, *cheema* (ant) *musuru* and *mala* (black) *musuru*. No detailed studies on their biology or honey production are so far reported. Neto (1949) emphasizing the need for further taxonomic work on the dammar bees reports of the remarks of H.S. Schwarz, a great specialist in the systematics of the dammar bees of the world, thus: "the number (of the species of the dammar bees) when the size of India is taken into consideration is notably small." In Malaysia 29 species of meliponinae (stingless bees) were recorded. India which has a larger geographic area and variety of vegetations and climates should have a larger number of species.

Pugh (1947) in the Assam region, gave a short account of the species of stingless bees found in the Khasi hills. He identified these as species of *Melipona*. These are probably *Trigona*, since no *Melipona* is reported to occur in India. An interesting feature of all the three species is their similarity in size and appearance to the dwarf bee. *Ngap siwor*, identified by Pugh as *Melipona tunneli* is different from the other two in possessing yellowish spots on the abdomen. It forms a long tunnel about 5 to 6 cm long at the opening of the hive that is usually located in tree trunks. Honey is collected twice or thrice in a year, and a colony may yield about two kilograms of honey. *Melipona khasiana*, locally known as *ngap hamang*, is slightly smaller than *M. tunneli*, and does not form the long tunnel at the nest entrance. Yield from these bees in one or two collections is about a kilogram. Both these species are kept in earthen pots that are placed in the open cells of houses in villages. *Ngap khyndew*, *M. terrestris* is the third species, that lives in underground cavities made by termites. Honey is also collected once or twice from their nests and a little over a kilogram of honey is got in a year. Dammar bees are very hardy and work in harsh weather conditions. In spite of their short foraging range, they can exploit several small flowered plants like weeds and produce honey. They can be kept in any receptacle that provides adequate space for nest building. They withstand rough handling and repair quickly the nest that is damaged in handling. They recover and grow quite rapidly after adverse conditions that deplete the colony strength. They need little care by the beekeeper. The colonies can be divided by simple division of the brood cluster. The part of the cluster without the queen, rears readily another queen and a full-fledged new colony gets established. These features make dammar bees ideal for keeping for honey production, or for pollination of agricultural and horticultural crops that have small flowers, and cannot be useful to other larger bees. Dammar bee honey is much in demand for its high medicinal value.

Stingless bees can play a great role in the economy of the country. They are known to pollinate crop plants with small or minute flowers like carrot and coriander. Even when several species of *Apis* are used for honey production, the stingless bees can be kept by farmers for honey production and for improved yields of their crops. The bees require little attention. They are harmless, unlike the honey bees, whose sting is quite painful. The hive for stingless bees is quite simple and does not demand any great craftsmanship. For these reasons at least, rearing of stingless bees deserves to be encouraged.

Reproductive isolation among species of the genus *Apis*

In the 1960s, research on reproductive isolation in honeybees started with the pioneering work on *Apis cerana* and *A. mellifera* of F. Ruttner. Since then, the number of recognised *Apis* species increased from four to nine, and data on reproductive isolation played a key role in this development. The different mating behaviours include: behavioural mating barriers (mating season, mating place, sexual signals, daily mating periods), copulatory barriers (size, genitalia, mating sign) and physiological barriers (sperm transfer, sperm storage) and postzygotic barriers (fertilisation, development, hybrids). Allopatric *A. mellifera* and allopatric populations of the other species had a uniform mating period during the afternoon hours. Sympatric honeybee species were separated mainly by different daily mating periods. The mating period differed between populations of the same species from different regions. The sequence of the mating periods, however, described from Sri Lanka, Thailand and Sabah (Borneo) followed the same pattern and showed a taxonomic and size correlation: the dwarf bees (*A. andreniformis* and/or *Apis florea*) occupied the first position shortly after noon. The next mating period was occupied by cavity-dwelling bees and at sunset, *A. dorsata* drones flew out for mating. In addition, in the honeybee species that have been studied, various non behavioural mating barriers have been demonstrated.

Commercial beekeeping

Among the four commonly-recognized species of *Apis*, only *A. cerana* and *A. mellifera* are kept commercially by man. Behavioural limitations of the dwarf and giant honeybees, particularly their practice of open-air nesting, prevents their being kept in man-made hives for reasonably long periods, while hiving colonies in specially-constructed containers is essential in that it enables the colonies to be manipulated.

In many parts of the world, including several countries in Asia, commercial beekeeping depends on moving the honeybee colonies to places where forage is abundant at certain periods of the year. Such migratory beekeeping often calls for the colonies to be moved several times a year, over distances which may range from a few kilometres to several hundred kilometres from the home base. This approach is practicable only when the colonies are in movable-frame hives, which can he transported without danger to the hives or the colonies. From the practical standpoint, therefore, beekeeping can be a dependable agricultural occupation only when the beekeeper can determine and control the number of hives he owns.

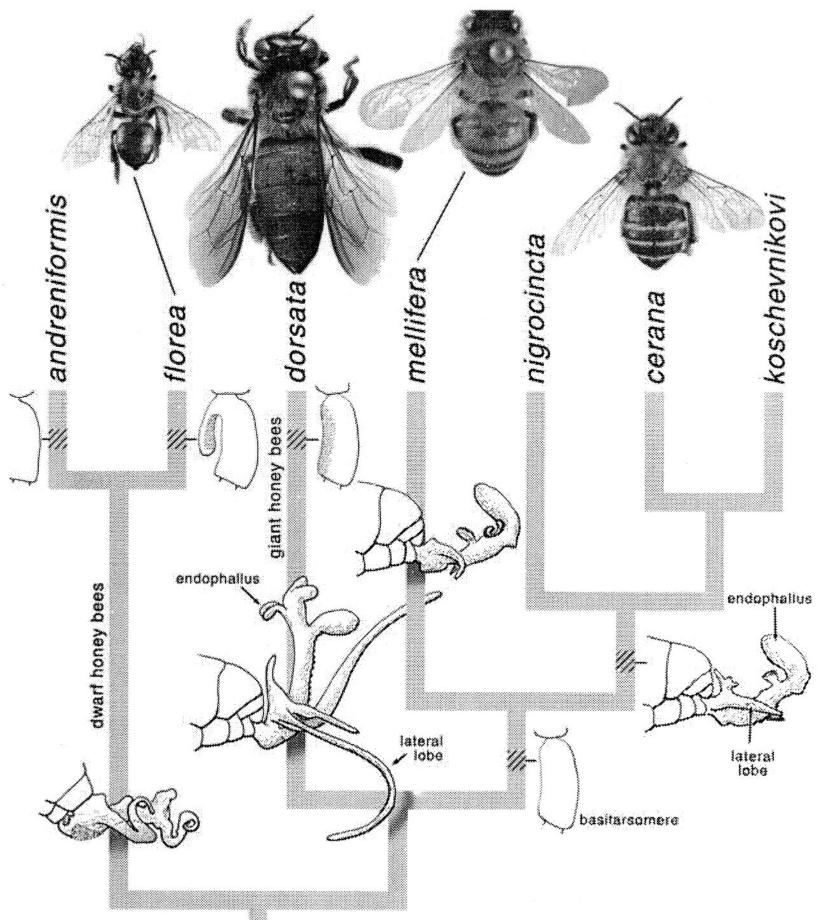

Figure 7. Relationships among species of recent honey bees, genus *Apis*, showing important variations in tarsomeres and male genitalia. Relationships from Engel and Schultz (1997)

Generally speaking, there are two possible approaches to the development of commercial beekeeping in Asia: the introduction of modern beekeeping with *A. mellifera* or the improvement of existing techniques for using *A. cerana*. Notwithstanding the difficulties involved in establishing new apiaries of the introduced colonies and in developing colony management techniques suitable to local conditions, *A. mellifera* colonies are generally more productive than those of *A. cerana* where forage is abundant, and the development of beekeeping with *A. mellifera* in Japan, the Republic of Korea, China and northern Thailand is based on this finding.

On the other hand, where forage is available only marginally, colonies of *A. cerana* survive better and can produce with lower management inputs than colonies of *A. mellifera*. It is the absconding behaviour of most, if not all, tropical races of *A. cerana* that creates a major obstacle to the development of beekeeping with this bee in rural areas in southern Asia. Since this behaviour is apparently triggered, at least to some extent, by an unfavourable hive environment, proper colony management may be able to provide at least a partial solution to this problem.

Thus, only through systematic research and development activities carried out locally is it possible to judge which of the two approaches to apicultural development should be adopted to suit the socio-economic situation, the vegetation pattern and the climatic conditions of each locality.

Chapter 6

Biology of the honeybee

Honey bees are social insects which live in colonies and can not survive on their own for long, but require the social setting of the colony. It is the colony that matters; where tasks are accomplished through division of labour. Every member works, not for itself, but for the benefit of the colony. For instance, during the summer, large quantities of food are gathered and stored by the workers, even though the individuals do not live long enough to feed on these reserves during winter. Defensive behaviour of individual bees is not for themselves but for the benefit of the colony. Since stinging mostly results in death, the use of the stinger is of no value to the individual bee. A honey bee colony typically consists of several thousand bees that cooperate in nest building, food collection, and brood rearing. Each member has a definite task to perform, but it takes the combined efforts of the entire colony to survive and reproduce. A colony normally has a single queen, fifty to sixty thousand workers at its peak, and several hundred drones during late spring and summer (Fig. 8) Individual queens, workers, and drones cannot survive by themselves. The social structure of the colony is maintained by the queen and workers and depends on an effective system of communication. The exchange of chemical secretions among members and communicative "dances" are undoubtedly responsible for controlling the activities necessary for colony survival. Division of labor within the worker caste primarily depends on the age of the bee but varies with the needs of the colony. Reproduction and colony strength depend on the queen, the quantity of food stores, and the size of the worker force. As the size of the colony increases up to about sixty thousand workers, so does the efficiency of the colony. It also includes a series of combs made up of six-sided cells on both sides for the rearing of young bees (brood) and for the storage of food-pollen and honey. Individual queens, workers, or drones cannot survive alone.

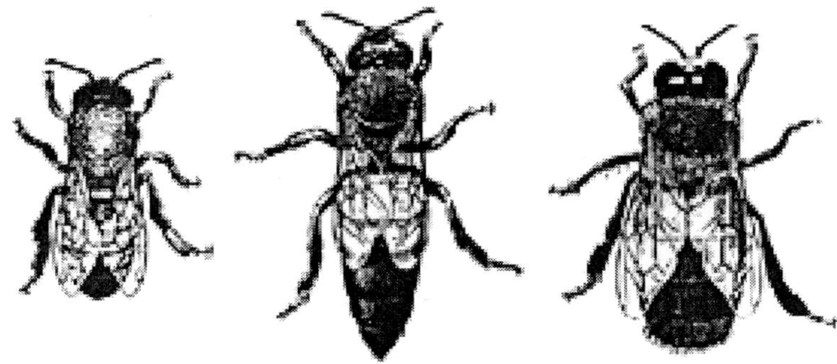

Figure 8. Three types of honey bees normally found in a honey bee colony: worker, queen, and drone

The Worker Bee

Worker bees have given up the functions of reproduction and egg laying, and passed on these tasks to the single queen of the colony. The worker bee has sex organs that are not fully developed. Her six-week life span in summer is devoted to carrying out the many tasks necessary for colony development and survival. Many duties carried out by the worker bee are the result of physiological changes that take place during the worker's life. The most important of these are the production and glandular secretion of royal jelly (brood food) and beeswax. In addition to numerous in-house duties, worker bees forage for nectar, pollen, water and propolis. They also serve as scout bees in finding these materials and in locating a future home site for a swarm.

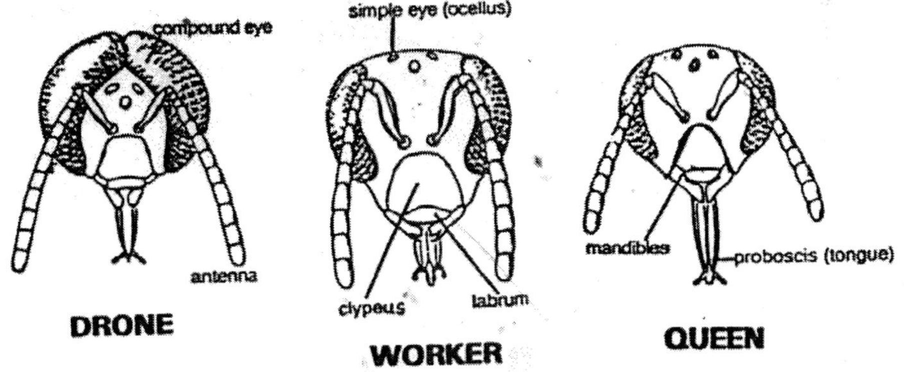

Figure 9(a). Comparison of heads of drone, worker and queen

Three distinct phases may be observed in the life of the worker bee: (1) The "nurse bee" phase lasts about a week. At first she helps in incubating the brood and preparing brood cells. Next comes the feeding of older larvae that are provided a mixture of honey and pollen. About three days later the special brood food glands in the head of the worker bee become active. The concentrated milky secretion from the glands called "royal jelly" is fed to the queen larva in its pure form while a mixture of pollen, honey and jelly is fed to the worker and drone larvae. (2) The "domestic" phase. For about one week, the young worker bee takes on various hive duties such as storing of honey, building and repairing comb, and cleaning the hive. During this period, the young worker bee makes its first orientation flights and may carry out guard duties at the hive entrance. (3) The "field" bee or forager. The bee is now about fourteen days old. Foraging may last two, three, or four weeks according to the amount of energy expended. At this final stage in life, at the age of 6-8 weeks, most worker bees will die in the field. In winter, bees live from fall to the next spring. Honeybees forage for four products which include (1) nectar, which is converted into honey; (2) pollen, which is the protein and fat portion of the bees' diet; (3) water; and (4) propolis, or bee glue. Propolis is a resinous material from the buds of trees. Bees use propolis to close small openings in the hive. When nectar is collected from flowers, it will be kept in the "crop" or honey sac where initial enzymatic conversions take place, while pollen and propolis are carried in the "pollen baskets" located on the hind legs. Workers are smaller than the queen and drones, but there number is much more. There number may range from a few hundred during winter and early spring to usually many thousands during summer when pollen and nectar are abundant. A strong colony at full strength can have close to 100,000 workers. These bees keep the colony going.

Life begins as a fertilized egg. Laid singly in cells, each egg is attached to the bottom of the cell and stands upright. Eggs hatch in about three days. Each larva is fed royal jelly for first three days and then on honey and pollen subsequently. Pollen and honey are not as rich as royal jelly, so the larva becomes a worker instead of a queen. The white grub-like larva molts (sheds its outer covering) five times during the six days. Just before maturity, house bees cap the cell. The larva then spins a cocoon and becomes a pupa. The adult emerges 12 days later. It takes about three weeks to mature from the egg to an adult bee ready to go to work. Workers' jobs change with their ages. Young bees, called house bees, do indoor duties. They produce wax and shape it into combs (structures of cells containing honey and brood) and use propolis (a gummy substance gathered from plants) to seal cracks or cover rough edges in the hive. They also fan their wings to ventilate the hive in summer, controlling temperature and humidity, and provide heat in winter. Some bees guard the hive to keep out raiders. Many produce honey and royal jelly. A lot of time is spent feeding brood and cleaning and repairing cells. House bees also feed the queen, the drones, and each other. Older workers, or field bees, gather nectar, pollen, and water. The average adult worker lives less than a month during the busy season; overwintering bees live several months.

Workers virtually do all of the efforts required to maintain function within a hive. During the latter part of their life, each will travel up to two miles in search of pollen, nectar and water. Each worker typically goes on ten food gathering journeys per day, each lasting approximately one hour. This heavy workload takes its toll; each worker lives for about a month prior to wearing out. Immediately after emerging from its pupal cocoon within one of the many brood cells, it immediately goes to work. During the first four days of its adult life, each worker is cleaned and fed by the other bees while its body hardens and it begins to produce substances in various glands. Activities during the next seventeen days include cleaning, feeding larvae, manipulating wax, processing honey, guard duty and air conditioning the hive by fanning. Any of these activities can be done at any time based on the needs of the colony. On day 21 the worker leaves the hive, and works for another 20 days, bringing in pollen, nectar and water before taking its final flight away from the hive and dying (Hooper, 1976).

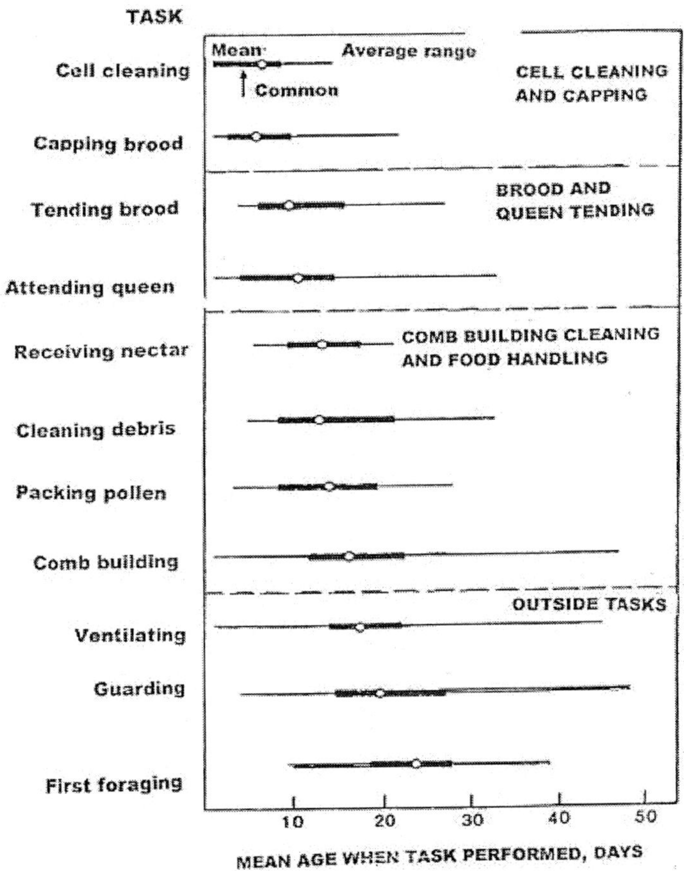

Figure 9b. Schedule of activity of worker has related to age of the bees.

There is considerable overlapping of the age groups engaged in the various duties (Figure 9b). When the age groups are not in normal balance, bees of any age can do the work necessary but not so efficiently. Bees under 3 days old and the field bees can feed the queen and raise brood or they can secrete wax and build comb even though their glands are not fully developed or they have degenerated from lack of use. Similarly, very young bees can forage for pollen and perhaps nectar when there are no field bees of normal age to do this work. Worker bees inherit many skills that they manifest purely on a behavioral basis.

They are skilled architects and craftsmen, qualified dieticians and nurses, proficient housekeepers, experts in heating and air conditioning, and fully qualified to police and defend their colony. Their architectural skill and craftsmanship is exemplified by the beauty of the honey comb, its structural strength, economy of material, and the rapidity with which they construct the uniform hexagonal cells. The building of comb is accomplished by first "plastering" the wax into approximate- mate position in the form of round cells, and then thinning down the wax walls to a uniform thickness to produce the hexagonal cells for strength and economy of wax. As dietitians they prepare one kind of food for the queen larvae and another for the worker and drone larvae. Each larva receives approximately 10,000 visits from the nurse bees during development. The hive is maintained immaculately clean at all times, and the guard bees with their stingers for armor protect the hives against all intruders.

Pollen, a plant protein source for the young, provides nitrogen, phosphorus, amino acids, and vitamins essential for development of these vegetarians. Pollen is collected in pollen baskets (corbicula) on the workers' rear legs. Nectar, obtained from floral nectaries deep within flowers, provides a pure carbohydrate source for all stages. Each worker fills her honey sac within her digestive system, increasing her weight by up to one half. Upon arrival at the hive, the worker regurgitates the contents of the honey sac to the younger workers within the hive. These younger workers receive the nectar, which is processed by enzymes within their honey sacs, and tipped into storage cells where it ripens for five days. At this point the substance becomes honey, and the cell containing it is capped for storage. Nectar from 5 million flowers is required to produce half kg of honey. Water, the final substance brought to the hive, is essential for hydrating all of the individuals within a hive and cooling it throughout the year. Approximately five gallons are required to hydrate and cool the colony each year (Hooper, 1976).

The worker bees are sexually underdeveloped females smaller than the queen but capable of laying small numbers of eggs under some conditions. Worker bees that lay eggs are called laying workers, their eggs, usually placed in worker cells, develop into undersized but functional drones. The presence of laying workers in a colony usually means the colony has been queenless for approximately two weeks. However, laying workers also may be found in normal "queenright" colonies during the swarming season and when the colony is headed by a poor queen. When a colony becomes hopelessly queenless, the ovaries of several workers develop and workers begin to lay unfertilized eggs. Development of the workers' ovaries is believed to be inhibited by the presence of brood, the queen, and her chemicals. Colonies with laying workers are recognized easily: only drones are reared in worker-sized cells; there may be anywhere from five to fifteen eggs per cell; and the eggs of a laying worker are smaller than those of the queen. In addition, laying workers scatter their eggs more randomly over the brood combs, and eggs are usually on the sides of the cell instead of at the base, where they are placed by a queen. Some of these eggs do not hatch, and many of the drone larvae that do hatch do not survive to maturity in the smaller cells. Drones that do mature are normally undersized. The activities of a laying worker are similar to those of normal workers. They consume pollen and honey as well as forage in the field.

Worker-bee larvae hatch from the eggs in 3 days, are fed royal jelly for 2-1/2 days, and then their diet is changed to include pollen and honey for 2-1/2 days. They are sealed in their cells for 12 days, during which period they spin a cocoon and transform from the larvae to the pupae, emerging as adult bees 20 days after the eggs were laid. The difference in the cell and food environment causes the worker bees to require 5 days longer to develop the queen, yet their life expectancy is only 5 weeks during the summer and a few months during the winter. Any worker larva under 24 to 48 hours old can be developed into a queen under the proper colony conditions that insures the nurse bees will construct a queen cell and feed royal jelly lavishly to the developing larva. The rearing queens for market is a highly specialized field of beekeeping. The worker bees differ markedly from the queen in many respects other than function, length of life, and behavior. Structurally they have a longer tongue for gathering nectar, modified mandibles (jaws) especially designed for comb building, special glands for secreting royal jelly, enzymes for the conversion of nectar into honey, and glands that function in communication; highly specialized leg structures for gathering and carrying pollen, four pairs of wax glands on the underside of their abdomen for the secretion of wax, and a straight barbed sting for the defense of the colony. The queen's sting is curved and smooth and is used only to destroy rival queens. The worker bees exhibit a well-defined division of labor based primarily upon their physiological age but modified to some degree by the needs of the colony. The physiological age of bees is similar to their actual age during the active season when the colony is raising brood and storing food. During dearth periods, especially in winter, a 60-day old bee may be younger physiologically than a 20-day old bee in summer.

Honey bees, like other insects, are cold-blooded and have a body temperature close to that of their environment. However, the honey-bee colony functioning as a single organism can maintain uniform hive temperatures under northern winter conditions identical with those in summer or in the tropics. Only recently has man accomplished this by developing elaborate heating and air-conditioning equipment. By clustering together, they generate and conserve heat, or they lower the temperature by evaporating moisture and establishing air currents through the colony to maintain a uniform temperature of 93°F. within the cluster, even though the outside temperature is at -50°F. or 120°F. Under low temperatures, the cluster temperature ranges from 45°F. on the surface to as high as 93°F. within when brood is being reared.

The most conspicuous characteristic dominant in honey bees is their great industry. Honey bees do not procrastinate by doing tomorrow what they can do now. They may fly 50,000 miles and visit 5,000,000 blossoms to gather enough nectar to produce one pound of honey, which is stored not for themselves but for the survival of the colony. The bees that gather this food do not live long enough to enjoy it. One bee, of course, cannot fly such a distance, yet the bees of a colony may store 5, 10, or even 20 pounds of honey in a day. They must gather 200 to 300 pounds of honey and 50 pounds of pollen (10 gallons) to meet the colony's needs each year. The beekeeper also expects to harvest a surplus of 100 or more pounds of honey for his efforts. The bees have to be industrious to gather so much food, rear so many young, build comb, air condition the hive, and perform all the other duties peculiar to the colony.

Activities of Worker Bees: Workers within a honey bee colony engage in various tasks, depending on their age and the needs of the colony (Figure 2). Division of labor by age exists within the worker caste. Bees less than 2 weeks old become involved in cleaning cells and feeding first the older larvae and then larvae of all ages. Workers function as nurse bees during the period that their hypopharyngeal glands are producing brood food.

Older house bees work with honey, pollen, wax, and propolis. Nectar-collecting field bees are met by house bees, usually near the entrance, and are relieved of their nectar loads. The conversion of nectar into honey requires both a physical and a chemical change. The physical change involves the removal of water, which is accomplished by externally manipulating nectar in the mouth parts and then placing small droplets on the upper side of cells and fanning the wings to increase air move ment and carry away excess moisture. (Nectar is 30 to 90 per cent water, but honey should have no more than 18.5 per cent water). The chemical change requires the addition to nectar of the en zyme invertase, which the bees produce in their salivary glands. This enzyme breaks the disac charide sugar, sucrose, into two monosaccharide sugars, glucose and fructose (Nectar may be nearly all sucrose or may contain no sucrose, but honey contains an average of only 1 per cent sucrose).

Pollen pellets are deposited in empty cells near the brood nest by the pollen-collecting workers. In the cells, the pollen undergoes a maturing process to what is commonly called bee bread. Details of the maturing process are not understood. When bees are about 12 to 15 days old, their wax glands become functional and comb building is possible. Wax scales are removed from between the ventral abdominal sclerites and positioned into place using both feet and mouthparts. Young house bees in the process of comb building hang in festoons and pass the wax scales from bee to bee.

Propolis-collecting bees also serve as propolis storage reservoirs. Propolis is not stored in combs or elsewhere, but is removed from the corbiculae of these field bees and used as needed. House bees fill cracks and cover rough parts with propolis. During their third week as house bees, workers take short flights for orientation and defecation. Hives painted different colors aid the bees with orientation and reduce the chance of young bees drifting between adjacent colonies. Some of the oldest house bees also serve as guards at the entrance. After approximately 3 weeks as house bees, the workers become foragers, gathering pollen, nectar, water, and propolis for the colony. This period of their lives also averages about 3 weeks. Most foragers collect nectar and pollen, but nectar is collected in greater quantities than pollen. Pollen collection tends to be an activity of younger foragers and nectar collection a function of older foragers. Water collectors may comprise 10 percent of all foragers, but this figure is much higher during periods of heat stress. Propolis collectors are quite rarely observed.

The schedule of worker bee activities is both flexible and reversible, depending more upon physiological age than on chronological age, and is altered according to the needs of the colony.

The first three weeks of her adult life, during which she is referred to as a house bee, are devoted to activities within the hive, while the remainder are devoted to field work, so that she is called a field bee.

Duties of the house bee

The duties of a house bee are -

1. cleaning the hive and the comb
2. feeding the brood
3. caring for the queen
4. making orientation flights
5. comb building
6. ventilating the hive
7. packing pollen, water, nectar or honey into the combs
8. executions
9. guard duty

Cleaning: The first activity of the worker bee on reaching maturity is to clean herself. She removes all unnecessary particles, grooms herself immediately and then crawls out of her cell. She takes in food and then starts cleaning the brood cells, employing both tongue and mandibles. The comb cells are cleaned to receive eggs laid by the queen who, before laying, examines the comb cell to satisfy herself that it has been properly cleaned. If she finds a cell that is not properly cleaned, she quickly rejects it. Other duties which may occasionally be necessary include removing dead intruders or dead bees from the hive, and removing debris and other objectionable material. Anything that is too large to carry is often dragged along and pushed outside, while dead snakes, wax moths or other carcasses too heavy to transport are encased with propolis brought in by field bees.

Feeding the brood: After three to five days, the worker bee starts to feed the brood. At this stage she is called a nurse bee. At first she feeds larvae more than three days old with a mixture of honey or nectar, pollen, small quantities of bee milk and some water. After a few days, she starts to feed the younger larvae (1-3 days old) exclusively on bee milk, which she produces in brood-food glands, also called milk glands or hypopharyngeal glands, located in her head.

Caring for the queen: The next work undertaken by the young worker is to provide for the needs of the queen bee. Whenever the queen needs food, she calls for it by stretching out her proboscis towards the mandible or mouth of the nearest worker. The workers are always anxious to satisfy her needs and make a circle or semi-circle around her. The queen contacts the nearest worker, and if she does not get as much as she needs, she approaches the next. This continues until all her demands are met. It is also the duty of the nurse bees to bathe her with their tongues and mandibles and to carry away her faeces.

Orientation flight: The orientation flight is not so much a house duty as an exercise for the young worker. She must learn how to fly, and she must know the vicinity, especially the location of the hive. She therefore first makes some short flights in front of the hive and in the immediate vicinity to acquaint herself with the environment, so that when in the near future she goes out to forage, she will be able to find her way back home.

Comb building: Comb building provides the needed "rooms" in the hive, in the form of hexagonal cells, for two main purposes: storing food and rearing brood. Beeswax, the material for the construction of comb, is secreted by the worker's wax glands, which are best developed and productive when she is 12-18 days old. The wax, which emerges from the glands as a liquid, hardens quickly and appears in the form of oval flakes similar to small fish scales, protruding from between the last four overlapping abdominal segments on the under-side of the worker's body. As we have already seen, the bee must consume large amounts of food (honey and nectar) to produce these wax flakes. Bees engaged in building combs usually hang themselves in festoons at or near the site of the building operation. There they hang quietly while their digestive organs transform the contents of their honey sacs into energy and beeswax. The wax is removed with the spines of the hind legs and is then manipulated with the mandibles to build the comb cells. Capping of comb cells is also the duty of comb builders.

Ventilating the hive: Temperature control is one of the important duties of the house bee. When the temperature is low, bees cluster to generate heat for themselves, but when it is high, some of them have to fan their wings to circulate air throughout the hive. The right temperature required is between 33° and 36°C, while the brood chamber requires a constant heat of 35°. Honey has to be cured in order to ripen, and this also requires the help of circulating air. According to Crane, 12 fanning bees positioned across a hive entrance 25 cm wide can produce an air flow amounting to 50-60 litres per minute. This fanning can go on day and night during the honey-flow season. The phenomenon is always at its peak in October in the high savannah and forest zones of Vest Africa.

Honey conversion and packing

It takes several bees to produce honey. No single honeybee completes the whole process. The forager brings a load of nectar to the hive and transfers it to a house bee, who proceeds to the empty or uncrowded part of the hive, where she rests and exposes the nectar to the air being fanned by the fanning bees. The air circulation helps reduce the moisture content of the nectar and thus aids sugar concentration. The house bee may load the nectar into the upper section of an empty cell or add it to the honey or nectar of a cell incompletely filled. The speed with which she manipulates the nectar depends on the intensity of the nectar flow. If nectar is abundant, the house bee may deposit her load quickly into a comb cell for later processing. The time required for the nectar to mature into honey depends for the most part on its original moisture content. Matured honey usually has over 80% sugar concentration. Ripening time is also determined by the quantity of the nectar: combs completely filled with nectar, even if strongly ventilated, may take as much as 36 days to mature.

Packing water, pollen and propolis: Other essential commodities which are brought in by the foragers and need the attention of the house bee are water, pollen and propolis. Water is required for cooling the hive, especially during the hot and dry season, when the atmosphere is very dry and temperatures are too warm for the bees' comfort. Water is mixed with honey and pollen and then fed to the older larvae, between 3-6 days old. Pollen is also packed to about three-quarters full in comb cells in the brood chamber, sometimes side by side with brood cells. Cells are never completely packed with pollen. Propolis is a resinous material collected from trees. It is difficult to unload, because it is gummy in consistency, and the house bees have to help the foragers to unload. The carrier holds firmly onto the walls of the hive, and the house bee removes the sticky gum from the hairy corbicula or pollen basket. Sometimes it takes more than three days to off-load a forager. The propolis is either stored or used immediately for the purpose required: to block holes and cracks in the hive, to repair combs, to strengthen the thin edges of the comb, or to make the entrance of the hive watertight or easier to defend. As already mentioned, propolis is also used to cover objectionable material in the hive and to embalm dead intruders such as wax moths, snakes, etc., too large to be removed. It is interesting to note that house bees are always eager to help unload the field bee that brings in material which the hive requires immediately. For example, when the weather is too warm and water is required to cool the hive, they will pay no attention to foragers bringing in nectar or propolis, who will have to wait until the heat situation is brought under control before they are offloaded.

Executions: Executions are a means of protecting the colony from hunger, disease and any catastrophic event. They may be performed to eliminate strange bees, to kill or drive away old and sick bees, to discourage other hive predators from entering the hive, to remove sick or unwanted unemerged brood, to eliminate useless drones, and to kill unwanted or strange queens.

Guard duty: Guard duty is the final activity of the house bee before she leaves the hive. By this time she has reached peak strength, is very energetic, and is best fit to defend the entrance of the hive, which is also the point of entry of the colony's enemies. The guard bee has the duty of inspecting all incoming foragers by smelling their odour. When satisfied, the guard allows the incoming bee to enter unmolested with her load. In most cases, foragers with loads to discharge are not intercepted unless the hive is greatly disturbed. After staying at the entrance for a while, the guard may fly out on patrol for some time before returning to the entrance. The guard bee is also responsible for watching any crack through which a robber bee or any other intruder might enter the hive. In an alerted hive, guard bees stand on four legs, their forelegs lifted and they antennae held straight, searching here and there. Any intruder, robber or other enemy first receives a frightening audible warning, followed by a sting; if he persists, the application of the alarm pheromone on the spot where the bee stings quickly summons more defenders. The scent helps other attackers to find the target and follow without delay. It has been observed that during the brood-rearing season, more guards are stationed at the hive entrance than during the peak of the honey flow.

The field bees: Activities involving flight may start from the third day after emergence from the brood cell, but the young worker begins her actual foraging activity later. Between the 18th and the 21st day, her hypopharyngeal and wax glands have become too weak to function, so that she cannot produce royal jelly to feed the queen and the young larvae, nor wax to build comb cells. But by this time she is in perfect condition to fly and knows the geography of the locality. She therefore starts field work, fetching nectar, pollen, propolis or water, but always concentrating her activity on the immediate needs of the colony.

Nectar gathering: Nectar, the sweet liquid secreted by plant nectaries, is collected by foragers, taken to the hive an turned over to the house bees for processing. The forager then returns to the flowers and collects more. The number of trips she makes in a day cannot be assessed precisely. It may vary from time to time for a number of reasons: the availability and accessibility of the nectar source, the quantity of nectar present, and the nectar requirement of the colony for the day. Sight and smell enable the bee to locate sources. She lands on the part of the plant that will support her and dips her stretched proboscis into the corolla of the flower. If there is nectar, she sucks it into her honey stomach. If there is none, she wastes no time before moving to the next flower. Some flowers have more nectar than others. Sometimes the bee can load enough by visiting one, two or three, but in plants with tiny flowers she can only get a full load by visiting hundreds. A fully loaded bee can carry 85% of her own weight. The time taken to complete a trip varies, but can reach 2 hours

Water collection: Bees consider water-carrying as one of their most important duties. They execute it regardless of what may be involved. If they need Water for the hive, they will resort to drastic methods to acquire it. In water-scarce areas, desperate bees sometimes attack farmers for their sweat, and clothes cannot be washed outdoors in the daytime for fear of molestation by desperate bees searching for water. Thirsty bees visit kitchens, bathrooms, toilets and all obscure humid places. They will land on any moist area, dip their proboscis and suck in water. Loading of water takes only a few seconds. The bee carries it to the hive and returns in a few minutes to reload if water is still present.

The scout bee: Foragers can take on scout duties as well. The scout bee locates food sources and passes on the information to other bees by a series of dance-like movements. She circles around and around, stamping her legs and wagging her abdomen; sometimes she stretches her proboscis, possibly to show the type of food she has found. The onlooking workers watch her dance, interpret it and act accordingly. It is believed that different dances show different types of information to be passed on. Another most important duty of the scout bee in a new swarm is to search for a suitable accommodation, while the rest of the swarm waits on a tree branch or in a small enclosure. On finding a suitable hollow or hive, she returns to the swarm and performs a characteristic dance to inform them about the find. When two or more bees make different finds, each scout dances especially vigourously in an attempt to win the support of the swarm.

Robber bees: All worker or foraging bees are thieves. They claim anything they like as their own property. They snatch honey away from honey harvesters from other swarms during the daytime, especially when the weather is sunny and bright. In the rich bee-zones where water is scarce, bees easily steal water from villagers. Robber bees visit other colonies' hives and try to take honey in order to store it in their own hive. The problem of hive robbing is not as serious in tropical Africa as in America and elsewhere. Only very weak colonies are sometimes robbed; usually it is abandoned hives that other colonies invade to take advantage of the honey stored in the comb cells. It is strange that bees often fail to take advantage of water or any sweet juice located close to the hive, but when it is placed further away (about 20 metres or more) they take it. This shows that the beekeeper should always watch his hives to avoid leakages of honey, for the leak will not be recovered by his own bees but by other bees from elsewhere, thus encouraging robbing.

During autumn, a reduction in brood rearing and an increase in pollen consumption result in a population of long-lived "winter" bees having increased fat bodies and protein reserves. The normal 6-week adult life of "summer" bees may be extended to several months in these "winter" bees. Similar extensions of life expectancy also occur when brood rearing is interrupted at other times.

The Queen

The honeybee queen is unique in being the only individual responsible for the reproduction of the colony population. She is surrounded by a group of attendants, as she goes about her single task of egg laying. During peak periods in spring she may lay as many as 1,500 to 2,000 a day. She is capable of egg production only and does not nurse her brood. It is essential for the queen to maintain a high rate of egg laying if the colony is to replace all the workers that die during the normal development of the colony. The egg-laying ability of the queen is key to success of the colony and ultimately to the beekeeper since a large population of worker bees is needed to optimize honey yields and pollinate crops. Unlike worker bees, a queen may live for 5 or 6 years. At any time, and especially later in life, the queen may falter in her egg-laying, which reduces brood development. The worker bees will construct queen cells in preparation to replace her. To reduce the risk of a slowdown in brood rearing or queen failure in the middle of production season, it is recommended to replace the queen every year or two.

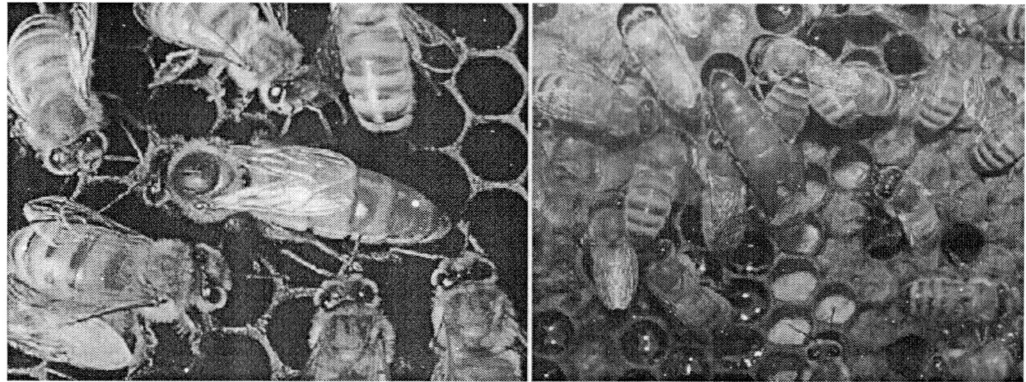

Figure 10. Queen honey bee surrounded by worker bees.

The queen is the longest bee in the hive but has the shortest wings. She is the mother of all the other bees. Her most important job is to lay eggs. Her productivity depends on the amount of food the workers bring in and the amount of brood space in the colony. A good queen does not waste any space. She lays a solid pattern of brood, meaning one egg in every cell. Few eggs scattered among many empty cells or several eggs per cell are signs of problems. The queen might be missing or old or parasites or disease might have weakened the colony. If it happens then there is need of examination of the colony to ascertain the problem. Worker bees usually rear new queens for one of three reasons: (1) the former queen left with a swarm; (2) the queen is laying increasingly fewer eggs; (3) the colony is overcrowded and has no place to expand. A colony that loses its queen suddenly is very upset but soon starts to rear a new one. Worker eggs or larvae less than three days old are raised in quickly built queen cells which hang vertically and are about the size and shape of a peanut shell. The queen develops from a fertilized egg the same as the worker bees. The egg hatches in 3 days, for 5 days the queen larva is fed a food secreted by the worker bees known as royal jelly. Then the larva is sealed in its cell where it spins a cocoon and goes through a number of transformations before emerging 7 days later as a fully developed queen bee.

A newly emerged queen stings the remaining queen cells in the colony and fights any other queens she finds. The former queen is killed if she is still in the hive. Usually she has already left with other bees in the colony. Six to eight days after emergence, the queen takes nuptial flights and mates high in the air with the male (drone). She may mate with up to ten or more drones, while flying. From these matings she is able to store 5 to 8 million sperm cells in a tiny organ called the spermatheca. She releases several sperm from the spermatheca each time she lays an egg destined to produce a worker or queen bee. When she places an egg in the drone cell, she does not release sperm to fertilize the egg. Once the mated queen commences egg laying-usually 8 to 10 days after she emerges-she never leaves the hive, except when the colony swarms. The different parentage has a number of benefits to the hive. Workers from different fathers begin to cool or heat the hive at different levels, leading to a more even temperature. Then she settles down and lays eggs. The queen is constantly attended and fed by the worker bees in the colony. These drones aren't hatched yet. It takes longer to hatch a drone than any other bee, and even when hatched, they aren't actually ready. It takes another week or so before they are ready to fulfill their dronely duty. Successful mating ends with the death of the drones.

After two to three years, the queen uses up all of her stored sperm and begins producing unfertilized eggs, which give rise to drones. Usually the workers raise one or more queens from the last of the fertilized eggs to replace the new queen (Jean-Propst, 1994). To maximize hive productivity, honey farmers replace the queen annually or every other year (Bishop, 2005). She will leave the hive only with a swarm (Swarming is the natural way by which colonies are established at new locations).

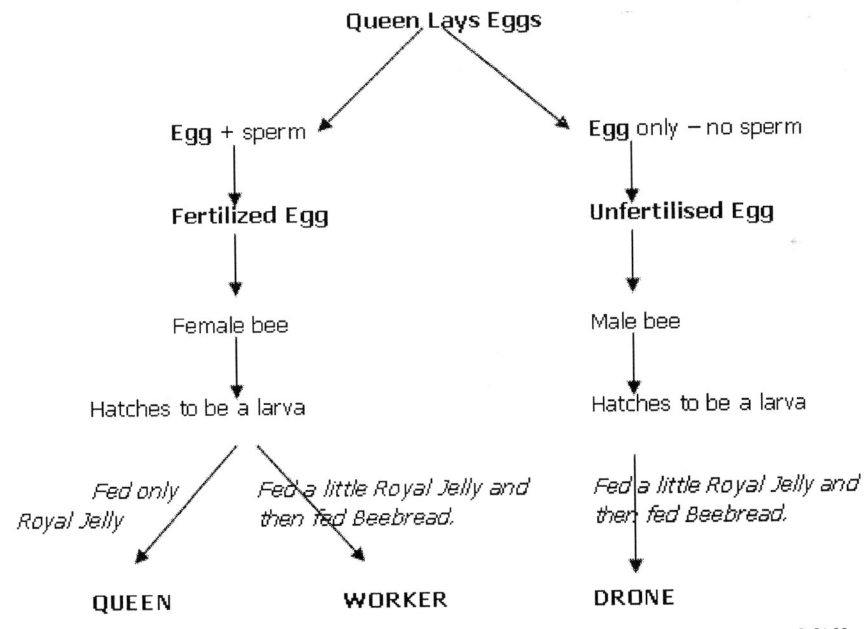

Figure 11. Impact of type of queen egg laying and food on development of different castes of bees.

The major role of queen centers around egg laying to insure the vast numbers of individuals required to maintain a hive. As stated earlier she may produce up to 1,500 eggs per day and 250,000 per year, and, under some circumstances, more than a million during her lifetime. The eggs laid in a single day may weigh more than her own body weight. She lays two kinds of eggs-those that she fertilizes and places in small hexagonal cells that develop into worker bees and those that she does not fertilize, which she places only in the larger hexagonal cells that develop into drone bees (Figure 11). Occasionally, she places fertilized eggs in round, peanut-shaped queen cells. These are temporary cells that hang down. The worker and drone cells lie in a horizontal position. Approximately 95 percent of the typical hexagonal cells of the honey comb are worker cells, 25 per square inch; the remainder is drone cells, 16 per square inch.

A queen is easily distinguished from other members of the colony. Her body is normally much longer than either the drone's or worker's, especially during the egg-laying period when her abdomen is greatly elongated. Her wings cover only about two-thirds of the abdomen, whereas the wings of both workers and drones nearly reach the tip of the abdomen when folded. A queen's thorax is slightly larger than that of a worker, and she has neither pollen baskets nor functional wax glands. Her stinger is curved and longer than that of the worker, but it has fewer and shorter barbs.

There is only one queen per colony, except during swarming preparations or supersedure. The qualities of the colony depend on the egg laying and chemical production of the queen. Her genetic makeup, along with that of the drones she has mated with, determines the quality, size, and temperament of the colony. The number of eggs the queen lays depends on the amount of food she receives and the size of the worker force capable of caring for the brood. When the sperm supply begins to be depleted, the workers prepare to replace or supersede her. The old queen and her new daughter may both be present in the hive for some time following supersedure.

The Drone

The drone is the male bee of the colony. It develops from an unfertilized egg and hence, is a haploid with only half the number of chromosomes. All his genetic characteristics originate from his mother, the colony queen. Drones cannot sting and do not perform any duties within the hive or gather nectar from flowers. The rearing and feeding of drones requires considerable resources from the colony. In the summer, a colony may raise 200-300 drones but in the autumn when the colony prepares for winter it will drive out the few remaining drones. The sole purpose of the colony to rear drones is to have one or more mate with a virgin queen from another colony and have the genes of the colony passed on. The drones are the largest bees in the colony, heavy bodied, and more or less rectangular in body shape. They are larger than workers but not as long as queens. A drone has large eyes that touch each other at the top of the head. Drones do not have stingers, pollen baskets on their legs, or glands for producing wax, and their mouthparts are too short to gather nectar. Moreover, they do not even do jobs they could like ventilating the hive. Their only function is to fertilize the queen, and they die in the process. Drones are banished from the hive before winter begins. While queens and workers develop from fertilized eggs, drones develop from unfertilized eggs. Drone cells are slightly larger than worker cells. This stimulates the queen to lay only unfertilized eggs in them. Drone eggs are also laid in worker cells that have become enlarged because of stretched or sagging combs. Small drones develop in worker cells if a queen gets old and loses her ability to fertilize eggs. They develop in 24 days from an unfertilized egg and thus have a grandfather but no father. They become sexually mature about 12 days after emergence and die instantly upon mating. The matings take place in flight, often several miles from the colony in which the queens and drones are produced. Like the pollen of flowers, there must be a great excess of drones to insure matings of queens that fly over a vast area. Only a small percentage of the drones fulfill their basic function. At the close of the honey harvest, the drones remaining in the colony are driven off the combs until they become weakened from starvation. They are then carried out of the hive by the worker bees to perish. A few drones may develop late in the season and overwinter with the colony. The drones are mainly reared and tolerated by the colony during spring and summer. Drones can be distinguished from workers and queens by their large size, rectangular abdomens, large conspicuous eyes, and noisy flight. All drones lack a sting, and have more eye facets than a worker (6,000-7,000 vs. 3,000-5,000).

Drones are capable of extracting honey four days after emergence, but prefer to be fed by workers. The role of the drones is largely to fertilize new queens. A group of drones follows each virgin queen on her early flights. Several males will mate with each virgin queen while flying, dying immediately after mating since his reproductive organs and the end of his abdomen break off, temporarily plugging the end of the queen's reproductive tract and abdomen

Under normal circumstances, drones live for about two months. Drones stay in the hive until about eight days old, then begin to take orientation flights. Flight from the hive normally occurs between noon and 4:00 p.m.

They do anything other than hang out in the local "Drone Congregation Area". No matter how many hives you have in an area, the drones select a particular area, and gather loosely there to watch for queens on mating flights. It is unknown how it is that drones from one season to the next choose the same congregation area, since none normally survive. Nor is it known how the queen, on her second flight from the hive, knows where the congregation area is. In the fall, the workers will drag the hapless drones from the hive, and refuse to allow them to return. Thanks, your services are no longer needed. During the summer though, the drone waits all day for a queen. His sole purpose in life is to procreate, but nature has played a cruel trick upon him. To mate is to die. The same mechanism that forms a worker's stinger is adapted to provide the drone's sexual organs. Bees die after stinging. Drones die after mating, his genitals left with the queen.

Activities of Drones and Queens: Drones take their first flights at about 8 days of age and are sexually mature at 12 days. Drones fly out on cleansing flights and orientation flights-both of short duration-and also on longer flights to congregation areas in search of a virgin queen. Drones do not forage and spend about three-fourths of their time at complete rest. Their normal lifespan is 8 weeks or less. Queens newly emerged from their cells are at first ignored but are later touched and licked by workers. This apparently helps prepare the virgin queen physiologically for her mating flight. Mating occurs in drone congregation areas-special locations in the air regularly visited by drones. These occur in the same places year after year.

A queen generally mates 6 to 10 days following emergence. She may go out on several mating flights, mating with several drones on each flight. Additional mating flights are taken until the spermatheca contains an adequate supply of semen (5 million to 6 million spermatozoa). If mating is delayed more than 3 weeks, there is a high risk of her becoming a drone-layer. Egg-laying usually commences within a week after mating, and a queen can continue to lay fertilized eggs throughout most of her life-usually 2 to 5 years. An old queen will not go out and mate again when her original supply of semen becomes depleted-she simply becomes a drone-layer. An old queen and her supersedure daughter sometimes coexist, thus contradicting the commonly accepted idea of one queen per colony.

The Brood Cycle: A honeybee has 4 life stages which include egg, larva, pupa and adult. Each life stage is very different from the others. A honey bee begins life as an egg. Bee eggs develop in the ovarioles or small tubes that make up the two ovaries of a queen. The egg is nourished and grows as it moves down this tube. When it is fully formed, it reaches the end of the ovariole, then moves through the oviducts into the vagina. The sex of the new bee is normally determined as the egg passes through the vagina. A lifetime supply of sperm (5 million to 6 million) is stored by each queen in the spermatheca, a little globular sac attached to the vagina. The queen controls the release of sperm with the so-called sperm pump. If an egg is fertilized, it will develop into a female bee, but if not fertilized, a male bee will result. The result is that male bees have only one set of chromosomes (haploid) acquired from the queen.

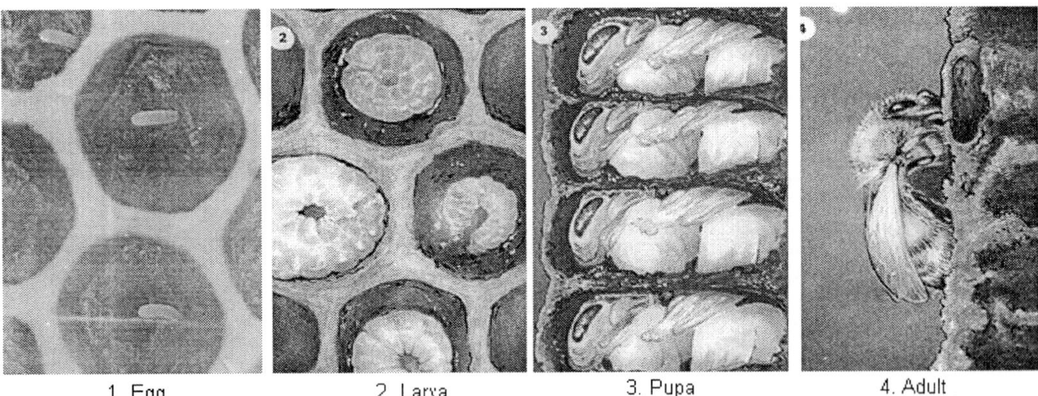

Figure 12 (a) Different stages in the life cycle of a honeybee

The queen bee attaches each egg to the base of an empty cell in combs that have been cleaned by workers. The honey bee egg is a smooth, white, sausage-shaped object about 1.5 ml in length. During the first day, the egg nucleus divides-if the egg is unfertilized; or if the egg is fertilized, the fusion nucleus or zygote divides. It is not until the third day that the embryo form (with head and body segments) can be seen within the egg. The head is present at the larger unattached end and the back (dorsum) is on the incurved (concave) side.

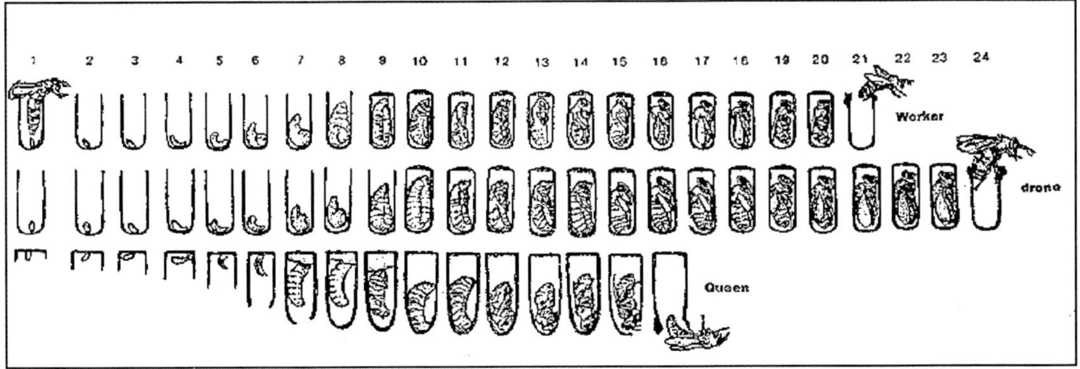

Figure 12 (b) Average development of times and stages for workers,drones and queens.

The first sign of hatching occurs when an egg is 72 to 84 hours old. Muscular contractions by the embryo cause a gentle, weaving motion that apparently results in a tiny hole being torn in the outer membrane (chorion). Fluid from within the egg soon emerges and covers the external surface. The embryo with its "tail" attached to the base of the cell continues to move about until its head also touches the base and an arch is formed. In this "croquet wicket" stage, the chorion evidently is dissolved. The larva then eases itself over against the bottom of the cell into the familiar C-shaped position). Honey bee larvae are fed a nutritious substance called royal jelly secreted by the brood-food glands (hypopharyngeal glands) of young workers. During the first 24 hours, worker larvae are fed lavish amounts of royal jelly by older nurse bees. During the second 24 hours, they get very little additional food and thereafter are cared for by nurse bees of all ages. Pollen and honey are present in the food of older worker larvae. Honey bees use two systems of feeding larvae. Young larvae are fed amounts excessive to their needs and older larvae are provided small quantities of food as needed. It has been estimated that 110,000 visits are made to a single bee during its egg and larval stages, 3,500 of these during the last 24 hours.

A female larva fed continuously on lavish amounts of royal jelly and provided a large, peanut-shaped cell will become a queen. Another larva given a mixture of honey and pollen during the latter half of its larval life and kept in a worker cell becomes a worker. The process that produces the complete expression of sexual characteristics in a queen has not been determined; however, it is considered to be caused by differences in both the quality and the quantity of the larval food provided.

Drone larvae grow larger than either workers or queens and, therefore, require more food. Food given to young drone larvae is nearly devoid of pollen and is milky-white, while that given to older drone larvae is a yellow-brown color and contains considerable pollen. The food given older drone larvae also is higher in pollen content than that given older worker larvae. Thus, both qualitative and quantitative differences distinguish the lar val food given queen, worker, and drone.

The developing honey bee larva is a helpless creature whose principal function is eating. Both the malpighian tubules and midgut are shut off from the intestine until a larva is nearly mature. In this way, body wastes are stored internally and the food surrounding each larva is protected from fecal contamination. The feces are expelled and pushed down to the bottom of the cell about the time the cocoon is made and after the larva has finished eating.

Table 12. Moults of the honey bee

	Workers		Queens		Drones	
Day	Stages	Moults	Stages	Moults	Stages	Moults
1	Stages	Moults	Stages	Moults	Stages	Moults
2						
3	egg		egg		egg	
4		hatching		hatching		(hatching)
5	1st larval	1st moult	1st larval	1st moult	1st larval	1st moult
6	2nd larval	2nd moult	2nd larval	2nd moult	2nd larval	2nd moult
7	3rd larval	3rd moult	3rd larval	3rd moult	3rd larval	3rd moult
8	4th larval	4th moult	4th larval	4th moult (sealing)	4th larval	4th moult
9						
10	gorging	(sealing)	gorging		gorging	
11						
12	prepupa	5th moult	prepupa	5th moult		(sealing)
13					prepupa	
14			pupa			
15						5th moult
16	pupa					
17			imago	6th moult emergence		
18						
19						
20					pupa	
21						
22	imago	6th moult (emerging)				
23						
24					imago	6th moult (emerging)

All castes of honey bees molt about every 24 hours during the first 4 days of larval life. When the ecdysis or molting occurs, the skin splits over the head and slips off the posterior end of the larva. This process normally takes less than 30 minutes. Each new larval stage (instar) is at first only slightly larger than the previous one, but it grows rapidly. The fifth larval instar gains about 40 percent of the total mature larval weight during days 8 and 9 (Table 12). By the end of the 8th day after the egg was laid, the cell containing the worker larva is capped. During the 9th day, the larva spins a cocoon using silk from a special gland in its head. On the 10th day, the larva stretches out on its back with its head toward the cell opening and becomes quiescent inside its cocoon. This stage usually is called the prepupa. The 5th molt, which occurs during the 11th day, reveals the pupal form-white in color and motionless. Color develops gradually, first in the eyes (13th day), then in the abdomen (15th day), legs (16th day), wings (18th day), and finally in the antennae (20th day).

Throughout this period, the pupa is encased in a thin outer skin which is shed in the 6th and final molt on the 20th day. Thus, legs, wings, and mouth parts are freed and the pupa becomes an imago (adult) which soon begins to chew its way out of the cell.

Because a bee egg hatches into a larva which becomes first pupa and then imago, bees are said to have a complete metamorphosis. The length of the egg stage (3 days) is the same for all three castes, but the larval and pupal stages are shortest for the queen and longest for the drone (Table 12, 13). As with most biological development, the duration may vary between individuals, and the data in Table 12 should be considered as close approximations. For example, hatching occurs between 72 and 84 hours, and workers may emerge as early as 19 days or as late as 23 days after the egg was laid.

Table 13. Developmental stages of the three castes of bees.

	Development duration of stage		
Stage	Queen	Worker	Drone
Days	3	3	3
Egg	5-1/2	6	6-1/2
Larval stage			
Pupal stage	7-1/2	12	14-1/2
Total developmental time	16	21	24

During its development, the honeybee undergoes four distinct stages of egg, larva, pupa, and adult, a process called complete metamorphosis. The queen is capable of laying fertilized eggs that produce worker bees or queens, and non-fertilized eggs that result in drones. During the first three days, the embryo inside the egg will develop rapidly. Then, just prior to hatching, the egg is provided a minute drop of bee milk. This applies to all brood regardless of caste. In worker brood, the larva is liberally fed glandular food or "royal jelly" for two or three days, followed by a honey and pollen diet. This change of food determines the worker caste. When the diet is not changed and remains royal jelly throughout the larval period, the larva will become a queen.

During the larval stage, five molts take place. After eight days, the larva is fully grown and fed, and at this point workers seal the cell with a porous capping. This is called the "capped brood" stage. The larva will spin a cocoon and undergo the process of pupation. The duration of the pupal stage depends on the caste.

Bee Life Stages

Typical of the most advanced insects, bees exhibit complete development or complete metamorphosis. This means that the young and the adults look very different and the diet of the young and the adults typically differ, preventing the parents from competing with their offspring for resources. The life stages are egg, larva, pupa and adult. Development from egg to new worker typically takes two to three weeks (Bishop, 2005).

Egg: Honey bee eggs are normally laid, one to a cell, by the queen. Each egg is attached to the cell bottom and looks like a tiny grain of rice. When first laid, the egg stands straight up, on end. However, during the three day development period the egg begins to bend over. On the third day the egg hatches into a tiny grub and the larval stage begins. The eggs are described as having an appearance similar to sausage-shaped poppy seeds. Each egg has a small opening at the broad end of the egg, the micropyle, that allows for passage of sperm. Hatching takes place three days after egg laying (Jean-Prost, 1994).

Larva: Healthy larvae are pearly white in color with a glistening appearance. They are curled in a "C" shape on the bottom of the cell. Worker, queen and drone cells are capped after larva are approximately 5.5, 6, and 6.5 days old, respectively. The period just after the cell is capped is called the prepupal stage. During this stage the larva is still grub-like in appearance but stretches itself out lengthwise in the cell and spins its cocoon. Larvae remain pearly white, plump and glisten during the prepupal stage. The larval stage lasts eight to nine days. Upon hatching, the larva is almost microscopic, resembling a small, white, curved, segmented worm lacking legs and eyes. For the first two days, all larvae are fed a diet of royal jelly. Beginning the third day, worker larvae are fed honey, pollen and water, while the larvae destined to become queens continue to receive royal jelly throughout their larval lives. Regardless of whether the larva is male or female, it molts five times during its larval stage (Jean-Prost, 1994). Care of the larvae is constant. Each larva receives an estimated 10,000 meals during this stage. Larval weight increases 5 1/2x during the first day, 1500x in six days. Larval stage durations vary: 5.5 days for queens (fertile females), 6 days for workers (sterile females), and 6.5 days for drones (fertile males) (Jean-Prost, 1994).

Pupa: After two days, prepupae begin to change from their larval form to adult bees. Healthy pupae remain white and glisten during the initial stages of development, even though their bodies begin to transform into adult forms.

Compound eyes are the first areas that begin to take on color; changing from white to brownish-purple. Soon after this, the rest of the body begins to take on the color of an adult bee. New workers queens, and drones emerge approximately 12, 7.5 and 14.5 days, respectively, after their cells are capped. The pupal stage is a stage of massive reorganization of tissues. Organs undergo a complete reorganization, while body changes from the wormlike larval body shape to the adult body shape with three distinct body regions. Pupation periods vary: queens require up to 7.5 days, drones require14.5 days, while workers require 12 days (Jean-Prost, 1994).

Table 14. Differences between three castes of honeybees

Queen	Worker	Drone
Number		
Only one queen present per olony under normal conditions	20,000-60,000 workers present per colony	Few hundred drones present during the period when they are required for mating
Sex		
It is perfect female	Imperfect females, reproductive structures not well developed with reduced ovaries.	Perfect males
Size		
Biggest in size	Medium in size	Bigger in size
Wing expanse		
Wings donot cover the entire abdomen	Wings cover the entire abdomen	Wings cover the entire abdomen
Sting		
Sting slightly curved & longer	They have sting called spines	Sting absent
Thorax size		
Thorax broad	Thorax short	Thorax broad
Abdomen		
Abdomen pointed V-shaped	Abdomen pointed V-shaped	Abdomen blunt
Pollen baskets		
Pollen baskets absent	Pollen baskets present	Pollen baskets absent
Wax glands		
Wax glands absent	Wax glands present	Wax glands absent
Food supply		
Dependent upon workers	Not dependent	Dependent upon workers for food
Development period		
16 days	21 days	24 days

Brood patterns: Healthy brood patterns are easily recognized when looking at capped brood. Frames of healthy capped worker brood normally have a solid pattern. Cappings are medium brown in color, convex and without puncture cappings.

Bee Development: All three types of honey bees pass through three developmental stages before emerging as adults: egg, larva, and pupa. While the developmental stages are similar, they do differ in duration. Unfertilized eggs become drones, while fertilized eggs become either workers or queens. Nutrition plays an important part in caste development of female bees.

Adult: Adult bees are either workers (sterile females), queens (fertile females), or drones (fertile males). A typical honeybee colony consists of 50,000-60,000 sterile workers, 500 to 1000 drones (fertile males) and one queen, the only fertile female in the colony and mother of the entire population of the hive (Bishop, 2005).

Workers and drones are reared in hexagonal-shaped cells which comprise the comb, but the queen is reared in an acorn-shaped cell, normally protruding vertically from the comb surface, with the opening at the bottom. Bees are perfectly equipped to collect pollen and nectar. They are covered with finely branched hairs that trap pollen as they visit flowers. While visiting flowers, the bees gather pollen from their hairs and store it in pollen baskets on their hind legs. A tongue-like portion of the mouthpart sucks up nectar.

Although man has managed bees for hundreds of years and carried them around the world, honeybees have not been "tamed" in true sense. Bees in the most modern apiary follow the same instincts as wild bees that live in hollow trees. Successful beekeepers anticipate and work with or around the bees' natural behavior.

Honeybees are social insects, living together in highly organized colonies. Each member has a specific job to do. A single honeybee cannot grow or survive by itself. The three distinct kinds of honeybees in a colony are queen, worker, and drone.

In many respects a honeybee colony is like a single animal. Individual bees and castes are like the cells and tissues. When one part is threatened, the whole colony reacts. If an essential segment of a colony becomes diseased or destroyed, the colony often can heal itself. It may divide and become two or more separate colonies. The colony also changes to survive different seasons. Let's follow the life of a colony through a year.

In mid to late summer, only small amounts of nectar and pollen are brought into the colony. Often no brood is being reared, so the colony does not grow. A fall nectar flow usually allows a small crop of young bees to carry through the winter. The colony needs honey for energy and pollen for protein, minerals, and vitamins to survive the winter and raise brood in early spring. Survival depends on a large cluster of young bees and a good food supply. If the cluster is too small, it cannot generate enough heat to survive the winter. Bees die if their body temperature gets much below 57°F. The colony must be able to make and save heat to survive in winter. Bees produce heat by digesting honey. They save the heat by bunching together in a tight cluster. The outer layer of bees is an insulating shell that traps the heat in the center of the cluster. The bees on the outer layers periodically change places with inside bees so that none of them become too cold. The cluster tightens or loosens depending on the temperature in the hive. Below 14°C, bees do not work in the hive. They do not even move to get honey that is not next to the cluster. If it stays cold for too many days straight, bees can starve even if honey is just a few inches away. The colony soon runs out of heat and freezes. Even if honey is within reach, they can freeze if there are not enough bees to produce some heat and save it. A large colony with plenty of food can keep the temperature at the center of the cluster around 32°C. This is warm enough to rear brood. They start doing this in late winter. As spring arrives, increasingly more brood is raised. As pollen and nectar are brought in, empty cells in the hive soon fill with brood and food.

Bees do not like to be crowded. If there is not enough room to add comb, some leave in a swarm. Colonies with plenty of space are less likely to swarm and will continue to grow. Beekeepers can keep healthy, productive bees by managing food and space wisely during the year. Honeybees were not significantly genetically selected by humans until recently because basic bee reproduction was not understood until 1845.

Chapter 7

Form and form function of the honeybee

ANATOMY OF HONEYBEE

The anatomy of honey is the assemblage of structural parts that enables the animal to do the things necessary for maintenance of its individual existence and the perpetuation of its kind. A complex combination of parts integrated into a finely tuned organism, the honeybee is capable of performing a broad range of athletic, graceful and purposeful tasks. It is very important to know the anatomy of honeybees to fully understand its behaviour unless and until, bee behaviour is understood properly one can not manage bee colonies efficiently.

Figure 13. External morphology of honeybee

A honey bee s made up of a series of hardened plates connected by membranes and covered in most cases regions by a dense pile of hairs (Figure 13). The external skeleton provides protection from the predators, prevents water loss and serves as a frame work for internal muscles attachment. The body of honey bee is divided into three regions, the head, the thorax and the abdomen each of which has number of segments. The head contains the mouth parts and serves as the major sensory region of the body having the eyes, antennae and sensory hairs. The thorax is made up of 3 segments each bearing one pair of legs and the posterior two segments bearing one pair of wings each. Thus, the thorax is the principal locomotory region of bee's body and contains powerful muscles for flight, walking and specialized functions such as pollen collection. The abdomen consists of seven visible segments and contains the bulk of the internal organs as well as a structure of great interest to beekeepers, the sting.

1. THE HEAD

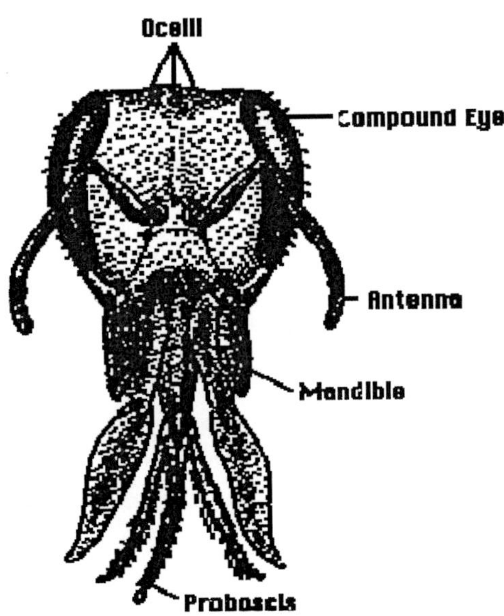

Figure 14. The head of worker honey bee with proboscis extended (Redrawn fiom Snodgrass, 1956).

The head is the center of information gathering. It is triangular in shape. It is here that the visual, gustatory and olfactory input are received and processed. Of course, food is also input from here. Important organs on or inside the head include: 1. Antennae, 2. Eyes, 3. Mouth parts, 4. Internal structures.

(a) **The eyes:** The seeing apparatus of the bee consists of a pair of compound eyes and three small simple eyes, called the ocelli. The compound eyes are composed of several thousands of simple light-sensitive cells, called ommatidia, which enable the bee to distinguish light and colour and to detect directional information from the sun's ultraviolet rays. These simple eyes called ocelli mostly help the bee determine the amount of light present in the environment. The eyes of the drone are larger by far than those of the worker or the queen bee, occupying a large proportion of the total volume of the head. They assist him to locate the queen as he pursues her during the mating flight.

(b) **The antennae** are a pair of sensitive receptors whose base is situated in the small socket-like membraneous areas of the head wall. They move freely in every direction. The antennae's functions are to feel or touch and to smell, and thus to guide the bee outside and inside the hive, to differentiate floral and pheromone odours, and to locate hive intruders. Another important function of the bee's antennae is to measure the flight speed

(c) The mandibles are a pair of jaws suspended from the head and parts of the bee's mouth. They are the strong jaws and help the bee to eat pollen for food, cut and shape wax, feed baby larvae and the queen, clean the hive, groom themselves, and fight They are also used to chew wood when redesigning the hive entrance. They also permit any activity requiring a pair of grasping instruments.

(d) The proboscis: The proboscis or the tongue functions mainly in drinking or lapping of nectar, honey, and water. It also plays a role in food exchange between the bees. Unlike the proboscis of all other sucking insects, that of the honeybee is not a permanent functional organ; it is improvised temporarily by assembling parts of the maxillae and the labium to produce a unique tube for drawing up liquids such as sweet juices, nectar, water and honey. The insect releases it when needed for use, then withdraws and folds it back beneath the head when it is not needed.

Mouthparts

Honey bees have a combined mouth parts than can both chew and suck (whereas grasshoppers can chew and moth can suck, but not both). This is accomplished by having both mandibles and a proboscis. The mandibles are the paired "teeth" that can be opened and closed to chew wood, manipulate wax, cleaning other bees, and biting other workers or pests (mites). The proboscis is mainly used for sucking in liquids such as nectar, water and honey inside the hive, for exchanging food with other bees (trophallaxis), and also for removing water from nectar. The workers can put a droplet of nectar between the proboscis and the rest of the mouth parts to increase the surface area, and slowly moving the proboscis back and forth.

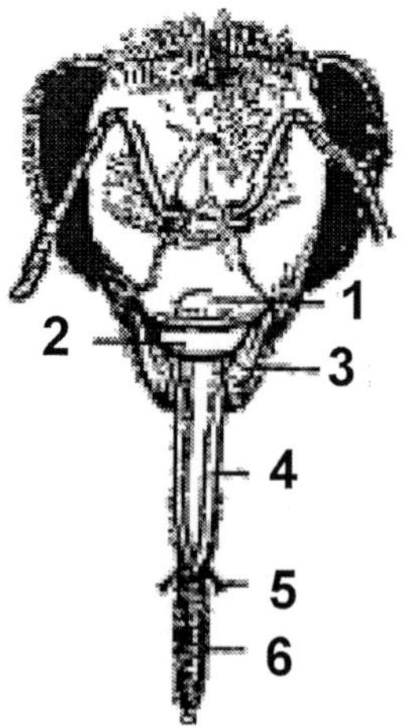

Figure 15. The mouth parts of the worker bee: 1 clypeus; 2 labrum; 3 mandible; 4 maxilla; 5 labial palpus; 6 glossa.

Internal organs

The main internal organs in the head are the brain and subesophageal ganglion, the main component of the nervous system, in addition to the ventral nerve cord that runs all the way through the thorax to the abdomen. The brain has a large area for receiving inputs from the two compound eyes, called optic lobes. The next largest input are from the antenna (antenna lobes). One important region in the middle of the brain is called the "mushroom body" because the cross section resembles two mushrooms. This area is known to be involved in olfactory learning and short term memory formation, and recently shown to be also important in long term memory formation in insects.

There are also endocrine organs attached to the nerve cords, very close to the esophagus (the food canal). One is called the corpora allata, which is the only source of one important hormone, juvenile hormone, which is involved in both the queen-worker differentiation, and also division of labor in workers. The other one is called corpora cardiaca (CC, the body near the heart), which is a neurohemal organ and stores and releases another hormone (PTTH, prothoracicotropic hormone). PTTH can stimulate the production of ecdysteroids, by a gland located in the thorax, the prothoracic gland.

Lastly there are exocrine glands inside the head also, most notably the mandibular glands, the hypopharyngeal glands and the salivary glands. Mandibular gland is a simple sac-like structure attached to each of the mandibles. In the queen this is the source of the powerful queen pheromone. In young workers the gland produces a lipid-rich white substance that is mixed with the secretion of hypopharyngeal glands to make royal jelly or worker jelly and fed to the queen or other workers. In old workers (foragers) the gland also produces heptanone, a component of the alarm pheromone. Similarly, hypopharyngeal glands produce protein-rich secretions when young (nurses), but produce invertase (an enzyme to break down sucrose into fructose and glucose) in foragers. The gland is consists of a central duct (which is coiled between the front cuticle and the brain) with thousands of tiny grape-like spheres (acini, singular: acinus). The secretion flows to the mouth through the long duct. The glands are large in size (hypertrophied) in nurse bees but become generated in foragers. There is also a pair of head salivary glands inside the head. The glands produce saliva which is mixed with wax scales to change the physical property of wax.

2. THE THORAX

The armour-plated mid-section of an insect, the thorax, supports two pairs of wings and three pairs of legs, and carries the locomotor, or "engine", and the muscles that control the movement of the head, the abdomen and the wings.

(a) The legs: Each pair of legs differs in size and shape from the other two pairs and is jointed into six segments, with a pair of claws at the tip which help the insect to cling to surfaces. The leg can be flexed at any of the six joints. Its primary function is to help the bee to walk and run, but various parts also serve special purposes other than locomotion. For example, the brushes on the inner surface of the fifth segment, (the tarsus) of the two front legs are used for sweeping pollen and other particles from the head, eyes and mouth parts. The same tarsi of the mid-legs serve as brushes for cleaning the thorax, while the spines found at the end of the fourth sections (tibiae) are used for removing the pellets of pollen and for cleaning the wings. Two important parts to note on the legs are the antenna cleaners on the front legs and the pollen baskets on the hind legs.

(i) The antenna cleaner, located on the inner margin of the tibia of the forelegs, consists of a deeply-cut semi-circular notch, equipped with a comb-like row of small spines. All three castes - drone, queen, worker -- have this cleaning apparatus.

(ii) Pollen baskets: The tibiae of the hind legs of the worker bee carry a special apparatus, called the corbiculae, or pollen baskets, which enables her to carry pollen into the hive. These pollen baskets, concave in shape, are surrounded with several long hairs which bind the contents into an almost solid mass, allowing the worker to carry the load safely home.

(b) The wings of the honeybee, like those of most insects, are thin, flat and two-layered. The front pair is much longer than the rear. The worker's wings are used both for flight and for ventilating the hive, while the drone and the queen use theirs for flight only.

Forewing- A Honey bee's forewing is typically larger than it's hindwing. Its main function is flight but it also is used as a cooling mechanism.

Hindwing- The hindwing's main function is also flight and can at times be attached to the forewings by hooks called hamuli so both pairs of wings can beat in sychrony. It is also used to fan away heat and cool the hive.

Legs- The main function of the bee's legs are for movement, however bees also use their legs to manipulate and carry pollen and propolis (a resin substance from trees). The hair on the legs help dust off pollen and other subtances. Bees also have a specialized structure for cleaning the antennae

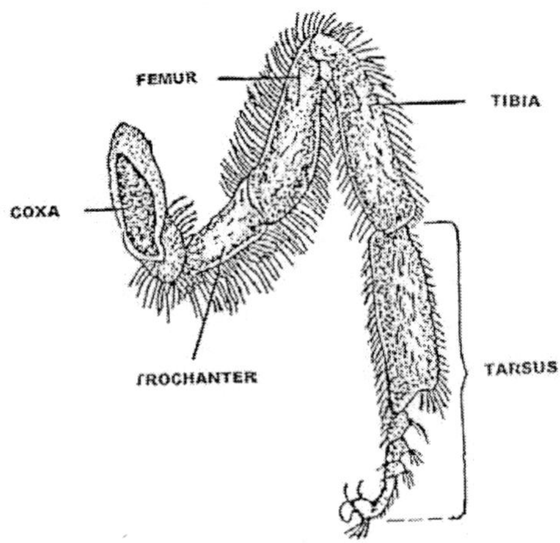

Figure 16. Outer view of middle leg of the worker honeybee.

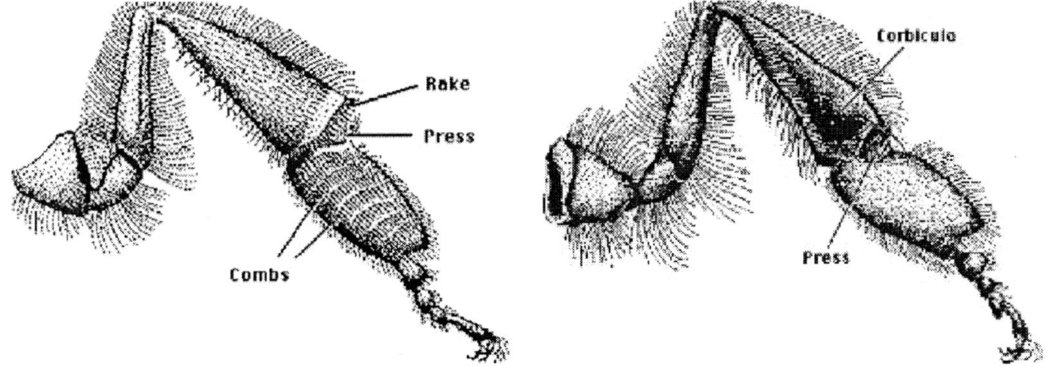

Figure 17. Hind leg of a worker bee: outer surface: **1** pollen basket; inner surface: **2** pollen press; **3** Pollen brushes.

Figure 18. Pollen carrying honey bee.

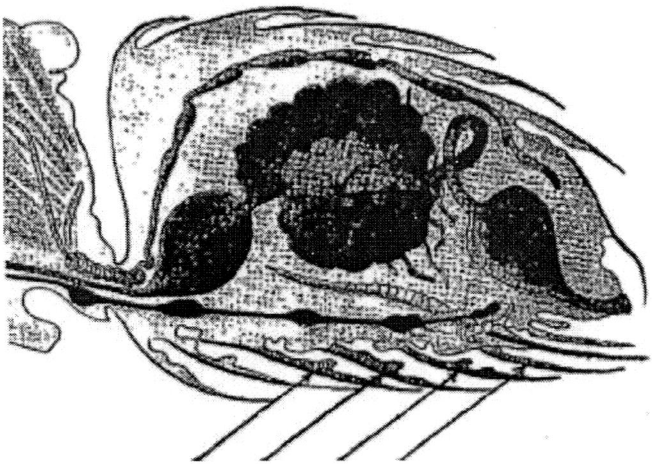

Figure 19. Wax glands (see arrows).

3. THE ABDOMEN

Like the thorax, the abdomen is armour-plated. It contains such vital parts as the heart, the honey sac, the stomach, the intestines, the reproductive organ, and the sting. As seen from the outside, only six segments can be observed, but the adult honeybee has nine, while the larva has ten. Most of the internal organs lie in abdomen along its length and collectively they form viscera. The anatomical systems involve the physiological functions of the body. These functions are; food, digestion, excretion, blood circulation, respiration nerve transmission and reproduction. In addition there are certain glands that are involved in specific functions and control muscles that aid in movement of body parts (Snodgrass & Erickson, 1992; Wigglesworth, 1984).

Digestive and excretory system

The digestive system consists of alimentary canal or the food tract (intestine). The mouth leads to pharynx which sort of sucking pump. It narrows into esophagus which passes through neck, thorax and peduncle. As it reaches the abdomen it swells into crop or the honey stomach. When expanded fully it may cover a greater part of abdomen. The honeybee occasionally fills it with nectar from the flowers or fresh water for drinking and sometimes with honey also. At the end of the honey stomach there is a narrow one-way valve which controls the onward passage of liquid food and pollen into proventriculus the (foregut). The honey stomach acts as a temporary store room for food and water most of which are regurgitated for processing in the liver; some is also passed on for digestion and personal use of the bee. The valve can not only retain food but also spores of *Nosema* protozoa, of Foulbrood bacteria and yeasts. After the proventriculus, follows the mid-intestine or the ventricular, where most of the digestion takes place. It is of thick tubular shape, having muscular layers on the outside and numerous folds on the inside, which greatly increases the digestive surface. On the inside it is lined with a thin layer of epithelial cells which multiply fast and provide the digestive fluids and enzymes. They act on carbohydrates and proteins of the pollen grain without affecting the spores of diseases causing micro-organisms that might reach there. The substances of digestion pass through various outer layers and reach the blood that circulates in the body cavity.

Some of the digested food is passed on to the narrow hind gut or the small intestine. In the anterior part there is further absorption of food and in the posterior part where it enlarges somewhat is the rectum. In rectum the waste of toxic substances are absorbed and are passed outside through the anus. During winter or prolonged bad weather when bees cannot fly out faeces are retained in the rectum for prolonged periods. At the junction of the midgut and the hind gut, protruding outward, there are about one hundred thread like malpighian tubules which float around in the blood. They are excretory in function and on absorbing waste chemical from the blood throw these into the hind gut to be excreted outside.

The surplus food is stored by honeybee in the form of fat body and it is meant for emergency use during their long flights. The fat bodies also contain glycogen, carbohydrates, hydrous iron oxide and even poisonous materials like uric acid.

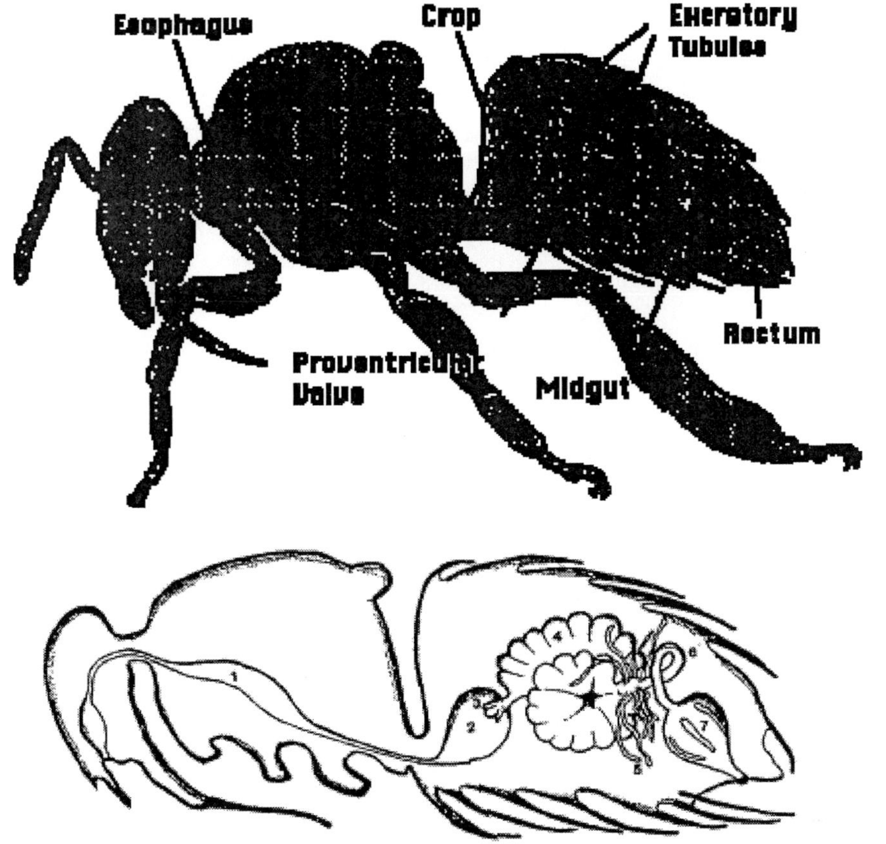

Figure 20. The digestive ansd excretory system of worker honeybee.1 oesophagus; 2 crop; 3 proventriculus; 4 ventriculus; 5 Malpighian tubules; 6 small intestine: 7 rectum.

Circulatory, respiratory and Nervous System

The body cavity of the honeybee is filled with pale yellow or amber coloured blood. The colour is due to haemocytes which resembles leucocytes of the vertebrates; the blood is not red because it does not contain hemoglobin of the blood cell of the mammals. The haemocytes are suspended in plasmic fluid through which the nutrients are transported to various tissues of body. The metabolic wastes are also carried in the blood.

Blood is circulated throughout the body by the pumping action of the heart which has five chambers arranged sequentially. It is located along the dorsal body wall of the abdomen and is carried forward to the head region though aorta. The flow of the blood is further regulated by the dorsal and ventral diaphragms in the body cavity.

In insects, the circulatory and respiratory systems are separate and the blood have only minor role in gas transport to and from the cells. The circulatory system is an open one with only a dorsal vessel and aorta to assist in blood circulation. The blood fills this body cavity so that the cells float freely in blood through vessels. The blood enters the heart through on way valve called as ostia and is pumped forward through aorta from which it is spills out into the head. Muscles attached to dorsal and ventral diaphragm are used to pump the blood through out the body and back to the heart. The main function of the circulatory system are transporting food from the midgut to the body cells removing waste material from cells and retuning it to the excretory organs lubricating body and providing defense against pathogens by means of blood cells that attack invading organisms.

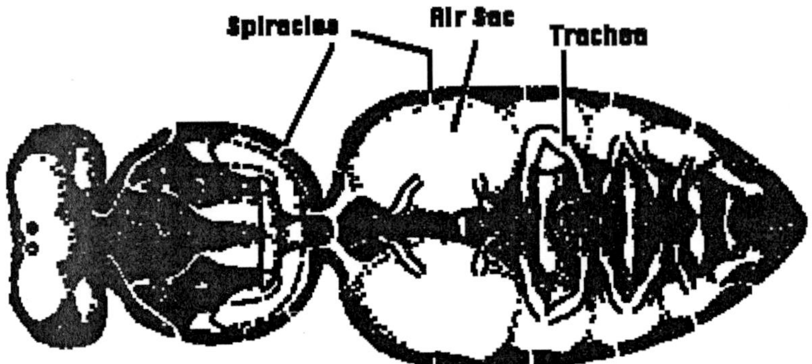

Figure 21. Respiratory system

Bees have no lungs for breathing rather they use a system of tubes which carry oxygen to and CO_2 away from the cells. The system of breathing tubes or tracheae is connected to the outside world by a series of holes in a cuticle called spiracles. When the bee is inactive gas exchange can operate simply by diffusion but during period of increased activity these pump their abdomens to increase gas exchange using expanded sacks of the tracheae as bellows (Bailey, 1954).

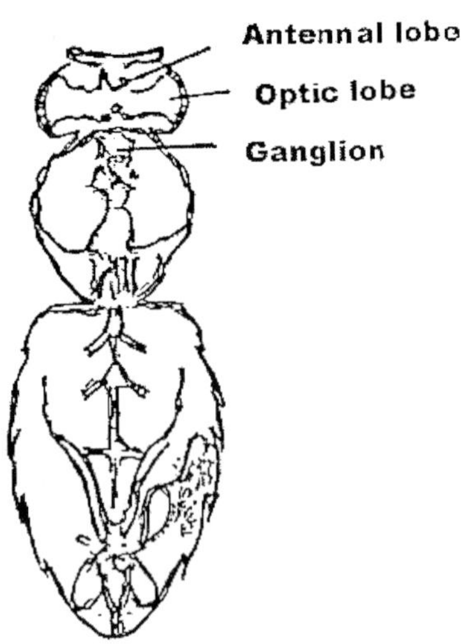

Figure 22. Nervous system of honeybee, showing the brain and the ganglia or nerve centers.

The cross structure of the nervous system is fairly simple one consisting of brain and seven ganglia are nerve centers at various junctions through out the body. Much of the nervous control exercised by bees is performed not by the brain but by these centers which provide local control over some of the bees musculature. For example, a beheaded bee can still beat its wings move its legs and sting although coordinated activity of these function is last.

The honey bee, has a very developed nerve and senory system. The bee uses the hairs on its body to determine touch or external pressure, by nerve receptors at the base of the hairs, especially the legs and antennae. The honey bee has no known olfactory organs. However, the antennae hairs allow the bee to smell or taste. The honey bee olfactory structures in the antennae are called plate organs. It has been estimated that the worker bee has 5 to 6,000 plates on the antennae flagella, 2 to 3,000 for the queen, and perhaps 30,000 on the drone. The best sensory organ the bee has is the eyes. The honey bee has 3 simple eyes on the top of the head and 2 compound eyes, one on each side of the head. Their eyes have external lenses to focus light and sub-lens, light sensitive retina connected to nerves in the brain. The simple eye has one lens for entire retina. The compound eye has many small lenses and retinas. It is supposed that an insect "sees" with a compound eye as many points of light as there are divisions of the eye, and thus gets a mosaic picture, many small parts to make a whole. There may be several thousand separate light receptive parts of the eye. It is known that insects respond quickly to movement of objects. Honeybees see color, shape, position of objects. They see most of the color spectrum. The central nervous system of the honey bee consists of a brain in the head above the pharynx, and a ventral nerve cord extending from head to end of abdomen. The brain is mostly the sensory center, receiving news from the eyes and antennae, and transmits the nervous impulses to the motor centers of the ventral nerve cord.

We know that if the head is removed from the honey bee it is deprived of these stimuli as well as the ability to eat.

Reproductive system

Drone and queens have specialized functions particularly those associated with reproduction. They do not perform such worker jobs as brood rearing, comb building and foraging and thus do not have many of the worker structures used for this tasks.

Drones

Drones have the only function of mating. The drones perform no other work for the colony consequently their work related structures are either reduced or obsolete. They have a very short proboscis, vestigial or very small cellulery gland no bags hypopharangial or nasonov glands small mandibles, no pollen collecting structures from the leg and no sting. In contrast orientation flight and mating related structures are highly developed. There compound eyes are much larger than the workers taking up much of the head and with up to 8600 facets as opposed to 6900 workers. The optical lobes of the brain are also much larger in drones and their antennae have 30,000 olfactory plate organs about 10 times the number found in workers. These antennae and eye characteristics are important in visual and olfactory orientation for finding and matting virgin females. Drones also have larger flight muscles and broader wings than workers. Both these adaptations are for mating flights. Mandible glands are small but may produce pheromone used for orientation. The drone reproduction system is designed to evert the genital apparatus into the queen during matting. The drone consists of an internal endophallus which is everted upon ejaculation and a pair of capulatory collaspers for griping the queen. Drones mate only ones and than die since much of the endophallus breaks off from the drone and is left in the queen during copulation. The sperms transport during mating are produced in the testes which reach their larger side during the pupal stage, the sperm production is largely completed by the time the adult drone is matured 12 to 13 days after emergence. The sperms are spores in the seminal vesicles until mating than they are ejaculated in a packet with mucus from the large mucus gland.

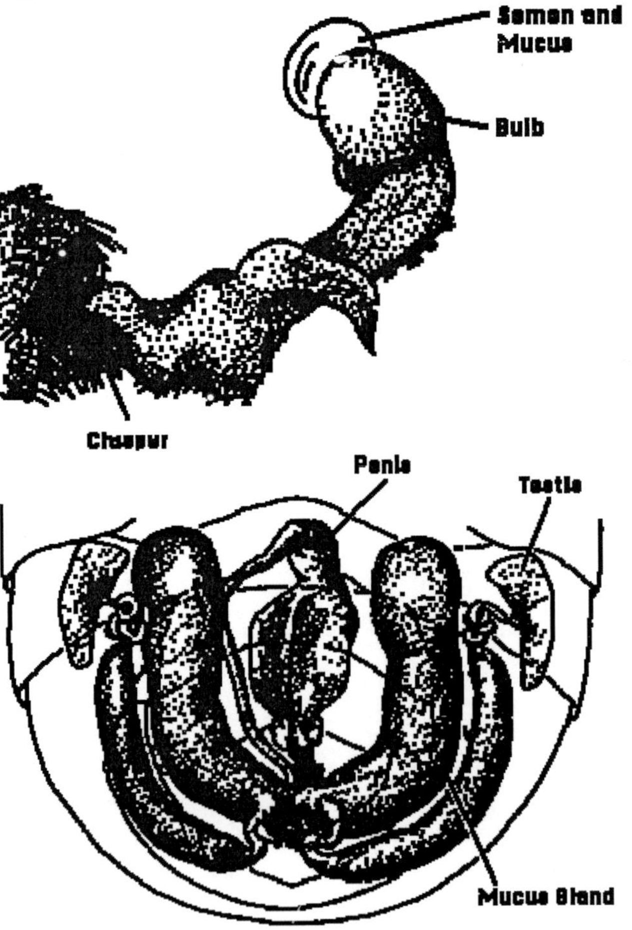

Figure 23. Reproductive system of drone (redrawn from Snodgrass, 1956)

Queens

Similarly, in queens many of the structures needed for performing different activities in worker bees are reduced. Queen lacks pollen collecting structures, proboscis is shorter and her wax, hypopharyngeal and nasonov glands are either absent or vestigial. Queens do have stings but are used in fight against rival queens but the barbs are small and ineffective, the associated plates are firmly attached so that they can be retracted after use. The queens do not die after stinging. Their poison sacs contain two to three times more times the volume of worker venom (Owen, 1978) possibly because the queen may need to sting many rivals in a short time. The queen also has some well developed glands for pheromones production, particularly the large mandibular glands and koshevnikov glands.

The queen's ovaries are much bigger than the workers because of her eggs laying function. Each of two ovaries consist of 150-189 egg producing ovarioles, whereas a worker ovary has only 2-12. These ovarioles can produce an unlimited number of eggs often as many as 1 million or more during the queen's lifetime. The eggs travel down the oviducts past the spherical spermatheca to the vagina, from where they are laid in cells. The spermatheca holds sperm from the drones which have mated with the queens early in her life, and the nutrients supplied by the spermathecal glands ensure that sperm will survive for many years.

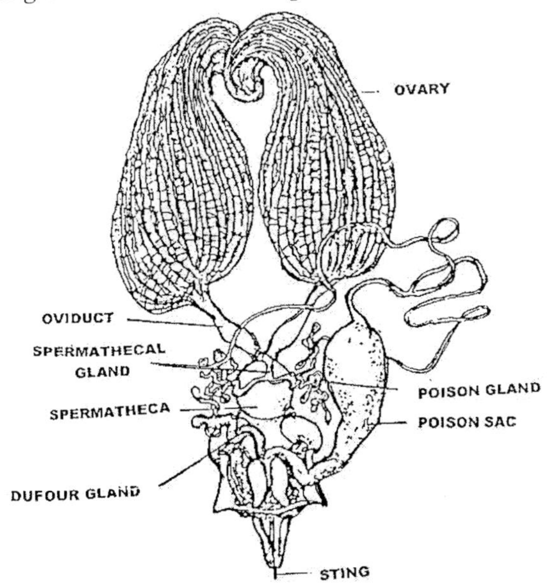

Figure 24. Reproductive system and sting of a mated queen (redrawn from Snodgrass, 1956)

When an egg passes down oviduct, it is probably pressed by a muscular flap, the valve fold, against the duct opening to the spermatheca (Ruttner, 1956). A small S-shaped pump and valve leading from the spermatheca to this duct opening allows the queen to draw a minute amount of sperm and seminal fluid into the duct, thus releasing only few sperms at a time (Harbo, 1979). This is important for a queen since she will be superceded and killed by the colony when she runs out of sperms. The spermatheca can hold up to seven million sperms (Dade, 1977) and it generally takes 2-4 year after mating before all of the sperms are used. Thus, it is the amount of sperm held in spermatheca, rather than the egg production which determines the natural life span of the queen. Workers have a rudimentary non functional spermatheca and also lack the various genital structures with which the queen can mate and accept from drones. Thus, workers can not mate or store sperms, and any worker laid egg can not be fertilized.

Figure 25. Male bees (drones) waiting for the queen to leave on the mating flight

Mating and fertilization of the queen

After several short orientation flights during the first week or so after emergence the virgin queen leaves the hive to find a drone congregation area. These places are often the same year after year, with many thousands of drones coming from different areas. The queen is able to fly distances of 10 to 12 km, sometimes crossing geographical barriers up to altitudes of 1000 m. As the queen reaches the congregation area, the "nuptial flight" begins. The drones fly slowly in a circular formation; the queen flies over the males, and slowly descends. Some tens of males are attracted by both visual cues and the queen's pheromones. Increasing their flight speed, the group of drones form a "comet-tail". Copulation occurs in flight, at a height of about 10 m and in few seconds (Fig. 25). After evagination of its endophallus in the queen's vagina and the transfer of sperm, the copulatory apparatus is torn from the body of the male, becoming the "mating sign" attached to the extremity of the queen's abdomen. The male then dies. The queen removes the "mating sign" with her legs and is then able to copulate a fresh. Over a few days, the queen may accomplish many nuptial flights; there may be no copulation at the time of the first nuptial flight. Usually, 8 to 10 copulations are needed to ensure life long fertility. Finally, the queen returns to the colony, often retaining the last "mating sign", which soon shrivels up and drops off. The sperm is stored in the queen's spermatheca, which contains a liquid for the first 20 days. After 20 days, if no spermatozoids are stored, the liquid in the spermatheca solidifies.

The drone mounts the queen, inserts his endophallus, and ejaculates his semen.

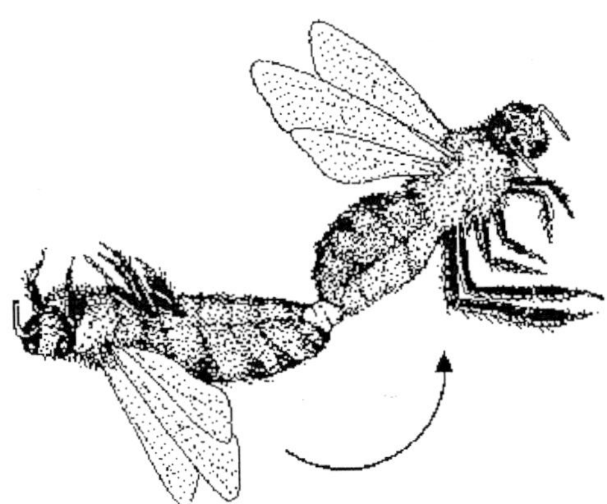

During ejaculation, the male falls back and his endophallus is ripped out of his body and remains attached to the queen.

Figure 26. Sequence of mating events in between a drone and a virgin queen (redrawn from Snodgrass, 1956)

Drones mounting later remove the previous drone's endophallus and lose their own through similar matings. The emasculated drones die very quickly with their abdomens burst in this fashion.

Of the 90 million sperm deposited by several males in the queen's oviducts, a mixture of about 7 million are stored in a special pouch in her body called the spermatheca.

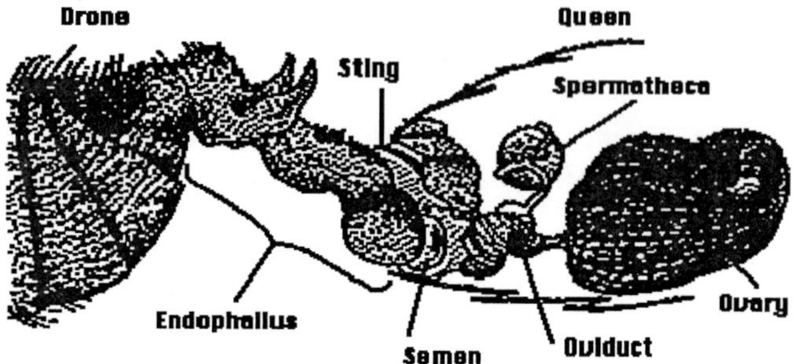

Figure 27. The position of the drones endophallus in the queens vagina during copulation. The sperm is being ejacyulated into queens oviducts as the drone flips backwards and begins to separate from the queen (redrawn from Snodgrass, 1956).

These sperm will be used, a few at a time, during the queen's life to fertilize her eggs. The one (or few) mating flight(s) in her first few days of life are her last. Should the old queen run out of sperm, a new queen will be produced through the intervention of the workers. Drone's eyes and antennae are specialized for seeing, following, and mating with a queen and most of the drones' internal organs are designed for flying fast enough to catch the queen or for delivering sperm cells into her oviduct. If the drone succeeds in this, it dies. Thus, all surviving drones are virgin.

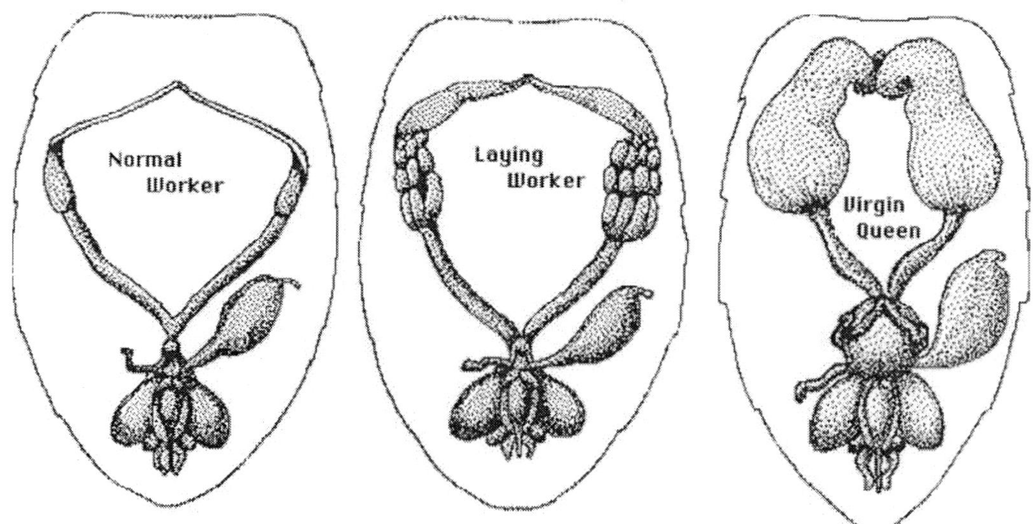

Figure 28. The diagrams included here show the various parts of a honeybee. The importance of the figures should be obvious with a little thought.

Young queen is not mated during the first 3 weeks of her life, she is unable to be fertilised. This is considered a major difficulty for apiculture in cold countries, where the climatic conditions may prevent nuptial flights within the required time. A normally mated queen stores enough spermatozoids for 2 to 3 years of continuous egg laying.

Oviposition

As the queen is mated in only one period during her whole life, she must retain the living sperm in her seminal receptacle. The duct of the seminal vesicle is supplied with a muscular sphincter and by relaxing or contracting the sphincter, the queen may or may not fertilise the eggs before laying:

(i) Fertilised eggs give rise to females (diploid individuals with 32 chromosomes).

(ii) Unfertilised eggs give rise only to males, which are haploid (n = 16), which means the multiplication of sexual cells without meiosis. It is the phenomenon of "parthenogenesis arrhenotoky". Egg laying may start approximately 2 days after mating, and proceeds regularly and continuously, except during periods of cold weather. The queen deposits one egg per cell, usually starting in the centre of the comb and continuing in an outward spiral pattern. The liberation of spermatozoids is dependent on the opening of the muscular sphincter of the spermatheca, itself controlled by a reflex triggered by the size of the aperture of the cell opening detected by the queen's first pair of legs. Fertilised eggs are deposited in the cells of workers or queens, and the non-fertilised ones in the drone cells.

Development of the brood

The brood is defined as the whole pre-imaginal period (eggs, larvae and pupae).

The eggs are whitish and elongated, measuring 1 to 1.5 x 0.5 mm. At deposition they are vertical on the cell bottom, attached by one end, then oblique, and finally become horizontal on the third day. The duration of egg incubation is three days. During the first three days after hatching, all the larvae are fed only with royal jelly (secretions of the pharyngeal glands of the nurse bees). Within three days, the weight of the larvae increases from 0.1 to 5 mg; i.e. by a factor of 50. Larvae that will become the future queens are fed ten times more often for the duration of their development, than are the larvae reared as workers. In addition, royal jelly contains secretions of the mandibular glands and more sugars, which enhances the quantity of food absorbed, by stimulating the appetite. The function of the corpora allata is also stimulated, particularly enhancing the secretion of juvenile hormone, responsible for the growth of the larva. After two and a half to three days, a worker larva is no longer capable of being reared as queen, even if it receives the same provisions as a queen. After the third day, the queen larvae still receive royal jelly exclusively, but of slightly different composition. The other larvae receive a mixture of nurse bee secretions, pollen and honey.

After the workers have covered the cells with a convex cap of wax (more convex for the male brood), the larvae cannot receive further food. Within 36 hours after cell capping, the larva spins a cocoon and defaecates. The last larval stage ends when the larva stops moving and stretches out on its back in the cell. After metamorphosis, the young adult emerges by chewing away the cell cap. The cell is then rapidly cleaned and reused.

Finally, there are two types of brood: (i) open brood = eggs and larvae; and (i) capped brood = last instar larvae, prepupae and pupae. If the regular spiral pattern of egg laying is followed, the open and capped brood is regularly distributed in concentric rings. When an irregular pattern is observed, interspersed with empty cells, it indicates an active destruction of eggs, larvae or pupae by the workers because of: (i) abnormalities of eggs or young larvae; (i) the presence of diploid males; and (iii) the removal of stored provisions when the queen needs additional cells in which to lay eggs. In addition, a pronounced irregularity of the brood pattern ("dispersed"; "in mosaics") may indicate the presence of a disease or toxic chemicals. Longevity of the individuals varies with castes (i) The queen: 3 to 5 years (ii) The workers: 4 to 6 weeks during summer time. Worker bees reared at the end of autumn can live until the following spring. (iii) The males: 1.5 to 2 months. They appear in spring, their numbers diminish during summer time and they disappear in the autumn.

Sting

The sting is similar in structure and mechanism to an egg-laying organ, known as the ovipositor, possessed by other insects. In other words, the sting is a modified ovipositor that ejects venom instead of eggs. Thus, only female bees can have a stinger. The sting is found in a chamber at the end of the abdomen, from which only the sharp-pointed shaft protrudes. It is about 1/8-inch long. When the stinger is not in use, it is retracted within the sting chamber of the abdomen. The shaft is turned up so that its base is concealed. The shaft is a hollow tube, like a hypodermic needle. The tip is barbed so that it sticks in the skin of the victim. The hollow needle actually has three sections. The top section is called the stylet and has ridges. The bottom two pieces are called lancets. When the stinger penetrates the skin, the two lancets move back and forth on the ridges of the Stylet so that the whole apparatus is driven deeper into the skin. The poison canal is enclosed within the lancets.

In front of the shaft is the bulb. The ends of the lancets within the bulb are enlarged and as they move they force the venom into the poison canal, like miniature plungers. The venom comes from two acid glands that secrete into the poison sac. During stinging, the contents of the alkaline gland are dumped directly into the poison canal where they mix with the acidic portion. When a honey bee stings a mammal, the stinger becomes embedded. In its struggle to free itself, a portion of the stinger is left behind. This damages the honey bee enough to kill her. The stinger continues to contract by reflex action, continuously pumping venom into the wound for several seconds.

The sting itself is a small, efficient, well engineered device that allows the bee to defend its own life and its nest, against predators. It is elaborated ovipositor modified as an instrument of defence, when supplied with venom, become a sting. The sting of the honeybee is contained within the large chamber in the end of abdomen enclosed between the tergal and sternal plates of segment VII. The sting chamber also contains the reduced and modified sclerites of segments VIII and IX and the apical anus bearing lobes, or proctiger, which represents the tenth segment.

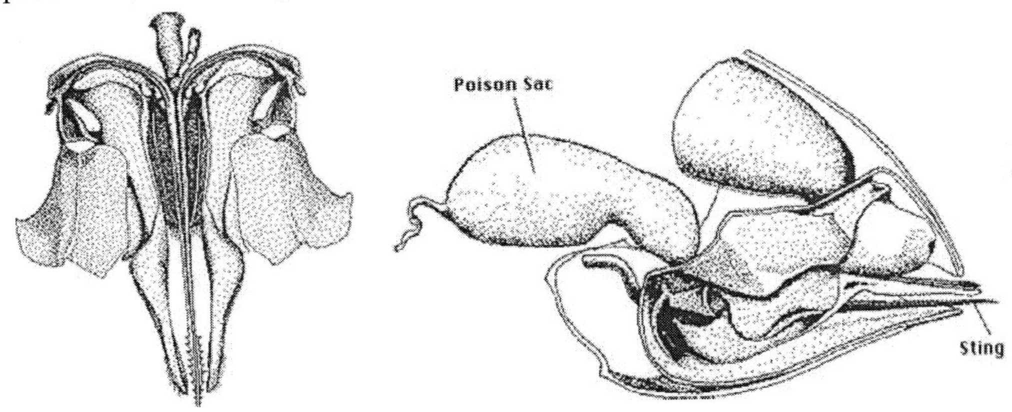

Figure 29. Side and front view of sting apparatus. Sting apparatus of the worker bee and cross section of sting shaft: 1 poison sac; 2 ramus; 3 quadrate plate; 4 oblong plate; 5 sheath, cross section of sting shaft; 6 stylet; 7 Lancets.

The action of stinging

When the sting is deployed, the bee bends its abdomen downward due to the actions of the muscles that connect the abdominal plates. The muscle set on the ventral side contracts thereby increasing the overlap between the sternite plates. On the dorsal side another set of muscles contract so that the membrane between the tergite plates is distended. This results in a bent abdomen, which in combination of the angular ventral movement of the sting shaft (in turn caused by the furcula muscle). This ensures that the sting shaft enters roughly perpendicularly into skin of the victim. Perpendicular penetration is the most efficient as venom is delivered more deeply and the path through the tough skin is shortest. The force from the bee's legs, the muscles of the abdomen and the effect of the backward raked barbs, as the lancets reciprocate alternately, all combine to produce a thrust that drives the penetration of the sting. Venom is pumped into the central canal by the reciprocating action of the lancets, each of which has an umbrella like collector and valve (within the liquid contained in the bulb) The pulses of venom are delivered through the canal and squeeze out between the rubbing faces of the two lancets. This is another feature that is of high efficiency as the venom comes into contact with a much larger area of tissue than would be the case if the sting were similar to a hypodermic needle. The venom also leaks to a small extent from the exposed portion of the sting shaft providing a pheromone alert at the earliest possible moment. This oblique drawing shows the action of the lancets sliding on the rails of the stylet. The grey outline is that of a needle tip and is to similar scale. The portion shown is approximately half a millimeter long and indicates that the sting will penetrate much more readily than a needle (and we all know how sharp needles area). Generally the bee pulls herself away, leaving the venom pouch and sting in the invader; the worker soon dies because of the abdominal rupture. Thus stinging for honeybees is an act of self-sacrifice. Unlike many other hymenopterans, honeybees are not likely to sting unless provoked.

The sting of the queen differs from that of the worker in several respects the plates of the basal apparatus are not of the same size or shape as in workers. The queen uses her sting only ob rival queens and do not suffer the consequences of losing it. The shaft of her sting is decurved beyond the base, the lancets have fewer and smaller barbs than those of worker's sting, but the poison glands are well developed and the poison sac is large.

Table 15. Different components present in poison glands of worker honeybee

Compound	% of Venom	Effects
Melittin	50%	Lysis of blood cells Release histamine Depress blood pressure
Phospholipase A	12%	Cell lysis Pain Toxicity
Hyaluronidase	& lt 3%	Spreading factor Hydrolyzes connective tissue
Acid phosphatase	& lt 1%	Induces allergic reactions
Histamine	& lt 1%	Itching Pain

Table 16. Some structural differences between worker, queen and drone

Characteristic	Worker	Queen	Drone
Sensory			
No. of facets of compound eyes	4000-6900	3000-4000	7000-8600
Optic lobes of brain	Medium	Small	Large
No. antennal plate organs	3000	1600	30,000
Relative ratio of antennal surface	2	1	3
Glandular			
Hypopharyngeal	Present	Vestigial	Absent

Mandibular	Large	Very large	Small
Head salivary (labial)	Large	large	Vestigial
Thoracic salivary (labial)	Large	Large	Small
Wax	Present	Absent	Absent
Nasonov	Present	Absent	Absent
Alkaline (dufour)	Reduced	Large	Absent
Koshevnikov	Reduced or absent	Present	Absent
Reproductive and sting			
Ovaries and testes	Reduced ovaries	Enlarged ovaries	Testes
No. ovarioles	2-12	150-180	None
Spermatheca	Rudimentary	Large	None
Sting barbs	Strong	Minute	No sting
Mouth parts			
Mandibles	Slender	Robust	Small
Mandibular groove	Present	Absent	Absent
Proboscis	Long	Short	Short
Leg and wing			
Pollen press and combs	Present	Absent	Absent
Pollen basket	Present	Absent	Absent
Wing sensilla	Medium	Fewest	most

Source: Data from Ribbands, 1953, Snodgrass, 1956; Michener, 1974, Dade, 1977.

Chapter 8

Beginning beekeeping and establishment of apiary

The art of beekeeping in easily maneuverable hives and extracting surplus stores of honey is called beekeeping. Honey is produced by the worker bees of a colony from the sweet and juicy nectar of wild and cultivated flowers. To know how the honeybees collect nectar from flowers in different localities and seasons is very exciting experience. This knowledge makes a person a good beekeeper. An insight into the remarkable social behaviour of the bees is necessary to get an idea of their proverbial industry, their tireless life and marvellous work they perform. Beekeeping has many attractions among them are the production of honey, royal jelly, propolis, bee pollen and beeswax that are highly valuable and have high market prices. But most importantly honey serves as a source of food with high nutrition value. In communities, where beekeeping is done for commercial purposes, it has led to self reliance through the innovation of local industries associated with the production of beekeeping equipment and bee products. Beekeeping provides an excellent opportunity for studying bee-pasture plants and the relationship of honey bees to flowers when bees are collecting pollen and nectar and when they are acting as pollinators. Beekeeping offers many income-producing opportunities, either as a hobby, as a part-time occupation, or as a fulltime commercial enterprise.

Importance of Beekeeping

Honeybees produce honey and other hive products (e.g. pollen, propolis, beeswax, royal jelly) for which there are unsatisfied local and export markets. Honeybees pollinate flowers, both wild and cultivated, and are crucially important to the environment and commercial crop production. The role the honeybee plays in ensuring the diversity of flora and fauna cannot be over emphasized. Honeybees play a more important role than mere honey producers. They assist in the production of a wide variety of commodities, such as apples, melons, cucumbers, almonds, flax, sunflowers, and clover. These and other fruits, vegetables, tree nuts, and field, seed, and forage crops require, or benefit directly from, bee pollination. For agriculture, the value of honeybees as pollinators far exceeds the value of the honey and beeswax they produce. In addition, honeybees are important to the production of plants that provide food and shelter for wildlife, control soil erosion, and beautify the environment. Since the honeybees forage on wide variety of resources, they may be used to monitor pollution of and/or, changes in the environment. In the future they may also be used to distribute fungicides to crops. Beekeeping supports ancillary industries, such as the manufacturers and suppliers of hives and equipment and publishers of beekeeping books and magazines as well as the retailers of products. They keep bees on a part-time basis and come from a wide variety of backgrounds crossing all areas of the social, religious, and political and gender divide. On average they keep five hives, but this varies according to circumstances, from one to over fifty per beekeeper.

Beekeeping Methods

Broadly speaking, beekeeping methods can be divided into two categories; mainly traditional and modern methods. The former is mainly practiced in societies in which beekeeping is done in small scale and in most cases done as a hobby. Another distinction between these two groups is on the equipment used.

Beginning beekeeping

A master beekeeper who worked with honey bees for more than 50 years often said, "Every time I look into a beehive, I learn something new about the bees, and I see another reason why I like the bees so much." Actual experience is the best way to learn about bees. An experienced beekeeper is your best source of information about honey bees. In fact, it is almost impossible to start working with honey bees without the advice of a helpful beekeeper. Plan to watch and study a beekeeper taking care of his or her hives. The more you see, the more you will understand. Ask questions. There is a lot to learn, but you have time to learn it all. Beekeeping can be a lifetime vocation and hobby. You may find it helpful to purchase some books and keep notes about what you are learning. You can also use your books to write questions that you think of so you will remember to ask them the next time you are working with the beekeeper. You need a basic understanding of bees and their activities before you actually take up the profession.

A basic part of beekeeping is understanding and accepting the fact that you are going to be stung from time to time. No matter how good a beekeeper you become, occasionally you will accidentally crush a bee. You may visit the hives when the bees are disturbed by a change in the weather, by hunger, or by something else beyond your control. As a result, you may be stung.

A few people have serious reactions to bee stings. They may have difficulty breathing after being stung or have some other very dangerous reaction. If you are ever with a person who is severely allergic to a sting, take them to a doctor immediately. It is very unusual for a person to have such a bad reaction to stings. For most people, the sting of the bee is a momentary discomfort that slows down. Be more careful when handling colonies if you are highly allergic to stings.

The only worker bee stings. When she pushes her stinger into the skin, it catches and pulls out of her body as she flies quickly away, causing her to die soon after. What she leaves in the skin is the barbed stinger attached to a poison sac. Often part of the bees intestine is still attached to the stinger in skin. You can scrape the stinger off the skin using a fingernail or hive tool. Then puff smoke from a smoker or rub dirt on the area of the sting. This covers the smell of the sting so other bees won't be disturbed. Although an ice treatment may reduce the swelling, there is really not much to do for it, except to get stung again. It seems that the more a beekeeper is stung, the less of a swelling reaction will result. So, there is some good in being stung; it will not be so bad when you are stung again.

Beekeeping is an integral part of agriculture because it is highly important in the pollination of vegetable, fruit, nut, and forage seed crops. Beekeeping has been an important part of human society from the prehistoric times when early settlers and farmers kept bees in primitive hives or cut down bee trees to get honey for home use and for sale. Honey bees have been selected and managed by man for many centuries. Nevertheless, they are still wild insects capable of living on their own without any assistance or special equipment. Probably because of this "resistance" to becoming domesticated, honey bees respond to beekeeping management practices in much the same way wherever they are kept. Management must always be adapted to fit local conditions and to take into account the somewhat variable behavior of different geographical strains, or races, of honey bees. Differences in climate and sources of nectar and pollen, even over short distances, will force you to adjust your beekeeping methods, especially their timing, to the area in which you live.

Beekeeping is continually changing, reflecting changes in cropping practices and agricultural land use. Commercial honey production now requires more extensive operations than in earlier days because nectar sources are more widely scattered. Fruit and vegetable growers, pressed by increasing production costs, are becoming more aware that the quality and quantity of insect-pollinated crops can be improved by renting bees for pollination. Providing bees for pollination is hard work, but it reduces the commercial beekeeper's dependence upon honey as the main source of income. After a period of decline, interest in keeping bees in urban areas has been renewed.

Learning to handle and manage bees is fun. Fortunately, bees will usually prosper if you make sure they always have enough hive space and enough food. By joining local beekeeping organizations, you can share your experiences with others and increase your pleasure from keeping bees.

For the beginner

It is best to start with four or five colonies. They will provide him with enough of honey for home consumption. Bees can also be kept on a large scale so as to make them sole source of income but its better in the begining to start on a small scale.

To be a successful beekeeper, one should acquire full knowledge of bee behaviour, live operations and acquaintance with nectar and pollen sources. One can learn handling of bees by seeing some one working with the bees. Sincere efforts and promptness in meeting the needs of bee colonies is the first and essential requirement in this profession. Bees face many hazards of diseases, enemies and pesticidal poisoning and therefore, beekeeper should be alert to safe guard bees against these problems

How to start beekeeping

Honey bees are truly wonderful creatures and keeping them can be a thoroughly rewarding and enjoyable hobby. Getting started though is not always easy. To ease your way and to make your introduction to the world of the honey bee as enjoyable as possible, consider taking the following steps:

(1) Gain Knowledge: Before becoming involved with live bees, often the best advice is to acquire sufficient knowledge of bee behaviour and management from literature such as bulletins, books and magazines available about beekeeping. Be prepared to read about beekeeping, the more reading you accomplish on the subject, the better informed you will be. There exist numbers of books on beekeeping, which can be found in many libraries. Read and reread books thoroughly. In selecting your books, realize that beekeeping is generally the same worldwide. There are some national and regional differences, however, in both equipment and methodology. Beginners will do well to stay with books written for their own area.

(2) Join an Association: Many regions have beekeeping associations or clubs, joining one of these can be a valuable resource. It may even be advised to join some time before you actually intend to keep bees. Join your local beekeepers' association. These groups welcome and encourage beginners. Beekeeping is much more difficult to learn in isolation, and beekeeper associations are a prime source of information about books and other publications, classes and workshops, and beekeeping events in the area. Many associations hold annual bee schools for beginners. There are a few appoaches to starting beekeeping, but the primary thing is to get as much information as you can before you start. Although keeping bees is mostly a routine once you know what you're doing, you need to know how to handle the bees to not annoy them too much, how to keep an eye out for problems in the hive, how to control swarming and how to keep the population of parasites like the varroa mite down.

(3) **Practical Experience**: The best way to gain knowledge on beekeeping is through practical experience. If possible locate a beekeeper in your area that may be willing to guide you along and gain sufficient experience by working with a successful beekeeper. Whenever possible ask lots of questions. Try to learn while working with him.

(4) **Attend a local beekeeping course**. The beekeeping training programmes are regularly offered by the state agricultural universities, beekeeping

organizations, departments of agriculture and Khadi Village Board Industries Commission. For beginners it is advisable to attend such beekeeping courses of at least one week to acquaint them, learn and use basic methods of their proper handling.

(5) **Start with new equipment**: Start with new equipment of standard design and dimensions. Used equipment or homemade equipment both have the potential to bring problems that the novice may not be equipped to recognize or handle.

(6) **Start with few colonies**: Start with few colonies and then increase their number with experience. A beginning beekeeper should start with at least two colonies but not more than four or five. With more than one colony, one is able to exchange brood, bees, and combs in case one of the colonies needs some help. With too many colonies one has to remain busy to keep them supplied with supers, and not be able to enjoy learning the details of the activities in any of them. The beginner can purchase established bee colonies from local beekeepers and organizations involved in beekeeping. Buy established colonies only after they have been certified as free of disease by the state Department of Agriculture or beekeeping organization, university etc. If full-sized colonies are purchased, one will lose the opportunity to watch the fascinating early development of the colony. Furthermore, the newly established small colonies are easier for the beginner to observe and manipulate than are the larger ones.

When to start beekeeping

Beekeeping can be started at any time of the year but it is always advisable to start beekeeping during spring when a good income of nectar and pollen is available. Though regional, climatic and floristic conditions vary from place to place so it is difficult to specify exact time for each location, yet the thumb rule is start beekeeping when enough blossoms are available for supply of food requirements for brood rearing. In plains colony may be started from September, October onwards, when sufficient flora for the development of bee colonies is available which becomes abundant during winter months. The colonies suffer greatly during summer. Therefore, a beekeeper should be well conversant of the needs of the colony and manage them accordingly. It must be borne in mind that the first year is a learning time for the beekeeper and a building time for the bees.

Selection of site for apiary

Both beginners and established beekeepers should select each apiary site carefully. Throughout the foraging season there must be nectar and pollen sources within a short distance of the hives. Pollen is essential for brood rearing, and nectar (honey) is the bee's basic source of energy. Large concentrations of floral sources are needed to produce a honey crop. Bees also need a source of fresh water so they can regulate the temperature of the hive, liquefy crystallized honey, and raise brood. If a water supply is not available within 1/4 mile of the hives, provide a tank or pan of water with a floating board or crushed rock for the bees to land on.

Bees are less irritable and easier to handle when located in the open where they can get plenty of sunshine. Shade from trees retards the flight of workers and hinders finding the queen and seeing eggs within the cells. A southern or easterly exposure gives colonies maximum sunshine throughout the day. The apiary is best situated near natural wind protection such as hills, buildings, or evergreens (Figure 30). Other requirements are dry ground and good air drainage. Avoid windy exposed hilltops or sites near the bank of a river likely to overflow and drown the bees. The apiary should not be located in woods or encourages development of such bee diseases as nosema and EFB in a damp bottom land, since excess moisture retards the flight of the bees.

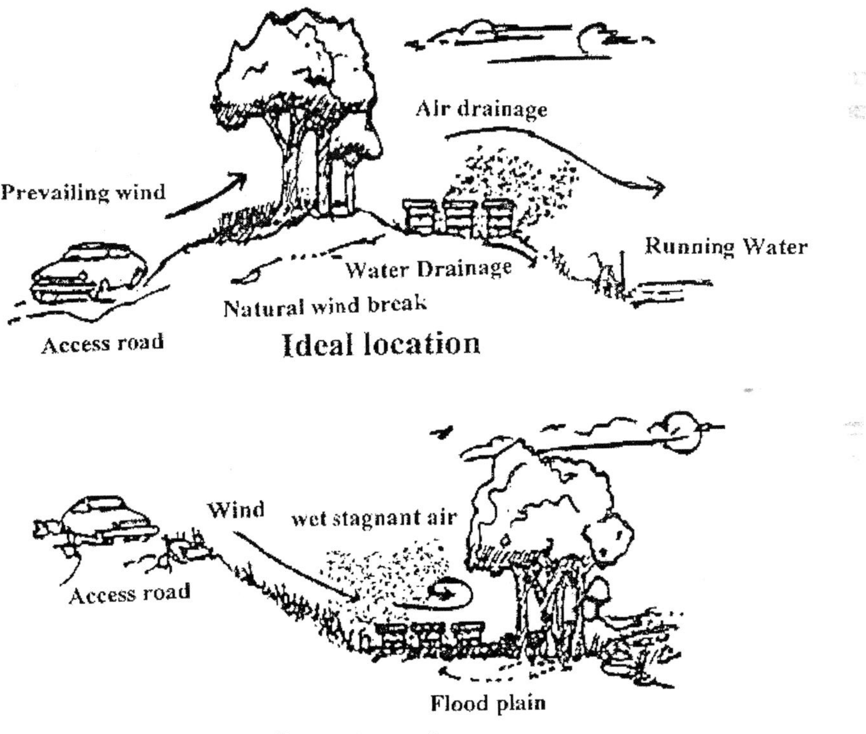

Figure 30. An ideal apiary location with a natural windbreak and good air drainage.

The location's accessibility is important, since many trips are made to the apiary each year in all kinds of weather. Avoid carrying equipment and heavy supers of honey to and from the apiary. Hives should be secluded from traffic, constant noise, and disturbance from animals and children. To discourage vandalism, place your colonies near a dwelling or area frequently visited yet screened from view if possible. Relative safety from repeated pesticide applications to colonies or their forage is also important. Acquaint yourself with the pesticides commonly used in the area. When practicable, place colonies away from fields that are treated routinely. When selecting sites for out yards, make inquiries to determine how many other beekeepers are operating in the area. If an area becomes overstocked with bees, there would be poor honey yield crop for everyone.

If possible, have your out yards near home. People tend to neglect out-apiaries that are located too far from headquarters. Soaring energy costs and efficient use of time should be included in each apiary site decision.

The economic venture of beekeeping starts with the establishment of apiary at the right place. Though honeybees can be kept anywhere but every location may not provide ideal conditions for setting up the apiary. The first and foremost thing to consider is the locality for establishment of apiary. It may be difficult to get an apiary site that will meet all the features required for ideal location, but there must be bee flora almost round the year. There are several things to look for as you attempt to locate the ideal apiary site as ginned below:

Floral resources

The bees depend upon nectar and pollen for their sustenance. The locality where apiary is to be established should have availability of bee flora round the year. Many flowers, fruits, vegetables, trees, bushes, and weeds provide abundant sources of pollen and nectar and bees will forage from areas in excess of 3 square miles around the hive to obtain them. There must be wild or cultivated plants which provide nectar and pollen and sufficient enough to produce surplus honey so that beekeeper can harvest good amount of honey and build up his colonies. One of the most important factor influencing the build up of colonies is the distance from food source. If the food source is nearer the bees will have to fly less, expend less energy and gather more food sources in a unit time, resulting in more brood production and higher honey yields.

Water sources

Honeybees require water for various activities in the hive which include: a) cooling of the colony during summer in tropical and subtropical areas, b) to dilute nectar of high concentrations to feed to larvae. Since the bees can not store water in the hive and they collect as and when required. Therefore, there is a permanent need of clean, hygienic and regular supply of fresh water. Avoid fresh running streams as the bees may be washed away while collecting water. If no natural source of water is available, then beekeeper should make arrangements for water near the apiary. A pan of water filled with gravel to keep the bees from drowning works well.

Set hives away from residential areas

Before setting up of apiary, talk with the neighbors if they are willing to allow the establishment of the same in their location. Getting answers to questions before you set up your hives can save a lot of headaches. Do the neighbors have young children who might inadvertently approach the hive? Are any of your neighbors allergic to insect stings? Are your bees likely to be attracted to your neighbors' properties for any reason in particular? Are there any penned or chained animals nearby? If answers to these questions don't make bee-keeping in town look enticing, then consider places in the countryside - in wastelands, pastures, farms, and orchards. Many people would love to have bees available to pollinate their crops; their fields could make an ideal site to place your apiary. Since apiary also requires regular monitoring, watch and ward and protection against theft and vandalism, it is better to have an apiary 200-300 meters away from residential areas and animal sheds. The swapping of tails by animals can cause bees to attack. The hives should be kept in an enclosure protected from grazing animals which might topple the boxes over. In some areas where attack of bear is more prominent hives are placed inside the electric fences.

Distance from main road

The apiary should be easily accessible in all the weathers and vehicles should reach near the apiary so that shifting of apiary during dearth periods, selling of honey and carrying of equipments is possible as and when desired. But it should not be located near the main roads with heavy vehicular traffic. The apiary site should not be located immediately near to the road side, as the higher vehicular traffic can create problems in smooth working of the bees and many bees may be killed due to striking against moving vehicles. Furthermore, exhaust and vibration emanating from vehicles is likely to disturb the bees and their developing brood. During nights the head lights of the vehicles attract the honeybees which may get disturbed and killed.

Distance from industrial areas and high tension wires

The apiary should not be located in areas with high tension electric wires, as their vibrations cause disturbance to bees. The sparking or occasional breaking of wires may cause hives to catch fire. Similarly, selection of apiary site near industries which pollute air should be avoided as the contaminants from these industries are likely to be carried to hive along with nectar pollen water and other resources collected by the bees, thereby killing the brood as well.

Distance from another apiary

The honeybees from different apiaries compete for food particularly during dearth periods or when floral resources are meager. The bees are likely to expand their foraging range beyond three kms when food resources are sparse or insufficient. Further more, the bees are likely to get attack of diseases if nearby apiaries are infected/infested. However, apiaries can be closer if food sources are abundant and apiary size is small.

Topography of the area

Establish apiary away from wet bottomland and stagnant air. It should be on upper side of slopes to improve air drainage away from hives. A flat surface with a general slope for proper drainage of water will allow easy and safe movement to beekeeper particularly when examining colonies and carrying apiary equipment during day to day manipulations. Marshy and low lying areas prone to flood and poor air circulation may be avoided. Sloppy terrains and peak of hills in hilly areas are also not good for establishment of apiary as the hives may be blown away with winds and or damaged due to land slides in rains. There must be open space to put the bee boxes in open sun during winter in subtropical areas, and shade to place the boxes under shade during summer to protect from heat.Therefore, the apiary should have open as well as shady area. A good mixture of sunlight and shade to protect the colonies from the intense heat of the late afternoon summer sun is very helpful. Colonies that receive either too much or too little sunlight produce less than those with a nice mixture. When setting out hives, avoid low spots where moisture and cold can accumulate.

Additionally, the chosen location should have a wind break to protect the bees from the cold prevailing winds of winter. Wind breaks also can shelter apiary from view. It is not necessary to inform the general public about the presence of your apiary. Even though many people are afraid to approach bee hives, yet many others are not. Unfortunately, theft of a whole group of hives and massive vandalism are not unheard of among beekeepers. Situate hives in hidden-away locations. The shrubs and bushes act as natural wind breaks during extreme winter and summer. The apiary site should be surrounded by hedges and bushes.

In areas with thick and dense vegetation flight of bees is hindered and there is reduction in bright sun light which affect the normal activities of bees. The thick vegetation also provides nesting sites for bee eater birds and predatory wasps. Moreover, the swarms that issue from hives in such locations may rest high up in the trees and thus would be difficult to catch. The dry grasses and vegetation can catch fire easily which may destroy the whole apiary, so the vegetation within the area of apiary should be cleaned.

Avoid areas prone to frequent pesticidal applications

Diseases and pests are not the only things that a beekeeper needs to worry about. Insecticides, if improperly applied (broadcast while bees are foraging, or allowed to run off into the bees' water supply, for instance) can lead to the death of forgers or an entire colony. Some insecticides leave residues that can be active for days after application and can either contact the bees directly or via mixing with nectar. Other insecticides are mistaken by bees for pollen, are gathered up, and delivered to the hive where they are fed to the brood. The end result is an insecticide kill. If the apiary is located in an area where widespread use of insecticides can be expected (fields, gardens, and orchards), check with those likely to apply them. High toxicity insecticides should not be sprayed on bee-pollinated plants during bloom. If this is to occur, be prepared to protect your bees by screening their entrance on the day of insecticide application. Provide additional protection from the sun so that the bees don't overheat and die from prostration. Beekeepers have certain rights under the law that will help protect both bees and beekeepers. Nonetheless, it is much wiser to be proactive than reactive when it comes to insecticides.

The arrangement of the apiary

The arrangement of the apiary is important to help meet the needs of the bees and to help make the beekeeper's work easier. In most areas of the tropics it is necessary to put the hives on stands to protect the bees from ants and toads. The stands should be at least 45 cm above the ground. The legs can be placed in cans, clay pots that contain used motor oil, or bands of grease can be placed around them to keep ants from the hives. Check periodically to see that the oil is not washed out of the cans by rainwater or that the grease is not covered with dust. Fresh ashes spread around the legs also keep ants away, but they must be replaced continually. Hives can also be elevated by placing them atop bricks or wooden beams treated with preservatives to avoid termites and prevent rotting. When setting out your hives tilt them so that rain won't come in through the entrance. Provide a ventilation hole near the top so that moisture doesn't accumulate within the hive. Damp hives leave colonies disease prone. Orient the hives so that their entrances face southeast or south, or any direction in between. Avoid placing a number of hives in a row as returning bees often become confused and drift back to the wrong hive. Drifting can be responsible for spreading diseases in an apiary, and will build up colony populations in the end hives at the expense of those more centrally located. Stagger the arrangement of your hives, or place them in a semicircle. Alternatively, place the hives among objects such as trees and bushes that can serve as landmarks. Some beekeepers that are forced to arrange their hives in rows paint the hives using varying light shades, or mark the entrances with different patterns, to help the bees distinguish one hive from another. Cut grass around entrances.

Cleanliness near hives

Keeping weeds cut in the apiary also reduces the ant problem. Tall weeds can provide bridges to the hives for the ants. A clean apiary also makes it easier to work around the hives. Protruding rocks and roots can cause the beekeeper to stumble or fall while working the hives. If there is no source of water for the bees within one kilometer, a container of water with floating sticks or protruding stones can be placed in the apiary. The sticks and stones are there to prevent the bees from drowning. The waterer should have a cover to prevent feces of flying bees from falling into the water. This helps to control nosema, a protozoan disease of bees which can be transmitted by contaminated water.

Density of bee hives in an area

Density of bee hives in an area is very important for beekeeping. Bees generally forage economically within a radius of 1 or 2 km. overstocking of an area with large number of colonies would result in reduced honey yields.

Generally, the honeybee colonies or apiary should be located at a distance of one km from another one for good economic returns, if sufficient bee flora is available. It is very difficult to assess the number of bee colonies in an apiary. Not more than 50 colonies must be placed in an apiary, otherwise problem of drifting and robbing will take place.

Arrangement of hives

Arrangement of hives in an apiary is the most important factor determining the success of beekeeping. Keep the hives in shade but too much of the shade is also not good for bees. The sunshine of early forenoon is most useful for bees. Only the bees be protected from severe heat of midday sun. If natural trees are not available some kind of shade boards be used for this purpose.

In the north we have cold winds of Himalayas. The beekeeper must provide protection to the bees against such prevailing cold winds. The wind should not pass directly to the hive through the entrance. For this purpose, hedges or a building in front of the apiary are useful. If no such facility is available, a fence of plank with one or two inches space left between each plank is erected in front of the apiary. The space between planks is essential because the fence with open spaces left, serve, to let the wind filter and break its force.

If the apiary is to be located in the town, the entrance of hives should be away from the nearby path, street or road so that bees in their flight are not hindered by passerby. Animals switching their tails are likely to be stung.

Traditionally, hives in apiaries have always been arranged in straight rows. It is much better to place the hives in some irregular pattern so that field bees are more likely to return to their own colonies. With hives in a straight row, foragers drift to the end hives and increase their populations at the expense of the colonies in the center of the rows. For convenience, the hives can be arranged in pairs about 6 inches apart. Pairs of hives can be separated from one another by several feet. A semicircle or U-shaped arrangement reduces drifting and makes it easy to handle the colonies with a hive loader.

The hives can be kept in rows of one hive each with a distance of 3 or 4 feet between each hive and 12 feet between each row or in groups of three or four hives each with the same distance having their entrances in different directions. The hives can also be kept in racks, but all this depends upon space available. One best decide the placement and adopt the method which suits best.

The hives should not be crowed together. Bees very well remember the situation of their own hive, they with surprising accuracy return from long distances to their correct hive, but when hives are placed quite close, their remains the danger of drifting. They may enter the wrong hives. For bees, it is not so important but for the queen bee sometimes it proves very dangerous, when she goes out for nuptial flight, she is destroyed if enters wrong hive. Furthermore, drifting can contribute to disease transmission. Do not keep the hives directly on the ground always use some kind of stands for this purpose.

Hives should be at least 45 cm apart and slightly tilted toward the entrance to aid the colony in removing residue that falls to the bottom. This also allows rainwater to run out. The placement of hives should allow the beekeeper to approach the colony and work it from behind. This is less disturbing to the colony since it does not interfere with the flight path of the foragers. It also allows the beekeeper a chance to smoke the colony before the guard bees at the entrance are alerted. All the grasses and weeds around the hives should be destroyed so that they may not prove an obstacle in the flight of bees. It is much better if a space of one meter radius all round the hive be always kept clean from all sorts of weeds and grasses. if different colours are used in front portions of hives, it also helps to recognize their own hive easily. The plot in which the apiary is placed should be well fenced to prevent the cattle and children from interfering the hives.

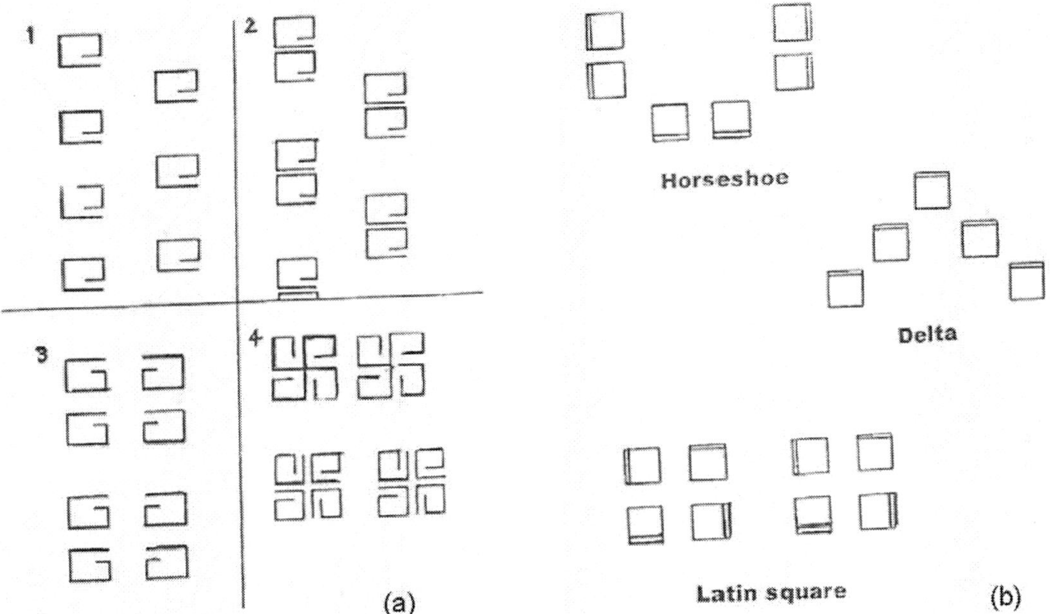

Figure 31 (a). Hive arrangement in an apiary 1- hives in rows of one each; 2 -hives in rows of 2 each having entrances towards the same side; 3- hives in rows of 2 each with having entrances in opposite directions; 4- hives in the group of 4 especially entrances in opposite directions **(b)** possible hive layouts within an apiary

Getting Bees

Sources of bees will vary depending on the region. In areas where beekeeping is practiced, the most practical way to obtain bees is to buy an established colony. If the colony is in a fixed-comb hive, it can be transferred into a top-bar or frame hive. The source of bees should be as local as possible and readily accessible to the farmer who wants to keep bees. Importing bees is not recommended. Not only is there the risk of introducing new diseases, pests, or undesirable bee strains into an area, but also the beekeepers involved can become dependent on that source.

Chapter 9

Beekeeping equipments

Introduction

Honey has a long and distinguished history in the human diet. For thousands of years honey hunters have plundered the hives of wild bees for their precious honey and beeswax – a practice still common today. Though honey and honey bees are known to the mankind since the prehistoric times, but the collection of honey was done by smoking away the bees and squeezing their combs till the end of the eighteenth century. The primitive hives like pot hives, logs of wood, pipes, baskets, clay pots and clay cavities fixed in walls of mud houses protect the bees from cold, wind or extremes of heat. The hive developed on the principle of bee space by L.L. Langstroth of USA revolutionized the whole concept of bee keeping from 1851 onwards. The salient points relating to bee space in moveable frame hive is that bees will propolise a space less than 6mm and will build brace or burr comb in a space greater than 9.525 mm. Bee space given for *Apis mellifera* (9.525 mm) by Langstroth has been modified depending upon the race of the Indian bee. Langstroth contribution to the industry was quickly followed by the invention of comb foundation mill by Johannes Mehring 1857, honey extractor by Frang von Hruschka in 1885 and many others. Beginning was made to keep bees in moveable frame hives in 1880's in India which occurred after the widespread adoption of Langstroth hive in the West. In South India, bee keeping with moveable frame hives started in the later part of the nineteenth century (1881-84). It was Rev. Father Newton who designed a small hive with movable frames for *A. cerana*. For nearly 50 years (1880-1930), a hive designed by Father Newton for *A. cerana* was commonly used all over India. However, the size of the worker bees of *A. cerana* and colony population are much smaller in Southern India as compared to those found in northern plains and higher altitudes. So the Newton bee hive proved unsuitable for *A. cerana* in northern states like Himachal Pradesh, Uttar Pradesh and Jammu and Kashmir.

The most widely used honey bees are the European *Apis mellifera,* which have now been introduced worldwide. Tropical Africa has a native *Apis mellifera,* which is slightly smaller than the European *Apis mellifera,* and is more likely to fly off the comb and to sting. They are also more likely to abandon their hives if disturbed, and in some areas the colonies migrate seasonally. In Asia there are three main native tropical species, *Apis cerana, Apis dorsata, and Apis florea; cerana* is the only species that can be managed in hives, but the single combs of the other two are collected by honey hunters.

There are three different kinds of bees in every colony: a queen, the drones, and the workers. The queen's job is to lay eggs, as many as several hundred in a day. These larva develop into drones, workers, or new queens, depending on how the workers treat them. Drones are the only male bees in the hive, and their main function is to mate with a virgin queen outside the hive. They die after mating. They have no sting, do not carry pollen, are unable to produce wax, and when resources are scarce they can be driven out of the hive to die. The all-female worker bees make up about 98 per cent of the colony, and they do almost all the work. They bring water, pollen, nectar, and propolis (bee glue) back to the hive, while some remain to guard the hive, and some clean it, build the wax comb, nurse the young, and control the temperature of the hive. Workers eat honey to produce heat in cold weather and fan their wings to keep the hive cool in hot weather. Their legs are specially equipped with pollen baskets, and they have glands that produce wax on their abdomens. The worker has a sting, but usually dies after stinging anything. A honey bee nest consists of a series of parallel beeswax combs. Each comb contains rows of wax with hexagonal compartments containing honey stores, pollen, or developing bee larvae (brood). To thrive and produce honey the bees need adequate supplies of nectar, pollen, and water. The combs are evenly spaced and are attached to the ceiling of the nest. The space between the faces of the combs is known as the 'bee space'; it is usually between 6 and 9 mm and is critical in maintaining optimal conditions within the nest, with just enough space for bees to walk and work on the surface of the combs while maintaining the optimum nest temperature. Bee space, dimensions of combs, and nest volume all vary with the race and species of honey bee. The bee-space is a crucial factor in the use of bee equipment, and honey bees cannot be managed efficiently using equipment of inappropriate size. Be careful! Most equipment is manufactured to the specifications of European bees.

Bees need a supply of food and water to live, and during dry periods the beekeeper may have to supplement natural sources. As a general rule, attempts to begin beekeeping should start with the area's existing bees, techniques, and equipment, which will all have been adapted for the local circumstances.

The modern hive the home of honeybees, has resulted from man's century old thoughts and searches. History knows of thousands of different shapes and forms of frame hives. The modern apiculture has accepted several types of hives. The hive is a wooden structure consisting of several parts, the names of which are self- explanatory. These parts as separated out of the hive (Figure 32) are as given below: first comes the stand, on which is placed the bottom board on which in turn is placed a wooden box called as brood chamber or the hive story. This chamber is filled with movable, parallel frames on which the bees build their combs out of the wax produced by them. On top as an upper story is the honey chamber which can also carry frames. Here, only honey is stored and there is no egg laying or brood rearing. This chamber is popularly known as 'super' and is covered on the top with an inner cover and the top cover or roof. There are three different types of hives *viz.:* (i) The Langstroth hive (ii) The Jeolikote hive and (iii) ISI hive. The Langstroth hive is bigger in size and carries ten frames each in thee brood and the super chambers. The Jeolikote and ISI hives are smaller than Langstroth hive. The only difference between the Jeolikote and ISI hive is that, the former carries seven frames and the later ten frames (the size of the frames is the same). The Langstroth hive is recommended for *Apis mellifera* and the other two hives may be used for *Apis cerana*. The dimensions of different hives are given in Tables 17-20.

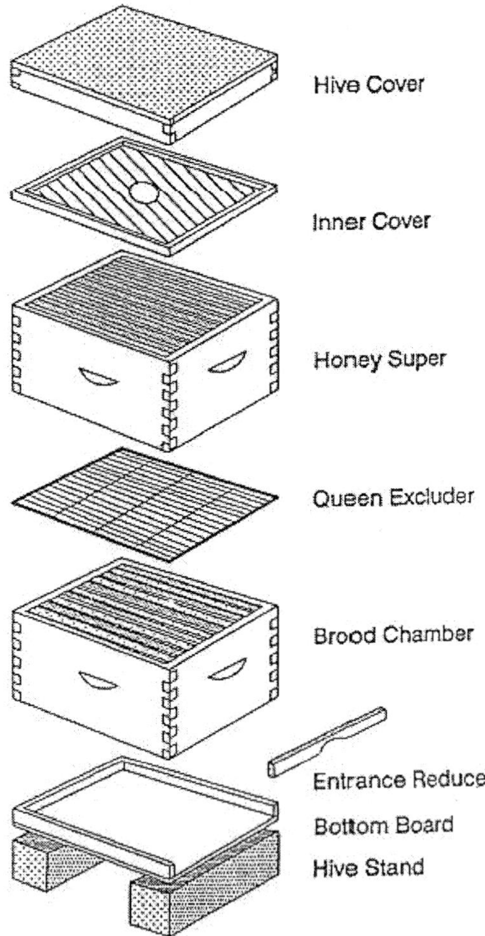

Figure 32. Diagram showing different parts of langstroth frame hive.

The Langstroth, Indian standard institute (ISI) and Newton frame hives are more in use than other types in India. While making the hive necessary details may be followed as given below:

Table 17. Dimensions of various parts of *Apis cerana* type A Beehives with ten frames (in mm)

Description	Distance between centres of two adjacent frames = 31 mm (bee space = 7 mm)			Distance between centres of two adjacent frames = 31 mm (bee space =8 mm)			Distance between centres of two adjacent frames = 31 mm (bee space = 9mm)		
	Length (L)	Breadth (B)	Height (H)	Length (L)	Breadth (B)	Height (H)	Length (L)	Breadth (B)	Height (H)
Floor board	361+2	356+2	50+2	361+2	366+2	50+2	361+2	376+2	50+2
Brood frame									
outside	230	-	165	230	-	165	230	-	165
inside	210+2	-	145+2	210+2	-	145+2	210+2	-	145+2
Brood chamber									
outside	286+2	356+2	172	286+2	366+2	173	286+2	376+2	174
inside	240	310	172	240	320	173	240	330	174
Super frame									
outside	230+2	-	85+2	230	-	85	230	-	85+2
inside	210+2	-	65+2	210+2	-	6+25	210+2	-	65+2
Super chamber									
outside	286+2	356+2	92	286+2	366+2	93	286+2	376+2	94
inside	240	310	92	240	320	93	240	330	94
Inner cover (crown board)	286+2	356+2	22	286+2	366+2	23	286+2	376+2	24
Roof (top)									
Outside	328+2	398+2	100+2	328+2	408+2	100+2	328+2	418+2	100+2
Dummy board	330+2	-	165+2	230+2	-	165+2	230+2	-	165+2
Division board	236	-	One end 182 other end 194	236	-	One end 183 other end 195	236	-	One end 184 other end 196

Source: Indian Standards Institution 1981.

Table 18. Dimensions of various parts of *Apis cerana* type A Beehives with eight frames (in mm)

Description	Distance between centres of two adjacent frames = 31 mm(bee space = 7mm)			Distance between centres of two adjacent frames = 31 mm(bee space = 8mm)			Distance between centres of two adjacent frames = 31 mm (bee space =9mm)		
	Length (L)	Breadth (B)	Height (H)	Length (L)	Breadth (B)	Height (H)	Length (L)	Breadth (B)	Height (H)
Floor board	361+2	296+2	50+2	361+2	304+2	50+2	361+2	312+2	50+2
Brood frame									
Outside	230	-	165	230	-	165	230	-	165
Inside	210+2	-	145+2	210+2	-	145+2	210+2	-	145+2
Brood chamber									
Outside	286+2	296+2	172	286+2	304+2	173	286+2	312+2	174

Inside	240	250	172	240	258	173	240	266	174
Super frame									
outside	230+2	-	85	230	-	85	230	-	85
inside	210+2	-	65+2	210+2	-	6+25	210+2	-	63+2
Super chamber									
outside	286+2	296+2	92	286+2	304+2	93	286+2	312+2	94
inside	240	310	92	240	258	93	240	266	94
Inner cover (crown board)	286+2	296+2	22	286+2	302+2	23	286+2	312+2	24
Roof (top)									
Outside	328+2	338+2	100+2	328+2	346+2	100+2	328+2	354+2	100+2
Dummy board	230+2	-	165+2	230+2	-	165+2	230+2	-	165+2
Division board	236	-	One end 182 other end 194	236	-	One end 183 other end 195	236	-	One end 184 other end 196

Source: Indian Standards Institution 1981.

Table 19. Dimensions of various parts of *Apis cerana* type B Beehives with ten frames (in mm)

Description	Distance between centres of two adjacent frames =31 mm (bee space = 8 mm)			Distance between centres of two adjacent frames =32 mm (bee space =9 mm)		
	Length (L)	Breadth(B)	Height(H)	Length(L)	Breadth(B)	Height(H)
Floor board	431+2	366+2	50+2	431+2	376+2	50+2
Brood frame						
outside	300	-	195	300	-	195
inside	280+2	-	175+2	280+2	-	195
Brood chamber						
Outside	356+2	366+2	203	356+2	376+2	204
Inside	310	320	203	356+2	376+2	175+2
Super frame						
Outside	300	-	105	300	-	105
inside	280+2	-	85+2	280+2	-	85+2
Super chamber						
outside	356+2	366+2	113	356+2	376+2	114
inside	310	320	113	310	330	114
Inner cover (crown board)	356+2	366+2	23	356+2	376+2	24
Roof (top)						
Outside	398+2	408+2	100+2	398+2	418+2	100+2
Dummy board	300+2	-	195+2	300+2	-	195+2
Division board	306	-	One end 213 other end 225	306	-	One end 214 otherend 226

Source: Indian Standards Institution 1981.

Table 20. Dimensions of various parts of *Apis cerana* type B Beehives with eight frames (in mm)

Description	Distance between centres of two adjacent frames =31 mm (bee space =8 mm)			Distance between centres of two adjacent frames =32 mm (bee space =9 mm)		
	Length (L)	Breadth (B)	Height (H)	Length (L)	Breadth (B)	Height (H)
Floor board	431+2	304+2	50+2	431+2	312+2	50+2
Brood frame						
outside	300	-	195	300	-	195
inside	280+2	-	175+2	280+2	-	175+2
Brood chamber						
outside	356+2	304+2	203	356+2	312+2	204
inside	310	258	203	310	266	204
Super frame						
outside	300	-	105	300	-	105
inside	280+2	-	85+2	280+2	-	85+2
Super chamber						
outside	356+2	304+2	113	356+2	312+2	114
inside	310	258	113	310	266	114
Inner cover (crown board)	356+2	304+2	23	356+2	312+2	24
Roof (top)						
Outside	398+2	346+2	100+2	398+2	354+2	100+2
Dummy board	300+2	-	195+2	300+2	-	195+2
Division board	306	-	One end 213 other end 225	300	-	One end 214 other end 226

Source: Indian Standards Institution 1981.

Langstroth ten- frame hive

(1) *Stand:* any four legged stand 15-25 cm high will do. Its upper dimensions should be such as to support the bottom board properly.

(2) *Bottom board:* it can be made either by taking a piece of wood 550 mm long, 450 mm broad and 22 mm thick, or by joining 2 wooden boards together nailing in position with wooden rods. Along each end of the longer side is nailed a wooden rod 550 mm long, 22 mm broad and 22 mm thick and another wooden rod 363 mm x 22 mm is nailed at the back. The front is provided with entrance rod which is 363 mm x 22 mm x 22 mm and this has an entrance 75 mm long and 22 mm deep in its middle. Two wooden blocks to be used for shortening the entrance, when necessary, should also be prepared, each block being 75 mm x 38 mm x 22 mm.

(3) *Brood chamber:* It is a rectangular box without top and bottom and is made of 22 mm thick wood. Its length on the outside is 406 mm and on

the inside 363 mm and its height is 238 mm. A rabbet 16 mm deep and 13 mm wide is cut along the entire of its width planks.

(4) *Frame:* consist of top bar, two side bars and a bottom bar dimensions for self spacing frame are given below

(i) *Top bar:* 475 mm long, 25 mm wide and 22 mm thick. It is cut to 9 mm thickness on both sides for a length of 25 mm. it has a groove, in the middle of its lower side for fixing the comb foundation sheet.

(ii) *Side bar:* each is made up of 9 mm thick wood and is 226 mm long. The upper part of each is 34 mm wide and lower part 25 mm wide. Each is cut from the middle portion at either end to accommodate the top and the bottom bars, respectively. There are 4 holes in each side bar for wiring the frame.

(iii) *Bottom bar:* 440 mm long, 19 mm wide and 9 mm thick. The outside measurements of the frame are 440 mm x 228 mm.

(iv) Two 15 mm staples should be driven in (to leave only 9 mm outside) to top bar on its opposite side, so that the frames stand 34 mm apart. One should make all frames either Hoffman or staple spaced type. Tinned wire of 28 gauge should be used in wiring the frame.

(5) *Super:* The dimensions of the super and the super frames should be same as those of the brood chamber and brood chamber frames, respectively.

(6) *Inner cover:* This is a wooden board to cover the brood chamber or the super as the case may be. It is 500 mm long, 406 mm broad and 9 mm length wood. It has 9 mm thick and 22 mm wide wooden bar nailed on to each of its four sides.

(7) *Top cover:* It is made up of 9 mm thick wooden board nailed to a rectangular frame 50 mm high, all covered over with a metallic sheet so as to make it impervious to rain water. Its inside measurements are 525 mm x 425 mm. It rests loosely over the hive.

Newton hive

Newton hive has the following parts:

(1) Stand: A log of wood about 10 cm in diameter and well soaked in solignum is buried deep into the ground. A length of about 20 x 30 cm is left above ground and a board 40 x 30 cm is fixed on its top with long nails and screws. The hive is placed on this platform on the log.

(2) Bottom board: It is a plank slightly wider and 25 mm longer than the brood chamber with beadings on three sides into which the hive body fits in tightly. The extension of the front serves as the alighting board.

(3) Brood chamber: It is a rectangular box without top and bottom and is made of 22 mm thick planks with outer dimensions 278 mm x 256 mm x 160 mm and 234 mm x 160 mm. Along the top of the front and rear planks a groove of 6 mm depth and 9 mm width is made for resting the frames and a clearance of about 6 mm is provided between the lower extremity of frames and the bottom board. The front plank has an opening 88 mm x 9 mm at its lower side to serve as an entrance.

(4) Brood frame: Self spacing (i) top bar breadth 22 mm, length 250 mm and thickness 3 mm (ii) side bar height 144 mm, width at top 28 mm, and (iii) inner length of frame 206 mm, inner height of frame 144 mm.

An extension of 3 mm is given on either side of the side bar and a clearance of 6 mm is effected when the frames are kept side by side. There are seven frames in the brood chamber.

(5) Super and super frame: It has the same length and breadth as the brood chamber, but its height is 78 mm. The dimensions of the super frame are those of brood frame but the internal height is 62 mm.

(6) Top cover: It has sloping planks on either side. An opening of 87mm square, fitted with wire gauge is made on the low ceiling plank to provide ventilation. Two holes in the front and rear planks of the top provide a clearance of about 6mmbetween the ceiling plank and the frames below.

For the manufacture of hives, light, well seasoned, good quality timber should be used. The wood should not have strong smell. Kail (*Pinus excelsa*), teak (*Tectona grandis*) and toon (*Toona ciliata*) are some of the woods suitable for the purpose. The hives should be preferably painted white or almunium on the outside to protect the timber from weathering agencies.

The British standard hive: It is also a big hive but smaller than the Langstrioth hive. It is of the same size as is used in England. The beekeepers in Darjeeling use it. The measurement of brood frame is 350 x 212 mm.

Jeolikote villager hive: It is of the medium size, most suitable to our country. It is popular in north India. It first came to use at Jeolijkote, the beekeeping training centre of Uttar Pradesh Government. It was developed by late Sh R.N. Mutto. Its dimensions are 300 x 175 mm brood frames and 300 x 175 mm super frames.

I.S.I hive: "A Type" Indian standard institute has approved two kinds of hive for our country. "A type hive" is of the same size as that of Newton hive in its measurement.

I.S.I hive: "B Type" It is the second type of hive approved by Indian standard institute. It is just equal to the villager size hive in dimension. Its brood frame is 300x195 mm size. This hive is manufactured in eight and ten frame sizes.

The Bee space

A major improvement in hive design was made in 1851 by Lorenzo Langstroth. He built a hive with frames that hung from the top ends of the hive, leaving a 3/8 inch space between all sides of the frames and the hive body (which is about 9.3 mm). The bees maintain a working space between the combs called a bee space. Bee space is the distance the bees reserve for their movement between combs. When bees encounter distances greater than a bee space, they use beeswax and build comb in that space. When distances are smaller than a bee space, they fill the space with propolis. The concept of bee space has made modern bee equipment and its easy management possible. Before bee space was integrated in equipment design, all interior hive components would become hopelessly glued together.

It is the most important thing to be kept in mind while manufacturing a beehive. Inexperienced people have many times spoiled their hives by not giving proper attention to this most important single matter. They either leave larger bee space or smaller than their bees actually need. The Indian standard institute has fixed this space 7, 8 and 9 mm for bees of different locality in the country. If this space is left less than the correct size, the bees will join up the two combs or the comb to side walls or the brood chamber to the super or inner cover and they cannot be removed independently for examination. If this space is enhanced, the same difficulty arises. The bees built the small combs or between the comb and side walls or between the brood chamber and the super. The situation also makes the manipulation or examination of the hive very difficult.

The hive protects the bees against rains, the heat of summer, snow and frost of the winter. To give the maximum protection, the hive must be made of strong and thick wood. It should be painted with an oil based paint so as to protect the wood against the deteriorating effect of winter.

A comb a collection of hexagonal cells made up of bees wax. These cells are horizontally parallel to the ground, their closed ends are back to back and open ends are towards the opposite direction, forming the two surfaces of the comb functionally, there are three type of cells.

1. The attachment cells with which the comb is attached to the top of a wooden frame.
2. The storage cells for storing honey and pollen.
3. Brood cells, for rearing the young ones. The brood cells are subdivided into:
   (a) The worker cells: The cells in which worker bees are reared. There are 17.6 such cells in 4 linear inches
   (b) The drone cells: The cells in which drone bees are reared. There are 22 such cells in 4 linear inches in the *Apis cerana* and comb and 19.3 cells in *Apis mellifera* comb.
   (c) The queen cells: The cells in which queens are reared. These are much bigger and look like ground nut shells

- A comb has three more or less distinct horizontal zones:
- The honey zone forms the top most portion.

- The pollen zone comes below the honey zone. These cells store the pollen.
- The brood zone is the lowermost portion. In this zone eggs and brood are reared.

Besides, hive there are number of other bee appliances which are used to have a good control over bees and also to provide them with necessary facilities. They include:

Equipment

Most of the equipment needed for small-scale beekeeping can be made at village level. It can be helpful to import basic equipment to serve as prototypes for local manufacturers. For practising on a large scale, some specialized equipment will probably need to be bought, such as honey gates, special filtering gauze, and gauges to determine honey quality

Timber used for making beehives

Bee hives are constructed mainly with seasoned timber such as teak, kail or toon and pine. The timber should be free from insect holes, dead knots, shakes, splits or cracks. The thickness of the wooden wall should be 22 mm. Hive bodies painted externally will last longer. The colour of the paint should be white, blue, yellow or green. White is generally preferred for its action of radiating heat from sun. Wood is the most generally preferred for hive construction. It offers durability, flexibility, easy handling and improves the colony efficiency in regulating hive interior temperature and humidity.

Types of Hives

Newton's hives BIS hives and Marthandam hives are suited for rearing Indian bees. Langstroth hives are suited for rearing Italian bees.

The Newton hive consists of a brood chamber, a super chamber and a sloping roof. The brood chamber measures inside 234 x 225 x 160 mm (outer-278 x 256 x 160) with a 240 mm wide and 360 mm long floor board, fitted at the bottom. This can be pulled in front to keep the brood chamber open from below for good ventilation. The front wall of the brood chamber has an entrance slit of 88 x 9 mm size at the lower edge in the middle, above the floor board. There are two 6 mm wide rabbets at the top on the side walls of the chamber inside to hold brood frames. The chamber holds 7 brood frames. The brood frame is about 250 mm long and 144 mm high, and has a 22 mm wide top bar. A bee space of a little more than 6 mm is maintained. The super chamber is like the brood chamber, but only 78 mm high. The top has inside a plank fitted to it a little above the bottom. The plank has a square hole of 90 mm sides that is fitted with a wire screen. When the top is in position, a 3 mm space is left above the super frames. On its front and back walls the top has ventilation holes fitted with wire screens. In view of its small size and a very small bee space, the Newton hive was found to be unsuitable in north Indian regions, where the bees are larger. Variations in the Newton hive were made to suit the local bee variety. In Jammu and Kashmir as also in the hills of Himachal Pradesh, the Langstroth hive has been used, because the bees, particularly in the Kashmir area are large and approach the size of the European bee. Although the 10- frame hives are used, the 6-frame type is quite common. The brood chamber is about 464 mm long and 368 mm broad in a 10-frame box, 235 mm broad in the 6-frame box. The outer dimension is 508 x 412 x 242 mm and that of inner is 464 x 368 x 242 mm. The brood frames are 448 mm long and 232 mm high. A bee space of 8 to 10 mm is kept between frames.

Based on the recommendations of Apiary Industry Sectional Committee, the Indian Standards Institution (ISI) bring out the standards for ISI Type A and ISI Type B bee hives and other equipment. The Type A is similar in hive and frame dimensions to the Newton hive, while Type B is similar to the Villager hive. For the large Kashmir bee, Langstroth type is found more suitable. The bee space for Type A hive was fixed as 7, 8 or 9 mm, while that for Type B was fixed as 8 or 9 mm. The important measurements for A and B types of hive now in use in India are given in Tables 17-20.

Depending on the size of the bee nest and the climatic conditions, the number of frames in both the I.S.I. type hives can be 4 (for nucleus colony), 8 to 10. The breadth of the roof, brood chamber, super chamber, and inner cover varies according to the number of frames. The dimensions of the super chamber and super frames are similar to those of the brood chamber and brood frame, except for the height. The super chamber is 93 mm in height, and the super frame is 85 outside and 65 ± 2 mm inside, in Type A box, while the dimensions in Type B box are 113 mm, 105, and 85 ± 2 mm, respectively.

Traditional hives

These are made from whatever materials are available locally: typically hollowed-out logs, bark formed into a cylinder, clay pots, woven grass, or cane. They are used to encourage bees to nest in a site that is accessible by the beekeeper. The bees build their nest inside the container, just as they would build it in a naturally occurring cavity. Eventually, the beekeeper plunders the nest to obtain crops of honey and beeswax. Bees may or may not be killed during this process, depending on the skill of the beekeeper. If the colony is destroyed, the hive will remain empty for a while. If there are plenty of honey bee colonies in the area, then eventually a swarm may settle in the empty hive and start building a new nest. Traditional beekeepers often own 200 hives, and expect only a proportion of these to be occupied by bees at any time. All the requirements for traditional beekeepers will be available locally, but beekeepers can be helped by the provision of protective clothing, smokers, and containers for the honey, and with help in locating markets for their products.

Movable frame hives

These are the hives used in industrialized countries and in some countries where beekeeping is a major industry. The objective of movable-frame hive beekeeping is to obtain the maximum honey crop, season after season, with the least disruption of the colony. These rectangular wooden or plastic frames have two major advantages:

- They allow easy inspection and manipulation of colonies.
- They allow very efficient honey harvesting because the honeycombs, within their frames, can be emptied of honey and then returned to the hive.

Frame hives must be constructed with precision. Frames are contained within boxes and each hive consists of a number of boxes placed on top of each other. Often a 'queen excluder', a metal grid with holes that allows only the smaller worker bees to pass through, is used to isolate the brood in the bottom-most boxes. The rest of the boxes will contain only honey.

Intermediate technology hives

Intermediate technology hives combine the advantages of frame hives with low cost and the ability to manufacture locally. The hive consists of a container with a series of 'top bars', on which the bees are encouraged to build their combs. These top bars then allow individual combs to be lifted from the hive by the beekeeper. The containers for the hives may, like traditional hives, be built from whatever materials are locally available. Top-bar hives can also be kept near the home and moved between flowering crops, enabling women to keep bees. The only items in the top-bar hive which need to be built with precision are the top bars themselves; they must provide the same spacing of combs within the hive as the bees would use in their natural nest. The natural comb spacing is the distance between the centres of adjoining combs, and this spacing will depend upon the species and race of honey bees which are being used. As a very general guide, *Apis mellifera* of European origin need top bars 35 mm wide, *Apis mellifera* in Africa need 32 mm, and *Apis cerana* in Asia need 30 mm.

The best way to determine the optimum width is to measure the spacing between combs in a wild nest of the same bees. The volume of the brood box should equate roughly with the volume of the cavity occupied by wild-nesting honey bees.

Other Beekeeping Accessories

Some commonly used equipment and implements are listed below (Figure 33).

1. Hive stand: This is used to keep the bee hive above the ground so as to protect the colony from termites, ants and other crawling insects, as also prevent soil moisture getting into the hive or facilitate ventilation from below the hive. Hive stand is a necessary accessory to the bee hive in modern beekeeping, particularly when frequent migrations are undertaken to farming areas. The stand is made of wood or iron tubing or angle iron. Any four legged stand of 15-25 cm high is sufficient. Ant wells of 15 cm in diameter kept under four legs prevent ants and other crawling insects entering the hive.

2. Comb foundation mill: This is a machine to prepare comb foundation sheet used in beekeeping to make bees build regular combs in frames, that are convenient to handle. J. Mehring of Germany made the first comb foundation in 1857. Comb foundation is made by passing plain sheets of beeswax between two rollers that have the regular 3-faced cell base pattern embossed on them. The patterns on the two rollers interlock properly, so that the 3-faced cell base on one roller matches with the base of each of the three cells on the other roller. The distance between the rollers is fixed in such a way that a thin foundation is made that is readily accepted by the bees. The rollers rotate on opposite sides. The rotation is done by a handle attached to the lower roller. The cell size in the cell base pattern varies according to the size of the brood cells.

3. Comb foundation sheet: It is a thin sheet of bee wax embossed with a pattern of hexagons of size equal to the base of the natural brood cells on both sides. The size of the hexagon varies with bee species. For *A. mellifera* there are 19 cells and for *A. cerana* 22- 23 cells/100 mm linear length. Bees consume about 6 kg of honey to produce 1 kg of wax. By providing comb foundation sheet, the energy of the bee is diverted for honey collection and thereby better and more honey harvest in honey flow season is possible. The sheet is fixed to the frames on fine wires threaded through holes in the side bars and stretched tight.

4. Embedder: A spur embedder or an electric embedder and transformer is used to embed wires into the comb foundation sheets. Electric wire embedder is useful to reinforce the comb and give extra strength to the comb.

5. Dummy or Division Board/ Movable wall: A wooden board of the size of a frame is used to reduce the size of the bee nest in a hive. It is placed inside the brood chamber like the normal frame. It prevents the bees from going beyond it. When a colony is weak, the bees have problem in maintaining brood nest temperature. Dummy board is used to reduce the free space in the hive that helps the bees in containing heat within the nest and also useful to protect them from enemies.

6. Feeders: Used for providing sugar syrup as feed to the bees during dearth period. A normal method of providing feeding is to keep a can with small holes punched on its lid. The can is filled with sugar syrup and kept over the frames in an inverted position. The syrup oozes out of the holes as the bees suck it. Another type is division board feeder that fits into the brood chamber like a frame and has space between two planks that are fitted to the frame of the size of the normal frame. Small sticks or grass or wire mesh is kept in the feeder to prevent drowning of the bees in the feed. Beeswax is coated inside to prevent leakage of the feed. The feeder is kept close by the side of the brood frames. These are commonly used in the *A. mellifera* colonies.

7. Queen Cage: This is used for transport of queen either with a few attendant worker bees, in packages. It is a cage made up of wood or wire gauge or plastic structure. This is useful for queen introduction.

8. Queen gate: It is a piece of queen excluder sheet. The holes in the sheet are large enough to allow free movement of worker bees in and out of the hive, but too small to allow queen's passage. Queen gate is used to arrest the queen bee in the hive, during seasons of food scarcity or during inclement weather, when there is a danger of bees deserting the hive. The queen gate is commonly used in beekeeping with the Indian hive bee.

9. Queen excluder: A flat metal screen with parallel slots of 4.0 to 4.2 mm width or an arrangement of thick wires like the cycle spokes arranged parallel to each other leaving a distance of 4.0 to 4.2 mm. The size of the worker abdomen is 3.2 to 3.5 mm and they move freely between the brood and super chambers. But the size of the queen abdomen is 4.4-4.5 mm and hence it is confined in the brood chamber itself. The screen is fixed in a wooden frame. The size of the queen excluder is the same as that of the brood chamber. This is placed above the brood chamber to prevent queen going into the honey storage area above, and laying eggs in the honey combs. Used mainly in beekeeping with *A. mellifera*, where the queen goes readily to the upper chamber, the queen excluder facilitates honey harvests during which super frames filled with honey are removed for extraction. In western countries, queen excluders are also used for propolis production. It is also used in producing royal jelly in queen rearing and in forming multi-queen colonies.

10. Queen cell protector: It is a cone shaped structure made of a piece of wire wound spirally. It fits around a queen cell. It is used to protect the queen cell, given from a queen right to queen less colony until its acceptance by bees.

11. Bee escape board or super clearer: This is a wooden or metal device which allows the bees to go through a self closing exit. Placed in position on a wooden board below the super, the bee escape facilitates removal of bees from the super frames for honey harvest. Porter bee escape is commonly used in India. It has metal spring strips that are attached on the opposite walls of a metal tray. Bees walk through the hole into the escape, and pass through the narrow opening left by the metal springs into the brood chamber below. The metal springs are close enough to discourage bee entry in the opposite direction.

12. Pollen trap: This is a device attached to the bee hive to remove pollen loads from the returning pollen foragers and to collect them in a tray. The trap works on the principle that the pollen loads attached to the forager bee on its hind legs get loosened and fall off, if the bee struggles to enter the hive through a narrow slot. The tray is withdrawn and the pollen loads collected for shade drying and preservation. The trap should not be kept continuously as the normal requirement of the pollen for colony is affected. This may be kept in alternate days or depending upon the requirement.

13. Drone trap: It is a rectangular box with one side open. The other side is fitted with queen excluder sheet. At the bottom of the box, there is a space for movement of worker bees. There are two hollow cones at the bottom wall of the box. Drone entering through the cones into the box get trapped. Drones are heavy honey eaters. The narrow end of the cone is wide enough to let the bees pass out but not large enough to attract their attention or reentry. This device is used at the entrance to reduce the drone population inside the hive.

14. Swarm trap: It is a rectangular box used to trap and carry the swarm. It is fixed near the hive entrance with one or two combs inside during the swarming season. This box traps and retains the queen only. But the swarm coming out from the hive reenters the hive and settles on the comb, since the queen is trapped.

15. Hive tool: This is a flat metal rod of varying design and size used commonly while handling bees in the hive. Hive tool has versatile uses; it is employed to separate the brood and super chambers, to loosen the frames for taking them out, or to scrape hive corners, frames and other hive parts to remove debris, burr or brace combs. It is made of tempered steel so as to make it strong enough to lift heavy supers, etc. The hive tool is usually 20 to 25 cm long. One end of the hive tool is flattened, broad and thin. The other end is bent at right angles at about a centimetre from the edge. This bend allows prying out the frames loose without disturbing the bees with jerks or shakes. A 3 to 4 mm wide hole is sometimes provided in the middle towards the broad end, and helps in prying out nails.

16. Smoker: A beekeeper uses a smoker to produce cool smoke to calm the bees while handling them. The smoker consists of a fuel box containing smouldering fuel (for example dried cow dung, hessian, or cardboard) with a bellows attached. The beekeeper puffs a little smoke near the entrance of the hive before it is opened, and gently smokes the bees to move them from one part of the hive to another. Bees respond to smoke by filling up on honey so that they are less likely to sting. The fuel is kept off the base by a perforated metal shelf above an airhole. The bellows are used to blow air into the fire box through two holes opposite each other. It is important that only smoke and no flames emerge from the smoker and that the fire should be put out immediately after use.

17. Protective clothing: Adequate protective clothing gives beginner beekeepers confidence, but more experienced beekeepers find that too much protective clothing makes it difficult to work sufficiently gently with the bees, and it is very hot. Always wear white or light-coloured clothing when working with bees - they are much more likely to sting dark-coloured clothing. It is most important to protect the face, especially the eyes and mouth; a broad-rimmed hat with some veiling will suffice. Individual items of clothing must be impermeable to bee stings, and every joint between them must be bee-tight; rubber bands can prevent bees from crawling up trouser legs or shirtsleeves. Some people find that a good way to protect their hands is to put a plastic bag over each hand, secured at the wrist with a rubber band.

a. Bee veil: It is worn over the face for protection against stings. It is particularly useful for a beginner, for protecting face from bee stings during the handling of bees. Some beekeepers feel that the Indian bees are mild and a bee veil is un-necessary. However, the stinging behaviour of the bees is unpredictable and sometimes factors other than the beekeeper may instigate the bees to sting. The number of stings is directly related to the number of bees lost in the colony. Use of the bee veil at all times is therefore advisable.

There is a great variety of types and designs of bee veils. In principle a bee veil should have a cloth hood with an arrangement to held a hat, and a screen in front of the face to allow free vision of the bees and the colony. The cloth portion protects the ears, head and neck, while the screen protects the lips, nose and eyes. Use of cotton cloth is advisable because other material may induce excessive sweating. The smell of sweat can also cause stinging. The screen is usually black nylon mosquito netting that makes it easier to see through. Other clothing may also be worn, including gloves, boots and overalls. White coloured clothing is advisable, as bees are much more likely to sting dark-coloured clothing.

b. Gloves: These are used while inspecting and handling colonies to protect hands and arms. Soft leather gloves with canvas gauntlets to the elbow are the best for use. Surgical gloves or thick canvas gloves can also be used, but under the tropical conditions in India these are quite uncomfortable for continuous use.

c. Overalls: White overalls are occasionally worn. Light coloured cotton materials are preferable since they are cooler and create less risk for antagonizing bees.

d. Boots: A pair of gum boots will protect the ankles and prevent bees from climbing up under trousers.

18. Swarm net: This is used to hold a swarm of bees and transport it to the apiary. A swarm net is quite useful when the swarm is located in difficult spots, where taking it into a hive directly is not possible. It consists of a mosquito net bag about 50 cm long and 40 to 45 cm wide, with one side open. The open side has a rope to close the bag after capturing the swarm. A thick wire ring of about 40 cm diameter is used at each end to keep the bag in the cylindrical shape that allows keeping the bees without crushing or harming them.

19. Bee brush: A soft camel-hair brush is used to remove bees off the honey comb before it is taken out for extraction of honey.

20. Decapping knife: Single or double edged steel knife is used for removing wax cappings from the honey comb before putting it in the honey extractor. In cooler climates like Kashmir, some sort of heating arrangement is made to that the blade remains warm and uncapping is facilitated. The blade can be heated by application of electricity or by steam circulation or simply by dipping the blade in hot water.

21. Frame gripper: A helpful tool for holding frames when inspecting the colony.

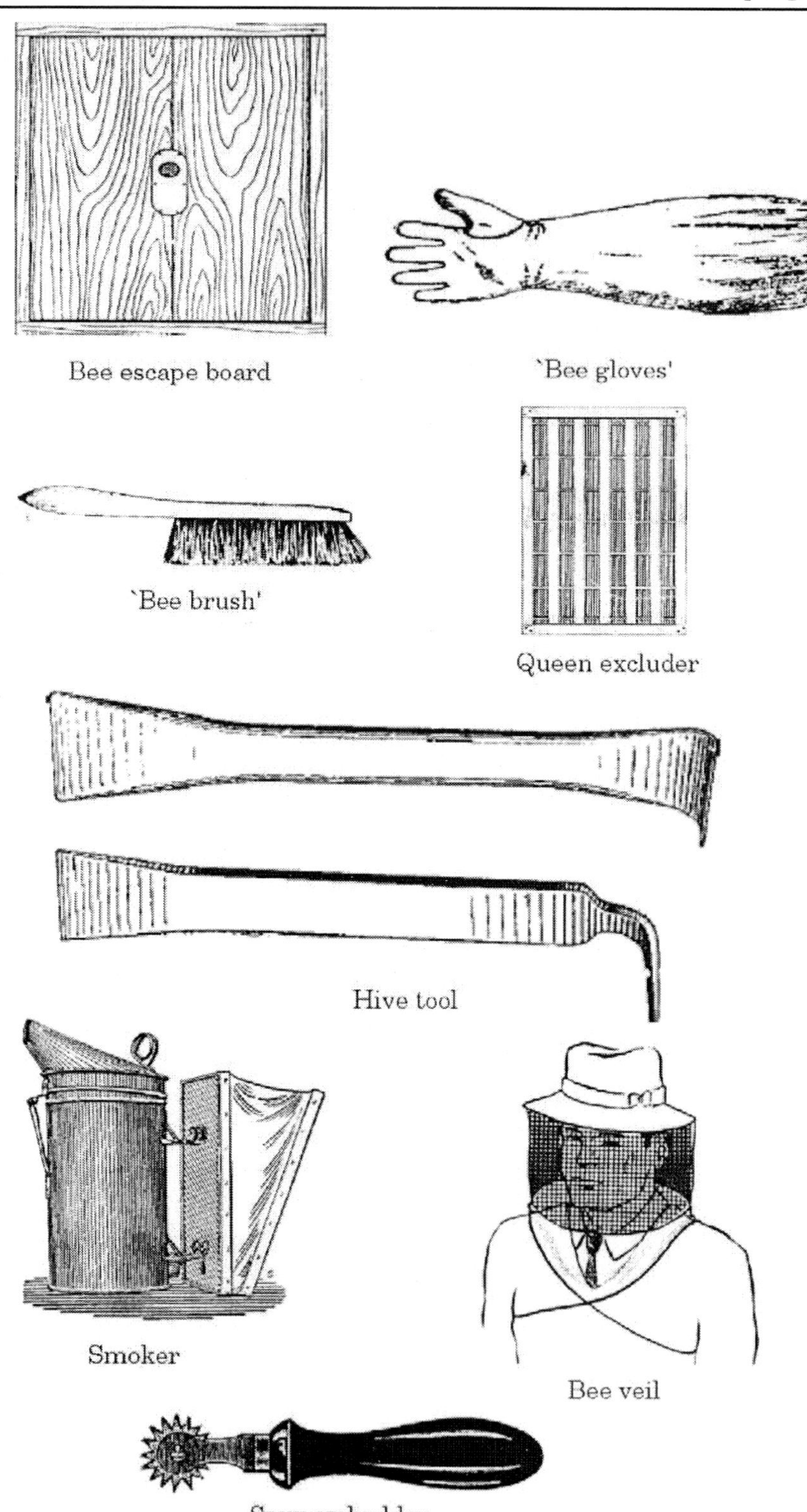

Bee escape board

`Bee gloves'

`Bee brush'

Queen excluder

Hive tool

Smoker

Bee veil

Spur embedder

Comb foundation mill

Honey Extractors

Wax sheet

Honey bee workers build combs as they add wax to the cell base imprints on the sheet of foundation

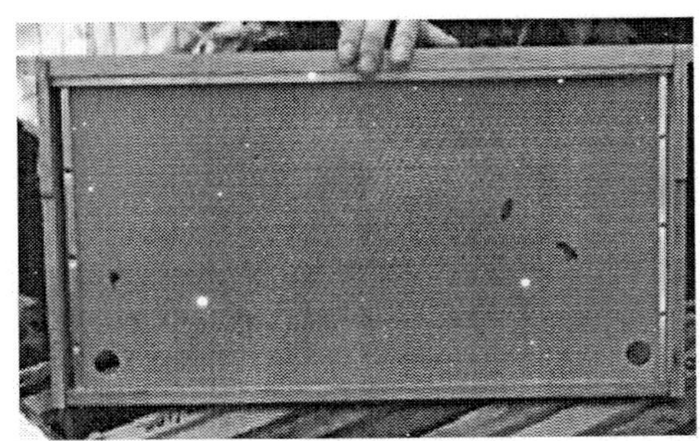

A sheet of comb foundation suspended in a wooden frame

Frame Gripper

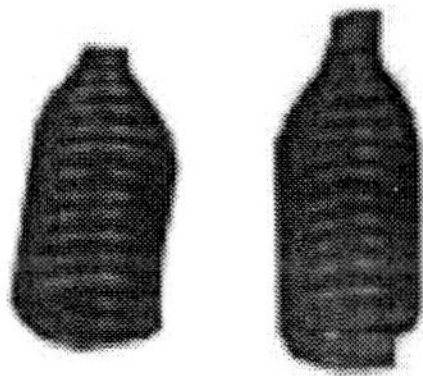

Queen Cell Protector

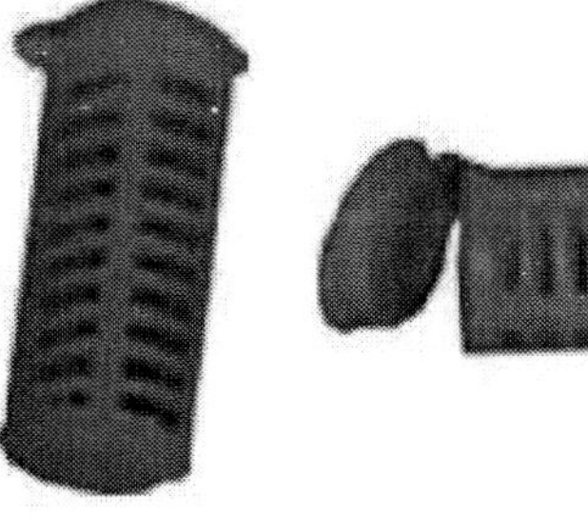

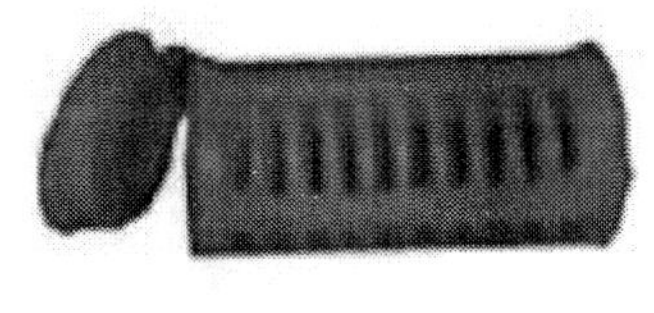

Queen Cage

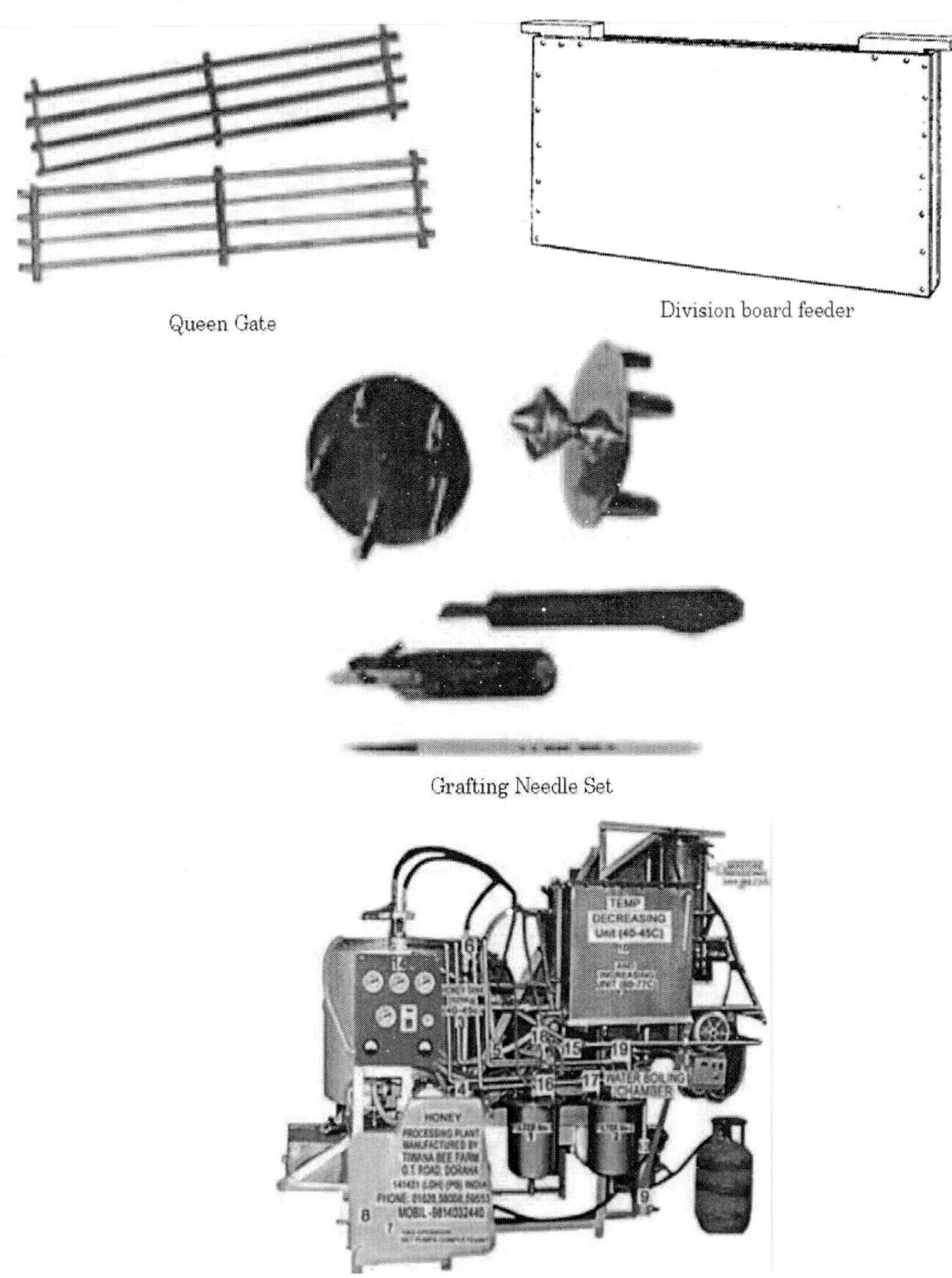

Queen Gate

Division board feeder

Grafting Needle Set

Honey Processing Plant

Figure 33. Bee keeping equipments and appliances

22. Honey Extractor: This is a metal drum or barrel with a gear mechanism to rotate a honey frame holder kept in it. Working of the honey extractor is based on the principle of application of centrifugal force to throw honey out of the combs, invented by Frang von Hruschka in 1885. Honey combs are kept in the extractor after removing the caps on the cells. After removal of honey, the combs can be re-used in the hive since the cells and the comb remain intact.

Harvesting honey and beeswax

Honey is harvested at the end of a flowering season. The beekeeper selects those combs which contain ripe honey, covered with a fine layer of white beeswax. These combs are usually the outside-most ones. Combs containing any pollen or developing bees should be left undisturbed. Honey will keep a long time if it is clean and sealed in an airtight container, but will deteriorate rapidly and ferment if it has absorbed water. Preventing this from happening is crucial in honey harvesting.

Harvesting the combs

Harvesting should be carried out in the evening or the early morning. Gentleness is the key to successful colony manipulation, so learn to carry out this process swiftly but calmly to avoid upsetting your bees.

1. Put on your protective clothing.
2. Get your smoker, brush or quill, knife or hive tool, and a rust- proof container in which to put the honeycomb.
3. Load your smoker, and puff some smoke gently around the hive for a few minutes. Wait a few more minutes, then puff smoke around the entrance holes.
4. After puffing the smoke, open the lid.
5. Knock the top bars to see which of them have combs; they will sound heavier than empty ones.
6. Use the knife or hive tool to remove the first bar from one end of the hive.
7. Puff smoke gently into the gap to drive the bees to the other side of the hive.
8. Start removing the bars one by one, until you get to the first comb, which will be white and new. It may be empty or it may contain some unripened honey. Replace it and leave the comb for the bees to develop.
9. Remove only the capped or partly capped combs, which will be quite heavy. Use a brush or feather to sweep any bees back into the hive.
10. Cut the comb off, leaving about 2cm for the bees to start building on again. Put the comb in a container and replace the top bar.

11. Carryon harvesting until you come across a brood comb,
12. which will be dark in colour and contain pollen too. Leave this honey for the bees.
13. Start the process again at the other end of the hive.
14. Close the hive carefully, replacing the lid.

Beekeeping Intermediate Technology Development Group

Honey extraction

The honeycomb can be simply cut into pieces and sold as fresh, cut comb honey. Alternatively, the honey and comb can be separated and sold as fresh honey and beeswax. It is important when processing honey to remember that it is hygroscopic and will absorb moisture, so all honey processing equipment must be perfectly dry. The most common traditional methods of honey extraction is squeezing. If your quantity of honey or financial resources are small, then squeezing the honey out by hand is probably the most viable option. The honey extracted by this method will have to be strained through several increasingly finer meshes to remove any bits of wax or debris, ending with something like muslin cloth. It is very important that this procedure be carried out hygienically, and that the honey is not left exposed to the air, where it will pick up moisture and deteriorate. Another good way of extracting honey from top-bar or movable frame hives is a radial or tangential extractor. This is a cylindrical container with a centrally-mounted fitting to support combs or frames of uncapped honey, and a mechanism to rotate the fitting (and the combs) at speed. The honey is thrown out against the side of the container and runs down to the bottom, where it is collected and then drained off with a tap. Most manufactured extractors are made to hold frames and have to be adapted to take comb from top bar hives. This is usually done by making wire baskets to hold the comb. The baskets can either lie flat horizontally, or be attached to the vertical frames and sit tangentially within the container. Top-bar combs in tangential extractors have to be spun twice, once on each side, to extract all the honey. The honey must be stored in airtight, non-tainting containers to prevent water absorption and consequent fermentation. If you want to sell your honey it would be helpful to add a label describing the source of the honey (for example sunflower, mixed blossom, tree honey), the country and district it was produced in, the weight or amount of honey in the container, and your name and address.

Beeswax extraction

The comb from which bees build their nest is made of beeswax. After as much honey as possible is separated from the combs, the beeswax can be melted gently over moderately warm water (boiling water will ruin the wax) and moulded into a block. Another option for processing the wax is a solar wax melter. This appliance is easy to make and consists of a wooden box with a galvanised metal shelf with a spout, a bowl or container that sits under the spout, and a glass or plastic cover. When placed in the sun the temperature inside the box will melt down a comb and the wax will flow into a container inside the box. Any honey that was left in the combs will sink to the bottom; it is usually used for cooking or beer making as its taste is spoiled somewhat by this process. Beeswax does not deteriorate with age and therefore beekeepers often save their scraps of beeswax until they have a sufficiently large amount to sell. Many beekeepers still discard beeswax, unaware of its value. Beeswax is a valuable commodity with many uses in traditional societies: it is used in the lost-wax method of brass casting, as a waterproofing agent for strengthening leather and cotton strings, in batik, in the manufacture of candles, and in various hair and skin ointments. Beeswax is also in demand on the world market. Beeswax for export should be clean and have been re-heated as little as possible.

Advantages and disadvantages of frame and top-bar hives

Advantages of the frame hive

a. The comb is fixed firmly to the four sides of the frame. This facilitates easy harvesting, and the beekeeper has little fear of damaging the comb.

b. The strength of the built-in comb also allows easy transportation, even over bad roads. It also affords easy control of a colony of bees without fear of breakage before the arrival at the new destination.

c. Honey is extracted by means of the centrifugal honey extractor, which makes it possible to remove the honey without damaging the comb. Empty combs are returned to the hive for the bees to refill with new honey, thus saving the insects from wasting time and energy to construct a replacement comb. Honey harvests are maximized, as the beekeeper can obtain several honey crops within the year.

d. During hive manipulations, very few bees are crushed between frames, whereas dozens of bees can easily be killed by careless handling of top-bars.

e. The hive is so designed (with queen excluder and supers) that the queen and brood are confined to the lower chamber. Supers contain only honey, and the lower brood chamber is undisturbed when honey is harvested.

f. Stealing a double- or triple-storey hive with a colony is a difficult feat for a thief. The Kenyan top-bar hive, on the contrary, can be carried away easily.

g. A swarm of bees can be hived with ease. Bees can easily pass through the numerous spaces between the frame and at the top of the hive.

h. Hive boxes can be stacked easily. This makes it easy to expand and contract the hive to meet the needs of the bee colony.

i. Drugs can be applied with ease through the openings.

Disadvantages of the frame hive in tropical Africa

a. A frame hive with two supers costs three times as much as a Kenyan 27-top-bar hive.
b. A high degree of craftsmanship is required to build the hive. Frame dimensions must be precise. Local village carpenters are not usually skilled enough for the job, and suitable tools for large-scale production of frame parts may not be available. Even if they are, it is never certain that the craftsmen have the patience to construct the hive correctly.
c. Wood for frame construction must be seasoned for at least a year. Very few carpenters or entrepreneurs can tie up their capital in this way.
d. The need to keep a stock of frames to replace those removed during the honey harvest creates an additional cost.
e. The need to import centrifugal honey extractors, decapping-knives, trays and other sophisticated equipment cannot be ruled out. In many countries, currency to order these from abroad cannot be obtained easily by local beekeepers.
f. A hive with supers is heavy and difficult to carry as a head-load; therefore, a vehicle may be required to move colonies if the need arises.
g. Because the frames do not fit together as the top-bars do, it is very difficult for the beekeeper to control the numerous bees which pass through the spaces between the frames and the top of the hive. This problem is very serious with the transitional long hive, which has as many as 30 frames arranged in a single-storey rectangular box. Such a beehive needs a special large smoker, and even such a smoker may not be able to produce enough smoke to "cool" the aggressive bees. The new beekeeper, upon seeing hordes of bees escaping, may run away, leaving the hive uncovered.

In the light of these serious problems, it is advisable for the beginner to start with the simplest type of movable-comb hive, which is, of course, the top-bar hive.

Advantages of the top-bar hive

a. This hive is cheaper and easier to produce than a frame hive. Any semi-skilled carpenter can make it. Only a few simple carpentry tools are required.
b. There is little or no need to import anything. All materials required can be obtained locally.
c. The hive can be opened easily and quickly. There is little or no need to employ a hive tool. Top-bars can be constructed to overlap the sides of the hive body slightly, and this makes it possible to use the thumb to pry up the top-bar.

d. Top-bars occupied by combs can easily be detected, so that the hive is opened from the empty side. This avoids crushing the bees between the top-bars when lifting the first comb.

e. Bees in the top-bar hive can easily be controlled when harvesting or inspecting the combs. The smoker puffs smoke through the opening created by the removal of one top-bar. Few bees can attack, since the beekeeper drives them away with smoke. When the top cover is removed from the transitional long hive, all the frames' 7-mm spaces are exposed, which permits numerous attackers to fly out and attack the beekeeper.

f. The top-bar hive is lighter to carry, even when the colony is inside.

g. More beeswax can be produced. Sales of beeswax increase the beekeeper's earnings and solve a great national problem. Beeswax is a multi-purpose industrial raw material required by factories and craftsmen.

h. There is no need to employ a queen excluder, which at the moment is not available. In practice, the bees keep their brood chamber separate from the honeycombs. Clean honey can be taken away, leaving brood combs undisturbed.

i. Honey production can be high. A well-managed hive with a good strong colony can produce between 50 and 120 kg of honey annually.

j. Honeycombs adulterated with pollen can be of high value. Pollen is a nutritious food supplement; the only way the nutrition is passed on is through honey harvested from such combs.

k. Only a few extra top-bars need be held in stock to replace worn-out or damaged bars.

In general, the top-bar hive is significantly cheaper and easier to use than hives with frames. The following disadvantages, however, cannot be overlooked.

Disadvantages of the top-bar hive

(a) A newly-constructed comb and all combs filled with honey must be handled with the utmost care. It is not advisable to move a top-bar hive, occupied by bees and combs, on lorries along bad roads full of potholes.

(b) Honey can only be extracted by destroying honeycombs, either by using the solar wax melter to dissolve the comb cells or by crushing them and squeezing out the honey. Bees have to build up new ones in their place, and this involves time, material and resources of the honeybees.

(c) Bees are often crushed between top-bars as the beekeeper rearranges the bars after removing them from the hive body. This problem can be serious when colonies are manipulated at night. When bees are crushed in this way, it is difficult to fix the last top-bar into place. Crushing bees is usually not a serious problem with frame hives.

(d) A top-bar hive is relatively easy to steal, as it is light and compactly designed. It is more difficult to steal hives and supers arranged one.

Chapter 10

Bee behaviour, Learning and Communication

Honeybees remain active throughout the year. In temperate areas which have prolonged winter and snow, the bees don't do much work. They eat honey and generate heat for warming themselves. They remain clustered together to remain warm. In lower hills or plains areas, where winter is mild they remain active in these months also and collect nectar and pollen from sarson flowers and also continue to rear the brood. In certain areas, where sarson is grown extensively, the colonies may collect and even store the extractable surplus from the source. The queen begins to lay eggs and brood rearing activity gears up resulting in spring colonies by March or April. Due to enormous brood rearing and increase in number of bees, the bees feel overcrowded and some of them may decide to desert, along with the queen. This is a natural phenomenon of the division of the colony called as swarming. Before swarming, it is necessary for a colony to produce a new daughter queen. for this purpose, workers build drone cells and the queen lays unfertilized eggs so as to produce males required for mating concurrently, with the production of drones, queen cells are raised, where she lays fertilized eggs, and the nourishment of future queen starts. As soon as the daughter queen is ready to leave the hive on warm sunny days. After flying for some time, the swarm settles on an interim clustering site may be nearby tree or bush. Some of the bees called scout bees move in different directions for the selection of a suitable nest site which could meet all the necessary requirements of colony such as protection, forage and water resources. When a new site has been discovered and a consensus arrived, the scout bees then lead the swarm to that site. However, in modern beekeeping such swarms are not allowed to escape but are captured and hived as new colonies.

In the parent colony, new queen on emergence, mates with one of the drones and starts egg laying. The normal activities continue in the parent colony and if the strength of bees increases, three or four such swarms may be issued in a season, each time old queen leaving with the swarm and a new queen taking up the control of the parent colony. When there is no more increase in number of bees, the colony decides not to issue more swarms and becomes stabilized, the queen that emerges the last becomes the mother of the colony and kills the remaining queen right in their cells. Normally February to April is the swarming season is also an index of good honey flow. Beekeepers should bc aware of the activity cycle orf the bees and adopt management practices.

Social behaviour of honeybees

The honey bee colony exhibits a combination of individual traits and social cooperation which is unparalleled in the animal kingdom. Honey bees are highly developed social insects. Their innate behavior has always fascinated the mankind. Their importance has promoted the mankind to undertake their perpetuation. Therefore, for efficient beekeeping, it is essential to understand their behavior.

Sequence of activities in the life of bees

It was believed that there existed some kind of scheme for assigning duties to various workers bees. This notion existed for hundred of years. Many researchers believed that queen bee is somehow active in organizing the activities or supervising the work that goes inside the hive. Though it is true that activities of the workers are affected by the presence of queen by the production of pheromones. However, there is a general agreement that activities argueably correlated with age of the worker bees. Honeybees have a well organized division of labour.

Division of labour

The diversity of bee activities and high degree of organization of activities is often referred to as division of labour. By looking at individual bee, one can ask questions. How does each bee know what to do next? Hcw to do and when and where the activity is appropriate. The answer is that each individual in a colony is slightly different from all other bees at a given moment, in terms of the combination of some of the genetic composition, internal stimuli and humuroal or physiological factors. Worker bees of different age groups perform different duties. During first three days after emergence the worker bees remain busy in brood area where they regulate the temperature up to 35°C. These bees are responsible for maintain ing the hygiene of the colony. They act as nurse bees and feed honey and pollen when they are about 4 days old. These nurse bees when 6 to 12 days old, secrete royal jelly from their mandibular glands develop which is used to feed the larvae destined to become queens. Again the duties change with age when the worker bees are of 14 days age, the wax glands develop and they start secreting wax to build combs. During this period, the bees may do any other activity as per the requirement of the colony. The worker bees start taking orientation flights during second week of their life. To acquaint themselves with the vicinity of the hive. The worker bees become foragers at the age of about 21 days after emergence. In general, first three weeks of workers life involves indoor duties and last three weeks outdoor duties. But this division of labour is not a hard rule and there can be any change in the schedule of activities depending upon the requirements of the colonies.

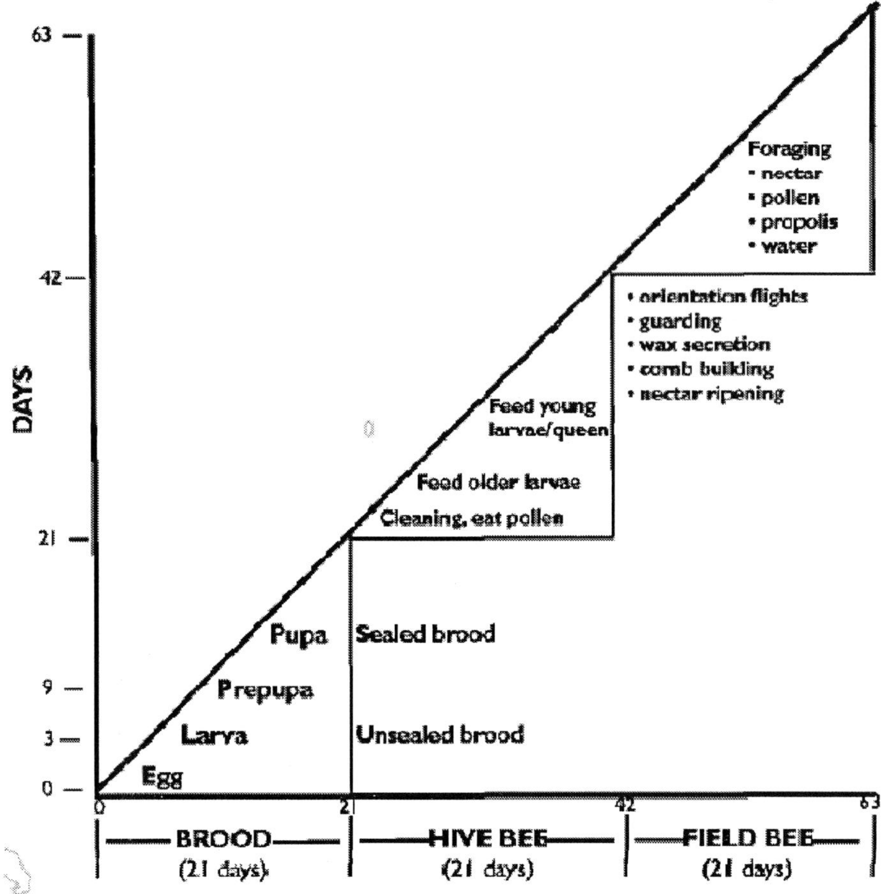

Figure 34. (a) Sequence of activities in the life of a worker bee (b)

Communication in honeybees

Honey bees are the wonderful creatures whose life activities have been topics of interest for early philosophers, such as Aristotle, Pliny, and Virgil. Though many others have studied bees throughout history, yet, most basic knowledge of their behaviour has been gathered only since the sixteenth century. Twentieth-century scientists have revealed the nature of honey bee sensory systems, behavior, communication, and population dynamics. Though, the honey bees do not possess the fine-grained retinal mosaics of humans, yet they can navigate with precision by optically scanning the sky's polarization patterns through their ultraviolet receptors. Moreover, they can easily perceive other colors and rapidly flickering patterns (as in flight at close range). Honey bees possess senses of taste and smell about as acute as those of humans and even more acute for some flavors and aromas. Other revelations are fast emerging about the flight muscle physiology of bees, their learning, memory, dances, chemical communication systems, and reproductive physiology.

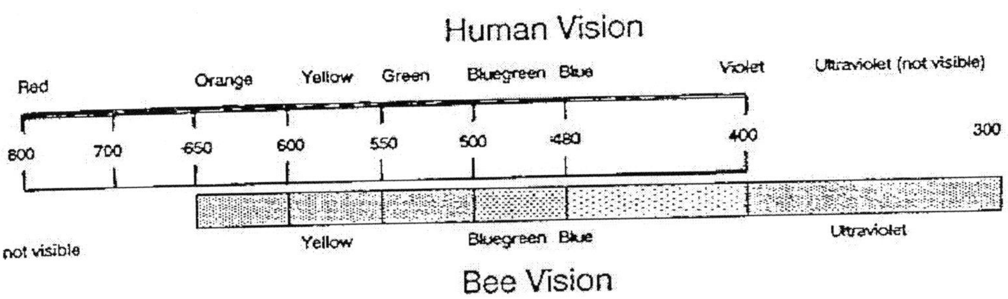

Figure 35 (a) A comparision of the visible spectrum of man and the beeredrawn from Gordon Townsend (1985).

Honey bees are unique because they generate and conserve heat, raise young, forage, communicate, swarm, store provisions, and pollinate or otherwise provide food for humans. They have fuzzy appearance (branched body hairs), robust body, and flattened hind leg. Honey bee colonies are perennial and nearly worldwide in distribution; they are distributed from the tropics to subarctic regions. This is possible because they hoard resources and can thermoregulate their nests, by fanning to evaporate water for cooling, by metabolizing honey to produce heat (0.1 cal per minute per bee), and by clustering to conserve the heat they produce. Unlike most other insects, bees build their nests out of their own body products, primarily wax and salivary secretions. In the summer, colonies managed by a beekeeper may consist of 40,000 to 60,000 adult bees. These will likely be distributed among the three castes as follows : 1 queen, 100 to 300 drones (up to 10,000 drones have been reported), 13,300 to 20,000 foraging workers, and 26,600 to 40,000 workers engaged in various hive duties such as brood rearing, comb construction, housecleaning, defense, and temperature regulation. The brood (eggs and young) in a typical colony includes 5,000 to 7,000 eggs, 7,000 to 11,000 larvae being fed, and 16,000 to 24,000 pupae in sealed cells. A single worker larva is fed approximately 1,300 meals per day for 5 days and grows to 300 times its original weight before pupation ensues (human growth is only 10 to 20 times original body weight). Three-fourths of a larva's weight is made up of fat. During brood rearing, the temperature of the brood nest is maintained at 32-35°C, usually 33-34°C. Since worker bees metabolize honey to generate the heat needed to warm the hive, honey must be present in the hive at all times. Colony humidity is also relatively constant at 48-50 +6% relative humidity. To accomplish this bees gather water and evaporate it in the hive when atmospheric humidity in the hive is low. However, most hive humidity, including that needed for cooling, is derived from the water content of nectar and from bee metabolism. When ambient relative humidity is high the bees are apparently able, by some means, to maintain a lower level in the hive. Bee behavior is instinctive and genetically programmed. Yet bees have plasticity in their central nervous system, as exhibited by their well-developed ability to learn and remember. Bees can learn in most sensory modalities (i.e., distance, sight, smell, taste, and touch). There is some evidence of a rudimentary ability to reason, such as associating flower type with reward. Bees are opportunists, as well, exploiting the most lucrative floral sources to save time and energy. Nevertheless, habituation (training) is an important part of bee behavior; bees can become accustomed to such things as frequent manipulation by beekeepers or the presence of floral sources.

Bees have evolved with numerous highly specialized sensory receptors that are unevenly distributed over the head, thorax, and abdomen. In function, the head is principally sensory for sight, smell, taste, and touch. The main portion of the brain within is a coalescence of three ganglia (nerve centers) that process much of the sensory information and initiate appropriate behavioral patterns. The thorax, tightly packed with well-innervated muscles, is the locomotory center for flight, walking, and sound production. The feet and legs are believed to bear important taste and touch receptors, as well. Certainly, the ability to build hexagonal wax cells of uniform size is dependent in part on sensory receptors on the legs. In the abdomen are found digestive, circulatory, reproductive, and defensive organs, along with the wax glands. Nerve and respiratory systems are distributed throughout the body. Since the insect is encased in a hard exoskeleton, it must of necessity have specialized external sensory areas. The bee's numerous external sensory receptors inform the central nervous system about the external environment, as well as about events within the body (such as the direction in which legs and wings are moving, whether the head is up or down, if the sting is in or out, or whether the bees is in contact with food or forage).

Honey bees communicate with one another both phonetically and kinetically. Tonal qualities of sounds produced by bees suggest that several moods" can be communicated. The best known of their communication systems is the dance language that provides information regarding the location of a particular source of forage. Information in this dance includes distance, direction, and quality and quantity of food. The recruitment dances instinctively performed by worker bees are in fact ritualized, miniaturized versions of the foraging trip the returning bee has just taken. The recruit bee learns and practices the trip in miniature and then leaves on her own (rather than being led, as with other social insects) to the source. Several other lesser known dances are performed by bees, and sound production by the dancing bees is an essential part of the dance activity.

Bees also communicate via the aroma and taste of the products they bring back to the hive, as well as by at least 31 pheromones (chemical compounds) they produce from glands in their bodies. Honey bees have no unusual nutritional requirements, only a balanced diet containing carbohydrates, fats, proteins, minerals, vitamins, and water. Even so, the logistics of honey bee foraging and food consumption are incredible. A typical colony may use up to 100 lb pollen for brood rearing each year; 4 million foraging trips are required to collect this quantity. A single bee may carry up to 5 million pollen grains in a single trip. Some pollen is probably consumed, but most is mixed with small quantities of honey or nectar and possibly salivary products and packed into cells adjacent to the brood nest where it undergoes a chemical change to a product called bee bread. This product is stored until consumed by adult bees for conversion into glandular larval food, a kind of "mother's milk" Bee bread is the principal food of the adult nurse bees. This nutrient was so named by primitive man because of its bready taste. Nurse bees eat and convert bee bread into at least two different glandular secretions, which are then fed to bee larvae. The diets of workers, queens, and drones are entirely dissimilar. Brood food recipes for each of the three castes involve differing ratios of the two glandular products (each is the product of bees of a different age); a quantity of honey is then added, the amount being caste dependent.

Cannibalism of eggs and larvae, which occurs during periods of food scarcity, can preserve the colony by conserving vital nutritional elements. Under ideal conditions and depending on the plant species involved, a colony may produce 90.7 kg (200 lb) or more surplus honey. The bees visit 2.5 to 500 million flowers for the nectar from which this quantity is produced. Approximately 3.6 kg (8 lb) honey are consumed in foraging and brood rearing to produce 0.45 kg (1 lb) surplus. Estimates of energy consumption indicate that foraging bees get about 11,265,100 km (7 million flight miles) to 3.8 L (1 gal) honey. Normally bees forage no more than a 3.2- to 4.0-km (2- to 2.5-mi) radius from the hive, covering 324 to 5062 ha (8,000 to 12,500 a) but may on occasion fly up to 16.1 km (10 mi) at a speed of 19.3 to 24 km (12 to 15 mi) per hour. The farther that bees must fly from a colony to forage for nectar, the more fuel they burn and the less efficient they are at producing a surplus of honey. The top speed of a worker bee is about 29 km (18 mi) per hour. As with the production of heat mentioned earlier, honey must be present in the hive to refuel foraging bees; without it foraging ceases

The fascinating social behaviour of honeybees is uniquely the result of millions of years of evolution through natural selection. It is not an exhibition of a high level of cleverness. Bees behave, or react to environmental stimuli, in a fairly predictable manner. Most of what they do and how they do it is in essence genetically programmed, and only through proper techniques of manipulation can beekeepers put the colonies to work.

A thorough understanding of basic honeybee biology is one of the most important requisites for success in beekeeping. Notwithstanding the fact that they have been kept by man for thousands of years, *A. cerana* and *A. mellifera* still retain all their natural biological characteristics. As compared with the other farm animals, they have not been truly domesticated, in that their life history, social organization, behaviour, physio-ecological characteristics and genetic composition have not been significantly altered by man. The honeybees of today are very much like their predecessors of millions of years ago. They live in a tightly knit social organization in which each individual member of each caste and sex plays a specific role in the survival of the colony and thus improves the chances of continuity of the species. An individual honeybee cannot survive for long when it is separated from its colony.

Bee behavior refers to what bees do as individuals and as a colony. By studying their behavior, one can learn how to manage them properly. Two practical discoveries of bee behavior were made by earlier workers which made our beekeeping of today possible. One was the discovery by Langstroth of bee space and another by G.M. Doolittle that large numbers of queens could be reared by transferring larvae to artificial queen cups. The discovery of the "language" of bees and of their use of polarized light for navigation has attracted considerable interest all over the world. Much has been learned about the behavior of insects, including bees, in recent years. Many bee behavior activities can now be explained by communication through pheromone. Recently, we have learned how certain bee behavior activities are inherited, and this information gives us a vast new tool to make the honey bee of our choice and produce specific strains for specific uses.

Bees communicate their floral findings in order to recruit other worker bees of the hive to forage in the same area. The factors that determine recruiting success are not completely known but probably include evaluations of the quality of nectar and/or pollen brought in. There are two main hypotheses to explain how foragers recruit other workers — "dance language" theory and the "odor plume" theory. The dance language theory is far more widely accepted.

Dance language

It has long been known that honeybees perform dances on their return to the hive, known as *round dance* and *waggle dance*. The nectar-laden bee dances on the comb in a circular pattern, occasionally crossing the circle in a zig-zag or waggle pattern. Aristotle in 330 BC, described this behaviour in his Historia Animalium. It was thought to attract the attention of other bees. In 1947, Karl von Frisch correlated the runs and turns of the dance to the distance and direction of the food source from the hive. The orientation of the dance correlates to the relative position of the sun. There is no evidence that this form of communication depends on individual learning.

Von Frisch performed a series of experiments to validate his theory. He was awarded the Nobel Prize in Physiology or Medicine in 1973 for his discoveries.

Honey bees are an excellent animal to study with regards to behavior because they are abundant and familiar to most people. An animal that is disregarded every day has very specific behaviors that go unnoticed by the normal person. Karl von Frisch studied the behavior of honey bees with regards to communication and was awarded the Nobel Prize for physiology and medicine in 1973. Von Frisch noticed that honey bees communicate with the language of dance. Honey bees are able to direct other bees to food sources through the round dance and the waggle dance. The round dance tells the other foragers that food is within 50 meters of the hive, but it does not provide much information regarding direction. The waggle dance, which may be vertical or horizontal, provides more detail about both the distance and the direction of the located food source. It is also hypothesized that the bees rely on their olfactory sense to help locate the food source once the foragers are given directions from the dances.

Another signal for communication is the shaking signal, also known as the jerking dance, vibration dance, or vibration signal. It is a modulatory communication signal because it appears to manipulate the overall arousal or activity of behaviors. The shaking signal is most common in worker communication, but it is also evident in reproductive swarming. A worker bee vibrates its body dorsoventrally while holding another honey bee with its front legs. Jacobus Biesmeijer examined the incidence of shaking signals in a forager's life and the conditions that led to its performance to investigate why the shaking signal is used in communication for food sources. Biesmeijer found that the experienced foragers executed 92.1% of the observed shaking signals. He also observed that 64% of the shaking signals were executed by experienced foragers after they had discovered a food source. About 71% of the shaking signal sessions occurred after the first five foraging success within one day. Then other communication signals, such as the waggle dance, were performed more often after the first five successes. Biesmeijer proved that most shakers are foragers and that the shaking signal is most often executed by foraging bees over pre-foraging bees. Beismeijer concluded that the shaking signal presents the overall message of transfer work for various activities or activity levels. Sometimes the signal serves to increase activity, when bees shake inactive bees. At other times, the signal serves as an inhibitory mechanism such as the shaking signal at the end of the day. However, the shaking signal is preferentially directed towards inactive bees. All three types of communication between honey bees are effective in their jobs with regards to foraging and task managing.

The Honey Bee Dance Language: components of the dance language

When an experienced forager returns to the colony with a load of nectar or pollen that is sufficiently nutritious to warrant a return to the source, she performs a dance on the surface of the honey comb to tell other foragers where the food is. The dancer "spells out" two items of information- distance and direction to the target food patch. Distance and direction are presented in separate components of the dance.

Distance

When a food source is very close to the hive (less than 50 meters), a forager performs a round dance (Figure 35b).

She does so by running around in narrow circles, suddenly reversing direction to her original course. She may repeat the dance several times at the same location or move to another location on the comb to repeat it. After the round dance has ended, she often distributes food to the bees following her. A round dance, therefore, communicates distance ("close to the hive," in this example), but not direction. Food sources that are at intermediate distances, between 50 and 150 meters from the hive, are described by the sickle dance. This dance is crescent-shaped and represents a transitional dance between the round dance and a waggle dance. The waggle dance, or wag-tail dance, is performed by bees foraging at food sources that are more than 150 meters from the hive. This dance, unlike the round dance, communicates both distance and direction. A bee that performs a waggle dance runs straight ahead for a short distance, returns in a semicircle to the starting point, runs again through the straight course, then makes a semicircle in the opposite direction to complete a eight figure circuit. While running the straight-line course of the dance, the bee's body, especially the abdomen, wags vigorously from side to side. This vibration of the body produces a tail-wagging motion. At the same time, the bee emits a buzzing sound, produced by wing beats at a low audio frequency of 250 to 300 hertz or cycles per second.

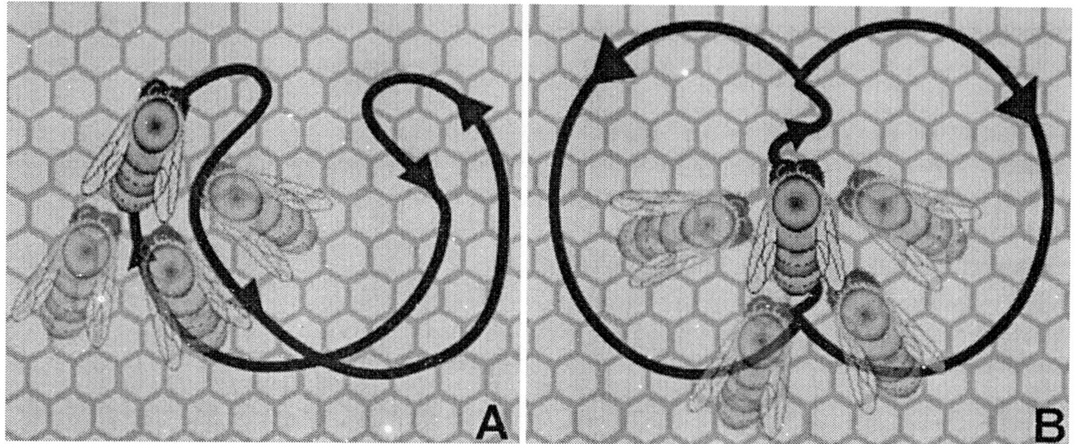

Figure 35b. (A) Round dance (B) Waggle dance

The Background

Honey bee dancing, perhaps the most intriguing aspect of their biology, is also one of the most fascinating behaviors in animal life. Performed by a worker bee that has returned to the honey comb with pollen or nectar, the dances, in essence, constitute a language that "tells" other workers where the food is. By signaling both distance and direction with particular movements, the worker bee uses the dance language to recruit and direct other workers in gathering pollen and nectar.

The late Karl von Frisch, a Professor of Zoology at the University of Munich in Germany, is credited with interpreting the meaning of honey bee dance movements. He and his students carried out decades of research in which they carefully described the different components of each dance. Their experiments typically used glass-walled observation hives and paint-marked bee foragers. First, they trained the foragers to food at sources placed at known distances from the colony. When the bees returned from gathering food from those sources, von Frisch and his students carefully measured both the duration and angle of the dances the foragers performed to recruit other bees to help gather food. Their findings led them to the concept of a dance language. Von Frisch's work eventually earned him the Nobel Prize for Medicine in 1973. The concept of a honey bee dance language, however, has had its skeptics.

Several scientists, among them Adrian M. Wenner, Professor Emeritus of Natural History at the University of California at Santa Barbara, have a different idea. They believe the dance exists, but they are not certain it communicates the location of a food source. These critics have argued that oral odors on a forager's body are the primary cues that enable the recruit-bees to locate new food sources. Many experiments have directly tested this alternate hypothesis and demonstrated the importance of oral odors in food location. In fact, von Frisch held this same opinion before he changed his mind and developed the theory of the dance language.

The biological reality probably lies somewhere between these two extremes. The most commonly accepted view is that recruits go to the area depicted in the dance, but then home in on the flower patch using odor cues. Indeed, researchers have built a robotic honey bee that is able to perform the dance language and recruit foragers to specific locations. But the robot is unable to properly recruit foragers to a food source unless it carries an odor cue on its surface. Nevertheless, it is clear that honey bees use the distance and directional information communicated by the dance language.

Buzzing occurs in pulse beats of about 20 milli seconds, delivered at a rate of about 30 per second. While several variables of the waggle dance relate to distance (such as dance "tempo" or the duration of buzzing sounds), the duration of the straight-run portion of the dance, measured in seconds, is the simplest and most reliable indicator of distance. As the distance to the food source increases, the duration of the waggling portion of the dance (the "waggle run") also increases. The relationship is roughly linear (Figure 36 a,b).

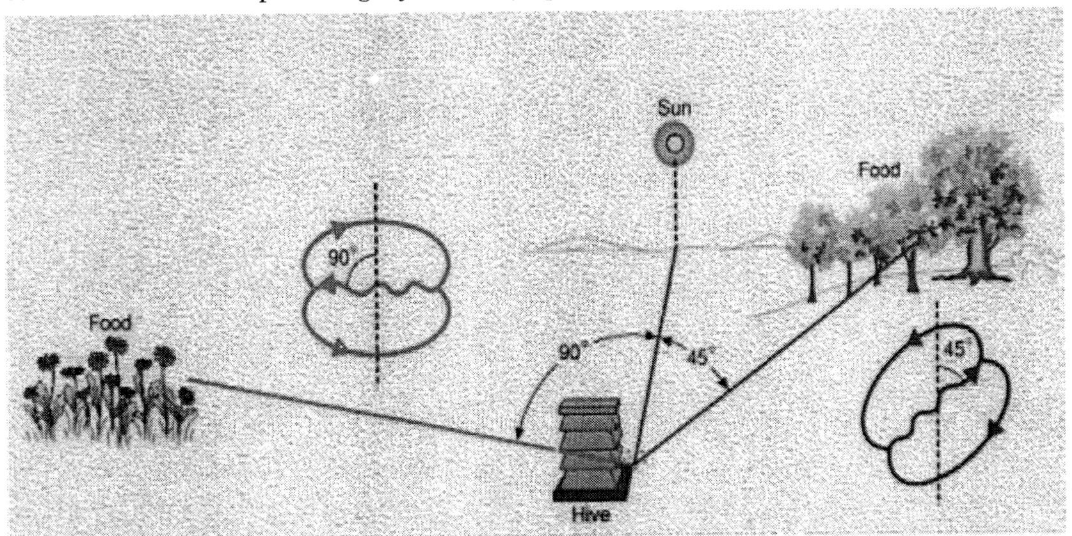

Figure 36 a. Relationship between the angle of the dance on the vertical comb and the bearing of the sun with respect to the location of food.

For example, a forager that performs a waggle run that lasts 2.5 seconds is recruiting for a food source located about 2,625 meters away.

Direction

Although the representation of distance in the waggle dance is relatively straightforward, the method of communicating direction is more complicated. The orientation of the dancing bee during the straight portion of her waggle dance indicates the location of the food source relative to the sun. The angle that the bee adopts, relative to vertical, represents the angle to the flowers relative to the direction of the sun outside the hive. In other words, the dancing bee transposes the solar angle into the gravitational angle. Figure given below gives three examples: A forager recruiting to a food source in the same direction as the sun will perform a dance with the waggle-run portion traveling directly upward on the honey comb.

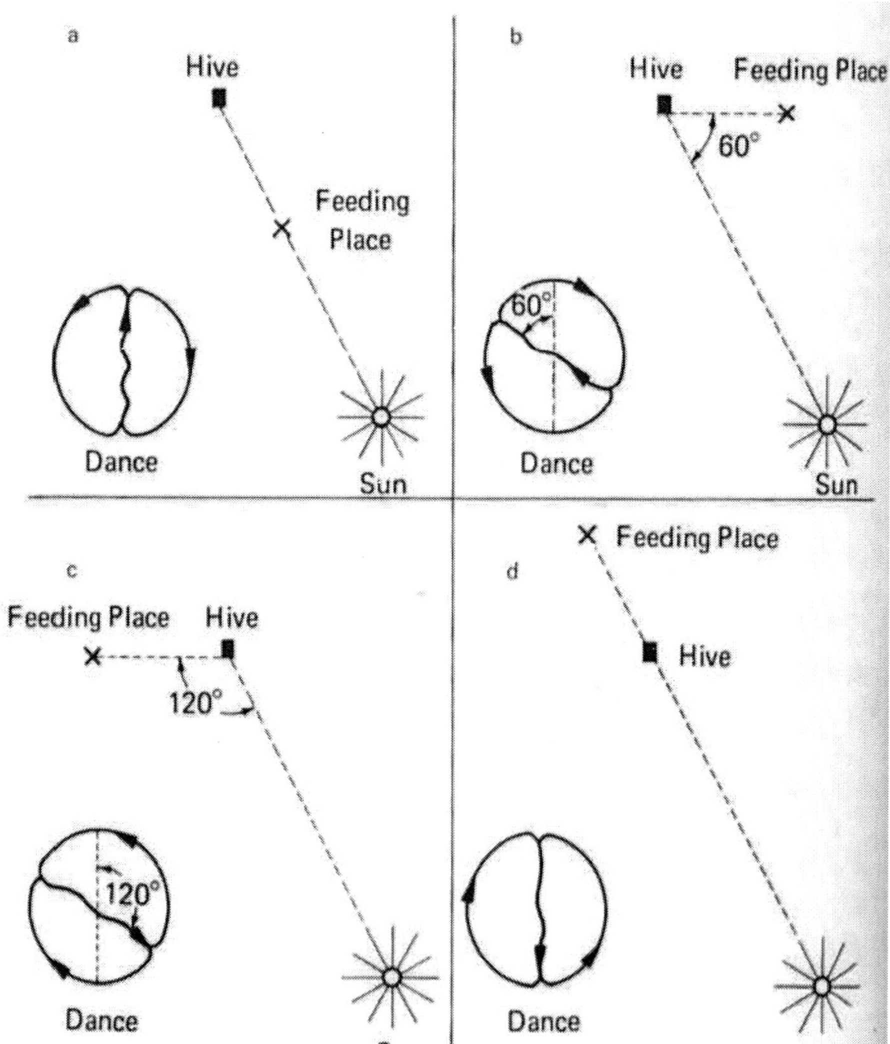

Figure 36 b. Relationship between the angle of the dance on the vertical comb and the bearing of the sun with respect to the location of food.

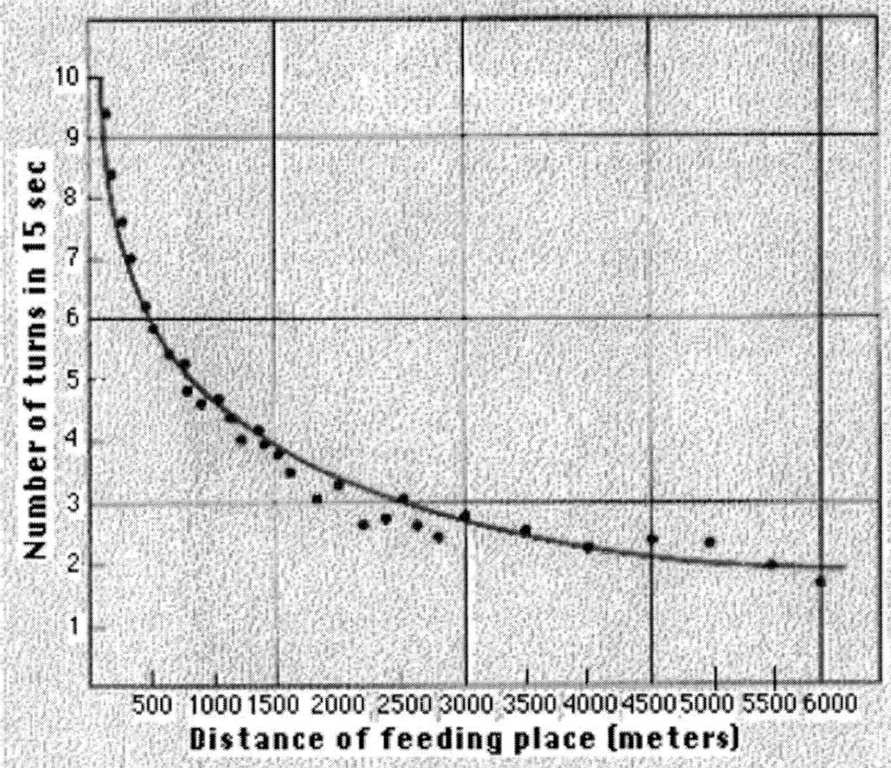

Figure 37. Number of turns per 15 sec in relation to distance of food source.

Conversely, if the food source is located directly away from the sun, the straight run will be performed vertically downward. If the food source is 60 degrees to the left of the sun, the waggle run will be 60 degrees to the left of vertical. Because directional information is given relative to the sun's position and not to a compass direction, a forager's dance for a particular resource will change during a day. This is because the sun's position moves during the day. For example, a food source located due east will cause foragers to dance approximately straight up in the morning (because the sun rises in the east), but in the late afternoon, the foragers will dance approximately straight down (because the sun sets in the west). Thus, the location of the sun is a key variable in interpreting the directional information in the dance.

The sun's position also is governed by geographic location and time of year. The sun will always move from east to west over the course of the day. However, above the Tropic of Cancer, the sun will move from southeast to southwest, whereas below the Tropic of Capricorn, the sun will move from northeast to northwest. Within the tropics, the sun may be located to the south or to the north, depending on the time of year. Thus, to translate the directional information contained in the honey bee dance, one must know the angle of the waggle run (with respect to gravity) and the compass direction of the sun, which depends on location, date, and time of day.

How do the bees calculate distance?

Von Frisch thought they measured the distance by assessing the amount of energy it took to get there. But the mechanism turns out to be quite different. The bees measure distance by the motion of images received by their eyes as they fly. Honeybees, like all insects, have compound eyes. These give little information about depth but are very sensitive to "flicker effect". It has long been known to bee keepers that honeybees respond better to flowers (A) moving in the breeze and (b) with complex petals. The importance of flicker effect can be demonstrated by training honeybees to visit food placed on cards with patterns. For example, the bees can distinguish any figure in the top row from any figure in the bottom row more easily than they can distinguish between any of the figures in either row.

Other features

Time Sense

When scouts remain in the hive for a long period, they shift the direction of the straight portion of the waggle dance as the day wears on (and the direction of the sun shifts). But they cannot see the sun in the darkened hive. Evidently, they are "aware" of the passing time and make the necessary corrections. The time sense of honeybees has long been known to people who have sweet snacks in their garden at a set time every day. Within minutes of the regular time, foraging bees arrive for their share of the jam. The speed of the bee's clock seems to be related to its metabolic rate. If normally punctual bees are (a) chilled (to lower their metabolic rate) or (b) exposed to an anesthetizing concentration of carbon dioxide. They arrive late to the picnic table.

Polarized light

Von Frisch also discovered that scouts (and foragers) don't actually have to see the sun to navigate. As long as they can see a small patch of clear blue sky, they get along fine. This is because sky light is partially polarized, and the plane of polarization in any part of the sky is determined by the location of the sun. Try it by rotating a pair of polaroid® sun glasses!

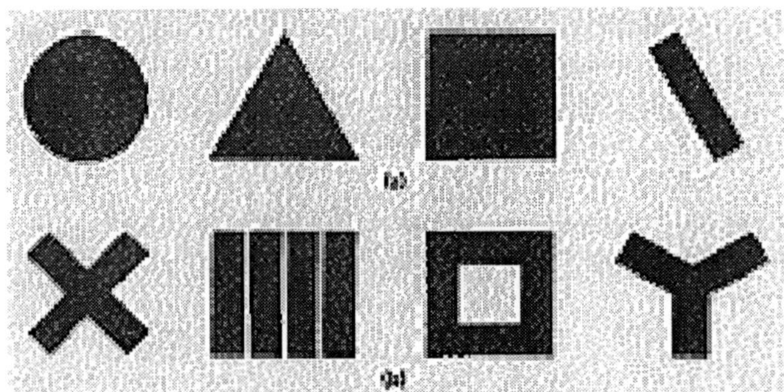

Figure 38. Different images to calculate duistance

Swarming

Before a new queen emerges, the old queen leaves the hive, taking many of the workers with her. The swarm usually settles somewhere, e.g., on a tree branch, while scouts go searching for a new home. Each scout that finds a promising site, returns to the swarm and dances on it just as though she had found food. Eventually, the swarm departs for the location promoted most vigorously.

Grooming

Workers have another type of dance — rapidly vibrating from side to side — that tells other workers that she needs help removing dust, pollen, etc. from hard-to-reach places on her body. Honey bee foragers communicate the location of rich forage patches to their nestmates via a well-characterized symbolic dance language (Lindauer 1985; von Frisch 1993). There also appears to be a relationship between dance tempo and distance of the forage source which varies widely between honeybee races (see von Frisch [1993]). A recent study (Esch *et al.*, 2001) suggested that these differences may be environmental rather than genetic in origin. However, Rinderer and Beaman (1995) studied the genetic basis for some of the observed differences in dance dialect. Their results strongly suggest that each of the transition points (round → sickle, sickle → waggle) is controlled by a single gene with multiple alleles (at least two).

The Controversy

Karl von Frisch won the Nobel Prize in 1973 for this discovery in spite of a challenge to his theory of the dance language thrown up by two scientists from California, Adrian Wenner and Patrick Wells. Their claim was that von Frisch's decoding of the dance language was a mere delusion and that bees really locate food discovered by their sisters by means of scent left by the original discoverer on the way to and at the location of the food. This counter argument cannot be dismissed out of hand because bees have a very well developed sense of smell. They are indeed capable of finding food discovered by their sisters by means of the scent of the discoverer on the way to and at the location of the food, under some circumstances. However, the question is not so much whether bees can ever locate a food source successfully by means of odour cues left by the discoverer. The question of interest is whether bees can ever successfully communicate using a symbolic, arbitrary dance language without the need for scent marks. The question of Interest Is whether bees can ever successfully communicate using a symbolic; arbitrary dance language without the need for scent marks of the discoverer along the way to and at the target location. This can only be tested by eliminating the scent marks of the discoverer and retaining only the information provided by the dancer. This was accomplished in an ingenious way by James L Gould of Princeton University. He made the bees "lie" and proved the existence of a true language!

Bees can tell lies

If a bee nest is removed from its normally dark cavity, the dancers need no longer use "up" to mean in the direction of the sun; they can as well aim their waggle runs in the direction of the food source using the sun directly as a reference point. The waggle runs can now be in the direction of the sun, away from the direction of the sun or 60° to the right of the sun and soon, depending on where the food is located. A nasty trick used by Gould was to place a powerful source of artificial light close to the nest. When you do this the bees then ignore the sun and confuse this light for the sun. In other words, they orient their dances with reference to this artificial sun. If the light is placed in a position very different from that of the real sun, bees use the real sun while foraging, and the artificial sun while dancing and thus communicate completely wrong information. In addition to their two compound eyes, honey bees have three simple eyes called ocelli on the back of their heads. These ocelli are used to sense the intensity of light and help bees to decide when to start and when to stop foraging. The ingenuity of Gould's experimental design was to paint the ocelli of the dancers so that they could not see the bright light.

They therefore use gravity for conveying orientation and let us say indicated 60° to the right of"up" because the food was located 60° to the right of the real sun.

However, the recruits (the dance followers) had normal unpainted ocelli and mistook the bright light to be the sun. Thus the recruits were expected to go to the misunderstood location, 60° to the right of the bright light instead of going to the true location, 60° to the right of the real sun. Gould of course was waiting at this "wrong" location (60° to the right of the artificial light, at the appropriate distance) where he was confident the recruits would go and sure enough he was rewarded by the arrival of the bees. Notice that the bees should have gone to the correct location, 60° to the right of the real sun if they had relied on the scent of the dancers on the way to and at the location of the food. Although von Frisch retained the Nobel Prize, even this elegant experiment did not set the controversy to rest.

The Robot Bee - A Novel Solution

The main problem is that we cannot claim to know exactly what it is that the dancer is telling the recruits. What we need is to be able to talk to bees in their own language and restrict communication to only those elements of the language that we have deciphered. Sounds impossible, does it not? Well, not quite. A. Michelsen and B.B. Andersen from Denmark and Wolfgang Kirchner and Martin Lindauer from Germany have constructed a mechanical "robot" bee that talks to the real bees through a computer, programmed by the scientists based on their idea of the bee dance language and, believe it or not, the bees understand! The robot bee was made of brass and was about the size of a real bee (Figure 5). It was coated with a thin layer of wax and made to sit in the nest among the real bees for about 12 hours and thus came to smell like the bees. It had wings made of razor blades which when vibrated produced acoustic signals similar to those A. Michelsen and B.B. Andersen from Denmark and Wolfgang Kirchner and Martin Lindauer from Germany have constructed a mechanical "robot" bee that talks to the real bees through a computer produced by real dancing bees. Two rods attached to the robot bee helped control its movements through a computer program. Most importantly a plastic tube near the model's head delivered, like the real bee, a drop of sugar solution from time to time. This was essential to prevent the real bees from attacking the model! Now by making the robot bee perform a waggle dance with a particular orientation of the waggle run, Michelsen and Co. thought that they could make the bees go wherever they wished them to go and sure enough they were right! This gives us confidence that our understanding of the bee dance language is sufficient to make the bees understand the true meaning of a message. This should lay the controversy to rest and vindicate the Nobel Laureate's theory but, knowing human nature; Wenner and Adrian are not going to be convinced. That won't be so bad really because it will motivate someone to do an even more ingenious experiment. Science, after all, thrives on controversy and a penchant for disbelieving others' findings.

Current questions

Although it is now quite clear that bees do decode the dances, odor does play a strong role in recruitment to food sources. It is appropriate to think of the dance as giving recruits a general idea of the direction and distance to the food source. Recruits then search in this area for sources matching the odors they have learned from the food carried by the dancing bee. Depending on the distribution of available food sources, the distance and direction information might be crucial in organizing a colony's food collection, or relatively unimportant. However, the relative importance of these two mechanisms in different habitats is just beginning to be investigated.

The angular scatter in the dance itself decreases with increasing distance indicated, as von Frisch reported. This change in scatter may be the result of changing duration of the waggle runs of the dance, but it also may be an adaptation to recruit bees to patches of more or less constant size at varying distances. This idea is supported by Seeley and Burmann's finding that the dances of scouts for nest sites,which are always single points rather than patches, have less scatter than those of nectar foragers. However, these results differ from those reported by Will Towne on the same issue, and this remains a question of current research. The evolutionary origin of the honey bee dance remains incompletely discerned. All species of *Apis* perform recruitment dances, though there are interspecific variations in a number of the aspects discussed earlier. The stingless bees (Meliponini), the bumble bees (Bombini), and the orchid bees (Euglossini) are the closest relatives of *Apis,* but the phylogeny of these different taxa within the Apidae remains controversial. Stingless bees are highly social and have a variety of mechanisms of recruitment that may provide possible antecedents to the dance language, but a determination of how the current form of the dance language might have arisen from these components must await both a greater understanding of recruitment mechanisms within the stingless bees and a more firmly established phylogeny within the family Apidae.

In conclusion, the honey bee dance language serves as a model of animal communication at all levels. It is one of the more intriguing behaviors in the animal kingdom and solidifies honey bees as one of the most interesting systems in biology.

CHEMICAL COMMUNICATION

Honeybee pheromones

Pheromones are chemicals used for communication among members of a species. A pheromone acting directly on the recipient's central nervous system and resulting in an immediate behavioural response is called releaser effect pheromone while resulting in physiological changes it is primer effect pheromone. Honeybee pheromones have releaser effect through sex attraction, alarm and aggregation, trial production, clustering and mutual recognition and primer effect through inhibition of reproduction. Pheromones of social bees (Free, 1987) and chemical world of honeybees (Winston, 1991) form a very good basis for understanding the honeybee pheromones. Knowledge on the gland producing the pheromones, chemical nature, mode of transmission and their influence on the colony behaviour is essential for their application in the management of honeybees. Most of the pheromone research done so far was focused on *Apis mellifera* while very limited attention was given for other honeybees and eusocial insects.

Pheromone Producing Glands

Glands producing pheromones are located mainly in the head and the abdomen. Mandibular glands in the head, tergite glands,, Nasonov gland, Koschevnikov's gland, sting sheath gland, Dufor's gland, wax glands and poison gland in the abdomen and Arnhart gland (tarsal gland) in the thorax are the important sites of pheromone production (Figure 39).

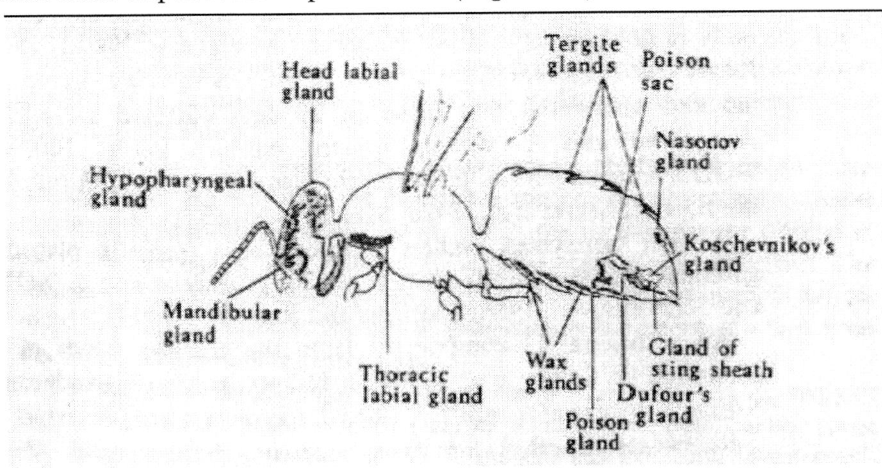

Fig. 39 Diagram of bee indicating location of pheromone-producing glands

Chemical Nature

A lot number of components have been identified from the pheromones of honeybees. In as early as 1973, 18 chemicals have been reported (Pain, 1973) to be produced by worker, queen, drone, brood and comb. The pheromones of queen and workers play a major role as indicated in Table 8.

Table 18. Pheromones produced by honey bees

Pheromone	Gland	Chemicals	Functions
Worker - produced			
Nasonov	Nasonov	Geraniol, Nerolic acid, Gernic acid, (E)-citral, (Z)- citral, (E–E) farnesol, Nerol	Orientation
Footprint	(Arnhart)	?	Orientation
Forage marking	?	?	Orientation at flowers
Alarm	Mandibular	2–Heptanone	Alarm and defense
Alarm	Sting	Isoamyl acetate, 2-onanol, N-butyl acetate, N-hexylacetate, Benzyl acetate, Isopentyl alcohol, N-octyl acetate, (Z)-11-eicosen-1-ol	Alarm and defense orientation
Recognition	?	?	Kin and/or colony recognition
Queen – produced			
Queen substances	Mandibular	9-keto-(E)-2-decenoic acid (9ODA) 9-hydroxy-(E)-2-decenoic acid (9HDA)	Inhibition of queen rearing, Inhibition of worker ovary development, Drone attraction, Worker attraction to swarms, Swarm cluster stabilization, Stimulate release of Nasonov pheromone, Stimulate worker foraging, Queen recognition
Olfactory	Koschevnikov	?	Worker attraction
Tergite	Tergite	?	Drone attraction and copulation Inhibition of worker ovary development Inhibition of queen rearing
Immature	?	?	Inhibition of queen rearing
Footprint	?	?	Inhibition of queen cup construction
Drone - produced			
Marking	Mandibular	?	Marking of congregation areas

Brood - produced			
Brood	?	?	Stimulate foraging Brood recognition, Inhibition of worker ovary development
Comb – produced			
Hoarding	?	?	Increase nectar storage

(? Indicates no known gland or chemical)

The honeybee scientists till date continue to pay much attention to solve many of the chemical communication mysteries. Oldroyd *et al.* (2002) discussed the egg marking pheromones of *A. mellifera* produced by Dufor's gland. Keeling *et al.* (2003) recently identified four additional queen-produced compounds that function synergistically with queen mandibular pheromone (QMP) in attracting workers to form a retinue around the queen. They are (1) methyl oleate (MO; methyl (Z)-octadec-9-enoate), (2) coniferyl alcohol [CA; (E)-3-(4-hydroxy-3-methoxyphenyl)- prop-2-en-1-ol], (3) hexadecan-1-ol (PA), (4) linolenic acid [LEA; (Z9,Z12,Z15)-octadeca-9,12,15-trienoic acid]. Hoover *et al.* (2003) discussed the influence of the above components in regulating the worker ovary. Among the various chemicals components of pheromones, 9 ODA [9-keto- (E)-2-decenoic acid] and 9 HDA [9-hydroxy-(E)-2-decenoic acid] are most thoroughly studied. These are the enantiomers of long chain hydrocarbons (Fig. 39).

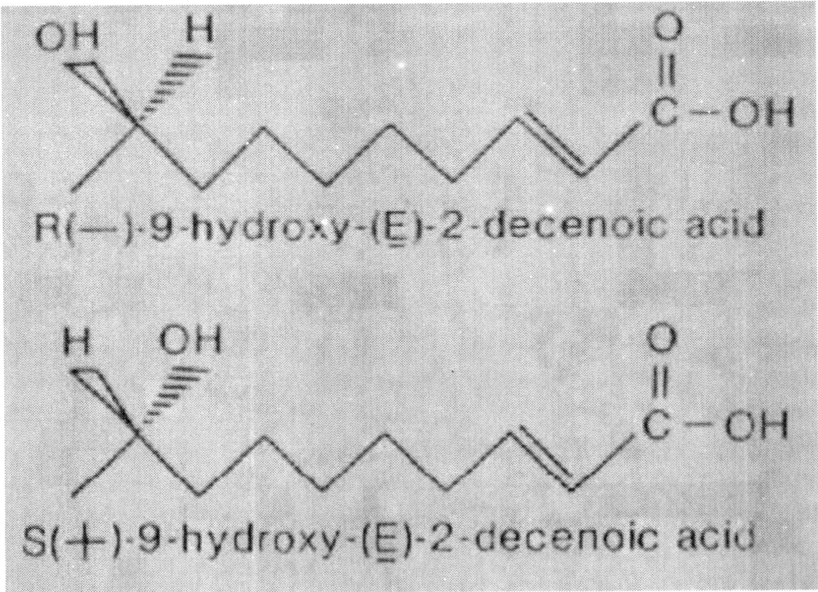

Fig 40. Enantiomers of 9-hydroxy-(E)-2-decenoic acid showing the two chiral forms of molecule

Mode of Transmission

There are a number of ways by which honeybee pheromones are distributed within the colony depending partly on the volatility of the pheromone concerned.

Highly volatile components of low molecular weight notable among which are the alarm pheromones and attractant pheromones are transmitted in the air. *Apis mellifera* and *Apis cerana* honeybee species whose colonies build several parallel combs and are kept in hives, live naturally in enclosed cavity or nests. The ability of bees to renew rapidly the air inside the nest or hive by fanning at the entrance and on the comb could help to remove the volatile pheromones when the need for them is past. Indeed, the sheltered conditions inside a nest would seem to favour communication by airborne pheromones. Airborne brood pheromones may help returning foragers to become aware of their colony's food requirements, and airborne queen pheromones could help all bees to be aware of her presence.

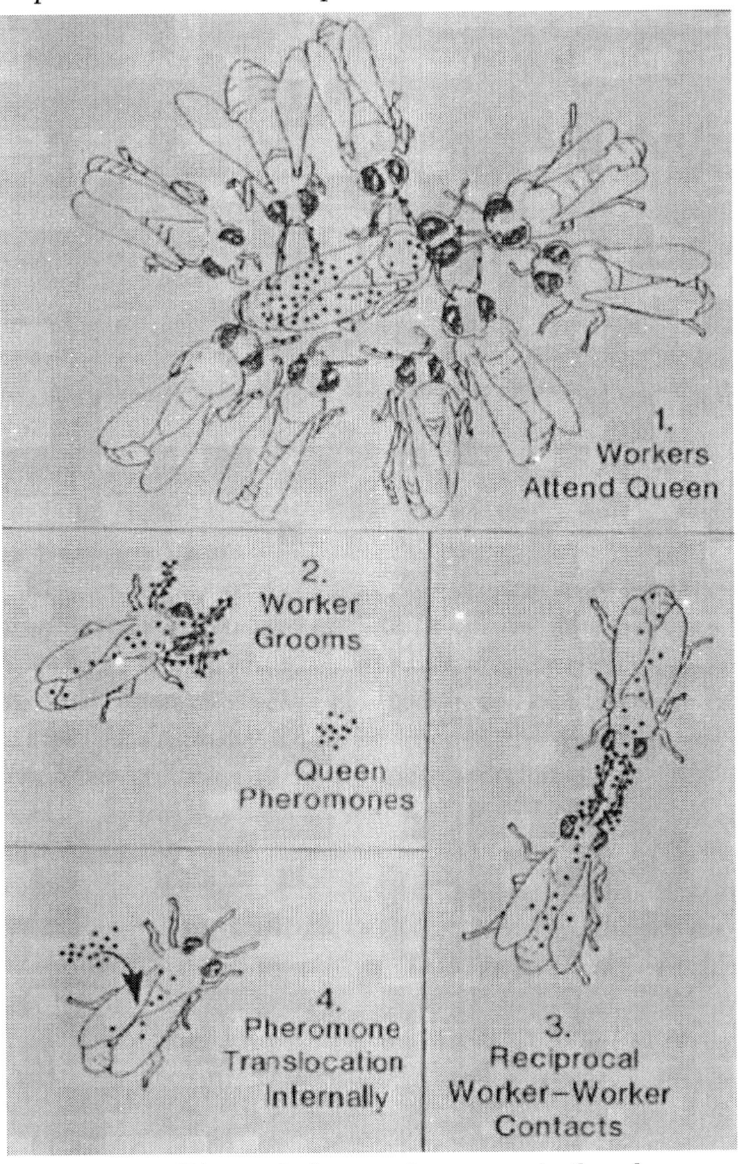

Figure 41. Dispersal of queen pheromones in the colony.

In contrast to the other honeybees, colonies of *Apis dorsata*, the great honeybee, and *Apis florea*, the dwarf honeybee, build only single combs and often nest in exposed situations. However, the curtain of worker bees, often several thick that screens the comb leaves a sheltered cavity beneath it in which the house-bees perform their activities and foragers enter to deposit their loads of nectar and pollen.

Other pheromones functioning within the nest are relatively non-volatile and are adsorbed onto the both surface where they are perceived by contact chemoreception. Long chain paraffins in the cuticular wax of honeybees tend to act as fixatives. Examples of such pheromones are those responsible for brood recognition, caste regulation and colony odour. They are passed from bee to bee either by direct bodily contact or in the food supply. They may also be deposited and perceived on the surface of the comb. Some transmission mechanism besides direct worker-queen contacts must operate, since workers not directly in contact with the queen nevertheless show the effects of pheromonal activity. Three transmission modes have been proposed: transmission by food exchange between workers, volatile transmission and transmission between the body surfaces of workers, particularly during antennal contacts. Of these, only the volatile transmission mode has been conclusively demonstrated. Although neither of the identified mandibular secretions are particularly volatile, both workers and drones outside the nest are attracted to queens, queen extracts and 9ODA and/or 9 HDA; workers are attracted during swarming and drones for mating. In both cases only airborne odors can explain the attraction.

Regarding the other two hypotheses, most evidence supports surface transport rather than transmission by food exchange. Since workers do exchange liquid materials frequently, the food exchange hypothesis is an attractive one. Studies showed that radioactively labeled sugar syrup can be distributed to most workers in a colony within hours. Although queen extracts presented in the drinking water showed some inhibitory effects on worker behaviour, extracts mixed with sugar syrup higher than 5-10% in concentration showed no effects. In contrast, there is abundant evidence for the surface transport hypothesis. The hypothesized movement of queen produced pheromones by worker messenger bees to other workers through surface transport and internal translocation is picturised in Fig. 40. First, radioactively labeled 9 ODA was found to move on immobilized workers from one body part to another, both externally and internally, through translocation, although some movement of radioactive material may have been due to labile tritium rather than 9 ODA. Second, a body part or other surface that has been in contact with a queen becomes attractive to worker bees. Third, there are specific olfactory receptor cells in the antennae specialized for 9 ODA perception. Fourth, behaviour of workers attending queens and following queen attendance strongly supports the idea of surface pheromone transport. Workers in the queen's retinue touch her with their antennae, particularly, and tongue. Most frequently, the antennal tips are lightly and quickly brushed over the queen, Fig. 40 although at times the entire antennae move over the queen's body in a flurry of activity. Following these contacts, workers generally groom their antennae with their forelegs, presumably passing pheromones between antennae, tongue and forelegs. Most significant, workers which have contacted queens in this manner move through the colony making reciprocated antennal contacts with other workers at high rates for approximately 30 minutes. A typical worker will exchange nestmate antennations with an average of 56 other workers during this time but only exchange food twice. Workers tend to perform these activities between 3 and 9 days of age; although workers of these ages do no specialize in queen contacts, they are particularly attracted to queen substances. These workers have been named 'Messenger bees'. It has been concluded that pheromones are distributed over the queen's body by her grooming, picked up by direct contacts between workers and the queen, and then transported throughout the colony by these messenger bees. Queen substances probably act on the worker's hormonal system when the pheromones are translocated internally from the worker's body surface. The injection of 9 ODA partly suppressed ovary development in an experiment probably owing to inhibited growth of the corpora allata, the gland that secretes hormones responsible for the stimulation of ovary development. The growth of neurosecretory cells may also be inhibited by queen substances. Pheromones used in the field are usually transmitted in the air. They are used by a swarm or migrating colony when it is clustering and probably also when it is in flight. Pheromones are also used to mark an enemy or a source of forage and attract other bees to it. Some of the pheromones used to mark flowers may be quite persistent.

PHEROMONES ON COLONY BEHAVIOUR

I. Queen Produced Pheromones

The major two components of the pheromone influencing the colony behaviour are 9ODA (commonly called as queen substance) and 9 HDA. These are mainly secreted by mandibular glands. Other pheromones may also be produced by abdominal tergite and Koschevnikov glands. Both 9ODA and 9HDA as well as the unidentified pheromones have been shown to have numerous biological functions, including inhibition of queen rearing and swarming, prevention of worker ovary development, attraction of drones for mating, attraction of workers to swarms, swarm cluster stabilization, stimulation of Nasanov pheromone release, induction of worker foraging, and queen recognition. However, although it is likely that queen-produced pheromones are involved to some extent in these functions, their effects are difficult to evaluate for a number of reasons. Research of these substances has utilized a wide range of techniques using field, within-colony, and laboratory caged bee assays. In addition, different concentrations, isomers and combinations of chemicals and extracts from bees have been used, and in some cases poor syntheses of these pheromones have resulted in the wrong chemicals or mixtures being produced. Also, factors such as worker and queen age and physiological condition are often not standardized and most colony functions are also controlled by non-chemical behavioural mechanisms in addition to pheromones. Nevertheless, queen substances are involved in at least some of the following six functions.

1. Inhibition of queen rearing and swarming

One of the first functions of 9 ODA and 9 HDA was the inhibition of queen rearing, which prevents reproduction by swarming or supersedure. Substances produced in the queen mandibular glands prevent workers from rearing queens and that these substances are more active in glands from mated queens than from virgins. The amount of queen substances necessary to inhibit a colony from rearing queens increased with the number of workers in a colony and naturally superseded or swarming queens contain less queen substance than mated and laying queens. It has been confirmed that both 9 ODA and 9 HDA act together to suppress queen rearing. Workers which have visited cells containing queen pupae move through the colony and make antennal contacts with other workers, indicative of pheromone transfer and the presence of queen pupae inhibits queen cell production; these effects are not found for queen eggs or larvae. Thus, immature queen pupae produce substances which prevent additional queen rearing, possibly but not necessarily 9 ODA and/or 9 HDA.

The footprint pheromone produced in the queen's tarsal glands also may be involved in the suppression of the first stage of queen rearing, associated with swarming, and cup construction. A queen walking along a glass surface deposits an oily, colourless trail at a rate of 0.8 mg of material/hr and this material evidently originated in the tarsal Arnhart glands. Colony over crowding may prevent the queen from moving along the bottom of combs and depositing the footprint pheromone. This, in combination with the mandibular gland secretion, may inhibit the construction of queen cups and thereby delay or prevent swarming.

2. Prevention of worker ovary development

Queen-produced substances prevent worker ovary development and egg laying. Only 9 ODA has been investigated for this effect and although it does inhibit worker ovary development, the inhibition is not as effective as the presence of a mated queen. Thus, other pheromones or behaviours associated with mated queens may be required for full inhibition, including substances from the queen's abdominal tergite glands in addition to brood-produced pheromones that are important for worker ovary suppression.

3. Attraction of drones for mating

The attraction of drones to queens for mating was due to 9 ODA and to a lesser extent 9 HDA. Drones may be attracted by queen pheromones at distances upto 60 m and 9 ODA may function at close range as an aphrodisiac to stimulate the mounting of queens by drones. Unidentified pheromones of the dermal tergite glands have been found to attract drones at distances less than 30 cm and increase their copulation activity. Attractive pheromones from the Koschevnikov glands may be released when the queen's sting chamber is opened prior to mating.

4. Attraction to swarms and their stabilization

The queen's mandibular gland secretions have three swarm-associated functions: they attract workers to the cluster, stabilize the cluster and aid in swarm movement to a new nest site. When workers issue from a nest in a swarm, they are attracted to the queen wherever she has alighted; 9 ODA appears to be more important for attracting flying workers, whereas 9 HDA seems to function more to stimulate alighting and clustering. Once the cluster has formed, both pheromones, but 9 HDA in particular, help to stabilize the cluster and prevent the workers from becoming restless and leaving the cluster prior to the swarm's movement to a new nest site. When the swarm lifts off to travel to the new nest, workers sense the presence of the queen through the 9 ODA odour and the swarm moves as a group only if the queen is present. For both cluster formation and swarm movement, Nasonov pheromones act in concert with the queen-produced substances.

5. Stimulation of Nasonov pheromone release and worker foraging

Exposure to 9 HDA stimulates workers to release Nasonov pheromones at nest entrances, but 9 ODA has no similar effect. This difference suggests that 9 HDA rather than 9 ODA may stimulate Nasonov release by workers which have recently settled on swarm clusters. Foraging for nectar can also be stimulated by the presence of a laying queen, queen extract or 9 ODA alone.

6. Queen attractiveness and recognition

Queen substances are important in worker recognition and attraction to their queen. There is little doubt that 9 ODA and probably 9 HDA attract workers and allow them to differentiate between fellow workers and queens; substances from the abdomen also have the same effect. The flowing flow chart (Figure 40) depicts the inhibiting and stimulating effects of queen pheromones.

II. Worker produced pheromones (Brood Pheromone)

In a honeybee colony, the ability of worker bees to recognize brood, its stage of development, sex and caste, is a prerequisite of brood care and progressive larval feeding. Cells containing brood are inspected at frequent intervals by house bees; presumably the stimuli they encounter release the appropriate behaviour patterns in physiologically and psychologically suitable bees.

In a honeybee colony, different diets are given to female larvae of different ages and castes, so workers providing the food must receive the appropriate signals, which are probably in addition to a more general brood-recognition pheromone. The amount of pheromone produced is likely to differ with a larval size; this may help determine the amount of food it is given.

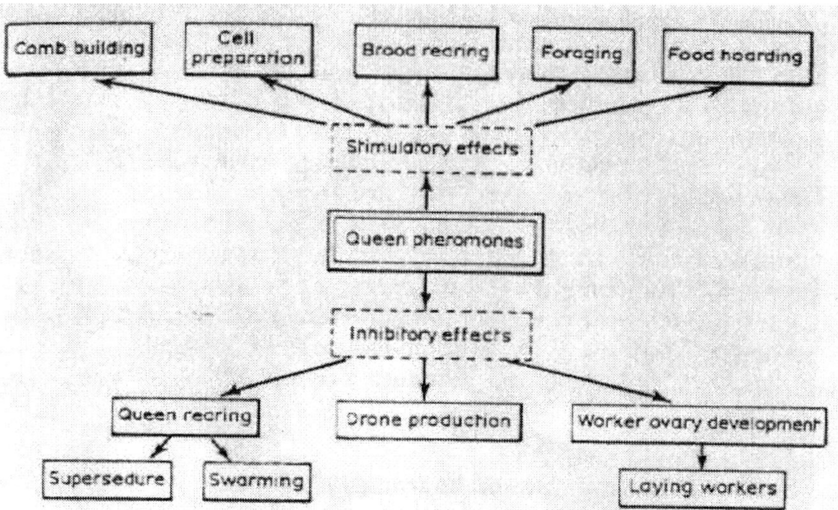

Figure 42. Flow chart exhibiting inhibiting and stimulating effects of queen pheromones.

The amount of pollen a colony collects increases with the amount of worker brood it is rearing. This may be partly because the amount of pollen collected directly influences brood rearing.

Honeybee larvae may be relatively more important than the queen in stimulating pollen collection in poor foraging conditions. The results of attempts to stimulate foraging by providing colonies with a larval extract are difficult to interpret and in general were unsuccessful; the extract was less effective than larvae when dispensed to colonies in poor foraging conditions, but in good foraging conditions it appeared to stimulate pollen collection by free-flying colonies with queens. The influence of brood pheromone on pollen collection is depicted below in Figure 41.

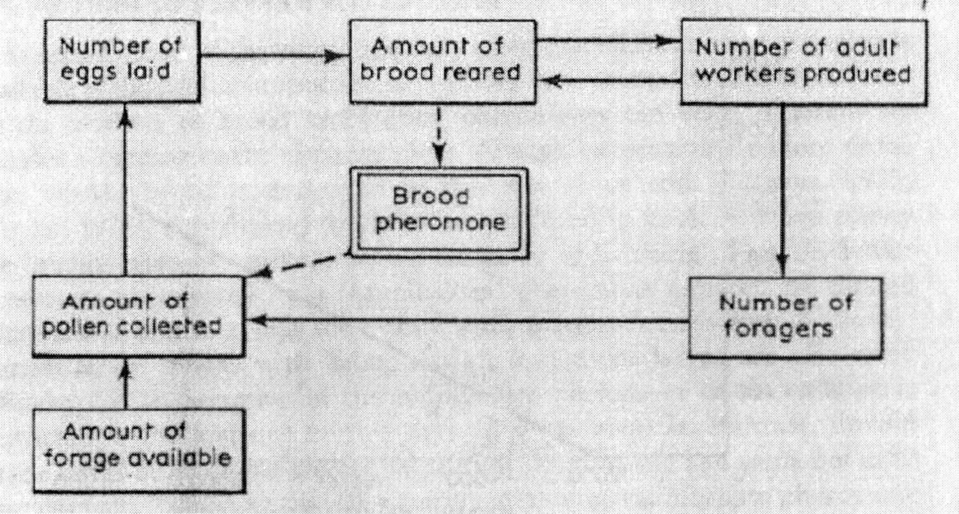

Figure 43. The influence of brood pheromone on pollen collection by honeybees.

III. Comb Pheromone

The wax secreted by bees and used by them to build comb in which food is stored and brood is reared could perhaps by itself be regarded as a pheromone. Furthermore, it is likely that pheromones incorporated into comb as it is built, or subsequently added to the cells may stimulate or regulate various activities of the colony. The stimulatory effect of brood, queen and comb pheromone are given in the figure below. Comb and the pheromones it contains may have stimulatory effects other than hoarding and foraging. The presence of worker comb stimulated the development of ovaries of queenless worker bees kept in groups of 150 in small cages. Subsequent attempts to demonstrate this in small combless colonies were frustrated because the bees consistently built some comb themselves. Colony defense is also encouraged by the presence of empty comb within the nest. The stimulatory effect of brood, queen and comb pheromone are given in the Figure 42.

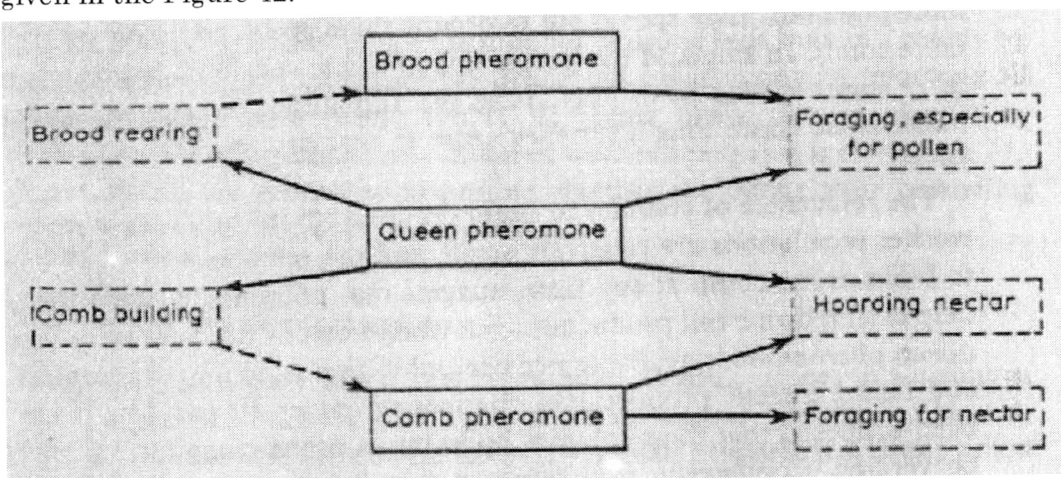

Figure 44. The stimulatory effect of brood, queen and comb pheromone

IV. Trail pheromones

Honeybee workers involuntarily deposit pheromones that attract others both when foraging and when entering their nest. Workers entering their hives deposit a persistent attractive material. The trial pheromone of workers from another colony was also attractive but slightly less so than that of workers from bee's own colony. The trail odour is often used to help bees orientation to a change in the location of the entrance to their hive. The floor and the inside walls of the hive or nest and the combs themselves are also marked with trail pheromone. Old comb accumulated with trial pheromone is more attractive to bees than the new combs. The trail pheromone is secreted either by tarsal glands (Arnhart's gland) or from all over the body. Honeybees mark the rewarding flowers with stable trail pheromone with an attractive odour while with deterrent pheromone in an unproductive food sources. Deterrence from unprofitable flowers should be short term only while they are replenishing their stocks of nectars. Pheromone signals used to mark flowers as attractive sources of forage should be longer term. Ultimate message given by the pheromones left on a flower will depend on the balance of the attractive and deterrent pheromones.

V. Nasonov pheromone

The Nasonov gland is present only in workers and released when swarm that is moving or forming cluster, marking the entrance to their nest and indicating a source of water on which they are foraging. Important chemicals identified in the glands are geraniol, nerol (E,E)- farnesol, (E)- Citral, (Z) Citral, Geranic acid and nerolic acid. Changes in pheromone amount appear to be associated with structural changes in the glandular cells.

VI. Alarm and Aggregation pheromones

Guard bees are usually present at the nest entrance and are few compared to the colony population. When danger is imminent, alarm pheromones serve the dual functions of being used to muster help and to direct the attack. The alarm pheromone is released by opening sting chamber and protruding her sting. Pheromone accumulates beneath the sting shaft setose membrane. Pheromone dispersal is aided by fanning of wings. Main component of the alarm pheromone of the sting gland is isopentyl acetate. Several more components mostly acetates and alcohols of low molecular weight have been identified. Alarm pheromones are unstable and do not persist for long in the region. Some of the less volatile oily components like alkanes and alkenes help in retaining the more volatile chemicals.

Mandibular glands produce 2-heptanone which acts as an alerting pheromone. This pheromone is mainly to repel the robber bees. Colonies differ greatly in the eagerness with which they defend themselves against intruders. These differences may be linked to the amount of alarm pheromones their bees produce or release, or to the threshold of response of their bees to alarm pheromone.

Nest and Nestmate Recognition

Honeybees use the odours emanating from their colony to help them locate the entrance to their nest on return from the field and use the odours emanating from individual bees to help them distinguish friend from foe. Foraging honeybees that have found the entrance to their hive after being temporarily disoriented release Nasonov pheromone at the hive entrance. Odours from inside the hive that release this behaviour include that of the comb, food stores, adult workers, drones and queen. Presumably all bees respond to these combined odours when returning home from field excursions and together they may be regarded as the hive or nest odour. Honeybees need to defend their food stores from robber honeybees of other colonies, especially at times of the year when little forage is available in the field. It is therefore of vital importance that guard bees should have a means of distinguishing between members of their own and other colonies. It has long been supposed that worker bees of the same colony share a common distinctive colony odour which is different from that of other colonies.

Mutual Recognition

When a colony rears a new queen without swarming the old queen may be killed by the workers immediately or she can remain alive for several months and mother and daughter may be found on the same comb without signs of antagonism. In all other circumstances in which more than one queen are present in close proximity, they are hostile to each other. When two virgin queens meet on the comb they will fight until one is killed and the first virgin queen to emerge in a queen less colony will sting and kill any others that are still in their cells. Queen use specialized sensilla for the perception of 9 ODA to discover new and potentially hostile queens in their colony.

Regulation of drone production

The production of drone comb and brood and the toleration of adult drones in a colony seem to be regulated by a balance between stimulatory and inhibitory pheromones. The below mentioned figure 43 depicts the regulation of drone production.

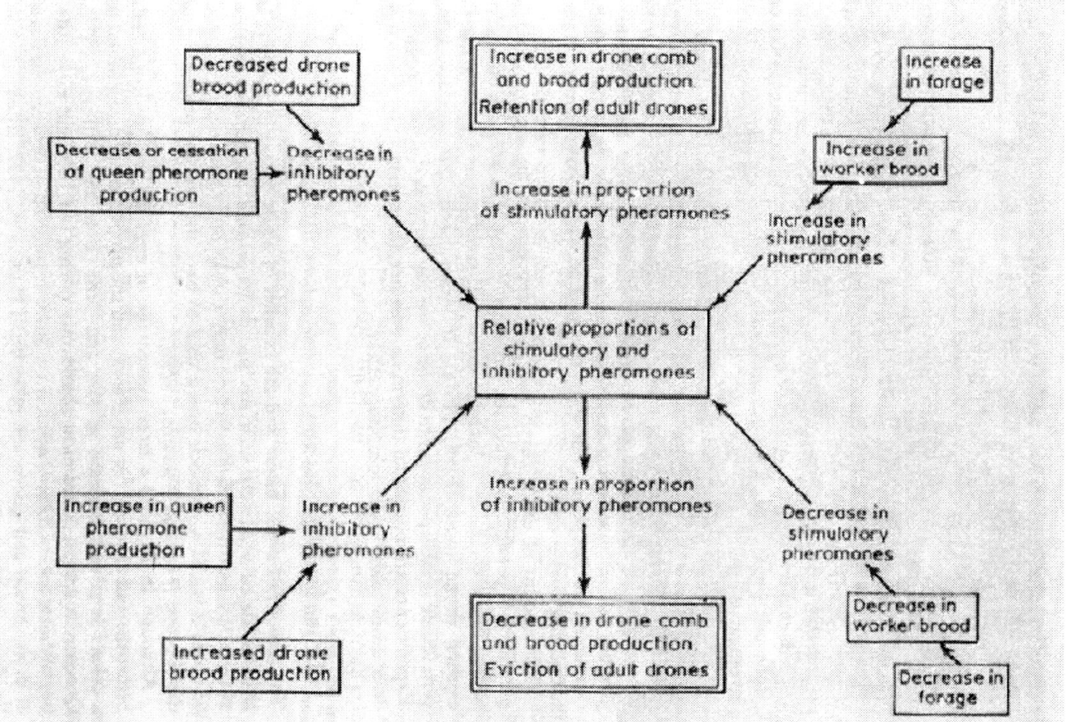

Figure 45. Depicts the regulation of drone production.

Probably a laving queen produces pheromone that inhibits the rearing of drone brood and the tolerance of drones. It seems likely that worker brood produces a pheromone that stimulates drone production and that the relative amounts of queen and brood pheromone determine the proportion of drone cells built, the amount of drone brood reared and the attitude of workers toward drones. Thus, when a queen is removed from a colony drone cells are built only if worker brood is present.

Beekeeping application

If synthetic queen pheromones were commercially produced, the same can be of great value for avoiding the time and labour invested for queen rearing and swarming. Introduction of worker brood with their inhibitory pheromone would prevent production of laying workers and their brood in queenless colonies and associated difficulties in introducing new queens. Foraging may be increased by brood pheromone possibly by providing the larval extracts or using a simple device to channel foragers from hive entrance to the brood combs. Drone population in a colony could be regulated by means of queen pheromones.

Synthetic mating pheromones could be used to survey the flight distribution pattern of drones of different genetic origins and further used to trap drones in sweep nets. Artificial mating station can be established by creating artificial drone congregation zone to avoid inbreeding. Increasing the brood or Identifying and applying the brood synthetic pheromone would enhance the foraging behaviour of bees.

Simple and inexpensive traps containing the Nasonov lure have proved to be an effective and competitive means of capturing honeybee swarms and moving to a particular site, trapping the undesirable colonies of Africanized bees, attracting the scout bees to potential nest sites and besides trapping the stray bees in locations where they are unwanted.

Alarm pheromones can be used in standardized tests to compare the aggressiveness of different colonies and so can be used as a tool in breeding a more docile strain. Application of synthetic alarm pheromone a few minutes before an insecticide application in the field might help to reduce any bee losses to acceptable levels. Probably it would only be necessary to apply the alarm pheromone to the borders of the field as foragers leaving or returning to the field would then become aware of it. Application of synthetic alarm pheromone at the hive entrance could repel robbers and ensure the defenders are alert. Colonies can rapidly become adapted to synthetic alarm components dispensed within their hives and as a result are less inclined to sting in defense.

In conclusion, honeybee pheromones are diversified and much complex in nature. The possibilities of using synthetic pheromones to control bee activities and increase their efficiency as crop pollinators and honey producers are enormously varied and challenging. There have been few advances in altering bee behaviour and so directing a colony's activities towards greater productivity.

These advances could eventually be achieved by appropriate synthetic chemicals, analogues to pheromone itself. A single pheromone has many functions and more than one pheromone is often involved in a single category of behaviour forcing the honeybee pheromone subject to be a highly complex situation and indicating the need for more thorough studies.

Temperature regulation

Honeybees, like all other insects, are unable to control their body temperature internally according to changes in the ambient temperature; for this reason they are referred to as "cold-blooded animals". However, although the individual bee cannot control its body temperature, but they are able to survive cold winters as a whole colony (Heinrich, 1993; Southwick, 1985, 1988; Moritz and Southwick, 1992). A populous honeybee colony can regulate the interior temperature of the hive, particularly within the area surrounding the developing brood. This is possible because at low temperatures they crowd tightly together in a winter cluster. The insulation by the tightly packed outer bees of the cluster, which reduces heat loss drastically, is a decisive factor in survival at low temperatures. In normal colonies, the brood-nest temperature is maintained at a remarkably constant 32-36°C. It is because of this ability honeybees have settled in a wide range of climates, from the tropics to the cool-temperate zones. By fanning their wings, evaporating the water film at the proboscises of the workers, and dispersing drops of water in empty cells, a honeybee colony can reduce its temperature markedly. When water is available, a colony of *Apis mellifera* can withstand an external heat of 70°C. When the external temperature is low, on the contrary, the bees reduce heat losses by clustering together, and the lower the temperature, the more compact the cluster. In addition, in order to generate more body heat, the worker bees will consume more food, especially honey: more heat is released as a result of the increased rate of food metabolism.

The survival ability of honeybee colonies during severe winter months depends on whether the colony has enough workers, adequately provisioned with food. Insulating the hive wall and decreasing the volume of the hive can also improve the effectiveness of the colony's thermal regulation. In the temperate regions, colonies of *A. mellifera* survive by forming clusters around the brood nest, the bees at the surface of the cluster and those within it changing their respective positions at intervals. In this manner, *A. mellifera* colonies can survive temperatures as low as -40°C. The regulation of brood-nest temperature is not confined to the European races of *A. mellifera* which further do not lose this behaviour characteristic on being transferred to tropical regions. Tropical honeybee species and races can also regulate their brood-nest temperature to a certain extent, but they are able to survive only mildly cold temperatures, generally not below 0°C. Mardan and Kevan (2002) found that in case of *A dorsata* capped brood died at temperatures above 36°C. Below 30°C adults tended to emerge deformed. Low temperatures delayed development. The optimal temperature for complete emergence of healthy adult workers was 34°C. Adult workers survived well from about 26 to 36°C. At 38°C they died within 5 days and at 45°C they died within 48 hours. Although syrup (1 sugar: 1 water W/W) consumption did not change but the water consumption rose rapidly above 38°C to over 3 ml/bee in 48 hours at 45°C. Nest temperature control is critical for survival of brood of *A. dorsata* and adult worker bees have tight constraints on their abilities to endure high temperatures. Water availability is vital for cooling the colony under hot, tropical conditions and rearing healthy brood.

It is well known that the brood nest of European races of the Western honeybee, *Apis mellifera* L., remains remarkably constant at about 34–35°C as a result of the thermoregulatory activities of the colony (Seeley, 1985; Moritz and Southwick, 1992; Heinrich, 1993). If capped brood is removed from the brood nest, it can be kept in incubators and the effects of different temperatures on emergence of adults and survival can be studied. In general, if the brood is incubated at temperatures even as low as 32°C, healthy adult worker bees emerge even though development is delayed (Jay, 1959), at 28°C and below brood dies (Himmer, 1927a, b). If the incubation temperature is elevated by even 1 to 2°C above 35°C, the emerging bees are deformed and mortality occurs; above 37°C, brood death is complete (Himmer, 1927a, b). As is true of most organisms, honeybee pupae have a smaller range of heat tolerance above their optimum than below. Adult worker bees of European races of *A. mellifera* are capable of withstanding very low temperatures if they can cluster (Allen, 1959), but as individuals they become torpid and immobilized at temperatures of 11°C and below (Goller and Esch, 1990, 1991). They freeze to death at about 0°C. The upper levels of temperature tolerance are less well understood. Colonies can withstand temperatures of 45–50°C for a few hours, but presumably the individual bees are striving to keep their internal temperatures below that level (Allen, 1959; Free and Spencer-Booth, 1962). Above 45°C enzyme activity and metabolism are likely adversely affected (Seeley, 1985). The Japanese race of the Asiatic hive bee, *A. cerana japonica*, has the remarkable capacity to cause the temperature in a ball of workers to rise to 46°C in the defense "cooking" of hornet invaders in its nest (Sasaki *et al.*, 1995; Ono *et al.*, 1995). The upper lethal temperature limit for this bee is 48-50°C (Sasaki *et al.*, 1995; Ono *et al.*, 1995). In clusters of European honeybees, temperatures of about 46°C have been recorded (Heinrich, 1993) but few studies have been made on the prolonged effects of high temperatures on worker honey bees.

Although open nesting honeybees, such as *A. florea* Fabr. and *A. dorsata* Fabr., are known to thermoregulate, it has been suggested that the effects of temperature on brood maturation may be less pronounced than in cavity nesting species, such as *A. mellifera* and *A. cerana* (Morse and Laigo, 1969). Nevertheless, Mardan and Kevan (1989) have suggested that excessive heat load on colonies of *A. dorsata* results in mass defecation flights (so-called "yellow-rain" (Seeley *et al.*, 1985)) as the bees dump excess heat and increase their efficiency of thermoregulation to maintain temperatures in the colony within tight tolerances. The brood nest temperatures for *A. dorsata* have been measured to fluctuate between 30 and 32°C in the Philippines (Morse and Laigo, 1969) and at about 33.5°C in India (Viswanathan, 1953) and Malaysia (Mardan, 1989). Evidently, the brood of *A. dorsata* is, like that of *A. mellifera*, highly sensitive to high temperatures. High temperatures would cause the death of adult workers and are less sensitive to low temperatures (Allen, 1959; Free and Spencer-Booth, 1962; Seeley, 1985; Heinrich, 1993 for *A. mellifera*).

Heat shielding is a recently identified mechanism used by worker honey bees (*Apis mellifera*) to help maintain constant hive temperatures. Only workers perform this behavior; drones actively avoided heated hive regions. Observations of marked day-old cohorts within brood comb regions indicate that heat shielding is performed by young bees to preferentially protect advanced stage larvae and pupae. As expected, the number of heat-shielders significantly increased with both the temperature of the heat source and the size of the colony. Of the young bees observed to perform the behavior, those aged 12–14 days were significantly more likely to heat-shield than expected. These data suggest that classifications of age-based tasks in honey bees should include heat shielding, and that the behavior is an adaptation designed to protect temperature sensitive

Honey bees have been shown to use temperature to thwart disease (Starks *et al.*, 2000) and predators (Ono *et al.*, 1995). But perhaps the most impressive temperature-related characteristic of honey bees is simple maintenance: honey bees (*Apis mellifera*) maintain colony temperatures throughout the year (Heinrich, 1980, 1985; Seeley, 1985).

Honey bees isometrically contract muscles to heat the colony and fan their wings to cool the colony-often spreading fluid in conjunction to induce evaporative cooling. Maintaining elevated and constant brood comb temperatures is clearly adaptive as it shortens brood development times and maximizes brood health (Heinrich, 1980, 1985; Seeley, 1985). An additional mechanism used by the honey bee to maintain constant colony temperature has recently been identified: heat shielding (Starks and Gilley, 1999). In response to localized temperature increases, worker bees shield the comb from the external heat source by positioning themselves on the hot interior region of the hive's walls. Heat shielding reduces the transfer of heat from outside to inside the hive (Starks and Gilley, 1999) and thus may serve to reduce incidences of over-heating. This behavior-placing the ventral surface directly on heated surfaces — appears to be similar, albeit with the opposite effect, to that reported by Bujok *et al.* (2002). These researchers observed honey bees with elevated thoracic temperatures position themselves directly over capped brood (i.e., over pupae or older larvae preparing for pupation) in order to elevate the temperature, or maintain elevated temperatures, on specific regions of the hive.

Precise control of nest temperature can be regarded as one of the major innovation in honeybee biology made possible by the evolution of their societies. From late winter to early autumn the annual period of brood rearing by honeybees, the temperature in the central, nursery region of each colony's nest is stabilized between 33°C and 36°C averaging about 34°C and usually varying by less than 1°C across a day. In case of capped brood, normal metamorphosis occurs only at the temperatures of 32-35°C. When colonies are without brood, from late autumn to mid winter, the temperature of the bees clustered inside this nest drops somewhat, but remain well above freezing. The core temperature of a broodless winter claster never falls below 18°C while that of tile cluster's mantle stays above about 10°C, even in the face of ambient temperature of -30°C or colder. The mechanisms underlying this impressive thermoregulatory ability comprise a set of tightly integrated behaviours and physiological devices whereby colonies regulate both the production of heat through metabolism and the loss of heat to the environment.

Honeybees need both to heat and to cool the nest, but in temperate climates at least, generating and preserving beat is the more pressing thermal demand. The ability of honeybee colonies to maintain a warm microclimate inside their nests is a direct extension of the adaptations of individual bees for flight. Being insects, bees fly by flapping their wings the most energetically demanding mode of animal locomotion and the flight muscles of insects are among the most metabolically active of tissues. The size: and metabolic rate of dying honeybees, combine to make a bee's thoracic temperature typically 10-15°C above ambient temperature of 15-25°C.

Thus in honeybees an elevated body temperature is an obligatory consequence of flight, worker honeybees must maintain thoracic temperature above about 27°C in order to fly, flight muscles cooler than this simply can not generate the minimum wing beat frequency and power output per stroke needed for take off and flight. Thus, while evolving biochemical machinery adapted to high temperatures, honeybees apparently also developed; the ability to conduct pre-flight warm-up, without which they would remain grounded once cooled below about 27°C. The motor pattern of bees warming up their flight muscles consists of simultaneous activation of the wing elevator and wing depressor muscles. These muscles therefore contract isometrically, producing much heat but few or no wing vibrations. This preflight warm-up behaviour evidently set the stage for the evolution of nest warming, since the mechanisms for warming flight muscles and heating nests are identical Whatever the precise pathway of control, it appears that bees have evolved the ability to generate heat as needed, for preflight warm-up, and that this ability provides the foundation for regulated heating of honeybee nests.

The full development of the honey bees' thermoregulatory skills, however, came only with the evolution of their societies. One of the links between group living and improved nest temperature control is simply the additive capacity for heat production when individuals group together. A colony of 20,000 bees presumably can generate heat some 20,000 times more powerfully than can a single bee. An equally important connection between socially and thermal homeostasis is the ability of the colonies to perform several tasks simultaneously rather than serially. Perhaps the greatest advantage in thermoregulation enjoyed by groups over individuals though, drives from their ability to crowd together into a cluster, through which individual bees greatly reduce the surface area exposed to the general environment and so dramatically reduce their heat loss.

The evolutionary histories of nest microclimate control and sociality are intricately interwoven for honeybees. On the one hand, as we have seen, group living enhances nest thermoregulation. On the other hand, once honey bee colonies evolved to a size of several tens of thousands of individuals, nest - temperature control evidently became essential for social life. With thousands of metabolically active adult and immature bees crowded together inside enclosed nest, a honeybee colony faces the threat of disastrous overheating when ambient temperatures rise above about 30°C. Sustained brood nest temperatures over 37°C, disrupt larval metamorphosis. Moreover, should the temperature inside a nest rise above about 40°C, the bees wax combs heavily laden with honey will soften and collapse. Also the adult bees in a colony can survive only a few hours at temperatures of 45-5°C, which is just 10-15°C above their optimum temperature for full activity (35°C). Thus honey bees, like most organisms, possess a smaller range of heat tolerance above their optimum than below it. The relatively low tolerance of heat by honeybees may reflect their not evolving enzymes more stable than arc normally necessary, since an enzyme which is stable at temperatures for above an organisms normal thermal range is probably too rigid to function efficiently at standard temperatures.

The adaptive significance of preventing nest overheating is clear, but what selective forces favoured the evolution of nest warming? The initial benefit was probably a reduction in maturation time for brood. Speedy brood development fosters rapid colony growth and thus quick recovery of full colony size whenever, a colony's population has declined, such as after winter, following swarming, or in the wake of heavy mortality from predation or disease. Colonies probably place a premium on rapid growth to full size for many reasons, including earlier swarming, superior defence against predators, and greater food collection. Various insect viruses fail to cause lethal infection when their hosts arc reared at temperatures no higher than those of the honey bees broodnest and the honeybees elevated nest temperature may also prove beneficial in resisting bacterial diseases.

The honeybee is originally a tropical insect whose range has expanded into cold-temperate regions through various adaptations, especially the ability to maintain a warm cluster throughout long, freezing winters. Because certain tropical races of *Apis mellifera* appear to lack this ability, some of the mechanisms of temperature control inside winter clusters arc evidently relatively recent adaptations, shaped during the honeybees' penetration into colder regions of the world.

The primary problem in thermoregulation faced by honeybee colonics is one of staying warmer than the surrounding environment. How intensively a colony heats its nest varies with whether or not the colony is rearing brood. If it is, then the control brood nest region is heated to 32-36°C. If the colony lack brood, then the bees "turn down their thermostats," maintaining the core and mantle of the clustered bees only above about 18°C and 10°C, respectively. These two temperatures are critical minima: bees' chilled below about 18°C can not generate the action potentials needed to activate their flight muscles, their source of heat, and bees cooled below about 10°C become immobilized, entering a sort of chill coma. Survival of such hypothermia depends on its duration, chilling to 10°C or colder kills most bees within 48 hours. Honeybees employ two general methods of improving heat retention in their nests, one a long term measure involves selecting a protective nest site which will tightly enclose the colony's combs and bees, smaller nest cavity's entrance opening and sealing unnecessary openings with sticky plant gums and resins. Below 7-10°C, bees become immobile due to the cold and above 38°C bee activity slows due to heat. Honey bees can tolerate temperatures up to 50°C for short periods.

The second general mechanism for reducing heat loss is the process of clustering, whereby thousands of bees in a colony press tightly into a compact, roughly spherical, mass. A two part internal organization is revealed. There is an outer zone consisting of several layers of densely packed bees sitting with their heads pointed inward. These workers form a mantle of insulation. They fill all the empty cells in combs and press together as closely as possible in the spaces between combs. Bees in the inner zone, in contrast, have room to crawl about, feed on the honey stores, fan with their wings and tend brood.

Cooling the nest

To counter overheating, colonies employ several mechanisms of nest cooling in a graded response, starting with simple spreading out of the adults in the nest and proceeding through fanning, water evaporation and finally partial evacuation of the nest. The fanning bees deploy themselves throughout the brood currents. Gangs of fanners face inward at the nest entrance and expel currents of warm air from the nest. Without entrance fanners there would probably be little exchange between the nest atmosphere and outside air through the small entrance opening.

High rates of air flow through the nest are created by these fanners. When cluster expansion and ventilation can not cool the nest sufficiently, the powerful cooling technique of water evaporation is also brought into play. Water for nest cells, smeared as a thin layer over the roofs of open cells or places as hanging droplets in theses cells. It may also be rapidly evaporated via "tongue-lashing" whereby been hang over the brood cells and steadily extend their proboscises back and forth.

A remarkable communication system is employed to regulate a colony's water collection. The fundamental complication in this operation is that the bees the sense the overheating and become "water sprinkler" are the young nurse bees in the central brood nest and not the forager bees which can gather water. Finally, under conditions of externally high temperature and high humidity, the bees may partially evacuate their nest and form a mass of hanging, clustered bees just outside the nest entrance.

Thermoregulation in Swarms

The total resources possessed by a honeybee swarm as it sets out to establish a new colony are the bees themselves and the load of energy rich honey carried inside each bee. Both of these assets are carefully conserved throughout the process of finding of bees when in a swarm. So gentle are they that a bee keeper can perform the seemingly amazing stunt of having a swarm cluster on his thus creating a "bees bear" with little of being sting. By refraining from stinging unless severely disturbed, swarm bee help preserve their fledgling colony's labour force. A second major mechanism for preserving the swarm's capital is through energy-efficient thermoregulation. The outmost bees in a swarm cluster are never allowed to cool below about 15°C and thus are protected from death due to chilling. Simultaneously, the bees in a swarm's core maintain a 30-40°C micro-climate deep inside the swarm. This pocket of brood nest-level warmth may help keep the warm's queen and nurse bees physiologically primed for the massive brood rearing ahead. It may also facilitate the scout bees' flights from the swarm and so speed the passage through this perilous phase of the colony life cycle.

So effective are swarms at retaining their own heat that thermoregulation by bees in the core consists almost exclusively of adjusting the expulsion rate of excess heat. Core bees rid themselves of their excess heat by forming open ventilation channels through the swarm and fanning to force conventional cooling.

Thermoregulation during Foraging

The honeybees ability to thrive in a vast range of environments, from temperate regions to humid tropical and hot desert habitats, reflects powerful temperature control not only by whole colonies inside their nests, but also by individual foragers out amidst the flowers. These bees must maintain their body temperature below the lethal upper limit of 45-50°C, and while flying, must keep their thorax temperature above 27°C, the lower limit for steady flight. The foraging bees manage to fulfill these requirements across the range of ambient temperatures from about 5°C to 45°C thus achieving considerable freedom from temperature limitations when collecting resources for the colony. The stop-and-go foraging is expensive in time and energy spent on warm-up, but can still be profitable when working on rich flowers, especially since once a bee reaches the flowers her behaviour will likely consist of intermittent flight between flowers, thus automatically affording her opportunity for warm up when perched on two flowers.

Honeybees do possess an effective cooling system for flight at high temperatures. The primary mechanism for cooling the thorax appears to be to keep the head cool by regurgitating fluid from the honey stomach and holding he extruded nectar triplet between the tongue and mandibles.

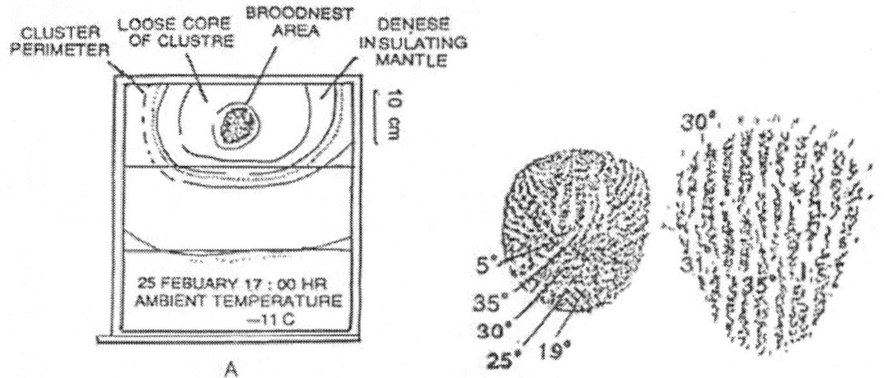

Figure 46. Thermoregulation in honey bees, A, Anatomy of a winter cluster of honeybees; B. Temperature (°C) regulation in a swarm cluster. Under cooler conditions workers cluster tightly, with little internal ventilation, and maintain a core temperature higher than at the surface. Under warmer conditions the cluster spreads out to cool the center, and surface and core temperatures are more uniform (Redrawn from Heinrich, 1981)

Foraging activities

Foraging activities provide necessary raw materials for the hive which include nectar, pollen, propolis and water.

Nectar

It is the chief source of carbohydrates energy for bees for their survival and various life processes. It is a sweet sugary secretion of the nectaries of flowers and other parts of the plant. The nectar is sucked up from the flowers and stored in their honey stomach. The forager bees unload the nectar load by distributing the load to the hive bees. Hive bees ripen the nectar by regurgitation of the drop of nectar to eliminate excess water by fanning process which also dries excess of moisture. During this process there is also enzymatic conversion of sugars. When the nectar has attained desirable concentration, it is stored in cells. Nectar load is usually between 15-30 mg in *Apis cerana* and 30-40 mg in *Apis mellifera*. Each bee may bring 10-20 loads full of nectar per day and each load may be collected from 50-500 flowers depending upon the quantity of nectar per flower.

Pollen

Pollen is an important component of diet, rich in proteins, lipids and is used for egg maturation or larval development. A load of pollen may vary from 10 to 30 mg and may be collected from 50 to 200 flowers depending upon the quantity of pollen per flower. A pollen load is usually collected in 10-20 minutes and as many as 47 loads may be collected each day by a forager.

Water

Water is required when nectars are highly concentrated or for dilution of honey for larval food when there is no nectar income. Water is also required by the bess to maintain the proper humidity levels in a nest as well as to regulate high temperature through evaporation by fanning of the bees. The evaporation causes cooling which regulates the nest temperature. Water foragers can bring a trip of water in 5 minutes and may make 50 such trips per day depending upon the needs of the colony.

Propolis

It is a sticky and resinous material gathered by bees from trees or other vegetation. It is used by the bees to reduce the hive entrance, stick down frames, proofing the walls of hives and cover unwanted debris which can not be pulled out by the bees. *Apis cerana* does not gather propolis but *Apis mellifera* bees are heavy propolisers. Propolis is generally disliked by beekeepers as (1) it sticks to the hands and clothing during warm weather (2) contaminant of beeswax (3) causes difficulty when removing frames. But uin case of migratory beekeeping propolis helps to hold hive parts together during migration.

Foraging range

Flight activity of honeybees is an index of their foraging potential for pollination and honey production. Bees recognize their hive and surroundings and can return to their hive after completing the field activities. If the hive is shifted by more than 3 feet from its original position; the bees returning from their foraging trips get confused and fail to locate new location. They will stay on the old spot and some of them may eventually die. The bees have a tendency to fly in one or two major lines of flight called as beeline. Bees generally prefer foraging near hive. The longer flights are generally avoided, because the food required for such an effort will be more than the nectar gathered. Bees generally visit flowers within a radius of 1 to 3 km. *Apis cerana* normally forages 1 to 1.5 km while as *Apis mellifera* may forage up to 3 km. but if the food sources are limited and the competition is more the foragers may go long distances. The foragers of *Apis cerana* have been reported to fly 600-1040 m while as that of *A. mellifera* from 3 to 14 km. the actual recognition of source of food or nectar is influenced by various factors such as colour, flavor and taste. For example, bees can distinguish green, blue, white, yellow, violet flowers but are insensitive to red flowers. The bees can also discriminate between different concentrations of sweetness and varying salt content in flowers. If flowers of different concentrations are blooming simultaneously, the bees will prefer those which have higher sugar content. Once the bees have started to work on a certain source for nectar or pollen, they continue to visit it and seldom visit more than one or two species on a single trip. Bees also drink water and they like stagnant water rather than fresh. If pinch of salt is added to the fresh water it also becomes attractive. It is again with the sense of smell, that bees can recognize their nest mates. if by mistake a bee of other colony intrudes, it is recognized and killed at once. Strangely enough, if the stranger is laden with nectar or pollen it is not driven away or killed. Surprisingly, drones of all hives are allowed free entry.

Working habits of field bees

It has been found that bees working on sugar solution feeders or flowers which supply food only during a particular portion of the day, seemed to be active only during that period while as rest of the time they remain idle in the hive. The memory of the bees enables them to exploit nectar and pollen from flowers without expenditure of any excess energy. Bees generally forage on single plant species; loads of pollen from more than one plant species may be collected. It is generally argued that that individual bee forages in a restricted area. The size of the area however depends upon the number of flowers in a area, their nectar and pollen content and competition by other foragers.

Speed of foraging activity or the rate at which bees visit flowers also depend upon the amount of nectar and pollen present. Type of flower, stage of development, climatic conditions and the number of competing insects. Generally, pollen collection trips are shorter than nectar collection trips.

Foraging activity also depends upon number of climatic factors such as temperature, light intensity, radiation, time of the day, dew drops on flower, concentration of nectar in flowers and wind velocity. Foraging activity is greatly influenced by temperature. The optimum temperature for active foraging lies between 13-38°C. In extremely high temperature of 43°C, nectar and pollen foraging ceases whereas water foraging continues. Beside, influencing the flight activity of honeybees, metrological factors also influence plant growth, flowering productivity and availability of nectar and pollen.

Speed of flight

In honeybees the speed of flight has been found to vary from 20.9 to 25.7 km per hour (average 24 km per hour) for loaded bees and from 10.9 to 29.0 km per hour (average 20 km per hour) for empty bees. This is perhaps due to the fact that bee, on outward flight does not make a beeline in all the cases for the source of supply. Bees do not like to work in a wind blowing over 24 km per hour.

Colony defence activities

The hive entrance is the point from which many enemies can enter. There are guard bees to check the intruders. Each of the guard patrols a particular area of the alighting board examining all bees entering the hive. It is very interesting to note that any forager from another colony that is loaded with nectar or pollen is allowed to enter without being checked/examined or attacked. Normally strange foragers are intercepted, examined and not allowed to enter the hive. Colony defense activity is one of the most important activities, not only from the point of view that it protects the bees from enemies but because stinging is involved which acts as a deterrents to keeping bees. Therefore, a better understanding of defensive behavior can greatly reduce the bee sting hazard and make beekeeping more pleasurable.

Chapter 11

Bee pasturage

Food of the Honeybees The natural diet of the bees consists of carbohydrates, proteins, vitamins, salts etc. The nectar of the flowers, which the bees collect and convert into honey, is the source of carbohydrates and vitamins. Pollen, the yellow powder in the flowers, is the main source of protein and is mixed with honey before it is fed to the larvae. The winged bees, when they are young, also feed on this mixture. Soon they become strong and secrete the royal jelly, which forms the food of the queen, and of the larvae in the early states of growth. These young bees thus act as nurse bees; but as they grow older they take up the field duties and feed on honey only. In a hive, the nurse bees are often seen going from one cell to another, feeding the young larvae.

For brood rearing, it is important that there are plenty of flowers in nature as the source of pollen and honey. For the growth of one larvae of honeybees into an adult bee, one cell full of honey and one cell of pollen is required. In other words, two frames of honey and pollen are required by the bees to raise one frame of brood.

Bees can also use the sugar syrup as food (sugar dissolved in an equal quantity of water). Sugar is offered to supplement honey resources or in the extreme case to save the weak colonies from starvation. Early in the spring, when the flowers are not in abundance bees can be stimulated to start brood rearing, and it should synchronise with the main honey flow, so that the bees can take best advantage of it. Under proper management, not more than one or two kilograms of sugar are needed to be used in a year/colony. A colony of normal size should have at least six to eight pounds of honey (two or three frames) in reserve. When the stores fall below this level, bees should be fed artificially. If sugar is given as a winter reserve, the syrup should be thick, prepared by mixing two parts of sugar with one part of warm water.

Collection of food in honeybees is a social enterprise and more then 10,000 of foragers may be engaged together in collecting nectar and pollen. The foraging is designed in a way to achieve high efficiency. The foragers sacrifice their individual foraging efficiency for colony efficiency. When they have discovered a rich source of food, they communicate the information through various types of dances to their hive mates. High foraging efficiency is achieved as a result of sharing information about location of rich food sources. Four important resources include nectar, pollen, water and resin. Nectar and pollen are diet, water is used for evaporative cooling of the nest in hot days and resin is used for plugging the unwanted openings.

Colonies managed for honey production rear 1,50,000 to 2,00,000 bees annually and consume 15 to 30 kg of pollen and up to 80 kg of honey. The number of trips required to pool these materials could be quite astonishing. To collect 20 kg of pollen, approx 1-3 million foraging trips are required. Each trip on an average measure 4-5 km of distance. Likewise, to collect 60 kg of honey, 3 million foraging trips are required. In brief, each colony can be thought of as an organism which weighs 1 to 5 kg (biomass of bees 7700 bees/kg), rears 1,50,000 bees, and consumes 20 kg of pollen and 60 kg of honey/year. To collect this food several million foraging trips are required and foragers fly about 20 million kilometers.

Bee flora

Honeybees are dependent upon flowering plants for their food requirement. The natural diet of honeybees consists of carbohydrates, proteins, vitamins, salts etc. The nectar of the flowers, which the bees collect and convert into honey, is the main source of carbohydrates and vitamins. Pollen the yellow powder in flowers is the main source of proteins and is mixed with honey before it is fed to the larvae.

Honey bees need pollen for their body growth, for production of food for the queen and drones, for nursing larvae to grow into adult bees and for general health and hygiene of their colony. They need nectar to meet their day-to-day energy requirements. In order to preserve nectar for future needs, it is converted into honey by partially digesting the complex sugar molecules and removing about 3/4th of the water in it. One of the major problems in beekeeping is the presence of floral dearth poeriods which result in dwindling and desertion of the colonies. Every locality has floral dearth periods of long or short durations. Beekeepers have to ensure regular supply of food to the colonies to run the industry efficiently.

Nectar is the sweet liquid which comes from floral and extra floral resources and is the raw material for honey whereas the pollen is the highly nutritious food for honeybees. The plants providing both these resources are collectively called as bee forage plants or bee pasturage. The period when good number of plants providing nectar and pollen are available to bees is called major honey flow period. When the amount of nectar to be collected is small is called as minor honey flow period. The days when there is no honey flow is available is called dearth period.

Efiicient beekeeping requires supply of both nectar and pollen in adequate quantities. Not all plant species are equally good for beekeeping. Ideally, a good beekeeping area is one in which honey and pollen plants grow abundantly and with a relatively long blooming season. Efficient beekeeping means managing colonies in such a way to maintain maximum colony popuation to coincide with major honey flow in an area to utilize the population for honey production and pollination. Evidently, the first step to start beekeeping in an area requires preparation of floral calenders to determine when honey flow occurs.

Floral calendars

A floral calendar for beekeeping is a time-table that indicates to the beekeeper the approximate date and duration of the blossoming periods of the important honey and pollen plants in his area. Beekeeping in any specific area cannot develop without an understanding of the calendar, and for migratory beekeeping, special calendars for the different foraging zones along the migration route are required. Preparation of a floral calendar for any specific area requires complete observation of the seasonal changes in the vegetation patterns and/or agroecosystems of the area, the foraging behaviour of the bees, and the manner in which the honeybee colonies interact with their floral environment. The preparation of an accurate, detailed calendar will therefore often require several years of repeated recording and refinement of the information obtained.

To determine when the honey flow occurs: (1) Make seasonal survey to identify the nectar and pollen yielding plants. (2) Make survey to record the flowering period of these plants. (3) Determine whether the plants are visited by honeybees for nectar, pollen or both, (4) determine whether bees could collect surplus honey from some abundant crops, (5) determine what are the nectar-secreting flowering plants besides the major crops of the area? (6) How long a dearth period if any lasts? (7) Examine weather records data and altitudinal variations.

The steps required are as given below:

1. Place several strong honeybee colonies in the area, inspecting the hives regularly and observing changes in the amount of food stored within the hive to determine whether it is depleted, stable or increasing. Any food gains or losses can be monitored accurately by weighing the hives.
2. Within the flight range, record the species of plants that the bees visit. Determine whether the plants are visited for nectar or for pollen. Pollen-foragers will have pollen pellets attached to their hind legs. To determine whether the bees visit flowers for nectar squeeze the abdomen of individual bees to obtain a drop of regurgitated nectar, tasting it for sweetness or measuring the nectar concentration with a hand refractometer.
3. Study the frequency with which the bees visit each flower species, in relation to changes in the level of the colonies' food stores. If there is a continuous increase in food stores, in direct response to the availability of the plants visited, the plants are good forage sources. When the food

stores remain stable, the plants can be depended upon to meet the colonies' daily food requirements, but they cannot be classified as major honey sources.

4. Carefully record all the changes in the blossoming of the plants visited. When the colonies begin to lose weight, the flowering season is finished for all practical purposes.
5. Once all the data on forage species have been assembled and repeatedly verified, they should be judged as they relate to the actual performance of the honeybee colonies. Make floral calendar, in the form of a circular chart, indicating the periods of availability of major nectar and pollen sources.

The suitability of beekeeping in a locality will depend upon answers to these questions. If nectar secreting plants are available in a large number, that is there are one or two major honey flow periods with minor honey flow periods during other parts of the year and the dearth period of not long duration then beekeeping can be successful in that area.

During the honey flow period a good colony of honeybees may collect from one to two kg of honey per day. It should be kept in mind that a few fruit trees or flowering plants or a vegetable farm or garden cannot sustain any number of bee colonies leaving alone surplus honey is hoped for, they must be able to take advantage of many across honey yielding pasturage.

April and May are usually considered to be the period of nectar flow. After honey flow, there comes the hot months of June and July when most of the colonies stop brood rearing in the lower hills and in the plains. The bees, which had put in hard work in collecting resources, die of old age and strength of the colony depletes. Some bees are also lost in their effort to keep the hives well ventilated and cool. The monsoon season starts in the first week of July and because of the heavy rains bees generally stay inside. Some of those, which go out for field duties, are lost in the heavy downpour and the strength of the colonies depletes further. This is a very difficult period for the bees.

The enemies of the bees such as wax moths, wasps and ants become active and make the colonies abscond. During July and August, pollen from maize becomes available and is used in brood rearing. The colonies thus further use their honey reserves. If such weak colonies are not fed on sugar, they have a tendency to leave hives and abscond. The shortage of food and insufficient number of field bees are the main causes of this behaviour. Furthermore, strong colonies acquire a tendency to attack the weak ones and ransack their poor honey reserves. This dearth continues throughout autumn and during these months, therefore, colonies need particular care and management.

The honey flow period, dearth periods etc. varies from one location to another and with altitudes. The flowering periods of even the same plant species vary in different geographical regions and agroclimalic zones. For instance, in the lower hills and valley areas of Himachal Pradesh the season for activities of bees starts right from the advent of the winter when sarson flowers are available in the fields. Then fruit crops provide flowers in the second week of February. They are followed by the flowering of Eucalyptus, Shisham (*Dalbergia sissoo*), Tun (*Cedrella toona*), and their flowers continue upto the end of April. After a gap of about one week, the soapnut (*Sapindus* sp.) starts blooming which provides the main honey flow. It continues upto the end of May and surplus honey is extracted in the end of May or the first week of June. Under Punjab conditions, surplus honey is stored from Eucalyptus, shisham, citrus, stone fruits, litchi and barseem flowers in March, April and May. Then follows the dearth period up to September and during this period the bees visit a number of wild flowers, and the blossoms of maize in the end of July. In the oilseed and cotton growing areas bees also store surplus honey from toria, cotton, arhar etc. in September and October.

In the higher altitudes of Himachal Pradesh winter is much colder and the bees are very much restricted in their activities. The stone fruits including pear, almond, plum, cherry, apple, apricot, etc. blossom in the spring and the bees are stimulated to rear their broods. In the beginning of May barberry flowers provide minor honey flow periods followed by soapnut in the end of the June. Early in August, maize inflorescences provide pollen for brood rearing to bees. The colony strength increases. At the end of August, *Plectranthus* blossoms appear which continue up to the end of October. This provides the main honey flow season. Extraction is done towards the end of the October, leaving sufficient honey reserves for the bees to tide over winter.

In Kashmir, the activity is instigated with the appearance of early blossoms from *Salix, Narcissus, Brassica, Pyrus, Prunus*, etc. in Feb.-March. During April-June blossoms of *Iris, Robinia, Corylis, Castanea, Aesculus, Alianthus*, Rose etc. provide the requirements at the peak of brood rearing activity and occasionally also lead to a spring honey crop.

While fruit blossoms, mustard and other early flowering herbs greatly help in encouraging brood rearing activity and in building up some reserves of nectar and pollen. Vegetable blossoms and other flowering annuals and trees, which appear during May and June, serve as valuable sources. July and August are the dearth periods. During August-September colonies are migrated to higher altitudes where pollen and nectar sources such as *Zea, Zinnia, Thyme, Mentha, Cyanoglossum, Balsam* etc. play an important role in gearing up the bee activity. Among the various shrubs providing nectar or pollen or both, *Plectranthus rugosus* Wall which flowers from mid-August to mid-October, provides for a major part of the honey flow in autumn, because of its abundance on many hillocks in Kashmir. *Saffron* (Crocus) blossoms during late October to mid-November, if availed are of immense value to bees for brood rearing even in late fall.

In Jammu, April provides the major honey flow period, June-July-August are dearth periods. During August-September colonies are migrated to higher altitudes to utilise *Plectranthus rugosus*. October onwards rapeseed and mustard crops are available up to February, followed by fruit blossoms etc. J&K state has varied climates, habitats, which support varied type of vegetation and cultivated crops. Broadly it may be divided into: 1. Subtropical zone 2. Intermediate zone 3. Temperate zone. The vegetation and blooming timimg vary from one region to another. Similarly, floral calendars of beekeeping plants are available in different states. However, a particular crop may be good bee plant in one locality but can be quite otherwise in another locality. Honey production from a crop in the same area may vary from year to year.

Information on different aspects of bee forage is essential for the efficient management of honeybee colonies. Management scheme for each apicultural region is closely correlated with the flowering of local honey and pollen producing plants as also the climatic conditions. Basic research in the area of forage ecology has been done and floral calendars for different regions have been prepared. On the basis of surveys potential beekeeping areas have been identified.

The most serious problem for Indian beekeeping has been the decline in flora due to deforestation and clearing of wastelands for extensive agriculture. Improvement of bee flora is not possible by individuals' efforts and a beekeeper has to adopt and adjust only to the cropping patterns of the area and forest wild flora available in the locality. Recently central and state Governments and local organisations have helped in expansion of planted areas of bee forage along highways, wastelands etc. To get good results, plantation of selective trees and shrubs is essential and this should be done on the basis of multiple use principle including bee forage as one of the uses. Flowers of many plant species are visited by bees for nectar and or pollen but relative importance depends on the quality and quantity of rewards available and also on the density of the plant species. Intensive research in this area has generated this type of information on many of the important flora. The knowledge accumulated can be made use of while planning plantations on the basis of accessibility of the potential bee forage areas and migration schedules can be worked out. Migratory beekeeping is practised by many commercial beekeepers in states like Himachal Pradesh, Bihar and south India but micro-regional survey of bee forage would be required for planning short and long distance migration schedules.

Assessment of areas for beekeeping

Productive beekeeping depends on good colony management and good beekeeping areas, and in order to promote it as a profitable agricultural occupation, areas with a good potential for beekeeping must be located and evaluated. One major problem in this respect is how to select sites for assessment. The following guidelines for the exploration and evaluation of potential beekeeping areas may be found useful:

1. Refer to lists of known major honey plants in other countries or regions with similar vegetation patterns, agro-ecosystems, climate and edaphic conditions; determine whether similar plants are to be found in the area under study.

2. The seasonal occurrence, in unusally high numbers, of wild (feral nests) of native honeybees can often indicate that there is ample forage in the area, at least during the period in question.

3. The mere presence of flowering trees and shrubs in limited numbers, or of a few hectares of land covered with good honey plants preferred by bees, does not necessarily indicate that the area has potential for commercial beekeeping.

4. Large scale beekeeping operations require large areas, usually hundreds or thousands of hectares of nearby land bearing good forage with high population densities. Good honey plants are characterized by relatively long blossoming periods, generally in terms of several weeks or months; high density of nectar-secreting flowers per plant or unit area; good nectar quality with high sugar concentrations; and good accessibility of the nectaries to the bees. The foraging land should be well proportioned, in terms of length and width, so as to promote foraging efficiency.

5. The supporting capacity of an area for honey production is best determined by monitoring weight changes in the bee colonies. Among other factors that affect the economic value of an area for beekeeping are average hive yields, prevailing honey prices in the area, as well as costs of colony management inputs.

6. The fact that a flower is brightly coloured or that it has a strong scent does not always indicate that it is good for bees, unless the fact is confirmed by the criteria set out above.

7. The large scale planting of honeybee forages has never been proved to be a profitable approach in terms of net economic return, except in integration with other agricultural activities, such as reafforestation, roadside plantings, animal pasture, etc.

Honey production and colony development are directly related to floral sources in the immediate area of the apiary. Major honey flows depend on a few plant species which yield nectar abundantly and are readily available. Besides the two or three main annual sources, there should be a great variety of minor plants yielding both pollen and nectar throughout the season to support the colonies between the main flows. Large acreages of flowers are needed to produce surplus honey. It is not economical to plant crops just for their nectar and pollen yields. An acre of blooming plants rarely provides surplus honey for more than one or two colonies of bees. The value of land not being used for other purposes can be increased, however, if it is planted to some nectar-yielding plants. Beekeepers should become familiar with the major floral sources of their area and when these species bloom. From the stand point of honey production, the most populous colonies produce the most honey. Thus, the essence of spring management is the development of strong colonies. Efficient management requires proper timing of colony development so that maximum populations will coincide with major nectar flows. Each area has its own set of plants and blossoming timings.

A colony of honey bees needs about 30 kg of pollen and 200 kg of honey in a year. Each flower provides but a few milligrams of pollen or nectar. Bees have therefore to visit hundreds of thousands of flowers. All the available flowers do not provide food of the required quality. Because of this, the bees which gather food have to select food plants from among a host of other plants. Availability of food plants for at least 8 months in a year is essential for bee colonies to survive.

Pollen contains proteins, carbohydrates, fats, minerals, vitamins, essential oils, and colouring materials. Of these the digestible protein is an important constituent that determines the bees' preferences to different plant species. In many plants pollen is released in the mornings. A few, like cucurbits, offer pollen in the evenings. Bees collect pollen when freshly produced. Bees adopt their foraging behaviour to suit the mode of presentation of pollen in the flowers. Pollen is packed into small balls on the third pair of legs. On returning to their hive, they seek a pollen store cell, drop the two pellets into it and resume their foraging.

Figure 47(A). Beekeeping map of J. & K. and Himachal Pradesh

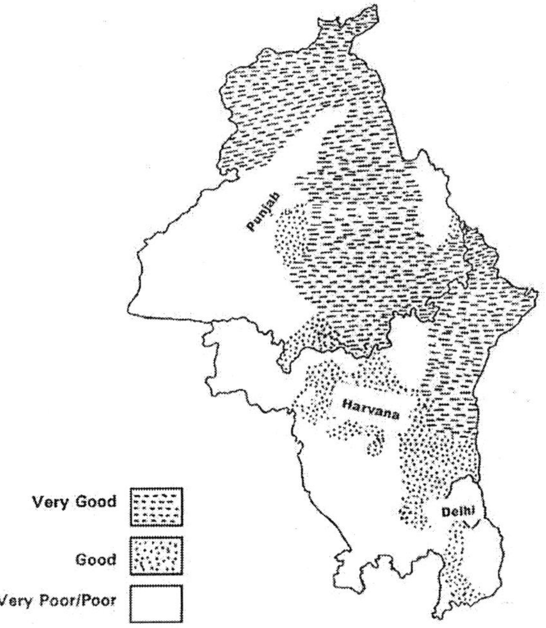

Figure 47(B). Beekeeping map of Punjab and Haryana,

Figure 47(C). Beekeeping map of Rajasthan, Gujarat, Uttar Pradesh, Madhya Pradesh, Bihar, Maharashtra and Orissa

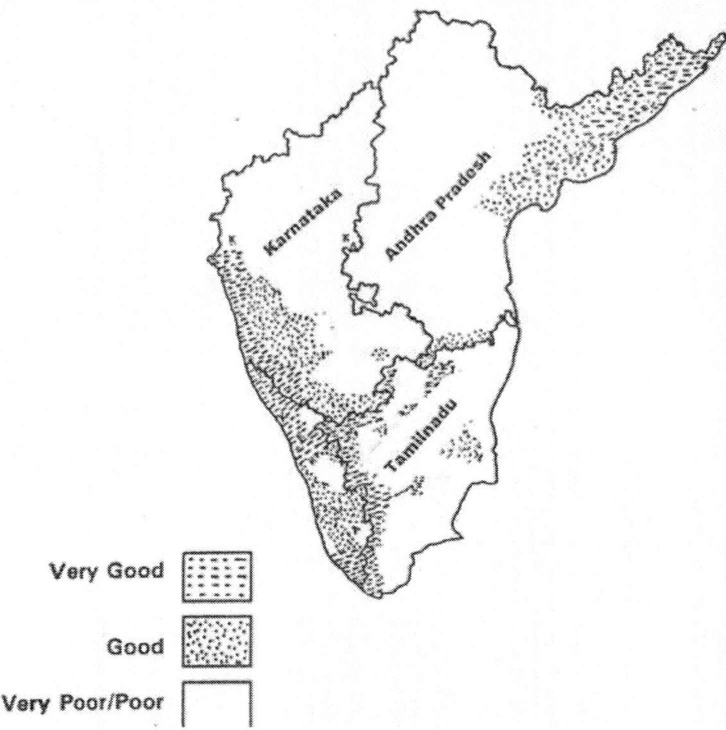

Figure 47(D). Beekeeping map of Southern Peninsula (Karnataka, Kerala, Andhra Pradesh and Tamil Nadu)

Nectar has sucrose, fructose and glucose, in varying proportions, essential oils, minerals and other material in traces. Bees prefer sucrose and fructose to glucose or an equal mixture of these sugars. The concentration of sugars in nectar varies. Bees prefer a concentration of about 35 per cent. Nectar is collected by sucking it from nectaries through proboscis. Salivary gland secretions are added to the nectar in this process. The nectar is then taken into the honey stomach, where some more digestive enzymes are added. On returning to the hive, nectar is regurgitated and passed on to the nectar seeking house bees through their proboscis. After similarly passing through several nectar seeking bees, it is finally deposited in a honey store cell. Some house bees take a small portion of this enzyme-enriched nectar and evaporate the water in it by making a thin film of the liquid in the mouth parts and exposing it to the wind currents created by fanning of wings. After the honey attains the required consistency, - about 80 per cent sugars and 20 per cent water - the honey cell is sealed.

Honeybees have closed links with the flora because adults and young ones live solely on nectar and pollens. Usefulness of a flora depends upon the quantity of food and energy harvested and the energy requirement of bees. Nectar is secreted by nectaries and is usually a reward for the visitors bringing about pollination. In general, the quantity of nectar secreted is directly related to the pollination requirements of the crop. Honey is made by bees from the nectar collected from floral and extrafloral nectaries. Sometimes honeydew is also an important source in certain localities, but the well being of bees is greatly dependant on the value of the flora. For the selection of the apiary site it is essential to know the plants, which provide nectar or pollen to bees. Data on different aspects can be gathered for couple of years to establish the potentials of an area for beekeeping. Scale colony provides useful information. Weight of the scale colony is regularly recorded and the changes in weight are correlated with the flowering plants, which are being visited, by bees and also the weather conditions. Flowers present nectar and pollen during specific time of the day, therefore, bee activity on the flora should be carefully recorded. Recording the bee activity at different day hours should give useful information. Melissopalynological studies are essential to ensure as to which plants are availed by bees. Pollen analysis of honey samples taken at different time of the year and comparing with reference slides give the exact information about the floral sources for bees in vicinity.

Bee flora should be studied from different angles to find out the value of bee forage though the total nectar production per flower is important. Some flowers secrete nectar only for one day and few others for short time and still there are flowers like *Schefflera wallichiana,* which continue nectar secretion for about a fortnight. The time for which nectar plant blossoms are another important point determining the value of flora. In some forages the duration between the start and end of flowering is very short, whereas in others like *Brassica* spp. there is a succession of flowering and it lasts for about a month. Trees present larger number of flowers as compared to bushes, shrubs and crop plants unless the latter are growing in large continuous areas. Concentration of nectar sugar gives an objective measure. Total sugars per flower per day is the sugar value which is normally estimated for such studies. Sugar value is the number of mg of sugar secreted by one flower over a period of 24 hr. However, total quantity of nectar produced and the amount available to, or harvested by bees give valuable information.

Nectar sugar concentration

The bees visit flowering plants to obtain nectar, which is the source of honey, as well as pollen. The concentration of sugar in nectar depends on several factors: the plant species and variety, the soil type, the time of day of collection, the temperature and relative humidity, etc. As a rule, plants with a higher sugar concentration in their nectar are more attractive to bees than those with weaker nectars, because in the process of making honey the bees are obliged to get rid of excess water in the nectar, so that in treating more highly concentrated nectars, they need to expend less time and energy.

When the excess water has been evaporated from the nectar and the enzymatic reactions in the conversion of nectar to honey have been completed, the honey is ready for storage, to serve as the bees' reserve of carbohydrates to cover the colony's energy requirements. In the broad sense, then, honey is the colony's energy reserve, all or part of which will be expended in the process of foraging. From the standpoint of the beekeeper, a colony is "productive" when it stores a surplus of honey, i.e. when it can collect and convert into honey more nectar than it consumes. The beekeeper harvests all or most of this "surplus" honey.

Nectar concentration in most bee forages varies between 20-50% but may be as low as 6% in *Bombax ceiba* L. to 15% in pear and as high as 79% in silver oak (*Grevillea robusta*). Nectar and sugar concentration (%) has been worked out for some flora in India. For example, *Grevillea robusta* had 79% sugars in the nectar 23 plants including *Tecoma grandiflora* had 14%, peach and pear 70%, *Brassica juncea* 52%, Barberis 48%, some citrus spp. 40-44%, *Sapindus detergens* 40%, Plectranthus 38%, *Cidrella cerata* 36% *Carvia callosa* 35% *Thelepaepale ixiocephala* 35-64% *Impatiens balsamina* 16-25% *Nephelium litchi* 61-78%, Plectranthus 26-54% *Woodfordia floribunda* 10-12% *Brassica campestris* var. *toria* 38.5-53.5% and onion 59-75.5%. Gupta found that there were considerable differences in the amount of nectar sugar production in the flowers of different cultivars of cauliflower and in the attraction of bees to them. Average nectar sugar contents varied from 0.035-to 0.150-mg/flower/24 hr. On the basis of such studies honey potentials per unit area are estimated for cultivated crop/fruit plants. For wild flora the density per unit area is estimated by sampling and honey potentials of the flora in the area are worked out. Honey potentials of forage are only estimates but it serves as a useful guide. Honey potentials of forage are liable to change. This is happening fast in developing agriculture in India where changes in land-use patterns have been very frequent. Under extensive agriculture vast wastelands are cleared for cultivation and this reduced the wild flora. Land may be put to urbanization or industrialization causing reduction in cultivated and wild bee flora. There are introductions and extension of cultivation of new crops like sunflower, safflower and other oilseed crops and this has substantially changed the scenario with regard to beekeeping potentials. There are many agricultural practices, which affect the bee flora. Mechanized agriculture reduces the weed plants, which may be serving as bee forage. The use of weedicides is another agricultural practice, which reduced the weed forage for bees.

The nectar yielding plants contributing to nectar/honey flow are specific to different areas and they have definite micro-regional habitats. Even in rich floral areas continuous succession of nectar yielding plants throughout the year is lacking. In some localities there is single surplus honey flow and in good areas 2 surplus flows may be available. In north-western hills of Himachal Pradesh and Jammu and Kashmir, the lower and mid hills present spring, early summer flow and in areas like Kashmir and parts of Himachal Pradesh, there is rich flow from *Plectranthus* in autumn. Bees face protracted dearth period in winters and only subsistence flora is available in rainy season but heavy downpours are hazardous to bees. In north Indian and Gangetic plains, major flora of *Brassica* is available from September through early February. This build up and surplus flow is followed by spring and summer surplus honey flow from *Eucalyptus, Dalbergia sissoo* and other trees and berseem is availed by bees till May. Hard summers are also no flora availability period but some weeds and crops present subsistence forage in rainy season. In Western Ghats and South India, there is medium to major flora available from October to May and important sources being jamun, hirda, carvi, soapnut rubber plant, Schefflera, etc. but June to September is a dearth period. Therefore, for beekeeping sub-tropics may have an inactive period of 1 or 2 months. In tropics it is dry season or excessive rainfall and in hills it is winter, which are troublesome to bees. Surplus flow season may vary from 1 or 2 and rarely 3 in a year.

Beekeeper, to maximize his honey crop, should have a thorough knowledge of the floral cycle, onset of major honey flow and dearth period. The bee colonies should be managed in a way so as to have maximum foraging strength to avail major flow and economical or minimum strength in dearth period.

Nectar Composition

Nectar is a solution and total solids in nectar are mostly sugars. There are only few exceptions where lipids are present in nectar. Chromatographic studies help to know the sugars in nectar. Monosaccharides, glucose, fructose and disaccharide, sucrose, are the common sugars in nectar. In general, flowers with tubular corolla secrete sucrose dominant nectar. In open flowers, such as *Brassica*, only glucose, fructose and sucrose are present. Bee's preference to nectar is also governed by the sugar balance and the nectars with equal amount of glucose, fructose and sucrose are preferred by bees. Minute amounts of other substances such as amino acids, minerals, essential oils, organic acids and other components usually comprise less than 0.03% of total dry weight. Essential oils impart characteristic aroma to nectar and honey and bees are attracted by this aroma to flower nectar. Solid particles in the form of pollen grains, yeast cells, fungal spores and bacteria can be found in small amounts.

Attempts have been made by Indian scientists to gather information on the nectar composition of bee flora. Sucrose, fructose and glucose make up to 60% of the total solids of *Thunbergia grandiflora* nectar. They also detected small amounts of aspartic acid, alanine, glycine, serine and valine by paper chromatography. *Moringa pterigosperma* Gaertn, nectar contained 0.90% of reducing and 11.81% of non-reducing sugars. Wakhle analysed nectars from 4 species; nectar of *Carvia callosa* contained fructose, glucose, maltose, raffinose and an unidentified sugar; nectar of *Thelepaepale ixiocephala* had only fructose, glucose and sucrose, whereas nectars of *Schefflera roxburghii* and *Grevillia robusta* had only fructose and glucose. Only these 2 sugars were observed by Mishra in *Woodfordia floribunda* nectar. Soapnut (*Sapindus emerginatus*) nectar had 85.5% sucrose and 7.25% each glucose and fructose. Sihag analysed the nectars of 44 plants visited by *Apis florea* and *Apis dorsata*. Nectar of *Tecoma stans* and all 11 cruciferous plants, *Althea rosea, Prunus persica, P. domestica* and *Petunia alba* contained glucose dominated sugars and sucrose and fructose were in very small fractions. Bahadur analysed nectar of 103 plants from 100 species and revealed that 54 plants had 3 sugars, *viz.,* sucrose (S), glucose (G) and fructose (F). In addition to these, 7 plants had 4 or more sugars, 6 having S+G+F+1 unknown sugar, one having S+G+F+2 unknown sugars. Forty plants had 2 sugars (29:S+G; 3:S+F; 8:G+F) and 2 had only one each (1:G; 1:S). In the nectar of these plants, 24 had dominant sugar as sucrose, 7 had S+G, 2 had G and one had G+F and there was no plants having either S+F or F dominant sugar. Six plants had balanced S+G+F sugars, 3 plants had S+G and there was no plants having G+F balanced sugars. In 90 plants amino acid was present, whereas in 15 plants, the amount was double than the remaining ones.

Factors Affecting Nectar Secretion

Honey flow from the same plant is not the same under varying weather, soil and vegetation habitat conditions. The nectar secretion in a plant is the function of specific features of the forage and other external factors. Factors related to a plant species are age of the flowers and the cultivar or varieties. In *Woodfordia floribunda* Salisb, the flowers continued to secrete nectar for 3 days; it was maximum on second day and minimum on first day of flower opening. The differences in nectar secretion have also been found in many plant species and differences with age of the flowers and cultivars have been reported in peach and cauliflower in India. Nectar secretion in flowers after opening is expected to be correlated with the degree of receptiblity of the stigma. Fertilization in flowers is also known to activate a feed back mechanism to switch off nectar secretion. After certain period of flower opening there is reduction in amount of nectar sugars. This happens because of reabsorption of nectar. Reabsorption only of nectar sugars takes place and not of water. Therefore, the reabsorption at the end of nectar secretion leads to lowered nectar sugar concentration. The total amount of nectar secretion over a period by a flower is more when it is periodically removed than in case of its non-removal by insects. Besides these, growth regulators have also been found to affect nectar volume and nectar sugars. The effect of GA3 was more pronounced in mustard and cauliflower than with other growth regulators.

Sunlight and temperature. Sunlight has a direct bearing on photosynthesis. Photosynthesis produces carbohydrates, which are secreted in nectar. Effect of sunlight and consequently of photosynthesis may not be immediate because stored carbohydrates do make good but ultimate effect is there. Temperature has direct relationship with nectar secretion. For every plant species there is a specific threshold temperature at which the nectar secretion is started and it increases when the temperature is optimum. This is the range of temperature at which enzymes responsible for nectar secretion are activated. On the other hand very high air temperature may result into water stress in plants and the waste stress causes more water loss than the uptake. The imbalance results into lowered nectar secretion through reduced sugar transport in the conducting tissues.

Relative humidity. Nectar is hygroscopic in nature and for this reason atmospheric humidity is inversely linked to nectar sugar concentration. After secretion nectar sugar concentration changes and attains equilibrium with the moisture in the air. Water is always lost from nectar unless the relative humidity is near 100%. High relative humidity causing reduction in sugar concentration can affect the attractiveness of the source to bees. Conversely, higher sugar concentration at lower relative humidity can affect the nectar column in the tubular flowers and bees may be unable to reach and imbibe the nectar. Water loss in the plant is also a function of relative humidity as also of air temperature.

Soil. Optimum soil moisture is essential for good plant growth. With soil water as a limiting factor the number of flowers is reduced and nectar secretion is also adversely affected. Balanced nutrient level in soil which supports good plant growth also favours nectar production. Phosphorus and potassium increase nectar production but high level of phosphorus reduces it. A balance between the 2 elements should be beneficial. Excessive application of nitrogen causes abnormal vegetative growth and comparatively lesser number of flowers are produced. Some of the bee forages have been briefly discussed in this chapter.

Bees exhibit floral constancy or fidelity. As long as food is available from a source, they visit that source exclusively and ignore others. Because of this behaviour, in any good vegetation with abundance of an useful nectar source, the beekeeper can get honey almost exclusively from that source. Such honeys are called unifloral or monofloral honeys, while honeys prepared from nectars of several sources is called multifloral.

Bee flora of India

In several countries, beekeeping is agriculture based. Clovers, orange and other citrus fruit trees, pomaceous fruit trees, sunflower, rapeseed, mustard and other cultivated crop plants are important as sources of nectar. Honey is also produced in plantation areas with trees like locust or forests of eucalypts. In the coniferous forests of Europe, bees make honey from the sweet excretions of aphids that infest coniferous trees. This is called honey dew honey.

In India beekeeping with the Indian honey bee has been by and large forest based. In recent years cultivated areas are increasingly used for honey production. A little over 70 per cent of this was under food grains including rice, wheat, jowar, bajra, maize, gram and other pulse crops. Over 24 million hectares were under oilseed crops like ground nut and mustards. Details of the areas at present under several other cultivated plant species are not readily available. Among such species are coconut, areca nut, red oil palm, date palm, cacao, mango, custard apple, jujube, cinnamon, clove, cashew, fodder legumes, coriander, cumin, dill seed, fennel, fenugreek, garlic, turmeric, ginger and other spice and condiment crops, road side plantations that contribute to honey production like eucalyptus, karanj, tamarind, gulmohr, peltaphorum and soap nut. Hedges and fence plants like mehndi (Indian privet), duranta, mulberry, justicia and jatropha, do also add to the bee forage value of farms and orchards.

Cereals in general are not very useful as sources of bee forage. However, jowar, bajra and maize are valued for their pollen, particularly during the kharif and rabi crop seasons, when natural sources are scarce. In the case of jowar, some varieties produce sugary exudation on the nodes. Leaves also secrete a thick sweet liquid when infected by rust fungi. In both cases bees collect the sweet substances. Among other crops which are not useful to honey bees are ground nut and sugar cane. The latter is occasionally useful as a supplementary source of carbohydrate food. This happens after the crop is harvested. Honey bees eagerly visit the freshly cut stumps for the sugary sap, oozing out of them.

Many pulse and oilseed crops are good sources of bee forage. Among the plantation commercial crops, coffee, orange and other citrus fruit, apple and other pomaceous fruit species, cardamom and rubber tree are important from the beekeeping point of view. The last mentioned is the single most largest source of nectar in India. Rubber plantations are found in southwestern and northeastern parts of India, where tropical humid climate prevails. Kerala, Tamilnadu, Karnataka and Tripura have large areas under rubber plantation. The nectaries on young leaves of rubber trees secrete nectar profusely in the refoliation stage, before the tree blooms. Next in importance is litchi tree. The entire north India from West Bengal to Jammu has large areas under litchi orchards that constitute an excellent source of nectar during March to May. A point to note here is the relatively low utility of garden plants. Several garden annuals, shrubs and climbers are not useful to honey bees, mainly because the showy flowers lost their reproductive function and produce no pollen or nectar. Some garden species do however contribute to the bee forage in the otherwise useless areas under gardens and parks. These include the railway creeper, ocimums, salvias, coleus, poinsettia, petunia, zinnia, phlox, and daisy.

Forest trees, ornamentals and avenue

Eucalyptus spp. (Fam.: Myrtaceae). These are evergreen trees and are indigenous to Australia. In Australia most species are bushes and shrubs but fortunately in India Eucalyptus spp. attain gigantic size. Different species and even varieties flower during different parts of the year but main bloom is available in spring. Eucalyptus plantations have expanded very fast on road sides, canals, even waste lands and thus presenting vast potentials to bees. Honey is light amber coloured and granulation is fairly quick.

Hirad (*Terminialia chebula* Retz; Fam.: Combretaceae). Hirad is distributed in the Western Ghats and in submountainous regions. The fruits have medicinal value and also a source of dyes and tannins. It is a major flora in beekeeping areas of Maharashtra. It flowers in April to June. Hirad honey is composed of 17% water, 35% glucose, 40-41% fructose, sucrose, maltose, melezitose, ash and it also contains calcium, phosphorus, iron, magnesium, sodium, potassium and silicon. Tartaric, citric, malic, succinic acids are also present. Honey is light yellow coloured; granulation is slow and has characteristic pungent aroma and tastes akin to fruit tannins. *Terminalia arjuna* (Rob.) Wight and Arn. has been widely planted in forests and as avenue plantation. It is also a nectar and pollen source but details of value as bee forage have not been worked out. *T. bellerica* (Gaertn.) Roxb. has also both nectar and pollen forage but no data on nectar and honey is available.

Jamun (*Syzygium cumini* Skeels; Fam.: Myrtaceae). Jamun is widely planted avenue-cum-fruit tree and naturalized as forests in many parts of India. It is a predominant flora of Western Ghats. It is reported to be a major flora in Bihar, Maharashtra, Punjab, Tamil Nadu and Uttar Pradesh. Its flowers are dirty white and bloom in April-May. The honey flow extends over a period of 2 or 3 weeks. Nectar-sugar concentration is very high up to 72% but may be as low as 9%. Jamun is also a pollen source. Chemical composition of honey has been given by Narayana. Fructose is high (43.30%) and glucose is 32.26%, sucrose, maltose, reffinose and melezitose are present in jamun honey. Protein (0.656%), dextrin (1.55%), ash (0.182%) and riboflavin, ascorbic acid, thiamine and niacin are also present in honey. It is light reddish-brown and has characteristic taste of jamun fruit. The honey does not granulate for years.

Karanj (*Pongamia pinnata* (L. Pierre); Fam.: Leguminosae). Karanj grows well in humid tropical area. *P. glabra* is grown as avenue and shade tree in drier climate. Leaves are used as fodder. It blooms in April-May and both nectar and pollen are availed from karanj. Data on nectar production, honey characteristics and honey potentials are not available but it is considered to be good flora by the beekeepers.

Moulsari (*Mimusops elegni* L.; Fam.: Sapotaceae). Moulsari is evergreen tree with whitish flowers. It has good nectar flow in May-June in Bihar and also in Uttar Pradesh. Bees also collect pollen from this flora. No information on nectar and honey from moulsari is available.

Phalsa (*Grewia asiatica* L.; Fam.; Tiliaceae). Phalsa shrub is grown for its fruits. It blooms from April to August and yields nectar to bees. The extent of its cultivation is less. *G. oppositifolia* Roxb. is a medium-sized tree which is planted for lopping its foliage for fodder, for its best fibre for ropes and for timber. It blooms in April-May and is a nectar source to bees. These species of Grewia are only minor sources.

Rubber (*Hevea brasiliensis* Muell. Arg.; Fam.; Euphorbiaceae). Rubber plant is deciduous tree with male and female flowers on the same inflorescence. Nectar is available to bees only from the extra-floral nectaries at the base of young buds. The nectar flow is there for about 2 weeks when leaves are young. Rubber is a major flora in Kerala where large plantations are grown. Honey potential is estimated to be 3 kg per tree. Rubber honey is clear straw coloured.

Schefflera wallichina Harms. (Fam.; Araliaceae). It grows as a strangler or small tree in forests or cardamom estates of Coorg area of Karnataka. Schefflera blooms in April-June and is a rich nectar source. Most honey in Coorg is obtained from this source. Pollen availability to bees is very low.

Shisham (*Dalbergia sissoo* DC.; Fam.; Leguminosae). Shisham's habitat is very varied, from lower hills to dry plains; normally planted on banks of canals or roadside as shade tree. This is a source of valuable timber especially for furniture. Pale-yellowish flowers are put forth in April. Corolla is tubular but narrow tube has nectar well up to the reach of bees. Wind blows flower from branches and nectar availability is adversely affected. It is a major nectar source for bees in many states especially Himachal Pradesh. Punjab, Haryana and Uttar Pradesh. Shisham honey is dark amber with moisture 18.75%, glucose 34.6%, fructose 39.1%, sucrose 1.04% and ash 0.18%. Honey flavour is strong and not attractive.

Silver oak (*Grevillea robusta* A. Cunn. ex R. Br; Fam.; Proteaceae). Silver oak is planted in coffee plantations and also as shade and avenue tree. It profusely flowers and is an important nectar source and nectar secretion is abundant with 17-79% nectar concentration. Honey is reddish-black with prominent flavour and it granulates rapidly.

Soapnut (*Sapindus* spp.; Fam.; Spindanceae). Soapnut is an avenue and forest tree; fruits are used as substitute to soap. *S. emerginatus* Vahl is a species found in Andhra Pradesh, Karnataka, Orissa and Tamil Nadu. The species flowers in October-December. It gives 20-25% of total honey yield in some parts of Andhra Pradesh. S. emerginatus nectar has 85.5% sucrose and 7.25% glucose and fructose.

Tamarind (*Tamarindus indica* L.; Fam.; Leguminosae). Tamarind is a large evergeen shade tree cultivated in many parts of India for pods which are used in curries. The tree has many other uses like fuel, timber, etc. It blooms in April-July and is a good nectar and minor pollen source in South India. Honey is rich golden and has sour flavour.

Toon (*Toona ciliata* M. Roem.; Fam.; Meliaceae). Toon grows as timber and avenue tree in lower hills and plains of northern India especially in moist soil. It is also lopped for fodder but is not to the liking of cattle. Toon is a major nectar source in Himachal Pradesh, Kashmir, Punjab and Uttar Pradesh. It is a minor source of pollen. Large number of trees has recently been cut and floral source has declined. Nectar sugar concentration varies from 26 to 72% in freshly opened to 48 hr old flowers. Average nectar sugar value is 2.38 mg/flower/day and nectar is secreted for 4 days. Honey is light amber in colour with pronounced flavour. Normally its honey is mixed with Dalbergia sissoo honey and is obtained in May.

Whayati (*Thelepeaple ixiocephala* (Benth.) Bremk; Fam.: Acanthaceae). Whayati is a tree of moist forests of Western Ghats of Maharashtra and Karnataka. Whayati flowers after every 8 years. It blooms in November-January and is a major nectar source. Sugar concentration of nectar is 35-46%. Honey contains 38% glucose and 39% fructose; sucrose, maltose, raffinose and melezitose are present. Sodium, potassium, calcium, magnesium, iron, phosphorus and silicon are contained in whayati honey. It is light yellow and granules rapidly but granulation is uniform.

There are many other avenue and forest trees which are good source of nectar and pollen but their number is small and at best serve as subsistence sources. These trees are bottle brush (*Callistemon lanceolatus* DC.), pride of India (*Lagerstroemia indica* L.), drum stick (*Moringa oleifera* Lam.), Indian laburnum (*Cassia* spp.), puna (*Ehretia acuminata* B.), siris (*Albizzia* spp.), willows (*Salyx* spp.) and chestnuts (*Aesculus* sp., *Castanea* sp.)

Fruits

Banana (*Musa* spp.; Fam.; Musaceae). Perennial herb, stem formed by leaf petioles. Flowers are large and monoecious. *Musa* spp. flowers throughout the year and is a medium to good source of nectar which has 25-30% sugar concentration. Banana flowers are also visited by bees for pollen which is in abundance. Banana plantations are common in many states of India.

Cashew (*Anacardium occidentale* L.; Fam.: Anacardiaceae). Cashew is cultivated in South India. It is an evergreen tree, flowers pink, small and fragrant. Devadason has advocated migratory beekeeping to avail cashew flow. It is also a pollen source; value of the flora and honey characteristics are not known.

Citrus spp. (Fam.: Rutaceae). These are *C. aurantifolia* (Christm) Swingle; *C. grandis* (L.) Osbeck; *C. limon* (L.) Burm.; *C. paradisi* Macfad.; *C. reticulata* Blanco and *C. sinensis* (L.) Osb. These citrus species flower during February-March. Nectar production and nectar sugar concentration is medium. Citrus spp. also serve as pollen source. Citrus honey has delicate flavour. In India Citrus mostly serves as a build up flora and surplus is not extracted anywhere, though large areas are under citrus and it is widely distributed.

Coconut (*Cocos nucifera* L.; Fam.: Palmae). It is grown in coastal regions. Flowers are small, monoecious. both male and female flowers have nectaries. It blooms in May-June and bees collect abundant pollen from staminate flowers in spathes.

Jujube (*Ziziphus mauritiana* Lam.: Fam.; Rhamnaceae). Indian jujube is cultivated in tropical parts of India, advantageously below 600 M.S.L. It tolerates severe heat and is drought resistant. It flowers during July to October and flowering is very protracted. The flora gives surplus honey when the colonies are strong but bees get nectar source when there is no other flora. Some pollen is also availed and therefore, it is very useful forage for bees. The jujube honey is yellow-brown with very sweet flavour.

Litchi (*Litchi chinensis* Sonner; Fam.: Sapindaceae). Litchi has become very popular in sub-mountainous regions for expensive fruits. It blooms in March and is a rich source of nectar to bees. Sugar concentration of nectar is high. Juice from damaged fruits is also availed by bees. Litchi honey is light golden coloured with very pleasing aroma.

Pome and stone fruits. Apple, pear, plum, peach, apricot, cherry, almond and their closely allied wild species are included in this category. They flower from

February to April and bees gather both nectar and pollen. These fruit trees have local importance and have good build up sources before surplus honey flow season. Pear nectar is very low in sugar concentration. Therefore, it is normally avoided in favour of other competing flora. Surplus honey is not gathered from these cultivated fruit trees. But wild cherry, *Prunus puddum* Roxb. flowers in October to November when no other flora is available in mid and lower hills of Himachal Pradesh. On an average a flower secretes 35ml nectar for 4 days with 3.47 mg nectar sugar per flower. Nectar sugar concentration varies from 12 to 18%. Chromatographic separation revealed glucose, fructose, sucrose and one unidentified sugar in the ratio of 39.6, 40.7, 12.3 and 7.5.

A point to note here is the relatively low utility of garden plants. Several garden annuals, shrubs and climbers are not useful to honey bees, mainly because the showy flowers lost their reproductive function and produce no pollen or nectar. Some garden species do, however, contribute to the bee forage in the otherwise useless areas under gardens and parks. These include the railway creeper, ocimums, salvias, coleus, poinsettia, petunia, zinnia, phlox, and daisy.

Table 19. Area (million hectares) under different crops in India providing bee forage

Crop	Area	Crop	Area
Jowar	14.5	Coffee	0.2
Bajra	10.4	Fruit Crops	1.7
Maize	6.0	Vegetable Crops	1.6
Gram	7.4	Rubber	0.5
Other pulses	17.0	Cardamom	0.1
Rapeseed, mustards	5.7	Tea	0.4
Other oilseeds excluding groundnut	10.0	Citrus fruits and guava	0.1
Cotton	7.4	Coconut	108.0
Jute	0.8	Areca nut	1.8
Mesta	0.2	Banana	0.2

Source: Anonymous, 1993

The National Commission on Agriculture assessed the beekeeping potential of the cultivated area in India in terms of the need of bee colonies for pollination of entomophilous crop plants. Many crop plants depend upon insects for transfer of pollen from one flower to another, resulting in seed and fruit set. It had estimated that nearly one-third of the total cropped area in the country was under insect-pollinated crops, where bees had a great role to play. Taking a minimum requirement of 3 colonies per hectare for effective pollination of these crops, the Commission came up with a figure of 150 million bee colonies. Deodikar and Suryanarayana (1977) made a similar estimate of the requirement of bee colonies for pollination of bee dependent crops in India. At the rate of 3 bee colonies per hectare they felt that the 40 million hectares under bee pollinated crops would need about 120 million colonies.

We may also assess the beekeeping potential of the cultivated vegetation, based on the honey potential of the component plant species. Unlike the species under natural vegetation, crop plants are usually found in dense populations. The agro-climatic conditions are often controlled by agronomic practices. Due to these reasons, the honey potential figures are often close approximations of the actuals.

Nectar and honey potential

The amount of nectar produced during a season per hectare of land is called honey potential. A plant producing large quantity of nectar with surplus honey is referred to as major source and if a plant provides nectar just for survival of the colonies is called minor source. As the honey is produced from nectar, the amount of honey produced per hectare of land from a plant species under optimal conditions is called as honey potential. Honey potential varies from species to species, with climate, soil and geographical conditions. Factors affecting nectar production such as temperature, humidity, soil moisture also influence the honey production potential of a crop.

Table 20. Honey production potential of some common bee plants

Plant species	Honey potential (kg/ha)	Remarks/factors
Brassica campestris	50-100	Needs good irrigation
Eucalyptus spp	1000-1500	Varies with age and size of the trees
Litchi chinensis	500-1000	Healthy plants
Dalbergia sissoo	200-300	Requires good rains during winter
Syzygium cumini	200-300	Large healthy trees
Azadirachta indica	100-200	Large healthy trees
Cedrella toona	300-500	Uniform plantation and age of trees
Sapindus mukorossi	300-500	Uniform plantation and age of trees
Berberis spp	100-300	Thick bushes
Terminalia chebula	100-300	Healthy trees and Uniform plantations
Prunus armeniaca	100-300	Uniform plantation and healthy trees
Malus domestica	15-20	Requires plentiful rains during winter
Trifloium alexandrium	300-500	Requires good irrigation
Gossypium spp	100-300	Healthy crop, good irrigation
Robinia pseudoaccacia	500-1000	Requires plenty of soil moisture
Acacia catechu	100-300	Requires plentiful rains before flowering and light rain during flowering
Trifolium repens	200-500	Requires good irrigation and soil moisture

Plectranthus spp	500-1000	Requires plentiful rains during rainy season and good soil moisture
Prunus cerasoides	100-300	Requires good soil moisture
Brassica napus	300-500	Requires good irrigation
Helianthus annuus	100-200	Requires good irrigation

Source: Joshi, 1992

Table 21. Bee flora, their nectar sugar concentration, sugar value and honey production potential capacity

Botanical name	Common name	Nectar sugar concent-ration (%)	Sugar value (mg/flower/24)	Honey produ-ction potential (kg/colony/season)	Reference
Allium cepa	Onion	59-75.5	-	-	Rao and Lazar, 1980
Bombax ceiba	Sambal	35.0	--	-	Parmar, 1995
Brassica campestris L. var. *toria*	Toria	38.5-53.5	-	-	Kapil and Brar, 1971
Callistemon lanceolatus	Bottle brush	23.2	1.21	-	Bhat *et al.*, 1987
Carvia callosa	Karvi	20-37	-	-	Phadke, 1962
Citrus aurantifolia	Malta	22.0	-	-	Srawan and Sohi, 1985
Dalbergia sissoo	Shisham	32	-	4-9	Singh, 1962
Eucalyptus spp	Safeda	66-67	-	-	Srawan and Sohi, 1985
Fagopyrum esculentum	Buckwheat	35-45	0.10-2.68	60-70	Cirnu *et al.*, 1977
				70-90	Avetisyan, 1978
Helianthus annuus	Sunflower	35-38	0.27	56-69	Peter, 1977
Litchi chinensis	Litchi	More than 62	--	27	Nair, 1981
Malus domestica	Apple	30-65	1-3	36	Lovell, 1977
Medicago sativa	Lucerne	40.2	1.13	-	Bhat *et al.*, 1987
Plectranthus rugosus	Shain	38	-	50	Shah, 1979
Robinia pseudo acacia	Robinia false kikar	34-59	0.76-4.0	80	Konstantinovic, 1977
Sapindus detergens	Ritha	40	--	-	Sharma, 1958
Syzygium cumini	Jamun	9-72	-	-	Satyanaryana, 1975
Trifolium alexandrium	Berseem	32.4	-	165	Petkov, 1977
Vitex negundo	Banah	32-42	0.39	-	Gupta & Thakur, 1987a
Ziziphus Mauritian	Ber	More than 40	-	-	Zmarlicki, 1980

Crane (1952) reported an average honey production of 150 kg per hectare of cultivated crop plants in temperate regions. Plum, apple and cherry trees produced small quantities of nectar. Honey yield from orchards of these plants ranged between 30 and 80 kg per hectare. An hectare of phacelia (*Phacelia tenacetifolia*), ivy (*Hedera helix*), lucerne (*Medicago sativa*) and lavender

(*Lavandula* sp.) yielded 200 to 1300 kg of honey. Buckwheat (*Fagopyrum esculentum*) can produce 90 to 500 kg, cotton (*Gossypium hirsutum*) 29 to 90, rape seed (*Brassica napus*) 35 to 100 kg, raya (*Brassica juncea*) 50 to 60 kg, white mustard (*Sinapis alba*) 20 to 110 kg, sesame (*Sesamum indicum*) 45 kg, and sunflower (*Helianthus annuus*) 13 to 140 kg honey per hectare of the crop (Crane 1975). Depending on the colony strength, health of the bees, atmospheric conditions and the type of management of the colonies, 5 to 50 per cent of the total honey yield of an area is available to the beekeeper as surplus or production from a colony.

Different mustard species have a honey potential of 100 kg per hectare, other oilseed crops 100 kg, cotton 50 kg, coffee 200 kg, fruit crops 200 kg, vegetable crops 25 kg, rubber 1000 kg, cardamom 100 kg, citrus and guava 100 kg, coconut 20 kg and banana 25 kilograms per hectare. Taking one-fourth of the potential as surplus production, the number of bee colonies per hectare of these crops varies from 0.2 colony in coconut orchards to 8 colonies in rubber plantations.

Agricultural crops are seasonal and provide bee forage for limited periods. Bee colonies cannot be sustained throughout the year in any cultivated area, unless it has an integrated intensive agriculture, agro-forestry and social forestry systems. During the gap periods between two crop seasons, bee colonies will have to be moved to forest areas. Thus, the beekeeping potential indicated for the cultivated vegetation becomes a part of the potential estimated for the natural vegetation.

Sources of bee forage in India

It is believed that the highly advanced insect groups like the bees and the flowering plants evolved during the course of the last 100 million years, by influencing each other due to their inter-dependence. Plants depend on anthophilous insects for pollinating their flowers resulting in fruit and seed set. Honey bees depend upon flowering plants for their food. In any area with good vegetation, the natural or cultivated plant species are in an ecological balance with the local insect population.

There are several thousands of species flowering plants indigenous to India. In addition, there are a few thousand plant species that have been introduced as crops of economic importance, or as ornamental plants. Not all these are useful to honey bees, and even those that provide bee forage vary in their value to beekeeping. There are also variations in the utility of a species depending upon the climate, soil and other factors in the place of its occurrence. In assessing the beekeeping potential of any location it is of prime importance to learn about the composition of the vegetation in the locality, bee forage value of individual species and the usual flowering periods of important bee forage plants.

Sequence of flowering of the component plant species in the area has to be observed to find out the seasons when a large amount of nectar is available, which helps in storage of surplus honey that can be harvested by the beekeeper.

This observation also helps to find out whether bee forage is continuously available for at least eight months in a year. Long gaps in the bee forage availability affects the growth of bee colonies. During such floral dearth seasons, particularly when pollen is not available, colonies become weak. Such colonies use up a major part of the forage available immediately after the dearth for recouping their strength and build up of worker bee population. Often the important honey season that follows the dearth period, is half way through before these colonies can start storing surplus honey. Colonies can be shifted between farms and forests, depending upon the availability of bee forage or its absence in any place. Thus both the natural and cultivated vegetations are complementary for proper growth of bee colonies and for efficiency in the honey industry (Deodikar and Suryanarayana, 1977) felt that the agricultural crops could give only seasonal forage for bees restricted to their flowering periods around September-October, and December-January in the case of annual summer and winter crops, and around March-April for most of the orchard and other arboreal species. In India, the dearth periods on farms and orchards in plains more or less synchronize with floral peak periods in nearby forests in hills. Deodikar and Suryanarayana, therefore, felt that the 120 million colonies that would be required for pollinating crop plants, could be accommodated in the 60 million hectares under reserved forests.

Knowledge of bee plant species is necessary for the beekeeper so that he can make an efficient use of these resources. In view of this there have been several observations on bee plants of different locations (Deodikar and Suryanarayana, 1977). Similarly many other workers have evaluated the honey plants useful for bees (Saraf, 1972; Singh and Singh, 1971; Atwal and Goyal, 1974; Pratap, 1997). Still there may be many plants important to honey bees in India which may have been left out because of the paucity of information on them.

In both stationary and migratory beekeeping, the beekeeper seeks to place his colonies in or near areas where a sufficient quantity of honey plants - be they crop or pasture plants, weeds, shrubs, forest trees, roadside planting, etc. exists, in season or throughout the year, within the economical flight range of the foragers. Planting special crops for bees is not likely to yield a good economic return: arable land will provide better returns if it is used for other agricultural purposes. Beekeeping is thus one of the rare forms of agriculture in which the planting of crops is not specifically required.

List of flowering plant species providing bee forage in India

Deodikar and Suryanarayana (1977) have provided a very comprehensive list of bee forage plants available in different parts of India. Where: Bee forage value: N = source of nectar, P = source of pollen; 1 = major and important source; 2 = source of normal importance; 3 = source of minor importance. Flowering period: Serial numbers of months given thus, 1 = January, 2 = February, 11 = November, 12 = December. Habitat: Phytogeographic type of the region or type of forest, where the species is generally found; under cultivation the species can be found in any type of region. Economic Uses: Commonly used terms are abbreviated thus, Fod. = fodder; Fl. = flower; Ft. = fruit; Ind. = industry; lvs. = leaves; Med. = medicinal; Orn. = ornamental; Veg. = vegetable

Acanthaceae

Acanthus ilicifolius L.;--; N3P2; 1-12; Marshes, mangroves; Shrub; Food.

Adhatoda vasica Nees; Basuti, bakas; N3P2; 4-11; Tropical, sub-tropical; Shrub; Medicinal; hedge, green manure.

Andrographis paniculata (Burm.*f.*) Wall.; Bhuilimb, chirata, kal-megh, nela vemu; N2P2; 8 - 12; Tropical, sub-tropical; common weed in waste lands and road sides; Herb; Medicinal, for snake bites, lvs. for drinks and as vegetable.

Asystasia spp.; Bassati, lavana valli; N2P2; 6-12; Tropical, sub-tropical; common weed during rainy season; Herb; Medicinal, lvs. as vegetable, ornamental.

Barleria spp.; Bajradanti, kuranti, etc.; N2P1; 11-3; Tropical, sub-tropical, moist and dry areas, also cultivated; Shrub; Medicinal, ornamental, stem as toothbrush, root edible used in drinks, fls. sacred.

Blepharis spp.; Bhangari; N2P1; 7-2; Tropical, sub-tropical, forest undergrowth, and along forest paths; Herb; Medicinal.

Carvia callosa (Nees) Bremk.; Karvi, garikalu; N1P1; 7-9; Tropical, sub-tropical moist evergreen forests; Shrub; Sticks for construction, aromatic oil in perfumery and as insect repellent.

Hygrophila auriculata (Schum.) Heine; Gormidi, neeti gobbi, oont katela; N1P1; 7-2; Tropical, sub-tropical, in ditches, ponds and tanks, rice fields; Tall herb; Medicinal, leaves and shoots as vegetable.

Justicia spp.; Bankas; N2P2; 8-10; Tropical, sub-tropical, temperate; Herb; Medicinal, hedge, ornamental.

Lepedagathis cuspidata Nees; Akhra; N1P1; 10-2; Tropical, sub-tropical, moist and dry areas; Shrub; Fodder.

Rungia spp. Rungia; N3P3; 7-8; Cosmopolitan, wild; Herb; Medicinal.

Phlebophyllum kunthianum Nees; Kurinji; N1P1; 8-12; Sub-tropical moist evergreen open areas; Shrub; Ornamental, sacred plant.

Ruelllia tuberosa L.; --; N2P2; 6-11; Tropical, sub-tropical; weed in open moist lands; Herb; Ornamental.

Strobilanthinae strobilanthus; N1P1; 8-10; Tropical, sub-tropical; Shrub; Insect repellent, ornamental.

Thelapaepale ixiocephala (Benth.) Bremk; Whayati; N1P1; 11-1; Tropical-sub-tropical semi-evergreen forests; Shrub; Insect repellent.

Thunbergia spp.; Thunbergia, chakra kedar, chotaganti; N3P2; 1-3; Tropical-sub-tropical semi-evergreen forests, mostly cultivated; Climber; Ornamental, medicinal.

Aceraceae

Acer spp.; Great maple; N2P2; 3-4; Sub-tropical, temperate; Tree; Timber, fuel

Agavaceae

Agave americana L.; Century plant; N3P3; 9-11; Sub-tropical; Shrub; Ornamental, hedge, fibre

Dracaena spp.; Dragon's blood tree, charlang; N2P1; 3-5; Tropical, evergreen forests; also cultivated; Shrub; Ornamental, medicinal, red resin, ft.

Polianthus tuberosa L.; Tube rose; P1 1-12; Cultivated; Herb; Ornamental, fls. fragrant

Aizoaceae

Trianthema portulacastrum L.; Galijeru, pathorchattu; N2P2; 1-12; Tropical, sub-tropical moist and farmlands; Herb; Medicinal, lvs and shoots as vegetable

Alangiaceae

Alangium salvifolium (L.f.) Wang.; Ankol, alangi, dhela; N2P2; 3-6; Tropical, sub-tropical dry forests, waste lands; Small tree; Medicinal, for snake bite, fruit, hedge, timber for musical instruments, basketry

Amaranthaceae

Achyranthes spp.; Adhajhara, agya, latjira, uttareni; N2P2; 1-12; Tropical, sub-tropical, in forest and agricultural lands; Herb; Medicinal, leaves and roots for scorpion and snake bites, leaves and seeds as vegetable, sticks as toothbursh

Aerva spp.; Bui, bhol-phuli, chatra, pindikura; N2P1; 7-3; Tropical, sub-tropical, gregarious in open forests and farms; Herb; Medicinal, leaves as vegetable

Allmania nodiflora (L.) R.Br. ex Wt.; --; N2P2; 7-2; Open sandy and agricultural areas; Herb; Medicinal, vegetable

Alternanthera spp.; Jalsanchi, kodukka cheppu, etc.; N2P1; 1-12; Ruderal, agricultural areas; Herb; Medicinal, veterinary medicinal, leaves and young shoots as vegetable, root juice repels bees

Amaranthus spp.; Amaranth, math, etc.; N2P1; 7-12; Cosmopolitan; wild and cultivated; Herb; Vegetable, minor food grain, medicinal, root for scorpion and snake bites

Celosia argentea L.; Cocks comb; N2P2; 1-12; Cultivated, runs wild; Herb; Ornamental, vegetable

Gomphrena celosioides Mart.; --; N2P1; 7-12; Road sides, agricultural lands; Herb; --

Pupalia lappacea (L.) Juss; --; N2P1; 7-1; Open forest lands, near human habitation; Shrub.

Amaryllidaceae

Crinum spp.; Spider lily, chengalva; N3P2; 5-10; Tropical, sub-tropical, moist areas, along stream and river banks, field bunds in rice fields, etc.; Stout herb; Ornamental, medicinal, veterinary medicine

Pancratium triflorum Roxb.; Kandad; N3P2; 5-7; Tropical, sub-tropical, marshy areas, along stream banks; Herb; Ornamental

Anacardiaceae

Anacardium occidentale L.; Cashew nut tree, kaju; N1P2; 1-5; Tropical, coastal sandy locations; cultivated; Tree; Cashew apple and kernel edible, shell oil Cardol for industry, timber, medicinal

Buchanania lanzan Spr.; Calumpang nut tree, chironji, sara; N2P3; 1-5; Tropical, sub-tropical; also cultivated; Tree; Fruit and seed edible; oil from kernel, tannin, medicinal

Holigarna spp.; Black varnish tree, kattucheru; N2P3; 1-3; Tropical, sub-tropical, evergreen or semi-evergreen forests; Tree; Sap used in varnishes

Lannea coromandelica (Houtt.) Merr.; Jhingan, karambal; N2P2; 2-5; Tropical, sub-tropical; Tree; Timber, gum, fodder

Mangifera indica L.; Mango; N2P3; 3-4; Tropical, sub-tropical; cultivated; Tree; Fruit, fuel, timber, medicinal

Rhus spp.; Sumac, triseng, sitha; N3P3; 2-6; Tropical, sub-tropical, temperate, also cultivated; Tree, shrub; Gum, tannin, ornamental, seed oil edible

Semecarpus anacardium L.*f*.; Marking nut tree, jidi, bhilawa; N2P3; 6-7; Tropical, sub-tropical, deciduous forests; Tree; Fruit, juice for marking ink, seed oil, medicinal, veterinary medicine

Apiaceae

Carum spp.; Garnyu,vamu, zeerakarpo; N1P2; 10-2; Tropical, sub-tropical, cultivated; Herb; Medicinal, fruit, plant edible

Coriandrum sativum L.; Coriander; N1P1; 3-8; Cultivated; Herb; Vegetable, condiment, medicine

Daucus carota L.; Carrot; N2P3; 3-5; Cultivated; Herb; Vegetable, fodder

Foeniculum vulgare Min; Fennel; N3P3; 8-9; Cultivated; Herb; Spice, medicinal

Heracleum spp.; Hogweed; N3P3; 5-7; Sub-tropical, temperate; Herb; Medicinal

Pimpinella monoica Dalz.; Lady's lace; N3P2; 9-11; Cultivated; Herb; Ornamental

Apocynaceae

Carissa carandas L.; Karonda; N2P2; 4-5; Tropical, sub-tropical, temperate; Shrub; --; Fruit used as vegetable and in pickles

Araliaceae

Schefflera spp.; Bethu, etc.; N1P3; 3-5; Tropical, sub-tropical; Liana; Wood, fuel, ornamental

Arecaceae

Areca spp.; Betel nut palm, etc.; N2P2; 1-12; Cultivated; *A. triandra* Roxb. in mangroves; Tree; Nut edible, wood, leaves for thatching, ornamental

Borassus flabellifer L.; Toddy palm; N1P1; 1-12; Coastal and delta regions; cultivated; Tree; Stem sap for drinks, sugar; wood, leaves for mats, etc., fruit

Cocos nucifera L.; Coconut palm; N1P1; 1-12; Coastal areas; cultivated in a variety of habitats; Tree; Oil, Kopra, fibre, wood, leaves for mats, etc.

Elaeis guineensis Jacq.; Red oil palm; N2P1; 1-12; Tropical humid areas; Tree; Oil, ornamental

Nipa fruticans Wurmb.; --; N2P3; 1-12; Mangrove swamps, tidal forests; Tree; Stem sap for drinks, sugar; leaves for thatching, etc.

Phoenix dactylifera L.; Date palm, Khajur; N2P3; 1-12; Tropical, arid regions; Tree; Fruit edible, leaves for fibre

Phoenix spp.; Wild date palm, etc.; N2P3; 5-7; Tropical, sub-tropical; *P. paludosa* in salt water mangroves; Tree; Fruit, leaves for thacthing, pith edible,wood, fuel

Asclepiadaceae

Asclepias curassavica L.; Milkweed; N2P2; 4-6; Cosmopolitan weed; Shrub; Medicinal, fibre

Asteraceae

Ageratum conyzodies L.; Ageratum; N3P3; 7-9; S. Tropical, temperate; Herb; Ornamental

Artemisia maritima L.; Mugwort; N3P3; 8-10; Sub-tropical, temperate; Herb; Medicinal, ornamental

Aster spp.; Star-wort; N3P3; 7-11; Cultivated; Herb; Ornamental

Bellis perennis L.; Daisy; N2P1; 6-10; Cultivated; Herb; Ornamental

Bidens spp.; Spanish needle; N3P2; 9-12; Cosmopolitan weed; Herb; Medicinal

Calendula arvensis L.; Marigold; N3P3; 5-7; Cultivated; Herb; Medicinal, ornamental

Callistephus chinensis (L.) Nees; Aster; N2P1; 5-10; Cultivated; Herb; Ornamental

Cardus onopardiodes Fisch; Musk Thistles; N2P2; 5-8; Tropical, sub-tropical, temperate; Herb; Fodder

Carthamus tinctorius L.; Safflower; N2P2; 1-2; Cultivated; Herb; Oil, dye, ornamental, fodder

Centaurea cyanus L.; Corn flower; N2P2; 2-4; Cultivated; Herb; Ornamental, medicinal

Chrysanthemum spp.; Chrysanthemum; N2P2; 5-9; Cultivated; Herb; Insect repellant, ornamental

Cichorium intybius L.; Chicory; N2P2; 5-8; Tropical, sub-tropical, temperate; Herb; Medicinal

Cirsium spp.; Field Thistles; N2P2; 5-8; Tropical, sub-tropical, temperate; Herb; Vegetable, medicinal

Coreopsis drummondii Torr. & Gray; Coreopsis; N2P2; 6-10; Cultivated; Herb; Ornamental

Cosmos spp.; Cosmos; N3P2; 9-11; Tropical, sub-tropical, temperate; also cultivated; Herb; Dye, ornamental

Cyathocline purpurea (Don) O.Ktze.; Bongagra, chotagunja; N2P2; 1-12; Tropical, sub-tropical moist shaded areas; Herb; Medicinal, fish poison

Dahlia spp.; Dahlia; N2P1; 6-10; Cultivated; Herb; Ornamental

Echinops echinatus Roxb.; Globe Thistle; N3P3; 5-6; Cosmopolitan; Herb; Medicinal

Eclipta prostrata (L.) L.; Bhangra, bhringaraj, guntakalagara; N2P2; 1-12; Tropical, sub-tropical, moist areas as under growth, and farm areas; Herb; Medicinal, leaves for scorpion sting, leaves as vegetable

Eupatorium spp.; Through-wort; N3P3; 7-9; Cosmopolitan weed; also cultivated; Herb; Medicnial

Guizotia abyssinica Cass.; Niger, kalatil; N1P1; 9-11; Cultivated; Herb; Oilseed, ornamental

Helianthus annuus L.; Sunflower; N1P3; 1-12; Cultivated; Herb; Oilseed, ornamental

Inula cappa; Inula; N2P2; 9-11; Temperate; Herb; Ornamental, medicnial

Lagascea mollis Cav.; Phuli, koriyapatta, phukri; N2P1;1-12; Tropical, sub-tropical, open forest and agricultural areas; Herb; Medicinal

Mikania scanens Willd; Mikania; N2P3; 3-4; S.Tropical; Climber; Medicinal

Parthenium hysterophorus L.; Parthenium; NP3; 1-12; Cosmopolitan weed; Herb; Medicinal

Senecio spp.; Ragwort; N3P3; 6-9; Cosmopolitan, also cultivated; Herb; Medicinal

Solidago canadensis L.; Golden rod; N2P3; 6-9; Cultivated; Herb; Medicinal, ornamental

Sonchus spp.; Sow Thistle; N3P3; 6-10; Cosmopolitan, wild; Herb; --

Sphaeranthus spp.; Bodasaram, sitadori; N2P1; 10-4; Tropical, moist areas, rice fields, wastelands; Herb; Medicinal, fruit as fish poison

Tagetus spp.; Marigold; N3P3; 6-10; Cultivated; Herb; Ornamental, medicinal

Taraxacum officinale Weber; Dandelion; N2P2; 3-8; Cosmopolitan, wild; Herb; Medicinal

Tridax procumbens L.; Ekdandi, gad-dichamanti, patherchotti; N2P2; 1-12; Trpical, sub-tropical weed in waste lands, cultivated fields, etc.; Herb; Medicinal, leaves for scorpion sting, veterinary medicine

Tussilago farfara L.; Coltsfoot; N3P3; 4-6; Tropical, sub-tropical; Herb; Vegetable, medicinal

Vernonia monosis DC.; N1P2; 3-5; Tropical moist evergreen forest; Tree; Timber, fuel

Ximinia spp.; Iron weed; N3P3; 6-9; Cosmopolitan, weed; also cultivated; Herb; Medicinal, ornamental

Zinnia spp.; Zinnia; N1P1; 6-10; Cultivated; Herb; Ornamental

Averrhoaceae

Averrhoa carambola L.; Carambola tree; N1P2; 3-5; Tropical, sub-tropical; cultivated; Tree; Ftowers edible, ornamental, amenity tree Avicenniaceae

Avicennia officinalis L.; White mang-rove, tivar; N1P2; 8-11; Mangroves, salt marshes, tidal forests; Tree; Wood for fuel, leaves. for fodder, detergent, medicinal

Balsaminaceae

Impatiens spp.; Balsam; N1P2; 6-9; Cosmopolitan, wild; also cultivated; Herb; Ornamental, medicinal

Barringtoniaceae

Barringtonia racemosa (L.) Spr.; Samudra kadappai; N2P2; 1-12; Coastal, evergreen. high lands, cultivated; Tree; Timber, fuel, medicinal, ornamental road side tree

Careya arborea Roxb.; Kumbha, kumbi; NP2; 2-4; Tropical, sub-tropical deciduous areas; Tree; Timber, fibre, medicinal, root, bark edible, fish poison

Couropita guianensis Aubl.; Cannon ball tree, nagalingam; NP3; 3-5; Cultivated; Tree; Ornamental, amenity tree, sacred, timber

Berberidaceae

Berberis lycium L.; Barberry; N2P1; 5-6; Sub-tropical, temperate; Shrub; Medicinal

Betulaceae

Alnus spp.; Alder, kunis, utis; N3P2; 9-11; Sub-tropical, temperate; Tree; Timber, dye

Corylus colurna Dence.; Hazelnut; P2; 3-5; Sub-tropical, temperate; Shrub; Seeds edible, fuel

Bignoniaceae

Cybistax donnell-smithii Rose; Gold tree, basant rani; NP2; 3-6; Cultivated; Small tree; Ornamental, road side tree, timber valued

Dolichandrone spp.; Vil padiri, nuruddi, kadalathi; N3P2; 3-6; Tropical, sub-tropical, deciduous forests; Tree; Timber, fruit flowers as vegetable, medicine, leaves. used for snake bite

Heterophragma quadriloculare (Roxb.) K.Schum.; Varas, bondgu, udumaram; NP2; 1-4; Tropical, sub-tropical forests; Tree; Timber, medicinal

Millingtonia hortensis L.; Indian cork tree, akas nim; P2; 7-9; Cultivated; Tree; Ornamental, avenue tree

Oroxylum indicum Vent.; Pampini, ullu; N3P2; 4-6; Tropical, sub-tropical deciduous forests; Small tree; Medicinal, tannin, dye, flowers, fruit as vegetable

Pyrostegia venusta (Ker-Gawl) Miers; Golden shower; NP3; 3-5; Cultivated; Climber; Ornamental

Spathodea campanulata Beauv.; Scarlet bell; NP2; 3-6; Cultivated; Tree; Ornamental, avernue tree

Stereospermum spp.; Padiri, padal, gajpipal; N3P2; 4-6; Tropical, sub-tropical deciduous forests; Tree; Timber, medicinal, dye, amenity tree

Tabebuia spp.; Basant rani; NP2; 3-6; Cultivated; Small tree; Ornamental, road side tree

Tecoma spp.; Chinese trumpet creeper; NP2; 3-9; Cultivated; Shrub, climber; Ornamental

Bischofiaceae

Bischofia javanica Bl.; Bishop wood; N3P3; 2-3; Tropical, sub-tropical, moist eveergreen, forest; Tree; Timber

Bixaceae

Bixa orellana L.; Arnatto, kun-gamapoovu; NP1; 2-4; Tropical, sub-tropical; often cultivated; Small tree; Seed dye, Arnotto, Orlean, in industry, medicinal

Bombacaceae

Adansonia digitata L.; Baobab, gorakhimli; N1P1; 4-6; Cultivated; Tree; Ornamental, bark yields fibre, paper pulp

Bombax ceiba L.; Red silk cotton tree; N1P1; 2-3; Tropical, sub-tropical; also cultivated; Tree; Road side tree, fibre, fodder, soft wood

Ceiba pentandra L.; White silk cotton tree; N1P1; 3-5; Tropical, sub-tropical; also cultivated; Tree; Avenue tree, fibre, soft wood

Boraginaceae

Cynoglossum spp.; Hounds tongue, etc.; N3P3; 5-9; Tropical, sub-tropical; also cultivated; Herb; Ornamental

Heliotropium indicum L.; Naga danti, bhurandi; N2P2; 1-12; Tropical, sub-tropical, ruderal and sandy areas; Herb; Medicinal, for insect and scorpion bites

Trichodesma spp.; Aundhi, gheema; N3P2; 1-12; Wild in moist forest and open lands; Herb; Medilcinal, veterinery medicine, leaves as vegetable

Brassicaceae

Brassica juncea (L) Czern.; Indian mustard; N2P2; 12-3; Cultivated; Herb; Vegetable, oilseed

Brassica napus L. var.glauca (Roxb) Schutz.; Mustard; N1P1; 12-3; Cultivated; Herb; Oilseed, vegetable

Brassica napus var *toria* L. Toria; N1P1; 12-3; Cultivated; Herb; Oilseed, vegetable

Brassica oleracea var *capitata* L.; Cabbage; N2P3; 5-7; Cultivated; Herb; Vegetable

Brassica rapa L.; Turnip; N2P2; 2-4; Cultivated; Herb; Vegetable, fodder

Cardamine spp.; Cuckoo flower; N2P2; 1-2; S.Tropical, Temperate; Herb; Medicinal

Eruca sativa Mill.; Rocket salad; N1P1; 12-3; Cultivated; Herb; Medicinal, oilseed, fodder

Raphanus sativus L.; Radish; N1P1; 2-3; Cultivated; Herb; Vegetable, medicinal

Burseraceae

Boswellia serrata Colebr.; White dammar, gugal, salai; NP2; 4-6; Tropical, deciduous, forests, also cultivated; Tree; Gum resin, incense, med., timber, paper pulp, veterinary med., fuel

Canarium strictum Roxb.; Black dam-mar, manda dhup; NP2; 3-5; Tropical, sub-tropical, evergreen forests; Tree; Resin aromatic, insect repellant, in industry, incense, timber.

Cactaceae

Opuntia spp.: Prickly pear, thor; N2P2; 4-5; Cosmopolitan, wild and cultivated; Shrub; Medicinal, fruit, fodder, hedge plant

Caesalpiniaceae

Bauhinia variegata Wt.; Geranium tree; N2P3; 2-4; Tropical, sub-tropical, temperate; Tree; Vegetable, fodder, ornamental

Bauhinia vahli Wt.; Camel's foot climber; N2P3; 3-5; Tropical, sub-tropical; Tree; Fibre, seeds edible, leaves used as plates, fodder

Cassia spp.; Indian Labernum; N2P2; 4-7; Tropical, sub-tropical; wild and cultivated; Tree; Ornamental, medicinal, timber

Caesalpinia spp.; American Surmach; N3P3; 11-4; Tropical, sub-tropical; wild and cultivated; Small tree; Timber, Tannin

Delonix regia Raf.; Gulmohr; N3P2; 5-6; Cultivated; Tree; Ornamental, soft wood

Parkinsonia aculeata L.; Jerusalem thorn; NP2; 3-6; Tropical, sub-tropical, cultivated; Small tree; Ornamental, avernue tree

Peltophorum pterocarpum (DC.) Baker ex Heyne; Copper pod tree; N2P1; 3-5; Cultivated; Tree; Ornamental, avenue tree, timber

Tamarindus indica L.; Tamarind; N1P3; 3-5; Tropical,sub-tropical; wild, and cultivated; Tree; Condiment, fruit, vegetable, seed yields starch, fodder, road side and amenity tree

Cannabinaceae

Cannabis sativa L.; Hemp; N2P2; 6-9; Tropical, temperate; wild, sometimes cultivated; Herb; Alkaloid, fibre, vegetable

Capparaceae

Capparis spp.; Kanthar, etc.; N3P3; 4-5; Tropical, sub-tropical; Tree, climber; fruit edible, medicinal

Crataeva religiosa Forst. Barna; N2P2; 4-5; Cultivated; Tree; Ornamental

Caprifoliaceae

Lonicera spp.; Honey suckle; N3P3; 5-8; Cultivated; Shrub, climber; Ornamental

Viburnum spp.; Vikurum; N3P3; 5-6; Tropical, sub-tropical; also cultivated; Shrub; Ornamental

Caricaceae

Carica papaya L.; Papaya; NP2; 1-12; Cultivated; Tree; Fruit, medicinal

Caryophyllaceae

Gypsophila cerastioides D. Don; Gypsophila; NP; 3-4; Cultivated; Herb; Ornamental

Cassythaceae

Cassytha filiformis L.; Sitamma savaralu; N2P1; 7-1; Tropical, sub-tropical forests; Parasitic climber; Medicinal

Casuarinaceae

Casuarina equisetifolia Forst.; Beefwood tree, suru, saravu; P3; 2-4; Tropical, evergreen; extensively cultivated, along sea coasts, etc.; Tree; Timber, poles, fuel, afforestation, wind break, amenity tree, tanning, dye

Celastraceae

Lophopetalum wightianum Arn.; Banati; N2P2; 12-2; Tropical, sub-tropical, fresh water swamps; Tree; Timber

Chenopodiaceae

Chenopodium album L.; White goose foot; N3P3; 3-5, 8-10; Cosmopolitan, wild and cultivated; Herb; Vegetable, fodder

Cleomaceae

Cleome gynandra L.; Gulabi tiwan, vaminta; N1P2; 1-12; Tropical, open moist areas near villages; Herb; Medicinal, for scorpion bites

Cleome monophylla L.; Velakkeerai; N1P2 ;7-12; Tropical, sub-tropical, open, and cultivated areas; Herb; Leaves as vegetable

Cleome viscosa L.; Kukka vami-nta, pilitilwan; N1P2; 7-1; Weed along road sides, and open places; Herb; Medicinal, shoots as vegetable

Clusiaceae

Calophyllum inophyllum L.; Alexandrian laurel; NP2; 3-4; Tropical, sub-tropical, temperate, moist areas; Tree; Timber, seed oil for cooking and industry, medicinal

Garcinia spp.; Amsul, kokam, kuki; NP1; 3-5; Tropical, sub-tropical moist forests, cultivated; Tree; Fruit, vegetable, medicinal, dye

Mammea suriga (Roxb.) Kosterm.; Surangi, nagkesar; P1; 3-4; Tropical moist evergreen forests; Tree; Timber, flowers scented, cosmetic, medicinal

Mesua ferrea L.; Iron wood, nagchampa; P1; 1-4; Tropical, sub-tropical evergreen forests, also cultivated; Tree; Timber, seed oil, flowers in cosmetics, fruit, ornamental

Cochlorspermaceae

Cochlospermum religiosum (L.) Aston; Galgal, konda gogu, gong; N3P2; 2-4; Tropical, deciduous forests; Small tree; Seed cotton, gum Katira, medicinal, veterinary medicine

Combretaceae

Anogeissus spp.; Axle wood, dhaura, etc.; N3P1; 12-1; 6-8; Tropical dry deciduous forests; Tree; Timber, wood for coal, fuel, tannin, gum, veterinary medicine, snake bites

Quisqualis indica L.; Rangoon creeper; P2; 1-12; Cultivated, running wild; Climber; Ornamental

Terminalia arjuna (Roxb.) Wt. & Arn.; Arjun, maddi; N1P3; 4-6; Tropical, sub-tropical, deciduous forests; Tree; Timber, shade tree, amenity, medicinal

Terminalia bellirica (Gaertn.) Roxb; Beheda; N2P3; 4-5; Tropical, sub-tropical, semi-evergreen forests; Tree; Medicinal, timber, dye, tannin

Terminalia chebula Retz.; Ink nut, hirda; N1P2; 5-6; Tropical, sub-tropical, moist, evergreen forests; Tree; Medicinal, timber, dye, tannin

Terminalia tomentosa Wt. & Arn.; Ain, maddhi; N1P2; 4-6; Tropical, sub-tropical, moist, evergreen forests; Tree; Timber, fuel, host for tassar silk worm

Commelinaceae

Commelina spp.; Venna veduru, etc.; N2P2; 6-12; Tropical, sub-tropical, moist grasslands, etc.; Herb; Leaves edible, medicinal

Cyanotis spp.; Idra, bon gasun; N2P1; 7-9; Tropical, sub-tropical, moist, exposed lands; Herb; Tubers edible, vegetable, medicinal

Murdannia spp.; --; N2P1; 6-12; Tropical, sub-tropical, moist forests, common in farmlands; Herb; Ornamental

Tradescantia spp.; Tradescantia; N2P1; 1-12; Cultivated; Herb; Ornamental

Convolvulaceae

Convolvulus arvensis L.; Convolvulus'; N3P3; 4-9; S.Tropical, sub-tropical, temperate; Weed

Cressa cretica L.; Dahba; N3P2; 3-9; Mangroves, saline marshes and coastal sand; Climber; Plant edible, vegetable

Cuscuta reflexa Roxb; Amar-Vel; P3; 7-10; Cosmopolitan, weed; Parasitic, climber; Medicinal

Ipomoea batatas Lam.; Sweet potato; N2P3; 8-11; Cultivated; Herb; Vegetable, medicinal

Ipomoea pulchella Roth; Railway creeper; N3P3; 8-11; More or less cosmopolitan; Herb; Medicinal, ornamental

Jaquemontia pentantha (Jacq.) G. Don; Jaquemontia; N1P1; 1-12; Cultivated; Climber; Ornamental

Porana volubilis Burm.; Porana; N1P2; 3-5; Tropical, sub-tropical; Climber; Fibre

Cordiaceae

Cordia dichotoma Forst.*f.*; Sebesten plum, bohar; N2P2; 3-4; Tropical, sub-tropical deciduous forests; Tree; Timber, fruit, vegetable, pulp used as gum, medicinal

Costaceae

Costus speciosus (Koen.) Sm.; Besemati, kenkand; N2P2; 6-12; Tropical, sub-tropical, deciduous forests; Herb; Medicinal, veterinary medicine, edible, for pickles, drinks, as vegetable, root for snake bite, for rabid dog or fox bite.

Cucurbitaceae

Citrullus vulgaris Schrad; Water-melon; N2P2; 4-5; Cultivated; Climber, Fruit, medicinal

Cucumis spp.; Cucumber, melon; N2P2; 7-9; Cultivated; Climber; Vegetable, fruit used in pickles, oil, medicinal

Cucurbita maxima L.; Pumpkin; N2P2; 4-6; Cultivated; Climber; Vegetable, seed pulp edible, medicinal

Lagenaria siceraria L.; Bottle gourd; N2P2; 5-7; Cultivated; Climber; Vegetable, medicinal

Luffa spp.; Ribbed gourd; N2P2; 5-7; Cultivated; Climber; Vegetable, medicinal

Momordica spp.; Bitter gourd, etc.; N1P2; 5-10; Cultivated; Climber; Vegetable, medicinal

Cyperaceae

Cyperus spp.; Bimbal, chin-choda, gane-chi, gathiyan, motha; P1; 1-11; Wet lands, in farming and forest areas; Herb; Medicinal

Dilleniaceae

Dillenia indica L.; Chalta, pedda kalinga; N3P1; 3-4; Tropical, sub-tropical, moist evergreen or semi-evergreen forests; Tree; Timber, fruit, vegetable, medicinal

Dillenia pentagyna Roxb.; Agachi, karmal; N3P1; 1-3; Tropical, sub-tropical, moist evergreen forests; Tree; Timber, fruit, vegetable, medicinal, leaves for polishing

Dioscoreaceae

Dioscorea spp.; Dukka penda-lam, khamalu, pitakanda, tamboli; P1; 2-9; Tropical, sub-tropical forests; also cultivated; Climber; Medicinal, tubers as vegetable, starch, flowers. as vegetable

Dipsacaceae

Dispsacus inermis; Teasel; N3P3; 7-10; Tropical, sub-tropical, temperate; Herb; --

Scabiosa speciosa Royle; Scabious; N2P3; 5-7; Tropical, sub-tropical, Temperate; also cultivated; Herb; Ornamental, medicinal

Dipterocarpaceae

Dipterocarpus spp.; Black dammar; NP2; 12-2; Tropical, sub-tropical moist evergreen forests; Tree; Timber, resin, medicinal

Hopea spp.; Iron wood,etc.; NP2; 1-5; Tropical, sub-tropical moist, evergreen forests; Tree; Timber, fuel

Shorea robusta Gaertn.; Sal, guggal; NP1; 2-3; Tropical, sub-tropical; often planted; Tree; Timber, resin, fruit, seed oil edible, used in industry, leaves used as plates, sacred

Vateria indica L.; Indian coral, dhup; NP1; 2-4; Tropical moist evergreen forests; Tree; Timber, oleoresin in industry and as ornamental beads, seed oil edible, used in industry

Ebenaceae

Diospyros spp.; Persimmon, Malabar ebony, etc.; N1P3; 1-5; Tropical, sub-tropical, also cultivated; Tree; Fruit, leaves edible, timber, medicinal

Maba spp.; Irambali, alli, etc.; N2P2; 2-5; Tropical, sub-tropical; Small tree; Fruit

Ehretiaceae

Ehretia acuminata R.Br.; Ivory wood; N1P2; 2-4; Tropical, sub-tropical; Tree; Timber, fodder, fruit medicinal

Elaeagnaceae

Elaeagnus angustifolia L.; Sarseng; N2P2; 12-2; Sub-tropical, temperate forests; Shrub; Fruit

Elaeagnus conferta Roxb.; Amguli, hari, kulangi; N2P2; 10-2; Tropical, sub-tropical forests; Shrub; Fruit, medicinal, fuel

Elaeagnus kologa Sch.; Wild olive, kulangi; N2P2; 12-2; Tropical, sub-tropical forests; Shrub; Fruit

Elaeagnus parviflora Wall. ex Royle; Ginwai; N2P2; 4-5; Sub-tropical, temperate forests; Shrub; Fruit, also in curries, medicinial

Elaeocarpaceae

Elaeocarpus spp.; Wild olive tree, rudraksha; N2P2; 9-10; Tropical, sub-tropical, temperate moist evergreen forests; Tree; Timber, fruit as beads in rosary, considered sacred and are worn in garlands

Muntingia calabura L.;Singapur or Burma cherry; N2P2; 1-12; Cultivated in tropical to temperate areas; Small tree; Fruit, shade and amenity tree, ornamental

Ericaceae

Gaultheria fragrantissima Wall.; Fragrant winter-green; N2P2; 5-6; Temperate; Shrub; Fruit, medicinal

Rhododendron spp.; Alpine rose; N1P2; 3-5; Temperate; also cultivated; Small tree; Flowers edible, used in drinks and preserves, medicinal

Erythroxylaceae

Erythroxylum monogynum Roxb.; Bastard sandal, devadaru; N1P2; 2-4; Tropical, dry or moist deciduous forests; Shrub; Timber, incense, fuel, wood tar as preservative

Euphorbiaceae

Aporusa spp.; Vittil, sarali; N2P1; 12-4; Tropical, sub-tropical, evergreen forests; Tree; Timber, fruit, dye

Croton bonplandianum Baill.; Mirchayia, galivana; NP2; 1-12; Cosmpolitan, ruderal, and road side weed; Herb; Medicinal

Drypetes roxburghii (Wall.) Huru.; Putranjiva; N3P2; 4-5; Tropical, sub-tropical, semi-evergreen forests; Tree; Timber, fruit used as ornamental beads, medicinal

Euphorbia antiquorum L.; Tridhra sehund; N2P2; 11-1; Tropical, cultivated; Shrub; Hedge plant, ornamental, medicinal

Euphorbia pulcherrima Willd. ex Klot.; Lal patti; N3P2; 1-12; Cultivated; Shrub; Hedge plant, ornamental, medicinal

Euphorbia royleana Boiss; Euphorbia; N3P3; 4-5; Cosmopolitan; Shrub; Medicinal, hedge plant

Excoecaria agallocha L.; Blinding tree; N1P2; 3-4; Mangroves, tidal forests, swamps; Small tree; Soft wood, fuel, fish poison, sacred for Hindus

Glochidion spp.; Kalchia, bho-ma, nirveti; NP2; 2-5; Tropical, sub-tropical, evergreen forests; Small tree; Medicinal

Hevea brasiliensis (HBK) Muell.-Arg.; Para rubber; N1; 2-4; Tropical moist areas; cultivated; Tree; Latex made into rubber, timber

Jatropha curcas L.; Physic nut; N2P2; 6-9; Tropical, sub-tropical, widely cultivated; Shrub; Ornamental, seed oil in med. and industry, hedge plant

Jatropha gossypifolia L.; Belly ache bush; N2P1; 1-12; Cultivated; Shrub; Hedge plant, ornamental, seed oil for industry

Macaranga spp.; Chandada, vatta, etc.; NP2; 1-3; Tropical, sub-tropical, evergreen forests; Tree; Wood for matches, paper pulp, gum, shade plant, medicnial, leaves used as plates

Mallotus spp.; Kamala dye tree, pitali; P1; 8-11; Tropical, sub-tropical, evergreen forests; Tree; Fruit yields dye, timber, seed oil in industry, medicinal

Manihot esculenta Crantz; Cera rubber, maracheni-kilangu; N2P2; 3-5; Tropical, sub-tropical, cultivated; Small tree, shrub; Medicinal, root tubers edible, ornamental

Melanthesa spp.; Devadari, jirual, owi,jhanjika; N2P2; 7-8; Tropical, sub-tropical, deciduous forests; Shrub, small tree; Medicinal, fuel

Phyllanthus acidus (L) Skeels; Star gooseberry; N2P2; 4-6; Tropical, sub-tropical; also cultivated; Tree; Fruit, medicinal, tannin

Phyllanthus emblica L.; Emblic myrabolan, amla; N3P2; 3-5; Tropical, sub-tropical; also cultivated; Tree; Fruit used as vegetable, in pickles and preserves, medicinal, tannin, fuel

Ricinus communis L.; Castor; P2; 5-8; Tropical, sub-tropical; often cultivated; Large shrub; Seed oil used in industry, medicinal

Sapium spp.; Chiran, makhani; NP2; 2-4; Tropical, sub-tropical, deciduous forests; Tree; Seed oil and fat in industry, medicinal, timber, fuel

Securinega virosa (Roxb. ex Willd.) Pax & Hoffm.; Dalme, helpi, balli chettu; N2P2; 7-2; Tropical, dry open forests, along coasts, etc.; Small tree; Medicinal, stupification of fish, blossom forecasts rain

Synadenium grantii Hk.f.; African milk bush; N2P3; 1-12; Cultivated; Shrub; Ornamental, hedge plant

Fabaceae

Atylosia spp.; Bajhala, rantur; N3P2; 2-7; 10-12; Tropical, sub-tropical, deciduous open forests; Shrub, climber; Medicinal, for snake bites, veterinary medicine

Butea monosperma (Lam.) Kuntze; Palas, dhak; N2; 5-9; Tropical, sub-tropical; Tree; Dye, medicinal

Cajanus cajan L.; Pigeon pea, tuar, kandulu; N1P2; 5-9; Cultivated; Shrub; Pulse, fodder

Cicer arietinum L.; Gram, sanagalu, chana; NP2; 2; Cultivated; Herb; Pulse, vegetable, fodder

Crotoloaria juncea L.; Sannhemp; NP3; 7-9; Cultivated; Under Shrub; Fibre, green manure

Dalbergia lanceolaria L.*f.*; Kuralada, patradi; N1P2; 3-5; Tropical, sub-tropical, deciduous forests; Tree; Green manure, medicinal

Dalbergia latifolia Roxb.; Rose wood, bete, devbija; N1P2; 1-3; Tropical, sub-tropical, semi-evergreen forests; Tree; Timber valued, sacred, medicinal, veterinery medicine

Dalbergia paniculata Roxb.; Passi, pachari; N1P2; 3-4; Tropical, sub-tropical deciduous forests; Tree; Medicinal, host for lac insect, low value wood and fuel

Dalbergia sissoo Roxb.; Sissoo, shisham; N1P2; 3-5; Tropical, sub-tropical; also cultivated; Tree; Timber, fuel, fodder, green manure

Dalbergia spinosa Roxb.; --; NP; 3-5; Mangroove highlands; Tree; --

Desmodium spp.; Chirota, sarivan, karabeeja N2P2; 8-3; Tropical, sub-tropical, forest undergrowth; Under shrub; Med., root antidote to snake venom, stem fibre for paper, repels bed bugs

Erythrina spp.; Coral tree, dadap, etc.; N2P2; 3-6; Tropical, sub-tropical; also cultivated; Tree; Shade tree, soft wood, paper pulp, fibre, medicinal

Flemingia spp.; Ajhar, makri, nemali padam; N2P2; 10-5; Tropical, sub-tropical forests as undergrowth; Shrub; Medicinal, fruit and tubers edible, veterinary medicine

Glycine max (L.) Merr.; Soyabean; N3P3; 7-8; Cultivated; Herb; Pulse, oilseed, fodder

Indigofera spp.; Indigofera; N2P2; 6-8; Tropical, sub-tropical, temperate; Shrub; Fodder

Lespedeza spp.; Lespedeza; N2; 6-7; Tropical, to temperate; Shrub; Soil reclamation, fodder

Medicago sativa L.; Lucerne, alfalfa; N1P2; 5-8; Cultivated; Herb; Fodder, medicinal

Melilotus spp.; Mellilot; N1P2; 3-7; Tropical, sub-tropical, temperate; also cultivated; Herb; Fodder

Phaseolus spp.; Pulses, Beans; N1P2; 7-8; Cultivated; Herb; Pulse, fodder, green manure

Pisum sativum L.; Pea, matar; batani; N2P2; 3-7; Cultivated; Herb; Vegetable, green manure

Pongamia pinnata (L.) Pierre; Indian beech, karanj, ponga; N1P1; 3-6; Tropical, sub-tropical, moist and dry areas; Tree; Seed oil in industry, timber, fuel, medicinal, fish poison

Pterocarpus marsupium Roxb.; Kino tree, bijasal, yegisa; N2P2; 3-5; Tropical, sub-tropical, deciduous forests; Tree; Timber valued, medicinal, gum in industry, veterinery medicine

Pterocarpus santalinus L.; Red sanders; N1P2; 7-8; Tropical dry forests; Tree; Timber valued, gum resin dye, medicinal

Robinia pseudacacia L.; Black locust, false acacia; N1P1; 4-6; Cultivated; Tree; Timber, shade and avenue tree, fuel, leaves edible

Sesbania grandiflora Pers.; Basna, agasthi, avisi; NP1; 2-4; Tropical, sub-tropical, often cultivated; Small tree; Soft wood, fuel, leaves, flowers. and fruit as vegetable, medicinal, shade tree in plantations

Smithia spp.; Barka, kaula, titalboti, etc.; N2P1; 3-9; Tropical, sub-tropical, grass lands, moist lands; Herb; Leaves as vegetable, medicinal

Sophora spp.; Pagoda tree, etc.; N2P2; 3-7; Tropical, sub-tropical, temperate, cultivated; Tree; Ornamental, timber, medicinal, twigs used as broom

Trifolium alexandrium L.; Egyptian clover, berseem; N1P1; 4-7; Cultivated; Herb Fod., green manure

Trifolium pratense L.; Red clover; N2P2; 4-7; Sub-tropical, temperate; Herb; Fodder

Trifolium repens L.; White clover; N1P2; 4-7; Sub-tropical, temperate; Herb; Fodder, green manure

Trigonella spp.; Methi; N2P2; 6-9; Cultivated; Herb; Condiment, vegetable, medicine, fodder

Vicia spp.; Field Beans; N2P2; 4-8; Sub-tropical, temperate; also cultivated; Climber; Food, Fodder

Vigna spp.; Ran shevga, halunda, etc.; N2P2; 3-5; Tropical, sub-tropical, moist and dry forests; Climber; Vegetable, pulse, medicinal

Wisteria sinensis DC.; Grape flower; N2P2; 3-5; Tropical, sub-tropical; Climber; Ornamental

Fagaceae

Castanca sativa Mill.; Sweet Chestnut; N2P2; 5-9; Sub-tropical, temperate; also cultivated; Tree; Timber, seed edible

Quercus spp.; Oak, phaniat, sulentong; N2P3; 4-7; Sub-tropical, temperate; also cultivated; Tree; Timber, medicinal

Flacourtiaceae

Flacourtia spp.; Sweet thorn, attak, kakai; N2P2; 9-12; Tropical, sub-tropical; Tree; Timber, fruit, fodder, fuel

Flindersiaceae

Chloroxylon swietenia DC.; Satinwood tree, bhera, porasu, billu; N2P2; 3-4; Tropical dry deciduous forests, also cultivated; Tree; Ornamental, timber, medicinal, veterinary medicine

Gentianaceae

Gentiana spp.; Gentiana, neelkanthi; N3P2; 4-7; Sub-tropical, temperate; also cultivated; Herb; Medicinal, ornamental

Nymphoides spp.; Water lily, baan; N3P2; 1-12; Tropical, sub-tropical, bydrophyte in ponds, etc.; Herb; Medicinal, ornamental, veterinary medicine, rhizomes edible

Swertia spp.; Swertia; N3P3; 8-10; Sub-tropical, temperate; Herb; Medicinal

Geraniaceae

Erodium spp.; Erodium; N2P2; 4-6; Sub-tropical, temperate; also cultivated; Herb; Ornamental

Geranium spp.; Crane's bill; N3P3; 5-9; Cosmopolitan, wild and cultivated; Herb; Ornamental, medicinal

Hippocastanaceae

Aesculus indica Colebr. ex Cambess; Horse chestnut; N2P2; 4-5; Sub-tropical, temperate climates; cultivated; Tree; Timber, kernel edible, detergent for woollen, medicinal, amenity tree

Icacinaceae

Apodytes beddomei Mast.; Ammemara; N1P2; 11-1; Tropical, sub-tropical evergreen forests; Tree; Timber

Iridaceae

Crocus sativus L.; Saffron, kesar; N2P1; 9-11; Temperate, cultivated; Herb; Stamens for flavour and colour, ornamental

Iris nepalensis Don; Iris; N2P2; 4-5; Temperate, also cultivated; Herb; Ornamental, medicine

Juglandaceae

Juglans regia L.; Walnut, akhrot; N2P2; 2-4; Sub-tropical, temperate; Tree; Kernel edible, dye, medicinal

Lamiaceae

Clinopodium vulgare L.; Summer savory; P3; 9-10; Temperate; Herb; Medicinal

Dysophylla spp.; --; N2P2; 11-8; Tropical, sub-tropical open moist areas; Herb, shrub; Ornamental, medicinal

Lamium spp.; Dead nettle; N2P3; 4-10; Sub-tropical, temperate; Herb; Medicinal

Leonurus cardiaca L.; Mother wort; N3P3; 6-9; Sub-tropical, temperate; Herb; Ornamental, medicinal

Mentha spp.; Mint; N3; 7-10; Sub-tropical, temperate; also cultivated; Herb; Essential oil, medicinal, condiment

Nepeta spp.; Catmint; N3P3; 5-8; Sub-tropical, temperate; Herb; Medicinal

Ocimum spp.; Basil, tulsi, etc.; N3P3; 6-9; Tropical, sub-tropical; often cultivated; Herb; Ornamental, medicinal

Origanum vulgare L.; Marjoram; N2P2; 7-9; Sub-tropical, cultivated; Herb; Medicinal, essential oil

Plectranthus spp.; Plectranthus, bapachi; N1P2; 8-12; Tropical, sub-tropical, temperate forests; Herb, shrub; Medicinal

Pogostemon spp.; Ban tulsi, hanbila, pangla, pach; N1P1; 7-2; Tropical, sub-tropical, forests, hedges and open lands; also cultivated; Shrub; Leaves yield essential oil, ornamental, medicinal, veterinary medicine, roots for drinks

Rabdosia coesta (Buch.-Ham. ex D.Don) Hara; Kalthunia, shain; N1P2; 8-11; Sub-tropical, temperate mixed open forests; Shrub; Fodder

Rabdosia rugosa (Wall. ex Benth.) Hara; Shain, sulah; N1P2; 8-10; Temperate dry forests, open hill slopes; Shrub; --

Rosmarinus offinalis L.; Rosemary; N2P2; 4-6; Cultivated; Herb; Ornamental, condiment, medicinal

Salvia spp.; Sage; N2P2; 7-10; Sub-tropical, temperate; also cultivated; Herb; Ornamental, medicinal

Stachys spp.; Woundwort; N2P3; 6-9; Sub-tropical, temperate; Herb; Medicinal

Teucrium spp.; Wood Sage; N3P3; 8-9; Sub-tropical, temperate; Herb; --

Thymus spp.; Thyme; N2P3; 5-10; Sub-tropical, temperate; Shrub; Medicinal, condiment, essential oil

Lauraceae

Actinodaphne angustifolia Nees; Pisa, pishia; N1P1; 11-2; Tropical, sub-tropical, moist evergreen forests; Tree; Seed oil in industry, fuel, leaves for tonic, medicinal

Alseodaphne semecarpifolia Nees; Naramamidi, arambai, nelthare; N1P1; 10-1; Tropical, sub-tropical, moist evergreen forests; Tree; Timber, medicinal

Cinnamomum zeylanicum Bl.; Cinnamon tree; N2P2; 12-3; Tropical, sub-tropical, moist evergreen forests; Tree; Bark spice, leaves and roots yield camphor, medicinal

Cryptocarya wightiana Thw.; Palai, gulmur; N3P2; 11-12; Tropical, sub-tropical, moist forests; Tree; --

Litsea spp.; Maida, chandana, narra alagi; N1P1; 1-10; Tropical, sub-tropical, evergreen forests; Tree; Medicinal, veterinary medicine, edible oil, timber, fish poison

Litsea polyantha Juss.; Meda; N3P3; 12-1; Sub-tropical, temperate; Tree; Medicinal

Persea spp.; Avocado, ladder wood, lapche phal; N2P2; 12-1; Tropical, sub-tropical, evergreen forests, also cultivated; Tree; Timber, fruit

Leeaceae

Leea spp.; Hansiadhapan; N3P2; 3-9; Tropical, sub-tropical; Shrub; Medicinal, fruit

Liliaceae

Allium cepa L.; Onion; N3P3; 5-6; Cultivated; Herb; Vegetable, condiment, medicinal

Allium sativum L.; Garlic; N3P3; 5-6; Cultivated; Herb; Vegetable, condiment, medicinal

Aloe spp.; Aloe, gheekumar, kathaldi; N2P2; 3-4; Cultivated, running wild; Shrub; Hedge plant, ornamental, medicinal, leaves edible

Asparagus spp.; Satawari, nilichedi, pilli teegalu; N2P2; 8-11; Tropical, sub-tropical, temperate forests; also cultivated Shrub, climber; Medicinal, ornamental, root tubers as vegetable

Asphodelus tenuifolilus Cav.; Piazi; N3P2; 7-10; Sub-tropical, temperate; Herb; --

Chlorophytum spp.; Chamksa, kulai, janjari; N3P2; 5-10; Tropical, sub-tropical, dry deciduous, moist, shady areas; Herb; Medicinal, plant, leavesand flowers as vegetable

Gloriosa subperba L.; Kala lavi, nagaser; NP2; 7-10; Tropical, sub-tropical forests; Twiner; Medicinal, leaves and rhizome for insect and scorpion bites, veterinary medicine

Linaceae

Linum spp.; Flax, etc.; N2P1; 1-12; Tropical, grass lands; also cultivated; Herb; Medicinal, oilseed for industry

Lobeliaceae

Lobelia spp.; Boomari, dhaval, nimante, gorak nal; NP2; 8-2; Tropical, sub-tropical undergrowth, moist places, open forests; Herb; Leaves as vegetable, medicinal, veterinary medicine, latex leech repellent

Loganiaceae

Buddleja spp.; Buddleia, agia chita; N2P2; 6-9; Tropical, sub-tropical, temperate forests; also cultivated; Shrub; Ornamental, leaves and fruit edible, used in drinks

Strychnos nux-vomica L.; Nux vomica, kuchila, mushti; N2P1; 2-4; Tropical, sub-tropical, deciduous forests; Tree; Timber, oil from wood, and fruit, medicinal, fruit, fish poison, dye, hedge plant

Strychnos potatorum L.; Clearing nut tree, nirmali; N2P1; 2-4; Tropical, sub-tropical, deciduous forests; Small tree; Timber, seed mucilage for clearing water, medicine

Loranthaceae

Dendrophthoe spp.; Banda, irkum, badanika; N2P2; 3-5; Tropical, sub-tropical forests; Partial stem parasite; Med.icinal

Viscum spp.; Badanika, chathew, raghumola; N2P2; 7-11; Tropical, sub-tropical forests; Partial stem parasite; Medicinal

Lythraceae

Lagerstroemia indica L.; Crape myrtle, Pride of India, chinna goranta; N2P2; 7-9; Tropical, sub-tropical, also cultivated; Small Tree; Ornamental, medicinal, timber

Lagerstroemia microcarpa Wt.; Benteak, nana; N1P2; 3-5; Tropical, sub-tropical deciduous forests; Tree; Timber, medicinal

Lagerstroemia parviflora Roxb.; Nandi, bondar; N1P2; 4-6; Tropical, sub-tropical deciduous forests; Tree; Timber, medicinal, dye

Lagerstroemia speciosa (L.) Pers.; Jarul, poo maruthu; N3P2; 3-4; Tropical, sub-tropical, moist forests; also cultivated; Tree; Ornamental, timber

Lawsonia inermis L.; Henna plant, mehndi; NP1; 3-9; Tropical, sub-tropical areas, cultivated; Shrub; Hedge plant, dye for cosmetics, industry, medicinal

Sonneratia apetala Buch.-Ham.; Keora, Kalinga; N1P2; 4-6; Tropical coastal and mangrove forests; Tree; Fuel, Fruit

Woodfordia fruticosa (L.) Kurz; Dhaiti, dauri, jargi; N1P2; 2-10; Tropical, sub-tropical, moist and dry forests; Shrub; Dye, medicinal, flowers, fruits used as vegetable, in drinks

Malpighiaceae

Aspidopterys indica (Roxb.) Hochr.; Bandi kota teega; N2P2; 8-12; Tropical, sub-tropical, semi-evergreen forests; Large climber; Ornamental

Hiptage benghalensis (L.) Kurz; Madhumalati, aneta, kurukkathi; N2P2; 2-4; Tropical, sub-tropical, evergreen forests; Liana type shrub; Fuel, ornamental, flowers fragrant, medicinal, veterinary medicine

Malvaceae

Abelmoschus esculentus L.; Lady's finger, okra; N3P3; 7-9; Cultivated; Herb; Vegetable, fibre

Abutilon spp.; Thutti, etc.; N2P2; 1-12; Tropical, sub-tropical, forests, open lands; Shrub; Medicinal, fibre

Althaea rosea Cav.; Hollyhock; N2P2; 6-10; Cultivated Herb; Dye, ornamental

Gossypium arboreum L.; Cotton; N3P3; 7-9; Cultivated; Shrub; Fibre, oil

Hibiscus cannabinus L.; Deccan hemp; N3P3; 7-8; Cultivated; Herb; Vegetable, fodder, fibre

Kydia calycina Roxb.; Pula, warang; NP2; 2-4; Tropical, sub-tropical deciduous forests; Small tree; Fibre cordage, paper pulp, stem mucilage to clarify sugarcane juice, medicinal

Malva sylvestris L.; Mallow; N2P3; 6-10; Cultivated; Herb; Medicinal, ornamental

Sida spp.; Bala, bari-yara, kherati; NP1; 9-12; Tropical, sub-tropical; weed, ruderal or in forest undergrowth; Herb, shrub; Medicinal, fibre, plants used as broom,veterinary medicine

Thespesia spp.; Indian tulip, paraspipal, bhendi tree; N1P2; 1-12; Tropical, sub-tropical, coastal districts; Tree; Fibre, cordage, timber, dye, medicinal, veterinary, medicine, road side and amenity tree

Urena spp.; Boriyal, nichardi; N3P2; 7-11; Tropical, open forest and waste lands; Under shrub; Medicinal, fibre

Martyniaceae

Martynia annua L.; Baghnakha, bichua, vinchi; N2P1; 8-11; Open moist lands in tropics and sub-tropics; Herb; Medicinal, veterinary medicine

Melastomaceae

Melastoma malabathricum L.; Jatrangi, kadalil, nekkane; N3P; 3-9; Tropical, sub-tropical, evergreen forests; Small tree, shrub; Host of a var. of silkworm, fruit, medicinal

Memecylon spp.; Alli, kasavu, kaya, etc.; N2P2; 1-9; Tropical, sub-tropical, moist or dry forests; Shrub, tree; Wood for excellent fuel, charcoal, walking sticks, timber, dye, ornamental

Meliaceae

Aphanamixis polystachya (Wall.) Parker; Harinhara, akawatong, chemmaram; N3P2; 4-5; Tropical, sub-tropical, evergreen forests; Tree; Timber, seed oil in industry, amenity tree, dye, fish poison, medicinal

Aglaia spp.; Karagil, nimbadi, madrasa; N3P2; 1-2; Tropical, sub-tropical, evergreen forests; Tree; Fruit aril edible, timber, medicinal

Azadirachta indica Juss.; Neem, margosa; N2P3; 5-9; Tropical, sub-tropical, temperate, wild, also cultivated; Tree; Timber, seed oil in industry, medicinal, insecticide, avenue and amenity tree

Cipadessa baccifera (Roth) Miq.; Nalbila, jial, pithamari; NP2; 1-3; Tropical, sub-tropical; Small tree; Fuel, medicinal

Melia spp.; Pesian lilac, hebbevu, bakain; N2P2; 1-3, 7-11; Tropical, sub-tropical, temperate, often cultivated; Tree; Timber, nuts as beads, fuel, medicinal, insecticide, young shoots edible, vegetable

Swietenia spp.; Mahogani; N2P2; 4-6; Tropical, sub-tropical, cultivated; Tree; Timber, amenity, roadside tree

Toona ciliata M.-; Cedrella; N1P2; 3-6; Sub-tropical; Tree; Timber, dye

Trichilia cinnaroides (Wt. & Arn.) Bent.; Korakkadi, kora; N3P2; 3-4; Tropical, sub-tropical evergreen forests; Small tree; Fruit

Xylocarpus granatum Koen.; Chapa, sisumar; N1P2; 3-5; Tropical, mangrove swamps; Tree; Medicinal

Menispermaceae

Stephania glabra (Roxb.) Miers; --; P1; 7-12; Tropical, sub-tropical forests; Climber; Medicinal, veterinary medicine

Tinospora spp.; Chintil, goluncha, gunge; N3P1; 8-11; Tropical, sub-tropical evergreen forests; Large climber; Medicinal, veterinary medicine

Mimosaceae

Acacia auriculiformis A. Cunn.; Australian acacia; N2P2; 1-5; Cultivated; Tree; Ornamental, avenue tree, timber, fuel

Acacia catechu (L.f.) Willd.; Cutch tree, khair; N1P2; 2-5; Tropical, arid or semi-arid forests; Tree; Cutchu from wood, gum

Acacia chundra Willd.; Son khair; N3P3; 2-5; Tropical, evergreen or semi-evergreen forests; Tree; Timber, Cutchu from wood, tannin

Acacia decurrens Willd.; Green wattle; N3P2; 3-8; Tropical, sub-tropical humid areas; Small tree; --

Acacia leucophloea (Roxb.) Willd.; Panicled acacia, safed kikar; N2P1; 8-11; Tropical, sub-tropical, deciduous forests; Tree; Medicinal, bark and pods edible as vegetable, timber, fibre

Acacia melanoxylon R. Br.; Black wattle; N2P1; 8-11; Sub-tropical, temperate, cultivated; Tree; Timber, tannin

Acacia modesta Wall.; Phulai; N2P2; 5; Sub-tropical, temperate; cultivated; Tree; Timber

Acacia nilotica (L.) Del. ssp. indica (Benth.) Bren.; Babul, kikar; N3P2; 1-9; Tropical, arid or semi-arid areas;Tree;Timber, medicinal, tannin, fuel

Acacia pennata (L.) Willd.; Biswal, iklonda; N2P2; 7-12; Tropical, sub-tropical forests; Tree; Medicinal, bark for snake bite, detergent, fish poison, tannin

Acacia senegal (L.) Willd.; Gum acacia; N2P2; 2-5; Tropical, arid regions; Tree; Gum

Acacia sinuata (Lour.) Merr.; Soap pod wattle, shikekai; N1P1; 2-5; Tropical, sub-tropical, evergreen, semi-evergreen forests; Climber; Pods used as soap, bark for dyeing and tanning, live fence, fuel

Adenanthera pavonina L.; Coral wood, bari ghumchi; N2P2; 2-5; Tropical, sub-tropical, semi-arid regions; Tree; Timber, dye, seeds used as jeweller's weights

Albizia spp.; Siris, etc.; N2P2; 4-5; Tropical, sub-tropical; Tree; Fodder, fuel, timber, pulp

Dichrostachys cinerea Wt. & Arn.; Virtuli, vadathali; N2P2; 6-11; Tropical, sub-tropical; Shrub; Stems as walking sticks, fuel

Entada pursaetha DC.; St. Thomas bean, gila; N3P2; 3-5; Tropical, sub-tropical, semi-evergreen forests; Liana; Seed for handicrafts, edible, medicinal, fish poison

Leucaena spp.; Subabul, etc.; N3P2; 3-10; Tropical, sub-tropical; cultivated; Tree; Fodder, timber, green manure, fuel

Mimosa hamata Willd.; Jheryihani, erra Sandra; N1P2; 11-12; Tropical, sub-tropical dry forests; Shrub; Medicinal, ornamental

Mimosa himalayana Gamble; Bramble stalk, kasari, korintha; N1P2; 8-12; Tropical, sub-tropical, savannah and open forests; Shrub; Medicinal, for snake bite

Mimosa pudica L.; Sensitive plant; N1P1; 7-11; Tropical, sub-tropical; weed; also cultivated; Herb; Ornamental

Parkia timoriana (DC.) Merr.; Supota, yongchak; N3P2; 10-12; Cultivated; Tree; Avenue tree, fruit and seed edible as vegetable

Prosopis chilensis (Mol.) St.; Misquite, deb babul, gunda bavar; N1P1; 9-3; Tropical dry evergreen forests from coast to foot hills; Small tree, shrub; Timber, used for fuel, charcoal, drought resistant, fruit edible in thirst

Prosopis cineraria (L.) Druce.; Chonkar, shami, jammi; N2P2; 9-4; Tropical dry evergreen forests, also cultivated; Tree; Timber, medicinal, bark, gum and shoots edible, sacred for Hindus and Sikhs

Samanea saman (Jacq.) Merr.; Rain tree; N2P2; 3-6; Cultivated; Tree; Road side, shade and amenity tree, fodder

Xylia xylocarpa Roxb.; Irul, pynikado; N2P2; 3-4; Tropical, sub-tropical, deciduous forests, often cultivated; Tree; Timber, dye

Moringaceae

Moringa oleifera Lam.; Drumstick tree; N1P2; 1-12; Tropical, sub-tropical; often cultivated; Tree; Leaves, flowers, fruits used as vegetable, medicinal, gum in perfumery, fruit pickled, seed oil edible

Moraceae

Morus spp.; Mulberry; P2; 3-4; S.Tropical; Tree (W/C); Source of food for silkworm

Musaceae

Musa spp.; Banana, kela; N3P3; 3-12; Cultivated; Herb; Fruit, vegetable, leaves as plates

Myristicaceae

Knema attenuata (Wall.) Warb.; Jathikai, ruktmaru; N2P2; 12-3; Tropical, sub-tropical evergreen forests; Tree; Medicinal

Myristica spp.; Nutmeg tree, etc.; N2P2; 2-4; Tropical, sub-tropical, ever-green forests; Tree; Timber, medicinal, spice, seed oil aromatic, in candle making, Flowers ornamental for garlands

Myrsinaceae

Aegiceras corniculatum (L.) Blanco; Halsi, narikandam; N3P2; 1-2, 7; Tropical, littoral forests; Tree; Fuel, bark and seed as fish poison

Myrtaceae

Callistemon citrinus(Curt.) Skeels; Bottle Brush; N2P2; 4-5; Cultivated; Small tree; Ornamental

Eucalyptus spp.; Eucalyptus; N1P1; 2-4, 10-12; Cultivated; Tree; Timber, fuel, paper pulp, amenity and avenue tree

Psidium guajava L.; Guava; N3P3; 5-6; Cultivated; Tree; Fruit, tannin, dye

Rhodomyrtus tomentosa Wt.; Hill goose-berry, kora ita, thavittoo; N2P2; 4-8; Sub-tropical, temperate forests; Shrub; Fruit, medicinal, branches as walking sticks

Syzygium cumini (L.) Skeels; Jamun; N1P2; 4-6; Tropical, sub-tropical; also cultivated; Tree; Fruit edible, medicinal, fodder, dye, tannin

Syzygium heyneanum (Duthie) Gamble; Kathjamti, sohayam; N1P1; 3-5; Tropical, sub-tropical; Small tree; Fruit, medicinal

Nelumbonaceae

Nelumbo nucifera Gaertn.; Lotus, padma, kamal; P1; 6-8; Lakes, ponds and tanks; also cultivated; Herb; Ornamental, medicinal, flowers, rhizomes, leaves, shoots as vegetable and salad, sacred to Hindus

Nyctaginaceae

Boerhavia spp.; Ghetuti, kha-pra, pandeku-sir, punarnava; N2P2; 1-12; Agricultural fields and open forest areas; Herb; Vegetable, medicinal

Nymphaeaceae

Nymphaea spp.; Water lily, boomb,saluk, kaluva, poyana; NP1; 1-12; Village ponds, lakes; also cultivated; Herb; Medicinal, fruit, rhizome, root stalk, shoots, leaves, seed edible, as salad or vegetable, sacred to Hindu

Oleaceae

Ligustrum spp.; Koli, lokhandi, pungan chedi; N2P1; 1-6; Tropical, sub-tropical evergreen, semi-evergreen forests; Shrub, small tree; Branches used for walking sticks, ornamental, medicinal

Linociera spp.; Kunde, masseri, satapala, verripogada; N2P2; 1-5; Tropical, sub-tropical, moist evergreen, deciduous forests; Tree; Timber

Olea spp.; Gaild, kahi, kunde, parjamb; NP1; 3-6; 12-1; Tropical, sub-tropical semi-evergreen, deciduous forests; Tree; Timber, medicinal, fruit

Schrebera swietenioides Roxb.; Gante, magalingam, mokkam; N2P2; 2-5; Tropical, sub-tropical deciduous forests; Tree; Timber, medicinal, veterinary medicine

Onagraceae

Epilobium spp.; Willow herb; N2P2; 6-9; Sub-tropical; Tree; Avenue tre

Oxalidaceae

Biophytum sensitivum (L.) DC.; Chui mui, lajvanti; N2P2; 7-12; Tropical, sub-tropical, agricultural lands; Herb; Medicinal

Oxalis spp.; Indian sorrel, etc.; N3P1; 2-6; Cosmopolitan; weed; also cultivated; Herb; Green vegetable, medicinal

Pandanaceae

Pandanus spp.; Ketaki, keora, mogali, tarika; P1; 7-8; Tropical, marshy areas, sea coasts and river banks, also cultivated; Shrub; Plants ornamental, medicinal, fibre for fish nets, perfume, leaves for thatching

Papvaraceae

Argemone mexicana L.; Pivla Dhotra; P3; 5-7; Cosmopoliltan weed; Herb; Medicinal

Papaver spp.; Poppy; N3P1; 3-5; Cultivated; Herb; Medicinal, ornamental

Pedaliaceae

Sesamum indicum L.; Sesamum; N3P3; 7-9; Cultivated; Herb; Oilseed

Peperomiaceae

Peperonia spp.; Luchipata, sumlako; P2; 7-11; Tropical, sub-tropical forests, in moist places; Herb; Medicinal, leaves for snake bite, twigs as vegetable

Piperaceae

Piper spp.; Betel leaf plant, vetta, pipali, miri, kalilmirach; NP2; 1-10; Tropical, sub-tropical, temperate, forests, moist shady places; cultivated in plantations; Climber; Leaves as masticant, medicinal, seeds spice, for flavouring curries, sacred plant, root for drink

Plantaginaceae

Plantago spp.; Plantago; N2P1; 3-9; Sub-tropical, temperate; Herb; Medicinal

Poaceae

Bambusa spp.; Bamboo, bans, veduru, etc.; NP1; 7-8; Tropical, sub-tropical, moist deciduous forests; Culms; Paper pulp; poles; housing, furniture, basketry, medicinal, leaves and shoorts as vegetable, seed as grain

Cynodon dactylon L.; Dub; P3; 5-9; Tropical, Sub-tropical, Temperate; Herb; Fodder

Dendrocalamus spp.; Bamboo, choya bans, karail; P1; 10-11; Tropical, moist deciduous forests; Culms; Shoots as vegetable, seeds as grain, housing, basketry, cordage, medicinal, paper pulp

Oplismenus spp.; --; N2P1; 10-11; Tropical, sub-tropical, moist areas as undergrowth; Herb; Fodder

Oryza spp.; Paddy, dhan, bhat, etc.; NP3; 9-11; Tropical, sub-tropical; cultivated; Herb; Cereal, fodder, thatching

Panicum spp.; Chamulu, samai, kungu, etc.; P1; 7-11; Tropical, sub-tropical, open hilly areas; Herb; Minor millet, medicinal, fodder, thatching

Sorghum spp.; Pearl millet, jawar, baru, cholam, etc.; N2P1; 1-12; Tropical, sub-tropical; cultivated; Herb; Millet, fodder, sugar, medicinal, leaves for snake bite

Zea mays L.; Maize; P2; 7-8; Cultivated; Herb; Food grain, starch, fodder

Polemoniaceae

Phlox drummondii Hk.; Drummond phlox; NP1; 1-12; Cultivated; Herb; Ornamental

Polygalaceae

Polylgala spp.; Meradu, nilkanti, rali, sherita; N3P2; 4-8; Tropical, sub-tropical, weed in open plances; Herb; Medicinal, roots used to vomit out poison, ornamental

Polygonaceae

Antigonon leptopus Hk. & Arn.; Railway or coral creeper; N1P1; 1-12; Cultivated, often running wild; Climber; Ornamental

Fagopyrum spp.; Buck wheat, banphapar, lappasak; N1P2; 6-9; Sub-tropical, temperate; cultivated; Herb; Minor food grain, vegetable, medicinal

Polygonum spp.; Polygonum, bhilamgori, nhachu, etc.; N2P2; 6-9; Tropical to temperate, generally in moist habitats; also cultivated; Herb; Medicinal, ornamental, tuber edible, vegetable, fish poison

Rumex spp.; Rumex, amb-avati, jangli palak, etc.; N2P2; 6-12; Tropical, sub-tropical, temperate, moist areas; Herb; Medicinal, leaves and shoots edible as vegetable, dye for woolen

Portulacaceae

Portulaca glandiflora Hook. Portulaca; N2P1; 6-9; Cultivated; Herb; Ornamental

Primulaceae

Primula spp.; Primula; N3P3; 5-9; Sub-tropical, temperate; Herb; Ornamental

Proteaceae

Grevillea robusta A.Cunn.; Silver Oak; N1P1; 3-6; Cultivated; Tree; Ornamental

Punicaceae

Punica granatum L.; Pomogranate; N2P1; 4-5; Cultivated; Tree; Fruit, Medicinal, Ornamental

Ranunculaceae

Caltha spp.; Marsh Marigold; N3P2; 4-7; Temperate; Shrub; --

Clematis spp.; Clematis; N2P2; 3-5; Tropical, sub-tropical, also cultivated; Climber; Ornamental

Delphinium spp.; Larkspur, bad-soh-plih-rit; N2P2; 2-3; Sub-tropical, temperate, also cultivated; Herb; Ornamental, medicinal

Naravelia zeylanica DC.; Rammam riub, vadak-kodi valli; N3P2; 2-4; Tropical, sub-tropical, moist deciduous forests; Climber; Medicinal, stems as toothbursh, ornamental

Ranunculus arvensis L.; Butter cup; N3P3; 5-6; Sub-tropical, temperate; Herb; Medicinal

Rhamnaceae

Scutia myrtina (Burm.) Kurz; Tuvadi, gariki, kuradi; N1P; 10-4; Tropical, sub-tropical, evergreen forests; Shrub; Hedge plant, fuel

Ventilago spp.; Bonga sarjom, surul; N2P3; 11-3; Tropical, sub-tropical scrub jungles; Woody climber; Dye, medicinal, fibre, cordage

Ziziphhus spp.; Chinese date, jujub tree, ber, bor, regu, toran; N1P1; 3-6; 12-2; Tropical, sub-tropical; deciduous forests; also cultivated; Tree, shrub; Fruit, fodder, fuel, tannin, host of lac insect, medicinal

Rhizophoraceae

Bruguiera gymnorrhiza (L.) Lamk.; Thoddu pon-na, sigappu kakandan; N2P2; 6-10; Mangrove swamps; Tree; Timber, wood for charcoal, fruit

Ceriops spp.; Ankandal, chiru kandal, goran; N2P2; 7-11; Mangrove, tidal forests; Tree; Tubers, leaves and fruit edible and used as vegetable, timber, tanning, fuel, dye

Kandelia candel (L.) Druce; Kandile; N2P2; 5-9; Mangrove swamps; Shrub; Wood for charcoal, fuel, dye

Rhizophora spp.; Kandal, daboja, etc.; N1P2; 4-9; 11-2; Mangrove swamps, tidal forests; Small tree; Fruit, timber, tanning, dye

Rosaceae

Eribotrya japonica Lindl.; Loquat; N3P3; 2-3; Cultivated; Tree; Fruit

Fragaria vesca L.; Fragaria; N2P2; 5-9; Temperate; Herb; Fruit

Malus domestica Mill.; Apple; N1P3; 3-4; Cultivated; Tree; Fruit

Malus pumile Mill.; Apple; N1P1; 3-4; Sub-tropical, Temperate; Tree; Fruit

Potentilla spp.; Silver weed; N2P3; 6-8; Sub-tropical, temperate; Shrub; Medicinal

Prinsepia utilis Royle; Bekhal; N1P2; 9-11; Sub-tropical, temperate; Shrub; Oil

Prunus armeniaca L.; Apricot; N1P1; 3-4; Cultivated; Tree; Fruit, oil

Prunus amygdalus Batsch.; Almond; N1P1; 3-4; Cultivated; Tree; Fruit, seed edible, oil

Prunus avium L.; Cherry; N1P1; 3-4; Cultivated; Tree; Fruit

Prunus cerasoides DC.; Wild Cherry; N1P2; 10-11; Sub-tropical, temperate; also cultivated; Tree; Fruit, ornamental

Prunus domestica L.; Plum; N1P1; 2-3; Sub-tropical, temperate; Cultivated; Tree; Fruit

Prunus pashia Buch-Ham; Wild pear; N1P1; 2-3; Sub-tropical, temperate; Tree; Friot, rootstock for pear

Pyrus persica L. Batsch.; Peach; N1P1; 2-3; Sub-tropical, temperate; cultivated; Tree; Fruit

Pyrus communis L.; Pear; N1P1; 2-4; Sub-tropical, temperate; cultivated; Tree; Fruit

Rosa macrophylla L.; Rose; N3P2; 3-5; Cultivated; Shrub; Ornamental, medicinal

Rosa moschata Mill.; Wild Rose; N2P1; 4-6; Tropical, sub-tropical, temperate; cultivated; Shrub; Ornamental

Rubus spp.; Raspberry, akhan, sohper; N1P1; 2-6; Sub-tropical, temperate; cultivated; Shrub; Hedge plant, fruit, ornamental, medicinal

Rubiaceae

Adina cordifolia Hk.f.: Haldu; N3P2; 5-8; Tropical, sub-tropical forests; Tree; Timber

Anthocephalus cadamba (Roxb.) Miq.; Kadamba; N3P2; 5-7; Tropical, sub-tropical, moist evergreen dry forests; Tree; Timber, Road side, amenity tree, paper pulp

Borreria articularis (L.f.) Will.; Burdmutta, kair kar, madanaku; N2P2; 6-2; Tropical, sub-tropical, moist areas, waste lands; Herb; Medicinal, leaves as vegetable

Canthium spp.; Kirna, palaga, woppe; N1P2; 12-8; Tropical, sub-tropical moist evergreen forests; Tree; Timber, fruit

Catunaregam spinosa (Thunb.) Tiruv.; Boebindi, gela; N1P1; 4-6; Tropical, sub-tropical, evergreen forests; Tree; Timber, medicinal, fish poison, fruit seed, vegetable

Coffea spp.; Coffee plant; N1P2; 3; Cultivated; Shrub; Roasted seed decoction stimulating drink

Hamelia patens Jacq.; Hamelia; N2P2; 3-9; Cultivated; Shrub; Ornamental

Hedyotis spp.; Khet papra; N2P2; 2-5; Tropical, sub-tropical moist and waste lands; Herb; Plant edible, used for drinks, dye

Hymenodictyon spp.; Kuthan, bhaulan; N2P3; 6-7; Tropical, sub-tropical, deciduous forests; Tree; Timber, medicinal

Knoxia sumatrensis (Retz.) DC.; Kampu rodda; N2P2; 6-12; Tropical, sub-tropical forests, ground flora; Herb; --

Pavetta indica L.; Pavattai, paphti; NP2; 2-5; Tropical, sub-tropical, moist or dry forests; Shrub; Root medicinal, flowers and shoots used as vegetable, medicinal

Mitragyna parviflora (Roxb.) Korth.; Kaim, kadam; N3P2; 3-5; Tropical, sub-tropical, forests; Tree; Timber

Mussaenda frondosa L.; Bedina; N2P2; 10-11; Tropical, sub-tropical; also cultivated; Small tree; Medicinal, ornamental

Wendlandia exserta (Roxb.); Chanlai; N2P1; 11-2; Tropical, sub-tropical; Small tree; Timber, fuel, ornamental

Rutaceae

Aegle marmelos (L.) Corr.; Bael tree, maredu, vilvam; N2P1; 3-6; Tropical, sub-tropical deciduous forests, also cultivated; Tree; Fruit, medicinal, timber, sacred for Hindus

Atlantia spp.; Malai narathai; N2P2; 11-1; Tropical, sub-tropical, moist or dry forests; Small tree; Timber, seed oil aromatic, medicinal

Citrus aurantifolia Swing.; Lime; N1P1; 3-4; Tropical, sub-tropical; cultivated; Tree; Fruit, essential oil, root-stock for mandarins

Citrus limettoides L.; Mitha; N1P1; 3-4; Cultivated; Tree; Fruit, oil

Citrus maxima L.; Changotra; N1P1; 3-4; Cultivated; Tree; Fruit

Citrus medica limonum L.; Lemon; N1P1; 3-4; Cultivated; Tree; Fruit

Citrus reticulata L.; Mandarin; N1P1; 3-4; Cultivated; Tree; Fruit

Citrus sinensis L.; Sweet Orange; N1P1; 3-4; Cultivated; Tree; Fruit, medicinal

Evodia lunu-ankenda (Gaertn.) Merr.; Kattu shanbagam; N2P2; 6-7; Tropical, sub-tropical; Small tree; Timber, soft wood

Glycosmis pentaphylla (Retz) Corr.; Anam, golugu; N2P1; 3-10; Tropical, sub-tropical, wild or cultivated; Shrub; Hedge plant

Limonia acidissima L.; Wood apple, velaga, kavath; N2P2; 2-3; Tropical, sub-tropical, also cultivated; Tree; Timber, fruit, medicinal

Murraya paniculata (L.) Jack.; China box, marchala; N3P2; 3-9; Tropical, sub-tropical forests, also cultivated; Shrub; Timber, twigs for toothbrush, medicinal

Toddalia asiatica (L.) Lamk.; Kanj, kattu nidogu; N2P2; 2-4; Tropical, sub-tropical; Shrub; Dye, medicinal

Zanthoxylum spp.; Hanjor, timur, tirphal, yer; N3P2; 3-6; Tropical, deciduous, semi evergreen forests; Tree; Medicinal, spice, aromatic, leaves and shoots as vegetable, ornamental

Salicaceae

Populus ciliata Wall.; Poplar, piplas, makal; N2P2; 2-4; Sub-tropical, temperate forests; Tree; Timber, soft wood, ornamental, avenue, tree, medicinal

Salix spp.; Willow; N2P2; 2-3; Tropical, sub-tropical, temperate; Tree, shrub; Timber, ornamental

Salvadoraceae

Salvadora spp.; Jal, pilu, maram uga; NP2; 3-5; Tidal forests; Tree; Medicinal, fruit, timber, plant ash as detergent

Santalaceae

Osyris spp.; Popli, natadike; N2P3; 10-12; Tropical, sub-tropical, deciduous forests; Small tree; Timber, fuel

Santalum album L.; Sandalwood tree, chandan; N1P2; 5-8; Tropical, sub-tropical deciduous forests; also cultivated; Tree; Wood for craft work, incense, perfumery, medicinal, sacred

Sapindaceae

Aesculus indica Colebr.; Horse chestnut; N2P2; 5-6; Tropical, sub-tropical; Tree; Nuts edible, timber, fodder

Euphoria longan (Lour.) Steud.; Longan tree, ashphal, umb; N1P3; 2-5; Tropical, sub-tropical, evergreen forests; Tree; Timber, seed aril edible

Lepisanthes spp.; Korivi, netota ritha; N3P3; 2-3; Tropical, sub-tropical, evergreen forests; Tree; Fruit

Nephelium litchi Comb.; Litchi; N1P2; 3-4; Sub-tropical, cultivated; Tree; Fruit

Sapindus spp.; Soapnut; N1P; 5-6; 10-12; Tropical, sub-tropical, also cultivated; Tree; Nuts used as soap, timber, medicinal

Schleichera oleosa (Lour.) Oken; Ceylon oak, kusum, sagade; NP1; 2-3; Tropical deciduous forests; Tree; Timber, seed oil edible, cosmetics, industry, host for lac insect, charcoal, fuel, medicinal

Sapotaceae

Madhuca longifolia (Koen.) Mac Br.; Mahua tree, ippa, iluppai; N1P2; 2-5; Tropical, sub-tropical deciduous forests; Tree; Timber, flowers and fruits edible, also made into liquor, seed oil edible, used in industry, medicinal, flowers for cultural use

Manilkara spp.; Cylon iron wood, bakula, pala chettu; NP2; 11-2; Tropical, sub-tropical evergreen dry forests, also cultivated; Tree; Timber, fuel, fruit, seed oil, medicinal

Xantolis tomentosa (Roxb.) Rafin.; Jastimadhu, palai,kantabar; N2P2; 2-5; Tropical, sub-tropical, deciduous forests; Tree; Timber, fruit as vegetable and in pickle

Scrophulariaceae

Bacopa monnieri (L.) Penn.; Brahmi, nir-brahmi, jalna-veri, sambrani chettu; N2P2; 1-12; Tropical, sub-tropical, moist and hydrophyte, common in rice fields, ponds, ditches, etc.; Herb; Medicinal, tonic for hair, shoots and leaves as vegetable

Paulownia tomentosa Steud.; Paulownia; N2P2; 3-5; Cultivated; Tree; Ornamental, timber

Russelia equisetiformis Schlecht & Cham.; Bright coral red; N2P2; 1-12; Cultivated; Shrub; Ornamental

Scoparia dulcis L.; Atisirsa, mithi patti, mashla; N3P2; 4-11; Tropical, sub-tropical moist places, forest paths; Herb; Medicinal, plant and seed for drinks

Scrophularia spp.; Figwort; N1P2; 7-9; Tropical, sub-tropical, temperate, weed; Herb, --

Sopubia delphinifolia (L.) G.Don; --; N2P2; 8-12; Tropical, sub-tropical, moist grass lands; Root parasite

Striga spp.; --; N3P2; 6-12; Tropical, sub-tropical, grass lands; Root parasite

Torenia indica Saldanha; --; N2P1; 8-11; Tropical, sub-tropical, open forests; Herb; --

Verbascum spp.; Bantamakhu, jakhmirir; N3P2; 1-12; Tropical to temperate, in moist shady places; Herb; Medicinal, veterinary medicine, fish poison

Simaroubaceae

Ailanthus spp.; Tree of heaven, maharukh, mattipal; N3P2; 2-4; Tropical, sub-tropical; often planted; Tree; Timber, soft wood, road side, amenity tree, gum, medicinal, veterinary medicine

Smilacaceae

Smilax spp.; Nimbetta, kukardara, ramdalim; N2P2; 3-5; Tropical, sub-tropical; Climber; Fruit and leaves as vegetable, medicinal, fruit for snake bite, fibre

Solanaceae

Acnistus arborescens; --; N2P1; 7-9; 11-2; Cultivated (imported from Brazil); Shrub; Ornamental, hedge plant, flowers fragrant

Capsicum spp.; Chillies; N2P3; 7-8; Cultivated; Herb; Vegetable, condiment

Datura stramonium L.; Thorn Apple; P3; 6-9; Tropical, sub-tropical, temperate; Herb; Medicinal

Lycopersicum esculentum Mill.; Tomato; N3P3; 3-10; Cultivated; Herb; Vegetable

Nicotiana tabacum L.; Tobacco; N3P3; 6-8; Tropical, sub-tropical; often cultivated; Herb; Tobacco, insecticide, medicinal

Petunia spp.; Petunia; N3P2; 1-9; Cultivated; Herb; Ornamental

Solanum spp.; Brinjal; N3P3; 6-8; Many species occur wild; some cultivated; Herbs, Shrubs; Vegetable, ornamental

Sterculiaceae

Dombeya mastersii Hk.f.; Dombeya; N2P2; 11-1; Cultivated; Shrub; Ornamental

Kleinhovia hospita L.; Kleinhovia, punaitekku; N2P2; 5-9; Cultivated; Tree; Ornamental, medicinal

Symplocaceae

Symplocos spp.; kaula, lodha, paralai, singan; N2P2; 10-5; Tropical, sub-tropical, temperate, evergreen, shola forests; Tree; Tree, timber, medicinal, dye, seed oil edible, pollen as cosmetic, silkworm host

Tamaricaceae

Tamarix spp.; Jhau, kori; N2P2; 5-7; 8-2; Tropical, sub-tropical riparian forests; Shrub; Fuel

Theaceae

Camellia spp.; Tea plant, singo, theyaku; N2P1; 12-3; Tropical, sub-tropical, temperate climates; cultivated; Shrub; Leaves for stimulating drink medicinal, ornamental, stems for walking sticks

Schima wallichii; Needle Wood; N2P2; 5-6; Sub-tropical; Tree; Paper pulp, medicinal

Thymeliaceae

Daphne oleides Scherb, Daphne, N3P3, 7-9, Sub-tropical, temperate, Shrub, Ornamental, medicinal

Lasiosiphon eriocephalus Dcne.,Rametha, nunju, nacchinar, N3P2, 10-3, Tropical, sub-tropical, open forests, Shrub, Fibre, medicinal, fish poison

Tiliaceae

Corchorus spp.; Jute plant, banpat; N3P2; 8-3; Tropical, sub-tropical, cultivated, running wild; Shrub, herb; Fibre, cordage in industry, leaves as vegetable, medicinal

Grewia spp.; Dhaman, udipe, tegali, bhemal, makad meva; N2P2; 12-9; Tropical, sub-tropical, temperate; also cultivated; Shrub, tree; Fruit, fibre, cordage, timber, fodder, medicinal, veterinary medicine

Triumfetta spp.; Anyotti, nichardi; N1P2; 8-9;11-2; Tropical, sub-tropical, moist or dry areas; Herb, shrub; Fibre, cordage, medicinal, leaves as vegetable

Typhaceae

Typha spp.; bochokand, jammu, ghabajariya; P1; 5-7; Marsh lands, lakes, ponds, moist ditches, along road sides and waste lands; Herb; Medicinal, stems for thatching, mats and brooms

Ulmaceae

Holoptelea integrifolia (Roxb.) Planch.; Indian elm, kanju, navili, papri; NP2; 1-6; Tropical, sub-tropical, deciduous forests; Tree; Timber, seed oil, leaves for scropion sting, medicinal, fish poison

Urticaceae

Pilea spp.; Guge, jealear; NP2; 9-11; Tropical to temperate, moist shady locations; Under shrub; Leaves as vegetable, medicinal

Pouzolzia spp.; Chipli, chaok daba, utkra; NP2; 2-12; Tropical to temperate moist shady locations; Herb, shrub; Stem and leaves as vegetable, medicinal

Urtica spp.; Chichru, kaniyali, pachoo, etc.; NP2; 8-2; Tropical, sub-tropical, deciduous forests; Tree, shrub; Shoots, leaves, flower twigs as vegetable, medicinal

Verbenaceae

Cithraxylum subserratum Swartz; Fiddlewood; N2P3; 3-5, 8-10; Cultivated; Tree; Timber, flowers fragrant, ornamental roadside and avenue tree

Duranta repens L.; Duranta; N2P2; 1-12; Cultivated; Shrub; Hedge, fence plant, ornamental

Gmelina arborea Roxb.; Gumhar, sivna; N2P1; 11-12; Tropical, sub-tropical dry forests; Tree; Timber, medicinal, fruit

Petrea volubilis L.; Purple wrath; N2P2; 3-4; Cultivated; Climber; Ornamental

Phyla nodiflora (L.) Green; Jalbutti, ninkhadi, bokkana; N1P3; 1-12; Tropical, sub-tropical, weed in moist cultivated fields; Herb; Medicinal

Premna spp.; Aganyo, kondamanga, masathekku; N1P2; 3-6; Tropical, sub-tropical, moist evergreen forests; Tree, shrub; Timber, leaves as vegetable and flavouring

Vitex spp.; Bannah Violet; N2P2; 5-6; Tropical, sub-tropical, temperate; Shrub; Hedge, fuel, medicinal, ornamental

Violaceae

Viola odorata L.; Sweet violet; N3P3; 6-8; Sub-tropical, temperate; cultivated; Herb; Perfumery, medicinal, ornamental

Vitaceae

Ampelocissus spp.; Adavidraksha becchuti, chundura, etc.; N2P2; 6-10; Tropical, sub-tropical forests, in moist areas; Climber; Medicinal, fruit, root for snake bite

Cayratia spp.; Mura palleru, ludra danka, kanipu tige; N2P2; 5-10; Tropical, sub-tropical, dry forests, along forest road side hedges; Climber, shrub; Medicinal, leaves, shoots and fruit edible, as vegetable, roots for snake bite, veterinary medicine

Cissus spp.; Vajravalli, etc.; N2P2; 1-12; Tropical, sub-tropical, moist and dry forests; Climber; Stem, leaves used as vegetable, medicine, veterinary medicine

Vitis vinifera L.; Grape vine; N3P3; 5-6; Cultivated; Climber; Fruit edible and used variously, medicinal

Zingiberaceae

Curcuma spp.; Trumeric, bon haldi, haldi, shindalvan; N3P2; 7-12; Tropical, sub-tropical, moist deciduous forests; also cultivated; Herb; Spice, condiment, medicinal, dye, costmetic, rhizome as vegetable, for insect stings

Hedychium spp.; Ban haldi, dumpa rashtramu, kapur kachri, rubiri; NP2; 6-12; Tropical, sub-tropical, moist, shady forests; often cultivated; Herb; Ornamental, root for snake bite, medicinal, aromatic

Zingiber spp.; Ginger, adrak, gada, jangli adrak; NP3; 6-12; Tropical, sub-tropical, moist, shady forest lands; often cultivated; Herb; Rhizome edible, spice, also for scorpion sting, and snake bite, medicinal

Zygophyllaceae

Tribulus terrestris L.; Chotigothru, kanti, palleru; N2P2; 1-12; Tropical, ruderal, in farm lands; Herb; Medicinal, fruit, vegetable.

Table 22. Bee flora of India

SN.	Scientific Name	Common Name	Flowering period	Source	States where found/ grown	References
1.	*Acacia* sp.	Acacia	5–7	NP	Punjab, Karnataka	Atwal *et al.* (1970), Diwan and Rao (1969) Deodikar (1970)
2	*Actinodaphne ngustifolia*	Pisa	10–3	N_1	Jammu and Kashmir	Phadke (1962)
3	*A. hookerii* Meissn.	Pisa	10–1	N_1	Maharashtra (West Coast), Karnataka	Thakar *et al.* (1962), Chaubal and Deodikar (1965)
4	*Adhatoda vasica* Nees	Maker	8-10	N_3P_1	Maharashtra	Chaubal and Kotmire (1980), Garg (1989)
5	*Aegiceras corniculatum* (L.) Blanco	Mangrove	3-4	N_3P_3	Maharashtra (Sagarmal), Himachal Pradesh, Uttar Pradesh.	Chakrabarti and Chaudhary (1972).
6	*Albizzia stipulata* Boivin	Oee	4-5	N	West Bengal	Sharma and Raj (1985), Garg (1989)
7	*A. lebbeck* Linn	Siris	4-5	NP	Himachal Pradesh	Sharma and Raj (1985)
8	*Allium cepa* L.	Onion	5-6	NP	Uttar Pradesh (Jeolikote)	Verma (1983)
9	*Antigonon leptopus* Hook & Arn.	Kaju	12-1	N_1	Kerala, Karnataka, Orissa	Devadason (1971), Holmes
10	*Azadirachta indica* A. Juss	Coral creeper	3-4	N_1P_1	Bihar	Naim and Phadke (1976)
11	*Bauhinia vahli* Wt. Arn	Neem	5-6	N_1	Uttar Pradesh	Kapil (1975), Kohli (1958)
12	*Berberis lycium* Royale	Taur	3-4	N_1	Himachal Pradesh	Sharma and Raj (1985)
13	*Bidens pilosa* L.	Kashmal	6-10	NP	Himachal Pradesh	Sharma and Gupta (1993), Thakur and Gupta (1987)
14	*Bidens* sp.	Spanish needle	1-12	NP	Himachal Pradesh, Tamil Nadu	Sharma and Gupta (1993), Chandran and Shah (1974)
15	*Bombax ceiba* L.	–	1-13	N_1P_3	Kerala	Nehru *et al.* (1984)

16	*Brassica campestris* L. var. *sarson*	Simbal	10-11	N_1P_2	Bihar, Maharashtra, Uttar Pradesh, Karnataka, Himachal Pradesh	Naim and Phadke (1976), Ramesh (1980), Anonymous (1959), Sharma and Gupta (1993), Parmar (1995)
17	*B. campestris* L. var. *toria*	Sarson	10-11 1-3	NP NP	Uttar Pradesh, Bihar West Bengal, Assam Himachal Pradesh	Kapil (1957), Parmar (1995)
18	*B. compestris* L. var. *toria*	Toria	11-12	N_1P_1	Punjab, Haryana, Himachal Pradesh	Sharma and Raj (1985)
19	*B. juncea* L. Cosson	Raya	12-2 3-4	N_2 N_2	Punjab, Haryana Jammu& Kashmir	Chaudhari (1977), Sharma
20	*B. nigro* L. Koch	Lahi (Kali Sarson)	9-3 11	P_2 N_1	Maharashtra Uttar Pradesh	Chaubal and Kotmire (1980) Kohli (1958), Ramesh (1980), Rawat (1980)
21	*B. oleracea* L. var. *botrytis*	Cauliflower	3-5 1-2	NP NP	Himachal Pradesh Punjab	Sharma and Gupta (1993)
22	*Butea monosperma* (Lam) Taub	Dhak	1-2	N_1P_1	Tamil Nadu	Khan (1976)
23	*Cajanus cajan* (L.) Millsp	Arhar	9-12	N_1	Bihar, Uttar Pradesh, Haryana Punjab	Naim and Phadke (1976) Srawan and Sohi (1985)
24	*Callistemon lanceolatus* DC	Bottle brush	7-8	N	Uttar Pradesh Punjab, Himachal Pardesh, Assam, Karnataka, Maharashtra	Bhat *et al.* (1987)
25	*Carvia callosa* (Nees) Brem	Karvi	8-10 7-9	N_1P_1 N_1P_1	Maharashtra Karnataka	Phadke (1964), Chaubal and Deodikar (1965) Diwan *et al.* (1964) Diwan and Rao (1972)
26	*Cannabis sativa* L.	Bhang	7-9	P	Himachal Pradesh	Sharma and Raj (1972)
27	*Carissa spinarum* Linn	Garna	4-5	N_1	Himachal Pradesh	Sharma & Gupta (1993)
28	*Cassia fistula* Linn	Amaltas	5-6	N_2P_2	Himachal Pradesh	Sharma & Raj (1985), Parmar (1995)
29	*Cicer arietinum* Linn	Channa	12-2 12-2 2	N_2P_2 NP NP	Maharashtra Haryana Uttar Pradesh	Chaubal & Kotmire (1980) Vasu (1967) Chaturvedi (1969)

						Chaudhary & Singh (1989) Srawan and Sohi (1985)
30	*Citrus aurantifolia*	Malta	12-1	N_2P_2	Punjab	Chaubal and Kotmire (1980), Srawanj and Sohi (1985)
31	*C. reticulata* Blanco Mausami	Mausami	3	N_1P_1 N_2P_2	Himachal Pradesh Kerala	Nehru *et al.* (1984)
32	*C. sinensis* (L.) Osb.	Santra/ Orange	3	NP	Punjab	Srawan and Sohi (1985)
33	*Cocos nucifera* L.	Coconut	1-12	NP_1	Karnataka, Orissa Andhra Pradesh	Anonymous (1959)
34	*Coffea arabica* L.	Coffee	4	N	Karnataka	Ramani and Bhumannavar (1990)
35	*Coriandrum sativum* L.	Dhania	1-3	N_2P_2	Maharashtra, Punjab	Chaubal and Kotmire (1980)
36	*Dalbergia sissoo* DC.	Shisham	3-4	N_1	Bihar, Uttar Pradesh, Himachal Pradesh, Punjab, Haryana	Chaudhrai (1977), Rehman and Singh (1940,41), Ramesh (1980), Parmar (1995)
37	*Ehretia acuminata* R. Br.	Punna	4	N_1	Himachal Pradesh	Rehman and Singh (1940)
38	*Elaeagnus umbellata* Thumb	Wild olive	4-5	N_2P_3	Punjab, Uttar Pradesh	Singh (1962), Chaudhari (1977)
39	*Eruca sativa* Mill	Taramira	1-5	N_1 N_1P_1	Himachal Pradesh, Punjab, Haryana, Himachal Pradesh, Uttar Pradesh	Gupta *et al.* (1986)
40	*Erythrina myosorensis*	Corel tree	5-6	N_1P_1	Tamil Nadu	Chandran and Singh (1974)
41	*Eucalyptus* spp.	Safeda	11-12,2-3	N_1P	Haryana, Punjab, Uttar Pradesh, Bihar, Karnataka, Himachal Pradesh	Srawan and Sohi (185), Sharma and Gupta (1993)
42	*Eugenis* spp.	Bhedas, Gudda Panneralu	2-4	N_1 N_1	Karnataka Kerala	Diwan *et al.* (1964)
43	*Fagopyrum esculentum* Moench.	Buck wheat	7-9	N_1	Jammu & Kashmir, Himachal Pradesh, Uttar Pradesh	Anonymous (1959) Saraf (1973)
44	*Fragaria vesica* L.	Strawberry	5-9	N	Uttar Pradesh (Jeolikote)	Singh (1979)
45	*Gossypium* spp.	Cotton	7	N P	Punjab, Haryana, Uttar Pradesh	Srawan and Sohi (1985)

46	*Grewia* sp.	Beol	5-7	N_1P_1 N_1	Karnataka Jammu & Kashmir	Diwan and Rap (1969) Deodikar (1970)
47	*Guizotia abyssinica* Cass	Niger	9-10	N_3P_3	Maharashtra, Andhra Pradesh, Karanataka	Chaubal and kotmire (1980)
48	*Halianathus annuus* L.	Sunflower	7-8	N_1P_3 N_1P_3	Maharashtra, Punjab, Haryana	Chaubal and Kotmire (1980)
49	*Hevea brasiliensis* Muell. Arg.	Rubber	3	N_1	Karnataka, Kerala	Srawan and Sohi (1985)
50	*Lagerstroemia entada*	Pride of India	7-9	N_1P_1	Karnataka, Kerala	Diwan and Rao (1969)
51	*Leucas aspera* Link	Tumba	10-11	N_1	Kerala	Holmes (1965)
52	*Litchi chinesis* Sonner.	Litchi	3	N_1	Punjab, Uttar Pradesh, Bihar, Himachal Pradesh	Chaudhari (1977) Nair (1983), Kohli (1958), Chaturvedi (1969), Naim & Phadke (1976)
53	*Litsea stocksii* Hook. f	Chirandi	9-10	N_1P_1	Maharashtra, Himachal Pradesh	Deodikar and Thakur (1953), Sharma and Raj (1985)
54	*Machilus* spp.	Kardel	2	N_1P_1	Tamil Nadu, Assam, Himachal Pradesh	Chandran and Shah (1974), Sharma and Raj (1989)
55	*Madhuca longifolia* (Koenig.) J.F. Macbr.	Mahua	2-3	N_1P	Bihar, Uttar Pradesh, Karnataka	Naim & Phadke (1979), Chaudhary and Singh (1989)
56	*Malus domestica* Borkh.	Apple	3-4	NP	Himachal Pradesh, Uttar Pradesh, Jammu & Kashmir	Singh (1962)
57	*Mangifer sativa* L.	Mango	1-4	N_1	Karnataka, Kerala, Himachal Pradesh	Anonymous (1959), Sharma and Raj (1985)
58	*Medicago sativa* L.	Lucern	8-9	N_2P_3	Maharashtra, Punjab	Chaubal and Kotmire (1980), Atwal *et al.* (1970), Chaudhari (1977)
59	*Melilotus* sp.	Clover/ Melilot	3-7	N_1P_1	Punjab	Chaudhari (1977)
60	*Mimusops elengi* L.	Moulsari	5-6	N_1P NP	Bihar Mahar- ashtra, Orissa Assam	Naim & Phadke (1976) Subramanium (1979)
61	*Moringa pterigos- perma* Gaetrn.	Drum stick tree	1-3	NP	Bihar	Nair and Singh (1974)

62	*Plectranthus rugosus* Wall-Benth	Shain	8-10	N_1P_3	Himachal Pradesh	Gupta *et al.* (1984)
63	*Polygonum* sp.	Polygonum	6-9	N_1P_1	Punjab	Chaudhari (1977)
64	*Pongamia pinnata* (L.) Pierre	Karanj	3	N_1P_1	Bihar	Naim & Phadke (1976)
				N_1P_1	Maharashtra	Chaubal and Kotmire (1980)
				N_3P_3	Karnataka, Kerala,	
				N_1P	Uttar Pradesh, Orissa, Haryana	Anonymous (1959)
65	*Prunus amygdalus* Batsch	Almond	2-3	NP	Himachal Pradesh	Sharma and Gupta (1993)
66	*P. armeniaca* L.	Apricot	2-3	NP	Himachal Pradesh	Sharma and Gupta (1993)
67	*P. domestica* L.	Plum	3	NP	Himachal Pradesh, Tamil Nadu	Sharma & Gupta (1993) Chandran & Shah (1974)
68	*P. puddum* Roxb.	Pajja	11-12	N_1P_1	Himachal Pradesh	Gupta *et al.* (1990), Reddy & Gupta (1987)
69	*Psidium guajava* L.	Guava	3-4	N_2P_2	Maharashtra, Himachal Pradesh	Chaubal and Kotmire (1980), Garg (1989)
70	*Pyrus communis* L.	Pear	2-3	NP	Himachal pradesh, Tamil Nadu	Sharma and Gupta (1993), Chandran and Shah (1974)
71	*P. pashia* Buch Ham	Kainth	2-3	NP	Himachal Pradesh	Sharma and Gupta (1993)
72	*P. persica* L.	Peach	2-3	NP	Himachal Pradesh	Sharma & Gupta (1993)
				NP	Punjab	Sharma & Sohi (1985)
73	*Robinia pseudacacia* L.	Robinia/kikar	3-4	N_1P	Jammu and Kashmir, Himachal Pradesh, Uttar Pradesh	Saraf (1973), Shah (1972), Shah and Shah (1976); Gupta *et al.* (1992)
74	*Rosa moschata* Mill	Wild Rose	4-6	N	Himachal Pradesh	Sharma and Raj (1985)
75	*Rubus ellipticus* Smith	Akhan	1-3	N_1	Himachal Pradesh, Uttar Pradesh, Jammu & Kashmir	Gupta and Thakur (1987b)
76	*Rumex* sp.	Rumex	5-6	N	Jammu & Kashmir	Deodikar (1970)
77	*Salix aegyptica*	Willow	3-4	NP	Kashmir	Shah (1978)
78	*Sapindus detergens* Roxb.	Ritha	5	N_1P	Andhra Pradesh, Karnataka, Orissa, Tamil Nadu	Sharma and Raj (1985)
79	*S. emarginatus* Vahl	Ritha	10-12	N_1P_3	Karnataka, Kerala	Krishnaswamy (1970)

80	*S. lauripolius* Vahl	Nauraykai	10-1	N_1P	Karnataka, Kerala	Anonymous (1959)
81	*S. mukorossi* Gaertn.	Ritha	4	N_1	Himachal Pradesh	Singh (1962), Goyal (1974), Sharma (1958)
82	*Schefflera wallichiana* Harms	Doddabettu/ Po Ngabettu	5	N_1P_3	Karnataka	Diwan and Rao (1972)
83	*Sesamum indicum* L.	Gingelly/Til	8-9	N_1P_3	Bihar	Nalim and Phadke (1979)
84	*Syzgium aromaticum* (L.) Merrill & Perry	Jamun	4-5	NP	Karnataka, Kerala	Anonymous (1959)
85	*Syzygium cumini* (L.) Skeels	Jamun	5	N_1P	Himachal Pradesh	Sharma and Raj (1985),
			5	N_1P_2		Satyanarayana (1975)
			2-4	N_1P_1		Nehru *et al.* (1984)
				N_1P_3		Naim & Phadke (1976)
				N_1P_1		Anonymous (1959)
				N_1P_2		Chandran & Shah (1974); Chaudhary and Singh (1989); Chaubal and Kotmire (1980); Chaudhary (1977)
				N_1P_3		
86	*Tamarindus indica* L.	Tamarind/ Imli	5-8	N_1P	Karnataka, Kerala	Anonymous (1959)
				N_1		Ramachandran (1973)
87	*Terminalia arjuna* (Roxb.) Wight & Arn.	Arjun tree	4-6	N_1P	Karnataka, Kerala, Uttar Pradesh, Maharashtra, Assam, Punjab	Anonymous (1959), Kohli (1958), Chaubal and Kotmirew (1980), Atwal *et al.* (1970)
88	*Terminalia bellerica* (Gaertn.) Roxb.	Behda/Thari	4-5	N_1P	Karnataka, Kerala	Anonymous (1959)
89	*Terminalia chebula* Retz.	Hirda/inknut	4-5	N_1	Karnataka, Kerala	Anonymous (1959)
			5-6	N_1	Maharashtra, Himachal Pradesh	Chaubal & Kotmire (1980), Sharma and Raj (1985)
90	*Terminalia tomentosa* Wight & Arn.	Mathi	4-6	N_1	Karnataka	Diwan *et al.* (1964)
				N_1P	Kerala	Anonymous (1959)
				N_3P	Maharashtra	Thakur (1976)
91	*Thelepaepale ixiocephala* (Benth) Bremk	Darmori/ Whayati	11-1	N_1P_1	Western ghats	Phadke, (1965)
92	*Toona ciliata* M. Roem	Tun	4	N_1	Himachal Pradesh	Goyal (1974),
				N_1	Uttar Pradesh	Gupta *et al.* (1990)
				N_1P	Jammu & Kashmir	Ramesh (1980)
				N_1P_3	Punjab	Saraf (1973)
				N_1P_1	Tamil Nadu	Chaudhari (1977)

93	*Trifolium alexandrinum* L.	Berseem	4-5	N_1P_1	Bihar Punjab, Haryana	Chandran & Shah (1974) Naim & Phadke (1976), Srawan & Sohi (1985) Shah (1953) Chaudhari (1977)
94	*Vernonia* Sp.	Iron weed	6-9	N_1	Karnataka	Satyanarayana (1975)
95	*Vitex negundo* L.	Banah	6-7	N	Himachal Pradesh, Uttar Pradesh	Garg (1989), Gupta and Thakur (1987a)
96	*Wendlandia notoniana* Wall.	Renda/Tiliya	2	N_1P_1 N_3P_3	Karnataka Maharashtra	Diwan *et al.* (1964)
97	*Woodfordia floribunda* Salisb.	Dhai	4-5	N	Himachal Pradesh	Thakur *et al.* (1962) Mishra *et al.* (1987)
98	*Ziziphus mauritiana* Lam.	Ber	7-10	N_3P_3	Maharashtra, Haryana, Orissa, Uttar Pradesh	Chaubal and Kotmire (1980) Chaudhary and Singh (1989)

N=Nectar
N1 = Major honey source
N2 = Medium honey source
N3= Minor honey source
P = Pollen
P1= Major pollen source
P2 = Mediuml pollen source
P3= Minor pollen source
1 – 12= Months of the calender year, January to December.

Table 23. List of important forage plants

Common name	Scienfific name	Value as nectar source	Value as pollen source
Skunk cabbage	*Symplocarpus foetidus*	------	minor
Elm	*Ulmus* spp.	-----	minor
Willow	*Salix* spp.	minor	major
Maple	*Acer* spp.	minor	major
Cottonwood	*Populus* spp.	------	minor
Dandelion	*Taraxacum officinale*	major	major
Fruit bloom (apple, peach, pear, cherry, plum)		minor	minor
Wild cherry	*Prunus* spp.	minor	minor
Yellow rocket	*Barbarea vulgaris*	minor	minor
Black locust	*Robinia pseuda-acacia*	major	minor
Tulip poplar	*Liriodeudron tulipifera*	major	minor
Brambles (Blackberry, raspberry)	*Rubus* spp.	minor	minor
Blueberry	*Vaccinium* spp.	major	minor
Huckleberry	*Gaylussacia* spp.	minor	minor

Cranberry	*Vaccinium macrocarpon*	minor	minor
Yellow sweet clover	*Melilotus officinalis*	major	major
White sweet clover	*Melilotus alba*	major	major
Blue weed	*Echium vulgare*	major	minor
Alfalfa	*Medicago sativa*	major	------
Corn	*Zea mays*	------	minor
Basswood	*Tilia* spp.	major	------
Alsike clover	*Trifolium hybridum*	minor	minor
White Dutch clover (Ladino)	*Trifolium repens*	major	minor
Smooth sumac	*Rhus glabra*	minor	minor
Vetch	*Vicia* spp.	minor	-------
Birdsfoot trefoil	*Lotus corniculatus*	minor	minor
Canadian thistle	*Cirsium arvense*	minor	minor
Wild carrot (Queen Anne's lace)	*Daucus carota*	minor	minor
Milkweed	*Asclepias* spp.	minor	-------
Cucurbits (cucumber, squash, pumpkin cantaloupe, watermelon)		minor	minor
Button bush	*Cephalanthus occidentalis*	minor	minor
Catnip	*Nepta cataria*	minor	------
Chicory	*Chichorium intybus*	minor	minor
Star thistle (spotted knapweed)	*Centaurea maculosa*	minor	------
Sunflower	*Helianthus* spp.	minor	minor
Purple loosestrite	*Lythrum salicaria*	major	minor
Thyme	*Thymus* spp.	major	minor
Smartweed	*Polygonum* spp.	minor	minor
Red clover	*Trifolium pretense*	minor	minor
Wild bergamont (horsemint)	*Monarda* spp.	minor	minor
Wild marjoram	*Origanum vulgare*	minor	minor
Fireweed (willow herb)	*Epilobium angustifolium*	minor	minor
Pepperbush	*Clethra alnifolia*	minor	minor
Clematis	*Clematis* spp.	minor	minor
Lima bean	*Phaseolus lunatus*	minor	------
Soybean	*Glycine max*	minor	minor
Japanese knotweed	*Polygonum* spp.	minor	minor
Spanish needle	*Bidens* spp.	minor	minor
Staghorn sumac	*Rhus* spp.	minor	minor
Buckwheat	*Fagopyrum esculentum*	minor	------
Goldenrod	*Solidago* spp.	major	major
Aster	*Aster* spp.	major	minor

Table 24. Some important Asian bee forage plants

Species	Family	Common Name
Aesculus turbinata	Hippocastanaceae	Japanese horse-chestnut
Astracalus sinicus	Leguminosae	Chinese milk vetch
Bombax ceira	Bombaceae	Silk-cotton tree
Brassica campestris	Cruciferae	Rape
Brassica spp.	Cruciferae	Mustard
Calliandra calothyrsus	Leguminosse	Calliandra
Castanea pubinervis	Fagaceae	Sweet chestnut
Ceiba pentandra	Bombaceae	Kapok
Cirsium spp.	Compositae	Thistle
Citrus spp.	Rutaceae	Orange, pomelo
Clethra barbinervis	Clethraceae	
Cocos nucifera	Palmae	Coconut
Croton spp.	Euphobiaceae	
Cucumis spp.	Cucurbitaceae	Cucumber, melon
Cucurbita moschata	Cucurbitaceae	Pumpkin
Diospyros kaki	Ebenaceae	Persimmon
Eriobotrya japonica	Rosaceae	Loquat
Eucalyptus spp.	Myrtaceae	Eucalyptus
Eupatorium odoratum	Compositae	Snakeroot
Euphoria longan	Sapindaceae	Longan, lam yai
Fagopyrum esculentum	Polygonaceae	Buckwheat
Gossypium spp.	Malvaceae	Cotton
Helianthus annulus	Compositae	Sunflower
Hevea brasiliensis	Euphobiaceae	Rubber
Ilex pedunculosa	Aquifoliaceae	Gallberry
Ilex rotunda	Aquifoliaceae	
Lespedeza spp.	Leguminosae	Bush clover
Ligustrum japonicum	Oleaceae	Privet
Litchi chinensis	Sapindaceae	Litchi
Malus pumice	Rosaceae	Apple
Medicago sativa	Leguminosae	Lucerne, alfalfa
Melilotus alba	Leguminosae	Sweet clover
Prosopis cineraria	Leguminosae	Mesquite
Prunus spp.	Rosaceae	Cherry, apricot, peach

Pyrus pyrifolia	Rosaceae	Pear
Rhus spp.	Anacardiaceae	Sumac
Robinia pseudoacacia	Leguminosae	Black locust
Salix spp.	Salicaceae	Willows
Sesamum indicum	Pedaliaceae	Sesame
Styrax japonica	Styracaceae	Snowball
Tilia japonica	Tiliaceae	Linden, lime
Tilia maximowicziana	Tiliaceae	Linden, lime
Tithonia tagetifolia	Compositae	Mexican sunflower
Trifolium pretense	Leguminosae	Red clover
Trifolium repens	Leguminosae	White clover
Tridax procumbens	Compositae	
Ziryphus jujuba	Rhamnaceae	Chinese jujube

Table 25. List of important agroforestry plants suitable for beekeeping

Plant species	Suitable region	Family	Common name	Flowering period	Reward	Other econiomic uses
Acacia auriculiformis Cunn. Ex.Benth	S	Leguminosae	Austratian Acacia	June-July	N_1P_3	Timber, fuel, gum, medicinal
A. catechu (Linn.) Willd	S	Leguminosae	Khair	June-July	N_1P_2	Medicinal, fuel, fodder for goat and sheep, gum, katha and dye
A. modesta Wall.	S	Leguminosae	Phulai	March-May	P_1	Shade, medicinal, fodder, fuel, soil conservation
A. Senegal (Linn.)Willd	S	Leguminosae	Babool	March-May	P_1	Hedges, shade, medicinal
Aegle marmelos (Linn.) Correa	S	Rutaceae	Bael tree	July-Aug.	N_2P_2	Fruits as food, fuel and timber
Aesculus indica Colebr.	T	Hippocastanaceae	Horse Chestnut	March-April	N_1P_2	Fodder, fuel, shade, amenity and medicinal
Albizia chinensis (Osb) Merr.	S/T	Leguminosae	Siran	April-May	N_3P_2	Fuel, fodder and timber, gum and tan from bark
A lebbeck (Linn)	S/T	Leguminosae	Siris	April-May	N_3P_2	Fodder, timber and fuel, gum and tan from bark
Anthocephalus cadamba Mig.	S	Rubiaceae	Kadamb	May-June	N_2P_2	Leaves as fodder, fuel and timber
Azadirachata indica A Juss.	S	Meliaceae	Neem	March-April	N_1P_1	Fodder, fertilizer, timber, fuel, medicinal leaves etc.
Bauhinia vahili W.&A.	S/T	Leguminosae	Taur	April-June	N_1P_3	Fodder, food and fuel

B.vanegata Linn	S/T	Leguminosae	Kachnar	March-April	N_2P_3	Agricultural implements, gum, fodder, food and fuel
Bombax ceiba Linn.	S	Bombacaceae	Samal	Feb.-March	N_1P_3	Match box, packing case, ply wood, food, fodder and fuel
Buchanania lanzan Spreng.	S	Anacard-iaceae	Chiraungi	March-April	N_1	Food, fodder, timber, oil etc.
Butea monosperma (Lam.)	S	Leguminosae	Palas	Jan.-Feb.	N_1P_3	Timber, fuel, food for lac insect
Cassia fistula Linn.	S	Leguminosae	Amaltas	April-May	N_2P_1	Tan from bark, fuel and medicine
Cordia dichotoma Forest.	S	Boraginaceac	Lasura	April-May	N_3P_3	Timber, fuel and medicinal fruit
Dalbergia sisoo Roxb.	S/T	Leguminosae	Shisham	March-April	N_1P_3	Fodder, fuel, timber, amenity, soil conservation
Delonix regia (Boj.) Rafin.	S	Leguminosae	Gulmohar	June-July	N_2P_2	Fuel, shade and ornamental tree
Dillenia pentgyna Roxb.	S	Dilleniaceae	Aggai	March-April	N_1P_3	Timber, jam and jellies from fruit
Ehretia acuminate R. Br.	S	Boraginaceae	Puna	March-April	N_1P_2	Medicinal fruits, timber and fodder etc.
Emblica officinalis Gtn.	S	Euphoriaceae	Amla	March-April	N_3P_3	Food, fruits, medicinal
Eriobotrya japonica Lindley (Thumb.)	S	Rosaceae	Lokat	Nov.-Jan.	N_2P_2	Food, fruit, amenity, liquor from fruit
Gmelina arborea Linn.	S	Verbenaceae	Gamhar	March-April	N_1P_2	Fodder leaf, timber, fuel, paper and pulp etc.
Grevillea robusta Linn.	S/T	Proteaceae	Silver oak	Nov.-July	N_1P_2	Timber, fuel, shade, amenity, wind brake etc.
Grewia oppositifolia Buch. Hamex Roxb.	S/T	Tilliaceae	Biul	April-June	N_1P_3	Fodder, fibre, fuel
Lagerstromia indica Linn.	S	Lythraceae	Saoni	June-Aug.	P_1	Ornamental, amenity, fuel etc.
Leucaena leucocephala (Lam) de wit.	S	Leguminosae	Soobabul	April-May	N_1P_3	Fodder, food and timber
Litchi chinensis Sonu.	S	Sapindaceae	Litchi	March-April	N_1P_3	Food, fruits and timber
Madhuca indica Gmel.	S	Sapotaceae	Mahua	Feb.-May	N_1P_3	Fodder, oil from seeds, timber and fuel
Mangifera indica Linn.	S	Anacardi-aceae	Aam	March-April	N_1P_3	Fruits, fuel, timber etc
Melia azaderach Linn.	S	Meliaceae	Drek	March-April	N_1P_3	Medicinal, seeds with insecticidal property
Mimusops elengi Linn.	S	Sapotaceae	Moulsari	May-June	N_1P_3	Fruits as food, fuel etc.
Moringa oleifera Zamk.	S	Morinagaceae	Sohjan	Jan.-March	N_1P_3	Roots, flowers and pods as foods, hedges, amenity, medicinal, oil from seeds
Pongamia pinnata (Linn.) Pierre	S	Leguminosae	Karanj	March-April	N_1P_3	Fodder, fuel, erosion control, roots and seeds as fish poison, medicinal
Prunus puddam Roxb	S/T	Rosaceae	Paja	Sep.-Dec.	N_1P_3	Fuel, fruits as food

Psidium guajava Linn.	S	Myrtaceae	Amrud	Aug-Nov.	N_1P_3	Fruits as foods, timber, tannin and medicinal leaves
Punica granatum Linn	S/T	Ltythraceae	Anar	April-May	N_1P_3	Seeds as food and medicinal, fuel
Robinia pseudoacacia Linn.	T	Leguminosae	False acacia	April-May	N_1P_3	Timber, oil, fuel and soil conservation
Salix spp.	S/T	Salicaceae	India Eillow	March-April	N_1P_3	Wood for cricket bats and conservation
Sapium *sebiferum* Roxb.	S	Euphor-biaceae	Pahari Shisham	June-Aug.	N_1P_3	Seeds yield fat for soap, black dye from leaves, fuel etc
Syzygium cumini Skeels	S	Myrtaceae	Jamun	Feb.-April	N_1P_3	Fruit as food, fodder, timber, fuel etc.
S.heyneanum Linn.	S	Myrtaceae	Bhedar	Feb-Marcn	N_1P_3	Fodder, timber, fuel, soil conservation, shade and wind brake etc.
Tamarindus indica Linn.	S	Leguminosae	Imli	April-May	N_1P_3	Leaves, flowers and pods used in curries, oil from seeds, fuel etc.
Toona ciliata M. Reen.	S/T	Meliaceae	Tun	March-April	N_1P_3	Medicine from bark, insectic-idal properties, timber, fuel etc.
Terminalia arjuna (Roxb.)W.& A.	S	Combret-aceae	Arjun	April-May	N_1P_3	Timber, fuel, soil conser-vat-ion, gum, dye and tan from bark
T. belerica Roxb.	S	Combre-taceae	Bahera	April-May	N_1P_3	Medicinal fruits, fuel, timber etc.
T. chebula Retz	S	Combre-taceae	Harda	April-May	N_1P_3	Medicinal fruits, fuel, timber etc.
Zizyphus mauritiana Lamk.	S	Rhamnaceae	Ber	June-July	N_1P_3	Fruits as food, fodder, fuel, host for lac insect, dye and tan from bark.

Where, N = Nectar, P = Pollen, N_1 = Major honey source. N_2 = Medium honey source, N_3 = Minor honey source, P_1 = Major pollen source, P_2 = Medium pollen source, P_3 = Minor pollen source, S = Subtropical, I = Intermediate, T = Temperate

Table 26. Bee flora in different agro- climatic zones of Jammu and Kashmir.

Places	January - February	March -April	May -June	July -August	September October	November - December
Subtropical (comprising outer plains and hills of Jammu, parts of Kathua and Udhampur)	*Brassica, Mangifera, Wendlandia*	*Bassica, Cassia, Sapindus, Toona, Citrus, Dalbergia, Mangifera, Phyllanthus, Opuntia*	*Carthamus, Indigofera, Sapindus, Dalbergia, Zizyphus, Cassia, Acacia, Grewia*	*Zea, Carthamus, Dipsacus, Cassia, Indigofera, Acacia, Zizyphus, Grewia*	*Sesamum, Zea, Carthamus, Dipsacus, Acacia, Zizyphus*	*Bauhinia*

Intermediate (comprising the submountainaeous parts of Doda, Udhampur, Poonch and Rajouri districts)	*Salvia, Salix, Viburnum, Rosa*	*Indigofera, Astragalus, Salvia, Nepeta, Polygonum, Allium, Spiraea, Geranium, Viburnum*	*Indigofera, Trifolium, Medicago, Salvia, Nepeta, Origanum, Polygonum, Fagopyrum, Rumes, Chenopodium, Artemsia, Dipsacus, Spiraea, Potentilla, Fragaria, Geranium, Veronica, Plectranthus, Berbaris, Aesculus, Crotolaria, Rhododendron, Viburnum, Prunus, Pyrus, Rosa, Rhus, Rubus, Lonicera, Strobilanthes*	*Indigofera, Trifolium Medicago, Polygonum, Fagopyrum, Rumex, Nepeta, Mentha, Origanum, Artemsia, Dipsacus, Inula, Potentilla, Geranium, Plectranthus,, Rosa, Dianthus*	*Trifolium Medicago, Polygonum, Fagopyrum, Rumex, Dipsacus, Geraniun, Olea*	*Olea*
Temperate (comprising whole of Kashmir and higher reaches of Poonch and Rajouri districts)	Whole areas snow bound	*Brassica, Trifolium, Taraxacum, Plantago, Spiraea, Veronica, Tulipa, Salix, Almond, Cherry, Rosa, Aesculus Acacia, Viburnum, Rubus, Acacia, Viburnum, Rubus, Juglans*	*Brassica, Cabbage, Raddish, Impatiens, Stellaria,Rubus Geranium, Lonicera, Spiraea, Potentilla, Indigofera, Polygonum, Fagopyrum,Veronica, Tulipa, Iris, Garden flowers, Aesculus, Salix, Acacia, Tillia, Fruit trees, Rosa, Viburnum,*	*Zea, Trifolium, Indigofera, Polygonum, Fagopyrum, Impatiens, Geranium, Delphinium, Helianthus, Potentilla, Plectoranthus*	*Zea, Trifolium, Polygonum, Polygonum, Stellaria, Geranium, Helianthus,Crocus, Plectoranthus*	*Crocus*

Vegetation mapping for beekeeping

Most important issues of concern in apiculture are the identification of important plant species and their times of flowering. Therefore, appropriate monitoring would enable us to map vegetation communities of value for beekeeping. Areas of high potential could be located and also difficulties of access that would make them uneconomical. The methods used for such type of studies i.e. monitoring/ mapping arc collectively known as remote sensing. The term remote sensing implies sensing an object trom a distance. In other words it is gathering of information about the objects and features without placing instruments (sensors) in contact with them.

Advantages of Satellite Remote Sensing

a. Synoptic view: Wide area can be covered by a single image/photo.

b. Repetitively: can collect the data of any area repeatedly.

c. Coverage of inaccessible areas like maintains, swampy areas, thick forests, deep sea etc., possible.

REMOTE SENSING APPLICATIONS

Information on the natural resources, both renewable and non-renewable, is extremely important for their judicious use and wise management. The renewable resources like agriculture, forests, water, marine resources require time for regeneration and replenishment. The remote sensing techniques have proved their potential in acquiring reliable comprehensive and timely data concerning the distribution, character and productivity. A knowledge of the present and potential resource is essential for formulating and executing management strategies for their proper utilisation. Table 27 shows the application potential of the data for various discipline.

Table 27. Application Potential: Land Surface.

1. Agriculture	Crop identification.
	Crop condition
	Crop disease
	Crop stress
	Crop yield
	Soil
	Crop water management
	Soil moisture
2 Forestry	Forest cover
	Density
	Forest fire
	Forest management
	Forest degradation

		Species
		Encroachment
3	Land use/Landover	Urban Landuse
		Land use/Landcover
		Land use changes
		Saline land wetland
		Wasteland
		Site suitability
4	Water resources	Surface water bodies
		Flood
		Soil moisture
		Water quality
		Snow cover
		Irrigation management
		Watershed management
		Wetland
5	Geology	Structural geology
		Lithological mapping
6	Environmental monitoring and impact assessment	
7	Disaster	Earthquake prone area
		Flood
		Forest Fire
		Damage assessment
		Land slides

7. POSSIBLE APPLICATION OF REMOTE SENSING TO APICULTURE

1. Beekeepers know the best honey plants and when they bloom, and they can sometimes determine quite accurately whether or not the year's flowering is likely to provide a crop. However, at the level of large cooperatives, or governments wishing to exploit the full potential of a larger area, such information may prove inadequate.
2. Aerial photographs can be four types: black and white, true colour, false colour or infra-red. Use of more than one type can provide a broader basis for conclusive interpretations, but colour prints can be reliable on their own.
3. Photographs should be taken during the peak of flowering, to enable different plant species to be identified by the use of tone and texture, and by the form of flowers, twigs and crowns. Features important for beekeeping should be selected, e.g. concentrations of important plants and area damaged by fire. Also geomorphologic features which might

hinder the development of a healthy beekeeping industry, e.g. lack of access roads, should be identified.

4. The distribution pattern of relevant features can be mapped with the aid of extrapolations from small-scale photographs, which need only a few prints to cover a large area.
5. Various classification techniques can be performed on Landsat imagery using a computerized interactive image processor, if the distribution of flowering plants, burnt areas, water bodies etc., has been previously identified using aerial photography.
6. A sample area is divided in to classes after a ground survey. Each class is then described in tertins of its reflectance characteristics.
7. Having divided the area covered by the imagery into different classes or themes, a final thematic map is then derived. Such a map for beekeeping potential would highlight distribution patterns of valuable species.
8. Knowledge of specific areas can provide clues to possible colony stocking rates. 'When maps and aerial photographs are available, supervised classification can be accomplished more accurately and gives more convincing results, enabling planners and concentrate human resources and investments in areas with higher economic potential.
9. A beekeeping model would sequentially consider aspects such as weather, landform, vegetation, and hydrology and then classify vegetation into types and flowering seasons and also identify fire pone areas.

CONCLUSIONS

Taking into the scenario of technology developments and application needs at micro level development, more IRS satellites are being planned for launch in future. This will provide enhanced data for vegetation application and will allow multiple crop identification. Satellite imagery can provide useful data at a fairly low cost, and can help in the evaluation of large land areas if some ground knowledge is available. Space imagery covering large areas indicates differences in biomass and other critical features, and it can help those with beekeeping knowledge to focus on areas of high potential.

Chapter 12

Management of honeybee colonies

Management of honeybee colonies requires their timely inspection and then meeting the requirements so that they can work effectively for honey production as well as pollination.

The Inspection of honeybee colony

One of the greatest advantages of modern hives is that a colony can be examined thoroughly to know the working of it and determine problems if any. The examination of the colony makes it easy to promptly meet its needs. However, the beginner must open the hive less frequently and only when it is absolutely necessary. Unnecessary interference and sometimes mishandling of the bees during inspection, makes the colonies to abscond. The bees donot like too much interference, as it upsets their normal working. They should therefore be, disturbed as little as possible. During the build up or honey flow period, the bee colonies need to be examined every week and during the off season only once or twice a month a colony may be examined when, the bees are busily engaged in nectar or pollen collection, In summer, they should either be examined in the morning or evening but not during day time. During winter, they should be examined in the middle of a warm day. A hive should not be opened on a windy or rainy day or at night when the bees are not flying in and out. In winter, the brood frames should not be exposed to that' cold too long and the hive should not be opened too often. During a visit to the colony the beekeeper should not be perspiring profusely or strongly smelling of alcohol. Before approaching the bees for inspection, he or she should put an overall, a bee veil, and if required gloves also. Then should proceed as follows :—

After lighting the smoker (Rags, or cotton waste, wood shavings, dried cow dung, maize cobs, pine cones, dry leaves etc make good fuel, for smoker) the hive should be approached from side avoid interfering with bee's flights. A few puffs of smokes should be given at the entrance. Lift the inner cover gently and blow a few puffs of smoke below. Take care and avoid quick jerky motions. Then the frames should be pried apart with the hive tool and taken out one by one. They should stand in a vertical position and preferably over the hive. For examining, the other sides of the combs, it should be manipulated as shown in Fig. 47. After examining, replace them in the brood chamber in the order in which they were found. During these manipulations queen should always be kept in mind and avoid crushing her. The frames should be handled gently and jerking and crushing of the bees must be avoided so that they are not provoked to sting. Because apart from being painful, the odour of the poison irritates the bees even more and makes them difficult to handle. If a bee stings you, the visible sting should be pulled out with the sharp edge of the hive tool or knife and not with the tips of your finger. After finishing the work, the hive should be closed at once. The cover should be put on in such a way as not to crush any bee. Keep the record of every colony and enter your observations in data ledger. This record will help you to tackle the needs of the colony.

The aim of the colony inspection is to determine whether the requirements of the colony are being fulfilled at that particular time which would lead to an increase in production.

When a colony is examined following points should be taken care of

1. Whether the bees have sufficient stores of nectar, pollen or require artificial feedings.
2. Whether the eggs are present and properly laid in a regular pattern. This will indicate that queen is well and there is no need to search for her. But if there is no egg, then search must be made to locate her. Some beekeepers have marked queens on thorax to make the visibility easy.
3. Examine for the presence of queen cells, destroy if any and crush the larvae or eggs. But if the colony is preparing for swarming, allow it to rear queens.
4. The presence of queen cells during off season indicate that either the queen is missing or old queen is not working properly and workers intend to supercede it (Supercedure means replacement of old queen with new prolific queen).
5. Determine whether enough empty combs are available for queen to lay eggs and for the bees to store honey. If sufficient combs are not available, more frames should be provided.
6. Examine if any enemies or diseases of the bees are prevalent. Take immediately precaution and" remedial measures if you notice any symptom of disease.

Uniting honeybee colonies

In beekeeping operations, sometimes there is need to unite the weak colonies which are unable to sustain independently. The chances of survival of weak colonies increase if such colonies are united and given one queen. In some circumstances, when a colony loses queen and has the tendency to become the laying workers. To overcome such a situation it becomes essential either to provide new queen or immediately unite with other colonies. Whatever may be the reasons, but uniting procedure remains the same as given below:

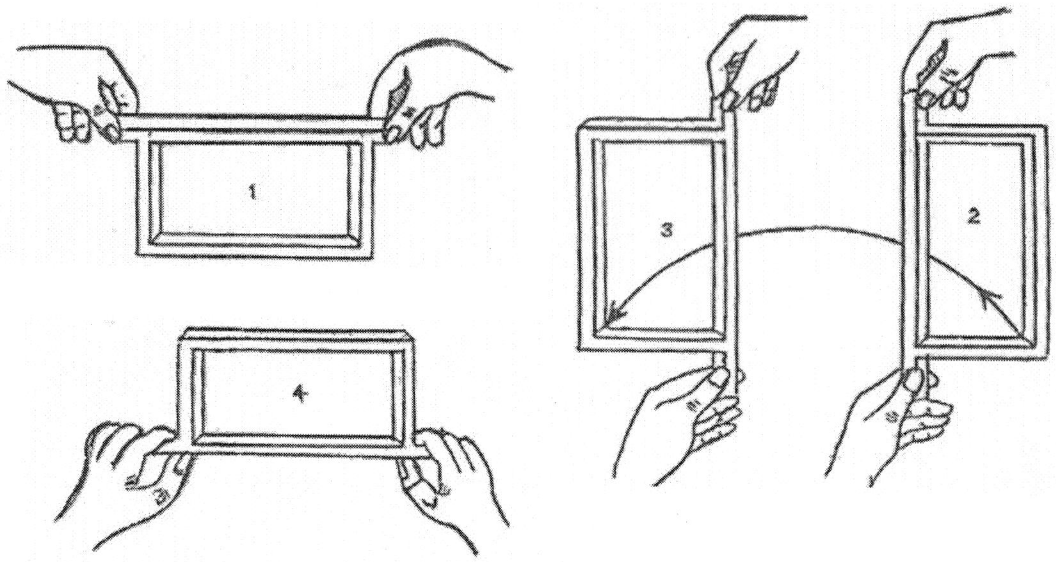

Figure 48. Manipulating a frame 1. Normal position 2 held at right angle 3. After revolving 180° and 4 the outer side of the frame.

Uniting colonies-Newspaper method

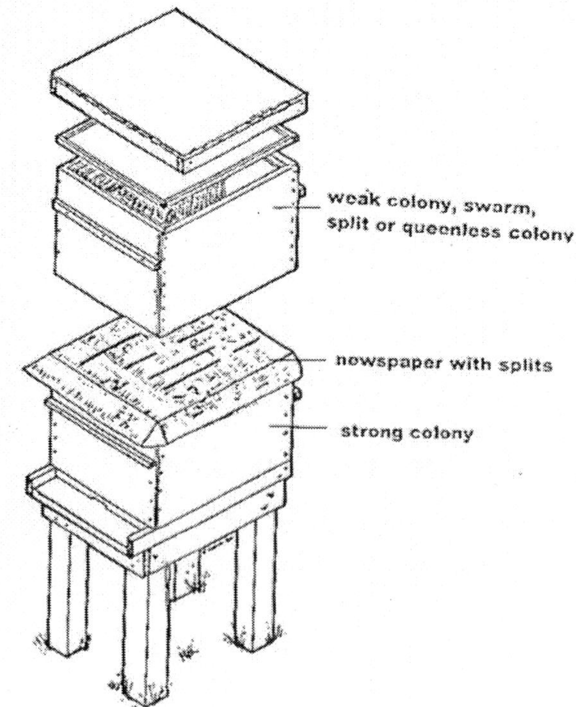

Figure 49. Uniting honeybee colonies by news paper method

Each colony has its own colony specific odours and it is very difficult to combine the two colonies unless their odour is mixed well. Any attempt to unite these colonies without mixing their odour result in infighting and deaths will occur on large scale. Therefore, first step will involve bringing the two colonies into contact with each other. If uniting is done abruptly, the field workers of the colony shifted will not recognize the new place and returning to their original place will persist. This problem can be overcome by moving a hive gradually at the rate of two or three feet per day, so that the field bees get accustomed to the changing position of their hive and will not drift back to the old site. When colonies are sufficiently close, one or two feet apart, they are ready for uniting. They can be united by following either newspaper method. Top cover is removed and the frames are covered with a piece of newspaper having a few holes made with a small nail the bottom, board of the upper colony is then removed and the brood chamber, is placed above the other colony, the newspaper forming a partition between the two. After a day or two, the odours of the colonies will mix and the bees will cut through the paper and will unite together, forming a single colony. After a few days all the frames can be placed in one hive and the upper chamber can be removed.

Alternatively, colonies can be united using smoke method. When the colonies to be united have been brought close to each other, both should be smoked heavily and thin sugary syrup scented with oil of peppermint or wheat flour sprinkled over them. The combs with the bees of the colony to be united should be altered with the combs of the other colony. More smoke and syrup or flour should be applied and the colony closed. The work of the queen may be checked up after three or four days.

It is better to unite a laying worker colony to several strong colonies by giving from one to two frames to each of them. If all its frames are united to one colony there is danger of latter's queen being killed by the laying workers.

Artificial feeding of honeybees

Honey is the best food for the bees and a beekeeper should always leave enough stores for them at all times. However, during dearth periods artificial feeding becomes necessary. When honey is not available, the best substitute is sugar syrup made from white crystalline sugar by dissolving 2 parts of sugar and one part of water or 1 part of sugar and one part of water. The use of gur, jaggery, molasses etc is not recommended as these substances ferment in the cells and also cause dysentery. The sugar syrup can be fed to the bees in small dishes with floating straws. The bees will not be dipped in sugar and feed while sitting on these straws. The feeding should not be done during the day time which may result in robbing. It should be done in the evening hours. During early spring, the feeding of the bees is done to stimulate them that season has started. The queen starts laying more and more eggs before the actual flora is available. This type of feeding is called as simulative feeding to prepare the colony to utilize the nectar flow season.

Feeding of sugar is must to survive the colonies during dearth periods, but if the feeding lasts a long time, the functional activity of bee is negatively affected, they winter more poorly and their productivity is greatly reduced. Experiments have shown that honey feeding is the best, honey mixed with sugar is of secondary importance and sugar alone as the lowest quality food. It is therefore recommended that while extracting the honey, sufficient honey stores be left for bees. If has been found that disturbance in animals nutrition diminishes their enzymatic activity, weakens the defence of organism and brings about various pathological changes.

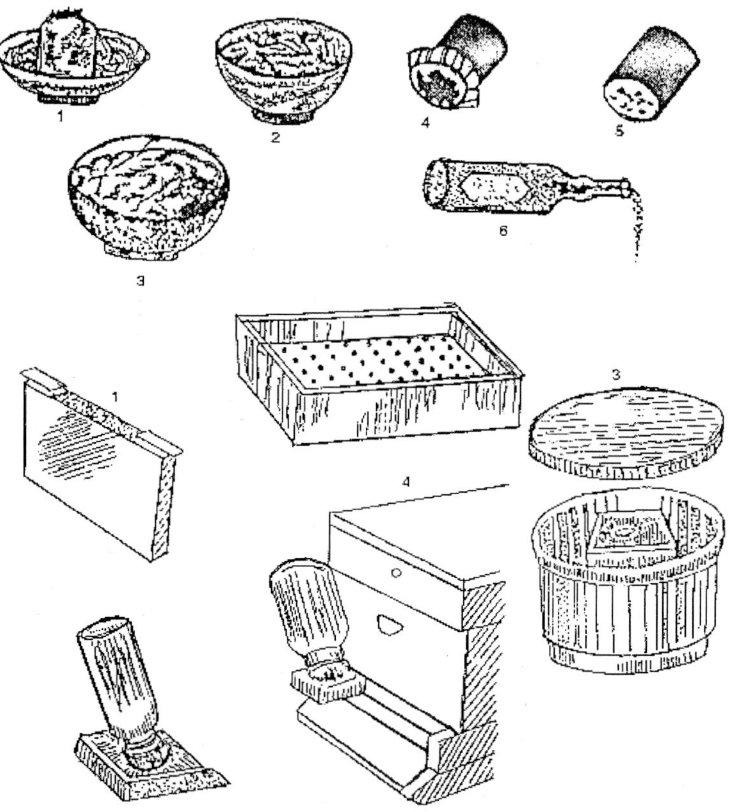

Figure 50. Feeding apparatus. upper indigenous feeders ; 1,2,3 cup feeder with cloth and glass inside. 4,5 tin feeders having cloth wrapped on the mouth and small holes in the bottom 6-bottle feeder lower-1 Division board feeder, 2 –flat feeder 3- rapid feeder 4- boardman feeder separately.

Pollen feeding

Pollen is a protein food. Brood rearing is impossible without it. Therefore there is a natural direct dependence between the arrival of pollen to the nest and the amount of its brood. Pollen like honey, must be available in adequate quantities to a wintering colony, otherwise spring dwindling occurs due to insufficient pollen stores. A wintering colony must replace its fall population with young bees and have a large active brood nest by the time natural pollen is available in late winter or early spring. Whenever there is a pollen shortage it can be provided in two ways

Pollen substitutes

Pollen substitutes are prepared using the following ingredients :

Ingredients	Quantity in g
Wheat flour	100g
Soyabean flour	100g
Skimmed milk powder	20g
Dried brewer's, yeast	25g
Sugarcane	60g
Fat or edible oil	50g
Sodium biocarbonate	50g
and cream of tartar (1 : 1)	5g

All the ingredients are mixed together and kneaded to a soft dough before being served to the bees.

Pollen supplements

Pollen is collected manually from flowers or by bees using pollen traps when available in abundance and is stored for future feeding. A pollen trap is used in front of the hive entrance. The bees while passing through the grid, loses the pollen she carries, the pollen pellets drop onto the grid and down into the pollen trap (Fig. 51) Without hindering in any way a colony's growth its preparation for winter, as well as its honey harvesting, one can obtain from it as much as 6 to 8 kg pollen within one season.

A pollen trap can be locally fabricated (Chhuneja and Goyal, 1996). It is made of kail Wood (*Pinus wallichiana*) and its construction details are given below (Fig. 51). It consists of a plastic strip of 1.5 mm thickness which serves as a vertical pollen trapping screen in which circular holes of 4, 5 mm diameter are made. The pollen balls in the corbiculae of foragers get detached when such bees pass through these holes. The plastic strip contains 5 rows of such holes (28 holes/row) with a working area of 28.2 x 6.0 cm. The balls detached from the corbiculae are collected into wooden pollen collecting tray kept below a 6 mesh wire grid on which the pollen foragers alight before entering the hive. The trap has two compartments. Lower compartment is comprised of a drawer type pollen collecting tray. It is enclosed from its back and sides by the trap body and from the front by a wooden strip attached to the tray for facilitating its easy removal. The tray (28.0 x 15.7 x 5.0 cm) is covered at its top by the wire grid (6-mesh) which serves as an alighting board for the bees, preventing the bees from taking away the collected pollen from the tray and also preventing the dead bees falling into the tray. The upper empty compartment is enclosed from back by the plastic trapping screen and from top and sides by the trap body. The trap body is in the form of trapezium the top plank with a little downward inclination towards the front. Its internal length is 28, 2 cm, whereas internal width is 17.5 cm. Its internal height at the rear end is 11. 4 cm, whereas it is 10.2 cm at its anterior end (upto the trap body). Beyond the trap body is a top plank extended by 5 cm in the form of an eaves so that the rain showers should not enter the trap and spoil the pollen. Leaving a space of 2-7 cm from the rear end and 8.7 cm from the bottom, one hole of 7 mm is made on each of its side plank. Such holes facilitate the exit of drones and the queen bee for nuptial flight. Leaving a space of 9.5 cm from the front, two wooden rods (17.5 x 2.22 x 2.22 cm) are fixed one each at bottom of each side plank so that these rods extend by 9.6 cm from the rear end.

For installing the trap on to the hive, the front wooden entrance rod is removed from the hive and the trap is fixed to the hive by inserting the extended portion of the rectangular insertion rods of the trap into the hive through the space so created after removing the entrance rod. Pollen trapping screen should be kept a little displaced from the exact position for the rest two days so that bees could pass through without dislodging their pollen loads and get accustomed to this new attachment otherwise bees show a considerable reluctance to, pass through the pollen trapping screen and remain disturbed for a long period of time. After the bees become used to pass through the trap, the screen is fixed to its exact position. The bees then readily accept it and start passing through the pollen dislodging holes. The pollen so collected can be fed to the bees during pollen scarcity periods. Yields per hive vary according to the plant source, colony strength, weather and type of the trap.

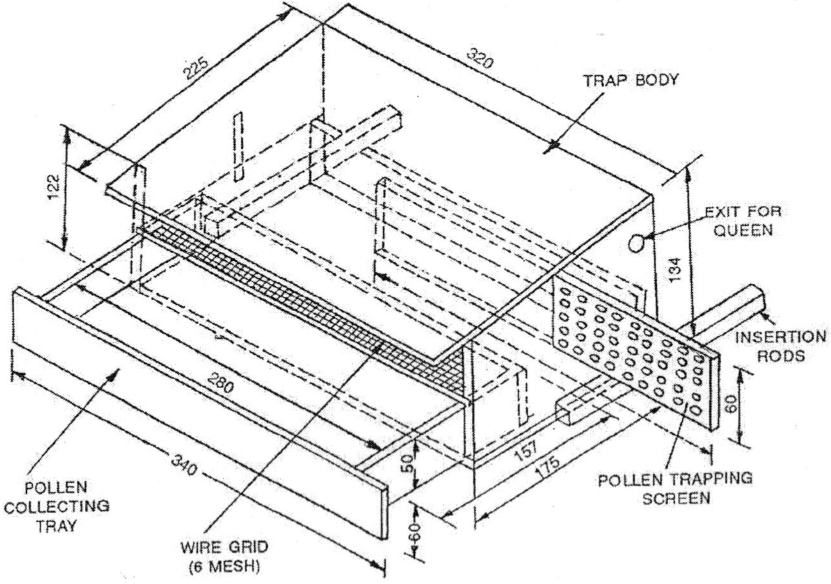

Figure 51. Pollen trap isometric view of apollen trap (all dimensions in mm) (redrawn from Chhuneja and Goyal, 1996).

The pollen is also collected manually from maize and other sources when abundant. During pollen shortage it is fed to the bees along with following ingredients :

Soyabean flour	3 parts
Pollen	1 part
Sugar syrup	2 parts
White of an egg'	1/20 th part

All these ingredients are mixed thoroughly and fed to the bees.

Pollen storage :- The collected pollen can be stored for long periods. Pollen may be stored for 3 days in refrigerator having + 4 degree centigrade temperature for long storage, its moisture content must be less than 20%. The lower the moisture content, greater will be its storage quality, and pollen may be dried in automatic thermoregulators maintained at 40-42 degree centigrade. When the pollen is ready that is its moisture content is reduced to 10-12%. It is immediately packed in polythene bags, milk cans, or glassware. It is very important to have the pollen tightly sealed, so that it does not contact air and absorb moisture during storage.

Pollen can retain its qualities if it is mixed with honey/sugar syrup and stored. When preserved with honey, the proportion is 2 parts honey to one part pollen; when preserved with sugar powder, the proporition of dried pollen is 1 : 1. Pollen so preserved is stored in some dark and cool place.

Moving bees : Flying bees return to their original location. Bee colonies can be moved by about a meter each day and by steps can be shifted to few meters in the apiary. If a colony is to be moved few hundred yards within few days, the best course is to move the colonies 4-5 km away keep them there for 5 to 6 days and then bring it back to the desired place. During this period, the bees will lose the memory of the earlier site in the apiary. Bees should always be moved in the evening or night.

Transporting bees: Transportation of hives to a new locality requires great care and precaution. The following procedure should be followed :

Move all the frames to one side of the brood chamber and press them tight. Fix one nail on each end against the last frame, so that they remain in position, screen the top of the hive with a piece of wire gauze and nail down all the movable parts making the hive completely bee proof. After dark, when all the bees have returned home, close the hive entrance with a piece of wiregauze the hive is now ready to be transported to any place. In hot weather or during a long journey, water should be sprinkled daily on the screen. There should be enough food for the bees to last during transportation or couple of days after transportation. Travelling in hot weather should preferably be done at night. On arrival, the hives should be placed at their new sites and their entrance screen removed. After sometimes when the bees have oriented themselves to new site, the top screens may also be removed and the colonies examined to remove any damaged or broken comb. A further examination may be conducted after a day or two to assess the working of the queens.

Comb space : It is very important that bees should always have sufficient comb space to work, for the queen to lay eggs and expand the brood area. The additional space is provided by giving the raised frames and if they are not available by offering frames fixed with comb foundation sheets. When the brood chamber is full, a super or honey chamber should be provided.

Increasing the number of colonies

Experience shows that the *Apis cerana* has a high swarming tendency and the number of colonies can be increased by hiving the swarms. Capturing the swarms and hiving them is tedious and risky job as some of the swarms may move far away from apiary site and get lost. It is, therefore, always advisable, that the strong colonies may be divided into two or three equal parts in the spring season when plenty of drones are available (Figure 52).

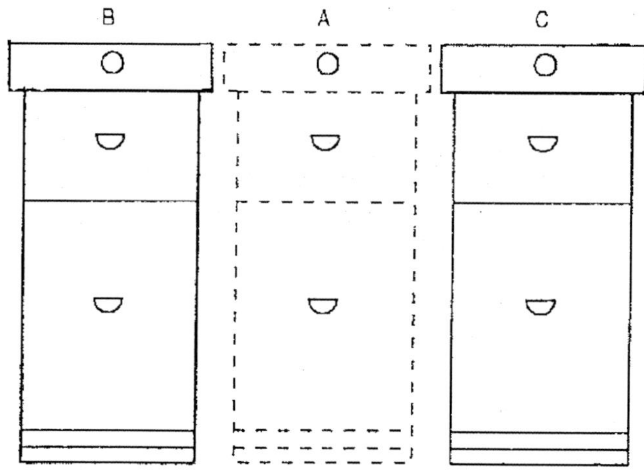

Figure 52. Dividing at the same place A- original colony; B and C–divided colonies.

A colony can be divided into 2 equal parts: Half of the frames (comb) (containing a minimum of 1 frame uncapped brood with one frame honey and pollen each) adhering bees and the old queen is established in a separate hive and moved to a distance of 4-5 km and kept there for a week or so and then brought back to the apiary. The queenless part is kept in the apiary at the site and is given a ripe queen cell or a mated/virgin queen.

In another method, one third of the frames covered with adhering bees along with old queen are removed and placed in a separate hive. This queen right portion is kept at the old site and the parent colony is shifted to a few yards away in the apiary and established there. The colony at the old site is augmented by some of the returning field bees from the parent colony. The queenless portion (parent colony) is given a ripe queen cell a virgin or a mated queen.

The European honeybees, *A. mellifera* swarms less and thus maintain strong colonies. In their case, division of the colony should be made immediately after the honey flow. The colonies can also be divided before the honey flow, without impairing their productivity. Two or three emerging brood combs, with adhering bees but without a queen are removed from a strong colony and placed in another hive. This new colony is shifted to a new site in a apiary and given a ripe queen cell. The older bees in the new colony will return to the parent colony, leaving behind the young and emerging bees. This method does not impair the productivity of the parent colony. The queen in the new colony will mate in due course and the colony may be strengthened by giving bees or brood from the parent. colony after the honey flow.

When the colonies are rearing brood at a faster rate than normal and there are chances of the 'swarming instinct' becoming dominant in the honey flow period, onc or two frames with capped brood along with the adhering young bees (as there is no fighting among young bees from different colonies) are taken each from several colonies and placed in a new hive in the apiary. A queen cell in a protector or a queen or a queen in an introducing cage is put in the new hive and work of the queen checked up after 10 to 16 days. This method helps to check swarming, but does not weaken the colonies. The bees are able to take full advantage of honey flow.

Method of queen introduction into a colony

The introduction of a queen is required either when there is no queen in the colony or when the old queen is replaced by a new one (Supercedure). So, the queen is the heart of the colony, it is most important organ. The stronger and more fertile she is, faster grows the colony. A queen with good inherited properties, upon mating with hybrid drones, will produce offspring of high productivity, the power of any colony is determined by the youth of its queen. If the old queen is to be replaced by a new queen, the colony is first made queenless, then new queen, is installed. If a colony has been queenless for some time and the bees have become laying workers then the new queen will not be accepted. Under such circumstances, the colony should be united with a normal one.

The first step towards introduction of queen is to destroy all the existing queen cells in the colony. A queen is then placed in a queen cage along with few attendants or worker bees. The entrance hole of the cage if plugged with candy prepared by kneading some sugar powder with a little honey. The cage is then hung in between two frames (one of the frames preferably be having brood) of the queenless colony to enable the queen to acquire the hive odour.

The bees will eat up the candy and in the meantime, the queen has acquired the odour of the hive. Her work may be checked up after about a week.

If the queen is not accepted and the bees are hostile towards her, she should be caged and suspended in the colony as before and should be again released after 24 hours or so. Some colonies especially, *A mellifera* are very senstive and resist requeening and thus take longer time to accept a new queen. So the queen be kept in the cage till the aggressive attitude of the bees towards her ceases and she is accepted. If the bees are building the queen cells in the presence of the caged queen in the colony, they should be broken. If possible all the brood combs from the colony may be removed and placed in another colony. This manipulation helps to bring about a speedy acceptance. If the queen is not introduced by using the above method, then a 'push-in cage' which consists of a wooden frame 20 x 13 cm, made of 1 cm thick wood on which piece of wire cloth with apertures of 3 mm square is used (Fig. 53). This cage is fixed on the emerging brood comb after shaking of the bees. The queen is placed on the comb and the cage is pushed in position over her, so that only newly emerging bees are inside with the queen, the cage is held in position by means of rubber bands. The cage provides plenty of room for the movement of the queen and she cannot be attacked by older bees of the colony. When the queen starts egg laying inside the cage, she may be released and at that stage she will accepted. This method works under all circumstances.

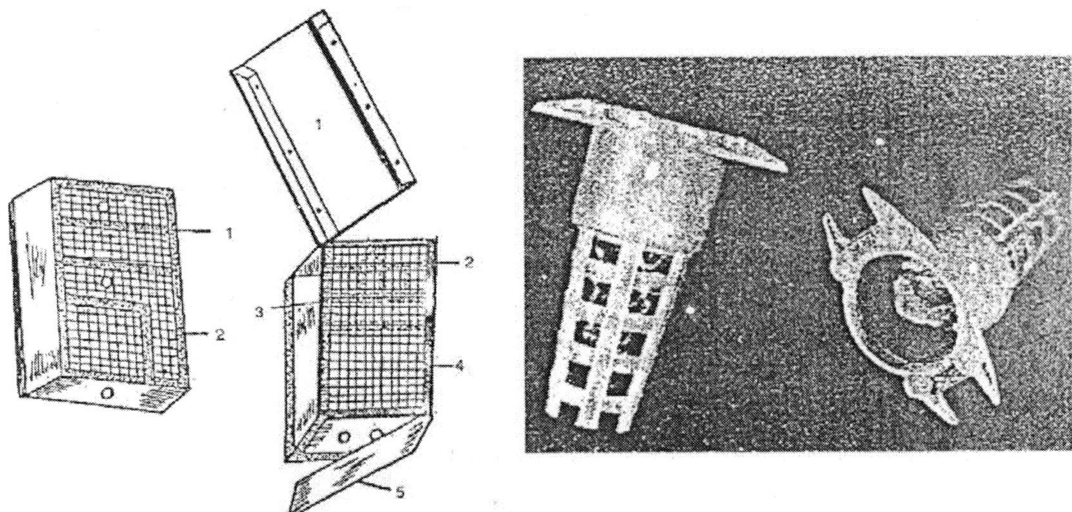

Figure 53 (a). Queen cages. Left –introduction cage 1-food chamber 2-bee chamber right mailing cage 1-cover 2-food chamber 3-water chamber 4-bee chamber 5-cover.
(b) queen cell protein

The Queen mating problems

The queen mating problems occur during certain parts of the year when drones are not produced by other colonies or they are scarce. Under these circumstances, virgin queen fails to mate and lay unfertilized eggs. Sometimes, the queens are also picked up and devoured by the birds while they are on nuptial flights and, thus, they fail to return. To overcome these problems, it is suggested that colonies should be divided or requeened only in the spring season when sufficient number of drones are available. In the beginning of spring season, the colonies should be given drawn combs with the maximum number of drone cells. Some surplus queens that get mated in spring may be kept in small colonies maintaining two or three frames for emergency requeening. The queen-mating yards should be established where there are no problems of bee eating birds. It may further be noted that drone of *A cerana* cannot mate with the queen of *A mellifera* and vice-versa. The colonies of honeybees must be carefully examined for the presence of Varroa mites. It has been found that Varroa mites attack drone brood, creating drone depression resulting in mating problems.

Robbing in honeybees and its control

Bees have strong tendency to search and collect sweet substances to their hives. This tendency often leads them to steal honey from weak colonies and especially when there is a dearth of bee flora. The robbers enter through holes, cracks and crevices other than the main hive entrance. Robbing once started is very difficult to control. In severe cases, the queen of the robbed colony is killed and many bees die in their life and death battles with the robbers. The colonies which are not strong enough to repel the invaders are robbed, ruined and ultimately abscond. Owing to the loss of hair in fighting, the robber bees become smooth, shiny and almost black. The Indian honebee *A cerena* is very prone to robbing. The European bee *A. mellifera* also robs *A. cerana* and *vice-versa* during dearth periods and it is very difficult to keep both the types of colonies together in a apiary. As soon as the robbing is detected, the entrance of colonies should be narrowed down to one bee space. Wet grass or an piece of cloth soaked in dilute carbolic acid or Kerosine, may be put near the entrance of the hive of the robbed colony. Efforts should be made to locate the colony from which the robber bees are coming. To identify, the entrances of all the colonies (except the colony being robbed) should be closed. Fine wheat flour or talcum powder may be sprinkled over the alighting board of the robbed colony. Thus marked bees could be easily located at the entrance of the colony, from which they come. After identifying the robber colony, the entrance of other colonies should be reopened. The robber colony or the robbed colony may be shifted some miles away.

It is much easier to prevent the bees from robbing than to fight it, when robbing takes the shape of an assault; to quench it is very difficult, and sometimes even impossible. Bees, invading the nests of others; may spread any infectious disease which they may be carrying on their bodies. But it is more common that invaders themselves get infected by drinking infected, honey or coming in contact with stranger's nest and foreign bees thus' bringing infections to their home.

Bee robbery is an extremely hazardous evil. To prevent it following precautions should be taken:

1. Do not examine colonies during dearth periods, if necessary proceed very quickly and do not keep the honey or other combs exposed for long.
2. Do not expose honey or any other sweet material in the apiary.

3. Feeding with sugar syrup is done in the late evenings, always inside the hive never outdoors.
4. Honey should be extracted in a bee proof room.
5. Minimize all chances of robbers gaining entry into the hive. Entrances can be reduced so that guard bees can defend effectively.
6. Do not drop even a single bit of honey when withdrawing it from the hives.
7. Try to keep the colonies strong with large stores of honey as the robber bees more frequently attack the weak colonies.
8. Do not open the nest completely at the time especially when there is no honey harvesting.
9. Badly robbed colony should be moved to a new place and an empty hive placed at its place.

Seasonal management of honeybee colonies

The management of honeybee colonies depends upon the climatic and floristic conditions in a locality. In view of the varied climatic and floral conditions of the country there cannot be one standard management that could be adopted for multiplication of colonies or for honey production. Management practices round the year in subtropical, intermediate and temperate areas are given in the Table 28.

Table 28. Beekeeping management practices round the year

S.N.	Name of the operation,	Outward symptoms	Sign And Subtropical areas	Intermediate	Temperate areas
1.	Removal of outer winter packing.	Warming up of weather and the blooming of earliest spring flowers such as peach, plum, apricot sarson and wild pear.	Generally no outer packing is given.	If packing is given remove it after the middle of February; clean the hives.	Outer packing should be removed in the end of February or early March and the hives should be cleaned.
2	Removal of inner winter packing.	The lengthening of days; further rise in day temperature; blooming of spring flowers.	The inner packing should be removed after the middle of the February or according to the prevailing weather conditions.	The removing of inner packing in the February or first week of March according to weather conditions.	Inner packing should be removed after the middle of the March according to the prevailing weather conditions.
3	Stimulative feeding.	The opening of spring flowers; increased bee and brood activity Abundance of nectar and pollen plants;	If need, stimulative feeding with 35% sugar syrup in the beginning of February to	Feed the bees in the middle of February with 35% sugar syrup to stimulate brood rearing and	Feed the bees in early March with 50% sugar syrup according to their needs. Give terramycin in sugar

			stimulate early brood rearing be given.	again the middle of March to ensure rearing.	syrup.
4	Control of the acarine disease.	Observe the symptoms, if K-shaped wings and crawlers are available, then acarine disease is present.	Segregate the diseased colonies and give them 'Folbex' treatments other colonies be treated in Feb. or March as a preventive measure against disease.	Segregate the diseased colonies and give 'Folbex' treatments Healthy colonies should be fumigated with 'Folbex in April as a preventive measure against the disease.	Segregate the diseased colonies and treat, them, with 'Folbex' treat healthy colonies with 'Folbex' in the middle of March or April as a preventive measure against the disease.
5	Strengthening of weak bee colonies.	Weak colonies covering less than three combs.	Strengthen the weak colonies by giving them sealed brood frames in Feb-March.	Unite exceptionally weak colonies with strong ones in the middle of February; strengthen weak colonies by giving sealed brood in March.	Weak bee colonies may be given sealed brood frames from strong colonies in March April. Exceptionally weak colonies may be united with strong ones in early March.
6	Providing comb space	Increased pasturage for bees; extensive brood rearing and overcrowding.	Give new empty frames for egg laying and give frames having comb foundation sheets on outer sides.	Make sure that queens are never short of the egg laying space; provide comb space whenever required.	Provide sufficient space for queen to lay eggs by adding more drawn combs or frames fitted with comb foundation sheets during the building up period in spring to relieve congestion in the hive.
7	Swarming and swarm control	Overcrowding the brood nest raising the drone brood and building of the queen cells. Bees may swarm soon after the first queen cell is sealed over.	Keeping hives ready for hiving the swarms. Control swarming in April. It is good time to start beekeeping for beginners.	Keep beehives ready for hiving the swarms after the middle of February. Control swarming in April. This is a good time to start beekeeping for beginners.	Keep bee hives ready to hive swarms in April. Control swarming after middle of May. This a good season to start beekeeping for begineers.
8	Honey flow and honey extraction.	The whitening of honey cells; the appearance of brace and burr combs; an increase in the weight of colonies.	Bees gather surplus honey from shisham citrus, Eucalyptus, litchi, Sarson, and berseem flowers.	Provide proper space for honey storage. If brood chamber is full provide super. Surplus honey is stored from	Surplus honey is stored from barberry, False acacia, clovers, wild rose horse chestnut etc. Control swarming during the honey flow. Provide

	Honey is generally extracted at the end of April of in May, depending upon the locality and opening of the flowers.	Tun, Puna, shisham, Eucalyptus, soapnut etc extract honey in May depending upon locality.	space for honey storage. Extract surplus honey in June. Leave enough honey for monsoon brood rearing.

Winter management

Proper management practices for successful over wintering of the colonies depend upon the following three important factors:

(1) Adequate population of young bees

(2) Sufficient stores of food

(3) Protection of colonies especially from moisture and cold.

(1) Adequate population of young bees

Prior to the commencement of winter each colony should have a large number of young bees. Young bees are not only capable of surviving the winter but are also essential in starting good brood rearing soon after the winter that is in early spring. Dwindling of colonies during and after winter is mainly due to the presence of proportionately greater number of old tired bees which had heavy work during autumn flow. These bees die soon; even those which may survive winter, collapse more rapidly in early spring. The rate at which new bees are produced in spring is slow compared to the heavy death of old bees. The condition of the colony becomes precarious if there are other factors such as inadequate stores, lack of proper protection from moisture, cold, infestation with diseases like acarine, nosema etc. An inferior queen will leave colony very weak and is incapable of surviving through winter. To have plenty of young bees before and after winter it should be seen that the entire colonies arc headed by young prolific queen capable of laying satisfactorily till the later part of September. Only such colonies have sufficient young bees for wintering. Every year or every alternate year a new queen should be introduced in a colony in spring or summer so that they can begin laying and produce enough number of young bees for a good winter cluster. Winter survival of weak colonies is very uncertain. It is therefore desirable to unite such weak colonies keeping the young queen in the colony.

(2) Stores of food in a colony

Sufficient quantity of honey stores should be left over in a colony for consumption during winter. One super of honey completely filled and scaled is adequate. Honey stores in brood frames and one super be an average per bee colony through winter. Where honey flow is very short when there is a little surplus honey gathered by the bees, one may have to feed sugar syrup to the bees for winter survival to substitute honey extracted. In such cases 500 ml by weight may be fed to bee colonies. Feeding should be given in the evening to avoid robbing. Depending upon the strength of the colonies 3 to 5 kg of sugar syrup (75 per cent sugar by weight) may be fed to them.

(3) Protection from the humidity and cold

Honeybees live in a thermal environment of their own and air condition their nests at temperatures between 32. 3 to 35.0°C. This is done at the cost of honey used for fanning in summer and for muscular movements to produce heat in winter. Insulation of their abode helps to reduce consumption of honey and save energy of the bees. At a temperature of 14-15°C bees start forming clusters, the first indications of formation of winter cluster and approach of winter is killing of drones. This is the time for assuring adequate stores in the colonies and their packing for protection from the cold. It has been established that if requirement under (1) and (2) above arc fulfilled, then even light packing is sufficient to ensure the winter survival of bees. Packing requirements further vary with the type of climate in a area. In plains, very light packing is sufficient but in hilly areas with severe cold a good packing is needed to protect the colonies from cold and keep away the moisture.

(a) Reduced entrance and adequate ventilation

Winter losses also occur due lo heavy packing. Condensation of water inside the hive disturbs the winter cluster resulting in the loss of the colony. An important factor to be borne in mind while packing the colonies is to provide a reduced entrance at the bottom and a small auxiliary entrance at the super chamber level. Prior to the packing of colonies a 13 mm hole should be bored to the super on anyone of the sides. The normal hive entrance should be reduced to a small slit of 5 cm wide and 1 cm high. The temperature around cluster is almost equal to the temperature outside the hive. Top entrance permits the bees to fly from the top of the cluster and also allows circulation of the air te help dissipate excess moisture and thus prevents its condensation. An inner cover with a feeder hole fitted with wire screen is placed over the super chamber and the hive top is placed over it.

(b) Winter packing

Colonies require heavy packing in mountainous areas and operation should be completed well before the first frost of the season (Brar *et al.*, 1987). A packing of chopped rice or wheat straw is placed in the surplus space of the brood chamber. This inner packing is done in brood chamber. This inner packing is further supplemented with an outer packing of straw mattings on all sides of the hive. These mattings are tied with a piece of string to protect them from flying away during a storm. If the hives are not under a roof and are lying out in the open, exposed to rain, the outer packing is further covered with a polythene sheet so that the packing does not become wet and soggy. Thermocole sheets can also be used for packing. Colonies thus packed, are not disturbed during the cold months and their condition is watched from the behaviour of bees at the entrance hole. If some thing, abnormal is suspected, the colony may be unpacked and examined on a warm sunny day.

Spring Management

Spring is the most important season for the bees. During this season they increase their brood rearing activity and build colony strength in preparation for the coming honey flow. When the spring comes, first the outer packing is taken out only when the weather warms up. After removing the packing, the colonies are examined on a warm sunny day. The bottom boards are cleared of all the debris. The condition of the queen is checked. If there is no queen in the colony, the latter should be united with another one. The honey stores are checked and the colonies having shortage of reserves are given sugar syrup. All the colonies should be fed when the spring flowers begin to appear. Artificial feeding is very important during this period to stimulate the bees to produce the brood and to time their activities. The bees which gather surplus honey during this period are generally old bees. For the new bees to reach from egg to adult stage it takes three weeks and for the first three weeks she attends indoor duties, it is only after three weeks that her field work begins. Thus, the colonies must be prepared for the main honey flow at least six weeks earlier. This is how the artificial feeding is helpful in timing the brood rearing, early in the spring. In a certain year, if the weather warms up earlier and the honey flow is expected earlier than usual, the bees can be stimulated likewise. When the brood rearing activity starts, extra comb space is provided, if necessary. Weak colonies should be given a few sealed brood frames taken out of the strong colonies. Soon, the colonies gain sufficient strength and the bees feel crowded. In certain temperate areas, inclement weather at the height of the brood rearing season in spring make the bees stay indoors. The suspension of the field activities, when the food requirements of the colonies are very high, lead to the death of many colonies owing to starvation. Under such circumstances, the beekeepers should feed the colonies, if there is shortage of food.

Swarming

This is the natural process of multiplication of colonies when the strength of the bees in the colony increases and bees feel over-crowing, they make preparation of swarming. When brood chamber is full of pollen, honey and there is overcrowding for queen to lay eggs, a fever or stimulus starts in the colony called as swarming fever. In this state, the queen cells are raised and new queens are reared. When the development is completed, these cells arc sealed and in the warmest part of a sunny day, the colony issues a swarm, taking the old queens along with it. It is generally the old queen and young bees who accompany the swarm. The swarm settles on it nearby tree or a bush and is caught. If the mother colony is weak, the swarm is returned to the same hive; otherwise it is put in a new hive, kept ready before hand, with raised frames or frames fitted with comb foundation sheets.

Catching the swarm

Catching a swarm looks rather difficult, but with little practice it turns out to be a very simple operation. The method is as under: Gently place the swarm basket over the swarm and move the bees with hand, pushing them in (Fig. 54). When the bees have entered the hollow of the basket, take it to the apiary and keep it in a dark, cool 'place. Late in the evening, take the swarm basket to the hive already prepared for the purpose. Remove the inner cover of the hive and place in a slanting position, slopping from the entrance hole down towards the ground. Shake the basket over this board. The bees will move up the slope and march into the hive through the entrance. They will soon occupy the frames and settle down. A brood frame from the parent colony and sugar syrup given at night will further help them to settle. A swarm caught in this way, if found small, weighing less than three pounds, should be united with the mother colony or ancther weak colony. Only a swarm from an average colony is obtained and kept as a separate colony. The weak colonies should not be allowed to issue swarms.

Prevention of Swarming

Swarming can be controlled by relieving congestion in the hive. More frames should be given, the old queen should be changed and shade should be provided. Provide proper ventilation as lack of ventilation also enhances swarming. Remove one or two frames of brood (sealed), transfer it to another colony to provide additional space in the mother colony. Care should be taken that frames that are provided should have minimum number of drone cells or may have full foundation sheets only. The production of morc drones also results in issuance of swarms. All queen cells should be destroyed after every 4th day. When the brood chamber is almost full, place a super over it. A few brood frames along with workers can be removed from the colony and placed in the upper chamber or given to other weak colonies. In their place, empty frames may be provided. If a colony swarms in spite of the precautions, the swarm that issues may be caught and introduced into a new hive. Before the honey flow, this swarm can be reunited with the mother colony or with another weak colony. It has been observed that one strong colony will give twice the quantity of honey, as yielded by two weak colonies collectively having the same number of bees.

Swarming generally occurs in the forenoon between 10 AM and 11.45 AM but occasionally it is noticed at other parts of the day. The bees come out in a rush and encircle the hive. When a fairly large number of bees have come out and the queen has accompanied, the swarm is issued. Other ways of prevention swarm is to use queen excluder which will allow the bees to pass but not the queen or clip the wings of the queen (Figure 53). If queen will not accompany no swarm will be issued. But in a bee colony if swarming impulse persists and is evident by raising regular queen cells, then best way is to divide the colony.

Management during honey flow

The honey flow commences after the swarming fever is over. The seasonal onset of honey flow can be known from the whitening of honey and of the honey cells and from an increase in the weight of bee hives the exact period varies from locality and the bee keeper himself has to find out the precise time when the bees store surplus honey. Not knowing the period or slight delaying in catching up with the bees can deprive a beekeeper of the valuable gift of honey that can be extracted. During honey flow, the beekeeper must keep the honey gathering instinct of the bees in mind. The bees would thus require additional combs for the storage of the honey. When the brood chamber becomes full, the honey frames should be removed and placed in the upper chamber of the super. The main honey now in the sub-temperate and temperate areas of the Jammu & Kashmir state occurs during April, May or up to the middle of June and in the higher hills in September and October. Depending upon the local climate, the altitude, the direction of the mountains, etc., the honey flow may be one or two weeks earlier or later and the beekeeper has to make adjustment likewise. In the subtropical areas, the honey flow may start as early as in the middle of march and may continue upto the middle of May, depending upon the locality.

Honey extraction

The extracting of surplus honey from the colonies is the ultimate aim of beekeeping. It depends upon the efficiency of a beekeeper that how much surplus honey and how many times he can extract honey from a colony. The beekeeper must also know exactly when to extract honey and how much of it to leave as food for the bees. When the honey flow has begun to slow down or when it has just stopped, it is the best time to extract honey and how much of it to leave as food for the bees. When the honey flow has begun to slow down or when it has just stopped, it is the best time to extract honey. The frames in which two thirds or three-fourth of the cells have been scaled over are taken out of the hives, leaving behind 3-4 kg of honey in an Indian colony and approximately double the quantity in the Italian Colony. The frames are removed in the evening and care is taken that three are no brood or bees on the frames. The frames are stored in a room in which bees cannot get in to rob honey and the extraction is done that very night.

The uncapping knife is heated in boiling water and with the sharp edge, wax capping cut from the tops of the cells on both sides of the frames (Fig. 54). Two such frames are placed at a time in centrifugal honey extractor. The machine is revolved vigorously for about two minutes and the honey flows out of the cells. The other sides of the frame are then turned and the machine is worked again. As the honey flows out of the machine it is strained through a piece of muslin cloth. Honey is stored in stainless steel drums, tins or glass bottles. The container should be airtight to avoid the absorption of moisture from the air and the consequent fermentation of honey. The empty frames are returned to the hives and all the appliances used in honey extraction are thoroughly washed with water so as to avoid robbing and fighting among the bees. As a precaution, the bees entrances of all the colonies are shortened on the eve of honey extraction. Pure honey has the property of granulation after sometime, particularly during the winter season. If honey is heated in a water bath at 1400F for half an hour and packed, it would not granulate so easily.

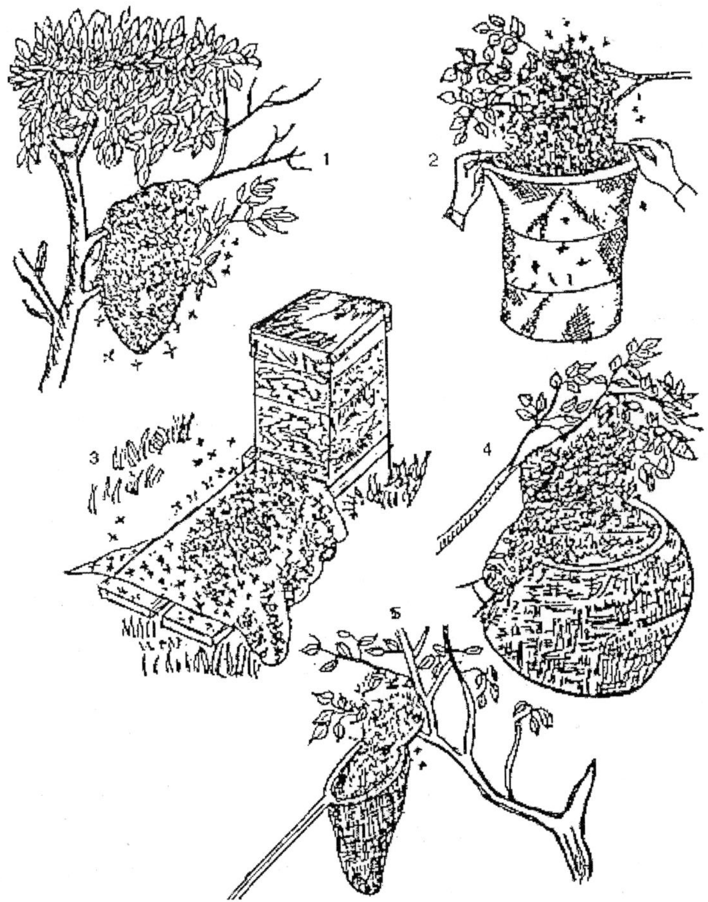

Figure 54. Catching and hiving a swarm. 1- swarm on a tree 2- capturing in a swarm bag 3- hiving a swarm 4-capturing swarm in a basket 5- capturing swarm in a swarm catcher

Summer Management

In most localities, honey flow season is followed by the summer dearth period. Bees adopt number of strategies to protect their stores as they throw out the drones because they arc useless for the colony. Unmanaged colonies stop brood rearing in order not to starve in future. The colony strength also declines considerably due to the death of old and decrepit bees who have put in strenuous work and exhausted due to honey gathering during honey flow season. In summer, besides defending the colonies against enemies, bees have to keep their hive cool and properly ventilated.

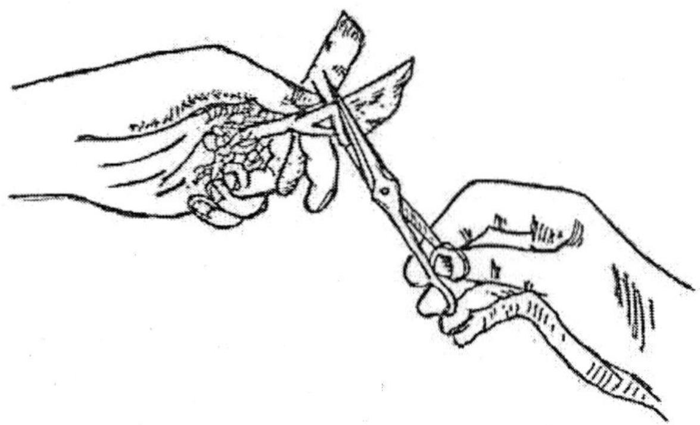

Figure 55. Clipping of the queen wings.

In unattended colonies, the bees mostly remain indoors, do little work and wait for the developing brood to emerge. The queen stops egg laying and if she persists' in depositing eggs, they are neglected by the bees and allowed to shrivel. When almost all the brood bas emerged and not enough is left, the colony absconds. Only infested combs and sometimes infested with wax moth are left. The deserting bees almost behave in a similar fashion as swarming bees except that these bees rise very high and it is rather difficult to make them settle down. Desertion or absconding is little known in *Apis mellifera* but the tendency is very strong in Indian honeybee *Apis cerana indica. Apis mellifera* colony would rather die because of starvation or due to attack of enemies but show little tendency to swarm.

The best way to save colonies during this period is to avoid broodlessness in colonies and stimulate them to rear brood. The strong colonies with sufficient stores would continue to rear some brood during summer. Colonies headed by young prolific queens will continue to lay eggs. Strong colonies are less troubled by enemies and are able to take advantage of subsistence floral sources. The colonies can not be kept in the open sun.

These should always be shifted to a place with thick shade. At some places, the temperature rises as high as 45°C-47°C. And gunny bags, moistened twice at noon and in the afternoon, can be spread over top covers. Ventilation of hive is to he ensured but holes, cracks and crevices also allow easy passage to enemies. The brood and super chamber can be raised slightly by putting shavings or splinters of wood in between. The space so made should not allow the bees to pass through to avoid robbing. Sufficient honey stores are must in a colony. Beekeepers should have complete information of subsistence flora that would be available to the bees and make his judgment about the stores to be allowed at a time of last honey extraction. The brood rearing in summer can also be boosted by feeding pollen substitute or pollen supplement, a continuous supply of which is required during dearth period since these are not stored in cells. Frequent inspection of colonies should be avoided. The above management practices during summer check the absconding of colonies. Infact, all efforts should be made to add to the comforts of bees. Some of the other operations include:-

1. Feed the bees with a thin sugar solution daily. One part of sugar and eight to ten parts of water. This also meets the water requirement of bees to maintain suitable humidity condition within the hive
2. Make provision of fresh water daily.
3. Some sort of wind break should also be provided to the bees so that the colonies be protected from the hot winds (Loo) of the day.
4. Keep the gunny bags or rice matting placed over the top covers of hives drenched with water to reduce the excessive temperature.

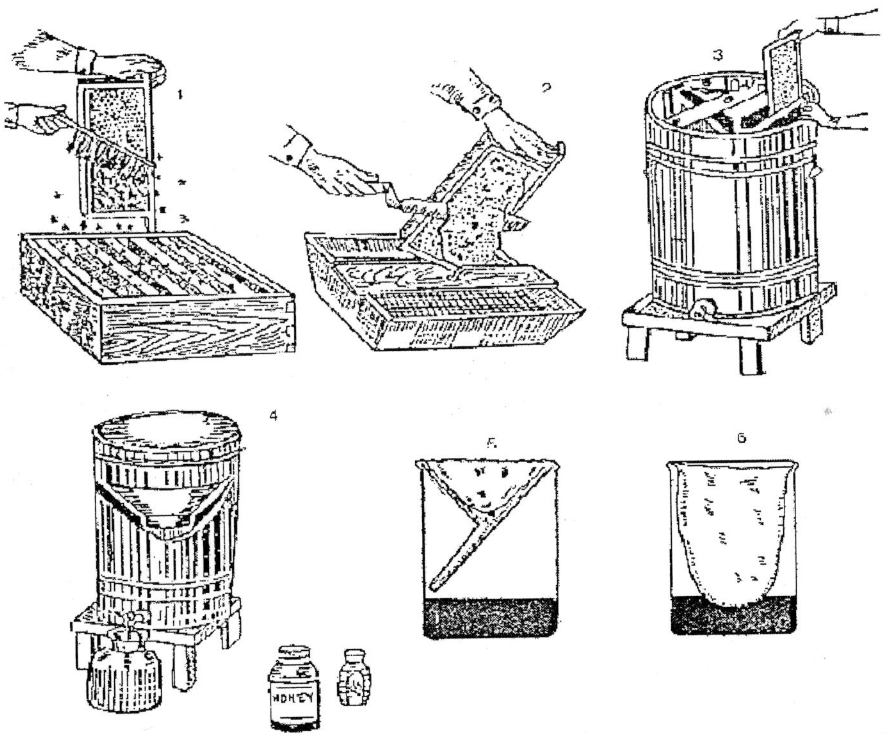

Figure 56. Honey extraction 1 Brushing off the bees 2- Uncapping the cells 3- Putting frame into the extractor 4- Straining and bottling 5- Straining honey (Correct method) 6- Straining honey (Wrong method) (Redrawn from Rawat, 1982).

Monsoon management

Honey flow season is mostly over by the end of May and colonies are required to be prepared for monsoon management. For survival of the colony the first thing to be seen after honeyflow season or during the last extraction of honey is that adequate quantity of honey should be left over in the brood and super combs for consumption by bees during the prolonged monsoon period of two months. The colonies should be well protected from rains, wind and enemies like ants, lizards, wasps, birds and other enemies. Mass desertions immediately after monsoon during September-October is a major problem for beekeeping. This is mainly due to the weakening of colonies, coupled with fungal growth, wax moth infestation on old combs uncovered by honeybees and general un-hygienic conditions. Removal of empty combs at periodic intervals during monsoon saves colonies from wax moth to a greater extent. The major wet spell of monsoon is generally over by mid or end of August and there arc intermittent sunny days. At this time weeds, grasses and many kharif crops like coriander, maize and oilseeds flower and bees start collecting pollen. Comparatively nectar is not freely available. On a clear sunny day the colonies should be quickly inspected and old black combs uncovered by bees or combs with fungus or wax moth infestation should be removed, even though such combs have some brood or stores. Wired frames fitted with comb foundation should be provided in place of the removed combs and 300 to 500 ml sugar syrup (50% sugar by weight) may be given to the colonies in the evening. With availability of pollen and simulative sugar keeping, bees start constructing new combs. This may be repeated at 10-12 days intervals complete comb renewal can thus be done by end of September. Queen lays eggs in such newly drawn combs and there would be adequate sealed and unsealed broods in the colonies. With this management, the problem of post-monsoon desertions cannot only be avoided, but the colonies can be made strong for multiplication or minor honey production.

Post-monsoon management

Depending upon the duration of forage availability from October to December, severity of winter and the dearth of January, there can be two alternate methods that could be adopted.

(a) If the honey flow period is short, the colonies may be developed for minor honey flow. Some strong colonies may show signs of swarming. But when the flow season is short it is not advisable to divide the colonies in view of the ensuing dearth periods. Sealed brood combs from such colonies may he removed to reduce their swarming impulse and given to comparatively weak colonies to make them strong as drawn out super combs, if available, may be given to the colonies at appropriate time. Alternatively, full comb foundation sheets may be fixed in super frames and these may be got drawn out for use during the major honey flow season. After extraction of any surplus honey in the minor honey flow, the drawn out empty supers may be preserved using Paradichlorobenzene.

(b) If post monsoon flow is long, with sufficient pollen and nectar and if the winter rains are not heavy or winter dearth is not very acute, few selected strong colonies may be strengthened further by giving them scaled brood frames and frames with pollen and honey stores from different colonies. This will induce early drone breeding and swarming impulse in them. Swarm cells from such colonies can be used either to increase the number of colonies or to requeen the colonies having old queens. In any case it has to be seen that the divided colonies with newly mated queens get sufficient time in the latter part of the flow for their development and for building up pollen and honey stores for the following dearth period. Colonies headed by new queens over winter well.

Control over desertion of bee colonies

Desertion is one of the problems faced by beekeepers. During spring period nectar and pollen are available to the bees in abundance and swarming and colony multiplication takes place during this period. After the spring, the major honey flow season commences which is followed by a prolonged dearth period and adverse environmental conditions for the bee colonies. This results in desertion of a large number of colonies. Desertions occur mainly during monsoon or post-monsoon period i.e. in August-September or in the beginning of October. This particularly occurs in areas where spring and summer are building up and honey flow periods and summer monsoon is prolonged dearth period.

In other regions, where September-October is a flow period, desertions occur in winter or during winter rains. The losses are to the tune of 40% and go upto 60% or more, particularly when buildup period or honey flow season fails. In either case about 50% of the bee colonies desert and the remaining 50% of the colonies though surviving the dearth are in a very weak condition. These colonies take a long time to develop and to come to swarming stage or to honey yielding stage. As a result of this situation the colony number in the country remains almost static and at times honey production goes down. It is therefore necessary to know fully the reason for desertion or dwindling of colonies and the precautionary measures necessary to avoid the colony losses due to desertion. The main causes of desertion or dwindling are as follows:-

1. **Starvation**: In the hills, where most of the beekeeping activity is concentrated, high rainfall and foggy conditions confine the bees to the nest for at least two months, the adequate stores of honey are not left in the colony by the end of honey flow season, the bees starve and the colony perishes. The surviving colonies try to desert during sunny days of dearth period.

2. **Unhygienic conditions in colonies**: Most of the brood combs used for brood rearing during spring period get old and black. After emergence of bees, these combs are not used as the queen stops egg laying after flow. Due to depletion in bee population, such, combs are not covered by the bees. Such old black brood combs generally get heavily infested with wax moth and fungus. As the strength of the colony dwindles, other pests, predators and enemies of bees like wasps, ticks, mites, ants, birds etc. further demoralize the bee colony. This also results in robbing and fighting among bee colonies and finally the colonies attempt to desert en-masse.

3. **Absence of young bees in colony**: At most of the places, flow period is followed by dearth period. By the end of flow period queen stops egg lying. Having worked vigorously, most of the worker bees get old, worn out and exhausted. These are of very little use after 2 to 4 months long dearth period.

4. **High temperature and humidity**: Water management of bee colonies is a very important factor, during summer, requirement of water of the bee cclonies increases. Large number of bees collect water for maintaining the temperature of the colony and for brood rearing. If fresh water is not available near the apiary, working of the colony is affected. Bees cannot tolerate excessive humidity and accumulation of water in the colony. This leads to unhygienic condition in colony and prevalence of brood diseases.

Desertions can be avoided to a great extent by proper management of colonies during pre-monsoon and post-monsoon period or during flow and winter dearth period. The following precautions may be taken to prevent desertion.

(1). Leave adequate quantity of honey stores in the colony. Do not extract all the honey. This will help in two ways: adequate stores are available for the prolonged dearth and bees are also not overstrained.

(2). During building up and flow period get 2 or 3 combs newly drawn by giving comb foundation strips.

(3). Ensure that all the colonies are headed by young healthy queens of less than two years age.

(4). Ensure that egg laying of the queen continues as late as possible in the flow period. This will ensure enough force of young bees, emerging from brood during dearth period, and which will be useful as foragers after the dearth period.

(5). Keep queen gate in position and undertake proper repairs of the bee hives if required to protect the bee colonies from the attack of wasps, reptiles etc.

(6). The colonies may be kept in shade for protection from direct rains, dusty winds. Avoid accumulation of water on floor board by keeping the colony in a slightly inclined position.

(7). Clean and dry the bottom board periodically. If possible use reversible bottom board. This may be useful to check the attack of wax moth in. colonies.

(8). In the month of August or when the severity of monsoon is reduced inspect the colony casually on a clear sunny day; remove old black and uncovered combs and give dilute (50:50) simulative feeding.

(9). A colony showing symptoms of desertion such as dull foraging, no sealed stores, no larval brood etc. may not be kept in the apiary. It may be kept in isolation to avoid nuisance to other settled colonies in the apiary. When one or two such attempt desertion, generally bees from other good colonies also join the deserting bee colonies, get mixed up and fighting starts, demoralizing settled colonies.

(10). When the severity of monsoon reduces, bees may start collecting pollen. If the rate of pollen collection increases, give frames fitted with comb foundation strips and heavy dose of sugar feeding. Continue this management if the bees have drawn out of the strips and the queen has started egg lying; till all the old black and uncovered combs are removed.

The above management practices will protect the colonies from pests and predators and induce bees to build new combs and collect pollen, have freshly drawn out good combs in the colony for the queen to lay eggs and have fresh young bees to take advantage of the post-monsoon/post-winter flow period. This will not only prevent desertions but will develop the colonies to multiplication or honey yielding stage.

Another way to prevent desertion is to migrate bee colonies to comparatively less rainfall area. Area having rainfall between 600 mm to 1200 mm are ideally suited for monsoon migration. These areas have arboreal, ground and agricultural plants and many, a time not only survival but multiplication of colonies or honey extraction is possible. If the forage is available for a prolonged period queen cells can be used for requeening and removal of old queens or for increasing the colony number. By the end of October or mid-November or by the end of February or early March strong and healthy colonies can be migrated back to their permanent locations.

Chapter 13

Migratory beekeeping – prospects and problems

There are frequent cases when a nectar plant which was focus of one's hope to get surplus honey-stores, fails suddenly or its blooming coincides with very hot weather or very rainy weather when conditions are not conductive for bee's flight activity. Some of the localities provide nectar flow for one season followed by dearth period during the other. Therefore, it becomes necessary to transport colonies to areas where sufficient nectar, pollen providing flowering plants are available. Such transportation of bees from one honey source to another is called migration.

Migratory beekeeping is a tremendous apicultural achievement. It permits gathering nectar where its quantities are the largest; consequently, one can get rid of "windows without honey flow" and to some extent become independent of the environment.

Migratory beekeeping is a very old technique in apiculture. Apiaries are transported not only to the sources of the main honey flow, but also to such nectar plants which are of secondary importance but help intensify young-bee rearing for the main honey flow, or provide for the bees-better preparation for winter. If the conditions are favourable one can gather surplus honey. It has been found that migratory beekeeping can bring three or four honey yields per year.

If a beekeeper has migrated apiary to a honey harvesting area at once, he has definitely appreciated the great advantages of this technique over the stationery ones. After that he will become a travelling beekeeper and will never miss the chance to gather honey flows some where nearby. Unfortunately, too many beekeepers do not travel with their bees. Some of them find it difficult to prepare their bees for transportation or regard the transportation very dangerous for their bees and are afraid they may perish enroute; others are too busy with their full time jobs to spare time for such migrations. Only full time beekeepers are able to get the benefits of migratory beekeeping. Those beekeepers which have very few number of colonies and cannot afford to migrate individually; they must follow co-operative migratory beekeeping. They can pool up their colonies and one of the trained beekeeper can look after the colonies at the migration site on wage basis. With migratory beekeeping, one can get an additional 30 or 40 kg of honey/hive. Migratory beekeeping not only increases the production of honey per hive but also enhances the chances of success in the division of colonies to a great extent. Migratory beekeeping can very safely be classifed as an economic proposition, because of the fact that total expenditure incurred on the whole migratory operations are much less than the profit received from it. Migratory beekeeping not only helps increased honey production, surplus honey stores for winter and build up of the colonies but also prevents the loss of colonies. Experience shows that more than 60% of the colonies suffer mortality due to floral death coupled with attack of wasps/birds and other predators in stationery colonies. Further more, weak colonies are more susceptible to wasp attack, bee diseases and abscond. Strong colonies on the other hand can very effectively defend against the wasp attack and have less tendency to abscond. Fruthermore, at migration sites there are no difficulties in protecting the bees in the forest or field, if the migrating beekeepers co-operate and take turn guarding the bees. With this kind of organized bee migration and a co-operative association no matter how small it may be the beekeepers can have pleasurable, memorable and highly advantageous beekeeping. Migratory beekeeping is one of the best forms of active recreation. It is particularly nice in summer time. One can fully enjoy the warm nights, rich in the aroma of flowers and fresh honey. Before preparing bees, for transportation, one must look for an appropriate area covered by a nectar-rich plant and verify the length of time the plant blooms. If two sites with equally good sources of nectar are available, the site with higher soil fertility should be selected. The plants growing on peat bogs, does not secrete much nectar while those on freshly burned areas secrete high yields of nectar. It is not desirable to locate one's apiaries close to large bodies of water or near the roadsides, especially when the sources are situated on the opposite side of the river, road. One will inevitably lose many his flying bees if these factors are neglected. The amount of precipitation within the period of nectar plant vegetation should also be taken into account. Rain is another most important factor if rains are normal, nectar secretion will be more.

It is highly desirable that next to the source of the main honey flow, or at least nearby, there are plants providing secondary nectar flows. In case the main nectar plant is not strong enough or its secretion lasts not the whole day or few hours, the bees will be able to work on nectar plants of secondary importance.

The qualities of the local nectar bearing flora and the expected honey flow is a most important factor determining the final outcome of the entire apicultural season.

When a beekeeper decides to migrate his bee hives to cultivated fields, he must get the permission of the farm management to accommodate his hives on their land and official grounds. It must be negotiated prehand, that no fields near the apiary will be sprayed with insecticides or only safer insecticides will be used when bees are not foraging. It is necessary to make a survey of the area where colonies are intended to be taken. A site suitable for an apiary is selected and all arrangements to locate colonies on site after transport are completed. Migration of colonies is done during the night time, when bees rest within the hives. All the colonies are inspected a day or two before migration. Sugar feeding is given where necessary.

If the beekeeper wants to shift his bees to a forest, he must address the local forestry authorities. The latter will also allocate the places for accommodating the apiaries. Such places are usually fixed in the forest skirt, near or among the nectar-bearing areas. The plot chosen should be dry, well heated by the sun, its vegetation should protect the hives against the sunshine during the hot day hours. Over crowdedness of colonies in a area should be avoided. It is better to divide large apiaries into group of 20 to 30 hives in a area. It has been found that the temporary division of apiaries into smaller units is a most successful technique for utilizing honey yielding areas to their utmost.

Transportation of bee hives

Transporting of colonies is not difficult if done systematically. The distance to which the colonies are migrated must be kept in mind. If they are to be moved more than 2 miles away from the apiary, then there is no problem as this distance lies outside their flying stage. But if they are to be moved less than 2 miles away from the apiary, we cannot move them all of a sudden. They should be either moved a yard daily in the evening towards the side we want to establish them or if the same is not possible they should be first shifted to a place which is at a distance of about 2 or 3 miles or more than this from both the places i.e. from the place where the colony is being shifted and where it is intended to be established (Figure 57). After keeping the bees for about a week to such a new site, they can be brought to the desired site without any hesitation. On reaching a new place, they become accustomed to the new site.

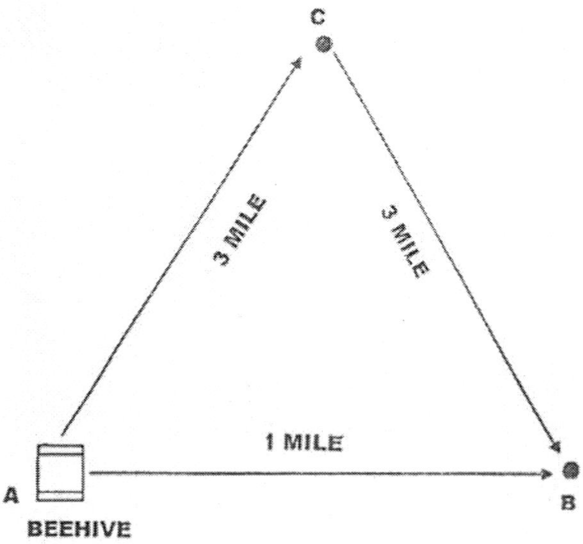

Figure 57. Shifting of colonies at short distances A-original place B-place to which the colony is to be shifted. C-place where the colony is to be shifted temporarily.

If the bees are to be moved to a considerable distance, they should be moved at lest 2-3 miles beyond their flight range. The bees should be moved when all the field bees have returned to their hive, either later in the evening or early in the morning.

Make sure that all the cracks or openings between the hive parts are closed by filled them with paper, rags, or similar materials so that the bees will not find their ways out during transit. The bottom board and hive parts must be fastened securely. Hive, staples are made to secure hive parts tightly so that that they may not shift position when raising and handling the colonies (Figure 58). In hot weather, the hives should be provided with a top moving screen to provide a clustering surface and air for the bees. When colonies are loaded on trucks or other vehicles each layer of hive should be separated by wood strips to provide a space between for free movement of air. Properly prepared colonies can be moved in the hottest weather. The truck should be kept moving, if stops are necessary it should not be parked in the sun. It is always advisable to transport colonies during night. After, all the colonies have been unloaded at the new location, the entrances can be opened and screens removed. On a bright hot day it is better to spray the bees with water to prevent rapid flight and drifting.

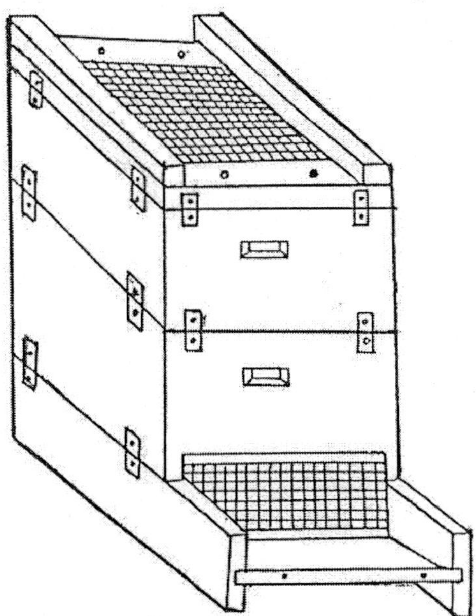

Figure 58. Hive parts stappled for migration

In the cool weather of early spring or late fall when bees are not flying, or after they have ceased flight for the day, colonies can be moved without any specific preparation as the bees make little attempt to come out of the their hives while the truck is in motion. If the weather is too cool for flight, the colonies may be left on the truck until morning before unloading. Sometimes if colonies are to be moved immediately at a short notice when the bees are working in the field, we can shift the colonies leaving one colony full of combs in the apiary so that bees returning from the field trips can be collected in this hive. It is preferable to reach the destination while it is still dark outside. Bees remain in their rest and come out of the hive to resume their usual activities in the following morning. On reaching the migratory site, the hives are distributed and kept in their selected spots. The entrance gate is kept open, removing the wire screen. They are left in that position for a day or two.

After about a day of migration, when the bees have got their acquaintance with the new location, the inspection work can be done after removing the screen covers and when the inspection work is over, the hives should be closed with their original cover. The broken combs should be replaced and all the bottom boards cleaned. Colonies are given sugar feeding to encourage foraging for pollen.

After the colonies resume normal working, necessary management is taken up for the purpose for which the migration is undertaken.

Migratory beekeeping

Because of temporal and spatial distribution of different bee plants, the availability of surplus flora or honey flow for bees varies in time and space. In any specific beekeeping area, only one or two honey flows are available and when honey flow ceases at particular site, the beekeepers move their apiaries to some other site within or outside their state, where honey flow from some other plant species can be available to the bees. The type of movement of honey bee colonies, several times a year from one region to the other, for getting multiple honey crops and increasing colony yields is referred to as 'migratory beekeeping' and is a common practice with many beekeepers (Mishra and Sharma, 1988). The need, types, some routes and useful considerations for migratory beekeeping, its prospects, problems and possible solutions are given below:

Migratory beekeeping- the needs

A beekeeper may need to move his honey bee colonies to long distance for various reasons such as to abridge the floral dearth periods, to exploit different bee floral sources/honey flows, to rent out his colonies for pollination of the crops/ orchard plants or to establish a new apiary at some new location. The migratory beekeeping in earnest is, however, an important component of commercial beekeeping undertaken not only to augment honey production but also to increase the colony strength either for exploiting the ensuing better honey flow to its maximum or to have an ease in the colony multiplication well in advance of some good ensuing honey flow.

Migratory beekeeping- the types

The migrations could be short distance ones when undertaken between different parts of the local forests, or between forests and adjacent farms when there is short duration floral dearth. Monsoon in the tropical and subtropical regions and winter in the temperate region along the Himalayas constitute the long periods of floral dearth when it is worthwhile to migrate the colonies to areas where bee flora is available (Suryanarayana and Rao, 1998).

Migratory beekeeping- some routes

Beekeepers of the Punjab shift their bee colonies near roads, railway tracts, canal banks etc. for availing Eucalyptus honey flow during February-March. There after some beekeepers who migrate the apiaries to Pathankot and Dehradun areas for litchi honey flow during April; to Hoshiarpur, Jalandhar, Amritsar, Ludhiana and Bathinda during Apirl-June on sunflower; to *Brassica campestris* var. *toria* growing belt in and around Amritsar, Taran-taran, Rajpura and Patiala areas during October-November; to Rajasthan (Ganganagar, Sanghria, Padampupr, Bhartpur, Karanpur, Alwar etc.) and Haryana (Hissar and Rewari) and Uttar Pradesh (Agra, Aligarh etc.) on *B. campestris* var. *sarson* and *B. juncea* during December-February. In J&K, watery honey yielding *Plectranthus* sp. blooms much earlier at higher hills and as we descend, the period of flowering in this wild bush is delayed by about 7-10 days for every 1,000 feet. So beekeepers get multiple honey crops by moving their apiaries from higher hills to lower hills periodically according to the blooming of *Plectrannthus* sp. Migration of apiaries from North Bihar and South Bihar to avail litchi, jamun and Karanj flows (Chand *et al.*, 1995); fom Himachal Pradesh (H.P) and Jammu & Kashmir (J&K) to Punjab to avail *Brassica campestris* var. *toria*, Egyptian clover and sunflower flows (Goyal and Rana, 1992), from Punjab to H.P. and J&K to avail *Plectranthus* sp; from Punjab and H.P to Rajasthan to avail *Brassica campestris* var. *sarson* and *B. juncea* flow and between various altitudes with in West Bengal to Sunderbans and in Southern states of Karnataka, Kerala and Tamil Nadu to rubber plantation, coffee, cardamom, cashew, cotton and *Eucalyptus* plantation is a standard management practice for honey production (Suryanarayana and Rao, 1998). Migration of honey bee colonies on rental basis to apple orchards as a joint/community approach for enhancing pollination is fast developing in Kulu region of Himachal Pradesh.

Migratory beekeeping-some useful considerations

While preparing the honey bee colonies for migration and practically migrating them, the beekeepers need to consider a number of factors discussed here after:

(i) Season: Migration of honey bees during winter is not a great problem. Exposure to cold simply has the effect of causing bees to consume more honey stores, produce heat and cluster together on the brood nest as they do in the winter. In cold weather, the bee colonies can be carted any where simply by closing their hive entrances with wire screen ensuring required ventilation especially if there is some warmth in the air. If the colonies are strong, full screened inner (migration) covers be nailed over the bee frames. In cold and rainy weather, the hives should be covered with a tarpaulin when being moved.

During summer or monsoon season, colonies should be migrated during night when it is cooler and all the bees are in the hives. The bees clustering outside can be made to drive inside the hive by giving a few puffs of cool smoke or light sprinkling of water over them. Enough ventilation must, however, be ensured by exchanging the inner wooden cover with full screened one. A small, piece of screen nailed on or thrusted into the entrance to keep the bees inside the hive will also facilitate the ventilation enroute. It is also wise to sprinkle a little water at the hive entrances and travelling (migration) screens at intervals if it is long distance migration but do not over do it, because if the hive becomes wet, inside temperature will rise instead of falling. It is worth emphasizing that it is not the cold but heat that injures bees while confined to their hives for migration. Infact very little harm ever results from chilling of the packed colonies while moving the bees (Manley, 1948). So the golden rule in moving bees is to provide ample ventilation, plenty of inner space and to keep them cool.

Making short time halts and temporary siting of the apiaries enroute on some suitable bee flora for a day or so is also advisable when long distances and several journey days, especially in summer season, are involved. It may not only lesson the problem of food dearth in the colony but will also avoid arousing of the absconding instinct among the bees due to unnatural/unfavourable conditions. For colony migration in summer or monsoon season, the bottom board of the hives can also be modified for providing extra ventilation. The bottom board can be divided into two parts by nailing a lengthwise bar inside its centre. One half is provided with a regular entrance for bees. Whereas the other half can be covered with a full wire screen to prevent the bees from landing on its floor. The portion of the entrance rod in front of this screened half is cut and totally removed so that only the air can pass freely though that screened half but bees can not escape.

(ii) Distance of migration site: Very long distance migrations of bee-yard during winter seasons at a stretch is possible provided the bees have sufficient food reserves. During cold weather, bees consume excessive food stores to produce heat and cluster together over the brood nest. During such weather, the hive entrance can simply be closed and bee carted. But if some warmth is expected in some region enroute, the entrance should always be closed with screen/perforated zinc sheet and hive body (bee frames) be covered with full sized travelling screen. Bees are killed very often by overheating and lack of aeration but very seldom by getting chilled.

During long distance colony migrations in hot weather, as already discussed above, the populous colonies may be divided to create more room for the expansion of the cluster. Alternatively, these colonies be shifted with supers in position to allow plenty of room for expansion of the cluster. The hive body and supers and course be stapled properly to avoid their sipping enroute (Manley, 1948).

During summer, it is worthwhile to have one or two halts/journey breaks for short temporary siting of the apiary for a day or two at some suitable place having some bee flora by easing the confined bees. Moving the colonies continually for more than 48hours often leads to their brood mortality and such colonies produce poor results particularly when migrated for pollination.

(iii) Number of hives: If the number of hives to be migrated is small, it will not be economical to migrate as per colony carriage/ transport charges will be much higher than when the beekeepers has full vehicle-load. Hence, full vehicle-load (number) of colonies be migrated to make it more economical. However, in case the number of colonies is small, and migration at all has to be done, make it a full truck-load by joining with fellow beekeepers who also intend to take up migration to the same or nearby areas.

To accommodate more colonies in the vehicle, and for quick dissipation of heat, the top and inner covers of the packed colonies are removed and replaced with travelling screens. Before stacking the other row over the lower low of the colonies to be migrated, rectangular wooden bars on near the anterior edge and another near the posterior edge of the travelling screen be placed. Same may be repeated for stacking the subsequent (upper) rows of colonies in the vehicle. This will check chocking of ventilation through travelling screens. Alternatively, some built in provison must be there so that bottom boards of the upper row of stacked bee hives do not jam the travelling screens of the lower row of the stacked bee hives. The left over spaces towards the end of the rows of stacked colonies may be jammed by inserting the removed top and inner covers.

(iv) Colony strength: Migratory beekeeping means a good deal of movements of hooey bee colonies and this is a job that seems to frighten the beginners. Bees get excited in confinement which in populous colonies leads to increased production of heat in the packed colony and suffocation of bees until there is sufficient room for expansion of the cluster and plenty of aeration.

If weather is very hot and the colony is populous but the hive is not spacious enough to allow expansion of the cluster, bees may very quickly smother and/or get suffocated even when the bee frames are covered with full wired travelling screen. For safety, the heat must be dissipated sufficiently and rapidly to keep the temperature of the interior of the hive low enough for the bees to be comfortable. The moment the temperatue rises high enough, the bees get upset and crowd all over the travelling screen, thereby preventing the heat from escaping. This may result in the rise of inside hive temperature higher than the melting point of beeswax, the combs might melt, the bees immediately discharge the gorged honey and in a matter of seconds they start dying leaving sticky mass of half-animate bedraggled (Manley, 1948). Hence, populous colonies may be shifted with an empty super, filled with drawn combs or empty frames, in position to allow plenty combs may be added for the expansion of the cluster in the hives before migration. If empty combs are not available, empty wooden frames alone or with well fixed comb foundation sheet can be added in between brood/bee frames to ease congestion and facilitate ventilation.

(v) Preparing and packing the colonies for migration: Extract the surplus honey, if any, a few days prior to the migration. All cracks and crevices in the

hive should be sealed or closed. Excessively broken hive parts should be replaced with new ones. The hive body, bottom board and inner cover should be fastened together by stapling/nailing. Always use two staples on each side of the juncture and slant the staples in opposite directions to prevent the hive parts from shifting/slipping. An alternative to the hive staples is to use metal or nylon travelling (migration) belts around the hive. It is, however, not wide to use rubber belts as being elastic in nature, the rubber belts can not withstand slipping of hive parts.

Sometimes, some beekeepers move bees at night with their entrances open and covering the whole vehicle with nylon netting so that bees are not confined to the hives any way. In such case, the bees cluster out side the hive and make unloading at destination very difficult. If unloading is done during day time, the bees will get confused and are lost. Moreover, if because of some reason or the other, the vehicle need to be stopped enroute, the leaking bees may get lost and can also pose serious problems to the passerby. Therefore, it is better to make the colonies bee proof but sufficiently ventilated before the same are loaded for migration.

(vi) Type of the vehicle. Generally the colonies should be migrated by using shockproof vehicle (vehicles with effective shockers), such as trucks, mini trucks four wheelers (tempo). However, sometimes, for short distance migration, beekeepers use their own tractor-trailers. While migrating the honey bee colonies in vehicles such as trucks, the jerking movements will be forward and backward hence, the length side of the hives should be kept parallel to the body of the vehicle (i.e. length of the hive should be perpendicular to the axle of the vehicle). While loading the colonies in a tractor-trailer where the jerking movements are side ways, the bee hives should be loaded with their length side parallel to the vehicle or the axe of the vehicle. Such type of loading will avoid striking of the bee frames enroute with each other, thereby avoiding the bee mortality due to to crushing during migration.

It is best to have engine of the vehicle running during the loading and unloading process. The vibrations caused by the engine have an effect on the bees which causes them to move up and away from the hive entrance. This makes it less likely for the bees to fly out of the hives during loading and unloading. For extra precaution, cover the sides of the vehicle with wire mesh and the entire load of beehives from the above with some type of nylon netting. Some extra provisions (ventilators) on the sides of the vehicles near its floor may also be built in to ventilate the bottom most beehives of the stacks loaded in to vehicle. If a short duration stop is necessary on the way, it is recommended that engine be left running unless it is for a long period of time (Ambrose, 1992).

Do not turn off the engine upon arriving at the destination site until unloading is finished. It is better to place the hives in the new location late in the evening or at night so as to reduce the tendency of some of the foragers to fly away from the hive and get lost in search of food before they have learnt the new location of the hive. After unloading, wait a few minutes to remove the entrance screens. New site location can be reinforced by placing a slanting wooden board against the hive entrance. This will tend to confuse the forgers and encourage them to take orientation flights and learn the new hive location. Avoid facing hives towards the walkway or areas of human activity.

(vii) Time of the day: If the whole of the apiary is to be shifted, it is better to move the bees in the evening or at night (when all the bees are inside the hive and temperatures is low) or during inclement or cold weather when the bees have not gone out for foraging (Ambrose, 1992). However, if a part of each colony (nucleus) purchased is to be shifted, it can be done even during day times as the field bees will return to the parent hive left behind and will not be lost. As already discussed above, the colonies at the new site should be opened preferably in the late evening so that bees leave the hive with prior orientation flights the following day and are not lost in an endeavor to forage.

(viii) Timing migration in relation to flowering of crops: Colonies should not be taken to crop needing pollination until it is in good bloom and can be considered the predominant species in the locality (Kumer *et al.*, 1998). It has been concluded that delay in shifting colonies to the crop until flowering period has begun, invariably increases pollination particularly when the crop has short flowering period and is less attractive to bees than the other crops in the area. The same is true for honey production. In either circumstances, few colonies should be released in the crop before successive days of flowering. However, it should be pointed out that if a great proportion of flowers of crop are open before colonies are taken to it, an important part of crop may fail to be pollinated as the receptivity to fertilize the flowers of many species diminishes. A flower produces/ releases most of its nectar and pollen at a time characteristic to its species. Some plant species produce nectar in morning hours and release pollen grains in afternoon or vice versa. For example in case of *Pyrus* and *Malus* (apple), most of pollen is released in the afternoon. In musk melon, pollen is released in early morning. If the colonies are to be shifted to these floral sources, they should be brought at the time when majority of the pollen is being released. If the colonies are sifted be brought at the time when the crop in question does not shed pollen, bees will approach the flowers of other crops that are releasing pollen during that time and the bees will then get conditioned to that crop which we do not intend to pollinate. Under these situations, hive entrances of the colonies should remain closed with wire mesh for the first day after transportation and should be opened at a time when the crop intended to be pollinated, releases the pollen. Following this type of practice, we can increase both the nectar collection and the pollination efficiency of the honey bees.

(ix) Placement or sitting the migrated stock: The migrated honey bee colonies should be placed away from the passage/walkways where human or domestic animals' movements are expected. If there is no choice but to site along passages, the hive entrances be kept in the opposite direction to the passages. If migration is for pollination purpose, the bee colonies be placed within the crop and should be evenly distributed over the area to harvest maximum pollination benefits and

should not be crowded at one place. Even for honey production, care should be taken that the site is amply away from any other migrated bee-yard so that overcrowding could be avoided.

The colonies should not be opened immediately after placement. Give time to the bees to calm down lest they should be aggressive and resort to stinging. Entrances of the colonies should preferably be opened late in the evening to avoid the loss of foragers in the field that are taking off to the crop for food immediately with out any orientation flights. The drifting behaviour which is extensive when identical hives are arranged is regular fashion is particularly undesirable because it results in undue weakening of some colonies while strengthening of others. This results in less nectar collection by the weaker colonies and lowers the average foraging potential for pollination. The drifting can be reduced by arranging hives, irregularly, placing them with their entrances in different directions, spacing them apart, putting them near the land marks and putting different coloured boards above hive entrance. The first two measures are particularly effective and whenever possible they should be applied when migrated particularly for crop pollination. New site/location can be reinforced by placing slating wooden planks against the hive entrance. This will tend to confuse the foragers and encourage them to orientation flight and learn then new hive location.

(x) Increasing attractiveness of the crop: When honey bee colonies are migrated for pollination purposes, many attempts need to be made to direct bees to a certain crop by simulative feeding with a little sugar syrup containing the scent of target crop intended to be visited by the bees. The techniques implies to immerse and/or crush the flowers of the crop in sugar syrup for some hours, then strain/sieve the flowers and feed the sugar syrup to the colony (Kumar *et al.*, 1998). Often fresh flowers of the crop are also provided for bees to stand on while lapping/licking the syrup. Bees, that take syrup communicate odour of target crop to other bees in the hive which then leave the hive to visit the communicated food source. Spraying the crops with 'bee cues' (bee lures) and use of reflected UV light on the crops have been reported to increase the crops' attractiveness to the bees.

(xii) Conditioning of the colonies: For pollination purposes, the pollen gatherers are more useful than nectar gathering bees if the purpose is the increased pollination of the crop on which the colonies are intended to be migrated. Therefore, it is beneficial to increase number of pollen gatherers for higher efficiency in pollination. Since the proportion of foragers that collect pollen increases with the amount of brood in the colonies, the colonies for pollination should contain plenty of brood. Simulative sugar feeding is useful for increasing colony's brood rearing and thus enhances its pollen foraging potential. Colonies can be induced to pollen collection by removing some of their pollen stores and giving them extra combs of young larval brood to be reared. Restriction on incoming pollen to the colonies with pollen traps is also helpful in increasing pollen foraging.

(xii) Anticipated profits: From the beekeeper point of view, one must be able to have good honey crop and the colony growth as well on the migrated crop, not only to offset the expenditure incurred on migration but also to have sufficient monetary and physical benefits. Good colony growth would help the beekeeper in selling the bees or would ensure him its division or he would better exploit the ensuing honey flow elsewhere owing to colonies' increased strength. Thus, to achieve above benefits the beekeeper must be conversant with the climatic conditions of the area, bee flora and all the more its relative utility to the bees. The beekeepers must be aware of the honey production potential of the bee flora on per unit area basis, honey collection potential of the honey bee colonies and the expected level to which his colonies would develop (Deodikar *et al.*, 1957; Kohli, 1958; Devanesan, 1971; Singh and Singh, 1971; Chakarbarti and Choudhary, 1972; Diwan and Rao, 1972; Chaubal and Kotmire, 1980; Gupta *et al.*, 1984; Mishra *et al.*, 1985; Srawan and Sohi, 1985; Bhat *et al.*, 1987; Brar *et al.*, 1989; Garg, 1989; Brar *et al.*, 1992a; Brar *et al.*, 1992b; Goyal and Rana, 1992; Gupta *et al.*, 1992; Brar *et al.*, 1994; Gotoria *et al.*, 1994; Chand *et al.*, 1995; Chhuneja *et al.*, 2001). The migration should be undertaken only when estimates ensure the required profits/benefits. In the Punjab, colonies under migratory beekeeping for eight months, yield on an average 52.93 kg honey as against 21.20 kg by stationary colonies (Gatoria *et al.*, 1994).

Migratory beekeeping-the prospects

As already discussed, migratory beekeeping has a great scope in commercial apiaries to maximize honey yield per colony, enhancing colony development, increasing crop yields, abridging the dearth periods and perpetuating the colonies through the harsh weather/seasonal conditions.

In Himachal Pradesh, migratory beekeeping with *A. mellifera*, between high hills (Rohru area), low hills (Nauni-Solan area) and plains (Karnal area) increased the honey yield per colony by more than 16 Kg with economic benefit of Rs. 584 per colony per year the stationary beekeeping (Goyal and Rana, 1992). Migratory beekeeping in the Punjab, increased honey yield by 149% over stationary ones (Gatoria *et al.*, 2000). Wakhle (1998) has also visualized the bright prospects of migratory beekeeping as this technology is becoming very popular with commercial beekeeping with *A. mellifera* and beekeepers in the Punjab, H.P., J&K and Bihar are obtaining 50-70 kg honey yield per colony from migratory beekeeping.

In Jammu and Kashmir, diverse climatic conditions of the state with rich floral diversity is most suitable for migratory bee keeping between plains and hills for producing 2-3 crops per year. The migratory beekeeping helps the beekeepers to enhance their honey yield by maintaining the colony strength, enabling them to have more honey harvests annually. Almost every area of the state has a potential for migratory beekeeping. For obtaining an autumn crop, migration of apiaries to the areas which abound in *Plectoranthus rugosus* has got to be effected. These migration sites in each district are as follows:

- Srinagar: Dara, Anderwan, Cherwan, Kulan, Raizan.
- Pulwama: Check (Khonmoh), Barten (Khrew), Machama, Nagbal, Latergam, Lam, Satura, Naristan (Tral).
- Anantnag: Lehan, Dagan (Salar), Warsar (Chatergul), Kapran (Kuker Nag), Aishmuqam and Kund (Qazigund).
- Budgam: It has a potential of only spring crop but for autumn, migration to some other districts has to be made.
- Baramulla: Patches of *Plectranthus* in Baramulla proper, which can sustain only meager number of colonies, Chontimula (Bandipur), Lagama (Uri).
- Kupwara: Kalaroos (Lolab), Gulgam, Awura.
- Doda: Nowgam, Thethar, Neel, Basti (Banihal), Shreekhi, Baija, Bahles, Chanoti (Bhaderwah), Assar, Baggar (Batote).
- Kathua: Bani, Hot, Bwindhi
- Udhampur: Sudmahadev, Basantgarh, Gool
- Rajouri : Thanamandi, Budhal, Drass Hattal, Lamberi
- Poonch: Mender, Mandi Loran, Bafallaz

Migratory beekeeping-the problems and suggestions to overcome them

Although migratory beekeeping has bright prospects, yet beekeepers face several problems too:

Due to lack of awareness in the farming fraternity about the utility of honey bees in increasing crop yields through bee pollination, farmers often do not allow or hesitate to allow the beekeepers to place their migratory apiaries in or around their fields. Many farmers have a misconception that sucking of floral nectar lead to lowering the vitality of flowers. There is need to educate the farmers about the benefits of pollination by the bees.

1. Oilseed crops such as *Brassica*, Sunflowers etc. constitute major utility bee flora. However, because of import of cheap vegetable oils under WTO regime, domestic cultivation of these oilseeds is now not more remunerative, resulting in drastic reduction in area under these bee forage crops. Area under many other non-oilseed bee plants is also declining. This has reduced the carrying capacity of the localities in relation with respect to the colony density and thereby increasing competition among migrated bee yards in those localities. Hence, crop diversification assumes a great significance. Increasing import duty on vegetable oils would be in favour of domestic cultivation of oils seed crops.
2. In India, there is no agency that, unlike those in Israel, registers the beekeepers and allots them the specific sites for colony migration so as to distribute the colonies evenly in the area and thereby reducing the above

referred competition. Proper census of honey bee colonies throughout the country and registration of beekeepers with some nodal agency is urgently required.

3. Because of decline in cultivated bee flora, the forest based beekeeping has great scope in the country. There is an urgent need to adopt multipurpose agro-foresty and plant avenue trees that would, in addition to other plant utilities, serve as good bee flora too.

4. At certain instances, even the forest department also do not allow the beekeepers to site the apiaries in their jurisdiction along side the canals, national highways, railway tracks or in the forest plantation itself. The matter needs a considered attention of all the concerned agencies.

5. While evolving new varieties of plants, which serve as good bee flora, due consideration must be given to its nectar/pollen production potential i.e. they should also be evaluated from beekeeping point of view before their release.

6. Migratory beekeeping also carries a risk of spreading potential bee diseases and enemies from one region to the other as there is no functional agency or any concept of 'Apiary Health Inspectors' who would enforce domestic quarantine need to be given special emphasis in the country. Recently, migration of honey bees colonies and unlawful import of queen bees and honey bee colonies across the border from Nepal-based commercial apiaries, ignoring all the quarantine procedures, by some private agencies have paved the way of spreading European foul brood Sac brood disease and *Varroa destructor* in *A. mellifera* in the northern India.

7. Excessive use of pesticides in the areas of colony migration by the farmers without any prior information to the beekeepers is another serious concern for the migratory beekeepers.

8. Unlike advanced countries, there are no specific vehicle-types and other allied equipment required for loading and unloading of colonies. Beekeepers also lack the honey house facilities, blowers to dislodge the bees off the honey frames, motorized/power driven radial honey ex tractors, automatic uncapping machines etc. In the Punjab, however, some commercial/migratory beekeepers modify their vehicles, used for colony migration, to a movable honey house by fixing a wire mesh all around and temporarily covering it at the top with nylon netting at the time of honey extraction. Alternatively, many beekeepers use large sized mosquito nets, specially fabricated for this purpose. Research and development of the equipment required for commercial beekeeping must be taken up by some engineering agency.

9. Poor adoption of scientific management technologies such as timely replacement of old/exhausted queen bees, supering the colonies is also a limiting factor in realizing the full honey potential of migratory beekeeping. Most of the commercial beekeepers extract the honey from

single chambered colonies development and lowering the quality of the honey too. In apicultural advancc countries colonies are taken to several supers. The lower most two chambers as brood nests and honey is extracted only from the supers above them.

10. Many a times, migratory beekeepers are harassed at the interstate borders. Since apiculture is not recognized as an agricultural activity, octroi has to be paid by beekeepers while moving their yield back home. The beekeepers are also harassed by octroi personnel while the formers are moving the required equipment to and fro for migratory beekeeping. Certain regulations must be made or identity cards must be issued for avoiding any harassment to the migratory beekeepers.

Chapter 14

Honey bee nutrition and supplemental feeding

Successful beekeeping is governed by a set of golden rules which include (1) Regular requeening — requeen any failing queen as necessary. (2) Comb replacement — replace old, dark and damaged combs, particularly in the brood nest. (3) Pest and disease surveillance — regular monitoring of brood. (4) Nutrition management — monitor food intake and stored honey and pollen. Though almost all the beekeepers attend to first three problems but ignore the nutritional aspect of bees which is not less important.

The natural diet of the adult honey bee is pollen and honey. However, honey bees need a range of elements to satisfy their nutritional requirements for normal growth and development. These elements include proteins (amino acids), carbohydrates (sugars), minerals, fats/lipids (fatty acids), vitamins and water. Honey bees collect three substances viz. water, nectar and pollen to satisfy their nutritional requirements. These materials are collected by field bees according to the needs of the colony and the availability of these substances in the field. Sometimes, however, when nectar is not available, bees collect sweet-tasting juices from overripe fruit and plant exudates. Also, certain insects secrete honeydew, which bees may collect and store as honey. During periods when no pollen is available, bees may collect powdery animal feed or spores from plants and store this material as they would pollen. This may have some food value but does not sustain brood rearing and is considered a poor substitute for pollen The quality and quantity of these materials available to field bees does not always match the requirements of the colony. Lack of one or more of these substances will potentially lead to a serious reduction in the population of the colony, reduced longevity of the bees, reduction in drone populations, increased disease susceptibility and ultimately, death of the colony. A beekeeper's skill is to be able to ascertain the nutritional status of a colony, predict what the floral conditions that are immediately, and in the future, available to it and determine a course of action to ensure his/her goals are achieved. These goals may include building populations prior to a major nectar flow, maintaining populations, or even allowing populations to reduce to a more sustainable level, for instance, during a prolonged drought period or over wintering when often colonies are not foraging to any large extent. Healthy colonies can resist the disease incidence efficiently. Nectar flows, for instance, stimulate hygienic behaviour in the brood nest so dead and dying adults and brood are removed more swiftly than if there was no fresh nectar stimulus.

The nutrition status of a colony can be manipulated so as to encourage a greater ratio of foraging bees to collect pollen than nectar. Pollen gathering bees are considered to be good for as far as pollination of crops is concerned. With no fresh nectar available in the field, bees will often be reluctant to forage for pollen. Strategically providing sugar syrup may, under the right circumstances, induce the colony to collect greater volumes of pollen than they would have without the beekeeper's intervention. The specialist beekeeping industry that produces mated queen bees for the export and domestic market has to pay particular attention to nutritional issues. If the availability of pollen in the field is scarce, then one of the first reactions by a colony will be to cease rearing drone brood. Any drone eggs and larvae will be neglected or even eaten by the nurse bees, and the queen will not be stimulated to lay drone eggs. This will potentially create a major problem with inadequate drones of the correct age available when large numbers of virgin queens are ready to mate. Poorly mated queens are said to be superseded earlier than usual and often the queen fails within months without the colony replacing her, eventually equating to the ultimate demise of the colony.

Nutritional Requirements: Honey bees require proteins (amino acids), carbohydrates (sugars), lipids (fatty acids, sterols), vitamins, minerals (salts), and water, and these nutrients must be in the diet in a definite qualitative and quantitative ratio for optimum nutrition.

Proteins and Amino Acids: Adult worker bees (1 to 14 days old) obtain dietary protein from pollen which workers collect and bring back to the hive; adult drone bees (1 to 8 days old) obtain dietary protein from food supplied by young workers which is a mixture of glandular secretions, pollen, and honey; and larval and adult queens obtain their protein from royal jelly secreted by young worker bees. Royal jelly also is fed to worker larvae less than 3 days old. Royal jelly is a secretion of the hypopharyngeal glands of worker bees normally 5 to 15 days of age. It is a creamy, milky white, strongly acid substance with a moisture content of 65 to 67% and rich in protein lipids, reducing sugars, vitamins B, vitamin C and minerals.

Honey bees require specific amino acids for normal growth and development, reproduction, and brood rearing. The protein and amino acid requirements of larval and adult queens are unknown, but we have a fairly comprehensive knowledge of the chemical constitution of their basic food, royal jelly. During the first 5 or 6 days of adult life, worker bees consume large amounts of pollen to obtain the protein and amino acids required to complete their growth and development. If young adult worker bees do not consume needed proteins, their hypopharyngeal glands (brood food glands) will not develop completely, and their royal jelly will not support normal growth and development of worker larvae or egg production in the adult queen. The requirement for protein decreases when worker bees discontinue nursing (between 10th to 14th day of adult life). Subsequently, the chief dietary constituent becomes carbohydrates obtained from nectars and honey.

Carbohydrates

Carbohydrates are abundant in the natural diet of the honey bee and are used mainly for the production of energy, but may be converted to body fats and stored. Some carbohydrates can be utilized by bees, some cannot, and some are toxic. Adult bees thrive on glucose, fructose, sucrose, trehalose, maltose, and melezitose, but they cannot use rhaminose, xylose, arabinose, galactose, mannose, lactose, raffinose, dextrin, or insulin. Differences in carbohydrate utilization between larvae and adults may be due to the absence of appropriate enzymes.

Lipids: Information on the nutritional need for dietary lipids (fatty acids, sterols, and phospholipids) in honey bees is fragmentary and inconclusive. Generally, lipids are used for energy, synthesis of reserve fat and glycogen, and for the functioning of cellular membranes. The lipid composition of adult bees differs from that of pollen. However, a phospholipid found in pollen also is found in the body tissue of adult bees. Another substance, 24-methylene cholesterol, also found in pollen, is the major sterol of the body tissue of adult queen and worker bees. Possibly, certain lipids have a significant role in the lubrication of food when it is ingested and prepared for absorption. All insects studied critically were found to require a dietary sterol; therefore, it is reasonable to assume the honey bee also requires this lipid.

Vitamins: When bees begin producing royal jelly for the young larvae and the queen, they need a diet high in vitamins. Nurse bees need following vitamin B complex for brood rearing: thiamine, riboflavin, nicotinamide (niacin, nicotinic acid), pyridoxine, pantothenate (pantothenic acid), folic acid, and biotin. Pantothenic acid is needed in worker-queen differentiation and nicotinic acid, in initiating brood rearing. In addition to these vitamins, ascorbic acid (vitamin C) also seems essential for brood rearing. In general, the vitamin needs of a honey bee colony are satisfied as long as the pollen stores are abun-dant in the hive or fresh pollen is available to bees in the field. Micro-organisms naturally present in the alimenatry canal of bees may provide vitamins, and other essential substances, which may make an otherwise unsuitable diet adequate.

Minerals: Minerals required in the diet of humans and other vertebrates (sodium, potassium, calcium, magnesium, chlorine, phosphorus, iron, copper, iodine, manganese, cobalt, zinc, and nickel) are needed by some species of insects. Pollens contain all these minerals, some of which are required by bees.

Water: Water is collected by bees and used primarily as diluent for thick honey, to maintain optimum humidity within the hive, and to maintain appropriate temperatures in the brood area. The amount of water required and collected by a colony is generally correlated with the outside air temperature and relative humidity, strength of colony, and amount of brood rearing in progress.

Ingestion and Digestion: Food enters the alimentary canal by way of the mouth and passes through the esophagus to the honey stomach. In the honey stomach hydrolyzing enzymes break down the principal sucrose of nectar to the simpler monosaccharides glucose and fructose present in honey. Immediately behind the honey stomach is the proventricular value or honey stopper. It retains the nectar load in the honey stomach, controls passage of food into the midgut or ventriculus, and prevents food substances in the midgut from returning to the honey stomach.

The midgut is a relatively large segment of the alimentary canal, where food is temporarily stored and most digestion occurs. The inner wall of the midgut is lined with a peritrophic membrane, presumably to protect the cells from damage by the gut contents.

The alimentary canal is completed by a short small intestine and a large intestine or rectum that comprises the hindgut where food digestion is completed. Undigested food residues are reformed into feces in the rectum and eliminated through the terminal anus. Passage of pollen through the alimentary canal of adult bees requires about 2 1/2 hours. Feces of adult bees contain almost intact, empty pollen grain shells. The complex foods ingested by bees must be broken down (digested) into simpler units before they pass through (absorbed) the gut wall into the hemolymph (blood) for ultimate assimilation and utilization. Digestion depends on the activity of enzymes. Enzymes are present in the secretions of the salivary, postcerebral, and hypopharyngeal glands and in the secretions of the midgut epithelial cells. In addition, digestion may be facilitated by the micro-organisms present in the alimentary canal. Compound sugars must be broken down by enzymes to simple sugars before they can be absorbed and utilized.

Bees apparently do not have the enzymes or micro-organisms needed to digest the complex carbohydrates (cellulose, hemicellulose, and pectin) in the outer wall of pollen grains. Enzymes gain access to food inside punctured pollen grains and also by dissolving the "soft germinal pore areas" with digestive enzymes. Enzymes that digest protein are abundant in the alimentary canal of the adult bee and are furnished almost entirely by the midgut and hypopharyngeal glands. Proteins are first broken down to peptones and polypeptides; and these, in turn, are hydrolyzed to amino acids. The lipid-splitting enzyme lipase is abundant in the midgut of adult workers and drones. In higher animals, lipids are digested by lipase or esterases into free fatty acids and glycerol. The fatty acids are made water soluble by neutralization with alkalies in the alimentary canal. Some insects produce enzymes that hydrolyze certain phospholipids (that is, ecithin and spingomyelin), but probably digestion of the esters and fatty acids usually results from the activity of bacteria. Certain lipids may be absorbed unchanged also.

Food absorption begins in the upper portion of the large intestine and is completed in the rectum, where water salts and other organic molecules are selectively absorbed. There are two pairs of rectal glands or pads on the sides of the rectum that function in water and possibly fat absorption.

Sources and Chemical Composition of the Natural Foods

Nectar

Nectar is the major source of carbohydrate in the natural diet of honey bees. It may contain 5 to 75% soluble solids (sugars) although most nectars are in the 25 to 40% range. The primary sugars are sucrose, glucose, and fructose. As nectar is manipulated and finally stored as honey, much of the sucrose is inverted to approximately equal parts of glucose and fructose. The sucrose is converted by the natural presence and addition by bees of enzymes to levulose (fructose) and dextrose (glucose). At the same time this "ripening process" is occurring, the moisture content of the nectar is reduced to somewhere between 12–21%. Occasionally with very thin nectars in areas that are experiencing high humidity, the moisture content may remain higher than 21%, which will lead to the fermentation of the honey. Once honey is ripe it is capped with beeswax and sealed in the comb. A normal-sized honey bee colony may use the nectar equivalent of 300 to 500 pounds a year.

Pollen

Pollen is eaten by adult bees and fed (via the mixture of glandular secretions, honey, and pollen supplied by nurse bees) to worker and drone larvae after they are 3 days old. Unlike young house bees, field bees do not require pollen in their diet. Stored pollen (bee bread) is consumed by nurse bees. Under natural conditions, pollen collected by bees is usually stored on the periphery of the brood area. In a colony rearing brood, pollen placed next to a comb full of eggs is consumed in 2 or 3 days; if placed on the periphery of larvae, it is used by nurse bees within 1 or 2 days but it may be stored for much longer periods of time. A normal-sized colony may consume 100 pounds or more pollen a year. Not all pollens are nutritionally alike; bees generally collect and utilize a mixture, and many individual pollens are nutritionally inadequate. It has been demonstrated that the longevity of worker bees is greatly enhanced when they receive a diet of high protein pollen, as compared to diets dominated with pollens with low protein levels. The amount of brood rearing is also reduced when colonies are experiencing a lower protein diet. The protein content of pollens varies from 10 to 36 per cent. Some pollen contains proteins that are deficient in certain amino acids required by bees. All the amino acids listed in Table 29a, except threonine, are essential for normal growth of the young adult bee. With the exception of histidine and perhaps arginine, they cannot be synthesized by bees and must be obtained from the consumed pollens or from some other appropriate protein source. An average pollen mixture contains lipids (fats) and the following minerals: calcium, chlorine, copper, iron, magnesium, phosphorus, potassium, silicon, and sulfur. Vitamins include ascorbic acid, biotin, vitamins D and E, folic acid, mositol, nicotinic acid, pantothenic acid, pyridoxine, riboflavin, and thiamine. Amino acid content is listed in Table 29a.

Table 29a. Amino acid content of average pollen expressed as percent of crude protein

CONSTITUENT	Average pollen (Crude protein, 26.3%) Per cent
Arginine	5.3
Histidine	2.5
Isoleucine	5.1
Leucine	7.1
Lysine	6.4
Methionine	1.9
Phenyalalamine	4.1
Threomine	4.1
Trypotophane	1.4
Valine	5.8

Supplementary feeding: Honey bees can be fed various foodstuffs to supplement inadequate supplies of pollen or honey. In early spring before pollen and nectar are available or at other times of the year when these materials are in short supply, supplementary feeding may help the colony survive or make it more populous and productive. As modern land-use practices reduce dependable nectar and pollen supplies, the need for supplemental food becomes more and more urgent. Colonies provided with adequate stores in the autumn may not need supplemental foods. However, if the spring weather is unusually cold and rainy, colonies may need supplemental foods for subsistence and continued brood rearing until nectar and pollen can be collected. A sudden curtailment of food when brood-rearing activities are in progress will result in reduced bee population.

In practice, beekeepers feed their bees supplemental foods to develop and maintain colonies with optimum populations for: (1) nectar flows, (2) pollination of crops, (3) autumn and spring divisions, (4) queen and package-bee production, and (5) overwintering. Supplemental feeding may also be of value for building up colonies after pesticide damage.

Protein supplements : Numerous kinds of plant and animal products have been fed to bees in attempts to find a substitute to replace pollen in their natural diet. None has been found that is a complete replacement for natural pollen. Certain protein foodstuffs, however, will improve nutrition and ensure continued colony development in places and times of shortage of natural pollen. Wheast, soybean flour, and several brewer's yeast products-fed singly or in combination-are palatable to bees and contain the quality and quantity of proteins and amino acids, lipids, vitamins, and minerals required for growth and development of individuals and reproduction of the colony. The yeast products and soybean flour formulations presented in this publication can be fed as a dry mix or moist cake inside the hive or as a dry mix in open feeders outside the hive. Bees are unable to collect yeast in its original dry state because of its large particle size; therefore, it must be fed as moist cake inside the hive.

The choice of sugar to use in protein diets depends partly upon the cost of sugar. Equally important is to use a sugar or sugar syrup that will not cause the moist cake to become hard in a few days when exposed to the warm and comparatively dry environment of the brood nest. Sucrose (cane or beet sugar), isomerized corn syrup, and type-50 sugar syrup with protein supplements produce cakes that maintain their consistency for long periods. Cakes prepared with honey maintain their consistency for a prolonged period. However, it is comparatively expensive, and may transmit bee diseases. The addition of 10 to 12 percent pollen to a supplement fed to bees improves palatability. The addition of 25 to 30% pollen improves the quality and quantity of essential nutrients that are required by bees for vital activity.

To prepare moist pollen supplement cakes for feeding inside the hive, first dissolve the granulated sugar in the volume of water indicated in the formula. Subsequently, add the brewer's yeast, torula yeast, soybean flour, or other suggested material to the sucrose syrup and stir thoroughly. Bee-collected pollen pellets should first be dissolved in water (one-third gallon of water for each pound of pollen pellets) since they do not readily soften in sucrose syrup. The pollen water suspension is then mixed with the sucrose syrup. More water may be necessary in formulae containing pollen. Each beekeeper should experiment with the formulations to determine the amount of water necessary. In humid areas, the suggested amount of water may be excessive.

A dairy yeast (*Saccharomyces fragilis*) grown in cottage cheese whey containing 54 to 60% protein, produced by means of a multiple enzyme-conve-rsion of corn syrup and composed largely of the simple sugars glucose (dextrose) and fructose (levulose) contains 77% solids by volume, equal parts of dextrose and levulose, 1.1 kilograms of dry weight/liter, 0.5% ash, specific gravity of 1.39 and a pH of 4.8 to 5.7.

Do not use pollen from diseased colonies or from an unknown or questio-nable source. Preferably, the beekeeper should get the pollen from his own disease-free colonies to avoid possible infection from pollen. The pollen should be stored in a freezer or dried and stored in air-tight containers. Dried pollen more than 2 years old loses much of its nutritional value. Pollen can be used to replace other protein material in any formula.

After thoroughly mixing the combined materials, the final product should be of a dough like consistency. It should then be divided into 1-1/2-pound cakes and wrapped in waxed paper to prevent the loss of moisture. If cakes are not fed on the day prepared, they can be held in a cool place for several days or in a freezer for several weeks without deterioration or loss of food value. Protein supplements can be fed any time inside the hive on the top bars or in division-board feeders.

When weather permits bees to fly, the materials can be fed in the open, in trays, boxes, tubes, or other open containers. When the supplement is supplied to bees as a moist cake inside the hive, it should be in close proximity to the unsealed larvae in brood combs where nurse bees have ready access to it. A dry mix of moist formulae can be prepared by replacing the water with an equivalent amount of sucrose. Dry mix is usually fed in open feeders. To protect it from rain or dew, the feed should be placed under a roof or hive cover where it is readily accessible to the bees.

Therapeutic drugs may be incorporated in the protein or carbohydrate foods for control of bee diseases. Drugs, however, should never be fed to a colony during or 5 weeks before a major nectar flow. All medicants that are presently recommended for the prevention or treatment of brood diseases and *Nosema* can be fed in pollen and carbohydrate (dry sugar or syrup) supplements.

Carbohydrate Supplements: Proper colony management should ensure adequate honey reserves or stores in the hive at all times, but feeding sugar may sometimes be necessary. Whenever the honey supply in the colony is low and nectar in the field is in short supply, or inaccessible due to adverse weather, the colonies should be fed sugar supplement. Brood rearing requires a large amount of honey and pollen. Cane or beet sugar, isomerized corn syrup, and type-50 sugar syrup are satisfactory substitutes for honey in the natural diet of honey bees. The last two are supplied only as a liquid to bees.

Preparation and feeding: Sugar Syrup.- For spring feeding, mix one part by volume (two parts for autumn feeding) of sugar with one part water heated to 50°C to 65°C. Isomerized corn syrup or type-50 sugar syrup.---Dilute syrup with an equal volume of water. Sugar syrup can be supplied to bees inside the hive by one of the following methods:

1. Friction-top pail -- Puncture several holes in the cover and invert pails on the top bars of the frames or over the hole in the inner cover only and place an empty hive and the hive cover over all.
2. Combs within the brood chamber-- When some of the combs are empty, sugar syrup can be poured directly into the cells with a sprinkling can.
3. Division board feeder -- A container that replaces a comb in the brood nest. A plastic bag type is also available.
4. Boardman feeder -- This feeder is widely used, especially by hobby beekeepers. Its capacity is smallm however, and the syrup, being outside, tends to cool down excessively at night. In addition, the air in the feeder may expand during the day, forcing the food out of the container faster than the bees can consume it, and its exposure to other bees may stimulate robbing.

When package bee colonies are established on empty combs or comb foundation, they should be fed thick syrup (two parts sugar and one part water) for 2 or 3 weeks. Colonies also are fed sugar syrup to stimulate brood rearing for queen and package-bee production to meet early shipping date schedules by producers.

Precautionary measures if robbing is a danger

1. Feed late in the day.
2. Disturb the bees as little as possible.
3. Reduce hive entrances.

Feeding dry sugar: As an emergency measure in late winter when it is too early to feed sugar syrup, bees may be fed dry sugar by placing a pound or two on the inverted inner cover. Some beekeepers increase the feeding space by providing a wooden rim on top of the inner cover.

Preparation and feeding of sugar candy: Mix one part sugar with one part water by weight and heat this mixture until it becomes the thickness of fudge (soft-ball stage). Pour the candy mix on waxed paper and allow to harden. Feed by placing the candy on the top bars directly over the brood nest and cluster.

Supplying Bees with Water: A supply of water must be available to bees at all times. A lack of it adversely affects the nutrition, physiology, brood rearing, and normal behavior. If a natural source is not within a half mile or less, a supply should be provided. Pan or trays in which floating supports-such as wood chips, cork, or plastic sponge-are present may be satisfactory.

The beekeeper that sees that his colonies always have adequate provisions of food and water is likely to have strong, productive colonies.

Chapter 15

Two queen colony system for high-honey yields and pollination

Maximum yield of honey is possible if colonies are very populous at the start of the of the honey flow. To produce such populations, multiple queens and two queen method of management has bee advised (Farrar, 1937, 1942, 1944; 1952, 1960 Cale, 1963). Colonies are maintained with two laying queens in one hive, with the two queens kept in different brood boxes by placing queen excluders between them. This management technique is thought to increase the honey bee population and, subsequently, the honey yield. In addition, if one queen dies, the colony survives without an interruption in brood production. It has been claimed that when one requires large bee volumes to maximize honey crops then one should use a two queen system. Such thinking arises because of the widely held view that a large unit will gather more honey when compared with two units of lesser size. Over the years the idea has become widely established to join up two small colonies into one big one or deliberately developing a large special unit having two laying queens in order to obtain large crops of honey.

It has long been stated that large colonies collect more than small ones and one big one will collect more than two small ones (Matheson, 1984) as given below:

Size of colony (× 1000)	Honey yield (kg)
10	4
20	14
30	23
40	32
50	41
60	50

If a graph is drawn of the results recorded by Matheson it will be apparent that there is a linear relationship between the size of the colony and the honey yield ranging from the smallest to the largest colony with 60k bees. Evidently, the most important factor in maximizing honey yields is the prevention of swarming throughout the active season.

In two queen system, the two queens are separated by the queen excluder so that the colony has two separate brood nests with a flight for bees in each nest.

Colonies of honeybees in the same area and with the same nectar resources, if properly managed will produce surplus honey crops exceeding what could be obtained with the same resources available by introducing a second queen into an established colony of honey bees for additional buildup of the worker bee force prior to the early honey flow. The replacement of the resident queen requires that the established brood is divided into two divisions and placed in separate brood chambers positioned side by side with the resident queen in one division and an introduced queen in the other division. A hive chamber containing drawn comb or comb foundation is placed over each of the chambers containing brood to provide space for brood expansion. Worker bees can move vertically and diagonally among the hive chambers. The two queen bees can move either vertically or diagonally between two hive chambers in a manner which prevents direct contact between them during the brood buildup.

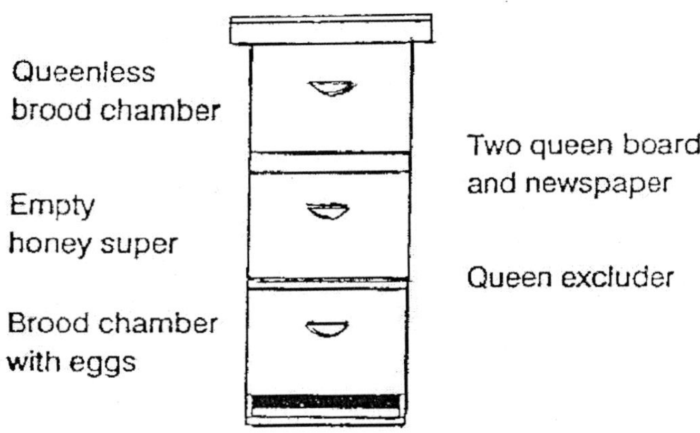

Figure 59. The two queen system of colony management.

This invention relates to the management of honey bee colonies for high honey yields, and more particularly to a method and system for two-queen operation of honey bee colonies with the eventual replacement of the resident queen by a more desirable queen.

Honey bee colonies are housed in hives which usually contain a plurality of hive chambers stacked vertically. Each hive chamber is essentially a rectangular box, without top or bottom, filled with removable frames within which the colony raises brood and stores pollen and honey. The removable frames provide a means of inspection and manipulation of the colony and of removal of honey in excess to the colony's needs. The lower most chambers, usually two, are used for raising brood and for storing honey and pollen for the colony's own use, and are known as "brood chambers." The hive chambers above the brood chambers are for the storage of honey which is in excess to the colony's needs and are known as "supers." A queen excluder, which usually is essentially a series of wires mounted parallel at a spacing which will allow the passage of worker bees but prevent the passage of queens which are larger, is usually employed between the upper most brood chamber and the lower most super to prevent brood rearing in the supers. A cover and a bottom board which provides an entrance and exit opening at the bottom of the hive assembly completes the basic hive structure. Each honey bee colony has a distinctive hive odor by which the colony recognizes its own members and intruder bees. The colony readily admits its own members but intruder bees are usually prevented from entering.

Honey bee colonies during an acitve period normally contain a single queen, worker bees, drones, and brood which is primarily worker bee brood. Two queens in the same colony almost never occur in a typical hive because if they contact each other they will contest for survival until one is destroyed. Depending on the season of the year, queen condition, brood space, colony condition, and other factors, queen brood may be present in preparation for swarming or for the replacement of a failing queen which is known as "supercedure." Modern management methods are designed to prevent natural swarming because the colony is left in a weakened condition after swarming and the swarm may be lost. Often a young, more desirable queen is introduced into an established colony to replace the resident queen before supercedure has started to insure the pedigree and control the temperment of the colony. Requeening as currently practiced requires the removal of the resident queen and gradual introduction of the replacement queen. A gradual introduction of the replacement queen is required for the queen odor and the colony odor to become unitized so that she will be accepted by the colony. This introduction period usually requires about 4 days, which is an interruption in brood rearing since the resident queen must be removed from the colony, and if done prior to a honey flow will result in several thousand fewer worker bees when they are needed most for the honey harvest. For this reason, requeening is often delayed until after the honey flow.

Two-queen operation of a honey bee colony provides a larger worker bee force for the honey flow than is provided by a single queen. The colony is reverted back to single queen operation at the appropriate time after the worker bee force is built up. Two-queen operation as currently practiced is may not be practical for commercial operations because of the large amount of labor required to effectively manage the colonies.

The method for two-queen operation of honey bee colonies involves the division of the colony population and brood into two parts with each division in separate hive chambers for purposes of introduction of the second queen. After the introduction has been accomplished, the brood chambers containing the introduced queen are placed above the brood chambers containing the resident queen and the other division of the colony population and brood. The two divisions are separated by newspapers which the bees tear away and intermingle or by a board with a screened opening to separate the bees but allow the mixing of the characteristic odors of the two divisions so that they will become indistinguishable from each other. If a board with a screened opening is used, it is removed after a certain time has elapsed. The vertical stacking of the brood chambers requires that the upper brood chambers and any supers can be removed for inspection and manipulation of the lower brood chambers. This is a difficult task since the heavy hive chambers are stacked to a height above the effective lifting height of people. It is an object of this invention to provide an efficient means of simultaneously requeening established honey bee colonies and operating the colonies as two-queen systems so as to provide larger worker bee forces for the main early honey flow than would be provided by a single queen in each colony. This invention combines the two operations, requeening and two-queen colony operation, into a single procedure which is expected to require substantially less labor than two-queen colony operation as currently practiced. Instead of the vertical stacking of the brood chambers containing the separate queens as currently practiced, this invention provides for side-by-side placement of double brood chambers, which requires less effort for effective management. Worker bees can travel vertically between either lower brood chamber and the brood chamber immediately above, and they can travel diagonally between a lower brood chamber and the brood chamber above the other lower brood chamber. The movement of the queens is limited to either diagonal movement or vertical movement between brood chambers. This control of the movement of the worker bees and the queens is provided by queen excluders and a beeway-isolator which is an element of this invention. The beeway-isolator is centered on the adjoining walls of the lower brood chambers, and they cover a portion of each lower brood chamber. When configured for diagonal movement of both worker bees and queens, the beeway-isolator provides for the isolation of the queens from each other and prevents them from contesting for survival, except when one queen is near the position where she was placed in the hive complex, and the other queen has made a complete traverse from the position where she was placed to a position in the hive chamber above the queen that has not moved significantly. If this occurs, the queens can contest for survival through the queen excluder. When the beeway-isolator is configured for the passage of worker bees only, the queens cannot physically contact each other. The beeway-isolator, in addition to providing for passage of the bees, provides for thorough intermixing of the colony odors so that neither division of the colony can develop its individual characteristic odor.

The colony is reverted back to a single queen colony at the appropripate time for the honey harvest by removing the resident queen and stacking the brood chambers in a conventional manner. If desired, the brood chambers may be stacked without removing the resident queen, in which case the two queens will contest for survival and one will be destroyed. Queen excluders are used to prevent the remaining queen from re-entering and depositing eggs in any hive chamber to be removed after the brood has emerged.

Sustainability of two-queen system

The establishment of a two-queen colony is based on the harmonious existence of two queens in a colony unit. Any system that ensures egg production of two queens in a single colony for about 2 months before the honey flow will boost honey production (Moeller 1956, 1961, 1976a, 1976b).

The population in a two-queen colony may be twice the population of a single -queen colony. Such a colony will produce more honey and produce it more efficiently than will two single-queen colonies. A two-queen colony usually enters winter with more pollen than a single-queen colony. As a result of this pollen reserve, the two-queen colony emerges in the spring with a larger population of young bees and is thus a more ideal unit for starting another two-queen system.

To operate two-queen colonies, start with strong over wintered colonies. Build them to maximum strength in early spring. Obtain young queens about 2 months before the major honey flow starts. When the queens arrive, temporarily divide the colony. Replace the old queen, most of the younger brood, and about half the population in the bottom section. Cover with an inner cover or a thin board and close the escape hole. The division containing most of the sealed and emerging brood, the new queen, and the rest of the population is placed above. The upper unit is provided with an exit hole for flight.

At least two brood chambers must be used for the bottom queen and two for the top queen. Two weeks after the new queen's introduction, remove the division board and replace it with a queen excluder. The supering is double that required for a single-queen operation, or where three standard supers are needed for a single colony, six will be needed for a two-queen colony.

When supering is required, larger populations in two-queen colonies require considerably more room at one time than is required for single-queen colonies. If a single-queen colony receives one super, a two-queen system may require two or even three empty supers at one time.

The brood chambers should be reversed to allow normal upward expansion of the brood area about every 7 to 10 days until about 4 weeks before the expected end of the flow, after which the honey crop on the colony may be so heavy as to preclude any brood nest manipulations. Thereafter, give supers as they are needed for storage of the crop. As the honey is extracted, the supers are returned to the hive to be refilled. They should never be replaced directly over the top brood nest, unless a second queen excluder is used to keep the queen out of them. The top brood nest may tend to become honey bound. If this occurs, reverse the upper and lower brood nests around the queen excluder. This puts the top honey-bound brood nest on the bottom board and the lighter brood nest with the old queen above the excluder.

There is no advantage in having a second queen when about a month of honey flow remains, because eggs laid from this time onwards will not develop into foragers before the flow has ended. However, entering the brood nest during the middle of the flow to remove one of the queens is impractical. Uniting back to a single-queen status can be done after the bulk of the honey is removed from the colony. By this time some colonies may have already disposed of one queen. When this happens, simply remove the queen excluder and operate the colony as a single-queen unit.

Improved Stock

Production of honey is one major criterion in selecting honey bee stock and breeding for improvement. Superior stock must also be reasonably gentle, not prone to excessive swarming, maintain a large but compact brood nest which winters well. It should ripen its honey rapidly, seal the cells with white wax, and use a minimum of burr comb. To obtain all the desirable characters in a superior stock, specific inbred lines from many sources must be selected and developed and then recombined into a genetically controlled hybrid. When this is done, hybrid vigor or heterosis usually results (Moeller, 1976).

Queens of common stock reared under favorable conditions and heading well-managed colonies probably will be more productive than poorly reared queens of superior stock. Queens of superior stock reared under favorable conditions will require a higher standard of management than is demanded of common stock. To realize the maximum benefits from improved stock, the beekeeper must provide unrestricted room for brood rearing, ripening of nectar, and storage of honey.

The queen breeder should produce the best queens possible to obtain the maximum benefits from improved stock and the honey producer receiving these queens should manage them in such a way that they can develop their maximum colony populations.

Merits and demerits of two queen systems are as given below:

Merits

1. Non producing colonies are eliminated
2. If one queen dies, the colony survives without an interruption in brood production.
3. Average honey yield increases from 35-60% in different seasons depending on the richness of the flora
4. Colonies enter the winter season populous and over winter successfully.

Demerits

1. Moving of two queen colony is difficult
2. Operation of this system is a time bound programme
3. Joining two small colonies is not always an option as they may be small because of failing/scrub queens or affected by disease.
4. Developing a two queen system with two fecund queens is also not a solution. It is particularly labour intensive and impracticable because of the vertical height of the unit if it is successful.
5. Furthermore, problems in managing colonies with *Varroa* is magnified even.

Chapter 16

Honeybee breeding, mass rearing of queens and artificial insemination

HONEYBEE GENETICS AND BREEDING

Information on genetics and reproductive biology is required for planned bee breeding and stock improvement programmes. It was hypothesized by Dr. Deodikar and co-workers that during the course of its trans-Himalayan migration, the primitive tetraploid Indian honey bee might have gradually differentiated into an advanced tetraploid *Apis mellifera*. All along these migratory routes, *Apis mellifera* differentiated further into a number of African, European and Sino-Japanes races. Cytogenetic studies of *Apis* spp. have shown that *A. cerana indica* has the same number of chromosomes as the European bee. The male has 16 chromosomes, which occur in 8 homomorphic pairs showing pronounced somatic association, it was inferred that the males though numerically haploid, may be closely approaching diploid condition genetically. The females show a close approximation to a tetraploid condition. This inference has been corroborated by chromosomes in *A. florea* and *A. dorsata* which have 16 chromosomes in female and 8 in the male. Reproductive biology is also worked out in Indian honeybee. The queens of the species can be instrumentally inseminated but difficulties like production of low quality semen; its separation from mucous and lesser concentration and activity of spermatozoa are encountered. Experiments on hybridization between two hive species have not proved successful. Intensive research is needed in reproductive biology and genetics of Indian honeybee and breeding of *Apis* spp. The present day *A. mellifera* stock has descended from varied and heterogeneous blood, therefore, there is a need to study about inbreeding depression with caution and verifying the facts. We need to be over cautious in importing more *A. mellifera* bees since it will run the risk of introducing diseases and enemies. Artificial queen bee insemination techniques has been standardised/practised in our country, therefore safest alternative would be to import semen and use for hybridisation as and when the need arises.

Almost everything that we eat or wear is the product of our domesticated plants and animals. Most of these living forms have been cared for and propagated by man since long before the beginning of written history. Domestication further implies bringing the growth and reproduction of the plant or animal at least partly under man's control. Man uses domesticated animals or their products for his own advantage, and he usually keeps them in or near his dwelling. From the variable wild species man has, by selection and reproduction, developed individuals that suit his purpose.

Domestic animals, such as chickens, cattle, sheep, pigs, and horses, have been selectively bred by human beings for thousands of years. Consequently, when modern breeding practices came into use, much selection had already been done. The honey bee races were not strongly controlled and bred by people and are much more variable than a breed of domestic animal.

Breeding and selection have therefore led to the establishment of types or varieties useful to man. Selection for usefulness and survival under man's care has frequently resulted in varieties incapable of survival without man's care. Man did not create these genetic forms. By controlled breeding and selection he merely succeeded in putting together or fixing these varieties from the genetic variability that existed in their wild ancestors.

The honey bee has a definite place in our modern world. Its products of honey and wax are useful to man, although perhaps not essential to all men. However, the pollination activities of bees affect the lives of most of the people of the world. Although, we may never succeed in taming the honey bee or in getting it to change its mating habits, it seems evident that the definition of domestication is sufficiently broad to include this most useful insect. The ways and means of improving the economic value of honey bees by controlling their growth and reproduction is discussed in this chapter.

The breeder of animals must decide which animals shall produce the next generation. His choice is determined by his knowledge of the variability that exists in the potential breeding population and the usefulness that he desires in the succeeding generations. The variability in the breeding population is due in part to differences in inherited factors and also in part to differences in environment. A knowledge of the science of heredity should make it easier to choose intelligently among the many animals in the potential breeding population.

Because of the social nature of the honey bee, the colony is the unit upon which selection must be based for most economically important characteristics. The individuals of a colony consist of a queen and her worker and drone offspring. The queen is mated before she begins egg laying, and her mate is thus the father of the workers in the hive. She may have more than one mate, and therefore the genetic variability of the colony may be due to several individuals. Each of these individuals can be influenced by environmental factors that are not always easy to evaluate. The bee breeders' problems in properly evaluating the genetic and environmental factors affecting each individual colony are much more difficult than those of other animal or plant breeders.

Although honey bees have been propagated by man for 5,000 years or more, there is little evidence that much progress in bee breeding has been made.

Comparable progress in the improvement of honeybees stock has not yet been achieved. In other words, honeybees appear to have been static in form for many centuries. Fossil evidence shows that honeybees attained the peak of its development about 150 million years ago, which would account for rigidity with, as a genetic type it is fixed. No superior breeds of bees have been established. The species has been subdivided into races, and all these races show considerable variability. Apparently they are the product of many generations of random matings within populations that have been partially isolated by geographic barriers. Man may have had some influence in fixing these races by selection that altered the frequency of certain inherited factors.

The aim of the bee breeder is to produce better or more profitable bees. This may mean more honey per colony or it may mean more efficient pollinators. The breeder must work with the reproductive individuals that are available. Since colonies of bees differ in many characteristics, the breeder has variability from which to select. This variability may be due to both genetic and environmental factors. The successful bee breeder must observe his colonies closely so that he can make proper allowances for environmental factors affecting the genetic variability in his breeding stocks.

His next problem is to mate the breeding individuals so as to obtain genetic improvement. From the variable colonies he attempts to unite the genes (the genetic factors) for good qualities from many stocks into one line or breed while eliminating inferior or less desirable qualities. In order to succeed he must have some knowledge of the science of heredity. Geneticists have accomplished this with many plants and animals, and therefore it is possible with bees.

Since the honey bee is an animal, it would appear that animal breeding methods should be used. The honey bee, however, is quite different from the economically important animals. Its method of reproduction is in a sense more like some plants. We shall attempt to describe these differences and likenesses and show how they affect methods of bee breeding.

The science of heredity is relatively new, having only become recognized as a science in the 20th century. Yet a great amount of real progress through selective breeding was accomplished in most of our economically important plants and animals long before Mendel, the father of genetics, was born. Most every breed of horses, cattle, hogs and chickens used in the United States today was established before the 20th century. Yet even today there are no fixed breeds of bees - we still have only races.

The fundamental biological principles of inheritance are the same whether we are breeding plants or animals. Therefore, much about bee genetics can be learned by studying the genetics of other animals and plants.

Early naturalists supposed that species were fixed and had no genetic variability. However, modern geneticists have shown that wild populations of one species of animal were not genetically uniform. They have also shown that selected breeding within these wild populations, with their existing genetic variability, can produce distinctly contrasting races or breeds within a few generations.

It seems probable that all the races of bees, as we know them today, developed from a single wild species of honey bee. In time they became dispersed over a large part of the Old World and were somewhat isolated into many small groups. There was, of course, some interbreeding between groups, but geographical barriers prevented a great many matings between groups. As a result a certain amount of inbreeding took place within each geographic group. By mutations, natural selection, and perhaps some human intervention in selection, the bees within each geographic area became different from those in other areas.

The bees from the various areas are called races. Probably the most geographically isolated race of bees is the Caucasian. A Russian author who studied this race in its native home divided it into six separate varieties. These varieties ranged in color from the banded appearance of the Italian to completely black. Other genetic characteristics within this and other races of bees are equally variable. The bees of one race are fertile when crossed with those of another race.

Why have not bee breeders done the same thing with bees? Many diverse races and strains from all areas of the world have been brought into the United States. There is thus great genetic variability in our bees. When we look around we find outstanding individuals (colonies). We have seen that successful breeders of other animals succeeded in establishing superior breeds of animals by inbreeding and selecting from among outstanding individuals and their descendants.

Bee breeders did try methods used by breeders of other animals but they did not succeed. It is true that they could not control matings so successfully as other animal breeders, but this was not the major cause of their failure. Studies indicate that a major cause for this failure was a lack of knowledge of sex determination in honey bees. Its importance to bee breeding will be explained in a later section.

The honey bee was not strongly selected by humans because basic bee reproduction was not understood until 1845. Without this understanding, very little could be done. In 1851, when this basic understanding was becoming widely accepted, Langstroth developed the movable frame hive. Suddenly beekeepers not only understood bee reproduction, they could also manipulate the hive and control the queen.

However, controlled mating was the only obstacle remaining in further development of breeding programme. Island isolation was one means, but it was of very limited value. Between 1860 and 1940, dozens of attempts were reported to induce queens and drones to mate in the confines of a jar, cage, tent, or greenhouse. Some claimed success, but the successes could not be verified or repeated. With the development of instrumental insemination as a practical technique in the 1940's, controlled bee breeding began.

Therefore, as people begin breeding bees, they enjoy the benefits of having a large and variable population with which to work. Breeders quickly discover that honey bees respond well to selection. In part, this is because human beings are just beginning to modify the bee through selection and controlled breeding.

Genetics of the Honey Bee: The family relationships within a colony of bees are different from other agriculturally important animals as a consequence of mating habits, social structure, and drones developing from unfertilized eggs. The honey bee colony found in nature is a complex family group, best described as a Superfamily. This superfamily consists of: (1) one mother queen, (2) several father drones present as sperm in a sperm storage organ (spermatheca) of the queen, and (3) the worker and drone offspring of the mother and fathers.

Within a superfamily are usually 7 to 10 subfamilies, that is a group of workers fathered by the same drone. Since all the sperm produced by a drone are genetically identical, each subfamily is composed of sisters that are more closely related than full sisters of other animals. Thus, workers belonging to the same subfamily, often called supersisters, have three-quarters of their genes in common by descent. They receive identical gametes from their father and, on the average, half-identical gametes from their mother.

Workers belonging to different subfamilies have the same mother but different fathers. They are half sisters and have one-quarter of their genes in common by descent. On occasion, brother drones mate with the same queen. In such instances, their subfamilies are related to each other as full sisters rather than half sisters. Through natural mating, such full sisters probably are uncommon.

Despite the complicated family structure, the basic principles of genetics still apply to bees. The chromosomes contain hereditary units called genes. The specific place on a chromosome where particular genes are found is called a locus. On rare occasions, a gene entering an egg or sperm has changed somewhat and will have a different effect than the original gene. The process of change is called mutation, and all the forms of a gene that might occur at a locus are called alleles.

Honey bee eggs hatch regardless of whether the are fertilized. The female bees-queens and workers-develop from fertilized eggs that contain 32 chromosomes. These 32 chromosomes consist of two sets of 16, one set from each parent. Hence female bees are said to be diploid in origin. The males (drones) develop from unfertilized egg which contain only one set of 16 chromosomes from their mother. Drones are thus haploid in origin This reproduction by the development of unfertilized eggs is called parthenogenesis

Since queens and workers have paired chromosomes, they carry two alleles for each gene, one on each member of the pair. If both alleles are of the same type, the condition is homozygous; if they are different, the condition is heterozygous. In some heterozygous circumstances, one allele will mask the expression of the other and is said to be dominant. The allele which has its expression masked is said to be recessive. Drones can carry only one type of allele because they are haploid; thus, they are called hemizygous.

At one time parthenogenesis was considered to be the basis of sex determination in bees. The theory was that a chromosome dosage effect occurred such that two sets of chromosomes resulted in females and one set resulted in males. While this is a reasonable explanation, this theory is now known to be untrue.

Research workers investigating apparent low egg viability in inbred lines discovered that sex in bees is determined by the alleles at a single locus. If an egg is a heterozygote at this locus, it will develop into a female. If it is homozygous or hemizygous, it will develop into a male.

The apparent nonviable eggs found in the inbred lines were diploid eggs homozygous at the sex locus. Worker bees selectively remove and destroy homozygous diploid larvae from the comb just after they hatch. Research efforts have been made to rear these diploid drones to maturity with the hope of producing diploid sperm and triploid queens and workers. However, artificial rearing is a difficult procedure; and the resulting diploid drones had reduced testes and produced very little sperm.

While sex determination is genetically complicated, other characteristics can be even more complicated. Different combinations of alleles at a locus result in different expressions of characteristics. Alleles at other loci also can affect a characteristic. All these different events result in complex genetic systems which produce a wide variety of character expression in bees. For example, worker bee response to isopentyl acetate (a component of alarm pheromone) was estimated to be influenced by at least seven to eight genes. This variety is some of the raw material necessary for the genetic improvement of bee stocks.

As with other animals, variety in bees is further increased by events that occur when a female (queen) produces an egg. During this time pairs of chromosomes in cells that are destined to become eggs exchange segments. Further in the process, the chromosome number of germinal eggs is halved. This process results in a haploid egg, with chromosomes having a new combination of alleles at the various loci.

The unfertilized egg will develop into a drone which will produce sperm. The processes of re-combination of alleles and reduction of chromosome number do not occur in drones. All the sperm cells produced by a drone are genetically identical. They are identical to each other, and they are identical to the chromosomes in the unfertilized egg that developed into the drone.

Communicating Pedigrees: Like other animal and plant breeders, bee breeders need a simple format for communicating ancestry and breeding plans. A pedigree format usually is a standardized diagram, simply showing a line from the father and one from the mother to one or more offspring. Because of the haploid drones, bee pedigrees are different.

To properly discuss a pedigree, two terms must be defined. These are gamete and segregation. An animal gamete is an unfertilized egg or a sperm cell containing half the chromosomes needed to produce a worker or queen. Segregation is the random sorting of paired chromosomes to produce gametes. In most animals, segregation occurs in the ovaries and in the testes. In bees, segregation occurs only in the ovaries of queens.

Therefore, in honey bees, all new gametes originate with a queen. We say "new" gametes because drones propagate only existing gametes. The drones then have two reproductive functions: first, they convert and extend the queen's female gamete (the single unfertilized egg that develops into a drone) into about 10 million identical male gametes (sperms). Second, they serve as a vehicle to move the propagated gametes to the queen (the act of mating).

In bee reproduction, then, the female progeny receive one gamete from the queen that produced the egg and the other gamete from another queen (via drone conversion of the gamete to a sperm cell). Thus, a bee pedigree contains only females. So instead of using the traditional circle to represent females and the square to represent males, the bee pedigree in figure 2 has only circles (ovals), for only females (or queens) need to be recorded.

Stock propagation and maintenance controlled mating: Some degree of controlled breeding has been practiced by queen producers for over 75 years. During that time, beekeepers had the capability of producing hundreds of queens from a selected colony rather than relying on natural supersedure or swarming. Thus, the female line was controlled.

Controlling the mating has been possible only by establishing isolated mating yards or through instrumental insemination. Isolated mating yards have two major shortcomings: (1) Absolute control of matings is difficult to achieve because a queen can mate with drones that are up to 5 miles away, and (2) one isolated mating yard is needed for every drone line used in a breeding program. Mating yards usually are not used for breeding stock, but rather for production of queens where absolute control of matings is not quite as critical. Controlled breeding through instrumental insemination has been well established since 1947 and has solved the controlled-breeding problem for bee researchers, but commercial queen producers rely primarily on natural mating.

Recently, a few commercial queen producers have tried instrumental insemination for mass production of queens. Frequently, their reason for using it has been to eliminate mating nucleus (small queen-mating colonies), rather than to make specific matings.

Instrumental insemination is, in itself, not a complicated process. Simply stated, it is a mechanical transfer of semen from the drone to the oviduct of the queen. This transfer can be made with any one of many designs of insemination stands and syringes available. All of them use carbon dioxide (CO_2) as an anesthetic to keep the queen still, use a device to hold the queen in position, and use some type of syringe to collect the semen and discharge it into the queen. Probably the most popular apparatus is that developed by Mackensen. Usually, one can become proficient at instrumental insemination, after practicing with 50 to 100 queens. Thereafter, the major problems and the major workload involve drone rearing, holding adult drones to maturity (about 2 weeks), queen storage, coordination of queen and drone production, queen introduction, and record keeping.

Instrumentally inseminated queens differ from naturally mated queens in their initial egg laying because instrumental insemination does not stimulate egg laying as does natural mating. If it were not for the carbon dioxide anesthetic given to queens during the following insemination, an instrumental inseminated queen would begin laying eggs no sooner than a virgin queen that was not allowed to leave the colony to mate. The usual procedure is to render queens unconscious with CO_2 for 10 minutes on each of the 2 days following insemination.

Germplasm Storage: Germplasm is the hereditary material that can produce new individuals. In honey bees, this includes eggs, sperm, and tissue that can potentially produce eggs or sperm. Since every breeding program needs to keep certain stock for current and future use, the problem becomes one of either storing it or continually propagating it.

At the present time, honey bee germplasm is kept primarily through propagation. Thus, germplasm is usually in the form of mated queens-their ovaries and the sperm in their spermathecae Alleles are lost gradually through inbreeding, so each generation of propagation reduces the variability of the germplasm slightly.

To avoid this loss and the labor involved in propagation, attempts have been made to store honey bee germplasm. Among the possible candidates for storage (eggs, larvae, pupae, virgin queens, sperm), sperm storage has been the most successful. Sperm stored less than 2 weeks at nonfreezing temperatures seems to be as viable as fresh sperm, but longer storage results in fewer sperm reaching the spermathecae. Although inferior inseminations result from sperm stored in liquid nitrogen (-196°C), nitrogen shows great promise for long-term storage where survival of the germplasm is the major concern.

Mutations : More than 30 specific visible mutations have been described in bees, and a number of these are maintained by research laboratories. Generally, these mutations produce a striking effect, and the majority has been easily observed by their discoverers. Many other mutations might occur in bees that also cause subtle changes yet to be observed. Known mutations affect the color, shape, and presence of eyes, the color and hairiness of bodies, the shape and size of wings, and nest-cleaning behavior.

Probably because of their distinctive appearance, most of the honey bee mutants so far collected had variations in color of eyes. Various shades of white, tan, and red have been described and about 20 still are maintained. In addition to their value as curiosities, these mutants have value as scientific tools. For example, by studying various colors of eye mutants, the biochemical pathway for the production of eye pigments in honey bees was determined.

In addition to contributing to work on eye pigment biochemistry, mutants have been used as tools to investigate a variety of other questions. Resistance to American foulbrood, mating behavior, sex determination, pollination activity, fertilization technology, sperm storage, population dynamics, longevity, and bioacoustics all have been explored with experimental designs utilizing bees identifiably different because of mutations they carry. Because of this history of usefulness and further potential applications, it is desirable for the scientific community to maintain a number of mutations. Newly discovered mutations may have special applications in science; therefore, beekeepers could help by reporting mutations they observe to a research laboratory.

Most mutations are recessive. Mutations, therefore, are often first observed in drones, for drones are haploid and do not mask recessive genes. A mutation might occur in a single drone in a colony or in many drones.

Gene pool: Across the world, bees are quite diverse. Time, mutations, and selection pressures have resulted in populations of bees called races, somewhat isolated from each other, that excel for various combinations of characteristics. These combinations of characteristics are finely tuned for survival in specific local environments. Worldwide, the races of bees form the gene pool or genetic base available to bee breeders for stock improvement.

The great virtue of our past imports is that we can breed highly desirable bees from the many varieties of stocks that we now have on the continent. Yet, some people still want to import more stock. Usually, the first thought that comes to mind when improved stock is desired is to import. It seems like a simple solution. Importing, however, should not be used as an easy substitute for a selection program. If used, importing would be only preliminary to a selection program, as an effort to expand the genetic base from which to select.

Thus, further stock imports are of questionable benefit. Such imports may result in spreading of many diseases and may themselves be undesirable, or may combine with local stocks to produce undesirable hybrids. Many past imports of "select stock" proved to be poor or even undesirable. The best way to improve bee stock, therefore, is to select from the plentiful gene pool already available.

Stock improvement: Using improved stocks of bees is an effective way to improve the productivity of a beekeeping operation. Regardless of the stock of bees used, basic operational expenses will remain much the same.

Success in improving bee stocks is a reachable goal. As we have seen, there is great variation in bee stocks available with the bee breeders. This variation is the raw material used by bee breeders. Working with the tool of selection, bee stocks can be molded to show high performance for desired characteristics.

Selection methods describing the desired stock: The first task of a bee breeder is to describe in rather specific terms what characteristics are desired in the bee stock to be produced. Most certainly a number of characteristics will be listed. Generally, desired characteristics will relate to the production needs of a group of beekeepers who are in similar localities or have similar needs. Desirable characteristics might include fast spring buildup, intensive honey production, and strong overwintering ability, disease resistance, and good handling qualities, heat tolerance and pollination activity.

A knowledgeable bee breeder will be careful to be only as specific in his stock descriptions as good information permits. Unless scientific proof is developed to the contrary, physical characteristics such as color, size of bees, and shape of wings are poor choices. Generally, if such characteristics are important, they will be selected and improved automatically along with more general characteristics such as honey production or disease resistance.

Overall, there is a need in the beekeeping industry for a number of bee stocks, each having a collection of characteristics economically important to different segments of the diverse beekeeping community. No one bee stock can possibly be universally acceptable, and attempts to produce such a stock would prove fruitless. Thus, communication between an individual bee breeder and the beekeepers using the breeder's stock is important. This communication will help the breeder decide which characteristics to emphasize in the breeding program.

Measuring superior breeding stock: Once the breeding goal has been established by describing the desired stock, choices need to be made as to how the various characteristics will be measured. Although more precise ways to evaluate colonies may be devised in the future, at present the bee breeder must choose his stock from the on-site performance of colonies established in apiaries.

Test apiaries should be established with colonies arranged in an irregular pattern, with the colonies spaced as far apart as possible. Apiary sites that have trees, shrubs, or other such landmarks are valuable. These various precautions will tend to prevent the drifting of field bees. Other management procedures also should be reasonably uniform so that all colonies have an equal opportunity to perform. Management procedures should conform reasonably well to the management procedures used with production colonies.

Test colonies will be evaluated for the various characteristics to determine which colonies will be used as breeding stock. In all cases, beekeeping judgment will be brought to bear on the evaluation. The power to select more accurately the best breeders, however, will be enhanced if each colony is given a numerical score for each characteristic being evaluated. This will require the keeping of extensive records on colonies. Obviously deficient colonies can be left out of the record keeping to ease the load. Such records are particularly important when different evaluations, such as honey production and overwintering ability, are made at different times.

Once all the colonies have been evaluated, which will take 1 to 2 years, depending on the characteristics to be improved, breeder colonies can be chosen. The scores given to a single colony for all the various characteristics can be added to provide a single numerical score for the entire colony. Such scores can then be compared to select the best colonies available. More emphasis can be put on one or another characteristic by adjusting the scores given for that characteristic. For example, honey production may be scored on a scale of 0-20, while temper may be scored on a scale of 0-10. This arrangement would be used if honey production was considered twice as important as temper.

Breeding methods line-breeding : The common method of breeding practiced by queen breeders is known as line-breeding. It can be defined as breeding and selecting within a relatively small closed population. The bee breeders' colonies constitute such a population to the extent that mismatings with drones outside their stock do not take place.

The general procedure in line-breeding is to rear queens from the best colonies. These queens are both sold as production queens and used to requeen the bee breeder's test colonies. The queens are allowed to mate with the drones present in the bee breeder's outfit at the time the queens are reared.

A number of variations can be made on this general procedure which would be of benefit. General control of drone brood in the majority of colonies, coupled with purposeful propagation of drones in a good number of more exceptional colonies, would improve the selection progress by controlling, to a limited degree, the male parentage of the stock. Of course, this procedure is used with the best success in areas where mating yards can be reasonably isolated.

In line-breeding, some inbreeding is inevitable. Its main effects are (1) fixation of characteristics so rapidly that effectiveness of selection for good qualities is reduced, (2) the stock loses vigor as a general consequence of inbreeding, and (3) the poor brood pattern from homozygous sex alleles. These effects can be lessened by using as many breeding individuals as possible for every generation.

To keep inbreeding at a minimum, one should rear queens from as large a number of outstanding queens as possible and requeen all the field colonies with equal numbers of queens from all the breeders. Each group of queen progeny is then considered a queen line and each year, after testing, at least one queen in each line is used as a grafting mother.

Despite these several precautions against in breeding, stock may begin to show a spotty brood pattern and other symptoms of inbreeding. When this occurs, new stock must be brought into the operation. At least 10 virgin queens from each of several promising stocks should be mated with drones of the declining stock and established in apiaries outside the mating range of the bee keeper's queen-mating yards. They should be evaluated there to determine which stock(s) combine best with the deteriorating stock. Once this evaluation is made, the preferred stocks can be established as new queen lines.

Hybrid breeding: When inbred lines or races of bees are crossed, the hybrid progeny often are superior to either parent for one or many traits. This phenomenon is called hybrid vigor or heterosis. Hybrid bees have more heterozygosity in their genome than do inbred or line-bred bees. This heterozygosity is thought to be the basis for hybrid vigor.

Hybrid-breeding programs in bees are considerably more complicated than line-breeding programs. At the very least, three inbred lines must be combined so that both queens and their worker daughters are hybrids. An inbred queen mated to inbred drones will produce hybrid workers. However, the egg-laying qualities of the inbred queen probably would be inadequate. Therefore, there is a need to mate hybrid queens to inbred drones so that both queens and workers in production colonies are hybrids.

Four-line hybrids also are possible and commercially available. Such a hybrid may involve lines 1, 2, 3, and 4 and could be combined in the following way: An inbred queen of line 3 artificially mated to drones of line 4 is used as a grafting queen to produce hybrid (3 x 4) queens. These are allowed to mate naturally and are used to produce drones. Queens of line 1 are then mated to drones of line 2 and hybrid virgin queens (1 x 2) are reared from the mating. Production queens are produced from a cross of virgin queens (1 / 2) mated to the drone progeny from the 3 x 4 queens. Colonies produced by this cross will be headed by two-way hybrid queens, which will be uniform in appearance, whereas the worker bees will be four-way hybrids and variable in appearance, unless the color markings of the parent lines are very similar.

Comparative tests of hybrids have shown their superiority. Increased productivity of 34 to 50 per cent over the average of line-bred strains has been reported. Segregation and random mating in the generations following hybridization are likely to result in queens that are no better than the average supersedure queen. Hybrids are an end product, and to make best use of them it is necessary to requeen every year.

Whatever the specific choice of breeding scheme, hybrid breeding requires the use of instrumental insemination and careful record keeping. As a consequence, few bee breeders have undertaken the entire operation of a hybrid program. However, many have become involved as producers of hybrid queens with the breeding stock supplied by an outside source.

Queen rearing

Drones, which are reared from the unfertilized eggs, are haploid and carry the genomes of their mother. The males and queen contribute to the heredity of the offspring workers. Under natural conditions a beekeeper cannot exercise control on parentage and drones from the vicinity can take part in fertilizing a queen. But good queens can be reared from better performing colonies so that the hereditary characters contributed by the mother can be improved to a greater extent. In an apiary many queens may be required for frequenting and colony divisions. Therefore, mass queen rearing is discussed here.

Requeening of some or all the colonies in an apiary becomes a feature of annual management. With some additional effort an element of multiplying a better stock can be introduced. All the colonies in an apiary are not equally superior performers. Beekeeper is concerned with higher honey production but honey production is directly correlated with amount of brood rearing as also the queen's egg-laying capacity, industriousness of workers, swarming and absconding tendency, frugal behaviour and disease resistance. Apiary records for these attributes are essential to select the best colonies for queen rearing. In the absence of records for all these attributes, only the honey yield records of colonies can serve an important criterion for selecting colonies for mass queen rearing. Superior colonies can also be induced to rear drones. Though the mating with these drones cannot be assured yet the chances can be increased by making these colonies to rear more drones.

Queen bee lays fertilized as well as unfertilized eggs. Both the workers and the queen develop from fertilized eggs. A larva from a fertilized egg can be reared into a worker or a queen. There seems to be no or very little genetic bias in caste determination. Worker and queen larvae are fed on glandular food for the first 2 or 3 days and there is hardly much difference in the quality of food of the 2 castes. Moreover, supply of food is more than the requirement. At about the age of three days, there is a shift in food to worker larva, whereas queen larva continues to get the nutritious glandular secretion food. Honey and pollen is mixed in the food of worker larva and also the feeding is progressive or in other words the developing larva is starved to some extent.

Larvae from fertilized egg up to the age of 72 hr can be reared into queens but the queens reared from larvae older than 72 hr may develop into intercastes. Diet is the major caste determining factor. Though the better food is responsible for higher JH titre in queen larvae which consequently affects the development of reproductive organs. Feeding of stored royal jelly is also not a perfect food for queen rearing which indicates that the royal jelly has some labile fraction.

A colony rears queens, only when certain favourable conditions are available. The bee colony must be crowded so that the bees feel the necessity of rearing queens under swarming impulse. This condition can be altered in case of queenlessness where the influence of queen or queen cell building and queen rearing is removed. For feeding queen larvae large number of young nurse bees are required because worker bees of the age of 6 to 12/13 days have well-developed and active food glands. Food glands of worker bees remain actively secretory if they get a continuous supply of pollen. Therefore, continuous income of pollen and nectar also creates favourable conditions for queen rearing. The queen rearing is possible if these prerequisites are naturally available or are artificially created by manipulations.

Queen rearing is inhibited by pheromone secreted by the mandibular glands of the queen and spread over the queen's body to be licked by the workers. Certain level of this pheromone would be required to suppress the queen rearing urge. The level of the queen pheromone available to the workers may be reduced when the colony population is very high. Such overcrowded and populous colonies rear queen cells under swarming instinct. The reduced amount of queen pheromone is available to the workers in a colony when the queen is old and is exhausting. The bees feel the necessity of superseding the old, failing queen and queen cells are raised on the face of the comb but are fewer in number. The queen can be reared from the eggs or the larvae. Failing queen, herself lays eggs in the supersedual cells raised by the workers. In case of sudden loss of queen, the queen cells are built under emergency impulse and these queens are reared from worker cells with larvae of certain age. Therefore, they are built on the face of the comb and the queens vary in development stages.

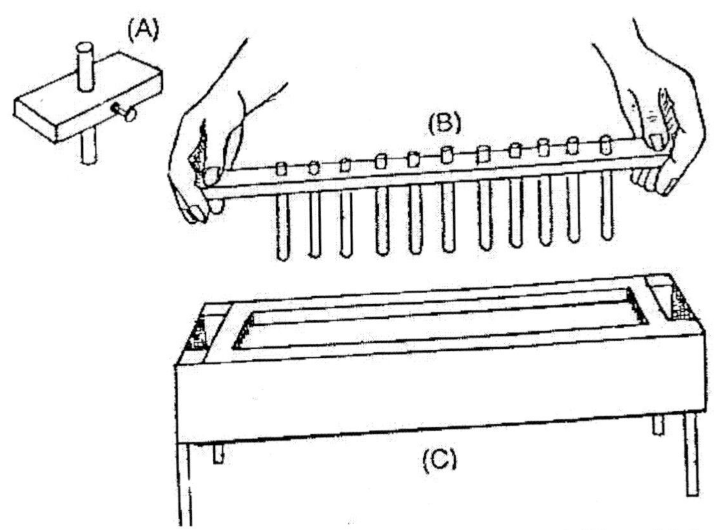

Figure 60. A former (left) for making an artificial queen cup and (B) multiple former (below) being dipped into the wax tank (C).

Queen rearing is triggered off when any of the above 3 conditions are available or artificially created. When few queens are required, a beekeeper can remove the queen from a colony and few emergency queens are raised by the bees. Modified swarm box method used for swarm control can be profitably used without removing the queen. As discussed under Damree method of swarm prevention, the queen with a brood comb, honey and pollen comb and remaining frames with empty combs or comb foundation is restricted to the lower chamber. All other frames with sealed and unsealed brood are kept in upper chamber with a queen excluder in between the two. The second super if already there can be placed in between the 2 chambers. Few queen cells are built above the queen excluder and the sealed queen cells can be carefully removed from the base and the emerging queens can be used.

Biological basis of queen rearing

The performance of the honeybee colonies depends on the inherited qualities of its queen. The queen transmits to the off springs her characteristics pertaining to longevity, industriousness, disease resistance, temperature, swarming, absconding and other behavior attributes. Thus, for successful and profitable beekeeping young freshly mated and pedigree queen bees are required because:

1. A young vigorous queen generally is more prolific and lays larger number of eggs and for longer duration as compared to the old queen
2. Moreover, colonies headed by young queens swarm less in comparison with those having old queens. In order to harvest large benefits, the

beekeepers, therefore, must maintain young and prolific queens in their colonies.

3. Besides queens also need to be replaced in case of sudden loss during manipulation, transpot1ation and due to the attack of enemies.
4. Reserve stocks of queens need to be maintained not only for routine replacements but also for rapid and commercial multiplication of honeybee colonies,
5. Young queen bees produced enmass are also required for breeding programs and for sale to diversify the beekeeping enterprise.

Selection of breeding stock

In India, the concept of bee breeding is by and large lacking. To achieve this objective, the quality queens need to be produced in large number. Generally the stock of the honeybees with the beekeepers is not of uniformly good quality from the production point of view. Individual colonies differ from one another in various attributes. Selecting best colonies for breeding is an art which depends upon experience and keen sense of observation on the part of the beekeeper. The past honey production if maintained by the beekeeper for individual colonies are helpful in making the right selection of colonies. Besides this, such factors as fecundity, longevity, industriousness, disease resistance, lack of swarming instinct etc. are also important for consideration during selection. On the basis of attributes listed below, the whole stock need to be screened and a few colonies are finally selected as breeder queens for queen cell production and stock multiplication (Laid law, 1979).

Important Attributes for Selection of Breeder Queen

Based on queen as well as its progeny/off springs.

1. Purity of queen
2. Compactness in brood rearing (spotted brood < 3.5%).
3. Surplus honey produced
4. Least swarming tendency
5. Gentleness
6. Quietness on the comb
7. Fecundity/prolificacy
8. Resistance to diseases/enemies
9. Comb raising
10. Winter/summer hardiness
11. Old queens working satisfactorily are not good for producing colonies but are indicative of other high vitality and longevity and are, hence, good for breeding daughter queens.

Conditions leading to queen cell construction

1. Swarming impulse (Impulse for natural increase)

If the population of a colony gets greatly increased, it leads to crowded conditions and insufficient availability of queen substance to the workers. Thus, swarming impulse gets dominated in the colony, and the workers start constructing the queen cells on the edges of the brood frames. The young larvae selected for queen rearing are considered to be well fed during this impulse and good quality/heavy weight queens are expected.

2. Supersedure Impulse (Impulse for replacement of aged queen bee)

This impulse is triggered off by the failure of old queen to produce sufficient quantity of queen substance. Under this impulse, a few queen cells are constructed, usually on the edges of brood area. This impulse also can not be created at will by the beekeepers for utilization in the mass production of queen cells on commercial scale.

3. Emergency impulse (Queenlessness)

Queenlessness impulse is characterized by the sudden absence of queen which deprives all the workers of the queen substance. The queen cells are, thus, constructed under emergency condition anywhere around the existing eggs/ larvae of suitable age by removing the partition wall of adjacent worker cells, thereby enlarging the cell for rearing the queen. Queen cells constructed under this impulse are not necessarily on the lower edges of the brood area. This impulse can easily be aroused simply by dcqueening the parent colony during breeding season and is utilized for commercial production of the quality queens from the selected stocks.

Prerequisites for queen rearing methods

For all the methods of mass queen rearing important operations are management of breeder colony, starter colony and cell builder or finisher colony.

Breeder colony: To get larvae from selected colony for grafting specific comb arrangement in the breeder colony is required. The hive body of the breeder colony is divided by inserting a piece of queen excluder vertically into two compartments so that one part contains six frames of honey and pollen and other part contains one frame of open brood and one frame of sealed brood. In between these two brood frames one empty, slightly dark coloured empty comb with worker cells is placed for the queen to lay eggs. The mother queen bee is confined to this three frame compartment of breeder colony. The colony is supplemented with sugar solution feeding or enhancing brood rearing by the nurse bees. After 3 days of deposition of eggs in the empty dark comb it is full of newly hatched larvae for grafting in the queen cell cups. Each time removing the larvae another empty comb is replaced for egg laying. After emergence of adult bees, that comb is replaced with another scaled brood comb to ensure the proper strength of nurse bees in the colony.

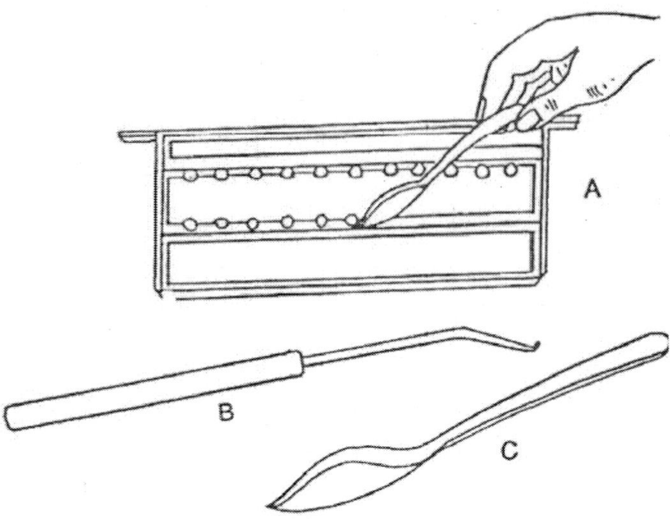

Figure 61. Artificial queen cups former fastened to a bar with wax. The bar can then be swiveled round so the cups hang vertically (B) grafting tool (C) how to bend a tea spoon to make it an efficient wax porerer.

Queen cell Starter colony

Usually this is a queen less colony with enough food stores (honey and pollen) and good number of nurse bees. This can also be prepared by uniting two colonies. This colony is regularly supplied with food, sealed and larval brood. Frame with grafted larvae is placed in between larval brood frame and pollen stores.

Cell builder or finisher colony

Cell builder colony is established 4 days prior to larval grafting by dequeening a 10 frame strong colony. The dequeened colony will raise a few queen cell from which the royal jelly is extracted to prime the artificial queen cell cups by discarding the royal larvae. Extra young bees or sealed brood frames are given to the colony to maintain a sufficient number of nurse bees. The comb arrangement in the cell builder colony is kept as suggested by Mackinley (1963)

HSEYCP

Where,

H = Honey comb

S = Sealed comb

E = Emerging brood comb

Y = Comb with young larval brood

C = Frame with queen cell cups/grafted larvae

P = Pollen comb

Cell finisher colony can also be strong double chamber queen right colony in which the queen is confined to the lower chamber by using a queen excluder and queen rearing frame is placed in the upper queenless hive body having the bee frame arrangement as described for queenless cell builder colony.

Methods for commercial queen rearing

There are several methods of mass queen rearing discussed by various authors Laidlaw and Eckert 1962; Laidlaw, 1975, 1979: Ruttner, 1983; Graham. 1992; Mishra, 1994, 1998) which are briefly described below?

1. **Miller method**: In this method comb foundation cut into deep v-shaped sections and reaching about two-third way down the frame is used/fixed. The wires except the upper one from such frames should be removed. The frame is then inserted into the middle of brood nest of the selected strong queen-right - colony). By feeding the colony the sugar syrup the bees are induced to work/raise comb cells on v-shaped comb foundation sheet for the queen to lay eggs in cells. When this has been achieved the frame is removed from the mother colony and is inserted in the centre of a cell builder colony. Cell builder colony should be very strong and be dequeened at least 24 hours before inserting the egg frame. All the eggs and young larvae should also be removed from the cell builder colony. The bees will build queen cells along the border of the given v-cut comb and after about 10 days sealed queen cells would be ready for transplanting in the desirous dequeened colonies. With this method even up to thirty queen cells are frequently produced on each comb under favorable conditions.

2. **Alley method**: The breeder queen is induced to lay eggs in the newly drawn comb. After 3 days of laying, the eggs will hatch and the comb is cut into strips of one row cells wide. The cell walls of the comb containing the larvae are shaved down to about 6 mm from the cell base. The next step is to destroy two out of every three cells alongwith their larvae (by puting a match stick in the cells) so as to provide sufficient space between the two royal cells. The strip of combs, containing young larvae, is then glued to the lower edge of a comb already trimmed in a semi-circle. This frame then given to the cell builder colony is prepared as mention in Miller methods. The workers will remodel the cell containing newly hatched larvae and raise them into queen cells.

3. **Smith method**: This method is a modification of the Alley method in which the stripes of cells containing eggs or very young larvae (1 to 2 days old) are glued by wax to the horizontal wooden bars and then the bars are fitted into frames parallel to the top bars. Such frames are then placed in the cell builder colonies.

4. **Hopkin method**: With this method a large number of queen cells are obtained. A fresh comb is placed in the brood nest of mother colony until it is filled up with eggs. Then it is removed and brought into a warm room and placed

flat on a table. Three out of every four rows of cells across the comb are destroyed upto the mid rib, leaving every fourth row intact. Two out of three eggs are destroyed in each row. This frame is then laid over the top bars of the frames in the brood chamber of cell builder colony with the prepared side facing downward. In order to provide space for bees to raise the queen cell properly, one empty wooden frame without wires is placed just beneath the laid over comb. The bees will start raising the queen cells. When the raised queen cells are sealed they are cut out with the sharp warm knife and grafted in to the dequeened colonies.

5. Deolittle or Grafting method: The most common method of queen rearing is the Doolittle or Grafting method (Doolittle, 1951). This method is commonly used by commercial queen breeders. In this method, the artificial queen cell cups are used. These cups are rinsed with royal jelly and selected larvae of the desired age (<24 h) are grafted into them. This method involves the following steps:

(i) Making the queen cell cups: For making queen cell cups, light coloured pure bees wax and a forming stick made of hard wood with the tip moulded to tile shape/size (Q-IO mm diameter) of the average queen cell are required. The wax should be melted in a water bath and held at temperature just above the melting point. Dip the forming stick into a weak solution of honey in water. Shake off the excess liquid and then dip the stick into the molten wax to a depth of 6-8 mm for a while and withdraw it and hold in air until the wax solidifies (Fig. 60). Repeat the process five to six times until a cup is obtained with a good base and tapering thin walls. Allow the wax to cool and then with a slight twist, the cup can easily be removed from the forming drip stick. The secret of obtaining successful cups is their thin walls especially at the open end.

(ii) Affixing cell cups: The next step in the process is to attach the cell cups to a cell bar made of wood, resembling the bottom bar of the frame but is suspended horizontally from about middle of the frame. This bar should be easily removable or rotatable to facilitate grafting of the larvae into attached cell cups to the wooden bar. It is, however, more convenient to attach the cell cups to flat pieces of soft wood and then these pieces are coated by dipping them in molten wax. Small wax blocks or cork pieces can also be used for affixing queen cell cups for easy removal upon sealing. The cell cup attached wood pieces; wax blocks or corks are glued to the wooden (Fig. 61).

(iii) Larval grafting: The next step in the process is the transference of the, young larvae to the cell cups. Younger larvae less than 24 hours are more easily accepted for queen rearing. Before larval grafting, the cell cups may be offered to a colony for an overnight for the bees to work on the cells and polishing in order to make them more acceptable. The larval grafting can be dry grafting (grafting the worker larvae of appropriate age in the queen cell cup without priming the cell cups with royal jelly); wet grafting (grafting the worker larvae after priming the queen cell cup with royal jelly) or double grafting (to ensure sufficient amount of royal jelly for nursing the queen. First grafted larvae are removed after 24 hours and then new young larvae of 12-24 h age are regrafted in to that cell cups). Remove the selected larvae of about 24 hours age, one at a time with a grafting needle and float them off on the royal jelly in the cell cups. Less than one day old larvae are very delicate, therefore, a great care. must be taken so that no physical damage is done to them during the transfer. While grafting the larvae, its original position should not be changed. The frame with the grafted larvae should be given to the queenless cell builder colony.

To ensure proper supply of nurse bees, queen right cell building colony consisting of two storeys with a queen excluder in between the two chambers can be used. The queen is confined in the bottom chamber and the emerging brood frames shifted to the top chamber where queen cells are being nursed and finished.

Double grafting is also practiced to ensure good quality queens. In double grafting the first grafts are removed after the cells have been started and second graft is given in the same cell cup. This is to avoid starving of the graft and the second graft will get food supply immediately.

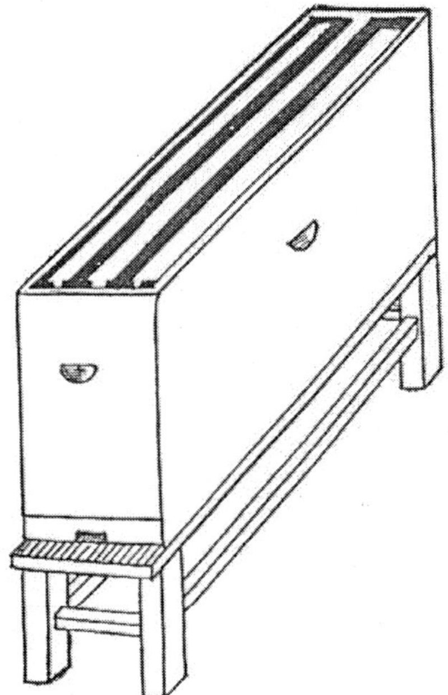

Figure 62. A three frame nucleus bee hive for queen rearing.

(iv) ***Queen cell construction***: In each method discussed in the proceeding text the main objective is to raise a large number of queen cells in the cell builder colony. The queen cells are sealed in about 10 days. Hence, it becomes imperative that the sealed cells before the emergence of the queen are transplanted to the mating nuclei or to the dequeened colonies demanding queen replacement.

7. **Karljenter Queen cell cup kit method**: Karljenter (1986) developed "Queen Cell Cup Kit Apparatus" for grafting worker larvae without touching. The queen bee is made to lay fertilized eggs in a plastic comb where comb cells are extended beneath to a removable brown plastic cell cups. After egg hatching, these plastic cell cups along with young larvae are removed with cell cup cap and

are fixed on to a cell bar of queen rearing frame with already fixed plug holder. The frame is then given to a queenless cell builder colony for nursing the gynes.

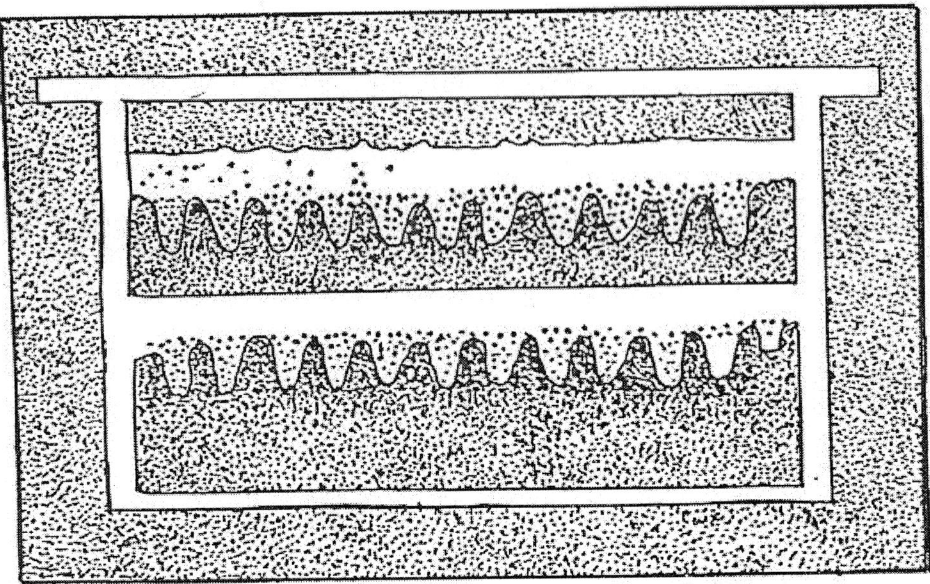

Figure 63. A grafting frame with cells steady for distribution to mating nuclei. In the lower row the fifth cell from the left is too small for use and the end cell has been rejected by the bees.

Transplanting queen cells: This is an important operation in queen rearing and the success lies in the accuracy and precision of this operation. Since a large number of queen cells are raised at a time, one should prepare in advance an equal number of nucleus colonies to which these sealed queen cells are to be transplanted. Nucleus hives/nucs should have sufficient number of worker bees of all ages, mature drones and sufficient quantity of food reserves so that the emerging young queen is attended well in the colony. Before actual transplantation of a queen cell, it must be ensured that the workers in the nucleus hives already have felt the absence of the queen atleast 24 hrs prior to the transplantation, otherwise, the chances of tearing off immature queen cells by the recipient colony workers can not be ruled out. Removing the queen cell from the cell builder colony and transplantation into the nucleus hive is an operation, which needs a good amount of experience. Individual queen cells are scrapped off with a sharp edged knife. The cells removed in this way should be trans-planted just near the brood area and it must be ensured that it is properly fixed on the raised comb in the recipient colony. The transplanted cells should be fixed in such a way and at a location (i.e. just adjacent to the brood area) on the frame that it is properly covered by the bees.

After the successful transplantation, the young queen emerges out from the cells within a couple of days. Alternatively, sealed queen cells are covered with queen emergence cages. After the queens emerge they are introduced into mating nuclei/queenless and queen may mate within a week's time. The mating nuclei should be placed in the mating yard before the emergence of queens from the cells. The emergence of the young queens should coincide with the peak period of drone population in the mating yard so that the successful mating of the virgin queen is ensured. The direction of the mating nuclei should not be changed/ disturbed till the mated queen starts oviposition

Relative advantages of different sizes of queen bee mating nuclei

1. Full frames sizes: free interchange of frames with regular hive bodies
2. Small sized frames/small nucs

a. Require lesser bees and brood for stocking/establishing nucleus

b. Easy to locate queen and feed

c. Easy to stock

Placing the mating nuclei: Entrance facing towards different directions. Keep their direction undisturbed till the mated queen starts oviposition.

Reserve stock of queen bees (Queen Bee bank)

At the end of the queen breeding and mating programme, when production of new colonies have been successfully achieved, the beekeeper should also maintain additional stock of queens. These mated queens are held in small nucleus hives or queens be caged in queen cage holding frame so that one always has a ready stock of selected and laying queens for use in emergency. A queen bank of about 5-10% of the number of the colonies in an apiary should be maintained.

Woyke (1971) found that body weight and number of ovarioles of virgin queens as a well as the number of spermatozoa in the spermatheca of naturally or instrumentally inseminated queens can be increased by rearing queens from brood grafted at the earliest possible age. This fact is of utmost importance in practical beekeeping for selection of better queens, the weight of virgin queens can be used as an index.

Artificial Mating

Artificial queen bee insemination technique is now available and mating in the open can be bypassed. There are distinct advantages of the technique. By artificial queen bee insemination method the parentage can be controlled. The virgin queens can be inseminated with the semen from desired drones and hence the technique is the only method for bee breeding but for commercial beekeeping the artificially inseminated queens are less suitable. The queen insemination can be possible at any time of the day and even in bad weather. But the disadvantage is that the operator should be experienced to achieve high success. Success of instrumental insemination also depends on the simplicity and reliability of insemination apparatus. Mackensen and Roberts were the first to develop it in

1948, which met these requirements. During the same year. Laidlaw also developed artificial insemination equipment. 'Mackensen and Roberts' apparatus is more widely used because of its simple construction and easy handling. Several improvements have since been made in this apparatus.

INSTRUMENTAL INSEMINATION OF QUEEN HONEYBEES

Apiculture at present is one of the world's significant minor industries. Even though man in most regions of the world is not dependent upon honeybees for wax and concentrated sugars as he once was, it is likely that he is becoming more dependent on them for pollination of fruit, vegetables and seed crops. This dependence springs from several agricultural practices that destroy wild pollinating insects.

Honeybees have contributed to the advancement of science in impressive ways. Published researches of Swammerdam, von Siebold, Boveri, D. Keilin, G.F White, Snodgrass and K.von Frisch have dealt with bees in one way or the other. The discovery of pathogenesis in bees by Dzierzon was an important advance in the knowledge of reproduction. Moreover, it may have been Dzierzon's counts of black and yellow drones from the unfertilized eggs of a hybrid queen that suggested the Mendel the new idea of classifying and counting the progeny of each mating. Such a technique applied by Mendel to more complicated diploid species, enabled him to make the first profound break in the study of heredity.

Although Mendel and other early investigators experimented with bees, the study of bee genetics was hampered until the technical genius of Lloyd R.Watson was brought to bear upon the problem of mating control. Watson devised artificial (instrumental) insemination for this insect in 1926. The Watson technique was utilized by other workers, and eventually greatly improved. Nolan, Laidlaw, Mackensen and Mackenzie and Roberts principally deserve great credit for improving the technique and making the apparatus and equipments commercially available. Apparently, the honey bee is still the only insect that can be inseminated instrumentally with high success.

Within recent years the technique of artificial insemination has been improved and used successfully in experiments on bee breeding. Thus a major barrier in our efforts to produce better breeds of bees has been successfully hurdled. Instrumental insemination of honey bee queen is one of the advanced technologies and essential tool for ensuring the selective mating for stock improvement (reviewed by Gatoria and Gupta, 2004). Need for the artificial/ controlled insemination in honey bees arose for controlling the parentage for stock improvement. Instrumental insemination is an essential tool for the research scientists requiring specific crosses and for the beekeepers involved in a bee breeding programme. Mating control in queen honey bee could be made possible only as per given: (i) Development of natural methods for controlled mating in isolated mating yards and (ii) Development of artificial insemination technology in honeybees through hand mating and subsequently through instrumental insemination.

History of controlled insemination in honey bees

Hand mating: Earliest attempt was made by Mclain (1887) who placed drops of semen upon vulva of a queen bee. Late efforts were made by many scientists (1917-1933) to introduce the copula Tory organs of drone into the queen's sting chamber but only Quin (1923) with the help of laid law claimed some success. The most extensive studies on manual mating were conducted by Muzalesvskii and Kozlovd (1933). They used ready to mate queens and claimed 10-15 per cent success. Subsequently, during 1950-1956, several persons experimented using different types of apparatus. However, Tryasko (1959) using modified method of hand mating, showed that only traces of sperms (less than 2% of the normal filling) were found in the spermatheca. Normal brood rearing was observed only in those cases where queens mated naturally afterwards without being noticed by the apiarists. All queens, that were not allowed to mate subsequently, produced only the drone brood.

Instrumental insemination

First attempt to inseminate the queen bee using instrumental insemination equipment was made by Huber (1738-1791) by introducing the drone semen into the vagina of the queen bee by means of a hair-brush pencil. Many other workers tried to use a syringe but without any success. The first successful attempt in instrumental insemination was that of Waston (1927) who used a microsyringe clamped in a micromanipulator. The queen was tied to a wooden block using several silken threads. The sting was opened by a pair of forceps held in hand. Later, Nolan (1932 and 1937) devised a simple apparatus for artificial insemination to introduce more semen into the vagina of the queen. Ladlaw (1944) discovered the role of the valve fold and reported that for successful insemination, the semen must be injected beyond the valve fold in the median oviduct. Laidlaw (1948) modified the apparatus by controlling the movements of needles and syringes with screws. Mackensen and Robert (1948) modified the apparatus of Nolan (1937) and obtained better results over those of latter. The former workers used two needle attached to two posts for opening with a probe and injected the semen in the median oviduct. Later, Mackensen (1948) designed a new diaphragm-type syringe. Vesely (1960) changed the shape of the tip of syringe and Ruttner (1964) that of the queen holder. Subsequently, several improvements were made to the 'standard model' of Mackensen-Roberts insemination apparatus and was tested by Ruther, Schneider Fresnaye (1974). Presently, a number of instrumental insemination equipments like Schely, Swiently, Kuhert-Laidlaw are in use.

Development in stimulating egg laying by inseminated queen bee: Laidlaw (1930) studied the utility of CO anesthesia to queen bee. Mackenson (1943-1947) discovered that two anesthesia of queen bee with CO would stimulate the queen to initiate oviposition.

Development in semen storage studies: Taber and Blum (1960) successfully demonstrated the storage of semen in vitro at temperatures above freezing point for longer timetable (1961) successfully shipped the semen between countries and continents. Poole and Taber (1970) successfully stored semen at 13-15°C for 35 weeks after mixing it with streptomycin and keeping it in flame sealed glass capillary tubes. Kumar and Gupta (2003) found that in semen stored at room temperature (23-32°C) in glass capillaries the sperm motility was 46% after 60 days of storage. Sawada and Chang (1964) reported the semen resists very low temperature but Lensky and Schinde (1967) reported that semen degenerates fast at temperatures below the freezing point. Ruttner *et al.* (1972) and Woyke (1973) demonstrated that *Apis cerana* Fab. queens may be inseminated by the same method as *Apis mellifera* Linn.

Instrumental insemination equipment

Good results in large-scale use of instrumental insemination depend upon the simplicity and reliability of the insemination equipment. The standard model of insemination apparatus consists of a heavy base supporting the two hook holding blocks with ball bearings, syringe holder and a queen holder. It is a specially designed manipulator, which facilitates both collection of drone semen and insertion of semen into the vaginal orifice of the queen. Semen is collected in the syringe; the assembly of which is shown. The parts of the instruments which have contact with the semen and saline solution must be sterilized to avoid problems with contamination. A house-hold pressure cooker is useful to sterilize the metal parts of the instruments. Plastic/rubber parts may be sterilized using alcohol.

MANAGEMENT OF QUEENS AND DRONE BEES FOR INSTRUMENTAL INSEMINATION

(I) MANAGEMENT OF QUEENS BEES

(a) Rearing and Maintaining Queens

In establishments where instrumental insemination (II) is practiced, an efficient and productive method of mass queen bee production must be adopted. Large and vigorous queens make the instrumental insemination procedure easier and only good queens justify the extra effort involved. Studies on mass queen rearing has revealed the use of artificial queen cell cups of 9.0 mm diameter and young worker larvae (<24h) from the selected breeder colony for grafting are efficient for queen rearing as compared to the older larvae (Gatoria *et al.*, 2000, 2004a; Singh *et al.*, 2001). Double grafting is often considered to produce large queens. The number of open queen cells in the cell finishing colony is an important factor in queen quality and it should never exceed 30. Queen Bees can also be reared using queen cup kit apparatus (Gatoria *et al.*, 2004 (b)).

(b) Emergence and Marking of the Queen

Sealed queen cells can be given individually be queen less nuclei but placing the sealed queen cells in an incubator at 35°C about 2 days before emergence should be preferred. A few hours after emergence, the queens are marked as per the international colour codes using coloured plastic disks or making colours white, gray, yellow, red, green and Blue (Table 29a). Clipping the distal part of one of the fore wings, i.e. left wing or right wing in alternate years, is also helpful for determining age of the queen bee.

(c) Maintaining the Queens after Emergence

Push in cage method for queen introduction is considered to be satisfactory. Maintaining queens in small cages is very laborious. Under favourable conditions of climate and nectar sources the storage for virgin for a short duration and instrumentally inseminated queens (for longer duration) is possible in special queen less nursery colonies-termed as queen banks. According to the studies conducted it has been found that up to 20 mated and laying queen bees could be kept and maintained in a strong dequeened queen bee bank colony, for over four months by frequent addition of ready to emerge brood from other colonies (Singh and Gatoria, 2003). To prevent uncontrolled mating, the virgin queens must be confined to the mating nuclei using a large piece of queen excluder. Caged queen must be inserted in the middle of the cluster between two combs. Entrance of the nuclei should be large enough so that it must not be clogged by the drone bees attempting to fly out. Virgin queens left free in nuclei should be inseminated at 5-6 days' age.

Table 29(b). International colour code for marking of queen bees

Year of queen production ending with	Colour accepted for marking queen bees
1 & 6 (say 2001 & 2006)	White or gray
1 & 6 (say 2002 & 2007)	Yellow
1 & 6 (say 2003 & 2008)	Red
1 & 6 (say 2004 & 2009)	Green
1 & 6 (say 2005 & 2010)	Blue

(d) Preparing Queens for Insemination

If a queen is old than 5 days and good flight weather prevails, she must be caught up before 1000 h or after 1600 h because if tried to catch it between 1000-1600h, she may try to escape for natural mating. After catching, queen should be placed in small cage with 5-10 attendant bees and left in the nucleus until insemination. If several queens are to be inseminated one after the other, they should be fed and provided with several worker bees and kept in a cabinet at about 25°C. Caged queens, just before insemination, can be permitted to fly against a window pane for short time to allow defecation.

(ii) Management of drones

(a) Rearing and Maintenance of Drones

Availability of sexually nature drones for instrumental insemination is also very important task. During early build up period place an empty drone comb in the brood nest of the selected colony for laying of drone eggs. Alternatively, queen bee can be confined to drone comb in an excluder cage or by using vertical queen excluder. Older drone combs in which several generations have been reared should not be used again and again. Drone rearing colony should have higher bee strength, abundant food stores, good population of nurse bees, and young worker brood on both sides of the drone brood comb.

(b) Obtaining Drones from worker Bees

When genotype of worker bees is to be exploited, about 500 young worker bees are obtained by shaking in a receptacle. A medium sized nucleus is formed on drone combs and its is fed profusely. The nucleus is then placed in an isolated location. After few days, the colony will become laying worker. The comb with drone eggs laid but without bees is then transferred to the middle of the brood nest of a queen right nurse colony with the queen prevented to reach that comb (using queen excluder cage) and lays eggs. The drones thus produced from this comb will conform to the genotype of laying workers.

(c) Maintenance of Adult Drones

Drones required for semen collection must be in apiary near to the insemination laboratory. For maintenance of these drones, following points must be considered.

1. Drone comb is placed in an incubator 2 days before emergence. At 2-3 day interval, the emerged drones are marked and released in a strong colony. Change the colony every 3 days to recognize the age of marked drone. These drones are then placed in a colony, which does not have its own drones. A drone rearing colony should not have more than 2000 drones. Making the drone nurse colony queenless will ensure better nourishment of the drones. In this colony ensuring free flight, continuous flight (by providing an overhead drone flight cage) and simulative feeding is advantageous.
2. Alternatively, the comb of the drone brood for emergence should be given to a colony or a strong nucleus made up on empty combs with shook swarm method over a queen excluder to remove its own drone bees if any. A mated queen several days before the emergence of the drones is also given. Gate of the hive is also provided with a piece of queen excluder to check the entry of alien drones.
3. If required, hold the drones with nurse bees outside the hive (in drone cages) from emergence to sexual maturity. The temperature must be more than 30°C else at low temperature, the migration of spermatozoa to seminal vesicles will be retarded. Ensure proper temperature, or migration of spermatozoa to seminal vesicles will be retarded. To ensure proper temperature, drone cages with drones can also be placed inside the colony in middle of the brood nest.

4. Store drones after emergence till utilization in drone cages/excluder cages in the colony itself. Each cage may hcld drones of one different origin marked with different colour. Supplement the young brood from other colonies around the drone cages and continuous feedings to the drones nursing colony are necessary for better nourishment of the drones. Thus, groups of drones of different genotypes can be kept in different cages in the same drone nurse colony.

5. If the drones caught have no chance to fly in the drone nursing colony, they can be permitted to fly in a drone flight cage measuring 50 x 40 x 40 for 10 minutes before semen collection. During this time the drones will defecate and after flight activity, they will evert much better and the risk of contamination of semen is also reduced.

6. For rearing drones off-season, create artificial swarming conditions by reducing the space and number of combs, liberal addition of bees from other colony, supplementary pollen and sugar feeding are also necessary. After egg laying dequeen the colony for ensuring better care of the drone brood under emergency impulse.

7. Two CO_2 treatments/anesthesia for 15-20 minutes (once 5-6 days after queen emergence, and again 2-3 days later) may stimulate the virgin queen to lay unfertilized eggs. Such drown layer queens be maintained on drone combs in strong colonies, as the drones reared in worker cells are smaller in size, producing less semen.

(d) Effect of the Flight Activity on the Sexual Maturity of Drones

The beginning of flight 7-10 days after emergence is not proof of sexual maturity of drones. Migration of mature spermatozoa into the seminal vesicle starts 6-9 days after emergence. Maximum aversion ability in drones is reached after 12th day of emergence. Thus to get mature drones for semen collection, drone rearing must be initiated at least 20 days prior to queen rearing. Reports regarding stimulatory effect of flight on ability to drone eversion are contradictory. But Mackensen (1995) found 10% increase in usable semen from free flying drones as compared to caged ones.

(e) Length of Life and Performance of Drones

Drone longevity depends upon rearing, nursing and seasonal weather conditions. Drone longevity has been recorded upto 21-70 days but proportion of long lived drones has been reported to be only 3-5%. Caging shortens the life of the drones. Eversion ability in old drones is also found between 76-95%.

(E) The Insemination Techique

Requirements

In addition to queen and drones, an insemination instrument, complete with a syringe, hooks accessories and a source of carbon dioxide for anaesthetization are required. A stereo microscope with a suitable cold-light source is also essential. The detailed list of the material required below:

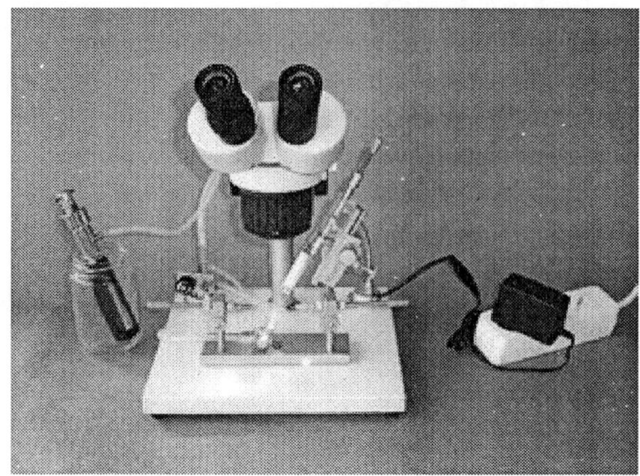

Figure 64. Schley's instrumental insemination device.

The necessary equipment consists of the insemination device, a syringe for the sperm transfer, a stereomicroscope, a cold-light-source and the anesthetizing device to immobilizing the queen bee.

Table 30. List of tools, chemicals and other material required

S.N	Items required	S.N.	Items required
1	Disinfectant (sterilium/alcohol)	15.	Incubator
2	Chloroform	16.	A pair of scissors
3	Boar sperm diluent or normal saline	17.	Forceps
4	Distilled water	18.	Stereo microscope (15-20 x) with 80-120 mm clearance
5	Towel	19.	Cold light source.
6	Cleaning paper	20.	Insemination apparatus
7	2 ml and 20m injection syringes	21	Queen marking
8	Cleaning tissue	22	Queen cage candy
9	Sterilized ear buds	23	Drone cages
10	Camel hair brush	24	Drone flight chamber.
11	Two petridishes	25	Mature drone bees (15-20 days old)
12	Two 500 ml beakers	26	Ready to mate queen bees (5-7 days old)
13	Autoclave	27	Queen introduction cages.
14	CO_2 cartridge/cylinder with regulator	28	Mating nuclei

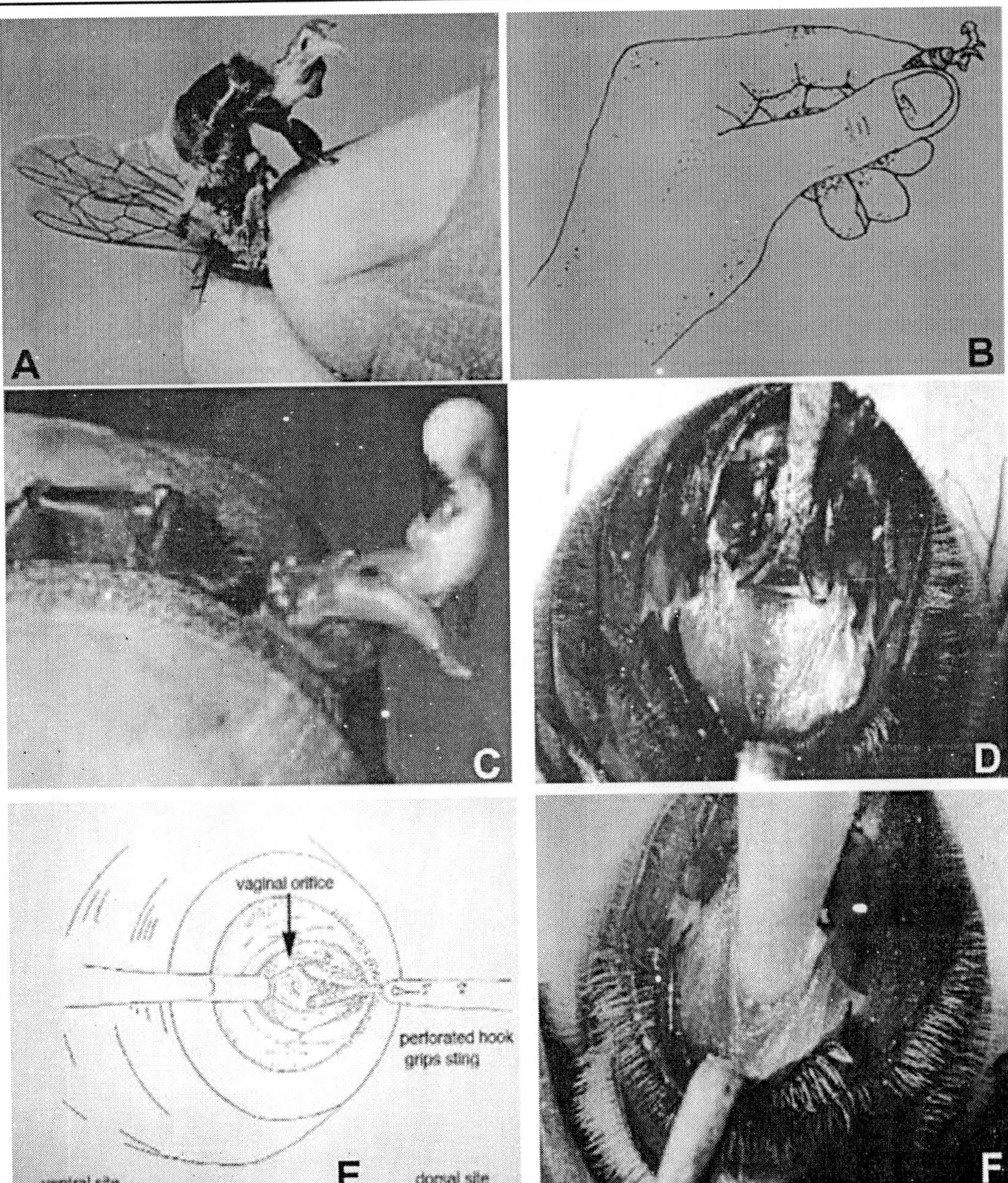

Figure 65. Pictures showing the *instrumental insemination* process (a) First, the head and thorax of the drone are crushed. (b) This causes the abdominal muscles to contract and turn part of the drone inside out (partial eversion of the endophallus). **(c) The partial eversion must be completed to expose the semen.** (d) The rest of the body is then crushed to turn the endophallus fully inside out. The semen from many drones is collected and combined. (e) Hooks are used to open the queen genetalia for insemination. **(f) The opened sting-chamber.** (g) The semen is injected into the queen genetalia (A newer syringe is slightly thinner)

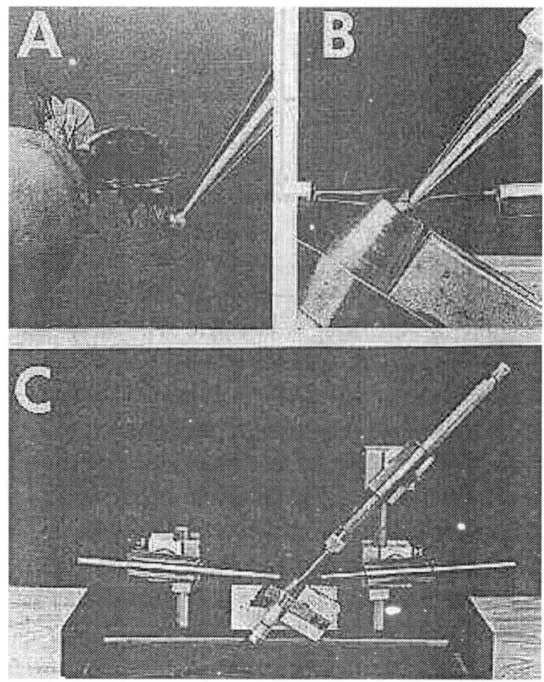

Figure 66. A shows semen collection from a drone. B shows the insemination of a queen who is in the holding tube of an instrument and anesthetized with carbon dioxide. C is a wider view of B

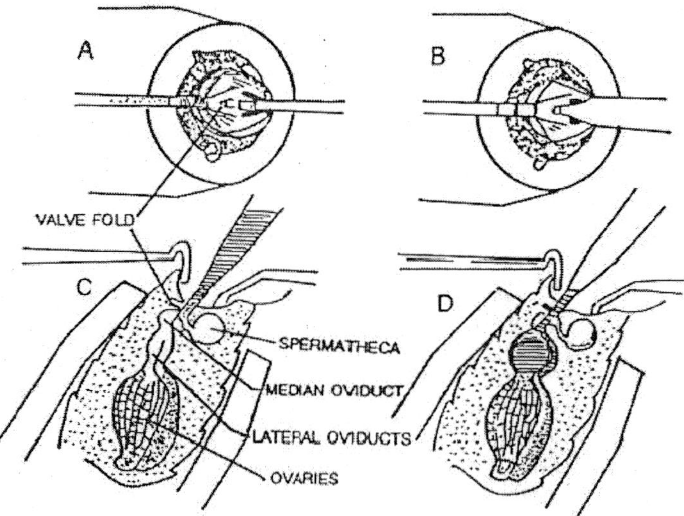

Figure 67. The insertion of the insemination tip (a) the operators view of a queen ready to be inseminated (b) the same view as (a) but showing the first stage of tip insertion. The tip has been inserted about 0.5 mm (about twice as far as the tip is wide). (c) A side view of (b). (d) final placement of the tip in the median oviduct. After the first stage of tip insertion (b and c) the tip is moved ventrally (left) about 0.5 mm to bypass the valvefold, and then inserted another 0.75 mm. the total insertion depth is about 1.25 mm. semen that was discharged from the tip has caused the lateral oviducts to swell (From Harbo,1985).

2. Filling of the syringe

The syringe is sterilized and filled with sterile saline solution. A common physiological saline solution (0.9% sodium chloride solution) can be used. Sterile water can be obtained in a glass ampoule from a scientific supply company or a pharmacy. Alternatively, distilled water run through a bacteria filter with a pore size of 0.2 mm can be used. An antibiotic is also added to prevent contamination/infection.

3. Attaching the Glass Tip to the syringe

The blunt end of the glass tip is pushed into the syringe end/sleeve and the tip is attached with the sealing tube/ squeeze-sealing. The functioning of the hydraulic system of the syringe can be easily tested with the movement of control knob.

4. Stereo Microscope

Magnification required for insemination is 15 x 20 x and a working space of 80-120 mm below the objective is required. Not all types cf microscopes are suitable to use with the insemination instrument. Hence, be sure that the microscope and instrument chosen must be compatible. A cold light source, which does not produce heat, facilitating proper illumination is necessary for insemination operation.

5. Eversion of endophallus of drone

The reproductive organs of the drone bee. To collect semen, already marked mature and healthy drones are used. Before actual collection of semen the drones are stimulated to evert their bulb of endophallus (penis) inside out so as to expose the semen. Drone eversion is obtained in two steps i.e. partial eversion and finally completer eversion to expose the semen on the bed of thick mucus on the surface bulb of endophallus.

The drone is stimulated by decapitation, which results in partial of full eversion. Alternatively partial eversion is obtained by subjecting the mature drone, caught from drone flight cage, to some chloroform fumes. Further for full eversion, grasp from anterior part of abdomen and slightly press to get semen alongwith mucus. Semen is light pinkish to cream coloured, appearing mottled or cloudy.

6. Semen collection

A small air space in the syringe is created by withdrawing some amount of saline using plunger of syringe. This will prevent dilution of semen with saline. The ejaculated drone (fully everted bulb of penis/end phallus) is brought near the syringe tip under the microscope. After making contact with semen, it is skimmed out form the more viscous mucus layer below. Avoid placing the tip of the syringe deep into the thick mucus layer. Also avoid drying of the semen at the tip by collecting a small amount of saline in the tip of the syringe between drones.

Repeat this process to collect the quantity of semen needed. Small amount of saline solution is drawn into the syringe every time when you stop semen collection to prevent sealing/clogging up of the tip by dried semen. Expel this saline solution out when you again start semen collection each queen is required to be given 8 micro liters of semen, (i.e. equivalent to 12.5 mm length of glass capillary). An experienced person can collect this much semen in just five minutes.

7. Placing the queen in the Holder

Place the queen into the cylindrical backup having a small hole on one end. When queen reaches the end of the tube with the small hole, she will backup into the holding tube which is immediately held next to this backup tube. The queen moving backward will protrude its abdomen out of the tapering end of the holding tube. The queen can be helped in this process by light blowing with mouth form the other end. The holding tube with the queens then put back on the queen holder of the instrument, which is connected to the carbon dioxide line. Flow of CO_2 is kept constant at one bubble per sec. The tube is kept in such a manner that the back of the queen is on the right side. The queen is ready for insemination when she becomes motionless.

8. Anaesthetization

Carbon dioxide through the tube connected with the queen holder is used to keep the queen anesthetized continuously during the insemination. A slow and steady rate of carbon dioxide bubbled through water adjusted at the rate of one bubble per second is generally used to keep the queen quiet during the process. The carbon dioxide treatment is also important to stimulate egg laying. Two treatments are necessary to initiate egg laying, one during the insemination procedure and a second one day before or one day after the insemination. Longer exposures should be avoided as this can shorten life expectancy of the queen bee.

9. Opening the sting Chamber

The reproductive organs of queen bee showing the position of valve fold (Figure 67). For injection of semen, the inseminating syringe must pass through the vaginal orifice and vaginal passage beyond the valve fold. As the anesthetized queen in the queen holder becomes quiet, sting chamber of the queen is opened with holding hooks. Ventral hook, by keeping to the left (ventral side of the queen, is first put in the ovipositor to stretch the ventral side (last sternite) of the abdomenting tip. The sting hook which is holed is used to thread the sting by keeping it to the right (dorsal) side of the queen bee. When the hooks are properly positioned, the tissue will stretch to form a large triangle. Within the triangle is 'v' shaped wrinkled tissues defining the vaginal orifice. Below this is the location of the valve fold, which is not readily visible.

10. Injecting the Semen

The whole process of opening of sting chamber of a queen bee and injection of semen is done under a microscope. The syringe tip having semen is passed just above and slightly right 'v' shaped wrinkled tissue i.e. the vaginal orifice'. To bye pass the value/fold, a slight zigzag motion of the syringe tip is used. This is essential to move the valve-fold for the passage of the semen. If the syringe is properly place there will not be movement of surrounding tissues when it is inserted further, in the vaginal office.

The semen can now be injected and it should not leak out if the surrounding tissue moves with the syringe tip, you have not bye-passed the valve-fold and the semen would leak out/back up.

(F) Post insemination care of the queen bee

Inseminated queen bee can be dropped between the combs while it is still anesthetized. But it is better to put it into the queen cage with candy and let the queen after insemination become active in the incubator. Upon activation, introduce the queen in the queenless colony or nucleus. The tendency of the queen for mating flight persists even after one insemination or one anesthesia with CO_2. Therefore, queen must remain confined in the colony until second insemination or anesthesia is completed. Queen guard should not be removed from the entrance until queen starts oviposition.

The first 24 hours after insemination is a critical period. Caged queens must always be near the brood nest because at lower temperature, transfer of injected spermatozoa to spermtheca is difficult. Hence, newly inseminated queens should be kept at brood nest temperature and be attended by nurse bees to facilitate sperm migration. Sperm migration is influenced by environmental conditions. Therefore, queens should not be subjected to cold or isolation.

Table 31. Number of spermatozoa in the vesicular seminales of 4 *Apis* species.

Origin of spermatozoa	numbers of spermatozoa per drone	numbers of spermatozoa adjusted to *A. mellifera* drone
A. mellifera	7.6 + 1.47 n = 10	1 drone: 7.6 + 1.5 b n = 10
A. cerana	1.0 + 0.11 n = 5	8 drones: 8.0 + 1.0 n = 10
A. dorsata	1.81 + 0.18 n = 5	5 drones: 7.8 + 0.3n = 5
A. florea	0.38 + 0.03 n = 5	20 drones: 7.1 + 1.3 n = 8

Queen bee inseminated with less than 5mm^3 semen can be maintained under the normal colony conditions. But queen bee inseminated with semen from single drone should be kept in nuclei where a lack of room restricts egg laying and thereby prolongs considerably the life and fertilized queen maintained in a strong queen bee bank/reservoir colony (Singh and Gatoria, 2003) for over four weeks.

Instrumental insemination of *Apis mellifera* vs *Apis cerana*

Woyke (1973) explored the possibility of instrumental insemination (II) in *Apis cerana* queens. He found that semen was difficult to be separated from mucous and average semen was only 0.20 mm^3 per drone as compared to 1.60 mm^3 in *Apis mellifera.* Concentration and activity iof spermatozoa is also lower in *Apis cerana* (4.655 million/ mm^3) of that found in *A. mellifera* (11 million). The penetration ability of *Apis cerana indica* spermatozoa into spermatheca is also lower than that of *A. mellifera* queens. Naturally mated *Apis cerana indica* queens had lower number of spermatozoa in the spermatheca (the highest number found 2.665 million) than *A. mellifera* queens (average 5.340 million). Nevertheless, the queens of *A.c. indica* must copulate with higher number of drones (15 in one flight or 30 in two) than do *A. mellifera* queens (average 8 matings in one flight). Woyke (1973) further found that artificial insemination of *A. cerana indica* queens with 1 mm^3 to 4 mm^3 of semen resulted in 411 thousands to 1,121 thousands of spermatozoa in their spermatheca. Queens inseminated with doses higher than 3 mm^3 of semen produced at the end of the season exclusively worker brood. Two or three inseminations with 3 mm^3 of semen each are recommended. For this purpose, 100 to 150 drones should be tried, out of which 40 to 60 may give the required quantity of semen for effective insemination of *A. cerana* queen. Comparative account of spermatozoa and their motility are given in Table 31 and 32. Taking *Apis mellifera* as a reference, spermatozoa from 8 drone of A. *cerana*, 5 drones of *A.dorsata* and 20 drone of *A. florea* is equivalent to one drone of *Apis mellifera* (Table 31). Similarly, motility of spermatozoa stored in *Apis mellifera* is more followed by *A. cerana, A. florea* and *A. dorsata*, respectively (Table 32). Crosses between the two species *Apis mellifera* and *A.cerana* have also been attempted but could not be successful.

Table 32. Number and motility of spermatozoa of 4 species stored in the spermatheca of an *A. mellifera* queen.

origin of spermatozoa	sperm numbers	motility after 3 days	motility after 4 weeks
A. mellifera	0.16 + 0.05	98.5 + 3.2%	96.9 + 4.6%
	n = 21	n = 13	n = 8
A. cerana	0.22 + 0.07	97.5 + 5.4%	93.8 + 5.4%
	n =16	n = 8	n = 8
A. dorsata	0.11 + 0.04	61.2 + 37.8%	26.0 + 37.2%
	n = 10	n = 5	n = 5
A. florea	0.15 + 0.06	83.4 + 22.5%	33.9 + 38.9%
	n = 16	n = 8	n = 8

Conclusions

It is true that II is indispensable in modern apiculture, but there are some problems associated with this process. Some times there is considerable loss of semen during insemination which results in the queen losing, rapidly her fertility becoming sooner or later a drone layer/breeder. In natural mating, the semen is introduced into the vagina and median and lateral oviducts from whence it is taken up into the spermatheca. Loss of the semen is avoided because of the closure of the vaginal orifice by a plug, or the mucous and the tip of the drone genetalia, the so called "mating sign". Once the semen is absorbed into the spermatheca the plug is removed by the worker bees and the queen subsequently commences egg laying. Furthermore, there is risk of queen survival. Queen loss after insemination can be a serious problem. Probably, the two most common causes are (1) a semen residue remaining in the oviducts for more than 1 or 2 days after insemination and (2) worker aggression when II queens are first introduced to colonies. Queen loss when first introducing an II queen to a colony is common because a young II queen that has never laid eggs is not accepted by the workers as readily as a laying queen. Evidently, there is no quick or easy way of producing improved honeybee. Indeed it would seem that nature is what rebellious so far as the honeybee is concerned and in view of this it is doubtful, if the production of so called "improved bee" is at all possible. Nature would seem to take the view that the honeybee as we have it today is perfect as a type; in other words, its development has reached the ultimate. In practice, it is found that performance between colonies of bees varies so first step in the improvement of bee stock is a critical evaluation of individual colony performance. The main criterion in evaluation is production level. Clearly, a type of bee that gives consistently higher yields, relative characters as vigour, longevity, nectar gathering ability, ease in taking to supers, ability to draw combs, resistance to disease etc. therefore, continual systematic selection together with maintenance of colony vigour is the only course open to the practical beekeeper.

Chapter 17

The use of honeybees for pollination

Flowers are the reproductive parts of a plant, where seeds are formed. For the cycle to begin, a pollen grain, which is often carried on an insect or by the wind, comes in contact with the stigma (or female part) of the same flower species. Fertilization takes place when pollen from the anther (male organ) unites with a female ovule, which forms the seed and fruit. The transfer of pollen from male to female sex organs is called **pollination.** All plants must be pollinated before seed (or fruit) will set. Pollen is transferred from the anthers to the stigma by wind, water, gravity, mammals, birds, humans and insects. If the transfer takes place on the same blossom or on another blossom on the same plant, it is called self-pollination.

Cross pollinated flowers must have transfer of pollen from one flower to the next. Even self-pollinated plants may benefit from cross-pollination with hybrid vigor, or more rapid and complete growth. Male sterility has been found in many crops and plant breeders exploit this characteristic to develop hybrids which are then dependent on cross pollination. Most flowers secrete nectar, a sugary, scented liquid to help attract insects to the flower. This is usually offered deep within the flower near the base where petals originate from around the ovary. In seeking nectar or gathering pollen, the insect accidentally brushes against anthers so that pollen grains are transferred to the stigma. Pollination is not a deliberate behavior but an accidental one performed by the insect as they collect food.

Pollination is an important step in the reproduction of seed plants: the transfer of pollen grains (male gametes) to the plant carpel, the structure that contains the ovule (female gamete). The receptive part of the carpel is called a *stigma* in the flowers of angiosperms and a *micropyle* in gymnosperms. The study of pollination brings together many disciplines, such as botany, horticulture, entomology, and ecology. Pollination is important in horticulture because most plant fruits will not develop if the ovules are not fertilized i.e. the transfer of pollen from the male part of the flower, the anthers, to the receptive female part, the stigma is necessary. The story may be summed up as getting the right pollen to the right place at the right time. In some plant species pollination is achieved by pollen grains carried in the wind. Most fruit species require some insect to carry pollen to the flowers. **Fertilisation** occurs when the pollen grains on the stigma germinate and grow down the stem of the stigma (the style). The sperms of the pollen unite with the ovules in the ovary of the flower and subsequently produce seed. Flowers may be fully pollinated but not necessarily fertilised because they may have received incompatible pollen. The pollen does not germinate or grow on the stigma or reach and fertilise the ovules. In such a case, unless the species is parthenocarpic, no fruit will result. Not all plants require pollination and fertilisation, as some are able to produce fruit parthenocarpically; that is, the fruit will develop without fertilisation of the flower and production of seed.

Two types of pollination may occur: self - and cross pollination.

1. Self pollination

It is the transfer of pollen from anther to the stigma of same flower or to the flowers of the same or other plants of the identical genetic material such as apple flowers.

2. Auto – pollination

In some plants pollen is automatically deposited on the stigma when the flower opens and this type of self pollination may be called auto pollination.

3. Cross- pollination

It is the transfer of pollen between plants which are not of identical genetic material i.e. pollen gametes are not genetically identical to the ovule gametes e.g. cross variety pollination of apples, cross inbred lines pollination for hybrid seed production. The most important and efficient carrier of pollen from anther to stigma is the honeybee. An understanding of the activities of honeybees is essential if the best return is to be obtained in fruit production.

4. Self fruitful

Self fruitful are capable of setting fruit with its own pollen. The pollen in such cases is heavy, sticky and for commercial crop production, it must be moved from flower to flower by bees.

5. Self unfruitful or self incompatible

Varieties which do not set commercial fruit/seed when self pollinated are called self unfruitful or self incompatible. Most apple varieties are in this category. There is a need to plant other varieties with compatible pollen so that cross variety pollination takes place.

6. Cross fruitful varieties

Varieties which set a commercial crop when properly interplanted for cross pollination. They have compatible pollen.

7. Cross unfruitful or incompatible varieties

This condition usually refers to a physiological incompatibility. The pollen may not send down a pollen tube to deliver the male gamete to the ovule or the embryo may abort after union of gametes. Some varieties have sterile (non viable) pollen. As a rule physiological incompatibilities exist between varieties with a close genetic relationships. There are varying degrees of incompatibilities which may be altered by different environmental conditions.

8. Monoecious plants

Monoecious plants are those plants in which staminate (male) flowers and the pistillate (female) flowers are separate but on the same plant e.g maize, cucumbers etc.

9. Dioecious plants

Dioecious plants are those plants in which staminate and the pistillate flowers are on different plants e.g. willow, poplars.

10. Gynoecious plants

The term gynoecious F_1 hybrids has been applied to certain cultivars of picking cucumbers and they are often referred to as "gynoecious" varieties. The cultivars have predominantly pistillate flowers.

Many of the fruit crops, legumes, vegetables, and oilseed crops depend upon pollinating insects for set of fruit or seeds. Inspite of the best agronomic practices-proper seed bed preparation, fertilization, soil moisture, cultivation, pest control, and harvesting methods bountiful harvests are not obtained if pollination is neglected.

Flower structure

In order to understand the entire pollination process, the knowledge of the flower structure and fruit setting is necessary. The flower has a simple basic pattern, but with seemingly infinite variations. Typically, the flower (Figure 68) is composed of the *sexual organs*, protected by delicate colorful *petals* that form a tube or crown like *corolla,* and which in turn are supported and partially protected by the usually green, more durable *sepals*, collectively called the *calyx.* The calyx and corolla combined are referred to as the *perianth.* There may be leaf like *bracts* just below the sepals.

The *male* part (or *androecium*) of the sexual organs are the *stamens,* which consist of the hair like *filaments* bearing the pollen-producing anthers on the extremities. At the appropriate time, these anthers *dehisce* or split open and disgorge the male element, the numerous microscopic and usually yellow grains of *pollen.* Evidently, there are four series of floral parts, usually arranged in whorls, each with its own set of functions.

The **sepals** (collectively the **calyx**) is the lowermost whorl functions to protect the *entire flower in the bud* as its other parts are developing. As the flower opens the calyx often folds out of the way.

The **petals** (collectively the **corolla**) constitute the second whorl of parts of a flower (painted red). Their primary function is to attract animal pollinators via a visual signal, such as color, pattern, and shape. Neither the calyx nor corollas are involved directly in production of gametes; however, their roles are critical to the success of the reproductive process in flowering plants.

The **stamens** constitute the third whorl of floral parts. Each generally consists of a stalk-like filament tipped by an anther. The anthers are is the structures in which the pollen grains are produced.

The female part of the flower (the **pistil**) consist of three parts: The **ovary**, the chamber that *contains* the **ovules/seeds**; the **style**, the tissue through which the **pollen tube cell** grows to the vicinity of the ovules; and the **stigma**, the receptive tissue that recognizes legitimate pollen and promotes the germination of the pollen grain.

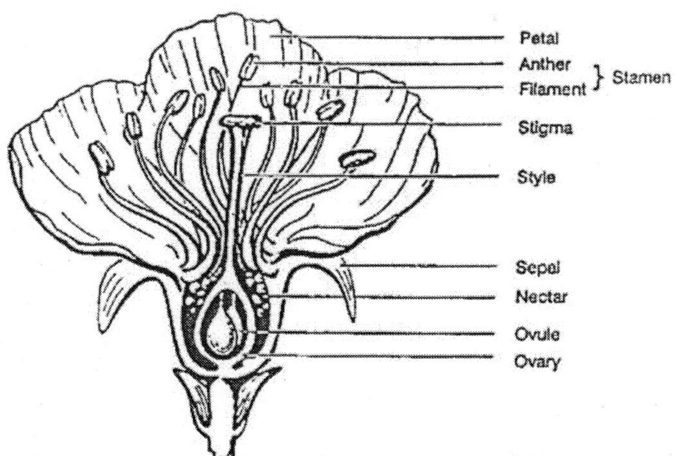

Figure 68. Typical structure of a generalized flower.

Nectaries and nectar secretion

Flowers frequently have one or more *nectarines.* Nectaries vary in size from microscopic to the 11-inch nectary of the orchid (*Angraecum sesquipedale* Thou.). The nectary is most often located within the flower, usually at the base of the sexual column inside the circle of petals. In cotton, however, there is a nectariferous ring just outside the base of the petals on the inner base of the calyx. Nectaries are also found outside the flower, on the stem or leaves. Nectar secretion within the flower usually starts about the time the flower opens and ceases soon after fertilization. Secretion of nectar on the stems and leaves is not influenced directly by flowering and may continue for several weeks.

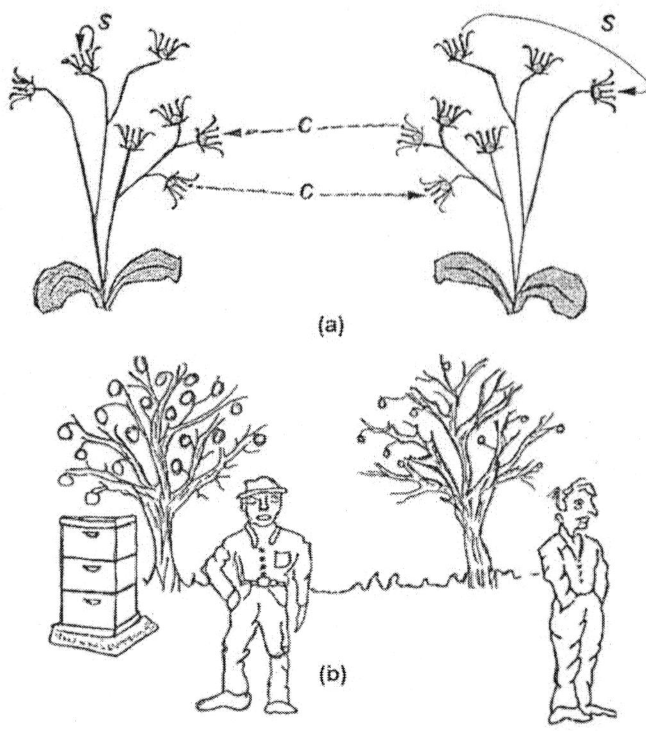

Figure 69 (a) Showing three methods of pollen transfer (b) model exhibiting impact of pollination on fruit set

The amount of nectar secreted varies from infinitesimal in numerous species to more than an ounce in the orchid *Coryanthes* spp. and in *Protea mellifera* Thunb, which natives in Africa reportedly remove and drink. Nichol (1952) reported that the nectar of the *Agave parryi* Engelm. flower stalk was gathered by Indians in the Southwest and used as a syrup. Numerous bee specialists have calculated the amount of nectar produced in the flowers of various crops. For example, McGregor and Todd (1952) calculated that the cantaloupe flowers on 1 acre produced 1.7 pounds of nectar in 1 day, whereas alfalfa flowers on 1 acre produced 238 pounds in 1 day.

Pollination and Fertilization

Pollination is usually considered in the combined sense of transfer of pollen and set of fruit or seed. In fact, it is not so, for this purpose two sets of factors are involved: (1) transfer of viable pollen from the anther to a receptive stigma, and (2) sprouting of the pollen grain and growth of the pollen tube down the style into the ovary, and ultimately the union of male nuclei of the pollen grain with female germ cells in the ovule that results in seed development. Pollination is of no value without fertilization.A plant may be spoken of as *self-fertile* or *self-compatible* if it can produce fruit without the need for the transfer of pollen to it from another cultivar. If the plant is not receptive to its own pollen, it is *self-sterile*. Even self-pollinating plants are frequently benefited by *cross-pollination*, the transfer of pollen from one flower to another. They may also benefit from having the pollen more thoroughly transferred and distributed over the stigma at the most receptive period. A plant is *cross-compatible* if it can normally be pollinated with pollen of another cultivar, but it is *cross-incompatible* if it is not receptive to pollen of certain cultivars.

Stigma receptivity

When the stigma is receptive to pollen, it is coated with a colorless, relatively tasteless stigmatic fluid. If viable, compatible pollen comes in contact with this moist stigma, it adheres, germinates, and sends a pollen tube bearing the tube nucleus and the two sperm nuclei down through the style into the ovary and, finally, into one of the ovules. *Fertilization* follows this *pollination* process by the sexual union of one of the two sperm nuclei of the pollen grain and the egg nucleus of the ovule to form the fertilized egg or zygote. Through this process of sexual union, a viable seed is formed that is capable of producing another complete plant.

In general, the sooner pollination can occur after a flower opens, the greater the likelihood that fertilization of the ovule and seed development will occur. As time elapses, the pollen may be lost to insect foragers, wind, gravity, or damage by heat, moisture, or drying out. Also, processes may set in that result in the shedding of the fruit.

Pollination also requires consideration of pollenizers. (The terms "pollinator" and "pollenizer" are often confused: a ***pollinator*** is the agent that moves the pollen, whether it be wind, bees, bats, moths, or birds; a ***pollenizer*** is the plant that provides the pollen). Some plants are **self-fertile** or **self-compatible** and can pollinate themselves. Other plants have chemical or physical barriers to self-pollination and need to be cross-pollinated: with these **self-infertile** plants, not only pollinators must be considered but pollenizers as well. In pollination management, a good pollenizer is a plant that provides compatible, viable and plentiful pollen and blooms at the same time as the plant that is to be pollinated.

Pollination and pollenizer

Pollination can be **cross-pollination** with a pollinator and an external pollenizer, **self-pollenization** with a pollinator, or **self-pollination** without any pollinator:

Cross-pollination (***syngamy***): pollen is delivered to a flower of a different plant. Plants adapted to outcross or cross-pollinise have taller stamens than carpels to better spread pollen to other flowers.

Self-pollenization (***autogamy***): pollen moves to the female part of the same flower, or to another flower on the same individual plant. This is sometimes referred to as self-pollination, but this is not synonymous with autogamy. Clarity requires that the term "self-pollination" be restricted to those plants that accomplish pollination without an external pollinator (example: the stamens actually grow into contact with the pistil to transfer the pollen). Most peach varieties are autogamous, but not truly self-pollinated, as it is generally an insect pollinator that moves the pollen from anther to stigma. Plants adapted to self-fertilize have similar stamen and carpel length.

Cleistogamy: pollination that occurs before the flower opens is always self-pollination. Some *cleistogamous* flowers never open, in contrast to ***chasmogamous*** flowers that open and are then pollinated. Cleistogamous flowers must of necessity be self-compatible or self-fertile plants. Other plants are self-incompatible. These are end points on a continuum, not absolute points.

Hybridization is effective pollination between flowers of different species of the same genus, or even between flowers of different genera (as in the case of several orchids).

Pollinating Agents

The process of pollination requires pollinators as agents that carry or move the pollen grains from the anther to the receptive part of the carpel. Faegri and van der Pijl (1966), in their classic book on pollination ecology, defined two major types of pollen dispersal: *Biotic pollination* in which the pollen dispersal agent is an animal (i.e., either an invertebrate or a vertebrate); and *abiotic pollination* where pollen is dispersed by an inanimate physical agent, such as wind or water. The various flower traits that attract different pollinators are known as pollination syndromes. Methods of pollination, with common pollinators or plants, are:

Biotic pollination

Invertebrates

- Beetles : Cantharophily
- Flies : Myophily
- Hymenopterans
- Wasps : Sphecophily
- Bees : Melittophily
- Ants : Myrmecophily
- Lepidopterans
- Butterflies : Psychophily
- Moths : Phalaenophily

Vertebrates

- Birds: Ornithophily
- Mammals: Therophily
- Marsupials: Metatherophily

- Placentals: Eutherophily
- Bats: Chiropterophily
- Rodents: Sminthophily

Abiotic pollination

- Self-pollination: Autogamy + Cleistogamy
- Wind-pollination: Anemophily
- Water-pollination: Hyphydrophily
- Gravity pollination

Historical background

The process of coevolution between flowering plants and pollinators has been proceeding for 225 million years [Leppik 1960; Anonymous 2003b]. It has been established from fossil record that insects were potentially effective pollinators when first flower appeared, which was small (5 mm) [Leppik 1960]. Gradual increase in flower size (10-12 cm) resulted about 140 million years ago [Smart and Hughes 1973]. Aristotle had the idea of relationship between bees and flowers but detailed floral structure and role of insect pollinators was described by Koelleuter Dobb in 1870 [Verma, 1990] and explained by Sprengel's theory [Sprengel, 1793], i.e. "every peculiarity of plant anatomy and physiology is related to the peculiarity of the structure and behaviour of the insects which visit and pollinate the flowers". Work of Darwin [1887] is helpful for understanding of plant perpetuation and vigour maintenance through cross-pollination.

There are about 250,000 species of angiosperms out of which 70% of plants rely on insect pollinators and 30% of our food comes from insect pollinated plants [Anonymous 2003a]. Angiosperms' relationship with insects as pollinators and birds as seed dispersers is source of its dominance [Regal, 1977, Anonymous, 2003a]. Grasses are also visited by many insects in tropical regions [Soderstrom and Calderon, 1971], which elsewhere are anemophilous. Insects are important transporters of pollens [Faegri and van der Pijil, 1978] and spores of some mosses [Erlanson, 1930] and fungi, especially rust fungi [Alexopoulos, 1952], having mutualistic relationship with each other, while birds are seed dispersers [Regal, 1977]. Insect pollinators belonging to important insect orders [Proctor and Yeo, 1973] depend upon floral characteristics. Important orders of anthophilous insects (feeding on flowers) are Collembolla and Arachnida [Kevan *et al.,* 1975], Coleoptera [Muller, 1883, Proctor and Yeo, 1973], Diptera [Thien, 1980], Lepidoptera [Kendall, 1981] and the most important one, Hymenoptera [Spradbury 1973, Keeler, 1981]. Role of honeybees as efficient pollinators of different cultivated crop plants is discussed by several authors [Kozin, 1976, Crane and Walker, 1984] because they can be managed in sufficient number and show flower constancy [Yazdani and Agarwal, 1997] especially in apple [Dulta and Verma, 1987, Verma, 1987].

Insect pollination permits a plant species to outcross with others of the same species [Dressler, 1968; Heinrich and Raven, 1972]. These pollinators are also responsible for long distance seed dispersal and ultimately reduced seed predation and vulnerability to epidemic diseases [Regal, 1977], high evolutionary rate [Levin and Wilson, 1976] and increase in seed or fruit yield [Deodikar and Suryanarayana, 1977]. The study of pollination ecology requires a detailed understanding of floral characteristics, pollinator's characteristics and environmental factors

About 80% of all plant pollination is biotic. Of the 20% of abiotically pollinated species, 98% is by wind and 2% by water. Flowers have evolved to effect pollination. Different flowers are adapted to suit a particular pollinator. There are just a limited number of pollination mechanisms that have evolved because there are just a limited number of ways pollen can be successfully transferred from one flower to another.

Table 33. Summary of flower types based on pollination mechanisms

Pollinator class	Colour	Scent	Nectar tube	Toughness
Insects	Yes	Yes	Yes	Not tough
Bees	Blue, yellow	Sweet smell	Nose size or Long body width	Not tough
Flies	White cream	Musty or off smell	Open cups	Not tough
Butterflies	Bright colours no	No - may stink	Long narrow	Not tough no Landing platform
Moths	White to show at night	Strong sweet	Long narrow	Not tough no landing platform
Birds	Red or green or yellow	No	Long wider	Tough, leathery, Plenty of nectar flowers clustered
Wind	no	No	No	Not tough Big anthers, Plenty of pollen. Stigmas feathery to catch the pollen.

Abiotic vectors

Wind: Wind pollination is the dominant type of abiotic pollination and is especially prevalent in several plant families, including the grasses (Poaceae) and sedges (Cyperaceae). Most gymnosperms (including pines, firs, spruces, etc.) are also wind-pollinated. Abiotic pollination is, with few exceptions, a wasteful process. The plant must release as much pollen as possible, given the low probability that wind or other physical forces can effect transfer of grains of that pollen to a receptive stigma.

To this end, anemophilous plants have morphological adaptations that increase pollen dispersal and capture. Modifications of the flowers include a highly reduced perianth (no need to be showy) and green to dark brown to reddish bracts and perianth. It has been suggested that dark red is important for the temperature conditions of the blossom, especially of the bud. The flowers of anemophilous species are often found above or outside the leaves which may increase their access to wind currents. Some species flower before the leaves are out, which allows maximum wind speed around the flowers.

Flowers of anemophilous species are generally unisexual (either monoecious or dioecious species) and are often temporally or spatially separated from flowers of the opposite sex. This separation not only prevents self-fertilization and increases outcrossing but prevents stigmas from being clogged by self pollen. One common arrangement of anemophilous flowers is for the female flowers to be located higher on the plant than male flowers so that pollen will not just fall down onto stigmas of the same plant.

Because of the high inefficiency of wind pollination, anemophilous species produce huge numbers of pollen grains. Filaments are frequently very long, causing the anthers to extend outside the surrounding perianth. Some species even have "explosive" anthers. The filaments are under strong tension in the bud stage and spring out, throwing pollen into the air, after the flower has opened. Anthers generally do not open unless the weather is favorable, i.e. warm and dry, because pollen is rapidly washed out of the air in rain.

Pollen grains of anemophilous species are usually very small (diameter of 10-20 mm), which increases their buoyancy. Some relatively large conifer pollen grains solve the problem of buoyancy by the addition of one or more air-sacs, which give a larger surface area without appreciably increasing weight. Great quantities of pine pollen have been found hundreds of kilometers away from the nearest forests. The typical rate of fall for wind pollen in calm air is of the magnitude of a few centimeters per second. Also contributing to buoyancy is the tendency of grains of anemophilous species not to adhere to each other but to be smooth and dry. Conversely, in entomophilous species, the pollen grains are usually highly ornamented and/or sticky, increasing the probability of their attaching to insects.

The female portion of anemophilous flowers has evolved to capture and utilize wind-born pollen. In contrast to the large number of pollen grains, the number of ovules per flower is generally rather low in anemophilous species; many species have but one ovule, so each flower only produces one seed. To capture wind-borne pollen, anemophilous flowers usually have greatly enlarged stigmas. For example, feather-like stigmas are often found in grasses and brush-like stigmas are found in cattails (*Typha* spp.), both of which increase the ability of the plant to "capture" pollen.

Several common plants use both anemophily and entomophily. Several species of *Plantago*, a common roadside weed, have inconspicuous, green flowers that produce massive numbers of pollen grains. These plants have been shown to be pollinated by wind in some circumstances, but they are also frequently pollinated by honeybees and flies. Other evidence comes from the loads of pollen-collecting bees; these frequently contain pollen from anemophilous species, sometimes exclusively.

Water: Another much rarer form of abiotic pollination is hydrophily, or water pollination, which can occur through a variety of mechanisms. In some flowers, the pollen is released into the water and floats to the water surface. The female flowers emerge onto the surface, receive pollen, and are then withdrawn back under the water. The most famous case is of *Vallisneria* in which the whole male flower is released instead of individual pollen grains; the pollen itself therefore does not touch the water surface. The female flowers create small depressions in the surface tension of the water. These depressions cause nearby floating male flowers to slide down allowing the anthers to contact the stigma of the female flower. In accordance with this effective mode of pollination, the number of pollen grains per male flower is drastically reduced.

Gravity: Pollination by gravity also called geophily is found in self pollinated crop plants. In this case pollen falls because of gravity on to receptive stigma of other flowers. However, gravity is highly unreliable and a rare and insignificant pollinating agent.

Biotic vectors

Biotic pollination is carried out by animals which act as the pollination agent or the pollen vector. The common pollinating animals include bees, wasps, butterflies, moths, birds, bats, and flies.

Most plant species have adopted one of two very different kinds of relationships with biotic pollen vectors. One option is to be a generalist and try to attract a wide variety of different pollinators. The other is to specialize (and often coevolve) with a single type of pollinator. Species that generalize can occur in a wide variety of habitats and survive conditions under which some of the pollinators cannot persist. On the other hand, they encounter a great deal of "foreign" pollen from other species, which can clog stigmas and prevent pollination e.g. *Dacus carota*. Flowers on these species are often grouped together in large showy, flat or gently rounded inflorescences. On their inflorescences one can usually find a motley crowd of insects-bees, wasps, flies and beetles of many kinds, and even some butterflies, although these are generally considered specialists.

Specialists can adapt to have very specific and highly efficient pollination mechanisms but are restricted to co-occurring with their pollinators. It is these mutualists that draw the most attention in pollination biology with their intricate and sometimes outlandish mechanisms for pollination.

Whether adapted as generalists or specialists, animal-pollinated plants share several characteristics. The pollen is sometimes larger than in anemophious species, is often sticky and/or highly ornamented with spines and bumps, and sometimes adheres in clumps of several grains. The number of pollen grains, sometimes expressed as a ratio with the number of ovules, is much lower in biotic pollination than in anemophilous species. Stamens are located so as to contact the pollinators, rather than to be exposed to wind. However, much of the floral structure is related to attracting specific types of animals to act as pollinators.

Table 34. Pollination syndromes as described by Fægri & van der Pijl (1979). Syndromes are named after their main pollinator taxon. They are considered to be "typical" representatives of the taxa: e.g. "bees" can include bumble-bees and beeflies. Only insect syndromes are considered, because vertebrate pollinators do not occur in central and northern Europe. Flowers belonging to the syndrome of carrion flies (*sapromyophily*) did not occur in the research area.

syndrome	flower morphology	colour	odour	primary attractants	sexual organ	remarks
beetles *canterophily*	actinomorphic, no special shape, large, shallow, often bowl shaped blossoms	dull nectar guides absent	strong fruity	mostly pollen, sometimes nectar: open, easily accessible	exposed	
flies *myophily*	actinomorphic, regular, simple, no depth effect	light or dull, whitish nectar guides frequently present	imperceptible	nectar: open, easily obtainable	exposed	
bees *melittophily*	zygomorphic, with great depth effect, mechanically strong, adequate landing facilities	lively, yellow or blue nectar guides present	weak, fresh	nectar: hidden, deep but less deep than moth flowers, abundant. pollen: open, (pollen source for bees often from simple flowers)	concealed	Syndrome of bumblebees, honeybees and larger (tropical) solitary bees. Most solitary bees in temperate Europe are "primitive" and comparable to e.g. beetles.
wasps (Vespidae)	not described	dull, brown	not described	nectar: accessible	exposed	Not considerd as a real syndrome
butterflies *psychophily*	actinomorphic, erect, radial, good landing facilities	vivid, including red simple nectar guides	weak	nectar: in ample quantities, hidden in tubes or spurs	exposed	
moths *phalaenophily*	actinomorphic or zygomorphic, horizontal	white or faint nectar guides absent	strong, nocturnal	nectar: in large quantities, deeply hidden in tubes or spurs	exposed or concealed	nocturnal anthesis

Attractants and rewards for biotic pollinators

Christian Sprengel (1750-1816) was one of the first to realize that bees and other flower visitors do not provide their pollen-carrying services free of charge. The plant must offer a reward in the form of something the pollinators need, either for their own survival or for that of their offspring. To secure the services of pollinators, the plants also must advertise these rewards. Flowers announce their rewards by being conspicuous in color, scent, size, or shape, making it easier for visitors to pick them out from their surroundings. Recognize that animals perceive flowers in many different ways, and certainly in different ways than do humans; what seems like an inconspicuous flower to us may be like a flashing neon sign to an insect or bird.

The most common rewards are nectar and pollen. Pollen may have been the original reward (bribe) by which flowers began attracting insects in the Cretaceous period some 140 million years ago. Because pollen is the male gametophyte, it is obvious that some pollen must be transported and not consumed for fertilization to take place, but flowers always produce much more pollen than is necessary. Pollen is a highly nutritious and well-balanced food material containing protein, sizable amounts of starch, sugars, fat or oil, minerals, antioxidants, and vitamins such as thiamin. It is also rich in free amino acids. There can be no doubt therefore about its value as a food for insects and mammals. Plants that offer pollen as a reward often produce it in huge quantities

In order to minimize the loss of viable pollen to pollinators, some plants produce, in the same flower, two types of anthers, normal ones, which produce healthy pollen, and others that yield sterile or at least less viable but still very nutritious and probably tasty pollen. Good examples of food stamens can be found in mullein (*Verbascum thapsus*) and partridge pea (*Cassia* spp.). In dayflower (*Commelina sativa*), the feeding anthers do not produce pollen, but offer a milky fluid instead.

Some orchids have hit upon still another way to offer imitation pollen. They produce "food hairs" on the lower lip of the flower, which easily fall apart into individual cells, rich in starch. Orchids cannot offer pollinators some of their pollen for food because, as, all the orchid pollen within a flower comes in neat little packages called pollinia and any attempt to sacrifice a few grains would sacrifice the whole pollinium.

Finally, it should be pointed out that not all the credit for pollen movement should be given to the insect pollinators. Pollen itself has been shown to "jump" from the stamens onto the insect, then from the insect onto a receptive stigma. It is known that electrostatic charges can build up on the body of a foraging honeybee and that these charges may reach a magnitude of several hundred volts. The bee is therefore flying at the center of its own highly charged electrostatic field. Most floral surfaces are well insulated, but the pistil is an exception: indeed a path of very low resistance leads from the stigma to the ground, almost like the earthing track of a lightning conductor. The result is that the bee's electrostatic field is attracted to the stigma, and with it the pollen.

Nectar is another important reward offered by many plants. Many scientists believe that nectar was already being produced before flowers first appeared on the scene. Primitive ancestral plants appear to have had many nectaries on their leaves, stems, and other non-floral areas. These nectaries may have been important for excreting excess sugars in maturing portions of plants. It is quite conceivable that these nectaries were simply incorporated into the structures we call flowers at the time those structures developed.

Nectar is an ideal substance to offer as a reward. It is easily produced by plants, it can be produced quickly as demand requires, and, being a sugar solution, it is readily digested and quickly assimilated as a source of energy. However, nectar has been modified in some species to reward specific pollinators. One critical factor is the amount of nectar, as flowers "want" to reward pollinators without satiating them. If too much nectar is provided, the pollinator has no need to forage further, and pollen will not be transported. On the other hand, flowers of different species, and even within species, are competing with one another for pollinators, and those that provide the most nectar are likely to attract the most pollinators.

Another way in which nectar can vary is in concentration and content. Bird-pollinated flowers usually have a fairly dilute nectar, with a sugar concentration between 20 and 30%. Birds simply cannot suck up a nectar that is too syrupy. On the other hand, horse-chestnut flowers may have a sugar concentration as high as 70%, with the results that bee visitors have to dilute it with their own saliva before they can suck it up.

The observation that certain nectars, and consequently some honey, can be poisonous has received a great deal of attention. Some of Xenophon's 10,000 soldiers were incapacitated on their epic homeward trek from Persia after consuming poisonous honey. In all likelihood, the culprit was *Rhododendron ponticum*. Other species are also known to produce poisonous nectar or pollen, although the reasons for doing so are not clear. Of course, nectar that is poisonous to one pollinator (or humans!) may not be poisonous to another.

Many other types of rewards are offered in the great diversity of types of flowers. Some flowers produce fatty oils (glycerides) to offer visiting bees instead of a sugary nectar. The bees that pollinate these species have specialized hairs to collect, transport, and temporarily store the somewhat sticky oil. In general, it is only the female bees that collect the oil, and it is they who mix the collected oil with pollen to produce a nutritious and protein-rich "bee-bread" for their larvae.

Other species produce resins and waxes, which can be valuable as nest-building materials, especially in wet-tropical areas. Some flowers may provide heat as a reward (Stanton and Galen, 1989). Figs are well known to provide environments for their pollinators in which to raise offspring, and a few orchids offer shelter from adverse conditions.

How do plants let the pollinators know that they offer these rewards? The primary attractants of flowers include color, size and shape, and scent. Plants use pigments contained in vacuoles to generate the huge variety of flower colors that we see around us. These pigments include anthocyanins (blues to reds), betacyanins (red), anthoxanthins (ivory to yellow), chlorophyll (greens), and carotenes (orange to yellow). Virtually the only purpose for floral color is to attract pollinators. Plants combine these pigments, along with special surface textures, to create the many additional colors, as well as shiny and velvety appearances. As we shall see below, these colors are often intended to attract specific pollinators, as pollinators differ greatly in the ability to see color and pattern.

Of course, many flowers produce scents, usually to attract pollinators (they sometimes smell nice to humans as well). These aromas must be manufactured at a cost to the plant. Many classes of pollinators have keen senses of smell. Their preferences vary from pollinator to pollinator, as does their color vision, and the scents given off by flowers reflect these differences. In many orchids, the scent is so specific that one, and only one, species of insect will visit each species of orchid.

Flowers of different species are distinguished more by their smells than by their colors, simply because there are many, many odor categories and relatively few colors. For this reason, odor may be even more important than color in establishing a phenomenon known as flower constancy. Flower constancy can be defined as the "loyalty" a pollinator displays towards the flower of just one plant species. Constancy is a learned behavior pattern and should never be confused with the inborn, specific pollination connection that may exist between a particular insect species and a particular species of flower. The pollinator exhibiting constancy may be perfectly capable of pollinating other flower species, but will stick to its original behavioral pattern even when flowers that have more nectar or pollen to offer are becoming available in the environment.

At first sight, such faithfulness may seem foolish, yet it can be advantageous for both the pollinator and the plant species to which it is loyal. The plant benefits in not having its pollen wasted on flowers of another species, and the pollinator, having the opportunity to concentrate all its efforts on one type of flower, soon learns to handle it with maximum efficiency and speed.

Pollination Syndromes

More than 100,000 different animal species - and perhaps as many as 200,000 - play roles in pollinating the 250,000 kinds of flowering plants on this planet. In addition to countless bees, wasps, moths, butterflies, flies, beetles and other invertebrates, as many as 1,500 species of vertebrates such as birds and mammals serve as pollinators, including hummingbirds, perching birds, flying foxes, fruit bats, possums, lemurs and even a lizard (gecko) (Ingram *et al.,* 1996).

Honey bees, birds, bats and insects play a vital role in pollination of most fruits and vegetables. With over 90% of all flowering plants and over three-quarters of the staple crop plants that we use for food rely on animals for pollination of the hundred or so crops that make up most of the world's food supply, only 15% are pollinated by domestic bees, while at least 80% are pollinated by wild bees and other wildlife (Prescott-Allen and Prescott-Allen, 1990; Ingram *et al.*, 1996a; Buchmann and Nabhan, 1996).

Hymenoptera

The hymenopterans are an abundant, highly specialised and highly successful order of insects, rivalling the dipterans and lepidopterans in number of species and probably exceeding both in variety of adaptive radiations (Riek 1970). This large group of insects includes two divergent suborders: the primitive, phytophagous Symphyta or sawflies; and the much more specialized and abundant Apocrita.

Symphyta

The suborder symphyta includes sawflies. Most sawfly species deposit their eggs in plant tissue and there is a marked tendency for the adults to visit the flowers of their larval food-plant (Proctor and Yeo, 1973). Sawfly mouthparts are very short, usually 2 mm long or less, so that nectar and pollen can be obtained only from blossoms in which these nutrients are well exposed. Species of the genus *Clarissa* (Pergidae) are commonly found on blossoms in early summer; they are particularly attracted to *Wahlenbergia* (Campanulaceae) blossoms, but have also been observed seeking nectar from eucalypt flowers (Riek, 1970).

Apocrita

The suborder Apocrita can be conveniently divided into three categories: "wasps", bees (Apoidea), and ants (Formicidae).

Pollination by bees *(Melittophily)* Among different groups of insects, bees are most important and have been described as efficient pollinators. The family Apidae contains the highly social bees which carry their pollen in specially adapted pollen-holding corbiculae on their hind legs. *Trigona* bees are stingless and their pollen and honey are stored in large wax pots, very different from the vertical combs of honey bees. There are about 30,000 known species of bees meeting to the same or the other pollination requirements. In India, many bee species have been recorded which include four honeybee species belonging to *Apis* genera, 28 *Bombus,* 52 *Halictus,* 25 *Nomia,* 19 *Xylocopa,* 44 *Megachile* and certain species of stingless bees (Abrol, 1991). Honeybees mainly, *A. cerana* and *A.mellifera* are managed in hives for improving pollination in majority of crops. There can be no doubt that, on a global scale, bees (especially honeybees and bumblebees) are the most important pollinators. It has been calculated that in Germany alone, honeybees pollinate about ten trillion flowers in the course of a single summer's day. To make 0.5 kg of honey, bees have to visit about ten million clover flowers, leading to the production of about 13.5 kilograms of clover seed. The pollen collected by a single honeybee colony may be more than 29 kilograms per year.

The "intelligence" of pollinators (i.e. their ability to perceive, discriminate between, and remember the characteristics of flowers) reaches a peak in bumble bees and honeybees. The bees have a large range of relationships with flowers, from the rather simple ones in parasitic bees to the incredibly complicated ones of honeybees and other social bees, in which communication and temporal variation play a large role.

Typical bee flowers are open in the daytime, have a minty fragrance, and offer their visitors nectar, pollen, or both. Bee eyes have trichromatic vision, and contain pigments sensitive to green, blue, or ultraviolet. Bees are blind to red. Thus, bee flowers are brightly colored (often yellow or blue), in order to attract insects from a distance, but are not red.

For close-range guidance of pollinators, bee flowers often possess nectar-guides in the form of detailed color or smell patterns (which may include special UV reflection patterns), which direct the pollinator to places where the flower "wants" the pollinator to go. Many species provide the bees with a landing platform, usually in the form of a broad lower lip on which the visitor can alight before pushing its way deeper into the flower. Thus, bee-pollinated flowers are generally bilaterally rather than radially symmetric. Because bees can visit many types of flowers, it appears that there has been some selection for pollen to be deposited on different parts of the visitor's body (e.g. the back, underside, or head), so as to increase the probability of the transfer of "correct" pollen.

The pollen grains of bee-pollinated species are sticky and spiny or highly sculptured, so that they cling easily to the mouth-parts, legs, or bodies of the visitors. Because one animal may carry thousands of pollen grains, it is not surprising that the ovary of the receiving flower contains a great many ovules. Getting to that pollen may not be easy, especially for bumblebee flowers. Many bumblebee flowers have petals or other floral parts that must be pried apart to allow entry into the flower. This arrangement prevents undesirable insects from taking pollen or nectar.

A number of bumblebee species "steal" nectar by piercing the corolla and removing nectar without properly entering the flower. These species generally do not pollinate the flower, and some plant species have adaptations that appear to be specifically designed to discourage robbers. These include hidden nectaries and thickened corollas. Some plant species have developed mutualistic relationships with ants, in which the plant produces small amounts of nectar on the outside of the flower, attracting ants. These ants, in turn, defend flowers from nectar robbers.

Coleoptera

The coleopterans, with over 300 000 species, constitute the largest animal order in the world. Coleoptera are considered to be the most primitive pollinators have gnawing mouthparts and seemingly primitive instincts, which generally preclude their effective participation in the pollination process. Beetles are unable to land precisely on blossoms and hence most anthophilous species are restricted in their visits to the simplest flowers, chiefly of the "dish and bowl" type (Leppik's, 1956) amorphic and haplomorphic blossom classes) characteristics of aminoid or fermenting fruit scent (e.g. *Magnolia*), in which pollen, nutritive tissues, or very exposed and easily accessible nectar is offered as the nutrient reward.

In such flowers, the beetles imbibe floral secretions (including nectar if present), eat pollen, and chew on floral parts or special food bodies. Many flowers, often specialized and not exhibiting the syndrome of cantharophily are visited and some beetles seem restricted to floral diets as adults. Flower visiting coleopterans have structural adaptations which include forward projection of the mouth parts by up tilting of the head and associated prolongation of the prothorax and the neck. Maxillary setae may also be elongated. These modifications allow the beetles to reach deeper nectarines and extract nectar more readily e.g Cerambycidae. In tropical American *Nemognatha* the maxillae are elongated to exceed the body length and may reach deep into such long tubular flowers as those of *Ipomoea*.

The importance of odour as a beetle attractant has recently been demonstrated in the Araceous species *Typhonium brownii* (Monteith 1973). Several authors (e.g., Grant & Grant, 1965; Faegri & van der Pijl, 1966) have suggested that beetle pollination may be secondary in some cases, i.e., "represented in taxa, the ancestors of which were pollinated by more highly developed insects" (Faegri & van der Pijl 1966).

Diptera

The dipterans with suctorial and lapping mouth parts are also considered to be primitive pollinators. In most families of Nematocera, the proboscis is short, but variable in form and flowers visited by these flies have readily accessible and exposed nectar in tubes that are short (e.g *Achillea, Senecio, Polygonum,* various crucifers). However, specialized pollination mechanisms exist between Nemaotcera and some highly evolved plants e.g mosquitoes and *Habenaria* orchids. Whether Nematocera represent an interm stage in the pollination mechanisms is still unresolved. The Brachycera contain a wider variety of flower visitors. They visit deeper tubed flowers. Among the Cyclorrhappa, the Aschiza include the most important anthophilous diptera, the Syrphidae tse flies feed on nectar and pollen of a wide variety of flowers worldwide. The members of small family Conopidae (Acalypterae) stand out as anthophiles. They have long proboscides, upto 6 mm in *Sicus* and 4 mm in *Conops*.

Pollination by Butterflies (*Psychophily*)

Butterflies are sun lovers that like to perch while feeding. They have long and slender probosces, can perceive a wide spectrum of colors, and have an excellent sense of smell. With few exceptions, they are nectar feeders. From these characteristics, one can easily deduce the characteristics of typical butterfly-pollinated flowers. They are open in the daytime, produce a good amount of nectar, possess a long, thin corolla tube (often with a "spur" or out-pocketing which contains the nectar), and are generally vividly colored (often red), although sometimes white. They also provide their butterfly visitors with a platform to land and walk on. The flat-topped inflorescences of *Verbena, lantana*, red valerian, milkweeds, and various composites provide excellent examples of butterfly-pollinated species. The spike inflorescence of purple loosestrife and some violets represent yet another type of butterfly-pollinated flowers.

Pollination by birds (Orinthophily) Birds that feed on nectar are highly specialized and feed on little else. Thus, they depend on flowers with large amounts of nectar available on a year-round basis. Plants in temperate areas cannot bloom year round and must therefore rely on migratory bird species which feed elsewhere during the winter. In the tropics, bird pollination of flowers is at least as important as insect pollination. At least 2,000 species of birds feed on nectar, flower-inhabiting spiders, and insects, or (rarely) pollen. Even some fruit-eaters among tropical birds will occasionally consume nectar. In New Zealand, which has no native bees pollination by birds is the rule rather than the exception.

Many of these birds are highly specialized to feed on nectar. Their bills are long and narrow, and their tongues are tube-shaped or have brush-like tips. They include hummingbirds, sunbirds, honeyeaters, honey creepers, and brush-tongued parrots. One can often determine their food-plants by looking at the pollen attached to their bills and feathers. This method can work even with museum skins and even in some cases with species that are now extinct.

Most of these birds also have very high metabolic rates, so their caloric intake per gram body weight must be high. Although few people who watch humming birds flit from flower to flower are aware of it, these birds are constantly on the verge of starvation. To meet their energy needs they must visit thousands of nectar-rich flowers every day. Night provides severe problems: some humming birds have solved this problem by essentially going into hibernation every night. Their temperature drops, and their rate of metabolism may go down to about one-fifteenth of the peak daytime value, so they save a great deal of energy.

Birds have excellent color vision and appear to favor red. In contrast, their sense of smell is very poor. These traits are reflected in the flowers they feed on. Bird-pollinated flowers have no odor. The amount of nectar produced can be quite large, up to a cupful a day in some cases. The colors of bird-pollinated flowers vary enormously. Many are red, but others are yellow or blue or almost any other color. Red flowers may be common simply because bees cannot see them well, so little nectar is lost to ineffective pollinators. To provide food during all active hours, bird pollinated flowers are generally open all day. Flower size and shape can vary, but corollas are often tubular (and are larger in diameter than butterfly flowers) and are sturdily constructed as a protection against the probing bills of their visitors. The ovules are usually below the perianth (that is, they have "inferior ovaries"), out of harm's way. Usually, stamens are quite numerous and often stick out of the flower, in which case they are brightly colored and quite strong.

The visiting bird is normally touched on the head or breast as it feeds. However, in some flowers the stamens are "explosive" and cover the bird with pollen. Hummingbirds, which are only found in the Americas, feed on the wing and so hummingbird flowers hang down or are downward-facing and lack a landing platform. In Asia and Africa, however, many of the flower-birds do not hover and, accordingly, the plants offer them a landing-platform or perch.

Fly and Beetle Pollination (Myophily and Cantharophily): Faegri & van der Pijl (1966) have suggested that a greater diversity of pollination methods is found among flies than among any other insect group and, since numerous dipteran taxa frequently visit flowers to ingest nectar (or rarely pollen), the order is important in pollination ecology. Both flies and beetles present a wide variety of pollinators and pollinator syndromes, making it difficult to generalize, but a number of flies and beetles should be specifically mentioned because they are not really adapted to flowers at all (although the flowers are adapted to them). These are the carrion, dung, and mushroom flies and beetles that are trapped by various flowers or inflorescences. Species in this group are primarily attracted by smell. Some are looking for food, others are looking for egg-laying sites. In either case, they are tricked by the flower odor into thinking that they have found their normal prey. They enter the flower, eventually figure out that they have been fooled, then move on.

Common traits of the flowers that use such pollinators include dull colors, often large, open flowers, (although some species instead have enclosures, which trap the pollinator for a time), and distinct odors, which are sometimes quite unpleasant to humans. The names of some of these flowers are can be equally unattractive. The world's largest "flower" (really it's an inflorescence) is the stinking corpse lily (*Raflesia* sp.).

The suborder Nematocera consists of at least 12 families known to contain anthophilous taxa. In most of these families the mouthparts are short (generally a millimeter or two long) and the flower-visiting taxa are thus restricted to visiting either flat or bowl-shaped blossoms (in which the nectar is well exposed and the stamens are very short), or very small tubular blossoms in which the nectar is easily accessible. According to Willis and Burkill (1895-1908), flower-visiting Nematocera are chiefly nectarivorous although pollen-collecting species have been reported in the Bibionidae, Mycetophilidae, and Scatopsidae. The Culicidae are exceptional among the Nematocera in that most taxa possess the piercing-and-sucking type of elongate mouthpart, which characterizes the typical "proboscis" of the true mosquitoes. It has only recently been realized that mosquitoes are very frequent blossom visitors with both sexes of many, perhaps all, species feeding on sugary substances such as honey-dew or nectar (Faegri and van der Pijl, 1966). The most important dipteran suborder in relation to pollination is the Brachycera exhibiting a gradual transition from very simple to more highly evolved blossom visitors, such as the Syrphidae, Conopidae, and Bombyliidae.

Ants and Pollination *(Myrmecophily)*: Morethan 15000 species of the Formicidae are known in the World. Although these insects are commonly anthophilous, many behavioural and general morphological characteristics preclude them from effectively participating in the pollination process. Pollen does not adhere readily to their hard and generally smooth bodies, and the diminutive size of these creatures allows them to avoid contact with the anthers and stigmas when they visit flowers in search of nectar. Also, the flightless worker ants are unlikely to facilitate cross-pollination, because most of their foraging time is spent crawling from one blossom to another on the same plant. Until recently, no beneficial role in pollination was attributed to ants: they were considered to be just thieves of nectar and pollen. At best, certain investigators were willing to concede that an "ant guard," provided with food by the plant in the form of extrafloral nectar, might keep potential flower robbers at bay. However, it is now recognized that there is a genuine ant-pollination syndrome. Because worker ants do not fly and therefore, do not expend many calories on traveling, the whole system is low-energy. The nectaries are small and produce a quantity of nectar so modest that larger insects are not interested. The flowers, likewise, are small, sessile, and close to the ground, and exhibit minimal visual attractions. They produce only small quantities of sticky pollen grains, so the ants are not forced into intensive self-cleaning activities that would remove the pollen from their bodies. Outbreeding is promoted because on each plant only a few flowers are open at the same time and also because these low-growing plants occur in groups with their branches closely intertwining. It is believed that this syndrome is most often found in hot, dry habitats, which are certainly among those favored by ants.

Jones (1972c) observed numerous "black jumper" ants *(Myrmecia pilosula)* on the inflorescences of *Prasophyllum alpinum* (Orchidaceae) (a species pollinated by the wasp *Pterocormus promissorius* (Ichneumonidae)); although pollinia were often observed on the jaws of these ants, their habit of cleaning their mandibles on the vegetative parts of the plant, before proceeding to other nectar-rich blossoms, prevented pollen exchange.

A genuine case of ant pollination, in the Australian flora, has been reported in *Microtis parviflora* (Orchidaceae). Jones (1975a) observed worker ants of three different genera *(Iridomyrmex* sp. (Dolichoderinae), *Meranops* sp. (Myrmeciinae); and *Rhytidoponera tasmaniensis* (Ponerinae)) feeding on the abundant nectar secreted by the labellum of this orchid species. All three ant species effectively pollinate the *Microtis* blossoms, but *Iridomyrmex* sp. is the most frequent pollinia-bearing visitor.

Although *Microtis parviflora* is an autogamous species, 75% of the flowers are ant pollinated within 3 days of anthesis (Jones 1975a). Bates (1979d) recently recorded a case of pseudocopulatory pollination in the South Australian orchid *Leporella fimbriata,* by winged male ants of the genus *Myrmecia iurens* group).

Pollination by Wasps *(Sphecophily)*: Wasps" possess unspecialized mouth-parts and hence are generally restricted to allophilic blossoms (Faegri and van der Pijl 1966), with open, easily accessible nectar. In the large genus *Ficus* (Moraceae), a unique situation prevails in which almost every fig species has a different stingless wasp species as a pollinator.

The wasps form a large and highly diversified group. Adults are generally predators or feed on carrion, and for that reason nectar is important to them only as a source of carbohydrate for their own energy needs. As pollinators, they don't begin to compare to honeybees. In fact, they may exert a strong negative influence on other pollinators by attacking honeybees, other wild bees, and butterflies. The question of whether or not there are flowers adapted specifically to pollination by wasps has been a hotly debated issue.

The one well-known relationship is that of figs and fig wasps. Gall wasps are specialized pollinators of figs (*Ficus* of which there are about 700 species, mostly tropical). Female wasps deposit eggs in specialized fig flowers, which are clumped together into an inflorescence. These eggs cause the flowers to change into galls, each with a larva inside. When a male larva matures and hatches, he chews his way out into the fruit, finds himself a female in the same inflorescence "prison," and mates. The female then enlarges the hole the male used to enter, crawls out of her gall, and leaves the fig in which she was born. In doing so, however, she must cross a region of male flowers near the fruit's entrance, which have just opened. Powdered with their pollen, she now makes her way to the young edible figs that have just formed on the same tree or on another one close by. These contain a different type of specialized flowers, which are quite hard and resist the female wasp's attempt to deposit eggs in them, but in the attempt, the flowers are pollinated. This fruit gradually matures and turns into the fleshy fruit we know as figs. Long before, however, the unsuccessful female has left the immature fig to continue her efforts in another fig of about the same developmental stage, but again in vain. She cannot satisfy her urge to deposit eggs in fig flowers until the type of flower in which she hatched again begins to develop in trees. These small, inedible flowers are generally found on the upper branches and are excellent incubators for the eggs and larvae of the female wasp. Thus the cycle starts all over.

Several authors (Ramirez 1974) have indicated that the specificity of the agaonid-fig association ensures that a given fig will produce seed only when its specific agaonid is present. Species of Agaonidae live exclusively in fig inflorescences where they act as pollinating agents. In *Ficus macrophylla* the winged females of *Pleistodontes froggatti* (Agaonidae) are the pollen vectors, whereas the wingless, highly modified males play no part in the pollination process. Scoliid wasps (Scoliidae) pollinate *Calochilus* orchids in eastern Australia (Gumprecht 1977). Jones and Gray (1974) have more recently shown that *Calochilus holtzei* is crosspollinated by male scoliid wasps and that these wasps are involved in pseudocopulatory pollination rather than the pseudoparasitic pollination referred to by van der Pijl and Dodson (1966). Self-pollination mechanisms are also well developed in *Calochilus* orchids (Jones and Gray 1974). Jones and Gray (1974) have noted that the various species will self-pollinate if insect pollination has not occurred within 3 days of anthesis. Thynnid wasps (Thynninae: Tiphiidae) appear to be extensively involved in pseudocopulation in Australian terrestrial orchids (Gumprecht, 1977). These parasitic wasps are well represented in Australasia and South America and consist of winged males and smaller, wingless, often ant-like females.

Pollination by Moths (*Phalaenophily*) Hawkmoths, also known as sphinx moths because of the peculiar position their caterpillars adopt when disturbed, are the nocturnal counterparts of hummingbirds. Like hummingbirds, hawkmoths normally feed on nectar, and, because they are usually quite large, fly at high speeds. In some ways, they operate rather like warm blooded animals, having higher energy requirements than most insects and thus high nectar needs. Some moths can consume a good teaspoon of nectar at a single sitting. Hawkmoths are highly specialized flower visitors, equipped with a long, thin, and very flexible proboscis, which is kept coiled up when the animal is not feeding but which can be stretched out to take nectar. The tongue is usually extended just as the moth reaches its floral target.

Not surprisingly, flowers that cater to hawkmoths open in the evening and are extremely fragrant. They are snow-white or light-colored, offer no landing platform and may have fringed petals--possibly for guidance. Many have both visual and olfactory nectar guides. The corolla tube is long and narrow, a feature that discriminates against other, short-tongued, visitors, and nectar is abundant. Typical hawkmoth flowers are evening primrose (*Oenothera biennis*), some tobacco flowers (Nicotiana spp.), and most honeysuckle (*Lonicera*) species.

Pollination by other invertebrates

Apart from the major invertebrates already listed, flower-visiting taxa have been recorded amongst the following invertebrate orders in various parts of the world (Miiller, 1883, Knuth, 1906, Faegri and van der Pijl, 1966, Proctor and Yeo, 1973): Collembola (pollen feeding), Dermaptera (pollen- and nectar-feeding), Dictyoptera (nectar-feeding), Hemiptera (often recorded at flowers), Mecoptera, Neuroptera, Odonata, Plecoptera (pollen-and nectar-feeding), Psocoptera, Thysanoptera (often found in flowers) and Trichoptera; snail (and slug) pollination (Malacophily) is known, but has not been reported in any Australian plant taxa. The pollinating effectiveness of many of these invertebrate groups is probably negligible and, as Faegri and van der Pijl (1966) state, "one hesitates to accept these allotropic animals as anything more than chance pollinators". Flower visiting amongst "other invertebrates" has rarely been reported in Australia, apart from the following records. COLLEMBOLA: spores and pollen have been found in the gut contents of members of the family Sminthuridae (Wallace and Mackerras, 1970). HEMIPTERA: Armstrong (unpublished data) recorded nectar-seeking Lygaeidae on *Aotus lanigera* (Fabaceae), *Nysius clevelandensis* (Lygaeidae) on *Helichrysum* sp. (Asteraceae), and members of the family Miridae on *Astrotricha floccosa* (Araliaceae) and *Dillwynia retorta* (Fabaceae). MECOPTERA: Symmington (1963) noted *Harpobittacus similis* (Bittacidae) visiting *Calectasia cyanea* (Calectasiaceae) blossoms, and Kenneally (1970) observed an undescribed species of *Harpobittacus* seeking nectar from the male flowers of *Diplopeltis huegelii* (Sapindaceae); Riek (1963,1970) noted//, *australis* obtaining nectar from *Leptospermum* (Myrtaceae) blossoms and Keighery (1974) recorded *H. similis* adults actively foraging in the floral heads of *Eryngium pinnatifldum* (Apiaceae) and *Podolepis lessonii* (Asteraceae). ORTHOPTERA: Clyne (in Key 1974) found that *Zaprochilus* and a closely related undescribed genus of "Zaprochilinae" feed on the pollen and nectar of flowers. THYSANOPTERA: Reed (1970) noted that *Liothrips* (Phlaeothripidae) frequently visits flowers and that *Isoneurothrips australis* (Thripidae), which is commonly found in eucalyptus blossoms, probably feeds on nectar; Churchill and Christensen (1970) recorded *Soneurathrips australis* (Thripidae) in *Eucalyptus diversicolor* blossoms; Mound (1974) vaguely noted that the genera *Odontotripiella* (Thripidae) and *Pseudanaphothrips* (Thripidae) are found in the flowers of Australian "herbs and shrubs"; and Endress (1979) recently reported that the genera *Steganthera* and *Wilkiea* (Monimiaceae) show a "normal carpellary pollination by small insects (e.g., Thysanoptera)".

Mammals (Therophily)

Pollination by mammals is called as therophily (Rourke, 1978). On the basis of functional aspects of pollination biology, two types of mammal pollination occur: non flying mammal pollination, involving nocturnal marsupials and placental rodents; and flying mammal pollination, involving nectar-seeking placental bats.

Flying mammals

Pollination by Bats: Bats constitute the order Chiroptera, the only mammals capable of true flight. The family Pteropodidae (suborder: Megachiroptera) contains the flower-visiting chiropterans. The genus *Pteropus* (Pteropodidae) contains the largest number of Australian flower-visiting species. *P. poliocephalus* is the most common but observed to feed on the nectar-rich blossoms of the following eucalyptus species: *Eucalyptus cloeziana, E. crebra, E.fibrosa* (listed as *E. siderophloia), E. grandis, E. intermedia, E. maculata, E. microcorys, E. moluccana, E. pilularis, E. resinifera,* and *E. siderophloia* (listed as *E. drepanophylla).* The spectacled flying-fox *(Pteropus conspicillatus)* is restricted to dense rainforest and swampy habitats regularly visits nectar rich *Eucalyptus* (Myrtaceae) and *Banksia* (Proteaceae) blossoms and is often seen during the day roosting in trees, close to its blossom sites.

Certain bats have become pure "flower animals"- especially those in the Macroglossinae or "big-tongues," found in southern Asia and the Pacific. Their protein requirements are met entirely by pollen, which they deliberately collect and consume in great quantities, along with nectar. Several New World species have the same life-style.

Of course, bats are most active at night. They have reasonably good eyesight but appear to be color-blind. They have a keen sense of smell, displaying a preference for odors that humans find definitely unpleasant: mouse- or urine-like, stale, musty or rancid, resembling butryric acid or sweaty feet. The sonar sense of flower-pollinating bats is not as well developed as is that of other bats.

Most flower-pollinating bats are small, sharp-snouted animals with long tongues that can be stuck out very far and have special projections (papillae) and, in some cases, a soft brush-like tip-devices that enable them to rapidly pick up the copious pollen and nectar soup that bat flowers offer. Normally, it is a bat's head that becomes dusted with pollen, transfer of the precious powder to the pistils of other flowers is no problem. Considering their extreme food specialization, it is not surprising that these bats are usually lacking teeth. The few teeth found in the males are used for fighting and not for eating. While foraging, most macroglossine bats hook themselves into the petals of the flowers (which they probably locate by smell) with their thumb-claws. Several of the New World bats hover like hummingbirds while feeding.

The typical bat flower is large, sturdy, and bell-shaped; has a wide mouth; and is either snow-white or drab in color. The drabness is probably correlated with the lack of color vision in bats, and white makes sense because it is conspicuous by contrast with the darker background. Food is offered in abundance. Bat-flowers have increased pollen production during the process of evolution in two ways: by increasing the number of stamens (up to 2,000 per flower in some species) and by making very large anthers, as in century plants (*Agave* spp.). Other bat flowers offer their visitors nectar or special food bodies in the form of succulent petals or sweet-tasting bracts.

Another adaptation of bat-pollinated plants is that flowers must be accessible to the relatively large flower bats with poor sonar abilities. To this end, the flowers dangle beneath the crown of trees, or they may be placed on the main trunk or the large limbs. Examples include the white saguaro cactus (Carnegiea) flowers and the century plant, the large flowering spike of which makes the flowers easily accessible high in the air. Some bananas (*Musa* spp.) are also bat pollinated. *Musa fehi*, for instance, is found on the Hawaiian Islands. Ethnobotanists have noted that, because bats do not occur in Hawaii, Musa was most probably introduced by man.

Non-flying mammals

Otto Porsch (1934) was the first to suggest that certain Proteaceae and Myrtaceae in Australia were pollinated by mouse-like marsupials; Since then very little new information has come to hand and pollination by non-flying mammals is largely ignored or given the very briefest mention in current summaries of pollination ecology (Faegri and van der Pijl, 1966, Proctor and Yeo, 1973). To date, no unequivocal evidence for regular pollination by non-flying mammals can be presented and all the available information relating to this phenomenon is, to quote Rourke and Wiens (1977), "either circumstantial, inferential, or anecdotal". The floral features of hooked-styled *Banksia* species appear to be adaptations for more efficient exploitation of both bird and mammal pollen vectors, rather than adaptations for exclusive pollination by mammals as proposed by Carpenter (1978) and others".

Marsupials (Metatherophily)

Australasia is the only region of the world in which marsupial pollination has been reported. The most highly adapted nectar-feeding marsupial is the honey possum *(Tarsipes spencerae),* a species confined in its distribution to the southwestern corner of Western Australia. The long prehensile tail of this mouse-like animal enables it to dangle from one blossom while feeding on another Its extremely etiolated, almost tubular snout is well adapted for nectar collection This new term is based on the name for the mammalian infraclass-Metatheria. and is used to probe nectar-rich blossoms in much plant species: *Adenanthos cuneata, Banksia* the same way as do the long-beaked honeyeating *attenuata, B. baxteri, B. occidentalis, B. sphaero*birds.

Rodents (Sminthophily)

The only rodent known to visit blossoms in Australia is the southern bush rat *(Rattusfuscipes:* Muridae) a shy nocturnal animal that lives on native vegetation, and hides in burrows during the day. Morcombe (1978a, b) was the first to suggest that this animal is attracted to nectar-rich *Banksia* blossoms; recently Carpenter (1978) have verified this suggestion, with their identification *of Banksia ericifolia* and *B. baxteri* pollen on the heads of trapped *Rattus* individuals. Rodents are known to visit blossoms in other parts of the world. In Hawaii, introduced rats are attracted to the fleshy bracts surrounding the flowers of *Freycinetia* (Pandanaceae) (Degener 1954), and in South Africa, the cryptic, geanthous species of *Protea* (Proteaceae) are regularly visited and pollinated by a variety of indigenous rodents (Rourke & Wiens 1977, Wiens & Rourke 1978).

Crops dependent upon or benefited by insect pollination: The following crops (Table) are dependent upon or benefited by insect pollination. Unfortunately, the information on their pollination needs is scanty or based upon earlier popular varieties. For some, insect dependence is absolute; for others, the benefit ranges from scant to great. For most of them, there is a great need for study of current varieties under different environmental conditions to determine the precise dependency on insect pollinators.

Table 35. Crops dependent upon or benefited by insect pollination ([1]For production of seed).

Fruit and nut crops	Vegetable crops	Forage crops	Oilseed crops	Others
Acerola	Artichoke	Alfalfa	Cotton	Buckwheat
Almond	Asparagus	Alsike clover	Flax Linseed	Cacao
Apple	Balsam pear	Arrowleaf clover	Peanut Guar	Cashew
Apricot	Broccoli Bean	Ball Clover	Rapeseed & Mustard Niger	Chicory
Blackberry	Brussels sprouts	Berseem clover	Soybean	Clove
Blueberry	Cabbage	Black medic	Safflower	Coffee
Chestnut	Cantaloupe	Broadbean	Sunflower	Kola
Chinese Gooseberry	Cardoon	Cicer milkvetch	Tung	Lupines
Coconut	Carrot	Crimson clover		Tea
Crabapple	Casaba melon	Crownvetch		Many garden flowers Coriander
Currant	Cauliflower	Kenaf		
Feijoa	Celeriac	Kidneyvetch		
Gooseberry	Celery	Kudzu		
Grape (some kinds)	Chayote	Lespedeza		
Grapefruit	Chervil	Mung bean		
Guava	Chive	Persian clover		
Jujube	Coriander	Pigeon pea		
Lemon	Cowpea	Red clover		
Litchi	Crenshaw	Rose clover		
Loquat	Cucumber	Sainfoin		
Macadamia	Dill	Scarlet runner bean		
Maney sapote	Eggplant	Strawberry clover		
Mango	Endive	Sweet clover		
Nectarine	Fennel	Sweet vetch		
Olive	Honeydew	Trefoil		
Orange (some kinds)	Kale	Vetch		
Papaw	Leek	White clover		
Papaya	Lima bean	Zigzag clover		
Passion fruit	Muskmelon			
Peach	Mustard			
Pear	Onion			
Persimmon	Parasley			
Pomergrante	Parsnip	Pea		

Plum	Pepper
Prun	Persian melon
Quince	Pimento
Raspberry	Pumpkin
Tangelo	Radish
Tangerine	Rutabaga
Temple orange	Squash
	Turnip
	Vegetable sponge
	Welsh onion

Many cash crops are actually self-sterile and require cross-pollination to produce seeds and fruit (McGregor, 1976; Free, 1993). But it is not only self-sterile varieties that benefit from cross-pollination, but self-fertile varieties also produce more and better quality seeds and fruits if they are cross-pollinated (Free, 1993). Logically, the increase in the cultivation of cross-pollinated cash crops will also increase the need for managed pollination.

Pollinators provide an essential ecosystem service that contributes to the maintenance of biodiversity and ensures the survival of plant species including crop plants. Two types of pollinators occur in nature. These include abiotic pollinators such as wind, water and gravity, and biotic pollinators such as insects, birds and various mammals. It has been estimated that over three quarters of the world's crops and over 80% of all flowering plants depend on animal pollinators, especially bees (Kenmore and Krell, 1998). Globally the annual contribution of pollinators to the agricultural crops has been estimated at about US$ 54 billion (Kenmore and Krell, 1998).

Insects are the most commonly occurring pollinators of many agricultural and horticultural crops. Different kinds of insect pollinators such as bees, flies, beetles, butterflies, moths and wasps are important pollinators of many crops. Among insects, bees are more effective pollinators than other insects because, unlike other insects, they are social and collect nectar and pollen not only to satisfy their own needs but to feed their young; their body hairs help transfer pollen from one flower to another; they show flower constancy and move from one flower to another of the same species; and many species can be reared and managed for pollination.

The value of bees as pollinator was first discovered by Koelreuter (1761), while Knight first noted the relation between plants and hybrid vigour. Darwin's concept of natural selection opened the way for a complete understanding of the correlationship of bees and flowering plants, with resultant appreciation of factors affecting mutual adaptation, speciation and distribution.

Honeybees (*Apis* spp.) are the,most important pollinators of agricultural and horticultural crops. Honeybees have attributes that makes them valuable for crop pollination. These can be marshaled in adequate numbers at desired places.

Their body parts are modified to effect pollination. Their body size and proboscis length suits them to forage many varied types of flowers. Their wide host range enable them to pollinate many types of crops. They work for longer durations and are also less effected by adverse climatic conditions as compared to other insect pollinators. Their foraging behaviour patterns are highly favourable to qualify them as most efficient pollinators of crops. In addition to pollination services, colonies can be managed to produce honey, bee wax, propolis, royal jelly etc. which are all saleable products

The honey bee is well adapted for pollination. Each colony contains thousands of individuals which are available to forage for the food (pollen and nectar) that's required to rear their replacements. Like other bees, their sense of smell, eyes, mouthparts, and numerous branched body hairs are ideally suited for finding food sources, sipping nectar, and collecting and distributing pollen. Honey bees visit a wide variety of flower types. In a single day, one bee makes 12 or more trips from the hive, visiting several thousand flowers. On each trip, it confines its visit to one plant species, collecting one kind of pollen. It has a complex communication system that enables the colony to find and collect food with maximum efficiency. These characteristics make honey bees most valuable agent for cross-pollinating crops.

They are relatively abundant and manageable. Bees from a colony will visit a large number of plants over a large area, collecting pollen and nectar, with individual bees visiting one species of flowers in the same location until the supply of nectar or pollen is exhausted. This loyalty or constancy is not found in some other social bees, which visit different plant species during the same trip in the fields. This indescriminate behavior reduces the effectiveness of the bee species as a pollinator because the pollens are mixed.

Honeybee's adaptations of as Pollinators

Honeybees have evolved numerous adaptations which make them the most efficient pollinators. Some of the more important ones are:

(1) **Social unit**: Because the entire colony overwinters, large numbers of bees are ready to forage in early spring, as compared to other bees such as the bumble bee, where only the queen overwinters.

(2) **Forked body hairs**: These hairs enable the bee to pick up pollen and move it around better than insects with simple, unforked hairs, or insects without any body hairs. Branched hairs are capable of carrying up to 5 million pollen grains. They have specialized mouth parts, honey sac, honey combs for storing nectar and pollen a efficient distribution over available territory.

(3) **Corbiculae** : The pollen basket: they have intricate pollen baskets. Pollen baskets are structures on the hind legs of worker honeybees which allow these bees to carry pollen back to the colony.

(4) **Antennae cleaner**: This is a structure on the front legs, which enables the bee to clean the antennae.

(5) **Communication from bee to bee**: By dancing, and other specialized systems of communication, scout bees can transmit information about a crop to other bees, thus greatly improving pollination activity

Honey bees are desirable pollinating agents for these reasons

a. Colonies of bees can be moved to crops needing pollination.

b. Each colony contains large populations of foragers to work crops within a narrow pollinating window.

c. Bees will usually work only one type of flower on each trip, (flower fidelity) not mixing pollen types. For example, when the honey bee flies out to gather nectar and pollen from an apple blossom, this is the only type of flower she will visit on this flight.

d. Crops can be sprayed with certain attraction pheromones.

e. Assuring the bees will work only the target crop. Although this handbook principally concerns honey bees and beekeeping, pollinating agents other than honey bees should be mentioned and their relative value considered.

The bees, especially honey bees, are the major pollinators of cultivated crops. The flowering plants and honeybees co-evolved long before the appearance mankind on the earth. It is a known fact that only where pollinators lived could plant evolves a mechanism of reproduction that involved pollination. A majority of 2,50,000 flowering plants on the earth has amazingly complex relationships with the bees and to other less important pollinators. The value of the honeybee as a pollinator is far greater than its value as a honey producer.

Benefits of honeybee pollination

As a result of cross pollination by bees, somatic reproductive and adaptive heterosis or hybrid effects occur in plant progeny. Such hybrid effects bring the following qualitative and quantitative changes in the economic and biological aspects of the plants

1. Stimulates germination of pollen on stigma of flower
2. Increase viability of seed embryo and plant
3. More nutritive and aromatic fruits are formed
4. Increases vegetative mass and stimulates faster growth of plants
5. Increases numbers and size of seed and yield of crops
6. Increases fruit set and reduces fruit drop
7. Increases nectar production in the nectaries of flowers
8. Enhances resistances to diseases and pests
9. Increases oil content of oilseed crops

Not all crops need pollination. Some can produce fruit without fertilisation of the flower. Some flowers are self pollinated, which means that pollen is transferred from the anther to the stigma of the same flower or flowers on the same plant variety. Although this transfer can be achieved by wind or rain, insect pollinators are the most effective.

Other flowers are cross pollinated. In these cases, the pollen is transferred from the anther of a flower on one plant variety to the stigma of a flower on a different plant variety. Plants requiring cross pollination usually can not produce fruit from their own pollen. Again, the most important and efficient carrier of pollen from the anther to the stigma for such plants is the honeybee.

Wind probably is the most important pollinating agent, insofar as it benefits our existence. Most of the forest trees, practically all the grasses and grains, with the exception of some that are completely self-pollinated, and many weeds are wind- pollinated.

The flowers of most wind-pollinated plants are either male or female. The male flowers produce an abundance of pollen to be carried by the wind. The female flowers usually have large stigmatic areas to receive the pollen. Corn is a good example of a wind-pollinated crop.

Birds of several different species feed upon nectar, pollen, or insects in some flowers and serve as pollinators. None is of significance in pollinating our cultivated crops. Their visits are confined largely to deep-throated, usually showy wild flowers.

Insects of many species visit flowers and pollinate them. These include bees, wasps, moths, butterflies, beetles, thrips, and midges. Bees are the most efficient and the only dependable pollinators, because they visit flowers methodically to collect nectar and pollen and do not destroy the flower or the plant in the process. Various species of bees, including the managed wild bees are highly efficient. An estimated 80 per cent of our insect pollination is done by bees.

Honey Bees and Pollination

Modern agriculture has come to depend greatly upon honey bees to fulfill its pollination needs. This insect has several valuable qualifications for this role. Beekeepers maintain honey bees at a high population level in most agricultural areas for the honey and wax they produce. The colonies easily can be concentrated when and where needed to satisfy pollination requirements, and by using techniques developed for honey production their numbers can be increased in a relatively short time. The honey bee is adapted to many climates and can successfully revert to its original wild state in most parts of the world, quickly becoming part of the natural reservoir of pollinators.

Beekeeping in Relation to Pollination

When growers conclude that their crops must be insect-pollinated but do not have sufficient pollinators to do the job, they may decide to rent colonies of honey bees from a beekeeper. There may be a local contractor with whom growers can deal-one who acts as contact agent on behalf of a few neighboring beekeepers.

Most of the bee rentals, however, are personal arrangements between growers and local beekeepers. Unfortunately, growers and beekeepers consider the rental arrangement from entirely different points of view.

Growers may consider only the fees they are paying and the potential value of the bees to their crops, along with having the beekeeper, the beekeeper's vehicles and crew, and the bees on their premises. Growers want to buy a service, pay for it, then go about their other duties.

Beekeepers consider the value of the bees to themselves before, during, and after the pollination contract is concluded. They weigh the advantage of the rental fee against the possibility of a reduced honey crop and the possibility of better forage than in their permanent location (Beekeepers are always looking for better locations). They also consider the constantly threatening adverse effect of pesticides, and the danger of damage to the bees and equipment in making the move. The bees usually are moved at night, and this is hard work with many chances for accidents on the roads or in the fields. When the colonies are moved from a location, another beekeeper may take it over. Exposure to bee diseases increases when colonies, particularly from many beekeepers, are concentrated in one area. And finally, problems associated with collecting the fee after the service is rendered frequently develop. These disadvantages discourage many beekeepers from renting bees to growers.

Pollinators qualifications

Constancy

The insect should show flower fidelity that is it must restrict itself to the blossom of the plants of one species at a time so that pollen grains left by it during its visits on the stigmas of the flowers may fertilize the ovules. Insects which flit from flowers of one plant species to those of another will cause little fertilization. However, too much specialization may limit the usefulness of the insect.

Thoroughness

The visitors should work their way into the flowers so that they come into touch with the essential organs namely, anthers of stamens and stigmas of carpels. Otherwise, they will not be able to transfer pollen from one flower to the other. If they take both nectar and pollen for food, they will be more effective as compared to those who visit the blossoms for nectar only. The contact of the latter class of insects with pollen grains is accidental and no sure results can be expected.

Working hours

Those visitors which start their work early in the day and continue late in the evening naturally put in more work quantitatively. Further, those visitors which work for comparatively long periods during inclement weather such as cloudy and cold days will be beneficial to those crops which put forth blossoms under such conditions.

Total population

It would be highly desirable to have big population of pollinating insects at the blossoming time of crop so that a maximum yield of the crop is obtained and the number of insects available at the time does not act as limiting factor. If the pollinators and their placement are under the control of man, it would be an added advantage.

Though certain species of insects have adapted to pollination of certain crops on a commercial scale, namely *Blastophaga* spp to figs or flesh flies-*Phorima* and *Lucila* to onions in breeding cages,most insects are general pollinators.

Over 25,000 species of bees are found in the world. These include honeybees, bumble bees, stingless bees and solitary bees. Bees are the most effective pollinators of crops and natural flora and are reported to pollinate over 70 per cent of the world's cultivated crops. It has also been reported that about 15 per cent of the hundred principal crops are pollinated by domestic bees (i.e. manageable species e.g. hive-kept species of honeybees, bumble bees, alfalfa bees, etc.), while at least 80 per cent are pollinated by the wild bees (Kenmore and Krell, 1998).

Honeybees as the most efficient and manageable pollinators

Many crops are partially or completely self-incompatible and cannot produce fruit or seed without cross-pollination of their flowers. Moreover, it is not only self-incompatible varieties that benefit from cross-pollination, but self-fertile varieties also produce better quality fruit and seeds if they are cross pollinated (Free, 1993). While other agronomic inputs, such as the use of manure, fertilizers, pesticides and irrigation are important, without cross-pollination desired crop yield and quality of harvest cannot be achieved.

Honeybees are the most widely known of all the bees because they provide honey, beeswax and other products and beekeeping is a prevailing tradition among mountain farming communities. They are the most efficient pollinators of cultivated crops because their body parts are especially modified to pick up pollen grains, they have body hair, have potential for long working hours, show flower constancy, and adaptability to different climates (Free, 1964, 1966; McGregor, 1976). Research has shown that pollination by honeybees increases fruit set, enhances fruit quality and reduces fruit drop in apple (Dulta and Verma, 1987). Among different species of honeybees, the hive-kept species (*Apis cerana* and *Apis mellifera*) are of special value because they can be managed for pollination and moved to fields/orchards where and when necessary for pollination. Pollination using honeybees is the most cost-effective method for pollinating apple and other fruit crops. Use of beekeeping is, therefore, the most promising method of cash crop pollination in the Himalayan region.

In fact, the main significance of honeybees and beekeeping is pollination, whereas hive products are of secondary value. It has been estimated that the benefit of using honeybees for enhancing crop yields through cross-pollination is much higher than their role as produces of honey and beeswax. Various estimates have been made to prove the economic value of honeybees in agriculture in developed countries. Recent estimates by Morse and Calderone (2000) show that the value of honeybee pollination to crop production in the US is US$ 14.6 billion. Similar estimates have been made for other countries. For example the value of honeybee pollination has been estimated at US$ 3 billion in EEC (Williams, 1992), and US$ 2.3 billion in New Zealand (Matheson and Schrader, 1987). Cadoret (1992) estimated that the direct contribution of honeybee pollination to increase farm production in 20 Mediterranean countries was US$ 5.2 billion per year – 3.2 billion in developing countries and two billion in other countries. Similarly, Chen (1993) estimated the value of honeybees to four major crops in China, including cotton, rapeseed, sunflower and tea, at US$ 0.7 billion.

Table 36. Expected increase in crop yield due to bee pollination

Crop	Pollination requirements	Expected percent increase in yield due to cross-pollination	No. of bee colonies needed per ha.
Almond	Highly cross pollinated	20	5 - 8
Apple	Highly cross pollinated	20	2 - 3
Citrus	Often cross pollinated	20	2 - 3
Coconut	Pollen transfer is essential	5	2 - 3
Grape	Often cross pollinated	20	2 - 3
Guava	Cross pollinated	10	2 - 3
Mango	Bee visit is helpful	3	2 - 3
Papaya	Pollen transfer is essential	10	2 - 3
Mustard rape	Mostly cross pollinated	20	3 - 5
Sesamum	Often cross pollinated	20	2 - 3
Sunflower	Cross pollinated	20	2 - 4
Cotton	Often cross pollinated	10	2 - 6

Honeybees are reported to play a vital role in enhancing the productivity levels of different crops such as fruit and nuts, vegetables, pulses, oilseeds and forage crops. A number of studies have been done to show the impact of honeybee pollination on different cash crops. However, the role of honeybees is not very well understood in the countries of the Himalayan region. Most of the research work has been done in developed countries of the world where honeybees are being used for the pollination of various crops. However, the limited research carried out in the countries of the Himalayan region has proved that bee pollination increases the yield and quality of various crops. Bee pollination increased yield and fruit quality in apple (Dulta and Verma, 1987; Gupta *et al.*, 1996), peach, plum, citrus, kiwi (Gupta *et al.*, 2000) and strawberry (Partap, 2000; Partap *et al.*, 2000). Bee pollination did not only increase the fruit set but also reduced fruit drop in apple, peach, plum and citrus (Dulta and Verma, 1987; Partap, 2000; Partap *et al.*, 2000). Reports have also indicated an increase in fruit juice and sugar content in citrus fruits (Partap, 2000). In strawberry, bee pollination reportedly reduces the percentage of misshapen fruits (Partap, 2000).

Studies have shown that honeybee pollination enhanced seed production and quality of seed in various vegetable crops such as cabbage, cauliflower, radish, broad leaf mustard and lettuce (Partap and Verma, 1992, 1994; Verma and Partap, 1993, 1994). These results confirm the usefulness of bee pollination and its role in increasing crop productivity and improving the quality of fruits and seeds (Table 36).

Table 37. Percentage Increase in yield of some crops due to bee pollination

Crop	Increase (%)	Crop	Increase (%)
Fruit crops		Fodders and legumes	
Apple	18.00-69.50	Alfalfa	23.00-19,733
Almond	50.00-75.00	Berseem	193.00-6,800
Apricot	5.00-10.0	Clovers	40.00-33,150
Cherries	56.00-1000	Vetches	39.00-20,000
Citrus	7.00-223.00	Birds foot	3.00-1000
Grapes	23.00-54.00	Miscellaneous crops	
Guava	12.00-30.00	Buck wheat	63.00-100.00
Litchi	453.00-10,246	Coffee	17.00-39.00
Plums	536-1,655	Cotton	2.00-50.00
Vegetable crops		Field beans	7.00-90.00
Cole crops	100.00-300.00	Oil seed crops	
Radish	22.00-100.00	Mustard	13.00-222.00
Carrot	9.00-135.00	Safflower	4.00-114.00
Turnip	100.00-125.00	Sunflower	21.00-3,400
Cucumber & squashes	21.00-6,700	Sesame	24.00-40.00
Onion	353.00-9,878	Niger	17.00-45.00
Cabbage	100-300.00	Linseed	2.0-49.00

Table 38. Summary of pollination management of different crops

Crop	Blooming period of the crop	Number of *A. mellifera* colonies/ha	Number of *A. cerana* colonies/ha	Time of placement of colonies
Fruit crops				
Almond	Mid-February to mid- March	5-8	10-12	5-10% bloom
Apple	April (7-10) days	5-8	10-12	5% bloom
Apricot	Mid - February (2-3 weeks)	2-3	4-6	5-10% bloom
Avocado	April -May	5-8	10-12	10-15% bloom
Cherry	February (7-10) days	2-3	4-6	5% bloom

Citrus	March-April	2-3	4-5	5-10% bloom
Kiwifruit	March-April	8-9	16-20	5-10% bloom
Litchi	March-April	2-3	4-6	5-10% bloom
Mango	February	2-3	4-6	5-10% bloom
Peach	February–March (3-4 weeks)	1-2	2-3	5-10% bloom
Persimmon	March-April (2 weeks)	2-3	4-6	5-10% bloom
Plum	February (1-2 weeks)	2-3	4-6	5% bloom
Strawberry	February to April (2 months)	>15	25	5-10% bloom
Vegetable crops				
Cabbage	February –March	5	8-10	10-15% bloom
Carrot	March-April	5-8	10-12	10-15% bloom
Cauliflower	March-April	5	8-10	10-15% bloom
Cucumber	June –September	1 for monoecious 8 for dioecious	2-3 for monoecious 12-16 for dioecious	10-15% bloom
Cucurbits (pumpkin, squash gourd)	June –September	5-8	10-12	10-15% bloom
Okra	June –September	1-2	2-3	10-15% bloom
Onion	April	5-8	10-12	5-10 % bloom
Radish	March-April	2-3	4-6	10-15% bloom
Turnip	February –March	2-3	4-6	5-10% bloom
Oilseed crops				
Mustard and rape	December-January February-March	3-5	5-8	10-15% bloom
Niger	August-September	3-5	6-8	5-10% bloom
Safflower	March-April	5	4-6	5-10% bloom
Sunflower	June	5	8-10	5-10% bloom
Spice crops				
Cardamom	March-April	2-3*	4-6	10-15% bloom
Chilli	July-September	2-3*	4-6	10-15% bloom
Coriander	February-April	2-3*	4-6	10-15% bloom

*no specific recommendation; (Source: Partap, 1999)

Scientific evidence confirms that bee pollination also improves the yield and quality of other vegetable crops such as asparagus, carrots, onion, turnips and several other crops (Deodikar and Suryanarayana, 1977). Recent experiments carried out in different parts of the northeast Himalayan region show that honeybee pollination does not only increase fruit set in rapeseed and sunflower but also increases the oil contents in these oilseed crops (Singh *et al.,* 2000).

The quality of pollination is determined by the number of colonies per unit area, strength of bee colonies, placement of colonies in the field, time of placement of bee colonies and the weather conditions. Experiences from pilot experiments have shown that the best results are achieved by placing strong bee colonies, having large amount of unsealed brood, free of diseases, at the time of 5-10 percent flowering in the crop (Free, 1993; Verma and Partap, 1993).

Table 39. Yields in cages without bees in percent of yields in cages with honeybees.

Crops	Percentage increase	Crops	Percentage increase
Fruits		**Legumes**	
Malus domestica	6 – 29	*Trifolium pratense*	1 – 34
Ribes nigrum	40 – 29	*Trifolium repens*	9 – 19
		Trifolium hybridum	0 – 8
Oil plants		*Trifolium incarnatum*	3 – 14
		Medicago sativa	0 – 68
Brassica napus (*)	50 – 72	*Melilotus alba*	0 – 6
Brassica napus (**)	35 – 51	*Onobrychus viciifolia*	0 – 14
Brassica rapa (*)	78 – 89	*Vicia villosa*	3 – 37
Brassica rapa (**)	39 – 69	*Vicia faba*	83
Sinapis alba	51 – 75	*Coriandrum sativum*	63
Helianthus annuus	70 – 86	*Fagopyrum esculentum*	14 – 53
Papaver somniferum	49 – 91	*Phacelia tanacetifolia*	12 – 62

Pollination Behavior of Honeybees

Honeybees are cold-blooded animals. They depend upon the surrounding temperature for their activity outside the colony. For pollination purposes, bee flights should be expected within the temperature range of 55' to 100 F. Below 55 few bees fly outside the hive, and when they do it is only for cleansing or orientation flights. As temperature increases, more bees search for water to cool the colony.

Strong winds slow bee flight activity and stop it completely when it exceeds 25 mph. Hot winds will also reduce nectar secretion, thereby reducing flower attractiveness. Hot, dry conditions also reduce the favorable climate for pollen germination.

Applying pesticides to crops near other plants that are attractive to bees should be done carefully. For example, mow dandelion in orchards before bloom to eliminate competition for bee visits and reduce the risk of pesticide injury.

Factors affecting bee populations

The number of colonies of honeybees and the number of bees in each hive can be influenced by bee diseases, pesticide sprays, poor nectar flows, and seasonal changes. A single, wide-scale pesticide application for control of a harmful pest can weaken or destroy many colonies. Drought or prolonged periods of rain, cool temperatures, and early frosts, all limit nectar production by plants. In these cases, the beekeeper may need to feed his bees

Manipulate Bees for Crop Pollination

A. One mistake often made by a fruit, vegetable, or seed producer regarding bee pollination is to assume that "the neighbor's bees" will provide sufficient pollination of his crop. There are several reasons why a producer who does this can be disappointed at harvest time with low yields:

1. Although honeybees from a single hive will fly 2 to 3 miles from the colony, the radius of greatest efficiency and greatest pollination is only 200 to 300 yards from the colony. The reason for this is mathematical, because the farther away from a colony a bee flies, the greater the number of acres it can potentially visit.

2. There are many flowers competing for the bee's attention; some may be more attractive or "profitable" to work than the crop needing pollination. Bees will not fly a great distance to forage on an unattractive crop.

B. By moving bees directly into a field, a grower can increase the flight activity in that area. However, there are several considerations he should make, and procedures he should follow:

1. Colony arrangement: Bees are most efficient when closest to the colony; therefore, to achieve the most efficient distribution of foragers, the colonies should be distributed at 300 to 500 yard intervals. Inexperienced foragers work closer to the hive, then disperse with age and experience.

2. Some crops or specific varieties of some crops are unattractive to bees. In such cases, a solution is to flood the area with bees. Sugar sprays and attractants have failed to increase pollination.

3. Provide water for bees to minimize wasted effort by foragers.

Protecting Bees from Pesticides

Until better methods of covering or otherwise protecting bees during insecticide sprays are developed for use under the humid conditions, the only safe way to protect honeybee colonies is to remove them from the area of the spray. This is difficult, or impossible in some situations, particularly if a large number of acres are to be sprayed or if many hives are involved.

It is suggested that beekeepers should cover hives with burlap during insecticide sprays and keep the burlap wet. Confining bees to the hive by screening the entrance and the top after removal of the inner cover might prevent overheating and minimize losses, as long as there is no direct hit by the insecticide on the open colony.

Minimum Colony Strength

Each hive, enclosing the colony or swarm of bees, is a family unit and consists of a laying queen, a few drones, and several thousand worker bees. Because the worker bees pollinate crops, it is important to understand how their populations change throughout the spring and summer. In early spring, a colony may have 10,000 to 20,000 workers, but by summer the colony should have reached a population of 40,000 to 60,000 workers. During these months the lower portion of the hive, called the brood chamber, is filled with eggs, larvae, and pupae called brood, which are reared in the cells of the comb. Generally, the larger the brood area, the greater the demand for nectar and pollen, thus the greater foraging activity by the bees. This in turn results in more pollination. Colonies used in pollination rental must have brood, and the amount of brood is a strong indication of the effectiveness of the colony for pollination.

Value of Bee Pollination to Agriculture

Unfortunately, the pollination mechanisms of some crops are poorly understood, particularly in some modern hybrids and new varieties. A few crops have been investigated and the pollination requirements are well known. Some scientists have estimated that the value of bee pollination is worth more than one billion dollars per year in the United States, when determined in value of crops involved. The contribution of bee pollination to the production of agricultural crops is considerable. Without this contribution, modern agriculture would be seriously affected.

Unmeasured Pollination Benefits

There are areas where it is almost impossible to place a dollar value on bee pollination. One is the contribution that pollination of plants makes to wildlife food production. Another is the natural beauty resulting from pollination of wildflowers. Many shrubs and countless annual plants reproduce by bee pollinated berries and seed, which provide food for birds and other animals. The beekeeper with a small scale operation is very valuable in providing honeybees for this unmeasured contribution to pollination, because there are many small time, hobbyist beekeepers spread throughout the state who increase food production for wildlife. Thus, it is difficult to place a dollar value on the enjoyment created by birds, wildflowers, and wild animals for the nature enthusiast.

Honey bees as pollinators

Domesticated species: There are two most important hive species. European honey bee, *Apis mellifera* L. is a native of old world except tropical Asia and introduced to most parts of the new world. It has a foraging range of 3 Km. The Indian hive bee, *A. cerana indica* F., a native of tropical Asia is prevalent in a wide region with a flight range of 1.5 Km. It is even a better pollinator than *A. mellifera* because of their longer foraging period and many other characters Sihag and Mishra (1995).

Wild species : Two other species, *A. dorsata* (rock/giant bee) and *A. florea* (little bee) are also native of tropical Asia and efficient pollinators. But these can not be managed for long time, as they do not live in artificial hives. Their foraging range is 2.5- 4.0 and 1 Km, respectively.

Non Apis Bees : Some other wild non-Apis bees are recorded from different countries on different crops serving as important pollinators (Crane and Walker, 1983).

Table 40. Important non-Apis wild bee pollinators

S.N.	Family	Important genera	No. of species
1.	Megachilidae	*Anthidium, Litburgus, Megachile* (leaf cutler bees), *Osmia*	20
2.	Anthophoridae	*Anthophora, Mellisodes, Nomada, Xylocopa*	13
3.	Apidae	*Bombus, Euglossa, Melipona, Trigona*	6
4.	Halictidae	*Aagapostemon, Dufournea, Halicutxis, Nomia* (alkali bees)	5
5.	Andrenidae	*Anderna, Panurginus, Pseudopanurginus, Perdita*	4
6	Melittidae	*Hesparapis, Melitta*	2
7.	Colletidae	*Colletes, Hylaeus*	2

Small amounts of pollen are collected by honey bees in tomato and they require buzz pollination. So bumblebees can do the better job than honey bees. However in closed conditions 2 colonies are required in an 800-1000 m^2 greenhouse (Houbaert and Jacobs, 1992).

Stingless bees: Stingless bee of the genus *Trigona* and *Melipona* (more than 50 species) with a wide foraging range on cultivated, fruits and other crops are important pollinators. In nature they are found in burrows, holes, cracks and crevices in the soil, trees and other dwellings. They have been successfully managed in small wooden boxes and used efficiently in crops like litchi. They resemble honeybees in behavior. They are especially effective in crops on small flowers.

Management of honey bees for pollination

Due to their morphological characters, social behaviour and management practices followed by our farmers, honeybees have become the most important and mostly the only pollinators of the crops. The uncertainty of populations and the difficulty in maintaining and using solitary bees, bumblebees and other pollinators' places all the onus of planned pollination on honeybees alone. The following points are most important to consider a colony for pollen foraging and getting desired pollination outputs.

1. **Foraging strength of colonies**: In order to get maximum honey and pollination benefits from a colony, it must be full-sized and populous not a growing one- as the brood/bee ratio diminishes in it, so that a greater proportion of bees is available for foraging. Matheson (1986, 1991) suggested that a colony

used for pollination should contain at least 7 frames with 60% brood (at least 25% in egg or larval stage) in the brood chamber, a young prolific queen, at least 12 frames covered with bees, sufficient empty combs for expansion and enough honey and pollen stores to sustain it. Separate colony strength standards were also recommended for field colony and orchard colony by Roubik (1995).

2. **Pollination requirement and concentration of colonies**: The number of colonies needed per hectare of crop will depend on local conditions including the number of honeybees and other pollinators already present, the size of the crop and the presence of competing crops of the same and different species like weed species. In Southwestern Haryana, the population of natural pollinators especially the solitary bees is so high that it suffices to great extent the pollination requirement of mustard crop. There is a difference in the number of colonies that could be profitably maintained for honey and the numbers necessary for pollination in a crop. In general, 3-5 colonies /hectare placed equidistant from each other within the crop are recommended. This number of colonies can be handled easily.

Table 41. Pollination requirement of different crops as given below.

Sr.	Crops	Pollination requirement Number of colonies/ha
1	Apple	2-3
2	Almond	5-8
3	Citrus	2-3
4	Sunflower	2-4
5	Turnip	2-5
6	Coconut	2-3
7	Cauliflower	3-5
8	Mango	2-3
9	Watermelon	1-5
10	Mustard	3-5
11	Pumpkin	1-2
12	Sesame	2-3
13	Guava	2-3
14	Onion	2-8
15	Lucerne	3-6
16	Cotton	2-6
17	Papaya	2-3
18	Grape	2-3
19	Sprouting brussels	5
20	Musk melon	1-5

21	White clover	0.2-1.0
22	Kiwi fruit	8
23	Berseem	3-4
24	Buckwheat	3
25	Carrot	6
26	Cherry	2-5
27	Citrus	2-5
28	Cucumbers	6-20
29	Hairy vetch	3.5-7.5
30	Pears	4-6

It is generally recommended to have around 2.5 colonies/ha but it will depend on many factors like concentration of flowers, their attractiveness, competing insects and crops, species, place, percentage of flowers open at a time, etc. and the number of colonies required may be increased or decreased accordingly.

3. **Foraging efficiency of colonies and their distance from the crop**: Honeybees are capable of foraging considerable distance from their hives but their efficiency is indirectly proportional to the distance covered. Generally foraging range is 2.5 km for *Apis mellifera*, 1.5 km for *A. cerana*, 3 km for *A. dorsata* and 1 km for *A. florea*. *Apis mellifera* have been observed to forage up to 11.3 km but foragers were concentrated within 0.6 km of their hives. The yields of the crops are more when the colonies are kept up to a distance of 0.5 km and decrease to almost half at a distance of 1.0 km and these impacts are even greater in poor season. The number of foraging bees on a crop diminishes with increase in the distance from the hive (Free, 1993). In general, placing the hive within 0.5 km radius increases the crop pollination. Recruitment of nest mates to the nearby sites is also greater as this information is more easily communicated. Colonies placed near crops collect more pollen and nectar, spend less time collecting load of pollen and nectar, the number of flights increases for both types with proximity to the floral source.

4. **Moving colonies to crops**: It is advisable to take as short migrations as possible because of the risks involved in long distance migration which result in killing of brood and low foraging efficiency. The settling and arrangement of colonies after migration is very important. Formal arrangement with identical hives should be avoided as it may result in drifting of foragers especially during their first flight which may result in weakening of some colonies and strengthening of others, lowering the mean foraging potential, honey production and pollination efficiency. The colonies should be arranged irregularly in different directions and spacing. They should be put near landmarks and windbreaks and different colored boards may be placed above the entrance.

5. **Time of moving colonies on crops**: The origin of food stores of a colony plays important role in determining the species as they did before moving them. But the predominance of one species at the new site results in changing their

foraging behavior. Therefore, colonies should be moved to the crop needing pollination only when it is sufficiently flowering. The colonies should be moved to a target crop when bloom is between 5-10% to influence the number of foragers that are turnaround on it. Care must be taken to avoid the blooming of too many flowers as an important part of the crop may fail to be pollinated. It is practical to move a few colonies to the crop at the beginning of flowering and the rest when more flowers have bloomed (Sale, 1983).

6. Replacement or Rotation : Colonies should be replaced or rotated with fresh ones when they begins to forage outside target crop. Colonies involved in replacement should be at least 2.4 kms apart or the bees may return to their former sites. This system is particularly useful, where the target crop e.g. Pear is relatively unattractive to bees.

7. Temporary placement: The flowers of crop generally present their pollen/nectar at a certain times of the day Thus confining the bees to their hive until the target crop is presenting its pollen / nectar appear to maintain bees, at least temporarily, on a crop.

8. Removing floral competition : Weeds or other non target crops should be eliminated or mowed when in flower, to avoid competition for foraging bees.

9. Pollen Dispensers: Pollen dispensers (pollen inserts) apply pollen to bees leaving hives so that they can cross pollinate when few pollinizers varieties are available. This may increase pollination efficiency of bees without necessarily maintaining more of them in a target crop. Dispensers may stimulates foraging activity and that may induce bees to forage for the type of pollen in the dispenser.

10. Disposable Pollination Units (DPU's): DPU's are small comb less colonies housed in inexpensive containers that are trucked or parachuted into target crops that arc inaccessible, and then destroyed or left to die when flowering is over

11. Arrangement of colonies: Arrangement of colonies in crop is very important to ensure uniform distribution of foraging bees which will depend on the plant species, amount of nectar and pollen available/unit area, weather conditions and physical features of the area like topographic gradients, wind direction, shelter, etc. Placing colonies in the middle of the crop increases foraging area (Free, 1993). Ideally colonies should be distributed singly over the crop which is always not feasible and practical. So, colonies should be kept in small groups of 4-5 at distances of about 200 m throughout the crop to provide sufficient overlap of foraging area between different groups. The colonies should face the direction of sun in the winter and should be sheltered during summer and rainy season. Windbreaks (natural and artificial) greatly benefit during chilly and windy conditions (Kumar *et al.*, 1998).

12. **Directing bees to crops**: Attempts to 'direct' the bees to the crops to be pollinated either by feeding them sugar syrup containing the scent of the crop or flowers immersed in the syrup or by spraying the crop with sugar syrup have met

with a limited success. Recently the use of "Bee attractants" (Bee-Q and Bee-here) or attractive odors present in the pollen on high value crops met with varied success. Their application though increased forager population but they may be diverted from the flowers and spend much time in searching the plant leaves and stems for pollen but of no avail (Patil *et al.*, 2000).

13. **Increasing the attractiveness of crops:** The more appropriate approach for improved pollination is to increase attractiveness of the crop itself to the bees, by developing plant strains with more accessible nectar or pollen, or developing strains producing more nectar. Unfortunately, this factor is not a priority with the breeders and the recent hybrids (of berseem and sunflower) produce very low amount of nectar, further reducing their attractiveness to the bees (Kumar *et al.,* 1998).

Onion can be honeybee-pollinated, mainly for seed production, though bees will fly over flowering onion crops to other crops in flower. Onion nectar contains about ten times more potassium than most other plants (3600 to 13000 ppm). Potassium can make the nectar unattractive (that is, repellent) to honeybees in high concentrations. Honeybees forage entirely for nectar, but become heavily dusted with pollen in the process (Kumar and Gupta, 1993; Prasad *et al.*, 2000). By use of attractants and sprayers bees can be directed to visit target crops. The bees perform recruitment dances and increase the numbers of bees visiting target crops. These includes : scented sucrose solution, food supplement sprays (eg. Beeline[(R)]), pollen attractants, pheromones and other chemicals

14. **Increasing the proportion of pollen gatherers**: Increasing the proportion of pollen gatherers, which are more efficient pollinators than the nectar gatherers in a colony, is always beneficial. To increase the proportion of pollen gatherers Goodwin *et al.* (1991) have suggested that the foragers may be stimulated to gather more pollen by manipulating colony by placing brood frames opposite to the hive entrance or by removing pollen stores from the colonies or by using pollen traps. Another method is by feeding sugar syrup to colonies which will increase its pollination efficiency by greatly increasing the number and proportion of pollen-gatherers.

15. **Using man-made devices to increase pollination:** Devices like “pollen dispensers" or "pollen inserts" have been used in high valued crops like fruit trees. They are placed at the hive entrance containing hand collected pollen of required compatible cultivar. The out-going bees are forced to carry this pollen. In sunflower the male sterile plants caged with a honeybee colony and a dispenser with pollen had similar percentage of flower set, seed yield and seed oil to the male fertile plants. Newer and more effective dispensers have been developed now using bee collected pollen (Sasaki, 1985). The simple "pollen enhancer" having row of soft nylon bristles at the entrance developed by Free *et al.* (1991) too are effective for the pollination of self in compatible tree fruits and hybrid seed production.

16. **Breeding honeybees for pollination**: Selection and use of Italian honeybees with long-tongue (6.8 mm) has been greatly reported for the

pollination of red clover. Mackensen and Nye (1969), Nye and Mackensen (1970) and Hellmich *et al.* (1985) have demonstrated the possibility of selecting honeybee colonies with high (2-15 times) or low levels of pollen collection for crops like alfalfa.

17. **Use of Synthetic pheromones**: Use of synthetic queen pheromones to stimulate foraging (especially pollen collection) and Nasanov pheromone whose two components (citral and geraniol) have been reported to increase honeybee foraging in onion (Woyke, 1981) and yield of apple (Ohe and Praagh, 1983) hold a good promise. Use of synthetic alarm pheromones, a few minutes before insecticide application may help in repelling them from the crop thereby reducing bee loss, and some other pheromones need further refinement. Repellents sprayed on the target crops (e.g. Carbolic acid, Acetic acid, Propionic anhydride, Benzaldehyde, Calcium chloride) reduce the bee forage and induce them to forage on then less attractive target crops. The mandibular glands pheromones Alarm pheromones can also be used.

Insect pollination is a requirement for the production of many crops, but in agroecosystem native pollinators are often too scarce to ensure adequate pollination. Large scale monoculture, necessary for economic production of mechanically cultivated and harvested crops provides no continual source of pollen and nectar necessary to maintain strong colonies. The indiscriminate and widespread use of pesticides highly toxic to bees either weakens or destroy many colonies or their use causes the beekeepers to relocate the colonies in profitable areas. Intensive agriculture has reduced honeybee population to a point where they are insufficient for pollination of commercial planting.

Pollination and orchard management systems

Pollinators

There are three types of major bee pollinators that may be found in an orchard-solitary bees, bumble bees and honey bees. There are several hundreds of different kinds of solitary bees. Their name indicates much about the biology of these bees. Each individual female bee builds its own nest in hollow weed stems, or makes nest burrows within the ground. In a typical nest the bee would line the fairly straight tube with leaf disks that she cuts from various plants. The bee would collect pollen sufficient to make a small ball in the nest section and then lays an egg onto the pollen. She then caps off the section and repeats the process several more times. Such nests may produce only one generation per year or some species may have several cycles in a season. Since each cell must have pollen for the young larva to develop, the foraging solitary bee is always a pollen collector. Some orchards have an abundance of solitary bees either because of favorable site locations or agricultural practices beneficial to their preservation. Many of our practices have discriminated against these bees because of reduced nesting sites (fence rows), large orchard blocks and by the use of pesticides during the year. Some species of solitary bee can be managed within an orchard system, for example the *Osmia* spp. With these bees the orchardist can provide nest blocks made out of wood. These blocks can be kept in a cooler and removed prior to the blossoming period.

Bumble bees have an annual nest that is started each spring by a queen that has overwintered in a freeze-protected location such as the leaf litter in a woodlot. During the fruit bloom period it is usually only the queen that is foraging. She is collecting pollen and nectar to produce workers that may number to a few dozen by late summer. The colony in the fall produces new queens and males that mate, and the new queens hibernate until the next spring. Bumble bees forage at lower temperatures than either the solitary bees or honey bees, and are less affected by increased wind speed. The technology for management of bumble bee colonies has advanced to where managed colonies are being used for glass-house pollination, and to a limited extent for fruit crops.

The European honey bee (*Apis mellifera* L.) is worldwide in distribution. The early colonists brought them to the New World and they spread throughout the country via swarms. Pollination of fruit trees was mainly provided by nearby colonies, either wild (feral) or managed, until recent years. With increased size of fruit orchards, more insecticide use, and fewer varieties within each orchard, there has been an increased need for planned pollination.

Foraging (pollination) by bees

Bees use pollen as their source of protein and nectar for energy (movement). The solitary bees are always pollen collectors. However, even a nectar collecting bee is almost always a pollinator if she is carrying compatible pollen. When bees clean their bodies to remove the pollen to feed to their young they still leave thousands of pollen grains on the branched hairs that cover them. Thus, when they move from flower to flower they accidentally pollinate the fruit. Most bees visit only one species of flowering plant on a foraging trip. This flower constancy favors effective pollination except that if such constancy would narrow the bees collection to a single variety within a self-incompatible crop, such as apple.

Most fruit crops have a good nectar sugar concentration and thus are competitive for bee visits. However, pears have a low sugar concentration and bees will sometimes leave pears for other crops that are in bloom at the same time.

Foraging trips from a colony are often limited by low temperatures. Honey bees and solitary bees do not forage much below 55°F (13°C). Solar radiation may enhance flight somewhat at these low temperatures. Bumble bees can elevate their body temperature by muscle activity and often forage below 55°F. Between 60° (15.5°C) and 90°F (32.2°C) the increase in available foragers is essentially linear. At the highest temperatures there is a reduction in foraging as more bees are used to circulate air in order to cool the hive.

Increasing wind speed has just the opposite effect on available foraging flight. When the wind speed exceeds about 25 mph most honey bee flight is stopped. Wind speeds can also influence pollination directly. For example, when the wind makes landing on a moving fruit blossom too costly in energy the bee will forage on the dandelion blossoms on the orchard floor where the wind speed is reduced.

Foraging within an orchard will depend upon the size of the tree and often the planting pattern. A honey bee will often find sufficient forage on a standard-size apple tree to complete a load of pollen (or nectar) on a single tree. This would prevent any cross pollination from occurring on that flight. If flower constancy extends to varieties then the problem of compatible pollen transfer is increased. The between-tree spacing is usually much less within a row than between rows, and bees will often continue down a row rather than cross to another row. This foraging pattern is caused by the bees in an attempt to maximize their energy gain, as the greater distance of flight between rows uses more flight energy and thus fewer calories to be gained.

Pollen compatibility and pollinating fruit crops

Many fruit crops are pollen self-compatible - tart cherry, peach, some plums, pear. Others are completely, or partially, self-incompatible. Examples of these are apple, almond and sweet cherry. Then there are some varieties of sweet cherry, or polyploid apple, that are also cross-incompatible.

The problem of self-incompatibility and the optimal fruit-set level becomes intertwined in determining the optimal concentration of pollinating bees. For example, in pollinating peaches we have a self-compatible variety along with a low fruit-set requirement. The number of honey bee colonies recommended is generally less than one colony/acre. When we move to a self-compatible crop such as tart cherries that has a high fruit-set requirement then we need morethan bees one colony/acre. A self-incompatible crop along with a low fruit set (e.g., apple) generally requires about one colony/acre. When we move to high fruit-set levels (35%-40% of blossoms) along with self-incompatibility (almonds and sweet cherries), pollination requirements are between two and three colonies/acre.

There may be an important disclaimer to these recommendations. Most research on fruit set was conducted prior to the loss of the wild (feral) colonies of honey bees due to the newly introduced parasitic mites. The combination of the tracheal mite and Varroa have essentially removed most feral colonies from the landscape. Estimates of the past abundance of these feral colonies has varied from about 50% to 80% of the pollinating population.

Orchard planting patterns

In pollinating self-incompatible crops the orchardist must consider the pollinizer variety, the timing of pollen dehiscence and the mix of pollen within the orchard. For example, in pollinating Delicious apples care must be given to selecting the pollinizer if the king blossom is to be set. The pollinizer variety should begin to bloom at least a few hours before the first flowers on the main variety.

Orchard planting patterns have been developed to take into consideration both the bees foraging patterns and also to maximize the yield of the main variety. Self-incompatible varieties may have pollinizer rows alternating with the main variety, or alternating with two or more rows of the main variety. Growers also may plant a pollinizer tree within the row to circumvent the bees tendency to work down the row. The difficulty with this pattern is that having the pollinizer within the row may conflict with pest management practices or in harvesting. One orchard plan has a pollinizer tree every third tree, every third row. This pattern breaks down into eight main-variety trees surrounding a pollinizer tree. In theory the pollinizer is only one tree away from a main-variety tree. However, at least two factors may be working against this orchard plan. The first is that there is still the problem of a greater distance between rows, and also the ratio of pollinizer pollen to the main variety may become too low under certain pollinating conditions. We often get a cool, wet pollination period. Then fruit set may be reduced since we just do not have enough pollinizer pollen transferred because of few available pollination periods.

Planned pollination

The first step in getting a fruit crop is having sufficient pollination. Without pollen transfer all other management practices are only meaningful for a future years crop. Adequate and sustained pollination helps stabilize fruit set from year to year. Currently it makes sense to rent sufficient colonies for pollination needs, as the number of wild colonies have also declined in recent years. There has been another change in orchards over the last few years. Standard-size trees have been replaced with dwarfing rootstock. There now may be four or five times the number of trees per acre as were previously planted. The blossom density has more than doubled. Yet many orchardists have not increased the number of colonies of bees. They are using the same number of colonies of bees as their fore fathers. This is true even though the orchard has increased block size, decreased the diversity of cultivars useful in cross pollination, removed the nesting sites for solitary and bumble bees, and planted dwarf-rootstock trees.

Monitoring pollination

It is difficult to give good guidelines for observing your trees to assess whether you are getting adequate visitation by bees to ensure pollination. The weather, time of day, intensity of bloom and the type of fruit crop all affect the number of bees that may be seen in a tree. Pollen dehiscence is usually early in the day and if weather is good bees will collect most of what is available in the morning. If you are examining your trees for bees, or honey bee colonies for flight, remember to consider the temperature and wind speed in your determination. We like to see one bee/minute on a dwarf fruit tree under average conditions. Remember, not every bee is carrying compatible pollen on crops such as apple. In apples each flower that is to be pollinated needs about five viable, compatible pollen grains on the stigma in order for that apple to grow and remain on the tree.

Pollination agreement (contract)

Fruit growers and beekeepers should draw up a written contract. If it is written and signed the agreement protects both parties. The contract should specify the size of the colony in number of frames of bees and brood that is minimum. There is a growing trend to pay the beekeeper on the total number of frames rather than colonies. This arrangement benefits both parties. The beekeepers benefit because they could move fewer colonies for the same price, and growers because there are more foragers per relative population in larger colonies.

Evaluating honey bee colonies for pollination

Honey bees are accurately described as indispensable pollinators.

1. The reasons for the success of honey bees as pollinators are seen in certain aspects of their biology. Honey bees forage for nectar and pollen from many thousands of plant species, so they efficiently pollinate a wide variety of crops we all consider important.
2. The second reason is that, while foragers from one hive may visit many species of plants in a given day, individual foragers display flower fidelity or constancy. When a forager begins collecting nectar or pollen from the flowers of one species of plant, she will continue to visit flowers of only that species for at least one foraging trip and more often for several days, or until the resource is no longer producing nectar or pollen. This is obviously important to the plant she visits as it requires pollen from a flower of the same species for pollination.
3. The third reason honey bees are so successful in commercial agriculture is that colonies are mobile. Hives are easily moved to locations where native pollinators may not occur in sufficient numbers to adequately pollinate a crop.

The most important question regarding pollination for the commercial purposes include:

1. How many colonies are necessary to insure maximum pollination of a given crops. This is a difficult question to answer because absolute pollination requirements are not established for most agriculturally important plant species.
2. Furthermore, bloom periods, bloom density, bloom attractiveness, blossom structure, competing bloom, and weather play major roles in determining how well honey bees will forage on and ultimately pollinate a given crop.
3. Finally, the quality of individual colonies need consideration in determining how many to use.

Colony size and efficiency. The older worker bees in a colony serve as foragers. As a colony grows in worker population, the proportion of bees old enough for foraging increases. As a general rule, smaller colonies send out a smaller percentage of bees as foragers. On the other hand, larger colonies send out not only more bees but also a higher proportion of the population as foragers. The greater value of larger colonies for pollination can be illustrated by the amount of honey produced by colonies of different populations. The ability of a direct result of the number of bees foraging and the amount of forage available. It is the foragers that pollinate the flowers they visit.

1. One colony of 30,000 bees produces 1½ times as much honey as the sum of two colonies with 15,000 bees each
2. One colony of 45,000 bees produces 1½ times as much honey as three colonies with 15,000 bees each.
3. One colony of 60,000 bees produces 1½ times as much honey as four colonies with 15,000 bees each.

Challenges in managed crop pollination

Although, both the need and the potential exist in the developing countries, the practice of managed pollination is practically absent. What to space of rearing and using species of bumblebees and solitary bees; even there is no practice of using hivebees such as *Apis cerana* and *Apis mellifera* and stingless bees even though beekeeping is a tradition throughout the developing countries. Develop-ment and use of insects other than hivebees in this part of world will take a long time and need major research and extension efforts before such insects can be reared and managed for pollination of crops in the country like India.

Like soil, water and nutrients, pollination is also a limiting factor in crop productivity. The declining agricultural productivity can be attributed to a number of factors, but pollination plays a crucial role. We may make use of plant husbandry techniques, such as the use of better quality seed and planting material and provide all agronomic inputs including, good irrigation, use of organic and inorganic fertilizers and biocides, but if there is no pollination, no fruit or seed will be formed.

The pollination problem is relatively new and needs due attention at this early stage. Since pollinator scarcity is the main factor responsible for inadeq-uate pollination, solutions to this lie in increasing the number of pollinators. This can be done by conserving populations of natural insect pollinators by promoting integrated pest management and making judicious use of chemical fertilizers and pesticides, however, the most practical and preferred solution to increase the number of pollinators would be by promoting manageable species of honeybees for pollination. There is need to formulate policies that include pollination as an integrated input to agricultural production technologies. Other challenges include strengthening research and extension institutions and human resources development.

Pollinator arrangement

Placement of hives in orchards

Spatial placement of hives in orchards is of utmost importance. The hives should be grouped in units of four to five which should be placed in the orchard or field at intervals of about 150 metres. The reason for this is two fold; it facilitates handling and off-loading of the hives and also creates areas of concentration of honey-bee populations. This increases the chance of bees visiting flowers of different clones or cultivars on the same trip, thereby increasing their cross-pollination effectiveness. The spacing of hives at 150 metre intervals throughout the crop is to ensure good coverage at all times. Bees donot forage at larger distances during inclement weather conditions.

Cover crops: The presence of cover crops which may include attractive plants such as clovers, capeweed and others. This results in a division of the field force and between the flowers and may even mean that the economic crop is completely ignored. This type of competition can be eliminated by mowing or spraying with a selective weed killer.

Location of the hives

The colonies must be placed at the site where there is least disturbance from farming operations. The site used in early spring when cold and excess moisture are avoidable may affect the colony development. To overcome this problem, ensure that the 'cold moisture laden air can move freely and naturally away from the hives.

The colonies should be strong, because within certain limits the stronger a colony the more forager bees it is able. to put into the field. The colonies must be headed by young vigorously laying queens. The colony should have both sealed and open brood when it goes into the orchard and it would be ideal if the sealed brood was far enough advanced to start emerging very shortly after arrival in the orchard.

Furthermore, the hive entrance should face the morning sun, and should be, protected from cold winds especially early in the season. In the case of summer pollination the hives should, if possible, be provided with some form of protection against the excessive noon temperatures. This shade will reduce the number of bees concerned in maintaining a reasonable hive temperature and will permit the colony to put more bees into the field. Trees serving as wind breaks should be planted around the orchard.

Role of bees in sustainable agriculture and Environment

The sustainable development of agriculture has necessitated the re orientation of the present crop production technologies (Free, 1993; Abrol, 1993; Tikoo and Abrol, 1994). Instead of making substantial use of chemical fertilizers, biocides, irrigation, facilitates and heavy machinery for yield enhancement, a shift towards biologically based agriculture will become necessary to increase food productivity. Emphasis in future should be on the full utilization of underutilized resources which are environmentally more friendly. One such resource is an increase in the yield of different cultivated crops through cross pollination by honeybees (Abrol, 1992; Free, 1993; Abrol, 1993). The vital role honeybees play in enhancing the productivity levels of different crops such as fruits and nuts, vegetables and pulses, oil and forage crops has often been underestimated especially in developing countries all over the world. The economic returns of increased crop production through cross-pollination by honeybees are difficult to quantify yet these far outweigh direct values of honeybees as producers of honey and other hive products. The value of bee pollination in crop production in USA has been estimated at US $ 20 billions per year (USDA - ASR, 1991). Evidently, there is a need to create awareness among policy makers, planners and aid agencies about promoting bees and beekeeping as an important component of present day strategies for sustainable agriculture and integrated rural development programme.

Many wild plants are dependent upon bees for continuation of their race which maintain the ecologically balanced ecosystem and maintain the environmental quality. Integration of beekeeping with agriculture will not only benefit the farmers with the direct benefits in the form of honey, bees wax, hive products and employment opportunities but indirectly benefit greatly through increased crop production. It has been established that almost all the agricultural crops depend upon pollinating insects for commercial seedlfruit production (Free 1993; Deodikar and Suryanaryana 1975, 1977; Rao *et al.*, 1980; Abrol 1985; 1986 a,b,c,d; 1987; 1988 a,b,c,d; 1989 a,b,c,d; 1990a,b; 1991a,b; 1993a,b). Insect pollination is extremely important as only 5% of flowers are self-pollinated and 95% cross-pollinated. Among 95%, 85% of pollination is brought about by insects. Evidently for increased crop production and sustainable agriculture it is essential to explore the possibility of planned bees or any insect pollination.

Throughout the world over, three thousand plant species are directly used as food either as seed-beans, peas and soyabeans; as fruits - apples, pears, almonds, cherries, tomatoes, egg plant and score of others as seeds and fruits - flax, coconuts, cotton, oil palm, olives, pea nuts, mustard and sunflower serve as source for edible oils. There is a still another category for example alfalfa, clovers lespedeza' 'which make a part of the diet of domestic animals as green fodder or hay and ultimately recycled to contribute indirectly to our diet as milk and its products, beef, pork, poultry and lamb.

Productivity can simply not include what is obtained as a food through plants and animals, it involves a system which needs be energised from time to time, either naturally or by man's efforts. Constant enrichment of soil nutritionally by some leguminous crops through nitrogen fixation is an example of a basic structure in productivity. Obviously pollination makes a critical junction not less important than other coaxial junctions and perhaps without which even the basic system may break down. On this functional analogy, it has been estimated that about one third of our total diet is dependent directly or indirectly upon crops, primarily pollinated by insects. Their role in maintaining ecological relationships has further been established in situations where absence of insect pollination has resulted in drastic reduction of many soil holding and soil enriching plants. Evidently, need for insect pollination has made the bees a vital component in crop production technologies and sustainable development of agriculture.

Honeybees by far are the most important agents of cross- pollination. Apart from direct benefits derived from honeybees in the form of honey and beeswax, indirect benefits realized by way of increased yields of certain farm crops and forest products due to their pollination services are immense. Compared to the value of honey and beeswax, monetary returns in the form of increased yields may be 20 times or even more. In case of highly cross-pollinated crops like almond, clovers, some varieties of apple, melons, any amount of fertilizer or cultural practices may not even give a fraction of yield unless pollinated by insects. There are some crops which exclusively depend upon bees for production of seeds/fruits.

Table 42. Crops dependent on bees for pollination

Category of crops/fruits	Name of the crop
Vegetables	Pumpkin, cucumber, bottle gourd ridgegourd, carrot, radish, cabbage, knolkhol,cauliflower,onion, soyabean,
Oilseeds	Sarson, toria, sunflower, niger, seasame, safflower, linseed
Pulses	Tur, urad, mung, beans, guar, pea, cowpea
Forage legumes	Lucerne, berseem, clovers
Fruit crops	Oranges, pears, apples, peach, plum, almond, cherry, persimon, strawberry, guava, pomegranate, Jamun, fig, craneberry, grapes, lemon, raspberry, blackberry
Other crops	Buckwheat, cotton, coffee, tobacco, sweet clover
Plants of forest importance	Toon, shisham, soapnut, wild raspberry, stain, Wild cherry, shain, *Euretia* sp., *Robina* sp., *Trifolium* sp., *Eupatorium* sp. *Azadirachta* sp, maple chestnut, eucalyptus, willow, linden, catalpa and magnolia etc.

The fact that the bee pollination enhances crop yields has been established (Free, 1993; Abrol, 1993; Deodikar and Suryanaryana, 1977). The increase in yield of different crops as reported through the bee pollination is presented in Tables 36 and 37.

Evidently, bee/insect should be encouraged for increased and sustainable agriculture production. In India, the total cultivated area is about 160 million bectares and at least one third (if not half) of the areas is under entomophilous crops which require insect pollination. At a very modest rate of 3 colonies per hectare we need 160 million colonies of honeybees but against this requirement we have just less than one million bees colonies at present. This would be possible only if the apiculture as a whole gets boost and number of colonies are multiplied and managed. Besides, it is very necessary to survey different agroclimatic zones to determine the distribution and abundance of pollinating insects so that pollinators specific to different zones are multiplied and managed. Regions with less abundance should be supplemented with apiaries.

Honeybee colonies are active throughout the year. They can be ofitably used both for honey production and pollination and can be aced in crops/orchards wherever and whenever required. Honeybees n also be trained to pollinate specific crops. The extracts of flowers the crop to be pollinated are fed to the bees in sugar syrup. The bees get conditioned to the aroma -of the crop, visit the flowers thereby effecting pollination. For pollination purposes, strong colonies of honeybees be maintained to coincide with the flowering of the crop.

Breeding bees for pollination of specific crops

Preference for specific crops is a well established phenomenon of bee behaviour. Plant breeders have been actively modifying crops for many purposes, generally associated with improved yield and quality or adaptations to mechanized agriculture. Apiculturists should be more intimately associated with breeding programmes involving bee pollinated crops. Breeding crops which are more attractive to bee has enormous; possibilities. The lines which have more nectar production potential and are attractive to bee should be selected. On the other hand some of crops such as alfalfa posses a special challenge as the nectar collecting honeybee rarely trips the florets to expose the genital column. This opened possibilities for breeding honeybees for specific crops.

Pollination requirements of crops

The number of colonies/ha recommended also varies with crop. It also been observed that placement of colonies influences the distribution of bees on the crop. Crops differ in their profitability of saturation pollincation. Blueberries show a direct correlation between the amount of pollination, the number of seeds, the rate of growth and the size and the yield of the fruit. Placement of honeybee colonies in or near the crop can be an important factor for different crops. Information of foraging areas of individual bees is useful to indicate best varietals interplanting arrangement in fruit orchards, row arrangement and cross-hybridization by bees and isolation distances required to prevent varietals contamination of bee pollinated seed crops.

Effect of insect pollination on environment

The benefits derived from insect pollination are not limited to cultivated crops. The absence of these beneficial insects would have a drastic effect on the non cultivated soil holding and soil enriching plants which would dosappear. Furthermore, without insect pollination to produce the fruits, berries other seeds on these non cultivated plants essential to a well balanced wild life population, many forms of wild life would disappear. Insect pollination is an essential link in our ecological global chain. It has been estimated that more than 1,00,000 species of wild plants depend upon insects for pollination and reproduction (Teale, 1957).

Efficient insect pollination is of utmost significance for continuation and propagation of the plants growing in wild which maintain the environmental quality. Among insect pollinated forests and trees, maple chestnut, eucalyptus, willow, linden, magnolia and catalpa should be mentioned. Pollination needs of many wild species to produce fruits and nuts that are eaten by birds and small mammal are unknown. If such crops were not available agricultural crops would receive more depredations. There should be economic studies to determine costs and benefits from the use of bees, so that both the grower and beekeeper can be assured of adequate profits. This research should be conducted by agronomists, entomologists, horticulturists and apiculturists. Therefore, for sustainable agriculture, research should not be confined to agricultural and horticultural crops but reproductive and pollination needs of forest plants must also be evaluated.

Management during pollination

The individual crops may vary in their requirement of bee colonies for pollination. The number of bees required depends on the number of flowers available in a given area for pollination and therefore, on the species of the crop, planting method, age of the crop, stage and duration of flowering etc. The number of bees available for foraging depends on the ecotype of the bees, management of the colony, competing flower species number of wild pollinators and weather conditions. Generally 3-9 bee colonies of good strength are needed for one hectare of crop. A large number of colonies help pollination, but may be unsuitable for either honey production or colony multiplication. Keeping this in view, optimum number of colonies per hectare have to be placed in the fields after migration. It is advisable to keep colonies near the plants to be pollinated. Colonies may be unloaded from the truck and distributed evenly all over the field. In northern regions, during winter, colonies can be kept in the open sunshine with entrance facing south-east. In more tropical regions shady cool places are necessary.

Efficiency in pollination depends largely on the management of colonies. The basic need for management of colonies for pollination is to keep the colonies in peak foraging activity, which is possible when there is peak brood rearing activity. Some of the management practices which help in this., are (i) provision of clean, new combs to provide adequate space for constant brood rearing, (ii) preventive methods against diseases or pests which divert the attention of colonies (iii) provision of a young newly mated vigorous queen etc. Dilute stimulative feeding (one part of sugar to two or three parts of water by measure) proves beneficial in making bees collect pollen. Thick syrup should, however, be avoided.

In order that colonies fix up their visits to the crop to be pollinated, it is advisable to migrate colonies to the field when the crop starts flowering. It is often very important to synchronize flower opening and bee activities. Colonies freshly migrated to a crop are more likely to visit the crop than those remaining there for a long time. Shifting of a few colonies at a time to the field, 2 to 3 times during the flowering of the crop, may help in thorough pollination. Visits to the crop by foragers can often be enhanced by feeding colonies with sugar syrup in which a few fresh flowers and pollen of the crop are soaked for some time. Such feeding is best given in the nights or early in the morning before bees go out.

It may not be uncommon when the crop to be pollinated is ignored by the bees, which may find other plants in the same place more attractive. When such a contingency arises, bringing new colonies to the field and conditioning them to the crop as above may be helpful. The earlier colonies can be shifted to a place at least 2-3 km away from the field for a few days and can be brought back with the above arrangements for conditioning them to the crop. A knowledge of the bee foraging behaviour and ecology of the crop helps in solving such problems.

The farmer's role in pollination of crop by bees can also be over-emphasized. Planned cropping pattern to avoid unnecessary competition between crops in attracting bees, removal of weeds, proper manuring, irrigation and other cultural practices for healthy growth of the crop, using less toxic insecticides or sparing use of insecticides and similar toxic chemicals, arc some of the responsibilities of the farmers for effecting efficient pollination.

Beekeeping is as much an art as a technique- It needs good deal of practical experience of some years. The busy farmer may not be able to achieve efficiency in a short time in good beekeeping. The forage gap periods and hot summers of plains may make it difficult for him to maintain bee colonies in adverse periods. He may initially prefer, therefore, to hire bee colonies for pollination when crops are in full bloom just as he hires sheep for manuring. Bee colonies are hired for pollination in western countries, especially in the United States. As 20% to 40% increase in many crop yields by bee pollination is possible, it may not be unfair on the part of the farmer to spare a portion of the increased yield to the beekeeper as his tees for pollination service.

During this periodic migration of colonies by a beekeeper the farmer himself can learn beekeeping technique and should try to have, gradually, a few of his own bee colonies rather than depend upon beekeepers in adjacent hills, wherever this is feasible. By this way a farmer can become self-reliant. He will have the satisfaction of getting some quantity of nature's precious sweet honey, from his own farm and to the same time will add to his crop-yield both in quality and quantity.

It is clear from above that beekeeping in addition to being a cottage industry to produce honey and beeswax, provide job opportunities, creasing economic returns of the small land holders, is an effective Id increasing farm productivity, maintaining environmental quality and staining balanced agriculture.

Utilization of bees for crop pollination

Man has been aware of importance of cross-pollination of flowers since hundreds of years, and hand pollination was the sole means to accomplish this task. The ancient European paintings and sculptures depict the pross of cross pollination in some plants. Darwin too highlighted the importance of cross pollination in plants. During the course of evolution, pollinators and flowers had parallel evolution. In the beginning the flowers were simple, open and actino-morphic and beetles were the pollinators. These beetles visited the flowers and the floral parts for food and in turn brought about pollination. This type of pollination was called dystrophic pollination. Later on, more complex flowers evolved along with the evolution of very important and most effective breeding barrier-the self-in compatibility and entomophily was the mode ofpollination.All other forms of pollination like anaemophily, hydrophily, geophily and zoophily are secondary derivatives of entomophily (Faegri and van der Pijl, 1979). Within the entomophilly, melittophily is considered to be the most effective for pollination of crops.

Apis has become more independent and consequently more reliable as a pollinator being active all the year round. *Bombus* on the other hand is very strong, more or less disastrous competitor to the native solitary bees. Social hymenopterans comprise the most versatile, the most active and consequently the best known pollinators. The large bees-social or solitary-are also stronger and can open floral doors that are closed to smaller and weaker insects. Their perception and ability to remember plant forms also seem to surpass those of most other pollinators.

Management of bee pollination of crops

Following groups of bees can be managed for pollination of crops:

1. Honey bees: Two species honey bees are effectively managed and utilized for pollination purpose. These are: *Apis mellifera* L. and *A. cerana* F. Honeybee species can effectively be utilized for pollination of crops because

(i) Honeybees are dependent on flowers for pollen and nectar as their food (Free, 1993).

(ii) The bees possess some morphological adaptations favourable for pollen carry over and transfer (Free, 1993).

(iii) The honeybees can be kept in the hives and are very easy to be managed (Singh, 1960; Dadant and Sons, 1992).

(iv) Due to their polylectic nature, honeybees visit large number of plants (Sihag, 1990a,b), therefore, they can pollinate a wide variety of crops (McGregor, 1976; Free, 1993; Sihag, 1995a).

(v) Their abundance on the crop can easily be manipulated (Jay, 1986).

(vi) Honeybees have better communication system for food sear-ching and gathering. When a honeybee forager begins foraging on a plant species, continues to do so until the resource gets exhausted. This behaviour of individual foragers has been termed as floral fidelity or constancy (Wells and Wells, 1983; Waser, 1986). This is very important for the plant species they visit for the effectivity of pollination.

(vii) Behaviour of honeybees can be manipulated by modifying the reward system of the plant/or nectar and pollen storage in the hive (Free, 1965, 1967; Baker, 1971; AI- Tikrity *et al.*, 1972) or colony's unsealed brood (Free, 1967; Cale, 1968).

(viii) Colonies of honeybees can be moved to a place of short pollinators supply (Jay, 1986; Mardan, 1995).

Due to their habits of making nest in the cavities these bees can be domesticated in the wooden hives. For effective pollination of a crop, following points are taken into account.

(i) Time of placement of honeybee colonies

When 5-10% of the flowering has initiated, honeybee colonies are placed on the crop.

(ii) Site of placement of honeybee colonies

The colonies are placed near the target crops so that they can actively move and bring about cross pollination and less energy is utilized in travelling:

(iii) Method of placement of honeybee colonies

Colonies are placed either in groups or are scattered, and further, either around the target crop or scattered in between as per the convenience of the beekeeper.

Types of colonies

Strong 8-10 frames standard colonies are placed for pollination purpose. The colonies should meet the following requirements.

Bee strength	8-10 frames
Unscaled brood	2 frames
Sealed brood	I frame
Honey	2 frames

The colonies should have a new mated queen and these should be free eases and pests. Other general and specific management practices honeybee colonies are also to be observed. Honeybees can be utilized for pollination of crops in all the agro-climates. Hence these are utilized most extensively world ever and for very diverse crops.

2. Bumble bees

These are the bees of temperate climates. These are one of the largest in size bear thick plum of hairs. These are mainly used for pollination of crops in green especially useful for crops with large size flowers e.g. tomato, brinjal especially valuable in pollinating those flowers in which their large size help transfer pollen while they are visiting the nectaries (e.g. *Gossypuim* spp. cultivars of tree fruits); in pollinating flowers with deep narrow corolla tubes from which only insects with long tongues can obtain nectar. In nature, these bees make nest in the deserted burrows of rats and voles. The same habit of the bee is exploited by man.

Accordingly, these bees can easily be domesticated in artificial wooden boxes with cotton swabs and artificial feed at the bottom. The over-wintering mated queens during winter months are kept in these boxes. They raise colony in the coming flowering season. These are also important pollinators of apple and alfalfa.

3. Stingless bess

These are the bees of the hot climates and are pollinators of diverse types of crops grown under such climates. In nature, these bees make nests in hollow tree trunks. The same habit of these bees have been exploited by man. The whole colony is removed from the tree trunks and is transferred to a wooden box especially designed for this purpose. These boxes are then placed near the target crop.

4. Solitary bees

Several species of solitary bees are now managed for pollination of crops. These include alfalfa pollinating bees of the genus *Megachile, Chalicodoma* and alkali bee *Nomia,* and cucurbit pollinating bee *Xylocopa, Megachile, Chalicodoma* and *Xylocopa* make nests in hollow stem tunnels. Accordingly, these bees have been exploited for pollination purpose using appropriate nesting materials. Nomia is a soil nesting bee. Its nesting beds are prepared near the alfalth crop.

In some crops like oil seed toria, brown sarson, taramira, turnip, caulifower, radish etc. there exists strong self-incompatibility. Here without cross pollination there will be no seed set. In such crops, for effective seed production, cross-pollination by honeybees is a must. In cucurbit crops and papaya, the flowers are unisexual. The pollen from male flower must reach the stigma of female flower for successful fruit set. In crops like carrot, fennel, coriander, cumin and onion, the flowers are protandrous i.e. they shed their pollen much before the receptivity of the stigma and self-pollen is therefore, useless in the these crops. Here too, cross-pollination by honeybees is essential for effective seed yield. In crops like alfalfa (lucerne) and berseem, if the flower is not tripped (forcibly opened) by the bees, there can not be pollen transfer and seed set in the flowers. In all such cases, bees are the agents which can be managed for pollination of their flowers. In all the cross-pollinated crops, seed yield due to bee pollination increases to many fold (McGregor, 1976; Free, 1993). Therefore, bee-pollination should be taken as an input in agriculture for higher seed yield.

Chapter 18

Management of wild bees for pollination

The term "wild bee" is used commonly for all bees except honey bees in the genus *Apis*. Bees generally are distinguished from other flying hymenopterous insects by their characteristic plumose body hairs. Bees are of many sizes, shapes, and colors. Some of the smallest bees, *Perdita*, are less than 3 mm; whereas the largest leaf cutter bee is over 80 mm. Almost the entire range of colors is found among the brightly marked bees, including many beautiful metallic species.

One can easily observe many species of bees actively visiting flowers for nectar and pollen or engaged in the processes of constructing nests. According to an estimate there are atleast 30,000 species of bees in the world. This number of species is more than all the fish, bird, and reptile species combined.

Most bee species construct either single or complex nests in the ground. Some make earthen, leaf, or resin nests on rocks and plants. Other bees make or utilize crevices in rocks or plant stems, insect borings, and plant galls for their nesting sites. Most bees live a solitary existence-each female after mating locates and builds her nest without the aid of other bees, and usually at a distance from her sister bees. However, some bees are quite gregarious and nest close to one another, sometimes in dense populations of up to a million nests in a few acres of soil. Some bees prefer to nest at the same site year after year, but others relocate their nests each season. A small percentage of wild bees are social or semisocial; that is, there is a division of labor among the bees occupying a single nest.

Value of wild bees as pollinators

One cannot easily determine the figure on the value of wild pollinators, simply because total impact on the environment is not known. Studies on the impact of each pollinator species on fruit or seed production of our major crops is almost nonexistent. The reproduction of wild flowering plants is often taken for granted to aid in maintaining soil moisture and fertility, and to provide food not only for wild life but for our domestic livestock as well. How many billions of dollars are these benefits worth? It is easy to document the value of crop species visited by bees, but here again the importance of wild bees as crop pollinators has been neglected. It has long been the general consensus that honey bees adequately pollinate crops and there is little need for wild bees. Unfortunately, it is not true since adequate research on the economic benefits of wild pollinators has not been done. Conversely, the research completed on the few wild pollinator species thus far studied has returned much compared with its investment costs.

The dependence on one species for crop pollination sometimes creates problems. It seems wise to make greater efforts to study, conserve, and try to manage as many species of wild bees as possible. There are several crops that are under pollinated by the honey bees, either because the bees are not physically adapted to pollinate them or the crops are not attractive to honey bees. Some of our most important crops, valued at billions of dollars, are in this category. These crops are alfalfa, soybeans, cotton, vegetable seed, and sunflowers, each of which is adapted to specific types of pollinators. Recent research on the utilization of several species of wild bees as crop pollinators is just beginning to indicate some of their economic benefits, e.g the alfalfa leafcutter bee. The alkali bee was the first wild bee to be utilized as a crop pollinator in the United States beginning in the early 1950's. Since that time, the alfalfa leafcutter bee and the blue orchard bee have been domesticated as crop pollinators.

Friese (1923) estimated that out of 20,000 species of bees (Superfamily; Apoidea), only four species of honeybees (now nine) and 300 species of stingless bees (Family: Meliponinae) live in the permanent perennial colonies. The majority of the bees are solitary where a female constructs a nest consisting of one or more brood cells stocked with nectar or pollen that provide food for the larvae that will emerge from the eggs she deposits just before the sealing of the cell. In general, two thirds of the bee fauna is comprised of the solitary bees (Michener, 1965; Linsley, 1958; Bingham, 1897; Batra, 1977).

Michener (2000) apprehended 16,325 species of bees, grouped under 425 genera.The taxa found in whole of the world were reorganized under 7 families. Still much needs to be known from different regions, more particularly, from the Oriental region of the world.

Unlike other commonly known insects, excluding honey bees that belong to genus *Apis*, bees have least attracted the attention of Indian taxonomists and biodiversity workers. No doubt various aspects on honey bees (*Apis*) such as their domestication, management and crop pollination have been considerably explored in India. However, this is not true for non-*Apis* bees. Most of them forage on wild flowerings located in the forests, buffer zones or more often that grow as weeds all along the cultivated crops. The wild flowerings infact constitute the primary resource for nectar and pollen for most of bees. Non-*Apis* solitary bees also visit cultivated crops in good populations but, as an alternate to wild flowerings.

In North India, Batra (1977) recorded 89 species of solitary bees out of the 97 species studied. The occurrence of major bee genera in India (Table 43) shows that bees belonging to family Megachilidae and Anthophoridae are most commonly distributed in India. Since India is a vast subcontinent with marked topographical and climatic differences, the climatic and florestic conditions vary from tropical and subtropical to subtemperate and temperate conditions, this is the reason that bee fauna from one region differs from the other. Most bees are distributed from valleys through plains to seashores. The Indian species of *Bombus* is generally restricted to higher elevations especially in the Himalayan ranges. Bingham (1897) recorded 24 species of Bumble bees from higher elevations of Kashmir; Himachal through Sikkim and Assam. Mani (1962) reported four species of bumble bees at elevations of over 4000 m at Himalaya. William (1991) has recorded 28 species of bumble bees from Kashmir including areas under illegal control of Pakistan. *Xylocopa* species and *Pithitis smaragdula* in the north western states of India remained confined to 914 m (Kapil and Dhaliwal, 1968) and 391m (Kapil and Kumar 1969; Kapil *et al.,* 1971) above the mean sea level. The other species are generally abundant in warmer, semi-arid areas, yet distributed in temperate and mountain hill ranges also. The number of species reported in each genera holds the following numerical order: *Halictus* 52, *Nomia* 25, *Bombus* 28, *Xylocopa* 19, *Megachile* 44, and others are still less (Bingham, 1897; Batra, 1977).

Table 43. Important non- *Apis* bee pollinators of some agricultural crops in India

Crop/plant	Bee species	References
Family : Leguminosae Alfalfa (*Medicago sativa)*	*Megachile bicolor, M. disjunta, M. flaviceps, M. femorata, M. lanata, Nomia oxybeloides, N. divisus, N. pusilla, Pithitis smaragdula, Xylocopa fenestrata*	Kapil *et al.*, 1974
	Megachile nana M. flaviceps M. femorata M.cephalotes	Kapil *et al.*, 1975
	Braunsapis spp	Kapil and Jain, 1980
	M. lanata, M. cephalotes M. cephalotes Pithitis smaragdula	Abrol, 1985
	M. flaviceps, M. nana	Abrol, 1986b
Berseem (*Trifolium alexandrium*)	*M. flaviceps, M. nana*	Abrol, 1986b
White clover (*T. repens*)	*Bombus asiaticus, B. albopleuralis*	Abrol, 1987
Red clover (*T. repens*)	*Bombus asiaticus, B. albopleuralis*	Abrol, 1987
Pigeon pea (*Cajanus cajan*)	*Megachile lanata, M. bicolor, M. flavipes, M. cephalotes, M. femorata*	Chaudhary and Jain 1978
	Megachile lanata, Xylocopa fenestrata, X. pubescens, M. bicolor, M. cephalotes	Abrol, 1985
Sunhemp (*Crotolaria juncea*)	*X. pubescens, X. fenestrata*	Kapil and Dhaliwal, 1968
	Megachile lanata, M. fasciculata, X. fenestrata	Grewal and Singh, 1978
	Megachile lanata	Abrol and Kapil, 1986

	M. bicolor	Abrol, 1987a
Pea (*Pisum sativum*) Family Compositae	*X. fenestrata X. pubescens Megachile lanata*	Kapil *et al.*, 1975
	Braunsapis spp.	Kapil and Jain, 1980
	X. fenestrata X. pubescens M. cephalotes M. flavipes	Abrol, 1985
	B. albopleuralis Bombus asiaticu Lasioglossum spp.	Abrol, 1987
Sweet potato (*Ipomoea batatas*) Family Convolvulaceae	*X. fenestrata B. albopleurali, Bombus asiaticus s*	Abrol, 1987
Egg plant (*Solanum melongena*)	*B. asiaticus*	Abrol, 1987
	X. fenestrata Ameigilla delicata A.subcosrulea Nomia caliphora Pithitis spp	Batra, 1967
Onion (*Allium cepa*) Family Liliaceae	*Nomioides spp*	Kapil *et al.* 1975
	Lasioglossum spp Nomioides spp X. fenestrata	Abrol, 1987
Field mustard (*Brassica campestris*) Family Cruciferae	*Nomioides Megachilids Andrenids Halictids*	Kapil *et al.*, 1971
	Andrena ilerda A.leaena	Abrol, 1986c
Rape (*Brassica napus*)	*Andrena ilerda*	Mohammad, 1938
	Halictids	Rahman, 1940
Raya (*Brassica juncea*	*Andrena ilerda*	Kapil *et al.*, 1971
	A.leaena Andrena ilerda	Abrol 1985, 1986
Taramira (*Eruca sativa*)	*A.leaena Andrena ilerda Colletes Halictus spp*	Kapil *et al.*, 1971
Cabbage and cauliflower (*B. oleracea*)	*Andrena ilerda Lassioglossum spp Pithitis smaragdula*	Batra, 1967
Raddish (*Raphanus sativus*)	*Anthophora spp Nomia spp., Lassioglossum spp Colletes spp*	Batra, 1967
Pumpkin and squashes (*Cucurbita spp*)	*X. fenestrata X. pubescens Halictus spp Nomioides spp*	Atwal, 1970
Smooth loofah (*Luffa aegyptica*)	*X. fenestrata X. pubescens P. smaragdula*	Kapil *et al.*, 1965-1970
Cucumbers (*Cucumis melo*)	*Nomia spp P. smaragdula Nomioides variegata Halictids*	Kapil *et al.*, 1965-1970
	Lasioglossum spp	Abrol and Bhat 1987
Cotton (*Gossypium* spp) Family Malvaceae	*Lithurgens attratus*	Batra 1977
Corriander (*Corraindrum*	*Nomioides spp Halictidae X. fenestrata*	Abrol 1985

sativum) Family Umbelliferae		
Saunf (*Foeniculum vulagre*)	*Halictis spp X. fenestrata*	Abrol 1985
Carrot (*Dacus carota*)	*Lasioglossum spp Sphecoides Hyleaus Nomioides Braunsapis Pithitis smaragdula*	Batra 1967
Jowain (*Traechyspermum ammi*)	*Andrena spp Nomioides Halictus spp Lasioglossum spp*	Batra 1967
Orange and lemon *Citrus* spp Family :Rutaceae	*Lasioglossum spp X. fenestrata*	Batra 1977
Guava (*Psidium guajava*) Family : Myrtaceae	*X. pubescens X. fenestrata Megachile lanata*	Batra 1977
Mango (*Mangifera indica*) Family :Anacardiaceae	*Xylocopa spp Megachile spp Nomia spp Lasioglossum spp*	Batra, 1967
Pomegranate (Punica granatum) Family :Punicaceae	*Nomioides Lasioglossum Halictus spp*	Batra, 1967
Family : Rosaceae Apples (*Pyrus malus*)	*Colletes nursei, Lasioglossum* spp., *Cattulum Osmia cornifrons*	
	Andrena spp., *Bombus haemorrhoidalis Halictus vacchalli Osmia* spp *Pithitis* spp *X.fenestrata Nomia* spp	Anon. 1986
Almond (*P. amygdalus*)	*Lasioglossum* spp., *Xylocopa valga*	Abrol 1987
Cherry (*P. avium*)	*Xylocopa valga Nomia* spp	Abrol 1987
Pear (*P. cumminis*)	*Xylocopa valga Nomia* spp	Abrol 1987

Table 44. Nesting behavior, nest acceptance and temperature tolerance of sub-tropical *Megachild* and *Xylocopinae* bees

Bee species	Crops pollinated	Nesting behavior	Nest tunnel accepted		Temperature for activity °C
			Diameter (mm)	Length (cm)	
Megachile haryanensis (M. nana)	Alfalfa	Leaf cutter bee	4.5-5.5	10-12	26-43
Chalicodoma rubripes (M. flavipes)	Alfalfa	Masson bee	5.5-6.5	10-12	28-43
C.lanata (M. lanata)	Pigeon pea	Masson bee	6.5-7.5	10-12	26-38
Chalicodoma cephalotes (M. cephalotes)	Pigeon pea	Masson bee Utilizes resin / animal faecal material for partitioning and closing the	5.5-6.5	10-12	27-41

		cells / tunnels			
Xylocopa fenestrata	Cucurbits	Carpenter bee	10-12	23-30	25-48
Xylocopa pubescens	Cucurbits	Carpenter bee	10-12	23-30	25-48
X. valga	Almonds, apples	Carpenter bee	10-12	23-30	5-30
X. aestuans	Cucurbits	Carpenter bee	10-12	23-30	10-47

Modified from Sihag (1994)

Wild bees and their management

The extensive need for insect pollination in modern agriculture has made bees a vital factor in crop production. To most people, the word bee suggests only the common honeybee, perhaps also bumblebees; but the bees are infact a superfamily of the order Hymenoptera, containing an estimated twenty thousand species. The true honeybee, *Apis mcllifera,* is only one among these many species, relatively few of which arc social in the usual entomological sense. They are infact a group of flower-visiting wasps termed as bees that has abandoned the wasp habit of provisioning nests with insect or spider prey and instead feeds its larvae with pollen and nectar, collected from flowers or with glandular secretions ultimately derived from the same sources. They are a large and diversified group, considered to be consisted of nine families (Michener, 1974). Except in certain Colletidae which carry pollen with nectar in the crop, the structures used for carrying pollen consist of scopal hairs having various locations and arrangements. One such group is the Masaridae, a family of wasps closely related to the Vespidae and in the superfamily Vespoidea. The other group of wasps which abandoned predation as a source of larval food was from an entirely different source than the masarids, namely the wasps of the superfamily Sphecoidea. From this group arose the bees. Bees entirely dependent on flowers for food, could not have arisen before the appearance of the angiosperm plants.

How early bees arose from the sphecoid wasps is unknown; it might have been as late as the middle Cretaceous since angiosperms also became the dominant vegetation in middle Cretaceous times. Primitive angiosperms had relatively shallow flowers, such as can be used as pollen and nectar sources by short-tongued insects, including many beetles, wasps, and the short-tongued bees. The first five families, listed as Colletidae. Halictidae. Oxaedae, Andrenidae and Mellitidae are characterized by usually short mouthparts and are often grouped as the short-tongued bees. The other four families-Fidelidae, Megachilidae, Anthophoridae and Apidae are classified as long tounged families, all are equipped with elongated glossae, maxillary galeae and basal segments of the labial palpi forming the sucking apparatus for taking advantage of nectar sources with deep tubular flowers. Flowers with deep corolla tubes probably arose in co evolution with their principal pollinators. The legume crops and their pollinators such as alfalfa-megachilid constitute an important example. There is close relationship between bumble bee species and their major plant hosts accordingly to their length. The worldwide distribution of these bees and their remarkable ability for proliferation are the results of their higher degree of adaptability (Stephen *et al.*, 1969, Michener, 1974).

In many colonies there are interrelations among individuals, such that behavior of one influences the behaviour or development of another. All these interrelations are termed social interactions. Feeding of a larva by a bee is an example of a social interaction. As indicated previously, colonies of bees range from those that seem almost insignificant-two or three bees in a burrow in the ground or in a hollow stem- to the large colonies of the honeybees. The kinds and amount of division of labour and communication among bees in colonies vary greatly. Species are often called solitary, communal, social, and so forth. Such terms are generally applied to the most complex type of organization attained during life cycle of the species. In the majority of species of bees each female makes her own nest, or sometimes several of them, without regard to the locations of other nests of the same species. Such bees mass-provision the ce lls by placing enough pollen and nectar in each to provide for the entire growth of a larva. After oviposition, the mother seals the cell and goes on to construct and provision another. Ordinarily she dies before her progeny mature and emerge from their cells; therefore, there is no contact between generations. Probably the majority of species of solitary bees have only a single generation per year, the adults emerging and flying about during a relatively brief season, sometimes only two or three weeks. Such species pass the rest of the year in the nest. The feeding stage of the larva is ordinarily brief, often only a few days, and most of the year is passed in the pupal stage or as young adults either still in their natal cells or in special hibernating or aestivating places. Some solitary bees, however, regularly have two generations per year, for example one in spring and another in the autumn, while others go through a succession of overlapping generations so that, except in the spring, all stages can be found at anytime during the warmer months of the year. Examples of North American bees having such life histories are *Megachile brews* and *Augochlora pura.* In the tropics, where there is no winter, some species of solitary bees breed continuously throughout the year.

Nest aggregations occur most commonly among bees that burrow in the soil. Aggregations of such burrows may vary from a few nests to dense clusters of nests like those of *Lasioglossum versatum* in Kansas. Some bees that make burrows in stumps or logs instead of soil also form aggregations. For example. one may find numerous nests of carpenter bees *(Xylocopa)* in a single post or building. while nearby none are found in similar posts or buildings. Bees. mostly megachilids. that construct exposed cells of mud and other materials brought to tbe site sometimes also form aggregations of nests: for example. *Chalicodoma muraria* sometimes covers large portions of walls in southern Europe and North Africa with masses of its cells. Communal quasisocial and semisocial groups are so similar superficially that the convenient collective term para social has been proposed for them. Parasocial colonies are simply any colony in which the adult bees consist of a single generation, unlike the eusocial forms in which two generations of adults are ordinarily present. A communal colony consists of a group off emales of the same generation using a single nest. each making, provisioning, and ovipositing in her own cells. In the enormous genus *Alldrefla,* most species are solitary, some nesting in aggregations. However, *Alldrella bucephaja* (Perkins, 1917) and *A. ferox* live in colonies that are probably communal. These are small colonies in which two to several females, usually of about the same age and of the same generation, cooperatively construct and provision cells. Morethan one bee working on a given cell as in communal groups, each female has enlarged ovaries and is mated, indicating that each is an egg layer. Some species of *Nomia* are possibly quasisocial, although knowledge of their colonies is inadequate (Batra, 1966a). The best known of such species is *N. capitata* from India. These are small groups which show cooperative activity and division of labor among adult females as in eusocial groups. Polygynous young colonies of other halictines (Vleugel, 1961) are often temporarily semisocial in that division of labor develops among gynes, one becoming the egg layer or queen, the others auxiliaries or workers. These are family groups each consisting of one adult female and a number of immature offsprings which are protected and fed by the adults. The mother leaves or dies before or at about the time that the young reach maturity. There is no division of labour among adults, as is found in semisocial and eusocial groups. Young colonies of *Bombus,* before workers are produced, are subsocial; the queen progressively feeds the growing larvae in a more or less subdivided common cell. However, the true social Hymenoptera, for which the word eusocial was coined by Batra (1966b), live in colonies which are family groups consisting of individuals of two generations, mothers and daughters. Usually in bees a eusocial colony contains only one queen and the bulk of the females are workers (daughters). Division of labor, with some individuals functioning as egg layers or queens and other as workers. that is, with more or less recognizable castes occurs in both the semi social and eusocial colonies but not in the other types of colonies.

Colonies are long-lived and sustained through periods of adversity by food for adult as well as larval consumption, stored in the nests. Integration within colonies is complex and involves a variety of behaviour patterns, pheromones, and physiological adaptations that would have no obvious function in solitary forms. Aggressive behaviour among individual of the same colony is rarely evident, nor is the egg eating that is often associated with such behaviour. Communication concerning food sources and at swarming time, concerning nest sites. is well developed in many of these bees. Larvae are fed at least in large part on glandular secretions of workers. Populations of colonies are commonly in the thousands (up to over 60000 for *Apis,* 180,000 for some species of *Trigona),* although some species often have colonies of only one or two hundred. Only *Bombus* and tbe highly eusocial bees store food in quantity outside of brood cells for use of adults and for transfer to larvae or brood cells as needed. In most groups of social insects interactions between adults and young (i.e., brood-eggs, larvae, pupae) are universal and important parts of the social organization. Exchange of food between larvae and adults is well known in ants and vespid wasps, and it has often been supposed that larval activity or secretions are of great importance in maintaining the social group. In most kinds of bees, however, there are no contacts between adults and young because the cells in which the young are reared are closed before the eggs hatch, each cell being mass provisioned with enough food to provide for the entire growth of the larva. Progressive feeding, which of course involves adult-larva contact from time to time during the growth of the larvae occurs among bees only in *Apis, Bombus* and most allodapines. Even the highly eusocial meliponines have mass-provisioned cells which, together with the cocoons spun by the mature larvae, completely enclose the immature stages for the whole developmental period.

The number of species of solitary bees is greatest in the warmer, more arid sections of the world, particularly in the semidesert areas as typified by those of western North America, North America, South Africa, Australia, northwestern Argentina, and South-central Eurasia. An abundant and diverse solitary bee fauna is common adjacent to mountainous areas where moderate rainfall conditions exist. The rich bee fauna found in mountains adjacent to arid or semiarid areas is only partially explained by the stratification into altidudinal zones. The varied soil type and exposures, rock niches, beetle holes in wood, and pithy-stemmed plants offer many diverse nesting niches. The world-wide distribution of bees and the remarkable proliferation of species attests to their high degree of adaptability. Numerous definitions have been proposed to distinguish between social and solitary bees, but recent information has shown that hard and fast distinctions cannot always be nude.

The biology and behaviour of the solitary bees has attracted the attention of an increasing number of research workers during the past 15 years: those exploring the value of solitary bees for pollination purposes (Bohart, 1972: Stephen, 1969) and those attempting to evaluate the significance of biological patterns as a supplementary tool for the determination of phylogenetic relationships among bees (Michener, 1974). In the Northwest about half of the *Megachile rotundata* larvae of the first generation pupate and emerge as adults in the late summer. In some seasons a small percentage of the progeny of the second generation emerge as a third generation, although they usually have little time for nesting before being killed by cold temperatures). Some *Megachile* and *Hylaeus* and some *Anthophora* have more than one complete generation and over winter as prepupae. Some single-generation species overwinter as adults in their netal cells. For example, *Osmia lignaria* usually emerges as an adult in April and lives for about three weeks, The type of life cycle described above is apparently an adaptation for early spring emergence, although some species exhibiting it (for example, *Osmia tera)* do not until late spring. It is interesting to note that another species with relatively late emergence. *Osmia californica,* usually has some individuals over wintering as prepupae. Other bees overwintering as unemerged adults include most *Osmia* and *Andrena* and some Anthophora, *Megachile,* and *Emphoropsis. Bumblebees* undergo a life history similar to that of halictines. The overwintered female is sole egg layer for several generations, which overlap broadly because egg laying is continuous. The over wintermg female (queens) which are distintly larger than the earlier generations of workers, are usually produced after the worker: brood ratio is favourable for intensive feeding of queen brood. The Xylocopinae overwinter as emerged adults, as do halictines and many apids. However, both sexes of Xylocopines overwinter in a dormant condition and mate in the spring (some mating is reported to take place also in the fall). The females usually overwinter in the natal nest with males from other nests often joining them.

In many megachilids, males appear more numerous but the exact causes leading to the variance are not well understood. In *Megachile rotumdata,* the ratio of males to females in large samples taken from different nesting population has been seen to vary from I : I to 10 : 1. The difference in ratios is directly correlated with the diameter of the tunnel in which they are nesting. Emergence of males in advance of females is evident in the alkali bee and in *Megachile rotundata,* as in most other species of bees that have been studied. This phenomenon, referred to as protandry, is a general rule among solitary bees and is interpreted a evolutionary adaptation, assuring the presence of males for mating with emerging females.

The availability of suitable substrates for nesting is one of the most common factors limiting the population and distribution of bee species. The principal type nesting micro environments include soil, wood, small and large cavities and even fully exposed surfaces. Species nesting in soil may select horizontal to gently sloping surfaces or vertical banks. The vertical surface may be bare or overhung with vegetation, or rarely with a grassy cover, and its exposure may provide maximum or minimum shade. Vertical banks are usually dry, but they may be moist in shaded gullies or dram ditches. The soil surface may be wet or dry, but appreciable moist is usually available where and when the cells are constructed. At least one observation has been made of *Anthophora* species tunneling in moderately hard sand-soil. *Ceratina,* some *Megachile.* Some *Xylocopa* and several genera of small sphecid wasps burrow in the soft pithy plant stems of plants such s raspberry, black-berry and sunflower in constructing nesting tunnels. A wide variety of nesting material is utilized by megachilids. Their nests may be found pockets or cracks in rocks attached to twigs or rock surfaces (some *Dianthidirim),* or in narrow crevices or almost any material *(Megachile rotundata).* Tendency of bees to renest in close proximity to their parent's nest is one of the main causes of gregariousness, and that selection of a peculiar soil condition has a minor effect. Most burrowing bees construct laterals from the main burrow with one or more cells arising from each lateral. Exceptions occur among species that construct a main burrow with one or more cells attached to the end, along the sides, or arranged within the main burrow in linear series. After the cells are provisioned and capped, the nest, or a portion of it, is plugged. Some species only plug the area immediately exterior of each cell, others completely backfill the laterals, and still other plug the entrance when the nest is completed. Most species that arrange their cells in line series construct complete cells, i.e. provide walls as well as top and bottom for each cell (most *Megachile. Anthidium).* In *Megachile,* the rounded base of each cell inserted into the concave apex of the cell below, resulting in a nearly intact, weakly differentiated column of cells.

The number of cells per nest ranges from one to several thousands. Most solitary soil burrowing species make only one nest, which contains as many cells as foraging conditions and the reproductive potential of the bees allow. Nest of *Nomia melanderi,* may contain from 5 to 24 cells, depending upon the availability of forage and the quality of the substrate in which they are nesting. Some species of *Osmia* normally select small pockets in rocks that accommodate only one cell, but the same specie may place several cells in somewhat larger holes. *Megachile rotundata* will accept tunnels which accommodate only a single cell but more commonly uses long tubular cavities in which it places as many as 20 cells. The reproductive potential of this species is even higher (up to 40 eggs), but there seems to be an upper limit to the number of cells it places in a single tube, independent of its length.

The world wide distribution of these bees and their remarkable ability for proliferation are the result of their higher degree of adaptability. Floral fidelity is the majoı attribute of these species making them sure of maintaining species characteristic within plant species. Based upon degree of association in between bees and the plant species i.e. the number of plant types visited for pollen collection (Linslay *et al.,* 1958), the bees are termed monolactic (visiting one species) oligolactic (visiting few related genera) or polylactic (visiting many types of plants). The solitary bee species are mostly oligolactic. Differences in seasonal adaptability, the pattern of origin vis-a-vis individual bee specialization basically determine bees abundance for pollen collection on a particular plant. The principle factor which determines the effectiveness of such pollinators for a particular crop or plant species depends upon the bee abundance, bee flight period, bee flight hours per day and the number of flowers visited pcr day. The factors which contribute to bee survival in nature and their propagation depends upon the availability of natural or man made nesting devices of preferred dimensions, abundance of natural parasitic, or predators, incidence of disease or pesticide poisoning, and the natural brood mortality during active or the dormant season. Most important is the synchronization of the bee flight period with the major blooming period of the crop. This is achieved through appropriate provisions of nesting devices and regulating development of adults so that there is synchrony in adults formation with crop blooming. Following are the characteristic features of such bee management programmes for crop pollination.

(i) Provision of appropriate nesting devices of brood cell formation.

(ii) Collection and safe storage of brood nest of cells at low temperature.

(iii) Checking/controlled emergence of parasites or removal diseased cells.

(iv) Incubation of cells at appropriate temperature to regulate formation

Following are the under mentioned few crops in which the wild bees have been successfully domesticated and managed for pollination.

(a) Pollination in alfalfa

Alfalfa *(Medicogo sativa)* is a perennial herbaceous protein rich legume. It is one of the most important fodder crop throughout the world. It is a cross plant and its flowers being typically papilionaceous has staminal column held inside keel. Tripping i.e. release of staminal column is considered to be a prerequisite, of cross pollination (Free, 1993). Most investigators have concluded that long tongued bees are not very important in alfalfa seed production particularly in dry or arid conditions because of the fact that the honey bee collect nectar from alfalfa flowers from the side of the keel, thus avoid the mechanical shock of the staminal column. Most of the workers has emphasized importance of certain non-*Apis* species in the pollination of its flowers. Among a variety of wild bees associated with alfalfa flowers, the investigations in (U.S.A.) and Alberta (Canada) showed that the leaf cutter bee *(Megachila rotundata)* and a alkali bee *(Nomia mellandari* could be exploited and propagated as alfalfa pollinator (Bohart, 1972).

A variety of non-*Apis* bees have been reported to be associated with some other crops in this region (Kapil and Jain 1980). They reported megachilid species associated with alfalfa under local conditions. The studies (Sihag, 1990) have suggested the inclusion of some of these bees *M. flavipas, M. femorata, M. lanata* and *M. cephalates* under the genus *Chalcodoma.* The major period of their activity for alfalfa pollination is in the April-May months. Their broods however undergo dormancy as mature larvae, first in the months of June-July and then in winters. Until mid March, there are many overlapping generations in the months of September to November. High day temperature and dry condition favour their foraging activity, brood formation viz-a-viz alfalfa pollination. Important non-Apis alfalfa pollinators are *Mellita leporilla. Atlthoph quadrifasciata* and few other species of *Alldrella* and *Megachile.*

The leaf cutter bee earlier called as *Megachile pacifica,* is a fast nest gregarious bee and it nest in any such tunnel available in wood above the ground. It has been found to accept the man made artificial nesting devices including tunneled wooden block, corrugated boards or the plastic tubes of appropriate tunnel size in diameter. Its native region is South-West Asia and it is spread to eastern USA around 1930. Today, it is successfully managed for alfalfa pollination in many states of USA and southern parts of Canada as well as many of the European countries. It is to, termed as million dollars bee as it involve the inputs and outputs of millions of dollars in making various mechanical devices and nesting materials and management techniques for alfalfa seed production.

The alkali bee is another fast nesting gregarious bee. It however, nests in big alkaline soil. Once established favourable site may produce as many as 2 lakh cells per acre. The best sites are bare or slightly vegetated tender silty loamy soils. Major disadventage in its management is that its testing sites or nests can't be easily removed, stored or transported, although there has been many refinements in developing artificial nesting sites. The most important step has been to keep the site puffy a little moist by providing plastic films several feet below the Alrface. Heated calli are sometimes employed to enhance pupation and emergence of adults to coincide with alfalfa blooming. In addition to rains moulded soil, rats, skunks, birds and variety of other parasites are its natural enemies. A temperature of 30°C is most appropriate for rapid pupation of the diapausing larvae.

Pithitus smaragdula, a green metallic bee is another important alfalfia pollinator. It is a small carpenter bee and it make nesting tunnels in cut pithy stems such as common reeds *(Erianthus munja).*

(b) Other crops

The wild bees are associated with many other crops also. The crops like clovers (Ladino clover, red clover, egyptian clover, white clover and sweet clover), soybean pigeon-pea sunhemp, broad bean, mustard, rape, coffee, papaya, cotton, sunflower safflower and apples are few other important crops which could be benefited from cross pollination. A variety of bumble bees belonging to the genus *Bombus* utilized for pollination of clovers. The bumble bees are semisocial bees as these initiate nesting as an individual female, but later establish a large size colony. *Osmia coerulescens,* a leaf cutter bee has also been reported to increase seed yields in red clover in USA. *Osmia corniifrons* has been successfully managed for apple pollination in northern and central Japan. It nest in bamboo and hollow reed. The stingless bees commonly termed as 'melipona' bees are another important non-*Apis* group of true social bees which resembles to *Apis* species as these form large colonies and yield sufficient honey for human consumption. The indigenous people of tropical America and Africa have for centuries managed various species of stingless bees for honey. Continuously changing climatic conditions and over exploitations of forests and barren lands for agriculture have been the major hurdles in natural propagation of wild bees. In north India alone there has been about 73% decline in their natural population during 1976-78 (Jain, 1993). Looking upon their utility and importance and their implications in preserving flora, there is a need to protect and conserve these bees for pollination.

Inspite of being represented in large numbers, non-*Apis* bee pollinators have-received little attention. The probable reason seems their unpredictable seasonal availability, lack of knowledge about their biology and host plant relationship. It was only in the mid sixties that scientific interests were generated to study and understand their life processes in India (Atwal, 1970). Failure of honeybees in the pollination of alfalfa which requires tripping (Kapil *et al.,* 1977; Kapil and Jain 1979, 1980) dwindling of bee colonies due to acute floral dearth in heavy monsoon during June to September and similar effects during severe winter in Himalayan ranges from November to February and prevalent bee diseases has created the necessity for exploration of alternate yet suitable non-*Apis* bee pollinators to augment crop yields in India. In addition to alfalfa, clovers and fruits like apple, pear, peach, almond and several other crops such as sunflower, hybrid tomato, cotton, onion, carrot and cucurbits (Kapil and Dhaliwal, 1968) can be future potential crops requiring the services of the non-*Apis* pollinators.

Table 45 summarizes the occurrence of various bee fauna within the modern territorial limits of our country. Around 92% of known species were recorded from northern (J & K, Punjab, Uttaranchal, U.P., Himachal Pradesh, Haryana) or western part (Rajasthan, except extreme north and eastern green parts and Gujarat) and remaining 8% were described from rest of the Indian regions. In other words, maximum investigations on biosystematics and floral relationships of bees have been made from the northern territories and a huge area of southern peninsula is left to be explored. The calculation concerning diversity of bees has revealed that a total of 633 species grouped under 60 genera are found in our country. Certainly, this is not a satisfying number for a huge area with enough of climatological variations, such as India besides, synonymies for many species is still pending to be worked out. The recent most biosystematic arrangement presented here will be further useful to the biodiversity, ecological and pollination-workers of the country.

A detailed historical account, regarding discovery of different genera and species, has not been described here. It has been written earlier by this author as well as many taxonomists of the world in prior publications. This author has recently (2003) presented a catalogue for the bee species distributed from Turkey and Cyprus in the west through all the south Asian countries upto Indonesia. Readers interested in a detailed species level taxonomy and their distribution, are advised to consult it.

The country has been subdivided under certain specific area based upon homogeneousness of ecological factors facilitating the presentation. The regions specified are:

(a) Extreme northern region: It includes high altitude of Himalayas ranging between 6-7,000 feet and more in J&K, Himachal Pradesh and Uttaranchal.

(b) Northern region: This indicates the medium and low altitudes area and, plains of Tarai at and south of Himalayan range i.e. including southern J & K, Punjab, Haryana, eastern Rajasthan adjacent to U.P. and plains of Bihar.

(c) Western region: Almost whole of Rajasthan (excluding eastern hilly-green region) and Gujarat.

(d) Western Ghats: Western hilly Maharashtra, Goa, western Karnataka and, otherwise specified Lakshadweep and Minicoy Islands.

(e) Central region: Whole of M.P., most of the Maharashtra, Chhattisgarh, Andhra Pradesh, excluding coastal area.

(f) Northeastern region: Includes hills of eastern Himalayan range, all the seven states of the northeast and Districts like Darjeeling and Siliguri in W. Bengal.

(g) Eastern region: Excluding hilly north, West Bengal, hilly South Bihar, Orissa and extreme coastal area of Andhra Pradesh.

(h) South-central: The region include Karnataka, and northern parts of Tamilnadu.

(i) Extreme southern region: Specified for Pondicherry, coastal Tamilnadu and Kerala and, otherwise not specified Andaman and Nicobar Islands.

DIVERSITY OF BEES IN INDIA

Table 45. Diversity of bees in India* (Gupta, 2003)

SN	Family	Subfamily	Tribe	Genus	No. of Species	Broad distribution in india
1.	**Colletidae** Lepeletier	Colletinae Lepeletier	-	1. *Colletes* Latreille, 1802	005	Northern region
		Hylaeinae Viereck	-	2. *Hylaeus* Fabricius, 1793	014	13 in Northern region and 2 in Southern region
2.	**Andrenidae** Latreille	Andreninae Latreille	-	1. *Andrena* Fabricius, 1775	034	Widely distributed in northern region
3.	**Halictidae** Thomson	Rophitinae Schenck	-	1. *Systropha* Illiger, 1806	002	Punjab & northern Rajasthan
		Nomiinae Robertson	-	2. *Nomia* Latreille, 1804	067	Widely distributed throughout India
			-	3. *Pseudapis* Kirby, 1900	002	Northern region
			-	4. *Steganomus* Ritsema, 1873	003	Northern and western Ghat regions
		Nomioidinae Börner	-	5. *Ceylalictus* Strand,1913	003	Discontinuous: extreme northern, coastal western and south-central region
			-	6. *Nomioides* Schenck 1867	003	Deserts of north-western region
		Halictinae Thomson	Halictini	7. *Halictus* Latreille, 1804	055	47 in hilly region of north extending upto north east and 8 in west, central and western ghats
				8. *Homalictus* Cockerell, 1919	006	North and north eastern region
				9. *Sphecodes* Latreille, 1804	017	11 in north and north eastern region and 6 in south Rajasthan & Gujarat
				10. *Thrinchosto-ma* Saussure, 1890	002	Northeastern region

4. **Melittidae** Schenck	Melittinae Schenk	Melittini	1. *Melitta* Kirby, 1802	001	North and north eastern region
5. **Megachilidae** Latreille	Fideliinae Cockerell	Pararho - phitini	1. *Pararhophites* Friese, 1898	002	1 in northwestern region and 1 in Gujarat
	Megachi-linae Latreille	Lithurgini	2. *Lithurgus* Berthold, 1827	014	Discontinuous distribution: northern, western coastal, extreme southern and northeastern region
		Osmiini	3. *Chelostoma* Latreille, 1809	001	Sikkim and allied hilly territories of north east, continue upto Myanmar
			4. *Heriades* Spinola, 1808	022	17 in Himachal Pradesh, Uttaranchal & north-eastern hills and 5 in southern Rajasthan & Gujarat
			5. *Hoplitis* Klug, 1807	010	8 in Northwestern region and 2 in wes- tern coastal region
			6. *Noteriades*, Cockerell, 1931	004	Himachal Pradesh
			7. *Osmia* Panzer, 1806	005	4 in Himachal Pradesh, Uttaranchal and 1 in eastern coast of southern region
			8. *Protosmia* Ducke, 1900	001	Northern (medium hilly) region
			9. *Pseudoh-eriades* Peters, 1970	005	Northwestern region (and eastern buffer zone of Thar)
			10. *Wainia* Tkalců, 1980	002	So far recorded from Lonavala [Maharashtra only
		Anthidiini	11. *Acedanthi-dium* Michener & Griswold, 2000	001	Northern (Himachal Pradesh & Uttaranchal only)
			12. *Anthidiellum* Cockerell, 1904	001	Northeastern region
			13. *Anthidium* Fabricius, 1804	004	Northern and northwestern hilly and xeric regions
			14. *Dianthidium* Cockerell, 1900	004	Northwestern xeric region
			15. *Eoanthidium* (Popov, 1950)	002	Northwestern xeric region
			16. *Euaspis* Gerstaecker, 1857	002	1 in northeastern region and 1 widely distributed in almost whole of India except, extreme southern region

			17. *Icteranthidi-um* Michener, 1948	002	Northwestern xeric area
			18. *Indanthidium* Michener & Griswold, 1994	001	So far recorded from Pune
			19. *Pachyanthi-dium* Friese, 1905	003	Discountinuous: Kangra, Bengal, Kerala, Bombay & Panchmarhi
			20. *Trachusa* Panzer, 1804	004	Discontinuous: north western region, Central region, Bombay
			21. *Trachusoides* Michener & Griswold, 1994	001	So far recorded from Appangala (Karnataka)
		Dioxyini	22. *Aglaoapis* Cameron, 1901	001	So far recorded from Bombay
		Megachilini	23. *Coelioxys* Latreille, 1809	032	Widely distributed all over country
			24. *Megachile* Latreille, 1802	105	Widely distributed all over country
6. **Apidae** Latreille	Xylocopinae Latreille	Xylocopini	1. *Xylocopa* Latreille, 1802	036	Widely distributed all over country
		Allodapini	2. *Braunsapis* Michener, 1969	006	Discontinuous: northern, parts of western and extreme southern region
		Ceratinini	3. *Ceratina* Latreille, 1802	014	Almost throughout India
	Nomadinae Latreille	Nomadini	4. *Nomada* Scopoli, 1770	013	Mostly confined to northern regions
		Epeolini	5. *Epeolus* Latreille, 1802	003	2 in north and north-eastern region and 1 in Central region
		Ammobatini	6. *Parammobato-des* Popov, 1931	001	Distinct territory not known [label indicates only India, pe rhaps extreme northern India]
			7. *Pasites* Jurine, 1807	001	Southern region
	Apinae Latreille	Ancylini	8. *Tarsalia* Morawitz, 1895	001	Western Ghats
		Eucerini	9. *Eucera* Scopoli, 1770	003	Northwestern region
			10. *Tetralonia* Spinola, 1839	015	13 in Northern region and 2 in Central region

			11. *Tetraloniella* Ashmead, 1899	002	1 in extremity of northeastern region and 1 in central region
		Anthophorini	12. *Amegilla* Friese, 1897	021	Widely distributed all over country
			13. *Anthophora* Latreille, 1803	010	Widely distributed all over country
			14. *Elaphropoda* Lieftinck, 1966	004	Mountains of extreme northeast region
			15. *Habropoda* Smith, 1854	014	Confined to northern to northeastern region
		Melectini	16. *Melecta* Latreille, 1802	002	Extreme northern region
			17. *Tetralonioidella* Strand, 1914	003	Southern slopes of Himalaya through northeastern region upto Indonesia
			18. *Thyreus* Panzer, 1806	006	Widely distributed all over country
		Bombini	19. *Bombus* Latreille, 1802	026	Exclusively along the high and medium range of mountains in Himalaya
		Meliponini	20. *Lisotrigona* Moure, 1961	001	Central region
			21. *Trigona* Jurine, 1807	003	Arid zone in western region and Northeastern upto Indonesia
		Apini	22. *Apis* Linnaeus, 1758	005	2 Widely distributed, 1 introduced and 1 confined to very high altitudes in Himalaya
006	**012**		**060**	**633**	

* Table contains the taxonomic categories found and known from India.

Family Stenotritidae Cockerell

The smallest family of bees that consists of only two genera namely, *Ctenocolletes* Cockerell with ten species and, *Stenotritus* Smith with eleven species. Both are exclusively found in Australia and, so far no species has been recorded from anywhere else in the world.

Family Colletidae Lepeletier

This is a medium size family subdivided into 5 subfamilies namely, Colletinae Lepeletier, Diphaglossinae Vachal, Xeromelissinae Cockerell, Hylaeinae Viereck and Euryglossinae Michener. Among them, subfamily Colletinae and Hylaeinae are the lone representatives of this family, known by one genus each, in India. The moderately populated genus *Colletes*, represented by 5 species, is restricted to northern region so as other identically populated genus *Hylaeus* with 15 species. It was interesting to note that 3 species of *Colletes* are more or less confined to middle range height (around 5000 to 6000') in southern slopes of greater Himalaya in Uttaranchal and Sikkim. All of them are quite limited in their distribution. On the contrary, species found in plains and those reaching upto Rajasthan in south, are comparatively widely distributed. The area of species distribution has been considered as Jammu & Kashmir, Punjab, Haryana, Himachal Pradesh, Uttaranchal, eastern Rajasthan and one species record exists from Sikkim and Pune.

Probably, majority of species of this genus are also distributed in intermediate area namely, union territory of Delhi, northern U.P. as well as the green northern Rajasthan. Practically there exists no ecological barrier between the noted territory and the probable area of distribution. These areas require intensive surveys. Species of *Colletes* have been collected digging deep burrows on sloppy surfaces of hard grounds in Sikkim. Grewal, *et al.* (1970a) described the life history of *C. nursei* from Punjab. Gupta (2003) also collected many females in Punjab while busy excavating burrows in sandy, bare and dry crop fields. Batra (1977a) made many ecological observations on nesting sites of *Colletes*. Batra (1968a) and Gupta and Yadav (2001) noted down different floral species visited by some species of this genus.

Genus *Hylaeus* in major contains those species which were earlier grouped under genus *Prosopis*. The name *Hylaeus* has priority thus accepted. *Prosopis* however, still is a subgenus of reorganized *Hylaeus* (Michener, 2000). It is a world wide genus having representatives on all continents. Most of the species collected from oriental region are yet to be assigned a subgeneric category. Species of our destined area have been collected from Kashmir, Punjab, Rajasthan, northern part of W. Bengal and Sikkim. Two out of 15 species known from our country were collected from southern extremities probably reaching upto Sri Lanka (Wijesekara, 2001). An intensive survey to the nation waits so that exact picture of its distribution would be marked on the map of our country. Species of this genus are known to visit some cultivated crops in Punjab (Batra, 1968a).

Family Andrenidae Latreille

It is a large family of bees subdivided under four subfamilies namely, Alocandreninae Michener, Andreninae Latreille, Panurginae Leach and Oxaeinae Ashmead. However, in our country this family is represented by subfamily Andreninae only having the single representative genus *Andrena*. Species of this genus can be marked all over northern India, more particularly plains of Punjab, Haryana, Himachal Pradesh, Rajasthan, Uttaranchal, U.P., Bihar and W. Bengal. Two species were collected one each from western coast of Maharashtra, the central Maharashtra and Assam. 34 species are known so far after judging synonymies. Major bulk of them show affinities for their western allies found in Pakistan, Afghanistan and, upto Turkey. On the contrary, eastern species are fairly distributed upto Myanmar and Malaysia. Precisely, most of species seem to have adapted for their specific eco-geographical environments in India but many more have their affiliates towards both neighbouring territories. An intensive survey of the remaining part of the country would yield several new species and their floral data, unknown so far.

Rahman (1940) briefly described a nest of *Andrena ilerda*, collected near a toria field. This is not yet known that species of *Andrena* excavate the tunnels themselves or they occupy pre-existing subterranean burrows. However, the highly branched, 50-60 cm. deep burrows have at least one cell at the end. Wain (1968) also recorded nests of *A. bellidoides* in the hills of western ghats. They were constructed gregariously. Grewal *et al.* (1970b) published the nesting behaviour of *Andrena leaena*.

Gupta (2003) concluded that species of this genus bear a good affection for the flowerings of cruciferous crops. The oil seed producing region all along the Shivalik and Aravali range harbours comparatively good population of many species. Mohammad (1935), Rahman (1940), Batra (1968), Atwal (1970), Bhalla *et al.* (1983), Kumar *et al.* (1988), Kumar *et al.* (1994) and, Gupta and Yadav (2001) have listed many species of *Andrena* collected on different cruciferous crops in the mentioned area. Abrol (1986b, 1988a, 1989) detailed the behavioural aspect of some *Andrena* species from Jammu & Kashmir.

Family Halictidae Thomson

This is a considerably large family having many semisocial bee species, commonly known as sweat bees. It is further subdivided into four subfamilies namely, Rophitinae Schenk, Nomiinae Robertson, Nomioidinae Börner and, Halictinae Thomson. The group is represented by 160 species in our country. Subfamily Rophitinae and Nomioidinae consist of a limited number of species with a restricted distribution whereas, subfamily Nomiinae and Halictinae are widely distributed.

The single representative of Rophitinae, genus *Systropha* was reported with two species limited to western Punjab and northern Rajasthan. Batra and Michener (1966) described one of them alongwith its subsoil formed nest and the larva. Identically, two representatives of Nomiinae, genus *Pseudapis* and *Steganomus* are confined to certain pocket area of the country. The third representative, genus *Nomia* with 67 species, is widely distributed all over India. Subfamily Nomioidinae is known by genus *Ceylalictus* and *Nomioides* both with three species each. Among them first one exhibits discontinuous distribution and the second is found in the arid region of north-west.

Subfamily Halictinae is further subdivided into tribe Augochlorini and Halictini. Among them Halictini has representation in our country and, of fairly wide occurrence. Two species of genus *Thrinchostoma* are known exclusively from the northeastern region of the country. In major its species were recorded from the eastern countries next to India. Genus *Homalictus*, found in similar territory, is comparatively widespread, reaching upto the north at the southern slopes of Himalayan all along the range from north to northeast. Genus *Sphecodes* is also represented in the area inhabited by *Homalictus* but in addition to its eleven species, six are found in the southern hilly zone of Rajasthan and at the tail end of Aravali hills in Gujarat. The species found in western zone have affinities with their western allies. Genus *Halictus*, represented with 55 species, is largely confined to the northern territories and at low level hills of central and western parts of the country. The aspect of taxonomy of the family needs intensive investigations for most of the area of our country besides, the bionomics and biology.

Works pertaining to the nesting biology of a few species of *Lasioglossum, Nomioides, Nomia* and *Halictus* appeared so far, were published during 1960s to 1980s. Among them significant contributions regarding social behaviour and nests of nomiine bees, specially observations made for several species of *Lasioglossum* and *Halictus*, were published by Batra (1964a, 1964b, 1965, 1966a, 1966b, 1966c, 1968b, 1970, 1971, 1995a, 1997a, 1997b etc.), Kukuk (1980), Batra and Bohart (1970). Same aspect was described for many other halictid bees of Indo-Malaysia regions by Sakagami, Ebmer and Tadauchi (1996) and, Sakagami and Ebmer (1987). Most of the halictids nest gregariously in the moist subterranean soil (resembling *Colletes*). However, their burrows were comparatively shallower culminating into 5-6 collateral branches. Each of which further bear around 1-3 or even 9 cells at the end. Precisely, a nest is a cooperative effort of as many as 20-25 females, suggesting their gregarious or semisocial behaviour. Nest tunnels were lined by mandibular secretions that provided evidence that they were self-excavated by females (Rahman, 1940; Sakagami and Michener, 1962, 1963; Sakagami and Wain, 1966).

A comparative account of the properties of nest-building secretions of *Nomia, Anthophora, Hylaeus* and other bees was presented by Batra (1972). Batra (1978b) further presented aggression, territoriality, mating and nest aggregation of some solitary bees that belong to Halictidae, Megachilidae, Colletidae and the earlier known family Anthophoridae.

Kumar *et al.* (1994) compared the pollination efficiency of bees by making comparison in the rate of visits made on toria in Himachal Pradesh. They identified *Halictus catullus* and *Halictus splendidulus* alongwith some other bees, on this crop. The regional studies made with regard to identification of species of various genera pollinating some selected crops, were published by Batra (1968) and Gupta and Yadav (2001).

Family Melittidae Schenck

This fifth family of bees is of rare occurrence in India. Michener (2000) further classified it into three Subfamilies, Dasypodainae Börner, Meganomiinae Michener and Melittinae Schenck. Among them Melittinae, including tribe

Melittini, with single Genus *Melitta*, known with single species [*M. harrietae* (Bingham, 1897)] reaches India. It is found in a very scanty population in the extreme north and north eastern hilly regions at the middle range altitude in Himalaya (Gupta, 2003). It has many allies in the hilly and arid countries, west and east to India. The family has numerous representatives in Africa and countries in the middle east. Probably, high range of Himalaya imposed a restrictive entry of erstwhile Palaearctic species into our country. Comments on nesting of a species of *Ctenoplectra* were recorded by Bingham (1897) from Tenasserim. This genus is now placed in tribe Ctenoplectrini, under subfamily Apinae of Apidae.

So far nothing is known about the nesting behaviour and pollination aspects etc about *Melitta* in India. Malyshev (1923) described them for some species known from Russian region.

Family Megachilidae Latreille

Until now it exists as the largest family of bees with regard to number of taxa known from India. Gupta (1993) consolidated 161 species of bees included in this family found in six states in northwest with their detailed taxonomy alongwith their flower records. Still work pertaining to synonymy for several taxa of the country, is pending besides, the multi-seasonal expeditions of the southern peninsular region. As a whole, 24 genera with 229 species are included in this family from India.

The taxonomic categories of this family have undergone a fair amount of shuffling by virtue of cladistic analysis made by Roig-Alsina and Michener (1993). This resulted into recognition of two subfamilies namely, Fideliinae and Megachilinae (Michener, 2000).

Subfamily Fideliinae is represented by a single tribe Pararhophitini in India, including genus *Pararhophites* with two species. The small bees of this genus love xeric conditions and were largely known in the western area beyond Indian limits. Its two species were collected from area adjacent to Baluchistan [Ferozpur, at the western border of Punjab, reaching upto Rajasthan] and second was found in Junagarh in Gujarat. These bees were collected while pollinating flowers of *Convolvulus* at both vicinities. The nests of one species were found in good aggregation made in subterraneous burrows in Baluchistan (McGinley and Rozen, 1987).

Subfamily Megachilinae includes five tribes and all have good representation in our country. The first, Lithurgini represented by genus *Lithurgus* Berthold (14 species), has discontinuous distribution. Species like *L. atratus* and *L. dentipes* have been of great concern with regard to their pollination of Cotton (Malvaceae) and, earlier many have commented upon their nests build in hollow sticks (Horne, 1870, 1872; Malyshev, 1930; Lieftinck, 1939).

Second tribe Osmiini includes numerous genera but most of them are known from the colder regions in the north and arid northwest. Genus *Chelostoma*, represented by a single species is an exception. It was initially recorded from Sikkim and is found upto Myanmar. On the contrary, Genus *Heriades* (22 species) is widespread in Himachal Pradesh, adjacent Punjab and extends southward all along Aravali hills upto Gujarat, in a scattered pattern. These small black bees have great affection for Compositae flowers (Gupta and Yadav, 2001). *Hoplitis* (10 species) also has more or less similar distribution pattern however, its two species were recorded at northern territory of the western coast. *Noteriades* (four species) and *Protosmia* (one species) are restricted to Himalayan range. Former is found at medium altitudes in Himachal Pradesh and the later is exclusively confined to high mountains in the extreme north.

After making taxonomic revisions, genus *Osmia* is reduced to five species found in India. Most of them are restricted to Himachal Pradesh and Uttaranchal at medium range mountains. Its one species was exclusively recorded from Karaikal located at south-eastern coast. Genus *Pseudoheriades* is quite widely distributed in Rajasthan and coastal Gujarat. The nests and immatures of one of its species were described from 'Moonj straws' (Gupta and Sharma, 1995: misident. as of *Heriades*). Recently many more bees were collected from western Rajasthan and Gujarat [yet to be named]. More particularly they were collected while busy pollinating Coconut at Somnath, Veeraval and Diu in Gujarat. Genus *Wainia* was exceptionally recorded from Lonavla in Maharashtra with two species.

Tribe Anthidiini has fair representatives in our country but mostly with a discontinuous distribution. Genus *Acedanthidium, Anthidiellum, Eoanthidium, Icteranthidium, Indanthidium, Trachusoides* are known with merely one or two species. *Euaspis* (=*Parevaspis*), the cleptoparasite, is almost cosmopolitan throughout the country. Its one species *Euaspis abdominalis* is found in the northeastern states extending upto Myanmar and Malaysia. The second, *Euaspis carbonaria* is found almost everywhere except the extreme southern region of the country. Both species are well known cleptoparasites of *Megachile.* Species of *Pachyanthidium, Trachusa* and *Dianthidium* are found in different pocket areas at different parts of the nation. However, most of the species are of prominent occurrence in arid zones of the northwest. A thorough expedition would yield many more species of this tribe.

Tribe Dioxyini, represented by a good number of taxa in the countries beyond the western limits of India, is known by a single genus *Aglaoapis* here. Its single species, *Aglaoapis brevipennis* was collected from Bombay around a century ago and until now not found anywhere else.

Tribe Megachilini, now consist of only two genera namely, *Megachile* that include the leaf cutting and mason bees and, its well known cleptoparasitic genus *Coelioxys*. Both are found almost everywhere all over the country (Gupta, 2003). Species of genus *Megachile* (exceeding 130 in number), have attracted many authors with regard to their nesting and pollination aspects.

The comments on nests of various leaf cutting bees were initially published by Horne (1868, 1870 & 1872), Maxwell-Lefroy and Howlett (1909), Bingham

(1897, 1898a, 1898b, 1908) and Dutt (1912). Until then, the interesting act of cutting leaf pieces with its mandibles, followed by their transportation to the nesting site by a female megachiline bee attracted enough attention all over the world. They use leaf cuttings, petals, mud, soil, pebbles and plant resins etc. for the construction of their nest chambers in self excavated or existing horizontal burrows in muddy walls or in subterranean soil. The nesting biology for a few Indian species was described by Pagden (1934), Malyshev (1930), Piel (1930) and, Wain (1968). Later Kapil, Grewal and Atwal (1970), Kapil *et al.* (1975), Choudhary and Jain (1978), Gupta and Sharma (1995) and, Gupta, Naval and Charan (2003b) included nesting behaviour, immatures and correlated aspects in their studies.

The family includes several good pollinators of Leguminosae, Compositae, Solanaceae and, several fruit crops etc. Rahman (1940) could identify very few species of Megachilini on Sarson and Toria. Batra (1968) identified several megachilids pollinating many cultivated and wild crops in Ludhiana. Mishra *et al.* (1976) and Batra (1976a & 1979a) presented some more works related to pollination of some cultivated crops from Himachal Pradesh and Punjab, respectively. Abrol (1986a, 1988b, 1988c) noted *Megachile nana* and *Megachile flavipes* (sic) on alfalfa and recorded their eco-pollination behavioural relationships. Kumar *et al.* (1993) recorded around 9% of bees belonging to *Megachile* and 28% belonging to *Ceratina*, on *Cichorium intybus* in Himachal Pradesh. Gupta (1993) presented the complete flower record of megachilid bees collected from six states in the northwest. Gupta and Yadav (2001) recorded a good number of megachilids from four cultivated crops in eastern Rajasthan and adjacent U. P.

Gupta (2003) during his studies with bees for more than previous two decades concluded some of the good and effective megachilid species such as: *Megachile bicolor, M. lanata, M. cephalotes, M. gathela, M. albifrons* and *M. creusa* for almost whole of the northern half of the country.

Family Apidae Latreille

Based upon the cladistic analysis of Roig-Alsina and Michener (1993), the reorganized Apidae consists of subfamilies Xylocopinae, Nomadinae and Apinae (Michener, 2000). Table 45 lists the three named subfamilies, their 12 tribes and corresponding 22 genera found in India. So far 192 species have been grouped under different generic categories. Still works related to synonymies of several species is pending and many of them would face taxonomic shuffling or would be synonymized. Species of *Xylocopa* (36 species and subspecies), *Ceratina* (14 species), *Amegilla* (21 species), *Anthophora* (10 species) and *Thyreus* (6 clepto-parasitic species), are widely distributed.

On the contrary, *Braunsapis* (six species, having discontinuous distribution), *Nomada* (13 species, mostly confined to northern region, upto the area of western arid Rajasthan), *Tetralonia* (15 species, out of them only two were recorded from M.P. and adjacent Maharashtra and rest are confined to north India), *Habropoda*

(14 species, all restricted to Himalayan range) and *Bombus* (restricted around or above the altitude of 3000 feet in Himalaya), are region specific.

A few genera namely, *Epeolus* (3 species), *Parammobatodes* (one species with doubtful occurrence in India), *Pasites* (one species), *Eucera* (3 species found in northwest), *Tetraloniella* (two uncommon species), *Elphropoda* (four species) and *Melecta* (with two species in extreme north, in Himalaya), *Tetralonioidella* (3 species all found throughout the Himalayan range upto east), *Lisotrigona* (one species known from M.P.) and *Tarsalia* (one species from Poona) come under the category of lesser known genera in the country. The social stingless bees belonging to genus *Trigona* (three species on record) were originally collected from all over the Himalayan range however, during past two decades, many more have been collected from several additional territories, by this author. They were Agra, Allahabad, Jamalpur (Bihar), Jabalpur and Pachmarhi (M.P.) and, recently during the summers of 2003, more specimens of this genus were collected at Dwarka and Somnath, i.e. at the costal Gujarat and, more as a surprise during September 2003, many more specimens of this genus have been collected on the flowers of *Techtona stans* (Bignonaceae) in Jodhpur.

Genus *Apis*, the honey bee, has five representative species in our country. Among them *Apis mellifera* the Italian honey bee, was introduced in India during sixties in the previous century. It ultimately 'merged' with *Apis indica* and now 'becomes' a cosmopolitan species by virtue of its adoption in artificial domestication programme by ICAR stations all over India. Engel (2000) has recently published the information concerning honey bees of India. *Apis dorsata* (with three subspecies *A.d. dorsata, A.d. laboriosa* and *A.d. bighami), Apis cerana, Apis florea* and *Apis andreniformis* are the other species. *Apis andreniformis* and *Apis dorsata binghami* are known with limited distribution records. The first of these was recorded from Khasia hills and second was collected from Sikkim and Meghalaya. *Apis dorsata laboriosa*, the giant Himalayan honey bee is confined to the high altitudes in northern region. Roubik *et al.* (1985) noted distribution and nesting of *A.d. laboriosa*. Batra (1996) described its biology and declared it as a good pollinator of apple at high altitudes in Garhwal (Uttaranchal). Qayyum and Ahmad (1967) described the biology of *A dorsata*. Thakar and Topani (1961, 1962) detailed the nesting behaviour of *A. dorsata* and *A. florea*. Sharma and Thakur (1999) put forward the morphometric characterization of *A. dorsata* and, Sharma *et al.* (2000) detailed the diurnal activity of *A. cerana* and *A. mellifera* on different flora.

Since early times the carpenter bees that belong to genus *Xylocopa* have been of great interest to mankind therefore, enough observations on nestings and immatures were made on various species of this genus. Horne (1872), Perkins (1899), Maxwell-Lefroy and Howlett (1909), Dover (1924), Dutt (1912), Kannan (1925), Iwata (1964), Ma (1938 & 1954) and, Beeson (1938) published comments on their nests and nesting behaviour. Sakagami and Yoshikawa (1961), Kapil and Dhaliwal (1968a, 1968b, 1969), Bhaskar and Gopinath (1975) further recorded their biology from woods, trees, logs and bamboos in northern India.

Green (1899) recorded interesting sleeping habits of *Crocisa ramose.* Malyshev (1925, 1936) described the nesting habits of several *Anthophora* and other solitary bee species. Schwarz (1939), detailed social species of *Trigona* found in Indo-Malayan region. Sakagami (1960, 1966) noted Ethobiological aspect of several *Allodape* and other genera of Apidae. Lieftinck (1955, 1957, 1962, 1964, 1972, and 1974) made remarks on ethobiology of many species that belong to former Anthophoridae such as *Crocisa, Xylocopa, Thyreus, Melecta,* and *Habropoda* etc. Iwata (1964) referred the egg of *Xylocopa* as the largest among all insects.

Kapil and Kumar (1969), Batra (1972, 1977, 1980), Pagden (1957), Batra, Sakagami and Maeta (1993), Reyes and Sakagami (1990), Reyes and Michener (1990), Batra and Norden (1996) and, Williams (1991, 1994, 1998) are the few references, dealt with nesting behaviour and immatures of some Indian species of *Ceratina, Anthophora, Amegilla, Braunsapis, Bombus* and *Trigona.*

Identically, thousands of references regarding pollination are available for this family of bees, from all over the world. However, when referring to Indian works, some of the important ones may be listed as follows:

Howard *et al.* (1920) first of all referred a number of pollinators, including honey bees, for several Indian crops. Work of Mohammad (1935) concerning pollinators of toria and sarson has been referred above. Similarly, role of *Apis indica* was studied by Latif *et al.* (1960) on toria and sarson. Several notes on the pollinators of leguminous crops including alfalfa were put forward by Batra (1977, 1991, 1994b, 1994c, 1995a). Sandhu *et al.* (1976) used *Ceratina (Pithitis) smaragdula* in India on alfalfa. Rai and Gupta (1983) published a useful note on the role of honey bees on the pollination of apple and pear. Abrol (1989, 1990, and 1992) described the ecology and behaviour of pollinators of strawberry and apple, including honey bees in major. Sharma and Gupta (1993) listed the flowering plants visited by *Apis mellifera* and *Apis cerana.* They concluded a total of 119 species of flowering plants in Solan (H.P.) among them 72 were visited regularly. Out of these 41 species provided both nectar and pollen, 22 nectar only and 9 species of plants were visited exclusively for pollen. Goyal and Gupta (1994) presented a detailed account of beekeeping with *Apis mellifera* in India. Batra (1995b) wrote on the use of solitary bees for blueberry pollination. Recently, Sharma and Gupta (2001) published the impact of bee pollination on sustainable production in apple orchards. Kumar and Singh (2001) recorded a preference for competing flora with sunflower for honey bees. Kumar *et al.* (2002) made observations on different modes of honeybee pollination and its effect upon the oil content in seeds of sunflower.

Non-*Apis* bees and future prospects

Honey bees (Genus *Apis* Linnaeus) have often been credited with pollination services that are actually performed by other bee species. Hundreds of entomophilous crops are now known that are very poorly pollinated by honey bees. The act in major is shared by non-*Apis* or the so called wild bees (Parker *et al.*, 1987). There are few estimates available for the value of non-*Apis* pollination. Estimations declared the value of wild bee industry was well over US $ 1 million per year in terms of expenditures and benefits in USA alone (Bohart, 1970, 1972). The benefits increased up to a range of US $18-40 million in 1981 and, the total involvement of money over crossed US$ 81 billion (Levin, 1983).

Recent technological advances in agronomic practices have focused primarily on improving yield, increasing the number of crops grown and increasing the area of harvestable crops. These advancements have been applied indiscriminately to the majority of crops and, in a very short duration, they have transformed farms into intensive monoculture systems. The positive results of these practices are impressive. The quality and quantity of food has increased, food costs have decreased, numerous fresh fruits and vegetables of high quality are available for much longer period, the quality and types of prepared food products have greatly improved and, the large labour force once required has been reduced, at the same time crop area have increased. On the other hand, the technical advances and intensive farming practices have evolved numerous negative impacts on crop pollination and non-*Apis* populations (Richards, 1993).

A number of conservation studies have concluded that clearing land of trees and increased cultivation have inadvertently eliminated many of the nesting sites previously used by non-*Apis* pollinators (Renner, 1996; Cane, 2001). Frequent applications of broad-spectrum weedicides and pesticides have been responsible for the rapid decline of pollinator numbers within agricultural areas (Batra, 1995c). Changing irrigation practices have had long-term negative effects on soil nesting pollinators. Overgrazing of rangeland and the use of herbicides have indirectly reduced the presence of pollinators by decreasing diversity of pollen-nectar resources and, by eliminating required plant resources that are utilized by various wild-non-*Apis* bees in nest construction (Batra, 1979b & 1979c). One of the consequences of an increased food supply for the world has been inadvertently depopulating both numbers and species of native pollinators within agricultural environment (Roubik, 2001). This situation must be addressed if our agricultural ecosystem has to be sustained.

Second interesting aspect of pollination with honey bee for several crops lies in the fact that after one or two morning forages honey bees are least interested in pollen collection. It prefers collection of nectar only and pollens are attached to the scopal bristles accidentally (Sharma and Gupta, 1993). In such events, honey bee enters and after drinking nectar moves out of the corolla tube so swiftly that a flower hardly receives any help in tripping. In the field, less than one percent of the self-tripped flowers produce seed and most of the non-tripped flowers fail to do so. Studies available with regard to many leguminous crops have reported several non-*Apis* bees as intentional trippers and they continue to do so for many full days (Zaleski, 1956). They deposit pollens in their nest chamber adjacent to their egg that hatch later as larva and feeds upon it. The process usually requires at least 3-15 days as per nest size of the species. Apparently, several wild-solitary bees are known for their effective pollination values abroad. Otherwise also, majority of non-*Apis* females are supported with dense brush of pollen collecting scopa and in larger area of the body so that maximum amount of pollen load may be collected in each trip to the field (Torchio, 1987).

The benefits we derive from native pollinators are believed to be increasing as the honey bee industry experiences continued difficulties from mites, Africanized bees and diseases, and as crops that are better pollinated by bees other than honey bees, are grown more intensively. To protect the native bee pollinators, two alternatives have been suggested, one is the preservation and management of habitats and another is artificial domestication and management of bee species (Stephen and Every, 1970; Williams *et al.*, 1991). Bohart (1970, 1972) quoted that honey bees are not entirely satisfactory in their use for the maximum output, in enclosures. On the contrary, non-*Apis* species when invited to artificial nesting devices, have been found much more effective pollinators. Apparently, a newer aspect has evolved which deals with "artificial domestication and management of wild bees for crop pollination (ADMP)".

In North America 3,500 species of wild-solitary bees have been recorded, commonly referred as "Pollen Bees" (Batra, 1994a). Alfalfa leafcutter bee (*Megachile rotundata*), blue orchard bee (*Osmia lignaria propinqua*), fuzzy foot bees (*Anthophora pilipes*) and, mustached bees (*Anthophora abrupta*) are some of the wild-bee cross pollinators which are successfully used in the artificial domestication and management programme, to enhance the crop pollination (Batra, 1982, 1991, 1993, 1994b). Robinson *et al.* (1989b) detailed several parameters applicable for the pollinators. In brief, such programmes intends to sustain a good population of pollinators close to the crops by providing them artificial nesting devices and, the normal principle usually understood was 'higher the number of pollinators higher will be seed yield'. Kapil, Grewal, Kumar and Atwal (1970) recorded some of the insect pollinators and noted their comparative abundance. However, after certain threshold limit population increase of the pollinator does not put any further positive impact upon the seed yield (Strickler, 1997). Instead, a consistency in increase in floral resource has to be maintained to sustain higher population of pollinator on the crop (Strickler, 1999; Strickler and Freitas, 1999), otherwise fear concerning escape of bee species would persist.

A few other successful programmes exist that has enhanced the number of native pollinators for fruit crops such as, using horn faced bees *Osmia cornifrons* in Japan and use of *Osmia cornuta* in Europe (Maeta, 1978; Torchio and Asineo, 1985; Maccagnani *et al.*, 2003) and its introduction and establishment in USA (Batra, 1978a). Identically, many species of *Megachile*, *Heriades* and *Osmia* are found in sufficient numbers and need intensive investigations in this direction because they are quite efficient visitors of several cultivated crops in northern India.

Field hut for nesting solitary bees

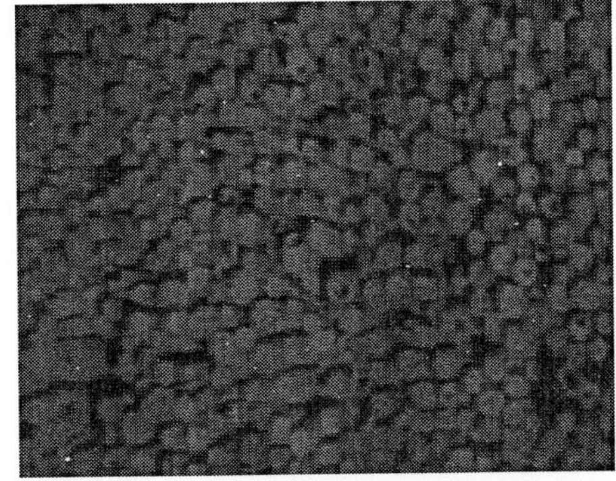

Pithy stems for *Pithitis* species

Castor stems for nesting of *Xylocopa* and *Megachile* species

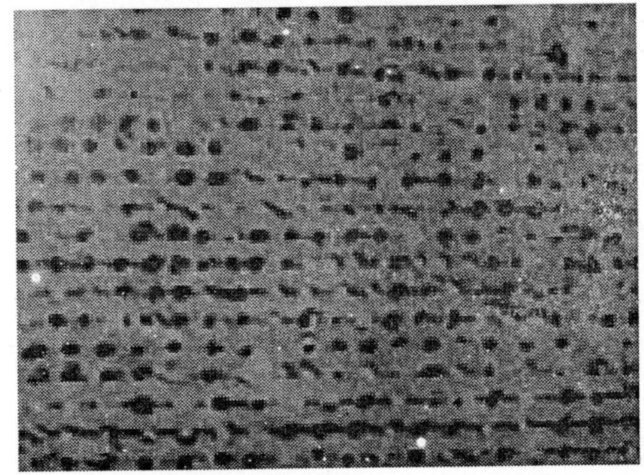

Soda water straws in wooden blocks for *Megachile* bees

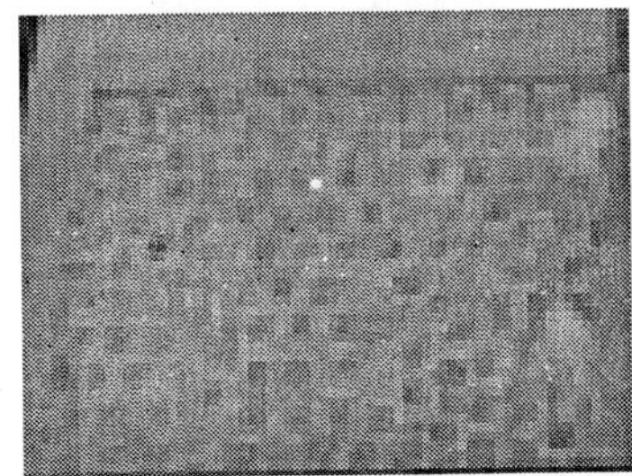

Soda water straws for *Megachile* bees

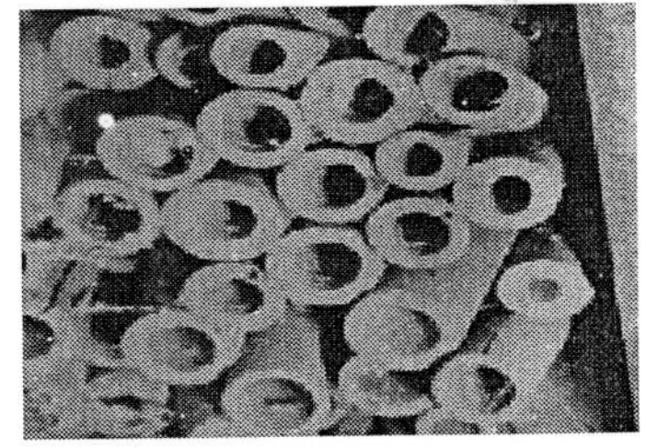

Bamboo stems for nesting of *Xylocopa*

Figure 70. Management of solitary bees.

Certainly, ADMP has emerged as a new entomological industry (Bohart, 1970). In India, the studies relevant to pollination of some more crops and the efficiency of several bee species were recently commenced by Kumar *et al.* (1994), Sharma and Gupta (2001) and, Kapila *et al.* (2002). Some of these workers have made significant contribution by using the criteria such as amount of carried pollen loads, number of visits made on flowers etc. for different bee species. Megachiline bees have the maximum area for the collection of pollen grains (beneath their abdomens) hence usually top the list with regard to carried pollen loads.

Gupta (2003) suggested that several correlated aspects should be undertaken in future such as, bee foraging studies coinciding with rotation of crops; population studies correlating different bee species on different crops; pollinator efficiency studies; effects of insecticides on various pollinator bee species; study of immatures, their developmental periodicities and, its variation under controlled environment, so that supply of broods as seeds may be implemented in practice, alongwith blooming of different crops; brood transfer techniques; studies relating parasites-predators and pests of adults and broods and, their mortality rate; impact of ecological factors on adults and broods; crop yield studies; development of nesting devices for different bee species so as to retain maximum population of bees on crops etc. These will ultimately help initiate artificial domestication and management programme in the country for many useful bee species to obtain better seed yields.

Chapter 19

Bee products

Honey is the most important primary product of beekeeping both from a quantitative and an economic point of view. It was also the first bee product used by humankind in ancient times. The history of the use of honey is parallel to the history of man and in virtually every culture evidence can be found of its use as a food source and as a symbol employed in religious, magic and therapeutic ceremonies, as the only concentrated form of sugar available to man in most parts of the world. The same cultural richness has produced an equally colourful variety of uses of honey in other products.

In India, honeybees are mainly known for the production of honey and to some extent wax. Besides, honey and wax, other valuable products such as pollen, royal jelly, propolis and bee venom can also be obtained from bees. Bees also help in increasing production of cross pollinated crops.

Honey is defined as natural sweet substance, produced by the honey bees from the nectar of blossoms or from the secretions of other parts of the plants, which honeybees collect, convert into honey and store in wax comb to ripen (Crane, 1975). Bees also collect sweet exudates of insects 'honey dew' and it is like flower nectar. Some plants also have nectarines other than on flowers. These nectarines also secrete liquids and bees do avail this extra floral nectar. Besides, bees can collect any sweet material like fruit juices, cane juices etc. Depending upon the source of sweet liquid and also the plant species in case of floral and extra floral nectar, the quality of honey varies. On this basis the honey can be classified as floral honey or honeydew honey. Though unifloral honeys are not common, yet honeys can be categorized on the basis of floral source such as litchi honey, beseem honey, *Brassica* honey, eucalyptus honey, sunflower honey etc. On the basis of colour honey can be categorised as light and dark coloured honeys. The honey from Giant honeybee combs is squeezed and it is out to contain more pollen grains and even the juice of some brood and the honey is normally called squeezed honey. The honey from combs of domesticated species is extracted with the help of honey extractors, so it is extricated honey. These honeys may remain in liquid form or may crystallize, and, have can be presented to consumers. As liquid honey or granulated honey. There are some other preparations for marketing honey which are not common in India. Small combs can be made to be raised and scaled honey in these combs is marketed as such as comb honey. Small pieces of combs are suspended in liquid honey packing and the preparation is marketed as chunk honey.

Colour Honey varies from very light to dark coloured and high mineral content were considered to attribute to dark colours. The colour of honey consists of melanodrin from the amino carbonyl reaction and flavoured pigments acquired from pollen by the nectar and honey, which is also a product of reaction of sugars (especially fructose) with amino acids. Caramelization of sugars due to acidic honey and the reaction of phenolic materials on iron honey content in honey or from containers or equipments, also contribute to colour. Honey also darken when heated or stored. Colour is also related to flavour and taste. Many light honeys are to mild flavour and dark honeys have more pronounced taste.

Chemical composition of honey

Chemical composition of honey is complex. Honey mainly contains sugars but acids, minerals, vitamins, enzymes and antibiotic substances are also present in small amounts. The solids in honeys vary with the floral source. Honeybees, seal the honey in comb cells after evaporating the excess moisture to reduce it to less than 20%. The average composition of honey is shown in Table 46.

Table 46. Average composition of honeys and ranges of values (White *et al.*, 1962)

Component (% except pH and diastase value)	Average	Standard deviation	Range
Water	<17.2	1.5	13.4 - 22.9
Fructose	38.2	2.1	27.2 - 44.3
Glucose	31.3	3.0	22.0 - 40.7
Sucrose	1.3	0.9	0.2 - 7.6
Maltose (reducing disaccharides calculated as maltose)	7.3	2.1	2.7 - 16.0
Higher sugars	1.5	1.0	0.1 - 8.5
Free acids (as gluconic acid)	0.43	0.16	0.13 - 092
Lactone (as glucolactone)	0.14	0.07	0.0 - 0.37
Total acid (as gluconic acid)	0.57	0.20	0.17 - 1.17
Ash	0.169	0.15	0.020 - 1.028
Nitrogen	0.041	0.026	0.000 - 0.133
pH	3.91	-	3.42 - 6.10
Diastase value	20.8	9.8	2.1 - 61.2

In honey sugars are the main component and account for 95 to 99% of honey dry matter. The majority of these are the simple sugars fructose and glucose which represent 85-95% of total sugars and 70 per cent of total solids. Generally, fructose is more abundant than glucose. This predominance of simple sugars and particularly the high percentage of fructose are responsible for most of the physical and nutritional characteristics of honey. Sucrose is most common disaccharide and varies between 0 to 5%. Exceptionally high sucrose (up to 20%) is present in *Carvia* honey or in honey produced by bees that had been fed with excessive sugars syrup. High sucrose is considered to be due to adulteration. A reducing disaccharide maltose is also present in higher amount that is 3-16%. This maltose may not be present initially but the quantity increases with storage and this probably by conversion of other sugars by the action of enzymes and acids. Small quantities of other sugars are also present. Presence of over 19 minor sugars in honey have been confirmed by using high performance liquid chromatography (HPLC). Other lesser common sugars present in honey are isomaltase, trehalose, genitiobiose, panose, turanose etc. These complex sugars account for 1 to 2% only. Though quantitatively of minor importance, their presence can provide information about adulteration and the botanical origin of the honey.

Water is quantitatively the second most important component of honey. Its content is critical, since it affects the storage of honey. Only honeys with less than 18% water can be stored with little to no risk of fermentation. The final water content depends on a number of environmental factors during production such as weather and humidity inside the hive, but also on nectar conditions and treatment of honey during extraction and storage.

Among the minor constituents **organic acids** are the most important and of these gluconic acid, which is a by-product of enzymatic digestion of glucose, predominates. The organic acids are responsible for the acidity of honey and contribute largely to its characteristic taste. Honey is known to be acidic due to the presence of tartaric, citric, malic, mccinic, formic, acetic, butyric acids. Gluconic acid is mot common. pH of Indian honeys ranged from 3.25 to 4.50 (Wakhle and Desai, 1991) as against the European honey with 3.20 to 4.50. Acidity of honey contributes to taste but is not evident due to great sweetness of honey. Honey gives sour or acidic taste due to increased acidity due to long storage or on fermentation.

Minerals are present in very small quantities, potassium being the most abundant. Minerals in honey vary widely and are grouped as ash. Generally dark honeys contain more minerals than light honeys (Kalimi and Sohonie, 1964). Minerals identified in honey are potassium, calcium, phosphorus, sodium, magnesium, manganese, copper, sulphur, silica, silicon zinc and iron and all these account for upto 1% of solids in honey. Honey contains much less sodium than potassium, hence, it is suggested by some workers to use Na/K ratio to detect honey adulteration with high fructose corn syrup.

Other trace elements include **nitrogenous compounds** among which the enzymes originate from salivary secretions of the worker honeybees. Enzymes are most important part of honey as these are added during the conversion of nectar into honey by bees. They have an important role in the formation of the honey. The most important enzymes reported. Indian honeys are invertase, diatase, glucoseoxidase, and catalase.

Traces of other proteins, enzymes or amino acids as well as water soluble vitamins are thought to result from pollen contamination in honey. Vitamins present in honey are thiamine (B_4), riboflavin (B_2), nictonic acid, Pentothenic acid, Niacin, vitamin K, folic acid, biotin, carotene, pyridoxin, and ascorbic acid. Very low amounts of vitamins have hardly any nutritional significance in honey.

Virtually absent in newly produced honey, hydroxvmethylfurfural (HMF) is a byproduct of fructose decay, formed during storage or during heating. Thus, its presence is considered the main indicator of honey deterioration.

Even though some of the substances responsible for honey colour and flavour have been identified the majority are still unknown. It is more than likely that honeys from different botanical origins contain different aromatic and other substances which contribute to the specific colours and flavours and thus allow to distinguish one honey from another. Similarly, it is very likely that, depending on their botanical origin, honeys contain traces of pharmacologically active substances. Some of them have been identified, such as those responsible for the toxicity of certain honeys, but for the majority of possible substances, scientific verification requires further studies.

The honey in north Indian hot plains have lesser moisture content as compared to honey from western and eastern ghats and areas with very humid climates. In these areas the moisture level in honey can be upto 23% and honeys are thinner. The moisture content may even change after extraction or during storage because honey is hygroscopic and absorbs moisture.

Generally, Indian honey contains more than 20% moisture as compared to 17-18% in European honey which is mainly due to low inversion and, evaporation capacity of native *Apis cerana* bees. Beekeepers also does not extract fully ripen honey from *A. mellifera* colonies. In case of honey produced by *A. dorsata* whose comb is generally exposed to the atmospheric high humidity and tropical high temperature and these not only increase the moisture content but also reduce the enzyme activity. The keeping quality, particularly granulation and fermentation of honey, mainly depend upon the moisture content. Hence great care is necessary to measure the moisture and decide the mode of processing for quality control of honey.

Physical Properties of Honey

Honey is a supersaturated solution of sugars. Most of the physical properties vary as per water content in honey. The physical characters include viscocity, density, specific gravity, refractive index etc. Besides chemical, physical properties are also used to lay down honey standards. These characters also give an inkling of purity of honey.

Viscosity: Honey is viscous liquid and the viscosity determines the flow ability. Resistance to flow is termed as viscosity. Viscosity depends mainly on the moisture content of honey and therefore, can be used to determine moisture content. Heating reduces the viscosity thus the heat treatment is given to handle honey. But in addition, the viscosity is also dependent on the protein contents which in turn is dependent on the source of nectar. Honey with higher protein contents to be more viscous. The viscosity of honey is of great practical importance to honey processors and packers. Viscosity of honey as high as 600 poise at 11.5 to 13°C reduces to 2.6 at 71.1°C. Honey of 13.7% moisture show viscosity of 420 poise and with 21.5% moisture it showed 13.6 poise (Munro, 1943). Gatoria *et al.* (1996) also found that Indian honey (Punjab) has the viscosity of 509 poise at 8.1°C for honey with 15.5. per cent moisture and it was 34.36 poise at 35°C.

Specific gravity: Ratio of weight of honey to the weight of the same volume of water is the specific gravity of honey. Pure honey should have the specific gravity of about 1.35-1.44. Specific gravity is also termed as relative density. This depends on temperature and moisture content. Specific gravity can be measured by different methods and it is customary to specify the temperature.

Refractive Index: Refractive index of honey is the ratio of the velocity of light in honey to that in air. Moisture in honey can be determined by using refractometers specific gravity and refractive index measurements can be used to know moisture content of honey.

Aroma and Flavour: Aroma to honey is imparted by nectar of flowers which is an important physical property for judging honey. Honey has an ability to acquire foreign odours, so proper storage is necessary to maintain the original aroma. Aroma or flavour is most important characteristic of honey, which depends on its plant source. Some of the honeys are known for its fine flavour like litchi, plectranthus, etc. Some have characteristic aroma like Jamun, sunflower, mustard etc. The raw honey may have fine flavour but processing of honey may change it if proper care is not taken because it is very sensitive to heat, storage and light intensity. Honey contains several flavour producing volatile organic compounds. The most common "honey flavour" compounds are phenyl-acetic alcohols and esters, which are found is most honeys (Cremer and Reidman, 1965). The taste of honey is attributed to the sugars, gluconic acid and proline, besides volatile constituents (Phadke, 1964).

Hygroscopicity: The ability to absorb moisture from surrounding atmosphere is hygroscopicity. In areas with high humidity, the honey if not stored properly tends to draw water from moist air. This happens in surface layer of honey which increases in moisture content. The hygroscopicity depends upon the sugar composition and moisture in honey. Moisture in honey is in equilibrium to particular relative humidity level.

Among all the sugars, levulose is most hygroscopic which varies in different honeys. Thus water content increased with increase in relative humidity (Martin, 1958). The moisture content of honey can be lowered, if exposed to air with lower relative humidity. Tropical honeys, particularly honey produced in coastal areas, like rubber honey of Kerala, contain moisture as high as 24.5% (Nair, 1980). Honey can be saved from absorbing moisture and rendering it liable to fermentation by proper storage of honey in very humid areas. Therefore, hygroscopcity has practical application to the beekeeper.

Optical rotation (Polarization): Due to presence of sugars in honey it has the property of rotating the plane of polarization of polar light. Generally honeys are levorotary (i.e., rotates the plane to the left), due to higher contents of levulose over dextrose (which is dextrorotatory i.e., rotates to right). Almost all India unifloral honeys are reported to be levorotatory (Phadke, 1962). This property also reflects the wide variability of the sugar contents of various honeys.

(i) Invertase: Invertase is for conversion of sucrose to simple sugars (glucose and fructose). Honey invertase is different from yeast invertase in that it transforms glucose from sucrose molecule, either to fructose and glucose or other sugars to form more complex products.

(ii) Diatase: Diatase converts starch to dextrins when added by the bees during ripening. Wakhle and Desai (1983) showed that though most stable among other enzymes, the diatase activity reduces on heating and also on storage at room temperature for two years. Diatase is very heat sensitive hence used as criteria to estimate heating of honey.

(iii) Catalase: Decomposes hydrogen peroxide in honey.

(iv) Phosphatase: Decomposes glycerophosphate.

(v) Glucose oxidase: Glucose oxidase added by the bees during ripening, oxidises small amounts of glucose to glucone, lactone and to gluconic acid and the acidity thus formed contributes to the stability or the ripening nectar against fermentation. In the reaction one molecule of hydrogen peroxide is produced for each molecule of glucose oxidised. Ripening of nectar is also stabilized by this hydrogen-peroxide and helps in preservation of honey. White *et al.* (1963) showed that most of the antibacterial activity or inhibine in honey is due to production of hydrogen peroxide. Wakhle and Desai (1991) have reported noticeable antibacterial activity in honey even in absence of glucose-oxidase.

Proteins and amino acids: Proteins are found in small quantities in different honeys. Proteins originate mostly from bees and some from the forage plants. Among the Indian honeys, Carvia honey contained as high as 1.20% proteins (Suryanarayana, 1981). Amino acid present in honey are cystine, aspargin, lysine, glycine, aspartic acid, glutamic acid, alanine, tyroxine, valine, methionine, leucine.

Pigments: Carotene, Chlorophyll, xanthophylls

Flavours and aroma substances: Apart from sugars these compounds include amino and other acids (especially gluconic): prolline. tannins: glucosidic and alkalolidal compounds, also specifically diacetyl (or other diketoalkane) and methyl anthranilate in orange honey. Hydroxymethyl furfural seem to be responsible for the characteristic honey aroma.

In honey, fructose in presence of acids produces hydroxy-methyl furfural (HMF). This reaction starts right from the ripening of nectar in the hive. The rate of production of HMF increases with temperature. Fresh honey from the hive contains no or very low HMF and 30 year old honey contained 200 mg/kg (Wakhle and Phadke, 1995). HMF has been considered as an indicator of the heat exposure to honey and possible adulteration with artificial sugar syrup that invariably contain HMF. Codes Alimentarius Commission (1969) had prescribed 40 ppm HMF, where as International Standards by Codex (1987) recommend to raise the HMF limit to 80 ppm. The revised Indian standards IS 4941-1994 has now specified limit of 80 ppm of HMF in honey, for the first time.

Uses of honey

Honey is used as traditional medicine in India whereas all over the world it is used as food. The nutritive value of honey has been compared with items like milk. fruits, jam, eggs etc. It is highly appreciated as food for infants and old persons. It helps to build hemoglobin of the blood. As it provides energy in a readily available form, honey is largely taken by sports persons. In an average Indian home, honey may be used in tea, coffee and milk in place of sugar. Honey-lemon drink is a welcome drink in summer. Honey when used in baking bread, cakes and biscuits, improves their flavour and enhances their keeping quality by retarding their drying up because it is hygroscopic and absorbs and retains moisture. It can be used on fruits, cereals, in salad dressings, canning and preserving. Honey is used as carrier in all Ayurvedic and Unani medicines. Honey is put to countless medicinal uses in Indian homes. Honey is a good luxative, a blood purifier preventive against cough and cold and relieves of sore throat. Honey is a remedy for tongue ulcers. Its regular use is recommended in severe cases of malnutrition with impaired digestion and stomach and intestinal ulcers. It contains number of nutrients required for balanced human diet (Table 47). Honey is also used for curing bums and wounds. Honey lemon juice drink is known to reduce obesity. Large amounts are used in skin and beauty lotions. Chewing gums have honey as a sweetening agent. In laboratory, honey has been found to stimulate plant growth and helps the rooting of cuttings. The ancient mead with 11% alcohol content made with fermented honey is still popular in some European countries. In African countries especially in Ethiopia. large quantities of honey have traditionally been fermented with fruits to make beer. The vinegar made with fermented boney bas added advantages of honey flavour and nutritive value.

Table 47. Nutrients in honey in relation to human requirements (Crane, 1980)

Nutrient	Unit	Average amount in 100 g honey	Recommended daily intake
Energy equivalent	kcal	304	2800
Vitamins			
A	I.U.	-	5000
B1 (Thiamin)	mg.	0.004 - 0.006	1.5
B2 (Riboflavin)	mg.	0.002- 0.06	1.7

Nicotinic acid (niacin)	mg.	0.11.- 0.36	20
B6 (Pyridoxine)	mg.	0.008 - 0.32	2.0
Pantothenic acid	mg.	0.02 - 0.11	10
Bc (Folic acid)	mg.	-	0.4
B12 (Cyanocobaltamine)	mg.	-	6
C (Ascorbic acid)	μg	2.2 - 2.4	60
D	mg.	-	400
E (Tocopherol)	I.U.	-	30
H (Biotin)	I.U.	-	0.3
Minerals	mg.		
Calcium	mg.	4 - 30	1000
Chlorine	mg.	2 - 20	
Copper	mg.	0.01 - 0.1	2.0
Iodine	mg.	-	0.15
Iron	mg.	- 3.4	18
Magnesium	mg.	0.7 - 13	400
Phosphorous	mg.	2 - 60	1000
Potassium	mg.	10 - 470	-
Sodium	mg.	0.6 - 40	-
Zinc	mg.	0.2 0.5	15

Some of the properties of the honey has been described in detail as given below

Digestive

Honey is said to improve food assimilation and to be useful for chronic and infective intestinal problems such as constipation, duodenal ulcers and liver disturbances. Honey is said to facilitate better physical performance and resistance to fatigue, particularly for repeated effort; it also promotes higher mental efficiency. It is therefore, used by both the healthy and the sick for any kind of weakness, particularly in the case of digestive or assimilative problems. Improved growth of non-breast fed newborn infants; improved calcium fixation in bones and curing anaemia and anorexia may all be attributed to some nutritional benefit or stimulation from eating honey.

Respiratory

Honey is a well known remedy for colds and mouth, throat or bronchial irritations and infections. The benefits, apart from antibacterial effects, are assumed to relate to the soothing and relaxing effect of fructose.

Skin and wound healing

Honey is used in moisturizing and nourishing cosmetic creams, but also in pharmaceutical preparations applied directly on open wounds, sores, bed sores, ulcers, varicose ulcers and burns. It helps against infections, promotes tissue regeneration, and reduces scarring also in its pure, unprocessed form. If applied immediately, honey reduces blistering of burns and speeds regeneration of new tissue. A cream, applied three times per day and prepared from equal parts of honey, rye flour and olive oil, has been successfully used on many sores and open wounds-even gangrenous wounds in horses

Eye disorders

Clinical cases or traditional claims that honey reduces and cures eye cataracts, cures conjunctivitis and various afflictions of the cornea if applied directly into the eye, are known from Europe, Asia, and Central America.

Medicine

Frequently, specific benefits of unifloral honeys are reported based on the traditional assumption that honey made from the nectar of a medicinal plant has the same or similar beneficial activity as the one recognized for the whole plant or some parts of it.

Diabetes

Frequently, claims are voiced that honey is good for diabetics. This is unlikely to find confirmation because of its high sugar content. However, it is better than products made with cane sugar. It revealed that insulin levels were lower when compared to the uptake of equal caloric values of other foods, but blood sugar level was equal or higher than in the other compared products shortly after eating. In healthy individuals, the consumption of honey produced lower blood sugar readings than the consumption of the same quantity of sucrose.

Ayurvedic medicine

Traditional, but well-studied medicinal systems as the ayurvedic medicine of India, use honey predominantly as a vehicle for faster absorption of various drugs such as herbal extracts. Secondarily, it is also thought to support the treatment of several more specific ailments, particularly those related to respiratory irritations and infections, mouth sores and eye cataracts. It also serves as a general tonic for newborn infants, the young and the elderly, the convalescent and hard working farmers

Other benefits

Honey is said to normalize kidney function, reduce fevers and help insomnia. It is also supposed to help recovery from alcohol intoxication and protect the liver; effects also ascribed to fructose syrups. Heart, circulation and liver ailments and convalescent patients in general improved after injection with solutions of 20 and 40% honey in water.

Energy source

As food, honey is mainly composed of the simple sugars fructose and glucose, which form the basis of almost all indications on how, when and why to use it. The main consideration is the fact that honey provides immediately available calories, from which it derives its energy value for healthy and sick people: quick access to energy without requiring lengthy or complicated digestive action. The same direct absorption also carries a risk of pathological sugar metabolism, such as diabetes and obesity.

Honey in infant feeding

In India and many other parts of the world honey is given to newborn babies during the first few days of life as a special tonic, particularly if they were born weak or prematurely. However, the consumption of honey by infants less than one year of age is not recommended by the US Food and Drug Administration. This recommendation is based on the correlation of some infant deaths with the ingestion of botulism spores (*Clostridium botulinum*) from honey. Though omnipresent in the environment and in many foods, the spores normally are not capable of developing in the intestines of adults or children. However, since very young infants often have less acidic digestive tracts and less competition from a bacterial flora as yet little developed, *Clostridium* spores may develop in their intestines. The toxin produced by the bacteria binds irreversibly to motor nerve endings and can only be overcome by the growth of new nerve endings. Some of the typical symptoms are mild paralysis (failure to thrive), moderate to severe paralysis, and fulminant (sudden and intense) paralysis which can lead to death without warning. Other milder symptoms can be constipation, listlessness, lethargy, diminished appetite or activity and lack of muscle

Non-energetic nutrients

Often honey is recommended because of its content of other nutrients like vitamins and minerals, but their quantity is so low that it is unrealistic to think they can provide any significant supplement in a deficient diet. Micronutrients in unprocessed honey can be assumed to be of the highest quality possible. Thus from a nutritional point of view, a synergistic balancing effect or one that unlocks the availability of other nutrients already present, is one of the more plausible yet untested hypotheses.

Topical applications

Topical applications under controlled conditions have shown accelerated wound healing in animals but also of various types of wounds, including post-operative ones in humans Similar, yet not equal, effects are obtained with the application of purified sucrose and special polysaccharide powders. External as well as internal wounds from operations become bacteriologically sterile within a few days and dry out. The simultaneous stimulation of tissue regeneration by honey reduces scarring and healing times. In addition, dressings applied with honey do not stick to the wounds or delicate new skins. In many tropical field hospitals, where antibiotics and other medicines are scarce, honey has been employed successfully for a long time.

Antibacterial activity

Antibacterial activity is the easiest to test and is probably the most studied biological activity of honey. In normal honey it is attributed to high sugar concentration and acidity (pH range 3.5 to 5.0). Yet, since also diluted honey has shown antibacterial activity, the active ingredient was attributed to an elusive substance generically termed "inhibin". Much of this activity was later attributed to hydrogen peroxide (H_2O_2) an enzymatic by-product during the formation of gluconic acid from glucose. The responsible enzyme, glucose oxidase is basically inactive in concentrated normal honey. Thus, in honey solutions with the right pH, antibacterial activity is largely due to the presence of hydrogen peroxide. The biological significance of such a mechanism arises from the requirement to protect immature honey (with high moisture content) inside the colony until higher sugar concentrations are achieved. However, it has been well demonstrated that most of the antibacterial activities of honey are lost after heating or prolonged exposure to sunlight.

OTHER USES OF HONEY

As a food

Honey is most commonly consumed in its unprocessed state, i.e. liquid, crystallized or in the comb. In these forms it is taken as medicine, eaten as food or incorporated as an ingredient in various food recipes.

Uniformly crystallized honey is attractive both visually and for its convenience of use. It is also less likely to ferment than badly crystallized or semi-crystallized honeys. Different storage temperatures in different climates, among other factors influence the crystallization and speed of re-liquefaction of honeys. Stored above 25°C, most honeys remain liquid or reliquify slowly, but lose much of their aroma in just a few months

As a food ingredient

The traditional use of honey in food preparations has been substituted in most cases by sugar and more recently by various sugar syrups in baked products, confectionary, candy, marmalades, jams, spreads, breakfast cereals, beverages, milk products and many preserved products. Honey can substitute all or part of the normal sugar in most products.

As an ingredient in medicines

The medicinal use of honey is probably its most widely known use, but such uses do not require special preparations. More common is the use of honey in herbal and other traditional extracts.

Products of honey fermentation

Honey is or was the only the most accessible source of fermentable sugars. The mead was obtained from honey. Honey vinegar can be produced from mead in the same way as wine vinegar. Mead can also be used as a base from which distilled alcoholic beverages can be produced. Such production is usually for home consumption only.

Others

The tobacco industry is estimated to use more than 2000 tons of honey annually to improve and preserve tobacco's aroma and humidity. Since tobaccos, at least in part, are valued according to the rate at which they dry, the importance of honey for the more valuable tobaccos can easily be understood.

The cosmetic industry uses honey as a skin moisturizer, softener and restorer of the skin's own moisturising factors in creams, soaps, shampoos and lipsticks. Because of its stickiness it can however only be employed in small quantities.

CRYSTALLIZATION OF HONEY

Crystallization is one of the most important characteristic for honey marketing. In temperate climates most honeys crystallize at normal storage temperatures. This is due to the fact that honey is an oversaturated sugar solution, i.e. it contains more sugar than can remain in solution. The crystallization results from the formation of monohydrate glucose crystals, which vary in number, shape, dimension and quality with the honey composition and storage conditions. The lower the water and the higher the glucose content of honey, the faster the crystallization. Temperature is important, since above 25°C and below 5°C virtually no crystallization occurs. Around 14°C is the optimum temperature for fast crystallization, but also the presence of solid particles (e.g. pollen grains) and slow stirring result in quicker crystallization. Usually, slow crystallization produces bigger and more irregular crystals.

TOXIC HONEYS

Honeys of some flowering species are reported to be toxic (White, 1975c and Kerkvliet, 1981) because of their content of active ingredients from nectars or honeydews which are noxious or toxic to humans. Although these honeys are not very common, they can be of particular importance in some localities. The plant families and genera from which a few species have been reported to produce toxic honeys include Ericaceae (*Rhododendron, Azalea, Arbutus, Andromeda,* and *Kalmia*); Solanaceae (*Datura, Hyoscyamus,* and *Atropa*; Compositae (*Senecio jacobaea* or ragwort); Lagnonaceae (*Gelseminum*); Ranunculaceae (Aconitum), some species of the genus Euphorbia in South Africa and the honeydew of *Coriaria arborea* from New Zealand (Crane, 1990). Bitter and off-flavoured honeys are produced from many more species.

It has been found that many plants produce nectar that is toxic or repellent to some floral visitors. Although secondary compounds in nectar and toxic nectar are geographically and phylogenetically widespread, their ecological significance is poorly understood. Several hypotheses have been proposed for the possible functions of toxic nectar, including encouraging specialist pollinators, deterring nectar robbers, preventing microbial degradation of nectar, and altering pollinator behavior. All of these hypotheses rest on the assumption that the benefits of toxic nectar must outweigh possible costs; however, to date no study has demonstrated that toxic nectar provides fitness benefits for any plant in addition to these adaptive hypotheses, Adler (2001) considered the hypothesis that toxic nectar provides no benefits or is tolerably detrimental to plants, and occurs due to previous selection pressures or pleiotropic constraints. For example, secondary compounds may be transported into nectar as a consequence of their presence in phloem, rather than due to direct selection for toxic nectar. Individuals are often simultaneously under selective pressures exerted by multiple interactions, including both mutualisms and antagonisms. For example, although plant-herbivore and plant-pollinator interactions are typically studied separately, most plants must attempt to attract pollinators while also escaping herbivores. Herbivores and pollinators can therefore both exert selective pressures for plant traits via direct effects on plant fitness (Schemske and Horvitz ,1988; Juenger and Bergelson, 1997; Strauss and Armbruster, 1997).

Selection for defense against herbivores may be influenced by selection for attracting pollinators, and vice versa (Strauss, 1997). Many traits that affect herbivory in plants are closely related to those affecting pollination. For example, floral resins that once functioned as defenses can be co-opted for pollinator rewards (Armbruster, 1997, Armbruster *et al.,* 1997), and pleiotropic effects of an allele determining floral pigmentation may influence vegetative resistance to herbivores (Simms and Bucher, 1996). If plant defensive compounds are present in floral tissues and are deterrent to pollinators, or if the cost of producing the compound results in less attractive floral structures, then pollinators may select against the production of these compounds (Detzel and Wink, 1993, Strauss *et al.,* 1999). Alternatively, plant secondary compounds could increase pollinator attraction if decreased herbivory improves floral displays or rewards (Karban, 1993, Lohman and Berenbaum, 1996, Juenger and Bergelson, 1997, Lehtila¨ and Strauss, 1997, Callaway *et al.,* 1999, Krupnick *et al.,* 1999, Strauss *et al.,* 1999, Mothershead and Marquis, 2000, Adler *et al.* in press). Thus, evolution of some plant traits may be constrained by opposing selection from herbivores and pollinators.

Secondary compounds that are associated with resistance to herbivory have been frequently documented in floral nectar (e.g., Baker and Baker, 1975, Baker, 1977, Guerrant and Fiedler, 1981), although nectar is usually studied in the context of pollination rather than herbivory. The general function of nectar, with its array of sugars and amino acids, is to attract pollinators and:or natural enemies of herbivores (Fægri and van der Pijl, 1979). Why, then, would secondary compounds, which are generally toxic or repellent, be present in a structure whose function is the attraction of mutualists? Although this phenomenon is widespread (Baker, 1977, 1978), it has received relatively little attention from ecologists. Integrating our understanding of multispecies interactions, such as plant-pollinator and plant-herbivore interactions, may shed light on traits, such as secondary compounds in nectar that previously seemed anomalous.

NECTAR SECRETION AND COMPOSITION

Nectar is about 90% sugar by dry weight (Luttge 1977); the other 10% consists of a myriad of compounds, including amino acids, lipids, antioxidants, mineral ions, and secondary compounds (Luttge and Schnepf, 1976, Baker, 1977). The chemical composition of nectar varies widely between species, and even between different types of nectaries within the same plant species (Davis *et al.*, 1998). Nectars are characterized by their ratio of sucrose: (glucose fructose), which is consistent within species but varies widely between species. Amino acids are virtually ubiquitous in nectar, and their composition ranges widely between species but is generally consistent within a species (Baker and Baker, 1982). Pollinator taxa have been correlated with both sugar ratios and amino acid composition across species, suggesting that there is selection for a characteristic "taste" that is recognizable to specific pollinators (Baker and Baker, 1982). Some nectars fluoresce under UV illumination while others do not; the color and intensity of fluorescence vary between more than within species, suggesting species-specific differences in compounds responsible for fluorescence (Thorp *et al.*, 1975). Finally, many types of secondary compounds have been found in nectar from different plant species in small but consistent amounts. The species-specific differences in nectar composition could be explained in two ways that are not mutually exclusive: (1) the secretory process in nectaries controls chemical composition and varies between species or (2) the constituents of nectar reflect the chemical composition of phloem, and phloem composition varies between species.

Table 48. Reports of nectar that is toxic or deterrent to floral visitors or humans. The compounds responsible for the deterrent or toxic effects of nectar were not identified.

Species	Family	Effects	Reference
Aesculus californica	Hippocastanaceae	toxic to bees	Eckert 1946, Mussen, 1979
Astragalus spp.	Fabaceae		
Cuscuta spp.	Convolvulaceae		
Cyrilla racemiflora	Cyrillaceae		
Gelsemium semperirens	Loganiaceae		
Kalmia latifolia	Ericaceae		
Solanum nigrum	Solanaceae		

Veratrum californicum	Liliaceae		
Zygadenus enesosus	Liliaceae		
Corynocarpus laeigata	Corynocarpaceae	toxic to honeybees	Palmer-Jones and Line 1962
Agauria spp.	Ericaceae	honey toxic to humans	Jaeger, 1961
Andromeda spp.	Ericaceae		
Kalmia spp.	Ericaceae		
Rhododendron flaum	Ericaceae		
Rhododendron ponticum	Ericaceae		
Paullinia australis	Sapindaceae		
Angelica triqueta	Apiaceae	toxic to bees	Bell 1971
Astragalus lentiginosus	Fabaceae	toxic to bees	Vansell and Watkins 1934
Azalea pontica	Ericaceae		Kebler 1896
Kalmia latifolia	Ericaceae	honey toxic to humans	
Camellia thea	Theaceae	lethal to honeybee larvae	Sharma *et al.* 1986
Erythrina fusca	Fabaceae	deterred ants	Feinsinger and Swarm 1978
Hippobroma longiflora	Campanulaceae		
Euphorbia spp.	Euphorbiaceae	honey bitter to humans; induced nausea	Pryce-Jones 1942
Ochroma lagopus	Bombacaceae	toxic to bees and other insects	Paula *et al.* 1997
Sophora microphylla	Fabaceae	toxic to honeybees	Clinch *et al.* 1972
Tilia spp.	Tiliaceae	toxic to bees and other insects	Crane 1977
Veratrum californicum	Liliaceae	toxic to bees	Vansell & Watkins 1933

Table 49. Reports of nectar containing secondary compounds

Species	Family	Secondary compound	Reference
Aesculus hippocastaneum	Hippocastanaceae	saponins	Schulz-Langner 1966
Atropa belladonna	Solanaceae	alkaloids	Detzel and Wink 1993
Brugmansia aurea			
Nicotiana tabacum			
Campanula rapunculoides	Campanulaceae	alkaloids	Baker and Baker 1975
Cucurbita pepo	Cucurbitaceae		
Cuscuta salina	Convolvulaceae		
Iris pseudocorus	Iridaceae		
Lotus corniculatus	Fabaceae		
Mimulus moschatus	Scrophulariaceae		

Nymphoides peltatum	Gentianaceae		
Rhododendron ponticum	Ericaceae		
36 of 66 species	various	Nonprotein amino acids alkaloids phenolics	Baker 1977
50 of 567 species	various		
191 of 528 species	various		
86 of 248 floral nectars	various	Nonprotein amino acids	Baker *et al.* 1978
86 of 248 floral nectars	various	Nonprotein amino acids	Baker *et al.* 1978

Toxic nectar: its nature and occurrence

There are many reports of nectar that is toxic or deterrent to animals, in which responsible compounds are not identified (Table 48). Most of these studies focus on honeybees or on humans poisoned by honey made from nectar of a specific plant. Several reports are anecdotal and describe bee death or narcosis following visits to flowers (Vansell and Watkins 1933, 1934, Pryce-Jones 1942, Eckert 1946, Jaeger 1961, Bell 1971, Crane, 1977). Some studies removed nectar from flowers and performed laboratory assays to demonstrate that nectar was the cause of toxicity (Palmer-Jones and Line, 1962, Clinch *et al.,* 1972, Berenbaum *et al.,* 1986, Sharma *et al.,* 1986, Paula *et al.,* 1997). Only one study offered both nectar and a sugar solution control in field tests; sugar solution was preferred over nectar by ants in two of four plant species (Feinsinger and Swarm, 1978). These studies suggest, albeit largely through description rather than experiments, that some plant species produce nectar that is toxic or deterrent to some floral visitors. More experimental studies comparing nectar and control solutions would strengthen this argument. Also, the emphasis on toxicity to honey bees, which are not the native pollinator for many of these plants, raises the question of whether native pollinators are as affected as introduced species. Numerous studies have demonstrated that secondary compounds occur in nectar without testing the effects of these compounds on floral visitors (Table 49). It is therefore not known whether these compounds occur in sufficient concentrations to have any ecological consequences. Thus these studies do not document nectar that is actually toxic, but identify plant species whose nectar might adversely affect floral visitors.

Future directions Much work remains in order to clarify the role of toxic nectar in plant ecology and evolution. Some major avenues of research are suggested below; these are intended only as jumping-off points for future studies.

Phylogenetic patterns and physiological mechanisms

Phylogenetic studies are needed to determine whether toxic nectar occurs concurrently with the evolution of novel secondary compounds or arises subsequently.

Is toxic nectar adaptive?

In order to assert that toxic nectar evolves in response to selection by any agent, it is necessary to demonstrate that this traits exhibits heritable variation. Only after all these points have been investigated can we assert that toxic nectar may have evolved in response to selective pressures exerted by ecological interactions.

In conclusion, toxic nectar is a widespread but poorly understood phenomenon. No study of toxic nectar has established that this trait benefits the plant. Clearly, answering this question is of central importance in evaluating the hypotheses reviewed above. If a benefit of toxic nectar is found, then more specific studies can address whether this is due to decreased nectar robbing, specialist pollinators, antimicrobial properties, altered pollinator behavior, or other reasons. If studies reveal that toxic nectar is not beneficial, then this would be strong evidence that toxic nectar is the result of previous evolutionary forces no longer acting on the plant, or a pleiotropic consequence of other traits such as resistance to herbivores. Integrating our understanding of multispecies interactions, such as plant-pollinator and plant-herbivore interactions, may shed light on traits, such as toxic nectar, that previously seemed anomalous.

HONEY HARVESTING AND PROCESSING

High colony yields are only possible with well populated colonies in areas with abundant nectariferous flora. The honey needs to be harvested before the bees can consume it for further colony development, but sufficient quantities have to be left to provide for the basic needs of the colony. However, the different management and harvesting techniques can influence the final quality of the honey

Processing

Uncapping is the first real step of honey processing. It consists of the removal of the thin wax layer that seals the honey cells. The wax caps can be sliced off with a sharp, thin, long knife or special knives heated by steam or electricity. Large numbers of frames are more rapidly processed with partially or completely automated uncapping machines which cut or chop the wax caps with blades, chains or wires. In comb harvesting the equivalent step is the comb selection and the removal of bees etc. Honey frame processing proceeds, after uncapping, to centrifugal extraction. Extractors range in size from a manual 2-frame model to motorized units extracting more than 12 deep supers at a time. Though honey can be extracted faster and more completely at higher temperatures, the combs will become softer and might break. Therefore, extraction temperatures should not exceed 30°C.

Purification

The next step is the removal of any impurities such as wax particles, other debris and air bubbles incorporated during extraction. There are two practical techniques: settling and straining. The first simply consists of leaving the honey in a suitably large container, so that impurities can separate according to their specific weight, i.e. air bubbles, wax particles, insect pieces and other organic debris float to the surface while mineral and metallic particles drop to the bottom. The surface scum can be removed carefully, or honey can be drawn off near the bottom for bottling without disturbing either surface scum or bottom sediment. Settling velocity varies with particle size (the smallest settle the slowest), container size and honey viscosity, i.e. moisture content and temperature.

At temperatures of 25-30°C settling is generally rather quick and can be completed in a few days. Tanks have to be well covered to avoid excessive contact with air. The process can be accelerated by letting honey flow through special buffer tanks prior to filling into the settling tanks.

Straining can be used instead of, or in addition to, settling. It is more frequently used in larger processing plants, where many tonnes of honey are processed every day and where it is therefore inconvenient and uneconomic to immobilize honey for as long as is required for settling. All the above purification methods can only be applied to liquid honeys. It is therefore preferable to use them immediately after extraction, when honey is still naturally liquid and at the right temperature. Extracted, cleaned or purified honey is ready to be consumed directly or to be included into other products.

Moisture content

Moisture content of honey is practically the most important quality parameter, since it affects storage life and processing characteristics Even though moisture can be removed after extraction, only completely ripe honey should be harvested, i.e. combs with more than 75% of the honey cells sealed. Post-harvest reduction of moisture content can be achieved by leaving honey supers in warm rooms at 30 to 35°C and circulating warm air through them. In industrial plants, vacuum driers are used at less than 45°C, similar to those for dehydrating fruit juices. Many other systems have been designed over the years, but honey should require such treatment only under exceptional conditions.

Prevention of fermentation

Fermentation is the only microbiological alteration to which honey is susceptible. Only osmophilic yeasts can grow in the high sugar concentrations, but their presence is ubiquitous in honey, nectar, hive interiors, dust and soils. Their rate of multiplication increases proportionally with increasing water content, up to a certain point. Below 18% moisture content there is little probability of fermentation, but even at concentrations below 17.1% the risk of fermentation cannot be completely excluded. This aspect of fermentation depends on factors such as the quantity of yeast and other growing factors - honey temperature and the distribution and availability of water following crystallization.

Appropriate and expensive cold storage but above all, careful production techniques, can prevent fermentation. If, after all precautions and care, honey cannot be harvested at less than 18% water content, excessive moisture should be removed. Another method is based on pasteurization and the destruction of the yeasts. The osmophilic yeasts found in honey die after only a few minutes of exposure to temperatures between 60 to 65°C. If the honey is heated and cooled quickly enough, with special heat exchangers feasible only on an industrial scale, very little damage occurs to the honey. Often these pasteurization treatments have two functions, the prevention of fermentation and the postponement of crystallization.

Relatively small quantities of honeys with high moisture content do not justify complicated and costly pasteurization, or drying.

Heating

Any application of heat has a negative effect on honey through the loss of thermolabile, aromatic substances, which is proportional to the temperature and duration of heating. The basic concept therefore is to heat the honey only to the lowest temperature and for the shortest period consistent with the desired technical objective.

Honey owes its distinctive characteristics not to the stable major compounds which can be found in any other sweet product such as sugar, molasses, syrup and marmalade, but to the multitude of minor components originating from the nectar and the bees themselves. Many of these substances which give honey its specific aroma, flavour and some of the biological activities are unstable over time and thermolabile, i.e. they are destroyed by heat. For the melting of crystallized honeys in large containers, thermoregulated rooms or water baths are maintained at temperatures between 35 and 50°C.

Packaging

The bottle or package should be leakproof and airtight so as to safely contain the product, but also present the product in an attractive form, enticing the consumer to buy it. The label, container shape and material or other packaging material should be chosen accordingly.

Storage

Storage containers for liquid or crystallized honey should be made either of glass or stainless steel or coated with food approved plastic, paint or beeswax. Nothing should be allowed to impart any odour to the honey. Particularly if used containers are recycled, care must be taken that they are absolutely clean and have not the slightest residual odour. Fermentation remains the major threat to unprocessed honey, whether it is liquid or crystallized. Therefore storage conditions have to prevent fermentation through either low temperature storage or by preventing further absorption of moisture.

Quality control

The quality control of honey has two principle purposes. To verify its genuineness i.e. to reveal possible adulteration etc., and to determine its quality in respect to the needs of the processor and the market.

Market outlook

The world-wide honey production in 1991, reached almost 1,200,000 tonnes. The increased production in the last 20 years, despite fluctuations in individual regions and countries is accredited to an increase in the number of hives and production per colony. The major producers are Russia, China, USA, Mexico, Argentina, Canada, Brazil and Australia. The major exporters are China, Mexico and Argentina, but the highest colony yields are recorded in Australia and Canada which have a favourable environment as well as highly developed colony management. The major consumers and importers are the industrialized countries led by Germany, Japan, USA and UK. The increased consumption over the last few years can be attributed to the general increase in living standards and a higher interest in natural and health products. Western Europe as a whole imported approximately 140,000 tonnes which is about 55% of consumption. The average EU per capita consumption of 600 g per year varies widely amongst individual nations, from Greece with 300 g per capita to Germany with 1,800 g per capita.

Bees Wax

Honeybees produce large quantity of beewax. Beewax is a true wax and secreted as liquid but solidifies when exposed to air and scales are formed. This is secreted by the four pairs of wax glands on the ventral side of the abdomen of worker bees of age 14-18 days and it is synthesized from reducing sugars of alimentary origin. It has been estimated that about 1.5 million wax scales are required to produce one kg of wax. Average amount of honey consumed in producing one kg of wax is 840 kg. Pure bee wax as found in the scales that bee secrete, is white, irrespective of whether the bees had fed on sugar syrup or dark honey but the colour to wax is impalled by the pigments of pollen (fat soluble carotenoid pigments).

Physical properties of bees wax Beewax is white when produced as scales by bees but its colour changes to light yellow or dark brown when comes in contact with pollen or brood. The melting point of bees wax varies from 6 1 -66°C as per various pharmacopoeis. At 15°C, its relative density is 0.958-0.970 g/cm and its electrical resistance ranges from 5x 10 to 20x 10 ohm (Crane, 1990). Its thermal conductivity coefficient is 2.5 x 10 Jcm/ Ccm. The saponification value of bee wax ranges from 85 to 100. Beewax is insoluble in water but is soluble in organic solvents such as ether, benzene, benzol, chloroform, turpentine oil and on warming in alcohol and fatty acids. Bee wax is resistant to many acids. Specific gravity of bee wax is 0.95.

The composition of beeswax

Bee wax is a complex substance. Pure beeswax from *Apis mellifera* consists of at least 284 different compounds. Not all have been completely identified but over 111 are volatile. At least 48 compounds were found to contribute to the aroma of beeswax. Quantitatively, the major compounds are saturated and unsaturated monoesters, diesters, saturated and unsaturated hydrocarbons, free acids and hydroxy polyesters. There are 21 major compounds (Table 50) each making up more than 1% of the pure unfractionated wax. Together they account for 56% of the wax. The other 44% of diverse minor compounds probably account for beeswax's characteristic plasticity and low melting point (Tulloch, 1980). The composition of wax from Asian honeybee species is much simpler and contains fewer compounds in different proportions (Phadke *et al.*, 1969, 1971; Phadke and Nair, 1970, 1973 and Narayana, 1970).

Complex esters of mono atomic alcohols and fatty acids form 70.4 to 74.7% of wax; other components are, free acids 13.5 to 15.0% saturated hydrocarbons 12.5 to 15.5%. Beewax is rich in vitamin A (4096 IU/100g). There arc slight variations in properties of waxes of different species (Phadke and Phadke, 1975).

Table 50. Composition of beeswax

Description	% of fraction	Number of components in fractoin	
		Major	Minor
Hydrocarbons	14	10 (5)	66
Monoesters	35	10 (7)	10
Diesters	14	6 (5)	24
Triesters	3	5	20
Hydroxy monoesters	4	6 (1)	20
Hydroxy polyesters	8	5	20
Acid esters	1	7	20
Acid polyesters	2	5	20
Free acids	12	8 (3)	10
Free alcohols	1	5	?
Unidentified	6	7	?
TOTAL	100	74	> 210

Uses of Beewax: Beeswax is considered safe for human consumption and has been approved as an ingredient in human food in the USA. (USA, 1978). It is inert, i.e. it does not interact with the human digestive system at all and passes through the body unaltered. This property is exploited in many medicinal preparations. Beewax is mainly used in candle, cosmetics and pharmaceutical, and bee industry. Comb foundation sheets arc being prepared by beewax. Wax is also used in cosmetics like lotions, creams, lipsticks, ointments, roughes etc. In pharmaceuticals it is used in ointments. Capsules, pill coatings etc. It is also used in deodorants. Wax is used for preparing polishes, furniture etc. Minor quantities arc used in adhesives, chewing gums and inks etc.

Wax collection and processing

There are several ways of collecting beeswax. In frame hive beekeeping, wax is collected from the cappings removed during honey extraction. This produces a very high quality, light coloured wax. Light coloured broken combs provide the best quality of wax, whereas old black brood combs yield the smallest proportion and lowest quality of wax. Different qualities of wax can be produced by separating new white honey combs from darker ones or from those with portions of brood. Since whole combs are harvested and crushed or pressed, the proportion of wax per kilogramme of honey (10-15%) is much higher than with frame hive beekeeping, where the yield is only 1-2%.

Before processing, all comb or wax pieces should be washed thoroughly to remove honey and other debris. Wax can be separated in solar wax melters, by boiling in water then filtering, or by using steam or boiling water and special presses. Wax should never be heated above 85°C. If wax is heated directly (without water) or above 85°C discolouration occurs. Therefore, wax always needs to be processed in water or in a water bath. Wax should not be processed in unprotected steel, iron or copper containers, since it will discolour from reaction with these metals. Direct exposure of wax to hot steam results in partial saponification.

Storage

Beeswax should only be stored in its rendered, clean form. Before rendering, it will quickly be attacked by wax moths, which are able to destroy large quantities of wax in short periods of time. Clean wax in large blocks is not attacked by wax moths. Storage should be in cool dry places and never in the same room with any kind of pesticide. Wax will slowly crystallize over time and as a consequence become harder, but this process is reversible without any damage, just as with crystallized honey. When melted or pressed with the rest of the wax it reverts to normal beeswax without any residues or impurities. Wax can be stored for very long periods of time without losing its major characteristics as items from Egyptian graves more than 2000 years old have shown.

Quality control

Beeswax, when sold in solid blocks should always both be clean and have the colour and odour characteristics. Quality standards for wax are set in most countries according to their pharmacopoeias. To detect adulteration, a number of tests may have to be conducted. The simplest is to determine the melting point, by measuring the temperature at which the first liquid wax appears during very slow heating. It should be between 61 and 66°C or preferably between 62 and 65°C. However, values within this range are not a guarantee of purity.

BEE POLLEN

Pollen grains are small, male reproduction units (gametophytes) formed in the anthers of the higher flowering plants. The pollen is transferred onto the stigma of a flower (a process called pollination) by either wind, water or various animals (mostly insects), among which bees (almost 30,000 different species) are the most important ones. The pollen collected by honeybees is usually mixed with nectar or regurgitated honey in order to make it stick together and adhere to their hind legs. The partially fermented pollen mixture stored in the honeybee combs, also referred to as "beebread" is the food given to honeybee larvae and eaten by young worker bees to produce royal jelly. Pollen has apicultural value in addition to pollination in 2 ways (1) as source of protein, fats and minerals in the honeybee diet and (2) as possible surplus product of the apiary. Pollen is collected by bees from flowers as pollen pellets and stored in comb. A bee colony can annually collect 15-20 kg of pollen. It has been estimated that one colony of *Apis mellifera* uses about 50 kg of pollen in a year as a source of proteins, fats, minerals and other substances. Proteins are essential for honeybees in the build up and repair of body tissues, egg production, and brood rearing.

Physical characteristics of pollen

Pollen grains range from 6 to 200 μm in diameter and all kinds of colours, shapes and surface structures may be observed. Most pollen grains have a very hard outer shell (sporoderm) which is very difficult or impossible to digest. There are, however, pores which allow germination and also extraction of the interior substances.

The composition of pollen

Composition of pollen changes from species to species, variation in absolute amounts of the different compounds can be very high. Protein contents of above 40% have been reported, but the typical range is 7.5 to 35%: typical sugar content ranges from 15 to 50% and starch content is very high (up to 18%) in some wind-pollinated grasses. Composition of pollen and bee-collected pollen however, has to be distinguished. Some average values for bee-collected pollen are shown in Table 51. The major components are proteins and amino acid, lipids (fats, oils or their derivatives) and sugars. The minor components are more diverse (Table 52). All amino acids essential to humans (phenylalanine, leucine, valine, isoleucine, arginine, histidine, lysine, methionine, threonine and tryptophan) can be found in pollen and most others as well, with proline being the most abundant. Many enzymes (proteins) are also present but some, like glucose oxidase which is very important in honey. have been added by the bees. This enzyme is therefore more abundant in "beebread" than in fresh pollen pellets.

Most simple sugars in pollen pellets such as fructose, glucose and sucrose come from the nectar or honey of the field forager. The polysaccharides like callose, pectin, cellulose, lignin sporopollenin and others are predominantly pollen components. After storage in the comb the further addition of sugars and enzymes creates beebread, through lactic acid fermentation.

Table 51. The average composition of dried pollen

	Bee-collected		Hand-collected
	%[a]	%[b]	%[b]
Water (air-dried-pollen)	7	11	10
Crude protein	20	21	20
Ash	3	3	4
Ether extracts (crude fat)	5	5	5
Carbohydrate			
Reducing sugars	36	26	3
Non-reducing sugars	1	3	8
Starch	-	3	8
Undetermined	28	29	43

Table 52. Minor components of bee collected pollen (Crane, 1990)

Flavonoids	At least 8 (flavonoid pattern is characteristic for each pollen type)
Carotenoids	At least 11
Vitamins	C, E, B complex (including, niacin, biotin, pantothenic acid, riboflavin (B_2), and pyridoxine (B_6)).
Minerals	Principal minerals: K, Na, Ca, Mg, P, S. Trace elements: A1, B, C1, Cu, I, Fe, Mn, Ni, Si, Ti and Zn
Terpenes	
Free animo acids	All
Nucleic acids and nucleosides	DNA, RNA and others
Enzymes	More than 100
Growth regulators	Auxins, brassins, gibberellines, kinins and growth inhibitors

In addition to these, other minor contents include vitamins, organic acids, flavonoids, carotenoids, enzymes, growth factor etc.

The physiological effects of pollen

The effects and benefits derived from pollen consumption have been described to be endless. Most of the major ailments reported to improve with pollen preparations (Table 53) are based on personal experiences of users rather than concrete medical evidence or other scientific investigation of claims. The only long-term observations on the medicinal effect of pollen are related to prostate problems and allergies. Several decades of observations and a few clinical tests have shown pollen to be effective in treating prostate problems ranging from infections and swelling to cancer. Supplementation of animal diets with pollen has shown positive weight gain and other beneficial effects for piglets, calves, broiler chickens and laboratory cultures of insect. Certain bacteriostatic effects have been demonstrated (Chauvin *et al.*, 1952) but this is attributed to the addition of glucose oxidase (the same enzyme responsible for most antibacterial action in honey) by the honeybee when it mixes regurgitated honey or nectar with the pollen. Therefore, this activity varies between pollen pellets and is much higher in beebread. A very slight antibacterial effect can also be detected in pollen collected by hand (Lavie, 1968). There is some evidence that ingested pollen can protect animals as well as humans against the adverse effects of x-ray radiation treatments.

Table 53. Benefits or improvements derived from the use or consumption of bee-collected pollen.

Improvements	Cures of benefits
Athletic preformance	Cancer in animals
Digestive assimilation	Colds
Rejuvenation	Acne
General vitality	Male sterility
Skin vitality	Anaemia
Appetite	High blood pressure
Haemoglobin content	Nervous and endocrine disorders
Sexual prowess	Ulcers
Performances (of a race horse)	

Properties of pollen: Pollen vary greatly in colour, shape and size. Pollen grains range from 6 to 200 microns in diameter. Pollen has very hard outer shell, which can not be broken or digested. However, its germlination and exit of most substances takes place through the pores of outer shell. Pollen composition varies from species to species and protein content generally ranged from 7 to 30% depending on species, however, up to 40% protein content has been reported. Main constituents of pollen given by (Crane, 1990) are as under:

Uses of pollen

Commercial application of pollen includes:

1. Plant breeding programme.
2. Fruit pollinatin (using pollen dispensers)
3. Studying and treating allergic conditions such as hay fever
4. Extracting certain components.
5. Production pollen supplements for feeding to bees.
6. Feeding human beings and domestic animals alone or with honey or royal Jelly.
7. Pollen harvesting is generally for last two applications.

Maximum use of pollen is for feeding the bees during death period when it is not available from the flowering plants. For preparing pollen supplements, 5-10% natural bee collected pollen is added to increase consumption. Small percentage of pollen in the diets of animals lead to increased weight gains and other useful effects. Pollen also has medicinal and prophylactic value. It is quite effective for treating hypertension when mixed with honey (I : I). It can also be used in complaints of nervous and endocrine systems. Pollen normalizes the activity of intestine especially in colistis or chronic constipation. improves appetite, and increases fitness for work. Pollen has been found useful in lowering blood pressure, increasing haemoglobin and erythrocyte content of the blood and useful in pernicious anemia. Presence of gonadotrophic hormones in the pollen of date palm contributes for treating sterility. Pollen has beneficial effect on the prostate gland. A pollen preparation 'Zernilton' is sold in Sweden as a prophylactic for complaints of the prostate and abdomens. Pollen is also used in cosmetic preparations with claims of rejuvenating and nourishing effects on skin.

Besides, pollen collected by honeybees reflects environmental pollution levels when examined for metals, heavy metals and radioactivity (Bromenshenk *et al.*, 1985) and can be used for monitoring pollution levels. Contaminants can be quantified and sampling may be cheaper than most standard methods currently in use. Attempts have also been made to use pollen-collecting honeybees for the identification of potential mining areas (Lilley, 1983). The same effect of accumulating aerial deposits and selective plant secretions of minerals beneficial when used to monitor pollution control becomes a hazard if pollen from heavily polluted areas is used for human or animal consumption.

Pollen collection

Extreme care should be taken that pollen is not contaminated by bees collecting from flowers treated with pesticides. During, and for several days or weeks after treatment of fields or forests in an area of several square kilometres (in a circle of at least 3-4 2 km diameter) around the apiary, no pollen should be collected. This is independent of the method of pesticide application. Even systemic pesticides have been shown to concentrate in pollen of, for example coconut (Rai *et al.*, 1977). Since a pollen pellet is collected from many flowers, even small quantities of pesticides per flower can be accumulated rapidly to reach significant concentrations.

Though pollen pellets are collected before they enter the hive, treatment of colonies for bee diseases, can contaminate the pollen pellets. Though, for example, cleaning of debris from the hive and bees regurgitating syrup, nectar or honey during collection of the pellets.

Pollen pellets are removed from the bees before they enter the hive. There are many designs of pollen traps some easier to clean and harvest, others more efficient or easier to install. The efficiency rarely exceeds 50%, i.e. less than 50% of the returning foragers loose their pollen pellets. Bees are ingenious in finding ways to avoid losing their pellets, like small holes or uneven screens and may even rob pollen from the collecting trays, if access is possible. Under some circumstances, pollen collection methods and regimes may interfere with normal colony growth or honey production (Dadant, 1992).

Pollen should be collected daily in humid climates but less frequently in drier climates. To avoid deterioration of the pollen and growth of bacteria, moulds and insect larvae, pollen should be dried quickly. Ants can remove considerable amounts from pollen traps and the losses can be up to 30% in temperate climates.

Quality control

Quality control of pollen is difficult and under most circumstances impossible. It is therefore very important that the buyer knows the supplier well and can trust him. A reliable supplier should have all necessary storage and processing facilities and use them. Furthermore, the production area, or processing centre, should be free of agrochemicals and industrial pollution.

Storage

Pollen, like other protein rich foods, loses its nutritional value rapidly when stored incorrectly. Fresh pollen stored at room temperature loses its quality within a few days. Fresh pollen stored in a freezer loses much of its nutritive value after one year. Longer, improper storage leads to the loss of a few particular amino acids, which cause deficiencies in brood rearing (Dietz, 1975). When dried to less than 10% (preferably 5%) moisture content at less than 45°C and stored out of direct sunlight, pollen can be kept at room temperature for a several months. The same pollen may be refrigerated at 5°C for at least a year or frozen to –15°C for many years without quality loss as tested by feeding to honeybee colonies and recording brood rearing rate (Dietz and Stephenson, 1975, 1980). Since sunlight, i.e. UV radiation, destroys the nutrient value of pollen, other more subtle characteristics probably suffer worse damage. Storage of dry pollen in dark glass containers, or in dark cool places, is therefore, a requirement.

Quality control

Only a few countries, such as Switzerland and Argentina, have legally recognized pollen as a food additive and established official quality standards and limits. The moisture content should not exceed 8% (controlled by vacuum drying at 45 mm Hg and 65°C). Other limits include a pH of 4-6, protein content of 15-28% Kjeldahl (N x 6.25) of dry weight, total hydrocarbons of 45-55% of dry weight and a maximum ash content of 4% of dry weight (determined at 600°C).

Since air pollutants and agro-chemicals have been shown to accumulate in pollen collected by bees pollen should originate from unpolluted areas with the lowest chance of contamination by agrochemicals, industrial pollution.

Royal Jelly

Royal jelly is a milky white substance secreted by hypopharyngeal glands of house bees of age 6 to 12 days. It is secreted to feed the queen bee throughout her larval and adult life, young worker and drone larvae. Royal jelly is a complex substance composed of water (57-70%), proteins (17-45%), sugars (18-52%), lipids

(3.5-19%), minerals (2-3%) and vitamins. The proteins are mainly amino acids which are alanine, arginine, aspartic acid, gultonic acid. glycine, isoleucine. lysine. methionine, phenyl alanine. tryptophane. tyrosine and serine. Carbohydrates in royal jelly are glucose, fructose. melibiose. trehalose. maltose and sucrose. Vitamins are A, D and C. Phosphorus, iron. copper, silicon and sulphur arc also present in royal jelly. Royal jelly makes the differentiation between worker and queen due to some other hormonal constituents and identification of these is still lacking.

Physical characteristics of royal jelly

Royal jelly is a homogeneous substance with the consistency of a fairly fluid paste. It is whitish in colour with yellow or beige tinges, has a pungent phenolic odour and a characteristic sour flavour. It has a density of approximately 1.1 g/cm^3 and is partially soluble in water. Aqueous solutions clarify during basification with soda. Viscosity varies according to water content and age. Stored royal jelly often develops small granules due to precipitation of components.

Table 54. A list of some effects of royal jelly on humans.

Applications	Description	References
Premature bebies and those with nutritional deficiencies of various origins	8-100 mg orally, improvement of general condition; increase in weight, appetite, red blood cells and haemoglobin	Malossi & Grandi, 1956 Prosperi & Ragazzini 1956 Prosperi *et al.*, 1956 Quadri, 1956
Elderly (70-75 years), anorexic, depressed and low blood pressure patients	20 mg injected every second day, improvements on all accounts 20 mg taken orally every second day, improvements as above	Destrem, 1956 Destrem, 1956
Psychiatry	Improvements of asthenia, nervous breakdown, emotional problems and counteraction of side effects of psychoactive drugs	Telatin, 1956
Chronic metabolism	Mixture or royal jelly, honey and ginseng, improvements in weight gain and psychological conditions, but changes of blood characterisics	Borgia *et al.*, 1984
Stimulating metabolism	Stimulating effects comparable to that by proteins, effect assumed to be due to activity of enzymatic complexes	Martinetti and Caracristi, 1956
Wound healing	5-30 mg/ml injected into burn blisters, improved regrowth of skin	Gimbel *et al.*, 1962

Uses

Royal jelly is very nutritious food for human beings as it increases vigour and vitality (Table 54, 55). Much research has been done on the use of royal jelly to treat disorders of the cardia-vascular system and gastrointestinal tract. Royal jelly normalizes metabolism, has a diuretic effect, can be used to prevent obesity and emaciation, builds up resistance to infections, regulates the functioning of the endocrine glands and is good for arteriosclerosis and coronary deficiency. Royaljelly is a tonic restoring energy, getting rid oftbe feeding of indisposition and improper appetite. Feeding royal jelly by human beings has shown an effect on physical output (resistance to fatigue), intellectual performance (greater learning capacity and better memory) and on their mental condition (greater self-confidence, feeling of well-being and euphoria). in other words, royal jelly appears to act as a general stimulant, improving immune response and general body functions.

Table 55. A list of properties, benefits and improvements attributed to royal jelly

Internal Use	External Use
Tonic	Skin conditions
Stimulant - physical performance, better memory, learning capacity and self-confidence	Epithelial stimulation and regrowth
General health improvement	Anti-wrinkle
Anorexia	Sebaceous secretion (fat secretions of skin glands) normalized
Increased appetite	
Skin conditions	
Sexual desire and performance	
Influenza	
Increased resistance to viral infections	
High blood pressure	
Low blood pressure	
Anaemia	
Arteriosclerosis	
Cholesterol levels	
Chronic and incurable disorders	

Antibiotic action has been proven against the following microorganisms: *Escherichia coli*, *Salmonella*, *Proteus*, *Bacillus subtilis* and *Staphylococcus aureus*. It shows one quarter of the activity of penicillin against *Micrococcus pyrogens* and is also fungicidal. This same antibiotic action of fatty acids is neutralized by raising the pH above 5.6.

Marketing of royal jelly

Royal jelly can be sold in its fresh state, unprocessed except for being frozen or cooled, mixed with other products, or freeze-dried for further use in other preparations. For larger industrial scale use, royal jelly is preferred in its freeze-dried form, because of easier handling and storing. Freeze-dried royal jelly can be included in the same products as the fresh form. A large amount of royal jelly is sold and consumed as it is harvested. In its unprocessed, natural state, it is preferred by most producers, because it does not require any special technology, and by consumers because of its unaltered "naturalness".

Unprocessed royal jelly is usually packaged in small, dark glass bottles of sizes that correspond to the duration of a "treatment" e.g. 10, 15 or 20 g. A tiny plastic spatula is usually included for the "correct" dosage of 250 - 500 mg. Special isothermal packaging (usually a moulded polystyrene box) is sometimes used to make the product look even more precious and protect it perhaps from brief temperature fluctuations

Storage

Royal jelly has a limited shelf-life. Information is, however, available on changes in composition due to long term storage, such as a higher acid titre, a large unsoluble protein fraction, less free amino acids, less glucose oxidase and others. Such changes make it appear likely that also biological activity is influenced by storage. Refrigeration and freezing delay and reduce the chemical changes. Although freeze-dried jelly is the most stable form of royal jelly, some changes still take place.

Royal jelly collection

Royal jelly is produced by stimulating colonies to produce queen bees outside the conditions in which they would naturally do so. A well-managed hive during a season of 5-6 months can produce approximately 500g of royal jelly. Since the product is perishable, producers must have immediate access to proper cold storage (e.g. a household refrigerator or freezer) in which the royal jelly is stored until it is sold or conveyed to a collection centre. The most rational and economic methods for large scale production are variations of the Doolittle method of queen rearing. Usually, the starter colony is omitted and cell cups, with transferred larvae, are directly introduced into the finisher colonies. Strong queen right colonies are preferred, in which the queen chamber is separated from the cell rearing chamber by a queen excluder. The only required adaptation is to shorten the cycle in the finishing colonies (3 days versus 10) before cells are removed for harvesting. For occasional and small scale production any other queen rearing method can be used. However, there are many queen rearing methods which differ only in hive design and the use of starter and/or finisher colonies.

The basic requirements are movable comb hives, preferably some queen excluders, queen cups (made from wax or plastic), a transfer needle, a spoon or suction device to remove royal jelly, dark glass vials and a refrigerator. Feeding with sugar syrup (1:1 in sugar/water) increases cell acceptance, even when flowers are available.

Individual queen cells should not contain less than 200 mg of royal jelly. Low cell content means that there are too many cells for the finisher colony or that the colony is not in a condition to provide for queen rearing. There are racial differences in productivity and specially selected strains can be obtained.

Bee venom

Bee venom is a clear liquid and this liquid remains in poison sac of bee which is attached to its sting. Bee venom has a sharp, bitter taste, an aromatic odor, an acid reaction with a specific gravity of 1.13. It dries quicidy at room temperature to 30-40 per cent of the original weight. Dried venom takes on a light yellow colour and some commercial preparations are brown, thought to be due to oxidation of some of the venom proteins.

Composition of bee venom: Composition of bee venom is complex and is composed of many active substances such as histamine, apamine, acithinase, hydrochloric acid, formic acid, orthophosphoric acid, sulphur, calcium, copper and magnessium sulphate. Water constitutes 88% of venom. The glucose, fructose and phospholipid contents of venom are similar to those in bee's blood. At least 18 pharmacologically active components have been described, including various enzymes, peptides and amines (Table 56). Venom from other *Apis* species is similar, but even the venoms from the various races within each species are slightly different from each other. The toxicity of *Apis cerana* venom has been reported to be twice as high as that of *A. mellifera*

Table 56. Composition of venom from honeybee worker

Class of molecules	Component	% of dry venom[a]	% of dry venom[b]
Enzymes	Phospholipase A_2	10-12	10-12
	Hyaluronidase	1-3	1.5-2.0
	Acid Phosphomonoesterase		1.0
	Lysophospholipase		1.0
	a -glucosidase		0.6
Other proteins and peptides	Melittin	50	40-50
	Pamine	1-3	3
	Mast Cell Degranulating Peptide (MCD)	1-2	2
	Secapin	0.5-2.0	0.5
	Procamine	1-2	1.4
	Adolapin		1.0
	Protease inhibitor	0.1	0.8
	Tertiapin[c]	13-15	0.1
	Small peptides (with less than 5 amino acids)		
Physiologically active amines	Histamine	0.5-2.0	0.5-1.6
	Dopamine	0.2-1.0	0.13-1.0
	Noradrenaline	0.1-0.5	0.1-0.7
Amino Acids	a-aminobutyric acid	0.5	0.4
	a-amino acids	1	
Sugars	Glucose & fructose	2	
Phospholipids		5	
Volatile compounds		4-8	

Uses

Bee venom has long been used in traditional medicine for the treatment of various kinds of rheumatism. Although venoms of the different honeybee species differ slightly, there have been reports of successful rheumatism treatment with *Apis dorsata* venom by Sharma and Singh (1983) and with *A. cerana* venom by Krell (1992). Bee venom therapy (Apitherapy) has many potential uses. The venom collected can be used for making intramascular injections for medical cares. Some interest has been expressed in introducing a pure bee venom product on the American drug market, to be used in *two* ways (1) for the desensitization of hypersensitive individuals and (2) for the treatment of rheumatoid arthritis. Ointment made by mixing apitoxin, vasaline and salicylic acid (1: 10: 1) can be applied on the affected areas. Bee venom has long been used to treat certain eye diseases. Bee venom is known *to* bring blood pressure down, by lowering down the cholesterol level and deposition in blood vessels. Patients with hypenension, improved soon after they began to work with bees where they were stung. Their headaches disappeared, fitness for work improved and blood pressure dropped to normal. Some people, however, may be allergic *to* bee venom and producc symptoms of diarrhea, vomiting etc.

The list of benefits to human beings as well as to animals is very long. Most of the reports of cures are of individual cases, though several unrelated patients have experienced the improvement or cure of similar ailments. The diseases and problems which have been reported by patients or doctors as improved or healed with bee venom therapy are listed below (Table 57).

Table 57. List of diseases and health problems improved or healed according to anecdotal reports

	Humans	
Arthritis, many types	Multiple sclerosis	Premenstrual syndrome
Epilepsy	Bursitis	Ligament injuries
Mastis	Some types of cancer	Sore throat
Chronic pain	Migraine	General immuno-stimulant
Decreases blood viscosity & coagulability	Dilates capillaries & arteries	Decreases blood cholesterol level
Neruoses	Rhinosinusitis	Endoarteriosis
Therosclerosis	Polyneuritis	Radicultitis
Infectious spondylitis	Neuralgia	Endoarthritis
Infect. Polyarthritis	Malaria	Intercostal myalgia
Myositis	Tropical ulcers	Slowly healing wounds
Thrombophiletritis	Cancer, temporary	Keratoconjunctivitis
Iritis	Iridocytis	Asthma[h]
Animals		
Arthritis		

Venom collection

The various trap designs stimulate bees by applying a mild electric shock through wires above the collecting tray. The trays are placed either between the bottom board and brood chamber at the hive entrance or in a special box between supers and the hive cover, when shocked, bees sting the surface on which they are walking. In some traps, this may be a glass plate or a thin (0.13 mm thick) plastic membrane, nylon taffeta or silicon rubber under which a collecting plate (preferably made of glass) or absorbent tissue receives the venom. Venom dries rapidly on glass plates and can be scraped off with a razor blade or knife. Absorbent tissue is washed in distilled water to extract the venom, which then should be freeze-dried. Collection on glass is generally easier and produces a product which is easier to store, ship and process. During handling of dry bee venom, protective gloves, glasses and dust masks should be worn to avoid any contact with, or inhalation of the highly concentrated venom.

Venom products

Bee venom may be sold as whole bee extract, pure lidquid venom or an injectable solution, but in either form the market is extremely limited. Most venom is sold in a dry crystalline form. For injections, the venom can be mixed at the time of injection with injectable fluids, such as distilled (sterile) water, saline solutions and certain oils, or it may be taken from prepared ampoules. Ampoules with set doses of ready-to-inject venom should only be prepared by certified pharmaceutical laboratories, because of the need to maintain stringent aseptic conditions and to measure the dosages very precisely.

There are creams available which include bee venom (e.g. Forapin and Apicosan in Germany, Apivene in France and Immenin in Austria) which are used for external application on arthritic joints.

Buying

The best way to buy bee venom is in the crystallised form, since it is more stable, impurities are easier to detect and adulteration is less likely. The colour of both crystals and powder should be a very light yellow.

Storage

Even dried bee venom should be stored refrigerated or preferably frozen and it should always be kept in dark bottles in the dark. All producers and buyers should closely observe these conditions. Dried bee venom can be kept frozen for several months, but should not be kept refrigerated for more than a few weeks. Liquid venom and diluted venom can be stored for similar periods if maintained in well sealed, dark glass containers.

Quality control

There are no official quality standards, since bee venom is not recognized as an official drug or as a food. Purity analysis may be carried out by quantitative analyses of some of its more stable or more easily measured components such as melittin, dopamine, histamine, noradrenaline or those for which contamination is suspected.

Caution

Collecting bee venom requires careful work with the highest degree of cleanliness, since the venom will be injected directly without further processing or sterilization. Protection of the collector against the disturbed bees and highly irritative dry venom is very important, too. When handling dry venom, laboratory gowns, gloves and face masks should be worn to avoid getting venom dust into the eyes and lungs. Using bee stings for self-treatment of various diseases can be risky, because allcrgies to bee venom can be developed quickly even after long periods of use. Since severe allergic reactions are possible, bee venom should not be casually included in any health or medicinal products.

Market outlook

Bee venom is a highly specialized product with only very few buyers. The market volume is relatively small too, although there are no comprehensive surveys. The main venom producer in the USA has produced only about 3 kg of dry venom during the last 30 years (Mraz, 1982) but there is a large producer in Brazil and more or less significant amounts are produced in many other countries. Prices in 1990 varied greatly between US$100 and US$200 per gram of dry venom (Schmidt and Buchmann, 1992). Prepared for injections or sold in smaller quantities, prices can be much higher.

Propolis

Propolis is a mixture of various amounts of beeswax and resins collected by the honeybee from plants, particularly from flowers and leaf buds.These resins are used by worker bees to line the inside of nest cavities and all brood combs, repair combs, seal small cracks in the hive, reduce the size of hive entrances seal off inside the hive any dead animals or insects which are too large to be carried out and perhaps most important of all, to mix small quantities of propolis with wax to seal brood cells. These uses are significant because they take advantage of the antibacterial and antifungal effects of propolis in protecting the colony against diseases. Propolis has been shown to kill the bee's most ardent bacterial foe, *Bacillus larvae* - the cause of American Foul Brood. The use of propolis thus reduces the chance of infection in the developing brood and the growth of decomposing bacteria in dead animal tissue.

The composition of propolis depends on the type of plants accessible to the bees. Propolis changes in colour, odour and probably medicinal characteristics, according to source and the season of the year. Foraging for propolis is only known with the Western honeybee *Apis mellifera*. The Asian species of *Apis* do not collect propolis. Only Meliponine or stingless bees are known to collect similarly sticky resinous substances, for sealing hives and constructing honey and pollen pots for storage.

Physical characteristics of propolis

The colour of propolis ranges from yellow to dark brown depending on the origin of the resins. At temperatures of 250 to 45°C propolis is a soft, pliable and very sticky substance. At less than 150°C, and particularly when frozen or at near freezing, it becomes hard and brittle. It will remain brittle after such treatment even at higher temperatures. Above 45°C it will become increasingly sticky and gummy. Typically propolis will become liquid at 60 to 70°C, but for some samples the melting point may be as high as 100°C. It has an aromatic odour, extremely brittle when cold, melts at 150°C, is insoluble in alcohol but readily dissolves in ether and chloroform.

The composition of propolis

More than 150 compounds have been isolated so far. It appears that with every new analysis, new compounds are found. Propolis resins are collected from a large variety of trees and shrubs. Each region and colony seems to have its own preferred resin sources, which results in the large variation of colour, odour and composition. A list of the major classes of chemicals occurring in propolis from different countries is given in Table 58. The major compounds are resins composed of flavonoids and phenolic acids or their esters, which often form up to 50% of all ingredients.

Table 58. The major compounds of propolis (Source : FAO, 1996)

Class of components	Group of components	Reference
Resins	45 to 55%	
	Flavonoids	Papay *et al.*, 1987 - Hungary
		Bankova *et al.* 1987 - Bulgaria
		Nagy *et al.*, 1989- Czechoslovakia
		Omar, 1989 - Egypt
		Greenaway *et al.* 1990a- UK
		Greenaway *et al.* 1990b - Austria, Ecuador, Germany, Israel, UK, USA
		Wang and Zhang, 1988 - China
		Mizuno *et al.* 1987 - Japan
	Phenolic acids and esters	Nagy *et al.*, 1985 - Hungary
		Wollenweber *et al.* 1987 - West Germany
		Bankova *et al.* 1991- Bulgaria, Mongolia
Waxes and fatty acids	25 to 35%	
	most are usually from bees wax, but many are of plant origin	Papay *et al.* 1987 - Hungary
Essential oils	10% volatiles	Petri *et al.* 1988 - Hungary

Pollen	5%	
	proteins probably from pollen: free amino acids (AA): 16 AA's at more than 1% of total AA's of which arginine and proline together make up 45.8%, 8 AA's occur in traces	Gabrys *et al.* 1986 - Polland
Other organics and minerals	5%	
	14 trace minerals of which Fe & Zn are most common, othe-rs e.g.: Au, Ag, Cs, Hg, La, Sh	Scheller *et al.,* 1989 - Polland
	Ketones	Bankavo *et al.* 1987 - Bulgaria
	lactones	Cueller and Rojas, 1987 - Cuba
	quinones	Cueller and Rojas, 1987 - Cuba
	steroids	Cueller and Rojas, 1987 - Cuba
	benzoic acid and esters	Greenway *et al.,* 1987- UK
	vitamins, only B_2	Greenway *et al.,* 1987- UK
	sugars	Greenway *et al.,* 1987- UK
General review		Walker and Crane, 1987- World Asis, 1989 - World Crane, 1990 - World Inoue, 1988 - Japan

Uses

Bees use propolis to seal cracks, openings, varnise brood cells, smoothen the hive, strengthen comb attachments and to seal intruders and object ionable objects' in the hive that are too large to throw out. Propolis is a local anaesthetic especially in dental mcdicine propolis has antibacterial and antifungal properties and is believed to cure burns, external ulcers and eczema in humans (Table 59). Many scientific tests have been conducted with a variety of bacteria, fungi, viruses and other micro-organisms. Many of the tests have shown positive control of the organisms by various extracts and concentrations of propolis. It has long been used in folk medicine for removing corns. Propolis ointment prepared with vaseline and sunflower and henbane oils in proportions of 1 : I and 1: 5: I was effective in necrobacillosis in cattle than other remedies. Propolis ointment prevents radiation skin relations. The diseases of upper respiratory tract and lungs (e.g. bronchitis and tuberculosis) were cured by inhalation of propolis. For an inhalation 60g of propolis and 40g of heeswax are put in 300-400 ml water in an aluminium vessel, which is stood in a large metal bowl of boiling water. Mixture is inhaled for 10 to 15 minutes twice a day over a period of two months.

Table 59. A list of microorganisms against which propolis or its extracts have been shown to have a positive effect.

Target organisms	Comments	References
Bacterial effects		
Bacillus larvae	causes American Foul Brood in honeybees	Meresta and Meresta, 1988
B. subtilis and others		Meresta and Meresta, 1985, 86
Bacillus de koch	tuberculosos	Karimova, 1975 Grange and Davey, 1990
Staphylococcus species	associated with pneumonia	Chernyak, 1973
Staphylococcus aureus	positive synergistic effect with action of 13 antibiotics against 10 strains	Kedzia and Holderna, 1986 Meresta and Meresta, 1988 Dimov *et al.* 1991
Streptococcus		Rojas and Cuetara, 1990
Streptomyces		Simuth *et al.*, 1986
S. sobrinus, mutans & ericetus	dental caries in rats	Ikeno *et al.*, 1991
Saccharomyces cerevisiae	brewer's yeast	Petri *et al.*, 1988
Escherichia coli		Simuth *et al.*, 1986
Salmonella and *Shigella*	review	Ghisalberti, 1979
Salmonela	potential use in salmonellosis treatment	Okonenko, 1986
Salmonella	reduction in pathological changes after Salmonella infections in mine	Okonenko, 1986
112 anaerobic strains	inhibitory effect on most	Kedzia, 1986
Giardia Lambia		Olariu *et al.*, 1989
Bacteroides nodosus	reduction of foot-rot in rams	Munoz, 1989
Klebsiella pneumoniae		Dimov *et al.*, 1991
reduced or no bacterial activity		Brumfitt *et al.* 1990
general	6 species of bacteria, major (4%) component - flavonoid, Cuba	Cuellar *et al.*, 1990
Fungicidal effects		
Candida albicans	weak effect by ethanol extracted propolis (EEP) no effect by aqueous extracted propolis (AEP)	Valdes *et al.*, 1987
	better effect in vitro	Petri *et al.*, 1988
	in comparison with 10 antibiotics EEP had best effect in synergism with natamycin and flucytosine	Holderna and Kedzia, 1987

Aspergillus niger		Petri *et al.*, 1988
Botrytis cinerea	in vitro EEP is fungicidal, but in vivo with strawberries has insignificant effect	La Torre *et al.*, 1990
Ascosphaera apis	chalkbrood pathogen in honeybee colonies	Kedzia, 1986 and Ross, 1990
6 fungi infections in humans	antifungal properties vary with different samples of propolis	Millet-Clerc *et al.*, 1987
Plasmopara viticola	ineffective, greater leave damage by *P. viticola* with 1% propolis treatment	Hofmann *et al.*, 1989
general	antifungal activity increased in presence of propylene glycol	Millet-Clere *et al.*, 1987 Milena *et al.*, 1989
Antiviral effects		
Herpes	Herpes 1 and 2 in vitro anti-herpes ointment patent	Sosnowski, 1984 Popescu *et al.*, 1985
Potato virus	EEP is effective, AEP less so	Fahmy and Omar, 1989
Influenza	reduced influenza mortality in mine with oral and injected propolis extracts	Maksimova-Todorova *et al.*, 1985 Neychev, *et al.*, 1988 Serkedjieva, 1992
Newcastle disease		Maksimova-Todorova *et al.*, 1985
general	review	Benkova *et al.*, 1988 Konig and Dustmann, 1989
Nematodicidal effect		
Ascaris suum	in intestines of guinea pigs, assessed to be effective through immunostimulation	Benkova, *et al.*, 1989

Chapter 20

Value added products from bees and beekeeping

The best known primary products of beekeeping are **honey and wax, but pollen, propolis, royal jelly, venom, queens, bees and their larvae** are also marketable primary bee products. While most of these products can be consumed or used in the state in which they were produced by the bees, there are many additional uses where these products form only a part of all the ingredients of another product. Because of the quality and sometimes almost mystical reputation and characteristics of most primary bee products, their addition to other products usually enhances the value or quality of these secondary products. For this reason, the secondary products, which partially, or wholly, can be made up of primary bee products, are referred to here as "value added" products from beekeeping. Many of the primary beekeeping products do not have a market until they are added to more commonly used, value added products. Even the value of the primary products may increase if good use is made of them in other products, thereby increasing the profitability of many beekeeping operations.

In some cases the traditional and early technological uses of primary bee products have been replaced by other (often synthetic products) because of better availability, lower cost and/or easier processing. But in regard to food or health products, there are no synthetic substances which can substitute for the wide variety of characteristics of primary bee products. Only when it comes to highly specialized applications and conditions, will synthetics sometimes out perform these unique and versatile products. In that sense, all products containing one or several of the primary bee products are value added products. Furthermore, the combination of several bee products synergistically increases their beneficial significance beyond their individual biological values.

Since monetary resources are limited in many societies the additional value cannot always be obtained in the form of higher prices, but may show itself in the form of preferred purchases. For the same reasons though, some products may not be able to compete against cheaper synthetic products. In such cases, the added value and cost may make a product unsuitable, unless other markets can be found.

Traditionally, honey is considered the major beekeeping product. Wax has played a considerable role in only a few parts of the world and propolis is even less known. However, with increasing knowledge about beekeeping and an awareness of the beneficial aspects of many bee products, the use and demand for other products is increasing. The inclusion of "natural" bee products in cosmetics, medicines and foods has improved consumer appeal. While such appeal is not always based on scientific evidence, more and more studies confirm at least some of the traditionally claimed benefits of primary bee products.

Honey in medicine

Traditionally honey has been used in folk medicine as in Ayurveda, Siddha, Unani, and other indigenous systems of medicine. Its super-saturated nature does not allow growth of bacteria and other microbes. It is thus anti-microbial and antiseptic. At the same time its nutritive property helps in healing of wounds and burns in a natural way. These and other properties of honey contribute to its utility as medicine. Consumption of honey for medicinal purposes is the highest in the pharmaceutical industry and in Ayurvedic medicine.

Honey in Ayurveda

Honey is commonly used in Ayrvedic medicine. The southern states are the largest consumers of honey for Ayurvedic medicine. The Arya Vaidya Sala, Kotakkal in Kerala consumed over 160 tons during 1985-86. Both forest and apiary honeys are used. But the contribution of the former is double that of the latter. Besides being cheaper, forest honeys are obtained from several medicinal trees and shrubs.

Honey as Food

Honey was the only sweet food in ancient India. It was valued for its nutritive value, giving energy, health, strength and vigour. It was considered as the food of the gods that made men immortal. After the discovery of the grasses that gave sweet juices, followed by the invention of sugar, the importance of honey as food began to decline. Beekeeping was already a closely guarded secret, and the craft was not pursued seriously because of the difficulties in producing honey and the comparative ease with which sugar could be prouduced. Forest honey was thin and fermented.

Because of its association with immortal gods, honey remained mainly as an essential food in religious ceremonies at the time of birth, marriage and death, as well as an item offered to the gods during worship. Social and cultural customs in all religious groups include use of honey.

Use of honey as food continued, however, because of its natural flavour, taste and its preservative properties. Honey has been used directly, and as an ingredient of other regular food items. It is used in cooking and baking.

Honey is now consumed for table and other household purposes. It is used in pickles, jams, jellies, marmalades and preserves because of its keeping quality. In cooking honey is used as a sweetener and flavouring agent in the preparation of sweetmeats like tarts, pancakes, pastries, pies and puddings. Other major household use of honey as food is in bakery, dairy products including yoghurts, and processed foods and confectionery.

Institutional consumption of honey as food is also significant. Hotels, health clubs, and restaurants serve honey on the table, as bread spread, in salads, or in their special food and drink preparations.

In 1985-86, about 1200 tons of apiary honey was sold in the retail organized market. Major Ayurvedic companies sold 350 to 400 tons of forest honey in this market under brand names like Baidyanath, Charak, Dabur, Phondaghat, Ramtirth, and Zandu. Over 55% of the production went in the unorganized rural sector for household consumption. Institutional consumption accounted for about 825 tons (Anonymous 1986).

Honey in Industry

About 10 per cent of the total production is used in Industry. This is largely due to the fact that honey is costly, and therefore adds significantly to the cost of production. In several end uses honey, constitutes one of the ingredients, and undergoes physical and chemical changes. The end product does not have the flavour or taste of honey. Naturally, use of honey in the industry is more for its medicinal properties, than for its nutritional value. Forest honey is usually preferred in the industry because of its low cost.

Food Products

Use of honey in mass production and fast-food industry has recently been increasing. Some products that use honey are processed foods like cereals, ready-to-eat mixes, health foods and baby foods, and dairy products like yoghurts, milk sweetmeats, creams, fudges and ice creams. Health food producing companies and ice cream producing factories have Kwality and Vadilal increased their sales by advertising the use of honey in their products.

Confectionery

During the past decade or so, use has been made of honey in confectionery like sweets, candies, toffees, eclairs and chocolates. Raolgaon Sweets, Nutrine Confectionery Products and Hindustan Cocoa Products are the major consumers for confectionery that accounts for about 100 tons of honey.

Bakery

Small quantities of honey are also being used in bakery products like special types of breads, biscuits, cakes and cookies. Britannia Biscuit Company is using honey in its popular brands of cookies and biscuits. The hygroscopic nature of honey keeps bread soft, while improving its keeping quality.

Fruit and vegetable preservation

This newly growing industry has good potential to use honey both as a preservative and as a flavouring medium. The fruit products industry that includes manufacture of jams, jellies, canned fruits, syrups, squashes, sauces and ketchups is just beginning to use honey in its products. The Muzaffarnagar Industrial Honey Producers Co-operative Society in Uttar Pradesh has been successfully marketing fruit and vegetable products developed by it with honey as an important ingredient.

Pharmaceuticals

Much of the honey used in industry is for manufacture of medicines like cough drops, losenzes or syrups, tonics and other medical formulations. Pharmaceutical companies like Cynamid, Proctor and Gamble, Roussel and Warner Hindustan consumed about 183 tons of honey in 1985-86. Interestingly, forest honey contributed to only 22% of the honey used in this industry. Dental creams and antiseptic eye lotions are newly developed products of this industry using honey.

Cosmetics

Honey has a mild bleaching action on skin, but at the same time protects it from infections and nourishes it. Because of these properties it is used in facials, shampoos and other cosmetics. Consumption in this sector has been negligible. With the recent perceptible shift towards herbal preparations, honey is increasingly used for manufacture of skin-care and hair-care products, as a natural therapeutic and medicinal ingredient. Jiwadaya Netraprabha Karyalaya, Bombay uses 10 to 15 tons of honey for preparation of the commonly used mascara for eye-care. Honey is also used in special herbal soaps and cosmetic powders and packs. The Muzaffarnagar Industrial Honey Producers' Co-operative Society in Uttar Pradesh and a few herbal products firms in Bombay introduced some cosmetic items that have honey.

Other Industrial Uses

Significant quantities of forest honey are used in incense sticks manufacture, perfumery, tobacco curing and cigarette manufacture. The Golden Tobacco Co. and other cigar and cigarette manufacturers use 5 to 10 tons of honey. The Ambika Chemical Works, Eluru, Andhra Pradesh uses over 10 tons of honey in the manufacture of their famous perfume and incense sticks that are even exported.

Beeswax

Beeswax is a product from a bee hive. Beeswax is secreted by honeybees of a certain age in the form of thin scales.

Uses as a product

Beeswax is used commercially to make fine candles, cosmetics and pharmaceuticals including bone wax (cosmetics and pharmaceuticals account for 60% of total consumption), in polishing materials (particularly shoe polish), as a component of modelling waxes, and in a variety of other products. It is commonly used during the assembly of pool tables to fill the screw holes and the seams between the slates. Beeswax candles are preferred in most Eastern Orthodox churches because they burn cleanly, with little or no wax dripping down the sides and little visible smoke. Beeswax is also prescribed as the material (or at least a significant part of the material) for the Paschal candle ("Easter Candle") and is recommended for other candles used in the liturgy of the Catholic Church.

It is also used as a coating for cheese, to protect the food as it ages. While some cheese makers have replaced it with plastic, many still use beeswax in order to avoid any unpleasant flavours that may result from plastic.

The burning characteristics of beeswax candles differ from those of paraffin. Beeswax has negative ionization, which binds particulate matter to clear the air. A beeswax candle flame has a "warmer," more yellow color than that of paraffin, and the color of the flame may vary depending on the season in which the wax was harvested. Bees wax is also an ingredient in moustache wax, and was used in the manufacturing of the cylinders used by the earliest phonographs.

Bee venom

Several patients suffering from phlebitis and thrombophlebitis noticed an increase in their skin temperature with a change in the blood circulation. However, good results were shown in all the sciatic nerve cases, and those with nerve and articular pains. In chronic inflammatory nerve affections the results were also very good, with ceasing of pains and a partial recuperation of movement. When combined with an oral therapy, such as vitamin therapy, good analgesic effects were obtained in cortisone-dependent patients suffering from rheumatoid polyarthritis. The bee venom is applied for 4-5 days, followed by a 2-3 day break. Treatment is then re-commenced again. Using this method no adverse effects were reported. Applying the venom topically provided a long-lasting effect and offers significant benefits in arthritic and rheumatic conditions

Royal jelly

Royal jelly is a type of bee secretion that aids in the development of immature or young bees. It is secreted by the heads of young workers and used (amongst other substances) to feed the young until they develop to the desired rank. If a queen is desired, the hatchling will receive only royal jelly as its food source, in order that she will become sexually mature and have the fully developed ovaries needed to lay more eggs for the hive.

Uses

Propolis is marketed by health food stores as a traditional medicine, and for its claimed beneficial effect on human health. Depending upon its precise composition it may show powerful local antibiotic and antifungal properties. Also it is generally efficient in treating skin burns. Claims have been made for its use in treating allergy; it may stimulate the immune system, but some warn that it should not be taken if the user is likely to have severe allergic reaction to bees.

Pollen

Bees collect pollen in the pollen basket and carry it back to the hive. In the hive, pollen is used as a protein source necessary during brood-rearing. In certain environments, excess pollen can be collected from the hive. It is often eaten as a health supplementary.

Chapter 21

Diseases and enemies of honey bees

Honeybees like all other creatures suffer from many diseases (dis+ease) and are attacked by various pests, predators and enemies. Man has been concerned about the diseases of honeybees for hundred and thousands of years. Aristotle (384-322 BC) in his **Historia animalium** written some 2300 years ago described certain disorders in honeybees in his work on the history of animals. The Roman writer, Virgil, also commented on honeybee diseases some 300 years later. However, none of the earlier observations could identify the disorders with certainty. Dzierzon (1882) recognised two kinds of foul brood diseases in honeybees. Chesnire and Cheyne (1885) started microbial investigations into disease causing organisms. Evidently, it became widely believed that presence or absence of any disease was a matter of presence or absence of pathogens and eventual disaster was thought to be certain.

In recent times due to fast modes of transport and movement of bees and their products from one location to another and from one continent to another has aggravated the problem of bee diseases in their distribution. Now any problem occurring in any part of the world in no time can assume global dimensions. Honeybees suffer from various viruses, bacteria, protozoa, parasitic mites, wax moths, predatory arachnida, wasps, birds and mammals.

In most instances, the appearance of disease is abrupt and instantaneous which play havoc with honeybee colonies in apiaries. The first line of defence against honeybee diseases warrants their continuous monitoring and recognition of early stages of attack/infection so that preventive measures could be initiated well in advance before any catastrophe occurs with interference as little as possible with their natural propensities. As with many forms of life where disease is detected at an early stage, control and care is much easier to achieve. If a disease has reached an advance stage before it is detected, the care of the colony is virtually impossible, then the colonies should be destroyed. This will avoid a weak colony being robbed out and avoid general dissemination of the problem. The important diseases, pests and enemies include:

1. Brood diseases : bacterial diseases, fungal diseases, viral diseases

2. *Adult bee diseases* : protozoan diseases, bacterial diseases and viral diseases.
3. Mixed infections
4. Non infectious disorders : neglected brood, chilled brood, over heated bees, genetic lethality, plant poisoning, pesticidal poisoning.
5. Honeybee mites
    - i. Parasitic
    - *ii.* Endoparasitic *Acarapis woodi*
        - *i.* Ectoparasitic *Tropilaelaps clareae, T. koenigerum, Varroa jacobsoni, V. destructor*
        - ii. Stored product mites
    - iii. Phoretic mites
6. Pests waxmothas, bee louse, hawk moths
7. Predators wasps, birds, small mammals.
8. Enmies: ants, toads, spiders, dragon flies, squirrels.

Brood diseases

Brood is the most important stage in the life of honeybees. Honeybees suffer from various brood diseases such as American foulbrood, European foulbrood, Sacbrood, stone brood, Chalk brood and other maladies of uncertain origin which are of concern for honeybees. Recent discovery of *Varroa destructor* has made beekeeping industry at cross roads over the globe. These diseases can affect one or more of the following six stages in the development of the adult worker bee or the emerged adult bee:

a. The embryo develops for 3 days in the egg.
b. When the larva hatches from the egg, it is fed continuously for the next 5 days, while it is growing in an open cell, by young adult bees or 'nurse bees'.
c. On day 8 the fully-grown larva is sealed in its cell by nurse bees and then spins a cocoon. About 2 days after it is sealed over, the larva lies on its back with its head towards the cell capping.
d. The quiescent larva changes within a loosened fifth skin to a propupa, and after 2 days of this phase it sheds the fifth skin to become a white pupa.
e. The pupa, now resembling an adult bee in shape, slowly darkens in colour, beginning with the eyes.
f. The pupa sheds its skin, and a few hours later (21 days after the egg is laid) the adult emerges from its cell.

Brood combs of healthy colonies typically have a solid and compact brood pattern. Almost every cell from the center of the comb outward contains an egg, larva, or pupa. The cappings are uniform in color and are convex (higher in the center than at the margins). The unfinished cappings of healthy brood may appear to have punctures, but since cells are always capped from the outer edges to the middle, the holes are always centered and have smooth edges.

Plate 1 (A). Healthy larvae. These are white and neatly curled at the base of their cells (B) Healthy capped brood (pupae). Note the regular brood pattern, even appearance and convex caps. Normal brood expands in concentric circles. Larvae in unsealed cells are plump and the color of mother-of-pearl. Cappings on normal worker brood are slightly arched and uniformly brown (C) A comb showing an irregular (scattered) brood pattern

By comparison, brood combs of diseased colonies usually have a spotty brood pattern (pepperbox appearance), and the cappings tend to be darker, concave (sunken), and punctured. The combs may contain the dried remains of larvae or pupae (called *scales*), which are found lying lengthwise on the bottom sides of brood cells. Sometimes scales are difficult to locate because of the condition of the comb. In such cases, scale material can easily be located using long wave ultraviolet or near-ultraviolet light. Exposure to wavelengths of 3,100 to 4,000 angstroms causes scale material to fluoresce. Some discretion must be used with this technique because honey and pollen may also fluoresce. The brood infections can be determined by observing the symptoms and comparing both healthy and diseased colonies (Table 60).

Table 60. Comparative symptoms of various brood diseases

Disease	Cause	Appearance of Broodnest	Appearance of Cappings	Dead Larvae	Color and Consistency of Larvae	Scales	Odor
American Foulbrood	*Bacillus larvae*- bacterium, sporeforming	scattered brood pattern	sunken, perforated, discolored, greasey appearance	flat on bottom of cell	light brown, dull white, dark brown, eventually coffee to dark brown, sticky to ropey	black-brown and rough, removed by bees wih difficulty; lies flat on lower side of cell	unpleasant glue-like

Chalk-brood	*Acosphaera apis*, a fungus	scattered	light or dark, convex, any perforated	most often in sealed or perforated cells	white and mouldy, later grey-black, hard and chalk-like	none	normal
Chilled brood	sudden or prolonged low temperature	few or many dead larvae in cells at edge of broodnest	light or dark sun-ken and discolored over time	mostly in unsealed cells	dark or black, dry quickly	remnants are removed by bees easily	normal, roten odor in severe cases
Drone brood in worker cells	unfertilized or laying wor-ker eggs in worker cells	predominantly drone brood	bullet-like	none or few	normal	none	normal
European foulbrood, advance infection	*Strepto-coccus pluton*, a bacterium	scattered brood pattern, often pepperbox in appearance	discolored, sunken, perforated	in unsealed and sealed cells, in twisted positions, sometimes stretched out on the ventral side of the cell	black-brown, viscous, slightly ropey and stickey	rubbery, black-brown and smooth, are removed by bees with difficulty	unpleasant, sour
European Foulbrood-Early infection	*Streptoco-ccus pluton*, a bacterium	scattered brood pattern	some discolored, sunken, perforated	in unsealed cells, in twisted positions; trachea system often visible	yellow and brown; remains granular	yellowish or light brown; easily removed by bees	sour
Healthy Brood		pattern of sealed cells	light brown color, convex cappings	none	plump, white, mother-of-pearl appearnace	none	none or fresh
Sacbrood	a virus	scattered, often with many unsealed cells	often dark and sunken, many perforated	most often with head raised	greyish to black, watery and granular; skin has a sack-like appearance	head predomi-nantly curled up; yellow-brown or dark grey; removed by bees with ease	none to sour
Stonebrood	*Aspergillus flavus*, a	affected cells have a	some perfo-rated and	in unsealed and sealed	green-yellow, hard	none	mouldy in advanced

	fungus	greenish, mouldy appearance	covered with a greenish layer	cells	and shrunken		stage
Varroa disease	*Varroa jacobsoni*, a mite	scattered brood pattern; infestation greatest in drone brood	discolored and sunken	in sealed cells when heavily infected	dead larvae decay; surviving adults are often deformed	none, dead larvae and pupae easily removed by bees	unpleasant, rotten in severe infestations

By contrast, the gross symptoms of most adult bee diseases are not unique. The inability to fly, unhooked wings, and dysentery, for instance, are general symptoms associated with many disorders. Symptoms of a contagious disease are sometimes mimicked because of unrelated factors. For example, a brood neglected because of a shortage of nurse bees will often die from either chilling or starvation. Disease symptoms can also be the result of a failing queen, laying workers, toxic chemicals, or poisonous plants/ Noninfectious Disorders.

Normal honey bee development

A healthy honey bee colony has three distinct types of individuals: queen, worker, and drone. The queen is an especially important individual in the colony, as she is the only actively reproductive female and generally lays all the eggs.

A healthy worker brood pattern is easy to recognize: brood cappings are medium brown in color, convex, and without punctures. Healthy capped worker brood normally appears as a solid pattern of cells with only a few uncapped cells; these may contain eggs, uncapped larvae, nectar, or pollen.

Healthy C-shaped larvae

It is important to be able to identify healthy brood stages. Healthy worker, queen, and drone larvae are pearly white in color with a glistening appearance. They are curled in a "C" shape on the bottom of the cell and continue to grow during the larval period eventually filling their cell.

Status of brood diseases in India

Apiculture in India is a very young science and bee research was initiated in early forties. Within short period, all the diseases and enemies of honeybees present in other countries have been recorded from India also. The two viral diseases Apis *iridescent virus* and *Thai sac brood virus* have played havoc with Indian honeybee *Apis cerana* during the last one decade. The available information is summarized under the following heads:

a. Brood diseases

b. Adult diseases

c. Pests, predators and enemies.

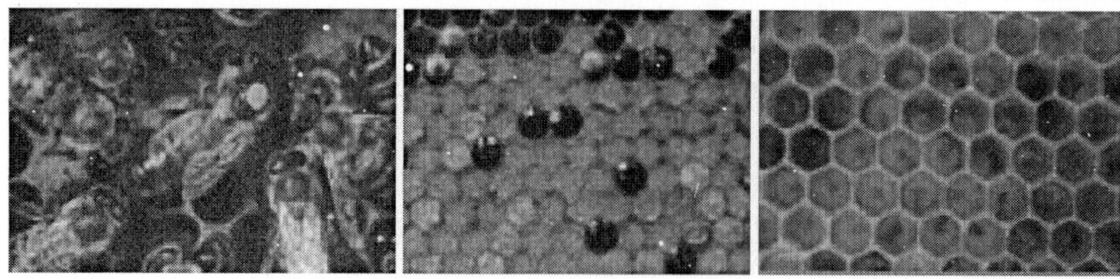

Plate 2. Showing a healthy bee colony

Brood diseases: These are generally easier to recognize as a group than adult diseases but are more difficult to control. A healthy brood comb typically shows a solid brood and compact brood pattern, whereas a comb from a diseased colony usually has a spotty brood also known as 'pepperbox' appearance. Also the cappings of healthy brood are uniform in colour and convex, whereas cappings of diseased brood tend to be darker and concave, and are frequently punctured. Sometimes the brood cells contain the dried remains of the larvae or pupae.

Brood diseases are either caused by bacteria or viruses or fungi. The occurrence of various brood diseases in India is summarized in Table 61.

Adult diseases

Most of the diseases of adult bees are difficult to diagnose, though inability to fly, unhooked wings and dysentery can be treated as general symptoms of of an unhealthy bee. Microscopic examination is often necessary for a definite diagnosis. the occurrence of various diseases in India is summarized in Table 61.

Table 61. Brood and adult diseases of honeybees, their causative agents and occurrence in India.

Disease	Causative agent	Bee species affected	Symptoms/colour of the brood	Locations
Brood diseases				
American foul brood (AFB)	*Bacteria, Bacillus* larvae	*Apis cerana indica*	Dull white dead brood becoming brown to white	Nainital (Uttar Pradesh)
Europen Foulbrood (EFB)	*Bacteria, Melissococcus pluton*	*Apis cerana indica*	Dull white dead brood turning yellow to dark brown	Mahabaleshwar Maharashtra Karnataka (around Castel rock near Goa border) Jammu, Punjab, Himachal
Sac brood disease (SBV)	Virus, *Morator aetatulus*	*A. mellifera*	Grayish or straw coloured becoming brown, grayish black or black or black head or black head end darker	Not recored from India so far

Thai Sac brood virus disease (TSBV)	Virus *Morator aetatulus* (Thai Strain)	*A. cerana indica*	Grayish or straw coloured becoming brown, grayish black or black or black head end darker	Whole India except Andhra Pradesh and Orissa
Chalkbrood disease (CB)	Fungus, *Ascosphaera apis*	*Apis cerana indica*	White chalklike mass sometimes referred to as mummy.	Not recorded from India so far.
Store brood (SB)	Fungs; *Aspergillus flavus*	*A. cerana indica* *A. mellifera*	The fungus forms a characteristic whitish yellow color like ring near the head end of the Infected larvae after death. Infected larvae become hardened and difficult to crush, hence called 'Stone brood'	Not recorded from India so far.
Adult diseases				
Nosema disease	Protozoan, *Nosema apis*	*Apis cerana*	Shining swollen abdomen	Uttar Pradesh, Himachal Pradesh, Jammu & Kashmir Punjab, Assam Orissa, Nagaland.
	Nosema ceranae	*Apis mellifera*	Shining swollen abdomen	Not recorded from India so far.
Amoeba disease	Protozoan, Malpighamoeba mellificae	*A. mellifera*	Cysts in Malphighan tubles	Not recorded from India so far.
Bee paralysis	Filterable virus	*A. mellifera*	Black hairless shiny bess	Not fecorded from India so far.
Septiceamia	Bacterium Pseudomonas apiseptica	*A. mellifera*	Destruction of connective tissue of legs, wings, and antennae	Not recorded from India so far.
Clustering disease	Iridescent bee virus	*Apis cerana indica*	Bees leave combs and form clusters on the wall of hive or outside the hive become sluggish, queen stops egg laying and drawlers appear around the colony.	Jammu & Kashmir, Himachal Pradesh, Punjab Maharashtra.

Table 62. Occurrence of various species of mites associated with honeybees in India

Mite species	Bee host	Occurrence in India
Acarapis woodi (Endoparasitic mite)	*Apis cerana*	Present
Varroa jacobsoni (Ectoparasitic mite)	*Apis cerana*	Present

Varroa destructor (Ectoparasitic mite)	*Apis mellifera*	Present
Euvarroa sinhai (Ectoparasitic mite)	*Apis florea*	Present
Tropilaelaps clareae (Ectoparasitic mite)	*Apis cerana; A. mellifera; A. dorsata*	-do-
Tropilaelaps clareae (Ectoparasitic mite)	*Apis cerana; A. mellifera; A. dorsata*	-do-
Tyrophagus longior (Provision mite)	*Apis cerana; A. mellifera*	-do-
Neocypholaelaps indica (Phoretic mite)	*Apis cerana; A. mellifera; A. dorsata*	-do-

Table 63. Occurrence of various pests, predators and enemies of honeybees in India

Category enemy/ animal	Disease/enemy	Causative agent	Bee species affected	Occurrence in India
Moths	*Achroia grisella*	Lesser wax moth	*A. cerana* *A.florea*	Present
	Galleria mellonella	Greater wax moth	*A. cerana* *A. florea* *A. mellifera* *A. dorsata*	Present
	Acherontia styx	Hawk moth	*A. cerana* *A. mellifera*	Present
Pseudoscorpion	*Ellingsenius indicus*	False scorpion	*A. cerana* *A. mellifera*	Present
Wasps	*Vespa orientalis, V. cincta, V. velutina, V. magnifica, V. mandarinia,V. tropica, V. ducalis, V. analis, V. asiatica, V. austriaca Vespa auraria, V. basalis, V. nursei, V. flaviceps, V. structur, Vespula vulgaris, V. germanica.*	All are predatory wasps	All the honeybee species	Present
Birds	*Meropes orientals*	Predatory bird	All the honeybee species	Present
	Meropes apiaster	Predatory bird	All the honeybee species	Present
	M. superciliosus	Predatory bird	all the honeybee species	Present

American foul brood

American foul brood disease (AFB) or American brood disease (ABD) is the most serious and infectious disease of brood honeybees. It occurs throughout the world and is responsible for considerable losses to bee and honey production. AFB is a disease of young larvae, causing their death, the loss of hives and if uncontrolled, decimation of apiaries. Uncontrolled infected hives act as a source of infection to other hives.

Cause

American foul brood is caused by a spore forming bacterium called *Paenibacillus larvae.* Young honey bee larvae become infected when they consume *P. larvae* spores in their food. The spores germinate in the gut; bacteria then move into the tissues, where they multiply enormously in number. Infected larvae normally die after their cell is sealed, and millions of infective spores are formed in their remains. These remains dry to form 'scales' which adhere closely to the cell wall and cannot easily be removed by bees. Consequently, brood combs from infected colonies are inevitably severely contaminated with bacterial spores. If the scales are not spotted and infected combs are subsequently used and distributed or moved from colony to colony during routine beekeeping management then infection has the potential to spread quickly.

The spores are very resistant to extremes of heat and cold, and to disinfectants. They retain their powers of germination for many years in honey, in old combs kept in store, or in derelict hives, skeps or boxes. Once a colony is infected the disease will usually progress until most of the brood is affected. The colony then becomes unable to replace the ageing adult bee population, causing it to become weakened, and finally to die out. The disease may develop for months before the colony succumbs, and death may occur at any time of the year.

Transmission

1. The beekeeper is the chief spreading agent of the disease. If combs honey or hive equipment are transferred from an AFB-infected colony to a healthy colony, it becomes infected.
2. Robbing and bee drift should be minimised to reduce spread. Interchanging of brood combs between hives and apiaries should be kept to a minimum. Swarms from infected colonies may also carry infection with them and become diseased after they are hived.
3. Second-hand bee equipment can remain infective for a long period, so it should not be used without rigorous heat sterilisation, e.g. inside of boxes should be charred with a blow-lamp and then scraped and painted.
4. Honey from unknown sources should never be used to feed bees. Second-hand honey tins should not be used if possible. If it is necessary, then bees should not have access to them.

Signs of American foul brood

1. Hive may show less than normal bee flight with dead bees on the bottom board. The colony may appear weak after opening the hive. Colonies become progressively weaker.
2. AFB generally affects only sealed brood. When infected larvae die within the sealed cell, the appearance of the cell cappings changes. A good way of remembering is that AFB = ***A*** (after sealing of the cell).

3. Wax cappings become sunken and perforated when adult bees nibble holes in them to try to remove the infected larva within. These perforations tend to be jagged and irregular in shape. Evidently, capped brood is uneven with puncture holes in the caps of brood cells.
4. Some cappings may become moist or greasy looking and slightly darker in colour than other cells.
5. At first only very few cells may show signs of disease, and the colony will appear normal in other respects
6. Eventually much of the sealed brood will become affected by the disease, causing a patchy or 'pepper pot' brood pattern.
7. There may then be an unpleasant smell associated with decomposition.
8. At the sunken capping stage the dead larval remains are light to dark brown in colour, and have a slimy consistency.
9. If a matchstick is inserted and slowly withdrawn, the remains can be drawn out in a brown, mucus-like thread or 'rope' 10- 30mm long. This is called the **'ropiness'** test and is a reliable test for the presence of AFB.
10. The ropy condition is followed by a tacky stage as the larval remains in the cell gradually dry up and the colour changes to dark brown.
11. The proboscis of dead pupae may sometimes remain intact, protruding upwards from the bottom edge of the cell.
12. Further drying leads to the final stage, which is a very dark brown, rather rough scale lying on the lower side of the cell and extending from just behind the mouth of the cell right back to the base.
13. The scales can be detected if the comb is held at an angle of approximately 15 degrees facing the light: they reflect the light from their rough surfaces and can easily be seen, even when their colour is almost the same as the comb itself. AFB scales can be readily detected in the field by holding the brood frame.
14. Over time, the larval remains in the cell will dry and harden into a dark brown leathery scale on the bottom side of the brood cell. A single scale contains millions of spores that remain viable for decades. Bees can not remove scales from cells.
15. Colonies with heavy infestation often display irritable behavior.

Laboratory Diagnosis

1. AFB is caused by *Paenibacillus larvae,* a spore-forming bacterium.
2. A microscope slide can be prepared by dissolving a small part of an AFB scale. Stir the scale with a toothpick in a droplet of water placed on a slide and apply a cover slip.

3. Under 400X magnification, the AFB spores are readily visible. AFB spores are characterized by being very slightly oblong, uniform in size and shape. The spores "jiggle" in a characteristic Brownian movement.
4. *P. larvae* is competitive and does not tolerate growth of other bacteria in the parasitized bee larva. As a result, most microscopic slides will show a predominance of *P. larvae* spores. This is not always the case with poor samples or those left in the collection bag for too long. In such case, secondary invaders such as moulds, will appear.

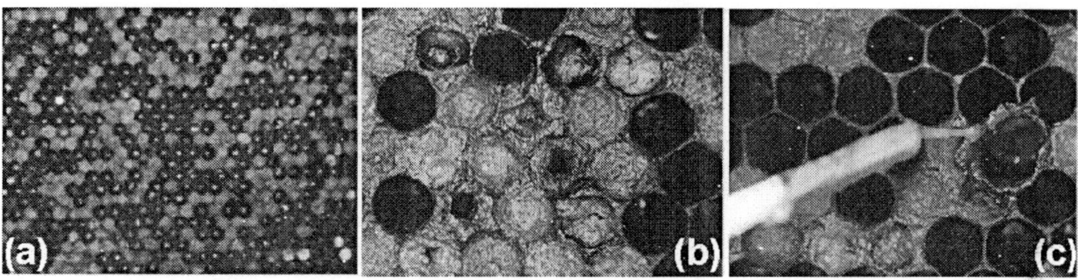

Plate (3a). Brood infected with American Foulbrood dies in the late larval or pupal stages, after capping of the cells. The brood pattern is spotty. Honey and pollen should not be present in the center of a healthy brood nest; **(3b)**. The cappings over cells with dead pupae are often sunken and dark. Cappings are often perforated by house bees; **(3c)**. Larvae and pupa killed by American Foulbrood first decay to a brownish, viscous mass and can be drawn out to a long thread that may snap back if pulled to far. The decaying brood stinks.

Control and Treatment

Methods of control of AFB using antibiotics that are used in some overseas countries are not effective, as they only serve to suppress signs of the disease without eradicating it and through frequent use allow the development of resistant bacterial strains.

1. Antibiotic-resistant AFB (r-AFB) has become established. Antibiotics must be used for treatment purposes only. Do not use antibiotics as a prophylactic (=preventive) measure.
2. Become thoroughly familiar with visual detection of brood diseases.
3. Inspect regularly, especially when disease has been reported in the area or after the colony has been placed in crop pollination.
4. For frames with suspect signs of brood disease, take a sample and send it to the laboratory for analysis.
5. When AFB has been confirmed, kill the bees and burn all the equipment. **Or:** Shake bees onto foundation and burn all the old equipment. Feed the bees with medicated sugar syrup at two week intervals until foundation has been drawn out.
6. Use antibiotics only as recommended. Never use the product after its expiry date, and follow preparation instructions carefully

7. Reduce the exchange of hive equipment between hives and apiaries.
8. Replace 20% of all brood frames each year so that after a few years, no brood frame is older than five years.
9. Don't leave used hive equipment exposed to foraging bees.
10. Apply hygienic management practices, including clean clothing, hive tools, and gloves.
11. Debris from inside the hive should be scraped off and burnt. Undestroyed hives and hive parts should be sterilized by scorching all the inner surfaces and edges to a dark-brown colour with a blowlamp. It is essential to repaint them

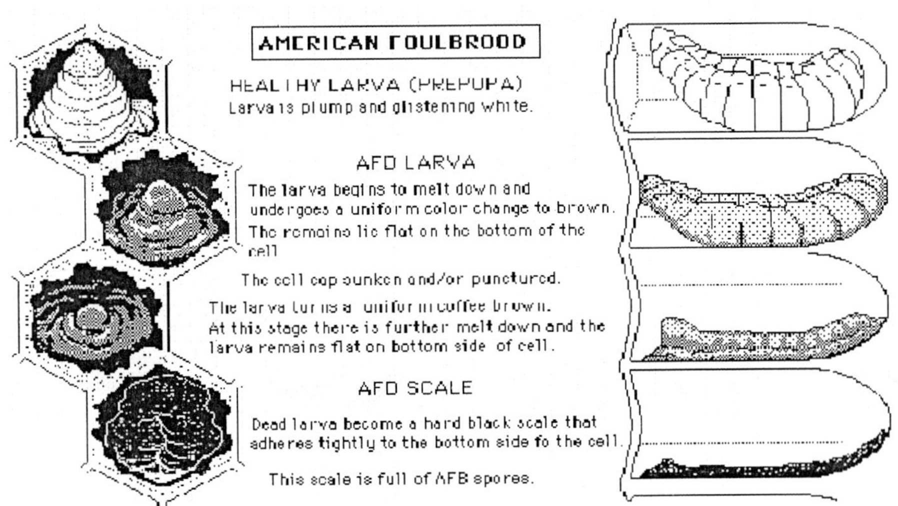

Figure 71. Showing different stages of AFB attacking larva

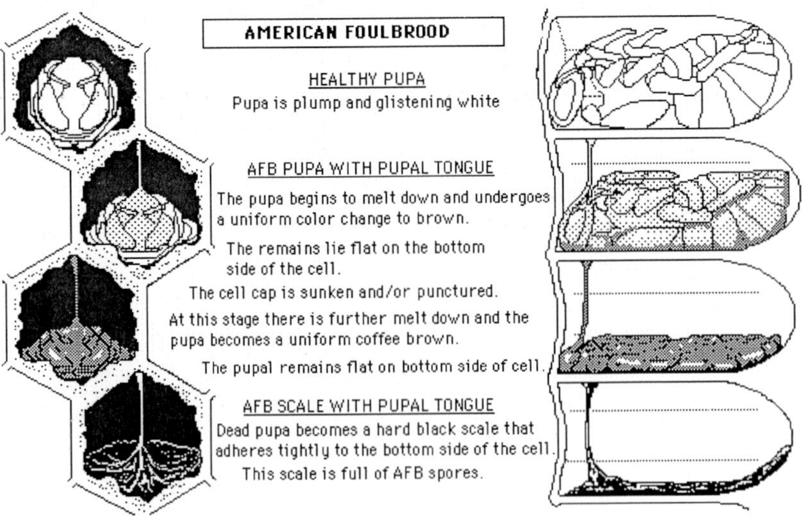

Figure 72. Showing different stages of AFB attacking pupa

European foul brood

Cause

European foul brood is caused by the bacterium called *Melissococcus plutonius* (formerly *pluton*). The bacterium *Melissococcus plutonius* (formerly *pluton*) (=*Streptococcus pluton*) causes the brood disease European foulbrood (EFB). *Streptococcus plutonius* (formerly *pluton*) was reclassified into the new genus *Melissococcus* by Bailey and Collins (1982a,b). *M. plutonius* (formerly *pluton*) is generally observed early in the infection cycle before the appearance of the varied microflora associated with this disease. The *M.* (formerly *pluton*) cell is short, non-spore-forming, and lancet shaped. The cell measures 0.5-0.7-1.0 m and occurs singly, in pairs, or in chains (Fig. 73). When stained with carbol fuchsin, the organism appears dark purple against a lighter background. Some distortion occurs during the fixing and staining process; this can be reduced by negative staining. *M. pluton* can also be detected using enzyme-linked immunosorbent assays (Pinnock and Featherstone 1984), polyclonal antisera (Allen and Ball 1993), or polymerase chain reaction (Govan *et al.*, 1998).

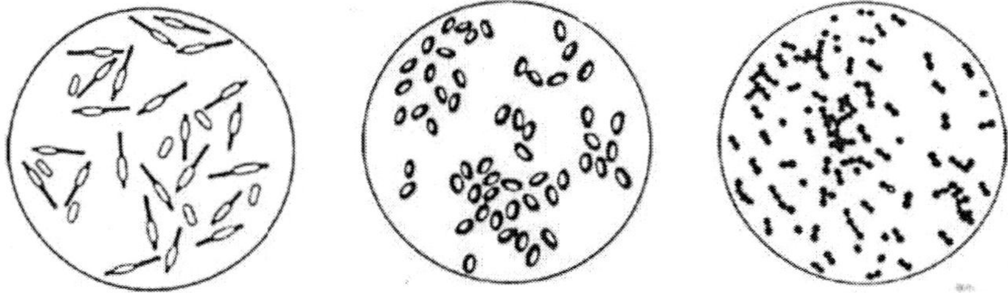

Figure 73. Bacteria associated with European foulbrood disease (not to scale). Top, *Paenibacillus alvei*. Middle, *Brevibacillus laterosporus*. Bottom, *Enterococcus faecalis*. (elongated in the direction of chain) are 0.5–1.0 m in diameter and are usually in pairs or short

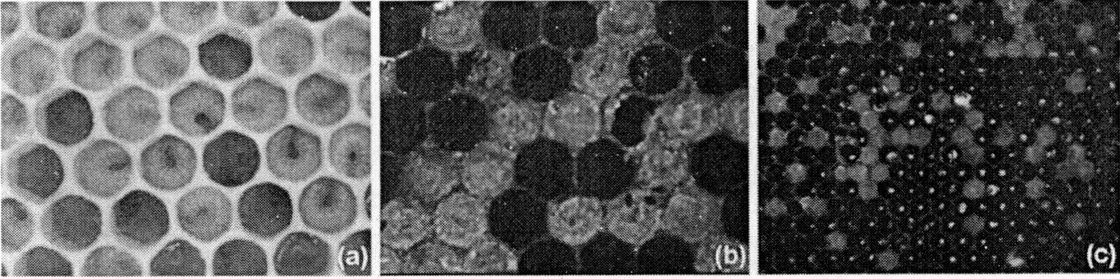

Plate 4. Trachea are visible in diseased larvae in unsealed cell ; (b) Advanced infection of European foulbrood. Dead larvae lie behind perforated cappings. The dead viscous mass is not as sticky as American Foulbrood, and can be drawn out only slightly; (c) Dead brood may be present in sealed and unsealed cells. The brood pattern is scattered as a result of the removal of sick and dead larvae from the broodnest by adult house bees.

This disease is considered a stress disease and is most prevalent in spring and early summer. This is the period when brood rearing is at its height. The earliest brood rarely is affected. European foulbrood frequently begins to disappear with a nectar flow and may disappear entirely for the balance of the year; or it may reappear during nectar dearths in the summer or fall. Occasionally, the disease remains active throughout the entire foraging season. Workers, drones and queens are susceptible, although various commercial bee strains differ in susceptibility.

The bacteria multiply in the mid-gut of an infected larva, competing with the larva for its food. They remain in the gut and do not invade the larval tissue; larvae that die from the disease do so because they have been starved of food. This normally occurs shortly before their cells are due to be sealed. Subsequently other species of bacteria may multiply in the remains of dead larvae as 'secondary invaders', such as *Paenibacillus alvei, Enterococcus faecalis, Brevibacillus laterosporus*, and *Lactobacillus eurydice.*

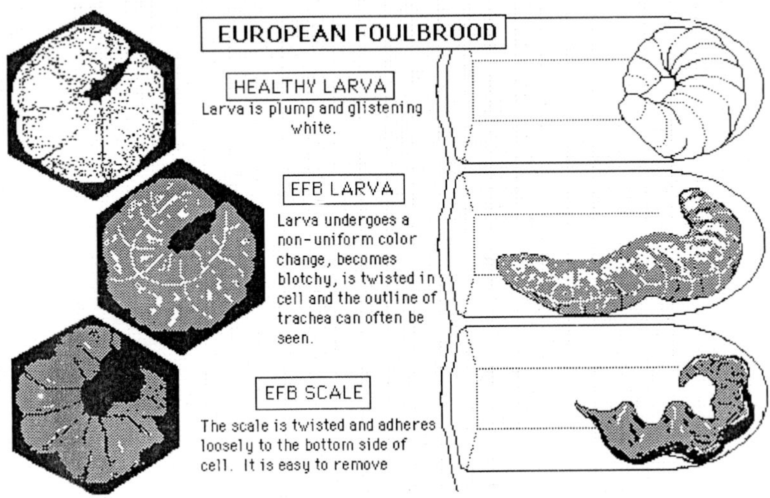

Figure 74 Showing different stages of AFB attacking larva

Progression of the disease

The development of the disease within a colony is complex, and still not fully understood. It appears that infection can develop over a period of months or years, debilitating but not killing the colony. During this time, signs of the disease may become more or less severe, or disappear altogether. Frequently there is a seasonal pattern, with signs becoming most obvious in late spring. This is thought to be because when there are many larvae relative to the number of nurse bees, larvae tend to receive less brood food overall, and those infected with EFB are more likely to suffer from starvation. At other times, larvae that are infected but receive an abundance of brood food may survive the infection, and develop into healthy adult bees. However, when such larvae pupate, they void their gut contents into the cell, contaminating the comb with millions of infective bacteria. Eventually the disease is likely to reach the stage where a high proportion of the brood is affected and the colony will be weakened and ultimately killed.

Transmission

M. plutonius does not form spores, but the organism often overwinters on combs. It gains entry into the larva in contaminated brood food and multiplies rapidly within the gut of the larva. Not all infected larvae die from the disease. Some develop normally and void or regurgitate the bacteria onto the underside of cappings, which then become the source of the disease. Since the honey of infected colonies is contaminated, EFB can be spread by robber bees or by the exchange of equipment among colonies and drifting bees.

Signs of European foul brood

European foul brood cannot be reliably identified visually, as the disease signs can easily be confused with various other brood abnormalities. The disease and its symptoms are highly variable, probably because several other types of bacteria are often present in dead and dying larvae. EFB usually does not kill the colony, but a heavy infection will seriously reduce population development. European foulbrood generally kills larvae two to four days old while they are still coiled in the bottom of the cells. Unlike American foulbrood, most of the larvae die before their cells are capped

1. EFB affects mainly unsealed brood, killing larvae before they are sealed in their cells. An easy way to remember is that EFB = ***E*** (early infection before sealing of the cell).
2. The EFB infected larva moves inside its cell instead of remaining in the normal coiled position characteristic of a healthy larva of the same age. When it dies it lies in an unnatural attitude – twisted spirally around the walls, across the mouth of the cell or stretched out lengthways from the mouth to the base. The dead larva often collapses as though it had been melted, turning yellowish-brown and eventually drying up to form a loosely attached brown scale.
3. The gut of an infected larva may be visible through its translucent body wall. It has a creamy white colour caused by the mass of bacteria living within it.
4. When a high proportion of the larvae are being killed by EFB, the brood pattern will often appear patchy and erratic as dead brood is removed by the bees and the queen lays in the vacant cells.
5. A very unpleasant odour may sometimes accompany severe EFB infection, depending on the presence of certain other species of bacteria in the remains of dead larvae.
6. A minority of infected larvae may die after the cell is sealed. In such cases, there may be sunken perforated cappings resembling AFB infection. However, the cell contents although brown and sticky cannot be drawn into a 'rope' as with AFB. Where larval remains dry to form scales, these are variable in colour, loose within the cell and somewhat "rubbery", unlike the hard black firmly attached scales of AFB.

7. A spotty pattern of capped and uncapped cells develops only when EFB reaches serious levels. Occasionally, pupae die from the disease. The most significant symptom of EFB is the non-uniform or blotchy color change of the larvae. They change from a normal pearly white to yellow, then brown, and finally grayish black. Larvae also lose their plump appearance and look under-nourished. Their breathing tubes or tracheae are visible as distinct white lines. In such cases, larval remains appear twisted or melted to the bottom side of the cell. Recently-dead larvae are rarely ropy. They form a thin brown or blackish-brown scale which can be easily removed from the cells, unlike American foulbrood scales. Bees remove scales under the stimulus because of strong hygienic behavior or due to a nectar flow or syrup feeding.

Field Diagnosis

1. European Foulbrood is much less serious than AFB. EFB shows up when the colonies have been under stress due to other diseases, colonies nearby, poor management and weather.
2. EFB is easily controlled with standard antibiotic treatments.
3. EFB affects bee brood much the same as AFB except that the disease affects open brood, i.e. the larvae die before they are capped.
4. Affected cells show discoloured larvae, often in twisted positions, with visible tracheal tubes.
5. The brood has a"sour" odour, distinctly different from AFB.
6. EFB scales are easily removed from the cell (compared to AFB scales).
7. When scales are detected, collect samples for laboratory examination. Although field analysis is often correct, accurate distinction between AFB and EFB can only be made through microscopic examination.

Laboratory Diagnosis

1. EFB is caused by *Melissococcus plutonius,* but the secondary invader *Bacillus alvei* is mostly observed when samples are examined microscopically.
2. Samples are prepared the same way as AFB samples.
3. At 400X, *B. alvei* is readily identified by its long spindle shaped spores.
4. The spores do not jiggle but float by in the solution.
5. Unlike AFB, EFB microscopic samples generally display a wide variety of microbes.

Control and Treatment

1. Inspect brood frames regularly and be familiar with field symptoms.
2. Remove all frames with significant numbers of affected cells.

3. If the disease threatens the survival of the colony, however, you may treat with an antibiotic at any time of the year. If treatment is done just before or during a honey flow, remove extracting supers before treating and do not use any honey produced during the season for human consumption. Spray or sprinkle antibiotics (oxytetracycline) dissolved in 250 ml of sugar syrup over the colony every 3-4 days for 10 days. Because most antibiotics are relatively unstable in honey or syrup solutions, they should be fed as a dust mixed with powdered sugar. Never medicate bees when any danger of contaminating the honey crop exists.
4. Requeening provides a distinct break in the brood cycle of the colony, allowing the bees to clean up existing disease. It may also provide new bees with better cleaning behaviour, i.e. less susceptible to disease.
5. Minimize robbing by preventing sugar spillage.
6. Apply hygienic management practices. Clean hive tools, smoker and gloves after inspection of each apiary. Clean clothes regularly.
7. Control of the disease by a husbandry method known as the *"shook swarm"* has also been shown to be effective and is an option available to beekeepers

Controlling foul brood

Both EFB and AFB are infectious diseases, and can spread without the intervention of the beekeeper by the natural processes of robbing, drifting etc.

Key strategies for controlling an outbreak of foul brood in bees include

1. Recognize the signs of foul brood

One should be able to distinguish between diseased and healthy brood and spot a diseasesd larva in a comb of several thousand. Send the individual suspect larvae to the laboratory for diagnosis

2. Use quarantine systems to avoid spreading disease

When colonies with signs of foul brood have been found and dealt with, there is still a significant risk that other colonies may be infected but not yet showing signs of disease. Then 'quarantine systems' need to be very effective in minimising the spread of infection between colonies while a foul brood outbreak is brought after control. These will also help minimise the scale of any new outbreaks that may subsequently occur.

Colony quarantine – avoid moving any combs, bees or equipment from one colony to another. It will be necessary to mark super frames and boxes so that they can be individually identified and returned to the same colonies after extraction. This is the most effective quarantine system, and the most appropriate for colonies, that are at particular risk – such as those that have been previously treated, and those that have had close contact with infected colonies – but involves significant effort to carry out on a large scale. This has worked very successfully to bring large outbreaks under control.

Apiary quarantine – avoid moving any bees, combs or equipment between apiaries, but allow some movement (e.g. super combs) within the apiary. This will not prevent spread within the apiary but involves less work than colony quarantine to implement on a large scale and helps prevent moving disease between apiaries.

Isolation apiaries

This keeps to a minimum any contact between diseased and healthy colonies, and makes it easier to operate quarantine systems appropriate to the level of risk in each apiary.

Disinfecting equipment – where it is necessary to move items between colonies, treat them to reduce the risk of spreading disease. Wooden hive parts can be made safe by scorching with a blowlamp. Hive tools, gloves, the smoker, etc. can be soaked in or scrubbed with a strong solution of washing soda or irradiating equipment is another option

Transfer colonies to new comb

The pathogens responsible for both AFB and EFB can exist in a colony's combs for long periods and remain capable of causing disease to develop. This is particularly true of colonies that have been treated against EFB with an antibiotic. A significant proportion of these colonies can suffer reoccurrence of disease within a year or so as a result of live bacteria remaining in the colony after treatment. Any method that removes contaminated comb from colonies and replaces it with new comb will be helpful in reducing the risks of disease. The more rapid and complete the transfer, the more effective it will be.

Shook swarm

The 'Shook Swarm' method aims to remove completely contaminated comb by transferring the colony to entirely new combs in one operation. This is done by shaking the adult bees into a clean hive fitted with frames of foundation during the season. The removed combs are then destroyed by burning.

Although this method can involve significant labour and expense, but recent research suggests that it is extremely effective at combating EFB and reducing subsequent recurrence of disease.

Integrated Pest Management

Integrated pest management, ('IPM') is a principal that has been widely used in agriculture, especially where it is desirable to keep chemical or medicinal inputs to a minimum. Significantly, no attempt is made to eradicate completely the pests or disease.

Instead, the aim is to keep them below the level where they cause significant harm by using a combination of controls applied at different times of the year.

More or fewer controls are employed depending on the levels of disease present. This is a much more effective approach than the alternative of waiting until pathogen numbers have reached a damaging level before applying controls, or applying the same controls each year regardless.

An IPM programme for foul brood control (with particular emphasis on EFB)

IPM is a principle that can be readily applied to control of many bee pests and diseases.

An IPM approach to foul brood control would aim to include:

1. A varied mix of controls working in different ways and at different times of year according to the level of risks
2. A mixture of prevention and intervention
3. Graded response depending on level of problem
4. Control at several points of the year makes it harder for the disease/ pathogens to reach
5. Harmful levels or threshold levels.
6. Use of management methods can reduce the need for antibiotic use.
7. Control strategies can be easily altered to reflect changing circumstances, infection levels

Foul brood control calendar is as given below (Table 64)

Table 64. Foul brood control calendar

	Jan	Feb	Mar	Apr	May	June	July	Aug	Sep	Oct	Nov	Dec
Sterilize supers and combs. Replace old wax with foundation	+	+	+								+	+
Routine comb change (3-4 older combs per brood box)			+	+	+	+						
Mark supers placed on individual hives				+	+	+	+	+				
Monitor for disease signs				+	+	+	+	+	+			
Shook swarm				+	+	+	+	+				
Antibiotic treatment				+	+	+	+	+	+			
Biological control treatment				+	+	+	+					
Colony destruction	+	+	+	+	+	+	+	+	+	+	+	+
Apiary/colony quarantine (barrier management)	+	+	+	+	+	+	+	+	+	+	+	+
Control robbing							+	+	+			
Extraction hygiene					+	+	+	+	+	+		

Table 65. Summary of brood signs causes and control

	Signs	Control
Normal brood	Uncapped: Pearly white, 'C' shaped larvae. Capped: Uniform brown colour, domed cappings.	None required
American foul brood *(Paenibacillus larvae)*	Affects only sealed brood sunken concave cappings, uneven brood pattern, 'pepper pot' or mosaic pattern, scales on bottom walls of open cells, brown decomposing larvae that 'rope' using matchstick test, moist dark perforated cappings.	Burn infected colonies and sterilise hives by scorching.
European foul brood *(Melissococcus plutonius)*	Affects mainly unsealed brood. Infected larvae discoloured yellow-brown lying in abnormal positions in cell with 'melted' appearance. Some dark sunken cappings may be present, but cell contents will not form a 'rope'	Infected colonies can be: "Shook swarmed", lightly infected colonies may be given with antibiotic and severe cases of EFB are destroyed as with AFB.
Chalkbrood *(Ascosphaera apis)*	Affects only sealed brood. Perforated cappings over cells containing hard white or mottled grey chalk like remains ('mummies').	No specific treatment. Keep strong colonies. Re-queen severely affected colonies.
Sacbrood *(Sacbrood virus SBV)*	Affects only sealed brood. Perforated cappings. Larvae become yellow-brown fluid filled sacs ('Chinese slipper'). Watery contents will not form 'rope'.	No specific treatment. Requeen severely affected colonies
Bald brood	Abnormal cell cappings over sealed brood. Affected cells have round hole in capping sometimes with a slight protrusion. Pupae have normal appearance. Signs of wax moth larvae may be visible in comb.	No specific treatment. Control wax moth infestation.
Drone laying queen or laying workers	Domed drone cappings over worker cells. Abnormally small drone pupae within cells. May be multiple eggs per cell. Unsealed brood may be neglected and dying.	Replace drone laying queen. Unite colony with laying workers to another colony. Shake bees out in front of the hive so that they return to other colonies slowly.
Chilled brood	Dead brood usually present in all stages. Unsealed brood turns very dark brown or black in colour.	Avoid conditions that prevent bees being able to care for brood.
Varroa infestation *(Varroa destructor)*	Signs vary. Sealed brood may be partially uncapped, dead pupae discoloured brown or black, watery or firm, but never ropy.	Control varroa infestation to low levels using appropriate treatment.

Chalkbrood

Cause

Chalkbrood, a fungal brood disease of honey bees, is caused by the spore-forming fungus, *Ascophaera apis*. The fungus *Ascosphaera apis* causes chalkbrood disease. It is a heterothallic organism and develops a characteristic spore cyst when opposite thalli (+ and –) fuse. Spore cysts measure 47–140 m in diameter. Spore balls enclosed within the cyst are 9–19 m in diameter, and individual spores are 3.0–4.0; 1.4–2.0 m (Figure 75a).

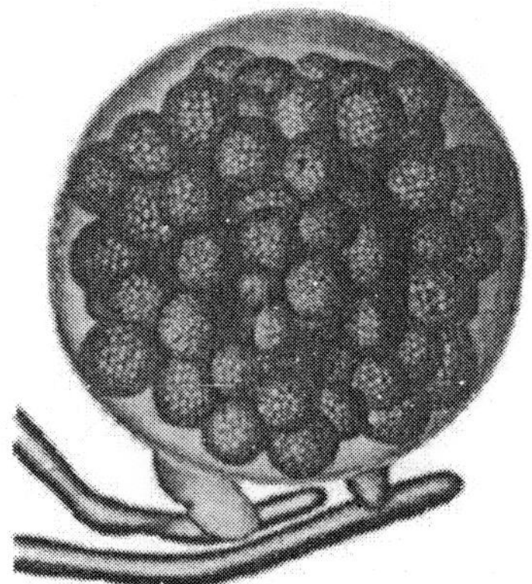

Figure 75a. Spore cyst of *Ascosphaera apis* containing spore balls, which in turn contain spores

Worker, drone, and queen larvae are all susceptible. Spores of the fungus are ingested with the larval food. The spores germinate in the hindgut of the bee larva, but mycelial (vegetative) growth is arrested until the larva is sealed in its cell. At this stage, the larva is about six or seven days old. The mycelial elements break through the gut wall and invade the larval tissues until the entire larva is overcome; this process generally takes from two to three days (Figure 75b).

Symptoms

Dead larvae are chalky white and usually covered with filaments (mycelia) that have a fluffy, cotton-like appearance. These mummified larvae may be mottled with brown or black spots, due to the presence of spore cysts or fruiting bodies of the fungus. Larvae that have been dead for a long period of time may turn completely black due to the spread of the fruiting bodies.

Diseased larvae can be found throughout the brood rearing season, but are most prevalent in late spring when the broodnest is rapidly expanding. Chalkbrood usually disappears or declines as the air temperature rises in the summer. Affected larvae are found on the outer fringes of the brood nest where sufficient nurse bees are unavailable to maintain brood nest temperature. Brood cells can be either sealed or unsealed. Infected larvae, stretched out in their cells in a lengthwise position, are removed by nurse bees two to three days after symptoms first appear. Dead larvae (mummies) are often found in front of the hive, on the landing board, or in a pollen trap. In strong colonies, most of these mummies will be discarded by worker bees outside of the hive, thus reducing the possibility of re-infection from those that have died from the disease.

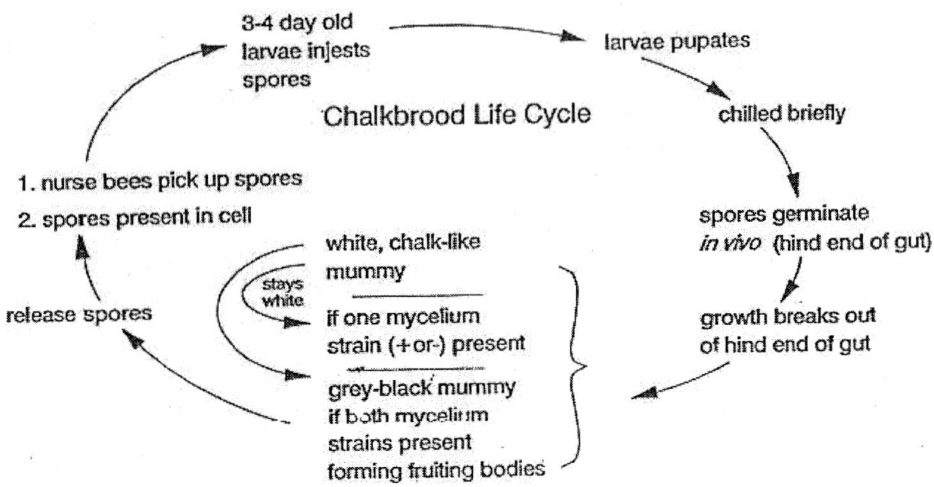

Figure 75(b). Showing life cycle of chalk brood.

Transmission

Spores remain virulent for years. Therefore, infected pieces of equipment, especially brood combs, are a reservoir for further infection. Be careful when buying used equipment since it is a possible source of disease contamination. Chalkbrood usually does not destroy a colony. When the disease is serious, however, it can prevent normal population buildup and surplus honey production. Research has shown that the spores are easily passed from bee to bee. Therefore, drifting and robbing bees are potential vectors of the disease. Both workers and queens taken from infected colonies can transmit infection to healthy colonies. Colonies fed pollen collected from infected colonies will also contract the disease. Chalkbrood infections are not always visible in the broodnest. Therefore, beekeepers who collect pollen to sell or to feed to their bees should check the pollen and pollen traps from each colony for the presence of whole or parts of mummies (the hardened remains of larvae infected with *A. apis*).

Field Diagnosis

1. Chalkbrood incidence increases in the fall and spring. Mummified larvae in front of the hive and on the bottom board are easily detected. Mummies on the bottom board may not necessarily indicate a serious problem, but confirm hygienic bee behaviour.

2. There is no control product available. High incidence of Chalkbrood mostly indicates poor hygienic behaviour and stress due to weather, poor management or diseases.
3. When there is a persistent Chalkbrood problem, replace the queen with one supplied by a reputable bee breeder.

Laboratory Diagnosis

1. Mummified larvae are generally white in colour. The mycelium of the fungus infiltrates the larval tissue that eventually hardens. The white colour is the result of asexual reproduction while sexual reproduction will produce black or grey coloured mummies.

Stonebrood

Cause

Stonebrood is a fungal disease. Several fungi belonging to the genus *Aspergillus* are associated with the disease, the most common being, *A. flavus and A. fumigatus.* However *A. flavus* is considered, by far, the most important.

Symptoms

Both larvae and pupae are susceptible. The disease causes mummification of the affected brood. Mummies are hard and solid, not sponge-like as in the case of chalkbrood. Brood infected by *A. flavus* become covered with a powdery green growth of fungal spores. The spores are found most abundantly near the head of the affected brood.

Treatment

This disease is considered of minor importance and is rarely encountered. No treatment is recommended. The bees remove the dead brood on their own and the colony normally recovers in a short period of time.

Viral Diseases

So far, the world over, 18 viruses have been found to infect honeybees. Out of these viruses, Thai sac brood virus and *Apis iridescent virus* have been found to cause heavy losses to bee industry in India. (Deodikar, 1971, a,b; Kshirsagar, 1982, 1983. Abrol and Bhat, 1990; Abrol 1992, 1993). The contiguity of land with Europe has been suspected to be the major factor for the spread of these diseases (Chahal and Gatoria, 1983).

Sac brood disease

Sac brood is caused by a virus. The infected larva dies and the tissue disintegrates into a brown watery solution held by the larval outer skin. The skin sac can be removed intact from the cell. The cell is often uncapped but may also be closed and the cap punctured similarly to AFB cells. Sac brood is a contagious disease which causes major damage to the bee-keeping industry. It is at an early stage of the attack, the diseased larva changes from its normal pearly white color to a grayish white. The larvae gradually darkens, turning from greyish to yellowish, with a black head. As the larva dries up, it becomes dark brown. The larvae in the cell shrink when they are attacked. At a later stage of the attack, some of the larva develop into a sac-like form containing liquid. This is the reason why this disease is called 'sac brood'. Larvae with sacbrood disease are easily removed intact from the cells, unlike those killed by American foulbrood. When removed and suspended from a toothpick or thin twig, the contents of the larvae are watery and the tough outer skin appears as a "sack" or bag of fluid which is filled with millions of sacbrood virus particles. The dried sacbrood scale lies flat, with the head end raised and darkened and the tail flat on the bottom side of the cell. The scales are rough and brittle and do not adhere tightly to the cell wall. As the dead larva decomposes, the outer skin becomes hard. The dead larvae can be removed more easily from their cells than live larvae. caused by a virus which attacks the larvae of the brood, causing their death. Sac brood attack reduces the number of larvae, thus later on reducing the population adult bees. Sac brood attack does not wipe out the colony but weakens it, making it susceptible to other pest attacks.

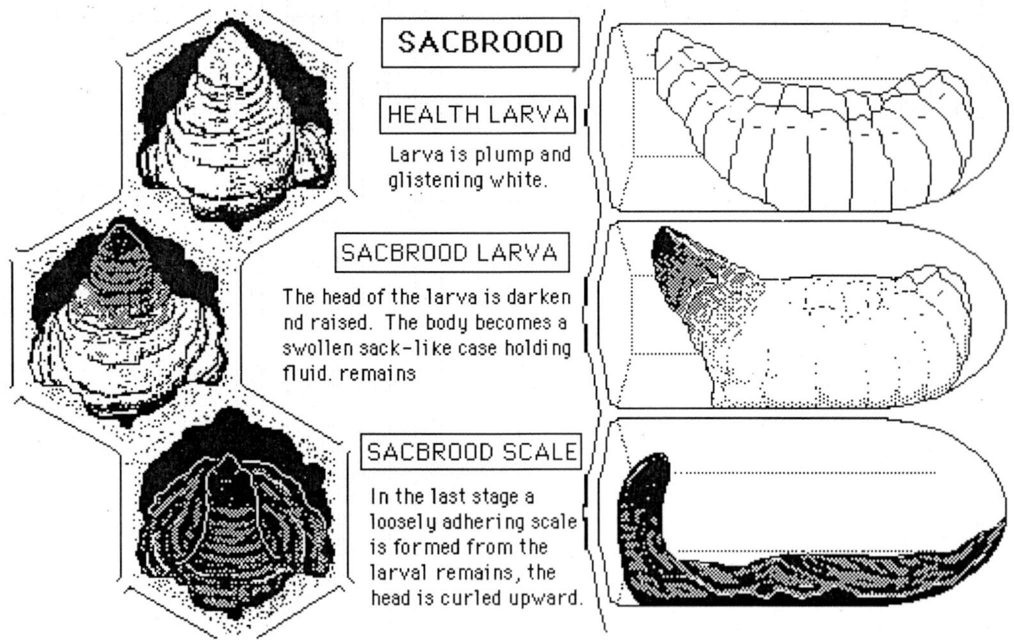

Figure 76. (a) Showing different stages of sacbrood virus attacking larva

Symptoms

1. The first symptoms of an attack is a change in the color of the larvae in open or closed cells of the comb. For closed cells, the presence of a small hole indicates an attack by the virus. The cappings over dead brood are

first punctured and later removed by the bees. Death usually occurs after the cell is sealed and the larva has spun its cocoon.

2. At an early stage of the attack, the diseased larva changes from its normal pearly white color to a grayish white. The larvae gradually darkens, turning from greyish to yellowish, with a black head. Larvae die in a stretched-out position with their heads raised.
3. As the larva dries up, it becomes dark brown. The larvae in the cell shrink when they are attacked.
4. At a later stage of the attack, some of the larva develop into a sac-like form containing liquid. This is the reason why this disease is called `sac brood'.
5. As the dead larva decomposes, the outer skin becomes hard. The dead larvae can be removed more easily from their cells than live larvae.

Eradicating the virus

Check the hive for sac brood once every two weeks. If sac brood is present, take the following actions. If the attack affects less than 20% of the comb, remove the comb and burn it. If more than 20% of the brood is infected, all adult bees must be destroyed by burning.

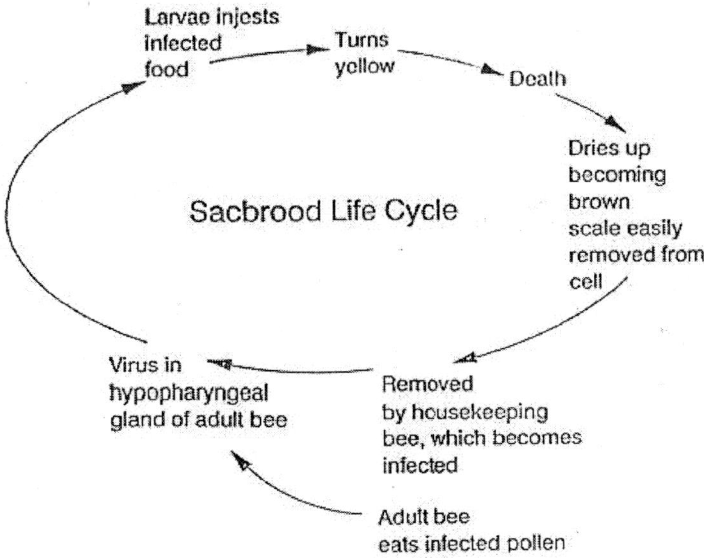

Figure 76(b). Showing life cycle of sacbrood

Transmission

Nurse bees are suspected of transmitting the disease by carrying the virus from cell to cell. It is also believed that robber bees spread the disease by carrying contaminated honey from colony to colony.

Management

There are no control products available. Since the disease is caused by a virus, no antibiotic is effective in preventing or controlling sacbrood. Strong colonies and regular requeening seem most effective in combating this disease. In a situation where 5-20% of the brood is infected, the colony can be helped to recover replacing the queen. This may be either: Self recovery, where the colony produces a new queen. The old queen is removed from the hive, and a new, healthy and fertilized queen is introduced, 21 days later. Once the new queen is introduced, sugar syrup should be added to speed up the development of the bee population. In a situation where less than 5% of the brood in a hive is infected, two colonies from two hives can be combined together. Sugar syrup as an additional food should be given to help the colony to recover rapidly.

Figure 1. Sac brood (sealed cells with punctures and affected larvae)

Figure 2. Larvae die with their heads raised. Dead larvae in perforated cells have begun to dry.

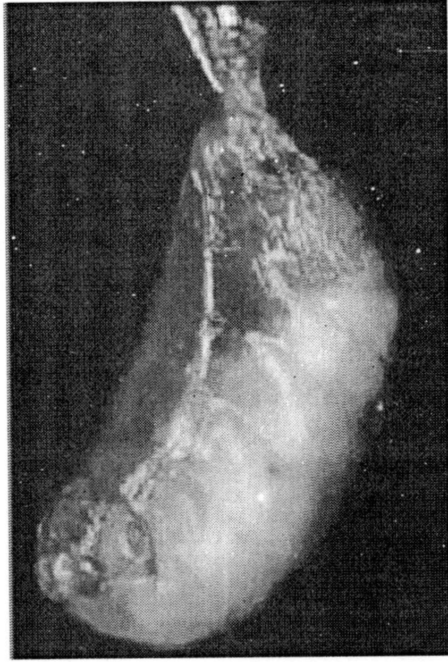

Figure 77. A layer of fluid insulates the pupa from the sac (prepupae in sack). The larvae itself hangs limply when taken from the sac, and is easily perforated to release fluid contents

Thai Sac brood

Thai Sac brood was first noticed in 1976 in Thailand from *Apis cerana* F. colonies. Bailey *et al.* (1982) established it as a new strain of sacbrood virus and named it as 'Thai sac brood virus' (TSBV). This virus spread to India via Malaysia and Burma (Kshirsagar, 1982, 1983; Kshirsagar *et al.*, 1982). After its first report in Thailand in 1976, the 'Thai sac brood disease' traversed to northern states such as: Utter Pradesh, Himachal Pradesh and to Jammu &

Kashmir by 1985-86 (Rana *et al.*, 1987, 1988; Shah and Shah, 1988; Joshi and Varma, 1985; Varma and Johsi, 1985; Abrol and Bhat, 1990; Abrol, 1992; 1993). The disease suddenly made its appearance in the far off south in late 1991 (Channabasavana, 1993; 1994).

The disease is caused by a virus (TSBV) which primarily infects the larvae of *Apis cerana* and is closely related to sac brood virus (SBV) which infects the *Apis mellifera* L., but the two are not identical (Bailey *et al.,* 1982). Thai sac brood virus shows certain physico-chemical and serological variations from SBV. It has been highly virulent in north-eastern and northern India, being responsible for heavy losses to bee-keeping industry, killing over 95% of *Apis cerana* colonies (Kshirsagar, 1983; Rana *et al.,* 1987; Abrol and Bhat, 1990; Abrol, 1992; 1993; Reddy and Reddy, 1994). Thai Sac brood virus is not known to infect *Apis mellifera* L. in India. Yang *et al.* (1988) isolated Chinese sac brood virus from *Apis cerana* colonies. This virus has properties very similar to TSBV with three close but well defined bands with molecular weight of 27000, 29000 and 39000. Similarities and differences between three types of sac brood virus diseases are presented in Table . Preliminary observations (Shah and Shah, 1988) have shown that larvae killed by TSBV are highly infective when they assume sac formation.

Symptoms: In case of infections in *Apis mellifera* by SBV mostly sealed larvae die after the cells are capped, whereas in *Apis cerana indica* late larval or mostly prepupal infections seem common, resulting in the death of larvae in the uncapped Cells. The examination of the diseased colonies show irregularly capped brood with sunken and faded caps. The dead larvae in such cells are observed with their heads turned up, partly across the cell openings. Such larvae when picked up with forcep show but entire larva/prepupa transformed into a 'sac' filled with fluid. Infected larvae turn pale yellow and finally brownish when dead.

Mode of spread: In an attempt of clearing the cells of dead brood, the young adult bees ingest infective virus and in turn contaminate healthy larvae, when these are fed by house cleaning bees. Even deserted swarms of infected colonies are capable of infecting newly developed young larvae reared in newly constructed combs. Drifting and robbing helps in spread of disease. Exchange of brood combs from infected to healthy colonies may spread infection.

Seasonal Occurrence: The infection generally takes place in the larval alimentary canal through ingested food. In Europe and Australia, the peak of the outbreak of SBV generally occurs in spring although its occurrence in other seasons is not very uncommon. But in India, as seen in Bihar, winter months seem too favourable for the outbreak. Its continued occurrence in other seasons, especially in the beginning of summer in north-eastern states is also recorded.

Thai sac brood diseases has appeared in epidemic form and wiped out 100 per cent bee stocks in some places. Kshirsagar and Phadke (1984), Shah and Shah (1988) have claimed that the disease has a four year cycle but this contention seems to be untenable since the disease is still present after 12 to 18 years in the areas of epidemic.

Studies made in Uttar Pradesh, India revealed that the 'Thai Sac brood disease' was more during winter months indicating thereby that low temperature plays an important role in the intensity of the disease and that the disease appears quickly when the brood rearing is at its highest (Varma and Joshi, 1985). There have been several records of the severity in almost all places where its incidence has been reported without specific quantification of loss sustained by the beekeepers. However, Abrol (1993) has reported that more than 13,000 colonies have been wiped out in Jammu and Kashmir. The losses in terms of pollination of fruit and other crops may be manifold.

Abrol and Bhat (1990) have stated that in nature the TSBV multiplies in adult bees without causing obvious disease. According to them TSBV is usually associated with certain aerobic bacteria which together increase the intensity of the disease.

DISEASES OF ADULT BEES

Apis iridescent virus

Another serious disease called as clustering disease is caused by *Apis iridescent* virus called as Iridovirus. *Apis iridescent* virus disease outbroke in mid fifties and again in the seventh decennial in Indian honeybee, *A. cerana indica* colonies and reports were published from different parts of the country (Mutto, 1956; Kshirsagar *et al.*, 1975; Singh, 1979; Shah and Shah, 1979; Mishra *et al.*, 1980; Bailey *et al.*, 1979; Bailey and Ball, 1978).

Symptoms: The disease appears during hot months but Mishra *et al.* (1980) correlated the appearance of the disease with dearth. The reports of disease incidence revealed that bees suffering from iridescent viral disease had reduced egg laying and brood rearing activity, the bees become sluggish and form clusters at the hive entrance and on the inner side of the hives many of which descend and crawl on the ground. The foraging discontinues resulting in depletion of honey stores, starvation of colonies and their dwindling/death. Verma ind Joshi (1985) found that incidence of *Apis iridescent* virus was maximum from March to April and in November and lowest in the February in the hills of Utter Pradesh, India.

Diagnosis: Infected adult bees can easily be diagnosed by illuminating their tissues, especially those in the abdomen and examining them with a hand lans, or in sunlight even with the naked eye. The exposed muscles under incident light under microscope look bluish, hence the name irridescent.

Shah and Shah (1988) reported that (AIV) infects various internal organs and is probably transmitted through glandular secretions in food. Mishra *et al.* (1980) fed vitamin B complex; vitamin B complex + antibiotics and yeast as a protein in sugar syrup to overcome the infection period. Yeast gave encouraging results in less severely infected colonies and authors believed that the virus multiplying in the fat bodies depleted the stored food, but the treatment restored the status.

Miscellaneous viruses, bacteria and fungi

Honeybee species are also attacked by various viruses bacteria and fungi. The world distribution of various viruses viz; black queen cell virus (BQCV); chronic bee paralysis virus (CBPO); Kashmir bee virus (KBV); acute paralysis virsus (APV); Chinese sac brood virus (CSBV); Thai Sac brood virus (TSBV); cloudy wing virus (CWV); deformed wind virus (DWV); Filamentous virus (FV); Apis iridescent (AIV); Arkanas bee virus (ABV); Japan bee virus (JBV); Slow bee paralysis virus (SBPV); bacteria viz. *Bacillis coaqulans*; *B. coryneforme, B. anthracis, Serratia marcescens* and bifid bacteria such as *Bifidobacterium asteroids, B. indicum, B. coryneforme, B. laterosporous; Bacterium Eurydice; Streptococcus apis, S. faecalis*, fungi viz. *Aspergillus flavus* and yeast like micro organisms causing melanosis; their symptoms, bee species affected and occurrence are presented. The data clearly reveals that these miscellaneous viruses, bacteria and fungi have rather limited distribution and occur in few countries of the world. However, they cause destruction of honey bee colonies in the apiaries, wherever they occur.

PROTOZOAN DISEASES

Nosema disease of honeybees

Introduction

Nosema is a serious disease of adult honeybees. *Nosema apis*, which causes nosema disease, is found world wide. It belongs to a unique group of spore-forming organisms known as Microspora, many of which are parasites of insects. *N. apis* is the most common cause of adult bee infection. Worker bees, queen bees and drones are all susceptible to infection by spores which can remain viable for considerable periods of time. Nosema disease has been reported from several parts of India. Kshirsagar *et al.* (1974) detected Nosema disease in *Apis carana indica* colonies in Jeolikote, Uttar Pradesh. Rahman and Rahman (1974) have recorded Nosema disease in *Apis mellifera* colonies in Assam. The disease causes serious losses wherever it occurs. Moffett and Lawson (1975) found that bees infected with *N. apis* consumed less oxygen than did normal bees. The infected bees also died earlier than healthy ones. Liu (1992) found that Nosemosis causes degeneration of oocyte in the ovary of the queen, the ovariole sheath gets wrinkled and ooplasm granules break down into the small spheres. Anderson and Giacon (1992) found that Nosemosis also reduces pollination efficiency of bees as the bees mainly collected nectar rather than pollen. Topolska *et al.* (1995) reported that in Poland Nosemosis was associated with bee virus infections.

Cause

The disease is caused by the microscopic protozoan, *Nosema apis*. Protozoa exist as single cells and some, including nosema, are parasites. Many form spores as part of their life cycle.

Life cycle

When spores of *Nosema apis* are swallowed by bees they germinate within 30 minutes inside the stomach. The organism then penetrates cells of the stomach lining. It continues to grow and multiply rapidly, using the cell contents as its food supply. Large numbers of spores are produced in the host cell within 6-10 days. The parasite may also penetrate and infect adjacent healthy cells. This spreads the infection further. During the normal digestive process of adult bees, healthy cells of the stomach lining are shed into the stomach. They burst open and release digestive enzymes. Cells infected with nosema are also shed in this way and burst, releasing nosema spores. Some of these spores infect other healthy cells of the stomach lining. Many pass through the intestines and are present in the faeces (excreta) of the bee.

Incidence and spread

Infection does not normally pass directly from infected bees to the next generation of bees. Instead, young bees become infected when they ingest spores during the cleaning of contaminated combs. Contamination of combs happens when adult bees defecate within the hive during long periods of confinement caused by inclement weather. During the summer months most colonies carry a few infected bees with little or no apparent effect on the colony. However, a number of spores may also persist on combs. As the weather in autumn changes, these spores, may initiate an outbreak of nosema. Losses of bees at this time of the year may be very heavy. Winter losses can also be heavy in some years. Infected bees confined in their hives due to inclement weather may defecate inside the hive during the latter part of winter soiling the combs and hive interior with excreta and spores. This, together with spores produced in the preceding autumn, causes infection in spring. Spring outbreaks usually begin in late August or September, when temperatures begin to rise. They may last until late spring or early summer. When the warm weather comes, the disease begins to decline due to improved flight conditions. The source of infection is largely removed because the bees are able to defecate outside the hive thereby reducing the contamination of combs.

Fortunately, serious nosema outbreaks do not occur every year. Research has indicated that the following conditions appear to be associated with serious autumn outbreaks : -heavy summer rainfall - an early autumn break in the fine weather about mid- March to early April - bees working grey box (*Eucalyptus microcarpa*), red mironbark (*E. sideroxylon*) and white box (*E. albens*). The exact reasons for these apparent relationships are not known. It is suggested that the heavy summer rainfall may be linked with the production of watery nectar in autumn flowering eucalypts which in turn could be related to the beginning of nosema. In these epidemics, strong colonies may be seriously weakened before the honey bee colony overwintering period begins. In fact, strong colonies may be reduced to the size of a nucleus in a matter of days. Those hives that survive the winter require a long build-up period before the population of adults of a colony reaches normal strength. Losses caused by nosema disease are not confined to areas of Victoria having the field conditions mentioned above., Spores of *Nosema apis* rarely occur in honey or pollen. Research reports indicate that honey bee workers can transmit nosema to queens in queen mailing cages, queen banks and mating nuclei.

Effect of nosema on bees

1. Hypopharyngeal (brood food) glands of infected nurse bees lose the ability to produce royal jelly and feed honey bee brood
2. Young infected nurse bees cease brood rearing and turn to guard and foraging duties usually undertaken by older bees
3. Life expectancy of infected bees is reduced. in spring and summer, infected bees live half as long as noninfected bees
4. Infected queens cease egg-laying and die within a few weeks. The queen-less colonies dwindle away in early spring.
5. Incidence of dysentery in adult bees may be aggravated by nosema disease. However, nosema is not the prime cause of dysentery.

Effects from *Nosema apis* on *Apis mellifera*

On colony level	On individual level
Increased winter mortality	Reduced lifespan
Reduced honey yield	Faster physiological ageing
Poor spring buildup	Atrophy of hypopharayngeal glands
	Supersedure of infected queens
	Aggravates dysenterea

Symptoms

Bees infected with nosema either show no symptoms, or none that are specific for this infection. Examination of adult bees using a light microscope is the only reliable method of diagnosing the presence of this disease. Many of the so-called 'symptoms' attributed to nosema disease may apply to other diseases or conditions of adult bees.

Infected colonies can lose population, sometimes at an alarming rate. Infected bees often die away from the hive and only a few sick or dead bees may be found near the hive entrance. The term 'spring dwindle' is often used to describe this condition. However, this should not be confused with the normal weakening of colonies caused by the natural dying-off of old, over-wintered bees in early spring. Sick or crawling bees outside the hive entrance, dead bees on the ground and excreta on hive components may be associated with nosema infection but may equally be caused by other abnormal conditions.

Clinical signs

The parasite invades the posterior region of the ventriculus.

Affected bees will present:

- Unconnected wings
- Missing hair
- Mortality
- Dysentery marked by brown faecal marks in the comb

Post-mortem findings

The size of the affected gut is two times the size of normal gut and the color is white instead of being light pinkish and the ventricle is white instead of being brown.

Differential diagnosis

Diagonsis

The microsporidium *Nosema apis* (Zander) is a protozoan parasite exclusive to the epithelial cells of the ventriculus of adult bees. Infection occurs by the ingestion of the spores in the feed or after the grooming of the body hairs of the bee. The polar tube of the spore is everted and penetrates the peritrophal membrane of the intestine, particularly in the posterior region of the ventriculus. The sporoplasm passes down the tube and enters the cytoplasm of the epithelial cells, where it reproduces. Autoinfections can occur at the same time as new infections. After a short interval, spores develop in large quantities. Infected bees are unable to fly and have been shown to be infected with up to 500 million spores.

The parasite is ubiquitous and multiplies at a specific rate throughout the year, with maximum numbers occurring during spring, coinciding with the increase in the brood. In winter, spores are rarely to be found, or are only found in heavily infected bees. The development stages necessary for hibernation are difficult to find microscopically because of their very small size.

Any inherent natural defence by a bee colony against a heavy infection with the parasite depends on the colony size as well as on the prevailing weather conditions during the early part of the autumn of the previous year. If these conditions are unfavourable, the overall life expectancy of the colony is reduced.

This may lead to the premature death of bees during winter or early spring. In a typical case of a colony being depleted because of a *Nosema* infection, the queen can be observed surrounded by a few bees, confusedly attending to brood that is already sealed.

The spores have a variable viability. In faecal droppings, spores may persist for up to 2 years, and in honey or in the carcasses of dead bees, they may be viable for 1 year. Mostly, the comb is infected. Spores may be killed by heating to a temperature of at least 60°C for 15 minutes. Fumes from a solution of at least 60% acetic acid will inactivate any spores within a few hours, depending on the concentration; higher concentrations are even more effective and will kill spores within a few minutes. Disinfection can be carried out, for example, by putting acetic acid solution into bowls or on to sponges that can soak up the liquid. Following disinfection after an outbreak, all combs should be well ventilated for at least 14 days prior to use.

1. Identification of the agent

In acute forms of infection, especially in early spring, brown faecal marks will be noted on the comb. At the entrance of the hive, both sick and dead bees will be seen, and during winter there will be an increased mortality rate in the colony. In milder infections, there may be no special signs except for colonies of weak appearance with large numbers of brood, but only a few adult bees.

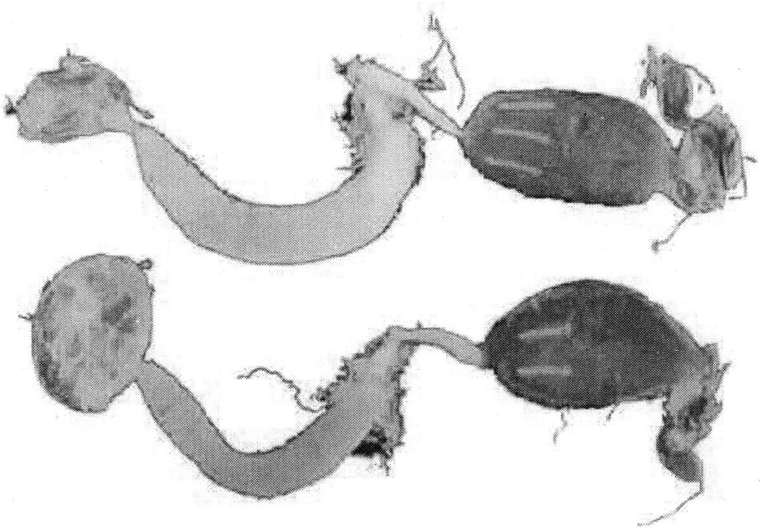

Figure 78. Nosema (comparasion of healthy and diseased honey bee gut)

To diagnose a *Nosema* infection, the posterior pair of abdominal segments are removed with a forceps to reveal the ventriculus, complete with the malpighian tubules, the small intestine and rectum. The ventriculus is normally brown but, following a *Nosema* infection, becomes white and very fragile. However, this appearance is given by other causes of intestinal disturbance, for example, feeding on indigestible food stores, such as syrup containing actively growing yeast. For a reliable diagnosis, a number of bees in a sample should be examined.

The only positive way of identifying nosema disease is through the dissection of adult bees. The hind gut and digestive tract of diseased bees are chalky white or milky white. Healthy bees, on the other hand, have amber or translucent digestive tracts. In addition, the individual circular constrictions of a healthy bee's gut are visible, whereas the gut of an infected bee may be swollen and the constrictions may not be clearly visible.

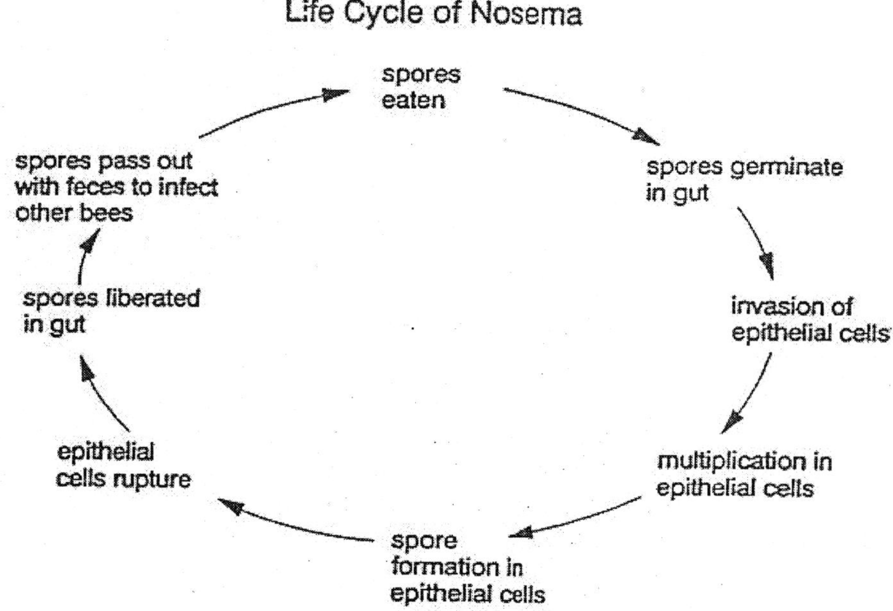

Figure 79. Life cycle of nosema

(a) Microscopy

A sample of about 20 dead bees is collected from a suspect colony. The abdomens are separated and ground up in 2-3 ml of water. Three drops of the suspension are placed on a slide under a cover-slip and examined microscopically at x400 magnification. The spores are about 5-7 μm long and 3-4 μm wide. They are completely oval with a dark edge. Their contents, consisting of nucleus, sporoplasm and polar tube, cannot be seen. Dyes are usually not necessary. *Nosema* spores must be differentiated from yeast cells, fungal spores, fat and calciferous bodies, and from *M. mellificae* cysts, which are spherical and approximately 6-7 μm in diameter In order to obtain accurate, reliable and meaningful quantification of levels of *Nosema* infections in honey bees, a standardised procedure must be used.

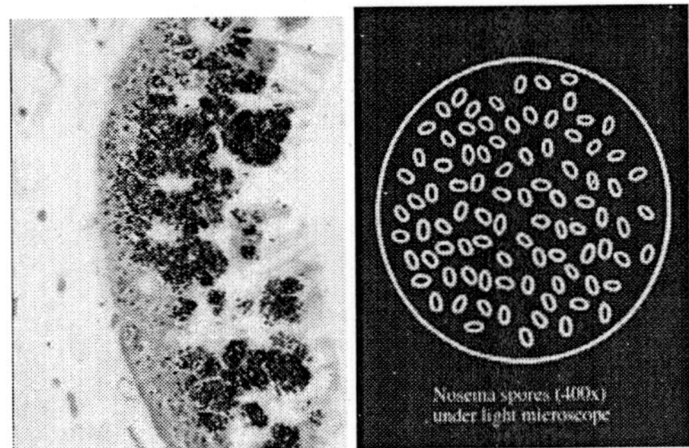

Nosema (nosema spores)

Nosema infection is most easily detected by the microscopic examination of macerated abdominal tissue for the presence of *Nosema apis* spores. The diagram shows nosema spores as they appear under the microscope. The other figure shows a section through the gut stained to highlight nosema

(b) Culture

There are no cultural methods for growing these organisms.

1. Serological tests
2. There are no serological tests available.

Control

Heat treatment

N. apis spores on combs and other hive equipment can be killed using heat treatment. The treatment involves heating the equipment at 49oC for 24 hours. This is best conducted in a room where the temperature is uniform and thermostatically controlled. Hot spots should be avoided, as higher temperatures may melt combs or cause them to sag. Ethylene Oxide (ETO) fumigation also eliminates Nosema spores when combined with the heat treatment as will acetic acid and heat. Treating equipment followed by treatment of bees with fumagillin is the best method to reduce Nosema losses and promote colony health.

Fumigation

Fumigation with acetic acid is effective, especially when the bees are transferred as early as possible in the season from contaminated equipment to fumigated equipment. An efficient method is to intersperse absorbent materials between piles of hive bodies containing the combs. Pour 150 ml of acetic acid (80% strength) onto the material between each box. The stacks should be left outside in a warm corner and protected from direct winds for about one week. It is also recommended that the material be aired for one day prior to use. Fumigation with ethylene oxide (ETO) has also been demonstrated to kill spores on combs (100 mg ETO/l for 24 hours at 37.8°C).

Chemotherapy

Fumagillin (Fumidil-B) is the only drug that has been found to be effective against *N. apis*. Fumagillin inhibits DNA replication of the microsporidian without affecting the DNA of the host cell. The activity of the fumagillin remains high in honey kept at 4°C for several years and for at least 30 days at 30°C It should be fed only in a sugar syrup with proper mixing and feeding carefully. The recommended feedings is 100 mg. fumagillin (about 1 teaspoon) to 1 gallon of sugar syrup (mix 2 parts sugar to 1 part water). Thymol (3-Hydroxy-*p*cymene) a constituent of the essential oil derived from thyme and many other plant species is effective in suppressing *Nosema* infection as an additive to dietary supplements used in commercial beekeeping.

Management practices

1. Protein deficiency is possibly the major cause of increased nosema levels. Colonies working late autumn and early winter flows are prone to developing high nosema levels when breeding becomes minimal as a result of protein deficiency. Putting bees on flora which provide high-protein pollens, before and directly after working a honey flow with low-protein pollens, will overcome any general protein deficiency
2. Maintain colonies with queens with good egg-laying potential. Colonies prepared for winter should have good populations of young bees ensure colonies have adequate supplies of pollen in autumn. This will also help to ensure good populations of young bees
3. Place the hives in a sunny position in the cooler months of the year. Choose apiary sites that have good air drainage and protection from cold winds. Avoid cool shady and damp sites. it has been found that the level of nosema infection in a colony can be reduced from about 85% to zero by placing the hive in a 'sun trap' where it obtains maximum sun and maximum shelter from cold winds maintain winter colonies in a minimum of hive space so they are compact and warm.
4. Remove supers of combs not required by the bees avoid colony stress which can be caused by excessive opening the hive and manipulation of combs, feeding and relocating colonies
5. Replacement of old, dark brood combs may help to lower the number of spores within the hive, although it will never totally eliminate the disease. Many beekeepers remove two or more old combs from the brood nest each spring, replacing them with sheets of foundation beeswax sheets available from apiary supply shops.

Warning

Acetic acid fumes will corrode the frame wire and nails in hive components. About five fumigations are possible before frame wires are completely corroded. Acetic acid is highly corrosive and contact with the skin should be avoided. Wash the area thoroughly if acid is spilt onto your person or any area where it is not required, such as concrete

CONCLUSION

Sound management practices will help reduce losses caused by nosema. Good management practices such as appropriate nutrition, young queens with populous hives, new comb rotation and placing hives in a warm sunny position over the autumn, winter and early spring all contribute to minimising nosema losses.

Nosema ceranae

Class: Dihaplophasea

Order: Dissociodihaplophasida

Suborder: Nosematiodea

Family: Nosimatidae

Genus: Nosema

Species: Nosema ceranae (Fries *et al.*, 1996)

Nosema ceranae is a microsporidian, a small, unicellular parasite that mainly affects *Apis cerana*, or the Asiatic honey bee. It causes nosemosis, also called nosema, see *Nosema apis* which is the most widespread of the adult honey bee diseases. The dormant stage of nosema is a long lived spore which is resistant to temperature extremes and dehydration.

History

Because of their economic importance, European honey bees (*Apis mellifera*) are found in almost all areas inhabited by humans (Goulson 2003). However, many factors may reduce the strength of honey bee colonies, including bee diseases (Ellis and Munn 2005).

Nosema is considered one of the most prevalent and economically damaging of honey bee diseases. Yet it often goes unnoticed because the causative agent, a microsporidium, is microscopic in size and therefore invisible to the naked eye, and because the disease rarely leads to the death of a diseased colony. At the beginning of the 20th Century (1909), the great German bee scientist Zander first described *Nosema apis* as 'the microsporidium responsible for Nosema disease'. Subsequently, all reports of microsporidia in honey bees, in both the western hive bee *Apis mellifera* and the eastern hive bee *Apis cerana*, were attributed to *Nosema apis*.

Disturbing developments

In 1995, Professor Ingemar Fries of the Swedish Agricultural University, Uppsala described a new microsporidium, *Nosema ceranae*, in indigenous honey bees *Apis cerana* (Fries *et al.,* 1996). *N. ceranae* and *N. apis* have similar life cycles, but they differ in spore morphology with *N. ceranae* having smaller spores. Bees die within 8 days after exposure to *N. ceranae* which is faster than bees exposed to *N. apis*. The foraging force seems to be affected the most. They leave the colony and are too weak to return, therefore die in the field. This leaves behind a small cluster and a weak colony. 'Colony collapse Syndrome 'in honeybee colonies seems to be a serious threat and possible cause could be *Nosema ceranae.* Symptoms of the problem are seen when strong colonies lose their workforce in a matter of weeks even though the hive has food stores, capped brood, and no outward appearance of a serious disease or parasitic condition. No build-up of dead bees occurs in front or inside the hive. Food stores are not immediately robbed by hives in the vicinity and attacks from wax moths and small hive beetles are noticeably delayed. The adult bees remaining in the small cluster are newly emerged and are reluctant to consume feed.

Professor Fries subsequently demonstrated in experiments that *Nosema ceranae* is infective for the western honey bee, *Apis mellifera* than *Apis cerana*. Evidently, it might be anticipated that the western honey bee would acquire *Nosema ceranae* if kept in Asia, where *Apis cerana* and its parasite *Nosema ceranae* are endemics. *Nosema ceranae* have shifted hosts from *A. cerana* to *A. mellifera.*

Symptoms

The symptoms are different from *N. apis*, where they do not seem to defecate inside the hive. Nonetheless, honey production is very low and colonies can collapse very readily. While the parasite is still susceptible to antibiotics, colonies seem to collapse before feeding fumagillin can rescue them. Contrary to the classic insidious form of Nosemosis:

1. Crawling and losses occurred during the whole year.
2. During winter colonies died within a very short time scale
3. Contrary to typical *Varroa* infestation damage, hives full of dead bees were observed.
4. In winter colonies in many apiaries were undertaking relatively strong cleansing flights, even at temperatures as low as 4°C.

Control

Management and control practices same as in case of *Nosema apis.* There is little advice on treatment but it has been suggested that he most effective control of Nosema ceranae is the antibiotic fumagillin as recommended for *Nosema apis.* Recently, beekeepers have questioned the efficacy of Fumagilin-B® and in some cases suggest treatment may lower colony strength. Reasons for reduced effectiveness include possible development of resistance in *Nosema* and/or poor application practices by beekeepers that allow this parasite to proliferate during winter.

Amoeba Disease

The problem of amoebae in honeybees was first detected by Maassen (1916) in Germany and Morgenthaler (1920) in Switzerland. Prell (1926) described and classified the protozoan from honeybees in Germany, naming it *Malpigha-moeba mellificae.* Amoebae have been discovered in Czechoslovakia, United states, England and wales, Soviet union, Yugoslavia and Venezeula.

Amoebae disease is caused by a protozoan (*Malpighamoeba mellificae*) that infects the Malpighian tubules of adult bees. It develops there first as an amoeba like individual and ultimately encysts. The epithelium of infected malphihian tubules may atrophy.

Multiplication

The cysts ingested by the adult bees presumably germinate within the intestine, possibly at the posterior end of the ventriculus where solid food particles accumulate. The infective amoebae probably enter directly into the malphigian tubules, which discharge into the posterior end of the ventriculus and apply themselves to the tubules of the epithelium. When it excysts the amoeba has a flagellated form that makes its way into the tubule where it changes into the trophic amoeba.

Spread

In temperate climates there is a sharp peak of infection around May in the northern hemisphere followed by an an abrupt decline, with infection becoming almost undectable after midsummer. Natural transmission of infection to the winter bees is is almost certainly by the remains of faecal contamination deposited on combs during the preceeding late winter and spring. *M. mellificae* is more frequently than can be accounted for by chance. The two parasites are independent, since they often occur alone, either in individual bees or even in colonies but they became associated because they are transmitted in the same way. *M. mellificae* forms only about 500000 cysts per bee and these taken about three weeks to develop, whereas *N. apis* forms upto 30 million spores per bee in about half the time. Therefore, *M. mellificae* spreads less easily than *N. apis*, usually only by severe dysentery. Accordingly, it is usually associated with the most severe infections by *N. apis* and with unusual mortality of colonies, but it is not a prime cause of such losses.

Since this protozoan is found in the Malpighian tubules of adult bees, diagnosis can be made only by removing and microscopically examining the tubules for the presence of amoeba cysts. The cysts measure 5–8 m in diameter. Malpighian tubules are long, threadlike projections originating at the junction of the midgut and the hindgut. They can be teased away from the digestive tract with a pair of fine tweezers. Once this is done, place them in a drop of water on a microscope slide and position a cover glass over them, applying uniform pressure to obtain a flat surface for microscopic examination. *M. mellificae* can be discerned using a high, dry objective and then changing to the oil immersion objective for more detail. Infection often occurs as a co-infection with Nosema, but may persist individually. It is rarely detected or identified as the sole cause of bee mortality. Amoebic infections may disrupt the normal water-balance physiology of infected bees. Due to the nature of infection, diagnosis can be only made by the removal and dissection of the tubules under a microscope. Tubule examination of infected individual will show the presence of amoebae cysts, which measure 5-8 μm in diameter.

Malphigamoeba mellificae has been reported from most European countries, Russia, New Zealand, USA and South America, England and Wales, Italy and Scotland.

Potential for Economic Loss: Slight to moderate. Co-infections with Nosema are likely to elevate the risk for economic loss of bees. The fact that amoebae rarely kill a colony, but causes a weekend condition reffered to as Spring dwindling or disappearing disease, suggests that amoeba disease like nosema causes economic losses.

Control Practices: Control of amoeba disease is currently based on hygiene and decontamination of equipment. No chemical control is recommended specifically for amoeba. Treatment of Nosema with fumagillin may help alleviate the compound effects of co-infections.

Other Protozoa

Gregarines

Four gregarines (protozoans of the order Gregarinida) are associated with honey bees: *Monoica apis, Apigregarina stammeri, Acuta rousseaui* and *Leidyana apis*. The immature stages, or cephalonts, average about 16-44 μm. Cephalonts are oval and consist of two distinct body segments; the posterior segment is larger. The mature stages, or sporonts, average about 35-85 μm and have a reduced anterior segment. Gregarines are found attached to the epithelium of the midgut of adult honey bees. To view gregarines, gently remove the midgut from the digestive tract of a suspect bee and place it on a microscope slide in a drop of water. The midgut can be separated from the digestive tract at the point of attachment with the proventriculus (honey stomach) and hindgut, using fine tweezers and a scalpel. Gently break open the midgut with fine tweezers and a probe, and place a cover glass over the resulting suspension. Gregarines can be seen using the low-power objective of a compound microscope.

Gregarines have been found in bees in Switzerland, Italy, Canada and South America, Lousiana but their incidence was very low. These organisms are not associated with any economically important diseases of the honey bees, and as such, there are no recommendations for their control. However, it is important to recognize their potential presence in honey bee tissues, especially when bees that have died from unknown causes are dissected.

Flagellates

The flagellates associated with honey bees are *Crithidia* (= *Leptomonas*) species. Flagellates have been found either free in the lumen or attached to the epithelium of the hindgut and rectum of adult honey bees (Fyg, 1954). Infection causes a dark spot or crust, easily visible on the intestinal wall on which the flagellates are attached to form a furry coating. Rounded forms of these protozoa occur and may be cyst stages. Queens are less commonly infected as compared to workers. Flagellates apparently of the same type, also occur in the rectum,either free in the lumen or attached to the epithelium.

The flagellates do not occur in bees less than 6 days old. In newly infected bees they move freely in the gut lumen, latter they clump into rosettes and adhere to the intestinal wall where the dark crusts appear in bees more than 16 days old. They are scare in winter bees. Flagellates vary in size from 5 to 30 m. Some appear as pearlike bodies with flagella; others are long threadlike forms or are round without flagellae (Lotmar 1946). To look for flagellates, remove the digestive tract of a suspect bee as described and place it in a drop of water on a microscope slide. Then, using fine tweezers and a scalpel, separate the hindgut and rectum at the point of attachment with the midgut. Macerate the hindgut and rectum, using the tweezers and a probe. Place a cover glass on the resulting suspension and observe under the high, dry objective of the microscope.

There is no evidence that the flagellates are pathogenic. They have been found commonly in Europe, scandinivia, Australia, where they were named *Crithidia mellificae*. They may multiply on artificial media and may not be bspecific to honeybees.

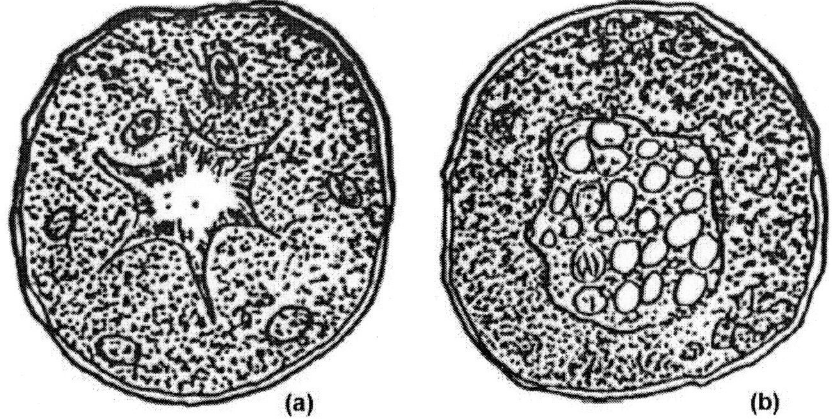

Figure 80. Cross sections of Malpighian tubules. (a) healthy tubule. (b) tubule containing cysts of *Malpighamoeba mellificae*.

Powdery Scale

Paenibacillus larvae subsp. *pulvifaciens* (=*Bacillus pulvifaciens*) (see appendix E) is the bacterium suspected of causing powdery scale disease. This disease is seldom reported because the incidence is low or, perhaps, because the average beekeeper is unable to identify it. A useful diagnostic characteristic is the scale that results from the dead larva. The scale is light brown to yellow and extends from the base to the top of the cell. The scale is powdery; when handled it crumbles into a dust. *P. pulvifaciens* vegetative cells measure 0.3–0.6-1.5–3.0 m. The spores are 1.0-1.3–1.5 m. The bacterium can be isolated on nutrient agar, but growth is more luxuriant on glucose agar. When first isolated, the organism produces a reddish-brown pigment that can be lost by subculturing. *P. pulvifaciens* closely resembles *P. larvae*, but the spores do not exhibit Brownian movement in the modified hanging drop technique. Also, *P. pulvifaciens* is distinguished by its ability to grow at 20°C on nutrient agar.

Septicemia

Pseudomonas aeruginosa (= *Pseudomonas apiseptica*) is the bacterium that causes septicemia in adult honey bees. This disease results in the destruction of connective tissues of the thorax, legs, wings, and antennae. Consequently, the affected bees fall apart when handled. Dead or dying bees may have a putrid odor. *P. aeruginosa* rods measure 0.5–0.8; 1.5–3.0 m. They are gram-negative and occur singly, in pairs, or in short chains. A bacterial smear and Gram stain (Appendix A) can easily be prepared after removing a wing from the thorax and dipping the wing base in a drop of water on a microscope slide. To isolate this organism, streak the base of a wing across Difco *Pseudomonas* isolation agar or *Pseudomonas agar* F. The optimum temperature for growth is 37°C. *P. aeruginosa* in culture is characterized by the excretion of diffusible yellow-green pigments that fluoresce in ultraviolet light (wavelength below 260 nanometers).

Septicemia can also be diagnosed by reproducing the disease symptoms in healthy, caged bees. This is accomplished by preparing a water extract (macerate the equivalent of one suspect bee per ml of water) and inoculating healthy bees in the thorax or dipping them in the water extract. Bees with septicemia die within 24 hours; they exhibit the typical odor and "break apart" symptom after about 48 hours.

Spiroplasmosis

Spiroplasma species is the bacterium that causes spiroplasmosis in adult honey bees. *Spiroplasma* is a helical, motile, cell-wall-free prokaryote that is found in the hemolymph of infected adult honey bees. The organism is a tiny, coiled, and sometimes branched filament 0.7–1.2 μm in diameter. Its length increases with age and ranges from 2 m to more than 10 μm (Clark, 1977, 1978a). *Spiroplasma* can best be seen in the hemolymph, using dark-field microscopy. They can also be seen by using the oil-immersion objective of a phase-contrast microscope. Hemolymph can be taken from adult bees by puncturing the intersegmental membrane directly behind the first coxae, using a fine capillary tube made from the tip of a Pasteur pipet. This organism can be cultured in standard mycoplasma broth medium (GIBCO) and in Singh's mosquito tissue culture medium with 20% fetal calf serum.

Reclassification of *Bacillus* spp. associated with Honey Bees

Bacillus spp. associated with honey bees have been reclassified as follows:

Bacillus alvei = *Paenibacillus alvei*

Bacillus apiarius = *Paenibacillus apiarius*

Bacillus laterosporus = *Brevibacillus laterosporus*

Bacillus larvae = *Paenibacillus larvae* subsp. *larvae*

Bacillus pulvifaciens = *Paenibacillus larvae* subsp. *pulvifaciens*

Plant Poisioning

The nectar and/or pollen of certain plants may contain compounds that are toxic to honey bees. Poisonous plants are only a problem in specific areas during the plant species' blooming period if the toxin(s) are nectar-borne. However, if the pollen contains the toxic compounds, the duration of poisoning may linger as long as the culprit pollen remains in the combs. There are few tell tale signs to distinguish between plant poisoning and pesticide poisoning, although pesticide poisoning tends to be more devastating and shorter in duration than plant poisoning. Areas with abundant poisonous plants should be avoided by beekeepers when the plants are in bloom. Signs of possible plant poisoning are dead adult or larval bees, hairlessness, queen supersedure, trembling, failing queens, discoloration, and larval mummification. Examples of poisonous plants affecting honey bees include *Cyrilla racemiflora* (southern leatherwood), *Aesculus californica* (California buckeye), *Asciepias* spp. (milkweed pollinia), *Gelsemium sempervirens* (Yellow jessamine), *Astragalus* spp. (Loco plants), and *Veratrum californicum* (False hellebore). Few, if any, cases of plant poisonings have been reported in New Jersey.

Non infectious disorders

Colony abnormalities : There are several abnormal colony conditions that are not caused by disease organisms but by environmental, management or biological factors. Some of these conditions can be as serious as a disease in their effect on colony health and honey production. The beekeeper should be familiar with both diseases and other abnormalities, in order to distinguish between the two and be able to correct the problem.

Overheated Bees

Bees can overheat during hot weather when they are confined in their hives without adequate ventilation or access to water, as can happen during the transportation of beehives between locations. Possible signs of overheating are a large accumulation of dead bees on the bottom board of a flight-restricted hive and/or worker bees crawling rapidly while fanning their wings. Colonies confined during transportation should have the solid hive cover replaced with a screen cover to allow for adequate ventilation. When protecting colonies for short periods from overhead pesticide sprays, thoroughly wetting the burlap cover may reduce overheating stress.

Genetic Lethality

Genetic abnormalities during brood development can also kill bees, usually without exhibiting symptoms of known diseases. However, drone brood from laying workers and failing queens often die with symptoms similar to European Foulbrood but in the absence of known pathogens. Inbreeding of queen stock may also result in the laying of non-viable bee eggs; these eggs are usually consumed by worker bees.

Chilled Brood

If larvae are underfed, or if adult bees cannot adequately maintain the hive temperature, some of the brood may become chilled and die. Brood killed by chilling turns grey and may resemble other conditions, such as sacbrood. Chilled brood will be removed from the colony by worker bees once colony homeostasis has returned. Working bees during cold or cool weather is one way beekeepers can cause chilled brood themselves. Providing adequate food reserves and working bees only during warm weather (above 60°F) are the best techniques to avoid the occurrence of chilled brood.

Mouldy Pollen

Stored pollen in comb is preserved when covered by honey. If the honey is removed and bees not present, the pollen will become mouldy. No disease is involved and the frame can be placed back into a strong colony for cleaning and use.

Starved Brood

Under periods of prolonged nectar and/or pollen dearth, larvae may be removed and/or consumed by adult bees. However, if there is a sudden and substantial reduction in the adult bee population, there may not be enough workers available to feed the larvae. In these cases, larvae may simply starve. Larvae crawling from their cells in search of food are a striking feature of starved brood. Larvae are most often the stage affected, however emerging bees may also starve if they were stressed as pupae and are not fed soon after chewing through their cell caps. These bees may die with only their heads protruding from the cell and have often their tongues extended. Starved brood may occur sporadically in particularly when rapid and massive bee kills resulting from pesticide use occur.

Gassed Brood

When colonies are killed off with hydrogen cyanide gas the uncapped brood will die, either from the cyanide fumes or from subsequent chilling or starvation. This dead brood is usually noticed when the beekeeper is sorting combs for brood chambers. It becomes dark and flattened, and will often superficially resemble larvae that have died from American foulbrood. When there is doubt, samples may be submitted to the provincial or state apiculture office for diagnosis. A small amount of gassed brood does not pose a problem to package bees, which will clean out any dead larvae in the combs. Gererally, however, the first brood chamber should not contain combs that are full of dead brood, as this will delay egg-laying and colony build-up. Such combs can be placed in the second brood chamber when the colony is stronger.

Scattered or Spotty Brood Pattern

A scattered or spotty brood pattern is often evidence of a failing queen. Instead of a solid pattern of eggs, young larvae or capped brood, many cells are empty. Spotty brood may also indicate the presence of a brood disease such as American foulbrood, European foulbrood, sacbrood or chalkbrood.

A spotty brood pattern is also very common in wintered colonies in early spring but is not necessary indication of a queen problem and it may result from a shortage of pollen. This problem generally corrects itself when pollen becomes available.

When a spotty brood pattern is observed, the bee keeper should examine the colony to check for other disease symptoms or signs of superseder. If suspenders cells are found, the problem is likely due to failing queen and she should be destroyed. The colony can then be requeened after all remaining queen cells have been removed and 3-5 days have elapsed. Alternatively, the remaining workers can be united with another colony.

Drone-Laying Queen

Normally a queen will only lay unfertilized eggs in drone cells. However, when she uses up the the sperm that have been stored in her spermatheca she will begin to regularly lay unfertilized eggs, even in worker cells. Sings of a drone -laying queen include a spotty brood pattern and the presence of protruding, rounded cappings on worker brood combs. Both are evidence of unfertilized eggs being laid in worker cells.

If the colony is still strong, the beekeeper may requeen it after removing the old queen and any queen cells. If the colony has dwindled in size, the bees may be united with another colony after removing the old queen. Otherwise the bees may be shaken on the ground and allowed to drift to other hives in the apiary after their own queen has been removed.

Multiple Eggs Per cell

In early spring when the space available for egg laying is restricted, a queen may lay several uniformly placed eggs in the bottom of a single cell. A normal worker will develop in a cell that originally contained several queen-laid eggs so no management action needs to be taken to correct this situation. This problem typically population expands and space available for egg laying increases. In contrast, a colony that has been queenless for several weeks, workers will lay eggs in large numbers. These eggs are not fertilized and will develop into drones. Signs of laying workers include a spotty pattern, drone brood in worker cells and multiple eggs per cell, laid on the sides rather than in the bottoms of cells.

It is difficult to find egg-laying workers because they are similar in appearance to other workers. Requeening a colony that has laying workers is not usually successful. The bees should be shaken onto the ground in the apiary and their hive removed. The bees will gradually drift to other colonies. The development of laying workers in a queenless colony can be delayed by adding comb with unsealed brood that has been removed from a healthy colony.

Queenlessness

Sign of a queenlessness colony include lack of eggs, larvae or all brood, and in most cases the presence of emergency queen cells on comb faces. In addition, the colony emits a perculiar, loud buzzing sound, and the workers run on the combs, often with their wings spread.

Depending on the time of year, a queenless colony may be requeened. This is best done with either a mated queen or a queen cell, but if neither is available, the emergency cells can be allowed to develop and hatch. If the colony has dwindled to a nonproductive size then it should be united with a queenright colony.

Dysentery

Honey bees normally defecate in flight. If they are confined over a long period, such as cold winter, or are feeding on poor quality food, the accumulation of indigestible matter in the hive occurs. Dysentery can cause premature death of honey bees, leading to weakening or death of colonies. Signs of a dysentery problem are fecal spotting on top bars, combs and entrances, especially in the early sping. Dysentery problems can be prevented by providing good quality feed in autumn. Only syrup made from refined white sugars should be fed. Burnt honey, honeydew, molasses and brown sugar should not be used. In addition, feeding pollen supplement before the bees are flying the spring or the fall may do more harm than good if the bees are subsequently confined to the hive for extended periods.

Pollen and Honey Shortages

Shortage of honey and pollen can occur at any time during the year. Proper management should prevent the occurrence of a food shortage, since shortage of pollen and honey will cause curtailment of brood rearing and may casue colony death. Honey shortages commonly occur in late winter and during nectar dearth periods throughout the beekeeping season. Signs of such shortage include the lack of stored honey in the hive. Symptoms of a starving colony include cessation of brood-rearing, slow-moving and trembling adult bees and dying and dead bees found head first in cells.

Colonies should have at least four to six frames of honey at all times during the beekeeping season, and far more at the beginning of the winter, to avoid slowing colony build-up. Honey stores may be moved with in the hive during early spring, or colonies may be fed with syrup when conditions are unfavourable. A starving colony should be fed immediately with sugar syrup. Pollen shortages may occur at any time during brood rearing. Occasional temporary shortages may occur when colonies are confined to their hives by bad weather during spring and early summer. However, a longer dearth of pollen can have serious effects on the colony population, especially when brood-rearing is at its peak and pollen is used as quickly as it comes in. At this time, a cessation of the pollen flow will cause brood starvation.

Signs of pollen shortage include a floral dearth period, lack of incoming pollen and combs devoid of stored pollen. Symptoms in the colony include: the presence of eggs but no uncapped brood; the presence of young larvae with no food provisions; a complete lack of brood; the presence of larvae and pupae on the bottom board and on the ground in front of the hive; and the presence of pupae that have been partly cannibalized. Pollen shortages may be corrected by feeding the colony with a high quality pollen supplement preferably before the shortage affects the colony. The beekeepers must become familiar with pollen and nectar plants and their blooming sequence and duration, in order to know when to expect dearth periods and thus be prepared for them.

Bald brood

The developing pupae are usually sealed in their cells under wax cappings 8-9 days after laying. Bald brood may be seen as small patches of normally developing larvae with uncapped or partially capped cells. These uncapped larvae will usually emerge as fully developed adults, although a few malformed adults may result from contaminants becoming deposited on the developing larvae. The most usual cause of bald brood is wax moth larvae (both the lesser (*Achroia grisella*) and greater (*Galleria mellonella*)) tunnelling below the surface of the comb. The larvae will perforate the cappings, which are then chewed down by the worker bees; sometimes these partial cappings have a raised lip protruding from the comb surface. Another cause of bald brood is genetic, where the worker bees do not cap the cells properly, either turning the cell edges inward or leaving a small hole in the centre of the capping.

Strong colonies of bees will reduce the effects of wax moth, and in the case of the genetic form of bald brood re-queening of the colony will usually resolve the problem.

Drone brood in worker cells

The larger domed cappings of drone brood can normally be seen throughout the height of the season, usually at the edges of the brood nest. However, there are disorders that sometimes cause drone brood to be reared in worker cells. This brood is very irregular, with extended cappings drawn out from worker cells to accommodate the larger drone larvae. The brood pattern will be poor with larvae of all stages of development throughout the comb. The surface of the comb may appear very uneven. There are two possible causes – a drone laying queen or laying workers.

Drone-laying queen (failing queen)

Queens lay two types of eggs, those that are fertilised and develop into worker bees, and unfertilised ones that develop into drones. The eggs are fertilised as the queen lays them, however, if the supply of sperm runs out or the queen is poorly mated or not mated for some reason, then only unfertilised eggs will be laid and these will develop into drones. It is usually older queens that become drone layers but it may also be apparent in younger queens that did not mate successfully. The best option in this instance is to re-queen with a young prolific, recently mated queen.

Laying workers

Under normal colony conditions worker bees do not develop functional ovaries, as their development is suppressed by pheromones produced by the queen. However, if a colony becomes queenless and the bees have no eggs or young larvae to rear a new queen from, workers will develop functional ovaries and start to lay unfertilised eggs, thus, they will develop into drones. The signs are very similar to those seen with a failing queen but there may also be multiple eggs per cell, often on the cell walls. Unlike colonies with a failing or defective queen, those with laying workers are very difficult to re-queen. The best course of action is usually to unite the colony with a stronger colony.

Half-Moon Disorder

Half-Moon Disorder (HMD) is a common non-infectious genetic/nutritional (pollen?) problem, the symptoms of which mimic EXACTLY those of EFB. HMD occurs everywhere. To a lack of adequate nutrition of the young queen prior to mating. This nutrition is directly related to the number of correct aged nurse bees in the mating nuc/hive. This causes faulty development of the queen's ovaries leading to larvae that are rejected by the feeding nurse bees whereupon they die of starvation. Larvae with EFB starve due to competition from bacteria in their gut. In HMD and EFB, the young larvae die at the same age from starvation. This results in Identical visual symptoms. Around the World it is almost certain that HMD is routinely diagnosed as EFB and hence no further explanation is ever sought. Only in New Zealand with its absence of EFB did the HMD mystery attract enough attention to enable it to be explained.

Neglected Drone Brood

This is not a disease but a condition, which can be confused with EFB during the discoloured larvae stage or AFB at the scaling stage. The cause is a drone laying queen or laying workers. Drone brood is raised on worker cells resulting in stunted and malformed drones. The colony is usually small and will have dwindled, the bees eventually neglect the drone brood in worker cells, which then die of starvation before sealing. They decompose and become yellow to brown. The decomposing larva becomes a brown watery mass (which does not rope) and eventually dries to a scale which can be removed by the bees.

Purple brood

Purple brood occurs when adult bees collect and use the pollen and nectar from *Cyrilla racemiflora* (titi, southern leatherwood). This "disease" is characterized by the blue or purple color of the affected larvae.

Paralysis

Aesculus californica (California buckeye) is probably the best known of the poisonous plants in the United States. Field bees exhibit symptoms similar to those of chronic bee paralysis; specifically, the bees are black and shiny from loss of hair and they tremble. Also, either the eggs do not hatch or the larvae die soon after hatching.

Honeybee mites

Honeybees are parasitized by many different organisms, and among some of the most serious pest are the parasitic mites. Mites are of particular importance to bees and beekeepers because of their potential to do great damage to both feral and managed honeybee colonies. The mites (Acari) that parasitize honey bees have become a global problem. They are threatening the survival of managed and feral honey bees, the beekeeping industry and, due to the role of bees in pollination, the future of many agricultural crops. Morethan 86 mite species have been recorded in association with *Apis* and their nests, but most are neutral or benign in nature. "Hive" mites fall into three ecological categories:

1. Parasitic.
2. Phoretic.
3. Stored product mites.

The five most economically important parasitic mites of honey bees are: *Varroa destructor*, *Varroa Jacobsoni*, *Euvarroa sinhai*, *Tropilaelaps clareae*, *Tropilaelaps koenigerum*, and *Acarapis woodi* (the tracheal mite) are the main pests, but about 100 mostly harmless mite species are associated with honey bees. Bee mites have greater dispersal potential than most other Acari, first through their hosts and then by humans who move bees primarily for commerce and pollination.

Table 66. Mites parasitizing bees, associated with the particular bee species.

Apis species	Mites	Source
andreniformis	*Euvarroa sinhai*	Delfinado-Baker *et al.*, 1989.
	Euvarroa wongsirii	Lekprayoon and Tangkanasing, 1991.
cerana	*Tropilaelaps clareae*	Delfinado-Baker *et al.*, 1989.
	Varroa jacobsoni	
	Varroa underwoodi	
dorsata	*Tropilaelaps clareae*	Delfinado-Baker *et al.* 1985.
	T. koenigerum	
florea	*Euvarroa sinhai*	Delfinado-Baker *et al.*. 1989.
	Tropilaelaps clareae	
koschevnikovi	*Varroa rindereri*	De Guzman and Delfinado-Baker. 1996
	Varroa jacobsoni	
laboriosa	*Tropilaelaps clareae*	Delfinado-Baker *et al.*, 1989.
	T. koenigerum	
mellifera	*Euvarroa sinhai*	Koeniger *et al.*, 1993.
	Tropilaelaps clareae	Delfinado-Baker *et al.*, 1989.
	Varroa jacobsoni	Anderson and Trueman, 2000
	Varroa destructor	
nigrocincta	*Varroa underwoodi*	Anderson *et al.*, 1997.
nuluensis	*Varroa jacobsoni*	Delfinado-Baker *et al.*, 1989.
	Varroa underwoodi	De Guzman and Delfinado-Baker, 1996.

Table 67. Mite pest of honeybees, hosts infestation and distribution

Mite	Host *Apis*	Colony/comb infestation	Distribution
Parasitic			
Varroa jacobsoni	*Mellifera,*	Worker and drone brood, queen cells during heavy infeatstion	Asia, Europe, Africa, S-America
	Cerana	Drone brood	Asia
Varroa destructor	*Mellifera,*	Worker and drone brood, queen cells during heavy infeatstion	Throughout the world except Australia
	Cerana	Drone brood	Throughout the world except Australia
Euvarroa sinhai	*florea*	Drone brood	Asia
Tropilaelaps clarea	*Mellifera,*	Worker and drone brood	Asia
	Cerana	Worker brood	Asia
	dorsata	Brood cells (No specific record)	Asia

T. koenigerum	*Dorsata*	Worker, brood and queen cells	Asia
	Laboriosa	Worker, brood and queen cells	Asia
	Mellifera	Worker, brood and queen cells	Asia
	Cerana	Worker, brood and queen cells	Asia
Acarapis woodi	*Mellifera*	*Trachea of adult bees*	Eurpe, S-America, Africa, Mexico, Asia
	Cerana	*Trachea of adult bees*	Asia
Acarapis externus	*Mellifera*	Neck region of adult bees	Eurpe, N-America and Pacific
Acarapis dorsalis	*Mellifera*	Thoracic groove of adult bees	Eurpe, N-America and Pacific
Non parasitic			
Neocypholaelapsi indica	*Mellifera*	Hive bottom debris, combs, on adult bees eucalyptus flowers	Asia
	Cerana	Combs and on bees	Asia
	Dorsata	Combs and on bees	Asia
N. apicola	*Cerana*	Worker brood cells	Asia

Parasitic bee mites

The parasitic mites suck the blood of honeybees. The numbers of parasitic mites are few but some cause serious diseaseas of honeybees. The attack by parasitic mites not only causes direct damage of honeybee colonies but indirectly increase the susceptibity to pathogens and viruses and they are

1. responsible for direct transmission of bacteria specific to mites in the act of sucking
2. secondary infection of bacteria/viruses entering through bite hole
3. Lowering of resistance and hence increasing the susceptibility to diseases causing organisms, pests and pathogens.

Three group of parasitic bee mites are of economic importance due to their destruction of honey bee colonies worldwide. They are the tracheal, *Varroa*, and *Tropilaelaps* mites. The tracheal mite sucks the heamolymph from the trachea of the bees as a result of which oxygen supply is reduced, bees become weak and die. The parasitic varroa mites, *Varroa jacobsoni* (Oudemans), is another serious pest of the honeybee (*Apis mellifera*). Varroa mites develop and reproduce in the sealed brood cells of honeybees where they feed on the hemolymph of bee pupae. It is an obligate parasite of bees and feeds externally on the immature stages of workers and the male reproductives (drones). Parasitized individuals may die or develop into weakened, crippled adults that are incapable of functioning normally. The impact of *Varroa* on agriculture has been significant. The number of managed colonies as well as feral colonies have been drastically reduced so they are no more effective as honey producers or pollinators. *Varroa* mites seriously reduce the numbers of drones therby, resulting in breeding depression in honeybees and many queens may remain un insenminated. This mite, *Tropilaelaps clareae*, feed only on bee brood. The mouthparts are stubby, with an apically bidentate fixed upper digit, and a longer, unidentate and pointed moveable digit. This piercing-grasping structure, is more suitable to piercing soft brood tissue, rather than the tearing-sawing type of *Varroa*. This implies that *Tropilaelaps* can feed only on soft tissues, such as honey bee brood. Symptoms irregular brood pattern, dead or malformed wingless bees at the hive's entrance, and the presence of fast-running, brownish mites on the combs, are diagnostic for *T. clareae*. Like *Varroa, Tropilealaps* is also considered as the primary cause for repeated failure of beekeeping in Asia.

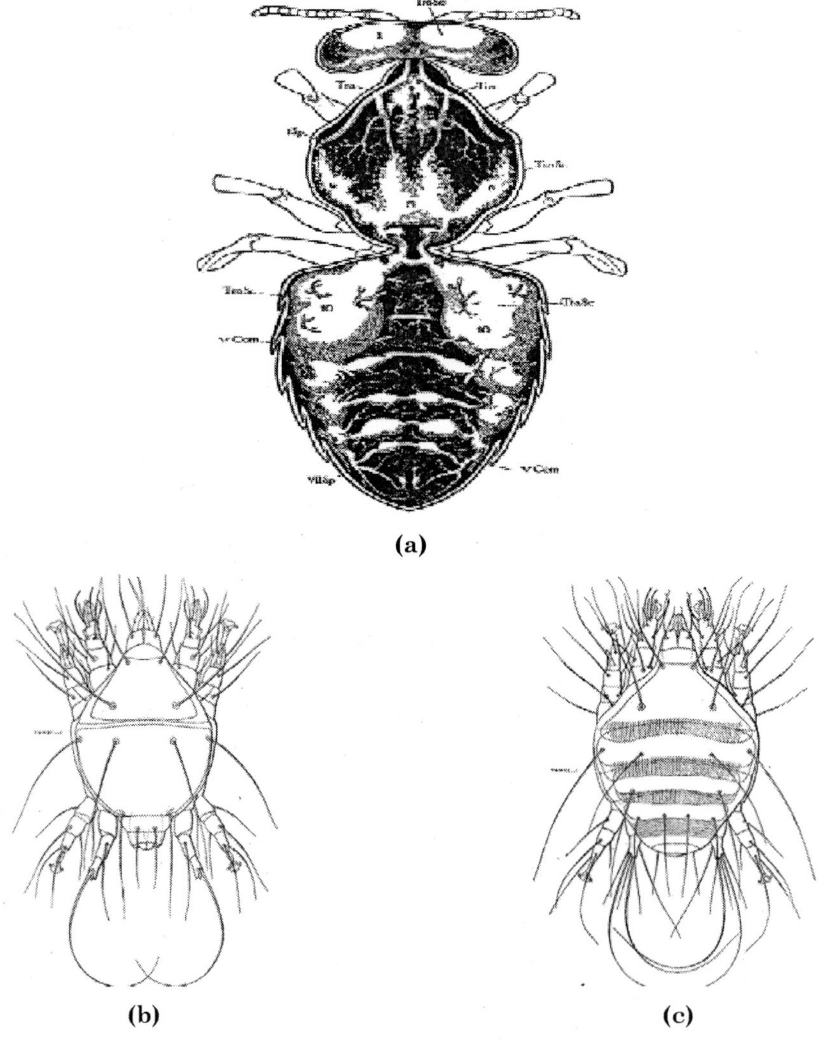

(a)

(b) **(c)**

Figure 81 (a). Internal Tracheal System of *Apis mellifera* (Snodgrass, 1956)
(b) Female, dorsal view (c) Male, dorsal view

Phoretic mites

Phoretic mites are flower or leaf feeding mites that use honeybees for transport from one plant to another and arrive accidently in a beehive. Among the many house guests are species that feed on old provisions and a few species that feed on other mites. Mites rarely feed on stored pollen in active hives, although large numbers of pollen feeding mites are often found in stored combs. The phoretic mite such as *Neocypholaelaps* reduces the pollen carrying capacity of the bees as they use bees for transports.

Stored product mites

Mites representing a variety of saprophytic and predatory species, are often found in the debris in the bottom of hives. The stored products astigmata wide spread contaminants of human food invade hive where they feed on stored pollen or honey or fungal contaminants thereof. They may become serious contaminants of the provisions. *Acaraus, Tyrolichus, Carpoglyphus* aand *Glyciphagus* are most common genera of these mites in bee hives.

THE TRACHEAL MITE (*ACARAPIS WOODI*)

The tracheal mite (*Acarapis woodi*) has been regarded a serious pest of honey bees in Europe and elsewhere. This mite was first described in 1921 in England, where it was believed to cause Isle of Wight disease. Today honeybee tracheal mite (HBTM) is known as acariosis or acarine; it is also called by several other common names around the world. This mite lives in the respiratory system of honeybees and shortens the life expectancy of an infested bee. A shortened life expectancy results in a smaller number of worker bees in a colony and lower colony productivity. As the name suggests, this mite spends most of its life cycle in the trachea of adult honeybees, except for brief periods when it will pass from bee-to-bee. The mites have an oval body shape and also possesses a hardened mandible that is used to feed on the blood of its host. The mite's small size is critical to its survival (females measure 120 to190 μm long by 77 to 80 μm wide males 125 to 136 μm by 60 to 77 μm. The tiny mites can hide under the flat lobe that covers the bee's first thoracic spiracle, which many individuals can thus occupy. Mites begin to disperse by questing on bee setae when the host bee is more than 13 days old, peaking at 15 to 25 days. *A. woodi* is vulnerable to desiccation and starvation during this time outside the host, and survival depends on the ambient temperature and humidity as well as on its state of nourishment. A mite can die after a few hours unless it enters a host. Mites are also at risk of being dislodged during bee flight and grooming.

Species diversity

External mites *A. externus, A. vagans* and *A. dorsalis*, all of which are morphologically similar to *A. woodi*, are often found on the thorax of healthy bees and can very easily be mistaken for *A. woodi*. It seems, however, that they do not cause any serious threat to bees or beekeeping. This method should therefore only be chosen if all that is required is a rough estimation of the degree of infection in a region. It is not suitable for determining a first outbreak.

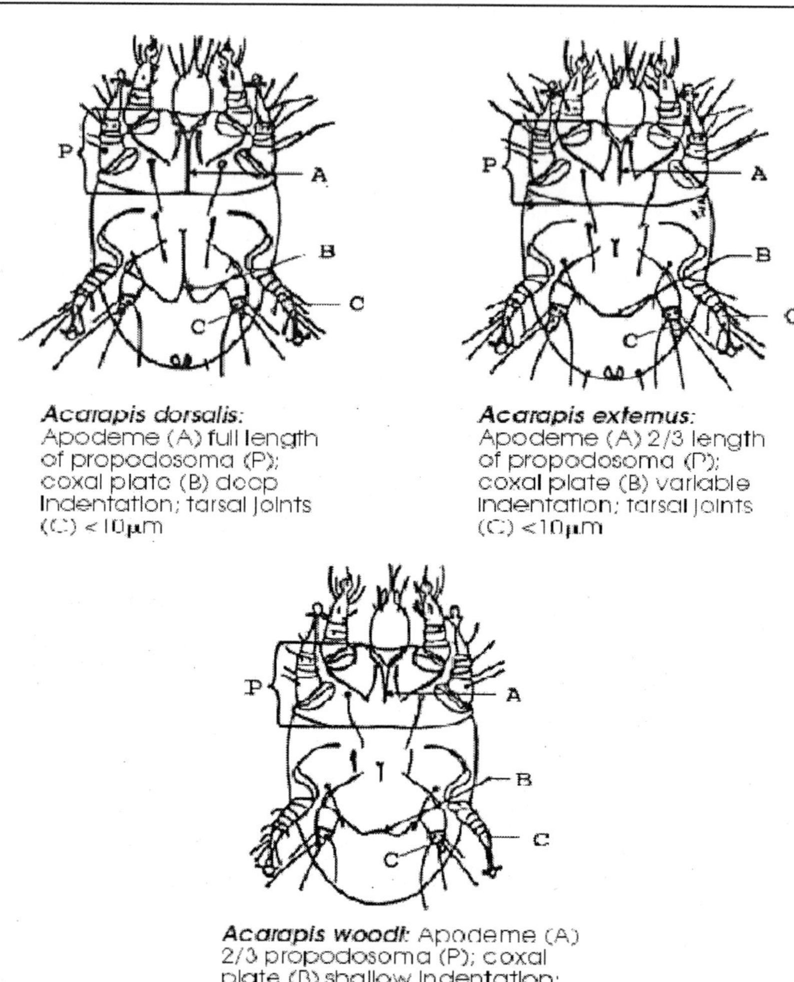

Figure 82. Morphological characters separating Acarapis species.

Three Acarapis species are associated with adult honey bees: *Acarapis woodi, A. externus*, and *A. dorsalis*. These mites are difficult to detect and identify because of their small size and similarity; therefore, they have frequently been identified by location on the bee instead of by morphological characters.

Only *A. woodi* can be diagnosed solely based on habitat, inside the breathing apparatus or tracheae of the honey bee. The position of other species on the host is a useful but not infallible characteristic. *Acarapis woodi* lives exclusively in the tracheae (it only leaves to transfer to other bees); *A. externus* inhabits the membranous area between the posterior region of the head and thorax or the ventral neck region and the posterior tentorial pits. Acarapis dorsalis is usually found in the dorsal groove between the mesoscutum and mesocutellum and the wing bases. Complete descriptions and illustrations are found in Delfinado-Baker and Baker (1982).

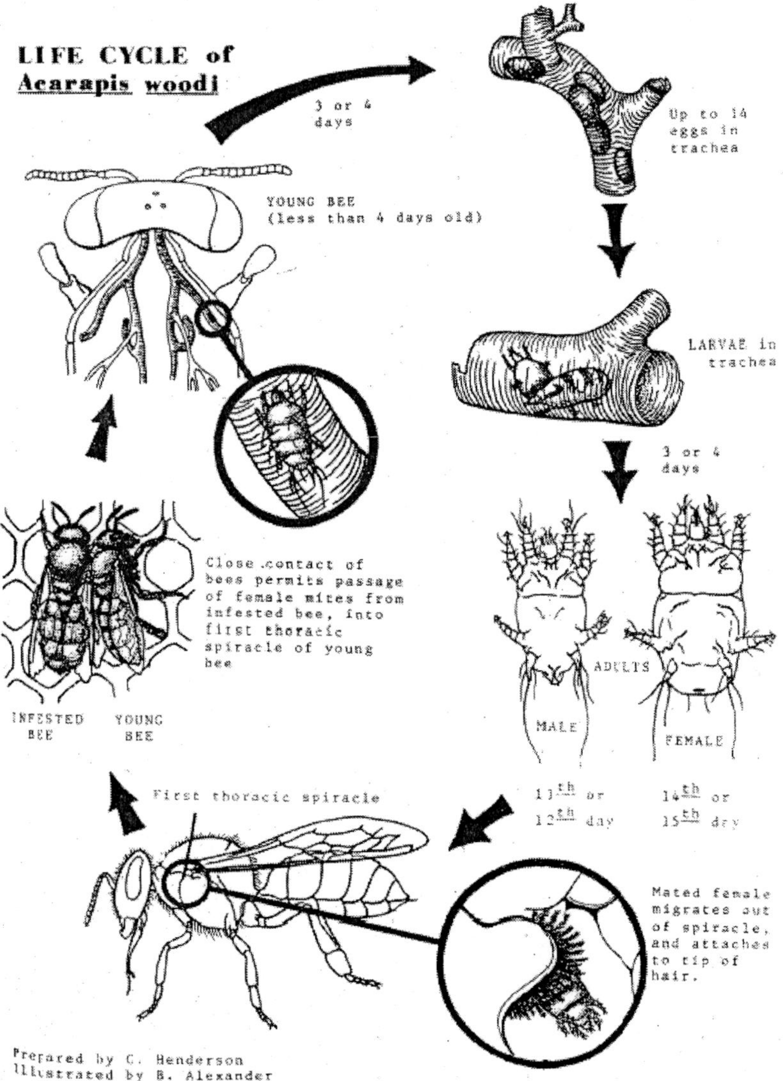

Figure 83. Showing life cycle of tracheal mite.

Life cycle

The HBTM has a four-part life cycle that consists of an egg, larva, resting stage, and adult mite. The female mite enters the tracheal tube of an adult honey bee through the first thoracic spiracle and takes up residence. She lays eggs which hatch, develop, and become adult mites within the trachea. These new adults are both male and female. Female mites leave the trachea soon after mating by passing through the thoracic spiracles of the bee. Once on the outside of the bee the mite will attach it self to the hairs of a passing bee and enter through the spiracles. Once inside the trachea, the female will begin to feed on the blood of the host, and will soon begin to lay eggs. The eggs hatch within 3 to 4 days inside of the trachea. The larva will develop in the trachea and will soon start the cycle all over again. Female mites prefer to migrate onto young bees, in particular bees less than 4 days old. The time from egg to egg, that is, the complete life cycle, is 14 days. Transfer to the new host takes place after the mite exits the tracheal tube through the spiracle. After exiting, she goes to the end of one of the surrounding hairs and waits for a passing bee as bees circulate in the hive. Within the tracheae all stages of development of the mite are found-eggs, larvae, and adult. The tracheal tubes can become very cluttered and eventually blocked. However, it is not clear that this blockage is what kills the bees. The flow of oxygen does not appear to be inhibited. Infested bees live and work normally, though their lives are shortened. The precise cause of death remains uncertain.

Economic Damage

The HBTM is responsible for significant colony losses throughout North America. Reports of losses as great as 90% were recorded just two years after initial discovery. Some of the earlier reports on tracheal mites suggested that the problem would be severe. in England during 1988/89 the losses were estimated in the range of 10,000 to 50,000 colonies and on national scale 300,000 to 1,000,000 colonies were lost. A heavy HBTM load causes diminished brood area, smaller bee populations, looser winter clusters, increased honey consumption, lower honey yields and, ultimately, colony demise. In temperate regions, mite populations increase during winter, when bees are confined to the hive, and decrease in summer when bee populations are highest. In subtropical climates, the cycle is similar, even though bees are not so onfined. Unfortunately, the introduction of varroa mites has overshadowed the impact HBTM has on bees.

Acarine disease can persist in a colony for years, causing little damage, but combined with other diseases, unfavorable conditions, scarcity of pollen and/or a poor foraging season, the disease can contribute to the death of a colony. Heavy mite infestations are most likely after seasons of poor honey yields and in situations when brood rearing is suppressed during the summer.

The foraging activity of worker honey bees at the hive entrance has been measured and used to study the effects of infestation by tracheal mites. No significant differences were found between infested and noninfested bees for the number of foraging trips, frequency of foraging trips, round trip times, frequency of pollen collection, or time between foraging trips. Nectar loads collected by forager honey bees infested with mites were compared with loads collected by noninfested foragers. The difference in mean honey sac volumes was not significant. Survival of worker bees as a function of mite infestation was determined by comparing, during a 6-week period, the ratio of infested to noninfested bees in subpopulations that were aged in common colonies. Stable infestation ratios throughout the experiment indicated no significant reduction in survival because of infestation. These results strongly indicate that mite infestations do not have a detectable economic effect on colonies when brood rearing is active and honey production is underway.

When the infestation is light or in its early stages, the bees behave as if not adversely affected. Infestation levels tend to decline in colonies when they are actively foraging. Colonies rarely show signs of infestation in summer or fall. Even the relatively few colonies with a large population of infested individuals in spring contain very few infested bees after a good season. The old infested bees from which mites are migrating are busy foraging during a nectar flow, so they become separated from the young susceptible bees and prevent most migrating mites from finding suitable new hosts.

Acarine disease shortens the lives of adult bees, affects flight efficiency, and causes a large number of crawling bees that are unable to fly. In extreme cases, colony populations often dwindle and, ultimately, the colony dies. Infested colonies may not develop normally and may exhibit symptoms of dysentery and exhibit an excessive swarming tendency. Often, however, severely infested colonies appear normal until their death during the winter. Colonies are most affected during winter confinement and early spring as with other stress diseases. Mite infestations are at a maximum in the early spring when the population is comprised of primarily older bees. Only old and heavily infested honey bees are killed by the mite. Whole colonies that have more than 30 percent of the individuals infested are often destroyed during late winter.

Symptoms

Tracheal mites are microscopic. They cannot be seen with the naked eye. Most of their life cycle takes place within the host bee, so positive diagnosis of a tracheal infestation can only be done after dissection and microscopic inspection. No one symptom characterizes this disease. An affected bee could have disjointed wings and be unable to fly, or have a distended abdomen, or both. Absence of these symptoms does not necessarily indicate freedom from mites. Positive diagnosis can be made only by microscopic examination of the tracheae; since only *Acarapis woodi* is found in the bee tracheae,this is an important diagnostic feature. A healthy trachea appears cream color or white. The trachea of a severely infested bee has brown or black blotches with crustlike lesions and is obstructed by many mites in different stages of development. The trachea must be examined carefully for the presence of mites. The trachea may not always be discolored when mites are present, and a cloudy or discolored trachea does not always contain mites.Some symptoms at the hive that suggest a tracheal mite infestation:

1. Bees crawling about at the hive entrance, unable to fly,trembling distendd abdomen.

2. Poor clustering in cold weather. If the colony dies, bees may be found randomly throughout the hive bodies, instead of in a cluster.
3. K wing--a phenomenon wherein the bees wings at rest are not folded along the abdomen but instead angle out to the sides, the two wings on each side forming the letter K with the bee's body. The K- shaped wings is due to chronic paralysis virus having syndrome of Type-I.
4. The chronic paralysis virus having syndrome of Type-II causes hairless black syndrome in bees. The bees are recognized by hairless, black shiny bodies and crawling in front of the hives.

Dynamics

The mite population is cyclical. There is a fall build up, a winter peak, and a summer crash. This is opposite to the normal honey bee population cycle in which there is a spring buildup, a summer peak, and a winter decline. A good queen bee will outproduce the mites in the spring and summer, getting ahead of them and easing the mite problem for the colony. Spring requeening helps boost the bee population further, and menthol treatment in the late season helps to control the mite's winter buildup. The greatest losses of both individual bees and of colonies are in the late winter and early spring. With shortened lives, larger numbers of individual bees die during the winter. This period then becomes even more stressful for the already weakened colony. It may die, or if it survives, it lacks vitality and may not be able to build up normally in the succeeding spring and summer.

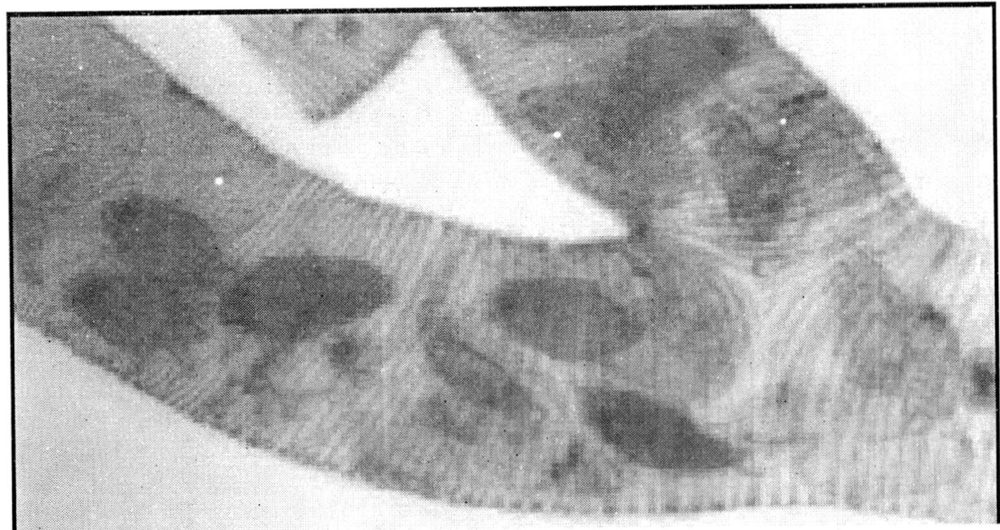

Figure 84. Honey bee trachea containing mites.

Diagnosing Acarapis woodi

Acarapis woodi can be diagnosed by following methods:

Method - 1

Pin the bee on its back and remove the head and first pair of legs by pushing them off with a scalpel or razor blade in a downward and forward motion. Using a dissecting microscope, remove the first ring of the thorax (tergite of prothorax) with forceps. This exposes the tracheal trunks in the mesothorax. When the infestation is light, it is necessary to remove the trachea. Place the trachea in a drop of lactic acid on a glass slide for clearing, and cover with a cover glass for examination at X 40-100 on a compound microscope.

Method- 2

Grasp the bee between your thumb and forefinger and remove the head and first pair of legs. Then with a scapel, razor blade, or fine pair of scissors, cut a thin transverse section from the anterior face of the thorax in such a way as to obtain a disk. Place the disk on a microscope slide and add a few drops of lactic acid. This makes the material more transparent and also helps to separate the muscle. With the aid of dissecting microscope, carefully separate the muscles, remove the trachea, and examine the preparations as in method 1. This is recommended for quick examination of a few bees.

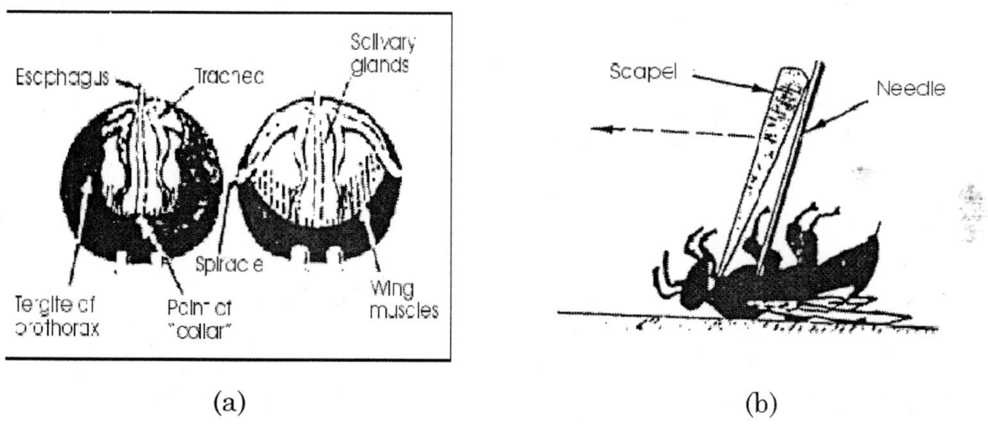

(a) (b)

Figure 85. (a). Location of the trachea in the thorax. (b) Positioning a bee for dissection.

Method- 3

Cut a few thoracic disks as described in method 2, place them on a slide, and add a few drops of 10% potassium hydroxide (KOH). Heat the slide gently for 1-2 minutes (do not boil), cover with a cover glass, crush the disks lightly, and examine microscopically. This procedure is advantageous when the bees have been dead for some time.

Method - 4

Prepare transverse-section disks from the thoraces of 50 honey bees as described in ethod 2, place them in 5% KOH, and incubate at 37°C for 16-24 hours. The KOH dissolves the muscle and fat tissue, leaving the trachea exposed. Then examine the disk-trachea suspension under a dissecting microscope. Remove suspicious tracheae from the disks and examine the tracheae microscopically (X 40-100). This procedure is recommended for large samples of bees.

Method - 5

Remove the heads, abdomens, wings, and legs from 20-200 thoraces and place them in a homogenizing jar with 25 mL of water. Homogenize three times for several seconds at 10,000 rpm, using just enough water to rinse the inside of the jar. Then strain the suspension through a 0.8-mesh sieve and rinse with water. The final volume of the filtrate should be about 50 mL. Centrifuge the filtrate at about 1,500 g for 5 minutes and discard the supernatant. Then add a few drops of lactic acid to the preparation, and allow it to stand for 10 minutes. Finally, place the sediment on a slide for examination. In this method, a microscope with an oil immersion objective is required to correctly identify *Acarapis woodi* because other mites associated with honey bees are morphologically similar. This technique is described by Cohn *et al.* (1979).

Method - 6

In the flotation method (Camazine, 1985), bees cannot be stored or killed in alcohol. For cleanest preparations, remove the head, wings, legs, and abdomen (saving only the thoraxes) of recently killed bees. This removal is easily done using one's fingers when the bees are frozen. Place 25-100 bees in a household blender with enough water added to cover the blades. Blend the preparation for no more than 15 seconds, just until the thoraces are broken apart (Blending for longer periods will pulverize the tracheae). Pour the resulting mixture into a series of test tubes (2-3 cm in diameter). Most of the denser muscle fibers and cuticular fragments fall to the bottom while the tracheae and air sacs float, forming a thin whitish layer on the surface of the water. Suction off this layer with a pipette, place on one or more slides, and cover with cover slips. Examine the slides under a compound microscope at X 100-250. Examine the slide in a systematic manner for darkened, blotchy, and discolored tracheae and for undamaged tracheae that may also contain mites and eggs.

Method - 7

In the modified methylene blue staining technique (Peng and Nasr, 1985), prepare transverse-section disks from the thoraces of 50 bees as described in method 2. Place the disks in a beaker of 8% KOH solution, and heat to boiling with continuous gentle stirring of the disks. Remove the solution from the heat and continue stirring until the soft tissues inside the disks are dissolved and cleared (about 10 minutes). Excessive stirring and heating will damage the specimens and subsequently reduce the color intensity of the mites. Recover the disks from the KOH by filtration through a perforated Tissue-Tek processing capsule. After filtration, cover the processing capsule with a lid, place in a beaker, and wash with tap water to remove the remaining KOH. After washing, transfer the processing capsule to a modified methylene blue staining solution (prepared by first dissolving 1% aqueous methylene blue and then adding sodium chloride to make a 0.85% sodium chloride solution). Immerse the capsule in that solution for 5 minutes and then in distilled water for 2-5 minutes; finally, rinse the capsule with 70% ethyl alcohol. Examine the disks for stained mites within the tracheae under a dissecting microscope at X 10-25.

Method - 8

Differentiation of live mites from dead mites (Eischen *et al.,* 1986) is the method of choice for evaluating chemicals used to control tracheal mites. Anesthetize live bees with carbon dioxide and remove the abdomens with a scalpel to prevent being stung during examination. Remove the head and first pair of legs of each bee by holding the bee on its back and gently pushing this section off with a downward and forward motion. Place each bee, held in this position, under a dissecting microscope, and remove the first ring of the thorax with fine forceps. This exposes the tracheal attachment to the thoracic wall, which is often the only location of mites in a light infestation. Remove tracheae that appear abnormal with tweezers and transfer to a glass slide containing a thin film of glycerol. Then dissect the tracheae using a pair of fine needle probes. Mites are considered dead if they do not move; also, dead mites often appear discolored and desiccated. Living mites have a translucent gray or pearl color and move within a few seconds after dissection.

Method - 9

For serodiagnosis, Ragsdale and Furgala (1987) produced an antiserum against extracts of *Acarapis woodi*-infested tracheae to be used as the primary antibody in a direct enzyme-linked immunosorbent assay (ELISA). Ragsdale and Kjer (1989) improved the ELISA technique, making it as reliable as dissection for the detection of *A. woodi*. Their ELISA is accurate, sensitive, reproducible, cost effective, rapid, and easy to use.

CONTROL

(I) CHEMICAL CONTROL

At the present time there are two products labeled for the control of tracheal mites. They are menthol, sold under the trade name of Mite-A-Thol TM and formic acid, which is sold under the trade name of Apicure TM. Both menthol and formic acid are naturally occurring products in honey but they should only be used in the labeled forms which are Mite-A-Thol and Apicure. Mite-A-Thol is a crystal form of menthol and Apicure is a formic acid preparation in a gel base. The misuse of either product can result in bee death and affect the flavor and saleability of the honey.

1. Menthol

(a) Timing

Menthol must not be applied to honey bee colonies during periods of honey production to prevent residues from occurring in honey or wax. In the spring, treatment must be discontinued no later than two weeks before the anticipated honey flow. Best results are achieved if menthol treatment begins in mid-May to early June (or as soon as ambient day time temperatures reach 21°C). Treatment can also be implemented after the honey crop it is removed and before fall feeding begins.

(b) Application

Shortening-menthol mix: an equal weight of menthol and shortening can be combined by adding menthol crystals to liquid shortening just at melting point (i.e. 65°C). Immerse sheets of corrugated cardboard (30 cm x 30 cm) into the shortening-menthol mixture until the cardboard it is saturated (approximately 20 g of menthol). The cardboard is then removed and cooled (can be stored in freezer until needed). One sheet of cardboard is placed on bottom board. Repeat once in 7-10 days.

(c) Safety Information

Menthol is volatile and should be used only in well-ventilated areas. Protective glasses must be worn to avoid eye contact.

Skin contact: wash with soap and water.

Ingestion: induce vomiting, consult physician.

(d) Storage

Store in tight containers in cool place (e.g. freezer).

(e) Caution

Menthol will disrupt colony activities and may cause temporary decline in brood production.

2. Formic Acid

(a) Timing

Do not apply formic acid when honey supers are in place to prevent unwanted residues in honey. Apply formic acid when outside temperatures are between 10-25°C. and leave hive entrances fully open. In spring, treatment must be discontinued at least two weeks before the anticipated nectar flow.

(b) Application

For two-story (bees covering 8-20 frames): Apply 30-40 ml of 65% formic acid directly onto the bottom board or onto absorbent paper (use three 15 cm square napkins or paper towels) placed either on the bottom board or on the hive top-bars. Re-apply at 5-7 day intervals, for a total of three treatments. For one-story (bees covering 4-10 frames): Use half of the quantities of formic acid indicated above, following the same method and timing of application.

(c) Precautions

Wear goggles and chemically-resistant gloves, apron and boots when handling formic acid.

Avoid breathing vapours

Formic acid will disturb colony activities and its application may result in a slight increase in bee mortality or queen loss, especially if applied when outdoor temperatures are near 25°C. When treating indoor wintered hives, delay treatment until queens have began to lay and daytime temperatures approach 20°C to avoid queen losses.

(d) Single Application Methods

Research is being conducted on slow-release, single application methods. Of the two products, Apicure is the product of choice because it controls both tracheal mites and varroa mites. In addition, Apicure is less dependent upon ambient temperature for its effectiveness than is Mite-A-Thol. The best time to use Apicure for tracheal and varroa mite control is in the late summer or early fall. Both Apicure and Mite-A- Thol are most effective when honey bee brood is at a minimum and a treatment prior to winter conditions will be most effective in keeping mite levels from becoming a serious problem. Apicure becomes ineffective at temperatures below 45°F. Mite-A-Thol (menthol) is just as effective in controlling tracheal mites as is Apicure but it does not control varroa mites and it is ineffective at temperatures of less than 60°F. The most effective time to use this product for tracheal mite control is prior to winter conditions when bee brood levels are at a low level so late summer or early fall treatments are best.

(II) NON-CHEMICAL CONTROL

Resistant bee stocks

Resistant bee stocks have shown to be effective in reducing the impact of tracheal mites on honey bees. There are several strains of bees that show promise. The beekeepers should select and multiply the stocks which show resistance to mites.

Use of vegetable oils

The use of these vegetable oils and sugar patties seems to interfere with the transfer (spread) of mated tracheal mites from their old bee host to a new bee host. The patties are most effective when used in the winter and early spring period. The use of resistant (tolerant) bee stocks and the vegetable shortening patties have shown to be effective in reducing the impact of tracheal mites on honey bees, but chemical controls may be necessary even with the use of these products.

Diagnosis

HBTM are not visible to the naked eye, making diagnosis difficult. Consequently, beekeepers often use unreliable bee stress symptoms, which include dwindling populations, weak bees crawling on the ground with disjointed hind-wings (called K-wings) and abandoned, overwintered hives full of honey. The only certain way to identify mite infestation is to dissect the tracheae of bees and visualize the parasites. Bees are collected in winter or early spring, when HBTM populations are highest; fewer mites are found in the summer, due to the dilution effect caused by the rapid emergence of many young bees.

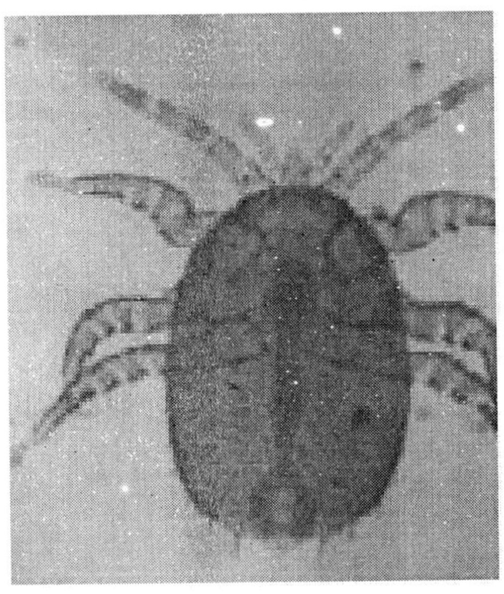

Tropilaelaps koenigerum

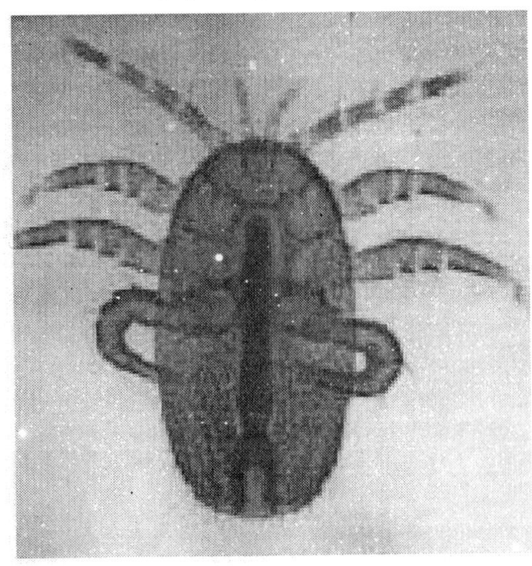

Tropilaelaps clareae

Figure 86. Showing Tropilaelaps koenigerum and Tropilaelaps clareae

THE TROPILAELAPS MITE (*TROPILAELAPS CLAREAE*)

Tropilaelapidosis is due to a haemophagous ectoparasitic mite *Tropilaelaps clareae.* After a short phoretic period on the adult bee, it enters the brood cell just before capping, where it reproduces. It causes a rapid decline of colonies of *Apis mellifera.* The *Tropilaelaps* mite is a parasite of brood only and causes brood mortality or reduced longevity of adult bees that survive the parasitised brood stage. It will breed and survive in bee colonies as long as brood is present. Its presence results in widespread losses of honeybee colonies causing serious economic hardship to apiarists and growers of those crops which require honeybee pollination to achieve viable production.

Epidemiology

The mite was first described on *A. mellifera* in the Philippines by Delfinado and Baker (1961). Later, it was described on the other species of the genus *Apis* (Table 68). This mite belongs in the mesostigmatic family Laelapidae, which has members that are mammalian parasites. *Tropilaelaps clareae,* originally obtained from a field rat from the Philippines, normally occurs on *A. dorsata*; this mite also parasitizes *A. mellifera.* It is also associated with other Asian honey bees, including *Apis laboriosa, Apis cerana* and *Apis florea.* Currently, *T. clareae* is restricted to Asia, from Iran in the northwest to Papua New Guinea in the southeast. A single, alarming report of this mite in Kenya has not been repeated. In India, the mite was responsible for the loss of 50% of the brood in *A. mellifera* colonies, introduced six years earlier. The same situation was observed in the Philippines. In other countries, where *A. mellifera* is native, *T. clareae* is considered a serious pest, making control treatments necessary. In 1982, *T. koenigerum* was reported as a new species of parasite infesting *Apis dorsata* in Sri Lanka. *T. clareae* has not been reported on the island. *T. koenigerum* has also been found in association with *A. laboriosa, A. cerana* and *A. mellifera* in Kashmir, and was recently detected in Nepal, Borneo and Thailand. Thought to be restricted to tropical or sub-tropical regions of Asia, their exact geographical range is unknown and is emerging.

There are currently two species of *Tropilaelaps mites* documented, *Tropilaelaps clareae* and *Tropilaelaps koenigerum. Tropilaelaps clareae* and *T. koenigerum* are serious mite parasites of honeybees. The primary host of the parasitic brood mite *T. clareae* Delfinado & Baker (Laelapidae: Acari) is the large Asian honey bee *Apis dorsata,* but the parasite is also associated with *A. laboriosa, A. mellifera, A. cerana* and *A. florea.* It is thought to be restricted to tropical or sub-tropical regions but its exact geographical range is unknown. The furthest west it has been found is in Khorasan and Sistan va Baluchestan provinces in Iran close to the Pakistan and Afghanistan borders, and the furthest east is Papua New Guinea.

The females of *T. clareae* are light-reddish brown and about 1.0 mm long x 0.6 mm wide, and the males are almost as large as the females. The life cycle and parasitism of *A. mellifera* is similar to that of *Varroa destructor. T. clareae* readily infests colonies of *A. mellifera* in Asia, particularly where colonies produce brood continuously. Adult female mites enter cells containing larvae where reproduction takes place within sealed brood cells. The mother mite lays three to four eggs on mature bee larvae 48 hours after cell capping. Development requires approximately 6 days, and the adults (including the mother mite) emerge with the hatching adult bee then search for new hosts. Mites move rapidly across the brood combs and are therefore easier to spot than *Varroa,* although they are much smaller. *T. clareae* has a shorter reproductive cycle than *V. destructor,* so when both mites are present in the same colony, *T. clareae* populations build up more rapidly.

T. koenigerum was first reported in 1982 as a new species of parasite on *A. dorsata* in Sri Lanka. It has also been found on *A. laboriosa, A. cerana* and *A. mellifera* in Kashmir. It has a similar life cycle to *T. clareae* and its biology and distribution is currently under investigation. *T. koenigerum* is smaller than *T. clareae,* with adult females being 0.7 mm long x 0.5 mm wide, oval and light brown, with males that are considerably smaller. Note that ongoing research in Asia may reveal the existence of other species of *Tropilaelaps mites.*

Parasitisation by these mites can cause abnormal brood development, death of both brood and bees, leading to colony decline and collapse, and can cause the bees to abscond from the hive.

Table 68. Distribution of *Tropilaelaps clareae* on its known *Apis* hosts (after Aggarwal, 1988)

Host	*A. dorsata*	*A. mellifera*	*A. cerana*	*A. florea*
Stage and cast	? Ad = adult bees; drB =drone brood; WB = worker brood.	?Ad = adult bees; drB = drone brood; WB = worker brood.	?Ad = adult bees; drB =drone brood	?Ad = adult bees;
Infested countries	India Philippines, Nepal Burma	India Philippines Burma Malaya, Java, Vietnam Pakistan Thailand Papua N.G. China, Taiwan Pakistan Afghanistan	India Burma Malaya Java Pakistan Papua N.G.	India

?Ad = adult bees; drB = drone brood; WB = worker brood.

Description

Tropilaelaps clareae is primarily a parasite of the bee brood; it is probably only phoretic on adult bees. Only the immature stages and adult females are haemophagous. The parasite reproduces on both drone and worker brood, although the drone brood is preferentially infested (ratio 3:l). The parasitism level can reach a maximum of 90% in the drone brood and in the worker brood. Typically 1 to 4 mated females enter a brood cell when it is approximately two thirds sealed. The latter authors distinguish five stages in the growth of the body size of female mites, correlated with the feeding behaviour, before egg-laying commences. The first egg is laid 50 hours after the cell has been capped and the majority of the eggs are laid before 110 hours. A female can produce up to six eggs (Feng, 1990). According to Kitprasert (1984), the mean duration of the mite progeny stages are: 1.05 days for the egg, 1.85 days for the larva, 2.11 days for the protonymph and 3.75 days for the deutonymph. Using these developmental times the first young adult mite appears about 18 days after the honeybee egg is laid, which is not in exact agreement with the maturation time of 16 days.

Males and females are produced in about the same proportion. When the bee emerges, by removing the cell capping, the adult female mites are released and start to move freely on the comb surface. The remaining few nymphal stages and the males in the cells do not survive after the bee has emerged. The adult female mites do not stay on the adult bees for longer than 1.4 days. More than 50% of the females are able to produce two viable offspring.

It was reported that in Thailand approximately 27% of the females entering brood cells did not reproduce and 2% of them give birth only to males. It has been found that only 18% of female mites were non-reproducing. Approximately 64% of the females produce one descendant and 33% produce two descendants. In colonies where *Varroa jacobsoni* is also present the ratio *V, jacobsoni* to *T. clareae* is 1:25, probably because only *T. clareae* produces viable progeny when it is in competition with *V. jacobsoni* (Burgett *et al.,* 1983).

Life Cycle

The life cycle and parasitism of *A. mellifera* by *Tropilaelaps* is similar to that of *Varroa destructor* although there are slight differences. Tropilaelaps has a higher reproductive rate than varroa as it has a shorter life cycle. This is because they have a faster development time and a shorter phoretic phase (non-reproductive transport phase, time spent on the adult bees) between reproductive cycles. The short life-cycle, as well as a very brief stay on adult bees, explains why populations of *T. clareae* increase faster than those of *Varroa* mites. Consequently, when both types of mite are present in the same colony *Tropilaelaps* populations build up far more rapidly than varroa, by a factor of 25:1 in favour of *Tropilaelaps*. When both *T. clareae* and *Varroa destructor* infest the same colony, the former may out-compete the *Varroa* mite. It has been reported that when both mite species are in the same cell, the reproduction of both mites declines. Adult mites enter cells containing larvae where reproduction takes place within sealed brood cells, particularly those of drones. The mites can reproduce in both worker and drone cells, but as with *Varroa* there is a preference for drone brood, may be almost 100% of drone brood parasitised. Typically, the mother mite lays three to four eggs on mature bee larvae 48 hours after cell capping, about one day apart. The eggs hatch after around twelve hours, then the larva goes through nymphal stages (protonymph, deutonymph) before reaching the adult stage. Once hatched, all stages of both female and male mites feed on the haemolymph (blood) of the developing bee, causing damage through feeding by depriving the developing bee of essential nourishment required for growth. Development from egg laying to the adult stage takes approximately 6 days. When the adult bee emerges, both adult male and female mites and the original invading mother mite exit the cell to search for new hosts. Up to 14 adult mites and 10 nymphal stages of mite have been recorded in a single cell. With *Varroa* infestations, immature females and the male mites die in the cell. Unlike the *Varroa* mite, Tropilaelaps cannot feed on adult bees because its mouthparts are unable to pierce the body wall membrane of the bees. The mites depend on the developing brood for food, and move from the adult bees to feed on the larvae as quickly as possible after emergence, so the phoretic stage is much shorter than that of *Varroa*, and may only be between 1-2 days. Phoretic survival on bees is quite short (only 1-2 days) because *Tropilaelaps* cannot pierce the integument of adult bees. The phoretic time for *Tropilaelaps* spp. is important in understanding the life cycle, and recent research suggests the period can be as long as 5-10 days. Development of the mite requires about 1 week. Gravid female mites (carrying eggs) will die within two days unless they deposit their eggs. *Tropilaelaps* is therefore less well adapted for survival in hives where there are long broodless periods.

Symptoms

Infestation by *Tropilaelaps* causes the death of many bee larvae (up to 50%), resulting in an irregular brood pattern and of which the cadavers that may partially protrude from the cells. Many malformed bees occur, with distorted abdomens, stubby wings and deformed or missing legs. Some of the affected bees crawl at the hive's entrance (1). In addition, perforated cappings are seen, the result of sanitation activities by the worker bees, which evict the infested bee pupae or young adults. Some infested colonies abscond, carrying the mites to a new location. Symptoms: Irregular brood pattern, dead or malformed wingless bees at the hive's entrance, and the presence of fast-running, brownish mites on the combs, are diagnostic for *T. clareae*. The faster development rate make *T. clareae* more dangerous to European honey bees than *Varroa* where they cohabit.

Spread and transmission

The adult female mite is the only stage responsible for the establishment and spread of infestation. A proportion of the adult female population remains in the colony where the mites can move with great agility, freely on the combs. Others are phoretic on the adult bee, often taking up a position between the thorax and the abdomen. *Tropilaelaps* mites are mobile and can readily move between bees and within the hive. However, to move between colonies they depend upon adult bees for transport through the natural processes of drifting, robbing, and swarming. Mites can spread slowly over long distances in this way. They are also spread within apiaries through distribution of infested combs and bees through beekeeping management. However, movement of infested colonies of *Apis mellifera* to new areas by the beekeeper is the principal and most rapid means of spread. The survival of the mite on worker bees maintained in an incubator at 35°C and 60% RH is a maximum of three days. Without food, mite survival is two days. The survival time does not seem to be closely correlated with the presence of adult bees, which is an argument in favour of mere phoresy. The main means of spread of the mites between colonies are robbing, drifting and absconding.

B. DIAGNOSTIC TECHNIQUES

According to Rinderer *et al.* (1994), if the population of *T. clareae* is allowed to develop unchecked the mite can rapidly cause the death of the colony. When the colony collapses, severely infested bee larvae and pupae are often seen at the hive entrance. Newly emerged adult bees often have vestigial or deformed wings or legs. The abdomen can also be malformed. The brood combs display an irregular pattern with dead or malformed immature bees. Pupae in infested cells often have darkly coloured spots, mainly on their extremities. At this stage, infestation in brood cells could be recognized by the adult bees. Queenless colonies have more severe infestations than queenright colonies. Differential diagnosis consists of distinguishing *T. clareae* from *T. koenigerum* although the latter has not been found on *A. mellifera*.

1. Identification of the agent

The first sign of an infestation by *T. clareae* is often the occurrence of large (almost 1 mm in length), red-brown, elongated mites on the combs or on adult bees. *Tropilaelaps koenigerum* is slightly smaller, only about 0.7 mm in length. Both species of *Tropilaelaps* can easily be recognised and separated from the *Varroa* mite using a x10 magnifying glass. The body of the *Varroa* mite is wider than it is long and it moves slowly, whereas the body of *Tropilaelaps* is elongated, with a heavily sclerotised holoventral or similar shield, and it is a fast-running mite.

(a) Mite collection

Methods to collect mites include an ether or sugar roll. Collect approximately 100-200 bees in a wide-mouthed jar with lid. Scrape the bees into the jar or use a modified vacuum to suck them in. Knock the bees to the bottom of the jar with a sharp blow; there should be about a 1-2 inch (2.54-5.08 cm) layer of bees on the bottom. Remove the lid and spray a 2-second burst with ether starter fluid. Alternatively, use enough 70% alcohol or soapy water to cover the bees; or add around 25 g (1 oz) powdered sugar (or flour). If using ether replace the lid and agitate or roll the jar for about 10 seconds; mites should stick to walls. If using soap or alcohol, agitate and then strain out the bees with a coarse hardware cloth or mesh strainer; mites will be in the liquid. If using sugar or other powder, put screening material (such as hardware cloth) on top of the jar and shake the mites on to white paper to count; repeat every 2 minutes. For a more accurate count, finish with an alcohol or soapy water wash to collect all the mites.

(b) Colony and brood examination

When monitoring honey bee colonies for the presence of *Tropilaelaps* (or *Varroa*), an examination of both drone and worker brood may provide an early indication of infestation. Mites can be observed inside capped bee brood by using a honey scratcher (with fork-like tines) to pull up capped pupae. The mites are clearly visible. The younger mite stages are whitish and may be almost motionless while feeding on their hosts' bodies, as their mouthparts and front legs are fixed to the cuticle of the bee host. The extent of parasitisation can be estimated by opening a predetermined number of brood cells; infestation rates are then calculated as per cent of capped brood containing live mites.

(c) Sticky board examination

A precise diagnosis can be made using a sticky board covered with a mesh, such as fly screen, that prevents the bees from removing the dislodged mites. The mesh must be large enough for mites to pass through. Make a sticky board with poster board, cardboard or other white, stiff paper coated with Vaseline or other sticky substance, or use a sheet of sticky shelf paper. Cut the paper to fit the bottom board of a hive. Cut a piece of hardware cloth or screen to fit on top of the sticky board. To keep the bees from cleaning off the board, fold under the outside edges of the screen to raise it off the board, and staple or tape in place. Leave the board in the colony for up to 3 days, collecting and examining the debris for mites. For faster mite diagnosis, smoke each colony adding 25 g (1 oz) pipe tobacco in the smoker. Puff the bees 6-10 times, close up the hive for 10-20 minutes. Pull out the sticky board after 10 minutes and count the mites. Acaricides are sometimes used to knock mites off bees and will appear on the sticky boards.

2. Serological tests

No serological tests are available for diagnosis.

Treatment

Many of the same acaricides used for *Varroa* will kill *Tropilaelaps*.

1. Strips of plastic-impregnated fluvalinate (Apistan T) or cimiazol (Apitol) trickled on to bees will kill mites.
2. Tobacco smoke in the smoker will cause mites to drop off bees.
3. Strips of filter paper, available in some countries as Folbex strips, are prepared by soaking in an aqueous solution of 15% potassium nitrate to which two drops of amitraz (usually 12.5%) are added. After the paper dries, the strip is ignited and inserted into the hive. The smoke causes many mites to drop off.
4. Use of plates or pads soaked with 20 ml of 65% formic acid. The pads are placed in the colonies, near the top. These last methods are not recommended, as they can harm both bees and humans.
5. Applicaion of 85% formic acid as acaricide has been found very efffectiive in controlling *T. clareae*. By means of a wick of gauze of a defined length inserted through a cork, 5 ml of formic acid is left to evaporate inside the colony for 14 days. The use of an absorbent plate impregnated with

formic acid is also effective and more successful than in the treatment of varroosis.

6. A biological method, consisting of caging the queen for 9 days and removing the sealed brood at the same time, is sufficient to eliminate the mite.
7. *Apis dorsata* bees have a natural defence mechanism against *Tropilaelaps* mites. *Tropilaelaps clareae* Delfinado & Baker (Acari: Laelapidae) is a natural brood parasite of Apis dorsata Fabr. a native honey bee species of South-East Asia. *A. dorsata* lives in the open, in single comb nests hanging on cliffs, tree branches or eaves of human buildings. *A. dorsata* colonies migrate regularly during the year and stop brood rearing in preparation for migration. This means that during migrations there is a period of broodlessness. *T. clareae* cannot survive more than 3 days on adult bees of *A. dorsata* because their chelicerae (mouth parts) are not specialized for feeding on adult bees. Thus the origin of infestations in colonies that have recently undergone migration is as yet unknown.
8. When *T. clareae* populations build up beyond the grooming capacity of *A. dorsata* workers, colonies may migrate which decreases the mite populations. In Malaysia a deserted comb of *Apis dorsata* contained as much as 1,060 *T. clareae.*
9. Simulaltaneous infestations by *T. clareae* and *V. jacobsoni* also occur in *A.mellifera* colonies. When multiple parasitism occurs in single brood cells, only *T. clareae* appears to be successful in producing viable progeny. *T. clareae* can successfully outcompete *V. jacobsoni* when both are present in the same brood cell.

VARROA JACOBSONI (OUDEMANS) : THE PARASITIC VARROA MITE

Origin and Distribution

The parasitic *Varroa* mite, *Varroa jacobsoni* (Oudemans), is another serious pest of the honeybee (*Apis mellifera*). The *Varroa* mite was first described on *Apis indica* from Java in 1904. It was first reported in association with *A. mellifera* in 1962, making this a relatively new host-parasite interaction. *Varroa* mites develop and reproduce in the sealed brood cells of honeybees where they feed on the hemolymph of bee pupae (Erickson, 1998). *Varroa* mites enjoy a near world wide distribution having been found on all continents. *Varroa* is currently the most important pest of honey bees world-wide. It is an obligate parasite of bees and feeds externally on the immature stages of workers and the male reproductives (drones). Parasitized individuals may die or develop into weakened, crippled adults that are incapable of functioning normally. Unfortunately, this parasite is not detectable in colonies of honey bees until there is a large number of mites present. Without effective control methods infested colonies dwindle in population or die.

The impact of *Varroa* on agriculture has been significant. The number of managed colonies as well as feral colonies have been drastically reduced so they are no more effective as honey producers or pollinators. However, sufficient data are not available with which we can assess the total impact. Another important consideration is the effect of *Varroa* on the production of male reproductives in populations, *Varroa* mites seriously reduce the numbers of drones produced by infested colonies. Queen honey bees mate with about 10 to 18 different drones while they fly at specific mating sites away from their colonies. The reduction in droe population can severely result in breeding depression in honweybees and many queens may remain un insenminated. *Varroa* is the greatest single threat to beekeeping worldwide.

Life History

Varroa is an external parasite of the honey bee. A mature, mated female enters a cell containing a worker or drone larva shortly before it is capped. The female secures herself in the bottom of the cell until the cell is capped and the immature bee pupates. Then, the mite opens a feeding hole in the cuticle of the developing bee. Approximately 60 hours after capping, the mite lays an egg that develops into a male. Then, at approximately 30 hour intervals, she lays additional eggs that develop into females. Mites go through a number of developmental stages: egg, larva, 6-legged protonymph, 8-legged deutonymph, and 8-legged adult. The male develops in approximately 3-7 days. The females develop in 5-9 days. Shortly after a daughter matures, she mates with the male. As additional females mature, the male mates with them in turn. Female mites emerge from the brood cell when the adult worker or drone emerges. Mites may leave on their own, either immediately before or after the bee emerges, or they may emerge on the bee. Males do not survive outside the brood cell. After emerging from the brood cell, mites spend a variable length of time on their adult host prior to entering a cell with a new immature host. Estimates of the number of reproductive cycles that an individual female mite goes through ranges from one to as many as seven. The adult worker bee emerges 12 days after the cell is capped. The adult drone emerges 15 days after the cell is capped. This means that a mite can produce more offspring if she has entered a drone cell because the longer development time of the drone allows time for more daughters to mature. The number of female offspring produced by a mite in a worker cell is between 1.0 and 1.7, while the number of female offspring produced in a drone cell is about 2.4. Mites capitalize on this difference, being found in drone brood between 2 and 30 times as often as in worker brood (when drone brood is available).

Adult female varroa mites are oval shaped and about 1.1 mm long and 1.5 mm wide, with a reddish-brown color. Males are smaller and paler in color. The adult female mites enter the brood cells in the hive just before they are capped by the bees. Once inside the cell the female will feed on the hemolymph of the larva, and will soon begin to lay eggs. The *Varroa* mite has a five-part life cycle that proceeds from egg to six legged larvae, to eight-legged protonymphs, to deutonymphs, to sexually mature mites in 6 to 10 days (BRL, 2001). When there is a shortage of brood available for the mites to feed on, adult bees serve as intermediate hosts. The mites will attach themselves between the abdominal head-thorax segments of the adult bee were they will feed on the bee's hemolymph.

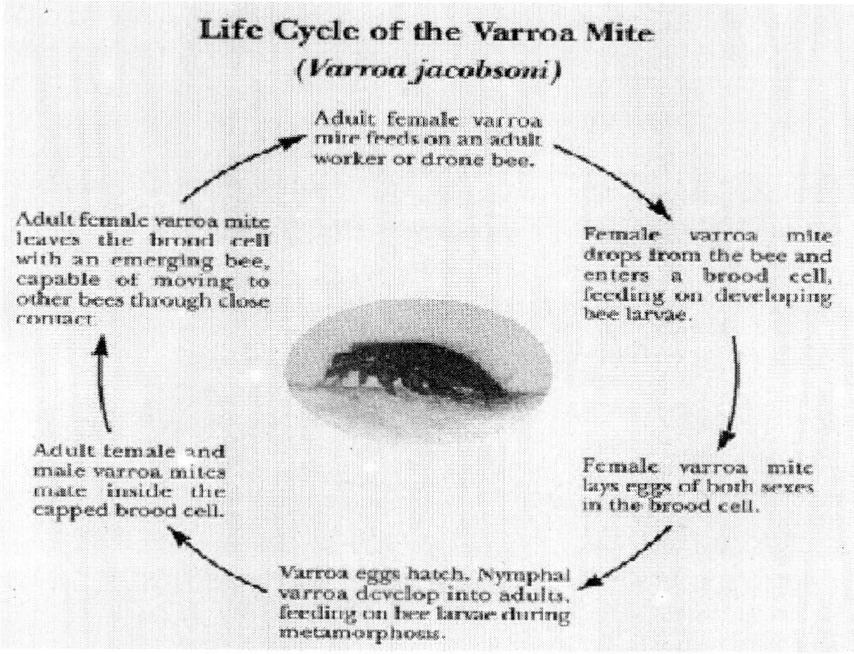

Figure 87. Life history of *Varroa jacobsoni*

Symptoms

Varroasis symptoms can be confused with other disorders, and even with pesticide poisoning. There are usually no obvious symptoms at low levels of infestation. Therefore, sampling using the ether roll method or the cappings scratcher method will be the only reliable way to detect the mites at low levels. As the infestation levels climb, emerging adult workers may be seen that have damaged wings, believed to be the result of a viral infestation associated with the mites. As the infestation rate climbs, more damaged workers will be seen, and bees may be seen crawling in front of the hive (although this could also indicate tracheal mite infestation). *Varroa* infestation causes a number of pathologies in the honey bee, including decreased lifespan, damaged wings, physiological abnormalities, and decreased body weight. Eventually, non-specific brood diseases begin to appear (see Parasitic Mite Syndrome below). If you observe any of these symptoms, you should immediately sample your colonies for mites.

Here are the most notable symptoms:

a. Pale or dark reddish-brown mites are seen on otherwise white pupae.

b. Colonies are weak with a spotty brood pattern and other brood disease symptoms are evident.

c. The drone or worker brood has punctured cappings.

d. Disfigured, stunted adults with deformed legs and wings are found crawling on the combs or on the ground outside. Additionally, bees are seen discarding larvae and pupae and there is a general colony malaise, with multiple disease symptoms. Because mite populations increase in proportion to the available bee larvae, *Varroa* can quickly overrun a colony and often colonies are dead by the fall. Interrupted brood rearing during the winter slows the population increase of varroa in temperate regions, but in warmer climates, colonies can be destroyed within months. Treated apiaries can still perish if the beekeeper is not diligent; reinfestation occurs due to the robbing of varroa-infested and weakly defended colonies.

Parasitic Mite Syndrome

Varroa is associated with a variety of brood diseases that characterize the end phase of the mite infestation. The symptoms associated with *Varroa* have collectively been designated parasitic mite syndrome. The deterioration in the brood usually occurs at moderate to high levels of mite infestation, althouh it occasionally occurs in colonies with Lower levels. The deterioration in the brood is believed to be a result of infection with a variety of pathogens, presumably viruses and bacteria. Although the symptoms superficially resemble AFB and EFB, these organisms have not been identified from infected larvae or pupae, and treatment with antibiotics such as terramycin does not eliminate this condition. From the time that colonies first exhibit symptoms of brood deterioration until the total collapse of the colony can be as little as 3 weeks. If a few cells of brood with disease are noticed, sample then your colony for mites immediately, and if mites are present, immediately remove any marketable honey and begin treatment. Procrastination at this stage insures the loss of colony.

Varroa's role in the transmission of pathogens is not well understood. Mites may introduce pathogens directly to developing pupae as they feed on them in the capped cell, or the feeding hole that the mother mite makes in the developing pupa could simply open a pathway for pathogens already present in the environment to enter the developing bee. Similarly, mites feeding on adults could introduce pathogens to their host, or they could simply open a pathway for pathogens to enter the adult host. Infected nurse bees may feed pathogens to developing larvae, and infected adults may transmit pathogens during trophallaxis. The exact relationship between *Varroa* and honey bee pathogens is not well defined and needs to be studied more thoroughly.

Transmission

Varroa infests new colonies in several ways. Moving brood among colonies for the purpose of strengthening or equalizing colonies is a common practice among beekeepers and a major source of transmission. When a beekeeper moves brood from an infested colony to another colony, mites are transferred with the brood. Robbing is also a significant source of transmission. Colonies that are allowed to become weakened by mites or disease are unable to defend themselves and are usually robbed by stronger colonies. In the process, the robber bees take home more than just a free load of honey. Swarms from infested colonies establish new nests with mites already present and are not likely to survive more than a year or two in the wild. This makes feral colonies prime sources of re-infestation for managed colonies. Drifting bees, especially in apiaries where colonies are kept close together, can also spread mites among colonies.

Identification and Detection

The adult *Varroa* mite is oval in shape with a width of 1.5 mm, a length of 1.0 mm, and 4 pair of legs. Typically, mites are light to dark brown in color, although immature stage may be nearly translucent. The bee louse, *Braula caeca*, is a wingless fly that is infrequently found in colonies in the northeast. The bee louse can be distinguished from *Varroa* by the fact that it has only 3 pair of legs, while the adult mite has 4 pair of legs.

Detection

Varroa can be detected in several ways.

1. The ether roll metod

Approximately 200-300 bees are collected from a comb and placed in a quart glass jar. A 1-2 second burst of an automotive starting fluid (ethyl ether) is sprayed into the jar. The jar is shaken vigorously for 10 seconds, then gently rolled along its long axis 2-3 turns. Mites, if present, will be seen adhering to the sides of the jar. This method generally detects about ½ of the mites actually present in the sample. Since mite levels are about twice as high on combs with brood as on combs with only honey, you maximize the chance of detecting mites in your colonies by sampling bees from the brood nest.

2. Removing some capped brood, preferably drone brood, with a cappings scrapper and examining the pupae for mites. This method has been found to be highly effective in detecting low levels of mites.

3. The sticky-paper collection device

A wooden frame is covered on one side with 1/8" hardware cloth. A piece of paper is attached to the other side of the frame and covered with a thin coating of a sticky material. The device is placed on the bottom board with the sticky surface facing up. The hardware cloth prevents bees from becoming stuck on the paper and from removing the mites. After a week, the device is removed and examined for mites. Mites can sometimes be seen on the adult bees or even walking on the comb, but this is more common when infestation rates are high and should not be used as a diagnostic method.

ECONOMIC THRESHOLDS

The best strategy for the use of any control measure is to use it only when the pest population reaches an economic threshold. Unfortunately, there are no established economic thresholds for *Varroa*. Evidently, colonies need to be sampled on regular basis. If *Varroa*, is detected initials treatment program and continue at regular intervals.

CONTROL OF VARROA

Optimizing a treatment program for a pest requires good information about the pest's population cycle. However, since *Varroa* is so virulent, there is no natural population cycle. Rather, once a colony is infested, the mite population simply builds up until the colony dies. Population cycles in managed colonies are a result of the constant battle between beekeepers applying treatments to control the mites and mite populations rebounding after treatment have been applied. Similarities in mite cycles from one beekeeper's operation to the next reflects similarities in treatment patterns and local environments. The object of an effective mite control protocol is to time the treatments so that the mite population never builds up to the point where it causes economic damage.

Cultural Controls

Other methods have been used to control mites, but most are too labor intensive and impractical in large apiaries. Used in combination with or in an integrated pest management (IPM) project, they may be helpful.

Smoke and Dropping Mites

Partial control in lightly infested apiaries can be obtained with tobacco smoke or smoke from other plant materials that cause mite knockdown. Smoke dislodges mites and can be used periodically to remove those that subsequently emerge from brood cells. A sticky board used in conjunction with smoke traps mites dislodged by the smoke. It has been found *Varroa* that dropped to the bottom board of a hive were more likely to remain there unless a bee passed within seven mm of it. Using a screen to separate fallen *Varroa* from bees may help keep mite levels lower.

Traps

Because *Varroa* prefer drones, combs of drone brood can be used to attract, trap, and remove mites by cutting out drone brood. Worker brood can also be removed. Drone brood can also be cut out of frames. Also, caging the queen of *A. cerana* for 35 to 40 days and separating the brood frames helped interrupt the brood/mite cycle in *A. cerana*. This method takes advantage of the fact that *Varroa* is found more often in drone brood than in worker brood. To turn this preference to the bees advantage, one or two full depth combs of drone comb are inserted into the brood nest during the beginning or middle of the drone rearing season. After the queen has laid eggs in the drone cells and the workers have reared and capped the larvae, the combs are removed from the colony and the drone brood is destroyed. This can be done by freezing the comb and then returning it to the colony for clean-up, or by placing the drone combs in a super and placing the super in an location where any emerging drones will not drift back into any colonies. An easy way to do this is to seal up the supers containing the drone comb so that the drones cannot escape.

The effectiveness of the drone trap method and the best time to use the drone trap method have not yet been determined, so, there is a need to find the best time for each location. The goal is to trap before the mites build up and become a problem, but as late as possible during the drone rearing season to produce the shortest possible interval between the time when the drone traps are removed from the colony and the end of the fall honey flow when colonies are treated with Apistan. Colonies that have been treated properly in the fall of one year, but not during the following spring, often start to exhibit high mite counts toward the end of the following summer. By trapping drones in May or June, the mite population is knocked down at a time that may prevent it from becoming a late summer problem. This may allow one to safely delay treating with Apistan in the fall until after the honey crop is removed from the colony.

Heat

The mite succumbs at or around 111.8°F (44.8°C), whereas sealed brood survives. Using a combination of these methods, along with a lactic acid treatment, mite numbers can be reduced considerably.

Resistant Bees

Selection of starins resistant to *Varroa* mites and their multiplication provides effective control.

Hygienic Behavior and Grooming

Bees will open capped brood cells and remove dead or dying brood. Such hygienic activity reduces the mite levels in untreated colonies, which require less chemical treatment to manage *Varroa*. Bee grooming (both autogrooming and allogrooming) has been observed in bees infested with mite. Grooming is an important component in mite reduction. However, grooming is highly variable in *A. mellifera*. Bees will remove mites from each other and some even kill them using their mandibles.

Length of Post-Capping Stage

The pupal period influences the number of mites completing development. Shortening this time results in fewer *Varroa* reaching maturity; if the capped cell stage is reduced by only six hours, fewer immature mites will become adults. Two African bee races have a heritable (worker) postcapping period of only 10 days, whereas European races require 11 to 12 days.

Brood Attractiveness

The larvae of European bees are highly attractive to *Varroa*; the ARS-Y-C-1 strain (*A. m. carnica*) was less attractive than other bee stocks. Differences in chemical components or levels in the brood may be the reason, but these possibilities have not been tested.

Low Mite Fecundity

At times, varroa mites do not reproduce, die without producing offspring, or get caught in the cocoons of bee pupae and die. Some of these traits are genetic. Mite fecundity can be classified into : (1) live mites that do not lay eggs, (2) live mites delayed in laying eggs, and (3) mites that die before oviposition. Mites with lower or no fertility were found to have fewer (or no) spermatozoa in their seminal receptacle. Queens selected for suppression of mite reproduction trait (SMRT) had reduced mite fecundity even after these queens were placed in susceptible colonies. This trait, used conjointly with hygienic behavior and other IPM methods, may help solve the bee/*Varroa* problem in the next decade.

Chemical control

The primary method of controlling *Varroa* had been the use of acaricides/ miticides. However, reports of resistance of the mites miticides are increasing. This emphasizes the need for developing not only new chemical controls for *Varroa*, but also a need to develop alternative methods to control the mites.

A final important consideration is that of contamination of the hive and hive products, especially beeswax, with miticides. Beeswax is an important world commodity which may become contaminated with miticide residues. There is a growing body of evidence that miticide contaminated wax used in the processes of queen rearing and the building of new comb for brood rearing is contributing to queen and colony viability problems.

Organic Acids

Formic acid kills *Varroa* and HBTM, but is temperature dependent and dangerous to humans. When used with absorbent paper over the top bars, the evaporating fumes kill HBTM and *Varroa*. Other organic acids, such as oxalic and lactic, are applied in sugar syrup trickled on bees. These acids require broodless bees and may cause bee mortality.

Essential Oils

Another approach is the use of volatile plant essential oils such as to control bee mites and other bee diseases. Many beekeepers are already experimenting with such "natural" products, but plant oils are complex compounds that may have unwanted side effects on bees and beekeepers, and could contaminate hive products.

VARROA DESTRUCTOR

Species: *destructor* (Anderson and Trueman)

The recent discovery of *Varroa destructor* and ability of their genotypes to reproduce both on the drone and worker broods of *A. mellifera* has led to their spread out of Asia, and resulted in heavy losses of *A. mellifera* colonies throughout the world. These mites have killed tens of thousands of honey bee colonies in different parts of the world in recent years. Of the two genotypes, the Korea genotype is the most widespread and common. It is found on *A. mellifera* in the UK, Europe, the Middle East, Africa, Asia, Canada, North and South America and New Zealand. The Japan/Thailand genotype has only been reported on *A. mellifera* from Japan, Thailand and the Americas. The chapter defines the problem caused by an important parasitic mite *V. destructor* and the discovery of virus in the mite body which has resulted in devastation of honeybee colonies across the globe.

The *Varroa* mite, *V. destructor* is considered the most serious pest affecting the Western honey bee *A. mellifera* (Fries *et al.,* 1994). Since the appearance of the mite on *A. mellifera*, honey yields have been reduced, and decreases in both apiary and feral honey bee colonies have reduced crop pollination and agricultural yields. The mite is found worldwide, excepts Australia. Recently, genetic studies have demonstrated the existence of genotypic variations among populations of the honey bee parasite, *Varroa jacobsoni* Oudemans, 1904. These studies also confirmed reproductive isolation of *Varroa* populations in Asia and this pest was renamed *V. destructor* Anderson and Trueman 2000. Past research on *V. jacobsoni* is probably referrable to *V. destructor* (Anderson and Trueman, 2000). Typical efforts to control the mite include biomechanical methods, organic acids and pesticides. Biomechanical methods that involve brood manipulation have proven to be inefficient or impractical for large-scale beekeeping operations. The misuse of pesticides and organic acids has resulted in mite resistance, bee health problems, and the contamination of honey bee products. Essential oils are good alternatives for control programs targeting *Varroa* mites. Several essential oils have shown acaricidal activity in screening tests, but it is necessary to prove their efficacy and bee compatibility in field trials.

Discovery of *Varroa destructor*

The recent shake-up in the taxonomy of *Varroa mites* were a scientific revolution. But for that fraction of the world's scientists who work on the parasitology of *Apis mellifera*, it may be as near as we get. For even though the news was non-controversial that *Varroa jacobsoni* is in fact a complex of at least two species, the discovery of *V. destructor* does have some far-flung implications for apiculture and bee science in general.

More recently, Anderson and Trueman (2000) published a new classification for *V. jacobsoni*, based on mt-DNA CoI gene sequences of *Varroa* mites collected in 32 countries. These authors described 18 different haplotypes of this mite.

Nine occur in the Malaysia-Indonesian region infesting *A. cerana* and include *V. jacobsoni* first described by Oudemans in 1904. Three other haplotypes of *Varroa* occur in the Philippines infesting *A. cerana*, but their taxonomic position is not yet established. Six haplotypes occur in mainland Asia parasitizing *A. cerana* and are reproductively isolated from specimens of *Varroa* found in the Malaysia-Indonesian region. Thus, they are considered a new species named *Varroa destructor*. Two haplotypes belonging to *V. destructor* transferred from *A. cerana* to *A. mellifera*. The Korean haplotype is the most widely distributed, affecting *A. mellifera* in Europe, the Middle East, Africa, Asia and the Americas. The Japan-Thailand5 haplotype is less common and is found in Japan, Thailand and the Americas. According to these discoveries, differences in mite haplotypes may explain bee tolerance to the parasite. It is likely that past studies on *V. jacobsoni* are applicable to *V. destructor* (Anderson and Trueman, 2000).

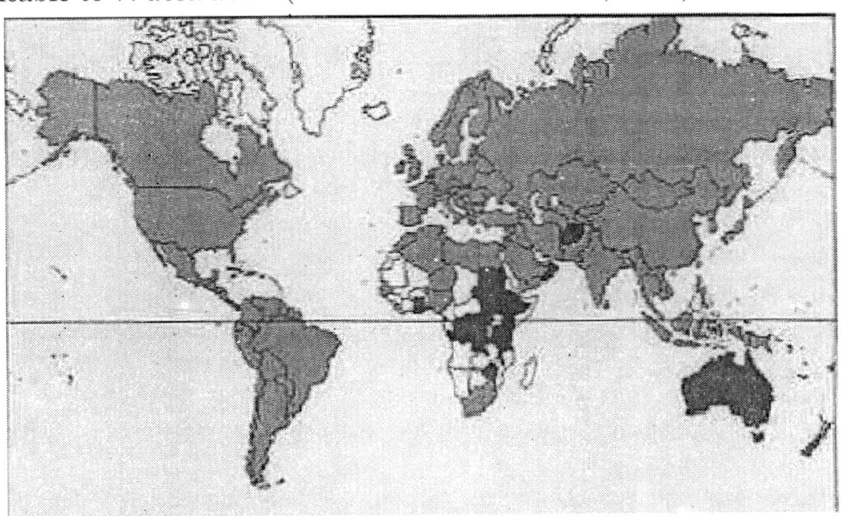

Figure 88. Map showing world distribution of *Varroa destructor* (Ellis & Munn, Bee World 86 (4): 88-101, 2000)

Species complex *Varroa* mites

Information gathered during the past five years has revolutionized our understanding of *Varroa* mite taxonomy, genetic diversity, host-relationships and epidemiology. Currently, there are four distinct species of *Varroa*; *Varroa jacobsoni* Oudemans, *V. destructor* Anderson and Trueman, *V. underwoodi* Delfinado-Baker and Aggarwal and *V. rindereri* De Guzman and Delfinado-Baker. The taxonomic status of three other *Varroa* genotypes has yet to be resolved (Anderson and Trueman, 2000). Each of these species and genotypes have been found to be natural ectoparasites on Asian honey bee species; *V. jacobsoni* on *Apis cerana* Fabricius mostly in India, southern mainland Asia and South-East Asia (Oudemans, 1904), *V. destructor* on *A. cerana* mostly in the western and northern regions of mainland Asia (Anderson and Trueman, 2000), *V. underwoodi* on *A. cerana* throughout Asia (Delfinado-Baker and Aggarwal, 1987), *V. rindereri* on *A. koschcvnikovi* v. Buttel-Reepen in Borneo (Koeniger *et al.*, 2002) and the three unresolved genotypes on *A. cerana* in the Philippines (Anderson and Trueman, 2000). The level of genetic variation among *V. underwoodi, V. rindereri* and the three unresolved Philippine genotypes has yet to be determined. However, a great deal of genetic variation has been reported among *V. jacobsoni* and *V. destructor* infesting Asian bees. To date, 15 different genotypes (or haplotypes – mites with distinct mtDNA CO-I gene sequences) of *V. jacobsoni* and eight of *V. destructor* have been reported, each only found on a specific geographic population of an Asian bee and each therefore named after the location (usually a country or Island) from which it was first discovered on its natural Asian bee host (Anderson and Trueman, 2000).

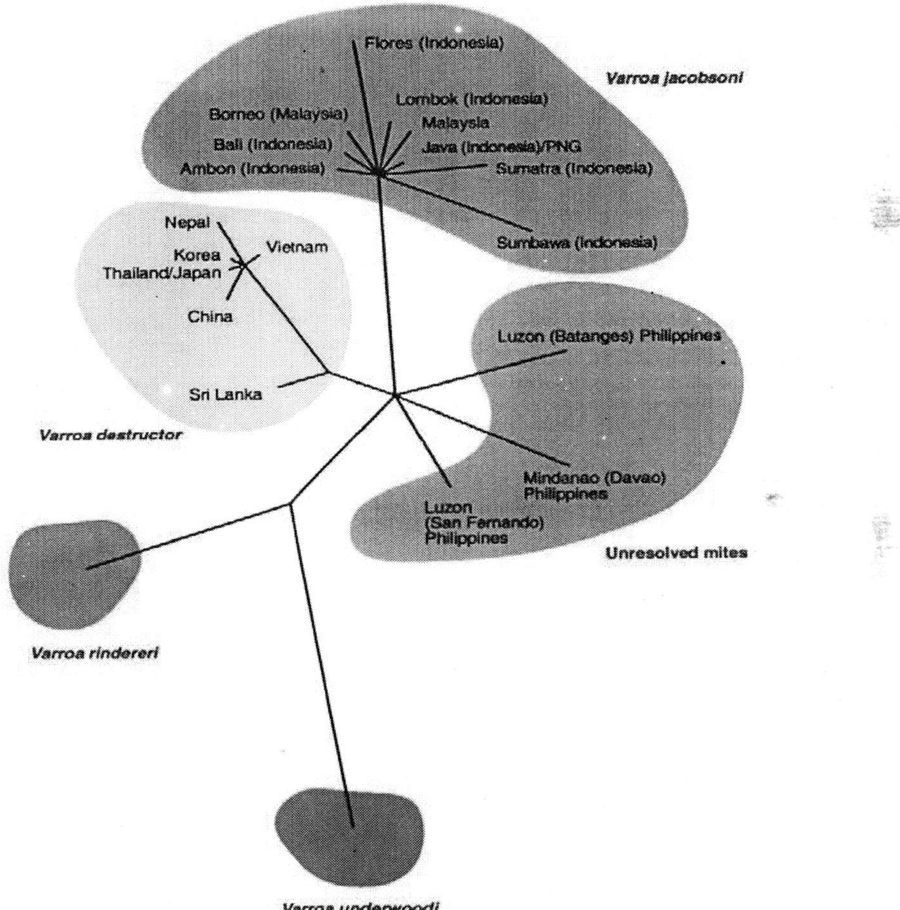

Figure 89. Species complex of *Varroa* mites

Hosts

Among the bees that serve as hosts of the *Varroa* mite are *Apis cerana, A. koschevnikovi, A. mellifera mellifera, A. m. capenis, A. m. carnica, A. m. iberica, A. m. intermissa, A. m. ligustica, A. m. macedonica, A. m. meda, A. m. scutellata,* and *A. m. syriaca.*

Description of *Varrroa destructor*

The infestation is caused by an ectoparasitic mite (*Varroa destructor* Anderson & Trueman), which sucks the blood of larvae, pupae and adult bees. It is reddish brown measuring 1.1 to 1.2 mm long and 1.5 to 6 mm broad. It has 4 pairs of legs. The female enters the cell with 4-5 days old larvae and lays eggs there. Life cycle is completed in 8 to 10 days in females and 6 to 7 days in males. It prefers drone brood over worker brood.

Honeybee mites have been extremely destructive to honeybees. In some counties in the UK, more than 90% of the beehives have been killed. The *Varroa* mite, *Varroa destructor*, is considered the most serious pest of the European honey bee, *Apis mellifera*. In the U.S., this ectoparasite causes approximately $160 million worth of damage annually. In different apiaries at Jammu and Kashmir the loss has been to the tune of more than 80%. Infestation ranged from 2-5 mites per brood cell.

Life-cycle

The entire life cycle of *Varroa destructor* mites occurs within the bee hive. *Varroa* is only able to produce offspring when honey bee brood is present in hives. The full life cycle is illustrated in Figure 90. It consists of a phoretic stage (transport phase) as a parasite on adult bees and a reproductive stage side the sealed brood cells. To breed, an adult gravid female mite (carrying eggs) enters occupied brood cells just before the cell is capped over, where she remains in the brood food under the larva until the cell is sealed. Mites prefer to breed in drone brood (10-12 times more frequently), but will also breed in worker brood. About four hours after capping she then starts feeding on the immature bee and establishes a feeding site on the developing bee that her offspring can feed from as they develop. Spots of white faeces or feeding sign of *Varroa* usually towards the hind end of the developing pupa can be seen. These spots will also be seen on the cell walls. Mated female *Varroa* enter drone and worker brood cells containing mature larvae just before the cells are about to be capped by hive bees. The female *Varroa* move to the base of the cell and submerge themselves in the larval food. When the cell is capped, the submerged mites move to the developing bee and begin feeding.

Individual females lay up to six eggs, beginning about 60-70 hours after the cell was capped and thereafter at intervals of about 30 hours. The first egg laid is male and all the others are female. Eggs are laid on the base of the cell and walls, and sometimes on the developing bee. Development of female *Varroa* from egg to adult takes about 8 to 10 days. The long interval between the laying of individual eggs means that mites of different stages of development may be seen in the one cell. Eggs hatch into a protonymph about 12 hours after they are laid. The protonymphs enter the larger duetonymph stage before reaching the final adult stage.

The single male *Varroa* mates with its sisters while they are in the brood cell. When the new adult bee emerges from its cell, the young *Varroa* females and mother mite also leave the cell, often on the emerging bee. The daughter mites feed on adult bees and after a short period will enter other brood cells to lay eggs. The cream-coloured males live for only a short time inside sealed brood cells and are never seen outside the cell. Each female lays 5-6 eggs, the first being a male followed by 4-5 female eggs laid at regular 30-hour intervals. The *Varroa* male emerges first, and the oldest daughter moults to adulthood 20 hours later. By laying only one male egg, *Varroa* increases the number of females that can reproduce at the next generation. Since the males do not survive outside the cell, females must be fertilized before the bee emerges from the cell, otherwise they remain sterile.

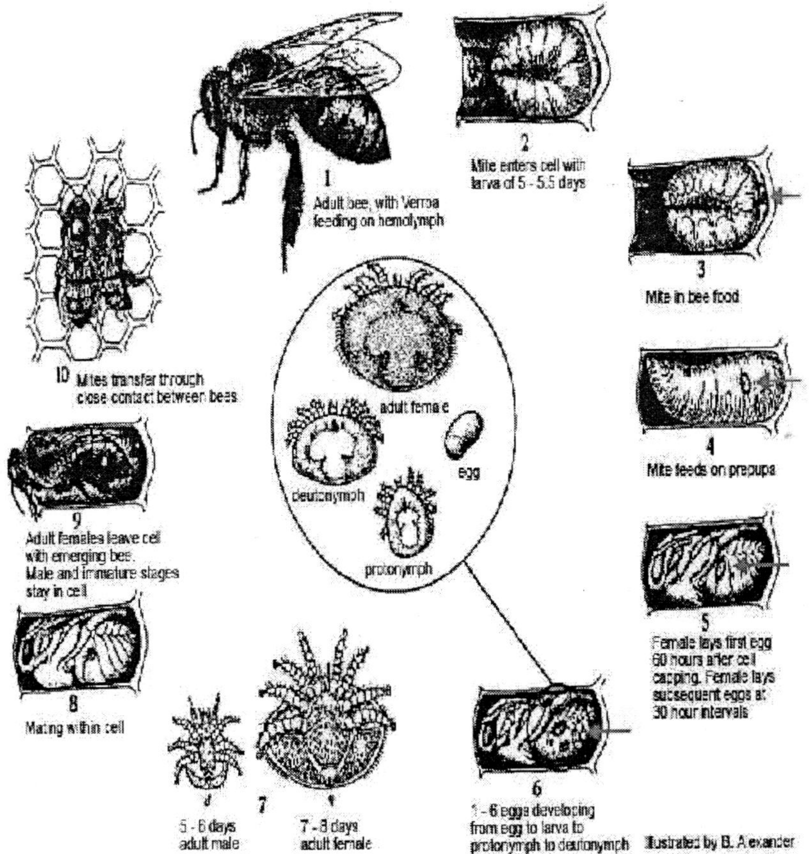

Figure 90. Showing life cycle of *Varroa destructor*

The duration of each reproductive cycle is limited by the development time of the bee, so not all reach maturity and mate by the time the bee emerges from the cell. Males and any remaining immature females die, unable to survive outside the sealed cell. With heavy infestation, two or more female mites may enter the same cell to breed, and female mites may produce more than one generation. Mature female mites leave the cell when the host bee emerges. Some of these may produce a second or third generation of mites by entering new brood cells. The success rate of reproduction (new mature female mites) in worker brood is about 1.7 to 2 but increases to between 2 and 3 in drone brood due to the longer development period. The development and status of a colony affects mite population growth, and depending on circumstances mite numbers will increase between 12 and 800 fold. This means that mite levels can return to a previous level 1 year after a high efficacy treatment has been applied.

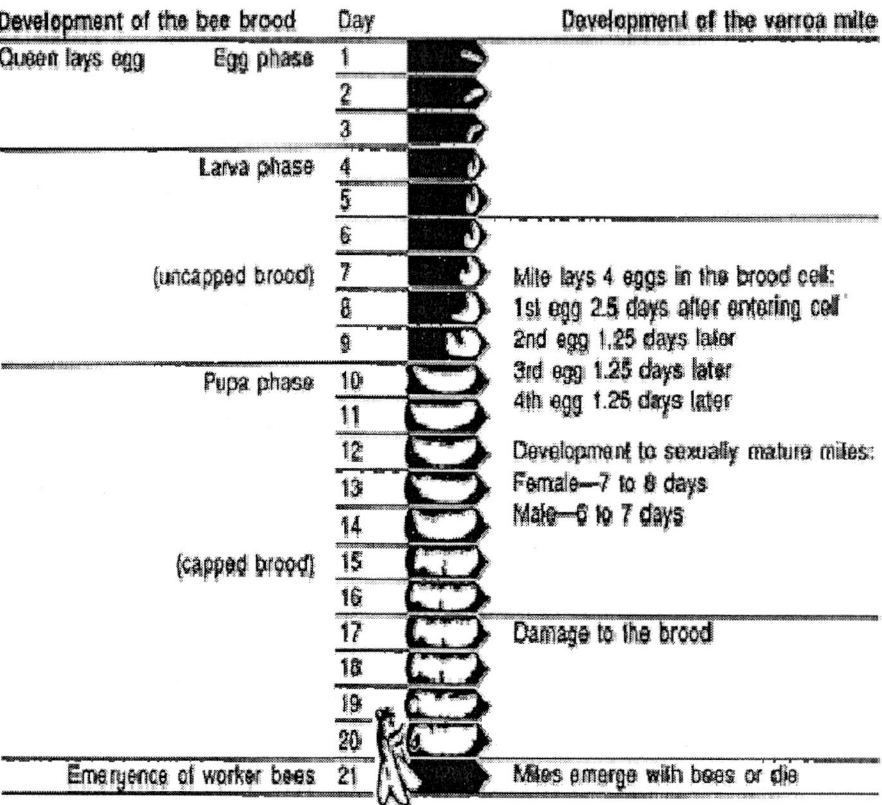

Figure 91. Showing life cycle of *Varroa destructor* in relation to its bee host *Apis mellifera*.

Varroa life cycle in a glance

1. Entering an uncapped cell
2. A fecund adult female *Varroa* leaves an adult bee and enters the uncapped cell of a five to five ½ day old larva of a worker or a drone, between 30 and 60 hours prior to capping. In heavy infestations, morethan one female mite enters the cell. Drone cells in particular are often infected by more than one female.
3. Inside brood food

4. Inside the cell, the mite submerges itself in the brood food, supplied to the developing larva by nurse bees. Oxygen is still available via channels called peritremes that protrude from the ventral surface of the mite through the semi-liquid brood food.
5. Feeding on the captive host: Shortly after the cell is capped the larva consumes the food, thereby releasing the *Varroa* mite, which then proceeds to feed on its captive host.
6. Laying Eggs: *Varroa* eggs are laid on the walls of the cell. The first egg is laid around 60 hours after operculation (when the bee larva has completed cocoon spinning) and subsequent eggs are deposited at approximately 30 hour intervals.
7. Larva grows six then eight legs: The six-legged larval stage of the mite develops within the egg during the first 24 hours and then develops into an eight-legged protonymph before hatching.
8. Reaching the adult stage: The protonymph gorges on the haemolymph of the bee pupa for 1½ - 2½ days before moulting to a deutonymph. The parasite continues feeding for another three to four days and then moults to the adult stage. Total development time from egg to adult is approximately 6 ¾ days for males and six days for females.
9. Adult males are small, globular and pale yellow/gray in appearance. Their mouth parts are specialised for spermatocyte transfer only so males are unable to feed. Mature females are larger and are red-brown in colour. In temperate climates, mites hatched in summer may live two or three months, while those hatching in winter or during broodless periods can survive for five to eight months, during which time they are not reproductive.
10. Mating: Both the male and the first female mite reach adulthood 10 days after operculation. Mating takes place between the male and the mature female mites in the cell.
11. Emerging from the cell: After mating, the mature female mites attach themselves to the young adult bee as it emerges from its cell. The male and remaining immature female mites stay in the cell and perish. Between one and three fecund female mites leave the worker bee cell, while three or four may emerge from drone cells.

Symptoms

1. The adult and developing stages of mites parasitise immature drone and worker bees within their cells
2. Colonies severely infested appear restless and weakened.
3. Only a few bees remain along with the queen
4. Adult mite can be seen on bee's surface.

5. Dead larvae, pupae, malformed workers and drones appear at hive entrance.
6. Spotty brood pattern.
7. Mites tear the integuments of the adult bees and suck the haemolymph.
8. "Parasite mite syndrome"
9. A parasitized pupa appears to have small, pale or dark reddish spot on its body.
10. White droppings are seen on the walls of empty cells.
11. Some larvae die in the pre-pupal stage with characteristic raised heads.
12. Reduced adult bee population in the infested colonies:queen supersedure, spotty brood are common.
13. Affected young larvae turn in to light brown colour
14. The brood fails to develop in to adults or malformed adults are formed.
15. Drone brood is more prone to attack but worker brood is also preferred.
16. Reduction in adult life span and loss of bodyweight.
17. The wounds inflicted by mites make the bees more susceptible to bacterial and viral diseases.
18. Feeding activity causes appreciable loss in haemolymph proteins
19. In heavy infestation damaged bees and pupae found on the bottom board
20. Immature bees with upto 6 mites develop normally; if parasitized by morethan 6 Mites they may be killed or deformed.
21. 'Acute bee paralysis virus' identified from *V. jacobsoni* infested colonies – the introduction of foreign proteins, such as digestive enzymes from mite in to the bees haemolymph can stimulate virus replication.
22. Crawling of affected bees with distorted and shortened wings in front of the hives.

Parasitic Mite Syndrome

Adult Symptoms

1. *Varroa* is present.
2. Crawling bees are seen.
3. Tracheal mites may be present.
4. Reduction in adult bee population
5. Evacuation of hive by crawling adult bees
6. Queen supersedure

Brood Symptoms

1. *Varroa* is present.
2. Brood pattern is spotty.
3. Symptoms resembling the foulbroods or sacbrood may be present.
4. Affected brood can be in any stage and anywhere on the comb.
5. Many symptoms are similar to American foulbrood, but there is no "ropiness," no typical odor and resultant scales are not brittle and easy to remove.
6. So predominant bacterial type is found and no known bee pathogen has been isolated from samples so far.
7. Affected brood can vary from C-stage larva to prepupa
8. Individual larva may appear
   a. twisted in the cell
   b. "molten" in the bottom of the cell
   c. light brown, gray to black in color
   d. watery to pasty consistency, resembling EFB, AFB, and sacbrood
9. Scales are not brittle and are easy to remove
10. No typical odor
11. No characteristic microflora

Mode of spread

- The close social contact between bees within the hive facilitates the transfer of *Varroa* mites from one host to another.
- Transfer of the mite between colonies of bees can occur in several ways:

Attachment to the bee in flight

Varroa mites attach themselves to the abdomen or thorax of the adult bees by gripping. Spines on their legs also entwine with hairs on the body surface of the bee. *Varroa* mites can achieve wide geographic distribution by securing themselves underneath or between the sclerites of the bee and being carried in flight.

Carried by a robber bee

- A robber bee that has been infested with varroa mites can transfer them to previously uninfested hives during the process of pillaging. Also a robber bee may become the unsuspecting host when stealing stores from an infected hive.

- The spread of the *Varroa* mite can also be accelerated by the following ways:
- Transport of hives by migratory beekeeping the transport of hives, used beekeeping equipment and queen bees by beekeepers is also a very effective means of transmission.
- Bees being moved between colonies

Drifting Bees

- *Varroa* could also be transmitted during swarming or by drifting bees. Drones especially can carry mites from one hive to another, sometimes over large distances.
- The spread of the *Varroa* mite can also be accelerated by the following ways:
- Transport of hives by migratory beekeeping
- Bees being moved between colonies
- Where social structure has already been weakened by *Varroa*. These hives are more vulnerable to robber bees, which pick up and then disperse the mites to their own and other colonies.
- Where social structure has already been weakened by *Varroa*. These hives are more vulnerable to robber bees, which pick up and then disperse the mites to their own and other colonies.

The mites are very mobile and can readily transfer between adult bees. *Varroa* can be spread between colonies and apiaries when hive components, infested brood and adult bees are interchanged during normal management apiary practices. Foraging and drifting bees, and swarms can also spread *Varroa*. In the case of foragers, mites can move from the bee to a flower and then hitch a ride with another bee or insect visiting the same flower. *Varroa* is not spread in honey.

DETECTION AND ESTIMATION OF *VARROA*

Diagnosis

Open drone or worker brood cells and remove the pupae with forceps and observe for mites on the body. A sheet of paper or a card board having a circle of grease on it is placed on the bottom board. Apistan under pressure is sprayed from the top of frame is used to fume the bees. The mites detach themselves and fall on the sheet, from where they can be picked and observed. Another easy method is to collect 200-300 bees in a plastic vial and kill them by using ether or ethanol. Once the bee die, mites detach from their body and fall into the liquid from where they can be picked and observed under a microscope.

Detection Methods

Sampling of *Varroa* mites

Field diagnosis

The mite *Varroa destructor* can be found on adult bees, on the brood, and in hive debris. The most severe parasitism occurs on the older larvae and pupae, with drone brood being preferred to worker brood.

Methods of Examining Adult Honey Bees

For a sample of adult honey bees, 500 to 1000 bees should be collected. This can be done by brushing honey bees off the comb through a large-mouthed funnel Individual honey bees can be examined with or without the aid of a hand lens. When the mites are moving about on a bee, they are fairly easy to detect; but once they attach themselves between segments, they are difficult to find. Mites can be detected and collected by three methods, as follows:

Shaking Method

Varroa mites can be dislodged by shaking the bees in liquids such as hot water, alcohol, detergent solution. Hand-shaking bees in alcohol for 1 minute dislodged about 90% of the mites and that mechanical shaking on a rotary shaker for 30 minutes removed 100% of the mites. The mites are collected by passing the bees and alcohol through a wire screen (8- to 12- mesh) to remove the bees and then sieving the alcohol through a 50-mesh screen or cotton cloth. The screen or cloth is then examined for mites.

Ether Method

This technique is a rapid and efficient detection method in the field and avoids the handling, time-consuming procedures associated with shaking adult bees in alcohol or other solvents. The bees (500-1,000) are collected in a jar and anesthetized with ether. The bees are then rotated in the jar for about 10 seconds. The majority of mites will have dislodged from their hosts and should be adhering to the inside wall of the jar. To complete the process, the bee sample is deposited on a white surface and spread around. This should cause any remaining mites to fall onto the white substrate.

Heating Method

Live adult honey bees can be shaken into a wire-based cage and placed in an oven over white paper. The bees are heated for 10-15 minutes at 46° - 47°C. Then *Varroa* mite if present, can be observed on the white paper

Methods of Examining Brood

To look for mites on brood, the pupae (preferably drone) are examined. *Varroa* mites can be easily seen against the white surface of worker or drone pupae after they are removed from their cells. It is suggested that a minimum of 100 drone pupae per colony be examined. The pupae can be collected by one of the following methods: The classic method of pupal collection is to uncap each cell and then remove the pupae with forceps or a hive tool.

Drone Brood Uncapping

Examination of honey bee brood

Varroa have a preference for drone pupae at the edge of the brood nest. If there are drone pupae use a pair of tweezers or a hive tool to remove individual pupae from their cells. Examine 50-100 of these for reddish-brown mites. When removing a pupa, carefully examine inside the brood cell, especially the base, for any mites. This is important, because *Varroa* may remain in the cell when brood is removed. Worker pupae should be examined if there are no drone pupae present in the hive.

Methods of Inspecting Hive Debris

Debris in a hive (such as wax particles, pollen, dead bees and brood, and mites) normally falls to the hive floor and is removed by house-cleaning bees during warm weather. This material can be collected and examined for the presence of *Varroa* mites as follows: The collection of hive debris can be facilitated by white construction paper on the hive floor. The paper is stapled under a wood (1/4-inch) and wire (8- to 12-mesh) frame, which protects the paper and debris from the bees. The paper is examined for mites, which can be easily seen against the white background. A magnifying glass or dissecting microscope can be helpful in locating the mites in the debris. Sticky boards or shelf paper (with the adhesive surface exposed) instead of construction paper will help hold the debris. The acaricides used to treat mite infestations can also be applied to bee colonies in combination with the paper method to detect.

Natural Mite Fall

The number of mites recovered from the hive floor debris can be a useful indicator of the mite population within a colony. The estimates are most accurate during the winter months when the colony is broodless or during the summer when the colony has an extensive brood nest with worker brood of all stages. If monitoring is carried out at other times the result will be less accurate and caution interpreting the results is required. To use this method the following should be taken into account:

1. Falling mites should be collected on a *Varroa* floor, or a tray that covers the hive floor and is protected by a mesh screen.
2. During the summer, mites falling over a period of approximately two weeks should be collected.
3. During the winter fewer mites will be falling, so a longer collection period of about 1 month is advisable. Mites can either be counted on the floor or separated from the debris before counting.

4. There should be no treatment in the colony while measuring natural mite drop.
5. Record both the number of fallen *Varroa* mites counted, and the number of days over which they were collected.

This method of monitoring is unreliable when the brood nest is rapidly fluctuating in size, e.g. when the colony swarms or is collapsing

Enter the mite count and the number of monitoring days into the *Varroa* calculator to get an estimate of *Varroa* mite numbers and a prediction of when treatment is required.

Sugar Shake Method

When *Varroa* are dusted with icing sugar, the fine granules stick to their pads (feet) and they are no longer able to grip the surface on which they cling. The dusting of adult bees with icing sugar causes mites to fall off the bee into the white sugar where they are more easily seen. A simple detection method using this fact is now used by many beekeepers throughout Australia. The method is described by the following points: obtain a large jar with a plastic or metal lid, drill 50-70, 3-4 mm holes in the lid and place a heaped tablespoon of icing sugar into the jar. Light a smoker and open a hive in the normal manner then shake about 300 bees from a comb of brood into the jar. It is helpful to first shake the bees onto a double thickness of newspaper and pour the bees into the jar. Place the lid on the jar and gently rotate the jar so all bees are dusted with sugar. Wait a few minutes, and rotate the jar a second time ensuring all bees are dusted. Shake the icing sugar through the holes in the lid onto a sheet of stiff white paper and gently shake the bees from the jar onto the ground in front of the hive entrance. Carefully examine the sugar, white paper, jar and lid for mites and insects. If you find any, carefully tip them into a small jar and place this in a cool position away from sunlight.

Sticky boards

Sticky boards or shelf paper (with the adhesive surface exposed) help hold the mites. Substances such as petroleum jelly, cooking sprays, oils have been used for preparation of sticky boards. Cover each IPM sticky board with Vaseline® (and 8 mesh hardware cloth screen if applicable). Insert sticky board above bottom board and under screen insert. Remove the board after 3 days. Count the mites and record.

STICKY PAPER TRAP

Sticky Paper: Place a sheet of white paper coated with cooking oil on the hive bottom and cover with #8 mesh screen. Check the "sticky board" daily for mites and replace when debris becomes excessive. To accelerate mite drop, place two Apistan® or CheckMite+® strips in the brood nest as directed by the label. Sticky boards are commercially available for this purpose.

Using Inserts

Bottom board inserts are commonly used in detecting mite infestations. These can be purchased, though many beekeepers make their own board. The inserts comprise two parts - a sheet of sticky paper or cardboard which covers the entire bottom board, and a screen that covers the sticky paper and serves to keep bees and larger hive debris off. The screen is raised slightly above the paper and is of a size that prevents the bees from passing through it. Mites, which are killed or stunned by appropriate hive treatment, Oxalic Acid aerosol, for instance, will fall to the bottom of the hive, pass through the screen, and be caught on the paper. The screen should be of fine enough mesh to pass mites while retaining the bulk of other debris, but at the same time the debris should not be allowed to accumulate and prevent the mites from falling through.

Measuring Mite Populations

The total number of Varroa mites in a colony of bees is determined by estimating the number of mites on adult bees and the number of adult mites within the capped brood cells. The procedure for these measurements is as follows:

Mites on adult bees

1. Weigh the entire hive, equipment and bees. We screen the colony's entrance during the night before weighing so that bees cannot leave during weighing.
2. Weigh the hive equipment without the bees. Brush all of the bees from the hive body and combs into an empty box before re-weighing the empty hive equipment. Do not shake the combs if they are to be examined for mite reproduction because immature progeny mites can be killed in a brood cell when the pupa is rattled against them.
3. The difference in these two weights is the weight of the adult bees in the colony.
4. Scoop *ca.* 1,000 bees from the box (after mixing) and put them into a pre-weighed jar. The jar is re-weighed, and the difference gives the weight of bees within the sample.
5. Wash each sample of bees with 75% ethanol. The bees are washed over a sieve that allows *Varroa* mites to filter through the mesh while retaining the adult bees. We rinse each sample until we get two consecutive washes that contain no mites. This procedure gives us an estimate of the number of mites per gram of bees. For example, if 30 mites were counted from 150 grams of bees, the estimate is 0.2 mites per gram of bees.
6. The total number of mites found on all adults bees is found by multiplying the total weight of bees by the mites per gram estimate. For

example, a colony with 3,000 grams of bees containing 0.2 mites per gram would have a total of 600 mites on all adult bees.

Mites in capped brood cells

1. Estimate the total area of capped worker brood cells in the colony. Use a 1 x 1 inch wire grid placed over the brood comb to estimate the total square inches of brood for each side of the comb. We only measure worker brood because we do not allow drone brood within our test colonies.
2. Convert square inches of capped brood into number of cells of capped brood. There are 23.6 worker-sized brood cells per square inch of capped brood. Multiply the total brood area by 23.6 to convert the area to number of brood cells. For example, 185 square inches of worker brood equals 4,366 cells.
3. Estimate infestation rate of capped brood. We select two brood combs from each colony to estimate the number of mites per 100 capped cells. We choose one comb containing young capped brood (prepupae and white-eyed pupae) and the second comb containing older brood (purple-eyed and tan pupae). We open 50 brood cells from each side of a brood comb and count the number of varroa mites within each cell. Brood cells are opened along a straight horizontal line that bisects the brood patch along its middle. Only foundress (or adult females that entered the cell) mites (and not progeny) are counted in this estimate. If 56 foundress mites are found in a total of 200 cells, we report 28 mites per 100 brood cells or 0.28 mites per brood cell.
4. Total mites in the capped brood is found by multiplying the total number of brood cells by the infestation rate. For our current example, the total mites in the capped brood cells is (4,366 cells) x (0.28 mites per cell), or 1222 mites.

Total mites = (mites on bees) + (mites in brood) = 600 + 1222 = 1822 mite

MANAGEMENT OF *VARROA* INFESTATION

Since some bee viruses are associated with *Varroa* mites and mites play a role in bacterial infections in bees, control of mites is not only important to prevent direct effects of their feeding behaviour, but also in controlling or combating these viruses and other opportunistic infections. The current options for the control of *V. destructor* include the application of various chemicals, organic acids and manipulative treatments. A variety of chemicals have been used to control *Varroa* mite populations in bee colonies. Among the chemicals commonly used are approved synthetic acaricides which specifically affect mites, such as, Tau-fluvalinate (Apistan®), flumethrin and coumaphos (Perizi). Organic acids (formic, lactic and oxalic acid vapour), essential oils and even simple compounds like sulphur are also in use. The chemicals are applied as vapours or by spraying. Thymol is a natural chemical found among many plant species, most notably in thyme, which is toxic to *Varroa* mites at doses relatively safe for honey bees. Two European products that use thymol as active ingredient are Apiguard®, produced by Vita Europe (England) and Apilife VAR® produced by Chemicals LAF (Italy). The higher surface area to volume ratio of the mites compared to that of the bees is thought to be one of the reasons why nonspecific toxic chemicals kill the mites faster than the bees. The continuous treatment of the beehives with these chemicals and compounds over a long period of time may reduce the mite populations, but several of these chemicals have clear disadvantages. Formic acid threatens the survival of the brood, young bees and the queens and acaricides fail to reach mites in capped brood. A way to circumvent these failures is to replace old queens when no capped brood is present in the colony. Mites have been found to develop resistance to the acaracide fluvalinate and if not used properly, the residues of the synthetic acaricides can accumulate and persist in honey and other bee products. In addition, some of the chemicals applied pose a health risk to the beekeepers.

Technical methods for the control of the *Varroa* mite are mainly used by small-scale beekeepers to minimise the use of non-specific acaricides. Most of the methods work by trapping mites in brood combs, which are then separated from the colony, or by causing mites to drop off adult bees by mechanical means. Some of these methods include the removal of infested drone brood before emergence, comb trapping of mites followed by destruction of the comb, and the use of open mesh floors, which prevent live mites that drop from the hive from returning. These methods are inexpensive and do not require the use of chemicals. Nevertheless, they are not sufficient when used alone. They are time-consuming, only effective in the case of moderate infestation and need high level of beekeeping skills.

There are a number of management techniques that can help in to keep *Varroa* mites under control. Most of these methods will not protect colonies by themselves, but they help to slow the rate at which the varroa population grows.

Measures against spreading bee diseases

The beekeeper and the honeybee are the two main agents that spread diseases among bees and between colonies and apiaries. Dead larvae, spores and dried scales transported for removal by the worker are sometimes dragged along the combs before they are disposed overboard. The beekeeper removes combs from a weaker colony and exchanges them with combs from a stronger colony. Sick and weak colonies are united. This transfer of bees and combs sometimes takes place from one apiary to another, thereby spreading diseases.

Further, honey contaminated with spores and parasites may be fed to a healthy colony, or the beekeeper may drop such contaminated honey-combs and bee products where they will be robbed by bees. Drones and workers straying into other colonies are also guilty of spreading diseases. The beekeeper must watch these thieves carefully and act in the interest of his own business. The following points are worth noting when there is an outbreak of disease:

1. The apiary must be kept clean. Honeycombs, wax, propolis and other hive products must not be thrown away near the apiary.
2. The beekeeper must not transfer infected combs from hive to hive or from apiary to apiary. Combs must be exchanged with great care.
3. Old hive parts, as well as used apiary equipment bought or acquired from doubtful sources, must be disinfected.
4. Unknown swarms should never be accepted when there is an outbreak of a bee disease. The beekeeper should set up a quarantine apiary four kilometres away from the nearest apiary, and make sure the swarm is disease-free before transporting it to the apiary.
5. Bees should never be fed with honey from a doubtful source.
6. If a colony dies of unknown causes, the hive should be closed pending an examination of a sample comb. The remaining stores in the hive should be protected from robber bees.
7. Robbing must be prevented. Place syrup or food for a colony inside the hive or in a properly designed feeder to prevent robbing.
8. Brood combs should be regularly inspected for signs of disease.
9. Hives should be spaced reasonably far apart. The beekeeper should try to arrange his hives so that it will be easy for every bee in the apiary to find its way into its own colony.
10. Isolation is a standard practice in many IPM programs. Locate apiaries 3-5 miles from other beekeeper's apiaries, to avoid contact with a number of problems, including Apistan resistant mites, as well as mites and diseases from other beekeepers colonies.
11. Maintain proper hygiene of the colonies. Do not discard comb and propolis in the apiary or exchange combs.
12. Removal of the drone brood limits the reproduction of *Varroa* mite.
13. In case of severe infestation, interruption of the brood cycle by caging the queen for 7 days at intervals is recommended so that the bees can remove infected brood. Learn to recognize healthy brood and bees.
14. Try not to squash bees when manipulating a colony.
15. Avoid robbing and drifting – don't spill sugar syrup or manipulate late in the season.
16. Assume second-hand equipment has contained a diseased colony. Do not use second-hand frames or combs.
17. Sterilize combs with 80% acetic acid and/or PDB.
18. Quarantine swarms – check for disease.
19. Consider keeping a separate hive tool and gloves in each apiary.

20. Survival of the fittest and adaptation applies to bees. Select bees resistant to mites.
21. Old honey bee brood combs are more infested by the mite *Varroa destructor* than are new brood combs
22. Prevent swarming as the swarms issuing from colonies are likely to serve as reservoirs for mites. Eventually, they will weaken and may be robbed by your bees. This will result in an increase in mite levels in your colonies. Follow an effective swarm prevention program to reduce the number of feral colonies in the vicinity of your apiaries.
23. Sterilize combs with 80% acetic acid and/or PDB.
24. Quarantine swarms – check for disease.
25. In case of severe infestation, interruption of the brood cycle by caging the queen for 7 days at intervals is recommended so that the bees can remove infected brood.

Key strategies for effective *Varroa* control

1. **Monitoring the infestation level of the colony:** This will indicate whether the mite population is building up to levels that will harm the colony. It will also indicate if the current method of control is not proving effective.
2. **Use a combination of methods**: The most effective control of varroa can be gained by using a combination of both biomechanical methods and chemical methods. These work in different ways and can be practised at different times of the year.
3. **Use approved varroacides such as Apistan® or Apiguard® :** These are proven to work and to be safe for bee and the user. It is also important to follow manufacturers' instructions. Incorrect use may result in residues in the hive products and it may promote the development of mite resistance.
4. **Use essential oil or organic acid treatments with great care:** If legal to do so, in rotation with registered acaricide products in a concerted Integrated Pest Management strategy.
5. **Use biomechanical methods :** Drone trapping and restricting queen movement can be a useful diagnostic and secondary control measure.
6. **Use a co-ordinated approach :** Developing a treatment programme with other beekeepers in the area will help reduce the likelihood of re-infestation.
7. **Treat only when necessary:** Treat only when necessary. Unnecessary treatments may promote mite resistance. Where possible, alternate between different types of varroacide. Avoid using

one type of product year after year. Never leave infested colonies unmanaged, as they will be eventually be killed by *Varroa*, and in the meanwhile they will re-infest the colonies of other local beekeepers who are trying to control the infestation. Where possible rotate the use of two or more unrelated varroacides. This is an effective strategy to slow the development of resistance. Avoid using the same varroacide year after year ahead.

Biological control

Biological control agents of pests are naturally occurring predators or parasites that will normally attack and kill a pest whilst sparing desirable organisms. Ongoing studies on the application of entomopathogenic fungi *Hirsutella thompsonii* and *Metarhizium anisopliae* (Kanga *et al.*, 2002), showed that they are pathogenic to *Varroa* mites at controlled conditions similar to those in a colony. Studies have also been performed to measure the pathogenicity of isolates of mitosporic fungi (Shaw *et al.*, 2002) and bacteria, such as *Bacillus thuringiensis*, to which the mites may be susceptible. The identification of pathogens harmful to *Varroa* but not to bees may open new avenues of controlling the mite in specific and environmentally friendly ways. In addition, the identification of biological control agents may have the potential for an effective, long-term and specific control method against *Varroa* mites.

The development of biological control of *Varroa* would provide an alternative to the use of acaricides, and could form part of an integrated pest management system. An effective pathogen would spread rapidly to maintain *Varroa* populations below a damage threshold giving long term or permanent control. A variety of microorganisms have been isolated from *V. destructor* mites that had apparent pathological symptoms, in an attempt to verify their pathogenicity. These include bacteria, viruses, *Rickettsiae*, fungi and parasitoids. Some like the bacteria *Bacillus thuringiensis* showed some disease symptoms, and like the rest of the microorganisms studied, it could affect other organisms besides the mites. No effective natural enemies able to cause wide population decline of *Varroa*, without harming other organisms in the process, have yet been identified.

Entomopathogenic fungi provide prime candidates for *Varroa* control since many species are active against acarines (mites and ticks) in nature. The potential of *Hirsutella thompsonii* and *Metarhizium anisopliae* as biological control agents of *Varroa* were evaluated in a laboratory setting and in observation hives (Kanga *et al.*, 2002). In the laboratory, the time needed to kill 90% of the mites (LT90) was 4.16 days for *H. thompsonii* and 5.85 days for *M. anisopliae* at 1.1 x 103 conidia mm-2, at a temperature (34.5°C) similar to conditions in a brood nest of a honey bee colony. The disadvantage is that the treatments did not significantly affect the mite population in sealed brood. *H. thompsonii* was apparently harmless to honey bee workers and brood, but Shaw *et al.* (2002) reported that four isolates of *M. anisopliae* caused significant mortality to honey bees.

The problem

The feeding habits of *Varroa* mites cause bees to suffer weight loss and physiological damage. In addition, the mite is also a vector of a number of pathogens, including bacteria, viruses and fungi, some of which also contribute to the killing of the bee populations. To date, no known biological control method is effectively able to eliminate any lethal pests in bee populations. The identification of a specific biological control agent against the mite will save bees and the livelihood of many beekeepers around the world. Therefore, it is crucial that potential agents are identified and investigated for their potential to control *Varroa* mites. Transportation of infested bee stocks and the beekeeping practices, including long distance transport of bees, contributed to the rapid spread of the mite and is a major threat to regions where the mite has not yet been seen.

GUIDELINES FOR THE USE OF CHEMICAL TREATMENTS

1. Oxalic Acid

- Oxalic acid should only be applied when the colony has no brood. Any open brood in the colony is likely to be killed by oxalic acid.
- Oxalic acid can be applied at cool temperatures, either through vapourization or trickling an acid-sugar syrup solution onto the bees.
- *Acid-sugar syrup solution*:
    - Prepare 1 litre of 1:1 sugar solution.
    - Add 35 g of oxalic acid crystals to the **warm** solution and stir gently until fully dissolved. The sugar syrup solution will have an acid concentration of 3.5%.
    - With a syringe or applicator, trickle 5 ml of solution directly onto the bees in each of the occupied bee spaces between frames in each brood box.
    - The maximum dose is 50 ml of acid solution per colony whether it is a nuc, single or multiple brood chambered hive.
- Vapourizer method:
    - Seal all upper hive entrances and cracks, and reduce the main entrance.
    - Smoke bees up from the bottom board.
    - Place 2 g of oxalic acid into the vapourizer. Insert vapourizer through the bottom entrance.

Timing of Application

- Remember that *Varroa* mites may be quickly re-introduced following a mite control treatment. Timing of treatment is therefore very important. When Apistan or Coumaphos is applied too early in the fall, the end of the

6-week treatment period may be at a time when there is still good flying weather, allowing for mite reintroduction.

- For fall treatment of Apistan or Coumaphos, select the end date of the treatment when the colony has little or no brood left.
- For many areas, the period of surplus honey comes to an end by mid-August. Immediately after honey removal, monitor the colonies for mites. It is recommended to use formic acid as a temporary control measure until strips can be applied later in the fall.
- Alternatively, an Apistan or Coumaphos treatment can be started after honey harvest in late summer, when mite levels demand treatment.
- To reduce the risk of resistance development, it is recommended to alternate between different control products. Experience has also shown that the efficacy of a product such as Apistan or Coumaphos can be re-established after a couple of years of non-use. (*Note that mites are not expected to develop resistance to formic or oxalic acid).*

2. Formic Acid

180 ml of 85% formic acid is filled in a bottle and placed in an empty space above the brood or adjacent to the brood. The bottle is corked in such a way so as to regulate 10 ml of the acid to evaporate daily.

- Effective against Varroa and tracheal mites (*Acarapis woodi*).
- Efficacy dependent on factors including size and condition of colony, time of year, humidity, temperature, etc. Efficacy of any one method may range from low to high.
- One effective method applied to a two-supered colony in the fall:
  - Remove lid and smoke bees off the top bars. Place paper napkins on the top bars and pour acid on the napkins. Prevent dripping, close the hive.
  - Each application equals 30-45 ml of 85% formic acid.
  - Apply three to four treatments, four to seven days apart.
  - Outside temperatures must be at least 12°C (55°F) in late afternoon.
  - Best results when there is no brood in the colony.
  - Mite drop can be monitored with sticky boards.
  - Formic acid treatments may increase risk of queen loss. Replace queen annually or bi-annually.

3. Lactic acid (15%)

Generally, 5ml of 15% lactic acid is sprayed directly on each comb face using a hand or back-pack sprayer. Two to four treatments per year are recommended.

4 **Sulphur** dusting @ 1 g per frame at weekly intervals is recommended.

5. THYMOL

Methodology : 0.25g /comb occupied by bees maximum (1.5-2 g/hive) on the top bar the frame every 4-7 days for 4-5 times. Or 1 gm of thymol in powder form mixed with 10-15 gm of wheat flour per colony may be dusted on infested frames at weekly intervals. Repeated treatments with 0.25 gm of thymol powder dust in passages between the combs can control upto 98% mites.

Time of treatment

After the honey removal

Main advantage

Effective treatment

Main disadvantage

1. Some variation in efficacy due to different parameters (weather, colony, colony strength etc.).
2. Damages on bee or brood under some condition.
3. Risk of queen loss.
4. Risk of hive abandonment.

Botanicals for Mite Control

Essential oils are secondary plant compounds which are used for the natural protection of these plants from pests and pathogens (Kevan *et al.,* 1999). This protection is provided through antifeedant and toxic properties. Several research reports show that essential oils have antimicrobial, fungitoxic, insecticidal and miticidal effects on various pathogens and pests under laboratory and field conditions.

In honey bees essential oils have been used for treating disorders, including parasitic mite infections and American Foul Brood. Laboratory and field tests have shown that they are 50% to 95% effective against *Varroa* mites and tracheal mites (Nasr and Kevan, 1999a, b). They also show minimal contamination of wax and honey in field trials. They are usually used in bee hives as fumigants, or mixed with sugar syrup for ingestion by bees. For *Varroa* control, the most commonly used essential oils are thymol, eucalyptus, and wintergreen. These oils are applied singly or as a mixture of different compounds to improve efficacy. Applied as fumigants, the effectiveness of thymol and other essential oils against *Varroa* mites depends greatly on temperature, time of the year, colony strength, and brood area. Due to the inconsistency and unreliability of essential oils for mite control, they cannot be used alone. However, their use does fit well into Integrated Pest Management (IPM) programs for alternating use with other control measures.

Non Chemical control

1. Queen arrest method

The queen is temporarily confined to a single brood frame or portion thereof. This method is labour intensive, slows down colony development and may only be suitable for the dedicated, small time beekeeper.

Physical Control (Traps and Oils)

- *Varroa* mites cling to their adult hosts and often loose their grip. When mites fall onto the bottom board they will climb up again and return to the bee cluster. The placement of a sticky board on the bottom board prevents mites from returning to the cluster. Sticky boards are commercially available or re-usable sticky traps can be easily constructed at home.
- Screened bottom boards allow mites to fall through, preventing them from crawling back up. The screened bottom board is a passive mite control device which has been reported to reduce mite levels by as much as 40%. Today, most beekeepers use screened bottom boards, with the additional benefit of improved air circulation in the hive.
- It has been reported that strips of cardboard dipped in mineral oil and suspended between brood frames offer limited *Varroa* mite control.

Controlling *Varroa* Mites with Walnut Leaf Smoke

Combinations of sticky traps and black walnut leaf smoke, could be useful in controlling *Varroa* mites. Walnut leaf smoke as an alternative to chemical treatment can be used to sontrol *Varroa* mites in honey bees. Mite counts in walnut-smoke-treated hives have been reported to be equal to or higher than the apistan treated and control hives. Evidently, using walnut leaves for mite control can improve the economics of honey production by reducing costs and increasing honey production.

Controlling *Varroa* Mites with Propolis

Propolis shows *Varroa* narcotizing and *Varro* acidal effects. Propolis extracted in 70% ethanol was found to be highly toxic, resulting in 100% narcosis of mites regardless of the contact time and concentration of propolis. However, the treatments with concentrated propolis solutions (e.g. 5% and 7.5%) led to higher mortality and little recovery rates from narcosis. The biologically active hydrophobic components of propolis have strong anti *Varroa* actions. The water soluble components of propolis are less active and constitute a minor proportion of its chemical makeup. Spraying honeybees with *Varroa* mites on their surface, with 10% propolis in 55% ethanol showed effective mite control. This is an indication that propolis solution can be used in the beehive interior as a varroacide.

Controlling *Varroa* Mites with Pollen traps

Pollen traps removed large numbers of *Varroa* mites from the bee population, but the removal was slower than chemical treatment so that the mites in pollen trap-treated colonies continued to reproduce. Pollen traps do have a value in *Varroa* mite IPM. Future studies should examine the effect of pollen traps on *Varroa* mite populations when the traps are used earlier in the year to prevent or reduce the chance that mite levels will reach economic threshold levels.

Controlling Varroa Mites with Neem oil

Topically applied neem oil did not result in direct mortality of *A. woodi* however, it offered significant protection of bees from infestation by *A. woodi.* Other vegetable and petroleum-based oils also offered selective control of honey bee mites, suggesting neem oil has both a physical and a toxicological mode of action.

A five percent neem oil spray killed 90 ± 6% of *Varroa* mites, three times morethan died in the control group, whereas thymol and canola oil spray killed 79 ± 8% and 65 ± 6% respectively. Colonies treated with thymol-oil had a significantly lower tracheal mite level (1.3 ± 7.5%) than untreated controls (23.3 ± 6%). None of the other treatments showed a statistically significant lower population. However, both neem and thymol-oil spray were detrimental to colonies exhibiting a 50% queen loss. Although effective for *Varroa,* however, formulations and application methods for these essential oils must be perfected to maximize control of parasites and minimize bee loss.

Controlling *Varroa* Mites with citric oil

D-limonene might be a good material to control mites. This substance is a readily available and relatively safe essential oil. It alongwith three others, all found in grapefruit (citral, linalool, citronella) and smoke from grapefruit leaves produced some tantalizing results for scientists. Citral was most effective, with a 72.8% *Varroa* knockdown.

OTHER PARASITIC BEE MITES

Most parasitic bee mites (with the exception of *Acarapis*) are in the tribe Varroini (or Group V) in the family Laelapidae.

Euvarroa

Euvarroa sinhai is a parasite of *A. florea,* occurring in Asia from Iran through India to Sri Lanka. The mite develops naturally on the capped drone brood but has been reared in the laboratory on *A. mellifera* worker brood. Development requires less than one week, and each female produces four to five offspring. Drones as well as workers are used for dispersal. The female mite overwinters in the colony, probably feeding on the clustering bees. Colony infestation by *E. sinhai* is somehow hindered by the construction of queen cells and its population growth is inhibited in the presence of *T. clareae* and of *V.*

jacobsoni. Transfer experiments confirmed that *E. sinhai* may survive on *A. mellifera* and *A. cerana*, emphasizing their ability to cross-infest exotic hosts. *Euvarroa wongsirii* parasitizes drone brood of *A. andreniformis* in Thailand and Malaysia. Its biology appears similar to *E. sinhai* and it can live for at least 50 days on worker bees outside the nest.

NON-PARASITIC BEE MITES

Three common suborders of mites associated with bees are the Astigmata, Prostigmata, and the Mesostigmata. Many astigmatic mites live on the hive's floor, feeding on bee debris, dead insects and fungi. *Forcellinia faini* (Astigmata), a common, whitish, slow-moving scavenger initially described in Puerto Rico, was abundantly collected in hive debris in northern and southern Thailand. A representative of the Prostigmata is the tarsonemid *Pseudacarapis indoapis* (Lindquist), a probable pollen feeder, which is apparently restricted to *Apis cerana*. *Melichares dentriticus* (Mesostigmata), a cosmopolitan predator on scavenger mites, is common in stored products, while members of *Neocypholaelaps* and *Afrocypholaelaps* live in flowers. They feed on the pollen of subtropical and tiopical trees and are phoretic on bees. Mites dispersing on *A. mellifera*, mostly as egg-bearing females, board and depart the bees via the tongue. The bees did not appear to be annoyed by these mites, nor were their foraging activities disrupted. One to 400 *A. africana* occurred on individuals of *A. cerana* in India. As bees return to the hive, mites disembark, roam on the combs and subsist on the pollen. *Melittiphis alvearius* has been found in hives and on bees in various parts of the world. Serological procedures demonstrated that, contrary to views formerly held, this mite is not a predator, but feeds on stored pollen.

PARASITIC MITES ON CTHER SPECIES OF *APIS*

All the honeybee species namely *cerana, dorsata, florea, mellifera, andreniformis, laboriosa, nigrocincta* and *nuluensis* have been reported to be attacked by several new mites. The relationship between bee nesting habitat and the genus or family of associated parasitic mites has been studied. The small bush-nesting species with single combs (*andreniformis* and *florea*) are attacked by *Euvarroa* spp. The bees that build multi-comb nests in caves and trees (*cerana, koschev-nikovi* and *mellifera*) are parasitized by *Varroa* spp., and the tree bees with single combs (*dorsata and laboriosa*) by *Tropilaelaps* spp.

NEMATODES

Mermithid nematodes have occasionally been found in honeybee workers, drones and queens. They are 10-20 mm in length, very thin whitish worms that parasitize many insect species. Mature adult worms swell in the soil and mate there, the eggs are laid on the wet grass by the active females or else the young larvae make their way there. Insects eat the grass, or take dew from it, and eggs are ingested or the young larval nematodes penetrate the insect cuticle. Queen honeybees probably receive nematode eggs that are brought in by bees collecting water. One queen whose ovaries contained no eggs was found to have an encapsulated nematode in her body cavity near the hind gut and ovary (Kramer, 1902). Nematodes probably have to go through a soil phase so they are not likely to multiply and spread within bee colonies. Their natural hosts are probably ground dwelling insects, including solitary bees and bumble bees. However, Vasliadi (1970) reported some 60% of many hundreds of honeybees he examined to be parasitized by mermithid nematodes in low lying regions of USSR. *Mermis nigriscens* has been found in honeybees in Switerzerland and an *Agamermes* spp. was found in honeybees in Brazil (Toumanoff, 1951). *Mermis albicans* has been found in worker, queen and drone honey bees in Europe (Fyg, 1959; Paillot *et al.*, 1949) and larval mermithids were found in worker bees in eastern USA by Morse (1955).

Nematode infections of bees are apparently accidental, and no report exists of the presence of large number of infected bees in a single colony. Milum (1938), the first to find a nematode in honeybee found *Mermis subnigrescens* Cobb in the abdominal cavity of one honey bee in the apiary at the University of Illinois AT Champaign, USA. Evidently, nematodes are not a major threat to beekeeping. Baur *et al.* (1995) exposed honey bee workers and brood to four entomopathogenic nematode species under conditions normally encountered in the hive by spraying nematodes onto combs. Mortality of adult bees exposed to any of the nematode species was less than 10%, and there was no evidence of nematode infection when dead adults were dissected. To assess the impact of nematodes on brood, smaller-size honey combs were placed in the second story (super) of a hive and large brood combs placed in the main section of the hive. Inconsistent results were found between these two experimental designs. The smaller honey combs sprayed with *Steinernema carpocapsae* contained the largest number of uncapped cells, those sprayed with *Heterorhabditis hacteriophora* or *S. riobravis* contained an intermediate number of uncapped cells, and control combs and those sprayed with *S. glaseri* contained the fewest number of uncapped cells. Large combs sprayed with *S. riobravis* contained more uncapped cells than controls or those sprayed with *S. carpocapsae*, although the differences were not significant. Their results do not support the hypothesis that high-temperature-tolerant species of nematodes are necessarily more infective to honey bees

However, *Sphaerularia bombi* a nematode (tiny worm) is found in bumblebee queens only and affect their behaviour. It belongs to Class: Secernent, Subclass: Diplogasteristeria, Order: Tylenchida, Suborder: Sphaerulariina and its scientific name is - *Sphaerularia bombi* It was first seen in 1742, and is one of the oldest known insect parasitic nematodes of Hymenoptera (*Bombus, Psithyrus, Vespula*). It is distributed in Canada, France, Germany, Netherlands, New Zealand, Norway, Sweden and USA

Control

No control measures are available to control nematode parasitism in honey bees. Such parasitism is not currently a problem. However, it might become one if nematodes are used for biological control of pest insects. It is, therefore, essential that like all organisms and chemicals, nematodes considered for control of pests should first be tested for their adverse effects on useful insects including the honey bee (Dutky and Hough, 1955; Cantwell *et al.,* 1972; Hackett and Poinar, 1973).

Insect enemies

A number of insects trouble honey bees which directly or indirectly inflict damage to honeybee colonies including adults, brood and stored food inside the beehives.

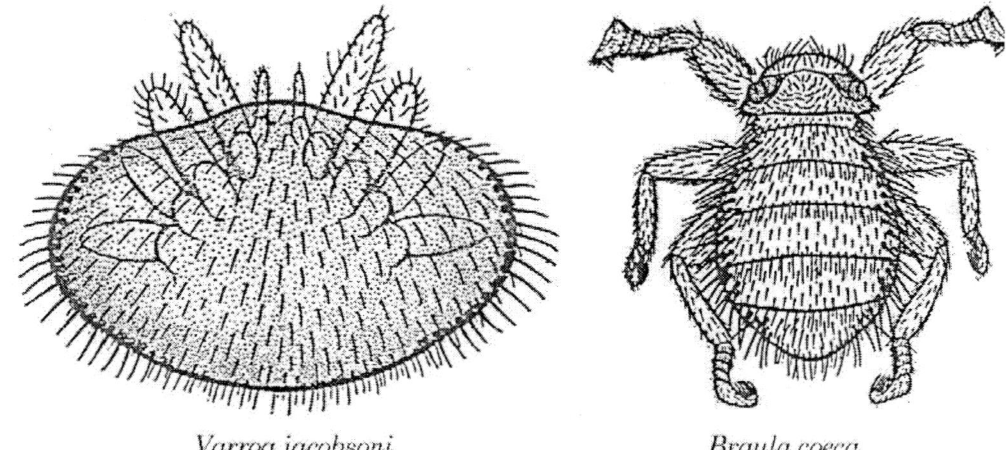

Figure 92. A comparison of the external morphology of the bee louse (*Braula coeca*) with *Varroa* mite (*Varroa jacobsoni*)

Bee Louse (*Braula coeca*)

Biology: *Braula coeca,* or the bee louse, is actually a wingless fly and an external parasite of adult bees. The adult lice are small (slightly smaller than the head of a straight pin), reddish brown, wingless flies. They first appeared in the US as "hitchhikers" on the bodies of imported queens. While several adult flies may live on a queen, usually only one lives on a worker. Bee lice seem to prefer nurse bees; only rarely do they live on drones. *Braula* move rapidly over the body surface, settling on the dorsal surface at the junction of the bee's thorax and abdomen. They remain there until a hunger response causes them to crawl up to the bee's head near its mouthparts. This movement seems to irritate the bee, causing it to regurgitate a drop of nectar. *Braula* then inserts its mouthparts into those of its benefactor and takes its food. Bees actively try to remove the lice. The louse lays its eggs on the cappings of honey storage cells during May through July. After oviposition, the adults die. Upon hatching, the young larvae burrow into the cappings. As the larvae grow, their tunnels lengthen and broaden; at this stage the infestation is easiest to observe. The larva pupates inside the tunnel after making a line of weakness in the wax to aid in its emergence as an adult. Soon after emergence, about twenty-one days later, the young adult crawls upon a bee. The diet of the larva appears to be wax and perhaps pollen grains incorporated into the wax by worker bees. In New Jersey, bee lice overwinter as adults and do not appear on queens until June.

Potential for Economic Loss: Very Slight (bees) to Moderate (comb honey appearance). *Braula's* damage to a colony of honey bees is limited. The amount of food taken by the larvae and adults is negligible. However, tunneling larvae can damage the appearance of comb honey. Honey production by strong colonies infested with bee lice appears to be little affected.

Control Practices: None. Little work has been done on control of *Braula*, and the measures that are suggested are antiquated.

Ants

The greatest natural enemies of the honeybee are all types of ants: driver, tailor, black, red, brown, large or small, all are dangerous to the hive. They eat sweets such as nectar, honey, sugar and the bee's body. In beekeeping, ants are one of the most widespread source of trouble for beekeepers. A great many species of ants attack honey bee colonies eat or carry off any comb contents, honey, pollen and brood. Weak or the small colonies are susceptible, but occasionally strong colonies are also lost after ant invasions. The harassment of honey bee colonies by ants causes aggressiveness leading to absconding of *Apis mellifera* and *A. cerana* colonies. However, much less is known about ants in relation to *A. dorsata* and *A. florea* colonies. Ants have been reported to inhibit pollination activity by reducing the foraging activity of honey bees in flowers.

Several species of ants attack honey bee colonies, but only a few of them kill a colony immediately, e.g., poneroid ants. These ants kill even larger animals that cannot get away. The poneroid ants belonging to subfamily Dorylinae are carnivorous and represent the savage or hunting stage in the evolution. Colonies include army, driver or legionary ants especially *Eciton burchelli* in South America and African driver ants *Anomma* and *Dorylus*. An army ants may contain up to 700,000 individuals which play havoc with honey bee colonies. Among myrmecoid ants (Subfamily Dolichodermae) is *Iridomyrmex humilis*, the Argentine ant. Evidently, the ants belonging to the Dorylinae and the Ecitioninae families are serious enemies of honey bees. Of the several species of ants, the Argentine ant, *Iridomyrmex humilis* Mary. is a serious pest of honey bees wherever it is found. The ants attack over a period of several days, and they are capable of destroying even strong populous colonies. May (1961) reported that over a period of six years in South Africa 160 colonies were lost due to predation by ants. *Iridomyrmex humilis* has also been reported attacking colonies of honey bees in Louisiana (Cockerman and Oertel, 1954), Florida (Robinson and Oertel, 1950), Rhodesia and Bermuda (Papadonpoulo, 1965). In Europe, the red wood ants (*Formica rufa* L.) have been reported to attack and destroy the honey bee colonies; a whole apiary consisting of 20 colonies, for example, was lost to these ants in Romania (Prosie, 1959). Similarly, in

Germany, bees in 26 hives were lost due to predation by *Formica rufa* (Muthel, 1959). Most of the ant species of genus *Camponotus* are particularly attracted to sweets. These ant species invade the hives and feed on stored honey. *Componotus abdominalis floridanus* Buckley has been reported as serious pest of honey bees in Florida (Walshaw, 1967).

In India, Singh (1962) recorded *Camponotus compressues* as a serious pest of honey bee colonies. Ants are highly organized and are the most successful of all groups of insects. They differ from bees and wasps in that they have one or two distinct humps between the thorax and abdomen, and in the existence of an entirely wingless worker caste is wingless and while this restricts the movement of ants, it has allowed them more efficiently to use the ground and plant substrates than do the winged bees and wasps. In spite of their small size their numbers and habits make them the most important of the invertebrate predators. Ants are most common in North America and their diversity increases as one moves from the temperate to the subtropical and tropical areas. In the lower latitudes, ants are more important, in terms of their total biomass and their effect on the environment. Ants are the most serious predators of honey bees in tropical and subtropical Asia. Being highly social, they attack hives en masse attacking virtually anything in them, the dead or live adult bees, the brood and the honey. Apiaries of *Apis mellifera* under ant attack become aggressive and difficult to manage, weak colonies most often than not abscond. Absconding is also the defense strategy of *A. cerana* against frequent ant invasion.

Many ant genera and species have been reported causing problems to both traditional beekeeping with *A. cerana* and to modern beekeeping with *A. mellifera*. Among the most frequently recorded species are the weaver ant (Occophylla *smaragdina*), the black ants (*Monomorium indicum, M. destructor, Oligomyrmes* spp., *Dorylus* spp.), the fire ants (*Solenopsis* spp.) and *Formica* spp. (Akratanakul, 1986). Available information on the diversity of ant species associated with honey bees is presented in Table 69. In India, not much work has been done on the ants. Ali (1991) reported that about 125 species of ants belonging to 7 families occur in Karnataka. Gadagkar *et al.* (1993) recorded 140 species belonging to 6 sub families and 30 genera from 12 localities in Uttara Kannada district of Karnataka. Abundance of ant species was highest in the genera, Pheidole (24 species), followed by *Monomorium* (17 spp.), *Crematogaster* (Family Myrmiemae) (14 spp.) and *Camponotus* (Family Formicinae) (12 spp.). But none of these studies recorded any relationship of ants with the honey bee. Singh and Naim (1994) reported *Tetraponera rufonigra* (Jerdon) as pest of honey bee *Apis cerana* during monsoon season. They found that attack resulted in complete destruction of 8.0 to 9.0% of colonies and partial destruction of 8.0 to 18.0% of the colonies. In general, ants interact with honey bees in three different ways: The Argentine ants, the driver or army ants are important predators attacking and destroying entire colonies of honey bees. The carpenter ants cause structural damage by tunneling in the wood of bee hives and nests (those in trees). A few species of ants complete with honey bees for sugar, e.g., wood ants.

Table 69. Ants associated with honeybees

Species/common name	Area	Reference
Camponotus noveboracensis Fitch	New York	Morse and Gary, 1961
C. pennsylvanicus DeGeer (Black carpenter ant)	New York,USA	Morse and Gary, 1961
Camponotus compressus	India	Abrol, 1997, Abrol and Kakroo, 1994, Burgi and Bhat, 1994
Camponotus sp	India	Gadagkar *et al.*, 1993
Crematogaster lineata Say	Missouri,USA	Burill, 1926
Dorylus labiatus Shuckard	India	Singh 1962
Formica fusca L.(Silky ant)	New York,USA	Morse and Gary, 1961
Iridomyrmex humilis (Red meat an, black sugar ant)	South Africa Australia	Buys, 1990 Woodward and Hones, 1991
I. peaninosum analis Aadre	Arizona, USA	Spanger and Taber, 1970
Lasius niger americanus Emery	Missouri, USA	Burill, 1926
Lasius niger Mayr	Englang,UK	Mace, 1927
Monomorium pharaonis L. (Pharaoh ant)	South Dakota,USA	Worden, 1917
M. indicum Morell	India	Singh 1962
M. destructor Jorden	India	Singh, 1962
Solenopsis geminata Fabricius (Fire ant)	Lousiana; USA	Cockerman and Oertel, 1954
S.invicta Buren (Red imported fire ant)	Lousiana, USA	Cockerman and Oertel, 1954
Tetraponera rufonigra (Jorden)	India	Singh and Naim, 1992 Ali, 1992, Sihag, 1991

Source: Abrol, 1998

Management of ants

Ants can be managed by following proper management practices as well as by chemical methods such as

1. **Reduction of nesting sites:** The area around the bee hives in the apiary should be kept clean by cutting the grasses, weeds, etc., and eliminating the rotten wood.

2. **Destruction of nests:** The ant nests near the apiary should be traced and destroyed by burning in situ as in case of weaver ants. In case of black ants which make underground nests, the colonies can be destroyed by fumigant and sealing them with mud. Pour boiling water or insecticidal soap down the anthills.

3. **Hive stands:** Use of hive stands provide effective control of ants that make the hive inaccessible to ants (Subhapradha, 1957). Hive stands should be 30 to 50 cm high to support the hives. The smearing of legs of stands with spent engine oil or grease provides effective control. Alternatively, the ants can be controlled by placing the bee hives on wooden stands with their legs in earthen cups containing water.

Termites

Termites are wood-infesting creatures and since most bee hives are made of wood, termites have to be listed as a hive pest. Termites are only after the wood – not bees or honey. Hives placed on the ground or bee equipment left lying around on the ground or stacked directly on the ground may be subject to termite infestation. If termites destroy the bottom board the bees may not have a bottom entrance and the colony could be more difficult to move. Termite ants are present in almost all tropical and subtropical areas. Termites destroy wooden structures in houses as well as our beekeeping equipment. It does not matter to termites whether bee boxes are empty or full of bees.

Control

Termites seek wood to feed upon and live in, so beekeepers need to avoid putting wooden equipment in direct contact with the ground. Active colonies on hive stands will usually be protected against termite attacks. Keep equipment stacks and spare equipment free from contact with the ground

WASPS

Members of the insect order Hymenoptera other than sawflies, ants and bees are often referred to as Wasps. There are over 2,000 varieties of wasps. These insects are medium sized (10-25 mm) and are readily distinguished by the bands of black and yellow or white on their abdomens. Several wasps species viz. *Vespa cincta, V. orienatlis, V. ducalis, V. mandarinia, V. velutina, V. analis, V. flaviceps, V. structor, V. vulgaris* and *V. germanica* have been reported preying honeybee from different parts of India.

Losses

On an average 20-25% of bee colonies are lost due to persistent wasps attack. The wasp's attacks usually coincide with the dearth periods when the bee forage sources, as nectar and pollen are scarce. Of all the *Vespa* species preying *Apis mellifera* L. and *Apis cerana* F., *V.cincta, V. velutina* and *V. basalis* are the most serious and cause heavy losses by feeding on adult bees, their brood and honey reserves. *Apis mellifera* is relatively more susceptible to wasps attacks than *Apis cerana* and predation often concides with flowerless dry season (July-Oct). The *V. mandarinia japonica* is the only hornet species which is known to attack en masse. First a single hornet kills a few bees and takes them to its nest to feed larvae; than it marks the site (eg. hive) with a pheromone from its Vender Vachit glands. When the three or more hornets have been attracted to the hive they attack en masses; a colony of 30000 of *Apis mellifrea* can be killed in three hours by 20-30 hornets. Wasps hover near hive entrance and catch returning/outgoing foragers and they also catch bees as they forage on the flowers. Adult bees, bee brood, honey, pollen etc. represent a vast storehouse of energy for wasps. The deserted colonies perish due to starvation as the wasps attack often synchronized with the dearth period. Wasps attack usually increases after the middle of August. But the *Vespa mandarinia* attack and crush honeybees one after another with their mandibles until all the honeybees are wiped out, then carry away all of the larvae. The hornets would enter the nest, kill the bees, and take their bodies home to their young. After a few successful trips, hornets rub a pheromone on the nest that signals other hornets to attack. So in autumn, most beekeepers think it is quite normal, that measures are needed to protect the honeybees from these natural enemies. They attach very often a wasp trap to the front of the hive entrance.

Management

Management of wasps requires an integrated approach involving cultural, mechanical, biological and chemical practices.

Management practices

(1) Strengthening of the bee colonies: Strong bee colonies can efficiently defend against the wasp attack. It is, therefore, always advisible to keep the colonies strong and vigorous with prolific queen.

(2) Mechanical destruction of the wasp colonies: Locate the nests or look for foraging wasps, install traps and check them at each visit. The other way is to track the foraging workers back to their nests. Mark the nests and destroy them at night when most of the wasps are inside.

(3) Discouraging or eliminating nests: Killing of early season foraging wasps which are nest founderess eliminates the problem to large extent. Early in the season, knocking down newly started paper wasp nests will simply cause the founding female to go elsewhere to start again or to join a neighboring nest as a worker.

(5) Physical killing of the wasps by flappers: Physical killing the wasps by fly flappers is one of the options for their management when populations are low. Flapping regularly for half an hour keeps the wasps away for at lest three hours.

(6) Baits /feeding attractants: Different baits can also be used to attract the wasps. Once they have accustomed to baits, they can be killed by mixing different chemicals in the baits. Some of the baits tested include: Cypermethrin + rotten fish/chicken; Cypermethrin + pear / apple / pumpkin / banana / pineapple; Cypermethrin + sweet candy; Fruit juice (Grapes juice fermented for three days); Mutton + 0.075 % Diazinon

(7) Aerial spraying of insecticides (e.g. Dichlorvos, Carbaryl 5%, Propuxor 1% etc.) effectively reduces the number of the foraging wasps. Subterranean colonies can be killed by pouring carbaryl 5% in to the external hole followed by plugging with cotton or cloth.

The best-known **chemicals for wasp control** are insecticides containing 0.5% propoxur, 2% malathion, or pyrethrins. Spraying is usually best after nightfall. The use of a red filter over a flashlight also permits the sprayer to see what they are doing without aggravating the wasps.

(8) To make an area **less attractive to yellow jackets:** do not litter, cover garbage cans with lids empty trash cans more frequently, caulk small cracks and holes in external walls, screen larger entrances such as vents, and eliminate puddle water.

(9) To control **ground wasps,** press a small cone made out of window screening over the hole. At the same time, place a shoebox over the cone. The workers will exit the nest, enter the trap and be unable either to leave the box or to re-enter the cone. The colony will die in a week or two. Stuff rags and dirt around the base of the trap to prevent escapees.

(12) Mass destruction of wasp nests through poisoning of brood

Capsule cup technique

Most of the wasp nests are either underground, hidden in forests, foliage, trees, rocks and or present in inaccessible areas. Their removal becomes a problem for beekeepers. The capsule cup technique can effectively be used to eliminate those nests (Mishra *et al.*, 1989). In this technique, wasps predating in the apiaries are captured, anesthegial and maintained in cages. Gelatin capsules (medicine capsules) are emptied and filled with poison such as fenitrothian. A cup filled with poison bait is fixed on to the thorax of a live wasp with quick fix, elfy or any other adhesive. The hind and middle legs were amputated upto the femur leaving the forelegs alone so that wasps could not shake off the poison load with the help of hind and middle legs. This technique, while retaining the body mobility, restricted the wasps from disturbing the poison on its back. Thus, the load could be safely taken by the wasps to the nests. The loaded wasps were then released to return to their nests. The wasps could carry a load of 100 mg which is immediately removed by the nest mates and gets distributed into the nest killing developing brood as adults.

Chemical fixation

In this technique a sticker is selected which would readily stick to the wasp and have less chances of being peeled off by the wasp. The nail polish has been reported to be the best sticker which had ideal consistency and did not come off when applied to the wasp's abdomen (Sharma *et al.*, 1986). No physical injury was noticed to the wasp on application of this sticker. The nail polish is applied to the tergal end soon after powdered CELPHOS (Aluminium phosphide) is fixed to it. On drying up of the nail polish the chemical got attached to the body of the wasp.

Wasps are collected in the apiary with the help of a butterfly net, caged and fed with dead bees and jaggery to avoid mortality owing to hunger and exhaustion in captivity. After the treatment, they are released to return back to their nests. To avoid shaking off the poison load, the hind and middle legs upto the femur are amputated, restricting the wasps from disturbing the poison on its back. Sharma *et al.* (1986) found that each wasp could easily carry 0.6 gm of CELPHOS and 3.0 gm of CELPHOS was sufficient to destroy the underground nest of *Vespa mandarina.* They further found that if the treatment is given during pre-monsoon period when the nest are being built and the colonies are less populous, it can help in destroying colony completely or making it to desert the nest.

(13) Topical application

Wasps can be effectively controlled by topically applying powdered Al_3Po_4 @ 0.68 g / wasp which was carried by wasp to their nest. It was found that 3.0 g of Al_3Po_4 was sufficient to completely destroy the colony of *V. mandarinia.*

Technique for collecting arboreal wasp nests

Wasps are serious enemies of honeybees and cause considerable damage. They kill the bees and their persistent attack results in absconding of the colonies. Morethan 25% of the colonies have been reported to be killed in mountainous areas (Adlakha *et al.,* 1975). Among various management techniques destruction of their nests near apiary is one of the most important weapon for their control. Though the underground nests could be easily destroyed by burning or spraying with insecticides but those on buildings, trees etc. at inaccessible places pose a serious threat for beekeeping. As they cannot be easily burnt for fear of forest fires or sprayed with the chemicals. A technique devised by Freeman (1973) is most useful for collection of such nests without any harm to beekeepers. The device could be successfully used to collect the arboreal nests complete with all their inhabitants. The successful device has a rigid rim within which the net was suspended by "Bull dog" type paper clips. Through the rim tape of the net a strong cord was threaded, attached to the rim at one end and at a point a few centimeters above the Y-Junction. The other end was passed through a steel eye at the junction and down the 9 m bamboo handle to act as a draw string. The action of pulling this string jerked the net from the clips away from the rim, securely enclosed the nest and tightly gripped the supporting branch around its irregular contours. A cleat was added at the hand end of the handle for securing the string. Branches and vegetation surrounding the nest to be collected were first cut back with a long handled pruner leading to the nest exposed at the tip of the branch to which was attached. The modified net was then carefully slipped over the end of the branch to draw string was pulled and secured.

A second person then cut the supporting branch near the nest and the captured nest was lowered to the ground. A thin wire net spreader (not figured) attached to the apex of the net rim was a useful addition that functioned to hold the net bag fully open while it was directed around the nest, thus minimizing disturbance to the wasps. Freeman (1975) could successfully remove nests from height upto 10 m additional higher nests could be collected by first climbing past way up the trees. Protective clothings including head net, was essential to avoid being stung.

WASP TRAPS

A number of wasp traps have been developed which differ in their effectiveness from one location to another and from one situation to another as given below: Glass Jar Wasp Trap, Double bait wasp trap, Lure wasp trap, Water Traps, Spur wasp trap, Soda bottles wasp trap (for details see Abrol, 2008)

Bio control of wasps

Rose *et al.* (1999) found that pathogens can be effectively used for long term control of social wasps. They found that wasps of the genera *Vespula*, *Vespa*, and *Dolichovespula* and their associated nest material contained 50 fungal, 12 bacterial, 5-7 nematode, 4 protozoan, and 2 viral species, although few have been confirmed through bioassay as pathogens of these wasp species. Despite few naturally-occurring host-specific pathogens and records of diseased colonies, wasps are susceptible to generalist insect diseases in bioassays. Fungi belonging to the genera *Aspergillus*, *Paecilomyces*, *Metarhizium*, and *Beauveria* have been confirmed through bioassay as Vespiniae pathogens, as have the bacteria *Serratia marcescens* and *Bacillus thuringiensis*, and nematodes *Heterorhabditis bacteriophora*, *Steinernema* (= *Neoaplectana)* sp., *S. feltiae*, *S. carpocapsae* and *Pheromermis vesparum*. Several of the pathogens listed here provide a resource from which inundative control agents might be developed, but none have potential as classical self sustaining control agents that can be transferred from generation to generation. As few studies have systematically searched for pathogens, it is likely other candidates suitable for use as control agents may be found.

DEFENCIVE BEHAVIOUR OF BEES AGAINST WASP ATTACK

Bees can handle invading wasps and kill them by 'baking' them alive (Abrol, 2006). The Japanese Honeybee's have been shown to exhibit thermal defense against the Japanese giant hornet, *Vespa mandarinia japonica*, preys bees. When a solitary hunter finds a nest, it marks it with a secretion from its van der Vecht gland. Other hornets in the area congregate to the area, and they begin a mass attack on the colony. While they are efficient at wiping out hives of the introduced European honeybee *Apis mellifera* (they are killed at rates as high as 40 per minute), the native Japanese honeybee, *Apis cerana japonica*, has an interesting defense against the predatory hornet. The Japanese honeybees can detect the hornet's secretion, and attack incoming hornets en masse. With approximately 500 honeybees surrounding the hornet in a tight ball, the temperature within the cluster rises to 47°C (117°F), which is above the upper lethal limit range of 44-46° for the hornet. This temperature is too high for the hornet, which quickly expires, but does not harm the honeybees. This temperature does not aversely affect the honeybees because their upper lethal limit is slightly higher, 48-50° (Ono *et al.,* 1987, 1995). The indigenous honeybee *A. cerana* also exhibits marked defence behaviour against predatory wasps. The bees starts simmering movements at the hive entrance and attack the incoming wasp enmasse forming a ball against the wasp which dies due to heat and suffocation. The balling continues for 5 to 10 minutes.

WAX MOTHS

Introduction

There are several species of moths regarded as pests of bee products. They include Greater Wax Moth - *Galleria mellonella;* Lesser Wax Moth - *Achroia grisella;* Fruit (pollen) Moth - *Vitula edmansae* and Mediterranean flour moth - *Esphestia kuehniella* and Indian meal moth-*Plodia interpunctella.* They belong to class insecta, order, Lepidoptera and family Pyralidae. These last 3 feed mainly on pollen and are less destructive as they do not make extensive webs in the wax combs.

Of all moths, two closely related moths, the greater wax-moth *G. mellonella* and the lesser wax-moth *A. grisella* are apiary pests in all parts of the world. Although, the greater wax-moth *G. mellonella* causes the greatest damage in apiaries which lead to material and financial losses every year followed by the lesser wax-moth *A. grisella* which also causes considerable damage, however, both have similar habits and can be controlled in the same way. Wax moth larvae are a very destructive pest and can quickly destroy stored beeswax combs. They tunnel and chew their way through combs, particularly brood combs and combs that contain pollen. Healthy, populous honey bee colonies do not tolerate wax moth larvae in the hive. However, larvae readily attack combs in hives in which the honey bee colony has died or where the colony has been severely weakened due to a number of causes. In such cases, combs left unattended by bees are vulnerable to attack by wax moth larvae. Wax moths are never the initial cause of colony destruction. The major damage caused by wax-moths is confined to stored combs and combs in hives which are weak or have died out. Under favourable conditions, the moth larvae can destroy good combs in a few weeks. It is not uncommon to find one or two wax-moth larvae even in a strong hive. These larvae seem to escape the attention of the bees and do not cause major damage. Individual larvae are often responsible for a condition called **bald brood** – a number of uncapped pupae forming a line 5 cm to 10 cm long, the result of a wax-moth larva tunnelling across the surface of sealed brood. These exposed pupae may become deformed and may be rejected by the bees. Moths can damage stored blocks of wax and foundation, though wax that has been thoroughly cleaned and stored in the light is usually safe from damage.

Moths usually enter colonies at night by evading the guard bees. Adult female wax-moths are capable of laying up to 1800 small, white eggs. These are just visible to the naked eye, and are laid on the side of combs or in cracks in the wood of the hive. After a few days these larvae hatch, crawl onto the comb, and begin their feeding activity. Newly hatched larvae feed on protein deposits in the form of pollen, larval excreta, pupal skins and some honey in wax. They damage or destroy the combs by boring through the cells as they feed on cocoons, cast skins, and pollen. As they chew through the wax, they spin silken galleries for protection. Combs are often reduced to a mass of webs and debris. Refined bees wax, such as foundation, is rarely damaged as it does not contain enough nutrients. Larvae develop quickly in old, dark combs, soon forming a mass of webbing. Final instar larvae often migrate from feeding sites to a suitable pupation position. They may chew into the timber and partially bury themselves before spinning a cocoon and pupating. If the moth population increases greatly, the colony may abscond.

Geographical distribution of wax moths

The geographical distribution corresponds reasonably with that of the bee. Distribution is limited by the inability of the pest to withstand prolonged periods of cold. This explains why wax moth problems are less acute in higher altitude locations or do not occur at all.

Biology of the Greater Wax Moth

Life cycle of wax moth depends upon the prevailing weather conditions and climate of an area extending from 6 weeks to 6 months. *Galleria* development goes through 4 consecutive stages-egg, larva, pupa and adult. This sequence is only interrupted if the temperature is too low or when there is no food.

A female moth lays 300-600 eggs, though some individuals may lay upto 1800 eggs per day. The eggs are spherical, smooth and pinkish to creamy white in colour 0.4 to 0.5 mm in size and are laid in batches of between 50 and 150. The eggs are laid by females by means of their ovipositor into crevices and gaps between hive parts in little used corners. They are not usually noticed unless specifically looked for. This puts them out of reach of the bees and prevents their destruction. The eggs hatch within 3-5 days when temperatures range from 29°C to 35°C. Hatching is delayed when temperatures are colder. For example, at18°C hatching commences about 30 days after egg laying.

The larva

At first, larvae are creamy white, but turn grey on reaching their fully grown size of up to 28 mm in length. After hatching, the small very active larvae tunnel in comb, lining their tunnels with silky web as they go. They move from comb to comb through a mass of webbing. Larval developmental time depends on temperature. The larval stage may last from 38 days to 5 months, depending on nutrition and environmental conditions. During this period, larvae may vary from 1/25 inch to 1 inch in length. After hatching, the young larva immediately searches for a comb in order to feed and to build the silk-lined feeding tunnels. Speed of growth is directly dependent on temperature and food supply. Under ideal conditions the larval weight can double daily during the first 10 days. The metabolic warmth, which is created by this rapid growth, can increase the temperature in the spun silk nests far beyond the environmental temperature. The larva feed in particular on impurities occurring in wax, such as feces and the cocoons of bee larvae as well as pollen. The larva also eats wax. Larvae, which have been reared exclusively on pure wax (foundation, fresh comb), do not complete their development. Dark, old combs that contain many bee larval cocoons are most at risk. When fully grown, the larva spins a rough silken cocoon, which is usually attached to the frame or inside of the hive. Frequently, the larva cements the cocoon inside a cavity chewed in the wood. Chewed frames are weakened and easily broken. Within the cocoon, the larva changes to the pupa and overwinters in the pupal stage. Under warm conditions, adults may emerge at almost any time of year.

Spinning

At the end of the larval stage, the larva spins a very resistant silk cocoon on a firm support, such as wooden frames, hive walls or in the comb storage chest. Frequently, the larva spins its cocoon in a hollow it had bored into the wood.

The pupa

In the cocoon, the larva changes into a pupa and then into the adult moth. These metamorphoses last from one to 9 weeks.

The adult Insect (imago)

Size and color of the imago vary considerably, depending on food composition at the larval stage and on the duration of the various developmental stages. Adult moths are pale brown to grey, usually about 20 mm long. The grey wings are often mottled and appear as "roof" or "boat" shaped when folded over the body. Females are larger than males. The females start laying eggs between day 4 and 10 after emergence from the cocoon. At dusk, the females attempt to enter the beehive to lay their eggs. If the colony is strong enough to repel the wax moth, they lay their eggs outside in cracks in the wood.

Biology of the lesser wax moth, *Achroia grisella*

The lesser wax moth is very common all over the world, except the colder regions. The reproduction of these moths comes in the form of ultrasonic signals sent out by the males in order to attract their female partners. The typical mating spot for these moths is in honeybee colonies, and the males can be seen there anywhere between six and ten hours in just a single night. Females are attracted to these ultrasonic signals in any one of three ways; peak amplitude, signal rate, and the asynchrony interval. Studies have been conducted that show these moths increase their signal rate when having to compete with others for the right to a local female, but due to the physical demands of an increased signal rate, its duration typically lasts only five to ten minutes.

The life cycle of the wax moth consists of five definable stages. The stages are egg, larva, spinning, pupa and adult. The larvae cause most of the damage to comb, the spinning stage causes the damage to woodwork and finally the adults cause further damage by mating and propagating the species.

Egg Stage

Eggs are laid in cracks between hive parts or in groups on the upper side of cells. Female lesser wax moths can produce up to 300 eggs and prefer to lay them in close proximity to used brood comb. Wax moth eggs hatch into larvae after five to eight days depending on ambient temperature. The eggs require a damp atmosphere to hatch. All stages of development are affected to a large extent, by temperature variation. Cool temperatures slow development and warm temperatures accelerate it.

Larval Stage

Lesser wax moth larvae are usually white with a brown head. They feed on combs, pollen and litter found on the hive floor. They are usually solitary, whereas greater wax moth larvae often congregate in large numbers. Freshly hatched larvae burrow into the comb towards the midrib. They are gross feeders and grow and feed for between one and six months depending on ambient temperature. When larval growth is complete they are 16 mm - 20 in length and look rather like a caterpillar. They have a reddish brown, dome shaped head, creamy white bodies with three segments that have a pair of legs and several other body segments, some of which have caterpillar style prolegs. The body colour changes slightly and progressively, with advancing age, to a light grey, occasionally with a pink or salmon pink tinge. The grub eats beeswax, but needs additional detritus within the comb structure (bee cocoons, feaces and pollen) to provide protein. All the while the larva is tunneling it leaves a fine silken tube and charcoal grey granular faeces behind it along the track that it has followed. The larvae is at it's most destructive in areas that are dark, warm and poorly ventilated. The larvae can detect each other somehow (maybe smell or Co_2 emission). If a box of combs that is infested with larvae is banged smartly on to paved ground and then removed, the larvae that are dislodged will set off, in a radial fashion, all in different directions like an expanding star. It may be due to the fact that by dispersing they both lessen competition between the individuals and explore as many as possible different habitats. Foundation is only partially attacked, usually around the edge where the wax fits into the groove in the wooden frame parts. This damage is caused by small larvae that have not found sufficient edible material, they stay small and as they are inadequately fed they often die before reaching adequate size to pupate.

Spinning

When the larva reaches full size (approaching it's last moult) it starts to spin a coarser silk, with which it makes a cocoon that is papery in texture and very strong. The silk used here is made from the material excavated from the surface on which the larva pupates. The colour of the cocoon is normally white if made on a softwood surface, but on the hardboard colour may be brown.

Pupal Stage

On completion of the cocoon, the larvae itself changes to the pupal stage. Wax moth pupae may hatch rapidly or take up to two months to change to the adult stage depending upon temperature. Pupation can occur within the comb or in the loose debris at the bottom of the hive, but most frequently it is firmly attached to the frame or hive woodwork, particularly in places where there is an internal angle in the woodwork. The cocoon is cemented into a boat shaped cavity the larva chews in the wood. It may be due to the fact that the cellulose mined from this cavity is converted into the silk that is used to make the cocoon and may account for the difference in silk texture. The pupae are about 12 mm long cigar shaped with the maximum width in the middle at about 3 mm. As the adult has no working mouth parts, it is not yet clear how it emerges from the cocoon. It certainly comes out of one end and leaves an almost circular flap, but how this is achieved is not known.

Adult stage

Adult wax moths cause no direct comb damage because their mouth parts are atrophied. They do not feed during their adult life. This moth is smaller than the greater wax moth and has a silver-grey to buff, slender body about 13 mm in length. Males do not rely solely on pheromones to find a mate they also use ultrasound. The moths, otherwise immobile can often be observed vibrating and this behaviour is perhaps involved with finding a mate. The vibration may be a by-product of emitting ultrasound. Infrasound may be used as well as ultrasound (infrasound will travel long distances in solid media). Wax moths fly mainly at night and during daylight they rest in dark places. They are reluctant to fly, preferring to run. This may be their energy saving strategy. Females appear like a clothes moth 10 mm to 15 mm in length, with wings folded at a shallow angle over the top of the body. The wings have scales and are a slate grey or dirty brown sometimes with a minor bronze tinge. The bronze colour becomes more apparent if touched with a finger. Female lesser wax moths can produce up to 300 eggs and prefer to lay them in close proximity to used brood comb. A component of the wax moth's female sex pheromone 'Nonanal' is also found in beeswax & may explain how wax moths find suitable wax rich places for laying their eggs.

Differences for the greater wax moth *Galleria mellonella*

Eggs up to 1800 in number are laid in batches of between 50 and 150. The eggs themselves are laid in cracks between hive parts in little used corners. Greater wax moth eggs are 0.5 mm across, olive shaped with a white/salmon pink hue. They are not usually noticed unless specifically looked for. Adult moths are pale brown to grey, usually about 20 mm long with wing span up to 40 mm.

The grey wings are often mottled and appear as "roof" or "boat" shaped when folded over the body. Adults are pinkish in colour. Grubs are pearly white initially, turning mushroom grey as they age. They are larger than *Achroia* larvae and more tapered at the back end. The cocoons are larger and the pupae inside them are a light golden brown. There is more silk produced that almost knits the frames together. Males are slightly smaller than females and have a scalloped leading edge to the wings. The greater wax moth flies in daylight even less than the lesser variety and if exposed to light will often scuttle away. Such dislike of light applies to the larvae of galleria as well as the adults (Fig. 93a,b).

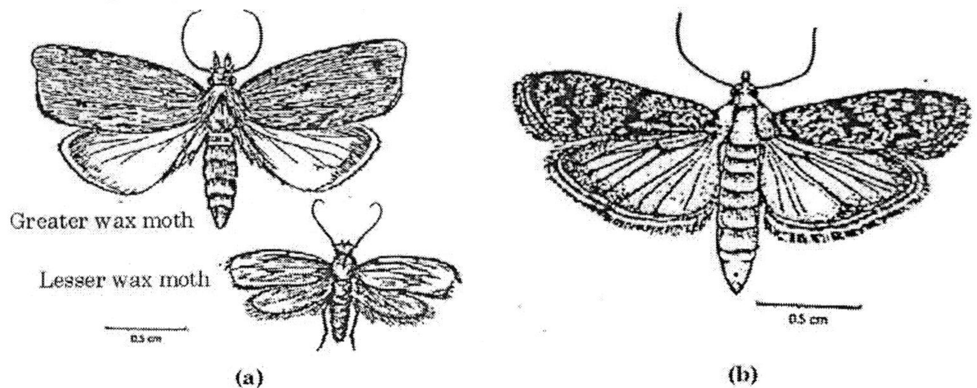

Figure 93(a). Comparative morphology of greater and lesser wax moths (b) dried fruit moth.

Lesser wax moth

This moth is smaller than the greater wax moth and has a silver-grey to buff, slender body about 13 mm in length. Lesser wax moth larvae are usually white with a brown head. They feed on combs, pollen and litter found on the hive floor. They are usually solitary, whereas greater wax moth larvae often congregate in large numbers.

Simultaneous infestations

In most cases, sooner or later, both the Greater and Lesser wax moths work in the degradation of the same colony. Their diet typically consists of honey, beeswax, stored pollen, bee shell casings, and, in some cases, bee brood. While tunneling through honeycombs attaining food, these moths are also protecting themselves from their main enemy, the honeybee. The size and voracity of the Greater wax moth larvae leave the lesser wax moth larvae very much in second place. The Greater consumes the combs and the lesser is relegated to survival on the remaining residues. The lesser has adapted well to this austere diet and can be found where the Greater is unable to grow. The Greater plays its part in the primary degradation of combs and when these have been reduced to a pile of debris formed of tiny wax fragments, larval excrement, cocoon remains and threads, the larvae of the Lesser take advantage of the remaining residues by reducing its volume, finally leaving a black, aromatic, granulate dust as the end product of their joint action. At the end of the summer the larval development starts to slow and progressively stops, almost completely, hibernate and pass the winter in their cocoons, bringing to an end the annual cycle, only to await the dawning of a new year in the coming spring. Keep wax moths at bay by maintaining strong, healthy colonies at all times, and be ever watchful for signs of wax moth eggs and larvae in your hives.

Conditions favouring infestation

Weak colonies are more prone to to the attack of wax moth than the strong colonies. Following are some of the factors which may cause the colonies to grow weak.

1. **Lack of food**. Generally the dearth periods due to the strong winter (in temperate climates) or extreme heat or a hot period coupled with rainy season, result in the physical scarcity of food or limitation of the bee activities, force the colonies to starve. Under such conditions, the colonies become weak and experience wax moth attack.
2. **Failing of queen**: During the starvation period or lack of pollen flow queen fails to sustain reproductivity and growth of the colonies, hence the colonies become weak.
3. **Queenlessness**: The queen may die due to several reasons or it is killed by the workers at the time of food scarcity or when it gets old.
4. **Pesticidal poisoning**: The application of pesticide in the field may render the colonies weak on account of heavy worker mortality through chemical poisoning.

Extent and nature of damage caused by wax moths

Adult wax moths cause no damage because their mouthparts are atrophied. They do not feed during their adult life. Only larvae feed and destroy combs. However, adult wax moths and larvae can transfer pathogens of serious bee diseases (e.g. foulbrood). In colonies infested with foulbrood. The feces of wax moths contain large amounts of *Paenibacillus* larvae spores. Newly hatched larvae may move to neighbouring honey bee colonies and can travel more than 50 metres. This has implications for the control measures and protection of apiary products and stored combs. Wax moth larvae are very active in warm weather, but become somewhat inactive in the extreme cold of winter. At the optimum temperature of about 32°C they reach full development in about 19 days after hatching. At cooler temperatures and when food is scarce, the larval period may extend to 5 months. The fully developed larvae spin silky cocoons which may be found in a mass of webbing in the comb or on the frame and internal surfaces of the hive. Larvae may form small depressions in the wooden hive components in which to spin their cocoons. Larvae can also bore through the wooden bars of frames. After spinning the cocoon, the larva commences the pupal stage which lasts about 14 days when temperatures are high but as long as 2 months at cooler temperatures. After emergence, adult moths mate and the life-cycle begins again.

The greater wax moth (*G. mellonella*) is the most serious and destructive insect pest of unprotected honey bee comb in warmer regions. Wax moths primarily infest stored equipment but will invade colonies whose worker bee population has been weakened by disease, queenlessness, failing queens, pesticide kills or starvation. However, many people outside of the beekeeping industry consider wax moths a beneficial insect because the larval stages are used as fish bait and for feeding insects in zoos. Newly hatched larvae are white but successive instars are medium to dark gray on the top with creamy white undersides. The larval head capsule is brown. Wax moth larvae prefer dark combs because they contain a variety of nutrients such as entrapped pollen and larval skins. The larvae grow rapidly and will migrate toward the edges of the frames or corners of the supers to spin a cocoon and pupate. Damage occurs as the larvae burrow into the comb feeding on the wax, larval skins, pollen and honey. As the larvae chew through the comb they spin a silk lined tunnel through the cell walls and over the face of the comb. These silk threads can tether emerging bees by their abdomens to their cells and they die of starvation because they are unable to escape from their cell. This phenomenon is termed galleriasis. In severe infestations, the wax comb, wooden frames, and sides of the hive bodies can be heavily damaged.

Furthermore, the wax moth larvae produce sufficient quantities of metabolic heat, temperatures as high as 25°C above the environmental temperatures are produced in the centre of the aggregation of larvae. Increased heat is detrimental to developing bee larvae, and workers have to spend a great amount of energy and time to lower down the increased temperature.

Heavy infestations by wax moth (along with lack of food, water and attack of diseases) causes absconding. Whenever, colony population decreases sufficiently to expose combs, wax moth larvae develop unhindered and gradually attack combs occupied by bees, causing the host colony to abscond.

Beehives

Honey bee colonies that have become weak and have low numbers of adult bees due to starvation, queen lessness, excessive swarming, disease, pesticide poisoning or neglect cannot effectively guard their hive against wax moth infestation. At first, combs left unattended by bees are subject to wax moth infestation and damage. As the colony weakens further, even combs with bees may be damaged. On occasions one or two wax moth larvae may be seen in a healthy, populous hive. These are mostly removed by the bees and very little, if any, damage occurs. Sometimes, the beekeeper may find one or two wax moth larva between the hive mat and the top bar of a frame. Again, these individuals cause little damage and may be removed by the beekeeper.

Bald Brood

Developing honeybee pupae are exposed when wax moth larvae partly remove the cell caps. Worker bees chew the remainder of the capping thereby exposing the heads of the pupae which continue to develop normally. The lines of bald brood follow the direction of the wax moth's travel. Some honey bee pupae nearing maturity may have deformed legs or wings. One of the causes of this deformity is a result of wax moth excreta affecting the final moult of the pupa before its emergence from the cell.

Galleriasis

Newly formed adult bees are sometimes unable to emerge from their cells. These bees are trapped by silken threads produced by greater wax moth larvae tunnelling at the base of the cells. The bees eventually die and are later removed by hive bees. This is a minor problem and is rarely seen.

Damage to stored combs and hive material

Damage will vary with the level of infestation and the time that has elapsed since the infestation first began. In time, stored combs may be completely destroyed and the frames and combs become filled with a mass of tough, silky web. In ideal conditions for wax moth development, a box (super) of combs may be rendered useless in about a week. Most damage occurs in the warm months of the year when wax moths are most active. Little if any damage is seen in the extremely cold winter period, because the larvae are relatively inactive. However, greater wax moth larvae have the ability to congregate together and create their own heat which is greater than the surrounding ambient temperature. When this occurs, considerable damage can still occur during the cooler months of late autumn and early spring. At the time of storage, combs that are apparently free of wax moth may contain eggs that will hatch later. They should be monitored at frequent intervals for signs of moth infestation unless treated as discussed later. Adult moths may also lay eggs in the external cracks and joints of stacked supers of combs. After hatching, the tiny larvae quickly tunnel though the joints to infest the combs. This explains why some combs that have been treated by freezing to kill all life-cycle stages of the wax moth become infested afterwards. Wax moth larvae prefer dark brood combs that contain some pollen. However, combs sticky with honey, white combs and combs containing honey are all prone to infestation. Larvae may damage frames and hives by chewing away the wood to make cavities for cocoons.

Damage to apiary products

Bee collected pollen

Wax moth larvae grow extremely well on a diet of fresh or dried pollen. Adult wax moths may deposit their eggs whenever pollen is exposed at any stage of collection, processing or packaging. The eggs can hatch at any time, even after packaging, making the pollen unsuitable for human consumption.

Comb honey

Honey sold in the comb must be protected immediately it is taken from the hive. Wax moth eggs present on the comb at the time of packaging will hatch soon after. The webbing and excreta produced by the larvae will render the comb unfit for sale.

Beeswax

Clean, refined beeswax including foundation is not readily subject to wax moth attack. Damage is usually minor and the larvae usually fail to grow to full size. Dirty, unrefined beeswax and slum gum is more readily infested.

Wax Moth Control

The methods employed in combating *Galleria mellonella* are generally effective against other moths identified as pests of bee products. The greater and lesser wax moths are serious pests of stored honey supers. The generally mild climate ensures active wax moth populations almost year around. Management of wax moth requires regular monitoring for signs of wax moth infestations which include webbing, debris, pupal cocoons, and tunnels in the combs. Stored equipment that contains comb is most susceptible to wax moth infestations. The various methods employed for the control of wax moths include:

Cultural control: Cultural control methods include the use of management practices and non- chemical methods as given below:

Management practices for wax moth control include: Strict vigilance by the beekeeper during the management of his colonies is more important to save the colonies from the pest. Following measures must be observed to combat the wax attack:

1. The most effective method for preventing wax moth damage in hives occupied by bees is to maintain strong colonies. The bees will remove the moth larvae and repair the damage as it occurs. Strong colonies can often protect themselves from wax moth intrusion, but weak colonies may not be able to control the infestation.
2. Weak colonies can be combined to make strong colonies which have a better chance of survival during a dearth period. The queen of one colony must be killed to combine the colonies. A strong colony can be divided again later during the build-up period to increase the number of colonies. During the dearth period, the objective is survival of bees, not survival of colonies. it is better to have one colony that survives rather than two that die out.
3. Maintain adequate food supply to maintain the strength of the colonies especially during dearth periods which may weaken the colonies.
4. Protection of colonies against disease and pests which may weaken the colonies.
5. Cleaning of hives fortnightly during the activity period of the wax moth.
6. Safety of colonies against pesticidal hazards.

7. Remove empty, unused comb from the hive during the dearth period. Put older, darker comb away from the brood nest so that it can be removed when it becomes empty.
8. Keep the bottom area of the hive clean of residue that drops. Wax moths can build up in the residue and move onto the comb. Slanting the hive a little in the direction of the entrance helps the bees to remove this residue before it builds up.
9. Do not leave comb lying around the apiary or stacked inside nearby shelters. This provides a good rearing site for the wax moth population of the area. Render unused wax into blocks instead of storing it as comb. The wax moth cannot complete its life cycle in blocks of pure beeswax.
10. Good beekeeping practice including scraping of burr comb and propolis from hive and frame woodwork will reduce the opportunity for wax moths of either species to become established.
11. Older comb is more susceptible than foundation or newer combs. This is another reason for embarking on a regular comb renovation program. Therefore, remove black old combs susceptible to wax moth attack.

Non-Chemical Control

Temperature extremes can be used as a non-chemical control measure for wax moth control.

Heat: All states of the greater wax moth are killed at a temperature of 115°F (46°C) for 80 minutes or a temperature of 120°F (49°C) for 40 minutes. Be sure to allow combs to reach the required temperature before measuring the exposure time. WARNING: Be careful not to expose honey combs to temperatures in excess of 120°F (49°C). Heat-treat only those combs having very little or no honey (combs softened at high temperatures may sag and become distorted). Heat-treat supers of combs only when they are in the normal, upright position. Provide adequate air circulation for the heat to be evenly distributed throughout the comb. Ventilating fans are useful for this purpose. Turn the heat off and allow combs to cool before moving the supers.

***Light*:** Combs with both surfaces exposed to light will have little if any wax moth damage. Supers of combs are stacked on specially constructed elevated insect-proofed wire floors so that light penetrates the stacks. It may be helpful to remove a comb from each super and space the remaining combs equally to allow penetration and even distribution of light. The tops of the stacks are covered with flywire to allow entry of light and prevent robber bees gaining access to honey in the combs.

Cold rooms: Wax-moth growth is restricted at temperatures below 18°C and death occurs at -6.7°C when experienced for 5 hours or longer. Commercial cold rooms are used to store frames safe from wax-moth damage. Cold is the best method of controlling wax-moths in comb honey. By holding the temperature down to 12°C, growth of the wax-moth is restricted and little damage can occur. The use of cool rooms to store combs and protect them from wax moths has become increasingly popular in the beekeeping industry. A temperature of 4°C will restrict wax moth activity.

Note: Combs stored at low temperatures become brittle and must be handled carefully to avoid damage. Turn the cool room off or remove stacks of combs from the cool room the day before use to allow the combs to reach room temperature. This practice will prevent damage during loading or transport.

Cold: A natural way of getting rid of wax moth is by freezing. It is also very effective since all life cycle stages are killed. The lower the temperature the quicker they wax moth stages are killed. This can be achieved in two ways. Firstly by storing combs in as cold a place as possible and hoping there will be a good period of freezing weather over the winter – a fortnight of hard frosts will do the trick nicely. However, since hard frosts do seem to be a thing of the past beekeepers can make certain by freezing combs themselves. Supers are frozen for a minimum 48 hrs. This kills all stages of both species of wax moth. The use of low temperatures can prevent the sagging problem which sometimes occurs when combs are treated with heat. Combs with honey and pollen can be treated by use of low temperatures without much danger to the combs. WARNING: Very cold honey combs are very brittle. The minimum temperature and exposure time needed to destroy all stages of the greater wax moth are shown in the Table 40.

Table 70. Showing range of temperature and exposure time needed to destroy all stages of the greater wax

Degrees Celsius	Degrees Fahrenheit	Time in hours
2	36	240
-7	20	4.5
-12.2	10	3.0
-15.0	5	2.0

Once the combs are treated, store them where no adult wax moths can get to them. Inspect combs monthly for any signs of infestation, especially if temperatures rise above 60°F.

Dessicating atmosphere: It is thought that wax moth eggs need a moist atmosphere to hatch. Therefore the addition of desiccants to the various methods would probably improve them as treatment against the wax moth.

Irradiation with gamma rays: Irradiation with gamma rays will kill all developmental stages including eggs, but costs are high and furthermore suitable storage container are required, that is totally moth proof, to house the combs after they have been sterilized.

Carbon dioxide: Carbon dioxide (CO_2) at concentrations above 95% can effectively control wax-moth. Carbon Dioxide (CO_2) can be used to exclude air, from comb honey that is intended for sale as well as for storing drawn comb in sealed containers or comb cupboards

Bee collected pollen: Collecting pollen from pollen traps every 2 or 3 days will help to minimise, but not prevent wax moth infestation. Drying and packaging of bee collected pollen should be done in a moth proof environment and the product frozen to kill all life-cycle stages of the moth. Allow sufficient time for the product to cool to the desired temperature before timing the commencement of the treatment. Freezing of pollen after packaging will ensure that the product is free of wax moth eggs and will not be at risk of infestation until such time as the package is open by the consumer.

Comb honey: This product may also be frozen to control all life-cycle stages of wax moth. However, some types of honey (for example, clover and lucerne) may candy after freezing. Allow time for the product to cool to the desired temperature before timing the commencement of the treatment. To some extent, freezing comb honey for 24-48 hours can eliminate the guesswork. Comb honey may also be held in cool rooms but this treatment will not kill wax moth eggs.

Apiary hygiene: Slum gum, old combs and scraps of beeswax are best collected and processed to form clean cakes of beeswax. These items left unprocessed and lying around the apiary or honey extracting facility can provide an ideal environment for development of wax moth. Combs and other items infested with wax moth are best treated or destroyed. This will significantly reduce the number of adult wax moths and help to reduce further infestation. A clean environment will always assist in minimizing the incidence of wax moth.

Control of wax moth in colonies : The bees themselves are the best control of wax moth in active bee colonies. It is not unusual to find an occasional wax moth adult or larva in a colony. They will be in out of the way places and in areas bees can't get to such as areas between top bars and inner covers. The bees may even have sealed the caterpillar off with a propolis fence. If many combs, especially darker combs that have had brood in them, or a weak colony is available, more wax moths and their damage may be evident. Beekeepers frequently state that wax moths are responsible for killing their colony. They are not capable of doing this. This may be due to the fact that the colony became weak, or more likely lost its queen, and the population dwindled to where there were too few adults to protect the combs. The adult female lays her eggs and the caterpillars hatch and grow. The caterpillar protected in its silken tunnel is hard for the bees to remove. Before the beekeeper discovers the weakened or queenless colony, the damage can accelerate. Under tropical climates, wax moths can completely destroy brood combs in a month. In addition to insuring active, populous colonies, keeping the hive clean and free of debris can help reduce wax moth damage. The bees need access to all parts of the hive. Don't neglect to remove the debris that accumulates on the bottom board or in cracks and crevices. Reasonable removal of burr comb and propolis will also help remove places where wax moths can become established.

Acoustics: Spangler (1988) stated that both lesser wax moth, *Achroia grisella* (F.) and greater wax moth, *Galleria mellonella* L. males produce sounds using tymbals located on their tegulae. Wing movement twists one end of a tymbal causing it to buckle and produce an ultrasonic pulse. Both sexes are equipped with tympanic ears that hear the high-frequency sound. *A. grisella* females use the sound to locate males prior to copulation. In contrast, female *G. mellonella* respond to the sound with wing fanning. This wing fanning sets off a more complex, three-step behavioral sequence that allows the females to locate males by male-produced pheromone. Techniques that make use of the moth-produced sounds to detect and control these pests of bee products include locating calling males with electronic detectors and using acoustically baited traps to capture receptive females could be an alternate strategy for their management.

Defencive behaviour of bees against wax moths

Wax moths *Galleria melonella* and *Achroia grisella* are also victims of biting and dragging by bees within colonies (Eischen *et al.,* 1986)). These behavioral responses, which occur inside the nest, involve the same principles of enemy identification that apply to intruders at colony entrances and may be behaviorally related to threats from outside the nest. Eischen *et al.* (1986) found that Africanized honeybee colonies were quicker, more persistent and more intense in their attack than our European colonies against adults of greater wax moth and their hive entrances. The intensity of attack was related to the number of bees guarding the entrance and not with the colony size. Africanized honeybee has more number of guard bees as compared to European honeybees.

Wax moth trap: Cushman (2000) has developed a wax moth trap which trapped both *A. grisella, G. mellonella* as well as a few species of wasps and other flying insects, but no bees. It is simply constructed from a 2 litre plastic drinks bottle. A 30 mm diameter hole is cut in the side of the bottle, just below the shoulder of the neck. The mixture in the trap as bait is made up of following ingredients: 1 Cup White Vinegar; 1 Cup Sugar (Any type) and 1 Banana Peel. The bottle is topped up with water until all the liquid represents 75% of the volume of the bottle. The mixture fermentation will take place, which may be attractive to wax moths. To use the trap just hang it up using a loop of baling twine, near to hives in the apiary. The wax moths are attracted, enter and just drown. Pheromone traps have been tried, but are only partly effective as several pheromones are utilised by the moths and other mechanisms (Ultra-sound) are involved in detecting the opposite sex.

CHEMICAL CONTROL

Chemical methods: Both wax moth eggs and larvae can almost always be found in colonies, but are routinely eliminated by worker bees and never build to injurious population levels. It is important to remember that large wax moth populations in bee colonies are generally the result of a bee colony being weakened in population for some other reason (starvation, pesticide poisoning, failing queen) which allows the moths to get established. Although, generally present in weakened bee colonies, wax moths are not the direct cause of a colony's demise. Traditionally, wax moth control in stored supers is accomplished by chemical fumigation. The availability and suitability of these chemicals, however, is constantly in flux. They are expensive, may require special training to use, and there is the ever-present chance the chemicals might find their way into the honey. Several chemical fumigants effectively used in the past are aluminum phosphide, methyl bromide, ethylene dibromide (EDB) and paradichlorobenzene (PDB). Unfortunately, only two (aluminum phosphide and PDB) remain legal, but their future is in doubt. Beeswax is similar in structure to many insecticides and often has an affinity for them. As a consequence EXTREME CAUTION should be exercised when using pesticides anywhere near a beekeeping operation. Alternatives to chemical fumigation have not been found to be practical in large-scale application, but may be useful in smaller outfits. These include the use of hot and cold temperatures, and fumigation with carbon dioxide. Fumigation with carbon dioxide (CO_2) is extremely dangerous, not because the chemical is inherently toxic, but because the user is at risk from suffocation.

Sulphur (sulphur dioxide, So_2): Burning of sulphur strips or spraying of So_2 from a pressurized vessel are the two main control methods using sulphur. This is still one of the most effective means against wax moths. It is highly volatile, not fat-soluble and therefore poses only a slight danger to bees, wax, and honey. After removing comb from the colonies, it is advisable to wait one or two weeks before treatment (So_2 is ineffective against eggs). For more safety, the treatment can be repeated after 2 weeks.

Acetic acid: Acetic acid vapor (80%) instantly kills eggs and moths. The larva, especially in the cocoon, is more resistant and must be exposed to the vapors for longer. For this reason, the combs must be treated immediately after removal from the colonies, before eggs can develop into larvae. Acetic acid can also be used to fumigate a stack of dry supers. This has the advantage that it will also clean other problems from the comb such as EFB bacteria or nosema spores but the disadvantage that it is corrosive to the frame nails.

Formic acid: Professional beekeepers in Europe successfully use *formic* acid against wax moths. The effects are comparable to that of acetic acid.

Paradichiorobenzene (PDCB) fumigation for stored combs: If supers must be stored in a warm room or basement, they may be protected by placing paradichlorobenzene (PDB) crystals on a small piece of paper on every fifth super in the stack, which should then be covered. PDB is heavier than air there is no need to put it at the bottom of a stack of supers/hive bodies. PDB works best above 70°F as it volatilizes to the gas state. It is non-explosive and non-flammable. To get the best fumigation, stack your hive bodies as tightly as possible, even taping cracks and broken covers. Combs removed for storage should not contain honey. The treatment must be continued at regular intervals throughout the winter. PDB kills adult and immature wax moths, but not eggs. The continuous presence of crystals within the stack not only repels moths and prohibits egg laying, but also kills any young larvae that hatch after the combs are placed in storage. Untreated combs should be inspected regularly for signs of infestation, especially if temperatures rise above 60°F and permit wax moth activity. Since beeswax comb can absorb the gas odor, supers should be aired after removal from storage before using them in the spring on bee colonies. *Caution:* Moth balls and crystals (naphthalene) should not be used to control wax moth.

For long storage, repeat application of PDB after crystals have vaporized (3 to 4 weeks in summer, longer in winter). A specially constructed fumigation room increases the effectiveness of PDB. Boxes should be aired for at least 24 hours before being placed on colonies. In high concentrations, PDCB can be toxic to bees. If several combs are put directly into the colony from a storage chest without airing, heavy damage may occur and can result in the death of the colony. Use 3 ounces of crystals for each stack of 5 full depth boxes or 8 half depths. Placing the crystals on a piece of cardboard or newspaper is preferred over putting the crystals directly on the top bars. Remember, as the gas is heavier than air so put the crystals at the top of the stack. Keep the bottom closed to help retain the fumigant in equipment stack. If the ambient temper-ature remains high, check the crystals every month or so and replenish as necessary.

Contamination of wax and honey by paradichiorobenzene (PDCB): PDCB is a highly volatile and lipophilic (easily soluble in fat and wax) substance. Beeswax can take up this material and a part of it may later migrate into honey. Honey analyses from Germany and Austria have shown that PDCB residues in honey are not rare. This applies to native as well as imported honeys. Even when measured values pose no problems as far as human toxicology is concerned, but detection of residues even in traces spoils the reputation of honey as one of the last natural products in the eyes of the public. Therefore, all beekeepers who are concerned about the quality of bee products are advised not to use PDCB and it is recommended that alternative control strategies be employed. Hohenheim (1992) analyzed 109 samples of honey and found that 51 samples contained PDCB residues in levels ranging from 3 to > 50 μg/kg.

Paradichlorobenzene in wax: The amount of PDCB stored in wax depends on the duration of exposure and the wax surface area. Foundation takes up PDCB more quickly than wax as a block (Table 71). Wax takes up PDCB like a sponge. The more PDCB crystals are added to combs, the longer it acts and higher the substance stored in the wax.

Table 71. Uptake capacity of a 1 kg wax block

Time Span	Paradichlorobenzene
After 1 Month	27.3g
After 2.5 Months	38.5g
After 9 Months	83.5g

Evaporation of PDCB from beeswax

Airing: Airing of combs over 1-2 days before insertion into the colony avoids visible damage to bees. Despite this, considerable amounts of PDCB may still be present in wax. Airing over several weeks is not enough to remove PDCB from wax completely. The amount and speed of removal are above all temperature-dependent. Thus, the considerably higher temperature in the colony causes PDCB evaporation from combs not previously aired enough. If these cells are now filled with honey, PDCB migrates slowly into the honey.

Melting old wax: When old comb is melted, the residues persist in the new wax. Examinations of wax carried out have shown that the majority of comm-ercial wax in Switzerland contained PDCB residues in the range of 5-10 mg/kg.

Stability of PDCB in honey: The use of PDCB poses several problems in honey as given below:

1. PDCB evaporates reluctantly from honey and only from the top most layer.
2. Honey cannot be aired as long as needed, since it attracts water and odors.
3. There is no possibility of significantly reducing paradichlorobenzene content of honey later.
4. Residues of PDCB in honey are not permitted and honey with any residue that is not normal is rejected.

Warning: Read and follow the safety directions on the label before use. Avoid inhaling fumes. Wear gloves and suitable protective clothing when hand-ling PDB to avoid contact with skin. Store unused crystals in sealed contai-ners away from food items. PDB is non-flammable and non-explosive.

Phosphine: Although phosphine (Phostoxin, Gastion, Celphide, Alphos Detia and Fumitoxin) will kill all stages of the wax-moth life cycle, however, reinfestation can occur in stored equipment if it is not securely sealed. Fumigate equipment out in the open when a gentle breeze is blowing towards an uninhabited area. No combs removed for storage should contain honey. Place these combs in supers in separate stacks up to five high. Each stack should have as a base a sheet of PVC (plastic) 0.25 mm thick or heavier. Fold up the edges of the PVC sheet and seal with PVC tape or clips. Securely seal all joints and cracks between the boxes. Place one tablet in a container (to catch residue) on the top bars of the frames in the top box. Cover the stack with a lid of PVC sheeting and seal with PVC tape.

After seven days, the wax-moths should be dead. If boxes are moved to a storage area, each stack must be resealed to prevent entry of adult moths. Check regularly for signs of reinfestation particularly during summer. All treated equipment must be ventilated thoroughly before use. This should be carried out in the open by allowing air to circulate through the equipment for five days.

Warning : Prominently display poison notices in the fumigation area throughout the exposure and ventilation periods and on all doors leading into rooms where boxes are stored. Phosphine is toxic to all forms of animal life. Avoid inhaling. Suitable gas respirators must be in hand when phosphine tablets are being used.

Withdrawal of ethylene dibromide (EDB): EDB has been banned for wax-moth control. It is a severe carcinogen and readily absorbed by beeswax and honey.

BIOLOGICAL CONTROL

Biological control includes the use of natural enemies, such as parasitic wasps and microorganisms such as bacteria, viruses and fungi.

Diseases and enemies of *G. mellonella*

Metalnikoff (1922) observed that *Galleria mellonella* was subject to natural epizootics due to certain viruses and bacteria. Significant mortality can be caused by the presence of microbes in the intestines or in the "blood" of caterpillars. Normally, caterpillars have a very scant microbial flora comprising only a *Micrococcus* spp. and a yeast (Toumanoff, 1951). In 1968, *G. mellonella* larvae infesting apiaries in Louisiana were found to be infected with a new strain of *Bacillus thuringiensis* Berliner (Barjac and Thomson, 1970). Spores of different strains of *B. thuringiensis* are now commonly used to control various species of harmful Lepidoptera. Among the Protozoa, *Coelogregarina* sp. and *Noserna galleriae* can cause fatal infections in *G. mellonella*. Some Hymenoptera specifically attack *G. mellonella*. In the Ichneumonidae group, *Eupelmus cinereus* Rondoni is a parasitoid of *G. mellonella* found in hives and in wax comb debris (Beljavsky, 1927). A polyphagous chalcidian, *Dibrachys boucheanus* Ratzb, also parasitises the caterpillars of *G. rnellonella*. A braconid Hymenopteran harmful to *G. mellonella, Apanfeles galleriae* Wilkinson (Singh, 1962) is found worldwide.

Natural Control: Wax moths can be controlled using bacteria or virus formulations as given below:

Bacillus thuringenisis

A natural microbial bacteria *Bacillus thuringenisis* (Certan®) discovered in 1911 is specific for wax moth. Commercial preparations based on *Bacillus thuringiensis* are supposed to be effective and environmentally acceptable. It is a microbial bacteria that can be kept alive by culturing in milk after macerating affected larvae. The bacterium produces spores containing a toxin and when the spores are ingested by the larvae, the toxin is freed and damages their intestinal walls. This results in the death of the larvae. There are different strains of this bacterium. The one used for bees is the Berlin strain. It is harmless for vertebrates (man, livestock) and bees, and leaves no residues in wax or honey. Bollhalder (2000) on the other hand found Mellonex®- as an effective biological control agent against wax moth which is very well tolerated by bee and brood. It is based on the natural bacterium *Bacillus thuringiensis* (Bt) and sprayed on the combs. Young wax moths are killed and the brood combs do not suffer any damage. Mellonex® is suitable as a biological control for wax moth, providing residue-free honey and wax production. It has no negative influence on the life span of the bees and the rearing of the brood in the recommended dosages. Therefore, Mellonex® can be used safely for the bees for an effective control against the wax moth on combs. Ahmad *et al.* (1994) tested three industrial preparations of *Bacillus thuringiensis* viz. Certan at the rate of 0.4, 0.5 and 1.0 ml; B-401 at the rate of 0.6, 0.75 and 1.0 ml and Dipel at the rate of 10, 20 and 30 g per 100 ml water against first and fourth instar larvae of greater wax moth, *G. mellonella* under laboratory conditions during July-August, 1988 and found that Certan at the rate of 1 ml per 100 ml water, B-401 at the rate of 1 ml per 100 ml water and Dipel at the rate of 30 mg per 100 ml water, respectively gave 73.3, 70.0 and 68.3% mortality of the first instar larvae and 43.33, 50.00 and 33.33% of the fourth instar larvae.

GmDNV (The Galleria mellonella densovirus)

The Parvoviridae family is a group of small viruses, of 250–280 Å diameter, that have single-stranded DNA genomes and are icosahedral. There are 60 protein subunits in each virion. An approximately 5.5 kb genome encodes the capsid protein and from one to three nonstructural proteins. Members of the Parvovirinae and Densovirinae subfamilies infect vertebrates and arthropods, respectively. The Densovirinae contain four genera distinguished by their genome organization and sequence similarities. Densoviruses, like other parvoviruses, are highly species-specific, possibly as a result of their extreme reliance on host function; this has prompted their use as highly specific pest-control agents. The larvae of the greater wax moth, *G. mellonella* (*Gm*), are parasitic on honey bee colonies and are frequently infected with *Gm* densovirus (*Gm*DNV). The *G. mellonella* densovirus (*Gm*DNV) capsid protein consists of a core-barrel motif, similar to that found in many other viral capsid proteins. The virus typically kills the host larvae within several days of infection (Belloncik, 1990; Simpson *et al.* 1998). Some densoviruses infect only the mid-gut and are transmitted through the intestinal tract, whereas others (e.g. *Gm*DNV) infect nearly all tissues except those of the mid-gut. Tal and Attathom (2005) evaluated the insecticidal potential of *G. mellonella* densovirus (*Gm*DNV) in third, fourth, and fifth instar larvae of the host, the greater wax moth, as a step toward the construction of a molecular vector for the introduction and expression of foreign genes in the larvae of these insects. Third instar larvae are most susceptible to *Gm*DNV. Viral RNA synthesis is more rapid in this stage and slowest in the fifth instar. Infection of prepupae by intradermic injection or by horzontal spread inhibited pupation. *Gm*DNV DNA is also infectious when introduced as a calcium phosphate precipitate. The two putative viral promoters were shown to be capable of driving the expression of the reporter gene chloramphenicol acetyltransferase (CAT) in DNA-injected larvae.

Apanteles galleriae

Wherever wax moths exist, we also find a wasp predator - a braconid wasp. Among the natural enemies *Apanteles galleriae* seems to the major predatory parasitic insect of *Galleria mellonella* larvae. It helps keep numbers down in an outbreak situation but is not effective enough for beekeepers and its commercial rearing operations are considered economically impractical. *Apanteles galleriae* Wilkinson 1932 (Hymenoptera : Braconidae) is a koinobiont, solitary, larval endoparasitoid of several lepidopterans including the pyralid wax moths, *G. mellonella* L., *A. grisella* Fabr., *Ac. Innotata* Walker, and *Vitula edmandsae* (Packard) (Watanabe, 1987; Shimamori, 1987; Whitfield *et al.*, 2001). Caterpillars of these host species are pests in beehives because they feed on pollen and generally destroy the combs. *Apanteles galleriae* adults feed on honey, fruit nectar, and host larvae in nature. It is a parasitoid of early-instar larvae of wax moths and emerges to spin its cocoon and pupate well before the host larvae reach full size, usually before the final instar. Whitfield and Cameron (1993) also reported *A. galleriae* to be a parasitoid of *Vitula edmandsae* (Packard), the common wax moth in North American *Bombus* colonies. Besides, there are a number of parasitic wasps which prey on larvae of the moths attacking bee hives. They are mostly members of the Family Braconidae and measure about 5 mm long (Table 72).

Table 72. Parasitoids associated with different moths attacking bee hives

	Galleria mellonella	*Achroia grisella*	*Aphomia sociella*	*Vitula edmandsae*
Braconidae				
Apanteles galleriae Wilk	+	+		
Apanteles hopliies (Ratz.)	+			
Apanteles lateralis (Hal.)	+			
Apanteles nephoptericis (Pack.)	+			+
Bracon brevicornis (Wesm.)	+	+		
Bracon bebetor (Say) Wesm.	+	+		+
Meteorus pulchbricornis (Wesm.)			+	
Meteorus salicorniae (Schm.)	+	+	+	
Microgaster deprimator (F.)	+		+	
Ichneumonidae				
Diadegma chrysostictum(Gmel.)	+			
Dolicbmitus messor (Grav.)	+		+	
Venturia canescens(Grav.)	+			
Pteromalidae				
Dibracbys cavus (Walk.)	+	+		

Nasonia vitripennis(Walk.)	+	
Chalcididae		
Pseudochalcis dircennae Bert.	+	
Eupelmidae	+	
Eupelmus cereanus Rond	+	
Trichogrammatidae	+	
Trichogramma evanescens Westw	+	+

Source: Patetta and Manino (1989).

Wani *et al.* (1998) found that parasitization of *G. mellonella* (Lepidoptera: Pyralididae) larvae by a larval endoparasitoid *A. galleriae* (Hymenoptera: Braconidae) leads to the precocious expression of premetamorphic behavior in the sixth (normally penultimate) instar host larvae prior to the parasitoid's emergence. They further found that the ecdysteroid titer in the hemolymph of parasitized sixth instar larvae (the last instar of parasitized larvae) was higher than that of unparasitized ones, and the high ecdysteroid concentrations induced premetamorphic behaviors such as wandering and cocoon spinning. However, the epidermis of the parasitized larvae was not pupally committed through this stage. The activity of JH esterase in the parasitized larvae remained low, and application of a JH analogue to these larvae caused the production of a larval-type cocoon. These facts suggest that the parasitization by *A. galleriae* induces precocious premetamorphic behaviors of *G. mellonella* larvae by changing host endocrine conditions without causing the typical larval-pupal metamorphosis.

It has been recognized that all females of several species of parasitoid wasps carry virus in their ovaries for successful host parasitization. Polydnaviruses (PDV) replicate into the ovaries of braconid and ichneumonid endoparasitoid wasps and are injected into their hosts during oviposition with a venom fluid (Stoltz and Vinson, 1979). It has been shown that in lepidopteras parasitized by endoparasitoid wasps, with the polydnavirus in their reproductive tract, viral gene expression is detected inducing immunosuppression and altering host development. These physiological and developmental alterations have been accompanied by major alterations in the protein content of host larval hemolymph beginning a few hours after host parasitization. These alterations may also occur in the late stages of wasp development inside the host or they may increase 4 post-parasitism and the expression continues for six more days.

Brochetto-Braga *et al.* (1995) detected the presence of a polydnavirus in the ovary of the microhymenoptera *Apanteles galleriae* parasitizing the larvae of the wax moth *Galleria mellonella*. They found that that polydnaviruses were capable of inducing significant physiological alterations in lepidoptera, which undoubtedly might be associated with suppression of host defense mechanisms, as suggested by several authors (Beckage and Kanost, 1993; Soldevila and Jones, 1994; Stoltz and Guzo, 1986; Strand *et al.*, 1992, Strand and Noda, 1991). They monitored alterations in the 6th and 7th instars of *G. mellonella* when associated with parasitoid wasps *A. galleriae*. However, for a better characterization of these alterations more detailed studies are needed.

Pheromone traps

Wax moth can cause problems to bee hives by laying its eggs in bee colonies. These eggs develop into caterpillars which consume young bee grubs. There is very little that can be done once the moth has layed its eggs in the colony. Use of the sterile male release technique has been shown to be a possible control strategy under test conditions but no program currently uses this methodology. There are traps available for stored product pests such as Indian meal and Mediterranean flour moths. They use synthetic sex attractants and live captured females to trap and eliminate the males. Fraser (1997) in Ontario studied the potential for attracting female greater wax moth using pheromone lures. Her lab results showed that there is considerable potential for trapping female greater wax moth. They can be attracted from a long distance with great effectiveness. In large commercial operations where individual examination of equipment is not feasible, or in small operations where an infestation can be pinpointed, a pheromone trap would be a useful monitoring tool.

The wax moth trap is a pheromone trap that uses lures to attract male wax moths. They are caught in the trap and are then not able to mate with females and this then reduces egg laying, which in turn reduces the caterpillar numbers. These traps can be used to monitor wax moth levels and help to reduce the overall number. Place the Wax moth trap out in May near the area where your bees are placed and replace the pheromone lure 10-12 weeks later. Each trap is supplied with two pheromone lures. Some studies have been conducted on the use of pheromone traps for capturing greater wax moth. It was found that greater wax moth male adults produce a sex pheromone in glands located on their forewings (Barth, 1937, Roller *et al.*, 1968). The pheromone was identified as a mixture of two aldehydes, nonanal and undecanal (7: 3) (Leyrer and Monroe, 1973). However, the response of females to the synthetic bait in laboratory tests was not as high as their response to live males (Finn and Pyne, 1977), and in field tests this mixture was practically inactive (Flint and Merkle, 1983). Two additional components, nonanol and undecanol, were found among volatiles collected from GWM males from Canada during their calling period. The ratio of the main components, nonanal and undecanal, of this population was found 1: 3 (Romel *et al.*, 1992). The ratio of these components in volatiles of greater wax moth from USA was 7: 3 (Leyrer and Monroe, 1973). Recently, Lebedeva *et al.* (2002) studied the composition of pheromone volatiles from calling males of the greater wax moth *G. mellonella* from six regions of Russia and found that the volatiles from calling males from all the regions contained nonanal and undecanal as the main components, but in different ratio for the males of greater wax moth from different regions. Hexanal, heptanal, octanal, decanal, undecanol and 6, 10, 14 -trimethylpentadecanon-2 were found as minor components also in different combinations. The structure of the ketone was proved by the comparison its mass-spectrum with spectrum of synthetic ketone. Further studies are needed for exploitation of pheromone traps for the management of wax moths.

Trichogramma wasps: Studies have shown that Trichogramma wasps could be used to control wax moths. Bollhalder, 1999) tested a biological control called *Trichogramma* which is a tiny stingless parasitic wasp. It is a parasite of many moths including the *G. mellonella* and is effective even under heavy wax moth infestation. *Trichogramma* wasp lays its egg in a *G. mellonella* egg and a wasp emerges instead of a moth. The wasp can only control the egg stage of the moth (*Trichogramma* wasps are solely egg parasites, meaning that they are ineffective on any stage of wax moths except eggs). *Trichogramma* wasps that are used commercially in green houses will control wax moth larvae, but they have to reach a particular size before the wasps will parasitise them, so some damage to combs is inevitable before control is gained. One or more species of parasitic (braconid) wasp naturally parasitize wax moth larvae. It oviposits single egg inside the larvae of *G. mellonella* and the developing wasps eat up the moth larvae.

Botanicals

Viraktamath and Basalingappa (2000) studied the efficacy of eight botanicals for management of greater wax moth under laboratory conditions and found that the larval mortality differed significantly and was dependent on the concentration and instar of the larva and the time interval. Seed extract of custard apple (*Annona squamosa*) gave significantly high mean mortality ranging from 94.08 to 98.29% in different instars and was the next best treatment to that of endosulfan (99.67-100%). Application with Indian privet (*Vitex negundo*), neem (*Azadirachta indica*) and sweet flag (*Acorus calamus*) showed high next mortality ranging from 98.29 to 91.98, 55.53 and 67.42 to 82.21%, respectively. The effects of pongamia (*Pongamia pinnata*) Clerodendron (*Clerodendrum inerme*) and tulsi (*Ocimum sanctum*) were moderate while (*Parthenium hysterophorus*) caused lowest mortality. High toxic levels of botanicals have been reported in the larvae of *Galleria mellonella* (Bolchi, 1979; Eischen and Dietz, 1987; Sylesha, 1787). Zaitoun (2007) examined the effects of ethanolic extracts of twenty one medicinal and health plants on the development of the greater wax moth *Galleria mellonella* and on honeybee workers and found that most of the extracts, prolonged the larval stage duration 2-40 days more than the control. Six extracts prolonged pupation period 2-5 days morethan the control. Extracts of *Abrus precatorius, Laurus nobilis, Petroselinum sativum* and *Plantago psyllium* had insecticidal effect against the moth; they killed 100 or 95% of the tested wax moths respectively without adverse effects on worker bees except in the case of *A. precatorius*. Worker honeybees were also affected adversely by few of the used extracts, the most poisonous was *Cicer arietinum*, followed by *Myristica fragrans* and *Raphanus sativus*. These extracts killed 80, 70 and 55% of the experimental bees, respectively. Some of the used plant extracts seem to act as insect growth regulators and toxicants and can be used effectively to control populations of wax moth. Exploration of botanicals for wax moth control opens a new area or investigation.

Beekeepers will never completely win the battle against wax moth. It is an insect well adapted for surviving around bee colonies. We need to be vigilant to not allow wax moth to take more than their share of drawn comb that the bees work so hard to produce. A summary of control methods is given in Table 73.

Table 73. Summary of methods for controlling wax moth

Method		Advantages (+) Disadvantages (-)	Procedure/Remarks
Technical		+ no residues	
- 	- Sorting comb		- supplementary measure - separate dangerous old comb from foundation and new comb
	- immediately melt old wax		- supplementary measure
	- storage in a cool, light, and airy place	+ simple	- Moths fear light and drafts; e.g. shed, porch; - Protect against weather, rodents and insects
Physical		+ no residues	
	- cool storage (<15°C)	+ effective - infrastructure, long term method	- cellar, cool place - good air circulation in comb stack
	- frost treatment	+ effective + kills all stages - expensive infrastructure	- 2 hours at -15°C or 3 hours at -12C or 4.5 hours at -7°C - strict period of frost
	- heat treatment	+ effective + kills all stages - infrastructure (warm air blower) - risk of wax melting	- 80 minutes at 46°C or 40 minutes at 49°C - good air circulation - accurate temperature control
Biological	- spores of *Bacillus thuringiensis* (B-401)	+ no residues + long-term effect (2-3 months) - average effect against the Lesser wax moth - expensive	- observe instructions - ensure good distribution on the combs - observe sell-by-date and storage conditions (living organisms) - if combs already infested, 1 x sulphur, then B-401 - ideal for the beekeeper with a few colonies
Chemical	- Sulphur	+ effective + good pollen conservation against molds - regular repeats - ineffective against eggs - fire danger	- treatment from above (SO2 heavier than air) - do not breath in vapors (respiratory and eye irritant) - burn in a small sulphur stove - treat every four weeks (in summer)

		- 1 strip per 100 liters (about 3 supers) - SO2 in spray can - 1 second (=2.5g SO2) per honey super or - 3-4 seconds per 100 liters hive volume - no fire danger
- Acetic Acid	+ effective + no problem residues + kills all stages + kills Nosema spores [10] - attacks metal parts - regular repeats - caution when handling	- treatment from above (vapors heavier than air) - do not breath in vapors, avoid contact with skin - 200ml acetic acid (60-80%) per 100 liters per hive volume [6;7;10;11] - in summer, treatment repeated 1-2 times with an interval of 2 weeks
- Formic Acid	+ effective + no problem residues + kills all stages - attacks metal parts - regular repeats - caution when handling	- treatment from above - do not breath in vapors, avoid contact with skin - 80ml formic acid (85%) per 100 liters hive volume - in summer, treatment repeated 1-2 times with an interval of 2 weeks
- Paradich-lorobenzine (PDCB)	+ Simple handling + effective - residues in wax and honey. - ineffective against eggs - toxic to bees at high dosages	- use cannot be recommended - aerate combs for 2-3 days before inserting into colony - treatment from above

BEETLES

Since beetles are the most numerous animals on our planet, it is little wonder that a few may be occasional bee pests. Some such as larger ground beetles may invade the colony or feast at the colony entrance. Reduction of the hive entrance or movement of the apiary location may be an effective control for these. Other beetles may live inside the shelter of a bee colony or infest stored equipment. Most are after stored pollen and bee bread and if the bees are strong enough they will keep the numbers of such beetles at a minimum. Stored equipment should be kept in tight stacks with PDB fumigation as for wax moth control. The small hive beetle, an accidental import from Africa, may be a serious problem.

SMALL HIVE BEETLE *(Aethina tumida)*

The small hive beetle (*Aethina tumida*) is the newest bee pest and a potential new threat to beekeeping. This beetle, indigenous to Africa, has recently spread to the USA, Australia and other countries where it has proved to be a devastating pest of European honey bees. There is a serious risk of its accidental introduction into other parts where it is still unknown. Mainly lives and breeds on its primary host the honey bee in colonies, stored comb and beekeeping equipment, but it can also survive and reproduce on certain types of fruit, particularly melons. The adult beetle is small (about 1/3 the size of a bee), reddish brown or black in color and covered with fine hair. The larvae are small, cream colored and similar in appearance to young wax moth larvae. You can differentiate the beetle larvae from wax moth larvae by examining their legs. Beetle larvae have three sets of legs just behind the head. Wax moth larvae, like all moth and butterfly larvae, have three sets of legs behind the head but in addition have a series of paired prolegs which run the length of the body. Prolegs are absent in beetle larvae.

LIFE CYCLE

Adult females lay their large egg masses on or near beeswax combs. In South Africa the eggs hatch in a few days, producing great number of small larvae. The larvae consume pollen and wax but also will eat honey, bee eggs and larvae. They complete their larval stage in 10 to 16 days and then drop to the ground where they pupate in the soil. Adults emerge from the soil in approximately 3-4 weeks. The females are capable of laying eggs approximately one week after emerging from the soil. They are good flyers and easily disperse to new colonies where they deposit eggs to begin a new generation.

Harmful effects of the small hive beetle

The SHB is considered to be a very minor economic pest of weak honey bee colonies and stored honey supers. It is a scavenger beetle, consuming dead colonies, in much the same way as wax moth. African bees have strong house cleaning and defensive traits, which include: preventing the beetles access to the colony by aggressively harassing them filling cavities where the SHB could hide with propolis removing SHB larvae and by confinement of beetles to 'propolis prisons'. Such behaviour limits SHB reproduction in African colonies and so keeps the beetle population down to manageable levels and below damage thresholds.

SHB and European Bees

Unlike African bees, European bees generally have fewer natural defences against the small hive beetle. Consequently, the beetle reproduces very successfully and the population grows much larger than in African colonies, and with far more harmful consequences. Weak colonies are at the greatest risk of infestation. Strong colonies will actively remove larvae (much as they can deal with wax moth larvae), but they are not able to deal with adult beetles due to their hard exoskeleton and their defensive behaviour. In the USA, colonies vary in their ability to resist the SHB. Scientists and bee breeders believe that bees that display these defensive traits and the ability to incarcerate beetles in propolis prisons should be selected for.

Damage to the colony

SHB larvae do the most damage in the colony, burrowing through brood combs and consuming the brood and stores. The level of harm to the colony depends on the number of beetle larvae present. Once present in large numbers, the survival of the colony is at great risk. Queens stop laying and colonies can quickly collapse. In heavy infestations, tens of thousands of SHB larvae may be present within the colony. In such cases there can often be up to 30 larvae per cell. Such large numbers can generate enough heat inside the hive to cause combs to collapse and subsequently for the colony to abscond. SHB larvae affect combs of stored honey and pollen and will also infest brood combs. During the feeding action by larvae an associated repellent sticky substance is laid down on the combs and this can result in bees abandoning the hive. When honeycombs are removed from colonies, bees then no longer protect the combs allowing the beetle larvae to feed uninhibited.

Honey spoilage

Defaecation of adult beetles and larvae in honeycomb causes the honey to ferment and drip out of cells. Affected combs become slimy and have a characteristic odour reminiscent of "rotten oranges". These combs are repellent to bees and can also cause absconding. It has also been demonstrated in quarantine facilities that small hive beetles can parasitise bumble bee colonies (*Bombus* spp.) causing very serious damage. This could have important ecological consequences if the beetles became established in the UK. However, it is not known if the beetles can find and infest these nests in the wild. Honey supers removed from colonies and stored before extraction are at great risk from the SHB, particularly if they are stored in a warm room, and contain pollen or brood. They can quickly be rendered useless for extraction and the honey within completely ruined and unfit for sale.

Potential impact of the small hive beetle on beekeeping

There is a serious risk that the SHB could be transported and introduced into other areas in swarms of bees or feral colonies inadvertently. Import regulations are one of the main defence against the introduction of the SHB (and other very serious bee pests and diseases) from overseas and it is absolutely essential that all beekeepers abide by them.

The SHB is well able to survive even in the colder climates where adult beetle can survive during winter in the bee clusters and can therefore survive in any location where bees exist. Light sandy soils are preferred and warm conditions are required for pupation and completion of the life cycle.

Figure 94. *Aethina tumida* (adult small hive)

CONTROL

Management Practices

Manage strong colonies -weak colonies are more vulnerable, because there are not enough bees to protect comb and defend the brood nest. Freezing at 10F for 24 hours, is reported to kill all life stages of the beetle. keep apiaries clean of all equipment not in use and extract honey as soon as it is removed from colonies. Destroy these beetles as soon as they are detected

Chemical Control: Chemical control for SHB is targeted at the adult stage of the pest. The placement of CheckMite+ ® strips underneath plastic corrugated cardboard is the only in-hive treatment approved for SHB in active bee colonies. Soil drenching with GardStar ® 40% EC (permethrin) is also used to kill SHB larvae as they burrow into the ground to pupate.

Biological Control: Various entomopathogenic fungi and nematodes have been evaluated for their ability to kill SHB larvae. No commercial products are currently available and it is likely none of them will represent viable alternatives to conventional pesticides from controlling SHB.

Flies

Flies are associated with honey bees in several ways. Some adult robber flies (Asilidae) are predators of adults bees. The bee lice (Braulidae) are restricted to honey bees; adults are ectoparasites of adult bees and their larvae of certain Conopidae, Tachinidae, and Sarcophagidae. Larvae of some Phoridae feed on honey bee larvae. The Drosophilidae, Phoridae, and Sarcophagidae each include a few saprophagous species whose larvae feed on dead bees as well as other dead organisms. Some of these saprophagous species have been incorrectly implicated as parasites. Although several of the dipterous pests of bees appear to be of only minor importance in their normal area of distribution, they have the potential of becoming pests of major importance in new areas where their own natural enemies, and other limiting factors, are not operative.

Minor Insect enemies

Thirteen of the 27 insect orders have members that are pests of honey bees, their comb, or their stored products. The Thysanura, or bristletails, are small, elongate insects with two or three tail like appendages at the year. Toumanoff (1939) indicated that large numbers of bristle tails may be present in colonies. Toumanoff and Borchert believed that silverfish seek the honey in a bee hive, but that they are not important enemies of the honey bee. Subhapradhs (1961) considered their presence unhygienic and believed that they should be destroyed. Their droppings can contaminate honey. The order Odonata includes two familiar types of insects, the dragonflies and the damselfies. Some of the larger species of dragonflies, also commonly referred to as mosquito hawks or darning needles, feed on honey bees. Foraging bees may wander into the clutches of several types of predatory insects such as praying mantids, assassin bugs or beetles. Such insects are not usually very numerous and none selectively feed on honey bees over other types of insects. Strong, healthy colonies can afford to suffer occasional losses to such pests without harm to the colony. There are several types of insects that may live for shorter or longer periods of time inside a bee hive or inside the inner cover of a bee hive. Roaches and earwigs are two good examples. Most of them do no detectable harm although the beekeeper may feel their presence unsanitary or unsightly. The Dermaptera, or earwigs, are long slender insects that have large forceplike tails. The adults have for wings but rarely fly. Earwigs are usually nocturnal and hid in cracks, crevices, under bark, and in similar places during the day. They are mainly scavengers and may be found in bee hives eat honey or the soft parts of injured and dying bees. Toumanoff (1939) indicated that one group of the true bugs, the Reduviidae or assassin bugs, are "dangerous" honey bee predators. They have been found in bee colonies and spread mites. The neuropterans are a diverse group of mostly predaceous insects. The familiar members of this order are the dobsonflies and fishflies. They are nocturnal and visit the bee hive. The beetle family of greatest consequence to beekeepers is undoubtedly the Meloidae, commonly called blister or oil beetles. Several species have been described as pests on honey bees. The Meloidae that are honey bee pests have a changing larval shape during this stage of the life cycle. Most beetle larvae, like the larvae of bees, butterflies, flies, and other insects with four life stages, simply grow in size through a serious of moults during larval life. The Meloe larvae change from very active, mobile larvae to inactive, grublike creatures. A beetle in this family, *Aethina tumida* (Murray), called the small hive beetle, has been labled as destructive as the wax moth *Galleria*. The beetle is only about 0.5 centimeters in length but both adult and larval stages cause damage in the hive. The spider beetle *Ptinus fur* (L.) will attack and damage beeswax comb in weak colonies in storage. The scarab beetles are a large family with diverse habits. Several members of this family have been recorded as bee pests. Singh (1962) described a tenerbrionid beetle, *Platybolium alverarium* (Blari), as occupant of the debris on the bottom boards of weak hives in India. This species, as well as species of *Bradymerus*, thrives in unhygienic hive conditions. Regular cleaning of the bottom board and inspection of empty combs will keep bee hives free of their invasion. Lundie (1952) and Smith (1960) identified two larger beetles as invaders of bee hives. *Diplognatha gagates* (Forster) and *Holplostomus fuligineus* (Oliver). Specimens lived quite well in captivity for over a year on honey and pollen. The strepsiptera, or twisted-winged parasites, are very small, unusual insects. There are two other reports of Strepsiptera attacking honey bees. Johnsen (1953) stated that he identified stylops on honey bees in Denmark. More recently, Adlakha and Sharma (1976) presented the strongest evidence for honey bee parasitism by styplops. This large moth is well known in the forest for entering hives between June and November. It makes a special sound which paralyzes the bees, and they may refrain from attacking it. The moth may then be able to load its stomach with honey. Sometimes the bees are able to catch the moth and break it into pieces. The dead *Atropos* may be found disposed of near the entrance of the hive.

Since the wing-span of the moth is as wide as 12 cm, *Atropos* cannot enter any hole which is only 8 mm in diameter. The use of hives with slot-like entrances should be avoided if the area is infested with this insect.

ARACHNIDS

SPIDERS

There are several types of spiders that may eat bees. The large web spinning spider will usually eat a bee that it can capture in its web. Some of the ground hunting spiders may also eat bees. Such spiders are seldom abundant and strong colonies should be able to suffer the occasional loss of a bee to a spider. The beekeeper will want to keep web building spiders from the immediate vicinity of his hives and out of potential flight Lanes of foragers.

SPIDERS AS ENEMIES

The knowledge of spider fauna in India and South Asia is very sketchy and the role of various spider species in killing the honey bee has been studied very little. It is well known that the common genera of spider have a universal distribution and the living habits of the important groups are similar to those studied in Europe and U.S.A. Langstroth believed that the common domestic spider Theridion that makes nests inside the buildings plays a useful role in killing moth worms of combs in storage. However, the orb weavers, crab spiders, and the funnel web spiders are known to catch or ensnare honey bees and eat them up. There is a difference of opinion of the amount of damage caused to the colonies. On the whole spiders are believed to play a minor role as enemies of honey bees.

Orb Weaves

Agriopidae, or the orb weavers, make large beautiful webs in gardens, and not infrequently in the apiaries. The spiders are quite large and have black and yellow or silvery and grey markings on the body. Drones and honeybee workers are caught in webs and the spiders suck their body juices. Some authors believe they prefer honeybee as prey, but quite often Diptera, flies, and blow flies, are seen caught in their snares.

Crab Spiders

The orab spiders belong to the family Thomisidae and their body is short and broad, somewhat resembling a crab, Most of the time they move sideways and backwards and lie in a bush near flowers. They do not make webs; and they catch flies and bees much bigger than their own size. The prey is quickly in sensitized by a bite and its body juices are sucked. The jumping spider *Salticus* can catch honey bee twice their size and hole onto them till the bee is dead. The spider may build a silken case as nest inside the hive covers. For keeping a check on them it is recommended to scrape off the nests periodically. The golden rod crab spiders feed actively on honey bees and prefer them over other prey.

Funnel-Web Spiders

The funnel-web spiders belong to the family Agelenidae. They make funnel shaped webs in neglected buildings and godowns. Some species prefer to make their nests in grass and scrub. These webs are very effective in ensnaring honeybees as prey. Web building spiders in buildings should be of little concern since bees don't normally fly into buildings. The spider constructs webs around in the apiary or in an empty hive. Once the web catches bees, the spider will eat them.

All webs found in or near the apiary should be destroyed. The hive should be cleaned and all webs found within it removed. Otherwise, the scout bees will be caught and eaten, and no swarm will ever take possession of the empty hive.

Pseudoscorpions

Pseudoscorpions, literally 'false scorpions', are small (most are about 1/5 inch long) and reddish or brown. They have oval, flattened bodies with two conspicuous pedipalps (pincer-like appendages). When they walk, they often appear crab-like. They can move as quickly backward as they move forward. Pseudoscorpions are a type of arachnid and are related to spiders, ticks, scorpions, and mites. Without their large pedipalps, pseudoscorpions look very much like ticks. Pseudoscorpions are also similar to scorpions but lack the long tail and stinger that is common to scorpions.

Pseudoscorpions live along with the bees in hives, in some cases both the adults and nymphs were found clinging on to the adult worker bees close to the neck. *Ellingsenius* is not phoretic on any other insect and it has never been observed in the absence of bees.

The role of pseudoscorpions within the ecology of beehives is still not well known. In attempts to study this relationship, it was found that pseudosco-rpions feed on small and sluggish micro arthropod fauna found in the beehives, which includes different species of mites, larvae of moths and other insects. The role of the pseudoscorpions as predators on insects and other arthropods of economic importance have been well recognized. There is a perfect synchroniz-ation between the pseudoscorpions and the bees, as they do not normally interfere with the activities of the honeybees, and honeybees seem to ignore the arachnids.

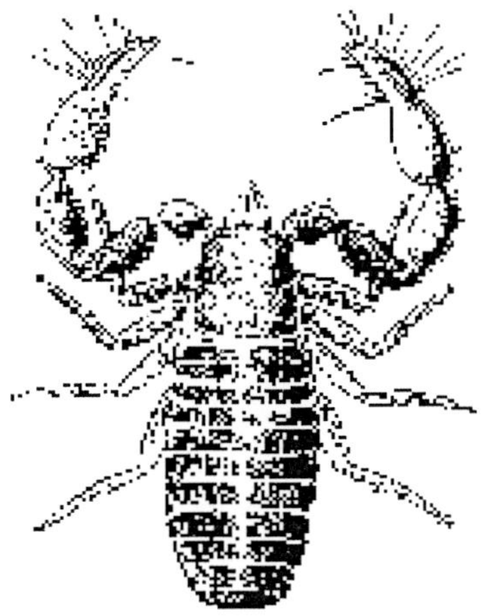

Figure 95. Showing pseudoscorpion.

The pseudoscorpions occupy restricted niches depending on the microclimatic conditions. They prefer habitats where the temperature is generally constant. Beehives maintain a temperature range of 33°C - 37°C, irrespective of the external temperatures. Also, beehives are usually dark with high humidity, whether they are in artificial moveable frame hives or in natural habitats such as hollows in tree trunks, under roof tiles or inside walls, which creates conducive conditions for pseudoscorpions. In the natural habitats the pseudoscorpions are seen in the crevices; in the moveable frame hives, they can be spotted on the floorboard between the frame of the spaces between the frame and the brood chamber walls.

The only time when pseudoscorpions interfere with behaviour of the bees, is when they sense their swarming behaviour. The pseudoscorpions sense this through the tactile setae present on the chela of the palps. It then immediately moves towards the bee and takes firm hold on one of its legs. In a majority of cases the hold is made at the region of the trochanter and very rarely have they been found hanging from the antennae. Workers, drones and queens were observed with these clinging arachnids. When the pseudoscorpion gets itself attached to the bee, the bee reacts by fanning its wings vigorously and rotating around itself, trying to shake off the pseudoscorpion, but since the hold is very firm it is not possible. Normally a bee can be seen with only one arachnid but when hiving takes place or during swarming or desertion, the last bees leaving the hive usually carry more than one arachnid. Bees carrying between two to eight pseudoscorpions on their hind legs are quite common in some cases. Swarming or desertion thus takes place with the pseudoscorpions attached to the bees, and even after the bees alight at a particular place, the pseudoscorpions retain their hold. Only when the bees become calm after the swarm settles down well in a new habitat, the pseudoscorpions leave the bees and retire to the convenient spots in the hive where they build their nests for brooding and for moulting. Less than 10% of the pseudoscorpions are not dispersed during swarming or desertion. The deserted seudoscorpions may live for about two to three months after which they die ultimately.

Are species of *Ellingsenius* pests or beneficial

On one hand anecdotal evidence from South Africa suggest that pseudoscorpions apparently controlled *Varroa*. From India the observations and photographs of *Varroa* being eaten, and the reports that other pests of bees were also eaten, strongly suggest that pseudoscorpions are beneficial to bees. Also, *E. indicus* never fed on bees or their younger stages, and observations in natural hives and test-feeding in the laboratory showed that neither the adult nor the larval stages of bees were harmed. In Nepal in 1999, the local beekeepers friends, but nobody knew why, although there was the tantalizing possibility that they assisted in keeping down the population of varroa, which were present in very low numbers.

On the other hand there are reports of pseudoscorpions causing harm to honey bees. In 1947 in a laboratory a pseudoscorpion tried to feed by applying its mandibles in between the joints of the leg of a bee. In 1954, when a bee was confined with *E. hendrickxi* from the Belgian Congo, the bee was killed. Also, the weight of *E. indicus* clinging to foraging bees in South India was reported in 1957 to be a hindrance, causing the general activities of the bees to decrease, and in a few colonies the number of pseudoscropions was so great that the colonies died. Pseudoscorpions (later identified as *E. fulleri*) were not a common pest of *A. mellifera* in Iran in 1984, but they had been collected from two weak colonies near Shiraz.

VERTEBRATE ENEMIES

1. MAMMALS

BEARS

Bear damage to apiary

Bears are a serious threat to beekeeping operations since they do a great deal of damage to hives and equipment. They normally visit the apiaries at night, smashing the hives to eat brood and honey. Once bears locate an apiary, they return again and again, and it becomes exceedingly difficult to control their marauding behavior.

How to protect bees

Use temporary or permanent electric fences to safeguard your hives. Be sure to keep weeds from shorting out the fence. Regularly maintain the chargers or batteries. Do not place fences next to trees which the bears can climb and then jump down inside the fence. Keep open, mowed areas on all sides around the hives. Do not locate hives in abandoned areas or close to wooded overgrown areas where bears have pathways to the hives. Do not place supplemental foods nearby as a distraction. This will attract or habituate bears and will be counter productive

Chemical controls

There are no effective chemical measures for protecting bees from bears or varmints.

Aversive conditioning

"Aversive conditioning" using emetic compounds has been tested as a means of repelling bears from beeyards. "Noisemakers, water guns, or bright lights may sometimes be used to discourage curious or non-habituated bears from people's backyards. Emetine hydrochloride, lithium hydrochloride, and alpha-naphthyl-thio-urea are among the compounds tested

MECHANICAL CONTROL METHODS

Non-Lethal Methods

Currently non-lethal methods in use are to deter black bears from these activities. Electronic fencing is a generally effective technique for protecting apiaries. The key to avoiding bear damage is a well-maintained electric fence. The fence must be put up before colonies are located in bear habitat, otherwise it often will not stop depredations. Such a fence must be dependable, relatively cheap to construct, and capable of operating in remote locations. Fences are totally ineffective if not installed and managed properly. Avoid a site with overhanging trees, as limbs might fall across the wires. It is also quite common for bears to climb trees and then drop down inside the fence. Control grass and weeds along the fence so that they will not contact the charged wires and short them out.

Non-electrified fences

Non-electrified fences consisting of heavy gauge woven wire panels 8 feet high with barbed wire brackets set outward at 250° will discourage most bears but are costly. and wired to a wooden pallet were effective in protecting hives.

Bright lights and loud music

Bright lights, loud music, noisemakers, and frightening devices have a limited effectiveness but may deter curious individual bears that are not focused on predation.

Trained dogs

Trained dogs may also be used to harass nuisance bears. Guard dogs are often effective in alleviating losses.

Propane cannons

Propane cannons have been successfully used to repel bears from crops. However, bears may become habituated to the noise and the period of aversion may be slight. The bears may also just move to another nearby field.

Live-trapping

Live-trapping and translocation is used to provide a temporary solution however, bears may return from substantial distances. Translocation also does not address the root cause of the problem, which may then recur.

Sport hunting

Sport hunting may occasionally be effective in reducing bear damage when the offending segment of the population can be targeted. In Massachusetts, bear hunting is popular in some areas in early September and hunters often hunt in or near corn fields, especially when natural foods are scarce. Crop damage permits are issued in some states to allow farmers to shoot offending animals.

Pepper spray

Oleoresin capsicum ("pepper spray") has been used since the 1960s as a sublethal inflammatory agent for personal defense and as a repellent. Pepper spray may be useful in a variety of field circumstances but variable responses by bears do occur. Spray should be directed into the bear's face at close range (<20 ft) so that the particles contact the animal's mucous membranes. Spray should be directed as a cloud rather than a stream so that noise and visual effects act as an additional deterrent.

Firearm

Fire arms may be useful in remote areas for rare instances of predatory attack or close chance encounter with a sow grizzly and cubs. Unless the shot is accurately directed, the bear may be wounded and continue its attack.

Small mammals

Skunks and other night-time foraging animals may find bee colonies an easy food source. The animal scratches on the outside of a hive and as guard bees come out to investigate the disturbance they are rolled with the animal's paw and then chewed. As long as only a few bees come out at a time the skunk (or opossum, or raccoon too, apparently) can avoid being stung. Continued feeding means continued disturbance and the bee colony may become very aggressive to the beekeeper.

Control: The best control for skunk, opossum or raccoon is to make the entrance less accessible or to increase the possibility of stings to the animal. Elevation of the entrance on a hive stand or large mesh wire ("chicken wire") before the entrance can be very effective.

Mice, Moles and Shrews

A serious pest problem of bee colonies is mice. Mice are serious pests of stored combs and active honey bee colonies during the fall and winter months. We may not immediately think of this tiny rodent as a predator, but it can cause serious damage to colonies and stored equipment just the same. In the winter, when the bees are clustered together, mice can enter the hive and are free to eat honey, wax and pollen. They may even build a nest. When it warms up enough to allow free movement of the bees, the mice will leave. These rodents chew combs and frames to make room for building their nests. Adult mice move into bee colonies in the fall and make their nest in the corner of the hive away from the bee cluster. Mouse urine on combs and frames makes bees reluctant to clean out these nests in the spring. Colonies in apiaries located in fields or at the edges of woodlots are especially vulnerable. Mice can successfully build a nest even in a strong colony. They move in and out of the colony while the bees are inactive, and their nests furnish additional protection. Their activity may disturb the bees, but the greatest damage is from the nest building of the mice.

Potential for Economic Loss: Slight to Moderate. If mice get in a hive during winter, they can cause moderate destruction to combs.

Management

Mice can be prevented from entering the hive in three ways. When new bottom boards are built, the top lip can be reduced in height to 3/8 of an inch. This leaves the bees with enough room to move in and out of the hive freely, but not enough space for mice to pass through.

Cultural Control: Chase away any mice found inside a colony, then remove the nest and restrict reentry. If comb chewing is extensive, replace the frames. When bees repair damaged combs, they replace worker-sized cells with drone comb. Combs in storage should be protected from mice by covering the top and bottom of each pile of supers with a queen excluder, wire screen, or outer telescoping lid.

Chemical Control: The use of poisons and traps is not needed if cultural control tactics are practiced.

Miscellaneous Large Livestock

Large animals like cattle, horses or goats are not pests but bees and livestock are best separated to avoid problems. The animals may attempt to scratch against colonies or they may knock them over accidentally as they browse. The bees may buzz around livestock, such as horses, making them nervous and less manageable.

Large livestock such as horse, sheep, or cattle, may cause damage to bee colonies by toppling hives. Some animals will scratch to remove flies or pests or become skittish around buzzing insects and as a result may inadvertently upset bee hives. To protect against these possibilities, colonies should be maintained outside areas of large livestock habitation or the apiary site should be fenced. Colonies should not be located so that the bee flight path is in an area where large livestock are located.

Control: Locate bee colonies outside enclosures fenced for livestock or fence in an apiary site for bees.

Primates

One mammal pest of the honey bee which seems to be becoming more and more of a problem is the human. Many authors list man as the greatest enemy of bees. Loss of entire bee colonies to bee rustlers can be listed as an example of human depredation. Honey crops may be taken, but more usually entire colonies are stolen.

Certainly humans can be destructive to bee colonies. As honey increase in value and as forage areas continue to decrease owing human population pressures, colonies are inevitably located in and around concentrations of people. Antisocial behavior, such as vandalism and colony theft, is likely to increase. Greater care in the selection of apiary locations and the moving of colonies from locations where trouble occurs are the best solutions. Out apiaries should be selected with care and should be visited regularly. The beekeeper's name and address should also be posted in each bee yard. Even with these precautions, vandalism and colony movement necessitated by human activities will undoubtedly result in the regrettable loss of valuable bee forage and crop pollination.

Three other primates, monkeys, boboons, and chimpanzees, have been referenced as pests of honey bees. Das Gupta (1945) noted that black faced monkeys of the genus Langur are a pest to Indian beekeepers. Subhapradha (1957, 1961) stated that monkeys may open hives and remove and destroy frames. Bee stings do not seems to drive the monkeys away. Punjabi (1961) indicated that monkey (and bear) damage in both the pain and hill sections of India is "noteworthy."

Subhapradha (1957) diagramed a hive stand to keep monkeys and ants away from the bees; it was made with solid pipe with a motor oil barrier for ant protection. Each hive is securely fastened to the hive stand by steel strapping over the top of the hive. The monkeys cannot remove covers or enter the hives when they are tightly fastened in this covers or enter the hives when they are tightly fastened in this manner. Although such straps may slow down colony manipulation, the extra precautions are worthwhile in areas where monkeys are a problem.

Botha (1970) stated that baboons will usually not tamper with well-constructed hives because of the stings they receive. However, they do raid and break up boxes set out for trapping bee swarms. Once they start such activities they may continue and raid large, established colonies. Baboons may tip colonies over repeatedly until combs and frames come apart and they frequently carry the frames away from the bee yard to consume the honey. Friedmann (1955) and Crane (1975a) related tales of baboons opening wild bee nests, after they had been guided to the nesting sites by a honeyguide, *Indicator indicactor*. According to Friedmann (1955) baboons (but never monkeys) may occasionally

Smith (1960) listed baboons and chimpanzees as pests of bees in the tropics. Deschodt (1969) described the action of a group of 20 chimpanzees raiding a wild bee nest in a granite wall crevasse. On large male climbed to the nest site, removed pieces of comb and threw them to the others on the ground. The stings did not drive the chimpanzees away, although there was much yelling and jumpi-ng about. The young chimps were the only members which did not participate as they snuggled their faces against the chests of their mothers to avoid stings.

Merfield and Miller (1956) watched chimpanzees use sticks as tools to obtain honey. The chimpanzees would poke a long stick down a hollow tree containing a wild bee nest. They would then withdraw it an consume the honey on the stick.

2. AMPHIBIANS

Frogs and toads

Frogs and toads are general insect feeders and they may include an occasional bee in their diet. Toads use the same methods as lizards, and will remain in the apiary if they can get bees to eat. The toad generally consumes only weak and dead bees, but if it can reach the hive, it will eat live bees as well. The toad does not pose as many problems as the lizard because it cannot climb. The best means of protecting hives against toads is therefore to install them at least 60 cm above the ground.

Beekeeping in tropical climates frequently suffers damage caused by the amphibians: toads (*Bufo melanostictus, Kaloula pulchra,* etc.) and frogs (*Ranlimnocharis, Rana tigrina,* etc.). The detection of this problem generally requires close observation: beekeepers are normally unable to observe intense predation by amphibians on honeybees in the daytime, when they are at work in their apiaries, because the heaviest attacks occur at night. Often the problem goes unrecognized until a substantial fall in colony populations is perceived.

One sign indicating that toads and frogs are preying heavily on the colonies is the presence of the predators dark brown droppings, scattered in front of the hive entrance. If the dry fecal deposits are spread apart (e.g. with a twig.), the remains of bee parts can be seen.

Continuous predation by toads and frogs, if not prevented, results in a loss of colony strength. While colonies with moderate or relatively large worker populations can withstand the predation and subsequently recover their full strength, weaker colonies are at considerable risk.

The attacking patterns of toads and frogs, if not prevented, results in a loss of colony strength. On arriving at the colony, the amphibians wait in the vicinity of the hive entrance, preying on passing bees. Colonies close to the ground provide easy access to the predators, which guard bees at the hive entrance are easy preys. If the attackers are small enough to squeeze through the hive entrance of a relatively weak colony, the outcome can be devastating.

Control: Although in some circumstances predations on honeybee colonies by amphibians cannot be over looked most beekeepers look on the problem as a minor one. Placing the hives on stands 40 to 60 cm high is usually sufficient as a protective measure. Where large numbers of the predators tend to congregate near an apiary, fencing it with fine-mesh chicken wire may be found necessary. Thus far, such other methods as trapping, baiting or poisoning have not been advocated.

3. REPTILES

Geckos, skinks and other lizards are among the most commonly found reptiles in tropical Asian jungles, woods, grasslands and even urban areas. Among the reptile species regularly recorded as present in commercial apiaries are the tokay (Gecko gecko), which can be as much as 35 cm in length, *Calotes* spp. *Acanthosaura* spp., and the skink *Sphenomorphus* spp.

Arboreal reptiles such as many geckos and skinks can attack bees either near the hive entrance or at the limbs of flowering trees visited by foragers. Smaller lizards, such as the gecko *Hemidactylus frenatus* attack the bees near the hive. In tropical areas, this type of predator frequently causes the sudden loss of the queen from a weak colony.

Lizards

Lizards, measuring about 25 cm from head to tail, are mostly found in backyard gardens, in villages and the outskirts of the city. The activity of the "home lizard" may cause great concern to the beekeeper. It sometimes stays very close to the hive or accommodates itself comfortably between the lid and the hive body, if it can find an entrance. From that convenient spot, it may feed indefinitely on the bees.

Even lizards not living near the hive will feed on the bees once they can locate the apiary. Although they prefer dead bees, they will eat live ones as well. A worker bee, acting as a scavenger, will pounce on an old, lazy or sick bee and try to tear the victim's wings, breaking them into pieces. While this action is in progress, the lizard will rush in and lick both of them up with its sticky tongue.

A serious lizard problem may lead to absconding. The simplest and most practical protective measure is there the bee hives are installed on a platform, with metal cones nailed on the legs about 70 cm above the ground, to prevent lizards from reaching the hives.

Snakes

Some snakes are known to eat bees. They do not cause much damage to the colony, but the beekeeper should always be careful to avoid being bitten by a poisonous snake near the hive.

4. BIRDS

Bee-Eaters *Meropidae*

The Bee-eaters must rank among the most delightful birds on Earth. They possess startling grace and beauty, well exemplified by Southern Carmine Bee-eaters. Carmine Bee-eaters are social birds breeding in colonies on river banks and foraging out over open country. On the other hand bee eater birds are of serious concern to apiculturists. Their role as predators of honeybees has long been realized. Aristotle recorded titmouse (Paridae) as predator of honeybees. Later, Virgil included, the green wood pecker, *Picus viridis*; the red backed shrike *Lanius collurio* and the blue titmouse *Parus caeruleus* as predators to honeybees. A number of bird families have been recognized as major and minor predators of honeybees.

Most of the earlier studies reveal that birds in fact are of little significance as enemies of honeybees, as they also feed on other insects like locusts. However, they do create serious problems during queen rearing operations. The king bird *Tyrannus Tyrannus* has been listed as the main culprit during queen rearing operations. In other parts of the world particularly Africa and Asia predatory birds result in severe losses to honeybees. Of the predatory birds those belonging to family Meropidae, the bee eaters which feed on adult bees, inflict tremendous losses. Interestingly, the birds of the family Indicatoridae are so much fond of honeybees and honey that they lead man and other animals to nests of the bees with the hope of getting food when nests are torn apart. Besides, other families of the birds such as wood peckers Picidae, the titmice, Paridae; the shrikes, Laniidae; the swifts Apodidae and the tyrant fly catchers Tyrannidae. The predatory birds could be classified into :

Major

Minor

Miscellaneous

Major predators include birds belonging to those families which obtain major part of their diet from honeybees e.g. Meropidae, Dicrusidae and Indicatoride.

Minor or occasional predators are placed in families, Picidae (woodpeckers), Paridae (titmice), Laniidae (shrikes), Apodidae (swifts), Tyranidae (fly catcher or king bird), Muscicapidae (Swallow), Ploceidae (Subfamily passerine–house sparrow and yellow throated sparrow).

Miscellaneous predators include such families of birds which occasionally feed on bees such as sparrows.

Beekeepers have traditionally feared the arrival of bee eaters, which eat the cultivated bees. Hives are especially vulnerable to attack when the weather is cool, since there are not as many insects flying about for hungry bee eaters to prey on. But the damage bee eaters do to hives is generally out-weighed by their control of harmful insects such as locusts, as well as the wasps and hornets that are themselves honeybee predators. Fry *et al.* (1988) reported that Bee-eaters will lively pursue any type of flying insect, however, honeybees predominate in their diet. Bee-eaters are almost strictly predators on flying insects. Once an insect lands the bee-eater ignores it, even if in plain sight. Bee-eaters are just programmed to catch things on the wing except one species that eats vegetable matter at all (the White-throated Bee-eater *M. albicollis* of West Africa).

The world range of the bee-eaters is nearly congruent to the native world range of the four species of honeybees. Fry *et al.* (1992) found that in the diet of 16 kinds of bee-eaters, Hymenoptera (ants, bees and wasps) comprised from 20% to 96% of all insects eaten, and honeybees formed on average about one-third of the Hymenoptera thereby indicating that bee eater birds have evolved in relations to honeybees, as their predators.

Bee-eaters are temperate and tropical Old World birds that catch insect prey in the air. There are 26 species of bee-eaters (Meropidae) native to temperate and tropical regions of Old World as given below:

Table 74. Bee-eaters (Meropidae)

Common Name	Scientific Name
Black Bee-eater	*Merops gularis*
Blue-headed Bee-eater	*Merops muelleri*
Red-throated Bee-eater	*Merops bulocki*
White-fronted Bee-eater	*Merops bullockoides*
Little Bee-eater	*Merops pusillus*
Blue-breasted Bee-eater	*Merops variegatus*
Cinnamon-chested Bee-eater	Merops oreobates
Swallow-tailed Bee-eater	*Merops hirundineus*
White-throated Bee-eater	*Merops albicollis*
Green Bee-eater	*Merops orientalis*
Boehm's Bee-eater	*Merops boehmi*
Blue-throated Bee-eater	*Merops viridis*
Blue-cheeked Bee-eater	*Merops persicus*
Madagascar Bee-eater	*Merops superciliosus*
Blue-tailed Bee-eater	*Merops philippinus*
Rainbow Bee-eater	*Merops ornatus*
European Bee-eater	*Merops apiaster*

Chestnut-headed Bee-eater	*Merops leschenaulti*
Northern Carmine Bee-eater	*Merops nubicus*
Southern Carmine Bee-eater	*Merops nubicoides*

Green bee eater

Species : *Merops orientalis*

Size : Sparrow

Small green bee-eater *Merops orientalis* 21 cms.

Field Characters: A grass-green bird tinged with reddish brown on head and neck. Central pair of tail feathers prolonged into blunt pins. Slender, long, slightly curved bill. Conspicuous black "necklace". Sexes alike.

Habits: inhabits open country, neighbourhood of cultivation, forest clearings, fallow land, gardens etc. launches aerial sallies after bees, etc. snapping them up into its bill and circling back gracefully on outstretched motionless wings to the perch, where the quarry is battered to death and swallowed.

Food: Insects chiefly Diptera (Flies) and Hymenoptera (Bees and Wasps).

Nesting: February - May. *Nest:* A horizontal or oblique tunnel ending in a widened egg chamber, dug in the side of an earth-cutting, borrow pit, river bank etc. *Egg:* 4 - 7, pure white, roundish ovals. Both sexes share in excavating nest-tunnel and feeding young.

Blue-tailed Bee-eater (*Merops philippinus*) is a widespread resident in India. **Size:** 23-25 cm. The **Blue-tailed Bee-eater**, *Merops philippinus* is a near passerine bird in the bee-eater family Meropidae. It breeds in southeastern Asia. It is strongly migratory. This species is sometimes considered to be conspecific with the Blue-cheeked Bee-eater, *M. persicus*.

This species, like other bee-eaters, is a richly-coloured, slender bird. It is predominantly green; its face has a narrow blue patch with a black eye stripe, and a yellow and brown throat; the tail is blue and the beak is black. It can reach a length of 23-26 cm, including the two elongated central tail feathers.

This is a bird which breeds in sub-tropical open country, such as farmland, parks or rice fields. It is most often seen near large water bodies. Like other bee-eaters it predominantly eats insects, especially bees, wasps and hornets, which are caught in the air by sorties from an open perch. This species probably takes bees and dragonflies in roughly equal numbers. The insect that are caught are beaten on the perch to kill and break the exoskeleton. This habit is seen in many other members of the coraciiformes order.

These bee-eaters are gregarious, nesting colonially in sandy banks or open flat areas. They make a relatively long tunnel in which 5 to 7 spherical white eggs are laid. Both the male and the female take care of the eggs. These birds also feed and roost communally. The call is similar to that of the European Bee-eater.

Chestnut-headed Bee-eater

The **Chestnut-headed Bee-eater**, *Merops leschenaulti* is a near passerine bird in the bee-eater family Meropidae. It is a resident breeder in southern Asia from India east to southeast Asia and Indonesia. Their area of distribution is western ghats, Ceylon, terai area near the himalayas upto 4000 elevation, Assam, Orissa, Bangladesh. The food and nesting habits are similar to those of green bee eater. This species, like other bee-eaters, is a richly coloured, slender bird. It is predominantly green, with blue on the rump and lower belly. Its face and throat are yellow with a black eye stripe, and the crown and nape are rich chestnut. The thin curved bill is black. Sexes are alike, but young birds are duller. This species is 18–20 cm long; it lacks the two elongated central tail feathers possessed by most of its relatives. This is a bird which breeds in sub-tropical open woodland, often near water. It is most common in highland areas. As the name suggests, bee-eaters predominantly eat insects, especially bees, wasps and hornets, which are caught in the air by sorties from an open perch. These bee-eaters are gregarious, nesting colonially in sandy banks. They make a relatively long tunnel in which the 5 to 6 spherical white eggs are laid. Both the male and the female take care of the eggs. These birds also feed and roost communally. The call is similar to that of the European Bee-eater.

RAINBOW BEE EATER

Order: *Coraciiformes*

Family: *Meropidae*

Genus & species: *Merops ornatus*

The rainbow bee eater is also known as the rainbow-bird, pin tailed bee eater, spine tail, berrin-berrin, gold-digger, and gold-miner. Like all bee eaters, rainbow bee eaters are very social birds. When not breeding, they roost together at night in dense under growth or large trees. The rainbow bee eater perches on power lines or branches of dead trees to search for food. It is always alert for flying insects and can spot a bee from 150 feet away. The bird immediately flies after it and almost always catches its prey. Once the bee eater catches a bee it usually removes the stinger even though the bird is immune to bee and wasp stings, which kill other small birds. Bee eaters eat several hundred bees and wasps a day. They cough up the indigestible portions in the form of pellets. When a flock of bee eaters spots a large swarm of small termites, it attacks the swarm and eats as many insects as possible in a short time. Key facts about birds include: **Length:** 7-8 in. Central tail feathers extend 3/4 in. beyond male's body, 3 in. beyond female's. **Wingspan:** Approximately 10 in. **Sexual maturity**: 1 year. **Mating**: Before and after rainy season in northern part of range; November to January in south. **Eggs**: 3-7, usually 4-5. Glossy white. **Habit:** Very social. Pair for life. **Diet:** Mainly bees and wasps. Also inchneumon flies, dragonflies, damselflies, moths, butterflies, and grasshoppers.

Many species of bee-eater are migratory, including the European Bee-eater *M. apiaster* that travels thousands of miles from European breeding grounds to west and southern Africa, and the Rainbow Bee-eater, the "Rainbow bird" of Australia that moves north to New Guinea and the Moluccas after nesting. An exception is the Little Bee-eater of tropical Africa that apparently makes only short local movements in response to the rainy and dry seasons. Because of their migratory nature they may feed in an apiary for a short period and then shift to another locality. The bee eater are particularly dangerous when they attack bees in an apiary in a flock of up to 250 birds.

Black drongo *Dicrurus adsimilis* 31 cms.

Field Characters: A slim and agile glossy black bird with long, deeply forked tail. Sexes alike.

Habits: A familiar bird of the open country, usually perched on telegraph wires, or attending on grazing cattle. From exposed look outs it keeps a vigilant watch for grasshoppers and other insects. It also rides on backs of grazing cattle and takes toll of insects disturbed by the animals' movements. Forest fires or fired grass patches attract number of drongos. Highly beneficial to agriculture by the vast quantities of injurious insects it destroys. Extremely territorial.

Food: Insects, flower nectar also occasionally small birds.

Nesting: April - August. *Nest:* A flimsy-bottomed cup of fine twigs and fibres cemented with cobwebs, in the fork at extremity of branch up in large trees preferably standing alone in open.

Egg: 3-5, mostly whitish with brownish red spots. Both sexes share all domestic duties and are bold in defense of their nest.

The swifts

They belong to family Apodidae which has world wide distribution with the exception of north most Eurasia and the Americas. The spine tailed swift *Chaetura dubia* is an important predator of honeybees in the Philippines and may cause limitations on beekeeping. In India, Ceylon, Burma and Bangladesh two species of the size of a sparrow are common. The palm swift (*Cypsiurus parvus*) is a sooty grey with a narrow deeply forked tail. The birds fly about gregariously in open country dotted with palmyra palms, with which they are inseparably linked. The birds roost and nest in these palms and also feed on nectar loving insects including honeybees that are attracted to the palm sap.

The house swift *Apus affinis is a* small smoky-black bird with white throat, white rump, short square tail, and long narrow wings. Sexes alike. Flying about gregariously near human habitations. Fond of ruined buildings and also dwelling houses – deserted or occupied – often in the midst of noisy cities. Flies about at great speed almost incessantly throughout the day, hawking flies. Food consists chiefly of dipterous insects. Nesting occurs during February - September. The nest consists of a round untidy cup of feathers and straw cemented with the birds' saliva, with a slit between wall and nest for entrance. Usually in colonies, plastered in angle of walls and ceilings in verandhas, porches etc. it lays 2-4, pure white, longish oval eggs. Both sexes share in building and feeding of young.

Shrikes

Laniidae (Shrikes) are a widespread group of medium-sized sit-and-wait predators. They have a strong, hooked bill (sometimes toothed), large head, longish tail, strong legs and sharp claws. Most Laniidae species are patterned crisply in gray, black, white, bay or chestnut. They feed on larger insects, grass-hoppers, rodents, reptiles and even small birds. They are considered to be the minor pests of honeybees during the active period of April–August. They are seen near the hives diving at honeybees and catching them. Two species are common in India Srilanka and Bangladesh. The black headed cuckoo shrike, *Coracina melanoptera*, is the size of a bulbul, is ashy grey with black head, wings, tail, and is whitish with bars on the underside. It is distributed throughout the plains and mountains of India upto 6000 ft hunting around in parties mixed with other insectivorous birds. Some species impale their prey on thorns or barbed wire for later retrieval. Most shrikes are birds of open country. All are generally solitary, monogamous, territorial birds. Sexes generally alike. The clutch size is 2-6 eggs; both parents care for the young, sometimes with helpers. The different species include Red-backed Shrike (*Lanius collurio*) - sparse passage migrant Burmese Shrike (*Lanius collurioides*) - sparse passage migrant Brown Shrike (*Lanius cristatus*) - widespread winter visitor Northern Shrike (*Lanius excubitor*) - wide-spread resident Rufous-tailed Shrike (*Lanius isabellinus*) - widespread winter visitor Southern Grey Shrike (*Lanius meridionalis*) - resident Lesser Grey Shrike (*Lanius minor*) - vagrant Long-tailed Shrike (*Lanius schach*) - widespread resident Grey-backed Shrike (*Lanius tephronotus*) - local resident and wides-pread winter visitor Bay-backed Shrike (*Lanius vittatus*) - widespread resident.

Tit mice or the grey tit

It is of the size of a sparrow, the grey tit *Parus major* Linnaeus belongs to the family Paridae and its earliest written record as a pest of honeybees dates back to the days of Virgil in Greece. More recently it has been observed as bee predator in Russia and Great Britain. The bird attack is more pronounced in winter months as they lure the bees out and catch them, sometimes even carry dead bees to feed their young ones. The Great Tit is very bright and lively, and is the largest of the Titmice family. It looks quite like the Blue Tit, but it is bigger, and has a black head, apart from the white cheek patches. It has a dark black stripe through the yellow feathers from its chin down its tummy to its tail.I t flies in a jerky scattered pattern. It is good at climbing branches. Unlike the blue-tit, it can often be seen on the ground, in search of food. It can be found in woods, hedges, gardens, parks and mountains, as long as trees are present. The Great Tit eats insects, oily seeds, nuts and fruit. It has a larger and stronger beak than other Tits, and so it can eat more acorns, sweet chestnut and hazelnuts. The

Great Tit builds a nest of moss, hair and feathers, in a hole in a tree, wall or in a water pump, flowerpot or letterbox. It will lay about eight eggs in April, May or June, and lays two batches of eggs each year. Great Tits are found throughout Britain, all year round. Some of them migrate, heading off to the sun in September or October. Many return in March. It length is 14 cm.

Woodpeckers

Woodpeckers are near passerine birds of the order Piciformes and family Picidae. They gained their english name because of the habit of some species of tapping and pecking noisily on tree trunks with their beaks. This is both a means of communication to signal possession of territory to their rivals, and a method of locating and accessing insect larvae found under the bark or in long winding tunnels in the tree. They are found worldwide except extreme north of Asia, Australia Madagascar and Oceanic Islands. Most of the species in family 214 species are migratory in nature and are considered only a nuisance near the apiaries and can be easily scared away. The green wood pecker *Picus viridis* is quite a serious pest of bees in Greece, Russia and Europe during the winter season when no other food is available. It makes hole with its sharp bill near the hive entrance to reach to the bees. The holes are further widened by mice and rats.

Three species of wood peckers are widely distributed in India, Ceylon, Pakistan, Bangladesh and Burma; the chestnut-rufous wood pecker, having black bars on the wings is *Micropternus brachyurus;* the golden backed wood pecker, *Dinopium benghalensehas* golden yellow upper plumage with margins of black aand a red crest and the yellow fronted pied wood pecker, *Picoides mahrattensis,* is smaller in size and is spotted black and white above with red spots on vent and abdomen. The wood peckers breed from March to August and make nests in hollowed out holes in tree stems and branches. They generally lay three glossy china white eggs and both sexes take parental care and domestic duties. They are seen genaerally pecking at the bark of trees for the hidden grubs and insects and only occasionally peck at beehives for catching bees.

Woodpeckers also use their beaks to create larger holes for their nests which are 15-45 cm (6-18 inches) below the opening. These nests are lined only with wood chips and hold 2-8 white eggs laid by the females. Because the nests are out of sight, they are not visible to predators and eggs do not need to be camouflaged. Cavities created by woodpeckers are also reused as nests by other birds, such as grackles, starlings, some ducks and owls, and mammals, such as tree squirrels.

Drongos

The king crowes or drongos belong to family Dicruridae, which is distributed throuout Asia. They are primarily carnivorous birds and feed on medium or large insects. These insect-eating birds are found in usually open forests or bush. Their value as insectivorous birds is well known but their harmful role in picking drones and queens near mating yards is sometimes disputed. Most are black or dark grey in colour, sometimes with metallic tints. They have long forked tails, and some Asian species have elaborate tail decorations. They have short legs and sit very upright whilst perched, like a shrike. They flycatch or take prey from the ground. There are three common species in India Ceylon, Bangladesh, Burma and Pakistan. The Racket tailed drongo is of the size of the myna *Dicrurus paradiseus* (Linnaeus, 1766). This bird is a rare resident. Locally called **Bhimraj** or **Bhangraj**, it is found in hill forests of India and in Haryana at Morni Hills, Saraswati and Kalesar. The Greater Racket-tailed Drongo is a resident breeder in tropical southern Asia from Kashmir, India and Sri Lanka east to Indonesia. This species is usually found in broadleaved forest. Three or four eggs are laid in a cup nest in a tree. These are aggressive and fearless birds, 32 cm in length, and will attack much larger species if their nest or young are threatened. This courageous drongo usually leads the mixed feeding flocks typical of Asian jungle habitats.

The adult Greater Racket-tailed Drongo has spangled metallic green-blue plumage, and a large bill. The tail is long and shallowly forked, with the shafts of the two outermost feathers greatly extended and ending in the rackets which give this species its name. There is a head crest, prominent in the Indian race, *D. p. grandis*, but much reduced in some other subspecies. The young bird is duller and uncrested. The Sri Lankan form, *D.p. ceylonicus*, has a smaller crest and has a long deeply forked tail without rackets. It is sometimes given specific status as *Dicrurus ceylonicus*

Two to four eggs are laid in a nest high in a tree. These are aggressive and fearless birds, given their small size, and drongos will attack much larger species if their nest or young are threatened.

The **White-bellied Drongo**, *Dicrurus caerulescens*, is a small Asian bird. The White-bellied Drongo is a resident breeder in India and Sri Lanka. This species is usually found in open forest and well-wooded habitats. Two to four eggs are laid in a cup nest in a tree. These are aggressive and fearless birds, 24 cm in length, and will attack much larger species if their nest or young are threatened.

The adult White-bellied Drongo has dark grey upperparts, and the tail is long and deeply forked. The Indian race *D.c. caerulescens* has a pale grey underparts from the head to the breast, and is white from the belly down. One Sri Lankan race, *D.c. leucopygialis*, has the white on the underparts restricted to the vent area and undertail coverts. The young bird is duller and have a brownish grey breast.

The White-bellied Drongo has short legs and sits very upright whilst perched prominently, like a shrike. It is insectivorous.

The Fork-tailed Drongo, *Dicrurus adsimilis*

The Fork-tailed Drongo is a common and widespread resident breeder inindian subcontinent, Africa south of the Sahara. These insect-eating birds are usually found in open forests or bush. Two to four eggs are laid in a cup nest in a fork high in a tree. These are aggressive and fearless birds, given their small size, and will attack much larger species, including birds of prey if their nest or young are threatened. The male is mainly glossy black, although the wings are duller. It is large-headed and has the forked tail which gives the species its name. The female is similar but less glossy. The bill is black and heavy, and the eye is red. The Fork-tailed Drongo is 25 cm long. It has short legs and sits very upright whilst perched prominently, like a shrike. It flycatches or take prey from the ground and is attracted to bush fires. The call is a metallic *strink-strink.*

There are number of drongos distributed in Himalaya (Fleming *et al.,* 1976; Salim Ali 1977). They include the black drongos (*Dicrurus adsimilis*); the grey drongos (*D. leucophaeus*),the croebilled drongos (*D. annectans*), the lesser racket tailed drongo (*D. remifer*) the hair crested drong (*D. hottentottus*) and the large racket tailed drongo (*D. paradiseus*)

Himalayan honey guides

Animals like honeybees secrete wax to construct their hives. Such a honeybee hive with its stored food in the form of honey, eggs, larvae and pupae is a rich and attractive source of food for many animals. But it is a food source most ably defended by worker bees that launch suicidal attacks on potential predators braving certain death as they firmly embed their stings in the bodies of their enemies. Animals in turn have developed elaborate structures and behaviour patterns to overcome these defences. One such predator is the honey badger equipped with a thick coat of fur and a very loose skin which keep the bee stings out of reach of most of its organs. But honeybee hives are at a height and honeybees operating in the forest canopy are not easily visible to the badger. They are however much more easily tracked by a bird called the honeyguide. The honeyguide on its own is too frail to attack a honeybee hive. It has therefore evolved a mutualistic relationship with the honeybadger. When a honeyguide locates a hive it looks for its ally, the honeybadger and freezes in a characteristic posture. The honeybadger then begins to follow it on the ground. The honeyguide flies in stages towards the hive with the badger following it all the way. The honeybadger then successfully attacks the hive and feasts on the honey, the eggs, the larvae and the pupae. But it cannot digest the wax which is left behind. The honeyguide feeds on the wax, for it has yet another ally, a protozoan lodged in its gut that helps it digest the wax. Thus have organisms continually evolved more and more complex structures and behaviour patterns as they have devised ever more diverse ways of getting at energy.

Honeyguides (Indicatoridae) long have attracted interest because some species have the habit of leading certain mammals, including man, to sources of honey (they eat beeswax, not honey). A honeyguide, guiding bird exhibit a unique pattern of behavior: the bird leads a ratel (honey badger) to a bees' nest by its chattering and flying ahead; attracts attention with wavering, chattering "tya" notes compounded with "peeps or pipes" sounds and flies toward an occupied hive and then stops and calls again as the same time make an eye-catching spreading tail, showing the white spots and a "bounding, upward flight to a perch".

In all the typical honey guides, the white outer tail feathers are either very short or are tipped dark, producing a distinctive white wedge in the tail.The white tail patches are important in honeyguide biology. All species of honeyguide — the 15 in Africa and 2 in Asia — eat beeswax and honeycomb. The flashing white tail patches stimulates young honeyguides to follow others to bee-hives in the woods, and, this feature is more visible in immatures. Also serves to establish immature honey guides as dominant when at the wax sources. In parts of Africa, humans follow "guide" birds to these honeycombs. The birds benefit when men break open the beehives to get the honey, and people benefit from following the birds to the hives. Some "guide" birds give distinctive chipping notes to encourage men to follow them to the hives, yet there are other parts of Africa where this does not happen at all (Short and Horne, 2002).

The large animals and local men in Himalayas are guided by these birds to the hidden *Apis cerana* colonies in the hollow of rocks. Their area of distribution includes Pakistan and Indian Himalayas, Garwal, Assam, Sikkim, Nepal and North Burma. They can eat beeswax from the combs of *Apis dorsata* on their own without the help of a symbiont.

Management

Bird menace could be reduced using following methods:

Netting. One of the most effective methods of bird management is their netting. If installed properly, the netting is barely visible from a distance and will offer a long-term solution to the damage problem. If the birds move to another area of the dwelling, that too will need to be netted.

Frightening Devices

Visual. Stationary model hawks or owls, fake and simulated snakes and owl and cat silhouettes are generally considered ineffective as repellents. Toy plastic twirlers or windmills fastened to the eaves, and aluminum foil or brightly colored plastic strips, bright tin lids, and pie pans hung from above, all of which repel by movement and/or reflection, have been used with some success, as have suspended falcon silhouettes, especially if put in place soon after the damage starts. The twirlers and plastic strips rely on a breeze for motion. Stretching reflective mylar tape strips across a damaged area, or attaching them to the eaves and letting them hang down (weighted or unweighted) is a recent alternative to aluminum strips. Large rubber balloons with owl-like eyes painted on them are included in the recent array of frightening devices used to scare woodpeckers.

Sound. Loud noises such as hand clapping, a toy cap pistol, and banging on a garbage can lid have been used to frighten woodpeckers away from houses. Such harassment, if repeated when the bird returns, may cause it to leave for good. Propane exploders (gas cannons) or other commercial noise-producing, frightening devices may have some merit for scaring birds at least for short periods. Around homes, portable radios have been played with little success in discouraging birds. Expensive high-frequency sound-producing devices are marketed for controlling various pest birds but rarely provide advertised results. High-frequency sound is above the normal audible hearing range of humans but, unfortunately, above the range of most birds too.

Table 75. Average number of bee eating birds visiting apiary beforev and after protection.

Techniques	Time of day			
	6-8	8-10	14-16	16-18
No protection control	44.40	21.10	55.00	77.13
Hawk m	9.7	3.0	16.3	5.34
Hawk model	16.00	8.00	4.5	7.5
Hawk shaped kite	19.00	8.5	9.7	23.00
Scare crow	6.5	1.5	1.00	5.00

Thirakhupt, 1989.

Repellents

Taste. Many chemicals that have objectionable tastes as well as odors have been tested for treating utility poles and fence posts to discourage birds. Most have proven ineffective or at least not cost-effective.

Odor. Naphthalene (moth balls) is a volatile chemical that has been suggested for woodpecker control. In out of door unconfined areas, however, it is of doubtful merit. It is unlikely that high enough odor-repelling concentrations of napthalene could be achieved, to effectively repel birds. Odorous and somewhat toxic wood treatments, such as creosote and pentachlorophenol, which are frequently used to treat utility poles and fence posts, do not resolve the bird problem.

Tactile. Sticky or tacky bird repellents such as Tanglefoot®, 4-The- Birds®, and Roost-No-More®, smeared or placed in wavy bands with a caulking gun on limbs or trunks where sapsuckers are working, will often discourage the birds from orchard, ornamental, and shade trees. These same repellents can be effective in discouraging birds if applied to wood siding and other areas of structural damage. The birds are not entrapped by the sticky substances but rather dislike the tacky footing. The tacky repellents can be applied to a thin piece of pressed board, ridged clear plastic sheets, or other suitable material, which is then fastened to the area where damage is occurring.

Toxicants

Toxicants have only rarely been used to protect bees from birds. Bird problems can be resolved without toxicants and none are registered for such use.

Trapping

Wooden-base rat snap traps can be effective in killing the offending birds. The trap is nailed to the building with the trigger downward alongside the spot sustaining the damage. The trap is baited with nut meats (walnuts, almonds, or pecans) or suet. If multiple areas are being damaged, several traps can be used.

Shooting

Where it is necessary to remove the offending birds and the proper permits have been obtained, shooting may be one of the quickest methods of dispatching one or a few birds. The discharging of fire arms is often subject to local regulations in residential areas. At close range, air rifles or 22-caliber rifles with dust shot or BB caps can be effective. Shotguns or 22-caliber rifles may be needed for birds that must be taken from greater distances. Considerable discretion must be used around dwellings. Bullets and shot can travel long distances if they miss their targets. With appropriate permits, shooting has been occasionally used to reduce woodpecker damage in commercial fruit and nut orchards.

Reflective tapes

Reflective tapes of different colours (1 m x 3.5 cm) fixed on string at an distance of 20-30 cms at a height of 5 m on two poles/stems have been found effective to repel the birds. Keep bee hives under thick canopy from where the bees can fly but the birds are hindered in their predatory flights.

Chapter 22

Breeding bees for disease resistance

Introduction

The internal environment of a honeybee colony (characterised by constant warm temperatures, high humidity and extensive contact between individuals) is highly conducive to the spread of parasites and pathogens (Milne, 1983). Honeybees are attacked by various brood and adult bee diseases which are caused by viruses, bacteria and fungi. They include bacterial diseases such as American foul brood caused by *Bacillus larvae* and European foul brood by *Paenobacillus plutonius*. Chalkbrood disease is caused by *Ascosphaera apis* and stonebrood by *Aspergillus flavus*. Thai sac brood and sac brood are viral diseases. Nosema, Amoeba, bee paralysis, septicemia are the adult bee diseases. Besides, ectoparasitic mites such as *Varroa jacobsoni*, *Tropilaelaps clareae* and endoparasitic mite such as *Acarapis woodi* are the serious enemies of honeybees. Chalkbrood, a highly infectious disease that affects honeybee brood, is caused by ingestion of the heterothallic fungus *Ascosphaera apis*. The disease is characterized by the presence of mummified larvae (Gilliam *et al.,* 1978; Mehr *et al.* 1976). While the disease is most apparent in summer, *A. apis* spores can survive throughout the year in honey, in stored pollen and in the gut of adult bees. These spores are highly resilient and may remain infectious for 15-38 years (Gilliam 1990). Frequent food and water sharing among nest mates contributes to the infectious nature of the disease. However, chalkbrood is also stress related and certain predisposing physiological and environmental conditions are required for the development of infection (Gilliam, 1978; Heath, 1982). Further, it is clear that some stocks are far less affected by *A. apis* than others (Gilliam, 1986).

Breeding as a means of combating Disease

It is now generally recognized that breeding plays a determining role in the fight against disease in both the animal and plant. Today we have at our disposal a bewildering number of prophylactics, which are on all sides assumed to be absolutely essential. These means however must be regarded as only extra aids, as so often they bring with them very undesirable consequences. There is moreover another point when the use of one of these measures has succeeded in cleaning up a disease; there is always the danger of re-infection. As a top priority of all serious practical breeding there must always be the breeding of a strain genetically immune or resistant to disease. This, of course, raises the question as to whether this struggle against disease by breeding is a possibility. On all sides this possibility has up to now been seriously doubted, admittedly less so recently than a few years ago. All living creatures are provided with the capacity of protecting themselves against disease. It is this capacity that we can influence by breeding. Here the honeybee is no exception. In her case breeding is not concerned with individuals but with colonies of varying size. Moreover, the strength of a colony during the year is subject to considerable fluctuation. Since all living things are provided with the means of protection against disease, in different degrees, why should the honeybee be an exception? Here practical experience must be the deciding factor. But before a definitive answer can be given, the problems involved must first be set out.

We have at our disposal an almost unending mass of material of a scientific and practical nature on this or that disease, its causes, treatment and prevention. There is however a corresponding dearth of information on the fight against disease by means of breeding and also of the possibilities open to us by this means. This lack of information is easy to explain. Only very few scientific institutes or experimental centres have the resources required for a project of this nature. To obtain positive comparisons, not only is there necessary a large number of colonies, but also a number of reference points which only different races and strains can provide. Where we are dealing with only one particular race or a pure-bred line, the only differences which appear are those of that particular race or line. We cannot obtain any enlightenment on the breeding or profitable possibilities of other lines, crosses or races, above all in the realm of fighting disease. In breeding animals and plants the decisive role in this fight is in every case provided by select racial crosses.

The honeybee is no exception to this although we have to deal with problems not met with in the animal and plant world. In the case of the bee we always have to deal with the community, not with isolated individuals. Although this fact has been mentioned it needs stressing here when we are concerned with breeding as a means to combat disease. As a result of multiple mating each colony is made up of groups of indeterminate numbers of half-sisters and super-sisters. Each group has a common mother but fathers of different provenance with different hereditary characteristics. The number of the individuals of each group of sisters and their influence on the unity of a colony can be very varied. Hence in a colony there is a conglomeration of hereditary dispositions which within prescribed limits, make their appearance and it is these which we have to take into consideration especially their reaction to disease. It is clear from this that the different groups of sister queens will not prove resistant or susceptible to a disease in exactly the same way. This becomes even more clear where random matings take place.

Resistance and immunity

Before we go on to deal with the different diseases and the possibilities of combating them by means of breeding, it is essential to give precise definitions of the words "resistance" and "immunity". The two words are often used indiscriminately and this can lead to misunderstanding. It is therefore essential to make a clear distinction between the two words - a distinction, incidentally, which corresponds to reality - otherwise there is the danger of endless confusion and quite false assumptions.

The word "resistance" means hardiness and an ability to overcome weakness, the concept is that of a struggle between two opponents in which the stronger is the winner. "Immunity", on the other hand, means the complete lack of susceptibility to a disease without any exception. Resistance therefore refers to a midway stage between the two extremes of susceptibility to disease and immunity. The practical beekeeper has to be content with a highly developed resistance allowing the bees to be proof against disease, for complete immunity. is unobtainable in our case. But this highly developed resistance fully answers all the needs of modern beekeeping as confirmed by wide practical experience.

Experiments which are made in the laboratory do not always correspond to life in nature and the corresponding reactions to it. This is a fact which is not always recognised and in some cases is ignored. In spite of all these doubts and the conflict between theory and practice in this sphere, the possibility of breeding as a means of combating disease is always open.

Resistance to chalkbrood does occur in some colonies, and resistant bees can be selectively bred (Gilliam *et al.,* 1983). Good nest cleaning - or "hygienic" behaviour, in which bees uncap and remove dead and diseased brood - inhibits the survival of the fungus *A. apis* (Gilliam *et al.,* 1983), and is correlated with resistance to chalkbrood (Gilliam *et al.,* 1988; Milne, 1983; Palacio *et al.,* 2000).

Although other mechanisms have been demonstrated, hygienic behaviour remains the primary mechanism of resistance to chalkbrood and other brood diseases, such as American foulbrood (Gilliam *et al.* 1988; Spivak and Gilliam 1993; Spivak and Gilliam 1998a; Woodrow and Holst 1942). Shah (2000) reported that magnitude of hygienic behaviour was significantly correlated with the amount of TSBV killed larvae indicating that hygienic behaviour confers some resistance to TSBV in *A. cerana*. Interestingly, different colonies in the same apiary exhibited different levels of susceptibilities to infection thereby suggesting that resistance to TS-BV not only depends upon hygienic behaviour but is a complex phenomenon involving other mechanisms as well.

Honeybees must contend with a number of parasites both inside and outside of their natural environment. Mites are of particular importance to bees and beekeepers because of their potential to do great damage to both feral and managed honeybee colonies. Mites have also developed resistance to acaricides. Further the use of chemicals is harmful to bees and bee products.

Honeybees are parasitized by many different organisms, and among some of the most serious pest are the parasitic mites. Hygienic colonies also have some resistance to infestation by the parasitic mite *Varroa jacobsoni* (Boecking and Spivak 1999). Chemotherauptic treatments are ineffective against viral diseases such as Thai sac brood virus (TSBV) (Abrol, 1993). Evidently, there is a need to select stocks of bees resistant to diseases and mites.

Hygienic behaviour

Honeybee hygienic behaviour first received attention as a potential mechanism for control of American foulbrood (AFB) (Park *et al.* 1937). In a series of elegant studies conducted during the 1960s, Walter Rothenbuhler and his colleagues investigated the genetics of AFB resistance and hygienic behaviour (reviewed in Spivak and Gilliam 1998b). These investigations demonstrated that the level of expression of hygienic behaviour was variable between colonies from different lines (Rothenbuhler 1964a), that hygienic behaviour was carried out by workers up to 28 days old (Thompson 1964), and that hygienic bees removed large numbers of dead pupae just as quickly as smaller numbers (Jones and Rothenbuhler 1964). It was also shown that the behaviour was innate to hygienic bees and not learned, and that mixed colonies composed of equal numbers of hygienic and non-hygienic workers still expressed the behavioural phenotype (Trump *et al.* 1967). Hygienic behaviour - Whenever a larva in a sealed cell dies from a brood disease, the cell must be uncapped and the diseased larva removed before the cell can be reused.

Different steps of hygienic behaviour include

Uncapping	:	Making of the initial hole in the sealed cell
Enlarging	:	Enlarging the initial hole
Removing	:	Removing the infected larva

Rothenbuhler's two gene model for hygienic behavior

He proposed that hygienic behaviour is controlled by two unlinked recessive loci – one governing the uncapping behaviour (u) and the other removal (r). In order to express the hygienic phenotype, an individual must be homozygous recessive at both the 'uncapping' and 'removal' loci (Rothenbuhler 1964a; Rothenbuhler 1964b). Rothenbuhler's two-gene model is considered a classic example of the Mendelian inheritance of a complex behavioural polymorphism. He further stated that hygienic behaviour is a complex trait that has a strong environmental as well as genetic component.

It has also been shown that factors such as worker age (Thompson, 1964) and nectar flow influence the expression of hygienic behaviour. A large environmental component for both uncapping and removal behaviour is also supported by estimates of their respective heritabilities, although these estimates have varied widely.

Some researchers have challenged the two-locus model favouring a more complex genetic basis for hygienic behaviour. Re-evaluation of Rothenbuhler's data prompted Moritz to suggest a three-locus model, involving a single locus (u) for uncapping behaviour but two loci (r1 and r2) governing removal (Moritz, 1988).

Measurement of hygienic behavior

Methods

Different tests could be made simultaneously to test the hygienic behaviour on each colony.

Test 1 Removal of dead brood

Place a 2- inch square metal template, whose area delineated 100 cells on one side of the comb containing unsealed larvae. Sprinkle dry ice on the larvae within the template area to kill them. Record daily observations on the number of dead larvae removed by the bees

Test 2 Uncapping and removal behaviour

Cut 2-square inches section of healthy brood containing 100 capped brood cells on each side. Place comb sections in a freezer to kill the brood (Newton et al. 1975). At the same time a 2- inch square section is cut from the brood comb of each colony and placed with a comb section containing frozen brood. The bees usually add sufficient wax to secure the test comb section in the comb within 24 hours. Daily observations could be made from number of cells uncapped and empty cells from which brood has been removed.

Test 3 Pin prick brood assay

The pin-prick brood assay, developed by Newton and Ostasiewski Jr. (1986) and modified by Gramacho and Gonçalves (1994), is used to measure hygienic behavior. This assay consists of piercing the sealed worker brood of the same age (corresponding to the pink eye pupal phase), using a number 1 entomological pin. A test comb with capped brood cells is removed from each colony; 100 brood cells in one area of the comb are perforated, and another 100 cells in the same comb used as control. After perforating the brood, the combs are returned to the colonies. After 24 h, the test combs are again taken to the laboratory to count the cells that had been cleaned by the bees. The degree of hygienic behavior (HB) are determined as a percentage of cleaned cells 24 h after piercing the brood:

$$HB = \frac{EC\,(24\,h) - EC\,(0h)}{CC\,(0h)} \times 100 - Z$$

where, EC (24 h) = number of empty cells 24 h after perforation; EC (0 h) = number of empty cells before perforation; CC (0 h) = number of capped cells before perforation, and Z = control correction factor

$$Z = \frac{Y}{A} \times 100$$

where, A = number of capped cells in the control before placing the comb into the colony for the cleaning test; Y = number of new empty cells, calculated as Y = C – B; C = number of empty cells after the cleaning test, and B = number of empty cells before the cleaning test.

Twenty colonies of honey bees *Apis mellifera* L., in an apiary are tested for their hygienic behavior. Each of the colonies are tested five times, during a 10-15-day period. A colony are considered hygienic if it removed a mean of 80% or more of the dead brood.

Classification of colonies

The colonies could be classified as resistant or susceptible depending upon good or poor hygienic behaviour

(a) The resistant colonies removed 100% dead brood, and

(b) The susceptible ones less than 70% of the dead brood.

Resistance to Parasitic Mites in Honey Bees

Honeybees must contend with a number of parasites both inside and outside of their natural environment. Mites are of particular importance to bees and beekeepers because of their potential to do great damage to both feral and managed honeybee colonies. The two species of mites that are responsible for most of the damage to bee colonies are the *Varroa* mite *Varroa. destructor* (earlier *V. jacobsoni*), and the honeybee tracheal mite *Acarapis woodi*. Bees have several methods to cope with mites. One method is genetic resistance. Several types of bees have been bred that shows a significant resistance to honey bee tracheal mites (HBTM) as well as possible resistance to *Varroa* mites. There may also be resistant strains of bees to *Varroa*, but it is difficult to gage the success of genetic experiments. This may be in part due to *Varroa's* shorter association with the European honey bee (Cobey, 1997). Hygienic behavior in bees may also be a method associated with resistance to *Varroa destructor* n.sp (Boecking *et al.*, 2000). Africanized bees have shown a greater resistance to *Varroa* compared to European honeybees. This may be due to differences in grooming and hygienic behavior (Moretto, 2000). It may also be due to the longer association of the Africanized bees and *Varroa*. The infestation of HBTM seems to follow a pattern (Cobey, 1997). Initial infestation is very damaging. Over time the bees seem to cope with the mites and soon resume normal behavior.

The honeybee tracheal mite (HBTM), *Acarapis woodi* is an internal parasite that infects the respiratory system of adult honeybees (*Apis mellifera*). This mite was first described in 1921 in England, were it was believed to cause Isle of Wight disease. Today HBTM is known as acariosis or acarine. As the name suggests, this mite spends most of its life cycle in the trachea of adult honeybees, except for brief periods when it will pass from bee-to-bee. These mites are extremely small and therefore are hard to observe and study. Females are 0.004-0.007 inches (143-174 microns in length and 77-81 microns in width) and males 0.003-0.005 inches (125-136 microns in length and 60-77 microns in width). The mites have an oval body shape and also posses a hardened mandible that is used to feed on the blood of its host. The entire life cycle of the mite is spent within the trachea of adult honeybees.

The HBTM has a four-part life cycle that consists of an egg, larva, resting stage, and adult mite. Female mites leave the trachea soon after mating by passing through the thoracic spiracles of the bee. Once on the outside of the bee the mite will attach itself to the hairs of a passing bee and enter through the spiracles. Once inside of the trachea the female will begin to feed on the blood of the host, and will soon begin to lay eggs. The eggs hatch within 3 to 4 days inside of the trachea. The larva will develop in the trachea and will soon start the cycle all over again. Female mites prefer to migrate onto young bees, in particular bees less than 4 days old.

The parasitic varroa mite, *Varroa jacobsoni* (Oudemans), is another serious pest of the honeybee (*Apis mellifera*). The *Varroa* mite was first described on *Apis indica* from Java in 1904. It was first reported in association with *A. mellifera* in 1962, making this a relatively new host-parasite interaction. *Varroa* mites develop and reproduce in the sealed brood cells of honeybees where they feed on the hemolymph of bee pupae (Erickson, 1998).

Adult female *Varroa* mites are oval shaped and about 1.1 mm long and 1.5 mm wide, with a reddish-brown color. Males are smaller and paler in color. The adult female mites enter the brood cells in the hive just before they are capped by the bees. Once inside the cell the female will feed on the hemolymph of the larva, and will soon begin to lay eggs. The *Varroa* mite has a five-part life cycle that proceeds from egg to six legged larvae, to eight-legged protonymphs, to deutonymphs, to sexually mature mites in 6 to 10 days (BRL, 2001). When there is a shortage of brood available for the mites to feed on, adult bees serve as intermediate hosts. The mites will attach themselves between the abdominal-head-thorax segments of the adult bee were they would feed on the bee's hemolymph.

Both HBTM and *Varroa* mites can have a tremendous impact on feral and managed honeybee colonies. These mites have caused massive losses in all parts of the world where they are known to occur. The mites are especially devastating in previously un-infested areas that contain apiaries and feral colonies that have not been exposed to these parasites. The mites take their toll on the honeybee colonies, especially when both mite species are present, by stressing the colony and exposing them to other disease. This stress can lead to poor brood production, declining colony population, and the eventual death of the colony. *Varroa* mites have been known to kill entire colonies in 12 months to 4 years depending on the level of infestation. This loss of colonies takes its toll on beekeepers that must deal with the cost to replace colonies and the cost of reduced honey yield. As of yet there a no effective and efficient methods that beekeepers can use to control mites. There are several chemical treatments available such as Apistan,

Fumidil-B, and menthol that will decrease colony mortality. Colony losses decreased with the use of most mite control agents. These methods are not guaranteed and alone are not enough to control mites in every situation. Beekeepers are looking for new ways to help control the mite problem, and they have turned to the bees themselves for the answer.

Honeybees are not completely helpless when it comes to defending themselves against these parasitic mites. They are naturally stringent and clean insects. Honeybees have developed several different ways to help them cope with the mites including hygienic behavior, natural resistance, and high levels of compatibility to name a few. These methods are not completely understood, and research is underway to shed more light on how the bees deal with the mites. This behavior is very important to the health and survival of the colony and for the individual bee. Worker bees are responsible for the maintenance of the hive and for the care of the brood. This involves upkeep of the comb and removal of dead or diseased brood. This is done by the uncapping of the cell and removal of the dead or diseased larvae. Some colonies also show a high level of grooming behavior that is effective in removing mites from the bodies of adult bees. Some colonies of bees have been shown to have a more pronounced hygienic behavior compared to other colonies. These colonies also show an increased resistance to several diseases. Hygienic behavior is a mechanism of resistance to at least 2 diseases of honeybees: American foulbrood, caused by the bacterium *Paenibacillus larvae* (Woodrow and Holst, 1942, Rothenbuhler, 1964); and chalkbrood caused by the fungus *Ascosphaera apis* (Massen ex Claussen) Olive and Spiltoir (Gilliam *et al.*, 1983, 1988) Spivak and Downey, 1998. Hygienic behavior also is one mechanism of defense against the parasitic mite *Varroa jacobsoni* Oudemans (Spivak and Downey, 1998). Hygienic behavior may also be a defense against HBTM as well as other diseases. The original host of the *Varroa* mite, *Apis cerana*, is known to remove mite infested pupae from the hive, and in doing so interrupts the mite's life cycle. *Apis cerana* is the original host of *Varroa jacobsoni*; Behavioural and physiological adaptations of *Apis cerana* keep infestations low (Peng *et al.*, 1987). They effectively remove mites from bodies by self- grooming and perform grooming dance to attract other bees to remove mites. They bite, puncture and drop the mites outside the hive. Furthermore, they are equally effective in removing mites from the brood cells. *Apis mellifera* on the other hand has a limited capacity to detect and recognise mites. Grooming behaviour is also infrequent. This hygienic behavior is also seen in other species of honeybees. In European honeybees (*Apis mellifera), A.mellifera carnica* and *A. m. ligustica* are two examples of colonies that have shown a high level of hygienic behavior. Colonies of *A.m. ligustica* Spinola specifically bred for hygienic behavior removed significantly more infested pupae than non-hygienic colonies (Spivak, 1996, Spivak and Downey, 1998).

It is not yet completely known how this hygienic behavior is passed on in the bee's genetic code. It has been shown that hygienic behavior is not learned from other bees, and that it is most likely controlled by the genes of the bee (Rothenbuhler, 1964). The effect that hygienic behavior has on control of mite populations within the colony is still unknown. It is clear that hygienic behavior does confer some level of defense against the *Varroa* mite, and possibly against the other pests as well. It has been shown that hygienic colonies have lower incidences of chalkbrood and American foulbrood (Boecking *et al.,* 2000). There are several lineages of honeybees that show an increased hygienic behavior along with a mild resistance to varroa mites. Through selective breeding, beekeepers and researchers may eventually develop a commercially available strain of honeybee that is resistant to parasitic mites and other diseases. This strain will also have other favorable characteristics such as good temperament, high honey yield, and good weather tolerance.

One honeybee in particular that displays many of the characteristics mentioned above is the Buckfast bee. This strain of bee has been undergoing selection since about 1919, and is known for its good honey production, mild temper, and its resistance to HBTM. This strain of bee has its beginnings in England, where at a monastery a monk by the name of Brother Adam is credited for developing the strain.

The Buckfast bee has been shown to have a clear level of resistance to HBTM, especially compared to other strains of bees. The resistance of the Buckfast strain is due chiefly to a lower infestation rate among bees in a colony, as compared with susceptible bees (Danka and Villa, 1996) (Slabezki *et al.*, 2000). How exactly the bees achieve this lower infestation rate is still unknown. The Buckfast bee is an example of how selective breeding can and will be used as a way to control mites and other diseases that effect honeybees.

Host Compatibility

The issue of resistance versus host compatibility is currently under debate. Weather or not certain strains of honeybees are truly resistant to parasitic mites or are merely expressing a level of compatibility is the issue at hand. Vandame *et al.* (2000) stated that the term "resistance", when used to describe the host-parasite relationship, is incorrect, as they suggest the term "compatibility" to describe the relationship. When a host is said to have low susceptibility it is in reference to characteristics of the host that prevent parasite development and to the incapacity of the parasite to adapt to the host's resistance mechanisms (Vandame *et al.,* 2000). Resistance is an active response by the host to counteract the parasite's development (Vandame *et al.,* 2000). The term compatibility refers to a situation in witch the parasite is permitted to develop, but not to the point were the host is permanently harmed.

The relationship with the European honeybee (EHB) and the varroa mite just might be an issue of compatibility. Africanized honeybees (AHB) have had a longer host-parasite relationship with the varroa mite compared to the European honeybee. As a result the Africanized bees show a higher level of compatibility with *Varroa* compared to the European bees. This is in large part due to the lower reproductive rate reached by varroa females in the worker brood cells of AHB compared to the reproductive rate among EHB (Medina and Martin, 1999;

Moretto and Mello, 2000). Africanized bees also have the ability to remove mites that are attached to adult bees by implementing grooming behavior (Moretto and Mello, 2000)

Breeding *Varroa* resistant strains of honeybees

Breeding honeybees for resistance to *Varroa* (*Varroa destructor*) has been a goal since *Varroa* began to move into the worldwide population of honeybees and at times this goal seems unreachable. However, we now know that resistance to *Varroa* does exist in honeybees. And it seems to exist in more than one form. The aim is to produce honeybees that are genetically resistant to *Varroa*, and we may define mite resistance in honeybees as a slower growth or a decline of the mite population in a colony. Although acaricides control *Varroa* mite in colonies of honeybees, use of chemicals endangers bee and hive products. Acaricide resistant strains and limitations of chemical control has lead to exploration of selecting *Varroa* resistant strains

The *Varroa* defence is normally measured indirectly by per cent number of dead mites falling on bottom board or by the removal of artificially killed brood (Hoffman, 1996). These methods are strongly affected by environmental influences. Colony level selection may not be particularly efficient for *Varroa* resistant breeding because significantly intra-colonial variations occur. Colony composition has different patrilines with a different distribution of behavioural response. Uncapping of most of the *Varroa* infested cells begin between 4-5 days after infestation.

Disturbance by the mite caused to the growing pupa results in release of some cues or signals. These cues are possibly perceived by the uncappers and result in initiation of opening. When freshly capped brood cells are infested and returned to the colony, the cuts resulting from artificial infestation are re-sealed by the bees instead of uncapping and removing the larva at this stage. The initiation of uncapping between 4-5 days is caused by the rhythmic behavioural activity of mites searching for a feeding site on the pupal body. The rise in these activities may cause the release of certain stimuli or cues by prepupae because of iritation or injuries as a result of feeding of mites. These cues may be perceived by some specialists which start uncapping. The possible preference of only a few bees to respond to the above has a genetic basis.

Conclusions

Honeybees have evolved elaborate and diverse defence mechanisms to protect their brood, adults and food against a broad range of pests ranging from viruses to vertebrates. One such mechanism is the hygienic behaviour. Dead or parasitized brood are removed from their cells by hygienic workers who detect, uncap and remove them providing protection against brood diseases and tolerance to ectoparasitic mites *Varroa jacobsoni*. The resistance trait could readily be increased through a breeding programme using a field assay (freeze killed brood). The selected queens could be reared from each of these colonies by grafting worker larvae, which could be tested and retested for their hygienic behavour. Further data is needed on the host-parasite relationship between the parasitic mites and the honeybee. This data will be important in the quest to develop a more resistant or more compatible strain of honeybee for the commercial market worldwide.

Chapter 23

Impact of inbreeding depression in honeybees

Inbreeding is the mating of two individuals which are more closely related to each other than the average population. It is well known from animal husbandry and plant cultivation that inbreeding often leads to reduction of viability. Inbred organisms seem to be more susceptible to diseases, give smaller yields and generally appear to have reduced fitness. The common phenomena known as inbreeding depression is a major problem in animal breeding. The reduction of performance can be explained in terms of a loss of genetic variability in the inbred genome (the genome is the sum total of the genes in an organism. Both haploid and diploid individuals have only one genome which is contained either in one or two basic chromosome sets). According to recent hypothesis, the within locus variability and the gene interactions resulting from hetrozygosity are responsible for the phenomenon of inbreeding depression.

Many attempts have been made to provide animal breeders with guidelines for optimized selections. Breeding schemes have also been proposed for honeybee selection. The schemes simulate the fate of genes in a given population. at present we do not know whether inbreeding depression is affected solely by genes. For example, non - heritable maternal effects on the egg, or heritable cellular components which are transferred via cytoplasm to the egg also might cause inbreeding effects. In the fruit fly (*Drosophila)* such non-genic effects have been observed, but information is scarce for other organisms.

In honeybees which have a haplo-diploid structure, the dynamics of inbreeding may be quite different from those in diploid live stock populations. From studies of isoenzymes we know that degree of variability seems to be much lower in haplo-diploid hymenopteran population than in diploid hymenopteran population.

A lack of variability in gene pool should provide evidence for a close average relatedness of all members of the population. Hence inbreeding as defined above might result in the inbreeding observed in highly variable population. Nevertheless, the effects of inbreeding depression are found in honeybees. Breeding activity, defensive behaviour comb construction in inbred colonies and morphological and physiological characters of inbred workers are all subject to inbreeding depression. Either the reduced heterozygosity in honeybee population is still sufficient to cause inbreeding depression or it s caused by sources other than genome such as cytoplasmic compounds of the queen transferred to the egg (i.e. maternal sources).

Inbred drones

Drone is a good material for investigating the occurrence of non genic inbreeding effects. Drones result from unfertilized eggs, and therefore, get only a single copy of the mother queen genome. During development they multiply their set of chromosomes per cell which results in highly polyploidy tissue (i.e. with many sets of chromosomes per cell). Only a few tissues such as sperms, remain haploid in drones. Hence, inbreeding depression resulting from a decrease of genetic variability, would have no effect in drones if only gene interactions are the source of inbreeding depression. On the other hand if drones show a depressive response to inbreeding, this must be due to factors other than genes.

Reproductive ability

Reproductive capability for an organism is very important in characterizing organism's fitness. For the drone duration and frequency of mating flights and the number of sperm produced can be good estimators of reproductive capacity. Studies by Moritz (1986) on the age of drone at first mating flight and the duration and number of mating flights per day, for drones at different levels of inbreeding in relation to weather conditions (Table 76). Showed that there are severe inbreeding effects in drones which affect their fitness, inbred drones showed a worse flight performance than non bred drones–they made fewer flights and these were of shorter duration also. The number of sperm produced by inbred drones decreased dramatically as the level of inbreeding increased.

Table 76. The effect of inbreeding on flight performance and number of sperm of drones

Level of inbreeding	Age at first flight	Total daily flight time	No.of flights per day	No. of sperm
0	100	100	100	8.8×10^6
25	98	98	94	6.3×10^6
34	88	87	85	4.8×10^6
47	80	68	66	-----
50	73	63	62	2.2×10^6

Source: Moritz (1986)

Physiological performance

Physiological performance of freshly emerged drones (Table 77) also revealed the decreased reproductive ability in inbred drones. In most cases, inbreeding effects were very drastic. In bred drones did not resist the cold as well as non inbred drones. The temperature at which chill coma commenced (i.e. at which locomotive activity ceased) increased from 14°C to 18°C as the level of inbreeding of the mother increased. Furthermore, oxygen consumption of individual freshly emerged, under extreme temperature of hot and cold was significantly lower in inbred drones than in non-inbred drones.

Table 77. The effect of inbreeding on physiological performance of freshly emerged drones.

Level of inbreeding	Chill coma temperature (°C)	Oxygen consumption (ulO_2)in 5 min. interval	
		20°C	40°C
0	14.7	8.5	40.0
25	15.0	8.3	34.9
34	16.4	8.2	31.2
47	--	6.2	22.0
50	18.0	5.2	19.1

Source: Moritz (1986)

Biochemical characters

To study the problem at a deeper level, Moritz (1986) investigated the activity patterns of some key enzymes in several metabolic pathways in drone haemolymph. He found two different type of results, exemplified by those for the enzymes malate dehydrogenase (MHD), isocitrate dehydrogenase (ICHD), glutamate pyruvate transaminase (GPT), and glutamateoxal acetate transaminase (GOT). Enzymes concerned with oxidative metabolism (MDH and ICDH) showed a significant lower activity in inbred drones than in non inbred drones whereas enzymes concerned with protein metabolism (GPT and GOT) had a higher activity in inbred drones than in non inbred drones (Table 78). These differences in enzyme activities which are most significant in freshly emerged drones and become less as the drones age, may be the signs of a severe metabolic disequilibrium in inbred drones.

Table 78. The effect of enzyme activities in drone haemolymph.

Level of inbreeding %	% enzyme activity in drone haemolymph			
	MDH	ICDH	GOT	GFT
0	100	100	100	100
25	91	96	110	120
34	80	97	117	137
50	57	63	129	143

Source: Moritz (1986)

Components of inbreeding depression in honeybees

Two effective components of inbreeding depression in the diploid honeybee or queen include, a genic one, resulting from a loss of genetic variability, and a non genic one as described above. The possible components of inbreeding depression in honeybees are shown in a schematic model (Figure 96) in the diploid individual inbreeding depression would result from maternal effects, plasmatic factors, and the genome-and interactions between these compounds. In haploid individuals there would be no direct effect genome.

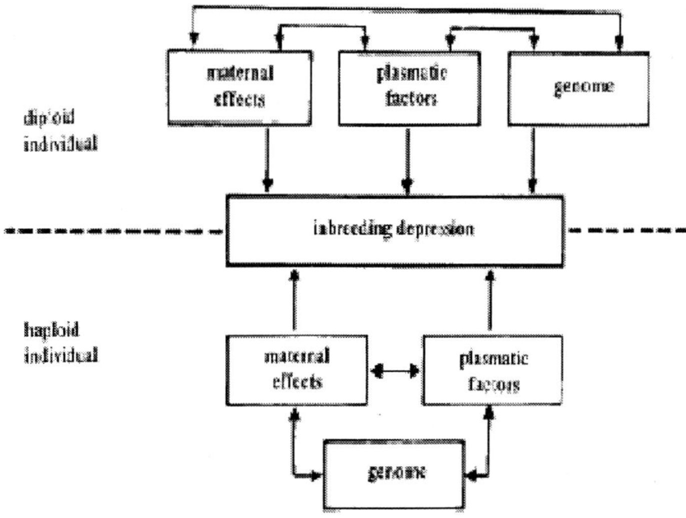

Figure 96. Factors affecting inbreeding depression in haploid and diploid individual

Comparison of workers and drones

Based on the hypothesis that there are two components of inbreeding, a comparison between inbred workers and inbred drones should enable us to quantify genetic and non genetic inbreeding depression. Moritz (1986) performed tests with inbred (47%) and non inbred, workers and drones using the enzyme sets mentioned above, however, this time the enzyme activity was tested on mitochondria isolated from flight muscle tissue. Mitochondria are the cellular organelle responsible for respiratory metabolism. They have their own DNA which is replicated independently of the cell DNA. All the enzymes tested are known to be located on mitochondria, and we therefore, expect more exact data from these organelles rather than from haemolymph samples.

In all cases workers showed a stronger inbreeding effect rather than drones (Table 79). This difference can be interpreted as the genic inbreeding depressive effect which is additional to the cytoplasmic effects. If for simplicity, we neglect the possible effect of interactions between nucleus and cytoplasm, we estimate that up to 83% of the inbreeding effects in mitochondria can be cytoplasmically determined.

Table 79. The effect of enzyme activities in drone haemolymph.

Level of inbreeding %	% enzyme activity in drone haemolymph			
	MDH	ICDH	GOT	GFT
0	100	100	100	100
47 (worker)	49	58	161	124
47 (drone)	67	75	119	115

Source: Moritz (1986)

Maternal or mitochondrial effects

A major component of non-genic inbreeding effects may be due to the inbred maternal genome. For example, the inbred maternal enzyme apparatus might not supply the egg with sufficient protein resources. Such non-heritable maternal effects, but also the effects of heritable plasmatic factors in honeybees, were investigated by making reciprocal crosses of inbred lines without bred lines. Table 80 shows the number of accepted (non-inbred) larvae after grafting. The viability of the larvae is strongly affected by whether the mother queen is inbred or outcrossed-larvae of inbred queens. Besides non heritable maternal effects other components in the cell also might effect inbreeding depression. Mitochondria isolated from various inbred lines showed *in vitro* lower enzyme activities than *in vitro* combination of mitochondrial complementation, enzyme activities were low in the inbred lines. Furthermore, this suggests that mitochondria themselves might cause phenomena comparable to inbreeding depression.

Table 80. Acceptance rate from grafted larvae of inbred and out crossed parents

Parent type (queen and drone)	No. of non-inbred larvae	
	accepted	rejected
Inbred x Outcrossed	86	78
Outcrossed x Inbred	114	49

Source: Moritz (1986)

This hypothesis would be applicable to honeybees provided that mitochondria are transferred in drone sperm during fertilization. Such a transfer is very likely as three to seven sperm (head and tail) enter the egg during fertilization. We know from other organisms that mitochondrial DNA is found quite frequently in the nuclear DNA and vice versa. To confirm this hypothesis, male mitochondria (i.e. those from paternal sources) will have to be located after fertilization. Up to now attempts to demonstrate any paternal mitochondrial gene flow in insects have failed-although this does not mean that none exists.

Moritz (1983) found that in flight muscle mitochondria of drones and workers (*Apis mellifera* L.) enzyme activities showed significant changes after inbreeding. The genome seems to have only moderate effect on this mitochondrial inbreeding depression. It is shown that mainly cytoplasmic factors are affecting the changes in enzyme activities observed after inbreeding.

Inbreeding depression in classical quantitative genetics is explained by an increase in homozygous loci in diploid organisms. The classical model must fail in haploid or hemizygous individuals because in these cases it is impossible to increase the degree of homozygosity.

Nevertheless, it was possible to find inbreeding effects in haploid individuals of haplo-diploid populations. Helle (1965) showed this phenomenon in an arrhenotokous mite (*Tetranychus urticae*) and Moritz in *Apis mellifera capensis* (1978) and in '*Apis mellifera cornica* (1981). Figure 96 shows the factors affecting inbreeding depression in haploid and diploid individuals. The genome has no influence on the inbreeding depression of haploids. Non-heritable maternal effects and heritable cytoplasmic structure should be the only causes of inbreeding depression in the case of haploidy.

The effects of maternal plasma on the egg development of *Apis mellifera* and *Apis cerana* was shown by Maul (1969). Biemont (1974) observed maternal inbreeding depression in *Drosophila melanogaster* on a zygotic level. The reduced allozyme pattem in cell plasma of the inbred mother could be an explanation for the inbreeding effects observed in her progeny.

Little is known about heritable cytoplasmic factors and their role in inbreeding depression, although such factors are quite common in eukaryotes (Sonneborn, 1957; Margulis, 1971). Some information is available about mitochondria, which are of great interest in this problem. Mitochondrial complementation (a heterotic phenomenon of isolated mitochondria) has been shown in maize (Sarkissian and Mc- Daniel, 1967), wheat (Sage and Hobson, 1973) and barley (McDaniel, 1969). Dzapo *et al.* (1973) reported the same phenomenon in mitochondria isolated from rabbit liver. Since heterosis is the inverse of inbreeding depression (Fischer, 1978) this mechanism could be an explanation for the inbreeding effects observed in haploid individuals. The activity of mitochondrial enzymes (Sacktor, 1953, 1954) from drone hemolymph samples is correlated to the inbreeding coefficient of the mother queen (Moritz, 1980). Thus it seemed worthwhile to examine *Apis mellifera* mitochondria and their role in quantitative genetics of haplo-diploid populations in general.

The enzyme activity changes observed after inbreeding were reported earlier (Moritz I980a, 1981 b) and seem to be characteristic of *Apis mellifera*. A disequilibrium in metabolism obviously arises after inbreeding. Further biochemical investigation will prove if there is a real compensation for a reduced Krebs cycle activity by an increased protein metabolism for energy production. From the present results one can draw the genetic conclusion that cytoplasmic structures may cause serious inbreeding depression

Inbreeding effects

Spotty brood cells and deterioration of the colony

A genetic trait is a characteristic that can be transferred from one generation to the next. Traits are determined by one or more genes. Genes are located on chromosomes. In humans, sex is determined by an entire chromosome. Human embryos that have two "X" chromosomes (XX) become females. Those which have one "X" and one "Y" chromosome (XY) become males. Sex determination in bees was first explained by John Dzierzon in 1845. He observed that virgin queens which were not permitted to go on mating flights produced only drone brood. From his observations, he correctly proposed that drones develop from unfertilized eggs while queens and workers develop from fertilized eggs. In the 1960s, J. Woyke established that sex was determined by a single gene and that there were multiple forms (alleles) of the sex-determining gene. When fertilized eggs are laid, normal females are produced if different alleles are contributed by the queen and drone (these offspring are called heterozygotes). When unfertilized eggs are laid (hemizygotes) normal males are produced. When queens mate with closely related drones, some of the fertilized eggs will receive the same allele from both parents (homozygotes). This union results in some of the fertilized eggs having two copies of the same gene. If hand fed by a scientist, they can be reared to adult drones. These larvae are cannibalized by the workers as soon as they hatch.

The beekeeper observes a spotty brood pattern, and the colony does not build up properly (Figure 97).

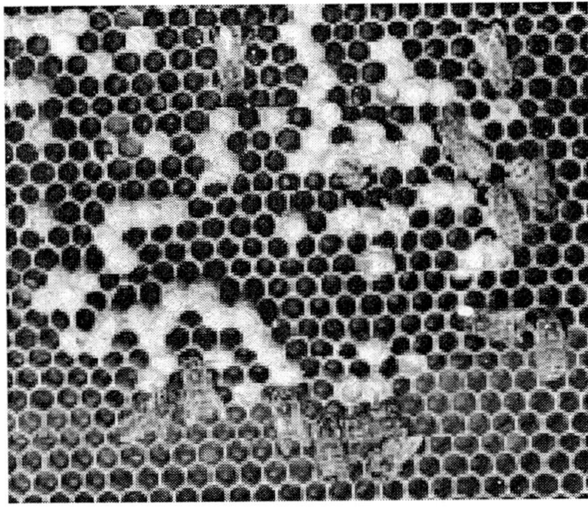

Figure 97. Inbreeding depression is characterized by a spotty brood pattern.

There are approximately 18 forms of the sex-determining genes in honey bees. Queens that mate with closely related drones produce spotty brood patterns due to the frequency of embryos that receive the same sex-determining allele from both their mother and father. Since queens mate with a dozen or more drones, the percentage of non-viable progeny is determined by the degree of inbreeding that occurs. A large population of wild honey bees existed in the United States prior to the introduction of the *Varroa* mite. This wild population served as a buffer to inbreeding by providing a genetically diverse component to the drone population. With the demise of the wild population of honey bees, beekeepers need to carefully guard against inbreeding if they rear their own queens. The most important precaution is to use multiple queen mothers and to select multiple lines. Infusing new lines into the population periodically will also reduce inbreeding. The loss of the wild honey bee population is a tragedy. A large part of the genetic variability available to beekeepers has been lost. Most of the honey bees in the United States are now the descendants of less than 500 breeder queens. The decrease in wild bees has both good and bad implications for bee breeding. Beekeepers should be able to make more progress in selecting open-mated stock. Without the competition from wild drones, it should be easier to maintain desirable traits. On the bad side, genetic variability has been lost, and beekeepers will need to increase their efforts to guard against inbreeding.

Does inbreeding level affect the hygienic behavior of honey bees?

The inbreeding is the basic method for maintaining the selection progress in the honeybee breeding. The nature of honeybees´ ability to resist the infectious diseases is strengthened by selecting for good hygienic behavior in their colonies. A question arises whether increasing the inbreeding level does not lower the hygienic reaction of bees. Cermak (1995) found that the effect of inbreeding on the hygienic behavior did not prove to be significant in the analysed selected population with a maximum inbreeding level of F = 31.4%. Neither inbreeding depression nor the opposite trend - heterosis in inter- line hybrids were apparent inside the breeding unit of Carniolan race. This makes possible the selection of bees for behavioral resistance against brood diseases without negative affecting of inbreeding depression. It is true for lower levels of inbreeding with F less than about 30%. An inbreeding depression in a population with closer inbreeding mating systems, i.e. higher inbreeding coefficients, cannot be excluded. The absence of a significant influence of inbreeding on hygienic behavior of bees may correspond with high heritability of the character. Low heritable characters are more negatively affected with inbreeding, and vice versa (Figure 98).

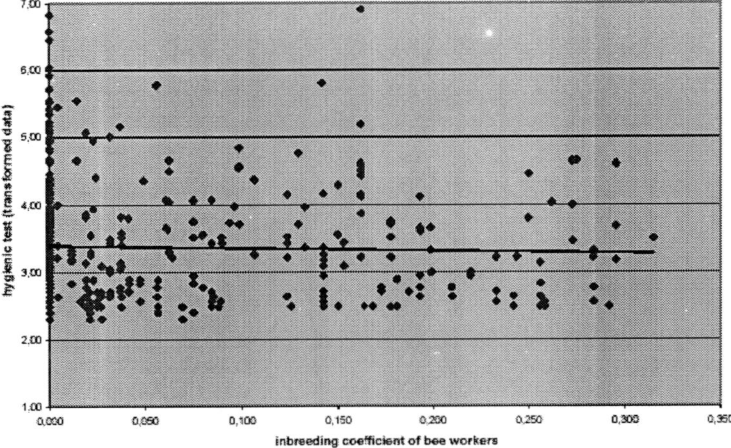

Figure. 98. Dependence of hygienic test on inbreeding level of bees hygienic test (transformed)

Breeding improved honey bees

Inbred and hybrid bees

Many beekeepers have produced excellent queen daughters of imported stock or daughters of some selected queen that they obtained from other beekeepers. The new stock queens usually mate with drones of the beekeeper's own stock. If the sex alleles of the two stocks are different, the eggs from most matings are high in hatchability.

The new stock may also be genetically different from the stock that the beekeeper already has. Thus the queen daughters of the imported queen produce hybrid progeny after mating with the drones of the beekeeper's own stock. These may be racial hybrids or hybrids between strains of bees. Such queens often produce excellent colonies with high-quality brood. The beekeeper then praises the imported stock apparently without realizing that his own drones contribute half to the genetic make-up of the workers in these superior colonies.

The story is often quite different after a few generations of back-crossing to the original drones or to drones produced by the hybrid daughter queens. Segregation and recombinations occur, and in the third or fourth year the queens and their progenies are frequently more variable than the original stocks. By this time the beekeeper concludes that the original imported stock was not so good after all or had "run out"; so he gets some other new stock for breeding purposes.

By this procedure the colony superiority may again be restored to a high level. It also falls again after a few years. To keep the quality of his bees up, the beekeeper continues to get new or different breeding stock every few years.

These results are best explained as due to hybrid vigor and the crossing of stocks having different sex alleles. The first-generation colonies were superior because they were hybrids - that is, they were headed by cross-mated queens whose worker progeny were hybrids between the imported and local stocks. By mathematical analysis of the action of genes in future generations of segregation and recombination after outcrossing, it can be shown that by selection alone the beekeeper is powerless to keep this stock at the high level of production and uniformity exhibited by the first-cross progeny.

Most of the complicated physiological characteristics, such as egg production and vigor are dependent for their maximum expression on the harmonious interaction of many genes and their alleles. Since these genes are located on many chromosomes and there is linkage of genes on the same chromosomes, the likelihood that certain factors conducive to vigor will be linked with injurious factors reduces the chance of obtaining maximum vigor by selection within inbred lines. Moreover, in breeding within a line the sex-allele problem again arises and selection is considerably weakened by a high proportion of low viability matings. In plant and animal breeding it has been found that certain hybrids are superior to the lines from which they are produced. The superiority of hybrid corn is unquestioned. Animal breeders have also obtained superior hybrids. Because a large proportion of the progeny is required for breeding purposes to maintain the lines, animal breeders have not used hybrids for extensive production. However, in hogs a modification of the controlled-hybrid program is often used. This is the criss crossing of three lines, using hybrid sows and purebred boars in each generation. The three-breed crossed pigs are more productive than those of the pure breed. Hybridizing inbred lines of chickens has become more popular and profitable within recent years.

Hybrid vigor is generally thought to be a characteristic of dominance. Dominant genes tend to have more favorable effects than their recessive alleles, and in a hybrid the dominant genes find expression at more loci than do the recessive alleles. If one inbred line is dominant (AA) at one locus and recessive at another (bb) and another inbred line has the reverse (aa and EB) characteristics, the hybrid will have one dominant gene at each locus (AaBb). If both loci contribute to vigor, then the hybrid will be superior to either inbred line. It is also thought that certain genes contribute more to vigor in the heterozygous than in the homozygous condition. An ideal breeding system is the production and maintenance of purebred but unrelated lines and the crossing of these lines for the production of superior hybrids. Not all- in fact, only a very few-of the inbred lines will give outstanding hybrids when crossed. The breeder, however, can perpetuate those lines that can be favorably used for the repeated production of superior hybrids.

Plant breeders can take advantage of hybrid vigor because of the large number of seeds produced in each generation and the relatively low cost of producing seed for the production of hybrids. Animal breeders, on the other hand, are not in this favorable position. The expense of producing and maintaining highly inbred lines together with the small number of offspring makes this practice unprofitable, except with chickens and to some extent hogs.

Honey bees more nearly approach the condition existent in plant breeding. Less than 1% of the daughter queens and drones of a queen is needed to perpetuate the line. Whereas only 4 to 10 queens are needed in each generation to insure the continuance of an inbred line, 3 or 4 out crossed queens can be used to produce 10,000 hybrid queens.

The low cost of producing and maintaining inbred lines of bees, together with the large number of offspring that can be produced from a single breeding individual, suggests that a bee-breeding program based on hybrid vigor is practical. The effect of the sex alleles on egg hatchability further indicates that other breeding plans, such as line breeding, are slow and expensive because too many poor-viability matings occur. To produce hybrid bees the bee breeder may cross different races, strains, or inbred lines of bees. Unless these bees are homozygous for the desired characters, the resulting hybrids will be variable. Furthermore, other crosses of the same races or strains will differ from each other. A sure method of having uniform hybrids is to have strains homozygous for the desired characters.

The fastest method of increasing homozygosity in a line of stock is to inbreed. Inbreeding is the mating of closely related individuals, such as parent-offspring, brother-sister, or cousins. Related individuals are likely to have many of the same inherited qualities, and mating of these individuals tends to fix these qualities in a homozygous, or pure, condition. Inbreeding is used chiefly for the one purpose of producing homozygosity, or genetic likeness of offspring.

Inbreeding within a population tends to separate the population into many distinct families. Each family becomes uniform within itself but distinctly different from other families. Selection between families of such inbred lines can then be made with more accuracy than selection between individuals. This is especially true for characteristics of low heritability, such as egg production and vigor. The inbred individuals become lower in average merit than non-inbreds. They are not nearly so vigorous or productive. Inbreeding is the severest test of the heredity worth of the individual that can be made, for it causes fixation of both good and bad traits.

Because of the mating habits of bees, it is necessary that all inbreeding matings be made by artificial insemination. After highly inbred lines are obtained it is necessary to control matings to insure keeping the stocks pure. Artificial mating is the only safe method known at present with bees.If the breeder makes very close matings, such as mother-son, brother-sister, or backcrossing to a queen, he can accomplish very little by individual selection. It is true that he can accomplish a great deal when selecting for color or other visible characteristics that are highly heritable. However, most of these characters have little economic importance. Characteristics of greater importance, such as brood production, vigor, or resistance to disease, are not visibly detectable in individual queens or drones. Selection for these traits is more effective after the lines are inbred. To determine traits in inbred lines, crosses are made to a line or lines known to possess certain inherited qualities. Thus the test of inbred lines for characteristics of low heritability is to cross all inbred lines to certain "tester" lines and compare the unknown inbred lines with each other. Selection then is between inbred lines.Before beginning to produce inbred lines the bee breeder chooses the stocks that he will use. He may desire to cross two or more of the selected non-inbred stocks before beginning to inbreed.

He should obtain a few queens each from a large number of sources and test all queens for one season under uniform conditions in a single yard if possible. In this way he can get first-hand knowledge of the potential breeding stocks. He may select some colonies for immediate inbreeding and perhaps allow the others to cross freely for use as future breeding stocks.

It has been shown in both plant and animal breeding that the best hybrids usually come from crosses of unrelated stocks of different origin. The breeder should therefore, select a number of different types of bees. He should select some black, some yellow, and some queens intermediate in color rather than all of one color type. He should also select some long, slender and some short, stubby queens. He should obtain some gentle bees and perhaps even one or two lines with a lot of evil temper! The superior hybrids are genetic heterozygotes. They are the result of crosses between stocks that are unlike genetically. Differences in type, color, or temper denote genetic differences for these characteristics and probably also for other characteristics that the breeder cannot readily see or measure.

There is some evidence that gentleness in bees is dominant over viciousness. In one of the experiments the hybrids of a cross between a vicious and a gentle line were gentle rather than vicious. The vicious line may bring into the hybrid the few genes that mean the difference between an average and a superior hybrid bee. By no means are we suggesting that we need viciousness in bees. This is merely an illustration of how genetic diversity might be obtained.

Once the stocks are chosen, close inbreeding is begun to fix the lines. The breeder should know which matings to make to obtain the desired inbreeding in the shortest time with the least expense. Figure 99 shows the percentage of inbreeding obtained in successive generations by the various systems of inbreeding possible in honey bees. The percentage of inbreeding is the percentage of heterozygous loci of the original selected individuals that become homozygous by inbreeding. Inbreeding has no effect on genes already homozygous in the lines; so we are concerned only with those genes that are not homozygous. Since the bee breeder cannot tell which genes were originally heterozygous and what effect each gene has, he can only measure the increase in purity of the stocks by the percentage of inbreeding.

With present techniques the two systems of mating that increase inbreeding fastest, backcrosses to a male and a mother-son, are not advisable economically. Loss of breeding individuals, and consequently of inbred lines, is high when these systems are followed.

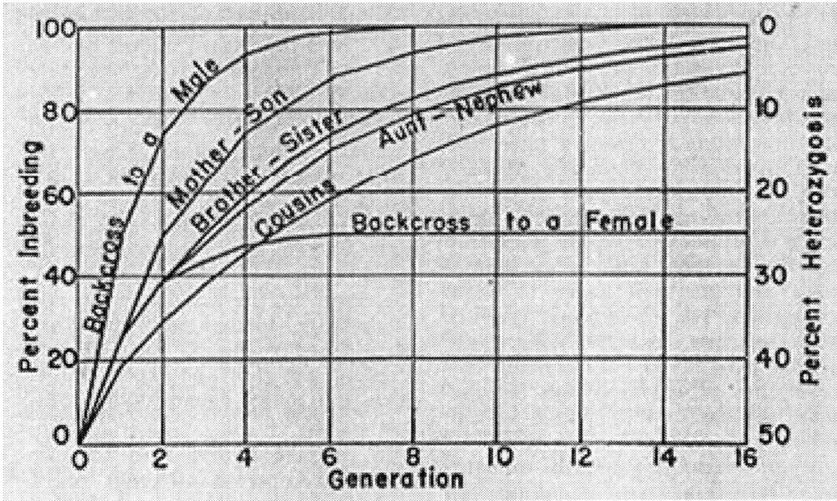

Figure 99. Percentages of inbreeding and of heterozygosis (assuming the initial value to be 50 per cent) in successive generations of various systems of inbreeding in honey bees. (From Crow and Roberts, 1950.)

The next most rapid method of increasing inbreeding is by brother-sister matings. This system is recommended, because at present it is the most practical of all systems shown. For the first two generations the systems of brother-sister and aunt-nephew matings and back-crossing to a female are identical in percentage of inbreeding. Because the aunt-nephew system depends upon the use of only one drone for each mating, this system is not so dependable for insuring the survival of an inbred line. Using only one drone to inseminate a queen often results in a poorly inseminated queen, who may turn into a drone layer before the next generation is produced.

Because drones mature more slowly than queens, backcrossing to a female produces inbreeding to 37.5% faster than brother-sister mating in time consumed per generation for the first two generations. Thus backcrossing to the original queen for the first two generations followed by brother-sister matings in all future generations is recommended for the production of inbred lines in the shortest time with the greatest chance of success in maintaining the inbred line's. Multiple drone matings may be made in all generations to insure good inseminations.

The bee breeder should know what inbreeding will do to his stocks. If he starts an inbred line by back-crossing and then makes brother-sister matings as illustrated in Figure 100, he may expect that each line will become more uniform as inbreeding progresses. Most noticeable, however, for the first few generations will be the quality of the brood.

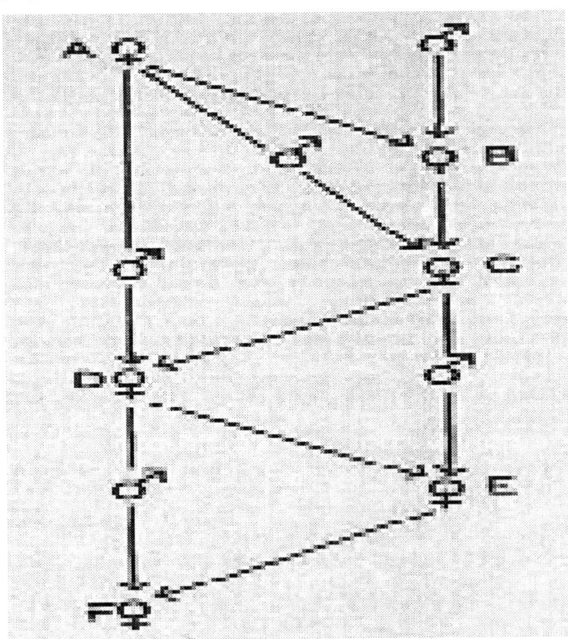

Figure 100. Arrow diagram of a recommended system of mating for inbreeding in bees. After two generations of back-crossing to a selected queen, the line is continued by brother-sister matings.

If queen B, a daughter of A, is mated to several drones that are sons of A, the egg hatchability of queen B will average 75%. A daughter of queen B, queen C, mated to sons of queen A will have egg hatchability that will average either 75 or 50%. If her egg hatchability is 50%, the line has been reduced to two sex alleles, and queens D, E, and F will also have eggs of 50% hatchability if mated as shown in the figure. If queen C has egg hatchability of 75%, then D may also be 75 percent but somewhere not far from E or F egg hatchability will probably drop to 50% and all future generations will remain at that level.

However, by selection it is possible to keep egg hatchability at 75%, but the breeder would be reducing the effectiveness of inbreeding slightly by selection against certain alleles. It is advisable to select the 50% egg-hatchability matings in the C or D generation and thus quickly reduce all inbred lines to two sex alleles, so that they will also have 50% egg hatchability. If this is done, an analysis of the sex alleles is more readily accomplished. By test crossing to identify the sex alleles in each line the breeder can then predict which crosses will give high egg hatchability in hybrids and which crosses will give inter-mediate or low egg hatchability.

Other than improving egg hatchability, the bee breeder can accomplish very little by selection while inbreeding. He can select queens and drones in each generation for color and general appearance, but selection for most other characteristics will be very inefficient.

In one season of inbreeding it is possible to produce a number of sister queens of the D generation and mate them to their brothers (drones produced by their mother queen C). These queens will be wintered, and the following year the breeder can make test crosses while continuing to inbreed the lines by brother-sister matings. It is advisable to test-cross the inbred lines at the E generation of queens (50% inbred). Brother-sister mating should be made each year after the first to insure continuing the inbred lines until they are selected in hybrid combinations.

Since egg hatchability in inbred lines with two sex alleles is only 50%, the inbreds do not develop colonies with sufficient populations to permit accurate evaluation of such economically important characteristics as honey production, swarming tendencies, and wintering. However, such qualities as tongue length, wing length, and temper may be measured in inbred workers. It is also possible to select for such characters in the queen as size and number of ovarian tubules. This selection in inbreds as inbreds is supplemental to rather than a substitute for selection between inbred lines when in crosses with other inbred lines.All crosses should be made artificially and the colonies tested under environmental conditions as nearly uniform as possible. The various hybrid crosses will differ, but the individuals of a given hybrid will be uniform as a group.

After it has been found that the colonies having (line 3 x line 4) hybrid queens mated to drones produced by (line 1 x line 2) hybrid queens are superior in production, then the breeding stock to produce this four-way combination can be distributed for commercial production. A commercial queen producer then needs only two artificially mated breeder queens from the breeding source to produce several thousand hybrid queens. One of these breeder queens is used to produce daughter queens that will head the 50 or more drone-producing colonies necessary to supply ample drones. An inbred queen of line 1, artificially mated to drones from inbred line 2, is the mother of the hybrid daughters needed to furnish drones. Since drones are produced parthenogenetically, those drone-producing queens can be naturally mated to the drones of the breeder's own stock.

An inbred queen of line 3 artificially mated to drones from inbred line 4 is used as the breeder queen. Her daughters will be hybrids (3 x 4). These virgins are naturally mated to drones produced by the (1 x 2) hybrid queens. The honey producer receives (3 x 4) hybrid queens mated to (1 x 2) drones, and the worker offspring in his colonies will be four-way hybrids - (1 x 2) x (3 x 4).

The Advantages and Disadvantages of Inbreeding

Inbreeding can rightly be regarded as the touchstone of selective breeding. It is the indispensable means of intensifying, fixing and standardizing the desirable characteristics and at the same time weeding out the factors we do not want. Inbreeding is employed in all forms of breeding both in the animal and plant world. In fact in plants self-pollination is the most intensive form of inbreeding, far more so than other means of propagation employed by Nature. On the other hand, where the honeybee is concerned nature uses every possible means of preventing inbreeding. This is clearly shown by the fact of multiple mating with drones at distances from the hive of up to ten kilometers.

As practical experience shows over and over again the most serious result of inbreeding is the progressive loss of stamina. This loss of stamina affects all the essential activities of the bees and even threatens the very existence of many colonies. The devastating losses about which we hear so much are due, largely if not entirely, to this loss of vitality. It is an insidious and illusive deficiency which always shows up mainly in unfavourable climatic conditions against which a weakened constitution has no resistance. Nature then takes control and weeds out the unfit. Lack of vitality shows up to in a reduced capacity for brood rearing, inability for self-defence and above all in an increased susceptibility to disease. While therefore, a carefully controlled inbreeding is undoubtedly an essential requirement it must be used with great care in breeding the honeybee. Experience has shown that without this care even the most productive strains of bees can. Inbreeding leads to reduction of variability which is discussed as one of the main reasons for inbreeding depression. In honey bees plasmatic and maternal effects also reduce the viability of inbred individuals (Moritz, 1983a). In addition to these factors the brood losses due to diploid drones reduce the viability of inbred colonies (Woyke, 1981) be ruined in a few generations.

Chapter 24

Quarantine control of honey bee diseases

QUARANTINE CONTROL OF HONEYBEE DISEASES

There have always been efforts to acquire races of bees which could serve as good honey producers, pollinators with better qualitative and quantitative characters. This has necessitated importing good bee stocks from one country to another. There are instances where these imports made by observing proper quarantine measures had contributed substantially to the economy but unwanted races of honeybees have caused serious concern for beekeeping and public health. Impotrtation of honeybees and other pollinators that are not indigenous to any country may endanger local honeybees and other pollinators. The inherent problems of accidental entry of parasites, pests and diseases and undesirable *Apis* spp to non existent areas, thereby threatening the existence of honeybees and other local pollinators have been reviewed by Gatoria *et al.* (2000) and Abrol (2001). The competition between existing honeybees and exotic species may lead to the elimination of indigenous species e.g. introduction of *Apis mellifera* eliminated *Apis cerana japonica* in Japan and the fate of *Apis cerana indica* in India is not different. Therefore, there is a need of legislative measures to control the entry of pests and diseases.

Quarantine

Quarantine is based on the principle that prevention is better than cure with the objective of reducing their risk factor against the introduction of exotic pests and diseases and if already present checks its further spread and multiplication. Quarantine can, therefore, be defined as a legal enforcement of measures aimed to prevent pests and pathogens from spreading, or prevent them from multiplying further in case they have already found entry and have established in a new restricted area. Though quarantine measures may not guarantee an everlasting protection against the entry of exotic pests but will certainly check or delay introduction of these unwanted organisms and their subsequent establishment in hitherto clean areas. The legislative measures such as international and national quarantine and modified campaigns need to be enacted to prevent the entry of new pests and diseases and their further spread. As a consequence various countries have promulgated bee laws/legislations to avoid the risk of introduction of bee diseases and other related problems.

Import of pollinators

Import of *Megachile rotundata* F. in Bulgaria from France and Canada proved to be effective pollinator of Lucerne (Dochkova, 1984). Populations of 1200-2000 female bees per hectare resulted in seed yield increases of 75-550 kg/ha. Aballay *et al.* (1986) reported the successful establishment of *Bombus ruderatus* (F.) imported from Newzealand into Chile for pollination of red clover (*Trifolium pretense* L.). Following the success of some earlier introduction of bumble bees for pollination of red clover in Newzealand, two bees *Megachile rotundata* and *Nomia melanderi* Cockerell were introduced as pollinators of Lucerne (Donovan, 1990). Similarly, Richards and Krunic (1990) imported *M. rotundata* bees from Canada for pollination of alfalfa (*Medicago sativa* L.) in Yugoslavia which also proved worthwhile. During 1996, a total of 2500 prepupae were imported. *M. rotundata* were imported from Newzealand to Australia. The immature stages of pollinators were screened for various diseases and reared under quarantine for adult emergence and foı later use as pollinators. Keven and Laverty (1990) cautioned in Canada that parasitic mite like *Varroa jacobsoni* Oudemans can survive on imported pollinators or on non honeybee insects that may come in contact with infested honeybees visiting these flowers in their native country. Hence, the import of pollinators is made through strict quarantine procedures after ensuring them from such parasites.

Import of honeybee packages, queen bees and spread of enemies and diseases

Varroa jacobsoni Oudemans

Lawrence (1989) reported the presence of *Varroa jacobsoni* in 17 states of USA. The united states department of agriculture's Animal and plant health inspection service (APHIS) imposed a quarantine to restrict the entry to non infested areas but it was found difficult to enforce. Rinderer *et al.* (1997) imported 100 queen bees from various apiaries in Russia with low incidence of *V. jacobsoni*. Prior to their use mite resistance research was undertaken and these imported queens were introduced in a quarantine apiary at Louisiana for further observations to ensure this consignment to be free of pests and diseases.

In order to reduce the risk of introduction of *V. jacobsoni* the import of honey bees from USA into Canada was prohibited in 1987 (Switzer, 1993). During 1988, Canada closed the border 'against import of queen bees from USA to prevent and slow down the introduction of *Acarapis woodi* Rennie and *V. jacobsoni*. The act was widely supported by Canadian beekeepers as it resulted in the annual saving of 3 million dollors on account of miticidal treatment (Winston, 1994).

During 1992, *V. jacobsoni* was detected in feral colonies of Asian honeybees (*Apis cerana* F.) on Dauan island adjacent to Papua New Guinea (Australian territory). The island was immediately declared as infested zone and adjacent areas as surveillance zones, monitored by Australian quarantine and inspection service (AQUIS). Schrader and Reid (1986) stated that it was because of strict quarantine rules that ensured the absence of American foulbrood, *Braula* and parasitic mites in New Zealand.

Acarapis woodi

Deodikar (1971) attributed the first reported acarine disease (*Acarapis woodi*) in 1950's in India to the importation of *Apis mellifera* and suggested that quarantine law should be made applicable to bees also. Kshirsagar *et al.* (1981) suggested the need of setting up of a quarantine procedure on offshore island in order to prevent the entry of honeybee diseases and enemies into the country.

Routine monitoring in Georgia (USA) detected *A.woodi* in April, 1985. The infested apiary was destroyed and quarantine of all apiaries within 3.2 km imposed (Hall *et al.*, 1987). There were no reports of occurrence of *A. woodi* in Finland until 1991. During summer of 1991 many queen bees were imported from USA with proper health certificates. When attendant workers of 9 queen from a batch of 300 queens were examined, 17 out of 52 bees were found to be infested with *A. woodi* (Korpela and Fakhimzadeh, 1991), thus putting a question mark even on the validity of health certificates.

Other mite species

Whitten *et al.* (1979) discussed the establishment of rigorous quarantine procedures in Australia to minimize the risk of the introduction of exotic diseases and parasitic mites.

Honeybee diseases

Protozoan diseases: About one quarter of a million mated queens were shipped to western Canada each year from united states (Jay and Dixon, 1982). However, 0.5-18.00% of imported queens were infected with *Nosema apis* Zanders. Later on there were reports of occurrence of *Nosema apis* and amoeba *(Malphiha-moeba mellificae* Prell) (Liu *et al.*,1987).

Viral diseases

Liu *et al.* (1991) reported the isolation of three Australian strains of Kashmir bee virus from bees in different parts of the Australia and cautioned the import of bees as the infected bees carry the virus in inactive phase for long.

Bacterial diseases

Bacterial diseases like American foul brood (Singh, 1961) from U.P. India and European foul brood in Maharashtra (Diwan *et al.*, 1971) from *Apis cerana* have been reported but thereafter neither in *Apis cerana* nor in *A.mellifera* bacterial diseases were found till 2000. Recently, European foul brood has been detected in Himachal Pradesh, Punjab and Jammu and Kashmir from *A. mellifera* (Anonymous, 1999, 2000a,b; Abrol, 1996) and in *A. cerana* in Jammu and Kashmir (Abrol and Smith,unpublished data) which could be consequence of illegal trafficking of the queen bees into India from across the border, thereby necessitating strict quarantine in India.

Predators

Brown (1979) has reported the ovccurrence of European wasp *Vespula germanica* L. nests in Sydney and New South Wales during 1978. The wasps were seen in large numbers in the vicinity of apiaries.

Introduction of unwanted races of honeybees

Africanized /killer bees

Accidental release of Africanized honeybees (AHB), *Apis mellifera scutellata* (*Apis mellifera adansonii* Laterille) in Brazil during 1957 is one of the most debated problem. Because of its defensiveness and aggressive temperament, it has created a serious public nuisance and has been designated as a killer bee. Stibick (1987) advocated very hard steps ranging from destruction of swarms to aerial spraying of insecticides. The AHB has spread throughout the south and central America and at present is heading northwards from southern U.S replacing European honeybee races in many countries (Cobey, 1999).

Cape bee problem

Allsopp and Creewe (1955) reported the invasion of AHB (*A. mellifera scutellata*) colonies by the cape bees (*A. mellifera capensis* Escholtz) and their usurpation of *scutellata queens*. This phenomenon was later on dubbed as 'cape bee problem'. The invasion of AHB colonies by cape bees resulted in dwindling and dying of 50,000-1,00000 bee colonies since 1991 from Transvaal and north cape region of Africa (Allsopp, 1993). During 1993, legislation was set up under agricultural pest act 1983 (Act No. 36) to prevent movement of cape bees into zone of AHB (Transvaal region). All cape bee colonies already in the region in that were eradicated under this act (Keetch, 1993)

The world's present honey industry would not exist if honeybees (*Apis mellifera*) had not been introduced from Europe into continents that had no native *Apis*, notably the Americas and Australasia. It is therefore, understandable that countries in the tropics just starting to develop their own honey industry look to the introduction of similar bees as a first step towards getting large honey crops. Though some honey-producing countries in Asia have met with a high degree of success in developing commercial beekeeping thanks to the introduction of colonies of productive European races of *Apis mellifera*. This is why foreign bees will continue to be imported to some countries and regions. Proper legal regulation and enforcement may help prevent the spread of disease; though it may not be sufficient. Providing quality stock and educating beekeepers about the risks of buying abroad is a necessary and long term solution. In any of these cases the following basic guidelines should be respected:

1. Before any attempt is made to introduce a new species or race of honey bee, it is important to study and evaluate the potentials and limitations of the native bees and their foraging ecosystems, including the availability of forage.
2. If introductions are found necessary; the entire process - alongwith all requisite precautions, including quarantine and interim colony inspection should be carried out by, or closely supervised by well-trained personnel.
3. Periodic evaluations of the current state of the local honey bee industry should be made to determine whether introductions continue to be necessary.
4. In countries or regions in which beekeeping with *Apis mellifera* has been reasonably well developed and there is no good technical reason to introduce more colonies or queens, the importation of new bee strains or races should be halted. It is preferable that beekeepers endeavour to improve their queen-rearing techniques and adjust their colony management schemes in order to obtain better yields.
5. All honey bee species native to Asia have their own diseases and enemies and it is unavoidable that introduced colonies will come into contact with bees from feral nests. Since inter-species transmission of certain diseases and parasites takes place reciprocally, it is of the utmost importance that the introduced colonies be disease-free and parasite-free. Since this is practically impossible, larger scale movements, or any scale of introduction, should either be prevented completely or reduced to a minimum, e.g. by producing sufficient and high quality local bee stock.

Those concerned may not even realize that they have their own valuable *Apis mellifera* or other *Apis* species. Introducing foreign bees can, however, have a negative result: it may prevent the successful development of apiculture and even destroy what already exists. Native bees are generally better adapted to the climate, changes in food supply and the pathogens present, etc. Importation of foreign bee species and races is therefore unnecessary. It has often been observed that honey bee colonies that have been introduced into new areas have had disastrous effects in spreading bee diseases and parasites.The subject of introducing exotic bees is thus a very important one.

Whether bees should be introduced from another country

1. Bees must never be introduced unless: *(a)* they are completely healthy and without parasites; *(b)* they have no characteristic that is disadvantageous in their use for beekeeping (such as inbreeding, poor performance. unacceptable behaviour characteristics such as 'aggressiveness'); *(c)* the characteristics of native honeybees have been adequately assessed. To these points must be added:

(d) no combs should ever be introduced, because some diseases can be transmitted by combs; the same applies to used beekeeping equipment. The above criteria must always be followed, in addition to points 2 to 8 below.

2. There is no apicultural objection, subject to restrictions in I above, to introducing bees to an isolated area **entirely without honeybees**, such as some Pacific islands.

3. Introduction of European-type *Apis mellifera* into an area **where no honeybees are native**, but where similar bees have previously been introduced (such as Central America), may be permissible. Such introductions may, however, result in hybridization with bees already there, which may or may not be desirable. In no circumstances should new bees be taken to a 'sanctuary' area where previously introduced bees have for some reason been isolated and now form a genetically valuable resource. An example is Kangaroo Island off South Australia, which has purer 'Italian' bees (*Apis mellifera ligustica*) than now exist in Italy or anywhere else.

4. Where **native *Apis mellifera* bees exist**, there is danger in importing foreign bees of the same species, because new genes may then contaminate the local gene pool, and a genetic breeding resource may be lost. This may not matter if the bees in the area are already very mixed as a result of many past introductions, but if there are isolated populations, genes valuable for breeding may disappear for ever. This would be true of parts of Turkey or of North Africa.

5. Where there are **native tropical *Apis mellifera*** (as in tropical Africa), any temperate-zone introductions may well not survive, because they lack several adaptations to life in the tropics

6. **Where *Apis cerana* is native** (in Asia), introducing *Apis mellifera* can have a variety of results. It may lead to the extinction of *Apis cerana*; this process is almost complete in Japan and is on the verge in India, for instance. Alternatively, competition between the two species may lead to the extermination of *Apis mellifera*, because *Apis cerana* is better at exploiting the local food resources. In other areas both may co-exist.

7. Where **any *Apis species* other than *mellifera* are present** (*cerana, dorsata, florea*) complications arise with the mating of queens of the introduced bees. because the sex pheromone that attracts drones to a queen is the same for all species. The presence of many drones of other species may prevent drones of the same species getting access to the queen and mating with her.

8. The results of introducing **tropical *Apis mellifera* from Africa** into Brazil in 1956 have been well documented. In no circumstances should tropical African bees. or Africanized bees from America be introduced elsewhere.

Points to consider if introductions are made

1. Point 1 in list above.

2. Bees should be introduced only under rigid quarantine conditions and permissions for importations should be granted only to institutions and

individuals that can maintain strict adherence to the quarantine requirements. For instance, the Director of Animal Quarantine in New South Wales, Australia, has described experimental quarantine procedures used there for imported honeybees.

3. Bees should be imported only from countries:

(a) that are free from serious bee diseases, and that have no bees with undesirable genetic characteristics;

(b) that maintains effective quarantine and disease control programmes;

(c) that can certify the health status of export shipments.

Pacific islands (including Hawaii) may well be the first choice, then New Zealand, and then certain careful suppliers in mainland USA and Canada.

4. The most common transport of bees is of a young mated queen in a cage with a few accompanying workers. The cage must be destroyed on receipt, and the workers (labelled according to the reference number of the hive into which the queen is introduced) sent to a competent authority for diagnostic examination. Only the queen should be allowed contact with bees or hive materials in the new area.

5. The best available way for introducing bees in bulk is to transport package bees, in new boxes that have had no previous contact with bees.

6. Insufficient information is available for comment on the introduction of productive *Apis cerana* into other tropical countries with less productive *Apis cerana*. Before any major introduction is undertaken, a pilot scheme should be carried out in an area as isolated as possible. The behaviour of the introduced bees under local conditions should be studied; also, if other *Apis* species are present, any robbing, competition in foraging, or other interactions between species should be looked for.

8. As far as is known, adult bee diseases are not transmitted by immature stages of honeybees, and for this reason methods have been worked out for transporting honeybee brood in an incubator, without comb, from one continent to another. Also, without risk of disease transmission, drone semen can be deep-frozen and transported/, and subsequently used for instrumental insemination of queens.

Successful introductions

Some successful introductions of European-type *Apis mellifera* have been made into territory of other *Apis mellifera* that are not too dissimilar from the native region of the introduced bees. Such importations have been on a large scale, so that hybridization between the wanted and unwanted bees was minimized, and importations have then been constantly repeated. An example is the introduction to Israel of *A. m. ligustica* (Italian bees) to replace the native *A. m. syriaca*, which are 'aggressive' and in other ways not easy to handle, and also not very productive. Dr. Y. Lensky of the Hebrew University of Jerusalem, Israel, has provided the following information :

Prior to the large-scale replacement of the native *A. m. syriaca*, observations on the introduction of Italian queens into *A. m. syriaca* colonies were made for several years. Introductions were successful when either mated queens or queen cells were placed into small *A. m. syriaca* colonies (2-3 frames covered with emerging workers). But the easiest way proved to be to split Italian colonies and to strengthen them with sealed brood from *A. m. syriaca* colonies.

Hybrid queens must constantly be replaced by Italian queens, because of the influx of *A. m. syriaca* drones across the frontiers with neighbouring countries. To preserve the Italian genes, every year they import several hundred Italian queens from California, USA, and Australia, and distribute these among queen breeders and beekeepers. The replacement of *A. m. syriaca* was achieved in Israel because of the persistent efforts of the beekeepers and of the Extension Service of the Ministry of Agriculture.

Introductions have also been made, and maintained, in certain regions of other countries, including the People's Republic of China, Taiwan, Burma, Thailand and India

Several countries of the world have their own quarantine regulations

Newzealand acts

The Act of August 31, 1922, entitled "An Act to regulate foreign commerce in the importation into the United States of the adult honey bee (*Apis mellifica)*" (referred to hereinafter as the Honeybee Act of 1922), prohibits the entry of honey bees from countries where diseases and parasites are known to exist that endanger the health of honey bees

The diseases, pests and germplasm specifically identified in the Honeybee Act and amendments, including regulations under the Federal Plant Pest Act entitled Exotic Bee Diseases and Parasites (Title 7 CFR Part 319.76) are as follows:

Exotic Bee Parasites

Acarapis woodi

Varroa jacobsoni

Tropilaelaps clareae

Euvarroa sinhai

Coelioxys spp.

Chrysis spp.

Exotic Bee Diseases

Aspergillus spp.

Bacillus spp.

Entomophthora spp.

Beauvaria spp.

Cordyceps spp.

Saccharomyces spp.

Because the protozoan *Nosema apis* is widespread in the United States, it is not considered an exotic disease.

Besides several honey bee hive products and used equipment involves the risk of spread of diseases which includes

(i) honey (which includes honeydew)

(ii) bee-collected pollen

(iii) royal jelly

(iv) propolis

(v) bees' wax

(vi) used beekeeping equipment

- hive parts, including comb, feeder and queen excluder
- pollen traps and pollen dispensers
- clothing and gloves
- smokers and hive tools
- honey extracting equipment
- containers used for transporting bees
- artificial insemination equipment
- *Varroa* floors

This risk assessment does not include live or dead honey bees, honey bee semen, or bee-collected pollen used for artificial pollination. For ease of interpretation, the decision-making pathway is outlined below.

Florida Laws

Florida legislated laws, also called statutes, are organized into chapters. The *Animal Diseases (Amendment) Bill 1997* repealed the *Apiaries Act 1928*. The provisions of the *Animal Diseases Act 1993* have replaced the *Apiaries Act* as the basis for legislative control of beekeeping in the ACT. Although there is no longer a requirement to be registered as a beekeeper with Environment ACT the disease aspect of beekeeping is administered under the *Animal Diseases Act*. As necessary, bee diseases can be declared as exotic or endemic diseases under the Act.

The following diseases have been declared Endemic Diseases under the Animal Diseases Act.

American Brood Disease
Chalk Brood Disease
European Brood
Nosema Disease
Stone Brood Disease

The following diseases have been declared Exotic Diseases under the Animal Diseases Act

Africanised Bee
Asian Bees
Asian Mite
Braula Fly
Tracheal Mite
Tropilaelaps Mite
Varroa Mite

The *Animal Diseases (Bees) Regulations* prohibits the keeping of honeybees other than in frame hives. The regulations also prohibits a person from exposing honey or honey comb, other than in a frame hive, in a way that honeybees may have access to it.

Registration of honey bees: Each person owning one or more hives of bees shall, annually, on or before the first day of October, make application to the State Entomologist for the registration of bees. The State Entomologist shall issue to such applicant a certificate of registration without fee. The State Entomologist shall (1) keep accessible to the public a record of the registration, the name and place of residence of the registrant and the definite location in the municipality where the bees are kept and (2) transmit a copy of such information to the town clerk of the municipality in which the registrant resides. Any owner of bees who fails to register as required by the provisions of this section shall be fined not more than five dollars.

Sec. 22-90. Inspection of bees for contagious diseases: The State Entomologist shall, to such extent as he deems necessary or expedient, examine apiaries and quarantine such as are diseased and treat or destroy them. No person or corporation shall sell, to be removed to another location, bees, brood comb, frames or hives that have been in use, with or without combs, until they have been inspected by an authorized inspector, who shall issue a certificate of health if they are found free of contagious disease. Any person violating any provision of this section shall be fined not more than fifty dollars.

Regulation for inspection and treatment of Honeybee related diseases : This regulation contains guidelines that authorized beekeepers, supervising beekeepers and bee inspectors must follow at inspections of bee families in relation to movement, migration beekeeping and changing of ownership, at suspect on about brood, posting of samples and treatment, advises and inspection in relation to AFB, EFB and stone brood.

Inspection of bee families: All bee families must be inspected in conjunction to movement, migration beekeeping and changing of ownership and also after movement out of a 2 km zone around bee yards with AFB or stone brood (and suspected attacks of AFB, EFB and stone Brood.

All brood frames in the bee family must be inspected, and also those, former used as brood frames, must be examined. The frames are examined for clinical symptoms (visible signs) of AFB, EFB and stone brood. It might be necessary to shuttle of the bees to have a good examination. Rubber gloves must be used when AFB or stone brood are suspected being present. They must be burned afterwards. Otherwise they can be cleaned by carefully use of water and soap. Leather gloves must not be used because they are difficult to clean.

Destruction: Bee families with clinical symptoms of foul brood may be killed if it is a wish from the beekeeper or the bee disease inspector determines that he families are for weak to be treated. The destruction takes place early in the morning or at a time where the bees are not flying. The bees are killed by use of cleaned gasoline. The entrance is tightened up. Then 2-3 dl cleaned gasoline is purged into the hive and the hive is closed.

Equipment and combs: Combs and bees from destructed families, together with leaders and tighting material must be burned. Furthermore, most old hives, not suitable for proper cleaning be burned. Burning can take place in the yard if permission is given, but it will normally be the best to transfer the material to a refuse disposal plant for burning. All combs from families with symptoms of clinical foul brood must be melted down. The process is done by steam. All build out and stored combs (both brood combs and honey combs) must likewise be melted down. If the beekeepers have more Yards, all build out combs must be melted down, unless the material from each yard is stored seperatly. When the combs are sent to the vax factory they must be packed up and marked clearly that they are from a yard with AFB.

All material not in use but used in the beeyard must be cleaned. The material is first scraped free of wax, then cleaned by using one of the following methods. Cleaning with a gas burner. Using a scrubbing brush together with hot soap water (1½ dl dishwashing chlorine soap in 10 litre water) followed with clean water flushing. Spraying with a 1%-solution of Virkon® S, is recommended which is an environment neutral disinfectant. The spraying is done with a water atomizer and there is used as much as needed to soak the material.

Canada

Canadian bee disease laws and regulations

1. Ist law governing bee diseases was passed by Ontario Legislature in 1897 and revised in 1906.
2. Quebec passed law in 1908 entitled "An act for the prevention and treatment of contagious diseases of bees"
3. British Columbia passed act in 1911, Manitoba in 1914 and New Brunswick in 1915.
4. In 1923, Ministry of Agriculture issued order restricting importation of bees; used beekeeping equipment; raw bee hive foods or products from Europe because of danger of introducing 'Isle of Wight disease'.

5. New Brunswick and Prince Edward Island enacted Foulbrood control act in 1926. Canadian Federal regulations.

The Canadian federal regulations in the Health of Animals Act also dealt with international movement of bees. The original regulation was adopted in 1923 amended in 1955 and 1974, which stated that no person shall import into Canada (a) honeybees on combs: used hive or used hive equipments : beewax unless accompanied by declaration that the beeswax has been liquefied or.: honeybees in combless packages unless the packages are accompanied by a declaration of the shipper that the food supplied to the bees and carried in the packages does not contain honey.

USA

Honeybee Act August 31, 1922 (Amended 1947, 1962 and 1972) restricted the importation of living adult honeybees to prevent the spread of diseases and parasites harmful to honeybees and the introduction of genetically undesirable germ plasma of honeybees except that honeybees may be imported into the united states:

(i) by USDA for scientific/experimental purpose.

(ii) From countries determined by Secretary of agriculture (a) to be free of diseases or parasites to honeybees and undersirable species or subspecies of honeybees (b) to have in operation precautions adequate to prevent the importation of honeybees from other countries where harmful diseases or parasites or undesirable species or subspecies of honeybees exist.

(iii) Honeybees semen may be imported into the United States only from countries determined by the Secretary to be free of undesirable species of honeybees and which have in operation precautions adequate to prevent the importation of such undesirable honeybees and their semen.

(iv) Honeybees and semen pursuant to the subsection (a) and shall be imported and other shall be intercepted and destroyed immediately.

Different states have adopted their own regulations which are as given below-

a. In 1877 California adopted regulations banning entry of diseased bees. The act was modified in 1883.

b. In 1906, 12 states passed regulations banning entry of foulbrood affected bees. Regulations of apiaries and permits system for interstate movement of bees and other materials were strictly enforced. In this act "Honeybees" meant all stages of honeybees and the germplasm of honeybees of the *Apis* except honeybees semen. Violations of this act resulted in a fine of $ 1000 and one year imprisonment.

India

(i) First act passed in 1906 under Sea customs Act of 1878 to stop the entry of Mexian cotton bollweevil.

(ii) Destructive insects and pests act No II passed on 3rd February 1914 restricting the entry of Mexican cotton bollweevil, unginned cotton seeds, sugar plants and seeds of cotton.

(iii) Amendments to this act were made from time to time and states were authorized to frame their own laws.

(iv) East Punjab Agricultural pests and noxious weeds Act of 1949 restricted the entry of cotton bollweevil, fruits and vegetables from Afghanistan, and import of potato seeds from areas infested with golden nematode.

(v) In Jammu & Kashmir state, Ladakh region has attack of codling moth *Carpocapsa pomonella* on apple and other horticultural crops. Its entry is restricted to other parts of the state and the country.

The Indian quarantine rules are not so elaborate and stringent as compared to those found in the United States and other countries. The government of India must adopt cautions approach and there is a need of strict quarantine measure as given below:

i. quarantine laws governing bee diseases

ii. Import restrictions on bee colonies, package bees queens etc.

iii. Regulations regarding import of beekeeping products.

iv. Domestic restrictions within state and interstate movement of bees.

v. Well equipped pathological laboratories for deterctions of bee diseases and their certifications.

Very recently, severe incidence of *Varroa destructor* mite has been reported from Italian honey bee (*Apis mellifera* L.) colonies in Punjab, Haryana, Rajasthan Jammu & Kashmir, and Himachal Pradesh resulting in great devastation of honeybee colonies. These have been identified to be Korean genotype of *Varroa destructor*. This genotype is highly damaging to *Apis mellifera* throughout the world. Previously *Varroa jacobsoni* has been reported from local bee *Apis cerana* in India which never reproduced in *A. mellifera* all these years. A little over two years ago, mites that have been referred to as *V. jacobsoni* on Asian bees (*A. cerana*) throughout Asia were found to be a complex sibling species. Over 22 genotypes have now been discovered in this species complex. Most belong to 2 species *V. jacobsoni* and newly named *V. destructor*. *Varroa* genotype gets its name from the location where it was first found on *A. cerana*. Only 2 species out of the 22 or so genotypes within the *Varroa* species complex on Asian bees have become pests of the European honeybee (*A. mellifera*) worldwide. These are the Korea and Japan genotype which have been found. More than 80% of the stocks have been wiped out. Probably illegal trafficking of black (*Apis mellifera carnica*) and hybrid queen bees and colonies from abroad in India could be the cause of introduction of *Varroa destructor* in India.

The import of new stocks of *Apis mellifera* into India may result in the the introduction of as yet unknown diseases, pests and Africanized bee blood problems, which are still not present in the country. There is a need of strict check on the import of bees of any kind whether intentional or otherwise. Only such checks can save Indian beekeeping and prevent another disastrous phase in its crisis ridden growth. This is not just about honey production, it is also about pollination and so this would affect all the pastoral industries and our fruit and vegetable markets as well," it's not just bees that could be at risk.

Conclusion

The greatest dangers in introducing new bees into an area are: (a) the introduction of diseases and parasites: European and American foul brood, and acarine and *Varroa* mites, have been gratuitously introduced in this way into various countries, to the lasting detriment of their beekeeping; (b) the loss for ever of genetic material that could be valuable for bee breeding; (c) the introduction of unsuitable bees, with a resultant reduction in the productivity of beekeeping. (d) the possible reduction of populations of native non-*Apis* bees (including stingless bees), through competition for food.

Chapter 25

Pesticidal toxicity to bees

Honey bees are required for the pollination of many vegetable and fruit crops. Without adequate populations of bees, the production of these and other crops would be impossible. In a world where people expect and demand more food and fiber each year, all branches of agriculture must continuously adapt and improve to meet this challenge. Farmers now find it essential to annually increase efficiency and production to remain in business and to show a profit. Practically every agricultural crop has insect pests that sometimes require treatment. Unfortunately beneficial insects such as honeybees are also susceptible to pesticides. The use of pesticides for pest control on the one hand and the role of honeybees (*Apis* spp.) for crop pollination on the other has become essential components of modern agriculture. Without either of the two, global food production would be seriously impaired. Unfortunately, these two practices are not always compatible, as honeybees are susceptible to many of commonly used pesticides (Johansen, 1977, MacKenzie and Winston, 1989, Russell *et al.*, 1998, Cunningham *et al.*, 2002, Sundararaju, 2003), used for the control of insect pests (Poehling, 1989, Stark *et al.*, 1995). The major constraint confronting pollinator-plant interaction is the indiscriminate and excessive use of pesticides for controlling insect-pests (Bisht *et al.*, 1980, Rana and Goyal, 1991, Zhong *et al.*, 2004). The loss of honeybees directly effect beekeeping through loss of honey production and indirectly the crop production due to inadequate pollination. Reduction of population of these beneficial insects due to insecticides, therefore, incurs significant environmental, ecological and economic costs (Bai and Reddy, 1977, Pimental *et al.*, 1980, Crane and Walker, 1983, Prakash and Kumaraswami, 1984, Kasana *et al.*, 1999, Khan and Dethe, 2004).

In India, 50 million hectare of area is under entomophilous crops, cross pollinated by different abiotic and biotic agents. About 90% pollination is carried out by insects, 85 per cent of which comprise the bees (Singh *et al.*, 1989). Honeybees constitute a major group of insect pollinators and their pollinating efficacy is manifested not only through increase in yield but also by the improvement of the crop quality through heterosis breeding (Melnichenko, 1976).

The bees earn about rupees 10 million to the national exchequer in terms of honey production and beeswax and it is expected that an additional crop yield worth rupees 90 million could be obtained due to pollination of crops (Sorthia and Chari, 1985). Levin (1984) estimated that the value of crops in the United States of America that benefited directly from the honeybee pollination approaches 20 billion US Dollars annually. The annual cost of crop loss due to insecticides poisoning of pollinating honeybees has been estimated at 135 million US dollar in the United States of America (Pimental *et al.*, 1980). Not withstanding with the absence of such estimates, the indispensability for a harmonious compromise between pest management and honeybee pollination of crops in India cannot be gainsaid.

Tropical and subtropical climate of India in particular presents suitable conditions for the outbreak and appearance of many pest problems. The pest problems have been further aggravated by the advancement in agricultural technology. Irrigated crops, intensive agriculture, introduction of crops and crop varieties and disturbing the indigenous and primitive cropping patterns have contributed in increasing the pest problem of crops. Reduction in uncultivated land, corners and bunds destroy nesting and hibernating places of wild pollinators and succession of nectar and pollen yielding flowers round the year is destroyed. Weedicides are used to control the weeds and hence lead to starvation of pollinating insects. The advanced agricultural technology has helped to destroy the agriculture cycle through indirect effect. There is also a prominent negative direct factor, i.e., the insect pollinators are killed by pesticidal usage in crop protection. There is increasing use of pesticides for the control of rodents, mites, insects, nematodes and fungal and bacterial diseases of crop plants. The loss by bee kill is direct, i.e., loss of honey production and indirect inadequate pollination of crops resulting in reduced productivity.

Integrated pest management in India has not yet shown desired results and blanket pesticidal applications are given. Most farmers apply large quantities of pesticides at regular intervals and in most cases the pesticides are non-selective coupled with untimely application. Unfortunately honeybees are susceptible to many pesticides used in pest control programmes. This problem is recently overshadowing all other problems in apiculture. Farmers in India have small holdings and hand sprayers/dusters are commonly used for treating small area each day. This results into a continuous threat of chemical poisoning to bees. Moreover, there is no coordination between the beekeepers and the farmers by any Government decree and therefore, measures to save bees cannot be taken.

Large number of killed bees are found in front of the hives or in the fields by insecticidal poisoning. It is not possible to quantify the loss in terms of food production or to assess the financial value of the bees killed. Even more important is the loss in future crop yields because a beekeeper whose bees are killed gives up beekeeping and others too are discouraged to take up beekeeping. Therefore, a balance sheet between the gains in crop yields by control of pests and losses due to decreased pollinator activity and honey production by bee kill should be worked out. While controlling pests the scientists and farmers are looking on to one aspect of the economic considerations in insecticidal applications. Our primary aim should be to assess how crop pests can be kept under control without killing insect pollinators and to ensure optimum pollination by these insects. Widespread destruction of beneficial insects (including pollinators) often occurs as a consequence of irresponsible and improper use of pesticides. It should be accepted that some loss is inevitable in certain circumstances and that a realistic aim should be an acceptably low level of loss rather than complete protection of bees.

Some insecticides have been screened in laboratory in India for their toxicity to bees. First study in this field was carried out by Cherian and Mahadevan with DDT and Gammexane against *Apis cerana indica.* Hameed allowed the worker bees of *Apis mellifera* to forage on cut flowers of mustard to which systemic and contact insecticides had been sprayed. On the basis of safety index formothion, Vamidothion, Dimethoate and Phosalone were considered to be relatively safer to bees. Contact toxicity to *Apis cerana* of insecticide applied as sprays was compared with Menazon by Kapil. Taking the LC50 for Menazon as 100, the comparative values for Endosulphan, Eormothion, Methyl demeton, Endrin, Dieldrin, Malathion, Parathion, Phosphamidon, Lindane, Phorate and Mevinphos were 1.17, 1.18, 14.00, 15.70, 17.79, 22.25. 26.01, 28.45, 36.99, 57.96 and 64.24 respectively. Singh tested 15 insecticides and reported that Menazon and Endosulphan were least toxic and were considered nontoxic to *Apis cerana* at their recommended doses. According to Thakur *et al.* (1981), Fenitrothion and Fenthion were highly toxic as compared to Endosulphan and Trichlorfon and Hinosan were moderately toxic as determined by residue film method. Comparative toxicity of organophosphates, chlorinated hydrocarbons and carbamates was worked out by Bai Attri who also assessed the contact and oral toxicities of some insecticides. Toxicity cf several organophosphates to *Apis cerana* was determined in the laboratory using topical application method. Determination of the kinetic parameters of the reactions by the authors showed that differences in anticho-linestrase activity were due mainly to differences in affinity rather than to different chemical structure of the compounds.

Cholinestrase inhibition by insecticides in Indian honeybee was studied by Dale and Bai who reported that signs of poisoning in *Apis cerana indica* were first observable when acetylcholinestrase inhibition exceeded 35% and death occurred at 96% or more inhibition.

Reddy reported the inhibition of magnesium activated adenosine triphosphate as the criterion to determine the degree of organochlorine insecticide poisoning to *Apis cerana indica*. Digestive amylase and protease of *Apis cerana indica* were inhibited to the same level by the insecticide poisoning from different groups of insecticides. Studies on the level of ions of amino acids in the haemolymph of worker bees of *Apis cerana indica* treated topically showed pronounced stimulatory effect with organophosphates, a relatively strong inhibitory action with chlorinated hydrocarbons and an intermediary effect with carbamate pesticides.

Bee poisoning or killing of bees from pesticides continuous to be a serious problem for beekeepers. Most bee kill occurs when pesticides are applied or allowed to drift on to blooming crops or weeds (Mayer, 2003). Most (99%) bee kills results from bees picking up the pesticides when foraging. The hazards of insecticidal application on flowering crops include- direct mortality, fumigative effects, repellent effect and toxicity of residues present on various floral parts and in nectar to the insect visitor. A highly toxic insecticide generally reduces the field force of a colony within a short period of time. Colonies may be reduced by one third to half in strength within 24-48 hours (Eckert and Shaw, 1960), thus adversely affecting both the production and marketing segments of the honey and beekeeping industry. Generally, fumigative action of insecticides used under field conditions is of much shorter durations then the effect of contact and stomach poison. A prolong repellent effect will deprive the flowers of pollination benefits of insect visits, while a short repellency will deter the insect pollinators from visiting the treated bloom for a brief period and thereafter, allow them to resume their foraging activity (with minimum residual hazards) without compromising with the yield potential of the crop.

Poisoning of insecticides to honeybees is generally more pronounced because of their long hours of working on the crop flowers for pollen and nectar collection, continuous working nature and long flights with pollen loads. Bee poisoning is more chronic in areas which lack of sufficient wild pollens and nectar plants to sustain the number of colonies required for crop pollination. Most bees are pastured in agricultural areas where they are subjected to killing from pesticides (Mayer and Johansen, 1983). The beekeeper has little or no control of when and what pesticides are applied in the areas of his bee forage (Caron, 2000, Mayer, 2003).

Conservation of honeybees for crop pollination is vital to agricultural production (Bisht *et al.,* 1983, Sorthia and Chari, 1985, Smirle *et al.*, 1990, Gupta, 1994., Dhuley *et al.*, 1996, Russel *et al.*, 1998, Kremen *et al.*, 2002).

Many colonies that are not killed outright may be weakened to such an extent that they are no longer effective as pollinators or honey producers, nor they can be divided to increase the number of colonies. This type of economic loss far exceeds the loss from colonies being killed out right by pesticides. Though the monetary loss to the beekeeping is large, yet the value of the seed and fruit lost through the lack of pollination is estimated to be 50-100 times greater (Atkins, 1975).

Pesticides application may also changes the physiology of nectar and pollen producing plants, change in attraction of bees to flowers, affect pollen viability, reduced pollen germination on contaminated stigma (Eaton, 1963, Church and Williams, 1978, Fell *et al.*, 1983, Kuriachen *et al.*, 1992, Sharma, 1993). All these effects of pesticides usages are serious to pollination potential and honey production.

The immature stages of honeybees are vulnerable to insecticidal poisoning that result in hidden damage to honeybee colony (Davis and Shuel, 1985,

Naumann and Isman, 1996, Shuel and Davis, 1998). Loss of brood and new bees as result of exposure of insecticides may be more deleterious than the loss of foragers (Singh and Kumar, 1998) because the latter could be replaced more quickly and provide less potential value to the colony than the emerging workers

The problem of bee poisoning is an old as about 1870s when Thompson (1881) detected accidentally killed bees by application of Paris green to pear trees in bloom and some sort of bee malady was found around beehives at that time. The intermediary period between 1870 and 1888 contributed further to the knowledge of the toxicity of Paris green and London purple. Brose (1888) demonstrated that these insecticides in sugar syrup repelled the bees to some extent but those fed on syrup were killed in 1-4 hours. Troop (1918), Hoskin and Harrison (1934) reported some inorganic compounds like arsenics highly toxic to honeybees. Likewise, Kingsmill (1917) reported accidental killing of bees to a great extent from Paris green and molasses bait mixture. Bourne (1927) reported that blossom treated with a mixture of lead arsenate, lime sulphur and nicotine sulphates were unattractive to bees.

Many major agricultural changes took place in the 1950's, shortly after World War II, when tractors replaced horses, chemical fertilizers replaced organic manure, aerial application of pesticides became common place, and farmers became increasingly conscious of business costs. At the same time, many farmers were encouraged to devote large acreages to the cultivation of a single crop, which necessitated the utilization of large quantities of synthetic fertilizers and pesticides to nourish and protect that crop. Consumers also came to expect all market fruits and vegetables to be completely free from insects and insect damage.

Thus, many growers found it advantageous to apply more and more pesticides each year. Unfortunately, some aspects of this agricultural modernization were not beneficial for beekeepers, whose needs were either frequently forgotten or ignored. Consequently, many honey bees were killed. To compensate, many commercial beekeepers had to keep larger numbers of honey bee colonies in a variety of locations to make up for losses from pesticides and to meet rising operating expenses. This interaction between the needs of crop farmers and the needs of beekeepers, coupled with frequent widespread application of the "newer" insecticides, such as parathion, proved devastating to thousands of colonies.

FACTORS INFLUENCING BEE POISONING

By nature, honey bees from a colony visit flowers over an area of several square miles. The intensity of visitation in any one part of the area is determined by the relative attractiveness of the flowers. The extent of damage to the colony by a pesticide application is influenced not only by the relative toxicity of the material, the number and methods of application, the time of day, and the weather conditions, but also by the number of bees from the colony visiting the flowers in the treated area, the type of food (nectar or pollen) they are collecting, the type of flowers the food is collected from, the season of the year the damage occurs, and even the influence of forage available to the bees for weeks before and after the application. Wild bees are also damaged by pesticides. Poisoning may result from contaminated food as well as from florets, leaves, soil, or other material used by the bees in nesting. The toxicity of a specific insecticide to honey bees and wild bees is not always the same, and even among wild bees some materials are more toxic to one species than to another. Different factors influencing bee poisoning are as given below:

Pesticide formulation: Dusts are highly hazardous to bees because of their tendency to drift to considerable distance and particles remain adhered to plant surface for long. So is the case with wettable powders which also remain unabsorbed on the plant surface for longer periods than emulsifiable concentrates. The emulsifiable concentrates are relatively safer (Kapil, 1970). The granular insecticides are safest. However, granular insecticides with systemic action may contaminate nectar and may result in losses to bees, foraging upon them. The insecticides with fumigant effect may also be hazardous to bees. Microencapsulated granules applied on flowers are sometimes collected by bees and stored in the hives, where they may be eaten by adult bees or fed to brood causing high mortality besides causing long term contamination of hive parts and hive products.

Selectivity of pesticides: There are some insecticides which have little effect on honeybees when applied as sprays e.g endosulfan, phosalone, pirmicarb, fluvalinate, or trichlorfon. High toxicity of carbamates to honeybees has been considered due to their very low level of phenolase enzymes. It has also been claimed that greater acetylcholinesterase concentration in young enables them to tolerate Malathion. Similarly, high tolerance of trichlorfon by honeybees has been correlated with relatively high pH of the body.

Period of application: Insecticides when applied to crops in flower may be hazardous to bees or when pesticides are applied to a non – flowering crop but having large number of attractive flowering weeds or hedges in the fields or in the adjoining fields. The bees are also affected if they pass through a field treated or sprayed with pesticides. In mango orchards, bees attracted to honey dew secreted by mango hoppers are killed in large numbers when insecticides have been applied to the trees.

Time of application: Little foraging occurs early in the morning or late in the evening. Application of pesticides during late evening or early morning provides relative safety. This avoids direct deposition of pesticides on the bee body and even residues on the treated surfaces are rendered less harmful, especially in case of short residual pesticides. Species variation has also been noticed. Whereas *Apis cerana* became active at 7-10°C ambient temperature, *Apis melli-fera* never became active unless the ambient temperature reached 13-16°C or above.

Attractiveness of crop: Some crops notably rapeseed and mustard are extremely attractive to bees which will forage from colonies upto 3 km away or more. These crops remain attractive until the very end of flowering. It remains attractive even in cool and dull weather when they not visit other crops.

Weather: Warm and sunny weather is conducive for foraging by bees. When there is a prolonged dull weather the foraging activity is reduced considerably and insecticides can be applied on the crop.

Temperature: Temperature is probably the most significant factor causing differences in the toxicity of pesticides. Immediate effects may be much greater at higher temperatures whereas, residual effects are likely to be less because the toxic materials breaks down more quickly.

Method of application: Aerial application of pesticides has been regarded as more hazardous than ground application. Bees get less time to escape the drift of insecticides. Systemic insecticides applied on the blooming crop may cause hazards to the bees. Fine sprays are safer than coarse sprays.

Colony strength: Populous colonies always suffer greater losses than small colonies, because more foragers are exposed to insecticides.

Age and body size of honeybees: Newly emerged bees are more susceptible to insecticides than older bees. Smaller bees likewise are more susceptible to insecticides because their body surface area is larger in relation to their body weight.

Distance of colonies: Honeybee mortality is inversely proportional to the distance of colonies from treated fields. Farther is the crop from colony,less likely is to attract large number of foragers.

How bees are exposed to pesticidal hazards

Many of the crop plants need cross-pollination. At the same time they are infested by pests even during flowering causing considerable losses which warrant the application of control measures. The pesticidal applications pose serious danger and eliminates large population of insect pollinators as well. Some of the crops benefiting from bees as well heavily attacked by pests include:

Cotton: It is the most dangerous crop for bees. As many as 15-20 insecticidal applications at shorter and regular intervals are recommended for the control of various cotton pests. The flowering continues for about 2 months and during this period insecticides are regularly applied for the control of many pests like bollworms, aphids, bugs, etc. Foraging bees are killed by these sprays. New generation of bees develop in 3 weeks. Insecticidal applications at shorter intervals than this kill more adult bees than can be replaced and ultimately the colonies die. But co-ordinated application of insecticides can minimize bee kill. (1) Flowering in cotton continues for about 2 months but flowers that set fruit appear within 3-4 weeks. Therefore, use of insecticides during this period should be reduced so that bees can be moved to the crop. (2) Nectar in flowers and extra floral nectaries is exhausted by mid day and very few bees are foraging in the afternoon when insecticides can be applied with reduced hazards to bees. (3) Air spraying has picked up for cotton. In such a situation the colonies should be located away from the flight path of the plane.

***Brassica* and vegetable seed-crops**: These are attacked by aphids, caterpillars and bugs during flowering and pod formation stage. These crops include oilseeds *Brassica*, seed crops of cauliflower, cabbage, raddish, turnip, carrot, fennel and coriander. In these crops too, the flowering is greatly extended, lasting for about 1-1½ months. These crops need insecticidal applications during flowering periods. But all these crops are also enthusiastically foraged by bees which are very useful pollinators of these crops. There is extensive pesticidal poisoning to bees on these crops. There are no specific recommendations to safeguard bees and only general guidelines to reduce bee kill can be followed, though Singh (1969) sprayed Endosulfon on mustard to control aphids at 0800, 0900, 1630 or 1730 hrs and found that Apis spp. foraged between 1030 and 1530 hr without any effect on foraging intensity and no bees were killed.

Sunflower: Its cultivation is gaining importance in India. Bees contribute much in increased crop production by pollination services but bee losses have been reported by insecticidal sprays for the control of aphids and caterpillars. In India Endosulfon was found to be less toxic to honeybees than Fenthion, Carbaryl or Parathion and seed set and yield were not affected since bee activity was not reduced in Endosulfon sprayed plots (Ramakrishna *et al.*, 1974; Bhattacharya *et al.*, 1982). Bees mostly forage in the forenoon and there is limited activity till early afternoon. Therefore, evening or late in the afternoon is appropriate time for chemical control operations.

Sesame: It is automatic self-pollinated but natural cross-pollination also occurs. Honeybees are very active on the flowers of sesame. The crop at flowering stage suffers from the attack of aphids, brown leafhopper, sucking bugs, white flies and caterpillars. Chemical application at blossoming would cause hazards to bees also.

Seed crops: Like lucern and clovers these are rich bee forages. Under semi-arid tropics the legume flowers usually close in the afternoon and it allows time for safe application of pesticides afterwards against caterpillar pests.

Pulses: Like soybean, cajanus and others are self-pollinated crops but yield increases by bee pollination have been observed. Considerable mortality of honeybees from insecticide poisoning is reported in some countries but lack of knowledge in India is due to non-monitoring of hazards.

Cucurbits : These require the control of fruit flies, pumpkin beetle and aphids when in flowering. Cover sprays of insecticides are given against these pests. Honeybees visit the flowers of melon and other cucurbits. Steps to minimize bee kill from these sprays are required.

Tobacco: In this case self-pollination is normal but honeybees and other insects visit the flowers for nectar, affecting some cross pollination. Aphids, whiteflies, thrips and caterpillars are the pests which may warrant insecticidal application during flowering which consequently would lead to bee hazards.

Flowering period in coffee is short and insecticidal applications can be avoided during coffee flowering. Coffee may be attacked by bugs, leaf miner and thrips during flowering. In case of outbreak during flowering, the crop should be treated when bees are not active and less persistent insecticides be used.

Pome and stone fruits: Apple, peach, plum, apricot and almond are attacked by caterpillars at blooming time. Insecticidal use has been suggested by economic entomologists against blossom thrips, though economic losses by thrips have not been ascertained. The recommendations are made in ignorance of insecticidal bee hazards. Therefore, caution is important so that the huge benefits from bee pollination are not reduced.

Other fruit crops. Insecticides can be applied for pest control at flowering-time in citrus, litchi, olive, grapes, coconut and cocoa. Care should be taken because they are also visited by bees for floral rewards.

Phtotoxicity to plants

The effect of an insecticide application may not be confined to damage to the pollinators of a distant crop or elimination of pollinators for the target crop. Another previously overlooked factor associated with the pesticide may be that it can detract from the plants' productiveness. Sedivy (1970) reported that only 10.5% of pollen grains germinated after they were dusted with Melipax (Melipax is a toxaphene like chlorinated camphene) as compared to 62.1% in the control pollen. When the pollen grains were treated with 0.3% Fribal emulsion, another apparently toxaphene like compound, only 28.2% germinated as compared to 81.5% of the control pollen. None of the grains treated with 0.7% Fribal emulsion germinated as compared to 79.0% of the control. Gentile *et al.* (1971) reported that the insecticide naled, at only 100 ppm, completely inhibited germi-nation of both tomato and petunia pollen. They also reported that azinphos-methyl, DDT, dichlorvos, dicofol, endosulfan, and Gardona R caused reduction in pollen germination and/or pollen tube elongation. Carbaryl and methomyl had little or no deleterious effect on pollen, and xylene was non injurious. The separation of the toxic or repelling effect of the presence of the insecticide on the plant from the possible less attractiveness of affected pollen is difficult, but the idea merits further examination, both from the effect of pesticides on the plants and on the pollinating insects.

Intensity of damage to bees by pesticides

Numerous surveys have been made to determine the extent of the losses of bees from pesticides. Levin (1970) stated that some 500,000 colonies were killed or damaged in the United States in 1967, of which 70,000 were in Arizona and 76,000 in California. Swift (1969) stated that losses in California in 1968 were even greater--83,000 colonies. Wearne *et al.* (1970) and Barnes (1972) concluded that the major problem confronting the beekeeping industry was bee losses due to pesticides-with which there is little disagreement by the beekeeping industry. All indications point to an annual loss by the industry in the neighborhood of

10% caused by pesticides alone. Few industries can tolerate such losses and survive. The effect of these losses on the adequacy of crop pollination is unknown.

Indirect effects of pesticides on bees

1. Pesticide application besides directly killing the bees, indirectly produced several indirect effects
2. Reduced foraging due to repellency or reduced foraging due to killing of foragers due to poisoning
3. Sub lethal doses can also influence other behaviours such as orientation, dance rhthym, dance velocity, walking speed, wing beat frequency
4. Physiological injury resulting into reduced longevity
5. Pesticide generally accumulate in combs as a result of absorption from stored pollen, honey/nectar and may cause chronic paralysis under stress conditions
6. Reduced egg laying and brood rearing due to small doses of pesticides
7. Amorphogenic effects on developing brood and delayed and abnormal development
8. Pesticide application changes physiology of plants affecting nectar, pollen production, pollen viability, consequently bee behaviour, nectar pollen collection honey storage and pollination

A schematic model exhibiting direct and indirect of pesticides on bee behaviour, ecosystem and crop productivity ig given in Figure 101a.

Factors influencing toxicity of pesticides to honeybees

Age and size of the bees: From the general principles, body size is likely to have a direct effect on the susceptibility of these contact pesticides. Smaller bees have a higher surface to volume ratio and contact poison will be more toxic to them than the larger bees.

Pesticides Involved - Basic Types and Classes

The following information is presented to help the beekeeper better understand pesticides and to successfully meet the challenge of pesticides killing honey bees.

Classes of Pesticides

The need of human beings to effectively control their environment is most evident in their agricultural pursuits. Modern farming covers large tracts of land under uniform planting, and this has made pest control mandatory. The evolution of pest control agents originated with natural products such as arsenicals, petroleum oils, and toxins derived from plants (nicotine and rotenone, for example). The advent of DDT, which was synthesized in a laboratory, heralded an era in which a mature chemical industry would screen synthetic chemicals for pesticidal activity. This effort spawned an impressive array of insect control agents. The selection of control chemicals is large. However, these materials can be grouped conveniently according to general chemical properties and modes of action.

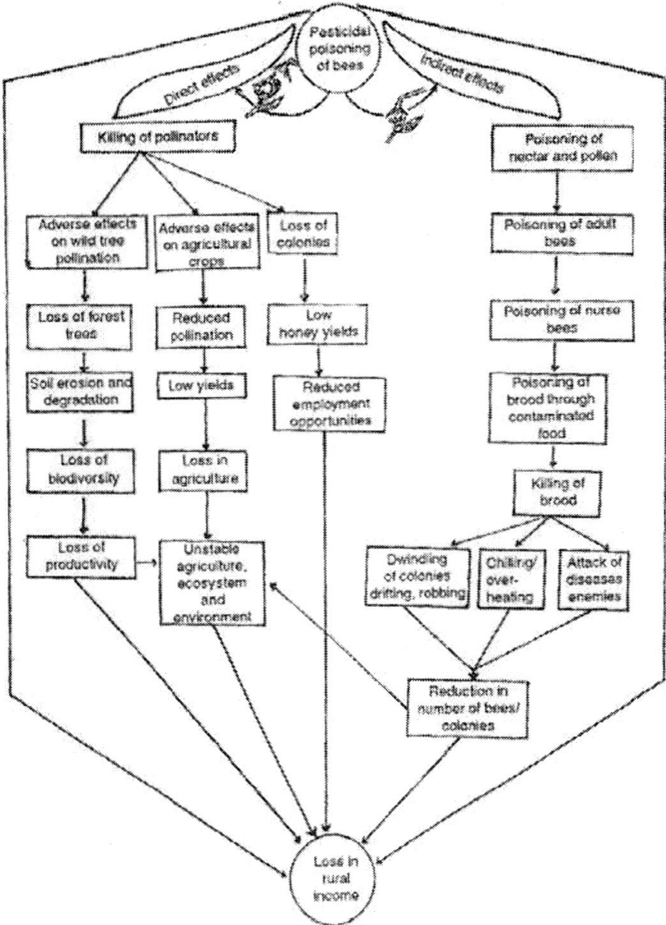

Figure 101 (a). Direct and indirect effects of pesticides on bees.

Insecticides

Insecticides affect bees in one or more ways as stomach poisons, as contact materials, and as fumigants. Arsenicals are typical stomach poisons, pyrethrum is a typical contact insecticide, and hydrogen cyanide, paradichlorobenzene, and carbon disulfide are examples of fumigants.

Botanicals – Only a small amount of our insecticides are derived from plants. These sources are cube, derris, nicotine, pyrethrins, ryania, sabadilla, and tephrosia. The bulk of this material is used in households and gardens, and, because of its inaccessibility to bees or the relatively minute amount used, it presents no hazards to pollinating insects. Sabadilla dust is sometimes used on citrus where it can create a bee poisoning problem.

Occasionally, bees are poisoned by feeding on nectar or pollen of certain plants, for example, California buckeye (*Aesculus californica* (Spach) Nutt.), locoweed (*Astragalus* spp.), or mountain laurel (*Kalmia latifolia* L.). Reaction of the bees to these plant poisons can usually be differentiated from those caused by most pesticides.

Inorganics - These pesticides include arsenicals, fluorides, mercury compounds, and sulfur. The method and limited use of the mercury compounds precludes their presenting a hazard to bees. Elemental sulfur alone or when used with other insecticides in the field, presents only a slight repelling action, although fumes from burning sulfur are highly toxic to insects. Fluorides are rarely used on a large scale and present no problem. In certain sections of Europe, fluoride compounds from smelters frequently cause bee damage. Whenever arsenicals are used they pose a serious threat to bees.

Organics – The chlorinated hydrocarbons, organophosphates, and carbamates vary in their toxicity to bees from relatively non hazardous to highly hazardous, depending upon the individual material or combination of materials.

Chlorinated Hydrocarbons

These include such important insecticides as BHC, toxaphene, and chlordane. The chemicals in this group are slowly reactive chemically, thus persistent in the environment. Biological degradation tends to be slow; hence, storage in fatty and muscle tissue causes these materials to become concentrated and enter our food chain. The mode of action of chlorinated hydrocarbons is still a subject of active research. They are classified as neuroactive agents which block the transmission of nerve impulses. Specifically, for example, DDT prevents the normal sodium-potassium exchange in the sheath of the nerve fiber-this exchange being the means by which a message is transmitted along the nerve. Because chemicals such as DDT are not very chemically reactive, it is felt that the mechanism of reaction with the sheath is not chemical, but rather that the size and shape of a DDT molecule may fortuitously permit it to fit into the proteins of the sheath. Such conceptualizations of the toxicological processes have promoted the search for new chemicals with better toxicological and environmental properties.

Organophosphorus Insecticides

They account for about 30% of the registered synthetic insecticides/acaricides. They possess the common characteristic of inhibiting the enzyme cholinesterase, which mediates the transmission of nerve signals. Hence, organophosphates also are neuroactive agents. As their name implies, these materials contain phosphorus, and as a group they include parathion, Systox, DDVP, and malathion. They are quite reactive chemically and are not regarded as persistent in our environment, unless they are microencapsulated.

Carbamate Insecticides

These also are inhibitors of cholinesterase and feature a nitrogen-containing unit known to chemists as a carbamate function. Members of this class of insecticide include carbaryl (Sevin), baygon, Furadan, landrin, and zectran. For the most part, these materials are easily biodegraded and do not constitute the residual hazard of the chlorinated hydrocarbon class of insecticides. Interestingly, cholinesterase inhibition tends to be reversible for mammals and insects alike. A sublethal dose can bring on the usual symptoms of nerve poisoning (tremors, loss of muscular control, incontinence, vomiting), but the poisoned animal will return to normalcy in a very short time.

Pathogens: bacteria, protozoans, and viruses: None of these that are currently recommended or that have been tested for biological control pose a hazard to bees (Cantwell *et al.* 1972).

Defoliants, desiccants, and herbicides

Most tests have shown this class of materials to be nonhazardous to bees, except for their removal of the food source from the plant; however, Morton *et al.* (1972) reported that paraquat, MAA, MSMA, DSMA, hexaflurate, and cacodylic acid were extremely toxic when fed to newly emerged worker honey bees at 100 and 1,000 ppm concentrations. Although newly emerged bees do not forage away from the hive, they consume food that others bring in. MSMA, paraquat, and cacodylic acid were also highly toxic when sprayed onto older bees in small cages (Moffett *et al.* 1972).

Diluents, synergists, and activators

There is little information on the influence of these agents on the toxicity of the primary pesticides on honey bees. Possibly different interpretations of the effects of certain pesticides may have been associated with the materials with which they were applied.

Fungicides

As used, the copper compounds, mercury compounds, pentachlorophenol, sulfur, and zineb have caused no trouble to bees. A wide variety of other synthetic chemicals may be applied to crops on which bees may be foraging. Herbicides and fungicides have bases for their activity which render them relatively much less toxic to honey bees. Still such materials are present in the biosphere of the honey bee, and little information is currently available dealing with the effects of these chemicals in combination with insecticides – a situation which occurs often under normal field conditions. Moreover, such materials as herbicides and nonconventional insecticides (such as insect sex attractants and insect growth regulators) to which bees are being increasingly exposed likely will be transferred to honey and stored pollen with, as yet, incompletely documented results

Sex lures, attractants, and other hormones

These usually cause no problems to bees, and their use near bees is generally welcomed. Occasionally, a few honey bees and bumble bees have been found in traps containing Japanese beetle lures (Hamilton *et al.*, 1970).

Biological control agents (parasitic and predatory insects)

Beekeepers would welcome biological control of harmful insects on crops because the control agents likely to be used would prey on the specific insects without harming bees. This would permit bees to forage with safety and effectively pollinate the crop.

Nonchemical Control

Along with the beneficial aspects of chemical pesticides come problems such as contamination of the environment and killing of beneficial insects, many of which are honey bees. To reduce agriculture's dependency on pesticides, a new concept of pest control has been developed called integrated pest management (IPM). Under IPM, all techniques and methods that are useful in controlling pests are used, including pesticides. However, a farmer applies a chemical pesticide only as a last resort. Primary reliance is on nonchemical controls, such as insect attractants, repellents, traps, insect-resistant plants, insect pathogens (disease), insect predators and parasites, time of planting, cultivation, time of harvest, sterilized insects, quarantines, and other practices. Not all of these techniques are utilized at the same time in controlling a pest; however, all of the control methods are considered non injurious to honey bees, except for chemical control. Where possible, beekeepers should encourage farmers to use IPM techniques

Toxicity of a pesticide

The toxicity of a specific pesticide is a composite of its physical and chemical properties, the method of formulation (description follows), and the inherent ability of the honey bee to deal with the material internally. If the pesticide is of high volatility (an example is the fumigant TEPP), then the chemical may be absorbed through the bee's spiracles or respiratory system. The physical properties of an insecticide and especially of its formulation would be largely responsible for the relative hazard from this mode of entry into the bee. Ingestion of contaminated pollen and nectar offers yet another route of entry. The alimentary tract may become altered or paralyzed, making feeding impossible, or the bee's gut may cease to function. The ability of an insecticide to contaminate nectar and pollen would again be a composite of the physical/ chemical properties of the material, its formulation, and the time of application of the spray relative to bloom.

How bee poisoning occurs

1. Bee poisoning generally occurs after a pesticide has been applied to crops or weeds which contain flowers or are providing secretions attractive to bees, e.g. from extra-floral nectaries.

2. The pesticide is applied directly onto bees foraging on the crop.
3. Bees fly to the treated plants and collect contaminated nectar and/or pollen.
4. Bees collect contaminated water on or near treated plants.
5. Bees forage on a cover crop associated with the treated crop, e.g. clover in an orchard.
6. Pollen collecting bees collect pesticide dust and/or contaminated pollen and return these to the hive.

Pesticides drift from their point of application onto flowering plants or across apiaries

Symptoms of bee poisoning

Bees come in contact with pesticides during foraging. There can be stomach poisoning to bees and also to brood when fed on contaminated pollen. Some pesticides may even cause hazards by fumigant action. After gaining entry to body, different pesticides having different modes of action. Some of the most important and common symptoms of pesticides poisoning in bees are as follows:

(i) Presence of large number of dead bees in front of bee hive.Individual bees that have been poisoned frequently are seen crawling on the ground near the entrance or twirling on their side in a tight circle. Others appear to be weak or paralyzed. These gross symptoms of poisoning vary with the type of pesticide and the degree of exposure. Foraging bees also may die in the field or on the flight back to the hive. Bees become paralytic, lose the power of orientation; legs, wings and digestive tract stop functioning and poisoned bees show unco-ordinated movements.

(ii) Abdomen becomes distended

(iii) Dead adult bees typically, but not always, have their wings unhooked and at odd angles to their body, their proboscis fully extended, and their hind pair of legs outstretched behind them.

(iv) Workers can get confused, forget their path and could not reach back to hives, hence their number is reduced. Bees may perform abnormal communication dances on the horizontal landing board at the hive entrance while under the influence of insecticide poisoning. Disorganized behavior patterns may lead to lack of recognition of affected field bees by guard bees and may harm them.

(v) Regurgitation of contents of gut is noticed particularly in case of organophosphate insecticidal poisoning. Stupefaction, paralysis, and abnormal activities of bees are commonly caused by chlorinated hydrocarbons and organophosphorus insecticides

(vi) Guard bees are also confused due to toxic effect of pesticides.

(vii) Brood chilling can occur due to reduction in the population of adult bees.

(viii) Dead brood can be seen inside the colonies if poisoned pollens are stored and fed to the brood.

(ix) Sometimes, the queens are also affected. Queens may stop egg laying or lay eggs in irregular pattern, there may be brood in only some of the cells of the, brood area as in case of the colonies suffering from foul brood. Sometimes queenlessness may develop. Queen cells may be raised and queens may be superseded in colonies which survive.

(x) If on crops, gamma-BHC (lindane) and endosulfan etc. from organochlorine group are used, the affected bees can not move properly and their legs adversely affected. They appear as though they had been chilled. Most of such bees die away from their hives.

(xi) If malathion, Dimethoate and other related insecticides from organophosphate group are used, a watery substance liquid comes out from their mouth due to which whole body of bees becomes wet, swelling of abdomen takes place and both pair of wings stick to the body of bees. The behaviour of such bees also changes and they die near their hives.

(xii) If insecticides of carbamate group such as carbaryl, carbofuran etc. are used for insect pest control, the affected bees become more angry and not able to fly properly. Most of the affected bees die near their hives. In such a bee colony, the queen stops egg laying. In extreme cases of pesticide exposure, the house bees also die. When the house bees die, the brood will show signs of neglect or poisoning and many, or all, immature bees still in the cells may die. Some pesticides, particularly systemic pesticides, have a less noticeable, but debilitating effect, resulting in an overall weakening of the colony. Signs are reduction in adult bee numbers and stages of the brood cycle or complete brood cycles missing. In severe cases, when insufficient numbers of adult bees remain, temperature and humidity control in the brood area is lost and brood is not fed. Brood die from chilling, overheating or starvation

By observing symptoms one can find out the group of insecticides sprayed on the crop in the vicinity of the apiary. In addition to this a scientist or a beekeeper as well as farmers should also know the degree of toxicity of different insecticides to the bees so that a person can spray such an insecticide which is quite effective against pests but least toxic to the bees.

Groups of insecticides based on their toxicity to bees

Toxicity is life property of a substance (insecticides) that causes any adverse effect in an organism. The toxicity may be due to single short term exposure period (acute toxicity) or due to repeated/continuous exposure within less than half life time of that animal (sub-acute toxicity) or repeated exposure over a period of at least half the life time of that organism (chronic toxicity). On the basis of toxicity to bees, the insecticides can be grouped into the following three Categories:

(i) More toxic insecticides

The insecticides which have adverse effect on bees even after 90 hours (about 4 days) comes under this category. Such insecticides are carbofuran, dichlorovous fenthion, monocrotophos, fenitrothion, lindane, malathion, carbaryl, methl-parathion, chlorpyriphos, dimethoate, phosphamidon, diaginon etc.

(ii) Less toxic insecticides

Those insecticides whose residues is less persistent in nature and remain toxic to bees upto 90 hours (Less than 4 days) belong to this category. Such insecticides are endosulfan and phosalone etc. Besides this, all systemic insecticides also fan in thus category if used properly. These insecticides should be sprayed after 3 PM and next day morning allow bees to go a little bit late in the fields on crop plants so that by that time most of the toxicant enters inside the plant system being systemic in nature resulting in minimum harm 10 the bees.

(iii) Non-toxic insecticides

There are some chemicals which are very effective for control of insect pests but DOD-toxic to bees. They are some bacteria (B.T. formulations), viruses (NPV) and insect growth regulators etc. Besides this, fungicides, weedicides and some plant growth regulators have also been found safe to bees.

Relative toxicity of pesticides

The kill of bees can be classified as:

<100 per day - normal die off rate

200-400 per day - low kill

500-900 per day - moderate kill

>1000 per day - high kill

Poisoning and developmental stages

Worker bees are those primarily affected by pesticides. The symptoms of poisoning can vary depending on the developmental stage of the individual bee and kind of chemical employed

Development of adult: It takes worker bees about twenty-one days to develop from egg to adult. During this process, each individual passes through a larval (feeding) stage followed by a pupal (transformation) stage. The larval stage is the most susceptible to pesticide poisoning during development.

House bees: These bees are emerged worker adults up to twenty-one days of age. They care for the brood, process pollen and nectar gathered in the field by older workers, and clean the nest. Eventually, they too will become field bees. House bees are usually poisoned by contaminated pollen which is collected in the field, brought back and stored in the hive. As house bees are killed, there are fewer bees to tend the brood and further decline in population results.

Field bees: These bees are workers twenty-one to approximately forty-two days of age. There appears to be no greater risk in bee society than to be a field bee. Should the insect avoid all the potential pitfalls due to predators like spiders, toads or skunks, it is still vulnerable at all times to the numerous pesticides applied in commercial agriculture, mosquito control, and home gardens. Most times, field bees are killed by contact with pesticides in the field, but other times they collect contaminated nectar and pollen and contribute to poisoning their sisters in the colony. If field bees are killed, then young bees are forced into the field earlier than normal, disrupting and thus disorienting the colony.

While foraging, field bees may range as far as two to five miles from a colony. They usually seek nectar and pollen systematically, not randomly, and once a food source is found, bees prefer to work that particular source to exhaustion before changing plants. This kind of resource partitioning by bee colonies accounts for the inconsistency observed many times between colonies undergoing pesticide poisoning in the same location. The bees are not all working the same plants and so some are affected more than others. Often it is those bees with established flight patterns located in an area before a pesticide is applied that are most damaged. Those placed in a field immediately after application are less affected by the pesticide because it takes some time for the bees to scout an area and locate food sources.

On the basis of mode of action the insecticides are classified as given below (Table 81).

Table 81. Classification of Insecticides on the basis of their mode of action

Group	Subgroup	Example
Physical poisons	-	Heavy mineral oils, inert dusts
Protoplasmic poisons	-	Heavy metals, e. g. Hg, acids
Metabolic inhibitors	Respiratory poisons	HCN, CO, H2S, rotenone, dinitrophenols
	Inhibitors of mixed-function oxidase	Pyrethrin synergists
	Inhibitors of carbohydrate metabolism	Sodium fluoroacetate
	Inhibitors of amine metabolism	Chlordimeform
	Insect hormones	Juvenile hormone analogues
	Inhibitors of chitin synthesis	Diflubenzuron

Neuroactive agents (nonmetabolic	Anticholinesterases	Organophosphorus Compounds, carbamates
	Effects on ion permeability	DDT analogues, pyrethroid
	Agents for nerve receptors	Ach*nicotinoids, Nereistoxin analogues, GABA* cyclodienes, HCH, Avermectins (Spinosoids) Octopamine* Formamidines
Hormone mimics	-	Methoprene
Stomach poisons	-	*Bacillus thuringiensis* toxin, Thiodicarb

Table 82. Pesticide Risk to Bees

Risk Rating	Chemicals	Remarks
High risk to bees foraging even 10 hours after spraying	Carbaryl, chlorpyrifos, diazinon, dimethoate, omethoate, methomyl, fenthion, methamidophos, methidathion, monochrotophos.	These should never be sprayed on flowering crops especially if bees are active and the crop requires pollination.
Moderate risk with some losses expected 10 hours after spraying	Acephate, demeton-s-methyl	
Some risk with low chance of losing bees 3- 5 hours after spraying	Endosulfan, dicofol, pirimicarb, petroleum oils, most pyrethroid chemicals, trichlorfon.	There is little risk of losing bees if these chemicals are sprayed in the evening when foraging has ceased.
No risk even if sprayed over foraging bees	*Bacillus thuringiensis*, propargite, oxythioquinox	

The relative toxicity of different insecticides to honey bees has been given in tables 1-6.

LD50 (μg/bee)	Classification
> 100	Virtually non toxic
11-100	Slightly toxic
2-10.99	Moderately toxic
<2.0 (ie., 0.001 to 1.99 μg/bee)	highly toxic

Application of these pesticides on blooming crops or weeds may cause severe damage to bees. Even after 10 hours of spray these pesticides are still very toxic to bees.

Table 83. Pesticides most toxic to bees (LD_{50} 0.001-1.99 μg/bee).

Acephate Fenthion*	Colep Naled D
Aldicarb G (applied 4 weeks before bloom) Heptachlor*	Crotoxyphos Omethoate
Aldrin* Isobenzan	Diazinon* Paraoxon
Aminocarb Lean arsenate	Dicapthon Parathion*

Azinphosethyl Malathion D*	Dichlorvos* Permethrin
Azinphosmethyl Malathion ULV	Dicrotophos Phenthoate
Benzene hexachloride* Methamidophos	Dieldrin Phosmet
Bomyl Methidathion	Dimethoate* Phosphamidon*
Calcium arsenate Methiocarb	Dinitrocresol Phoxim
Carbanolate Methomyl D	Dinoseb Propoxur
Carbaryl ULV (over 0.4 kg/ha) Methyl-carbophenothion	EPN Pydrin (over 0.1 kg/ha)
Carbofuran F Mehyl parathion*	Ethyl-methyl Guthion Pyramat
Carbophenothion D Mevinphos	Famphur Resmethrin
Chlorpyrifos* Mexacarbate	Fenamiphos Sulfotep (p)
Chlorthion Monocrotophos	Fenitrothion* Sulprofos
	Fensulfothion Thionazin

* There is an Indian Standard for one or more formulations of these pesticides (India Standards Institution, 1979).

1 For minimal hazard to honeybees, the insecticides below should NOT be applied on blooming crops or weeds. Their residual toxicity is usually high even after 10 hours.

Table 84. Pesticides very toxic to bees[1]

Malathion EC*
Naled WP
Phorate EC
Pydrin (0.1 kg/ha or less)

* there is an Indian Standard for one or more formulations of this pesticide (Indian Standards Institution, 1979).

1For minimal hazard to honeybees, the insecticides below should be applied only during late evening. Their residual toxicity is usually low with in 8 hours.

Table 85. Pesticides very toxic to bees[1]

Amidithion Kroneton	Disulfoton EC Propoxur MA
Aramite D Larvin	Endosulfan* Propyl thiopyrophosphate
Binapacryl* Leptophos	Endrin* Rotenone D
Carbaryl ULV (0.5 kg/ha or less) Lethane 384 special	Ethion Sabadilla
Carbophenothion Malathion MA	*ethyl-DDD Solvigran
Chlordane* Malonoben	Fenchlorphos Strobane
Chlorfenvinphos Menazon	Fenthion G or MA Temephos*
Chlorpyrifos or chlorpyriphos MA Methomyl S, SP	Fonofos TEPP
Coumaphos Methoxychlor	Formetanate Tetrachlorvinphos
DDD Methyl-demeton	Garrathion Tetram
DDT* Naled EC	Heptachlor G Thanite (isobornyl thiocyanate)
Demeton Nissol	Isobornyl thiocyanate Thioquinox
Dichlorfention or dichlorfenthion Oil sprays (superior type)	Isodrin Toxaphene*

Dichlorvos MA Oxamyl	Isolan Tranid
Dieldrin G Oxydemeton-methyl*	Isopropyl parathion Trichlorfon*
Dilan Phorate G	Malathion EC*
Dimetilan Phosalone*	Naled WP
Dinobuton Phostex	Phorate EC
Dioxathion Pirimicarb	Pydrin (0.1 kg/ha or less)

* there is an Indian Standard for one or more formulations of this pesticide (Indian Standards Institution, 1979).

1 For minimal hazard to honeybees, the insecticides below should be applied only during late evening. Their residual toxicity is usually low with in 8 hours

The following pesticides should be applied in the late evening when bees are not foraging in the field. bee hives should not be directly exposed to these insecticides. For minimal hazard to bees, dose, the timing and methods of application are very important.

Table 86. Pesticides least toxic to bees1 (LD_{50} 2.0-10 μg/bee).

Allethrin Fenazaflor	Chlorfensulphide Propargite
Amitraz Fenbutatin-oxide	Chlorobenzilate Propoxur G
Bromopropylate Fenson	Chloropropylate Pyrethrum*
Butoxy thiocyanodiethyl ether Fensulfothion G	Cryolite Quinomethionate
Carbaryl G Genite 923 or Genitol 923	Cyhexatin Rotenone EC
Carbofuran G Hydroprene	Dicofol* Ryania
Chlorbenside Lime-sulphur or lime-sulfur*	Dienochlor Schradan
Chlordecone Malathion G	Diflubenzuron Sodium fluosilicate baits
Chlordimeform Mirex G	Dikar Sulphenone
Chlorfenethol Nicotine sulphate	Dinex Sulphur or sulfur*
Chlorfenson Oxythane	Dinocap Tetradifon
	Disulfoton G

* there is an Indian Standard for one or more formulations of this pesticide (Indian Standards Institution, 1979).

1 The insecticides below can be applied at any time with reasonable safety to honeybees. Their toxicity is usually low with direct application

These can be used around bees with a minimum injury. They should be applied during late evening, night or early morning. These products can be used on fields or near hives with minimum damage to bees; in fact, a few of the listed acaricides can he used to control bee mites within the hive.

Table 87. Pesticides Non-Toxic (LD_{50} above 11μg/bee)

A. Insecticides and acaricides

acaraben (chloro-benzilate)	Ethodan	Omite
	Fundal	OMPA (schradan)
Allethrin	Galecron (chlorophenamidine)	Ovotran (ovey)
Aramite		Phostex

Bacillus thuringiensis	Heliothis virus	phrethrin
	Kelthane (dicofol)	Rhothane (TDC)
cryolite	Kepone	rotenone
Delnav (dioxathion)	methoxychlor	ryania
Dessin	Mitox (chlorbenside)	sabadilla
Dilan	Morestan	Sulphenone
Dylox (trichlorfon)	Nemagon	Tedion (tetradifon)
Eradex	Neotran nicotine	toxaphene

B. Fungicides

Arasan (thiram)	Cyprex (dodine)	Manzate (maneb)
Benlate (benomyl)	Dexon	Mylone
bordeaux mixture	dichlone	Parzate (nabam)
copper oxychloride sulfate	Difolatan	Phaltan (folpet)
	Dithane M-45	Polyram
copper sulfate (monahydrate)	(folcid)	sulfur
	Glyoxide (glyodin)	Thynon (dithianon)
cuprous oxide	Karathane (dinocap)	Zerlate (ziram)

C. Herbicides

amitrol	Eptam (EPTC)	picloram
Ammate (ammonium sulfamate)	Folex (merphos)	Planavin
	Herbisan (EXD)	Princep (simazine)
atrazine	Hyvar (bromacil)	Randox (CDAA)
Banvel (dicamba)	Igran (terbutryne)	Sinbar (terbacil)
Betanal (phenmedipham)	IPC	Stem F-34 (propanil)
	Karmex (diuron)	
Caparol (promytryne)	MCPA	TOK (nitrofen)
Casoron (dichlobenil)	Milogard (propazine)	Trysben (2,3,6-TBA)
	monuron	Vegedex (CDEX)
dalapon	NPA	2,4-D
DEF	paraquat	2,4-DB
diquat		2,4,5-T

Table 88. Common insecticides and miticides and their relative risk to honey bees*

Active ingredient	Trade Names	Risk Class	LD50	Residual
Acephate	Orthene	I	1.2	1 day to > 3 days
Aldicarb	Temik	I	0.35	>1 day to >2 days

Azinphos methyl	Guthion	I	0.43	2 days to >5 days
Bacillus thuringiensis	Biobit, Cutlass, Dipel, Javelin, Thuricide	III	NA	< 2 hours
Carbaryl	Sevin	I	1.5 to 26.5	<2 hours to 12 days
Chlorpyrifos	Dursban, Lorsban	I	0.11	5 hours to 6 days
Cyhexatin	Plictran	III	NA	<2 hours
Cypermethrin	Ammo, Cymbush	III	NA	<2 hours to > 3 days
Diazinon	Diazinon	I	0.37	<1 day to 2 days
Dicofol	Kelthane	III	NA	<2 hours
Dicrotophos	Bidrin	I	0.3	1 day to 1 days
Diflubenzuron	Dimilin	III	NA	<2 hours to 6 hours
Dimethoate	Cygon, Defend, Rebelate	I 0.19 <2 hours to >3 days	I	0.19 <2 hours to >3 days
Disulfoton	Di-syston	II	6.12	<2 hours to 7 hours
Endosulfan	Thiodan	II	7.8	<2 hours to 1 day
Ethion	Ethion oil	III	NA	<2 hours
Fluvalinate	Mavrik	III	65.8	<2 hours
Fonofos	Fonofos Dyfonate	II	8.68	<2 hours to 6 hours
Formetanate hydrochloride	Carzol	II	9.2	<2 hours to 2 hours
Lindane	Lindane	I	NA	>2 days
Malathion	Cythion, Malathion	I	0.73	<2 hours to 5 days
Methamidophos	Monitor	I	1.37	4 hours to 1 day
Methidathion	Supracide	I	0.24	1Day to 3 days
Methomyl	Lannate	I	1.29	<2 hours to >1 day
Methoxychlor	Marlate, Methoxychlor	III	NA	<2 hours
Methyl parathion	Penncap-M	I	0.11 to0.24	<1 day to >7 days
Mevinphos	Phosdrin	I	0.3	<2 hours to <5 hours
Naled	Dibrom	I	0.49	2 hours to >1 day
Oxamyl	Vydate	II	10.3	<2 hours to 12 hours
Oxythioquniox	Morestan	III	NA	<2 hours
Parathion	Parathion	I	0.18	10 hours to >1 day
Permethrin	Ambush, Permethrin, Pounce	I	0.16 12	<2 hours to >3 days
Phorate	Thimet	II	10.25	<2 hours to 5 hours
Phosmet	Imidan	I	1.13	8 hours to >3days
Profenofos	Curacron	II	3.46	<2 hours to 9 hours

Propargite	Omite, ornamite	III	NA	<2 hours
Sulprofos	Bolstar	II	7.22	<2 hours to >1 day
Thiodicarb	Larvin	II	7.08	<2 hours
Trichlorfon	Dylox, Proxol	III	NA	<2 hours to 6 hours

1 Table modified from Johansen & Mayer (1990) and Delaplane (1993).

2 Risk classes; I=highly toxic to honey bees, II=moderately toxic to honey bees, III = relatively nontoxic to honey bees. The risk class is closely associated with the LD50.

3 LC50 = the Lethal Dose required to kill 50% of the test honey bees, expressed in microbgrams per bee. The smaller the LD50, the more toxic the product.

4 Period of residual toxicity to honey bees after application. Evening applications of products with residuals of 8 hours or less generally cause only moderate harm to bees, even if the LD50 is small. For example, mevinphos is very toxic to bees, but because it has a short residual, it is fairly safe for early evening applications.

* Never spray during bloom periods unless it is absolutely necessary. If treatment is unavoidable, choose a product with a high LD50 and short residual. If a more toxic chemical is required, choose a residual under 8 hours and spray at night

Table 89. Pesticides recommended as relatively safe for honeybees (Indian Standards Institution, 1973)*

Pesticide	Use class	Cotton	Maize	*Brassicas*	Vegetables	Fruit	Others
Amidithion	3						X
Carbophenothion	3						X
DDT	3	X			X	X	X
Demeton methyl	3	X	X	X	X	X	X
Disulfoton	3,4						X
Endosulfan	3	X	X		X		X
Endrin	3	X				X	X
Ethion	3						X
Fenchlorphos	3						X
Isolan	3						X
Menazon	3			X	X	X	X
Methoxychlor	3						X
Naled*	2,3						X
Phorate	2,3						X
Ryania	4						X
Temephos	3						X
Toxaphene	3		X		X	X	X
Trichlorfon	3		X				X

*The recommendations apply to the crop specified, and for time and method of application stated in the publication

Figure 101 A schematic model exhibiting impact of pesticides on adult honeybees as well as brood.

Protection of Bees

Several management practices are known that may help protect bees from pesticide-caused mortality.

How Bees Can Be Protected From Pesticide Poisoning

Do not apply any insecticide unless the crop is so heavily infested and the insect in question has reached the economic threshold value (ET-value). If necessary, then

1. **Read the pesticide label carefully.** Read the pesticide label. Pesticides and formulations which pose a special hazard to bees are required to include a notification on the label.
2. **Use less toxic compounds**: Choose an insecticide of low toxicity to bees that will provide the needed pest control.
3. **Types of formulation and their toxicity to bees**
    a. **Use less toxic formulations:** Mostly dust, wettable/water dispersible liquid and granular formulations of insecticides are used to control insect/pests of various crops. Select the safest formulation of the pesticide that is available for the intended use. "Drifting" of the pesticide from the target pest and/or crop to areas frequented by bees should be minimized.
    b. **Dust formulations:** These formulations if used on crop result in maximum toxicity to bees and other pollinators as compared to other formulation. Dusts almost always drift more than other formulations and are generally more dangerous to bees than are sprays or granular application
    c. **W.P./W.D.P. Formulations:** These formulations are less toxic to bees as compared to dust formulation but definitely more toxic in comparison to other formulations of insecticides. **(i) Liquid formulations :** Of course liquid formulations are less toxic to bees as compared to dust and wettable or water dispersible powder formulations but they are less toxic as compared to granular formulations. **(ii) Granular formulations:** Among the insecticidal formulations which are commonly used to control pests of various crops, it is the granular formulation which is least toxic to honey bees visiting flowers to collect nectar and pollen. Further, ground applications are always safer than aerial applications.
    d. **Microencapsulated pesticides:** Microencapsulated pesticides present a very distinct and serious threat to honeybees. The particle size of this formulation is very similar to that of pollen,

and adult honeybees may carry this pesticide back to the hive, where it will be combined with pollen that is being stored in the hive. This pesticide will not kill the adult bees that collected it, but it will kill the immature stages of the bees and the young adult nurse bees that feed the brood. Bees have little protection from this product.

1. **The mode of pesticide application:** The mode of pesticide application is also important, particularly when it comes to drift. Aerial applications are generally more dangerous than applications by ground equipment, because of the location of target pests and/or crops to foraging bees or beehives. Never apply a pesticide directly over beehive.
2. **Do not treat crops in bloom:** Whenever possible, do not treat crops in bloom. If treatments are needed during bloom, choose a short-residual material. Remove weed blooms in orchard ground cover and in field edges before spraying. Flowering weeds may be removed by mowing or with an herbicide.
3. **Minimize spray drift:** Minimize spray drift onto adjacent crops or other plants in bloom. Honey bee hives should not be placed next to fields or orchards that are likely to be treated with pesticides toxic to bees. A small number of hives may be protected from pesticides by covering the colonies with wet burlap for a period of one to two days. In some cases it may be practical to move hives to a less exposed site. Beekeepers should inform farmers of the location of hives.
4. **Use pesticides only when needed:** Foraging honey bees, other pollinators, and insect predators are a natural resource and their intrinsic value must be taken into consideration. Vegetable, fruit, and seed crop yields in nearby fields can be adversely affected by reducing the population of pollinating insects and beneficial insect predators. It is always a good idea to check the field to be treated for populations of both harmful and beneficial insects.
5. **Apply pesticide when bees are not flying:** Bees fly when the air temperature is above 55-60°F and are most active from 8 a.m. to 5 p.m. Always check a field for bee activity immediately before application. Pesticides hazardous to honey bees must be applied to blooming plants when bees are not working, preferably in the early evening. Evening application allows time for these chemicals to partially or totally decompose during the night (Figure 101b).
6. **Do not contaminate water:** Bees require water to cool the hive and feed the brood. Never contaminate standing water with pesticides or drain spray tank contents onto the ground, creating puddles.
7. **Identify attractive blooms :** Bees also visit flowers of some weeds; hence before spraying remove such weeds from fields which might attract bees. In many instances bees have been killed even though the

crop being sprayed was not in bloom. Many times these attractive blooms can be mowed or otherwise removed, although mowing can result in destroying other beneficial insect habitat or force destructive insects into the crop being cultivated.

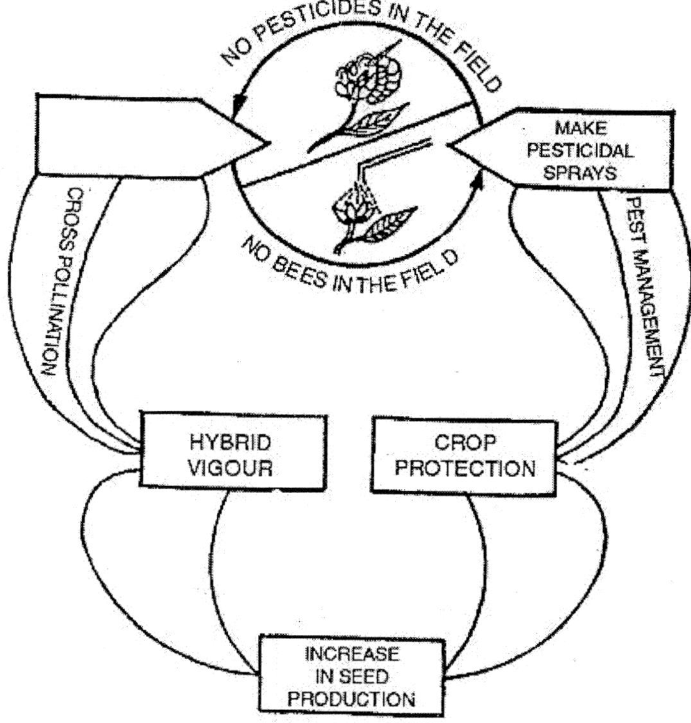

Figure 101 (b). A schematic model exhibiting bee behavioure with pest management

8. **Inform presence of apiary**: If your bees are located in any area where pesticides are commonly used, then identify yourself as a beekeeper to any neighbors who may use pesticides. Explain to growers the importance of your honeybees in the pollination of crops being grown in nearby fields so that they may consider the value of bees in pollination before applying any pesticides that may kill the pollinating insects. Do not place apiaries in areas that are used to grow crops that require heavy and frequent usage of pesticides.

9. **Use integrated pest management (IPM) to reduce pesticide hazard**. IPM is an important approach to controlling insects, mites, weeds and diseases of crops. Whenever possible, it is wise to use such integrated pest management practices viz., recognition of existing biological controls and evaluating their value before applying any pesticides; using pesticides that are less harmful to beneficial organisms; timing pesticide applications so they are less harmful to beneficial organisms. Biological control organisms may be effectively controlling pests even though they are not recognized. Even pests that occur

regularly, such as spider mites on field corn or green bugs on sorghum, may be effectively controlled under some conditions. Before applying pesticides, fields should be examined for the activity of natural enemies. If large numbers are present, pesticide use may be deferred or avoided.

10. **Use of repellents**: Chemical repellents have been studied for many years. The repellent is added to the pesticide before field application and is intended to discourage bees from visiting plant's until the pesticide becomes relatively nontoxic. Field tests showed several compounds to have repellency, but more research is needed before they are used commercially by farmers.

11. **Disposing of unused pesticides**: Follow proper precautions in disposing of unused pesticides and pesticide containers. Be particularly careful not to contaminate wter with pesticides, as the water may be collected by bees and result in bee kills. Pesticide dusts and small granules should not be left open or carelessly thrown any were because bees are likely to collect such dusts during acute pollen dearth period.

12. **Notify beekeepers:** Notify beekeeper that has beehives near an area to be treated with a pesticide so that they may attempt to protect their bees. In the areas of bee-keeping, there should be good coordination between the bee-keepers and crop/fruit vegetable growers so that the farmers may inform well in advance to the bee-keepers so that either they remove the colonies or should keep the bees confined inside the hive during the application. While confining the bees, due attention should be paid to the following points: (a) **Proper space:** While confining the bees in the hive, sufficient and proper space for all the bees including foragers should be made available. (b) **Proper ventilation**: Proper ventilation must be provided at the top or sides of the hive, Ventilation screen should have as large a mesh as possible (c) **Shade**: Shade is usually provided by the use of local materials. It must not hinder the flow of air past the hives. (d) **Covering hives**: Covering the hives with wet clothes or gunny bags is very useful because evaporation of water helps to reduce the rise in colony temperature. If pesticides applied from the air, then cover whole of the hive. (e) **Water availability**: Provide water inside the hives for bees. The bees take it and spread out in the hive where it evaporates and thus reduces the temperature. (f) **Minimize period of confinement**: The confinement must of course continue as long as the pesticide near the hive retains unacceptable toxicity and its duration can satisfactorily be reduced by application of less persistent insecticide. If there is no store in the hive in that case pollen supplement and sugar syrup may also be provided. If there is adverse effect on bees including brood due to pesticidal contamination of pollens, in that case bee-keepers can over come this difficulty by prevention of pollen being stored in the hive and provision of a safe pollen supply inside the hive. Provision of pollen cakes in the

hive during this period greatly reduces the collection of toxic pollen. Cake of pure pollen were more effective than pollen supplement made with soybean flour.

Management of poisoned colonies

Shift colonies away from foraging range from the source of poisoning. Keep bees warm by removing excess supers.

1. If a pesticide carried by the bees infiltrates the combs and contaminates the nectar or pollen, or both, then the combs may need to be replaced before a colony will recover. Remove contaminated pollen stored in combs by dipping the combs in water and washing by slight shaking or the entire comb melted and replaced with wax foundation. Before washing or melting combs, however, a sample of the pollen, wax, and honey should be analyzed chemically to determine the amount of pesticide residue present, if any.
2. Provide sugar syrup and pollen substitute. Feed colonies inside the hive with a 1:1 water: sugar syrup until recovered. Loss of field bees results in a lack of fresh nectar and water being brought into the hive. Add frames of sealed brood and adult bees from healthy hives, if require. Moreover, colonies weakened by insecticide treatment should be fed sugar syrup and pollen cake to stimulate brood production as an aid to population recovery. Frequently, weak colonies must be united to save the remaining bees and brood or a queenless package of bees added to the damaged colony to strengthen the population.
3. If above mentioned points are kept in mind, the bee-keepers can protect their bees to a great extent from toxic effects of insecticides.

Managing pesticide drift

Serious honeybee poisonings can be caused by pesticide drift. During warm periods, large numbers of honeybees may mass on the outside of the hive to help control hive temperatures. These exposed bees can be destroyed easily by drifting pesticides. Drift control is vital during every pesticide application. Several techniques can be used to reduce the possibility of drift:

1. use pesticides that have low volatility;
2. use formulations that resist drift and volatility;
3. use low pressures during spraying;
4. use nozzles which reduce formation of small spray particles;
5. use high water volumes during application;
6. apply pesticides close to the crop or soil surface;
7. avoid applying pesticides when the temperature is high;
8. avoid applying pesticides during windy conditions;

9. use drift reducing adjuvants.

Certain formulations of pesticides can help reduce drift. For example, low-volatile acid and amine 2, 4-D formulations have less potential for drift than ester formulations. Dust formulations drift much more readily than most sprays. Granular formulations are relatively less likely to drift.

Droplet size of pesticides during application is extremely important in determining the potential for drift. The ability of particles to drift increases greatly as the particle size decreases (below).

Distance water droplets drift while falling 10 feet in winds of 3 miles per hour.

Droplet	diameter (microns) classification	Particle Drift distance (ft)
30	Cloud	500
100	Mist	50
200	Drizzle	16
500	Light rain	7

The various spray nozzles and application equipment produce a wide range of droplet sizes. There are several techniques that will reduce the number of the smallest particles while still giving effective coverage.

Application pressures are important in determining the sizes of droplets that are formed. As pressure increases, the number of fine particles also increases. Drift can be reduced by reducing sprayer pressures during application.

Nozzle construction can also affect the number of small particles that are formed during spraying. Nozzle tips that produce larger droplet sizes help reduce drift. For example, larger nozzles can be used at lower pressures to get the same volume (Gallons per acre, or GPA) as smaller nozzles operated at higher pressures. Use of higher GPA applications are an alternate means of achieving adequate crop coverage with minimal pressures.

"Thickening" or "drift control" adjuvants can be added to the spray mixture to reduce drift. These compounds can increase the percentage of larger droplets which are formed but do not completely eliminate small droplets.

The weather conditions during application have a great effect on pesticide drift. Air movements, both horizontal and vertical, cause pesticides to move away from where you are spraying. The higher the wind speed, the larger the amount of pesticide that will be carried away. Pesticides should never be applied during high wind conditions (greater than 10 mph). This is particularly important when wind direction is likely to move drifted pesticides onto nearby sensitive crops or other sensitive areas. Drift to sensitive areas often can be avoided by spraying when the air is moving away from these areas.

Drift may also increase when warming air near the soil rises. Applications should be done at times when air and soil temperatures are most similar, often during early morning and late evening. At this time, vertical air movements are lowest.

If the air near the soil surface is cooler than the air above, an "inversion" exists. Small spray particles remain suspended in the cool air during temperature inversions, and the particles do not settle readily onto soil or plants. Later the suspended particles move out of the crop on winds and drift. Pesticide applications should be avoided during inversions.

Temperature and humidity can affect pesticide drift. When the temperature is high and humidity low, particles evaporate most rapidly. This evaporation causes droplet sizes to decrease and drift more readily. Volatile pesticides also evaporate more rapidly with high temperatures. Pesticides should be applied when the temperature is cool.

Height and orientation of sprayer nozzles can also affect drift. Distance and time for spray droplets to reach plants or soil is directly related to the height at which a pesticide is released. Sprays should be released as near the target as will permit adequate coverage. Sprays should also be directed so droplets are propelled downward to reduce the distance of droplet fall. Vapor drift of soil-applied pesticides can be reduced by properly sealing the soil after application. This often involves proper soil incorporation of the pesticides during application.

Avoiding pollution of ground and surface waters

Contamination of ground and surface waters is an imminent hazard of using pesticides in agriculture. This potential must be a major consideration in planning for pest control on cropland and other agricultural areas. Elements that enter into water pollution by pesticides are:

1. proximity of the treated area to surface waters;
2. proximity of the treated area to drinking water wells or aquifers;
3. depth of the water table at the treated site;
4. soil conditions that increase the potential for the pesticide to leach into ground water;
5. the hazard of the pesticide as a potential contaminant of ground waters;
6. conditions during application that affect pesticide drift into surface waters;
7. crop management practices that minimize pesticide leaching;
8. precautions during application to avoid leaching or direct ground water contamination.

Irrigate in a manner that reduces pesticide movement. High rates of irrigation can increase the amount of pesticide leaching. Excessive irrigation can also cause run-off and erosion. Particular care should be given when irrigating shortly after a pesticide application, since the pesticide is in the highest concentration at this time.

Honey bee indemnity program

Some Governments have enacted national legislation to partially repay each beekeeper for pesticide-killed bees. Beekeepers who exercised reasonable precautions to avoid pesticide damage but still lost bees could apply for indemnity payments as per the provisions made during 1967. The main goal of this program was to aid the bee industry in remaining financially stable and to ensure that enough strong colonies would be available to pollinate agricultural crops nationwide. To accurately acess the loss, beekeepers should maintain detailed records of their colonies, noting by date such items as colony condition, population size, syrup and pollen feeding, and honey production. The more detailed the records, the easier it is to establish the true magnitude of a loss and receive reasonable comp-ensation. Be prepared to manage the hives for queen failure or supersedure prob-lems which may occur a number of weeks after the pesticide problem occurred.

Plant Poisoning

Poisonous plants can be a problem under certain conditions in limited areas. If a plant's nectar is poisonous, the symptoms of plant poisoning are limited to the blooming period of the plant. However, if the poison is in the pollen, the symptoms may linger as long as the pollen remains in the combs. There is no clear-cut method for differentiating between plant poisoning and pesticide poisoning. The effects of plant poisoning are usually more gradual and last longer than the effects of pesticide poisoning. Plant poisoning usually occurs in the same geographical area at the same time each year, whereas pesticide poisoning is indiscriminate. For a good review of poisonous plants, Barker (1978). Some examples of plant poisoning are listed below and in table 90.

Purple Brood Purple brood occurs when adult bees collect and use the pollen and nectar from *Cyrilla racemiflora* (titi, southern leatherwood). This "disease" is characterized by the blue or purple color of the affected larvae. **Paralysis** *Aesculus californica* (California buckeye) is probably the best known of the poisonous plants. Field bees exhibit symptoms similar to those of chronic bee paralysis; i.e., the bees are black and shiny from loss of hair and they tremble. Also, either the eggs do not hatch or the larvae die soon after hatching.

Milkweed Pollinia The pollen of milkweed (*Asciepias* species) is produced in pollinia (coherent pollen grains) that are attached in pairs by a slender filament. When removed from a flower, the pollinia resemble a wishbone with pollen masses hanging from the ends. Honey bees become ensnared in the thin pollinia attachment and free themselves by pulling the pollinia from the flower. Honey bees often become seriously encumbered and unable to effectively fly or crawl because of the structures that remain attached to their body parts.

Table 90. Comparative symptoms in honey bees poisoned by toxic chemicals and selected plants

Source of poison	Stages most affected	Effect on adult	Effect on brood	Effect on colony
Toxic Chemicals	Adult	Field bees die in or near hive. Nurse bees may also die. Queens usually not affected.	Usually few larvae killed. symptoms of starvation may be evident if adult population reduced severely.	Weakened or killed. Many dead bees near hive.
California buckeye (*Aesculus californica*)	Young brood	Emerging young workers often defor-med, pale, Some hairless and tremble. Queens lay eggs at reduced rate, cease, or become drone layers.	Eggs normal at outset; later fail to hatch or all are drone eggs. Larvae die soon after hatching and disappear. Little or no capped brood; if present, scattered.	Weakened or killed. May be many dead bees near entrance. Supersedure of queen may fail.
Yellow jessamine (*Gelsemium sempervirens*)	Larva, pupa, and young adult	Young workers affected and soon die. Old adults appear normal.	Pupae die in cells and become mummified.	Slightly to severely weakened.
Loco plants (*Astragalus* spp.)	Adult and pupa	Field bees die. Some become black and tremble. Queen may die.	Many cells contain dried pupae.	Population dwin-dles. Colony may die.
False hellebore (*Veratrum californicum*)	Adult	Many field bees die between plants and hive. Adults die in curled state. Queens not affected.	No effect	Field population lost.
Southern leath-erwood (*Cyrilla racemiflora*)	Larva	No effect.	Many blue or purple larvae die in cells when nearly mature.	Slight to severe weakening.

Source: Modified from Burnside and Vansell (1936)

Adey et al. (1986) has given the following guide to minimize bee toxicity

Quick guide to minimize bee toxicity

↓

Has the infestation reached the economic injury level?

↓

Is there any effective –non chemical method for pest control?

↓

Chose the pesticide carefully

↓

Read the label in detail

↓

Are any plants in bloom in /around the crop to be treated?

↓ Yes ↓ No

Are bees foraging in /around the crop to treated

↓	↓
Yes	No
↓	↓
Can the bees be moved prior to treatment?	Treat
↓	↓
Yes	No

↓

Move bees, treat use non persistent pesticides

Apply when bees are not foraging

Chapter 26

Biotechnological potential of honeybees

India has long been well known for its glorious tradition of beekeeping and honey hunting, rich and diversified bee flora and fauna and suitable climate for beekeeping almost throughout the year. India is one of the leading countries of Asia in beekeeping research and training. Of the various problems and researchable issues, bee health is one of the major constraints in the development of beekeeping in India. The traditional methods of disease control have failed and use of new drugs/chemicals have resulted in the development of resistance in disease causing organisms and contamination of bee products. Evidently, there is a need for some innovative or biotech interventions to produce resistant and productive strains of bees. Since *Varroa destructor* has now been discovered in India (Abrol *et al.,* 2006) and it is too late to eradicate it. Should these pests establish themselves in India, agri-horti crops and beekeeping industry would only remain a distant dream. Efforts are therefore needed to tackle the menace of *Varroa* depredations on war footing to evolve management strategy for the control of this species and also to evolve strains which are tolerant to mite attack. *Varroa* was originally a parasite of the Asian honey bee (*Apis cerana*), which has evolved defensive mechanisms against this mite. However at some stage during the last century, *Varroa* 'jumped' to a new host species, the European honey bee (*Apis mellifera*) which have no defensive mechanisms against *Varroa*, and *Varroa* normally causes the death of European honey bee colonies it infests. Evidently, there is head of biotechnological interventions which include:

1. Recent advances in genetics and related disciplines has made it possible to move from traditional methods of controlling genetic changes in organisms to such as controlled breeding programme and selection, to the ability to directly manipulate the DNA, the genetic code itself. One major focus is on the area of biotechnology known as recombinant technology, locating or assembling specific genes and placing them in chromosomes of an organism. Methods of inserting genes into chromosomes are presently the major block to the

application of technology developed with *Drosophila* to other insects. Recent attempts are being made to apply this technology to honeybees and are limited to *Apis mellifera* only.

2. Biotechnology related research with honeybees is being conducted in USDA. The goal of this research is to develop the technology to locate, manipulate and transfer genes in honeybees. The first gene to transfer will be one which can be identified easily so that verification of a successful transfer is simplified. Once a reliable transfer method is developed other genes that will result in an improved stock can be transformed.

3. One class of genes that is being evaluated as the most promising for honeybees are the genes coding for proteins known as cecropins. These genes are from insects and produce proteins that have very strong bactericidal and fungicidal effects, even at low concentrat-ions. Thus, by transferring a single gene into honeybees, resistance to American foul brood (AFB), European foul brood (EFB), and chalkbrood might all be obtained simultaneously. It is possible that honeybees contain cecropins but in some cases they are either induced or turned on late when the larva is already infected and fail to provide acceptable level of resistance. Therefore, by examining honeybee DNA to determine the sequence of cecropin proteins it might be possible to modify the expression of this gene in honeybee larvae so that the cecropin protein is present to respond to the challenge of various diseases equally in *Apis mellifera* and *Apis cerana.*

4. Another area of biotechnology related research which is emerging is that of resistance to mites. *A. cerana* is the natural host of *Varroa jacobsoni,* but in some ways it is resistant to the effects of this mite to keep the levels very low. As an alternative to chemical control methods it would be desirable to have the same résistance in *A. mellifera* as is present in *A. cerana.* Since these two species do not hybridize the use of recombinant – DNA methods is the only solution to transfer this resistance. However, in order to make this transfer, the mechanism of resistance must be identified, its genetic bases determined, the specific controlling genes located, the desired genes removed, and thus gene transferred.

5. The genetic diversity of *A. mellifera* is well established. In case of *Apis cerana,* only three species have been recognized from India. All these subspecies seem to have different geographical populations/ sub-groups or ecotypes (Verma *et al.,* 1992; Abrol and Khan, 2002) which can be used for improvement through selective breeding along with suitable management practices.

6. Similarly, in *Apis cerana,* the transformation for the cecropin gene is an obvious choice. An important problem for, *A. cerana* is a virus

> disease, the so called Thai sacbrood. A disease caused by virus would not be affected by the cecropin protein. Identification of a gene in *A. mellifera* which conferred resistance to this sac brood virus would be important to *A. cerana* and would only be available through recombinant DNA technology. *Apis cerana* has significant potential to be improved by recombinant DNA technology. As it is resistant to diseases and parasites and has spectacular potential for genetic improvement to increase economic production.

India is blessed with atleast three native species of honey bees namely, *Apis dorsata, A. florea and A. cerana* and one exotic species i.e. *A. mellifera.* A stingless bee, *Trigona iridipennis* also occurs throughout India. *A. dorsata* is the most dominant and highly efficient pollinator of several cross pollinated crops. However, it is difficult to manage its bee colony because of ferocious nature and migratory behaviour. *A. florea* though is mild but its honey yielding capability is very low and it is also migratory. *A. cerana* can yield considerable quantity of honey, but it is highly erratic in its behaviour and absconds often. It is highly susceptible to wax moth attack. Thai sac brood virus disease has become a serious problem and caused havoc in the entire South India since early nineties. Though *A. mellifera* is highly productive, is highly susceptible to foul and pests like mites, birds and wasps.

Behavioural traits, honey yielding ability, temper and pollinating efficiency vary greatly among honey bee species. Interestingly these traits also vary among the colonies within the same species. Recently three cavity nesting honeybees closely resembling *A. cerana* viz., *A. nuluensis, A. nigrocincta* and *A. koschevnikovi* have been reported from Philippine and Indonesian Islands. Tribal people in some parts of Tamil Nadu are familiar with the existence of red honey bees (*A. koschevnikovi*?). Such reports are not yet scientifically confirmed and documented. From *dorsata* group, *A. dorsata binghami, A. dorsata. breviligula* and *A. laboriosa* have been described. *A. andreniformis* which resembles *A. florea* is known from Sri Lanka, China, Philippines and Indonesia. Occurrence of *A. andreniformis* was reported from North East India but not from South India From *mellifera* group; as many as 24 subspecies have been described.

In literature, *A. cerana* is reported having four subspecies with discontinuous distribution which indicates many lacunae in our knowledge about this species. Do these inter-specific and intra-specific variations indicate the existence of more than three species of honey bees and more subspecies/ecotypes within each species in India? This question has remained unanswered even today mainly due to lack of systematic and holistic studies on the diversity of the fauna of honey bees in India. Because of these limitations in our knowledge we have not been able to enhance the honey yield and pollinating efficiency of these honey bee species. Recently molecular studies of *A. cerana* also have been initiated in China, Japan, Philippine and other Asian countries. But no such studies have been made so far in India.

Evidently, there is a need for biotech interventions for bridging research gaps existing in the field of bee science using both conventional and biotechnological tools to document the biodiversity of honey bees in India. The studies can help to understand clearly about the morphological and genetic diversity of Indian honey bees with which the present stock of the bees can be improved to increase their honey yield as well as their pollinating efficiency in future

The Indian honey bee species comprising *Apis dorsata, A. florea* and *A. cerana* are likely to be a complex mixture of sibling species and ecotypes among themselves. A thorough understanding of morphometric, inter- and intra-colony genetic diversity (microsatellites) and phylogeny (mitochondrial genes) is expected to pin down the species composition and geographic mapping of Indian honey bees. This exercise expectedly can aid in further breeding disease resistance, rearing for better honey yields and pollination leading to better crop productivity. The key questions include: (1) Whether *Apis dorsata, A. florea* and *A. cerana* occurring in India are monotypic or polytypic species with sibling species, subspecies and ecotypes? (2) What is the spread of morphometrical, anatomical and genetic variation in Indian honey bee species? (3) What are the inter and intra colony genetic relatedness in Indian honey bee species. (4) How to distinguish the subspecies and ecotypes of each honey bee species? (5) What are their desirable biological traits and how are they distributed? (6) Where are the voucher reference collections of honey bees available for further studies?

Chapter 27

Honeybees as bioindicators of ecosystem health

Honeybees as bio-indicators

Honeybees are a good biological indicators because they indicate the chemical impairment of the environment they live in through two signals residues present in their bodies or in the bee hive products (in case of pesticides) and other contaminants like heavy metals and radionuclides) that may be detected by means of suitable laboratory analyses.

Several ecological and morphological factors make the honey bee a reliable ecological detector. It is an easy to breed, almost ubiquitous organism, with modest food requirements; its body is covered with hairs which make it particularly suitable to hold the materials and substances it comes into contact with it is highly sensitive to most plant protection products, revealing when they are improperly spread through the environment (e.g. during flowering during wind etc.); it has very high rate of reproduction and relatively short average life span, causes the colony to undergo rapid, continuous regeneration; its great mobility and flying range allows a vast area to be monitored; it is efficient sampler of the environment (numerous inspections per day). Furthermore, almost all environmental sectors such as soil, vegetation, water, air are sampled by honeybees.

A variety of materials are brought by honeybees into the hive through foraging (nectar, pollen, honeydew, propolis and water) and stored. Porrini *et al.* (2003) at the university of Bolgna has been studying the use of honeybees as bioindicators to detect pesticides, heavy metals, and radionuclides in many areas.

Monitoring of pesticides

Honeybees are highly sensitive to pesticides. The number of dead bees in front of the hive is the most important variable to be considered for these contaminants. Many bees directly stuck by the insecticides will not have the enough strength to return and die in the field or during their return flight. In case of dangerous compounds, the insect act as an indirect indicator i.e. not sensitive but exposed, and will provide us with the information in the form of residues. Other bees marginally hit gathering nectar pollen will eventually die in front of the hive.

Heavy metals present in the atmosphere can deposit on the hairy bodies of bees and can be brought back to the hive with pollen, or they may be absorbed together with the nectar of flowers or through the water or honeydew. The lead in the urban environment was found in higher quantities inside the bee than on the bees surface.

Monitoring of pesticides with honey bees, is an extremely important technique not only for proving potential bee poisoning risks by the use of pesticides, but also for determining the degree of environmental contamination due to plant protection products. In fact, because of its morphological and ethological features (such as its wide area of patrol and its intense foraging activity), the honey bee can be considered an excellent bioindicator. In many cases, pollution caused by abuse or by erroneous application of pesticides could not be proven without the help of honey bees (Porrini *et al.*, 2003).

Monitoring techniques, chemical and palynological analyses, and data processing enable us to characterize areas, to indicate periods of major bee poisoning risk, and to identify the most frequently used pesticides and the crops treated. The studies with honey bees reveal the type of plant protection management applied to the area under investigation and allow us to prove the application of molecules not permitted under certain circumstances or even forbidden. Between 1983 and 1986 in particular, the analysis of 581 dead bee samples revealed which compounds were most widely used in that period in cultivated fields, above all in northern Italy (Celli *et al.*, 1988a, 1988b).

In those years, bee poisoning incidents were caused primarily by treatments carried out in orchards, vineyards, on seed crops and by pollution due to drift on spontaneous plants. The most frequently found pesticides were phosphorganics (dimethoate, parathion, azinphos- methyl, methyl-parathion, diomethoate and metamidophos), followed by carbamates (carbaryl) and clororganics (endosulfan). Dithiocarbamates were almost always found in the dead bees samples, together with other products, insecticides for the most, which were the ones to blame for the bees' death. These include dimethoate, a compound serving a huge variety of purposes but often misused. Pollution by pesticides can be detected promptly and continuously at low costs by using bees as bioindicators, since they are able to signal immediately and unambiguously the incorrect use of pesticides. Anyway, some of the new insecticides do not induce high bee mortality but, even at low doses, they may cause severe behavioural changes, which could damage the entire family. They are also difficult to detect with chemical analysis and thus our tools (mortality and residues) are not always able to reveal them.

Brood and adult bees are directly or indirectly (through the process of royal jelly secretion by hypopharyngeal glands) fed with bee bread and can be exposed to pesticide residues for various times. Many research teams have worked on quantifying the amount of pollen needed to rear a bee larva. It mainly depends on season, pollen species, and conditions under which the larvae are reared. For example, it has been shown that worker honey bee larvae needed 86 mg of maize pollen for their complete development (Babendreier *et al.*, 2004).

Pollen supply also is involved in bee resistance to pesticide exposure. Amount and quality of pollen ingested in the first days of life affected the pesticide sensitivity of young and older bees (Wahl and Ulm, 1983). The most frequent residues (i.e., imidacloprid, 6-chloronicotinic acid, and fipronil) were searched with very low LODs compared with other pesticides.

The main question to be addressed remains, Are the doses found in pollen dangerous for bees? Let us consider this question for imidacloprid.We would like to know how much contaminated pollen an adult bee should eat to reach the LD_{50} quantity. If we rely on the average content, we have found in this study (1.2 g/kg) that the consumption of 33 g of pollen by one individual would be needed to meet the oral LD_{50} (0.04 g per bee). A certain amount of time would be needed for one worker to eat this quantity of pollen, which then would be stocked in cells in the form of bee bread. Our lack of knowledge on active ingredient fate and interactions between molecules stocked into hives remains unexplored.

Moreover, it has been shown that LD_{50} values would not be sufficient to assess the adverse effects of a pesticide. Very small quantities of active ingredients can lead to subtle effects at various levels of bee physiology and behavior. These effects are more difficult to detect, but they also may affect bee populations.

Contamination by pesticides ranged from 50%. Coumaphos and tau-fluvalinate residues were the most concentrated of all residues (mean concentrations were 925.0 and 487.2 g/kg, respectively). Fipronil and metabolite contents were superior to the limit of detection in 16 samples. Residues of Fipronil were found in 10 samples. Nine samples contained the sulfone compound, and three samples contained the desulphonyl compound. Residues of imidacloprid and 6-chloronicotinic acid were found in 69% of samples. Imidacloprid contents were quantified in 11 samples with values ranging from 1.1 to 5.7 g/kg. 6-Chloronicotinic acid content was superior to the limit of quantification in 28 samples with values ranging from 0.6 to 9.3 g/kg.

The use of honey bee *Apis mellifera* (Hymenoptera : Apidae), as a tool for monitoring environmental pollution is founded upon several ethological features such as high rate of reproduction, great mobility, large flying range, and numerous flower inspections per day. It is also founded on morphological characteristics: the honey bee body is covered with hairs that collect various particles and increase by this means the contact of the insect with its environment (Porrini *et al.,* 2002). Apicultural matrix analysis such as honey, wax, bees themselves, or pollen can provide useful indications of the diffusion of pesticides within the environment.

Because honey is a product of human consumption, its contamination has been studied in many countries such as Portugal and Spain (Blasco *et al.,* 2003), Germany (Wallner 1999), Belgium (De Greef *et al.,* 1994), Greece (Thrasyvoulou and Pappas, 1988, Balayannis and Santas, 1992), and Switzerland (Bogdanov *et al.,* 1999, Balayannis, 2001).

Wax and honey bees also have been subjected to various analyses to detect different types of contamination. Chemical contamination through *Varroa* treatments (*Varroa destructor* Anderson & Trueman [Acari: Mesostigmata]) has been found to be a route of wax contamination by coumaphos (Tremolada *et al.* 2004), bromopropylate (Hansen and Petersen, 1988), and fluvalinate (Tsigouri *et al.*, 2004). In addition, numerous studies have reported the use of honey bees to monitor environmental radionuclide contamination (Haarmann, 1997, Barisic *et al.*, 2002) or heavy metal contamination (Porrini *et al.*, 2002). Less commonly, honey bees have been used as bioindicators to detect the presence of phytopathogenic microorganism in the environment (Porrini *et al.*, 2002). There are three main purposes for monitoring bee products: consumer health protection, international commercial competition, and better product quality (Yakobson, 1996). Although pollen loads are a product for human consumption, they currently are subject to no legislation, leading to very limited data. For experimental purposes, pollen loads collected from traps set on honey bee colonies have occasionally been used to monitor environmental pollution: radioactive contamination after the Chernobyl accident (Barisic *et al.*, 1994) or new insecticide uses for particular crops, such as sweet corn (Erickson *et al.*, 1997) or oilseed rape (Kevan *et al.*, 1984, Fries and Wibran, 1987). Pollen also has often been analyzed for the presence of pesticide residues in bee mortalities (Johansen and Brown, 1972, Waller *et al.*, 1984, Kubik *et al.*, 1999). Samples were either collected with the aid of pollen traps or more often by hand, directly from flowers. To assess the contamination resulting from pesticide uses in agriculture and their consequences on beehive health, a field survey was initiated in France in 2002. This investigation was aimed to evaluate the impact of the bee weakness phenomenon in French apiaries.

The new agriculture which we so much hope for, will have to strive for that the application of these molecules in our cultivated fields will decrease and, for this purpose, the honey bee can be of great help. In any case, let us remember that the honey bee is killed by pesticides when they are incorrectly diffused in the environment (cultivated fields or private gardens) either in a qualitatively or quantitatively wrong way, that is by not applying the recommended dosage or by not obeying the accepted technical norms of use (culling of spontaneous plants, absence of wind, absence of honeydew).

Monitoring of radionuclide

Tonelli *et al.* (1990) analyzed the samples of honeybees,wax and pollen after the Chernobyl incident and demonstrated that pollen was the most efficient indicator of atmosphere radionuclide contamination. On a typical day, bees from a hive containing about 10,000 bees make more than 200,000 trips a day.

Honeybees and Landmines

People and bees have a long and mutually beneficial history. Ancient cave paintings in Spain depict a woman harvesting honey. The Egyptians moved bees on barges up and down the Nile originating near current-day Afghanistan. One species of honey bee, *Apis mellifera,* now lives all over the world, with the exception of the Antarctic and far Arctic regions. In every community and country, bees are kept for the honey and wax that they produce, and for the crops that they pollinate.

Honeybees are widely recognized as our most beneficial insects. Not only do they provide us honey and wax but they also serve as pollinators for many important crops. Now these industrious bees are being trained to serve us in still another quite remarkable way. Bees aren't busy just for the sake of being busy. The constant foraging for pollen by honeybees enables the hive to exist. This industriousness also has been proposed as a means to locate land mines. Trained bees can act as biological detectors of these explosive devices, and lidar is being investigated as a way to pinpoint the location of the insects as they momentarily hover over and thereby mark a land mine.

One of the saddest leftovers from warfare is the wide distribution of land mines. Worldwide, thousands of innocent civilians are killed each year by these mines; still more are crippled. In Cambodia alone today there are an estimated 35,000 amputees who were injured by mines, many of them innocent children. The mines are also a continuing daily threat in Afghanistan, Angola, Bosnia, Chechnya, Croatia, Iraq, Mozambique, Nicaragua, Somalia, and dozens of other countries. Morethan 90 countries, unfortunately including the United States, have manufactured and distributed 200 million antipersonnel landmines in the last 25 years.

Land mines are inexpensive weapons, but their cost to the world community is much higher. As many as 20,000 people a year are killed or injured by land mines. Mines can render roads, farms and other land useless. They sometimes remain active decades after a military conflict has ended. The United Nations estimates there are more than 100 million active mines and new mines are buried more quickly than old ones are safely unearthed.

Identifying and clearing minefields is itself a life-threatening activity and a wide range of techniques is being implemented. Although humans with metal detectors remain a common method, huge machines, chain mats dragged by helicopters and trained dogs have also been employed. And here is where the bees come in. They don't have noses, but they do have a strong sense of smell. A group of entomologists at the University of Montana led by Jerry Bromenshenk have spent several years developing techniques to employ honeybees in landmine and biological weapon detection.

According to Bromenshenk, "A honeybee's body has branched hairs that develop a static electricity charge, making it an extremely effective collector of chemical and biological particles, including pollutants, biological warfare agents and explosives. They also inhale large quantities of air and bring back water for evaporative cooling of the hive." Thus an individual hive has tens of thousands of foragers out sampling local air, soil, water and vegetation. Examination by the scientific team of a number of returning bees provides initial information about areas where materials of concern are to be found and appropriate relocation of hives can further zero in to areas of a few hundred meters. But even this detection is now being refined.

The scientists have devised methods to train the bees by the same kind of reward techniques (formally called operant conditioning) that are employed by dog trainers. The reward provided the bee is food which is associated with the particular target being sought -- the smell of the chemicals in land mine explosives in this instance.

"All land mines leak explosives into the environment. If bees go into that area they'll then collect the explosive signatures that are coming off the mines, they'll bring it back to the hive and we can detect it using various chemical sensors." Bees bring pollen back to the hive on their back legs after a morning of foraging. Testing of pollen, bees and honey can help detection of land mines. Samples of the plants can also reveal how much explosive is actually coming out of the mine and getting into the environment.

Limitations and Agricultural Benefits

Bees do not fly at night, during heavy rain or wind, or when temperatures drop to near or below freezing. As such, the use of bees is seasonal in temperate climates. Bees are active year-round in tropical regions. These limitations are not unique. Dogs do not perform well in wind, rain or frozen ground, and dogs and handlers usually do not work in the dark. Unlike a dog, bees do not need to bond to their handlers, and they work whenever weather conditions permit. At a weight of one-tenth of a gram, bees are not going to trip wires or set any mines off. Their wide foraging range offers the possibly of greatly speeding up survey times, while also increasing handler safety. Bee colonies can be established along the perimeter of the minefield, not in the minefield. With no leash to hold, the beekeepers can stay well clear of the dangerous area.

Future

We do not expect bees to replace dogs. Rather, we anticipate that bees, in passive and active modes, could be used in concert with dogs and other methods to reduce the time and expense of area reduction and landmine field surveys. Like dogs, bees are able to recognize multiple substances concurrently at very low concentrations. As with any vapor sensing system, bees cannot find a mine that is not leaking. We also see the need for additional research to define the performance of mine-detecting bees, taking into consideration environmental factors that influence the amount of chemical signatures.

Chapter 28

Beekeeping and subsidary industries

A subsidary or an ancillary unit is defined as a unit which produces parts, components, sub-assemblies, and tooling for supply against known or anticipated demand of one or more large units manufacturing/assembling complete product and which is not a subsidiary to or controlled by any large unit in regard to the negotiation of contracts for supply of its goods to any large unit. This shall not, however, preclude an ancillary unit from entering into an agreement with a large unit giving it the first option to take formers output

INDUSTRIES NECESSARY FOR APICULTURE

Supply industries

There are a number of specialist industries which operate solely (or predominantly) to supply requisites to the beekeeping industry. While not quantified as a part of the beekeeping industry, their existence is certainly a part of the indirect economic impact generated by the industry. This "supply" sector includes:

(i) **Beehive manufacturers** - while only small in number, companies manufacturing hives for the industry tend to specialize in this activity.

(ii) **Extractor/uncapping machinery** - any company skilled in the production of stainless steel equipment for the food industry would be capable of manufacturing the machinery required to "uncap" combs, and extract honey.

(iii) **Packers' equipment** - bottles and bottling equipment are required by packers. In addition, drum manufacturers provide the special galvanized drum, made with side bung that is not generally available from other sources for producers and packers. A gradual change to plastic drums is taking place, with intermediate bulk containers, of pallet size being used which hold more honey than do the traditional drums.

(iv) **Heat source** - every honey producer requires a steam or hot water boiler to generate steam, and hot water, for processing honey and wax. While not requiring a specialized manufacturing activity, the beekeeping industry generates a demand for such equipment.

(v) **Transport/handling equipment** - all commercial beekeepers must purchase trucks and utilities for transporting and servicing hives. Many also own front end loaders for loading hives, on pallets, and loading drums or use other forms of mechanisation.

(vi) **Other equipment** - beekeepers also have a need for other equipment such as electric generators and mobile extracting units. While the first two items (as for heat sources) do not require a specialized manufacturing activity, expenditure by the beekeeping industry can provide an important source of demand.

(vii) **Quality Assurance** - beekeepers and packers are increasingly introducing quality assured premises and equipment for handling honey, as a food product for human consumption. Again, this is not necessarily a specialised activity unique to the industry, but represents an additional demand for existing services.

It is clear that a considerable amount of other economic activity is generated as a result of the activities of the honey bee industry.

INDUSTRIES DEPENDENT ON APICULTURE

Major products of beekeeping are: honey, beeswax. In addition, specialized segments of the industry concentrate beekeeping activities towards the production of: queen bees, package bees - the provision of specialist paid honey bee pollination services for horticultural and agricultural industries. Queen bees and package bees are sold to other beekeepers both within the and overseas. Honey and beeswax are still produced by queen and package bee producers, but these products are not the prime goal of their beekeeping activities. Other minor products from the industry include: - pollen, royal jelly, propolis, bee venom. The economic dimensions of the industry, and the direct and indirect impacts that beekeeping are as given below:

Industry by sector

The overall apiary industry should be considered in terms of a number of sectors, as follows:

(i) Honey

Honey is the prime output of commercial beekeepers, and is produced by bees from plant nectar. The major producers are Russia, China, USA, Mexico, Argentina, Canada, Brazil and Australia. The major exporters are China, Mexico and Argentina, but the highest colony yields are recorded in Australia and Canada which have a favourable environment as well as highly developed colony management. The major consumers and importers are the industrialized countries led by Germany, Japan, USA and UK. The increased consumption over the last few years can be attributed to the general increase in living standards and a higher interest in natural and health products. Western Europe as a whole imported approximately 140,000 tonnes of honey which is about 55% of consumption. The average EU per capita consumption of 600 g per year varies widely amongst individual nations, from Greece with 300 g per capita to Germany with 1,800 g per capita.

In general, light-coloured honeys bring the highest price and dark ones are most frequently used for industrial production. Mild flavoured honeys are preferred, but characteristically flavoured honeys bring top prices in some countries. Large honey packers usually prefer honeys with a low tendency to crystallize. Some unifloral honeys such as Hungarian Black Locust honey bring twice the price of regular, multifloral honey. Small shipments into Switzerland of unifloral honeys such as lavender honey, in most cases already bottled, bring much higher prices. Local prices in most developing countries are higher than the international market prices and prices in neighbouring countries with less honey production or favourable exchange rates may sometimes be quite attractive.

Expansion of markets with honey-containing products should be considered on a national level or for across-the-border trade. Consumer education and of course, spending power will probably be the most important factors influencing the possibility of expanding local markets or for increased product diversity. The examples given in this chapter might serve as ideas for possible modification and adaptation to individual circumstances.

(ii) Beeswax

Beeswax is a substance secreted by the worker bees. It is recovered by beekeepers primarily from honey comb cappings, and also from cull combs and wax pieces. Beeswax is used in certain pharmaceutical and cosmetic preparations, as a base for polishes and some ointments, for candles - and for comb foundation for beekeeping. It has the highest melting point of natural waxes, and can be sold in either the raw or refined form. Commercial beeswax is generally refined for sale by a manufacturer of apiary products.

The cosmetics and pharmaceutical industries have no complete substitute for beeswax. At least small quantities will always be needed to maintain quality and specific characteristics. Like honey prices, prices for beeswax may vary considerably from place to place.

Markets and prices for products made from beeswax vary widely from country to country. In these industries, beeswax forms only a minuscule part both of the manufacturing process and of the final product. It is used in Candle making, Skin Creams, **Grafting wax for horticulture**, Polishes and varnishes, Paste furniture polish, Liquid furniture polish, Spray polish, Floor Polish, Shoe polish, cream type. Cravons, Leather preserves, Waterproofing textiles and paper, Paint, Wood preservative, Swarm lure, Veterinary wound cream, Adhesive lotions.

(iii) Live bees

The production of queen bees, and of entire colonies of bees, is the main diversification available to beekeepers. The queen bee industry is dependent on the existence of a profitable honey industry and on an export market to buy queens at a period when little or no sales. For example, In Australia the demand for queen bees is estimated at around 155,670 per annum - at an average price of $9/queen, this represents a farm-gate value of around $1.5 million. This is a conservative figure because export sales - estimated by industry sources to be $0.75 million - are not recorded separately and have not been added. Live bee exports is a potential growth area for the beekeeping industry, as further markets develop.

Package bees and nucleus colonies are other forms of live bee production, and are sold both within the country and overseas. Again, data on total value of production for this sector of the industry is not available, and has been estimated on the basis of known production. The total value of this sector has been assumed to be $2.25 million, which is almost certainly an under-estimate, but which has been used as a conservative minimum. In India where great potential of beekeeping exists marketing for live bees can be much more.

(iv) Other products

In addition to honey and wax, active beehives are also a source of other products. These include:

Royal Jelly - a milky white smooth jelly secreted by nurse bees, used to feed developing queen larvae and young worker bee larvae. The production of royal jelly is a very specialized procedure, and floral conditions must be ideal before production can be considered. Royal jelly is used as tablets, or mixed into creams and shampoos. Royal jelly can be sold in its fresh state, unprocessed except for being frozen or cooled, mixed with other products, or freeze-dried for further use in other preparations. The fresh production and sale can be handled by enterprises of all sizes since no special technology is required. In its unprocessed form it can also be included directly in many food and dietary supplements as well as medicine-like products or cosmetics. For larger industrial scale use, royal jelly is preferred in its freeze-dried form, because of easier handling and storing. Freeze-dried royal jelly can be included in the same products as the fresh form. The production of freeze-dried royal jelly requires an investment of at least US$ 10,000 for a freeze-dryer, sufficient production volume and an accessible market for the raw material or its value added products. Products containing royal jelly should be specially marked or packaged in order to distinguish them from similar products without it.

As dietary supplement

Royal jelly belongs to a group of products generically described as "dietary supplements" These are products which are consumed not for their caloric content nor for pleasure, but to supplement the normal diet with substances in which it might be lacking.

As ingredient in food products

A mixture of royal jelly in honey (1-3% royal jelly) is probably the most common way in which royal jelly is used as a food ingredient. Among the advantages of this product are that no special technology is required and the honey masks any visible changes in the royal jelly. The final product is pleasant-tasting and it provides the beneficial effects of both products.

As ingredient in medicine-like products

In medicine-like formulations royal jelly is generally included for its stimulatory effects. However, it is also used to solve specific health problems. A variety of formulations are available, often containing ingredients otherwise used to alleviate particular afflictions. or as medicine.

As ingredient in cosmetics

Except in Asia, probably the largest use of royal jelly is in cosmetics. Royal jelly is included in many dermatological preparations, but mostly in those used for skin refreshing, and skin regeneration or rejuvenation. It is also used in creams or cintments for healing burns and other wounds.

Others

The only other known uses for royal jelly are in animal nutrition. In particular, royal jelly has occasionally been used (fresh or freeze-dried) to stimulate race horses. For experimental purposes it is also used as a food for rearing mites and insects.

Royal jelly collection

Royal jelly is produced by stimulating colonies to produce queen bees outside the conditions in which they would naturally do so (swarming and queen replacement). It requires very little investment but is only possible with movable comb hives.

Propolis - a by-product of the bee hive. It originates as a gum secretion gathered by bees from a variety of plants, and can vary in colour depending on the plant species of origin. Propolis has remarkable therapeutic qualities, and is much sought after in some countries for the treatment of a range of human ailments, and for cosmetic purposes. It is used by honey bees as an antiseptic to varnish the interior of honey comb cells used by the hive to rear young brood, to seal cracks in the hive from the winter chill, and for general hive cleanliness purposes. The market for raw material and secondary products containing propolis will probably continue to grow as they find more acceptance in medicinal uses and as more cosmetic manufacturers realize their benefits and marketing value.

Bee venom: Collected by stimulating bees with a mild electric current. The venom is processed, and used in the preparation of pharmaceutical materials. It can be used to detect hypersensitivity or allergic reaction to bee stings. Bee venom is a highly specialized product with only very few buyers. The market volume is relatively small too, although there are no comprehensive surveys. The main venom producer is the USA which has produced only about 3 kg of dry venom during the last 30 years (Mraz, 1982) but there is a large producer in Brazil and more or less significant amounts are produced in many other countries. Prices in 1990 varied greatly between US$100 and US$200 per gram of dry venom (Schmidt and Buchmann, 1992). Prepared for injections or sold in smaller quantities, prices can be much higher. However, the beekeeper often does not get this price. The prevailing prices in European and Asian markets are generally slightly lower.

Pollen: Pollen can also be harvested by beekeepers, at a rate of around 7-10 kgs per hive per year. Pollen is used by bee colonies as a source of protein, but harvesting pollen by the beekeeper requires detailed knowledge of resources, hive management, species flowering variations and timing, and hive response to different honeys and pollens. Pollen is collected via specialized traps fitted to the hives, and must be processed rapidly after collection (usually via freezing or drying) to avoid excessive moisture absorption and fermentation. Many beekeepers harvest pollen to feed back to their hives during periods of natural pollen deficiency. Dried pollen prices in the USA range from US$5 to 13 per kg wholesale and US$11 30 per kg retail (*American Bee Journal*, 1993). Encapsulated pollen or pollen tablets sell vials of 50 to 100 units and retail at prices of up to US$900/kg, at least in Italy and the bulk pollen consumer market seems to be growing in industrialized countries, but pollen tablets are still a common feature of health food stores and command an excessively high price. Encapsulation and extraction of pollen lend themselves easily to small scale manufacturing and result in safer consumer products.

Most of the buyers and large scale sellers of pollen are also honey traders. Crane (1990) however, reports that a lot of commercial pollen is not bee collected, but machine-collected from certain wind pollinated plants which release very large quantities of dry pollen. At least in industrialized countries and those with increasing numbers of health conscious consumers, pollen consumption is likely to increase further. On the other hand, there seems to be a wide market for reasonably priced, encapsulated pollen and tablets.

Paid pollination services

Some beekeepers receive payment for placing hives in close proximity to flowering crops, according to contractual arrangements with farmers. For example, rates for pollination services in inland Australia varied between $25 and $35 per hive in 1996, with variations between crops. It has been estimated that at least \$2.9 million is received by the industry in this way, based on total payments received for pollination services in Tasmania (Gifford, 1989) and multiplied up to an Australian figure by numbers of hives.

Similar concept is picking up throughout the world including India. In Himachal Pradesh, India this practice has already been started and is likely to be followed in other states as the awareness about pollination benefits is realized by the farming community.

Evidently, to ensure the country's self-sufficiency in foodstuffs, to receive foreign currency from excess production, the stabilization of rural populations by complementary activities of both a financially rewarding and environmental nature, and there is no doubt that beekeeping fits perfectly within this framework and hence, efforts are required to popularize and increase beekeeping still an enormous potential waiting to be tapped.

Chapter 29

Handling, Processing, Storage and marketing of Honey

Handling, Processing and Storage of honey

Floral and extra-floral nectar is the raw material for honey production by bees, whereas honeydew is not available in sufficient quantity to serve as a raw material for honey production in India. Honey is the purest when stored in combs by bees is to bring to the table the same purest honey. But honey in combs is to be extracted, by the man himself, though some are unavoidable. Even during the process of gathering and production by bees, the beekeepers while use superior combs does some harm. The Indian consumers are not absolutely conscious about the delicate flavours and aroma but during handling and processing, it should be our aim to preserve them. Granulation (Crystallization) is the natural property of honey and all honeys sooner or later granulate. Uniformly granulated honey is preferred by the honey consumers in most parts of the world but ill informed consumers in Indian brand crystallized honey as adulterated. The nectar contains some pollen grains which get incorporated in it which gives haziness to the honey. In dry climates of Indian plains the air is heavily dust laiden particles of which are also incorporated into the honey during extracting. Uncapping of combs is required for extracting honey and by centrifugal extraction some wax particles are also added into the honey. The major honey producing sates in India are coastal and the areas have higher relative humidity. In such areas the honey contains higher moisture content and handling of honey at high relatives humidity conditions further aggravate the problem. Such honey with higher moisture content are liable to ferment.

Except in few cases i.e. Rubber, Litchi and *Brassica* we in India normally have mixed honeys. These mixed honey will create problems for marketing in other countries and even in India when greater awareness is created amongst the domestic consumers.

The honey produced in India is subject to high temperature in tropical and subtropical climates. Honey has to be handled and stored at higher atmospheric temperatures, therefore, changes in HMF values, enzyme contents, colour of honeys are frequent. Because of the international market standards even the genuine honeys from our country can be branded as bad.

Most of the problems highlighted here above can be reduced, if not overcome by careful handling, minimum processing and proper packing and storage. Consumer are also becoming more quality conscious and the principles from extraction to marketing of honey are briefly discussed in this chapter.

Handling of honey before processing

Producing good quality honey starts in the apiary itself. Beekeepers apply some chemicals in the bee colony for the control of diseases and enemies. Feeding of antibiotics is not a routine practice with Indian beekeepers since foul broods have not caused any concern as yet. In case any such treatments are to be given then feeding antibiotics should be stopped much before the honey flow. Ectroparasitic brood mite, *Tropilaelaps clareae* is a regular problem in *mellifera* colonies. Though formic acid can be sued for the control of the mite yet sulphur dusting is more commonly used. Unsystematic use of the these and other chemicals can lead to adding these to the honey.

Beekeepers, unmindful of the implications are dark brood combs or brood combs with pollen stores in the supers. But this practice causes the honey to contain more pollen content and honey produced in dark brooded combs is darker in colour. Therefore, care should be taken to use only super-honey combs in super chambers for honey gathering. Beekeepers will have to standardize the practice of using combs only for supers and not interchanging them with brood combs.

Beekeepers, in India are not particular about using queen excluders during honeyflow. Without queen excluders in most of the cases especially in A. mellifera the fertile queen goes to the super chamber and lays eggs. Sometimes all the super combs contain brood and sealed honey. The beekeepers is tempted to extract this honey. The unsealed brood combs and the brood body may rupture due to spinning movements of honey extractor under centrifugal force. This obviously means production of less than extracted honey since straining of honey by cloth bag would further render the honey bad.

Bees take 2-3 weeds of storage, ripening and sealing of honey in the comb. Ripenning and sealing can be aided by increasing ventilation by slight shifting or lifting of super chamber. Because of non-adherence to specific standards of honey in the market, beekeepers many a times do extract honey even from unsealed combs. In such cases doubtlessly the honey would contain higher moisture content. The above mentioned problems are under the control of man and careful beekeepers can produce relatively better quality honey.

For extracting honey uncapping of sealed combs can be done with different types of knives. The most commons used uncapping knives are manually dipped in hot water steam heated and electrically heated hand operated knives have found limited use with commercial beekeepers. During uncapping a good proportion of honey is removed along with the cappings. There are various methods such as gravity straining followed by pressing, draining boxes, etc. for getting this honey from the cappings by squeezing through a cloth bag. The honey separated from cappings by any of the method is likely to have greater amount of foreign particles, with air bubbles and heat treatment would render it darker. Therefore, mixing it with main batch of honey is not desirable. This honey should be processed and stored separately.

Moisture content

India has greatly varied climate and honey produced in different regions contain different levels of moisture. Honey produced in north Indian and Indo-Gangetic plains normally contain low moisture i.e. below 19% or so and more normally this is around 17-18%. Such honey pose less problems. But in some areas as in coastal regions the relative humidity is high and honey produced in these areas has high moistsure content; sometimes it goes upto 25-26% but moisture content above 20% is very normal. The honey is highly hygroscopic, therefore, moisture contents in honey can increase or desease even after honey extraction when it is exposed to atmospheric humidity. Equations, as given below, have been worked out in other countries between the honey moisture and relative humidity.

Per cent moisture in honey	15.9	16.8	18.3	20.9	24.2	28.3	33.1
Relative humidity	50	55	60	65	70	75	80

Water contents in honey can be reduced even before extraction. This is possible by blowing dry air at about 35-38°C over the honey combs in honey house/room. The reduction is very fast in unsealed combs but slower in sealed combs. This process is able to reduce moisture by about 0.5% in 12 hours. This hot air treatment increases the Hydroxy methyl furfural values in honey. But handling honey at low humidity conditions after extraction is a practical solutions and is possible in honey extraction rooms. The honey with higher water content can also be exposed in hot/dry air in open containers. It has been found that 1.2% reduction in moisture can be obtained when dried at 32-35°C. The honey with greater moisture content rises to the top of the storage tank and denser honey moves downward to bring tighter the lower layers.

Processing

The thumb rule for honey processing, "Least processing is the best processing". With any kind of processing the man is bound to deteriorate the quality, yet some minimum processing becomes essential and three main steps are given below:

(a) Heat treatment

(b) Straining

(c) Filtration

(a) Heat treatment

(i) To increase flowability: Heat treatment is an essential step before processing to increase flow ability of honey in cooler climates. But experience under Indian climatic conditions show that flowability is not a problem. In other countries with temperate climate this heat treatment is given in double jacketed storage tank or in pipes through which the honey flows from a tank or extractor to the storage tank.

(ii) To kill yeast cells: There are many species of yeast which are present in honey. These are mainly the species of Zygosaccharomyces and Saccharomyces and even the species of the former are not only sugar-tolerant but can also multiply in honey. These yeast are responsible for fermentation of honey if the moisture content is higher or if the honey granulation is not uniform. Honey should not contain more than 18% moisture to avoid fermentation and the maximum limit is 19%. Honey with moisture content above 20% is always in danger of fermentation. Yeasts can be destroyed by rapid heating and cooling of honey. Heating at 63°C for 2 min., 69°C for 1 min. or 71°C instantly can destroy all years.

(iii) Heat Treatment and granulation: There are number of factors which influence granulation but most important is that honey is a supersaturated solution of granulating sugars. Granulated honey is not acceptable to the common consumer. They honey to be sold in liquid form is processed by heating to 60°C for 5 – 7 min. or 71°C for 2-3 min. This dissolves all crystals and remove incorporated air. This heat treatment also destroys yeasts, therefore, granulation is delayed.

(iv) Temperature treatment for reduction of water content: In the processing of honey, heat treatment is also recommended in India, to reduce moisture content. Honey is recommended to be sprinkled on a hot revolving drum before and filtration in a processing plant. Heat treatment is given to achieve the above purposes but heating honey effects flavours and aroma, HMF value, enzymes and colour. These will be discussed alongwith the effect of storage temperature under the relevant section.

(b) Straining

Honey is passed through screens of different mesh to remove foreign particles. The temperature of honey should be around 35°C for straining since this temperature reduces the viscosity and straining is rapid. Straining is done below the surface of honey and straining area should be a large as possible. Heating before straining may not always be required under Indian conditions.

For small scale straining the honey from a container can is made to pass through a pipe of which the flow may be controllable. The temperature of honey is also raised to 35°C when it passes through this pipe. Honey is made to pass through a straining bag. For large scale screens of different meshes, one inside the other, is used. Honey passes from one screen to the next and is drawn off after passing through the finest one. I.S.I (BIS) specifications provides for the use of 180 u strainer in the first of the three settling tanks.

Honey processing plant specified by Bureau of Indian Standards consists of

(i) Plateform: A suitable graded Plateform to place the storage tanks so as to leave a clear vertical distance of about 150 cm between the outlet of storage tank and that of the spiral honey processing unit in order to maintain an adequate rate of inflow into the spiral processing unit.

(ii) Hot water bath

(iii) Processing Unit: This unit has two cylinders, outer and inner. A strip is welded around the outer side of the inner cylinder to give 2 x 1 cm ascending honey channel in between outer and inner cylinders. Inlet pipe is connected to the cock of the storage tank with a plastic tube. In the outlet pipe, provision is made for recording the temperature of outgoing honey from the outlet pipe of the double-walled spiral cylinder does not exceed 71°C and the minimum temperature is not less than 60°C.

(iv) Settling tanks: There are three settling tanks with stopping bottom, interconnected with inlet and out-let pipe. A strainer of 180 u is provided in the first settling tank. Stop cock are provided with the tanks for filling the processed honey in drums or bottles.

(c) Filteration

Honey is sometimes filtered under pressure to remove pollen, colloids and small air bubles which are left after straining from honey. Filtered honey remains liquid for a longer period than unfiltered honey and can be useful when granulated honey is not acceptable to the consumers. Filter aid made of diatomaceous earth is used for filteration in some countries like U.S.A.

Storage

There can be several types of damages that can be caused to honey during storage. Fermentation is the one when water content is higher and yeast cells are present. During storage, if the temperature is higher then HMF values are bound to increase. Colour of honey becomes darker with storage, flavours, aroma and enzymes are reduced. Therefore, appropriate storage aims at preventing these losses.

Shelf life of honey is normally specified and it is two years in India. The most important group of constituents of honey after the monosaccharides and sucrose are the reducing disaccharides (maltose etc.). As a result of enzyme activity their amounts are likely to increase during storage. On storage, delicate flavour of honey are lost and only strong flavours tend to survive but the honey does not go bad.

Presence of minerals is generally responsible for colour. The compounds which give fine delicate pleasing aroma are with boiling point and go very easily.

Components with higher boiling point are responsible for characteristic honey aroma in general. Aroma constituents are alcohols, ketones, aldehydes, acids and their esters. There is loss of aroma even during extraction because honey comes out in small droplets by centrifugal force.

Honey have different colours and these can range between white to nearly dark brown. Some with green or red tingue. Mustard, *Acacia* and *Plectranthus* honeys are white, sunflower and litchi honey are deep golden yellow and greenish-brown respectively. Honey from berseem and *Dalbergia* (Shisham) are dark brown. Light honeys command higher price and is more in demand in world market but lighter honeys are not liked by Indian consumers. Honeys with delicate flavours are always light, whereas darker honeys have a strong flavour but exceptions are there. Colour is also considered to be an indicator of quality. Honey becomes darker during storage, more rapidly at higher temperature. Contamination with metals also darkens honey. Colour of honey and their descriptions have been standardized in various countries and this is done with P fund colour grader but this remains to be done in India. Reduction in storage temperature lowers its darkening rate.

Honey contains enzymes which come from glands of the workers bee or they can be of plant origin. Invertase, diastase and glucose oxidase are important ones. Catalase and acid phosphates may also be present which are of plant origin. Enzymes derived from glands convert complex sugars into simple sugars. Invertase continues even after honey extraction but is inactivated by heating. Diastase is added to honey from food glands and it also comes from plants. It breaks down starch but for bees it is important because it helps in pollen digestion. Significance of this enzyme for quality determination is that it is more heat sensitive. In some countries low level of this enzyme is considered as an indication that thoney has been heated. But this is a bad standard for genuine honey since honeys from many sources like eucalyptus and citrus have little diastase. Honeys from nectars with higher sugar content have lower enzyme and diastase levels since worker bees quickly store the nectar in combs with less of manipulation. Enzyme loss can be reduced by storing honey at lower temperature since at 71°C half of the invertase is destroyed in 40 min. and half of the diastase in 4 1/2 hours.

Hydroxymethyl furfural increase with temperature. Every 10°C increase in temperature speeds up HMF production by 4 ½ times. Same amount of increase occurs at 70°C in few hour, at 60°C in 1 day, at 50°C in 4 days, at 40°C in 20 days in 100 days at 30°C. Higher HMF is an indication of heating honey. In Europe the upper limit is 40 ppm. But in India HMF is likely to cross the limit even if stored at room temperature or even when the honey is left in combs after flow. Thus there is a strong need for appropriate storage conditions if our honey has to find market outside India.

Storage temperature for honey have great significance to reduce the above losses in quality, avoid fermentation and delay granulation. Storage of honey at 0°C is expensive but storing at this temperature for 5 weeks can prevent subsequent granulation. A temperature range of 10-18°C is most favourable for granulation, where as temperature below 10°C greatly retards granulation. Honey tends to ferment if stored at 11-20°C, is less likely to lead to fermentation at 21-27°C but enzymes are destroyed and HMF is produced. Above 27°C, there is no fermentation but spoilage in terms of enzymes, flavour and aroma, colour, HMF etc. is rapid. Keeping this and above, quality losses in view it seems best to store honey at 0°C for 5 weeks and later on for longer storage at or below 11°C.

Honey Containers: Tinned metal containers are very common for bulk storage of honey, whereas waxed cardboards have also been in use. In recent past, there has been a trend toward the use of plastics but the plastic should be of very good food grade quality. The basic principle is that honey containers should not impart any odour to honey and there should also not be any contamination with metals since metals of the containers can react with honey. Steel or galvanized containers are used after protecting either with tinning or baked on lacquer. All honey storage containers should be moisture proof, with tight fitting lids. Small packing can be done in glass jars with air-tight, screw lids whereas wax fiber containers pick up some moisture at high humidity. Good quality plastic containers which are opaque and clear are popular among consumers but most plastics are brittle and there are complaints of their imparting odour to honey.

Table 92. Average value of multifloral honeys from four zones of India (Source: Phadke,1967)

	East Zone Assam W.B. Bihar	West Zone M.S. & Mysore	North Zone, Pb. U.P. & J&K	South Zone Kerala, Tamil Nadu	All India (Av.)
Specific gravity	1.392	1.392	1.397	1.40	1.399
Direct polarization	- 206	- 10.9'	-10.9'	-3027	-2021'
Moisture %	21.46	20.96	19.98	20.34	20.89
Total diss. Solids %	76.81	76.87	77.50	78.19	77.57
Reducing sugar %	68.91	69.97	72.78	72.88	70.24
Non-reducing sugar %	3.19	3.97	2.01	1.99	3.27
Levolose %	36.01	36.72	38.04	38.77	36.48
Dextrose %	32.31	33.00	35.03	34.07	33.39
L/D ratio	1.136	1.120	1.084	1.137	1.097
Dextrins %	2.050	1.960	2.025	1.771	1.966
Acidity %	0.189	0.180	1.174	0.159	0.180
Ash %	0.210	0.175	0.196	0.157	0.187
Proteins %	0.603	0.531	0.485	0.527	0.556
Undetermined %	2.399	2.011	2.149	1.875	2.184

Table 93. Physico-chemical composition of honeys from different flora (Source: Phadke, 1967)

	Floral Source	Karvi	Wha-yati	Khar-war	Pisa	Bura-mbi	Akhra	Jam-bul	Hirda	Gela	Rub-bers*
1	Moisture	18.33	19.11	18.25	19.37	18.86	19.57	18.15	17.13	16.09	24.05
2	Specific gravity	1.141	1.405	1.409	1.405	1.407	1.409	1.415	1.42	1.425	1.378
3	Direct polarization	+2	18'	-112'	-2.6'	-0.54'	-230'	-148'	-0.36'	-148'	-212'
4	Total dissol-ved slides	79.75	79.00	79.75	78.25	79.75	79.00	80.00	81.25	82.25	82.25
5	Total sugars	77.82	78.67	79.13	73.50	76.24	77.18	78.25	79.06	79.89	–
6	Reducing sugars	66.46	78.12	75.77	69.40	71.44	75.38	75.81	75.236	77.80	72.87
7	Non redu-cing sugars	11.36	0.55	3.36	4.06	4.80	1.80	2.44	3.70	2.09	1.71
8	Levalsoe	38.29	39.80	41.49	35.71	31.17	40.18	39.72	41.30	42.54	35.14
9	Dextrose	28.17	28.32	34.58	33.73	32.27	35.20	36.09	34.06	35.26	35.98
10	Ratio L/d	1.36	1.03	1.19	1.06	1.21	1.14	1.10	1.21	1.20	–
11	Dextrins	2.14	1.08	0.50	1.98	1.37	1.58	1.15	1.49	1.89	–
12	Acidity (as formic)	0.173	0.112	0.156	0.118	0.093	0.138	0.063	0.037	0.092	0.127
13	Ash	0.298	0.138	0.128	0.478	0.113	0.175	0.198	0.014	0.167	0.216
14	Proteins	0.817	0.579	0.79	1.135	0.389	0.546	0.485	10.61	0.419	0.138
15	Undeter-mined	1.92	1.89	2.00	2.38	1.39	1.43	1.85	1.62	1.66	–
16	Colour (visual)	P Gree-nish	Light	Light	Dark	Light	Light		Amber	Light	Pale
	Amber	Yellow	Yellow	black	amber	yellow	Yellow	Amber			

Table 94. Minor constituents of honey.

Constituents	Karvi	Wha-yati	Akhra	Eura-mbi	Pisa	Jamun	Hirda	Gela	A. Dorsata	A. Florea	Trig-ona
Sugars						+					
Unidentified				+		+		+		+	+
Raffinose	+	+		+	+	+		+			
Melezitose	+	+		+	+	+	+	+		+	+
Maltose	+	+		+	+	+	+	+	+		
Unidentified	+	+	+	+	+	+	+	+		+	+

Sucrose	+	+	+	+	+	+	+	+	+	+	+
Glucose	+	+	+	+	+	+	+	+	+	+	+
Fructose	+	+	+	+	+	+	+	+	+	+	+
Amino acids											
Cystine					+						+
Aspargine					+					+	+
Alanine					+						
Lysine					+						
Glycine					+						
Asp. Acid		+	+		+	+					
Glut acid		+	+		+	+					
Alanine					+		+	+	+	+	+
Tyrosine	+	+	+	+	+	+	+	+	+	+	+
Valine	+										
Methionine											+
Leucine	+			+	+	+	+	+	+	+	
Organic Acids											
Tartric			+	+							
Citric	+	+	+	+	+	+	+	+			
Malic	+	+	+	+	+	+	+	+	+	+	+
Succine	+	+			+		+	+	+	+	+

Marketing of bee products

Marketing of any product involves four components viz. Product, Price, place and Promotion i.e. 4 P's. Typical marketing strategy requires that product be considered first, because the customer's perception of the product has great bearing on the P's. However, in India the objective essentially is to provide remunerative prices to the beekeepers and to present the honey to the consumers at reasonably nominal price since it is important as also for expanding local/ domestic market. Each component of marketing is as discussed below:

1. Price

Honey price in any country or in the world is influenced by the following factors: (a) Production conditions and domestic demand. (b) Quality and type of Honey (c) Availability of competitive products. (d) Tariff and trade barriers and Government support programme. In general, the average yield of *Apis mellifera* is about 15 kg per hive per year in case of stationary beekeeping but the yield as the high as 50-70 kg is obtained by migratory beekeeping. However, in case of *Apis cerana,* the yields are lower. In most of the good beekeeping states of India, *Apis mellifera* is being used for commercial beekeeping.

Table 95. Prices paid to beekeepers for unprocessed-bulk purchases of their different type honeys (Source : Mishra, 2000)

1.	*Brassica* (white light amber)	35-40
2.	*Eucalyptus*-amber	40-45
3	Sunflower-golden	40-45
4.	Litchi-light amber	45-50
5.	Rubber-dark amber	45-50
6.	Mixed	35-40
7.	Egyptian clove-dark amber	35-40
8.	Forest (organic) honey (From domestic bees)	40 -45
9.	Squeezed (wild bees)	20

Table 96. Prices paid to beekeepers for their unprocessed-bulk purchases of different types of honeys) in some other countries (Source: Mishra, 2000)

USA	(Cennts / lb)
California	
Clover white	75
Cotton / alfalfa/ buckwheat	60
Mixed-extra light amber	62-63
Mixed-light-amber	60-63
Illinois	
Clover- White	80
Ohio	
Clover-white	74
Washington	
Alfalfa-white	65
Clover-white	75
Mint-amber	65
Argentina (East Coast)	
Clover-White	72-75
Mixed-extra white	74-75
Mixed-white	73-74
Mixed-extra light-amber	70-72
Mixed light amber	63
Mexico	
Mixed-light amber	70

The consumers in India pay Rs. 100 to 150 for one kg pack of honey. This means he pays about five times the price paid to beekeepers. In USA also the situation is almost the same. The retail prices of bottled honey in different markets in USA vary between three to five dollars per 2 lbs. Thus, consumers pay about 5 to 7 times the price paid to the beekeeper.

The domestic demand for honey in India is expanding. Many middle and upper class strata of Indian population like to have honey at their table, though most population still buy honey only for medicinal purposes. The use of honey for industrial users is also increasing. It is very much expected that domestic demand will continue to grow and domestic market would be very large. There can be following channels for honey marketing.

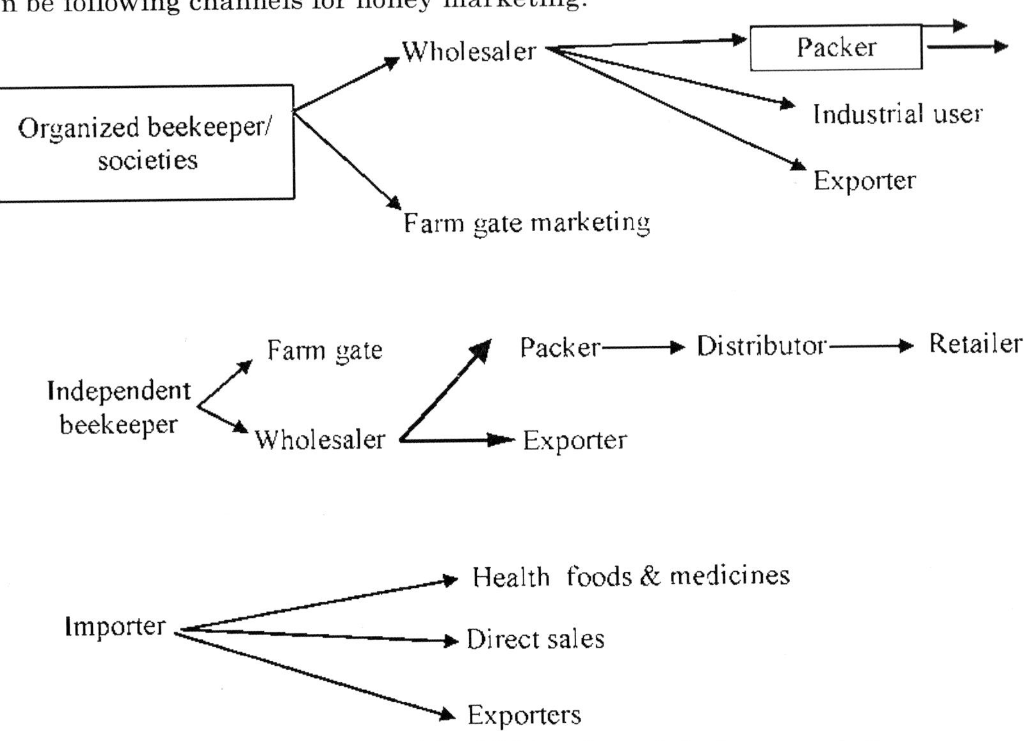

Figure 102. Schematic model exhibiting channels for honey marketing.

Individual beekeepers venture to market their own produce which can be referred to as "FARM GATE" marketing. Even quite a few of these purchase honey from the neighbor beekeepers for marketing. The farm gate marketing is traditional one but unsatisfactory because of problems of product quality, pricing, promotion and distribution. It is also not possible to handle large quantities of honey. Insertion of middlemen such as distributors, stockiest and retailers has an adverse effect on the income of the producer, but brings increased sales and overcome uncertainty and resistance to Farm Gate quality. Local prices in most developing countries are higher than international prices and prices in the neighboring counties with less honey production or with favourable exchange rates may sometimes be quite attractive.

In India, there is a scope for organizing the beekeepers into cooperatives. There is some successful example of cooperative beekeeper's societies in south India, especially the state of Kerala. With these societies the beekeepers get remunerative prices for their honey produce. In Kerala, the rubber honey is sold for more than Rs. 40 per kg even when the honey in terms of aroma, flavour etc. is not as good as some other honeys in the North. In contrast, the beekeepers in North India get Rs. 30-45 per kg for their produce. It is also important to bring out that with such a system the consumers get honey at nominal prices. In most of the South Indian states, the retail price of bottled honey is around Rs.100 per kg.it may also be noted that honeys old in retail in such states/areas is near 100 per cent from the societies and very little or no branded honey from wholesalers and packers is sold in markets. There is also a possibility for a honey project to enter in such as Punjab State Beekeepers Federation to procure honey from beekeepers at (fixed) remunerative prices and handle for marketing.

As far as international market is concerned, Germany, U.S.A and Japan are the major honey importing countries besides some other European Countries like Spain, UK, Italy, France and Belgium etc. World market is looking for new exporters because of declining acceptability of Chinese honeys. New Zealand and Australia are entering in the group of exporters. Importers are also looking towards India for honey. Some price fluctuations are influenced by market variations in producer as well as consumer nations and of course by currency fluctuations. The exports oscillates around one dollar. Average export prices by major exporters are giving. It can be seen that large honey importing countries are obtaining honey for less than a dollar per kg. However, countries which import honey in limited quantities are paying higher prices. From the statistics given in the tables it is clear that at present, it is rather difficult to get more prices than just over one US dollar per kg in the international market. We must be clear that it is buyer's market and not seller's market. Another point should also be clear to every one that honey prices in domestic market in India will sooner o later be controlled by international market and India cannot be in isolation in this era of open market. China had been offering honey for import by other countries at very low prices. This made the American beekeeping Federation and American honey producers association to agitate, alleging unfairly priced imports of Chinese honeys that were injuring domestic Industry. Consequently US Govt. Finalized "Honey Suspension Agreement' which was effective up to year 2000.

As yet in India there are no trade barriers and tariffs on honey. Instances have come to notice where exporters from other countries like China are offering honey to Indian importers at less than Rs. 30 per kg. This is important issue and our Government may have to impose some barriers and tariffs on honey imports to safe guard our industry.

There are couple of houses which are exporting honey from India and during 1999-2000 the total export was around 2500 tons. Exporting honey is important and essential to overcome the honey marketing problems in some of the provinces. But Indian exporters can expect to get 1 to 1.5 dollars a kg in the international market and they cannot afford to buy honey for more than Rs. 40 per kg or so from the beekeepers/wholesalers. In the international market the prices are static for the last about 20 years. Moreover, international market is buyers market and controlled by few countries. Therefore, incentives are required to be given to the exporters to compete in the world market. However, there are certain countries which are not large importers, but pay higher prices. Our exporters can fetch anything between 2 to 4.8 dollars per kg and such countries can be explored for export of our honey. This is also required to avoid competition for export of USA, Japan and Germany.

Indian beekeepers can expect to get marginally higher prices for their produce if the domestic market keeps on expanding but once the industry achieves about full growth and the marketing is systematized then the prices in India would be at par with those in international market.

PRODUCT

The product which is offered in the market is very important in marketing. Here product quality, packaging and brand are the major factors. In our domestic market, the consumers, as yet are not very much conscious about the quality and they need just he liquid honey since crystallized honey is considered to be the adulterated and bad. Our consumers have little appreciation for aroma and flavour; however amber coloured honeys are always preferred. The situation is more so with the buyers who purchase honey in small quantities. However, the consumers, especially those who purchase it for table purpose are getting enlightened about quality. Therefore, for the domestic market also, the producers packers and wholesales will have to be particular about the quality. Storage by beekeepers at collection centres and by wholesalers/packers in appropriate containers and in appropriate conditions would be very necessary.

In the international market the quality of honey in general, determines the price class e.g. table grade or industrial grade. Such parameters as moisture content, cleanliness, off flavors, homogeneity and transparency are major considerations. Some impoters require low HMF. The countries in European Union are particular about pesticidal residues and honey consignments are required to be accompanied by a sanitary certificate vis-a vis test report. Some countries also insist on sanitary certificate to the effect that the honey has been produced from disease free colonies.

Table 97. Outline of quality control measures to be taken by a typical honey processor on honey prior to processing

Parameters	Control method	Limits
Container	Direct observation	Adequate material and condition
Homogeneity of lot	Direct observation	Apparent homogeneity according to observable characteristics in whole Shipment.

Impurities	Direct observations of honey. Surface in containers filtration, or polarized light test.	Presence of limited impurities such as bee and wax particles
Organoleptic Characteristics	Organoleptic analysis on an average sample	Absence of defects such as strange odours and tastes, fermentation, overheating or otherwise characteristics. Correspondence to the samples from producers and to foreseen standards for end product classification.
Colour	Optical comparison with fund meter or according to Lovibond	Correspondence to producer sample and foreseen standards for product type
Moisture content	Refractometer measurements	Less than 18.0% for top grade (less that 21% max. limit)
HMF	Colorimetric method (codex Alimentarius, 1989)	Less than 10 mg/kg for top grade(40 mg/kg is maximum limit)
Microscopic	Quantitative and qualitative pollen analysis (Louveaux and geographic origin. *et al.,* 1978)	Correspondence to declared botanical and geographic origin.
Others	Official methods	According to legal limits.

Source: Mishra (2000)

There is inadequacy of laboratories in India for getting the honey tested for quality parameters. Large enterprises in developed countries have their own laboratories. While smaller manufacturers can perform only simple measurements and have to rely on outside laboratories for detailed analysis. In India some Govt., Semi-Govt., or Bee Board etc. Shall have to plan and provide a network of testing and certification mark *et al.* Offering quality honey starts from the producer itself. The flowers, places of production, and timing of collection will affect the quality of single flower honey in their colour, smell, and palatability. Therefore, enough attention should be given to the quality of products in production, purchase and processing.

Table 98. Beekeeping methods which may have negative effects on the quality of honey.

Beekeeping method	Possible damage to honey
Location of hives in densely urbanized or industrialized zones or areas otherwise Subjected to strong environmental including agricultural pesticide use.	Contamination of honey with noxious or toxic residues, possibly damaging to human health, or with sugars not of nectar or honey dew origin.
Inappropriate use of antibiotics and other drugs or chemicals to treat or prevent honey-bee diseases substances or control pests.	Contamination of honey with the same substances.
Use of organic chemicals like naphthalene, ethylene dibromide or paradichlorobenzol for comb protection during storage and treatment against wax months.	Contamination of honey with the same substances.

Use of chemical repellents during honey harvesting.	Contamination of honey with the same substances.
Inadequate use of smoke by quantity or type of combustion material.	Smoky odour and other flavours o f honey and contamination with microscopic soot.
Use of old dark combs and/or brood combs.	Honey of darker colour, comb odour, higher acidity and faster aging.
Use of combs with residual honey from a previous year.	Honey high in yeasts and possibly faster fermentation; premature crystallization of susceptible liquid honeys, contamination of unifloral honeys
Harvesting of incompletely sealed combs particularly during the nectar flow.	Excessive moisture content in honey.

Source: Mishra (2000)

For packaging, the honey is needed to be delicately and carefully handled during purification and processing so that the purest honey produced by the bees is not deteriorated. Our packers please note that “LEAST PROCESSING IS THE BEST PROCESSING”. Any fiddling by the beekeepers, or packers etc. will not deteriorate the quality rather than improving it. So, our aim should be to offer the same purest honey from combs to the consumers. Minimum processing of honey may be resorted to. A summary of the plan for processing would be as follows:-

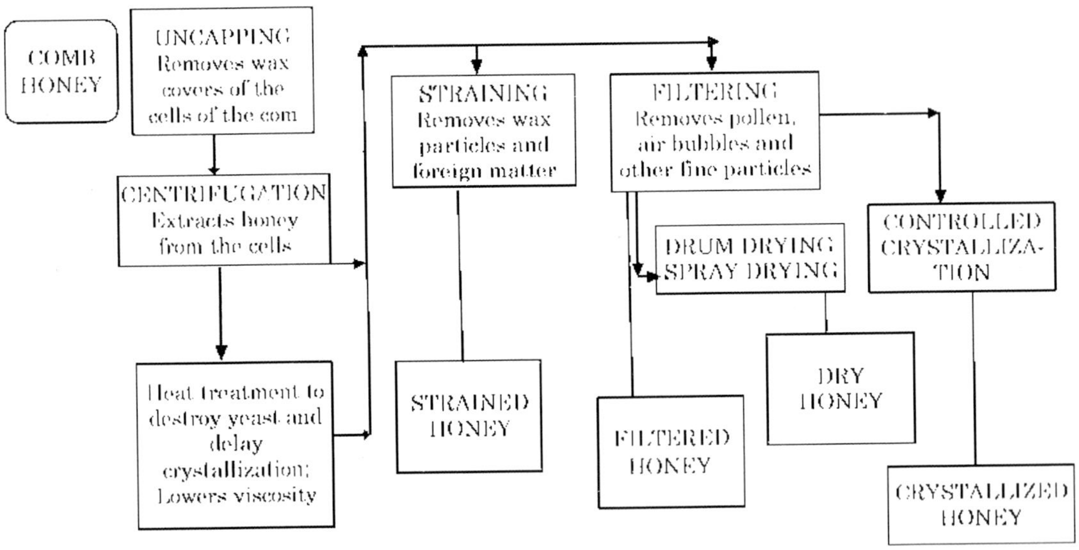

Figure 103. Schematic model exhibiting a summary of the plan for processing.

Packaging

The bottle or package should be leak proof and airtight so as to safely contain the product, but also present the product in the attractive form, enticing consumers to buy it. The label, container shape and material should be chosen accordingly. Labels have to provide all legally required information and preferably the lot number to help the producer track down any problem. In India, it should be relevant to give information on crystallization. Though packaging does not improve the product as such but very well adds to the value of the product.

The International market usually trades honey in 300 kg metal drums and only a small percentage of the market is traded in retail containers. The later is mostly between neighboring countries and within Europe.

The **brand** is important and this should be one which has earned good will by way of quality vis-à-vis price. Unbranded or un-popularly branded honeys are not readily accepted in the market. The brand name should necessarily distinguish the product as bee honey and more so from hive bees.

III. Places

Honey must be available to consumers at the right time and at the right place to encourage purchase. Place decisions are concerned with the location of marketing facilities and the selection of specialists, including transportation and storage agencies, distributors stockists and retailers.

In the International market high class honey should be light in colour, fragrant smelling, palatable with good transparency and not apt to crystallize. But there can be exceptions. Litchi honey is darker in colour, but smells good. It is regarded as good honey and well received in Japan, Hong Kong and Macao. Similarly crystallized *Brassica* honey finds market in Japan. Therefore, for different types of honeys the traders should find appropriate place for marketing.

IV. Promotion

The consumption of honey in India is very low. Widespread promotional efforts are required to expand domestic market. Some advertisements are given only by private houses such as Dabur (India). More aggressive approach, on the lines of promotional efforts for dairy and poultry products both by public and private sectors would be essential to expand the domestic market. India will need 5 lakh tons of honey for domestic consumption. If the present average per capita consumption increases from 3-8 gms to about 500 gms per year. It is worth mentioning that in U.S.A and Germany the honey consumption is one and 1.8 kg per capita per year respectively. It will also be the advantage of the beekeepers and honey traders if the domestic market expands. Since higher prices can be fetched in domestic market than in the international market.

Chapter 30

Genetically modified crops and beekeeping

Genetically modified, "transgenic" plants (GM plants) carry genetic information derived from other organisms (bacteria, plants or animals), or have been modified by genetic engineering to provide them with new properties. Genetically modified (GM) plants manifest new traits via the production (expression) of novel proteins encoded by inserted transgenes (DNA). Many insect resistant crops carry genetic information originally derived from the soil bacterium *Bacillus thuringiensis.* This gene enables the bacteria or the GM plant to produce a protein with insecticidal activity. Therefore, the crop plant can protect itself against insect pests. There are numerous variants of the gene, encoding different proteins active in a highly specific manner against defined groups of insects. For example, cotton modified to contain a Bt (*Bacillus thuringiensis*) gene and expressing Bt insecticidal protein in its leaves and buds will be protected from caterpillar attack. Both the transgene DNA and the novel protein in such plants could be considered "GM material". Cry3B is one of these proteins with activity against beetles.

Bees and plants are inseparable and depend upon each other for their survival and perpetuation of race. Bees collect pollen and nectar from plants as food for their colonies. They may also collect honeydew (plant sap that has been ingested and then excreted by sucking insects such as aphids) for food and plant resin to make propolis with which to seal up cracks in the hive. With GM plants, there is a possibility that transgene DNA and/or novel proteins may be present in the plant parts and secretions collected by bees.

Since bees are primary agents for cross pollination of several crops travelling considerable distances and potentially causing crossing of GM and non-GM crops. A hive may contain up to 50,000 bees and individual bees may visit up to 100 flowers on each trip out. Hive products intended for human consumption and derived from the plant parts collected by bees include honey (from nectar or honeydew, and containing traces of pollen), pollen and propolis. Honey can be contaminated by GM pollen, leading to potential economic losses for beekeepers. Evidently, there is a need to analyse whether Cry3B proteins have negative effects on honeybees. Other products, such as pure royal jelly, bee venom and beeswax, are secretions from the bees themselves and would not be expected to contain any plant matter.

GM plants occurring in hive products

The relevance of this to the production and marketing of bee products will depend on the tolerance limits for GM material in food. GM plants are now being grown commercially in thirteen different countries. In 2000, the estimated global area planted with these crops was 44.2 million hectares (James, 2000). Ninety nine percent of these crops were grown in the USA, Canada, Argentina and China. Herbicide-tolerant soybean was the most common GM crop, occupying 59% of the global area, followed by Bt corn (maize), herbicide-tolerant canola (oilseed rape), herbicide-tolerant corn, herbicide-tolerant cotton, Bt/herbicide-tolerant cotton, Bt cotton and Bt/herbicide-tolerant corn. In addition to these commercially available varieties, many other plants can now be genetically modified. Field trials are under way overseas for field crops (e.g. wheat, rice, barley, tobacco), flowers (e.g. roses, carnations), trees (e.g. poplar, spruce, sweet gum), oil crops (e.g. sunflower, peanut), grasses, sugar crops (beet and cane), fruits (e.g. apple, cranberry, grape, melon, strawberry) and vegetables (e.g. tomato, potato, broccoli, carrot, eggplant, lettuce, pea). Of the commercial crops, white clover is the most important for honey production. Trees grown for shelter (e.g. willow, hawthorn) and for fruit (e.g. apple, pear) and field crops (e.g. *brassicas*, sunflowers) can also be useful sources of forage for bees. Bees are also used to pollinate crops for fruit, vegetable and seed production. There have also been reports of bees visiting plants such as corn, grasses and potatoes in the absence of better sources of forage and of the occurrence of traces of pine pollen in honey. Since pollen is a plant tissue composed of cells capable of protein synthesis, it is reasonable to expect to find transgene DNA in pollen grains and also novel protein, if the transgene's promoter permits it. Table 99 presents currently available data on novel protein expression levels in pollen from GM plants. Thus, if modified, some of these plants might represent sources of GM materials that could be collected by bees. Compared to leaves, which are about 2% protein, pollen has a high protein content, estimated to be between 8 and 40%, and for this reason transgenes may be expected to express novel proteins at reasonable levels in pollen. Crops which could be genetically modified and their relevance to honey bees are listed in Table 100.

Table 99. Expression of novel proteins in pollen of GM plants

Plant	Novel protein encoded by transgene	Promoter	Expression level of novel protein in pollen	Expression level as % of total soluble protein (estimated)[1]	Reference
Maize	Bt toxin	maize pollen specific & PEP[2] (leafspecific) promoters	260 - 418 ng/ mg (of total soluble protein)	0.026 –0.0418	Kozeil *et al.*, 1993

Cotton	Bt toxin	Cry 1Ac CaMV 35S	0.6 μg per g fresh weight	0.00024	Greenplate, 1997
Oilseed rape	Oryzacystatin I	CaMV 35S	Nil	0	Bonadé Bottino *et al.*, 1998
Oilseed rape	Bowman-Birk trypsin inhibitor	CaMV 35S	Nil	0	Jouanin *et al.*, 1998
Oilseed rape	nptII (kanamycin resistance)	NOS	1.61 ng/mg of total protein	0.000161	Anon A, 1998
Cotton	Bt toxin	Cry1Ac CaMV 35S	11 ng /g fresh weight	0.0000044	Anon V, 2001

Values expressed as a proportion of fresh pollen weight in the original reference have been converted using the assumption that fresh pollen is 25% protein. 2 Phosphoenolpyruvate. 3 Cauliflower mosaic virus 35S promoter

Table 100. Crops which could be genetically modified and their relevance to honey bees.

Crops for which GM varieties might become available	Role of honey bees in relation to crop Potential hive products from crop	Potential hive products from crop
Crops		
White clover	Very important nectar source and pollen source to a lesser extent; bees required for seed production	Honey; pollen
Maize	Bees may collect pollen if no other forage available	
Potato	Bees may visit, but flowers have no nectar, little pollen	
Vegetable brassicas	Good source of nectar and pollen (to a lesser extent); bees required for seed production when crop not self fertile	Honey; pollen
Cotton	Source of pollen and nectar	Honey; pollen
Apples	Source of pollen and nectar (to a lesser extent); bees required for adequate pollination	Pollen; propolis?
Eucalyptus	Ornamental species are good source of nectar and pollen (to a lesser extent)	Honey; pollen; propolis?

Is GM Honey safe?

Honey contains pollen grains from plants that bees have visited, and can therefore be contaminated by GM pollen. It is not known what effect honey containing GM pollen could have on people consuming it. For example, the novel proteins produced by GM crops can also be present in their pollen are already cases of people who are allergic to honey, and this has been linked to pollen in the honey. Genetic engineering is imprecise and unpredictable. Genes are inserted from organisms which have never been eaten as food, and so new proteins are introduced into the human and animal food chains. There is concern that these could cause allergic reactions or other health effects.

There are between 20,000 and 80,000 pollen grains in a portion (10g) of honey. This means that honey containing GM pollen could pose a potential health risk. The researchers who studied this problem concluded that if GM pollen contained novel toxins or proteins it "could pose problems, not only to man who consumes honey as a food, but also to bee populations which rely on pollen as the sole source of protein"

It has been found that bees can pass proteins from nectar unchanged into honey. If GM crops produce novel proteins or toxins in the nectar as well, this may further threaten the safety of honey produced from GM crops. GM plants may also contain genes which provide resistance to commonly used antibiotics such as ampicillin. There is concern that these could be passed on to the bacteria that live in humans and animals. DNA from pollen has been found to be able to survive in honey for seven weeks. It may be that this could be a route for such gene transfer. In spite of the risks, there has been almost no safety testing of GM pollen, either for humans or for bees. Even the food safety tests of GM crops have been limited to short-term tests on animals.

Bees and gene pollution

Honey bees commonly forage up to two km from the hive, but oilseed rape fields are such an attractive source of nectar that bees may travel at least five km to get to them. When the oilseed rape was in flower, it made up 70% of the pollen that the bees carried back to the hive. One bee returning to the hive had 60,000 oilseed rape pollen grains stuck to its body. As the bees brush past each other in the hive any GM pollen is spread throughout the colony and taken out again by other bees. Not only are bee hives near to GM oilseed rape fields likely to become contaminated with GM pollen, but the bees may spread GM pollen to non-GM crops several miles away. It is likely that in the future farmers will be growing oilseed rape for the 'GM-free' market. In such cases, contamination of the crop could cause financial loss to the farmer. At the moment it is unclear who would be held liable for this, and whether the beekeeper might be held responsible as well as the farmer who grows the GM crop.

Impact on Beekeepers

As it stands, GM crops pose a serious threat to beekeepers and honey production. Honey contaminated with GM pollen will either have to be disposed of safely or sold as a GM product. Either of these options is likely to cause financial harm to beekeepers.

Impact on Wild Bees

Wild bees are vital for the survival of many of our wild plants, as well as being important for crops. Some species of wild plants are dependent on wild bees for their pollination and survival. Several species of bumblebees and solitary bees are also important pollinators for food crops. The decline of bumblebees and solitary bee populations has been linked to modern intensive farming. The widespread use of herbicides and the increased intensity of farming has removed the habitats that wild bees make nests in, and reduced the numbers of wild plants which they use for food. The widespread use of GM herbicide-tolerant crops is likely to threaten further wild bee populations.

It is clear that growing GM crops pose a serious threat to beekeepers and honey production. Issues around food safety, liability, the cost to beekeepers and the threat to wild bee populations have yet to be resolved. Vital questions need to be answered and there needs to be a full public debate on the acceptable uses of genetic engineering.

Bt insect protection of crops and honey bees

Genetically modified crops are being blamed for the declining number of bees? Though scientists from France have allegedly determined that *Baccillus Thuringiensis*, a bacteria normally found in the soil, and which has been widely used to develop various genetically modified crops, is behind deaths of honey bees. However, there findings have not been supported by many researchers. There is strong evidence that the production of specific insecticidal proteins from the soil bacterium *Bacillus thuringiensis* (Bt) in crops to control targeted caterpillar pests and beetles does not pose a risk to honeybees.

There is extensive information on the lack of non-target effects to diverse groups of beneficial insects including honey bees and other pollinators from Bt microbial preparations that contain Bt proteins. Bt proteins are ideal for use in organic production and in Bt crops because they bind specifically to receptors on the mid-gut of sensitive caterpillar pests and have no deleterious effect on beneficial/non-target insects under the conditions of use, including predators and parasitoids of targeted caterpillar pests and honeybees. Extensive studies on honeybee safety assessments on all insect-protected crops, including Bt corn and Bt cotton have shown that the Bt proteins in these crops have no adverse effect on the honeybee. Specific studies involving Cry1Ab provide strong evidence of the safety of MON 810 Bt corn to the honeybee (similar studies have been conducted with other Bt proteins in genetically modified crops). The EPA concluded that based on the weight of evidence there are no unreasonable adverse effects of the Cry1Ab protein expressed in MON 810 Bt corn to non-target wildlife or beneficial invertebrates.

There is extensive information on the lack of non-target effects to diverse groups of beneficial insects including honey bees and other pollinators from Bt microbial preparations that contain Bt proteins. The Bt proteins produced in Bt corn and Bt cotton are present in microbial products used in agricultural systems to control targeted pests. Bt proteins are extremely selective and are toxic only to specific pests. A generalized mode of action for Bt proteins includes ingestion of the protein crystals by insects, solubilization of the crystals in the insect midgut and proteolytic processing of the released Bt protein by enzymes, and binding of the partially digested "activated" protein to specific high-affinity receptors on the surface of the midgut epithelium of target insects. Bt proteins are ideal for use in organic production and in Bt crops because they bind specifically to receptors on the mid-gut of sensitive caterpillar pests and have no deleterious effect on beneficial/non-target insects.

Safety assessment of Bt crops

EPA evaluated studies of potential effects on a wide variety of non-target organisms that might be exposed to the Bt protein, e.g., birds, fish, honeybees, ladybugs, parasitic wasps, lacewings, springtails, aquatic invertebrates and earthworms. Such non-target organisms are important to a healthy ecosystem, especially the predatory, parasitic, and pollinating insects. These risk assessments demonstrated that Bt proteins expressed in Bt crops do not exhibit detrimental effects to non-target organisms in populations exposed to the levels of Bt proteins produced in plant tissues. To illustrate how the different Bt proteins produced in Bt crops are evaluated for safety to the honeybee, two representative studies are described below for the Cry1Ab protein produced in MON 810 Bt corn. These studies with Cry1Ab protein were conducted with the trypsin-resistant core because this is the insecticidally-active portion of the Cry1Ab protein. Specific studies designed to assess the potential for adverse effects to developing larval and adult honeybees are described below.

Honeybee Larva

The primary route of exposure for honey bee larvae to the Cry1Ab protein is ingestion of pollen collected by foraging adults from genetically modified plants. Therefore, honey bee larvae were exposed to Cry1Ab protein in their natural diet by including a maximum hazard dose (20 parts per million in distilled water mixed with honey) in developing brood cells. This maximum nominal concentration of 20 ppm was approximately 100 times greater than the maximum expected Cry1Ab protein level in MON 810 pollen. In addition to this treatment group, a negative control group was treated with distilled water. Another control group was treated with heat-attenuated (inactivated) Cry1Ab protein (20 ppm), and one set of larvae received no treatment (untreated control). At least 50 bees (1 to 4 days old) were in each replicate, and there were three replicates for each group. The treatments were administered to each larval cell through an electronic micro-applicator, which delivered 5 microliters (?L) of the test diet.

There were no statistically significant (P>0.05) differences in honeybee larval survival to adult emergence among the four treatment groups. The mean adult survival rates after emergence ranged from 91.7% to 96.0% across all groups, including the controls and Cry1Ab-treated groups. This study demonstrates that honeybee larvae were not adversely affected after being exposed to Cry1Ab protein at a concentration of 20 ppm in their diet.

Adult Honeybee

Adult bees reared in bee hives were immobilized using CO_2. The test diet was prepared by mixing the appropriate amount of the insecticidally-active Cry1Ab protein with a honey-water (50-50) syrup to a concentration of 20 parts per million (?g protein/g diet; ppm). The negative control group was fed the same diet with the exception that no Cry1Ab protein was added to the honey-water mixture. A second control group was fed heat-attenuated (inactivated) Cry1Ab protein at the same concentration (20 ppm) as the treatment group. A fourth test system was an empty cage to measure the amount of diet loss due to evaporation. All diets were presented to the bees in a 6 ml shell vial inserted through a cork in the holding cage lid. Three replicates of four test groups of at least 40 adult honeybees were selected and placed in each holding cage. Two observations were made the first day and were made daily for the duration of the 9-day study. At the time of the daily observation, the test diets were replaced with fre

Adult honeybees exposed to the Cry1Ab protein in a honey-water solution for 9 days at a concentration of 20 ppm showed no signs of treatment-related mortality or toxicity. At the end of the testing period, the mortality percentage was calculated for each group. Mortality in the treatment and the negative control groups was 16.20% and 22.28%, respectively. The heat-attenuated control group mortality was 32.59%. Mortality showed a sharp increase in all three groups from days 6 through 9. At the termination of the test, the highest mortality was observed in the group that was fed the heat-attenuated Cry1Ab protein diet, while the lowest mortality was observed in the group that was fed the Cry1Ab protein diet. The mortalities in the treatment group are not considered to be treatment-related because the two control groups showed a higher percentage of mortality over the same time interval. There was no significant statistical difference ($P>0.05$) in mortality patterns between any of the groups.

The studies concluded that based on the weight of evidence there are no unreasonable adverse effects of the Cry1Ab protein expressed in MON 810 Bt corn to non-target wildlife or beneficial invertebrates. They reported no measurable deleterious effects were observed in submitted studies of the Cry1Ab protein administered to honey bee larvae, honey bee adults, parasitic wasps, Ladybird beetles, green lacewings, Collembola (springtails), and Daphnia.

Arpaia (1996) found that the development of the larvae, the pupal weight and the number of bees in the Bt-exposed colonies were not affected. The conclusion of these experiments was that for honeybees no negative effects of GM crops producing the Cry3A protein are to be expected. Very similar experiments have also been conducted for numerous other Bt proteins, for example for Cry1Ab produced in MON810 maize plants (Babendreier *et al.,* 2005). In no instance, negative effects of Bt proteins on the health of honeybees have been published (Malone *et al.* 1999, Malone & Pham-Delègue 2001, 2002, Keil *et al.* 2002, Malone 2004, O'Callaghan *et al.* 2005, Sanvido 2006). Also other aspects of the honeybee's development were investigated. Several scientists looked at possible effects of Bt protein on the hypopharyngeal glands of worker bees, where protein rich food for the brood is produced. Neither for Cry1Ba (Malone *et al.* 2004) nor for Cry1Ab (Babendreier *et al.* 2005, 2006), detrimental effects of the purified insecticidal proteins were observed.

Honeybee's flight activity likewise was unaffected by Cry1Ba (Malone *et al.* 2001). In one case, a possible effect of Cry1Ab on the foraging activity was reported. However, this single experiment was carried out in winter, without direct comparison to untreated bees, and has not been repeated since, so that it is difficult to derive firm conclusions from this observation (Ramirez-Romero, 2005). Even the gut flora of honeybees, which is important to their health, was tested for its sensitivity against Cry1Ba; no effect was observed (Babendreier *et al.* 2007). Taken together, no evidence for risks to honey bee's health from Bt proteins used in GM crops has been reported in the scientific literature so far.

Potential impacts of GM plants on bee health

Although not directly related to the presence of GM material in bee products, there may be impacts of GM plants on bee health that could affect on the bees' ability to produce honey, pollen and propolis. There is a growing body of published research on the impacts of GM plants and novel proteins on bee health (Table 101). This research has recently been reviewed (Malone and Pham-Delègue, 2001; 2002; Pham-Delègue *et al.*, 2002). These reviews are summarized below. GM plants may have direct or indirect effects on bees. Direct effects are those that arise when a bee ingests a novel protein expressed by a GM plant. Indirect effects may arise if the process of introducing the transgene into the plant results in inadvertent changes to plant phenotype affecting its attractiveness or nutritive value to bees.

Table 101. Effects of novel proteins and GM plants on bees (Malone and Pham-Delègue, 2001).

Novel protein or GM plant	Type of experiment	Results
Bt toxins (lepidopteran active)	Larval survival	Not toxic
	Adult survival (in lab & in colony)	Not toxic
	Adult food consumption	No effect
	Adult flight activity (protein fed to colony)	No effect
Bt toxins (coleopteran active)	Larval survival	Not toxic
	Larval survival, pupal weight (protein fed to colony)	No effects
Bt-corn (lepidopteranactive)	Larval development, adult survi val, foraging frequency (in field)	No effects
Serine protease inhibitors	Adult survival (in lab and in colony)	High concentrations reduce survi-val by a few days; low concentratio-ns have no effect

	Adult digestive proteases	Inhibition of some proteases
	Adult flight activity (protein fed to colony)	Flight activity begins a few days earlier (when fed a high concentration)
	Olfactory learning response	One inhibitor offered in sugar reward reduced ability to learn; others did no
	Larval survival	High concentrations reduce survival
Cysteine protease inhibitors	Adult survival	No effect
Cysteine protease inhibitor-expressing oilseed rape	Foraging behaviour	No effect
Chitinase	Adult survival	No effect
	Olfactory learning response	No effect
	Foraging behaviour (sugar feeder with chitinase added)	No effect
Chitinase-expressing oilseed rape	Foraging behaviour	No effect
a-1,3 glucanase	Adult survival	No effect
	Olfactory learning response	No effect
	Foraging behaviour (sugar feeder with a-1,3 glucanase added)	No effect
Biotin-binding protein (avidin)2	Adult survival	No effect
	Adult food consumption	No effect
	Larval development and survival	No effect
Herbicide (glufosinate) resistant oilseed rape (pat gene)	Larval and adult survival, foraging behaviour (in colony)	No effect

Source: Brødsgaard *et al.*, 2001.

Direct Effects of Novel Proteins on Bees

Potentially, the ingestion of a novel protein expressed in pollen or perhaps occurring in nectar, resin or honeydew from a GM plant may affect bee behaviour, development or survival, or it may have no effect. As pointed out above pollen represents the most likely vehicle for the expression of novel proteins. Adult bees consume significant quantities of pollen during their first week after emergence and they have received the most research attention to date. Bee larvae also ingest pollen, especially during the later instars, but their food is composed in large part of glandular secretions from nurse adult bees.

Many experiments have been conducted in which bees are fed with purified novel proteins at concentrations estimated to approximate or to exceed likely pollen expression levels. Novel proteins with insecticidal properties aimed at making GM plants pest resistant have been the most thoroughly tested. There have also been trials conducted with small colonies of bees and potted flowering GM plants in glasshouses or under mesh in the field. Tests have assessed food consumption by adult bees, adult bee survival, olfactory learning and foraging behaviour in adult bees, larval bee development and survival. Results to date are briefly summarised in Table 101. Supporting references may be found in reviews by Malone and Pham-Delègue.

Indirect Effects of GM Plants on Bees

Indirect effects of transgenic plants on bees may occur when genetic modification results in an unexpected change in the plant's phenotype. Insertional mutagenesis is one such change. In this case, the random positioning of the transgene in the plant'sgenome interferes with a gene or suite of genes needed for a "normal" phenotype. For example, an insertional mutagenesis event that resulted in plants without flowers would have a definite negative impact on bees. Less obvious changes, such as alterations in nectar quality or volume would be harder, but not impossible, to detect. Effects due to insertional mutagenesis will vary among different lines of plants derived from separate transformation events and can easily be eliminated by line selection. Pleiotropic effects represent a second type of inadvertent phenotypic change. In this case, it is not the position of the transgene, but its product, which interferes unexpectedly with a biochemical pathway in the plant to create a phenotypic change. Such changes would occur in all lines of the transgenic plant and could not be remedied by line selection. Nectar analyses of GM oilseed rape plants have suggested that some modification events may lead to phenotypic changes that could influence bee behaviour (Pham-Delègue *et al.*, 2002). One line of GM oilseed rape expressing a chitinase produced more nectar of higher sugar concentration than the corresponding control line, as did one herbicide-resistant line compared with its control.

In brief, so far, no evidence for direct or indirect damages to bees by currently approved GM crops has been reported in the scientific literature. This is the result of numerous experiments in the laboratory or the field, in some of which the exposure towards GM crops or their products possible under natural conditions has been strongly exaggerated. Changing agricultural practices, like they are possible e. g. with herbicide tolerant crops, can result in changes to the number and seasonal composition of weeds within the fields – which is also the case for conventional agricultural practices, and not a direct result of the genetic modification. For mobile insects like bees, which have a large natural habitat far exceeding single fields, no negative consequences of such changes are to be expected.

However, strategies to minimize the presence of GM material in bee products may include:

1. Separating GM and non-GM crops (effectiveness will depend on bee flight distances)

2. Screening GM crops with bee-proof mesh (not practical except on a very small Scale)
3. Using bee attractants such as a sugar syrup spray on non-GM crops
4. Using bee repellents such as some pesticides on GM crops
5. Using GM plants where the transgene is not expressed in pollen, or the transgene
6. Occurs only in chloroplasts, or where pollen or flower formation is blocked
7. Removing pollen grains from honey by filtering after harvest.

Chapter 31

Diagnosis of Honey Bee Diseases

Introduction

Inspection for bee disease is an important part of beekeeping. Beekeepers must be able to recognize bee diseases and parasites and to differentiate the serious diseases from the less important ones. The purpose of this chapter is to acquaint readers with laboratory techniques used to diagnose diseases and to detect and identify parasites, pests, and other abnormalities of the honey bees.

Methods of Diagnosing Disease

Techniques of Microscopic Examination

Most bee diseases can be diagnosed by observing the associated microorganisms with a light microscope. The following techniques are commonly used to prepare microscope slides for examination:

Modified Hanging Drop

The modified hanging drop technique (Michael, 1957) can be very useful for differentiating diseases of the brood. Residue from a suspected cell is first mixed with water. Then a drop of this suspension (smear) is placed on a cover glass. The suspension used should always be dilute and only slightly turbid. The smear is dried and fixed under a heat lamp, or the smear can be air dried and heat fixed by passing it rapidly through a Bunsen burner flame two or three times. The fixed smear is stained with carbol fuchsin or a suitable spore stain for 10 seconds. Enough stain should be placed on the cover glass to cover the entire smear. The excess solution A: 0.3 g basic fuchsin (90% dye content), 10 ml ethyl alcohol (95%); solution B: 5 g phenol. 95 ml distilled water. Mix solutions A and B. Stain is then washed off with water. While the smear is still wet, the cover glass is inverted with the smear side down and placed onto a standard microscope slide previously coated with a very thin layer of immersion oil. The slide is gently blotted dry and examined with a microscope using the oil immersion objective. Results: Organisms that are not heat fixed are caught in areas where pockets of water have formed in the oil, and the organisms usually exhibit Brownian movement (see section on American foulbrood).

Simple Stain

This method relies solely on differentiating the bacteria by morphology. Place a drop of the suspension directly on a microscope slide. Heat fix and stain the smear as described in the previous section. Carbol fuchsin, methylene blue, and safranin are examples of stains that can be used. Air-dry or gently blot the stained smear. Place a drop of immersion oil directly on the smear. No cover glass is necessary. Examine the slide, using the oil immersion objective. Results: Organisms are uniformly stained and easily distinguished.

Gram Stain

The Gram stain is a standard microbiological method that can be substituted for the simple stain. Briefly, the procedure is as follows: A fixed smear is stained with crystal violet, immersed in iodine solution, decolorized in ethyl alcohol, and counterstained with safranin. Results: Gram-positive organisms are blue; Gram-negative organisms are red.

Wet Mount

The wet mount is especially useful for examining fungi or protozoa. Macerate a portion of the sample in water. Place a drop of the suspension on a microscope slide, and carefully drop the cover glass on it to minimize air pockets. No stain is required. The wet mount is usually examined with the dry objectives of a microscope. Results: Organisms refract light and are therefore visible on the slide. A phase-contrast microscope may be helpful, especially if an oil immersion objective is required.

Microinjection Techniques

To diagnose some diseases or to determine toxic levels of sample materials, it may be necessary to feed or inoculate larvae, pupae, or adult honey bees. Michael (1960) developed a technique using a microinjector equipped with a syringe and a 30-gauge needle. The microinjector can be calibrated to repeatedly deliver uniform inoculating volumes as small as 1 µl. The apparatus can be used to introduce material orally into the midgut (ventriculus) of a larva or to feed individual adult honey bees. The microinjection technique can also be adapted to direct injections into the body cavity of larvae, pupae, and adults.

Collecting Honey Bee Larvae and Pupae

Honey bee larvae 3 to 5 days old are readily obtained by removing a brood frame containing the desired larvae from a colony and placing it horizontally above a towel-lined tray in an incubator at 34°C. Within a few hours, the larvae crawl from their cells and drop to the tray below. Pupae can be easily obtained by collecting 5-day-old larvae as described above and incubating them in petri dishes until pupation occurs. This method of collecting larvae and pupae in large numbers saves considerable time and labor. It also eliminates the damage that can occur when attempting to remove these immature forms from cells by mechanical means.

Oral Introduction

Larvae

Honey bee larvae as young as 3 days and weighing as little as 25 mg can be force-fed by carefully inserting the needle through the mouthparts and into the esophagus. When the actuating lever is depressed, a predetermined volume of material is propelled through the esophagus and into the midgut, with no physical damage to the larva. After feeding, the larvae are placed in petri dishes lined with filter paper and incubated at 34°C.

Adults

Individual adult honey bees can also be fed known volumes using the microinjector. Adult bees are collected and held in a cage for about 4 hours without food. The material to be fed, at the final concentration, should be mixed into a sucrose solution to make it attractive to the bees. The microinjector is first actuated to produce a known volume of liquid at the tip of the needle. Then a bee from the cage is grasped by the wings, held up to the drop of liquid, and allowed to feed. Cold temperatures can be used to slow the bees and make them easier to handle. Avoid the use of carbon dioxide as an anesthetic to aid in handling bees, because they are reluctant to feed after exposure to carbon dioxide and their longevity is reduced. After feeding, the bees are placed in small cages with a supply of sugar syrup and held in an incubator at 34°C.

Direct Injection

Care should be taken to insert only the tip of the needle into the hemocoel and to not exceed inoculating volumes of 2 μl per bee.

Brood

Injections are usually confined to 4- to 5-day larvae. The larva is held gently between the first and second fingers and the thumb, and the larva must be absolutely parallel to the needle. After the integument is punctured by the needle under gentle pressure, the inoculum is expelled directly into the dorsal blood vessel; the needle is then withdrawn in a slow, steady movement. Any excessive pressure on the larva by the fingers, particularly when withdrawing the needle, must be avoided to prevent bleeding. Pupae can be inoculated dorsally between the third and fourth abdominal segments or through the propodeum of the thorax. After injection, the larvae or pupae are placed in petri dishes lined with filter paper and incubated at 34°C. If excessive bleeding has occurred, it can be seen on the filter paper.

Pupae can also be left in brood combs and inoculated in the head capsule (Wilson, 1970). The pupal head is exposed by removing the cell cap, and the needle is inserted between the ocelli or through the clypeal sclerite.

Adults

Adult bees can be injected either through the propodeum of the thorax or dorsally through the intersegmental membrane between the third and fourth abdominal segments. Adult bees should be carefully subjected to carbon dioxide anesthetic before and during the process of injection. When bee longevity is a factor in the test, cold temperature can be used as an anesthetic. After injection, the bees are placed in small cages with a supply of sugar syrup and held in an incubator at 34°C.

Removal of Digestive Tract

Intact digestive tracts that have been removed from adult honey bees are very useful for the detection of protozoan diseases. The digestive tracts can be easily obtained by removing the head of the bee to free the digestive tract, grasping as much of the stinger as possible with a pair of fine tweezers, and then with a steady, gentle pull withdrawing the entire digestive tract. Freshly killed honey bees are required for this procedure.

Brood diseases

Brood combs from healthy colonies typically have a solid and compact brood pattern. Almost every cell from the center of the comb outward contains an egg, larva, or pupa. The cappings are uniform in color and are convex (higher in center than at margins). The unfinished cappings of healthy brood may appear to have punctures, but since cells are always capped from the outer edges to the middle, the holes are always centered and have smooth edges. In contrast, brood combs from diseased colonies usually have a spotty brood pattern (pepperbox appearance), and the cappings tend to be darker, concave (sunken), and punctured. Also, the combs may contain the dried remains of larvae or pupae (scales), which are found lying lengthwise on the bottom side of brood cells. Sometimes scales are difficult to locate because of the condition of the comb. Scale material can be easily located using long-wave ultraviolet or nearultra-violet light. Exposure to wavelengths of 3100-4000 angstroms will cause scale material to fluoresce. Some discretion must be used with this technique because honey and pollen may also fluoresce. Symptoms of various brood diseases are summarized in Table 102.

Symptoms of a contagious disease are sometimes mimicked because of an unrelated factor. For instance, often brood that is neglected because of a shortage of nurse bees will die from either chilling or starvation.

Table 102. Comparative symptoms of various brood diseases of honey bees

Symptom	American foul brood	European foul brood	Sacbrood	Chalkbrood
Appearance of brood comb	Sealed brood. Discolored, sunken, or punctured cappings	Unsealed brood. Some sealed brood in advanced cases with discolored, sunken or punctured cappings.	Sealed brood. Scattered cells with punctured cappings. often with two holes	Sealed and unsealed brood
Age of dead brood	Usually older sealed larvae or young pupae.* Lying lengthwise in cells.	Usually young unse-aled larvae; occasi-onally older sealed larvae. Typically in coiled stage.	Usually older sealed larvae. Length wise in cells.	Usually older larvae. Lengthwise in cells.
color of dead brood	Dull white, beco-ming light brown, coffee brown to dark brown, or almost black.	Dull white, becoming yellowish white to brown, dark brown, or almost black.	Grayish or straw colored becoming brown, grayish black, or black; he-ad & end darker.	Chalk white. Sometimes mottled with black spots
Consistency of dead brood	Soft, becoming sticky to ropy.	Watery; rarely sticky or ropy. Granular.	Watery and granular; tough skin forms a sac.	Pasty
Odor of dead brood	Slightly to pron-ounced putrid odor.	Slightly to pene-trating sour.	None to slightly sour.	Yeast like, non-objectionable
Scale characte-ristics	Lies uniformly flat on lower side of cell. Adheres tightly to cell wall. Fine, threa-dlike tongue of dead maybe pres-ent. Head lies fl-at. Black in color.	Usually twisted in cell. Does not adhere to cell wall. Rubbery. Black in color.	Head prominently curled towards center towards center of cell. Does not adhere tightly to cell wall. Rough texture. Brittle. Black, in color.	Mummified. does not adhere to cell walls. Brittle. Usually chalky white in color. Sometimes black fruiting bodies are present

*__Bold text__ indicates the most useful field characteristic

Symptoms can also be the result of a failing queen, laying workers, toxic chemicals, or poisonous plants (See section on Noninfectious Diseases).

Bacterial Diseases

American Foulbrood

Bacillus larvae is the bacterium that causes American foulbrood disease (AFB). *Bacillus larvae* is a slender rod with slightly rounded ends and a tendency to grow in chains. The rod varies greatly in length, from about 2.5 to 5 microns (μm), and is about 0.5 μm wide. The spore is oval and approximately twice as long as wide, about 0.6 by 1.3 μm. When stained with carbol fuchsin, the spore walls appear reddish purple and quite clear in the center. The spores may form clusters and appear to be stacked. Approximately 2.5 billion spores are produced in each infected larva. If the larva has been infected for less than 10 days, the vegetative cells are present, and some newly formed spores may be seen.

The modified hanging drop technique can be very useful for differentiating American foulbrood from other brood diseases. In areas of the smear where pockets of water are formed in the oil, the spores of *Bacillus larvae* exhibit Brownian movement. This is an extremely valuable diagnostic technique because the spores formed by the other *Bacillus* species associated with the known bee diseases usually remain fixed. It is important to note that Brownian movement can be affected by slide preparation; also, debris and other bacteria can exhibit this motion. Therefore, Brownian movement must not be used as the sole criterion for diagnosis but must be considered together with the characteristic morphology of the spores and the gross larval symptoms. If microscopic examination is not conclusive, cultural tests can be made using the same suspension.

Cultivation of *Bacillus larvae*

Thiamine (vitamin B) and some amino acids are required for the growth of Bacillus larvae. Routine culture media such as nutrient broth will not support the growth of this organism. Good vegetative growth occurs on Difco brain heart infusion fortified with 0.1 mg thiamine hydrochloride per liter of medium (BHIT) and adjusted to pH 6.6 with HCl, but sporulation does not occur. Satisfactory growth and sporulation occur on the yeast extract, soluble starch, and glucose media recommended by Bailey and Lee (1962). The medium can be liquid, semisolid (0.3% agar), or solid (2% agar). For more information on sporulation, see Dingman and Stahly (1983).

Bacillus larvae spores also reproduce in the hemolymph of honey bee larvae, pupae, and adults when artificially introduced by injection (Michael, 1960, Wilson and Rothenbuhler 1968, Wilson, 1970).

To culture *Bacillus larvae*, we prepare spore suspensions by mixing diseased material (scales) with 9 ml sterile water in screw-cap tubes (We use cotton swab applicators to remove and transfer the scales from).

There are no reliable methods for making plate counts of *Bacillus larvae* because fewer than 10 percent of the spores will produce visible growth on the media presently available (Shimanuki, 1963). By calibrating our methods using spore and plate counts, we have determined that a minimum of 100 B. larvae spores are required to produce visible growth on BHIT.

Diagnostic Tests for *Bacillus larvae*

Holst Milk Test. The holst milk test (Holst, 1946) is a simple test based on the fact that a high level of proteolytic enzymes is produced by sporulating *Bacillus larvae*. The test is conducted by suspending a suspect scale or a smear of a diseased larva in a tube containing 3-4 ml of 1% powdered skim milk in water.

The tube is then incubated at 37°C. If B larvae is present, the suspension should clear in 10-20 minutes. It should be noted that this test is not always reliable.

Nitrate Reduction. *Bacillus larvae* reduces nitrate to nitrite (Lochhead 1937). The nitrate reduction test can be performed on a medium such as BHIT, which contains potassium nitrate (1-2 mg/L of medium). After growth has occurred, the addition of a drop of sulfanilic acid-alphanaphthol reagent produces a red color if nitrate has been reduced to nitrite. Diagnosis should not be based on this test alone but on this test along with larval gross symptoms, bacterial morphology, and growth characteristics of the bacterial colony (Table 103).

Table 103. Differentiation of Bacillus species in honey bee

Species	Brownian movement*	Catalase production	Nitrate reduction	Growth on nutrient agar
Paenibacillus larvae	+	–	+	–
Paenibacillus alvei	–	+	–	+
Brevibacillus laterosporus	–	+	+	+
Paenibacillus pulvifaciens	–	–	+	+

*In modified hanging drop technique.

*In modified hanging drop technique.the comb to the tubes). The suspension is heat shocked at 80°C for 10 minutes (effective time) to kill nonsporeforming bacteria. A sterile cotton swab is used to evenly spread a portion of the suspension (approximately 0.2 mL) over the surface of BHIT agar plates, which are then incubated for 72 hours at 34°C. Individual colonies are small (1-2 mm) and opaque; however, if large numbers of viable B. larva spores are inoculated, a solid layer of growth will cover the plate.

Catalase Production. A drop of 3% hydrogen peroxide is placed on an actively growing culture on a solid medium. Most aerobic bacteria break down the peroxide to water and oxygen and produce a bubbly foam, but *Bacillus larvae* is almost always negative for this reaction (Haynes, 1972).

Fluorescent Antibody. The fluorescent antibody technique requires the preparation of specific antibodies stained with a fluorescent dye. Rabbits are injected with pure cultures of *Bacillus larvae*, and the active antiserum is collected and stained with a fluorescent dye. This fluorescent antiserum is mixed with a bacterial smear on a slide and allowed to react. The excess anti-serum is washed off the slide, and the slide is then examined with a fluorescence microscope. *B. larvae* appears as brightly fluorescing bodies on a dark back-ground (Toschkov *et al.*, 1970, Zhavnenko, 1971, Otte, 1973, Peng and Peng, 1979).

Viability Test for *Bacillus larvae*

One method of controlling AFB is to destroy the viability of *Bacillus larvae* spores in contaminated bee equipment. This can be accomplished by gamma or electron beam irradiation or by fumigation with a sterilant gas such as ethylene oxide. Assessment of the efficacy of these methods should be based on the number of spores remaining viable in a test sample of brood comb containing at least 10 AFB scales.

A spore suspension is prepared from the sample comb by mixing 10 scales in 10 ml sterile water. Since each scale contains about 2.5 billion spores (Sturtevant, 1932), 1 ml of the suspension should contain 2.5 billion spores. A portion of the suspension (0.2 ml = 500 million spores) is spread onto solid BHIT plates as previously described and incubated for 72-96 hours.

The results of the viability tests are reported as the approximate number of viable spores on a per-scale basis. If no colonies form on the medium, the results are recorded as <100 viable spores per scale; 1-9 colonies are recorded as <1,000 per scale; 10-99, <10,000 per scale; over 100 colonies per plate, >10,000 viable spores per scale; and when the plate is completely overgrown with colonies, the results are reported as no detectable reduction of viable spores.

Terramycin (Oxytetracycline) Resistance Tests

Isolates of *Bacillus larvae* can be screened for sensitivity to oxytetracycline based on the size of inhibition zones on agar plates. A spore suspension of the B. larvae isolate to be tested is spread on solid BHIT as described previously. A disk (BBL Sensi-Disc) containing 5 µg oxytetracycline is then placed on the plate, and the plate is incubated at 34°C for 72 hours. The zones formed by sensitive strains usually average 50 mm in diameter. Alternatively, oxytetracycline incorporated into liquid BHIT will inhibit the growth of sensitive strains of *B. larvae* at concentrations as low as 12 µg/l of medium (Gochnauer, 1953). Care is necessary in making these tests, and adequate control strains should be included.

Any substantial reduction of the zone size or an increase in the concentration of oxytetracycline required to prevent growth of *Bacillus larvae* in liquid medium would be evidence of the development of resistant strains (Gochnauer *et al.*, 1975). However, when interpreting the results of the tests, the effects of growth rates should be considered. Because strains often grow at different rates, one may falsely conclude that a strain is resistant. No resistance of *B. larvae* to oxytetracycline has yet been reported.

Detection of *Bacillus larvae* Spores in Hive Products

Honey. Occasionally it may be necessary to examine honey for the presence of *Bacillus larvae*. Due to the high concentration of carbohydrate and other natural bacteriostatic substance(s) in honey, the examination of honey requires special considerations. The classical method (Sturtevant, 1932, 1936) is to dilute the honey 1:9 with water, centrifuge the mixture to concentrate the spores in the sediment, and then examine the sediment microscopically for the presence of spores.

Shimanuki and Knox (1988) have developed the following technique to detect *Bacillus larvae* spores in honey. Honey to be examined is heated to 45°C to permit easier handling and to decrease viscosity for more uniform distribution of any spores that may be present. Twenty-five milliliters of honey is placed in a 50-

ml beaker and diluted with 10 ml of sterile water. The diluted honey is then transferred into a 1.75-inch (44-mm) dialysis tube. The open end is tied after the tube is filled, and the tube is submerged in running water for 18 hours or in a water bath with three to four water changes in that period. Following dialysis, the contents of the tube are centrifuged at about 2,000 g for 20 minutes. The supernatant is carefully removed with a pipet to leave approximately 1 ml of residue. This residue is resuspended in 9 ml of water in a screw-cap vial and heat shocked at 80°C for 10 minutes to kill nonsporeforming bacteria. Next, 0.5 ml of the suspension is spread onto a plate of BHIT agar. The plate is incubated at 37°C for 72 hours and examined for colonies of *B. larvae.* Difficulties can sometimes occur when honey samples contain other sporeforming bacteria that may completely cover the surface of the plate.

Since approximately 100 *Bacillus larvae* spores are required to produce visible growth on BHIT, this technique can demonstrate the presence of *B. larvae* spores in samples that contain a minimum of 80 spores per ml of undiluted honey (25 ml honey X 80 sporesl/ml = 2,000 spores/dialysis = 2,000 spores in 10 ml or 200 spores/ml; 0.5 ml = a 100-spore inoculum). Lower spore levels can possibly be detected by the use of larger honey samples or a second centrifugation to further concentrate the spores.

Pollen. *Bacillus larvae* spores can also be recovered from bee-collected pollen pellets by physically removing bits of AFB scale. A series of sieves of different sizes is helpful. If scales are not detected, one may pass a water-pollen suspension through No. 2 filter paper, centrifuge the filtrate, and culture the pellet as described above (Gochnauer and Corner, 1974).

Beeswax. We have had some success in recovering spores morphologically similar to those of *Bacillus larvae* by melting beeswax in boiling water, removing the beeswax cake after cooling, and centrifuging the water at 2,000 g for 20 minutes. The sediment is then examined microscopically for the presence of spores. Spores have also been recovered from contaminated beeswax by chloroform extraction (Kostecki, 1969). However, in both cases, positive identification of the spores is not possible because the recovery techniques render the spores nonviable.

European Foulbrood

Melissococcus plutonius (= *Streptococcus pluton*) is the bacterium that causes European foulbrood disease (EFB). *Melissococcus pluton* has been reclassified into the new genus Melissococcus (Bailey and Collins, 1982a and b). However, *Melissococcus plutonius* has still not been adequately described to be accepted for the current editions of "Bergey's Manual of Determinative Bacteriology."

Melissococcus plutonius is generally observed early in the infection cycle before the appearance of the varied microflora associated with this disease. The *M. pluton* cell is short, nonsporeforming, and lancet shaped. The cell measures 0.5-0.7 by 1.0 µm and occurs singly, in pairs, or in chains.

Cultivation of *Melissococcus plutonius*

Melissococcus plutonius can be isolated on a medium developed by Bailey (1959). The medium consists of 1 g yeast extract (Difco), 1 g glucose, 1.35 g potassium dihydrogen phosphate (KH2PO4), 1 g soluble starch, 2 g agar, and distilled water to make 100 ml; the pH is adjusted to 6.6 with potassium hydroxide (KOH) and the mixture is autoclaved at 10 lb per square inch (116°C) for 20 minutes. It has been found that the addition of cysteine (0.1 g per 100 mL) improves the multiplication of *M. plutonius* (Bailey and Collins, 1982b).

It is difficult to isolate *Melissococcus plutonius* on artificial media because of its growth requirements and the competition from other bacteria. Also, once isolated, identification of *M. plutonius* is difficult due to its pleomorphic nature in culture. *Melissococcus plutonius* is best isolated when few if any other organisms are present. According to Bailey (1959), it is best to dry smears of diseased larval midguts on a slide. A water suspension of this material or a suspension prepared from larvae (apparently healthy, infected, or dead), cappings, etc., can be streaked on freshly prepared Bailey's agar medium. Or decimal dilutions of these suspensions can be inoculated into molten Bailey's agar medium (45°C) and poured into plates (Bailey, 1981). The plates are incubated anaerobically at 34°C. The "Gas Pak" (BBL) Anaerobic System, including a disposable hydrogen and carbon dioxide generator, is used to obtain anaerobic conditions. Small white colonies of *M. plutonius* should appear after 4 days.

Serodiagnosis

Pinnock and Featherstone (1984) developed an enzyme-linked immunosorbent assay (ELISA) for detecting *Melissococcus plutonius*. Using this technique, they were able to demonstrate the presence of *M. plutonius* even in apparently healthy honey bee colonies.

Organisms associated with European Foulbrood

Some organisms do not cause European foulbrood, but they influence the odor and consistency of the dead brood and can be helpful in diagnosis. These secondary invaders include the following:

Bacillus alvei: The bacterium *Bacillus alvei* is frequently present in cases of European foulbrood disease (EFB). It is a rod 0.5-0.8 µm wide by 2.0-5.0 µm long. Spores measure 0.8 by 1.8-2.2 µm. Like *Bacillus larvae*, the spores may be clumped and appear stacked. The sporangium may be observed attached to the spore. Typical strains of *B. alvei* spread vigorously on nutrient agar and may show "motile colonies"; free spores may lie side by side in long rows on the agar. The growth of this bacterium produces an unpleasant odor.

Bacillus laterosponis (= *Bacillus orpheus*): Rods of *Bacillus laterosporus* measure 0.5-0.8 µm by 2.0-5.0 µm, and the spores 1.0-1.3 by 1.2-1.5 µm (figs). An important diagnostic feature is the production of a canoe-shaped parasporal body that stains very heavily along one side and the two ends, and remains firmly adherent to the spore after lysis of the sporangium. The clear portion with the finely outlined wall is the spore. *B. laterosporus* grows moderately on nutrient agar, becoming dull and opaque, and spreads actively if the agar surface is moist. Growth on nutrient agar with 1% glucose added (glucose agar) is thicker and may become wrinkled.

Enterococcus faecalis (= *Streptococcus faecalis* = *Streptococcus apis* = *Streptococcus liquefaciens*) : Ovoid cells (elongated in the direction of chain) are 0.5-1.0 μm in diameter and are usually in pairs or short chains. This organism resembles *Melissococcus pluton* and may exhibit Brownian movement when the modified hanging drop technique is used. Growth occurs on nutrient agar usually within 1 day. Colonies are generally smaller than 2 mm; they are smooth and convex, with a well-defined border. When magnified, the colonies appear light brown and granular.

Bacterium eurydice (= *Achromobacter eurydice*). There is no standard description of *Bacterium eurydice.* White (1912) described this organism as a small, slender Gram-negative rod with slightly rounded ends, occurring singly or in pairs and measuring 0.5-1.4 μm long by 0.4-0.7 μm wide. According to White (1920), *Bacterium eurydice* is best isolated by plating the midgut contents of infected larvae on glucose agar and incubating at room temperature. Growth is slow and never luxuriant, and colonies are convex, smooth, and glistening. However, later researchers, who were unable to isolate *B. eurydice* as described by White, used the name *Bacterium eurydice* for a Gram-positive bacterium isolated from diseased larvae. Therefore, mention of the name in the literature causes confusion. This organism is not included in the current editions of Bergey's manual.

Bacillus apiarius: The bacterium *Bacillus apiarius* is rarely encountered and may or may not be legitimately associated with EFB. Rods are 0.6-0.8 μm in diameter and often less at the poles. Special diagnostic features include the ridged, thick, rectangular spore coat and the stainable remnants of the sporangium, which remain attached for a considerable time. Growth can occur on Sabouraud dextrose medium.

Powdery Scale

Bacillus pulvifaciens is the bacterium that causes powdery scale disease. Powdery scale disease is seldom reported, perhaps because the average beekeeper is unable to identify it. A useful diagnostic characteristic is the scale that results from the dead larva. The scale is light brown to yellow and extends from the base to the top of the cell. The scale is powdery; when handled, it crumbles into a dust.

Bacillus pulvifaciens vegetative cells measure 0.3-0.6 μm by 1.5-3.0 μm. The spores are 1.0 by 1.3-1.5 μm. The bacterium can be isolated on nutrient agar, but growth is heavier on glucose agar. When first isolated, the organism produces a reddish-brown pigment that can be lost by subculturing. Bacillus pulvifaciens closely resembles *Bacillus larvae,* but the spores do not exhibit Brownian movement in the modified hanging drop technique. Also, *B. pulvijaciens* is distinguished by its ability to grow at 20°C and by its growth on nutrient agar.

Fungal Diseases

Chalkbrood

Ascosphaera apis is the fungus that causes chalkbrood disease. *Ascosphaera apis* is a heterothallic organism and develops a characteristic spore cyst when opposite thallic strains (+ and -) fuse. Spore cysts measure 47-140 µm in diameter. Spore balls endosed within the cyst are 9-19 µm in diameter, and individual spores are 3.0-4.0 µm by 1.4-2.0 µm.

Chalkbrood disease can be easily identified by its gross symptoms. An affected larva becomes overgrown by fluffy cottonlike mycelia and swells to the size of the cell.

If only one strain (+ or -) of mycelium is present, the larva dries into a hard, shrunken, white chalklike mummy- thus the name chalkbrood. When the + and - mycelia are present in a diseased larva, spore cysts can form, and the resulting mummies appear either mottled (black on white) or completely black. In heavily infected hives, mummies can be found at the hive entrances or on the bottom boards. Mummies can sometimes be detected in brood cells by tapping the comb against a solid surface. This easy removal of larval remains also differentiates chalkbrood from other brood diseases.

Ascosphaera apis grows luxuriantly on potato dextrose agar fortified with 4 g yeast extract/L. Growth and sporulation also occur on malt agar but less profusely and with no aerial hyphae; this facilitates subculturing and microscopic examination. Cultures have a characteristic fruity odor similar to that of fermenting peaches. The optimum temperature for growth is 30°C.

Ascosphaera apis can be easily isolated from newly infected larvae or fresh mummies. These can be placed directly on the medium and incubated. New mycelial growth is usually visible within 24 hours. Small blocks of agar containing mycelia can be transferred to new plates to obtain pure cultures and isolates of the + and - strains. *A. apis* can be isolated from old mummies by placing them on water agar (agar with no added nutrients), incubating them, and transferring the new mycelial growth to a nutrient medium. Difficulties sometimes occur because *A. apis* may fail to grow or may be overgrown by other fungi, which can contaminate old mummies.

If only one strain (+ or -) is isolated, a fluffy cottonlike growth will eventually cover the plate. When both the + and - thalli are isolated, spore cysts form throughout the culture. The + and - thalli are morphologically identical. They can be distinguished by inoculating isolates on opposing sides of a plate. When opposite thalli grow together, a line of spore cysts forms at the juncture.

Stonebrood

Stonebrood is usually caused by *Aspergillus flavus*, occasionally *A. fumigatus*, and sometimes other Aspergillus species. These fungi are common soil inhabitants that are pathogenic to adult bees, other insects, mammals, and birds. The disease is difficult to identify in its early stages of infection. The fungus grows rapidly and forms a characteristic whitish-yellow collarlike ring near the head end of the infected larva. A wet mount prepared from the larva shows mycelia penetrating throughout the insect. After death, the infected larva becomes hardened and quite difficult to crush-hence the name stonebrood. Eventually, the fungus erupts from the integument of the insect and forms a false skin. At this stage, the larva may be covered with green powdery fungal spores. The spores of *Aspergillus flavus* are yellow green, and *A. fumigatus* spores are gray green. These spores can become so numerous that they fill the comb cells that contain the affected larvae. Stonebrood can usually be diagnosed from gross symptoms, but positive identification of the fungus requires its cultivation in the laboratory and subsequent examination of its conidial heads. *Aspergillus* spp. can be grown on potato dextrose or Sabouraud dextrose agars.

Viral Disease: Sacbrood

Morator aetatulas is the virus that causes sacbrood disease. It is the only common brood disease that is caused by a virus. Since sacbrood-diseased larvae are relatively free from bacteria, laboratory verification is usually based on gross symptoms and the absence of bacteria. Positive diagnosis requires the use of a special antiserum. Affected larvae change from pearly white to gray and finally black. Death occurs when the larvae are upright, just before pupation. Consequently, affected larvae are usually found in capped cells. Head development of diseased larvae is typically retarded. The head region is usually darker than the rest of the body and may lean toward the center of the cell. When affected larvae are carefully removed from their cells, they appear to be a sac filled with water. Typically the scales are brittle but easy to remove. Sacbrood-diseased larvae have no characteristic odor.

Mixed Infections

Bacillius larvae produces a potent antibiotic that eliminates competition from other bacteria typically associated with honey bee larvae. For this reason, American foulbrood and European foulbrood are rarely found in the same colony, except in cases where AFB is just becoming established in colonies that already have EFB.

It is not unusual to find chalkbrood and sacbrood on the same comb or on a comb with larvae infected with AFB. However, no single larva has been found to be infected with more than one disease. This is an important point to remember when selecting a sample for disease diagnosis.

Chapter 32

Conclusions

Honey and beekeeping have a long history in India. Honey was the first sweet food tasted by the ancient Indian inhabiting rock shelters and forests. He hunted bee hives for this gift of god. India has some of the oldest records of beekeeping in the form of paintings by prehistoric man in the rock shelters. With the development of civilization, honey acquired an unique status in the lives of the ancient Indians. They regarded honey as a magical substance that controlled the fertility of women, cattle, as also their lands and crops. The recent past has witnessed a revival of the industry in the rich forest regions along the sub-Himalayan mountain ranges and the Western Ghats, where it has been practiced in its simplest form.

Beekeeping plays a crucial role in the present context of commercialization of agriculture and liberalization of economy. It covers entire scope of honeybee resources, bee products, beekeeping practices, pollination services and their interface with business systems and environment integrity.

Resources and Potential

There are at least four species of true honey bees and three species of the stingless bees. Several sub-species and races of these are known to exist. In recent years the exotic honey bee has been introduced. Together these represent a wide variety of bee fauna that can be utilized for the development of honey industry in the country.

India has a potential to keep about 150 million bee colonies that can provide self-employment to over 6 million rural and tribal families. In terms of production, these bee colonies can produce over 1.2 million tons of honey and about 15,000 tons of beeswax. Organized collection of forest honey and beeswax using improved methods can result in an additional production of at least 120,000 tons of honey and 10,000 tons of beeswax. This can generate income to about 5 million tribal families.

Potential for honey industry in India

India has immense potential for development of beekeeping. It has been estimated that the Indian forests could provide shelter and food to over 120 million bee colonies. Even if we consider reduction in the forest area in recent years, due to deforestation, etc., the country can still hold over 100 million bee colonies, providing self-employment to over 10 million rural and tribal families. In terms of production, these bee colonies can produce over 700,000 tons of honey and about 30,000 tons of beeswax. If the potential that exists in the country is taken into account, then bee-keeping has never taken off. We do not have even half a million of the colonies at present. Ninety per cent of the potential remains untapped

India has been called a 'land of honey'. The country has some of the oldest records of honey industry in the form of prehistoric paintings depicting hives and the earliest attempts to harvest honey.

Problems with beekeeping in India

The eighties has been a turbulent period for the honey industry in the country. On the one hand, beekeeping with the indigenous bee faced severe constraints and has been left to fend for itself. The Thai sac brood disease that came into the country from the neighbouring Southeast Asian countries, prevailed during this period and delivered a body-blow to the industry. On the other hand, introduction of the exotic bee brought in a new set of problems. Beekeepers in several states that suffered from the viral disease were confused as to the merits of *mellifera* beekeeping, and in the absence of authentic information and know-how, fell prey to unscrupulous elements.

Apiculture industry in India received a serious setback due to the appearance of the acarine and European foulbrood diseases in the indigenous bees for the first time in 1956-57 and 1961. The appearance of these diseases was attributed to the importations of the exotic bees from Europe and America. Chahal and Gatoria (1983) tried to argue that the spread of the diseases in the native bees was not from *Apis mellifera*. They considered that the *Acarapis* mites causing the disease either evolved in the Indo-Malayan region or might have come from the temperate Alps into India through natural spread. Similarly, they suggested that the geographical continuity of Asia with Europe might have served as a natural route for the spread of the European foulbrood disease in India. Verma (1990) feels that epidemics such as the acarine disease cannot spread from ecologically different regions such as the temperate Alps or the tropical Indo-Malayan region.

The case of the spread of Thai sacbrood disease in India raised several controversies as to the source of infection and causes of its spread within the country. The disease was first recorded in Nagaland and Meghalaya in 1978. It spread to Bihar in 1980, Uttar Pradesh and further westwards during the 1980s till it reached Jammu and Kashmir.

The sac brood disease took a heavy toll of bee colonies in the northern states from Nagaland to Jammu and Kashmir. The loss to the honey industry was colossal in Bihar, Uttar Pradesh, Himachal Pradesh, and Jammu and Kashmir. In Punjab, the few bee colonies kept by beekeepers and government organizations, perished after getting infected by the sacbrood virus.

The beginning of the last decade (1991) saw incidence of *Apis iridiscent* virus, and other viral diseases, nosema, acarine, European foulbrood and sacbrood diseases in *Apis cerana indica* in different parts of the country.

Inspite of the vast potential of resources in India the desired level of development of beekeeping has not been acheived. There have been several constraints in the development of beekeeping which include

Constraints in development of beekeeping industry

There have been many problems in the further expansion of *A. mellifera* which include:

The honey resources and beekeeping areas have not been fully exploited.

1. Non - availability of essential materials and bee colonies at the proper time.
2. Lack of bee training facilities and bee nurseries.
3. Lack of disease investigation facilities.
4. No organized system of marketing bee products.
5. Problem of migration to hilly areas due to disturbances in the state.
6. Indiscriminate use of pesticides.
7. Rising cost of beehives and lack of diversification of bee products.
8. Using the correct species
9. Improving the available bee stocks
10. Promotion of mass planting of bee flora
11. Lack of proper management of bee colonies
12. Promotion of migratory beekeeping
13. Extension and Education
14. Beekeeping equipment, design and manufacture
15. Nutritional requirements of bee
16. Diversification of bee products
17. Banking and insurance facilities
18. Regional bee research centres
19. Lack of consumer awareness of honey and its products
20. Beekeeping not being considered as an employment generation activity

21. Lack of sufficient financial help from government and lending institutions for the development of beekeeping
22. Poor quality control for the production of honey
23. Pricing structures for honey
24. Future action plan for development of beekeeping at industry level

Extension

1. Manufacturing, stocking and distribution of bee boxes, honey extractors and other beekeeping equipment.
2. Qualitative improvement in worked out areas.
3. Survey of new potential areas
4. Establishment of demonstration cum training centres
5. Establishment of bee nursery units
6. Registration of beekeepers

Research

1. Standardization of beekeeping equipment
2. Standardization of bee management methods for different zones of India.
3. Protecton of bees from pesticides etc.
4. Maintenance and conservation of all the species of honeybees and assessing their utility as pollinators
5. Improving the strains of honeybees by selection from local stocks, imbibing in them diseases resistance and high honey yielding qualities
6. Assessment of crops and other flowering plants for their degree of utility to honeybees
7. Preparation of list of multiple tree species regionwise having staggered flowering and useful to honeybees during their flowering
8. Co-ordinating with agro forestry, farm forestry social forestry department for introduction of multipurpose plant species useful to honeybees.
9. Quarantine organizations may be developed.
10. Standardization and preservation of other bee products may be taken up.
11. Training and education
12. Syllabi of schools, colleges must include topics on apiculture as graded course.
13. trainings must be organized for unemployed youths and hobbyists]

Marketing

1. Marketing of other bee products like beeswax, propolis, bee collected pollen etc to be encouraged.
2. Quality control of bee products should be assured.
3. Export market for bee products be explored
4. Marketing organization and marketing channels may be established to avoid excessive stocking of honey in any region.

At pollinator end

1. Biodiversity, characterization and taxonomy, multiplication, conservation and domiciliation of pollinators
2. Protocol for multiplication of pollen vectors
3. Prepare bee maps in the country with respect to the distribution, availability of different bee species in different corps. scenario or staturs of bees in time and space.
4. A data base on pollinators, their abundance, behaviour, floral rewards and pollinator plant interaction needs to be generated.
5. The role of honeybees in crop pollination needs to be given due importance.
6. Non – *Apis* bees, pollinators such as stingless bees, carpenter bees, bumblebees, megachilids and many other bees need to be promoted for crop pollination.
7. In addition to cross pollination, the role o9f insects in self pollinatede crops need to be worked out.
8. Role of insects in increasing crop yields qualitatively and quantitatively.
9. Use of pollinators for seed/hybrid seed production using CMS lines and its economics.
10. Isolation distances for pure seed production assume greater significance while using pollinators for seed production to reduce contamination and increase the purity of the seed.
11. Use of pollinators for garden and green house crops
12. Bee lure, attractants, repellents to be used to increase the attractiveness to the crop or repel the bees from insecticidal treated crop.
13. Safety of pollinators from pesticides, diseases, parasites and predators and from adverse climatic conditions.
14. Fast and safe multiplication /domiciliation of pollinators both for green hose and open fields.

15. Need to extend technology through demonstration, farmers participatory trials and HRD by organizing on different aspects of insect pollinators for scientists.
16. Need for development of infrastructural facilities.

Conclusions

Evidently, the country with vast resources of bee flora has great scope for its further expansion if the confronting problems are solved in right perspective.

Glossary

Abdomen-the posterior or third region of the body of the bee that encloses the honey stomach, stomach, intestines, sting and the reproductive organs.

Absconding swarm-an entire colony of bees that abandons the hive because of disease, wax moth, or other maladies.

Acarapis woodi-a mite, called the Tracheal mite, which infests the bees' breathing or tracheal system; sometimes called Acarine Disease, this refers to bees that are heavily infested with the Tracheal mite.

Acidity- the state of being acid or sour; the acids in honey, called organic acids, including gluconic acid, formed by the enzyme glucose to produce the acid and hydrogen peroxide.

Adulterated honey- any product labeled "Honey" or "Pure Honey" that contains ingredients other than honey but does not show these on the label. (Suspe-cted mislabeling should be reported to the Food and Drug Administration.)

Afterswarm- swarms which leave a colony with a virgin queen, after the first (or prime) swarm has departed in the same season; afterswarms are also referred to as secondary or tertiary swarms.

Africanized Bee-a term used indiscriminately to describe the African honey bee *Apis mellifera scutellata* (formerly *A.m. adansonii*) or its hybrids; an African bee released in Brazil and known for its volatile nature, its aggressive behavior may cause concern to the non-beekeeping public.

Alarm odor-a chemical (iso-pentyl acetate) substance released near the worker bee's sting, which alerts other bees to danger; also called alarm pheromone.

Alighting board-A small projection or platform at the entrance of the hive.

Allergic reaction-a systemic or general reaction to some compound, such as bee venom, characterized by itching all over (hives), breathing difficulty, sneezing or loss of consciousness.

American foulbrood-a brood disease of honey bees caused by the spore-forming bacterium, *Bacillus larvae* and characterized by a ropy or gummy condition of affected larvae. It is the most widespread and destructive of the brood diseases, afflicting queen, drone, and worker larvae alike. Adult bees, however, are not affected by AFB. *Bacillus larvae* occurs in two forms: vegetative (rod-shaped bacterial cells) and spores. Only the spore stage is infectious to honey bees.

Anaphylactic shock-constriction of the muscles surrounding the bronchial tubes of a human, caused by hypersensitivity to venom and resulting in sudden death unless immediate medical attention is received.

Anther-from the Greek anthros (flower), referring to the pollen-bearing portion on top of the stamen or male part of a flower.

Antenna (pl –ae)-one of two long segmented sensory filaments located on the head of the bee, which enable bees to smell and taste.

Apiary (pl-ies)-the location and total number of hives (and other equipment) at one site; also called bee yard.

Apiculture-the science and art of raising honey bees.

Apis mellifera-a native European bee that is kept for its honey and wax in most parts of the world, has developed into several races differing in size, color, disposition, and productivity, and has escaped to the wild wherever suitable conditions prevail; subspecies include: *a.m. ligustica* (Italian), the most common domesticated bee; *a.m. caucasia* (Caucasian); *a.m. carnica* (Carnio-lan) *a.m. mellifera* (German black); and *a.m. scutellata / a.m. adonsonii / a.m. intermissa* (African).

Automatic uncapper-automated device that removes the cappings from honey combs, usually by moving heated knives, metal teeth, or flails.

Bacillus larvae-the bacterium that causes American foulbrood

Balling-refers to the action of worker bees surrounding a queen who is unacc-eptable, they are trying to kill her by pulling her legs, wings, and by stinging and suffocation; the bees form a small cluster or ball around this queen.

Basket extractor-a honey extractor that spins out one side of the frame at a time.

Bee blower-a gas or electrically driven blower used to blow bees from supers full of honey.

Bee bread-pollen collected by bees and stored in wax cells, preserved with honey.

Bee brush:-soft brush or whisk (or handful of grass) used to remove bees from frames.

Bee cellar-an underground room used for storing bee hives during long cold winters; difficult to use as constant temperature and humidity must be maintained to ensure colony survival.

Bee diseases-diseases affecting adult larval honey bees, not all of which are infectious (such as dysentary); important diseases are American and European foulbrood, highly infectious larval diseases.

Bee escape-a device constructed to permit bees to pass one way, but prevent their return; used to clear bees from supers or other uses.

Bee Go-a chemical, such as benzaldehyde, repellent to bees and used with a fume board to clear bees from honey supers.

Beehive-a box or receptacle with movable frames, used for housing a colony of bees.

Bee metamorphosis-the three stages through which a bee passes before reaching maturity: egg, larva, and pupa.

Bee space-a space big enough to permit free passage for a bee but too small to encourage comb building, and too large to induce propolizing activities; measures ¼ to 3/8 inch (9.5mm).

Bee suit-a pair of coveralls, usually white, made for beekeepers to protect them from stings and keep their clothes clean; some come equipped with zip-on veils.

Bee tree-a tree with one of more hollows occupied by a colony of bees.

Bee veil-a cloth or wire netting for protecting the beekeeper's head and neck from stings.

Bee venom-poisonous matter secreted by honeybees, used chiefly in defense and communicated by stinging; the poison is secreted by special glands attached to the stinger of the bee [Middle English *venom, venum, venim,* from Old French *venim,* from (assumed) Vulgar Latin *venimen,* alteration of Latin *venenum* drug, poison, magic potion, charm; akin to Latin *venus* love, sexual desire]

Beeswax-a substance that is secreted by bees by special glands on the underside of the abdomen, deposited as thin scales, and used after mastication and mixture with the secretion of the salivary glands for constructing the honeycomb.

Beeway super-the shallowest or section super used with wooden section boxes to make comb honey; has a built-in beeway or bee space.

Benzaldehyde-a colorless nontoxic liquid aldehyde C_6H_5CHO that has an odor like that of bitter almond oil, that occurs in many essential oils (as bitter almond oil and peach-kernel oil) and is usually made from toluene; used to drive bees out of honey supers, but is used chiefly in flavoring and perfumery, in pharmaceutical preparations, and in synthesis (as of dyes) -- called also *artificial bitter almond oil* [German *benzaldehyd,* from *benz-* + *aldehyd* aldehyde]

Black scale-refers to the appearance of a dried down larva or pupa which died of a foulbrood disease.

Boardman feeder-a wooden or plastic device that fits into the entrance of a bee hive and holds a quart jar that can be filled with syrup or water.

Bottling tank-a plastic or stainless steel tank holding 5 or more gallons of honey and equipped with a honey gate to fill honey jars.

Bottom board-the floor of a bee hive.

Brace comb-a bit of comb built between two combs to fasten them together, between a comb and adjacent wood, or between two wooden parts such as top bars.

Braula coeca-the scientific name of a wingless fly commonly known as the bee louse

Brood-immature stages of bees not yet emerged from their cells; the stages are egg, larvae, pupae.

Brood chamber-the part of the hive in which the brood is reared; may include one or more hive bodies and the combs within.

Brood diseases-diseases that affect only the immature stages of bees, such as American or European foulbrood.

Brood nest-the part of the hive interior in which brood is reared; usually the two bottom supers.

Buckfast hybrid-a strain of bees developed by Brother Adam at Buckfast Abbey in England, bred for disease resistance, disinclination to swarm, hardiness, comb building and good temper.

Burr comb-small pieces of comb made as connecting links between combs or between a frame and the hive itself; also called brace comb.

Cage shipping-also called a package, a screened box filled with 2 to 5 pounds of bees, with or without a queen, and supplied with a feeder can; used to start a new colony, or to boost a weak one.

Candy plug-a fondant type candy placed in one end of a queen cage to delay her release.

Capped brood- immature bees whose cells have been sealed over with a brown wax cover by other worker bees; inside, the non-feeding larvae are isolated and can spin cocoons prior to pupating.

Capping melter-melter used to liquefy the wax from cappings as they are removed from honey combs

Cappings-the thin wax covering over honey; once cut off of extracting frames they are referred to as cappings and are a source of premium beeswax.

Capping scratcher - a fork-like device used to remove wax cappings covering honey, so it can be extracted.

Carbohydrate-a food (organic compound) composed of carbon, hydrogen, and oxygen with the hydrogen:oxygen ratio frequently 2:1, as in water.

Carnolian bees-a grayish race of honey bee *Apis mellifera carnica* named for Carniola, Austria but originating in the Balkan region; while they are gentle and do not propolize, they tend to swarm more than other races.

Castes-the three types of bees that comprise the adult population of a honey bee colony: workers, drones, and queen

Caucasian bees-a black race of honey bee *A. mellifera caucasica*, originating in the Caucasus mountains; gentle but tend to propolize excessively.

Cell-the hexagonal compartment of a honey comb.

Cell bar-a wooden strip on which queen cups are placed for rearing queen bees.

Cell cup-base of an artificial queen cell, made of beeswax or plastic and used for rearing queen bees

Chalkbrood-a disease affecting bee larvae, caused by a fungus *Ascosphaera apis*, larvae eventually turn into hard, chalky white "mummies".

Chilled brood-immature bees that have died from exposure to cold; commonly caused by mismanagement.

Chimmey effect-the tendency for bees to fill only the center frames of honey supers; happens when bees are given too much room too fast.

Chromosome-a group of nuclear bodies (from the nucleus) containing genes; responsible for the differentiation and activity of a cell, and undergoing characteristic division stages such as mitosis.

Chunk honey-honey in the comb, but not in sections, frequently cut and packed into jars then filled with liquid honey.

Clarifying-removing visible foreign material from honey or wax to increase its purity.

Cluster-a large group of bees hanging together, one upon another

Cocoon-a thin silk covering secreted by larval honey bees in their cells in preparation for pupation.

Colony-the aggregate of worker bees, drones, queen, and developing brood living together as a family unit in a hive or other dwelling

Comb-the wax portion of a colony in which eggs are laid, and honey and pollen are stored.

Comb, drawn-wax foundation with the cell walls drawn out by the bees, completing the comb.

Comb foundation-a commercially made struc ture consisting of thin sheets of beeswax with the cell bases of worker cells embossed on both sides in the same manner as they are produced naturally by honey bees.

Comb honey-honey in the wax combs, usually produced and sold as a separate unit, such as a wooden section 4-1/2" square, or a plastic round ring.

Conical escape-a cone-shaped bee escape, which permits bees, a one-way exit; used in a special escape board to free honey supers of bees.

Creamed honey-honey that has been pasteurized and undergone controlled granulation to produce a finely textured candied or crystallized honey which spreads easily at room temperature.

Crimp-wired foundation-comb foundation into which crimp wire is embedded vertically during foundation manufacture

Cross-pollination-The transfer of pollen from the anther of one flower to the stigma of another flower of the same species.

Crystalize-See Granulate.

Cut-comb honey-Comb honey cut into various sizes, the edges drained, and the pieces wrapped or packed individually

Dearth-a period of time when there is no available forage for bees, due to weather conditions (rain, drought) or time of year.

Decoy hive-a hive placed to attract stray swarms.

Demaree-the method of swarm control that separates the queen from most of the brood within the same hive.

Dequeen-to remove a queen from a colony

Dextrose-also known as glucose (grape sugar), it is a simple sugar (or monosaccharide) and is one of the two main sugars found in honey; forms most of the solid phase in granulated honey.

Diatase-a starch digesting enzyme in honey adversely affected by heat; used in some countries to test quality and heating history of stored honey.

Disease resistence-the ability of an organism to avoid a particular disease; primarily due to genetic immunity or avoidance behavior.

Dividing-separating a colony to form two or more units.

Division-See split.

Division board feeder-a wooden or plastic compartment which is hung in a hive like a frame and contains sugar syrup to feed beesc

Double screen-a wooden frame, 1/2 to 3/4 inch thick, with two layers of wire screen to separate two colonies within the same hive, one above the other. An entrance is cut on the upper side and placed to the rear of the hive for the upper colony

Double story-referring to a beehive comprised of two deep supers, one for brood and one for honey.

Drawn combs-combs with cells built out by honey bees from a sheet of foundation

Drifting-the movement of bees that have lost their location and enter other hives; common when hives are placed in long straight rows where returning foragers from the center hives tend to drift to the row ends.

Drone-the male honeybee which comes from an unfertilized egg (and is therefore haploid) laid by a queen or less commonly, a laying worker.

Drone brood or drone comb-brood, which matures into drones, reared in cells larger than worker brood.

Drone congregating area (DCA)-a specific area to which drones fly waiting for virgin queens to pass by; it is not known how or when they are formed, but drones return to the same spots year after year.

Drone layers-a drone laying queen or laying workers.

Drone laying queen-a queen that can lay only unfertilized eggs, due to age, improper or no mating, disease or injury.

Drumming-pounding on the sides of a hive to make the bees ascend into another hive placed over it.

Dwindling-the rapid dying off of old bees in the spring; sometimes called spring dwindling or disappearing disease.

Dysentery-an abnormal condition of adult bees characterized by severe diarrhea and usually caused by starvation, low-quality food, moist sur roundings, or nosema infection

Eggs-the first phase in the bee life cycle, usually laid by the queen, is the cylindrical egg 1/16in (1.6 mm) long; it is enclosed with a flexible shell or chorion.

Electric embedder-a device allowing rapid em bedding of wires in foundation with electrically produced heat

Entrance reducer-a notched wooden strip used to regulate the size of the bottom entrance.

Escape board-a board having one or more bee escapes in it; used to remove bees from supers.

European foulbrood-an infectious brood dis ease of honey bees caused by *streptococcus pluton*.

Eyelets, metal-a small metal piece fitting into the wire-holes of a frame's end bar; used to keep the reinforcing wires from cutting into the wood.

Extracted honey-honey removed from combs by means of a centrifugal force; the combs remain intact.

Feeders-various types of appliances for feeding bees artificially.

Fermenting honey-honey which contains too much water (greater than 20%) in which a chemical breakdown of the sugars takes place producing carbon dioxide and alcohol; caused by naturally-occurring osmophylic yeasts of the genus saccharomeyces (formerly zygosaccharomyces).

Fertile queen-a queen, inseminated instrumentally or mated with a drone, which can lay fertilized eggs

Fertilized-usually refers to eggs laid by a queen bee, they are fertilized with sperm stored in the queen's spermatheca, in the process of being laid.

Festooning-the activity of young bees, engorged with honey, hanging on to each other and secreting beeswax.

Field bees-worker bees which are usually 21 or more days old and work outside to collect nectar, pollen, water and propolis; also called foragers.

Flash heater-a device for heating honey very rapidly to prevent it from being damaged by sustained periods of high temperature

Flight path-usually refers to the direction bees fly leaving their colony; if obstructed, may cause bees to become aggravated.

Follower board-a thin board used in place of a frame usually when there are fewer than the normal number of frames in a hive.

Food chamber a hive body filled with honey for winter stores.

Forage-natural food source of bees (nectar and pollen) from wild and cultivated flowers.

Foreign matter-in honey, unusually high amounts of wax, bee bodies, pollen grains, or other objectionable debris.

Foundation, wax-thin sheets of beeswax embossed or stamped with the base of a worker (or rarely drone) cells on which bees will construct a complete comb (called drawn comb); also referred to as comb foundation, it comes wired or unwired.

Foundation, wired-comb foundation which includes evenly-spaced vertical wires for added support; used in brood or extracting frames.

Foulbrood, American-a malignant, contagious bacterial disease affecting bee larvae caused by a spore-forming bacteria bacillus larvae.

Foulbrood, European-a serious, infectious larval disease of honeybees caused by melissoccoccus pluton formerly streptococcus pluton), a spore-forming bacteria.

Frame-four pieces of wood forming a rectangle, designed to hold honey comb, consisting of a top bar, two end bars, and a bottom bar (one or two pieces); usually spaced a bee-space apart in the super.

Fructose-see levulose.

Fume board-a devise used to hold a set amount of a volatile chemical (a bee repellent like bee go) to drive bees from supers.

Fumagillin-bicyclohexyl-ammonium fumagillin, whose trade name is fumadil-b (abbot labs), is a whitish soluble antibiotic powder discovered in 1952; it is mixed with sugar syrup and fed to bees to control nosema disease.

Fumidil-b-the trade name for fumagillin, an antibiotic used in the prevention and suppression of nosema disease.

Fume board-a rectangular frame, the size of a super, covered with an absorbent material such as burlap, on which is placed a chemical repellent to drive the bees out of supers for honey removal

Gloves-leather, cloth or rubber gloves worn while inspecting bees.

Glucose-see dextrose.

Gluconic acid-see acidity

Grafting-removing a worker larva from its cell and placing it in an artificial queen cup in order to have it reared into a queen.

Grafting tool-a needle or probe used for trans ferring larvae in grafting of queen cells

Granulate-the process by which honey, a super-saturated solution (more solids than liquid) will become solid or crystallize; speed of granulation depends of the kinds of sugars in the honey.

Guard bees-worker bees about three weeks old, which have their maximum amount of alarm pheromone and venom; they challenge all incoming bees and other intruders.

Gum-a hollow log beehive, sometimes called a log-gum (appalachia), made by cutting out that portion of a tree containing bees and moving it to the apiary; since it contains no moveable frames, it is therefore illegal.

Hay fever-an allergic condition that afflicts many people; caused by various plant particles, airborne fungal spores or pollen.

Hive-a manmade home for bees including a bottom board, hive bodies, frames enclosing honey combs, and covers.

Hive body-a wooden box containing frames.

Hive stand-a structure serving as a base support for a beehive; it helps in extending the life of the bottom board by keeping it off damp ground.

Hive staples-large c-shaped metal nails, hammered into the wooden hive parts to secure bottom to supers, and supers to super before moving a colony.

Hive tool-a flat metal device with a curved scraping surface at one end and a flat blade at the other; used to open hives, pry apart and scrape frames.

Hoffman self-spacing frame-frames that have the end bars wider at the top than the bottom to provide the proper spacing when frames are placed in the hive.

Honey-a sweet viscid material produced by bees from the nectar of flowers, composed largely of a mixture of dextrose and levulose dissolved in about 17 percent water; contains small amounts of sucrose, mineral matter, vitamins, proteins, and enzymes.

Honeydew-an excreted material from insects in the order homoptera (aphids) which feed on plant sap; since it contains almost 90% sugar, it is collected by bees and stored as honeydew honey.

Honeybee-the common name for apis mellifera (honey bearer), a highly social insect, order hymenoptera (membranous wings); correctly printed as two words.

Honey color-measured by a pfund grader, honey colors are classified between water white to white, to amber to dark amber (7 gradations).

Honey extractor-a machine which removes honey from the cells of comb by centrifugal force.

Honey flow-a time when enough nectar-bearing plants are blooming such that bees can store a surplus of honey.

Honey gate-a faucet used for removing honey from tanks and other storage receptacles.

Honey house-a building used for activities such as honey extraction, packaging and storage.

Honey plants-plants whose flower (or other parts) yields enough nectar to produce a surplus of honey; examples are asters, basswood, citrus, eucalyptus, goldenrod and tupelo.

Honey pump-a pump used to transfer honey from a sump or extractor to a holding tank or strainer

Honey sac-also called honey stomach, an enlargement at the posterior (back) end of a bees' esophagus but lying in the front part of the abdomen, capable of expanding when full of liquid such as nectar or water.

Honey stomach-an organ in the abdomen of the honey bee used for carrying nectar, honey, or water.

Honey sump-a clarifying tank between the extractor and honey pump for removing the coarser particles of comb introduced during extraction.

Honey supers-refers to hive bodies used for honey production.

Hornets and yellow jackets-social insects belonging to the family vespidae. nest in paper or foliage material, with only an overwintering queen. fairly aggressive, and carnivorous, but generally beneficial, they can be a nuisance to man. hornets and yellow jackets are often confused with wasps and honeybees. wasps are related to hornets and yellow jackets, the most common of which are the paper wasps which nest in small exposed paper combs, suspended by a single support. hornets, yellow jackets and wasps are easy to distinguish by their larger size, shiny hairless body, and aggressiveness. honeybees are generally smaller, fuzzy brown or tan, and basically docile in nature.

Hypersensitive-a condition in which reactions to any environmental stimulus are life-threatening; such as honey bee venom.

Imperfect-not fully formed, such as a worker, considered an imperfect female.

Increase-to add to the number of colonies, usually by dividing those on hand. see split.

Infertile-incapable of producing a fertilized egg, as a laying worker.

Inhibine-antibacterial effect of honey caused by an accumulation of hydrogen peroxide, a result of the chemistry of honey.

Injections, desensitizing-a series of injections given to persons with allergies (such as bee venom) so they might build up an immunity.

Inner cover-an insulating cover fitting on top of the top super but underneath the outer cover, with an oblong hole in the center.

Insecticide-any chemical that kills insects. Inspectors, state-persons usually employed by state agriculture departments to inspect colonies of bees for diseases and pests.

Instrumental insemination-the introduction of drone spermatozoa into the genital organs of a virgin queen by means of special instruments

Invertase-an enzyme in honey, which splits the sucrose molecule (a disacch-aride) into its two components dextrose and levulose (monosaccharides).

Isomerase-a bacterial enzyme used to convert glucose in corn syrup into fructose, which is a sweeter sugar; called isomerose, is now used as a bee feed.

Italian bees-a common race of bees, apis mellifera ligustica, with brown and yellow bands, from italy; usually gentle and productive, but tend to rob.

Langstroth, L. L.-a philadelphia native and minister (1810-95), he lived for a time in ohio where he continued his studies and writing of bees; recognized the importance of the bee space, resulting in the development of the movable-frame hive.

Larva, capped (pl-ae)-the second developmental stage of a bee, ready to pupate or spin its cocoon (about the 10th day from the egg).

Laying workers-worker bees which lay eggs in a colony hopelessly queenless; such eggs are infertile, since the workers cannot mate, and therefore become drones.

Leg baskets-also called pollen baskets, a flattened depression surrounded by curved spines located on the outside of the tibiae of the bees' hind legs and adapted for carrying flower pollen or other dusts.

Levulose-also called fructose (fruit sugar), a monosaccharide commonly found in honey that is slow to granulate (such as robinia or locust honey); chemical formula is like glucose, but has it's carbonyl group in a different place.

Mandibles-the jaws of an insect; used by bees to form the honey comb and scrape pollen, in fighting and picking up hive debris.

Maternal-from the mother's side of the family.

Mating flight-the flight taken by a virgin queen while she mates in the air with several drones.

Mead-honey wine

Midnight hybrid-a combination of the caucasian and carniolan races.

Migratory beekeeping-the moving of colonies of bees from one locality to another during a single season to take advantage of two or more honey flows.

Migratory cover-an outer cover used without an inner cover that does not telescope over the sides of the hive; used by commercial beekeepers who frequently move hives.

Moisture content-in honey, the percentage of water should be no more than 18.6; any percentage higher than that will allow honey to ferment.

Movable frames-a frame constructed in such a way to preserve the bee space, so they can be easily removed; when in place, it remains unattached to its surroundings.

Moving board-a framed screen that fits over the top as a hive cover; used to move bees in hot weather to provide sufficient ventilation to keep bees from suffocating.

Natural honey-unfiltered and unheated honey.

Nectar-a liquid rich in sugars, manufactured by plants and secreted by nectary glands in or near flowers; the raw material for honey.

Nectary glands-special nectar secreting glands usually found in flowers, whose function is to attract pollinating insects, such as honey bees for the purpose of cross pollination, by offering a carbohydrate-rich food.

Nectar guide-color marks on flowers believed to direct insects to nectar sources.

Nectaries-the organs of plants which secrete nectar, located within the flower (floral nectaries) or on other portions of the plant (extrafloral nectaries).

Newspaper method-a technique to join together two strange colonies by providing a temporary newspaper barrier.

Nosema disease-a widespread adult bee disease caused by a one-celled spore-forming organism nosema apis; it infects the gut lining.

Nuc, nuclei, nucleus-a small colony of bees often used in queen rearing.

Nurse bees-young bees, three to ten days old, which feed and take care of developing brood.

Observation hive-a hive made largely of glass or clear plastic to permit observation of bees at work

Osmotic pressure-the minimum pressure that must be applied to a solution to prevent it from gaining water when it is separated from pure water by a permeable membrane; in honey, its ability to absorb water from the air or other microscopic organisms, about 2000 milliosmols/kg.

Outer cover-the last cover that fits over a hive to protect it from rain; the two most common kinds are telescoping and migratory covers.

Outyard-also called out apiary, it is an apiary kept at some distance from the home or main apiary of a beekeeper; usually over a mile away from the home yard.

Ovary-the egg producing part of a plant or animal.

Ovule-an immature female germ cell, which develops into a seed.

Oxytetracycline-an antibiotic sold under the trade name terramycin; used to control american and european foulbrood diseases.

Package-see shipping cage.

Package bees-a quantity of adult bees (2 to 5 pounds), with or without a queen, contained in a screened shipping cage.

Paralysis-a virus disease of adult bees which affects their ability to use legs or wings normally

Parthenogenesis-the development of young from unfertilized eggs laid by virgin females (queen or worker); in bees, such eggs develop into drones.

PDB (Paradichlorobenzene)-a white crystalline substance whose vapors are heavier than air and are used to fumigate wax moths in stored hive bodies.

Piping-a series of sounds made by a queen, frequently before she emerges from her cell.

Play flights-short flights taken in front and in the vicinity of the hive by young bees to acquaint them with hive location; sometimes mistaken for robbing or swarming preparations.

Poison sac-large oval sac containing venom and attached to the anterior (front) part of the sting; stores venom produced by the poison gland, and its primary ingredients are peptide and mellitin.

Pollen-the dust-like male reproductive cells (gametophytes) of flowers, formed in the anthers, and important as a protein source for bees; pollen is essential for bees to rear brood.

Pollen basket-see leg basket.

Pollen cakes or pollen pellets-the cakes of pollen packed in the leg baskets of bees and transported back to the colony.

Pollen insert-a device inserted in the entrance of a colony into which hand-collected pollen is placed. As the bees leave the hive and pass through the trap, some of the pollen adheres to their bodies and is carried to the blossom, resulting in cross-pollination.

Pollen substitute-a food material which is used to substitute wholly for pollen in the bees' diet; usually contains all or part of soy flour, brewers' yeast, wheast, powdered sugar, or other ingredients.

Pollen supplement-a mixture of pollen and pollen substitutes used to stimulate brood rearing in periods of pollen shortage

Pollen Trap-a device for collecting the pollen pellets from the hind legs of worker bees; usually forces the bees to squeeze through a screen mesh, which scrapes off the pellets.

Pollen tube-a slender thread-like growth, containing sperm cells, which penetrates the female tissue (stigma) of a flower until it eventually reaches the ovary; there the sperm cells unite with the ovule.

Pollination-the transfer of pollen from the anthers to the stigma of flowers.

Pollinator-the agent that transfers pollen from an anther to a stigma: bees, flies, beetles, etc.

Pollinizer-the plant source of pollen used for pollination.

Porter bee escape-introduced in 1891, the escape is a device that allows the bees a one-way exit between two thin and pliable metal bars that yield to the bees' push; used to free honey supers of bees but may clog since drone bees often get stuck.

Prime swarm-the first swarm to leave the par ent colony, usually with the old queen.

Proboscis-the mouthparts of the bee that form the sucking tube or tongue

Propolis-plant resins collected and modified by bees; used to fill in small spaces inside the hive.

Propolize-to fill with propolis, or bee glue; used to strengthen the comb and seal cracks, it also has antimicrobial properties.

Protein-naturally occurring complex organic substances, such as pollen; composed of amino acids, the building blocks of protein.

Pupa (pl-ae)-the third stage in the development of the bee during which it is inactive and sealed in its cocoon; the organs of the sealed in its cocoon; the organs of the larva are replaced by those which will be used as an adult.

Queen-a fully developed mated female bee responsible for all the egg laying of a colony; recognized by other bees by her special pheromones (odors).

Queen cage-a special cage in which queens are shipped and/or introduced to a colony, usually with 5 or 6 young workers called attendants, and a candy plug.

Queen cage candy-candy made by kneading powdered sugar with invert sugar syrup until it forms a stiff dough; used as food in queen cages.

Queen cell-a special elongated cell resembling a peanut shell in which the queen is reared; usually over an inch in length, it hangs vertically from the comb.

Queen clipping- removing a portion of one or both front wings of a queen to prevent her from flying

Queen Cup- a cup-shaped cell hanging vertically from the comb, but containing no egg; also made artificially of wax or plastic to raise queens

Queen Excluder - a device made of wire, wood or zinc (or any combination thereof) having openings of.163 to.164 inch, which permits workers to pass but excludes queens and drones; used to confine the queen to a specific part of the hive, usually the brood nest.

Queenright- a colony that contains a laying queen.

Queen substance - pheromone material secreted from glands in the queen bee and transmitted throughout the colony by workers to alert other workers of the queen's presence.

Rabbet-a narrow piece of folded metal fastened to the inside upper end of the hive body from which the frames are suspended.

Races of bees-the four most common races of *apis are mellifera, cerana, dorsata* and *florea*; other newly discovered races are currently under investigation.

Radial extractor-a centrifugal force machine to throw out honey but leave the combs intact; the frames are placed like spokes of a wheel, top bars towards the wall, to take advantage of the upward slope of the cells.

Raw honey-see natural honey.

Requeen-to introduce a new queen to a queenless hive.

Rendering wax-the process of melting combs and cappings and removing refuse from the wax.

Resmethrin (sbp-1382)-a synthetic pyrethroid insecticide used to kill diseased honey bee colonies

Reversing-the act of exchanging places of different hive bodies of the same colony; usually for the purpose of nest expansion, the super full of brood and the queen is placed below an empty super to allow the queen extra laying space.

Robbing-the act of bees stealing honey/nectar from the other colonies; also applied to bees cleaning out wet supers or cappings left uncovered by beekeepers.

Ropy characteristic-a diagnostic test for american foulbrood in which the decayed larvae form an elastic rope when drawn out with a toothpick.

Round sections-sections of comb honey in plastic round rings instead of square wooden boxes.

Royal jelly-a highly nutritious, milky white glandular secretion of young (nurse) bees; used to feed the queen and young larvae.

Sacbrood-a brood disease of bees caused by a filterable virus which interferes with the molting process; the dead larva resembles a bag of fluid.

Scout bees-worker bees searching for a new source of pollen, nectar, propolis, water, or a new home for a swarm of bees.

Screened ventilated board-a framed screen used to cover the top of a hive being moved in hot weather.

Sealed brood-see "capped brood."

Sections-small wooden (or plastic) boxes used to produce comb honey.

Self-pollination-the act of a single flower, or flower from the same plant, pollinating itself.

Self-spacing frames-frames constructed so that they are a bee space apart when pushed together in a hive body.

Self-sterile-the inability of a flower, such as a fruit tree, to be fertilized within its own variety; it is only fertilized by pollen from another variety.

Settling tank-a large capacity container used to settle extracted honey; air bubbles and debris will float to the top, clarifying the honey.

Skep-a beehive without moveable frames, usually made of twisted straw in the form of a basket; its use is illegal in the u.s.

Slatted rack-a wooden rack that fits between the bottom board and hive body. Bees make better use of the lower brood chamber with increased brood rearing, less comb gnawing, and less con gestion at the front entrance.

Slumgum- the refuse from melted combs and cappings after the wax has been rendered or removed;usually contains cocoons, pollen, bee bodies and dirt.

Smoker-a metal container with attached bellows which burns organic fuels to generate smoke; used to control aggressive behavior of bees during colony inspections.

Solar wax melter or extractor-a glass-covered insulated box used to melt wax from combs and cappings using the heat of the sun.

Sperm cells-the male reproductive cells (gametes) which fertilize eggs; also called spermatozoa.

Spermatheca-a small sac connected with the oviduct (vagina) of the queen bee in, which is stored, the spermatozoa received in mating with drones.

Split-to divide a colony for the purpose of increasing the number of hives.

Spur embedder-a device used for mechanically embedding wires into foundation by employing hand pressure

Starline Hybrid-an italian bee hybrid known for vigor and honey production.

Stigma-receptive portion of the female part of a flower to which pollen adheres.

Sting-an organ belonging exclusively to female insects developed from egg laying mechanisms, used to defend the colony; modified into a piercing shaft through which venom is injected.

Sting sac-see poison sac.

Straining screen-a metal or plastic screen through which honey is filtered; also serves as a base for other, finer screening material.

Streptococcus pluton-bacterium that causes european foulbrood.

Sucrose-principal sugar found in nectar.

Sugar syrup-feed for bees, containing sucrose or table (cane) sugar and hot water in various ratios.

Super-a receptacle in which bees store honey; usually placed over or above the brood nest; so called brood supers contain brood.

Supering-the act of placing honey supers on a colony in expectation of a honey flow.

Supersedure-rearing a new queen to replace the mother queen in the same hive; shortly after the daughter queen begins to lay eggs, the mother queen disappears.

Surplus honey-any extra honey removed by the beekeeper, over and above what the bees require for their own use, such as winter food stores.

Swarm-a collection of bees, containing at least one queen that split apart from the mother colony to establish a new one; a natural method of propagation of honey bees.

Swarm cell-queen cells usually found on the bottom of the combs before swarrning.

Swarming-the natural method of propagation of the honey bee colony.

Swarming season-the time of year, usually mid-summer, when swarms usually issue.

Terramycin-an antibiotic used to prevent american and european foulbrood. See oxytetracycline.

Tested queen-a queen whose progeny shows she has mated with a drone of her own race and has other qualities which would make her a good colony mother.

Thin super foundation-a comb foundation used for comb honey or chunk honey production which is thinner than that used for brood rearing.

Thorax-the central region of an insect to which the wings and legs are attached.

Top bar-the top part of a frame.

Transferring-the process of changing bees and combs from common boxes to movable frame hives.

Travel stains-the darkened appearance on the surface of honeycomb caused by bees walking over its surface.

Uncapping knife-a knife used to shave off the cappings of sealed honey prior to extraction; hot water, steam or electricity can heat the knives.

Uncapping tank-a container over which frames of honey are uncapped; usually strains out the honey which is then collected.

Unfertilized-an ovum or egg, which has not been united with the sperm.

Uniting-combining two or more colonies to form a larger colony

***Varroa jacobsoni* (*V.destructor*)**-an external mite parasite on honeybees.

Veil-a protective netting that covers the face and neck; allows ventilation, easy movement and good vision.

Venom allergy-a condition in which a person, when stung, may experience a variety of symp toms ranging from a mild rash or itchiness to anaphylactic shock. A person who is stung and experiences abnormal symptoms should consult a physician before working bees again.

Venom hypersensitivity-a condition in which a person, if stung, is likely to experience an aphylactic shock. A person with this condition should carry an emergency insect sting kit at all times during warm weather

Virgin queen-an unmated queen bee.

Warming cabinet-an insulated box or 100m heated to liquefy honey.

Wasp-a close relative of honey bees, usually in the family vespidae; they are carnivorous, some species preying on bees (see also, hornet).

Wax-see beeswax.

Wax glands-the eight glands located on the last 4 visible, ventral abdominal segments of young worker bees; they secrete beeswax droplets.

Wax moths-usually refers to the greater wax moth, galleria mellonella whose larvae bore through and destroy honeycomb as they eat out its impurities.

Wax scale-a drop of liquid beeswax that hardens into a scale upon contact with air; in this form it is shaped into comb.

Wax tube fastener-a metal tube for applying a fine stream of melted wax to secure a sheet of foundation to an un-grooved frame.

Wind pollinated-plants whose flowers manufacture light pollen (and usually no nectar) which is released into the air to fall by chance on a receptive stigma; examples include the grasses (corn, oats) and conifers (pines).

Windbreaks- pecially constructed, or naturally occurring barriers to reduce the force of the (winter) winds on a beehive.

Winter cluster-a tight ball of bees within the hive to generate heat; forms when outside temperature falls below 57 degrees f.

Winter hardiness-the ability of some strains of honeybees to survive long winters by frugal use of stored honey.

Wire, frame-thin 28# wire used to reinforce foundation destined for the broodnest or honey extractor.

Wire cone escape-a one-way cone formed by window screen mesh used to direct bees from a house or tree into a temporary hive.

Worker bees-infertile female bee whose reproductive organs are only partially developed, responsible for carrying out all the routine of the colony.

Worker comb-comb measuring about five cells to the inch, in which workers are reared and honey and pollen are stored.

Bibliography

Abbott, W.S. 1925. A method of computing the effectiveness of an insecticide. *J. Econ. Ent.* 18: 265-267

Abrol D.P. 2002. Potential of beekeeping in Jammu and Kashmir. *Popular Science Magazine*10 (1): 33-35.

Abrol, D.P, Sharma, D and Monobrullah, M. 2005. Abundance and diversity of pollinating insects visiting peach and plum flowers and their impact on fruit production. *SKUAST Journal of Research* 4(1): 38-45

Abrol, D.P. 1986a. Ecophysiological adaptations between pollinating bees and their flowers. *Environ. Ecol.,* 4 : 161-162.

Abrol, D.P. 1986a. Time and energy budgets of alfalfa pollinating bees *Megachile nana* and *Megachile flavipes* Spinola (Hymenoptera, Megachilidae). *Proceedings of the Indian Academy of Sciences, Animal Science* 95(5): 579-586.

Abrol, D.P. 1986b. Significance cf environmental cues influencing pollination activity of *Andrena ilerda. Environ. Ecol* 4 : 169-171.

Abrol, D.P. 1986b. Wing beat frequencies of *Andrena ilerda* and *Andrena leaena* (Hymenoptera, Andrenidae). *Annals of Biology* (Ludhiana) 2(1): 98-99.

Abrol, D.P. 1986c. Wing beat frequency of *Andrena ilerda* and *A.leaena* (Hymenoptera : Andrenidae) *Ann. Biol.* 2 : 98-99.

Abrol, D.P. 1986d. Differential dilution of nectar by sub-tropical bees.*Environ. Ecol.* 4:188-189.

Abrol, D.P. 1986f. Flight range and significance of wing hooks in *Megachile femorata* Smith (Hymenoptera : Megachilidae) *J. Anim. Morphol Physiol.,* 33:106-112

Abrol, D.P. 1987. Activity time budgets and pollination potential of *Apis cerana indica* F. workers. *Environ. Ecol.* 5:757-759

Abrol, D.P. 1987. Analysis of environmental factors affecting foraging behaviour of *Megachile bicolor* F. on *Crotalaria juncea* L. *Bangladesh J. Agric. Res.* 12 : 5-14.

Abrol, D.P. 1987. Influence of thermal and energetic constraints on the pollination activity of *Xylocopa pubescens* Spinola (Hymenoptera : Anthophoridae) *Environ. Ecol.* 5: 90-93

Abrol, D.P. 1987. Interrelationships and path coefficient analysis of environmental factors influencing pollination activity of *Apis dorsata* F. *Trop. Ecol.* 28 : 147-154

Abrol, D.P. 1988. Ecology and behaviour of three bee species pollinating loquat (*Eriobotrya japonica* Lindley). *Proc. Indian Nat. Sci. Acad.* 54B : 161-163.

Abrol, D.P. 1988. Effect of climatic factors on the pollination activity of alfalfa pollinating subtropical bees *Megachile nana.* Bingh and *M. flavipes* Spinola (Hymenoptera: Megachilidae). *Acta Ecologia, Oecologia Generalis* 9 (4) : 371-176

Abrol, D.P. 1988. Environmental factors influencing pollination activity of *Apis mellifera* L. on *Brassica campestrisL. var toria J. Indian Inst. Sci.* (Biol. Sci.) 68 : 49-52.

Abrol, D.P. 1988. Foraging range of subpopical bees, *Megachile flavipes, Megachile nana* (Hymenoptera: Megachilidae) and *Apis florea* (Hymenoptera: Apidae). *J.Indian Inst. Sci.* (Biol. Sci) 68 : 43-47.

Abrol, D.P. 1988. Pollination studies in almond (*Prunus amygdalus* L.). *Curr. Sci.* 57(7): 397-398.

Abrol, D.P. 1988. Significance of thermal and visual constraints on the pollination activity of *Andrena ilerda* Cameron. *J. Anim. Morphol Physiol.* 35 (1) : 7-10

Abrol, D.P. 1988a. Ecology and behaviour of three bee species pollinating loquat (*Erio-botrya japonica* Lindley). *Proceeding of the Indian National Science Academy Part B. Biological Science* 54(2/3): 161-164.

Abrol, D.P. 1988b. Effect of climatic factors on pollination activity of alfalfa-pollinating subtr-opical bee *Megachile nana* Bingham and *Megachile flavipes* Spinola (Hymenoptera: Megachilidae). *Acta Oecol. Gen.* 9(4): 371-378.

Abrol, D.P. 1988c. Foraging range of subtropical bees, *Megachile flavipes, Megachile nana* (Hymenoptera: Megachilidae) and *Apis florea* (Hymenoptera: Apidae). *J. Indian Inst. Sci.* 68(1/2): 43-48.

Abrol, D.P. 1989. Energy intake and expenditure in *Andrena ilerda* and *A. leaena* (Hymenoptera : Andrenidae). *Korean J. Apic.* 4 (2) : 41-47

Abrol, D.P. 1989. Morphometerics and wing stroke frequency of some bees *Proc. Indian Natn. Sci. Acad.* 55B:369-376.

Abrol, D.P. 1989. Studies on abundance, diversity, behaviour and importance of native pollinators for crop production. *Korean J. Apic.* 4 (2) : 25-40

Abrol, D.P. 1989. Studies on ecology and behaviour of insect pollinators frequenting straw-berry blossoms and their impact on yield and fruit quality. *Tropical Ecology* 30: 96-100.

Abrol, D.P. 1989. Studies on ecology and behaviour of insect pollinators frequenting stra-wberry blossoms and their impact on yield and fruit quality. *Trop. Ecol.* 30(1):96-100.

Abrol, D.P. 1989. Wing beat frequency of alfalfa pollinating bees *Megachile nana* Bingh and *M. flavipes* Spinola (Hymenoptera: Megachilidae). *J. Anim. Morphol. Physiol.,* 36: 191-194.

Abrol, D.P. 1990. Conservation of pollinators for promotion of agricultural production in India. *J. Anim. Morphol Physiol.*, 38(1x2) : 123-139

Abrol, D.P. 1990. Ecology and behaviour of a carpenter bee *Xylocopa valga* Cerstacker pollinating fruit blossoms. In: *Social Insects and the Environment* (eds.) G.K. Veeresh, B. Malik and C.A. Virakthamath, Proc. 11th International Congress of IUSSI Banglore, India, *Oxford and IBH Publishing Co., New Delhi, India,* pp. 488-489.

Abrol, D.P. 1990. Energetics of nectar production in some apple cultivars as a predictor of floral choice by honeybees. *Tropical Ecology* 31: 116-122.

Abrol, D.P. 1990. Energetics of nectar production in some apple cultivars as a predictor of floral choice by honey bees. *Trop. Ecol.,* 31:116-122.

Abrol, D.P. 1990. Energy intake and expenditure in *Xylocopa fenestrata* F. and *X. pubes-cens* Spinola (Hymenoptera : Anthophoridae). In: *Social Insects and the Environment* (eds.) G.K. Veeresh, B. Malik & C.A. Virakthamath, Proc. 11th International Congress of IUSSI, Bangalore, India, Oxford and IBH Publishing Co., New Delhi, India.

Abrol, D.P. 1990. Studies on 'Thai sac brood virus' affecting indigenous honeybee *Apis cerana indica* F. colonies- prospects and future strategies. *J. Anim. Morphol. Physiol,* 37:101-108

Abrol, D.P. 1990. Studies on pollen loads and pollination efficiency of some bees *J. Anim. Morphol Physiol.,* 37(1&2):147-150

Abrol, D.P. 1991. Foraging strategies of honey bees and solitary bees as determined by nectar-sugar components *Proc. Indian Natn. Sci. Acad.,* 57B:127-132.

Abrol, D.P. 1991. Morphometrics and wing stroke frequency of bumble bees. *J. Indian Inst. Sci.* (Biol. Sci.), 71B: 365-372

Abrol, D.P. 1991. Path analysis of environmental factors influencing daily flight activity of *Apis dorsata* F. *Acta Ecologia,Oecologica Generalis,*,12(6):619-826.

Abrol, D.P. 1991. Pollination of brinjal flowers *(Solanum melongena* L.) by bumble bees.

Abrol, D.P. 1992. Bioenergetics in bee-flower interrelationships - An analysis of foraging behaviour. *Korean J. Apic.*, 7(1) : 39-66.

Abrol, D.P. 1992. Diel pattern of *Apis cerana indica* in *Asparagus officinalis* Linn. *Sci. & Cult.* 58(1&2): 25.

Abrol, D.P. 1992. Energetics of nectar production in some strawberry cultivars as a pred-ictor of floral choice by honey bees. *Journal of Biosciences* (Bangalore) 17(1): 41-44.

Abrol, D.P. 1992. Energetics of nectar production in some strawberry cultivars as a predictor of floral choice by honey bees. *J. Biosciences,* 17:41-44.

Abrol, D.P. 1992. Foraging ecology of *Xylocopa valga* Gerstacker (Hymenoptera: Anthophoridae) visiting almond flowers. *Sci. & Cult.*, 58(3&4): 71-72.

Abrol, D.P. 1992. Foraging in honeybees *Apis cerana indica* F. and *A. dorsata* F. (Hymenoptera: Apidae)- Activity and weather conditions. *J. Indian Inst. Sci.* (Biol. Sci.), 72: 395-401

Abrol, D.P. 1992. Pollination biology-pollinator plant interaction *Korean J.Apic.* 7(2): 130-155

Abrol, D.P. 1993. Ecology and utilization of insect pollinators frequenting pear flowers and their impact on fruit production. *Korean J. Apic.* 8 (1): 18-21.

Abrol, D.P. 1993. Energy intake and expenditure in *Megachile Femorata Smith* (Hymenoptera: Megachilidae). *Trop. Ecol.,* 34 (2): 173-180.

Abrol, D.P. 1993. Honeybee - virus infection and immunity. *Korean J. Apic.* 8 (2) : 116-118

Abrol, D.P. 1993. Insect pollination and crop production in Jammu & Kashmir. *Curr. Sci.* 65(3): 265-269.

Abrol, D.P. 1993. Intervarietal differences in pollinator attractiveness and fruit set in almond cultivars. *Pollination in Tropics. Proc. Intl. Symp. Pollin. Trop.* 1993, India. (eds. G.K. Veeresh, Umashankar & K.N. Ganeshaiah). IUSSI, Bangalore, 120-122.

Abrol, D.P. 1993. New pollinator bee. *Curr. Res.* 22 : 130.

Abrol, D.P. 1993. Oxygen consumption of foraging bees. *J. Anim. Morphol Physiol.* 39:27-42.

Abrol, D.P. 1993. Possible immune mechanism in indigenous honeybee *Apis cerana indica* F. against thai sac brood virus. *J. Anim. Morphol Physiol.* 39 : 165-166.

Abrol, D.P. 1993. Varroatosis in honeybees and its control. *Curr. Res.* 1993, 22: 128-129.

Abrol, D.P. 1994. Comparative ecology of European and Asian honeybee species in sub-tropical and temperate climates. Beekeeping and honey resources in subtropical-temperate transition climates of Jammu. *Korean J.Apic.* 9 (1) : 11-20.

Abrol, D.P. 1994. Ecology behaviour and management of social wasp *Vespa velutina* Smith (Hymenoptera: Vespidae) attacking honeybee colonies. *Korean J. Apic.* 9(1):5-10.

Abrol, D.P. 1994. Seasonal activity population dynamics and behaviour of bee eater birds attacking honeybees (Meropidae: Coraciformes). *Indian J. of Ornithol.* 32 : 153-159.

Abrol, D.P. 1995. Energetics of nectar production in some almond cultivars as a predictor of floral choice by honeybees *Apis cerana indica* F. and *Apis mellifera* L. (Hymenoptera: Apidae). *Proc. Indian Nan. Sci. Acad.*, B 61 (4) : 285-290.

Abrol, D.P. 1995. Foraging ecology and behaviour of an alfalfa pollinating bee *Megachile haryanensis* Rahman (Hymenoptera : Megachilidae) *Mysore J. Agric. Sci.* 29:113-119

Abrol, D.P. 1995. Haemocytes of honeybees *Apis mellifera* L. and *Apis cerana* F. (Hymenoptera : Apidae) and their alteration by mite parasitosis. *Indian Bee. J.* 57 (1):6-7.

Abrol, D.P. 1995. Important plant species for apiculture in Jammu region. *J. Anim. Morphol Physiol.*, 42 (1&2): 57-64.

Abrol, D.P. 1995. Studies on abundance, diversity and behaviour of insect visitors of Okra (*Hibiscus esculentus* L.) (Moench) blossoms and their importance as pollinators. *Korean J. Apic.* 10 (1):1-5.

Abrol, D.P. 1996. A contribution to the floral biology of *Acacia modesta* Wall and is importance as a source of forage to honeybees. *Indian Bee J.* 58 (3) : 134.

Abrol, D.P. 1996. Are *Xylocopa* species new hosts of the parasitic mite *Tropilaelaps Tropilaelaps clareae* (Acari : Laelapidae) *Insect Environ.* 2 (1) : 16

Abrol, D.P. 1996. Beekeeping in integrated rural development. *Indian Farming* 46(2):17-22.

Abrol, D.P. 1996. Foraging activity of insect pollinators especially honeybees visiting apple flowers and their impact on fruit production. *Current research* 25: 196-199.

Abrol, D.P. 1996. Incidence of *Varroa jacobsoni.* Oudemans in Jammu, India. *Indian Bee J.* 58 (1): 15-17.

Abrol, D.P. 1996. Seasonal activity, flower relationships and biology of *Xylocopa valga* Gerstacker (Hymenoptera : Anthophoridae) *Korean J. Apic.* 11(1):13-20.

Abrol, D.P. 1996. Status and prospects of beekeeping in Jammu and Kashmir. In: R.C. Mishra (ed.) Proceedings of the National Beekeeping Experience Exchange Confer-ence pp. 72-78. *Punjab Agricultural University* Ludhiana, India.

Abrol, D.P. 1996. Sunflower pollination: abundance and diversity of pollinating insects and their effect on seed yield. *Indian Bee J.* 58 (2): 60-63.

Abrol, D.P. 1996. *Vorroa jacobsoni* Oudemans as a carrier of micro-organisms. *Insect Environ.* 2(1): 20-21.

Abrol, D.P. 1997. Bees and Beekeeping in India. Kalyani Publishers Ludhiana, India, 450p.

Abrol, D.P. 1997. Diseases of Honeybees in India vis-à-vis world overview. In: R.C. Mishra and Garg, R. (eds.) *Perspective in Indian Apiculture* Agrobotanica Publishers, Bikaner, Rajasthan, pp 304-333.

Abrol, D.P. 1997. Effect of different feeding doses and durations on brood rearing and honey storage in *Apis mellifera* L. Colonies. *Insect Environ.* 3 (2): 44-45.

Abrol, D.P. 1997. Honey and diabetes. *Popular Science* 5(1): 19-20.

Abrol, D.P. 1997. Honeybee diseases and their management. Kalyani Publishers Ludhiana, India, 200.

Abrol, D.P. 1997. Impact of insect pollination on carrot seed production. *Insect Environ.* 3(3):61

Abrol, D.P. 1998. Ant pests of honeybees and their management.*Indian Bee J.* 60(1): 22-26

Abrol, D.P. 1998. Effect of climatic factors on population dynamics of *Andrena flavipes* pollinating apple flowers. *Trop. Ecol.* 39(1) : 143-147

Abrol, D.P. 1998. Food acquisition efficiency as a determinant of honeybees foraging decisions. *Indian Bee J.* 60(3): 133-136

Abrol, D.P. 1998. Foraging ecology and behaviour of the alfalfa pollinating bee species *Megachile nana* (Hymenoptera : Megachilidae) *Entomologia Generalis* 22(3): 233-237

Abrol, D.P. 1998. Honeybee diseases in India. In Souvenir FAO – Workshop sustainable Beekeeping Development and All India Honey Festival (Apiexi, 98), 1-5 August Dharwad Karnataka, India, pp. 11-15.

Abrol, D.P. 1998. Mite pests of *Apis mellifera* L. in Jammu region, India. *Indian Bee J.* 60 (2): 99-100

Abrol, D.P. 1998. Mites associated with bumblebee, *Bombus haemorrhoidalis* Smith (Hymenoptera: Apidae, Bombini). *Indian Bee J.* 60 (3): 153-154.

Abrol, D.P. 1998.Environmental factors influencing flight activity of honeybees, *Apis mellifera* L. and *Apis cerana* F. (Hymenoptera: Apidae). *Indian Bee J.* 60 (2) : 71-75

Abrol, D.P. 1999. Cultivation of *Plectranthus rugosus* Wall in lowland areas. *Insect Environ.* 5(1):34.

Abrol, D.P. 1999. Pollination importance value of bees. *Insect Environ.*, 5(3):140-141.

Abrol, D.P. 1999-2000. Energy budget versus plant pollinator interactions. Journal of Palynology 35-36: 17-36.

Abrol, D.P. 2000 Madhumakhi Palan – Sidhant avum Vidhian. Kalyani Publishers Ludhiana, India, 181 p.

Abrol, D.P. 2000. Bear attack on beehives *Insect Environ.* 6(1): 33.

Abrol, D.P. 2000. Beekeeping with *Apis cerana* in Jammu and Kashmir: present status and future prospects. *Bee World,* 82(3):145-148

Abrol, D.P. 2000. Floral biology and foraging behaviour of honeybees visiting Guava *(Psidium guajava* L.) flowers. *Insect Environ.,* 6(1) : 23

Abrol, D.P. 2000. Insect pollinators of kiwifruit (*Actinidia deliciosa*). *Indian Bee Journal* 62(3&4) : 55-57.

Abrol, D.P. 2000. Madhumakhi Palan – Kab Aur Kaise. Mata Vaishno Publications Jammu, India, 100 p.

Abrol, D.P. 2001. A beekeeper - with a difference. *Indian Bee J.* 63(3&4): 77-78.

Abrol, D.P. 2001. *Acacia modesta. Bee World* 82(2):90-91.

Abrol, D.P. 2001. Import-export policy of honey bees and other pollinators. Indian Bee J. 63 (3&4): 55-58.

Abrol, D.P. 2002 Beekeeping – An Indian Perspective.Vinod Publishers Pacca Danga Jammu. 350 p.

Abrol, D.P. 2002. Utilization of waste materials for outdoor wintering of *Apis mellifera* L. colonies. *Bioresource Technology,* 81:159-161

Abrol, D.P. 2003. *Antigonan leptopus Bee World* 84(3): 130-131.

Abrol, D.P. 2003. Energy intake and expenditure in *Apis mellifera* L. and *Apis cerana* F.

Abrol, D.P. 2003. Foraging behaviour of *Apis florea* F. on *Portulaca grandiflora* L *Sci.& Cult* 69: 289-290.

Abrol, D.P. 2004. *Abelia grandiflora* L. *Bee World* 85: 2004

Abrol, D.P. 2004. Introduction of *Apis mellifera* L. in Jammu and Kashmir- present status and future prospects *Bee World* 85: 37-40.

Abrol, D.P. 2004. Path analysis and correlation coefficient of environmental factors influencing foraging behaviour of four honeybee species pollinating litchi flowers (*Litchi chinensis* Sonn. *International Journal of Horticultural Science* 10,(1): 87-92

Abrol, D.P. 2004. Path coefficient analysis of environmental factors influencing flight activity *Apis florea* F. and seed yield in carrot (*Dacus carota* L.) *International Journal of Horticultural Science* 10 (1): 77-82

Abrol, D.P. 2005. Foraging ecology of insect pollinators visiting cherry flowers and their impact on fruit production. *Korean Journal of Apiculture.* 20(2): 159-168.

Abrol, D.P. 2005. Pollination Energetics. *Asia Pacific Entomol.* 8(1):1-12.

Abrol, D.P. 2006. Foraging Behaviour of Bees as Influenced by Quality and Quantity of Rewards from Flowers *J Asia -Pacific Entomol.* 9(2):1- 4.,2006.

Abrol, D.P. 2006. Two decades of development in apiculture in Jammu & Kashmir. *Korean Journal of Apiculture.* 21(2): 145-156.

Abrol, D.P. 2006a. Defensive behaviour of *Apis cerana* F. against predatory wasps. *Journal of Apicultural Science.* 50(2): 53-60.

Abrol, D.P. 2006b. Diversity of pollinating insects visiting litchi flowers (*Litchi chinensis Sonn.*) and path analysis of environmental factors influencing foraging behaviour of four honeybee species. *Journal of Apicultural Research and Bee World* 45(4): 180–187 © IBRA.

Abrol, D.P. 2006c. Factors influencing flight activity of *Apis florea* F, an important polli-nator of *Daucus carota* L. *Journal of Apicultural Research and Bee World* 45(2): 2–6.

Abrol, D.P. 2007 Differences in sugar metabolism among honey bees, *Apis mellifera* and *Apis cerana* potentially impact floral attractiveness. Journal of Apicultural Research *and Bee World* 46(1): (In press).

Abrol, D.P. 2007. Honeybee and Rapeseed- A pollinator plant interaction In: Advances in Botanical Research: Rapeseed Breeding Vol 45. (Ed.) Dr. S.K. Gupta, Elsevier London, U.K (in press).

Abrol, D.P. and Andotra, R.S. 1997. Contact toxicity of some pesticides to honeybee *Apis millifera* L. *Pest Management Econ. Zool.* 5 (1): 47-51.

Abrol, D.P. and Andotra, R.S. 2000. Contact and oral toxicity of some pesticides to honeybee, *Apis mellifera. Indian Bee Journal* 62(1-2): 29-34.

Abrol, D.P. and Andotra, R.S. 2000. Effect of pesticidal residues on the mortality of honeybee *Apis mellifera* L. *Korean Journal Apiculture.* 15(2):111-120.

Abrol, D.P. and Andotra, R.S. 2001. Field toxicity of pesticides to honeybee *Apis mellifera* L. foragers. *Korean Journal of Apiculture* 16(1) 19-26.

Abrol, D.P. and Andotra, R.S. 2001. Impact of pesticides on the foraging activity of honey-bee *Apis mellifera* L. on *Brassica campestris* var. *toria. Korean Journal of Apiculture,*16(2): 89-94.

Abrol, D.P. and Andotra, R.S. 2003. Relative toxicity of some insecticides to *Apis mellifera* L. *J. Asian Pacific Entomol.* 6(2): 235-237.

Abrol, D.P. and Ball, B.V. 2006. New record of European foul brood (EFB)-A bacterial disease of honeybee *Apis mellifera* L. in Jammu, India. *Journal of Research, SKUAST–J*: 5(2):256-260,

Abrol, D.P. and Bhat, A.A. 1987. Influence of atmospheric conditions and floral rewards on diurnal activity pattern, flower visitation rates and pollinating effectiveness of *Apis cerana indica* Fabr. foragers. *Res. Dev. Report.* 4 (1):13-15

Abrol, D.P. and Bhat, A.A. 1989. New record of *Xylocopa valga* Gerstacker (Hymenoptera : Anthophoridae) from India. *Curr. Sci.* 58 (1): 41 pp.

Abrol, D.P. and Bhat, A.A. 1989. Plants of apicultural importance in Kashmir 1. *Asparagus officinalis* L. *Indian Bee J.* 51(3) : 121pp

Abrol, D.P. and Kakroo S.K. 1998. Studies on seasonal activity and control of predatory wasps attacking honey bee colonies. *Indian Bee J.* 60 (1): 15-19

Abrol, D.P. and Kakroo, S.K. 1996. Studies on seasonal activity and control of wax moths (*Galleria mellonella* L. and *Achroia grisella* F.) attacking combs of four honeybee species. *Mysore J. Agric. Sci.* 30 (4) : 365-373.

Abrol, D.P. and Kakroo, S.K. 1997. Observations on concurrent parasitism.by mites on four honeybees species in India. *Trop. Agric.* 74 (2): 150-156

Abrol, D.P. and Kakroo, S.K. 1998. Pests, pathogens and predators of honeybees *Apis cerana* in Jammu, India. *Indian Bee J.* 60(3): 147-150

Abrol, D.P. and Kapil R.P. 1986. Nectar dilution pattern of bees in semi-arid environments. *Curr.Sci.,* 56 (13): 681 pp.

Abrol, D.P. and Kapil R.P. 1996. Studies on abundance and importance of insect pollina-tors for oilseed production. *J. Insect. Sci.* 9(2) : 172-174

Abrol, D.P. and Kapil, R.P. 1986a. Factors affecting pollination activity of *Megachile lanata* Lepel. *Proc.Indian Acad Sci.* (Anim. Sci.), 95 : 757-769.

Abrol, D.P. and Kapil, R.P. 1988a. Pollination studies in *Crotalaria juncea* L. *Sci. & Cult.,* 54 (7) : 243-244.

Abrol, D.P. and Kapil, R.P. 1988b. Foraging range of *Andrena ilerda and A.leaena* (Hymenoptera : Andrenidae) *Res. Dev. Reporter* 5 (1 x 2) : 65-67.

Abrol, D.P. and Kapil, R.P. 1994. On homing ability and pollinating effective-ness of bees. *Mysore J. Agric. Sci.* 28 : 249-252.

Abrol, D.P. and Khan A.A. 2002. Characterization of ecotypes of Indian honeybee *Apis cerana* F. in the northwest Himalayas *Korean Journal of Apiculture.*17(1): 13-20,

Abrol, D.P. and Kotwal, D.R. 1996. Honeybees foraging on *Calotropis procera* – a new host plant in semi-arid areas. *Insect Environ.* 2 (2) : 42-43.

Abrol, D.P. and Kotwal, D.R. 1997. Honeybee foraging on *Plectranthus rugosus* Wall. *Insect Environ.* 3 (3) : 60.

Abrol, D.P. and Mishra, R.C. 1997. Beekeeping equipments in India vis-à-vis world scenario: Retrospect and Prospects. In : R.C. Mishra and Garg, R. (eds.) *Perspective in Indian Apiculture.* Agrobotanica Publishers, Bikaner, Rajasthan. pp 387-312.

Abrol, D.P. and Putatunda, B.N. 1995. Discovery of the ectoparasitic mite *Tropilaelaps koenigerum* Delfinado – Baker and Baker (Acari : Laelapidae) on *Apis dorsata* F., *A. mellifera* L. and *A. cerana* F. in Jammu and Kashmir, India.*Curr.Sci.* 68 (1): 90.

Abrol, D.P. and R.C. Sihag 1997. Pollination energetics. In: R.C. Sihag (ed) *Pollination Biology* Series Rajendra Scientific Publishers, Hisar, Haryana. Vol I, pp 75-97

Abrol, D.P. and Rajiv Kumar 2000. Impact of pesticides on the brood rearing activity and foragers of honeybee, *Apis mellifera* L. visiting *Brassica campestris* L. var. *toria. Indian Bee Journal* 62(1-2): 19-27.

Abrol, D.P. and Rajiv Kumar 2000. Toxicity of a neem (*Azadirchta Indica* A. Juss.) insecticides to immature stages of honey bee *Apis mellifera* L. *Indian Bee Journal* 62 (3x4):47-53.

Abrol, D.P. and Rajiv Kumar 2001. Toxic and morphogenic effects of pesticides on the honeybee *Apis mellifera* L. brood. *Korean Journal of Apiculture.* 16(2): 95-112.

Abrol, D.P. and Sharma, D. 2007. Morphogenic and toxic effects of some pesticids on honey bees, *Apis mellifera* L. brood. *Journal of Research SKUAST- Jammu* (In Press).

Abrol, D.P. Bhat, A.A. and Khan, A.R. 1989. Influence of behaviour of *Apis cerana indica* Fabr. on cross pollination in apple. *Indian Bee J.* 51(3) : 103-104.

Abrol, D.P. Kakroo, S.K. and Putatunda, B.N. 1994. Stored product mite *Tyrophagus longior* (Gervais) with hive bees in India. *Curr. Sci.* 66 (2): 105

Abrol, D.P. Kapil, R.P. and Sihag, R.C. 1988. Physiological adaptations of subtropical bees in relation to energy intakes. *Res. Dev. Reporter,* 5: 68-79.

Abrol, D.P. Khan, M.S. and Monobrullah, M. 2002. Competition of fruit trees with a weed, *Lepidagathus incurva* for their pollination by *Apis* spp. in Jammu. *Indian Bee J.* 64(1&2): 55-56.

Abrol, D.P. Kumud, Mathur, S. and Mathur, R.B. 1989. Metabolic expenditures of some stored product mites as determined by their rate of oxygen consumption. In *Progress in Acarology* (eds.) G.P. Channabasavanna & C.A. Viraktamath, Oxford and IBH Publishing Co., New Delhi, India, pp. 271-273.

Abrol, D.P., Anderson Denis L., Kaul V., Sharma, D., Bhagat R.M., Hafeez, A. and Raghuraman, M. 2006. *Varroa destructor* Anderson & Trueman - A new threat to beekeeping in India. *Journal of Research, SKUAST–J*: 5(2):215-222,

Abrol, D.P., Bhagat, R.M. and Sharma. D. 2005. Mass rearing of *Apis cerana* F. *J. Asia Pacific Entomol,* 8(3): 309-317.

Abrol, D.P., Bhat, A. and Khan, A.R. 1990. Factors affecting pollination activity of honey-bee and fruit set in almond. In: *Social Insects: An Indian Perspective* (eds.) G.K. Veeresh,

A.R.V.Kumar & T. Shivashankar, IUSSI Indian Chapter, Bangalore Oxford and IBH Publishing Co., New Delhi, India, pp 50-53.

Abrol, D.P., Kitroo, A., Bhat A.A., Sharma B. and Khanna, Y.P. 1998. Interspecific varia-tions in haemolymph cation contents in *Apis mellifera* L. and *Apis cerana* Fab. of different age groups. *Indian Bee J.* 60 (3): 157-158

Abrol, D.P., Masoodi, M.A. and Bhagat, K.C. 1988. Bee flower interaction, parallel evoluation and adaptations with reference to *Apis cerana indica* F. pollinating loquat (*Eriobotrya japonica* Lindley) flowers in Kashmir. *Environ. Ecol.* 6 (3) : 737-736

Abrol, D.P., Sharma, D. and Bhagat, R.M. 2006. Apiculture and forestry: Management of resources for sustainable agriculture and ecosystem. In: Arora, A. and Sharma, V. (ed.) *Resource Management for Sustainable Agriculture.* pp 52- 57.

Abrol, D.P.1994. Factors affecting population dynamics of *Andrena flavipes* pollinating apple flowers. *Korean J. Apic.* 9 (2): 131-136

Abushady, A.Z. 1949. Races of bees. In : The Hive and the Honey Bee (R. Grout, ed.), Dadant and Sons, Hamilton. pp. 11-20.

Acolat, L. 1953. Les materiaux des nids de guepes. *Annals Scientifiques Universite d Besancon 8:* 39- 43.

Adalakha, R.L. and Verma, T.D. 1976. Scrab beetles and *Apis mellifera* colonies in India. *American Bee Journal* 116: 53.

Adam, Br. 1983. In: Search of the Best Strains of Honey Bees (2nd ed.). Northern Bee Book, England.

Adamson, A.M. 1943. Enemies and diseases of honeybees in Trinidad. Trinidad and Tobago agricultural society proceedings 43: 37-53.

Adey, M., Walker, P. and Walker, P.T. 1986. Pest control safe for bees. A manual and directory for tropics and subtropics, pp. 224.

Adlakha, R.L. 1975. Wasps a notorious bee enemy in India. *American Bee Journal* 115:55.

Adlakha, R.L. Sharma, O.P and Dhaliwal, H.S. 1974. Wasps as serious enemies of honey bees. *Farmer and Parliament.* 13 : 25.

Adler, L.S. 2001. The ecological significance of toxic nectar. - *Oikos* 91: 409–420.

Ahamad, R., N. Muzaffar and Q. Ali 1983. Control biologico de la pollila de la cera *Achroia grisella* y *Galleria mellonella* L. (Lep.: Pyralidae) mediante la reproducción del parasito *Apanteles galleriae* Wilkinson (Hym.: Braconidae) en Pakistan. Apiacta Int. Tech. Mag. Apic. Econ. Inf. 18: 15-21.

Ahmad F. U. Partap; S.R. Joshi and M.B. Gurung, 2002. Please do not steal our honey. Bees for Development Journal 64: 9.

Ahmad F; S.R. Joshi and M.B. Gurung, 2003. The Himalayan cliff bee *Apis laboriosa* and the honey hunters of Kaski. Indigenous honeybees of the Himalayas (Volume I). ICIMOD, Kathmandu. 52p.

Ahmadi, A.A. 1984. Incidence of honeybee (*Apis mellifera*) diseases and parasites in Southern Iran. *Bee World* 65(3): 134-136.

Akhurst, R.J. 1980: Morphological and functional dimorphism in *Xenorhabdus* spp., bacteria symbiotically associated with insect pathogenic nematodes *Neoaplectana* and *Heterorhabditis. Journal of General Microbiology 121:* 303-309.

Akratanakul, P. 1986. Honeybee diseases and their enemies in Asia-A practical guide, FAO Bulletin, Commonwealth institute.International Bee Research Institute London pp. 56.

Akratanakul, P. and Burgett, M. 1975. *Varroa jacobsoni*: A prospective pest of honey bees in many parts of the world. *Bee World* 56: 119-121

Akre, R.D. 1991: Critique. Wasp research: strengths weaknesses, and future directions. *New Zealand Journal of Zoology 18:* 223-227.

Akre, R.D. and MacDonald, J.F. 1986. Biology, economic importance and control of yellow jackets. *In:* Vinson, S.B. *ed.* Economic impact and control of social insects. New York, Praeger Publications. Pp. 353-412.

Akre, R.D., A. Greene, J.F. MacDonald, P.J. Landolt and H.G. Davis 1980. An Introduction to the Study of Insects. Saunders College Publishing, Fort Worth. Sixth edition.

Alexopoulos, C.J. 1952. "Introductory Mycology", Wiley, New York.

Alfonsus, A. 1891. Der Feind der Bienennlaus. *Deutsche lliustierte Bienenzeitung* 8: 503-506.

Alfonsus, A. 1922. An enemy of mites in the beehive. *Bee world* 4:2-3.

Ali, A.D., Barky, N.M., Abdelatif, M.A. and EJ-Sawaf, S.K. 1973. The control of greater wax moth,*Galleria mellonella* L. by chemicals, I. Z. Angew. Entomol. 74: 170-177.

Ali, T.M.M 1992. Ant fauna of Karnataka -II IUSSI Newsletter 6(1&2):1-9.

Ali, T.M.M. 1991. Ant fauna of Karnataka -I IUSSI Newsletter 5(1&2):1-8.

Allen, M.D. 1959. Respiration rates of worker honeybees at different ages and temper-atures. *J. Exp. Biol.* 36, 92–101.

Allsopp, M., Govan V. and Davison S. 1997. Bee health report *Varroa* in South Africa. *Bee World* 78: 171-174.

Allsopp, M.M., Kasimates, E.J., Swart, D.J. and Dipenaar-Schoeman, A.S. 2003. No evidence that Pseudoscorpions are controlling *Varroa* mites in South African honey-bee colonies. *South African Bee Journal* 75(2): 61-66.

Alpatov, V.V. 1948. The races of honeybees and their use in agriculture (In Russian). Sredi, Prirody 4, 1-65.

Alt, G. L. 1980. Bears, beehives, and beekeepers. *Gleanings in Bee Culture* 108: 137– 140.

Alt, G. L. 1988. Bear season results. *Pennsylvania Game News* 59(4):43– 45.

Al-Tikrity, W.S.; Benton, A.W., Hillman, R.C. and Clarke, W.W. Jr. 1972. The relationship between the amount of unsealed brood in honeybee colonies and their pollen collection. *J. Apic. Res.* 11: 9 -12.

Ambrose, J.T. 1992. Management for honey production, pp. 601-655, In: Grahm, J.M. (Ed.). Hive and the honey Bee, Dadant & Sons, Hamilton, lllionis, U.S.A., 1324 pp.

Amos, J.M. 1972b. Beekeeping in Virginia. Virginia extension publication 372.

Anderson D.L. 1994. Non-reproduction of *Varroa jacobsoni* in *Apis mellifera* colonies in Papua New Guinea and Indonesia, *Apidologie* 25, 412–421.

Anderson D.L. 2000. Variation in the parasitic bee mite *Varroa jacobsoni* Oud. *Apidologie* 31, 281– 292.

Anderson D.L. and Fuchs S. 1998. Two genetically distinct populations of *Varroa jacob-soni* with contrasting reproductive abilities on *Apis mellifera*, *J. Apic. Res.* 37, 69-78.

Anderson D.L. and Trueman J.W.H. 2000. *Varroa jacobsoni* (Acari: Varroidae) is more than one species, Exp. Appl. Acarol. 24, 165–189.

Anderson D.L. and Trueman J.W.H. 2000. *Varroa jacobsoni* is more than one species. *Exp. Appl. Acarol.* 24: 165–189.

Anderson D.L., Gibbs A.J. and Gibson N.L. 1998. Identification and phylogeny of spore-cy st fungi (*Ascosphaera* spp.) using ribosomal DNA sequences, *Mycol. Res.* 102:541-547.

Anderson D.L., Halliday R.B. and Otis G.W. 1997. The occurrence of *Varroa underwoodi* (Acarina: Varroidae) in Papua New Guinea and Indonesia, Apidologie 28, 143–147.

Anderson DL, Gibson NL. 1998. New species and isolates of spore-cyst fungi (*Plectomycetes; Ascosphaerales*) from Australia. *Aust Syst Bot* 11:53-72.

Anderson DL, Trueman JWH. 2000. Varroa jacobsoni (Acari : Varriodae) is more than one species. Expt Appl Acarology 24:165-18.

Anderson, D.L. 2000. Variation in the parasitic bee mite *Varroa jacobsoni* Oud. *Apidologie 31*: 281-292.

Anderson, D.L. and H. Giacon 1992. Reduced pollen collection by honey bee (Hymenoptera: Apidae) colonies infected with *Nosema apis* and sacbrood virus. *Journal of Economic Entomology* 85:47-51.

Anderson, L.D., and E.L. Atkins. 1966. "Research on the effect of pesticides on honey bees." *Amer. Bee Jour.* 106 (6): 206-08.

Anderson, L.D., E.L. Atkins, F.E. Todd and M.D. Levin. 1968. "Research on the effect of pesticides on honey bees." *Amer. Bee Jour.* 108 (7): 277-79.

Anonymous 1959. Bee flora of Karnataka and Kerala. *India Bee* J. 21 (7/8):90-92

Anonymous 1973. Mysore migrates bee colonies. *Indian Bee* J. 35(1/4): 71

Anonymous 1985. *Propagation of Bee Plants.* Tech. Bull. 20, Central Bee Res. Inst., Khadi and Village Industries Commission, Pune,

Argo, V.N. 1927. *Braula coeca* in den Vereinigten Staaten. *Archiv für Bienenkunde* 8:48-52.

Arias M.S., Tingek S., Kelitu A. and Sheppard W.S. 1996. *Apis nuluensis* Tingek, Koeniger and Koeniger, 1996 and its genetic relationship with sympatric species inferred from DNA sequences, *Apidologie* 27, 415–422.

Arnhart, L. 1923. Zur Entwicklungsgeschichte der *Braula coeca* Nitzsch. *Zool. Anz.* 56: 193-97.

Arpaia, S. 1996. Ecological impact of Bt-transgenic plants: 1. Assessing possible effects of CryIIIB toxin on honey bee (*Apis mellifera* L.) colonies. Journal of Genetics and Breeding 50:315-319.

Atkins, E.L. 1975. "Injury to bees by poisoning." In: *The Hive and the Honey Bee*, Rev. Ed. Hamilton, IL: Dadant & Sons. 740 pp. (see esp. p. 683, table 3).

Atkins, E.L. 1975. Injury to honey bees by poisoning. In The hive and the honey bee. p. 663-696. Dadant & Sons, Hamilton, Ill.

Atkins, E.L. and L.D. Anderson 1976. "Honey bee visitation patterns on some agricul-tural crops and their utilization in timing pesticidal applications." Univ. Calif. Coop. Ext., From: The U.C. Apiaries, May-June, 1976.

Atkins, E.L., D. Kellum and K.J. Neuman 1975. "Toxicity of pesticides to honey bees." *Univ. Calif., Div. Agric. Sci. Leaf. 2286.*

Atkins, E.L., D. Kellum, and K.W. Atkins. 1978. Encapsulated methyl parathion formu-lation is highly hazardous to honey bees. *Amer. Bee Jour.* 118 (7): 483-85.

Atkins, E.L., L.D. Anderson, and E.A. Greywood. 1970. Reseaech on the effect of pesticides on honey bees 1968-69. Parts I and II. *American Bee Journal* 110: 387-389; 426-429.

Atkins, E.L., L.D. Anderson, D. Kellum, and K.W. Neuman 1976. Protecting honey bees from pesticides. University of California, Division of Agricultural Science, Leaflet 2883.

Atkins, E.L., L.D. Anderson, II. Nakakihara, and E.A. Greywood 1975. Toxicity of pestic-ides to honey bees. University of California, Division of Agricultural Science, Leaflet 2286.

Atkins, E.L., R.L. MacDonald, and E.A. Greywood-Hale 1975. Repellent additives to reduce pesticide hazards to honey bees: Field tests. *Environ. Ent.* 4 (2): 207-10.

Atkins, E.L., R.L. MacDonald, T.P. McGovern, M. Berosa and E.A. Greywood-Hale 1975. Repellent additives to reduce pesticide hazards to honeybees: Laboratory tests. *Jour. Apic. Res.* 14 (2): 85-97.

Atwal, A.S and Goyal, N.P. 1974. *Apis mellifera* turns indifferent to Shain (*Plectranthus rugosus*). *Everyday Science* 19:25-26.

Atwal, A.S. (ed.) 1970. *Insect pollinators of crops.* Punjab Agricultural University Press, Ludhiana, 116 p.

Atwal, A.S. 2000. *Essentials of beekeeping and pollination.* Kalyani Publishers, Ludhiana, India. 393p.

Atwal, A.S., Bains, S.S. and Singh, S. 1970. Bee flora for four species of Apis at Ludhiana. *Indian J. Ent.* 32 (4): 330-334.

Babendreier D, Kalberer N, Romeis J, Fluri P and Bigler F, 2004. Pollen consumption in honey bee larvae: a step forward in the risk assessment of transgenic plants. *Apidologie* 35:293-300.

Babendreier D, Kalberer NM, Romeis J, Fluri P, Mulligan E, Bigler, F 2005. Influence of Bt-transgenic pollen, Bt-toxin and protease inhibitor (SBTI) ingestion on development of the hypopharyngeal glands in honeybees. *Apidologie* 36:585-594.

Babendreier D, Romeis J, Bigler F, Fluri P 2006. Neue Erkenntnisse zu möglichen Auswirkungen von transgenem Bt-Mais auf Bienen. Forschungsanstalt Agroscope Liebefeld-Posieux ALP, Schweiz.

Bailey J, Scott-Dupree C, Harris R, Tolman J, Harris B 2005. Contact and oral toxicity to honey bees (Apis mellifera) of agents registered for use for sweet corn insect control in Ontario, Canada. Apidologie 36:623-633.

Bailey L, Ball BV. 1991. *Honey Bee Pathology.* Academic Press, London.

Bailey L, Collins MD. Reclassification of *Streptococcus pluton (*White) in a new genus *Melissococcus,* as *Melissococcus pluton* nom. rev.; comb. nov. *J Appl Bacteriol* 1982;53:215- 217.

Bailey L, Gibbs AJ, Woods RD. Sacbrood virus of the larval honeybee. *Virology* 1964;23:425-429.

Bailey L. 1959. The natural mechanism of suppression of *Nosema apis* (Zander) in epizootically infected colonies of the honey bee, *Apis mellifera* Linnaeus. *J Insect Pathol* 1 : 347-50.

Bailey L. 1963. Infectious diseases honeybee, Land books London.

Bailey L. *Infectious Diseases of the Honey Bee.* Land Books, London, 1963.

Bailey, L. and B.V. Ball 1991. Honey Bee Pathology, 2nd Ed. Academic Press Inc., San Diego, USA.

Bailey, L., Ball B.V and Perry, J.N. 1983. Association of viruses with two protozoal pathogens of the honey bee. *Ann Appl Biol* 103:13-20.

Bailey, L., Gibbs, A.J. and Woods, R.D. 1963. Two viruses from adult honey bees (*Apis mellifera* Linnaeus). *Virology* 21: 390-395.

Bailey, L., Gibbs, A.J. and Woods, R.D. 1964. Sacbrood virus of the larval honey bee (*Apis mellifera* Linnaeus). *Virology* 23: 425-429.

Bailey, M.E., Fieger, E.A and Oertel, E. 1954. Paper chromatographic analyses of some southern nectars. *Gleanings in Bee Culture* 82 (7 & 8).

Baker, H.G. 1977. Non-sugar chemical constituents of nectar. – *Apidologie* 8: 349–356.

Baker, H.G. 1978. Chemical aspects of the pollination biology of woody plants in the tropics. – In: Tomlinson, P.B. and Zimmerman, M.H. (eds), Tropical trees as living systems. Cambridge Univ. Press, pp. 57–82.

Baker, H.G. and Baker, I. 1975. Studies of nectar-constitution and pollinator-plant coevolution. – In: Gilbert, L. E. and

Baker, H.G. and Baker, I. 1982. Chemical constituents of nectar in relation to pollination mechanisms and phylogeny. – In: Nitecki, M. H. (ed.), Biochemical aspects of evolutionary biology. Univ. of Chicago Press, pp. 131–171.

Baker, H.G., Opler, P.A. and Baker, I. 1978. A comparison of the amino acid complements of floral and extrafloral nectars. *Bot. Gaz.* 139: 322–332.

Baker, R.J. 1971. The influence of food inside the hive on pollen collective by a honeybee colony. *J. Apic. Res.* 10:23-26.

Balayannis, P. and Santas, L. 1992. Dissipation of malathion and fluvalinate residues in honey. *Journal of Apicultural Research* 31(2): 70-76.

Ball, B.V. 1989. *Varroa jacobsoni* as a virus vector. In: *Present status of varroatosis in Europe and progress in the Varroa mite control* (Ed. R. Cavalloro), Udine Italy, Office Official Publications EC, Luxemburg, pp. 241-244

Bambara, S.B. and J.T. Ambrose 1981. The parasites of the greater wax moth, *Galleria mellonella* observed in North Carolina. *Am. Bee J.* 121: 104-105.

Barker, G.M., Goh, H.H., Lyons, S.N. and Addison, P.J. 1991: Comparative pathogenicity to Argentine stem weevil of *Beauveria bassiana* from various hosts. Proceedings of the 44th New Zealand Weed and Pest Control Conference. pp. 214-215.

Barker, R.J. 1977. Some carbohydrates found in pollen and pollen substitutes are toxic to honey bees. *Journal of Nutrition* 107:1859-1862.

Barlow, N.D., Beggs, J.R. and Moller, H. 1998: Spread of the wasp parasitoid *Spheco-phaga vesparum vesparum* following its release in New Zealand. *New Zealand Journal of Ecology* 22(2):205-208.

Barlow, N.D., Moller, H. and Beggs, J.R. 1996. A model for the effect of *Sphecophaga vesparum vesparum* as a biological control agent of the common wasp in New Zealand. *Journal of Applied Ecology 33:* 31-44.

Barnes, G. 1972. Beekeepers continue to suffer pesticide losses. Ark. Dept. Com., *Apiary Bd. Bul.* 9(1): 1-2.

Barrett, E.C. and curtis, C.F. 1982. Introduction to environmental remote sensing. Nowich, U.K. Fletcher and Sons; New Yoirk, USA. Chapman and Hall, 2nd ed.

Barrett, P. 1996. The immigrant bees1788-1898, a cyclopaedia on the introduction of European honeybees into Australia and New Zealand. Peter Barrett: Springwood, NSW, Australia: 18 pp.

Barth, R. 1937. Bau und Funktion der Flugel drusen einiger Mikrolepidopteren. Untersuhungen an den Pyraliden: Aphomia gularis, *Galleria mellonella, Plodia interpu-nctella, Ephestia elutella* und *E. kuhniella. Z.Wiss Zool.* 150: 1-37.

Batra, S.W.T and Bohart, G.E. 1970. Brood care in halictid bees. *Science* 168: 875.

Batra, S.W.T and Michener, C.D. 1966. The nest and description of a new bee, *Systropha punjabensis* from India. *Jour. Kans. Entomol. Soc.* 39: 650-658.

Batra, S.W.T and Norden, B.B. 1996. Fatty food for their brood: How *Anthophora* bees make and provision their cells. *Memoirs of the Entomological Society of Washington* 17: 36-44.

Batra, S.W.T. 1964a. Behavior of the social bee, *Lasioglossum zephyrum*, within the nest (Hymenoptera: Halictidae). *Insectes Sociaux* 11: 159-185. [also *in Proceedings of North Central Branch of the Entomological society of America* 19: 74-75. 1964.]

Batra, S.W.T. 1964b. Organisms associated with *Lasioglossum zephyrum* (Hymenoptera: Halictidae). *Journal of the Kansas Entomological Society* 38: 367-389.

Batra, S.W.T. 1965. Behavior and ecology of the sweat bees (Halictidae) of India. *Year book American Philosophical Society for 1965*: Pp. 299-300.

Batra, S.W.T. 1966 b. Social behaviour and nests of some nomiine bees in India. *Insectes Sociaux* 13(3): 145-153.

Batra, S.W.T. 1966a. Nests and social behaviour of halictine bees of India. *Indian Journal of Emtomology* 28: 375-393.

Batra, S.W.T. 1966c. The life cycle and behavior of the primitively social bee, *Lasioglossum zephyrum* (Halictidae). *Dissertation Abstracts* 25(10). (Abstract).

Batra, S.W.T. 1968a. Crop pollination and the flower relationships of the wild bees of Ludhiana, India (Hymenoptera, Apoidea). *Journal of the Kansas Entomological Society* 40: 167-177.

Batra, S.W.T. 1968b. Behavior of some social and solitary halictine bees within their nests: a comparative study (Hymenoptera: Halictidae). *Journal of the Kansas Entomological Society* 41: 120-133.

Batra, S.W.T. 1970. Behavior of the alkali bee, *Nomia melanderi*, within the nest. *Annals of the Entomological Society of America* 63(2): 400-406.

Batra, S.W.T. 1971. Nesting habits of halictine bees (Halictidae: Hymenoptera). pp. 103-108. *In* A.S. Atwal (ed.), *Insect Pollinators of Crops*.

Batra, S.W.T. 1972. Some properties of the nest-building secretions of *Nomia, Antho-phora, Hylaeus* and other bees. *Journal of the Kansas Entomological Society* 45: 208-219.

Batra, S.W.T. 1976a. Comparative efficiency of alfalfa pollination by *Nomia melanderi, Megachile rotundata, Anthidium florentinum* and *Pithitis smaragdula* (Hymenoptera: Apoidea). Journal of the Kansas Entomological Society 49(1): 18-22.

Batra, S.W.T. 1976b. Nests of *Ceratina, Pithitis* and *Braunsapis* from India (Hymeno-ptera: Anthophoridae). *Oriental Insects* 10: 1-9.

Batra, S.W.T. 1977. Bees of India (Apoidea), their behaviour, management and a key to the genera. *Oriental Insects* 11(3/4): 289-324.

Batra, S.W.T. 1978a. Establishment of *Osmia cornifrons* in the United States. *Cooperative Plant Pest Reporter* 3: 166.

Batra, S.W.T. 1978b. Aggression, territoriality, mating and nest aggregation of some solitary bees (Halictidae, Megachilidae, Colletidae, Anthophoridae). *Journal of the Kansas Entomological Society* 51(4): 547-559.

Batra, S.W.T. 1979a. Osmia cornifrons and Pithitis smaragdula, two Asian bees introduced in the eatern United States for pollination. Proceedings of the IVth Inter-national Symposium on Pollination Md. Agriculture Experiment Station Special Miscellaneous Publication 1: 307-312.

Batra, S.W.T. 1979b. Insects associated with weeds of the northeastern United States: Quick weeds, *Galinsoga celiata* and *G. parviflora* (Compositae). *Environmental Entomology* 8: 1078-1082.

Batra, S.W.T. 1979c. Insects associated with weeds in the northeastern United States. II. Cinquefoils, *Potentilla norgegica* and *P. recta* (Rosaceae). *Journal of the New York Entomological Society* 87: 216-222.

Batra, S.W.T. 1980. Nests of the solitary bee, *Anthophora antiope* in Punjab, India. *Journal of the Kansas Entomological Society* 53: 112-114.

Batra, S.W.T. 1982. Hornfaced bees for efficient pollination of small farm orchards. Pp. 116-120. *In*: H. Kerr & L. Knutson (eds.), *Research for Small Farms*. USDA Miscellaneous Publications No. 1422, 301 pages.

Batra, S.W.T. 1991. Bees and pollination in south Asia. Invitational chapter in Annals of Entomology Festschrift honoring M. L. Roonwal, ed. P. sen Sharma. 9(2): 7-9.

Batra, S.W.T. 1993. India's buzzy biodiversity of bees. *Current Science* 65(3): 277-280.

Batra, S.W.T. 1994a. Diversify with pollen bees. *American Bee Journal* 134(9): 591-593.

Batra, S.W.T. 1994b. Shaggy fuzzyfoot bees. *Pomona* 27(3): 57-59.

Batra, S.W.T. 1995a. The evolution of "eusocial and the originof "pollen bees". *The Maryland Naturalist* 39(1-2): 1-4.

Batra, S.W.T. 1995b. Solitary bees for blueberry pollination. Pp. 69-72. In *Proçeedings of the 7th Southeastern Blueberry Conference, Savannah, GA.*

Batra, S.W.T. 1995c. Bees and pollination in our changing environment. *Apidologie* 26: 361-370.

Batra, S.W.T. 1996. Biology of *Apis laboriosa* Smith, a pollinator of apples at high altitude in the Great Himalaya Range of Garhwal, India (Hymenoptera: Apidae). *Journal of the Kansas Entomological Society* 69: 177-181.

Batra, S.W.T. 1997a. Fruit pollinating bees of the Garhwal Himalaya, U.P., India. *Acta Horticulturae* No. 437: 325-328.

Batra, S.W.T. 1997b. Biology of *Lasioglossum (Evylaeus) matieanense* (Bluthgen) [Hymenoptera: Halictidae], the predominant pollinating bee in orchards at high altitude in the Great Himalaya of Garhwal, U.P., India. *Proceedings of the Entomological Society of Washington* 99: 162-170.

Batra, S.W.T., Sakagami, S.F. and Maeta, Y. 1993. Behaviour of the Indian allodapine bee *Braunsapis kaliago,* a social parasite in nests of *B. mixta. Journal of the Kansas Entomological Society* 66(3): 345-360.

Batuev, Y.M. 1979. New information about virus paralysis. *Pchelovodsto* 7: 10-11.

Baur M.E., Kaya H.K., Peng Y.S. and Jiang, J. 1995. Non-susceptibility of the honey bee, *Apis mellifera* (Hymenoptera: Apidae) to steinernematid and heterorhabditid nematodes. *Journal of Nematology*. 27 :(3) 378-381..

Baxter J., Eischen F., Pettis J., Wilson, W.T. and Shimanuki, H. 1998. Detection of fluvalinate resistant varroamites in US honey bees, Am. Bee J. 138, 291.

Baxter, J R; Ibarra, J; Wilson, W T; Arther, R G; Kellerby, J D; Stewart, J 1999. Amitraz or coumaphos efficacy tests in Guatemala for control of *Varroa jacobsoni* in honey bees. *Southwestern Entomological Society*(Dec.): 309-313.

Beck, R. 1937. Mermis thread worm (Nematode) in wasp (Vespa vulgaris). Entomologist's Record and Journal of Variation 49: 65.

Beeson, C.F.C. 1938. Carpenter bees. *Indian Forester* 64: 135-137.

Beetsma, J., de Vries, R., Emami-Yeganeh, B., Emami-Tabrizi, M. and Bandpay V. 1989. Effects of *Varroa jacobsoni* Oud. on colony development, workerbee weight and longevity and brood mortality. In: *Present status of Varroatosis in Europe and progress in the Varroa mite control: proceedings of a meeting of the EC expert's group,* (Ed. R Cavalloro), Udine, Italy, 28-30 November 1988, Commission of the European Communities, Luxembourg, pp 163-170.

Beggs, J.R., Alspach, P A., Moller, H., Toft, R.J. and Tilley, J.A.V. 1992. Impacts of the parasitoid Sphecophaga vesparum on colonies of the common wasp (Vespula vulgaris). Proc. 41st Annual Conference Entomological Society of New Zealand.

Beggs, J.R.; Toft, R.J.; Malham, J.P., Rees, J., Tilley, J.A.V., Moller, H. and Alspach, P. 1998. Can the abundance of an introduced social wasp *(Vespula vulgaris)* be reduced enough for conservation gains? *New Zealand Journal of Ecology 22:* 55- 63.

Beier, M. 1937. The Pseudoscorpions of New Zealand, Norfolk and Lond Howe. New Zealand Journal of Zoology. 3:199-246.

Beier, M. 1976. Some remarks on *Ellingsenius indicus* Chamberlain (Pseudoscorpionidea). Annals and Magazine of Natural History 10(20): 633-634.

Belajvsky, A.G. 1927. The Enmies of bees.Mysti, Leningrad.

Bell, C.R. 1971. Breeding systems and floral biology of the Umbelliferae, or evidence for specialization in unspecialized flowers. – In: Heywood, V.H. (ed.), The biology and chemistry of the Umbelliferae. Academic Press, pp. 93-107.

Berenbaum, M.R. 1989. Nits, Gnats, and Nibblers. Univ. Illinois Press, Urbana.

Bertholf, L.M. 1925. The moults of the honey bee. *Journal of Economic Entomology* 18: 380.

Bertholf, L.M. and Pilson, J.E. 1941. Studies on toxicity to honeybees of acid lead arsenate, calcium arsenate, phenothiazine and cryolite. *Jour. Econ. Ent.* 34: 24-33.

Berube, C. 1999 Himalayan ceranaid development to promote *Apis cerana* beekeeping in Nepal. Part 1. *American Bee Journal.*139 (10):707-710.

Berube, C. 1999 Himalayan ceranaid development to promote *Apis cerana* beekeeping in Nepal. Part 2. *American Bee Journal.* 139 (10):784-7787.

Betts, A.D. 1912. A bee hive fungus, *Pericystis alvei,* gen. et sp nov. Annals of botany 26: 795-799.

Betts, A.D. 1939. Press mirror. *Bee world* 20:15-17.

Bhalla, O.P. 1971. Wasp enemies of honeybees.farmer and parliament 7(3):21-24.

Bhalla, O.P., Verma, A.K. and Dhaliwal, H.S. 1983. Insect visitors of mustard bloom (*Brassica campestris* var. *sarson*), their number and foraging behaviour under mid-hill conditions. *Journal of Entomological Research* 7(1): 15-17.

Bhaskar, V. and Gopinath, K. 1975. Nocturnal pollination in *Antirrhinum majus* by *Xylocopa rufescens. Journal of the Bombay Natural History Society* 72(2): 602-604.

Bhat, N.S.; Kandoria, J.L. and Goyal, N.P. 1987. Quantity and quality of nectar in some south Indian bee plants. Indian Bee J. 49:28-29.

Bingham, C.T. 1897. *The fauna of British India, including Ceylon and Burma. Hymen-optera. Vol. I. Wasps and bees.* 579 pp. Taylor & Francis, London [Reprint 1975, Today and Tomorrow's, New Delhi].

Bingham, C.T. 1898a. On some species of Indian Hymenoptera. *Journal of the Bombay Natural History Society* 12: 115-130.

Bingham, C.T. 1898b. Note on *Eumenes conica* Fabr. and *Megachile disjuncta* Fabr. and their parasites, Chrysis fuscipennis Brulle and Parevaspis abdominalis Sm. Journal of the Bombay Natural History Society 12: 585-588.

Bingham, C.T. 1908. Notes on the aculeate Hymenoptera. *Records of Indicn Museum* 2: 347-368

Boch, R, 1956. Die Tänze der Bienen bei nahen und fernen Trachtquellen. *Z Vergl Physiol* 38:136–167.

Boch, R, 1957. Rassenmäßige Unterschiede bei den Tänzen der Honigbiene (*Apis mellifica* L.). *Z Vergl Physiol* 40: 289–320.

Boecking, O. and Ritter, W. 1993. Grooming and removal behaviour of *Apis mellifera* intermissa in Tunesia against *Varroa jacobsoni. Jour. Apicultural Research* 32: 127–134.

Boecking, O. and Spivak M. 1999. Behavioral defenses of honey bees against *Varroa jacobsoni* Oud., *Apidologie* 30: 141–158.

Boecking, O., Bienefeld, K. and Drescher, W. 2000. Heritability of the varroa-specific hygienic behavior in honey bees (Hymenoptera : Apidae*). Journal of Animal Breeding and Genetics.* 117(6): 417-424

Boesch, C. and Boesch, H. 1990. Tool use and tool making in wild chimpanzees. *Folia Primatol* 54: 86-99.

Bogdanov, S. 1991. Criteria for the quality of beeswax. *Schweizerische-Bienen-Zeitung* 114(12): 689-694.

Bohart, G.E. 1970. Commercial production and management of wild bees-A new entomological industry. *Bulletin of the Entomological Society of America* 16(1): 8-9.

Bohart, G.E. 1972. Management of wild bees for the pollination of crops. *Annual Review of Entomology* 17: 287-312.

Bohart, G.E. 1972. Management of wild bees for the pollination of Crops. *Ann. Rev. Ent.* 31 : 49-65.

Bohart, R.M. and R.C. Bechtel 1957.How to Know the Insects. Wm. C. Brown Company Publishers, Dubuque. Hymenoptera of the World: an Identification Guide to Families.

Agriculture Canada. Research Branch. Bollhalder, F 1998. Thymovar for varroa control. *Schweizerische Bienen Zeitung* 121: 148-151.

Bond, V. 1947. How can we keep out ants. *American bee journal* 87:544.

Borchert, A. 1966, Die Krankheiten und Schadlinge der Honigbiene. Hirzel Verlag Leipzig [Diseases and pests of the honey bee

Borror, D.J., C.A. Triplehorn and N.F. Johnson 1989. The social wasps of California (Vespinae, Polistinae, Polybiinae). *California Insect Survey Bulletin* 4:73-101.

Bourdon, V.J., Lecompte, M., Leclercq, M. and Leclercq, J. 1975: Presence de noradr-enaline conjuguee dans Fenvelope du nid de *Vespula vulgaris* Linne. *Bulletin de la Societe Royale des Sciences de Liege 44: 414-A16.*

Bowen-Walker, P.L., Martin, S.J. and Gunn, A. 1999. The transmission of *Deformed wing virus* between honey bees (*Apis mellifera* L.) by the ectoparasitic mite *Varroa jacobsoni* Oud. *J Invert Pathol* 73: 101-106.

Bower, B. 1988. The brain in the machine. Science News 134 (Nov. 26):344.

Brady, J.R. and Maehr, D.S. 1982. A new method for dealing with apiary-raiding black bears. Proc. 36th Annu. Conf. Southeast. Assoc. Fiah and Wildl. Agencies, pp. 571-577.ues such as electric fencing (Gilbert and Roy 1975).

Brar, H.S., Gatoria, G.S. and Jhajj, H.S. 1992 a. Seasonal variations in honey stored by *Apis mellifera* L. colonies under different agro-climatic regions of the Punjab (India). *J. insect Sci.* 5(2): 141-143.

Brar, H.S., Gatoria, G.S. and Jhajj, H.S. 1994. Quantitative and qualitative evaluation of Eucalyptus as a source of nectar to honey bees. *Indian Bee J.* 56:28-30.

Brar, H.S.; Gatoria, G.S. and Chahal, B.S. 1989. Bee flora of Punjab, its relative utility and calendar of availability to honey bees. *Indian J. Ecol.* 16(2): 159-163.

Brar, H.S.; Jhajj, H.S. and Gatoria, G.S. 1992b. Abundance and activity of flower visiting bees or pigeon pea (*Cajanus cajan* L.) (Mill sp.) and role of *Apis melifera* L. in its pollination. Indian Bee J. 54(1-4):76-80.

Brewer, S.M and McGrew, W.C. 1990. Chimpanzee use of a tool set to get honey. *Folia Primatol* 54:100-104.

Brødsgaard, C. J.; Hansen, H.; Hansen, C. W. 1997. Effect of lactic acid as the only control method of *Varroa* mite populations during four successive years in honeybee colonies with a brood-free period. *Apiacta* 32, 81–88.

Brodsgaard, C.J., Jensen, S.E., Hansen, C.W. and Hansen, H. 1999. Spring treatment with oxalic acid in honeybee colonies as varroa control. *DIAS Report, Horticulture* 6: 16.

Brothers, D. J. and Carpenter, J. M. 1993. The taxonomy of the masarid wasps, including a monograph on the North American species. *University of California Publications in Entomology* 1: 369- 464.

Bruce, W.A., Hackett, K.J., Shimanuki, H. and Henegar, R.B. 1990. Bee mites: Vectors of honeybee pathogens? In: Proc. Intl. Symp on recent Research on Bee Pathology, September 5-7, 1990, Gent, Belgium.

Bujok B, Kleinhenz M, Fuchs S, and Tautz J, 2002. Hot spots in the bee hive. *Naturwissenschaften* 89:299–301.

Bunsen, J D 1992. Experimental Studies on the Control of the Mite *Varroa jacobsoni* Oud, an Ectoparasite of the Honey Bee (*Apis mellifera* L.), with Materials of Natural Origin Dissertation, Justus-Liebig University, Giessen, Germany.

Burges, H.D. 1977. Control of the wax moth *Galleria mellonella* on beecomb by H-stereo type V *Bacillus thuringiensis* and effect of chemical additives. *Apidologie* 8: 155-168.

Burges, H.D. 1978. Control of wax moth: Physical and biological methods. *Bee World* 59: 129-138.

Burgett M., Titayavan M. 1993. Brood thermoregulation by the giant honey bee (*Apis dorsata* F.), *Nat. Hist. Bull. Siam Soc.* 41, 93–98.

Burgi, M.R and Bhat, U.P.1994. Impact of urbanization on ant fauna in Bangalore. *IUSSI Newsletter* 8(1&2):10.

Burill, A.C. 1926. Ants that infest bee hives. *American bee journal* 66:29-31

Butani D.K. 1950. An *Apis dorsata* colony in New Delhi. *Ind. Bee J. 12*: 115.

Butler C.G. 1954. The world of the honeybee, Collins, London.

Butler, C.G. 1975.The honey bee colony life history. In The Hive and the Honey Bee, p.39-74. Dadant & Sons, Hamilton, Ill.

Butler, G.G., Finney, D.J. and Schiele, P. 1943. Experiments on the poisoning of honey-bees by insecticidal and fungicidal sprays used in orchards. *Ann. Appl. Biol.* 39: 143-150.

Buttel-Reepen, H. von. 1906. Beiträge zurSystematik, Biologie sowie zur geschichthchen und geographisdien Verbreitung der Honigbiene. Mitt. Zool Museum Berlin 3:117-201.

Buys, B. 1990. Relationship between argentine ants and honeybees in South Africa. In: Applied Myrmecology a world perspective. Vander Meer, R.K., Jafee, K and Cedana, A.J. (Eds). Westview Press, Inc., Boulder, USA. pp 519-524.

Buza, L. 1978. Control of *Varroa* disease in Hungary. *Apiacta* 13: 176-177.

Byers, C.F. 1930. A contribution to the knowledge of Florida Odonata. Biological Science Series Volume 1 Number 1 university of Florida Gainesville.

Çakmak, I., Aydın, L., Camazine, S., and H. Wells. 2002. Pollen Traps and Walnut-Leaf Smoke for Varroa Control. *American Bee Journal* 142, 367-370.

Calderone NW and RM Turcotte 1998. Development of sampling methods for estimating levels of *Varroa jacobsoni* infestation in colonies of *Apis mellifera. J Econ Entomol* 91: 851-863.

Calderone, N W. 1999. Evaluation of formic acid and a thymol-based blend of natural products for the fall control of *Varroa jacobsoni* (Acari: Varroidae) in colonies of *Apis mellifera* (Hymenoptera: Apidae). *Journal of Economic Entomology* 92: 253-260.

Calderone, N W; Nasr, M E. 1999. Evaluation of a formic acid formulation for the fall control of *Varroa jacobsoni* (Acari: Varroidae) in colonies of the honey bee *Apis mellifera* (Hymenoptera: Apidae) in a temperate climate. *Journal of Economic Entomology* 92(3): 526- 533.

Calderone, N W; Spivak, M. 1995. Plant extracts for control of the parasitic mite *Varroa jacobsoni* (Acari: Varroidae) in colonies of the western honey bee (Hymenoptera: Apidae). *Journal of Economic Entomology* 88(5): 1211-1215.

Calderone, N.W. 2005. Evaluation of Drone Brood Removal for the Management of *Varroa destructor* (Acari: Varroidae) in Colonies of the Honey Bee Apis mellifera L. (Hymenoptera: Apidae) in the Northeastern USA. *J. Econ. Entomol.* 98: 645-650.

Cale, G.H. 1963. Management for honey production. *In* The Hive and the Honey Bee, Dadant & Sons, Inc. Hannibal, Missouri, Pp. 249-302.

Cale, G.H. 1968. Pollen gathering relationship to honey collection and egg laying in honeybees, *Am. Bee J.* 108:8-9.

Cale, G.H. and Rothenbuhler, W. C. 1975. Genetics and breeding of the honey bee. In the hive and the honey bee, p. 157-184. Dadant & sons (ed.) Journal printing co., Carthage, ill.

Calis J.N.M., Boot, W.J., Beetsma, J and van den Eijende, H.P.M., de Ruijter, A and van der Steen, J.J.M. 1999. Effective biotechnical control of *Varroa*: applying knowledge of brood cell invasion to trap honey bee parasites in drone brood. *J Apic Res* 38: 49-61.

Calis, J.N.M. 2001. Parasite-host interactions between the *Varroa* mite and the honey bee: A contribution to sustainable *Varroa* control. PhD Thesis Wageningen University, the Netherlands.

Calis, J.N.M., Boot, W.J and Beetsma, J. 1999. Model evaluation of methods for *Varroa jacobsoni* mite control based on trapping in honey bee brood. *Apidologie* 30: 197-207.

Calis, J.N.M., Fries, I. and Ryrie, S.C. 1999. Population modeling of *Varroa jacobsoni*. *Apidologie* 30, 111–124.

Camazine, S and Liu, T.P. 1998. A putative Iridovirus from the Honey bee mite *Varroa jacobsoni* Oudemans. *J Invert Pathol* 71: 177-178

Cameron, S.A. 1993. Multiple origins of advanced eusociality in bees inferred from mitochondrial DNA sequences. *Proceedings of the National Academy of Sciences, USA*, 90: 8687-8691

Cane, J.H. 2001. Habitat fragmentation and native bees: a premature verdict? *Conservation Ecology* 5(1): 3.

Cantwell, G.E. and Smith, L.J. 1970. Control of the greater waxmoth *Galleria mellonella* in honeycomb and comb honey. *Am.Bee J.* 10: 141-143.

Cantwell, G.E., Jay, E.G., Pearman, G.C. Jr. and Thompson, I.V. 1972. Control of greater wax moth Galleria mellonella L. in comb honey with carbon dioxide. Part 1. Laboratory studies. American Bee Journal 112:302-303.

Cantwell, G.E., Lehnert, T. and Fowler, J. 1972. Are biological insecticides harmful to the honey bee? *Amer. Bee Jour.* 112: 255-258, 294-296.

Capolongo, F.; Baggio, A.; Piro, R.; Schivo, A.; Mutinelli, F.; Sabatini, A. G.; Colombo, F.; Marcazzan, G. L.; Massi, S and Nanetti, A. 1996. Trattamento della varroasi con acido formico: accumulo nel miele e influenza sulle sue caratteristiche. *L´Ape nostra amica* 8, 4–11.

Caron, D. 1992 Pseudoscorpions *Bee Culture* July,389.

Carpenter, J.M. 1982. Superfamily Vespoidea. in Hymenoptera of the World. H. Goulet and J.T. Huber [Eds.] Agriculture Canada.

Carpenter, J.M. 1982. The phylogenetic relationship and natural classification of the Vespoidea (Hymenoptera). *Systematic Entomology* 7:11-38.

Carpenter, J.M. 1987a. The common names of social wasps. *Bulletin of the Entomological Society of America* 26: 126- 130.

Cavalloro, R. *et al.,* 1988. Current situation of varroatosis infestation and control. *Proc. Meet. €C-Experts'* Group, Udine, 1988, R. Cavalloro (ed.). C.E.C., Luxembourg, pp. 13-54.

Chahal, B.S., Brar H.S., Jhajj, H.S and Gatoria, G.S. 1986. Status of bee diseases and ectoparasitic mites in *Apis mellifera* L. in Punjab, India. *Indian J. Ecol.* 13(1),46-51

Chakrabarti, K. and Chaudhary, A.B. 1972. Wild life biology of the Sunderbans forests: honey production and behaviour pattern of the honeybee. *Sci. Cult.* 38 (6): 269-276.

Chand, H., Singh, R. and Yazdani, S.S. 1995. Migratory beekeeping with Apis mellifera, a profitable business in Bihar. *Indian Bee J.* 57(3):147-148.

Chandler, D., Sunderland, K.D., Ball, B.V. and Davidson, G. 2001. Prospectiv biological control agents of *Varroa destructor* n.sp., an important pest of the European honey bee, *Apis mellifera. Biocontrol Sci Tech* 11: 429-448

Chandran, K. and Shah, F.A. 1974. Beekeeping in Kodai Hills (Tamil Nadu, India). *Indian Bee J.* 36 (1/4): 1-8.

Charriere, E and Imdorf, A. 2002. Oxalic acid treatment by trickling against *Varroa destructor*: recommondations for use in control Europe and under temperate climate conditions. *Bee World* 83: 51-60.

Charriere, J., Imdorf, A. and Bachofen, B. 1998. Trials of 5 formic acid dispensers. *Schweizerische Bienen Zeitung.* 121(6): 363-367.

Charriere, J., Imdorf, A. and Fluri, P. 1998. Was kann von der Oxalsaure gegen die *Varroa* erwartet werden? *Schweizerische Bienen Zeitung* 121(8): 505-506.

Chaturvedi, P.L. 1969. A tip for migratory beekeepers of Kumaon region. *Indian Bee J.* 31:48-51.

Chaubal, P.D. and Deodikar, G.B. 1965. Morphological characterization of pollen grains of some major honey yielding plants of the Western Ghats (India). *Indian Bee J.* 27(1): 1 -28 + 2 plates.

Chaubal, P.D. and Kotmire, S.Y. 1980. Flora calender of bee forage plants at Sagarmal (India). *Indian Bee J.* 42(3):65-65.

Chaudhari, R.K. 1977. Bee forage in Punjab plains (India). I. Pathankot and adjacent villages. *Indian Bee J.* 39(1/4):5-20.

Chaudhary, J.L and Singh, B.P. 1980. Bee pasturage and a pattern for seasonal management for Amethi, Uttar Pradesh. *Indian Bee J.* 51(3):109-111.

Chaudhary, J.P. and Jain, K.L. 1978. Nesting and foraging behaviour of a mason bee, *Megachile lanata* (Megachilidae: Hymenoptera). *Indian Journal of Entomology* 40(4): 405-411.

Chaudhary, O.P. 1998. Role of pollinators in seed production and its quality. In: *Seed quality assurance,* CCS Haryana Agricultural University, Hisar, 86-109.

Chen Y. 1993. Honeybee diseases and their prevention and control, in: Chen Y., Li Z., Qiu Z., Qiao Y., Yan H., Jin Z., Wang S., Wang, Y., Qiu J., Huang W. (Eds), Apiculture in China, Agric. Pub. House, Beijing, China, pp. 52–59.

Chhuneja, P.K., Gatoria, G.S., Aulakh, R.K., Singh, J. and Garcha, S. 2002. Surveillance of *Apis mellifera* Linn. colonies for bee diseases and enemies in the Punjab (India). Paper presented at 6th International Conference and World Apiexpo-2002 organized by AAA, at Bangalore (India):133-134.

Chitwood, B.G. and M.B. Chitwood. 1950. Introduction to Nematology. University Park Press, Baltimore. 334 pages.

Christian, P., Carstens, E., Domier, L., Johnson, J., Johnson, K., Nakashima, N., Scotti, P and van der Wilk, F. 2005. Genus *iflavirus* and Family *Dicistroviridae.* In: *Virus Taxonomy.* The eigth report of the international committee on taxonomy of viruses (Fauquet, C.M., Mayo, M.A., Maniloff, J. Desselberger, U. & Ball, L.A. eds). pp 779-788.

Christian, P.D and Scotti, P.D. 1998. The Picorna-like viruses of insects. In: *The insect viruses,* (L.K. Miller and L.A. Ball, eds), pp. 301-336. Plenum Press, New York, N.Y.

Christine Y.S., Peng, Xinsheng Zhou and Harry K. Kaya 2002. Virulence and site of infection of the fungus, *Hirsutella thompsonii,* to the honey bee ectoparasitic mite, *Varroa destructor. Journal of Invertebrate Pathology, 81(3): 185-195*

Chun, L. 1965. Beekeeping in China. *Glean Bee Cult.* 93:38-39.

Clapperton, B.K., Alspach, P.A., Moller, H. and Matheson, A.G. 1989: The impact of common and German wasps (Hymenoptera: Vespidae) on the New Zealand beekee-ping industry. *New Zealand Journal of Zoology 16:* 325-332.

Clark, T.B. 1977. *Spiroplasma* sp., a new pathogen in honey bees. *Journal of Invertebrate Pathology* 29:112–113.

Clark, T.B. 1978. Honey bee spiroplasmosis, a new problem for beekeepers. *American Bee Journal* 118: 18–19, 23.

Clements, J.F. 1991. Birds of the World: A Check-List. 4th ed. Ibis Publishing, Vista, CA.

Clinch, P.G., Palmer-Jones, T. and Forster, I.W. 1972. Effect on honey bees of nectar from the yellow kowhai (*Sophora microphylla* Ait.). – N.Z. J. Agric. Res. 15: 194–201.

Cobey S, 1997. Honeybee tracheal mite resistance: Are resistance stocks developing and can they be maintained? *Am. Bee Journal* 137(10) : 738-741

Cockerman,K.L. and Oertel, E. 1954. Control of ants in apiaries. *Glean Bee Culture.* 82: 348-349.

Colwell, R.N. (Ed.) 1983. Manual of remote sensing vol. I & II. American Society of Phoogammetry.

Conte, Le Y; Colin, M; Paris, A; Crauser, D 1998. Oil spraying as a potential control of *Varroa jacobsoni*. *Journal of Apicultural Research* 37(4): 293-294.31

Corbet, P.S. 1963. Biology of dragonflies. Quadrangle books, Chicago.

Crane, E. 1968. Mites infesting honeybees in Asia. *Bee world.* 49:113-114.

Crane, E. 1975. Honey: A comprehensive survey. Heinemann. London, pp. 726. Crane, E. 1990. Bees and Beekeeping. Science, practicc and world resources. *Comstock Publ. Ithaca,* NY, USA.

Crane, E. 1976, Apiculture in tropical climates. Full report on the first conference on apiculture in tropical climates. International Bee Research Association, London, U.K.

Crane, E. 1977. Dead bees under lime trees. *Bee World* 58: 129–130.

Crane, E. 1979. Fresh news on the *Varroa* mite. *Bee World* 608: 8. Research Association, London.

Crane, E. and Walker, P. 1984 "Pollination Directory for World Crops", International Bee Research Association, London.

Crane. E. and Walker, P. 1983. The impact of pest management on bees and pollination. Published by Tropical Development and Research Institute. London. 232 pp.

Cristobal, V.L. 1937. Novedades Entomologicas: *Apanteles galleriae*. Labor C. Est. Univ. Nac. La Plata 20: 1-8.

Crumb, S.E., Ende, P.M and Bonn, A.E. 1941. The European earwig. United States department of agriculture. Technical Bulletin, No. 766.

Current Science, special issue on remote sensing for national development 61(3&4), Aug. 1991.

Currie, R W 1999. Fluvalinate queen tabs for use against *Varroa jacobsoni* Oud.: efficacy and impact on honey bee, *Apis mellifera* L., queen and colony performance. *American Bee Journal* 139(11): 871-876

Dadant and Sons (Ed.) 1978. The Hive and the Honey bees. Dadant & Sons, Hamilton, Illinois.

Dafni, Amots; Kevan, Peter G.; and Husband, Brian C. (2005). *Practical Pollination Biology*. Enviroquest, Ltd.

Danielyan, S.G and Nalbandhan, K.M. 1971. The hungarian oil beetle *Meloe hungarus* as a dangerous parasite. *Pchelovodstvo* 5:32.

Danka, R.G and Villa, J.D, 2000. Inheritance of resistance to Acarapis woodi (Acari: Tarsonemidae) in first-generation crosses of honey bees (Hymenoptera : Apidae). *Journal of Economic Entomology*, 93(6):1602-1605.

Darwin, C. 1887, "The different forms of flowers on plants of the same species", John Murry. London.

Das, B.P and Gupta, V.K. 1989. The social wasps of India and the adjacent countries. *Oriental insect monograph No.*11, 292pages

Davidson, G, Phelps K, Sunderland K.D, Pell, J.K, Ball BV, Shaw KE, Chandler D. 2003. Study of temperature-growth interactions of entomopathogenic fungi with potential for control of Varroa destructor (Acari: Mesostigmata) using a nonlinear model of poikilotherm development. *J Appl Microbiol.* 94(5):816-25.

Davidson, G., Pell, J. K., Ball, B. V., Shaw, K. V and Sunderland, K. D. 2000. Fungal biocontrol of Acari. *Biocontrol Science and Technology,* 10, 357- 384.

Davis, H.G.; Zwick, R. W.; Rogoff, W.M.; McGovern, T.P.; Beroza, M. 1973: Perimeter traps baited with synthetic lures for suppression of yellow jackets in fruit orchards. *Environm-ental Entomology 2:* 569-571.

De Buysson, H. 1900. Damage of Forficula auricularia to bee hives. Bulletin de la Societe Entomologique France. 1900.183.

De Greef, M; De Wael, L and Van Laere, O 1994. The determination of the fluvalinate residues in the Belgian honey and beeswax. *Apiacta* 29(4): 83-87.

De Guzman L.I., Delfinado-Baker M. 1996. A new species of *Varroa* (Acari: Varroidae) associated with *Apis koschevnikovi* (Apidae: Hymenoptera) in Borneo. *Int. J. Acarol.* 22, 23–27.

De Guzman, L.I, Delfinado-Baker, M. 1996. A scientific note on the occurrence of Varroa mites on adult worker bees of *Apis nuluensis* in Borneo. *Apidologie* 27:329–30

De Jong D, D De Andrea Roma and Goncalves, L.S 1982. Comparative analysis of shaking solutions for the detection of Varroa jacobsoni on adult honey bees. *Apidologie* 13: 297-306.

De Jong, D. 1988. *Varroa jacobsoni* does reproduce in worker cells of *Apis cerana* in South Korea. *Apidologie* 19: 241-243.

De Jong, D. 1997. Mites: Varroa and other parasites of brood. In: *Honey Bee Pests, Predators and Diseases* (Morse, R.A. and Flottum, K., eds.). A.I. Root Co., Medina, OH, USA, pp. 279-327.

De Jong, D., de Jong, P. H and Goncalves, L. S. 1982a. Weight loss and other damage to developing worker honey bees from infestation with *Varroa jacobsoni*. *J Apicult Res* 21: 165-167.

De Jong, D., Morse, R.A., and Eickwort, G.C. 1982. Mite pests of honeybees. *Ann. Rev. Entomol.*, 27,229-252.

De Miranda, J. R., Drebot, M., Tyler, S., Shen, M., Cameron, C. E., Stoltz, D. B. and Camazine, S. M. 2004. Complete nucleotide sequence of Kashmir bee virus and comparison with acute bee paralysis virus. *J. Gen Virol* 85: 2263–2270.

Delaplane K.S. 2001. *Varroa destructor*, revolution in the making. *Bee World* 82, 157–159.

Delfinado, M. D and Baker, E. W. 1974. *Varroidae*, a new family of mites on honey bees Mesostigmata: Acarina. *J Washin Acad Sci* 64: 4-10.

Delfinado, M. D. 1963. Mites of the honey bee in Southeast-Asia. *J. Apic. Res* 2: 113-114.

Delfinado,M.D and Baker, E.W. 1976. Note on hypoi (acarina associated with bees and wasps (Hymenoptera*)*. *J.New York Entomol. Soc., 84,*76-90.

Delfinado-Baker M and Aggarwal, K. 1987. A new *Varroa* (Acari: Varroidae) from the nests of *Apis cerana* (Apidae). *Int. J. Acarol.* 13, 233–237.

Delfinado-Baker M, Baker EW, Phoon ACG. 1989. Mites (Acari) associated with bees (Apidae) in Asia, with description of a new species. *Am. Bee J.* 122:416–417

Delfinado-Baker M, Underwood, B.A and Baker, E.W. 1985. The occurrence of *Tropilaelaps* mites in brood nests of *Apis dorsata* and *A. laboriosa* in Nepal, with description of the nymphal stages. *Am. Bee J.* 125:703–6

Delorme, R. and Dacol, L. 1989. Les acaricides: mode d'action et résistance. *Annales A.N.P.P.* - Colloque sur les acariens des cultures, Montpellier, 24-25-26 Octobre 1989, 2, 91-1 13.

Denholm, C. H. 1999. Inducible honey bee viruses associated with *Varroa jacobsoni*. PhD thesis. Department of Entomology & Nematology, Rothamsted Research, Hertford-shire, UK and School of Life Sciences, Keele University, Staffordshire, UK.

Denmark, H.A., Sanford, M. T and Fasulo, T.R. 2000. *Varroa jacobsoni* Oudemans (Arachnida: Acari: Varroidae). Publication number EENY-37. University of Florida, USA 40p

Deodasan, A. 1972. Rubber plantation and beekeeping. *Indian Bee J.* 34(1/2):38-39.

Deodikar, G.B. 1970. A preliminary survey of Beekeeping in Kashmir. *Indian Bee J.* 32:12-22.

Deodikar, G.B. and Suryanarayana, M.C. 1977. Pollination in the service of increasing farm production in India. In: P.K.K. Nair (Ed.), *Advances in Pollen Spore Research,* 2: 60 - 82.

Deodikar, G.B. and Thakar, C.V. 1953. A pollen study of major honey yielding plants of Mahabaleshwar hills. Poona, India: Village Industries Committee, Bombay State 8 pages.

Deodikar, G.B.; Thakar, C.V. and Phadke, R.P. 1957. Higher nectar concentration in floral nectaries of silver oak, *Grevillea robusta*. *Indian Bee J.* 19(7/8): 84-85.

Detroy, B.F. and E.R. Harp. 1976. Trapping pollen from honey bee colonies.11.8. Department of agriculture production research report no.163,11 p.

Devadason, A. 1971. Migratory beekeeping in Kerala. *Indian Bee J.* 33(3/4):35-38.

Dietz, A. 1975. Nutrition of the adult honey bee. In: The Hive and the Honey bee, Chapter v. Dadant & Sons, Hamilton, Ill.

Dippenaar-Schoeman A.S and Harvey M.S. 2000. A check list of the pseudoscorpions of South Africa (Arachnida: Pseudoscorpiones). *Koedoe* 43: 89-102

Diwan, V.V. and Rao, G.V. 1972. Altitudinal variation in bee forage of West Coorg. *Indian Bee J.* 33(3):39-50.

Diwan, V.V. and Rao, S.K. 1969 Foraging behaviour of *Apis indica* F. on the flowers of *Synadenium grantil* Hook F. *Indian Bee J.* 31 (1):22.

Diwan, V.V; Suryanaryana, M.C. and Thakar, C.V. 1964. Beekeeeping Potentialities of the Castle Rock area. I. Preliminary Surveys. *Indian Bee J.* 26:4-15.

Donovan, B. 2001. N.Z Pseudoscorpion. The *New Zealand Beekeeper*.6(1):14

Donovan, B.D. 1991: Life cycle of *Sphecophaga vesparum* (Curtis) (Hymenoptera: Ichneumonidae), a parasitoid of some vespid wasps. *New Zealand Journal of Zoology* 18: 181-192.

Donovan, B.J. 1983. The common wasp is here. *New Zealand Beekeeper 180:* 9-10.

Donovan, B.J. 1989. Potential enemies of the introduced wasp parasitoid *Sphecophaga vesparum* (Hymneoptera:Ichneumonidae) in New Zealand. *New Zealand Journal of Zoology* 16: 365-367.

Donovan, B.J. 1993. Problems caused by immigrant German and common wasps in New Zealand, and attempts at biological control. *Bee World 73:* 131- 148.

Donovan, B.J. and Read, P.E.C. 1987: Attempted biological control of social wasps, *Vespula* spp., (Hymenoptera:Vespidae) with *Sphecophaga vesparum* (Curtis) (Hymenoptera:Ichneumonidae) in New Zealand. *New Zealand Journal of Zoology* 14: 329-335.

Donovan, B.J., Moller, H., Plunkett, G. M., Read, P. E. C and Tilley, J. A. V. 1989. Release and recovery of the introduced wasp parasitoid, Sphecophaga vesparum vesparum (Curtis) (Hymenoptera: Ichneumonidae) in New Zealand. *New Zealand Journal of Zoology* 16: 355-364.

Doolittle, G. M. 1889. Scientific queen-rearing. 169 P. T. G. Newman & son, Chicago.

Doull, K. M. 1977. Tucson pollen supplements. *American Bee Journal* 117:266-297.

Doupe, R. 1921. Irish bee notes. *Bee World*.2:8.

Dover, C. 1924. Some observations on the bionomics of *Xylocopa aestuans* Linnaeus (Apidae). *Transactions of the Entomological Society of London* 73: 144-149.

Drescher, W. 1975. Bienennutzung in Tansania. Aug. Dtsch, Imkerztg. 9, 117-122.

Dressler, R.L. 1968 "Pollination in euglossine bees", *Evolution*, 22, 202-210.

Du praw, E.J. 1960. Research on the honey bee egg. *Gleanings in Bee Culture* 88(2):104-111.

Dulta, P.C. and Verma, L.R. 1987 "Role of insect pollinators on yield and quality of apple fruit", *Indian J. Hort.*, 44, 274-279.

Dyer F.C. and Seeley, T.D. 1994. Colony migration in the tropical honey bee *Apis dorsata* F. (Hymenoptera: Apidae). *Insectes soc. 41*: 129–140.

Dyer, F.C and Seeley, T.D, 1991. Nesting behavior and the evolution of worker tempo in four honey bee species. Ecology 72:156–170.

Dyer, F.C, 1991. Comparative studies of dance communication: analysis of phylogeny and function. In: Comparative studies of dance communication: analysis of phylogeny and function (Smith DR, ed). Boulder, CO: Westview Press; 177–198.

Dyer, F.C. 2002. The biology of the dance language. *Annu Rev. Entomol.*47, 917–949.

Dzierzon, J. 1845. Gutachten uber die von hr. Direktor stohr im ersten und zweiten kapitel des general-gutachtens aufgestellten fra gen. *Bienen-Zeitung (Eichstätt)* 1:109-113, 119-121.

Eckert, J.E. 1946. Injury to bees by poisoning. In: Grout, R.A. (ed.), The hive and the honey bee. Dadant and Sons, pp. 570–582.

Eckert, J.E. and West, M.C. 1948. Chlordane bees and ants. *American bee Journal.* 88: 584.

Edwards, R. 1977. Trapping wasps (*Vespula* spp.) and bees (*Apis mellifera*) at a sweet factory. International Union for the Study of Social Insects. Proceedings of the 8th International Congress. pp. 300.

Edwards, R. 1977. Trapping wasps (*Vespula* spp.) and bees (*Apis mellifera* L.) at a sweet factory. *Proc. VIIIth International Congress Wageningen-The Netherlands* pp 300.

Edwards, R. 1980. Social wasps their biology and control. East Grinstead, Rentokil Limited.

Eischen F. 1988. Varroa control problems: some answers. *Am. Bee J.* 138, 107–108.

Eischen, F.A. 1998. Trials (and tribulations) with formic acid for varroa control. *American Bee Journal* 138(10): 734-735.

Ellis, J.D. and Munn, P.A. 2005. The worldwide health status of honey bees. *Bee World* 86:88-101.

Ellis, M. 2000. Using powdered sugar to detect varroa. *Beekeeping and Honey Bees Newsletter, University of Nebraska*(2): 1-3.

Elzen P. J., Eischen F., Baxter J.R., Pettis J., Elzen G.W and Wilson W.T. 1988. Fluva-linate resistance in *Varroa jacobsoni* from several geographic locations. *Am. Bee J.* 138, 674–676.

Elzen P.J. and Westervelt D. 2002. Detection of Coumaphos resistance in *Varroa destructor* in Florida. *Am. Bee J.* 142, 291–292.

Elzen, P J; Baxter, J R; Spivak, M and Wilson, W T.1999. Amitraz resistance in varroa: new discovery in North America. *American Bee Journal* 139(5): 362.

Engel, M.S. 1999. The taxonomy of recent and fossil honey bees (Hymenoptera: Apidae: *Apis*). *Journal of Hymenoptera Research* 8: 165-176.

Engel, M.S. 2002. The honey bees of India, Hymenoptera: Apidae. *Journal of the Bombay Natural History Society* 99(1): 3-7.

Erickson, E.H., Atmowidjojo, A.H and Hines, L. 1998. Can we produce *Varroa*-tolerant honeybees in the United States? *Am. Bee Journal* 138(11): 828-832

Esch, H. 1960. Über die Körpertemperaturen und den Wärmehaushalt vonn*Apis mellifica. Z. Vergl. Physiol.* 43, 305-335.

Esch, H. 1988. The effects of temperature on flight muscle potentials in honeybees and cuculiinid winter moths. *J. Exp. Biol.* 135, 109-117.

Evans, H.E. 1969. Three new Cretaceous aculeate wasps (Hymenoptera). *Psyche* (Cambridge) 76, 251-261

Evans, H.E. 1985. The Pleasures of Entomology. Smithsonian Inst. Press, Washington D.C.

Eveland, J.F. 1973. Population dynamics, movements, morphology, and habitat characteristics of black bear in Pennsylvania. M.S. Thesis, Pennsylvania State University, University Park, Pennsylvania. 157pp.

Faegri, K. and van der Pijil, L. 1978 "The Principles of Pollination Ecology", 3rd ed., Oxford/ New York/ Toronto, pp. 244.

Fægri, K. and van der Pijl, L. 1979. The principles of pollination ecology, 3rd ed. – Pergamon Press.

Fahn, A. 1988. Secretory tissues in vascular plants. *New Phytol.* 108: 229–258.

Fakhimzadeh, K. 2000. Potential of super-fine ground, plain white sugar dusting as an ecological tool for the control of varroasis in the honey bee *(Apis mellifera*). *American Bee Journal* 140(6): 487-491.

Farrar, C.L. 1937. Influence of colony populations on honey production. *Journal of agricultural research* 54:945-954.

Farrar, C.L. 1942. Nosema disease contributes to winter losses and queen supersedure. *In Gleanings in Bee Culture* 70:660-661, 701.

Faucon, J.P., Drajnudel, P. and Fleche, C. 1995. Decrease in Apistan efficacy used against varroosis in the honey bee *(Apis mellifera)*. *Apidologie* 26(4): 291-296.

Fauquet, C.M., Mayo, M.A., Maniloff, J. Desselberger, U. and Ball, L.A.(Eds.) 2005. Virus Taxonomy: The eigth report of the international committee on taxonomy of viruses. Elsevier Academic Press.

Fay J.M. and Carroll, R.W. 1994. Chimpanzee tool use for honey and termite extraction in Central Africa. *Am J Primatol* 34:309-317.

Feinsinger, P. and Swarm, L.A. 1978. How common are ant-repellant nectars? – Biotropica 10: 238–239.

Fenster, C.B., W.S. Armbruster, P. Wilson, M.R. Dudash and J.D. Thomson 2004. "Polli-nation syndromes and floral specialization". Annual Review of Ecology and Syste-matics 35: 375-403.

Fernandez-Garcia, M A; Riol-Melgar, M J; Herrero-Latorre, C; Fernandez-Garcia, M I 1994. Evidence for the safety of coumaphos, diazinon and malathion residues in honey. *Veterinary and Human Toxicology* 36(5): 429-432.

Finn, W.E and Payne, T.L. 1977: Attraction of greater wax moth females to male produced pheromones. *Southwest Entomol.* 2: 62-65.

Fittkan, E.J. and Klinge, H. 1973. On the biomass and trophic structure of the central Amazonian forest ecosystem.*Biotropica* 5:2-14.

Fletcher D. 1978. The influence of vibration dances by worker honeybees on the activity of virgin queens, J. Apic. Res. 17, 3–13.

Flint, H.M and Merkle, J.R. 1983. Mating behavior, sex pheromone responses and radiation sterilization of the greater wax moth (Lepidoptera: Pyralidae). *J.Econ. Entomol.* 76: 467-472.

Floris, I; Satta, A; Mutinelli, F and Prandin, L 1998. Efficacy of winter applications of oxalic acid against *Varroa jacobsoni* Oudemans in beehives in the Mediterranean area (Sardinia, Italy). *Redia* 81: 143-150.

Fox-Wilson, G. 1946. Factors affecting populations of social wasps, *Vespula* species, in England (Hymenoptera). *Proceedings of the Royal Entomological*

Francki, R. I. B., Fauquet, C. M., Knudson, D. L and Brown, F. 1991. Classification and Nomenclature of Viruses. Fifth Report of the International Committee on Taxonomy of Viruses. Springer, Wien New York (*Arch Virol* [Suppl] 2)

Free J.B and Spencer-Booth Y. 1962. The upper lethal temperatures of honeybees, *Entomol. Exp. Appl.* 5, 249–254.

Free, J. B. 1987. Pheromones of Social Bees. Chapman and Hall Ltd.,

Free, J. B. and Spencer-Booth, Y. 1960. Chill-coma and cold death temperatures of *Apis mellifera*. *Entomol. Exp. Appl.* 3, 222-230.

Free, J.B. 1965. The effect on pollen collection of feeding honey bee colonies with sugar, syrup. *J. Agric. Sci.* 64: 167-168.

Free, J.B. 1967. Factors determining the collection of pollen by honey bee foragers. *Anim. Behav* 15: 134-144.

Free, J.B. 1993. *Insect pollination of crops (2nd edition),* Academic Press, London, 684 pp.

Free, J.B., Paxton, R.J. and Waghwchoure, E.S. 1991. Increased the amount of foreign pollen carried by honeybee foragers. *J. apic. Res.* 30: 132-136.

Freeman, B.E. 1973. A technique for collecting arboreal social wasp nests. *Ann. Entomol. Soc. Am.*66 (1):916-917.

Frey, T. 1914. Some common bee enemies in India. *Gleaning in bee culture.* 87: 360-361, 363.

Fries, I and Hansen, H. 1993. Biotechnical control of *Varroa* mites in cold climates. *Amer Bee J.* 133: 435-438.

Fries, I. 1991. Treatment of sealed honey bee (*Apis mellifera*) brood for control of *Varroa jacobsoni*. *American Bee Journal* 130, 313–314.

Fries, I., Feng, F., Silva, A.D., Slemenda, S. B and Pieniazek, N. J. 1996. *Nosema ceranae* n. sp. (Microspora, Nosematidae), morphological and molecular characterization of a microsporidian parasite of the Asian honey bee Apis cerana (Hymenoptera, Apidae). European Journal of Protistology 32: 356-365

Fries, I., Martín R., Meana, A., García-Palencia P and Higes, M. 2006. Natural infections of Nosema ceranae in European honey bees. Journal of Apicultural Research. In press.

Fries, I.; Camazine, S. 2001. Implications of horizontal and vertical pathogen transmission for honey bee epidemiology. *Apidologie* 32, 199 214.

Fries, I; Munn, P; Jones, R 1997. Organic control of varroa. *Varroa! Fight the Mite*. London, IBRA: 16-21.

Frisch K. von 1923. Über die, Sprache" der Bienen, eine tierpsychologische Untersu-chung, *Zool. Jb. (Physiol.)* 40, 1–180.

Frisch, K. von. 1967. "The Dance Language and Orientation of Bees." Cambridge, MA: Harvard University Press.

Fry, C.H. 1984. The Bee-eaters. T. & A.D. Poyser, Calton, England.

Fry, C.H., Fry, K and Harris, A. 1992. Kingfishers, Bee-eaters & Rollers. Princeton Univ. Press, Princeton, N.J.

Fry, C.H., S. Keith, and E. K. Urban. 1988. The Birds of Africa. Vol. 3. Academic Press, London.

Fuchs S., Koeniger N and Tingek S. 1996. The morphometric position of A. nuluensis (Tingek, Koeniger and Koeniger) within cavity-nesting honey bees, *Apidologie* 27, 397–406.

Fuchs S., Long L.T and Anderson, D.L. 2000. A scientific note on the genetic distinctness of *Varroa* mites on *Apis mellifera* L. and on *Apis cerana* Fabr. in North Vietnam. *Apidologie* 31, 459–460.

Fujiyuki, T., Takeuchi, H., Ono, M., Ohka, S., Sasaki, T., Nomoto, A and Kubo, T. 2004. Novel insect Picorna-Like Virus identified in the brains of aggressive worker honey bees. *J Virol* 78: 1093-1100.

Fyg, W. 1959. Normal and abnormal development of in the honeybee. *Bee World* 40:57-66.

Gadagkar, R Nair, P Chandrasekhra, K and Bhat, D.M. 1993. Ant species richness and diversity in selected locations in Western Ghats, India.*Hexapoda* 5(2):79-94.

Gal, H; Slabezki, Y; Lensky, Y 1992. A preliminary report on the effect of origanum oil and thymol applications in honey bee *(Apis mellifera* L.) colonies in a subtropical climate on population levels of *Varroa jacobsoni*. *Bee Science* 2(4): 175-180.

Gal, H; Slabezki, Y; Lensky, Y. 1992. A preliminary report on the effect of origanum oil and thymol applications in honey bee *(Apis mellifera* L.) colonies in a subtropical climate on population levels of *Varroa jacobsoni*. *Bee Science* 2(4): 175-180.

Gambino, P and Thomas, G. M. 1988. Fungi associated with two *Vespula* (Hymenoptera: Vespidae) species in the Eastern San Francisco Bay area. *Pan-Pacific Entomologist 64:* 107-113.

Garg, R. 1989. Bee flora of Paonta Valley. *Indian Bee J.* 51(3):113-114.

Garridoa C., Rosenkranza P., Paxtonb R.J and Gonçalvesc L.S. 2003. Temporal changes in *Varroa destructor* fertility and haplotype in Brazil, *Apidologie* 34, 535–541

Gary, N. E. 1967. A method of evaluating honey bee flight activity at the hive entrance. *Jour. Econ. Ent.* 60 (1): 102-05.

Gates, J 1997. Controlling Parasitic Mites in Honey Bee Colonies. Vernon, BC, British Columbia Ministry of Agriculture, Fisheries and Food. 26pp.

Gatoria, G. S., Baldev Singh and Jhajj, H. S. 1996. Need to revise or separately fix the standards for *mellifera* apiary honey. A case study. National beekeeping experience exchange conference. 29.31 May 1995, Punjab Agricultural University, Ludhiana.

Gatoria, G.S., Aulakh, R.K., Chhuneja,P.K., Singh,J and Kamaldeep. 2004. Evaluation of queen cup kit apparatus for mass queen rearing of *Apis mellifera* Linn. Queen bee under Punjab (Indian) conditions *Insect environment.* 10(3):101-103.

Gatoria, G.S., Chhuneja, P.K., Singh, J and Singh, B. 2004. Mass queen bee rearing in *Apis mellifera* Linn. queen bees. *Indian Bee J.* 66(1-2):41-51.

Gatoria, G.S; Jhajj, H.S and Ratual, H.S. 1994. History of introduction of *Apis mellifera* L. and development of beekeeping in the Punjab (India). *Indian Bee J.* 5(1-2):8-15.

Gatoria,G.S., Singh, B and Singh, L. 2000. Effect of diameter of queen cell cups on mass rearing *Apis mellifera* linn. queen bees. *Indian Bee J.* 62(1-2):59-61.

Gauss, R. 1970. Beitrag ziir Kenntnis von Parasitoiden bei aculeaten Hymenopteren. *Zeitschrift fur Angewandte Zoologie 65:* 239-244.

Gayger, J. and Dustmann, J.H. 1985. Rückstandsuntersuchungen von bienenproduten wax, honig und pollen. Archiv für lebensmittel-hygiene, 36: 93-96.

Gelbe, R., Kurscheid, D. and Madel, G. 1987. The ectoparasitic bee-mite *Varroa jacobsoni* - investigations on the morphology of mouthparts and histology of salivary glands. *Zentralblatt für Bakteriologie Mikrobiologie und Hygiene Series A- Medical Micro-biology Infectious Diseases Virology Parasitology* 265: 554-554.

Gentile, A.G., Gallagher, K.J. and Santner, Z. 1971. Effect of some formulated insecti-cides on pollen germination in tomato and petunia. *Jour. Econ. Ent.* 64: 916-919.

Gerig, L. 1985, Der Schweizerische Bienenvater, Veriag Sauerlander, 16. Aufl.

Gerstell, R. 1939. The growth and size of Pennsylvania black bears. *Pennsylvania Game News* 10(8):4–7.

Ghosh, R.C., Ball, B. V., Willcocks, M. M. and Carter, M. J. 1999. The nucleotidesequence of *Sacbrood virus* of the honey bee: an insect picorna-like virus. *J Gen Virol* 80: 1541-1549.

Gilbert, B.K. and L.D. Roy. 1975. Prevention of black bear damage to beehives using aversive conditioning. DepCIR766 "Florida Black Bear-Beekeeper Conflict: 1981 Beekeeper Survey," *American Bee Journal*, Vol. 122, No. 5: pp. 372-375

Gill, R. A. 1991. The value of honeybee pollination to society. *Acta Horticulturae* No. 288: 62-68.

Gilliam M. 1990. Chalkbrood disease of honey bees, Apis mellifera, caused by the fungus, Ascosphaera apis: Review of past and current research. *Fifth International Colloq-uium on Invertebrate Pathology and Microbial control,* 398-402.

Gilliam, M. 1978. Fungi. In: Honey bee pests, predators, and diseases, R. A. Morse, ed., Comstock Pub. Associates a division of Cornell University Press, Ithaca, N.Y., 78-101.

Gilliam, M. 1986. Infectivity and Survival of the Chalkbrood Pathogen, Ascosphaera-Apis, in colonies of Honey-Bees, Apis-Mellifera. *Apidologie*, 17(2): 93-99.

Gilliam, M. Chalkbrood disease of honey bees, *Apis mellifera*, caused by the fungus, *Ascosphaera apis*: a review of past and current research. *5th International colloquium on invertebrate pathology and microbial control*, Adelaide, Australia, 398-402.

Gilliam, M., Taber III, S. and Bray Rose, J. 1978. Chalkbrood disease of honey bees, *Apis mellifera* L.: A progress report. *Apidologie*, 9(1): 75-89.

Gilliam, M., Taber III, S. and Richardson, G.V. 1983. Hygienic behavior of honey bees in relation to chalkbrood disease. *Apidologie*, 14(1): 29-39.

Gilliam, M., Taber III, S., Lorenz, B.J. and Prest, D.B. 1988 Factors Affecting Development of Chalkbrood Disease in Colonies of Honey Bees, *Apis mellifera*, Fed Pollen Contaminate *Ascosphaera apis*. *Journal of Invertebrate Pathology*, 52, 314-325.

Glare, T.R., Harris, R.J and Donovan, B.J. 1996. *Aspergillus flavus* as a pathogen of wasps, *Vespula* spp., in New Zealand. New Zealand journal of zoology 23: 339-344.

Gochnauer, T.A., Furgala, B., Shimanuki H. 1975. Diseases and enemies of the honey bee. In: The Hive and the Honey Bee, revised edition. Dadant Hamilton, Ill., USA. 615-621.

Godfray, H.C.J. 1994. Parasitoids: behavior and evolutionary ecology. New Jersey, Princeton University Press, 473p.

Goetze. C. 1964. Die Honigbiene in nhturlicher und künstlicher Zuchtauslese. I. Systemat Zeugung and Vererbung. Parey Verig. Hamburg.

Goller F and Esch H. 1991. Oxygen consumption and flight muscle activity during heating in workers and drones of *Apis mellifera*., *J. Comp. Physiol.* B, 161, 61–67.

Goller F., Esch H. 1990. Comparative study of chillcoma temperatures and muscle potentials in insect flight muscles. *J. Exp. Biol.* 150, 221–231.

Goodwin, R.H., Houten, A. and Perry, J.H. 1991. Feeding sugar syrup to honeybee colonies to improve kiwifruit pollen collection: a review. *Acta Hort.* 288: 265-269.

Gotwald, W.H. 1986. The beneficial economic role of ants. In: Economic impact and control of social insects. Vinson S.B. (Ed.) Praeger, New York, USA, pp 290-213.

Gould JL, 1975a. Communication of distance information by honey bees. J Comp Physiol 104:161–173.

Gould JL, 1975b. Honey bee recruitment: the dance-language controversy. Science 189: 685–693.

Gould JL, 1982. Why do honey bees have dialects? Behav Ecol Sociobiol 10:53–56.

Gould, J.L, 1974. Honey bee communication. *Nature* 252:300–301.

Gould, J.L. and C.G. Gould. 1988. The Honey Bee. Scientific American Library, W.H. Freeman, New York.

Goulson, D. 2003. Effects of introduced bees on native ecosystems. Annual Review of Ecology, Evolution, and Systematics 34:1-26.

Goyal, N. P. & J. K. Gupta, 1994. Development of beekeeping with *Apis mellifera* L. in India: present status and future scope. *Indian Bee Journal* 56(3/4): 127-131.

Goyal, N.P. and Rana, B.S. 1992. Economics of migratory beekeeping between hills and plains of north-west India. Bee J.54(1-4):19-25.

Goyal. N. P. (1974) *Apis cerana indica* and *Apis mellifera* as complementary to each other for the development of apiculture. *Bee Wld* 55(3): 98-101

Graham, J.M. (ed). 1992. The hive and the honeybee. Dadant and sons, Inc, Hamilton, Illnois, USA.

Greatti, M; Iob, M and Barbattini, R. 1992. Effectiveness of spring treatments with lactic acid and formic acid against *Varroa jacobsoni*. *Apicoltore Moderno* 83(2): 49-58.

Green, E. E. 1899. Sleeping habits of a bee, *Crocisa ramose*. *Ent. Monthly Mag.* 35: 214.

Gregorc, A; Jelenc, J. 1996. Control of *Varroa jacobsoni* Oud. in honeybee colonies using Apilife-VAR. *Zbornik Veterinarske Fakultete Univerza Ljubljana* 33(2): 255-259.

Grewal, G. S., A. S. Atwal, S. Kumar and Kapil, R. P. 1970b. Nesting behaviour of *Andrena leaena* Cam. *22 Internat. Beekeeping Congr.* 1969: 422-424.

Grewal, G. S., Atwal, A. S., Kumar, S. & Kapil, R. P. 1970a. Some observations on the nest architecture of *Colletes nursei* (Cam.). *Indian J. Ent.* 32: 184-186.

Grewal, J. S. and Sidhu, A. S. 1978. Insect pollinators of some cucurbits in Punjab. *Indian Journal of Agricultural Sciences* 48:79-83.

Grobov, O. F. 1976. Varroasis in bees: Varroasis, a honey bee disease. *Apimondia*

Grobov, O.F. 1977. Varroasis in bees, In: Varroasis a honeybee diseases, Bucharest, Apimondia, pp. 48-49.

Guerrant, E. O., Jr. and Fiedler, P.L. 1981. Flower defenses against nectar-pilferage by ants. *Biotropica* 13 (Suppl): 25–33.

Gupta, G.A. 1967. *Varroa jacobsoni*: a mite pest of *Apis indica*. *Bee Wld* 48(1)17-18.

Gupta, J.K. and Dogra, G.S. 1987. Importance or *Robinia pseudacacia* as a bee forage. In: Social Foretry for Rural Development, ed. P.K. Khosla and R.K. Kohli, Solan: ISTS Publication, pp. 215-217.

Gupta, J.K. and Thakur, R.K. 1987a. Nectar of *Vitex negundo* and foraging intensity of honeybees. In: Proc. First Symposium on Social Insects held at U.A.S., Bangalore, 1987. pp. 54-59.

Gupta, J.K. and Thakur, R.K. 1987b. Nectar sugar production and flower visitors of bramble, *Rubus ellipticus* Smith (Rosaceae) at Solan, India. Apidologie 18:223-230.

Gupta, J.K.; Kumar, J. and Kumar R. 1990. Observations on the foraging activity of honeybees and nectar production in 'Toon', *Toona Ciliata* Roem. *Indian Bee J.* 52:34-35.

Gupta, J.K.; Mishra, R.C. and Kumar, J. 1984. Plectranthus as forage for *Apis cerana* indica F.and Apis mellifera L. Apidologie 15(1):75-82.

Gupta, J.K.; Mishra, R.C. and Kumar, J. 1992. Evaluation of false acacia, *Robinia pseudacacia* L. as honey source in Himachal Pradesh. *Apiacta* 27(1) 8-12.

Gupta, J.K.; Reddy, M.C.M. and Kumar, J. 1990. Pattern of nectar secretion in wild cherry, *Prunus puddum* Roxb. And the associated foraging behaviour of *Apis cerana* and *Apis mellifera* L. *Apidologie* 21(1):11-16.

Gupta, J.K; Kumar, J. and Mishra, R.C. 1984. *Plectranthus* as forage for Apis cerana indica. F. and Apis mellifera L. Apidologie 15(10):75-82.

Gupta, R. K. 1993. *Taxonomic studies on the Megachilidae of north-western India.* Reprinted 1999. Scientific Publishers (India). v+294 p., 1032 figs.

Gupta, R. K. 2002. An updating bibliography of the bees of the world. URL: http://geocities.com /beesind. Second updated issue released on 31 July, 2003.

Gupta, R. K. 2003. An annotated catalogue of the bee species of the Indian region. URL: http:// geocities.com/beesind2. [93 genera including 1284 species]

Gupta, R. K. and Sharma, S. L. 1995. Description of the nest and immature stages of *Heriades tolawasensis* Sharma & Gupta (Hymenoptera, Megachilidae, Osmini). *Entomon* 20(1): 35-39.

Gupta, R. K. and Yadav, S. 2001. Apoidean species composition on *Crotalaria juncea* L., *Cajanus cajan* (L.), *Helianthus annus* L., and *Brassica compestris* L. var *sarson* Prain in eastern Rajasthan, India (Hymenoptera). *Opuscula zoologica fluminensia* 198: 1-10.

Gupta, R. K., Naval, R. K and Charan, S. K. 2003. Nesting biology of the solitary bee, *Megachile (Pseudomegachile) creusa* Bingham (Hymenoptera, Megachilidae). *Mellifera* 3(5): 11-18.

Gurs, S. and Abrol, D.P. 1995. Mite pests of honeybees and their management. *Indian Farming* 1995, 45: 28-30.

Gurung, M.B.; F. Ahmad; S.R. Joshi and C.R. Bhatta, 2003. The value of *Apis cerana* beekeeping for mountain farmers in Nepal. *Bees for Development Journal* 69: 13.

Gutpa, J.K., Thakur, R.K. and Kumar, J. 1986. Nectar of wild olive, *Elaeagnus umbellata* Thumb. and foraging intensity of different visitors. *Indian Bee J.* 48:40-42.

Guy. R. D. 1972. Commercial beekeeping with African bees. *Bee World* 53, 14-22.

Guzmán, L. and Delfinado Baker, M. 1997. A new Species of Varroa (Acari: Varroidae) Associated with *Apis koschevnikovi* (Hymenoptera: Apidae) in Borneo. *Int. J. Acarol.* 22: 23-27.

Guzman, R. F. 1984. Preliminary evaluation of the potential *of Steinernema feltiae* for controlling *Vespula germanica. New Zealand Journal of Zoology*

Hackett, K.J and Poinar, G.O Jr. 1973. The ability of Neoaplectina carpocapsae Weiser (Steineremematidae Rhabditoidae) to infect adult honeybees(Apis mellifera Apidae: Hymenoptera). American Bee Journal 113:100.

Hadisoesilo S., Otis G.W and Meixner M. 1995. Two distinct populations of cavity-nesting honey bees (Hymenoptera: Apidae) in South Sulawesi, Indonesia." *Journal of the Kansas Entomology Society* 68 399–407.

Hall, D.R., Cork, A. and Lester, A. 1987. Identification and role of female pheromone. *J. Chem.Ecol.* 13: 1575-1589.

Hall, E. R. 1981. The mammals of North America. 2nd edition. John Wiley & Sons, New York, New York. 1181pp.

Hameed, S.F. and Singh, B. 1989. Effect of *Varroa jacobsoni* Oudemans infestation on foraging and brood rearing in *Apis indica* Fabr. *Indian Bee J.* 51(4):135-36.

Hamilton, W.J. 1939.American mammals their hives, habits and economic relations. McGraw – Hill, New York.

Hanley, A.V, Huang, Z.Y and Pett, W.L. 2003. Effects of dietary transgenic Bt corn pollen on larvae of Apis mellifera and Galleria mellonella. *Jour. Apicultural Research* 42: 77-81.

Harbo J. R. and Harris, J. W. 1999. Heritability in honey bees (Hymenoptera: Apidae) of characteristics associated with resistance to *Varroa jacobsoni* (Mesostigmata: Varroidae). *Journal of Economic Entomology* 92 (2): 261-265.

Harbo J. R. and Harris, J. W. 1999. Selecting honey bees for resistance to *Varroa. Apidologie* 30: 183-196.

Harbo J. R. and Hoopingarner, R. A. 1997. Honey bees (Hymenoptera: Apidae) in the United States that express resistance to *Varroa jacobsoni* (Mesostigmata: Varroidae). *Journal of Economic Entomology* 90: 893-898.

Harbo, J. R. 1995 Observations on hygienic behavior and resistance to chalkbrood. *American Bee Journal*, 135(2): 828.

Harcourt, S. J.; Harris, R. J.; Rose, E. A. F.; Glare, T. R and Nelson, T. L. 1997. The potential of Beauveria bassiana for the control of common and German wasps (*Vespula vulgaris* L. and *V. germanica* F.) in New Zealand. Proceedings of 4th international workshop on microbial control of soil dwelling pests.

Harris J. W. and Harbo, J. R. 2000. Changes in reproduction of *Varroa destructor* after honey bee queens were exchanged between resistant and susceptible colonies. *Apidologie* 31: 689-699.

Harris J. W. and Harbo. J. R. 1999. Low sperm counts and reduced fecundity of mites in colonies of honey bees (Hymenoptera: Apidae) that are resistant to *Varroa jacobsoni* (Mesostigmata: Varroidae). *Journal of Economic Entomology* 92 (1): 83-90.

Harris, R. J and Rose, E. A. F. 1999. Factors influencing reproductive strategies of the vespid parasitoid *Sphecocphaga vesparum vesparum* (Hymenoptera: Ichneumonidae). *New Zealand Journal of Zoology* 26: 89-96.

Harris, R.J and Read, P.E.C. 1999. Enhanced biological control of wasps. *Science for conservation* 115, 39pp.

Hart, T and R. Nabors. 2000. Pollen traps and drone pupae destruction as a method of varroa control. *Am. Bee J.* 140: 151.

Hassel, M.P. and Southwood,T.R.E. 1978. Foraging strategies of insects. *Ann. Rev. Ecol. Syst.* 9: 75-98.

Haydak, M. H. 1970. Honey bee nutrition. *Annual Review of Entomology* 15:143-156.

Haydak, M.H. 1963. Unwanted guests of honeybee colonies. *American Bee Journal.*103:129-131.

Haydax, M. H. 1968. Nutrition des larves d' abeilles. In chauvin,r. Traite de biologie de l'abeille, vol. I, biologie et physiologie generales, p. 302-333. Masson et cie, paris.

Heath, L. A. F. 1982. Development of chalk brood in a honeybee colony: a review. *Bee World*, 63: 119-135.

Heinrich B, 1980. Mechanisms of body-temperature regulation in honeybees, *Apis mellifera. J. Exp. Biol.* 85:61–87.

Heinrich B, 1985. The social physiology of temperature regulation in honeybees. In: Experimental behavioral ecology and sociobiology (Hölldobler B, Lindauer M, eds.), Sunderland, Massachusetts: Sinauer; 393–406.

Heinrich B. 1993. Hot-blooded insects: Strategies and mechanisms of thermoregulation, Harvard University Press, Cambridge, MA, USA.

Heinrich, B. 1981. The mechanisms and energetics of honeybee swarm temperature regulation. *J. Exp. Biol.* 91, 25-55.

Heinrich, B. 1993. The Hot-Blooded Insects, Strategies and Mechanisms of Thermoregulation. Berlin, Heidelberg, New York: Springer.

Heinrich, B. and Raven, P.H. 1972)"Energetics and pollination ecology", *Science,* 176, 597-602.

Hewitt. Goodfrey, R.1929.South African pseudoscorpions of the genus chelifer. *Annals of the Natal Museum* 6 (2):305-336.

Heyndrickx, M., K. Vandemeulebroecke, B and Hoste, *et al.* 1996. Reclassification of *Paenibacillus* (formerly *Bacillus*) *pulvifaciens,* as a subspecies of *P. larvae,* with emended descriptions of *P. larvae* as *P. larvae* subsp. *larvae* and *P. larvae* subsp. *pulvifaciens.* International Journal of Systematic Bacteriology 46:270–279.

Hicheri, L. K. (1978). *Varroa jacobsoni* in Africa. *Apiacta* 13: 178.

Hicheri, L. K. 1978. *Varroa jacobsoni* in Africa. *Apiacta* 13: 178

Hicks TC, Fouts RS, Fouts DH. 2005. Chimpanzee (*Pan troglodytes troglodytes*) tool use in the Ngotto Forest, Central African Republic. *Am J Primatol* 65:221-237.

Higes, M., R. Martín, and Aránzazu, M. 2006. *Nosema ceranae,* a new microsporidian parasite in honeybees in Europe. *Journal of Invertebrate Pathology* 92:93-95.

Higes, M; Meana, A; Suarez, M; Llorente, J 1999. Negative long-term effects on bee colonies treated with oxalic acid against *Varroa jacobsoni* Oud. *Apidologie* 30(4): 289-292.

Himmer A. 1927a. Ein Beitrage zur Kenntnis des Wärmehaushalts im Nestbau sozialer Hautflüger. *Z. Vergl. Physiol.* 5, 375–389.

Himmer A. 1927b. Der soziale Wärmehaushalt der Honigbiene. II. Die Wärme der Bienenbrut, Erlang. *Jahrb. Bienenkd.* 51, 1–32.

Hock, R. J and A. M. Larson. 1966. Composition of black bear milk. Journal of Mammalogy 47:539–540.

Holldobler, B and Wilson,E.O.1990. The ants. Springer verlag.

Holmes, A. 1980. Bin och boidling. Stockholm, Sweden: LTs Forlag 586 Pages (In Swedish)

Holmes, F.O. 1965. An apiary in India. *Glean. Bee Cult.* 98(2):117,120

Holst, E.C. 1945 An antibiotic from a bee pathogen. *Science* 102:593-594.

Hoover, S. E. R., Keeling, C. I., Winston, Mark. L. and Slessor, K. N. 2003. The effect of queen pheromones on worker honeybee ovary development. *Naturwissenschaffen,* 90: 477-480

Hoppe, H 1990. Control of Varroa with a Thermal Treatment in Combination with Wintergreen Oil. Dissertation. Justus Liebig University, Giessen, Germany

Hoppe, H and Ritter, W 1989. Use of heat and wintergreen oil for treatment of varroatosis. *Tierarztliche Umschau* 44(11): 712, 715-717.

Horne, C. 1868. Notes on the habits of some Indian hymenopterous insects. *Proceedings of the Entomological Society of London* 1868: xlix.

Horne, C. 1870. Notes on the habits of some hymenopterous insects from the North-west Provinces of India. With an appendix, containing descriptions of some new species of Apidae and Vespidae collected by Mr. Horne: by Frederick Smith, of the British Museum. Illustrated by plates from drawings by the author of the notes. *Transactions of the Zoological Society of London* 7: 161-196.

Horne, C. 1872. Notes on the habits of some hymenopterous insects from the northwest provinces of India. *Trans. Zool. Soc. London* 7: 161-196.

Hornitzky MAZ, Clark S. 1991. Culture of Bacillus larvae from bulk honey samples for the detection of American foulbrood. *J Apicult Res* 29:199-205.

Hornitzky, M.A.Z and Nicholls, P.J. J. 1993. Medium is superior to sheep blood agar and brain heart infusion agar for the isolation of *Bacillus larvae* from honey samples. *J Apicult Res*;32:51-52.

Hornitzky, M.A.Z and Wilson, S.C. 1989. A system for the diagnosis of the major bacterial brood diseases of honeybees. *J Apicult Res.* 28:191-195.

Houbaert, D. and Jacobs, F. J. 1992. Accelerated fruit setting of tomatoes by insect pollination. In: *Bees for pollination. Proceedings of an EC workshop,* Brussels, Belgium, 2-3 March 1992: 207-218.

Howard, A., Howard, L.C. and Khan, A.R. 1920. Studies in the pollination of Indian crops. *Mem. Dep. Agric. India (Bot. Ser.)* 10(5): 195-220.

Huang Wei-Fone, Jing-Hao Jiang, Yue-Wen Chen and Chung-Hsiung Wang. 2007. A Nosema ceranae isolate from the honeybee *Apis mellifera Apidologie* 38 30-37.

Huang, M. T. F. and Gorman, C. M. 1990. Intervening sequences increase efficiency of RNA 3' processing and accumulation of cytoplasmic RNA. *Nucl Acids Res* 18: 937-947.

Huang, W.F., Jiang, J.H., Chen, Y.W and Wang, C.H. 2007. A *Nosema ceranae* isolate from the honeybee *Apis mellifera. Apidologie* 38, 30-32

Huang, Z.Y, Hanley, A.V, Pett, W.L, Langenberger, M and Duan, J.J. 2004. Field and semifield evaluation of impacts of transgenic canola pollen on survival and development of worker honey bees. *Journal of Economic Entomology* 97:1517-1523.

Humber, R. A. 1992. Collection of entomopathogenic fungal cultures. Catalog of strains, 1992. USDAARS- 110.

Hunt, G.J, Guzmán-Novoa, E and Page, R.E Jr, 1997. Genomic mapping of honey bee defensive behavior. In: Apiculture for the 21st century (Hoopingarner R and Connor L, series eds). Cheshire, CT: Wicwas Press.

Hunt, G.J, Guzmán-Novoa, E, Fondrk, M.K, and Page, R.E, 1998. Quantitative trait loci for honey bee stinging behavior and body size. *Genetics* 148:1203–1213.

Hunt, G.J, Page, R.E, Fondrk, M.K, and Dullum, C.J, 1995. Major quantitative trait loci affecting honey bee foraging behavior. *Genetics* 141:1537–1545.

Hyngstrom, S. E. 1994. Black bears. Pages c5–c15 *in* S.E. Hygnstrom, R. M. Timm, and G. E. Larson, editors. Prevention and control of wildlife damage. University of Nebraska Press, Lincoln, Nebraska.

Hywel-Jones, N. 1995. *Cordyceps sphecocephala* and a *Hymenostilbe* sp. infecting wasps and bees in Thailand. *Mycological Research 99:* 154-158.

Imdorf A, Charriere, J.D Maquelin, C Kilchenmann, V and Bachofen, B 1995. Alternative Varroa Control. Federal Dairy Research Institute. Apicultural Department. Liebefeld, Switzerland.

Imdorf A. J-D Charriere, C Maquelin, V Kilchenmann and B Bachofen. 1996. Alternative Varroa Control. *Amer. Bee J.*136: 189-193.

Imdorf, A.; Bogdanov, S.; Kilchenmann, V and Maquelin, C.1995. Apilife VAR: a new varroacide with thymol as the main ingredient. *Bee World* 76, 77–83.

Imdorf, A; Bogdanov, S; Ochoa, R I; Calderone, N W (1999). Use of essential oils for the control of *Varroa jacobsoni* Oud. in honey bee colonies. *Apidologie* 30(2-3): 209-228.

Imdorf, A; Charriere, J; Bachofen, B. 1997. Efficiency checking of the *Varroa jacobsoni* control methods by means of oxalic acid. *Apiacta* 32(3): 89-91.

Imms, A.D. 1942. On *Braula coeca* Nitzsch and its affinities. Parasitology 34: 88-100. Nitzsch CL. 1818. Die Familien und Gattungen der Thierinsekten (Insecta epizoica); als prodromus einer naturgeschichte der selben. *Magazin Ent.* 3: 261-316.

Iselin, W.A. 1968. A dermestid, *Trogoderma glabrum* infests comb honey. *Annals of the entomological society of America* 61:1621.

Ishay, J.; Bytinsky-Salz, H and Shulov, A. 1967. Contributions to the bionomics of the oriental hornet (*Vespa orientalis* Fab.). *Israel Journal of Entomology 2:* 45-106.

Iwata, K. 1964. Egg giantism in subsocial Hymenoptera with ethological discussions on tropical bamboo carpenter bees. *Nature & Life in southeast Asia* 3: 399-431.

Jack. R. W. 1916. Parthenogenesis amongst the workers of the Cape honeybee: Mr. G. W Onion's experiments. *Trans. Ent. Soc.* London, 396-403.

Jaeger, P. 1961. The wonderful life of flowers. – E. P. Dutton & Co.

James B. Whitfield, Sydney A. Cameron,Santiago R. Rami´Rez, Kari Roesch,

James M. Carpenter and Lien Phuong Thi Nguyen. 2003. Keys to the genera of social wasps of South-East Asia (Hymenoptera: Vespidae) *Entomological Science* 6 (3): 183.

Jay S.C. 1959. Factors affecting the laboratory rearing of honeybee larvae (*Apis mellifera* L.), M.Sc. dissertation, Ontario Agricultural College of the University of Toronto, Guelph, Ontario, Canada.

Jay, S.C. 1986. Spatial management of honeybees on crops. *Ann. Rev. Ent.* 31; 49-65.

Jaycox, E R. 1963. Confinement of honeybee colonies to avoid pesticide losses. *J. Apic. Res.* 2: 43-49.

Jaycox, E R. 1964. effect on honey bees of nectar fromsystemic insecticide-treated plants. *J. econ. ent.* 57: 31 - 35.

Jaycox, E. R., Guynn, G., Rhodes, A. M. and Vandemark, J. S. 1975. Observations on pumpkin pollination in Illinois (USA). *American Bee Journal* 115(4): 139-140.

Jitendar Kumar, Mishra, R.C. and Thakur, S.S. 1998. Use of honey bees for increasing agricultural yields, pp. 188-211, in Mishra, R.C. (Ed.) 1998. Perspectives in Indian Apiculture, Agro Botnica, Bikaner (India), 412 pp.

Johansen, C. A. 1966. Digest on bee poisoning, its effects and prevention. *Bee World* 47:9-25.

Johansen, C. A. 1969. The bee poisoning hazard from pesticides. Wash. Agr. Expt. Sta. Bul. 709, 14 pp.

Johansen, C. A. 1977. Pesticides and pollinators. *Annual Review of Entomology* 22:177-192.

Johansen, C. A. Coffey, M. D and Quist, J. A. 1957. Effect of insecticide treatments to alfalfa on honey bees, including insecticidal residues and honey flavor analyses. *J. econ. ent.* 50: 721 - 723.

Johansen, C. A., M. D. Levin, J. D. Eves, W. R. Forsyth, H. B. Busdicker, D. S. Jackson, and Butler, L. I. 1965. "Bee poisoning hazard of undiluted malathion applied to alfalfa in bloom." *Wash. Agric. Exp. Stn., Cir. 455.*

Johansen, C.A 1979. Honeybee poisoning by chemicals, signs, contributing factors, current problems and prevention. *Bee Wld.* 60(3): J09-127.

Johansen, P.1953. Chafers and stylops as enemies of the honeybee.*Nordisk Bitidskrift* 5:98-100.

Johansson, T. S. K and Johansson, M. P. 1977. Feeding honey bees pollen and pollen substitutes. *Bee World* 58(3):105-118.

Johansson, T. S. K., and Johansson, M. P. 1976 and 1977. Feeding sugar to bees. *Bee World* 57(4):137-142; 58(1):11-18; 58(2) :49- 52.

Jones, GD and SD Jones 2001. "The uses of pollen and its implication for Entomology". Neotropical Entomology 30 (3): 314-349.

Jones, R. L and Rothenbuhler, W. C. 1964. Behavior genetics of nest cleaning in honey bees II. Responses of two inbred lines to various amounts of cyanide-killed brood. *Animal Behaviour* 12: 584-588.

Joshi, R.N.1992. Sthanantarit Madhu Makhi Palan.In: Madhu Makhi Palan, pp 65-69. Dr. Y.S Parmar University of Horticulture and Forestry, Solan (Himachal Pradesh), India.

Judson, M.L. 1990. Redescrption of the bee associate (*Ellingsenius fuller* (Hewett and Godfrey) (Arachinda: Chelonethi Cheliferidae) with new records from Africa Europe and middle east. *Journal of Natural history* 24:1303-1310.

Juenger, T and Bergelson, J. 1997. Pollen and resource limitation of compensation to herbivory in scarlet gilia, *Ipomopsis aggregata. Ecology* 78: 1684–1695.

Justus, E.R. 1991. Beekeeping industry in Tamil Nadu (India): Present status, problems and prospects. *Indian Bee J.* 53 (1-4):1-8.

K von Frisch. 1965. Bees - Their Vision, Chemical Senses, and language. Revised Edn.

Kahn, H. K.; Jayaraj, S and Gopalan, M. 1993. Muscardine fungi for the biocontrol of agroforestry termite *Odontotermes obesus* (Rambur). *Insect Science & its Application* 14: 529-535.

Kaiser, H. 1987. Biology, ecology and development of the European parasitoid of yellowjackets *Pheromermis vesparum,* new species (Mermithidae, Nematoda). *Zoologische Jahrbuecher Abteilungfuer Systematik Oekologie und Geographie der Tiere 114:* 421— 449.

Kanbar G and Engels, W. 2005. Communal use of integumental wounds in honey bee (*Apis mellifera*) pupae multiply infested by the ectoparasitic mite *Varroa destructor*. *Genet Mol Res.* 4(3):465-72.

Kanga, L. H. B., James, R. R and Boucias D. G. 2002, *Hirsutella thompsonii* and *Metarhizium anisopliae* as potential microbial control agents of *Varroa destructor*, a honey bee *Journal of Invertebrate Pathology,* 81(3):175-184

Kannan, K. 1925. Xylocopa and Hyperechia in India. Proceedings of the Entomological Society of London 1925: 12-13.

Kapil, R. P and Dhaliwal, J. S. 1968a. Defense of nest by the female of *Xylocopa fenestrata* Fab. (Xylocopinae, Hymenoptera). *Insectes Sociaux* 15(4): 419-422.

Kapil, R. P and Dhaliwal, J. S. 1968b. Biology of *Xylocopa* species. I. Seasonal activity, nesting behaviour and life cycle. *Journal of Research, Punjab Agricultural University* 5(3): 406-419.

Kapil, R. P., Chaudhary, J. P and Jain, K. L. 1975. Biology and utilization of insect pollinators for crop production. Sec. Ann. Reptr. Dept. Zool. H. A. U. Hissar: 50 p.

Kapil, R. P., Grewal, G. S. & Atwal, A. S. 1970. Biology and nesting behaviour of *Megachile lanata* Fabricius. *22 International Beekeeping Congr.* 1969: 449-458.

Kapil, R. P., Grewal, G. S., Kumar, S and Atwal, A. S. 1970. Insect pollinators and their comparative abundance. *In*: A. S. Atwal (ed.), *Insect pollinators of crops.* Final report of the PL-480 Project No. A7-Ent-19. At P. A. U. Ludhiana.

Kapil, R.P. 1957. The length of life and the brood rearing cycle of the Indian *Bee World.* 38(10):258-263.

Kapil, R.P. and Dhaliwal, J. S. 1969. Biology of *Xylocopa* species. II. Field activities, flight range and trials on transportation of nests. *Journal of Research, Punjab Agricultural University* 6(1): 262-271.

Kapil, R.P. and Kumar, S. 1969. Biology of *Ceratina Binghami* Cockerell (Ceratinini: Hymenoptera). *Journal of Research, Punjab Agricultural University* 6(2): 359-371.

Kapil, R.P., Putatunda, B.N. and Aggarwal, K. 1985. Studies on ectoparasitic mites of *Apis* species. Final Technical Report (Department of Science and Technology project). Deptt of Zoology, HAU, Hisar, India. 191p

Kapila, R. K., Singh, B., Sharma, J. K., Lata, S and Thakur, S. P. 2002. Effect of insect pollinators on seed yield and its quality in radish (*Raphanus sativus* L.). *Seed Research* 30(1): 142-145.

Kashkovrkei, U.G 1958. An enemy of the apiary. *Priroda,* Moscow 47:120.

Kassimatis E., Dippenaar-Schoeman A.S. and Harvey M.S. 2001. The pseudoscorpions of South Africa (arachnida: pseudoscorpiones). *Spider club of southern Africa Newsletter* 16(2): 20-22.

Kastinger C, and A Weber 2001. "Bee-flies (Bombylius spp., Bombyliidae, Diptera) and the pollination of flowers". Flora 196 (1): 3-25.

Kasturirangan, K *et al.* 1996. Indian Remote Sensing Satellite (IRS)-IC: The beginning of a new era. *Current Sci.* 70(7):45-46.

Katie E. Shaw, Gillian Davidson, Suzanne J. Clark, Brenda V. Ball, Judith K. Pell, David Chandler and Keith D. Sunderland 2002. Laboratory bioassays to assess the pathogenicity of mitosporic fungi to *Varroa destructor* (Acari: Mesostigmata), an ectoparasitic mite of the honeybee, *Apis mellifera Biological Control, Volume 24(3):266-276*

Kauder, S. E and Racaniello, V. R. (2004). Poliovirus tropism and attenuation are determined after internal ribosome entry. *J Clin Invest* 113: 1743-1753.

Kebler, L. F. 1896. Poisonous honey. – Am. Pharmaceut. Assoc. Proc. 44: 167 174.

Keeler, K.H. 1981 "Function of *Mentzelia nuda* (Loasaceae) post-floral nectarines in seed defense", *Am. J. Bot.,* 68, 295-299

Kelley-Tunis, K. K.; Reid, B. L.; Andis, M. 1995. Activity of entomopathogenic fungi in free-foraging workers of *Camponotus pennsylvanicus* (Hymenoptera: Formicidae). *Journal of Economic Entomology* 88: 937-943.

Kendall, D.M. 1981 "Bionomics of *Euxoa auxilliaris* Grote (Lepidoptera: Noctuidae) in the rocky mountains and comparison with two resident species of alpine moths", MBS thesis. Univ. Colo., Boulder, pp. 88.

Kevan, P. G., Laverty, T. W and Denmark, H. A. 1991. Association of *Varroa jacobsoni* with organisms other than honey bees and implications for its dispersal. *Bee World* 71: 119-121.

Kevan, P.G., Chaloner, W.G. and Savile, D.B.O. 1975 "Interrelation-ships of early terrestrial arthropods and plants", *Palaeontology,* 18(2), 391-417.

Kevan, S. D., Nasr, M. E. and Kevan, P. G. 1999. Feeding menthol to honeybees (Hymenoptera: Apidea): entry and persistence in haemolymph without causing mortality. *The Canadian Entomologist* 131: 279 - 281.

Khan, M.S.A. 1948. Some important nectariferous plants and pollen sources of Bhopal State, Central India. *Indian Bee J.* 19(11/12);107-108.

Khanbash, M S; Taleb, A M B. 1998. Evaluation of some chemicals to control Varroa mite (*Varroa jacobsoni*) during the presence of brood in honeybee colonies. *University of Aden Journal of Natural and Applied Sciences* 2(1): 13-20.

King, C. C. 1964. Effects of herbicides on nectar secretion. *J. Apic. Res.* 3: 5-9.

Kistner, D. H. 1979. Social and evolutionary significance of social insect symbiotes. In: Herman H R (ed) Social insects. Vol 1 Academic Press,New York,USA pp 2-244.

Kitroo, A and Abrol, D.P. 1996. Studies on pollen carrying capacity and pollination efficiency of honeybees visiting litchi flowers. *Indian Bee J.* 58 (2) : 55-57

Kitroo,A and Abrol, D.P. 1996. Effect of flowering phenology, bloom density and floral rewards on the attractiveness of insects visiting litchi flowers. *Indian Bee J.* 58 (1) : 1-5

Klee, J., Besana, A. M., Genersch, E., Gisder, S., Nanetti, A., Tam, D. Q., Chinh, T. X., Puerta, F., Ruz, J. M., Kryger, P., Message, D., Hatjina, F., Korpela, S., Fries, I. & Paxton, R. J. (2006) Widespread dispersal of the microsporidium Nosema ceranae, an emergent pathogen of the western honey bee, *Apis mellifera. Journal of Invertebrate Pathology.* Under review

Kleespies, R. G., Radtke, J and Bienefeld, K. 2000. Virus-like particles found in the ectoparasitic bee mite *Varroa jacobsoni* Oudemans. *J Invert Pathol* 75: 87 90.

Kleinschmidt, M. G. 1972. Insecticide formulations and their toxicity to honey bees. *Journal of Apicultural Research* 11:59-62.

Kmitowa, K. 1982: Characteristics of strains of *Paecilomycesfarinosus* (Dicks.) Brown et Smith. *Polish Ecological Studies* 5:419-431.

Knowlton, C.F. 1947. Boxelder bug nymphs feeding on dead honeybees.*J. Econ. Entomol.*40:915.

Kobayasi, Y. 1941. The genus Cordyceps and its allies. *Science Reports of the Tokyo Bunrika Daigaku Section B 5:* 53-260.

Koenig, W. D. 1981. Reproductive success, group size, and the evolution of cooperative breeding in the Acorn Woodpecker. *Amer. Naturalist* 117:421-443.

Koenig, W. D., and R. L. Mumme. 1987. Population ecology of the cooperatively breeding Acorn Woodpecker. Monographs in Population Biology 24. Princeton Univ., Princeton, N. J.

Koeniger G., Koeniger N., Anderson D.L., Lekprayoon C., Tingek S. 2002. Mites from debris and sealed brood cells of *Apis dorsata* colonies in Sabah (Borneo) Malaysia, including a new haplotype of *Varroa jacobsoni*, *Apidologie* 33, 15–24.

Koeniger G., Koeniger N., Mardan M., Otis G.W and Wongsiri S. 1991. Comparative anatomy of male genital organs in the genus *Apis. Apidologie* 23 :539–552.

Koeniger N and Fuchs S. 1975. Zur Kolonieverteidigung der asiatischen Honigbienen. *Z. Tierpsychol.* 37, 99–106.

Koeniger N and G. Koeniger, 1980. Observations and experiments on migration and dance communication of *Apis dorsata* in Sri Lanka. *J. Apic. Res. 19*: 21–34.

Koeniger N, Koeniger G, de Guzman, L.I and Lekprayoon, C. 1993. Survival of *Euvarroa sinhai* Delfinado and Baker (Acari, Varroidae) on workers of *Apis cerana* Fabr., *Apis florea* Fabr., and *Apis mellifera* L. in cages. *Apidologie* 24:403–10

Koeniger N. 1996. The 1996 special issue of Apidologie on Asian honeybee species. *Apidologie* 27:329–30

Koeniger N., Koeniger, G and Delfinado-Baker, M. 1983. Observations on mites of the Asian honey bee species, Apidologie 14, 197–204.

Koeniger, N and Fuchs, S. 1988. Control of *Varroa jacobsoni* Oud. in honey bee colonies containing sealed brood cells. *Apidologie* 19: 117 -130.

Koeniger, N and Fuchs, S. 1988. Control of *Varroa jacobsoni*: current status and developments. In: *Africanized honey bees and bee mites* (Needham, G. R. *et al.* eds). Ellis Horwood Limited, Chichester, pp 360-369.

Koeniger, N. 1976. Interspecific competition between *Apis florea* and *Apis mellifera* in the tropics. *Bee World* 57: 110-112.

Koeniger, N. 1982. Interactions among four species of the genus *Apis. Proc. Congr. IUSSI,* 9th (Boulder), 59-64.

Koeniger, N., Koeniger G., and Wijayagunasekera, N.H.P. 1981. Observations on the adaptation of *Varroa jacobsoni* to its natural host *Apis cerana* in Sri Lanka. *Apidologie* 12, 37–40.

Kohli, N. 1958. Bee flora of Northern India. Indian Bee J. 20 :7-8, 31-32,61-62, 83-85, 106-107, 127-128,113-134,150-151,178-179,192, 193.

Koivulehto, K. 1976. *Varroa jacobsoni*, a new mite infesting honey bees in Europe. *Br Bee Journal* 104:16-17.

Korpela, S., Aarhus, A., Fries, I., & Hansen, H. 1992. *Varroa jacobsoni* Oud. In cold climates: Population growth, winter mortality and influence on the survival of honey bee colonies. *J Apic Res* 31: 157–164.

Kozin, R.B. 1976)"Pollination of Entomophilous Agricultural Crops by Bees", Amerind Publishing Co. Pvt. Ltd., New Dehli.

Kraus B. and Page R.E. Jr 1995. Population growth of *Varroa jacobsoni* Oud. In Mediterranean climates of California. Apidologie 26: 149–157.

Kraus, B 1991. Lactic acid treatment as varroasis therapy: intermediate report on winter treatment. *Biene* 127(8): 427-430.

Kraus, B 1992. Further results on lactic acid application as treatment for varroatosis. *Apidologie* 23(4): 385-387.

Kraus, B 1992. Lactic acid treatment as varroasis therapy: further studies. *Biene* 128(1): 5-11.

Kraus, B and Berg, S 1994. Effect of a lactic acid treatment during winter in temperate climate upon *Varroa jacobsoni* Oud. and the bee *(Apis mellifera* L.) colony. *Experimental and Applied Acarology* 18(8): 459-468.

Kraus, B and Page, R E, Jr. 1995. Effect of vegetable oil on *Varroa jacobsoni* and honey bee colonies. *Bee Science* 3(4): 157-161.

Krishnaswamy, S.V. 1970. Economic effects of application of Farrar's method under the conditions of Serbia (Yugoslavia). *Proc. XXVI Int. Beekeep. Congr.* : 525-526.

Krishnaswamy, S.V. 1981. A preliminary study of bee flora and its calendar around beekeeping observation centre, Jakhanahalli (Chikmagalur Taluk), Guthi (mudigera Taluk) and at the migration site at Banur, Chikmagalur district, Karnataka. *Indian Honey* 4(13):12-13.

Krishnasway, S.V. 1970. Soapnut trees: a nectar source. *Indian Bee J.* 32(3/4):83.

Kristof, L.J. 1879. Ueber einheimische, gesellig lebende Wespen und ihren Nestbau. *Mitteilungen des NaturwissenschaftUchen Vereinesfuer Steiermark* 15: 38-19.

Krombein, K. V., Hurd, P. D., Jr., Smith, D. R. & Burks, B. D. 1979. *Catalog of Hymenoptera in America north of Mexico.* Smithsonian Institution Press, Washington, D. C. 2735 pages [In 3 Volumes.].

Kshirsagar, K.K. 1971. A wasp trap to control predator wasps in the apiary. *Indian Bee Journal* 33:55-57.

Kubik, M.; Nowacki, J.; Michalczuk, L.; Pidek, A and Marcinkowski, J. 1995. Penetration of fluvalinate into bee-products. *Journal of Fruit and Ornamental Plant Research* 1, 13–22.

Kukuk, P. F. 1980. Diurnal activity of a primitively social bee, *Lassioglossum zephyrum* within the nest. *J. Kans. Entomol. Soc.* 53(1): 149-156.

Kumar, J. and Gupta,J.K. 1993. Nectar sugar production and honey bee foraging activity in three species of onion (Allium species). *Apidologie* 24(4): 391-396.

Kumar, J., Gupta, J.K. and Dogra, G.S. 1988. Discovery of the ectoparasitic mite *Varroa jacobsoni* on *Apis mellifera* in Himachal Pradesh India. *American Bee Journal* 128(2):124.

Kumar, J., Gupta, J.K. and Dogra, G.S. 1988. Discovery of the ectoparasitic mite *Varroa jacobsoni* on *Apis mellifera* in Himachal Pradesh India. *American Bee Journal* 128(2):124.

Kumar, J., Mishra R.C. and Thakur. S.S. 1998. Use of honeybees for increasing agricultural yields. In:Perspectives in Indian Apiculture (Eds. R.C. Mishra and Rajesh Garg), Agro Botanica, Bikaner, India. pp 187-211.

Kumar, J., Mishra, R. C and Gupta, J. K. 1993. Foraging activity and abundance of insect visitors to chicory (*Cichorium intybus* L.) in mid-hills of Himachal Pradesh, India. *Indian Bee Journal* 55(3/4): 31-36.

Kumar, J., Rao, K. V. K and Gupta, J. K. 1994. Pollination efficiency of bees visiting blossoms of *Brassica campestris* L. var. *toria* in mid-hills of Himachal Pradesh, India. *Indian Bee Journal* 56(3/4): 202-206.

Kumar, M and Singh, R. 2001. Preference for competing flora with sunflower for foraging of honeybees. *Journal of Applied Biology* 11(1/2): 70-71.

Kumar, M., Chand, Hari, Singh, Ramashrit and Ali, M. S. 2002. Effect of different modes of honeybee pollination on oil content in seeds of sunflower (*Helianthus annus* L.). *Journal of Entomological Research* 26(3): 219-221.

Kumar, R and Gupta, J.K 2003. Collection and storage of semen of *Apis mellifera* L. and effect of storage on sperm viability. *Pest Management and Economic Zoology*. 11(2):207-209.

Laidlaw, H. H. Jr. 1977. Instrumental insemination of honey bee queens. In pictorial instructional manual. 144 p. Dadant & Sons Inc. Journal printing co., Carthage, Ill.

Laidlaw, Harry H.Jr. 1979. Contemporary queen rearing. Dadant and Sons Inc. Hamilton. Illinois.

Laidlaw, Harry H.Jr. 1994. Artificial insemination of the queen bee (*Apis mellifera* l). Morphological basis and results. *J. MorphoL.*74:426-465.

Laidlaw, Harry. H. Jr. and Eckert, J.E. 1962. Queen rearing. University of California Press, Berkely, California

Laigo, F. M and Morse, R. A. 1969. Control of the bee mites, *Varroa jacobsoni* Oudemans and *Tropilaelaps clareae* Delfinado and Baker with cholorobenzilate *Philippine Entomol* 1: 144-148.

Laird, M. 1991. Australasian Invertebrate Pathology Working Group Newsletter XI: 74.

Langtroth, L. L. 1862. The hive and the honey bee. 409 P. C. M. Saxton, New York.

Lanzi, G and Rossi, C. 2003. Characterization of the *Deformed Wing virus* from the honey bee. Unpublished. GenBank Accession.

Lanzi, G and Rossi, C. 2003. Characterization of the *Deformed Wing virus* from the honey bee. Unpublished. GenBank Accession [AJ489744].

Larch. L.C.L 1876. Grafting queen cells. *Gleanings in Bee Cult.* 4 (3) : 48.

Larson, B.M.H, Kevan, P.G and Inouye, D.W. 2001. "Flies and flowers: taxonomic diversity of anthophiles and pollinators". Canadian Entomologist 133 (4): 439-465.

Latif, A., Qayyum, H. A and Ul-Haq, M. 1960. Role of *Apis indica* in the pollination of toria and sarson. *Bee World* 4(11-12): 283-286.

Leat, N., Ball, B., Govan, V and Davison, S. 2000. Analysis of the complete genome sequence of *Black queen-cell virus*, a picorna-like virus of honey bees. *J Gen Virol* 81: 2111-2119.

Lecomte, J.; Bourdon, V.; Damas, J.; Leclercq, M.; Leclercq, J. 1976: Presence de noradrenaline conjuguee dans les parios du nid de *Vespula germanica* Linne. *Societe Beige de Biologie 170:* 212-215.

Lecq, T. 1933. The dung beetles.L' Apiculteur 77:138-140.

Lekprayoon, C and Tangkanasing, P. 1991. *Euvarroa wongsirii*, a new species of bee mite from Thailand. *Int. J. Acarol.* 17:255–58

Leoncini, I, Le Conte Y, Costagliola, G, Plettner, E, Toth, A.L, Wang M, Huang, Z, Becard, J.M, Crauser D, Slessor, K.N., Robinson, G.E 2004. "Regulation of behavioral maturation by a primer pheromone produced by adult worker honey bees". Proceedings of the National Academy of Sciences of the United States of America 101 (50): 17559-64.

Leppik, E.E. 1960 "Early evolution of flower types", *Lloydia.*, 23, 72-92.

Levin, M. D. 1983. Value of bee pollination in US agriculture. *Bulletin of Entomological Society of America* 29: 50-51.

Lieftinck, M. A. 1939. Uit het leven van *Lithurgus atratus*, een Indisch houtbijtje. *De Tropische Naturae* 28: 193-201.

Lieftinck, M. A. 1955. The carpenter-bees (*Xylocopa* Latr.) of the Lesser Sunda Islands and Tanimbar (Hymenoptera, Apoidea). *Verhandlungen der Naturforschenden Gesellschaft in Basel* 66: 5-32.

Lieftinck, M. A. 1957. Revision of the carpenter-bees (*Xylocopa* Latr., subgenus *Maiella* Michener) of the Papuan region (Hymenoptera, Apoidea). *Nova Guinea* (N.S.) 8: 325-376.

Lieftinck, M. A. 1962. Revision of the Indo-Australian species of the genus *Thyreus* Panzer (=*Crocisa* Jurine) (Hym., Apoidea, Anthophoridae) Part 3. Oriental and Australian species. *Zoologische Verhandelingen, Leiden,* no. 53: 1-212, pls. 1-3.

Lieftinck, M. A. 1964. The identity of *Apis aestuans* Linnaeus (Hymenoptera, Apoidea) and related Old World carpenter bees (*Xylocopa* Latr.). *Tidjschrift voor Entomologie* 107: 137-158.

Lieftinck, M. A. 1972. Further studies on old world melectine bees, with stray notes on their distribution and host relationships. *Tijdschrift voor Entomologie* 115(7): 253-324, 2 pls.

Lieftinck, M. A. 1974. A review of the central and east Asiatic *Habropoda* F. Smith, with *Habrophorula,* a new genus from China. *Tijdschrift voor Entomologie* 117(5): 157-224.

Lillesand, T.M. and Kiefer, R.W. 1979. Remote sensing and image interpretation John Wiley and sons, New York.

Lindauer M, 1985. The dance language of honeybees: the history of a discovery. In: The dance language of honeybees: the history of a discovery (H;auolldobler B and Lindauer M, eds). Sunderland, MA: Sinauer; 129–140.

Lindauer, M. 1956. Communication among the honeybees and stingless bees of India. *Bee Wld. 38*: 3–14, 34–39.

Lindauer, M. Communication Among Social Bees. Atheneum, New York. London. 218p.

Linsley, E.G and Macswain, J.W. 1941. The bionomics of *Ptinus californicus* a predator in the nests of bees. Southern California Academy of Science Bulletin 40:126-137.

Liu B, Xu CG, Yan FM, Gong RZ 2005. The impacts of the pollen of insect-resistant transgenic cotton on honeybees. Biodiversity and Conservation 14: 3487-3496.

Liu T. 1996. *Varroa* mites as carriers of honey bee chalkbrood. *Am Bee J* 136: 655.

Liu, T. P. 1991. Virus-like particles in the tracheal mite *Acarapis woodi* (Rennie). *Apidologie* 22: 213-219.

Liu, Y.H., Chen, S.L., Liu, Y.H, Chen SL. 2001. Influences of transgenic insect-resistant plants on honeybees. Entomological Knowledge. 38:258-262.

Lodesani, M., Olombo, M. & Spreafico, M. 1995. Ineffectiveness of Apistan treatment against the mite *Varroa jacobsoni* Oud. In several districts of Lombardy (Italy). *Apidologie* 26:67-72.

Lodesani, M., Pellacani, A., Bergomi S., Carpana, E. and Rabitti, T. 1992. Residue determination for some products used against *Varroa* infestation in bees. *Apidologie,* 23: 257-272.

Lodesani, M; Bergomi, S; Pellacani, A; Carpana, E and Rabitti, T 1990. A comparative study on the efficacy of some products for controlling Varroa, and determinations of their residues.*Apicoltura* 6: 105-130.

Lodesani, Marco and Costa, C.2005. Limits of chemotherapy in beekeeping : development of esistance and the problem of residues *Bee World* 86: 102-109

Loglio, G and Pinessi, E 1991. Use of wheat flour for ecological control of varroa disease. *Apicoltore Moderno* 82(5): 185-192.

Loglio, G. and Plebani, G. 1992. Valutazione dell'efficacia *dell'Apistan. Apicolt Mod,* 83: 95-98.

Luttge, U and Schnepf, E. 1976. Organic substances. – In: Luttge, U. and Pitman, M. G. (eds), Transport in plants II. B: tissues and organs. Springer-Verlag, pp. 244–277.

Luttge, U. 1977. Nectar composition and membrane transport of sugars and amino acids: a review on the present state of nectar research. *Apidologie* 8: 305–319.

Ma, Tsing-chao, 1938. The Indian species of the genus *Xylocopa* Latr. (Hymenoptera). *Records of the Indian Museum* 40: 265-329.

Maa T.C. 1953. An inquiry into the systematics of the tribus Apidini or honeybees (Hym). *Treubia* 21, 525–640.

Maa, Tsing-chao, 1954. The xylocopine bees (Insecta) of Afghanistan. *Videnskabelige Meddelelser fra Dansk Naturhistorisk Forening I Kjøbenhavn* 116: 189-231.

Maccagnani, B., Ladurner, E., Santi, F and Burgio, G. 2003. *Osmia cornuta* (Hymenoptera, Megachilidae) as a pollinator of pear (*Pyrus communis*): fruit- and seed-set. *Apidologie* 34(3): 207-216.

MacDonald J. F. and Matthews, R. W. 1975.The social wasps (Hymenoptera: Vespidae) of Indiana. *The Great Lakes Entomologist* 22:155- 175.

MacDonald J. F. and Matthews, R. W. 1984.Vespula squamosa: a yellow jacket wasp evolving toward parasitism. *Science* 190:1003- 1004

Mace, H.1927. Modern beekeeping Wyman and Sons,London,U.K.

Mackensen, O. 1943. Effect of carbon dioxide von initial oviposition of artificially inseminated and virgin queen bee. *J.Econ. Entomol.* 40:344-349.

Mackensen, O. 1951.viability and sex determination in the honey bee ("*Apis mellifica*" l.). *Genetics* 36(5):500-509.

Mackensen, O. and K. W. Tucker. 1970. Instrumental insemination of queen bees. U.S. Department of agriculture, agriculture handbook no. 390, 28 p.

Mackensen, O. and Nye, W.P. 1969. Selective breeding of honeybees for alfalfa pollen collection: Sixth generation and out crosses. J. api. Res. 8:9-12.

Mackensen, O. and Nye,W.P1966. Selecting and building honeybees for collecting alfalfa pollen. *J. Apic. Res.* 5:79-86.

MacRoberts, M. H., and MacRoberts, B. R. 1976. Social organization and behavior of the Acorn Woodpecker in central coastal California. *Ornithol. Monographs* 21.

Maehr, D. S. 1983. Black bear depredation on bee yards in Florida. *Proceedings of the Eastern Wildlife Damage and Research* 16:87–88.

Maehr. D. 1982. "Beekeeping Enters the Solar Age," *American Bee Journal*, Vol. 122, No. 4, pp. 280-281.

Maeta, Y. 1978. Comparative studies on the biology of the bees of the genus *Osmia* of Japan, with special reference to their managements for pollination of crops. *Bulletin of the Tohoku National Agricultural Experiment Station* no. 57: 1-221 + errata 1-7. [In Japanese, English summary.]

Maggenti, A.R. 1981. General Nematology. Springer-Verlag, New York. 372 pages.

Malham, J. P., Rees, J. S., Alspach, P. A., Beggs, J. R., Moller, H. 1991: Traffic rate as an index of colony size in Vespula wasps. *New Zealand Journal of Zoology* 18: 105-109.

Malone, L.A 2004. Potential effects of GM crops on honey bee health. Bee World 85: 29-36.

Malone, L.A, Burgess, E.P.J and Stefanovic, D 1999. Effects of a Bacillus thuringiensis toxin, two Bacillus thuringiensis biopesticide formulations, and a soybean trypsin inhibitor on honey bee (Apis mellifera L.) survival and food consumption. Apidologie 30 (6): 465-473

Malone, L.A., Burgess, E.P.J., Gatehouse H.S., Voisey, C.R., Tregidga, E.L and Philip, B.A. 2001. Effects of ingestion of a Bacillus thuringiensis toxin and a trypsin inhibitor on honey bee flight activity and longevity. Apidologie 32:57-68.

Malone, L.A., Pham-Delègue, M.H. 2001. Effects of transgene products on honey bees (Apis mellifera) and bumblebees (Bombus sp.). Apidologie 32:287-304.

Malone, L.A., Pham-Delègue, M.H. 2002. Using proteins to assess the potential impacts of genetically modified plants on honey bees. Honey bees: estimating the environmental impact of chemicals p. 290-311, Editors: Devillers, J.; Pham-Delègue, M. H. Publisher: Taylor & Francis, London, UK

Malone, L.A., Todd, J.H., Burgess, E.P.J and Christeller, J.T. 2004. Development of hypopharyngeal glands in adult honey bees fed with a Bt toxin, a biotin-binding protein and a protease inhibitor. Apidologie 35 (6): 655-664.

Malyshev, S. I. 1923. The nesting habits of *Melitta leporina* Pz. Izvestiya Leningradskovo nauchnovo Instituta imeni P. F. Leshaft [*Bulletin de l'Institut Leshaft*] 6: 1-7. [In Russian, English summary.]

Malyshev, S. I. 1925. The nesting habits of *Anthophora* Latr. *Trav. Soc. Natur. Leningrad, Se Zool. Physiol.* 55: 137-183.

Malyshev, S. I. 1930. Nistgewohnheiten der Steinbienen *Lithurgus* Latr. *Zeitschrift für Morphologie und Ökologie der Tiere* 19: 116-134.

Malyshev, S. I. 1936. The nesting habits of solitary bees. A comparative study. *Eos, Madrid* 11: 201-309, pls. III-XV.

Manley, R.O. B. 1948. Moving bees, pp. 212-218 In: Honey Farming, Faber and Faber Ltd, London, 289 pp.

Mardan M and Kevan P.G. 1989. Honeybees and "yellow rain", *Nature* (London) 341, 191.

Mardan M.1989. Thermoregulation in the asiatic giant honeybee *Apis dorsata* (Hymenoptera: Apidae), Ph.D. dissertation, University of Guelph, Guelph, Ontario, Canada.

Mardan, M and Kevan P. G.2002. Critical temperatures for survival of brood and adult workers of the giant honeybee, *Apis dorsata* (Hymenoptera: Apidae). *Apidologie* 33: 295–301

Mardan, M. 1995. Varied pollinators for sub-tropical Asian Crops. In: Pollination of cultivated Plants in the Tropics by Roubik, D.W. (Ed. FAO (UN), Rome, pp. 142-148.

Maria H. Corrêa-Marques, Luis Medina Medina, Stephen J. Martin and David De Jong (2003) Comparing data on the reproduction of *Varroa destructor*. *Genet. Mol. Res.* 2 (1): 1-6

Marie-Pierre Chauzat, Jean-Paul Faucon, Anne Claire Martel, Julie lachaize, Nicolas Cougoule, and Michel Aubert 2006. A Survey of Pesticide Residues in Pollen Loads Collected by Honey Bees in France. *J. Econ. Entomol.* 99(2): 253-262 (2006)

Marin, M. 1978. World spread of *Varroa* disease. *Apiacta* 13: 163-166.

Marletto, F., A. Patetta, and A. Manino. 1991. Further tests on varroa disease control by means of periodical drone brood removal. *Apicolt. Mod.* 82: 219-224.

Martignoni, M. E.; Iwai, P. J. 1986. A catalogue of viral diseases of insects, mites, and ticks. Fourth edition. *United States Department of Agriculture, Forest Service, General Technical Report PNW- 195.*

Martin S.J. 1994. Ontogeny of the mite *Varroa jacobsoni* Oud. in worker brood of the honeybee *Apis mellifera* L. under natural conditions. *Exp. Appl. Acarol.* 18: 87–100.

Martin S.J. 1998. A population model for the ectoparasitic mite *Varroa jacobsoni* in honey bee (*Apis mellifera*) colonies. *Ecological Modelling* 109: 267–281.

Martin S.J. 2001a. The role of *Varroa* and viral pathogens in the collapse of honeybee colonies: a modeling approach. *J. App. Ecol.* 38: 1082–1093.

Martin S.J. 2001b. *Varroa destructor* reproduction during the winter in *Apis mellifera* colonies in UK. *Exp. Appl. Acarol.* 25: 321–325.

Martin S.J., Holland K. and Murray M. 1997. Non-reproduction in the honeybee mite *Varroa jacobsoni. Exp. Appl. Acarol.* 21: 539–549.

Martin, E.C. 1970. Pesticides and honey bees. Mich. Coop. Ext. Serv. Bul. E-678, Farm Sci. Ser., 8 pp.

Martin, S. J. 2004. A simulation model of biological control of social wasps (Vespinae) using mermithid nematodes. *New Zealand Journal of Zoology*. 31: 241–248

Martin. E.C 1958. Some aspects of hygroscopic properties and fermentation of honey. *Bee WId.* 39(7): 165-178.

Matheson A. 2000.*Varroa* discovered in New Zealand. *Bee World* 81:43-44.

Matheson, A.G. 1980. Breakthrough in wax moth control in bee combs and comb honey. *New Zealand Beekeeper* 11: 15-18.

Matheson, A.G. 1986. Colony standards for Kiwifruit pollination. New Zealand Beekeeper 189:20-22.

Matheson, A.G. 1991. Managing honeybee pollination of Kiwifruit (*Actinidia deliciosa*) in New Zealand – a review. Acta Hort. 288: 213-219.

Matsumura, F. 1975. Toxicology of insecticides. Chapters 3, p. 47-103; 4, p. 105-163. Plenum Press, New York.

May, A.F.1961. Beekeeping in South Africa. Haum, Pretoria, S.A.

McGinley, R. J and Rozen, J. G., Jr., 1987. Nesting biology, immature stages and phylogenetic placement of the Palaearctic bee *Pararhophites* (Hymenoptera: Apoidea). *American Museum Novitates* 2903: 1-21.

Mcgregor, S. E Caster, A. B and Frost, M. H., Jr. 1947. Honeybee losses as related to crop dusting with arsenicals. *Ariz. Agr. Expt. Sta. Tech. Bul.* 114: 221-246.

Mcgregor, S. E., and Vorhies, C. T. 1947. beekeeping near cotton fields dusted with DDT. Ariz. Agr. Expt. Sta. Bul. 207, 19 pp.

McGregor, S.E. 1976. Insect pollination of cultivated crop plants. Agriculture Hand Book No. 496, Agriculture Research Service, United States. Department of Agriculture, Washington, 411p.

Mcindoo, N. E., and Demuth, G. S. 1926. effects on honeybees of spraying fruit trees with arsenicals. U.S. Dept. Agr. Dept. Bul. 1364, 32 pp.

Melathopoulos, A.P., Winston, M.L., Whittington R, Smith, T., Lindberg, C., Mukai, A and Moore, M. 2000 Comparative laboratory toxicity of neem pesticides to honey bees (Hymenoptera: Apidae), their mite parasites *Varroa jacobsoni* (Acari: Varroidae) and *Acarapis woodi* (Acari: Tarsonemidae), and brood pathogens *Paenibacillus larvae* and Ascophaera apis. *J Econ Entomol.* 93(2):199-209

Melathopoulos, A; Winston, M; Whittington, R; Higo, H; Le Doux, M. 2000. Field evaluation of neem and canola oil for the selective control of the honey bee mite parasites *Varroa jacobsoni* and *Acarapis woodi. Journal of Economic Entomology* 93(3): 559-567.

Michelsen, A and Andersen, B. B. 1989. Honeybees can be Recruited by a mechanical model of a dancing Bee. *Naturwissemcluiften,* 76: 277.280.

Michelsen, A. B. B., Andersen, J., Storm, W., Kirchner, H and Lindauer, M. 1992. How Honeybees Perceive Communication Dances, Studied by Means of a Mechanical Model. *Belum. Ecol. Sociobiol.* 30: 143-150.

Michelsen. A 1993. The Transfer of Information in the dance language of Miller, Julie Ann (April 23, 1983). "Do Bees Plan Ahead Intelligently? Science News 123 (17): 271.

Michener, C.D. 2000. *The Bees of the World.* The John Hopkins University Press, Baltimore and London. xiv + 1-913 pp.

Milani N. 1995. The resistance of *Varroa jacobsoni* Oud to pyrethroids: a laboratory assay. *Apidologie* 26, 415–429.

Milani N. 1999. The resistance of Varroa jacobsoni Oud. to acaricides, *Apidologie* 30, 229–234.

Milani, N. 1999. The resistance of *Varroa jacobsoni* Oud. to acaricides. *Apidologie* 30(2-3): 229-234.

Milani, N. and Barbattini, R. 1989. Treatment of varroatosis with Bayvarol strips (Flumethrin) in Northern Italy. *Apicoltura* 5: 173-192.

Milne, C. P. J. 1983. Honey Bee (Hymenotpera: Apidae) Hygienic Behavior and Resistance to Chalkbrood. *Annuals of the Entomological Society of America* 76: 384-387.

Milne, C. P. J. 1985. Estimates of the heritabilities of and genetic correlation between two components of honey bee (Hymenoptera: Apidae) hygienic behaviour: Uncapping and removing. *Annals of the Entomological Society of America* 78(6): 841-844.26

Milner, R. J and Staples, J. A. 1996. Biological control of termites: results and experiences within a CSIRO project in Australia. *Biocontrol Science and Technology 6:* 3-9.

Mishra, R. C., Dogra, G. S and Gupta, P. R. 1976. Some observations on insect pollinators of apple. *Indian Bee Journal* 38(1-4): 20-22.

Mishra, R.C and Rajesh Garg (Eds). 1997-98. Perspectives in Indian Apiculture, Agro Botanica, New Delhi. 412p.

Mishra, R.C. 1995. Honey bees and their management in India. ICAR, New Delhi. 168p.

Mishra, R.C. and Sharma, S.K. 1998. Technology for management of *Apis mellifera* in India, pp. 131-147, In Mishra, R.C. (Ed.) Perspectives in Indian Apiculture, Agro Botanica, Bikaner (India), 412 pp.

Mishra, R.C., Gupta, J.K and Kumar, J. 1985. Nectar sugar production in different cultivars of peach, *Prunus persica* L. *Indian Bee J.* 47 :37-38.

Mishra, R.C.; Gupta, J.K. and Kumar, J. 1987. Nectar secretion, amount and type of nectar sugars, insect foraging on *Woodfordia floribunda*. *Apidologie* 18(2):173-178.

Mishra, R.C Kumar, J and Gupta, J.K. 1989. A new approach to the control of predatory wasps (*Vespa* spp.) of the honeybee *Apis mellifera* L. *J. Apic. Res.*28:126-130.

Moeller, F.E. 1956. Behavior of nosema-infected bees affecting their position in the winter cluster. Journal of economic entomology 49:743-745.

Moeller, F.E. 1961. The relationship between colony populations and honey production as affected by honey bee stock lines. 20 p. U.S. Department of agriculture, production research report 55.

Moeller, F.E. 1976. Two-queen system of honey bee colony management. 11 p. U.S. Department of agriculture, production research report 161.

Moffett, J. 0., Stoner, A and A. L. Wardecker. 1977. The wardecker waterer. *American Bee Journal* 117(6): 364-365; 378.

Moffett, J. O., Morton, H. L and Macdonald, R. H. 1972. Toxicity of some herbicidal sprays to honey bees. *J. Econ. Ent.* 65: 32-36.

Mohammad, A. 1935. Pollination studies in toria (*Brassica napus* L. var. *dichotoma* Prain) and sarson (*Brassica campestris* L. var. *sarson* Prain). *Indian J. Agric. Sci.* 5: 125-154.

Moller, H., Plunkett, G. M., Tilley, J. A. V., Toft, R. J and Beggs, J. R. 1991. Establishment of the wasp parasitoid, Sphecophaga vesparum (Hymenoptera: Ichneumonidae), in New Zealand. *New Zealand Journal of Zoology* 18: 199-208.

Moller, H.; Wharton, D. A and Tyrrell, C. 1991. *Pheromermis* nematodes as biological control agents for the reduction of wasp abundance in New Zealand. *University of Otago Wildlife Management Report No. 21.*

Momot, J. P and Rothenbuhler, W. C. 1971 Behavior genetics of nest cleaning in honey bees VI.Interactions of age and genotype of bees, and nectar flow. *Journal of Apicultural Research* 10: 11-21.

Morandin, L.A and Winston, M.L. 2005. Wild bee abundance and seed production in conventional, organic, and genetically modified canola. Ecological Applications 15 (3): 871-881.

Moretto, G., Gonçalves, L.S and De Jong, D. 1993. Heritability of Africanized and European honey bee defensive behavior against the mite Varroa jacobsoni. Rev. Bras. Genet. 16: 71-77.

Moretto, G., Gonçalves, L.S. and De Jong, D. 1991. The effects of climate and bee race on Varroa jacobsoni Oud. infestations in Brazil. Apidologie 22: 197-203.

Moritz R.F.A., Kryger P., Koeniger G., Koeniger N., Estoup A., Tingek S. (1995) High degree of polyandry in *Apis dorsata* queens detected by DNA microsatellite variability. Behav. Ecol. Sociobiol. 37, 357–363.

Moritz, R. F. A. and Southwick, E. E. 1992. Bees as Superorganisms: an Evolutionary Reality. Berlin, Heidelberg, New York: Springer.

Moritz, R.F.A and Southwick, E.E. 1992. Bees as superorganisms: An evolutionary reality, Springer- Verlag, Berlin.

Moritz, R.F.A. 1988. A reevaluation of the two-locus model for hygienic behavior in honeybees (*Apis mellifera* L.). J. Hered 79: 257–262.

Morse R.A. and Flottum K.(eds). 1997. Honey Bee Pests, Predators and Diseases. 3rd edition The A. I. Root Company, Medina, OH

Morse R.A., 1970. Annotated bibliography on Apis dorsata. International Bee Research Association, Gerrards Cross, UK. 25 pp.

Morse R.A., Laigo N. 1969. Apis dorsata in the Philippines, *Monogr. Philippines Assoc. Entomol.,* 96 p.

Morse R.A.1978.Honey bee pests, predators and diseases. Cornell University Press

Morse, R.A and Gary, N.E. 1961. Insect invaders of the honeybee colony. *Bee World* 42:179-181.

Morse, R.A. 1955. Larval nematode recorded from honeybee *Apis mellifera* L. *Journal of Parasitology.* 41:553.

Mortan, S. J. 2001. The role of varroa and viral pathogens in the collapse of honeybee colonies- a modelling approach. *J. Appl. Ecol.* 38(5):1082-1093.

Morton, H.L., Moffett, J.O and Macdonald, R.H. 1972. Toxicity of herbicides to newly emerged bees. Environmental Ent. 1: 102-104.

Mothershead, K and Marquis, R. J. 2000. Fitness impacts of herbivory through indirect effects on plant-pollinator interactions in Oenothera macrocarpa. – Ecology 81: 30–40.

Mozes-Koch R., Slabezki Y., Efrat H., Kalev H., Kamer Y., Yakobson B.A. and Dag A. (2000) First detection in Israel of fluvalinate resistance in the varroa mite using bioassay and biochemical methods, Exp. Appl. Acarol. 24, 35–43.

Mugnai, L.; Bridge, P. D. and Evans, H. C. 1989: A chemotaxonomic evaluation of the genus Beauveria. Mycological Research 92: 199-209.

Muheto, R.N. and Wix, P. 1984. Remote sensing for natural resources evaluation: its value to apiculture. Proc. Third International Conference on Apiculture. In Tropical Climates.

Muller, H. 1883 "The Fertilization of Flowers", Macmillan, London, pp. 669.

Murthy V.A, Venkataramanan, R. 1985. Contribution to the biology of pseudoscorpions *Ellingsenius indicus* Chamberlain, *Indian Bee Journal* 47:34-35.

Murthy V.A, Venkataramanan, R. 1986. *Ellingsenius indicus* (Arachinda: Chelonethi) as a tool to the assessment of the seeding nature of honeybee Apis cerana indica colony as a new habitat Indian Bee Journal 48:54-55.

Mussen, E.C. 1979. Buckeye poisoning. – U.C. Apiaries: Univ. of California Cooperative Extension: 1–4.

Muthel, H. 1955. Ants can overwhelm apiary and destroy 26 colonies. Die Bienenz 8:202-203.

Mutinelli, F; Baggio, A; Capolongo, F; Piro, R; Prandin, L; Biasion, L (1997). A scientific note on oxalic acid by topical application for the control of varroosis. *Apidologie* 28(6): 461-462.

Muttoo, R. N. 1956. Facts about beekeeping in India. Bee World 37, 125—130.

Muttoo, R.N. 1953. Bee flora of Jeolikote (II). Indian Bee J. 15 : 49 - 54.

Muzzaffar, N. and Ahmad, R. 1986. Studies on hornets attacking honeybees in Pakistan. *Pakistan Journal Agricultural Research* 7(1):59-63.

Naim, M. and Phadke, K.G. 1976. Bee flora and seasonal activity of Apis cerana indica at Pusa (Bihar). Indian Bee J. 38(1/4):13-19.

Nair, K.S. 1980. Pbysio-chemical characteristics of rubber honey in India. Khadi Gramodyog. 26: 358-363.

Nair, K.S. 1981. Beekeepers of Muzaffarpur (Bihar). Indian Bee J. 43(2):47-49.

Nair, P.K.K. and Singh, K.N. 1974. A study of two honey plants, Antigonum leptopus Hook and Moringa pterigosperma Gaer. Indian J. Hort. 31:375-379.

Nanetti, A.; Stradi, G. 1997. Oxalsäure-Zuckerlösung zur Varroabekämpfung. Allgemeine Deutsche Imkerzeitung 31, 9–11.

Nanetti, A.1999. Oxalic acid for mite control - results and review. Coordination in Europe of research on integrated control of Varroa mites in honey bee colonies, Merelbeke, Belgium, Commission of the European Communities. pp. 6-11.

Nanetti, A; Stradi, G 1997. Varroosis: chemical treatment with oxalic acid in sugar syrup. Ape Nostra Amica 19(5): 6-14.

Nasr ME, Otis GW, and Scott-Dupree CD, 2001. Resistance to Acarapsis woodi by honey bees (Hymenoptera: Apidae): divergent selection and evaluation of selection progress. J Econ Entomol 94:332–338.

Nasr, M. & P. G. Kevan. 1999a. Diseases and parasites, eradication or management: what strategy for honey bee health? Bee World 80: 53-54

Nasr, M. and Kevan, P.G. 1999. Diseases and parasites, eradication or management: what strategy for honey bee health? Bee World 80: 53-54

Nasr, M. and Kevan, P.G. 1999. IMP for diseases and paraistes in honeybees. Pest Management News 10 (4): 3-12.

Nealis, V.G. 1986. Responses to host kairomones and foraging behavior of the insect parasite Cotesia rubecula (Hym.: Braconidae). Can. J. Zool. 64: 2393-2398.

Nehru, C.R.S.; Jayarasthnam, T.K. and Levi Josephy, P.M. 1984. Studies on seasonal bee forage in rubber plantation. Indian Bee J. 46:23-25.

Nelson, J. A. 1915. The embryology of the honey bee. 282P. Princeton University Press, Princeton.

Neumann P., Moritz R.F.A and Van Praagh J. 1999. Queen mating frequency in different types of honey bee mating apiaries. J. Apic. Res. 38, 11–18.

Neumann, P. Radloff, S. E. Pirk S. W.W., and Hepburn, R., 2003. The behaviour of drifted Cape honeybee workers (*Apis mellifera capensis*): predisposition for social parasitism? Apidologie 34, 585–590

Newton, D.C., Cantwell, G.C. and Bourquin, E.P. 1975. Removal of freeze killed brood as an index of nest cleaning behaviour on honeybee colonies (*Apis mellifera* L.). American Bee Journal 115: 388, 402, 406.

Nickle, W.R. 1991. Manual of Agricultural Nematology. Marcel Dekker, Inc., New York. 1035 pages.

Nielsen, R.A and Brister, C.D. 1979. Greater wax moth: behavior of larvae. Ann. Entomol. Soc. Am. 72: 811-815.

Nikiel, A.H.1972. Invading carpenter ants in beehive. Glean Bee Culture100:330

Nixon. M. 1982. Preliminary world maps of honeybee diseases and parasites. Bee World 63(1): 23-42.

Nixon. M. 1982. Introduction of non-native bees to new areas Bee World, 63: 50-53.

Nolan, W.J. 1932. Breeding the honeybee under controlled conditions.USDA Technical Bulletin No. 326.

Nordstrom, S. 2003. Distribution of deformed wing virus within honeybee (Apis mellifera) brood cells infested with the ectoparasitic mite Varroa destructor. Exp. Appl. Acarol., 29, 293-302

Nye, W.P. and Machensen, O. 1970. Selective breeding of honeybees for alfalfa pollen collection: with tests in high and low alfalfa pollen collection regions. J. api.Res.9:61-64.

O'Callaghan M, Glare TR, Burgess EPJ, and Malone, L.A. 2005. Effects of plants genetically modified for insect resistance on nontarget organisms. Annual Review of Entomology 50: 271-292.

Oertel, E. 1971. Honey prices since 1880. *American bee journal* 111(2):50-51.

Ohe,W. von der and Praagh, J.P. van. 1983. Versuche zur Duftlenkung in Obstanlagen. Nordwestdeustsche Imkerzeitung 35:100.

Oldroyd B.P., A. Smolenski, S. Lawler, A. Estoup and R. Crozier, 1995. Colony aggregations in Apis mellifera. Apidologie 26: 119–130.

Oldroyd B.P., A.J. Smolenski, J.M. Cornuet, S. Wongsiri, A. Estoup, T.E. Rinderer and R.H. Crozier, 1996. Levels of polyandry and intracolonial genetic relationships in Apis dorsata (Hymenoptera: Apidae). Anls. Ent. Soc. Am. 89: 276–283.

Oldroyd B.P., Rinderer T.E., Wongsiri S. 1992. Pollen resource partitioning by *Apis dorsata, Apis cerana, Apis andreniformis,* and *Apis florea* in Thailand. J. Apic. Res. 31: 3–7.

Oldroyd BP, Goodman, R.D, Hornitzky MAZ, Chandler D. 1989. The effect on American foulbrood of standard oxytetracycline treatments for the control of European foulbrood of honey bees (*Apis mellifera*). Aust J Agric Res.; 40:691- 697.

Oldroyd, B. P. 1999. Coevolution while you wait: Varroa jacobsoni, a new parasite of western honey bees. Trends Ecol Evol 14: 312-315.

Oldroyd, B. P., Ratnicks, F. L. W. and Wossler, T.C. 2002. Egg marking pheromones in honey bees Apis mellifera. Behav. Ecol. Sociobiol. 51: 590- 591

Oliveira, P.E, Gibbs, P.E and Barbosa, A.A 2004. "Moth pollination of woody species in the Cerrados of Central Brazil: a case of so much owed to so few?". Plant Systematics and Evolution 245 (1-2): 41-54.

Ono M., Igarashi T., Ohno E., Sasaki M. 1995. Unusual thermal defence by a honeybee against mass attack by hornets. Nature 377, 334–336.

Ono, M. Okada, I and Sasaki, M. 1987. Heat production by balling in the Japanese honeybee, *Apis cerana japonica* as a defensive behaviour against the hornet, *Vespa simillima xanthoptera* (Hymenoptera: Vespidae). Experientia 43:1031-1032.

Orosi-Pal, Z. 1975. Varroa in America. Mehezet 23: 123. (In Hungarian). Oudemans, A. C. (1904). Acarological notes XIII. Entomologische berichten uitgegeven door de Nederlandsche Entomologische Vereeniging 1: 169-174.

Orr, J.C., J.J. Obrycki, R.V. Flanders. 1992. Host acceptance behavior of Dinocampus coccinellae (Hym.: Braconidae). *Ann. Entomol. Soc. Am.* 85: 722-730.

Oudemans A.C. 1904. Note VIII. On a new genus and species of parasitic acari, Notes Leyden Mus. 24, 216–222.

Owens, C. D. 1959. Shade for bees. *Amer. Bee J.* 99: 481 - 482.

Owens, C. D. and Benson, C. E. 1962. Confining honey bee colonies wlth burlap. Amer. Bee J. 102: 260 - 262.

Pagden, H. T. 1934. Biological notes on some Malayan aculeate Hymenoptera III. J. Federated Malay States Mus. 17(3): 487-492.

Pagden, H. T. 1957. Notes on communal flight activities of stingless bees. Malayan Nat. J. 12: 15-19.

Pahdke, R.P. 1965. Nectar concentration in Thelepaepale ixiocephala Bremk. Indian Bee J. 27(2):73-76.

Palmer Jones, T. 1960. Effect on honey bees of some chemical weedkillers. N.Z. J. Agric. Res 3: 485-490.

Palmer Jones, T. and Forster, I. W. 1958. agricultural chemicals and the beekeeping industry. New Zeal. Jour. Agr. 97: 298 - 304.

Palmer K.A., Oldroyd B.P. 2000. Evolution of multiple mating in the genus Apis, Apidologie 31, 235–248.

Palmer K.A., Oldroyd B.P., Franck P. and Hadisoesilo S. (2001) Very high paternity frequency in Apis nigrocincta, Insectes Soc. 48, 327–332.

Palmer-Jones, T. and Line, L. J. S. 1962. Poisoning of honey bees by nectar from the karaka tree (Corynocarpus lae6igata J. R. et G. Forst). N.Z. J. Agric. Res. 5: 433–436.

Panda, P. and Padhi, J. 1995. Beekeeping in Orissa. Indian Bee J. 57(1):20-21.

Panda, P., Padhi, J., Nanda, U.K. and Mohandi, N.C. 1989. Record of V. jacobsoni, Euvarra Sinhai on Apis cerana indica F. at Bhuvneshwar, Orissa, India. Indian Bee J. 51:59.

Pandey, R.S. 1967. Varroa jacobsoni: a new mite infesting honeybee (*Apis indica*) colonies in India.Bee World 48:16-17.

Papadonpoulo, P.1965. Enemies of bees (1). Rhodesia Agriculture University 6:2-3.

Papp L. 1984f. Family Braulidae, p. 178-80. In: Soós Á, Papp L. eds., Catalogue of Palaearctic Diptera. Volume 10. Clusiidae--Chloropidae. Akadémiai Kiadó, Budapest. 402 p.

Parker, F. D., Batra, S. W. T. and Tepedino, V. J. 1987. New pollinators for our crops. Agricultural Zoology Reviews 2: 279-304.

Parrish, M. D.; Roberts, R. B. 1984. Toxicity of avermectin Bl to larval yellowjackets, Vespula maculifrons (Hymenoptera: Vespidae). Journal of Economic Entomology 77: 769-772.

Partap, U.1997. Bee flora of Hindukush Himalayas-Inventry and management. ICIMOD Nepal. 297p

Patetta, A; Manino, A (1988). Examination of the action on honey bees of chemicals used in the control of Varroa jacobsoni. Apicoltore Moderno 79(3): 109-114.

Patil, B.S.; Viraktamath, S.; Lingappa, S.; Giraddi, R.S.; Parameshwarappa, K. and Bhat, A.R.S. 2000. Effect of Bee-Q and Bee- here on pollination and yield of sesamum. Insect Environment 5(4): 151.

Paula, V. F., Barbosa, L. C. A., Demuner, A. J. et al. 1997. Entomotoxicity of the nectar from Ochroma lagopus Swartz (Bombacaceae). Ciencia e Cultura 49: 274–277.

Pellett, F. C. 1938. History of American beekeeping. 393 p. Collegiate Press, Ames, Iowa.

Pelton, M. R. 1982. Black bear (*Ursus americanus*). Pages 504–514 in J. A. Chapman and G. A. Feldhamer, editors. Wild mammals of North America. The John Hopkins University Press, Baltimore, Maryland.

Pelton, M. R., A. B. Coley, T. H. Eason, D. L. Martinez, J. A. Pederson, F. T. Van Manen, and K. M. Weaver. 1999. American black bear conservation action plan. Pages 144–156 in C. Servheen, S. Herrero, and B. Peyton, compilers. Bears. Status survey and conservation action plan. IUCN/SSC Bear and Polar Bear Specialist Groups. IUCN, Gland, Switzerland and Cambridge, United Kingdom. 309pp.

Peng C.Y.S., Fang Y.Z., Xu S.Y and Ge L.S. 1987. The resistance mechanism of the Asian honey bee *Apis cerana* Fabr. to an ectoparasitic mite, *Varroa jacobsoni* Oudemans, J. Invertebr. Pathol. 49, 54– 60.

Peng CY, Zhou X, Kaya HK. 2002. Virulence and site of infection of the fungus, Hirsutella thompsonii, to the honey bee ectoparasitic mite, Varroa destructor. 2: J Invertebr Pathol. 81(3):185-95

Percy, A.P, K.K. Kshirsagar and K. Chandran 1968. Further observations on phoretic mite *Neocypholaelaps indica* (Evans). Indian Bee J. 30, 47-48.

Perrott, D.C.F. 1975: Factors affecting the use of mirex-poisoned protein baits for control of European wasp (*Paravespula germanica*) in New Zealand. New Zealand Journal of Zoology 2: 491-508.

Petch, T. 1924. Studies in entomogenous fungi. Transactions of the British Mycological Societv 10: 28-80; 244-271.

Petch, T. 1932a: A list of entomogenous fungi of Great Britain. Transactions of the British Mycological Society 16: 170-178.

Petch, T. 1932b: Cordyceps on Hymenoptera. Transactions of the British Mycological Society 16: 219- 221.

Petch, T. 1934: Notes on entomogenous fungi. Transactions of the British Mycological Society 10: 161- 193

Pettis, J S; Shimanuki, H and Feldlaufer, M F 1998. An assay to detect fluvalinate resistance in Varroa mites. American Bee Journal 138(7): 538-541.

Pettis, J S; Shimanuki, H and Feldlaufer, M F 1998. Detecting fluvalinate-resistant Varroa mites. American Bee Journal 138(7): 535-537.

Pettis, Jeff S. 2004. A scientific note on Varroa destructor resistance to coumaphos in the United States Apidologie 35 : 91–92

Phadke, K. G., Bisht, D. S and Sinha, R. B. P. 1966. Occurrence of the mite *Varroa jacobsoni* Oudemans in the brood cells of the honey bee, *Apis indica F Indian J Entomol* 28:411-412.

Phadke, R.P. 1962. Physico-chemical compositions of major unifloral honeys from Mahableshwar (Western Ghats). Indian Bee J. 24(7/9):59-65.

Phadke, R.P. 1964. Nectar concentration in Carvia Callosa Bremk. Indian Bee J. 26(2): 22-25.

Phadke, R.P. and Phadke, R.S. 1975. Physio-chemical characteristics of waxes from Indian honeybees. II. Viscosity and refrective index. Indian Bee 1. 37 (1/4) : 15-18.

Pham-Delègue M.H, Jouanin L, Sandoz J.C 2002. Direct and indirect effects of genetically modified plants on the honey bee. Honey bees: estimating the environmental impact of chemicals p.312-326, 2002. Editors: Devillers, J.; Pham-Delègue, M. H. Publisher: Taylor & Francis, London, UK

Piccirillo GA, De Jong D. 2003 The influence of brood comb cell size on the reproductive behavior of the ectoparasitic mite Varroa destructor in Africanized honey bee colonies. Genet Mol Res. Mar 31;2(1):36-42.

Piccirillo, G. A. and Jong, D. D. 2004 Old honey bee brood combs are more infested by the mite Varroa destructor than are new brood combs Apidologie 35 : 359-364

Piel, O. 1930. Nidification de Megachile monticola Smith, chez Xylocopa rufipes Sm., pictifrons Sm. Notes d'Entomologie Chinoise, Musée Heude, fasc. 5: 108, 2 pls.

Poinar, G. O. Jr. 1981. Distribution of Pheromermis pachysoma (Mermithidae) determined by paratenic invertebrate hosts. Journal ofNematology 13: 421-424.

Poinar, G. O. Jr.; Lane, R. S.; Thomas, G. M. 1976: Biology and redescription of Pheromermis pachysoma (von Linstow) n. gen., n. comb. (Nematoda: Mermithidae), a parasite of yellowjackets (Hymenoptera: Vespidae). Nematologica 22: 360-370.

Poinar, G. O., Jr. and P. A and Van der Laan. 1972. Morphology and life history of Sphaerularia bombi. Nematologica 18:239-252.

Poinar, G.O. Jr and Ennik, F. 1972:.The use of Neoaplectana carpocapsae (Steinerne- matidae: Rhabditoidea) against adult yellowjackets (*Vespula* spp., Vespidae: Hymenoptera). Journal of Invertebrate Ecology 7P: 331-334.

Polhemus, M. S., J. L. Lush, and W. C. Rothenbuhler. 1950. Mating systems in honey bees. Journal of heredity 61(6): 151-155.

Popa, A. 1980. Agriculture in Lebanon. *Ameri Bee Journl* 120: 336-367.

Porrini C., 2002. Use and abuse of plant protection products in agro-ecosystems: biomonitoring with honey bees.- Proceedings "Agroecology: toward a new agriculture for Europe" (Celli G., Radeghieri P., Bazzocchi G., Eds). Bologna, Italy, December 14-15, 2001, 41-45

Prasad,R.; Chand, H. and Singh, R. 2000. Effect of honeybee, *A.mellifera* pollination on the yield and germination of onion, *Allium cepa* seeds. *Shashpa* 7(1): 53-55.

Proctor M, Yeo, P. and. Lack, A 1996. The natural history of pollination. Harper Collins, London.

Proctor, M. and Yeo, P. 1973. "The Pollination of Flowers", Collins, London, pp. 418.

Prosie, V.1959.The red wood ant: a dangerous pest of bees. *Apic Bucaresti* 32:89.

Pryce-Jones, J. 1942. Some problems associated with nectar, pollen and honey. – Proc. Linn. Soc. Lond. 155: 129–174.

Punjabi, A.A. and Saraf, S.K. 1969. Occurrence of *Varroa jacobsoni* Oudemans in Kashmir. *Indian Bee J.* 31:44.

Putatunda,B. N. and Abrol, D.P.2003. Mites associated with bees in Jammu and Kashmir, India. *Zoo's Print Journal*. 1 (2):1021-1024.

Qayyum, H. A. and Ahmad, N. 1967. Biology of *Apis dorsata. Pakistan J. Sci.* 19(3): 109-113.

Quinby, M. 1864. Mysteries of beekeeping explained. ed. 8, 393 p. C.M. Saxton, New York.

Rademacher, E; Polaczek, B and Schricker, B 1995. Control of varroosis using formic acid in applicator. (Part II). *Pszczelnicze Zeszyty Naukowe* 39(1): 133-142.

Radetzki, T. (2001) Oxalsäure Verdampfung im Feldversuch mit 1.509 Völkern.Rosenkranz, P. (1999). Honey bee (*Apis mellifera* L.) tolerance to *Varroa jacobsoni* Oud. in South America. *Apidologie* 30, 159–172.

Rahman, K. A. 1940. Insect pollinators of toria (*Brassica napus* Linn. var. *dichtoma* Prain), and sarson (*Brassica campestris* Linn. var. *sarson* Prain) at Lyallpur. *Indian Journal of Agricultural Sciences* 10: 422-447.

Rai, K. M. and Gupta, B.P. 1983. Role of honey bees and other insects as pollinators in apple and pear. *Indian Bee Journal* 45(2/3): 56-57.

Ramachandran, S. 1973. Beekeeping in South India Bull. Dept. Agric Madras No. 37:78 pages 2nd. Ed.

Ramani, S. and Bhumannavar, B.S. 1990. Floral biology and Melittophily of two species of coffee. Nat. Seminar on Poll., HAU, Hisar 1990. Abs. pp. 6.

Ramesh, B. 1980. A visit to the cultivated forest area of Peopalpada, U.P. *Indian Bee J.* 43(3):86.

Ramierz, W 1987. VII Brazil Congress: the latest on *Varroa* jacobsoni. *Newsletter for Beekeepers in Tropical and Substropical Countries* 10: 11.

Ramirez, W 1994. Conformation of the ambulacrum of *Varroa jacobsoni* and mite control with dusts. *American Bee Journal* 134(12): 835.

Ramirez-Romero R, Josette Chaufaux J, Pham-Delègue MH 2005. Effects of Cry1Ab protoxin, deltamethrin and imidacloprid on the foraging activity and the learning performances of the honeybee Apis mellifera, a comparative approach. Apidologie 36:601–611.

Ramsey, F.W.1946. Chasing ants from hives. *Glean Bee Cult.*74:546.

Rangarajan, A.V. 1964. Protataetia aurichalcea F. (Cetonidae), a new beetle feeding on pollen stores of *Apis indica* F. *Indian Bee Journal* 26:26.

Rangarajan, A.V. 1971. Occurrence of the mite *Varroa jacobsoni* Oudemans on the Indian honey bee in South India. *Sci. and Cult.* 37(6):292.

Rath W. 1999. Co-adaptation of *Apis cerana* Fabr., and *Varroa jacobsoni* Oud., *Apidologie* 30, 97– 110.

Raven, P. H. (eds),1972. Coevolution of plants and animals. Univ. of Texas Press, pp. 100–140.

Rawat, B.S. 1980. Anand Singh Mehta: an ideal commercial beekeeper. *Indian Bee J.* 42(3):89-90.

Reaumur, R.A.F. de. 1740. Memoirs Pour Servir a l'Histoire des Insectes, 5. Paris.

Reddy, M.C.M. and Gupta, J.K. 1987. Wild Cherry, *Prunus Puddum* Roxb., A honey plant. In: Social Forestry for Rural Development (Eds. P.K. Khosla and P.K. Kohli). I.S.T.S pp 207-214.

Reed, D. K., Hall, I. M., Rich, J. E. and Shaw, J. G. 1972. Birefringent crystal formation in citrus red mites associated with noninclusion virus disease. *J Invert Pathol* 32: 170-175.

Reed, D. K., Tashiro, H and Beavers, J. B. 1975. Determination of mode of transmission of the citrus red mite virus. *J Invert Pathol* 26: 239-246.

Rees, S., Coote, J., Stables, J., Goodson, S., Harris, S and Lee, M. G. 1996. Bicistronic vector for the creation of stable mammalian cell lines that predisposes all antibiotic-resistant cells to express recombinant protein. *BioTechniques* 20: 102-110.

Regal, P.J. 1977 "Ecology and evolution of flowering plant dominance", *Science,* 196, 622-629.

Rehm, S. M and Ritter, W. (1989). Sequence of the sexes in the offspring of *Varroa jacobsoni* and the resulting consequences for the calculation of the developmental period. *Apidologie* 20: 339-343.

Rehman, K.A. and Singh, S. 1940. Beekeeping in Punjab. Publ. Dept. Agric; Punjab: 23 pages.

Rehman, K.A. and Singh, S. 1941. Nectar and pollen plants of the Punjab. *Indian Bee J.* 4(3/4):32-35.

Remsen, J. V., Jr., J. S. Luther, and D. Roberson. 1976. A Ringed Woodpecker *Celeus torquatus* in Colombia. *Bull. Brit. Ornith. Club* 96: 40.

Renner, S. S. 1996. Effects of habitat fragmentation on plant pollinator interactions in the tropics. Pp. 339-361. *In*: D. M. Newbery, H. T. Prins & N.D. Brown (eds.), *Dynamics of Tropical Communities.* Blackwell Scientific, Cambridge, UK.

Reyes, S. G and Michener, C. D. 1990. Observations on a parasitic allodapine bee and its hosts in Java and Malaysia. *Tropical Zoology* 3: 139-149.

Reyes, S. G and Sakagami, S. F. 1990. A new socially parasitic *Braunsapis* from India with notes on the synonymy of *Braunsapis mixta* (Hymenoptera: Xylocopinae: Allodapini). *Journal of the Kansas Entomological Society* 63(3): 458-461.

Reznik, S.Y., D.G. Chernoguz, K.B. Zinovjeva. 1992. Host searching, oviposition preferences and optimal synchronization in *Alysia manducator* (Hym.: Braconidae) a parasitoid of the blowfly, *Calliphora vicina.* Oikos 65: 81-88.

Rhoads, S. N. 1903. The mammals of Pennsylvania and New Jersey. Privately published, Philadelphia, Pennsylvania. 266pp.

Ribeiro, J.C. 1994. Estudo do relacionamento entre *Galleria mellonella* (Lepidoptera: Pyralidae) e o parasitóide *Apanteles galleriae* (Hymenoptera: Braconidae). Dissertação de mestrado, UNESP, Rio Claro, SP, 52p.

Richards, K. W. 1993. Non-*Apis* bees as crop pollinators. *Revue Suisse de Zoologie* 100(4): 807-822.

Richez, P. and Le ContYe. (19 95). Efficacité thérapeutique d'une nouvelle formulation (Apivar) destinée au traitement de la varroase de l'abeille. *Abeilles et Fleurs,* 807: 12-1 3

Rinderer T.E., B.P. Oldroyd, C. Lekprayoon, S. Wongsiri, C. Boonthai and R. Thapa, 1994. Extended survival of the parasitic mite *Tropilaelaps clareae* on adult workers of *Apis mellifera* and *Apis dorsata*. *J. Apic. Res. 33*: 171–174.

Rinderer T.E., Oldroyd B.P., Wongsiri S., Sylvester H.A., de Guzman L.I., Potichot S., Sheppard W.S., Buchmann S.L. 1993. Time of drone flight in four honey bee species in south-eastern. Thailand. *J. Apic. Res.* 32 : 28–33.

Rinderer T.E.,Oldroyd B.P., Wongsiri S., Kuang B., Sylvester H.A., de Guzman L.I., Kuang H., Dong X., Zhai W. 1996. Comparative nest architecture of the dwarf honey bees. J. Apic Res. 35 :19–27.

Rinderer, T E; Guzman, L I d; Lancaster, V A; Delatte, G T; Stelzer, J A; de Guzman, L I (1999). Varroa in the mating yard: I. The effects of *Varroa jacobsoni* and Apistan R on drone honey bees. *American Bee Journal* 139(2): 134-139.

Rinderer, T. E. 1977. A new approach to honey bee breeding at the baton rouge usda laboratory. *American Bee Journal* 117(3):146-147.

Rinderer, T. E.; Stelzer, J. A.; Oldroyd, B. P.; Tingek, S. 1998. Levels of polyandry and intercolonial genetic relationships in *Apis koschevnikovi*". *Journal of Apicultural Research*, 37 (4): 281-288.

Rinderer, T.E and Beama, L.D, 1995. Genic control of honey bee dance language dialect. *Theor Appl Genet* 91:727–732.

Ritter W., Perschil F and Vogel R., 1992. Vergleich der Wirkung verschiedener Methoden zur Bekampfung von Wachsmotten. *Adiz* (1), 11 - 13 [Comparison of the effect of various methods for combatting wax moths.]

Ritter, W 1986. Varroatosis in the honey bee, *Apis mellifera*, and its control with Perizin [coumaphos]. *Veterinary Medical Review* 1: 3-16.33

Ritter, W and Rath, H 1988. Experiments with mite resistance to varroacidal substances in the laboratory. European Research on Varroatosis Control, Proc. Meet. EC Experts' Group, Bad Homburg, Germany, Balkema, Rotterdam. October 1986. pp. 157-160.

Ritter, W. 1981. Varroadisease of *honeybee Apis mellifera Bee world,* 62: 141-153.

Ritter, W. 1988. Medications registered in Western Europe for varroatosis control. *Apidologie,* 19: 113-116

Ritter, W; Ruttner, F 1980. New methods for the treatment of varroa disease. Formic acid - laboratory and field test. *Allegmeine Deutsche Imkerzeitung* 14(5): 151-153.

Robinson, F.A and Oertel, E 1950. Chlordane for control of argentine ants. *American bee journal* 90:406-407.

Robinson, W. S., Nowogrodski, R and Morse, R.A. 1989a. The value of honeybees as pollinators of US crops Part I. *American Bee Journal* 129: 411-423 & 477-487.

Robinson, W. S., Nowogrodski, R and Morse, R. A. 1989b. Pollination parameters. *Glean. Bee Cult.* 117: 148-152.

Roig-Alsina, A and Michener, C. D. 1993. Studies of the phylogeny and classification of long-tongued bees. *University of Kansas Science Bulletin* No. 55 124-162.

Romel, K.E., Scott-Dupree, C.D and Caster, M.H. 1992. Qualitative and quantitative analysis of volatiles and pheromone gland extracts collected from *Galleria mellonella* (L.) (Lepidoptera: Pyralidae). *J.Chem. Ecol.* 18: 1255-1258.

Rose E. A. F. Harris,R. J and Glare, T. R 1999. Possible pathogens of social wasps (hymenoptera: vespidae) and their potential as biological control agents. *New Zealand Journal of Zoology.26:* 179-190

Rosenkranz, P and Engels, W. 1985. Konsequente Dronenbrutentnahme, eine wirksame biotechnische Massnahme zur Minderung von Varroa-Schäden an Bienenvölkern. *Allgemeine Deutsche Imkerzeitung* 21, 265–271.

Rosenkranz, P. 1986. Factors affecting *Varroa* reproduction in colonies of *Apis mellifera*: a comparison of European and Africanized honey bees. *Varroa Workshop,* Feldafing, Germany, p. 16.

Rothenbuhler, W. C. 1964a Behavior genetics of nest cleaning in honey bees I. Responses of four inbred lines to disease-killed brood. *Animal Behaviour* 12: 578-583.

Rothenbuhler, W. C. 1964b. Behavior genetics of nest cleaning in honey bees IV. Response of F1 and backcross generations to disease-killed brood. *American Zoologist* 4: 111-112.

Rothenbuhler, W. C., J M. Kulincevic, and W. E. Kerr. 1968. Bee genetics. *Annual Review of Genetics* 2:413-438.

Rothenbuhler, W.C. 1964. Behavior genetics of nest cleaning in honey bees IV. Response of F1 and backcross generations to diasease-killed brood. *Am Zool* 4:111-123.

Roubik D.W., S.F. Sakagami and I. Kudo, 1985. A note on distribution and nesting of the Himalayan honey bee *Apis laboriosa* Smith (Hymenoptera: Apidae). *J. Kans. Ent. Soc. 58*: 746–749.

Roubik, D. W. 2001. Ups and downs in pollinator populations: when is there a decline? *Conservation Ecology* 5(1): 2-5.

Roubik, D. W., Sakagami, S. F. and Kudo, I. 1985. A note on distribution and nesting of the Himalayan honeybee *Apis laboriosa* Smith. *Journal Kansas Entomological Society* 58(4): 746-749.

Roubik, D.W. 1995. Pollination of cultivated plants in the tropics. FAO agricultural services bulletin No. 118., FAO, Rome. 196p.

Roy, D.B, Bohan, D.A, Haughton, A.J, Hill, M.O, Osborne, J.L, Clark, S.J, N, Rothery P, Scott, R.J, Brooks, D.R, Champion, G.T, Hawes, C, Heard, M.S and Firbank LG. 2003. Invertebrates and vegetation of field margins adjacent to crops subject to contrasting herbicide regimes in the Farm Scale Evaluations of genetically modified herbicide-tolerant crops. Philosophical Transactions of the Royal Society of London Series B-Biological Sciences 358 (1439): 1879-1898.

Ruijter, A d; Eijnde, J v d (1991). Field experiment to determine the efficacy of Bayvarol strips on Varroa mites in bee colonies and to determine the effect of the medicament on colony development in the months following application. *Veterinary Medical Review* 61: 30-35.

Ruijter, A. de and Eijnde J van den 1988. Field experimento determine the efficacy of Apistan on varroa mites in bee colonies and the effect on spring development of *treated colonies. Proceedingo f €C Experts'group* Meeting, Udine, Italy, 28-30 Nov. 1988, pp. 331-337.

Runner. F., and Maul. V. 1983. Experimental analysis of the reproductive interspecies isolation of *Apis mellifera* L. and *Apis cerana*. Fabr. *Apidologie* 14, 305-327.

Ruttner F., 1988. *Biogeography and Taxonomy of Honeybees.* Springer- Verlag, Berlin. 284 pp.

Ruttner F., Tassencourt L and Louveaux J. 1978. Biometrical-statistical analysis of the geographic variability of *Apis mellifera* L., Apidologie 9, 363–381. Spivak M., Gilliam M. (1998a) Hygienic behaviour of honey bees and its application for control of brood diseases and varroa: Part I, *Bee World* 79, 124–134.

Ruttner, F and Hesse. B. 1981. Rassenspezifische Unterschiede in Ovarentwicklung and Eiablage von weiseilosen Arbeiterinnen der Honigbiene *Apis mellifera* L. *Apidologie* 12,159- 183.

Ruttner, F. 1977. Interim report on the cause of *Varroa* infection. *Die Bienen* 13: 353-354. (In German). Samsinak & Haragsim, 1972

Ruttner, F., Ritter, W. and Götz, W. 1980. Chemotherapie der varroatose über die *hämolymph der biene. Allgemeine Deutsche Imkerzeitungt,* 1.4 (5): 160-16.

Ruttner, P and Kauhausen. D. 1985. Honeybees of tropical Africa: biological diversi-fication and isolation. *Proc. Int. Conf. Apic. Trop. Climates, 3rd Nairobi, Kenya* 45-51.

Ruttner. F. (ed.) 1983. Queen rearing. Apimondia Publishing House. Bucharest Romania.

Ruttner. F. 1977. The problem of the Cape bee (*A. m. capensis* Escholtz): parthenogenesis size of population evolution. *Apidologie* 8, 281—294.

Ruttner. F. 1980. *Apis mellifera adami* (n.ssp.), die Kretische Biene.*Apidologie* 11, 385—400.

Ruzicka, A. 1991. Secondary infections with *Varroa* infestation of honeybees, especially viruses. *Benenvater.* 112:8-9.

Sabins, F.F. Remote sensing: principles and Interpretation W.H. Freeman and Compnay, New York, pp 441.

Sagara, N and Kobayashi, T. 1979. *Hebeloma spoliatum* appeared from abandoned nest-chambers of *Vespula lewisii,* a ground wasp. *Transactions of the Mycological Society of Japan 20:* 266-267.

Sagara, N.; Kitamoto, Y.; Nishio, R and Yoshimi, S. 1985. Association of two *Hebeloma* species with decomposed nests of vespine wasps. *Transactions of the British Mycological Society 84:* 349-352.

Sakagami S. F., Matsumura, T. and Ito, K. 1980. *Apis laboriosa* in Himalaya, the little known world largest honeybee (Hym., Apidae). *Insecta Matsumurana,* New Series 19, 47-77.

Sakagami S.F., Matsumura, T. and Ito, K. (1980) *Apis laboriosa's* largest honey bee (Hymenoptera, Apidae), *Insecta Matsumurana* 19, 47–77.

Sakagami, S. F and Ebmer, P. A. W. 1987. Taxonomic notes on oriental halictine bees of the genus *Halictus* (subgenus *Seladonia*). *Linzer Biologische Beiträge* 19: 301-357.

Sakagami, S. F and Michener, C. D. 1963. Nest architecture in halictine bees. *Symposia Genetica et Biologica Italica* 12: 266-274.

Sakagami, S. F and Wain, F. L. 1966. *Halictus latisignatus* Cameron: A polymorphic Indian halictine bee with caste differentiation. *Journal of the Bombay Natural History Society* 63: 57-73.

Sakagami, S. F and Yoshikawa, K. 1961. Bees of Xylocopinae and Apinae collected by the Osaka City University Expedition to Southeast Asia 1957-58, with some biological notes. *Nature and Life in Southeast Asia* 1: 409-444.

Sakagami, S. F. 1960. Ethobiological peculiarities of the primitive social bees, *Allodape* Lepeltier and allied genera. *Insectes Sociaux* 7: 231-249.

Sakagami, S. F. 1966. Comparative ethology of Apidae. *Japanese Society of Systematic Zoology,* circular no. 35: 1-6 [In Japanese].

Sakagami, S. F. and Michener, C. D. 1962. *The nest architecture of the sweatbees (Halictinae): A comparative study of Behavior*. University of Kansas Press, Lawrence, viii + 135 pages.

Sakagami, S. F., Ebmer, P. A. W and Tadauchi, O. 1996. The halictine bees of Sri Lanka and the vicinity, III. *Sudila* (Hymenoptera, Halictidae) Part I. *Esakia* 36: 143-189.

Sale, P.R.1983. Kiwifruit culture. Wellington, New Zealand: Government Printing Office.

Samsinak, K and Haragsim, O. 1972. The mite *Varroa jacobsoni* imported into Europe. *Vcelarstvi* 25: 268-269.

Samson, R. A. 1974: *Paecilomyces* and some allied Hyphomycetes. *Studies in Mycology 6:* 1-117.

Samson, R. A.; Evans, H. C and Latge, J. P. 1988: Atlas of entomopathogenic fungi. Berlin, Springer-Verlag. *Science: Biomedical Technology, Microbiology, Parasitology and Infectious Diseases 12:* 264.

Sandhu, A. S., Sidhu, B. S and Mehdiratta, P. D. 1976. Note on the use of *Pithitis smara-gdula* (Fab.) as a pollinator for the breeding of lucerne. *Indian Journal of Agricultural Science* 46(7): 347-349.

Santas, L 1989. Different tests with Folbex-VA against the Varroa disease during winter in Greece. *Proceedings of the XXXIst International Congress of Apiculture*: 312-316.

Santas, L. A. 1979. Problems of honey bee colonies in Greece. *Apiacta* 14: 127-313.

Santas, L.A., and D. M. Lazarakis, 1984. Using drone brood in the control of the varroa disease of bees in Greece. *Entomologia Hellenica* 2: 63-68.

Sanvido O, Stark M, Romeis J and Bigler F 2006. Ecological impacts of genetically modified crops. Experiences from ten years of experimental field research and commercial cultivation. ART-Schriftenreihe Nr. 01, Agroscope Reckenholz-Tänikon Research Station ART, Switzerland.

Saraf, S.K. 1973. Honey flow in Kashmir Valley. *Indian Bee J.* 35 (1/4):32-35.

Saraf, S.K.1972.Bee flora of Kashmir. *Indian Bee Journal* 34:1-10.

Sasaki M., Ono M and Yoshida T. 1995. Some biological aspects of the north-adapted eastern honeybee, *Apis cerana japonica*, in: Evan P.G. (Ed.), The Asiatic hive bee: Apiculture, biology, and role in sustainable development in tropical and subtropical Asia, Enviroquest, Ltd., Cambridge, Ontario, Canada, pp. 59–78.

Sasaki, M. 1985. Utilization of pollen loads for artificial pollination of fruit trees. The 30th Internat. Api. Cong. Apimondia, Nagoya, Japan, 1985, 390-393.

Satyanarayana, I. 1975. Sugar concentration in nectar from *Syzgium Cumini* Skeels. *Indian Bee J.* 37(1/4): 21-24.

Schenk, E. 1938. O apicultor brasileiro. Guia completo de Apicultura no Brasil, 7th ed., 320p.

Schmidt-Bailey J., S. Fuchs and R. Buchler. 1996. Effectiveness of drone brood trapping combs in broodless honeybee colonies. *Apidologie* 27: 294.

Schneider S.S., Stamps J.A. and Gary N.E. 1986. The vibration dance of the honey bee. I. Communication regulating foraging on two time scales, *Anim. Behav.* 34, 377–385.

Schuster, H 1997. Comparison of different methods for varroa treatment with formic acid. *Imkerfreund* 52(7): 4-12.

Schwarz, H.F. 1939. The Indo-Malayan species of *Trigona*. *Bulletin of the American Museum of Natural History* 76: 83-141.

Seal, D.W.A. 1957. Chalkbrood disease of bees. *Newzealand Journal of Agriculture*. 95:562.

Seeley T.D. (1985) Honeybee ecology: A study of adaptation in social life, Princeton University Press, Princeton, N.J, USA.

Seeley T.D. 1992. The tremble dance of the honey bee: message and meanings, Behav. Ecol. Sociobiol. 31, 375–383.

Seeley T.D., Nowicke J.W., Meselson M., Guillemin J and Akranatakul P. 1985. Yellow rain, *Sci. Am.* 253, 128–137.

Seeley T.D., Seeley R.H., Akratanakul P. 1982. Colony defense strategies of the honeybees in Thailand, Ecol. Monogr. 52, 43–63.

Seeley, T.D. 1995. "The Wisdom of the Hive: The Social Physiology of Honey Bee Colonies." Cambridge, MA: Harvard University Press

Seltner, B. 1950. Larvae of Meloe as bee pests. *Der Imkerfreund* 5:56-58.

Shabanov, M 1984. *Varroa jacobsoni* in honeybee colony as a carrier of micro organisms. *Acta Microbiologica Bulgarica*. 15,78-81.

Shabanov, M. (1984). Role of *Varroa jacobsoni* Oud. in the bee family as a carrier of micro-organisms. *Acta Microbiol Bulgarica* 15: 78–82.

Shah F.A and Shah, T.A. 1976. An unknown adult honeybee disease. *British Bee J.*108:38

Shah F.A and Shah, T.A. 1979. More on *Apis* irrideescent virus. *Indian Bee J.*41(3&4): 24-25

Shah, F.A and Shah, T.A 1991. *Vespa velutina* a serious pest of honeybees in Kashmir. *Bee World* 72 (4):161-164.

Shah, F.A. 1972. False acacia-a promising bee plant of Kashmir. *Indian Bee J.* 34(1/2):34-35.

Shah, T.A. 2000. Hygenic behaviour in relation to thai sac brood virus disease in *Apis cerana*. *Indian Bee Journal*. 62(1-2): 35-40.

Sharma P.L. 1958. Sugar concentration of nectar of some Punjab Honey plants. *Indian Bee J.* 20(7).: 85-91.

Sharma, D and Abrol, D.P 2005.Studies on contact toxicity of some insecticides to honeybees *Apis mellifera* L. and *Apis cerana* F. *J. Asian Pacific Entomol.* 8(1):113-115.

Sharma, D and Abrol, D.P.2004. Studies on abundance and diversity of different pollinators visiting *Brassica campestris* var. *toria* and their impact on seed production. *Indian bee journal 66(3-4):* 109-113.

Sharma, H. K and Gupta, J. K. 1993. Diversity and density of bee flora of Solan region of Himachal Pradesh (India). *Indian Bee Journal* 55(1/2): 9-20.

Sharma, H.K and Gupta, J.K. 2001. Impact of bee pollination on sustainable production in apple orchards. *Pest Management and Economic Zoology* 9(1): 45-48.

Sharma, H.K., Gupta, J.K. and Rana, B.S. 2000. Diurnal activity of *Apis cerana* F. and *A. mellifera* L. on different flora during spring and honey flow period. *Pest Management and Economic Zoology* 8(2): 151-154.

Sharma, O. P., Raj, D. and Garg, R. 1986. Toxicity of nectar of tea (*Camellia thea*) to honeybees. *J. Apicult. Res.* 25: 106–108.

Sharma, O.P.Thakur, A.K and Garg, R. 1986. Control of wasps attacking bee colonies. *Indian Bee Journal*.47:27-29.

Sharma, P.L. 1958. Studies on seasonal activities of *Apis indica* F. at Lyallpur. *Indian Bee J.* 19(3/4):20-23.

Sharma, Vinita and Thakur, M.L. 1999. Morphometric characterization of giant honey bee *Apis dorsata* Fab. (Hymenoptera: Apidae) from Doon Valley. *Annals of Forestry* 7(1): 125-135.

Shaw, F.R. 1941. Bee poisoning: a review of the more important literature. *J. Econ. Ent.* 34: 16 - 21.

Shaw, K.E., Davidson, G., Clark, S.J., Ball, B.V., Pell, J.K., Chandler, D and Sunderland, K.D. 2002. Laboratory bioassays to assess the pathogenicity of mitosporic fungi to *Varroa destructor* (Acari: Mesostigmata), an ectoparasitic mite of the honey bee, *Apis mellifera*. *Biol Control* 24: 266-276.

Shimanuki H. 1981. Controlling the greater wax moth. USDA publication

Shimanuki H., Calderone, N.W. and D. Knox. 1994. Parasitic mite syndrome: the symptoms. *Amer Bee J.* 134: 827-828.

Shoemaker, H.W. and French, J.C. 1921. The black bear of Pennsylvania. Times Tribune Company, Altoona, Pennsylvania. 92pp.

Short, L. 1969. An apparently melanistic Hairy Woodpecker from New Mexico. Bird Banding 40: 145-146.

Short, L.L., and J.F.M. Horne. 2002. Family Indicatoridae (Honeyguides), pp. 274-295 *in* del Hoyo, J., Elliott, A., and Sargatal, J., eds., Handbook of the Birds of the World. Vol. 7. Lynx Edicions, Barcelona, Spain.

Shuel, R.W. 1952. Some factors affecting nectar secretion in red clover. *Pl. Physiol.* 27:95-110.

Sihag, R.C. 1988. Incidence of *Varroa, Euvarroa* and *Tropilaelaps* mites in the colonies of honeybee, *Apis melifera* L. in Haryana (India), *Amer. Bee J.* 128:212-213.

Sihag, R.C. 1990a. Seasonal management of honey bee (*Apis mellifera* L.) colonies in Haryana (India). *Indian bee J.* 52: 51-56.

Sihag, R.C. 1990b. Ecology of European Honeybee, *Apis mellifera* L. in semiarid sub-tropical climates. I. Melliferous flora and over seasoning of the colonies *Korean J. Apic.* 5(1): 31-43.

Sihag, R.C. 1995a (Ed.) Pollination Biology. Pollination, Plant Reproduction and crop seed Prdouction. Rajendra Scientific Publishers, Hisar 210p.

Sihag, R.C. and Abrol, D.P.1986.Correlation and path coefficient analysis of environ-mental factors influencing flight activity of *Apis florea* F. *J. Apic. Res.* 25: 202-208

Sihag, R.C. and Mishra, R.C. 1995. Crop pollination and Apis cerena. In: The Asiatic hive bee (Ed. P.G. Kevan). Enviroquest, Ltd., Canada. pp 135-141.

Simms, E. L. and Bucher, M. A. 1996. Pleiotropic effect of flower color intensity on resistance to herbivory in *Ipomoea purpurea*. *Evolution* 50: 957–963.

Simpson, J. (1961). Nest climate regulation in honey bee colonies. *Science* 133, 1327-1332.

Singh G and Singh, G 1971. *Plectranthus rugosus* Wall., the major honey plant in Kashmir valley. *Indian Bee Journal* 33:58-59.

Singh J.N and Venkataraman, T.V. 1948. Pseudoscorpions in the hives in India. *Indian Bee Journal* 10:6.

Singh, B and Gatoria, G.S. 2003. Maintaining queen bee reservoirs/banks.*Proc. interna-tional workshop on conservation and management and honey festival (Apiexpo)* 2003 (October 13-18,2003):38.

Singh, B., Gatoria, G.S and Singh, L. 2000. Effect of priming the queen cell cups on mass queen rearing in *Apis mellifera* Linn. Colonies. *Indian Bee J.* 63(3):68-71.

Singh, G. and Singh, G. 1971. *Plectranthus rugorus* Wall., the major honey plant of Kashmir valley. *Indian Bee J.* 33(3/4): 58-59.

Singh, S. 1962. Beekeeping in India. Indian Council of Agricultural Research, New Delhi, India, 214 pages.

Singh, S.; Kumar, J and Mishra, R.C. 1992. Nectar Sugar production and insect foraging in different cultivars of almond (*Prunus amygdalus* Batsch.) *Haryana J. Hort. Sci.* 21(3-4):135-140.

Singh. Y. 1979. Pollination activity on strawberry at Jeolikote (Distt. Nainital, India). *Indian Bee J.* 41:17-19.

Siracusa, L.D, Russell, L.B, Eicher, E.M, Corrow, D.J, Copeland, N.G and Jenkins, N.A, 1987. Genetic organisation of the agouti region of the mouse. Genetics 117:93–100.

Sistrank, T.W. 1956. Bee defence against ants.*Glean Bee Cult.* 84:681.

Skaife, S.H. 1921. The life-history of the bee louse. *South African Bee. Journal* 1: 38-39.

Skillman, G. J. 1979. Queen bee imports-a trial run. *Anim. Quarant.* 6(2): 12-16

Skinner, A. J., M. E. Nasr and P. G. Kevan. 1998. Impacts of tracheal mites on the thermoregulation and overwintering of honey bees. Presented at Entomological Society of America Annual Meeting Las Vegas, Nevada. (8 - 12 November, 1998) and Presented at Entomological Society of Ontario Annual General Meeting, Sudbury, Ontario (October 1998).

Smit, I.B Jr, Caron, D.M. 1985. Distribution of the beelouse, *Braula coeca*, in Maryland and worldwide. *American Bee Journal* 125: 294-96.

Smith D.R., Hagen R.H.1996. The biogeography of *Apis cerana* as revealed by mitochondrial DNA sequence data. *Journal of the Kansas Entomology Society* 69: 294–310.

Smith, D.R., Villafuerte L., Otis G and Palmer M.R.2000. Biogeography of *Apis cerana* F. and *A. nigrocincta* Smith: insights from mtDNA studies", *Apidologie* 31: 265–280.

Smith, F.G. 1958. Beekeeping observations in Tanganyika 1949-1957. *Bee World* 39, 29-36

Smith, I.B Jr. 1978. The bee louse, *Braula coeca* Nitzsch, its distribution and biology on honey bees. *Unpub. MS thesis, University of Maryland.* 111 p.

Smith, K.E, Wall, R, French, N.P. 2000 The use of entomopathogenic fungi for the control of parasitic mites, Psoroptes spp. *Vet Parasitol.* 92(2): 97-105.

Smith, M.V.1962. A portable incubator for transporting honeybee brood. *J. apic. Res.* 1 : 33-34

Smithers, C. N and Holloway, G. A. 1978. Establishment of *Vespula germanica* (Fabricius) (Hymenoptera: Vespidae) in New South Wales. *Australian Entomological Magazine 5*:55-59.

Snodgrass E. 1975. The anatomy of the honey bee. In the hive and the honey bee, p. 75-124. Dadant & Sons, Hamilton, Ill.

Soderstrom, T.M. and Calderon, C.E. 1971. "Insect pollination in Tropical rain forest grasses", *Biotropica,* 3, 1-16.

Southwick, E. E and Southwick, Jr., L. 1992. Estimating the economic value of honeybees (Hymenoptera: Apidae) as agricultural pollinators in the United States. *Journal of Economic Entomology* 85(3): 621-633.

Southwick, E. E. 1985. Allometric relations, metabolism and heat conductance in clusters of honey bees at cool temperatures. *J. Comp. Physiol. B* 156, 143-149.

Spangler, H.G and Taber, S.III. 1970. Defensive behaviour of honeybees towards ants. *Psyche*77:184-189.

Spivak, M and Downey, D.L 1998. Field assays for hygienic behaviour in honey bees (Hymenoptera: Apidae). *J Econ Entomol* 91: 64-70.

Spivak, M and Gilliam, M. 1998a Hygienic behaviour of honey bees and its application for control of brood diseases and varroa - Part II. Studies on hygienic behaviour since the Rothenbuhler era. *Bee World*, 79(4) : 169-186.

Spivak, M and Gilliam, M. 1998b Hygienic behaviour of honey bees and its application for control of brood diseases and varroa. Part I. Hygienic behaviour and resistance to American foulbrood." *Bee World* 79(3): 124-134.

Spivak, M and Reuter, G. S. 1998. Honey bee hygienic behavior. *American Bee Journal* 138(4): 283-286.

Spivak, M. and Gilliam, M. 1993 Facultative expression of hygienic behaviour of honey bees in relation to disease resistance. *Journal of Apicultural Research* 32(3/4): 147-157.

Spradbery, J. P. and Huppatz, R. J. 1993: Use of *Metarhizium* for wasp control. Report of Research 1991-1993. Australia, CSIRO Division of Entomology.

Spradbery, J.P. 1973. Wasps: an account of the biology and natural history of social and solitary wasps. Univ. Wash. Press, Seattle. 408 p. Dedicated to Roger Akre, who first showed me yellow jacket.

Spradbury, J.P. 1973 "Wasps. An account of the biology and natural history of social and solitary wasps", Univ. Wash. Press, Seattle, pp. 408.

Spreafico M., Eordegh F.R., Bernardinelli I., Colombo M. 2001. First detection of strains of *Varroa destructor* resistant to coumaphos. Results of laboratory tests and field trials, *Apidologie* 32, 49–55.

Sprengel, C.K. 1793. "The Secret of Nature in the Form and Fertilization of Flowers Discovered", 4 Sets (in German).

Spurr, E. B. 1991. Reduction of wasp (Hymenoptera:Vespidae) populations by poison-baiting; experimental use of sodium monofluroacetate (1080) in canned sardine. *New Zealand Journal of Zoology* 18: 215-222.

Spurr, E. B. 1993. The effectiveness of sulfluramid in sardine bait for control of wasps (Hymenoptera: Vespidae). *Proceedings of the New Zealand Plant Protection Conference 46:* 307-312.

Spurr, E. B. 1995. Protein bait preferences of wasps (*Vespula vulgaris* and *V. germanica*) at Mt Thomas, Canterbury, New Zealand. *New Zealand Journal of Zoology* 22: 281-289.

Spurr, E.B. 1996. Carbohydrate bait preferences of wasps (Vespula vulgaris and *V. germanica*) (Hymenoptera:Vespidae) in New Zealand. *New Zealand Journal of Zoology* 23: 315-324.

Srawan, B.S. and Sohi, B.S, 1985. Phyto sociological studies on *Apis mellifera* L. and *Apis cerana* F. in Punjab, India. *Indian Bee. J.*47 :15-18.

Stacey, P. B and W. D. Koenig. 1984. Cooperative breeding in the Acorn Woodpecker. *Scientific America* 251: 114-121.

Standifer, L N and J. P. Mills. 1977. The effects of worker honey bee diet and age on the vitamin content of larval food. *Annals of Entomological Society of America* 70(5) :691-694.

Stanghellini M.S., J.T. Ambrose and J.R. Schultheis. 1997. The effects of honey bee and bumble bee pollination on fruit set and abortion of cucumber and watermelon. *American Bee Journal* 137(5): 386-391.

Starks, P.T and Gilley, D.C. 1999. Heat shielding: a novel method of colonial thermoregulation in honey bees. *Naturwissenschaften* 86:438–440.

Starks, P.T, Blackie C.A and Seeley, T.D. 2000. Fever in honeybee colonies; *Naturwissenschaften* 87 229–231

Stein, K. J.; Wrensch, D. L. 1988. The pest status of yellowjackets in Ohio (Hymenoptera: Vespidae). *Great Lakes Entomologist 21:* 83-89.

Steinhaus, E. A and Marsh, G.A. 1962. Report on diagnoses of disease in insects 1951—1961. *Hilgardia 33:* 349-390.

Stephen, W. A. 1968. A beekeeping problem in Vietnam and India. *Bee World* 49: 119-120.

Stephen, W. P and Every, R. W. 1970. Nesting media for the propagation of leaf cutter bee. Oregon State Univ., Corvallis [*Coop. Extension with U.S. Dept. Agric.*] *Fact Sheets No.* 175: 1-2.

Stokes, D.W.1983. A Guide to Observing Insect Lives. Little, Brown and Co., Boston.

Strickler, K and Freitas, S. 1999. Interactions between floral resources and bees in commercial alfalfa seed fields. *Environmental Entomology* 28: 178-187.

Strickler, K. 1997. Seed and bee yields as a function of forager populations: Alfalfa pollination as a model system *Journal of the Kansas Entomological Society* 69(4): suppl., (1996): 201-215.

Strickler, K. 1999. Impact of flower standing crop and pollinator movement on alfalfa seed yield. *Environmental Entomology* 28(6): 1067-1076.

Stroo, A. 2000. "Pollen morphological evolution in bat pollinated plants". Plant Systematics and Evolution 222 (1-4): 225-242.

Sturtevant, A.P. 1932. Relation of commercial honey to the spread of American foulbrood. *J Agricb Res;* 45:257-285.

Subhapradha, C.K 1957. A Hive stand of proved utility. *Indian Bee Journal.* 19:150-151.

Subhapradha, C.K 1961. Enemies and pests of bees encountered in this area. *Indian Bee Journal* 23:16-19

Subiah, S and Mahadevan,V. 1957. The life history asnd buiology of the reduviid *Acanthaspis siva* Dist., a predator of Indian honeybee *Apis indica. Indian journal of veterinary science.*27:117-122.

Subramaniam, K. 1979. The role of Forest Department in developing beekeeping in Western Ghat areas of Maharashtra. *Indian Bee J.* 41(3/4):91-93.

Sumpter, D.J.T and Martin S.J. 2004. The dynamics of virus epidemics in *Varroa* infested honey bee colonies. *J Anim Ecol* 73: 52-63.

Suryanarayana, M.C. 1966. Studies on flora of coorg with emphasis on bee forage plants. I. General ecological and botanical features. *Indian Bee J.* 28(2):59-75.

Suryanarayana, M.C. and Rao, K.S. 1998. *Apis cerana* F. for Indian apiculture and its management technology, pp. 66-108. In Mishra, R.C. (Ed.), Perspectives in Indian Apiculture, Agro Botanica, Bikaner (India), 412 pp.

Suryanarayana, M.C.; Mohana Rao, G. and Lakshmi, K. 1981. The sandie tree a nectar source for *Apls aralia* and *Apis mellifera. Bee World.* 74:] 08-11 0.

Susanna Messinger,Olivia Mariko Taylor, and Dustin, C. 2001. Review of the *apanteles* species (hymenoptera: braconidae) Attacking lepidoptera in *Bombus fervidobombus* (Hymenoptera: Apidae) colonies in the new world, with description of a new species from South America. *Ann. Entomol. Soc. Am.* 94(6): 851-857.

Swadener, S.R and Yonke, T.R. 1973a. Immature stages and biology of Apiomerus crassipes (Hemiptera: Reduviidae).*Annals of Entomological Society of America* 66:188-196.

Swadener, S.R and Yonke, T.R. 1973b. Immature stages and biology of Sinea complexa (Hemiptera:Reduviidae). *Annals of Entomological Society of America* 66:123-136.

Swain, P.H. and Devis, S.M. 1978. Remote sensing: The quantitative approach, McGrew Hill Int. Book Company.

Swift, J.E. 1969. Unexpected effects from substitute pest control methods. Biological impact of pesticides in the environment. Symposium, Aug. 18-20, Oreg. State Univ., Corvallis, 16 pp.

Szabo, T.I; Szabo D.C, 2000. Attempts to reduce the varroa jacobsoni populations in honeybee colonies: research report for 1999. *Am. Bee Journal* 140(8): 654-658.

Taber,S. III and Blum, M.S. 1960. Preservation of honeybee semen. *Bee World* 33:53-62.

Taber. S. (1961) Successful shipments of honeybee semen. *Bee Wld* 42(7): 173-176

Taccheo Barbina, M and Paoli, M D 1994. Degradation in the laboratory, and residues in honey and wax samples from field trials, of flumethrin. *New Perspectives on Varroa*. A. Matheson. Cardiff, IBRA: 91-96.

Takahashi, Jun-ichi and Nakamura, J. 2003. A scientific note on levels of polyandry of 2 queens of the Himalayan giant honeybee, *Apis laboriosa. Apidologie* 34 : 191–192.

Tanada, Y. and Kaya, H. K. 1993. Insect pathology. New York, Academic Press.

Tarpy D.R., Nielsen D.J. 2002. Sampling error, effective paternity, and estimating the genetic structure of honey bee colonies (Hymenoptera: Apidae), *Ann. Entomol. Soc. Am.* 95, 513–528.

Taylor. O.R. 1977. The past and possible future spread of Africanized honeybees in the Americas. *Bee Wld* 58(1): 19-30

Tewarson, N.C.; Singh, A.; Engels, W. 1992. Reproduction of *Varroa jacobsoni* in colonies of *Apis cerana indica* under natural and experimental conditions. *Apidologie* 23, 161–171.

Thakar, C.V and Tonapi, K.V. 1961. Nesting behaviour of Indian honeybees I. Differen-tiation of worker, queen and drone cells in the combs of *Apis dorsata* Fab. *Bee World* 42: 61-62.

Thakar, C.V and Tonapi, K.V. 1962. Nesting behaviour of Indian honeybees II. Nesting habits and cell differentiation in *Apis florea* Fab. *Indian Bee J.* 24: 27-31.

Thakar, C.V.; Diwan, V.V and Salvi, S.R. 1962. Floral Calendar of major and minor bee forage plants in Mahableshwar Hills (Western Ghats). *Indian Bee J.* 24(4/6):35-38.

Thakur, R.K. and Gupta, J.K 1987. *Melliferous* properties of Wild flora: *Berberis lycium* Royal. In: Proc. Int. Workshop on Agroforestry for rural needs, held 22-26 Feb., 1987, Vigyan Bhawan, New Delhi, India.

Thakur, S. S. and Kashyap, N.P. 1997. Development and testing of a trap for *Vespa magnifica. Indian Bee Journal* 59(2):83-85.

Thakur, S.S. and Kashyap, N.P. 1996. Wasp control by lure trap in apiary. *Insect Environment* 2:6-7.

Thien, L.B. 1980 "Patterns of pollination in primitive angiosperms", *Biotropica,* 12, 1-13.

Thompson, V.C. 1964. Behavior genetics of nest cleaning in honey bees III. Effect of age of bees of a resistant line on their response to disease-killed brood. *Journal of Apicultural Research*, 3, 25-30.

Thorp, R.W., Briggs, D.L., Estes, J.R. and Erickson, E.H. 1975. Nectar fluorescence under ultraviolet radiation. *Science* 189: 476–478.

Tikoo, R.K. and Abrol, D.P. 1994. Apiculture in agroforestry to boost rural economy. In *Agroforestry Systems for degraded lands*, Vol. I : Proceedings of the International Conference on sustainable development of degraded lands through Agroforestry in Asia. Singh, D: Pathak, P.S. and Roy, M.M. (eds.) Oxford and IBH Publishing Co. Pvt. Ltd., New Delhi, pp. 507-513.

Tingek S., Koeniger N. and Koeniger G. 1996. Description of a newcavity-dwelling species of Apis (*Apis nuluensis*) from Sabah, Borneo with notes on its occurrence and reproductive biology (Hymenoptera, Apoidea, Apini), *Senckenbergiana Biol.* 76, 115–119.

Tingek S., Mardan M., Rinderer G., Koeniger N. and Koeniger G. 1988. Rediscovery of *Apis vechti* (Maa, 1953): the Sabah honey bee. *Apidologie,* 19, 97–102.

Todd, F.E and Mcgregor, S.E. 1961. Insecticides and honey bees. U.S. Dept. Agr. Yearbook 247-250 pp.

Todd, F.E and Tuft, T.O. 1952. Toxicity of several new insecticides to honey bees. *Jour. Econ. Ent.* 45: 466-469.

Toft, R.J. and Rees, J.S. 1998. Reducing predation of orbweb spiders by controlling com-mon wasps *(Vespula vulgaris)* in a New Zealand beech forest. *Ecological Entomology 23:* 90-95.

Torchio, P.F and Asensio, E. 1985. The introduction of the European bee, *Osmia cornuta* Latr. into the US as a potential pollinator of orchard crops and a comparison of its manage-ability with *Osmia lignaria* propingua Cresson. *Journal of the Kansas Entomological Society* 58(1): 42-52.

Torchio, P.F. 1987. Use of non-honey bee species as pollinators of crops. *Proceedings of the Entomological Society of Ontario* 118: 111-124.

Torikata, T. 1931. The nesting habits of *Sphex flammitrichus* Strand and *Megachile monticola* Smith. *Mushi* 4: 1-9.

Toumanoff, C. 1939.The enemies of bees. Imprimerie d' Extreme-Hanoi.

Toumanoff, K.1951. Les maladies des abeilles (The disease of bees). La Revue Francaise d Apiculture. Numero Special 68.

Trochet, J., J. Morlan, and D. Roberson. 1988. First record of the Three-toed Woodpecker in California. W. Birds 19: 109-115.

Trouiller J. 1998. Monitoring Varroa jacobsoni resistance to pyrethroids in western Europe, Apidologie 29, 537–546.

Trouiller J. 2001. Monitoring *Varroa* resistance to pyrethroids (1994-2001), *Bee Craft,* 15–17.

Trump, R.F., Thompson, V.C and Rothenbuhler, W.C. 1967. Behavior genetics of nest cleaning in honey bees V. Effect of previous experience and composition of mixed colo-nies on response to disease-killed brood." *Journal of Apicultural Research*, 6, 127-131.

Tutin, C.E.G, Ham, R. and Wrogemann D. 1995. Tool-use by chimpanzees (*Pan t. troglodytes*) in the Lopé Reserve, Gabon. *Primates* 36:181-192.

Uçkan, F. and A. Guel 2000. Effects of host species on some biological characteristics of *Apanteles galleria*e Wilkinson (Hymenoptera, Braconidae). *Turkish J. Zool.* 24: 105-113.

Underwood B.A. 1986. The natural history of *Apis laboriosa* Smith in Nepal, M.S. thesis, Cornel University, Ithaca, New York.

Underwood B.A. 1990a. The behavior and energetics of high-altitude survival by the Himalayan honey bee, *Apis laboriosa,* Ph.D. Thesis, Cornell University, Ithaca, USA, 144 p., A. A. 1212/91.

Underwood B.A. 1990a. Time of drone flight of *Apis laboriosa* Smith in Nepal, *Apidologie* 21, 501– 504.

Underwood B.A. 1990b. Seasonal nesting cycle and migration patterns of the Himalayan honey bee *Apis laboriosa, Natl. Geogr. Res.* 6, 276–290.

Underwood, R.M. and Currie, R.W. 2003. The effects of temperature and dose of formic acid on treatment efficacy against *Varroa destructor* (Acari : Varroidae), a parasite of Apis mellifera (Hymenoptera: Apidae) *Exp Appl Acarol.* 29(3-4):303-13.

Underwood, R.M. and Currie, R.W. 2005. Effect of concentration and exposure time on treatment efficacy against *Varroa* mites (Acari: Varroidae) during indoor winter fumigation of honey bees (Hymenoptera: Apidae) with formic acid. *J Econ Entomol.* 98(6):1802-9.

Urdaneta-Morales, S. 1983. Some biological characteristics of a *Crithidia* species from the yellow-jacket wasp, *Vespula squamosa* (Hymenoptera: Vespidae). *Revista Brasileira de Biologia 43:*409- 412.

Van der Geest, L.P.S and Van der Laan, P.A. 1971. Insect pathogens available for distribution. *In:* Burges, H.D. and Hussey, N.W. *ed.* Microbial control of insects and mites. New York, Academic Press. Pp. 733-739.

Vandame R., Colin, M.E., Morand, S. and Otero-Colina, G. 2000. Levels of compatibility in a new host-parasite association: *Apis mellifera/Varroa jacobsoni. Canadian Journal of Zoology* 78(11): 2037-2044

Vansell, G.H and Watkins, W.G. 1933. A plant poisonous to adult bees. *J. Econ. Entomol.* 26: 168–170.

Vansell, G.H and Watkins, W.G. 1934. Adult bees found dying on spotted loco. *J. Econ. Entomol.* 27: 635–637.

Vasliadi, G. 1970. Pchelovodstova 11:16-17. (Cited by Bailey, L and Ball, B.V. 1991).

Vasu, H.D. 1967. Studies on beekeeping at Delhi. I. Possibilities of beekeeping and the plants visited by bees for pollen, nectar or both at Delhi. *Indian Bee J.* 29:63-64.

Vedova, G., Lodesani, M. and Milaini, N. 1997. Develop of resistance to organophosphates in *Varroa jacobsoni. Ape Nostra Amica* 19(1): 6-10.

Velitchkov, V. and Natchev, P. 1973. Investigation about the *Varroa jacobsoni* disease - Oud. in Bulgaria. In: *Proceedings of the XXIV the Apiculture Congress*, Buenos Aires, Argentina. pp 375-377.

Velkov, V.V, Medvinsky, A.B., Sokolov, M.S. and Marchenko, A.I. 2005. Will transgenic plants adversely affect the environment? Journal of Biosciences 30: 515-548.

Verma, L.R. 1987 "Pollination ecology of apple orchids by hymenopterous insects in Matiama-Narkanda temperate zone", Final report, Ministry of Environment and Forests, Govt. of India, pp. 118.

Verma, L.R. 1990 "Beekeeping in Integrated Mountain Development", Oxford & IBH Pub. Co. Pvt. Ltd., New Dehli/Calcutta, pp. 23.

Verma, L.R. 1990. Beekeeping in Integrated Mountain Development. Economic and Scientific Perspectives. Oxford & IBH Publishing Co. Pvt. Ltd., New Delhi,

Verma, S. and Ruttner, F. 1983. Experimental analysis of the reproductive isolation of *Apis mellifera* and *Apis cerana. Apidologie* 14: 41-57.

Vesely, V., Machova, M., Hessler, J., Hostomska, V. and Lenicek, J. 1994. Reduction of fluvalinate residues in beeswax by chemical means. *Journal of Apicultural Research* 33(3): 185-187.

Vidano, C. and Onore, G. 1971. The Mediterranean *Potesia opaca* (Fabricius) (Coleoptera: Scarabeidae), a chafer injurious to bee colonies. *L' Apicoltore Moderno.* 62:169-182.

Vinson, S.B. and G.F. Iwantsch. 1980. Host suitability for insect parasitoids. *Annu. Rev. Entomol.* 25: 397-419.

Viswanathan, H. 1953. Temperature readings in *A. dorsata* combs, *Indian Bee J.* 12, 72.

Von Frisch K, 1993. The dance language and orientation of bees. Cambridge, MA: Harvard University Press.

Von Helversen D., M.W. Holderied, and O. Von Helversen 2003. "Echoes of bat-pollinated bell-shaped flowers: conspicuous for nectar-feeding bats?". Journal of Experimental Biology 206 (6): 1025-1034.

Wafa, A.K. 1956. Ecological investigation on the activity of the oriental hornet, *Vespa velutina* with special reference to its damage to colonies of honeybees. Cairo Univ. Faculty of Agriculture. Bulletin No. 103, pp1-35.

Wain, F.L. 1968. Notes on the biology of a new bee, *Andrena bellidoides* LaBerge (Hymenoptera: Apoidea) from western India. *Entomologist* (London) : 101-104.

Wakhle, D.M. 1998. Beekeeping technology production, characteristics and uses of honey and other products, pp. 150-186, In: Mishra, R.C. (Ed.), Perspectives in Indian Apiculture, Agro Botanica, Bikaner (India), 412 pp.

Wakhle, D.M. and Desai, D.B. 1983. Enzymes in Indian honeys, Second International Conference on apiculture in Tropical climates, New Delhi: 703-710.

Wakhle, D.M. and Desai, D.B. 1991. Estimation of antibacterial activity of some Indian honeys. *Indian Bee 1.* 53 : 80-90.

Wakhle, D.M. and Phadke, R.P. 1995. Need to abolish date of expiry marking on AGMARK honey as HHF content in honey-not harmful. *Indian Bee J.*

Waller, G.D. 1969. Susceptibility of an alfalfa leafcutting bee to residues of insecticide on foliage. *Jour. Econ. Ent.* 62: 189-192.

Wallner, K. 1995. Secondary effects in the control of varroa mites: the situation concer-ning residues in several bee products. *Bienenvater* 116(4): 172-177.

Wallner, K. 1995. The use of varroacides and their influence on the quality of bee prod-ucts. *American Bee Journal* 12, 817–821.

Wallner, K. 1999. Varroacides and their residues in bee products. *Apidologie* 30(2-3): 235-248.

Wallner, K. 1999. Varroacides and their residues in bee products. *Apidologie* 30, 235–248.

Walshaw, E.R. 1967. The raiders. *American Bee Journal.*107:14-15.

Wang, J.D., Zhang, W.H., Wang, Y.M and Lim, H.R. 1985. Biological study on several kinds of wasps preyed upon honeybees and their control. *Journal Fujian Agricultural College* 14(7):43-51.

Waser, N.M. 1986. Flower constancy: definition cause and measurement. *Amer. Nat.* 127(5): 593-903.

Watkins, M. 1997. Resistance and its relevance to beekeeping, *Bee World* 78, 15–22.

Watson, L.R. 1927. Controlled mating of honeybees. *American Bee J.* Hamilton, Illinois.

Way, M.J and Khoo, K.C. 1992. Role of ants in pest management. *Annual Review of Entomology* 37:479-503.

Wearne, R.A. *et al.* 1970. Bee losses--the impact on pollination- honey production. U.S. Dept. Agr. Ext. Serv., 12 pp.

Weaver, N. 1950. Toxicity of organic insecticides to honeybees: stomach poison and field tests. Jour. Econ. Ent. 43: 333-337.

Webster, T.C., K.W. Pomper, G. Hunt, E.M. Thacker and S.C. Jones 2004. *Nosema apis* infection in worker and queen *Apis mellifera. Apidologie* 35: 49-54.

Weiser, J. 1977. An atlas of insect diseases. The Hague, W. Junk.

Wells, H. and Wells, P.H. 1983. Honeybee foraging: optimal diet, minimal uncertainty or individual constancy behaviour. *J. Anim. Ecol.* 52: 829 -836.

Wenner, A.M and Wells, P.H. 1990. Anatomy of a Controversy- The Question of a 'language' Among Bees. Columbia University Press, New York.

Wenner, A.M. and Bushing, W.W. 1996. *Varroa* mite spread in the United States. *Bee Culture* 341-343.

White, 1.W.; Subers, M.H. and Schepartz, A.I. 1963. The identification of inhibine, the antibacterial factor in honey, as hydrogen peroxide and its origin in honey glucose oxidase system. *Biochcm. Biophys Acla.* 73: 57-70.

Whittington, R., Winston, M., Melathopoulos, A. and Higo, H. 2000. Evaluation of the botanical oils neem, thymol and canola sprayed to control *Varroa jacobsoni* and *Acarapis woodi* in colonies of honey bees. *American Bee Journal* 140(7): 567-572.

Wienands, A. 1988. Synopsis der weltweit gegen die varroa-milbe der honigbienen eingesetztenen präparate. *AIlg. dtsch. Imkerztg.*, 22: 31 3-351. 86

Wigley, P.J and Dhana, S. 1988. Prospects for microbial control of the social wasps, Vespula germanica and *V. vulgaris.* Australasian Invertebrate Pathology Working Group newsletter IX: 17.

Wijesekara, A. 2001. An annotated list of bees (Hymenoptera: Apoidea: Apiformis) of Sri Lanka. *Tijdschrift voor Entomologie* 144: 145-158.

Williams, I.H., Simpkins, J.R and Martin, A.P. 1991. Effect of insect pollination on seed production in linseed (*Linum usitatissimum*). *Journal of Agricultural Sciences* 117(1): 75-79.

Williams, J.L. 1976. Status of the greater wax moth, "galleria mellonella," in the United States beekeeping industry. *American Bee Journal* 116(11): 524- 526.

Williams, P.H. 1991. The bumble bees of the Kashmir, Himalaya (Hymenoptera: Apidae, Bombini). *Bulletin of the British Museum of the Natural History* (Entomology) 60:1-204.

Williams, P.H. 1994. Phylogenetic relationships among bumble bees (*Bombus* Latr.): a reappraisal of morphological evidence. *Systematic Entomology* 19: 327-344.

Williams, P.H. 1998. An annotated checklist of bumble bees with an analysis of patterns of description (Hymenoptera: Apidae, Bombini). *Bulletin of The Natural History Museum* [London] (Entomology) 67: 79-152.

Wilson, W.T. and Collins, A.M. 1993. Failure of formic acid to control *Varroa jacobsoni* in a hot climate. *American Bee Journal* 133(12): 871.

Winston, M.L and Scott, S.C. 1984. The value of bee pollination to Canadian apiculture. *Canadian Beekeeping* 11: 134-135.

Winston. M. 1987. The Biology of the Honey Bee. Harvard University Press, Cambridge, Massachusetts.

Wojcik, W. and Georgis, R. 1988. Infection of adult western yellowjackets with desiccated *Steinernemafeltiae* (Nematoda). *Journal of Invertebrate Pathology 52:* 183-184.

Womeldorf, D.J., E.L. Atkins, and P.A. Gillies. 1974. "Honey bee hazards associated with some mosquito abatement aerial spray applications." *Calif. Vector News* 21(9): 51-55.

Wongsiri S., Limbipichai K., Tangkanasing P., Mardan M., Rinderer T.E., Sylvester H.A., Koeniger G. and Otis G. 1990. Evidence of reproductive isolation confirms that *Apis andreniformis* (Smith 1858) is a separate species from sympatric *Apis florea* (Fabricius 1787). *Apidologie* 22 47–52.

Wongsiri, S., Lekprayoon, C., Thapa, R., Thirakupt, K., Rinderer, T., Sylvester, H., Oldroyd, B. and Booncham, U. 1996. Comparative biology of *Apis andreniformis* and *Apis florea* in Thailand. *Bee World* 77: 23-35.

Woo, K.S. 1992. New honeybee mite *Varroa underwoodi* on *Apis cerana* in South Korea. *Honeybee Sci.* 13, 173–174.

Woo, K.S., Cho, K.S. and Cho, Y.H. 1993. Control effects of Apitol against *Varroa jacobsoni* Oud. in Korea. *Korean Journal of Apiculture* 8(1): 48-55.

Woodrow, A.W. 1942. Susceptibility of honey bee larvae to individual inoculations with spores of Bacillus larvae. *J Econ Entomol* 35:892- 895.

Woodrow, A.W. and Holst, E.C. 1942. The mechanism of colony resistance to American foulbrood, *J. Econ. Entomol.* 35, 327–330.

Woodward, M. and Jones, W. 1991. Ant control in apiaries. *Bee Briefs* 8(2):2-5.

Worden, S.H. 1917. Caged bee and queen murdered by ants. *Glean Bee Cult.*45:795.

Woyke J, Wilde J. and Reddy, C.C. 2004. Open-air-nesting honey bees Apis dorsata and Apis laboriosa differ from the cavity-nesting *Apis mellifera* and *Apis cerana* in brood hygiene behaviour. *J Invertebr Pathol.* 86(1-2):1-6.

Woyke J., Wilde J. and Wilde M. 2001. A scientific note on *Apis laboriosa* winter nesting and brood rearing in the warm zone of Himalayas. *Apidologie* 32, 601–602.

Woyke J., Wilde J. and Wilde M. 2003a. Periodic mass flights of *Apis laboriosa* in Nepal, *Apidologie* 34, 121–127.

Woyke J., Wilde J. and Wilde M. 2003b. Flight activity reaction to temperature changes in *Apis dorsata, Apis laboriosa* and *Apis mellifera. J. Apic. Sci.* 47, 31–38.

Woyke, H.W. 1981 Some aspects of the role of the honeybee in onion seed production in Poland. *Acta Hort.* 111: 91-98.

Woyke, J. 1971. Correlations between age at which honeybee brood was grafted, characterstics of the resultant queens and results of insemination. *J. Apic Res.* 10(1): 45-55.

Woyke, J. 1973. Instrumental insemination of *Apis cerana indica* queens. *J. Apic Res.* 12: 151-158.

Woyke, J. 1973. Reproduction and artificial insemination of *Apis cerana indica* Fabr. *Indian Bee Journal* 33:10-20.

Woyke, J. 1976. Population genetic studies on sex alleles in the honeybee using the example of the Kangaroo Island bee sanctuary. *J. apic. Res.* 15(3/4):

Woyke, Jerzy, Wilde, Jerzy and Wilde, Maria 2004. Temperature correlated dorso-ventral abdomen flipping of *Apis laboriosa* and *Apis dorsata* worker bees. *Apidologie* 35: 493–502

Wright, M. 1944. Some random observations on Dragonfly habitat with notes on their predaceousness on bees. *Journal of the Tennessee academy of science* 19:295-301.

Yamagiwa, J., Yumoto, T., Ndunda, M. and Maruhashi, T. 1988. Evidence of tool-use by chimpanzees (*Pan troglodytes* schweinfurthii) for digging out a bee-nest in the Kahuzi-Biega National Park, Zaïre. *Primates* 29:405-411.

Yang, X. and Cox-Foster, D.L. 2005. Impact of an ectoparasite on the immunity and pathology of an invertebrate: Evidence for host immunosuppression and viral amplification. *Proc Nat Acad Sci* 102: 7470–7475.

Yazdani, S.S. and Agarwal, M.L. 1997 "Elements of Insect Ecology", Narosa Publishing House, Mumbai/Calcutta, pp. 210.

Yue, C. and Genersch, E. 2005. RT-PCR analysis of *Deformed wing virus* in honey bees (*Apis mellifera*) and mites (*Varroa destructor*). *J Gen Virol* 86: 3419 - 3424.

Zacarin, Giuliano G., Gobbi, Nivar and Chaud-Netto, José. 2004. Host preference of *Apa-nteles galleriae* Wilkinson (Hymenoptera: Braconidae) for *Galleria mellonella* (L.) or *Achroia grisella* (Fabricius) (Lepidoptera: Pyralidae). *Neotrop. Entomol.*, 33(1)65-70

Zaleski, A. 1956. Lucerne investigation. II. Pollination and seed setting in Lucerne strains. *Journal of Agricultural Sciences* 48: 236-244.

Zander, E. 1909. Tierische parasiten als krankenheitserreger bei der biene. Müchener Bienenzeitung 31:196-204

Zander, E. and Weiss, K. 1964. Das Leben der Biene. Ulmer, Stuttgart

Zhang, S., Bock, F., Si, A., Tautz, J. and Srinivasan, M.V. 5 2005. "Visual working memory in decision making by honey bees". *Proceedings of the National Academy of Sciences of the United States of America* 102 (14): 5250-5.

Zhang, Z. 2000. Notes on *Varroa destructor* (Acari: Varroidae) parasitic on, honeybees in New Zealand, Syst. Appl. Acarol. (Special Publications) 5, 9–14.

Zhi-Qiang, Zhang 2000. Notes on *Varroa destructor* (Acari: Varroidae) parasitic on honey-bees in New Zealand. *Systematic & Applied Acarology Special Publications* 5: 9-14

Zoberi, M.H. 1995. *Metarhizium anisopliae,* a fungal pathogen of *Reticulitermes flaviceps* (Isoptera: Rhinotermitidae). *Mycologia* 87: 354-359.